CW01151895

Proceedings of the Second International Conference on the Future of ASEAN (ICoFA) 2017 – Volume 2

Rizauddin Saian · Mohd Azwan Abbas
Editors

Proceedings of the Second International Conference on the Future of ASEAN (ICoFA) 2017 – Volume 2

Science and Technology

🕊 Springer

Editors
Rizauddin Saian
Universiti Teknologi MARA Perlis
Arau
Malaysia

Mohd Azwan Abbas
Universiti Teknologi MARA Perlis
Arau
Malaysia

ISBN 978-981-10-8470-6   ISBN 978-981-10-8471-3   (eBook)
https://doi.org/10.1007/978-981-10-8471-3

Library of Congress Control Number: 2018932546

© Springer Nature Singapore Pte Ltd. 2018
This work is subject to copyright. All rights are reserved by the Publisher, whether the whole or part of the material is concerned, specifically the rights of translation, reprinting, reuse of illustrations, recitation, broadcasting, reproduction on microfilms or in any other physical way, and transmission or information storage and retrieval, electronic adaptation, computer software, or by similar or dissimilar methodology now known or hereafter developed.
The use of general descriptive names, registered names, trademarks, service marks, etc. in this publication does not imply, even in the absence of a specific statement, that such names are exempt from the relevant protective laws and regulations and therefore free for general use.
The publisher, the authors and the editors are safe to assume that the advice and information in this book are believed to be true and accurate at the date of publication. Neither the publisher nor the authors or the editors give a warranty, express or implied, with respect to the material contained herein or for any errors or omissions that may have been made. The publisher remains neutral with regard to jurisdictional claims in published maps and institutional affiliations.

Printed on acid-free paper

This Springer imprint is published by the registered company Springer Nature Singapore Pte Ltd. part of Springer Nature
The registered company address is: 152 Beach Road, #21-01/04 Gateway East, Singapore 189721, Singapore

Preface

Following the ASEAN Vision 2020, it analyses the issues faced by ASEAN countries which are different and yet make ASEAN a competitive entity through partnerships. On the 30th anniversary of the ASEAN, all ASEAN leaders agreed to the establishment of the ASEAN Vision 2020, which is the formation of a peaceful, stable and dynamically developed region while maintaining a community of caring societies amongst Malaysia, Indonesia, Singapore, Brunei, Vietnam, Thailand, the Philippines, Myanmar, Laos and Cambodia. Following the ASEAN aspiration, Universiti Teknologi MARA Perlis took the initial steps to organise conferences and activities that highlight the role of the ASEAN. The Second International Conference on the Future of ASEAN (ICoFA) 2017 is a programme organised by the Office of Academic Affairs, Universiti Teknologi MARA Perlis, to encourage a more comprehensive integration amongst ASEAN members. This book comes in two parts—volumes 1 and 2—and is useful for those who conduct research on business, social sciences, science and technology. Volume 2 looks at how science and technology impact the future of ASEAN. As an addition, it is also valuable for researchers worldwide who want to gain more knowledge about ASEAN countries.

Perlis, Malaysia
Rizauddin Saian
Mohd Azwan Abbas

Contents

Landscape Design for Residential Area 1
Mohd Ramzi Mohd Hussain, Izawati Tukiman and Ismawi Hj Zen

A Review on Tropical Green Roof Maintenance Practice of High-Rise Residential Buildings 11
Shafikah Saharuddin, Natasha Khalil and Alia Abdullah Saleh

Cycling! A Way Forward 27
Emy Ezura A. Jalil, Lau Siong Heng, Tan Song Jun and Fong Sim Ling

A Review Factors Affecting Residential Property Price 37
Wan Nur Ayuni Wan Ab Rashid, Thuraiya Mohd
and Lizawati Abdullah

Tensile Properties and Thermal Characteristics of Linear Low-Density Polyethylene/Poly(Vinyl Alcohol) Blends Containing 3-(Trimethoxysilyl) Propyl Methacrylate 47
Razif Nordin, Hanafi Ismail, Zuliahani Ahmad and Ragunathan Santiago

A Review Analysis of Accident Factor on Road Accident Cases Using Haddon Matrix Approach 55
Nur Fatma Fadilah Yaacob, Noradila Rusli and Sharifah Norashikin Bohari

GIS Efficiency Analysis on Traffic Congestion for Emergency Responses in Alor Setar, Kedah 67
Ummu Syira-Ain Redzuwan, Sharifah Norashikin Bohari, Noradila Rusli
and Nursyahani Nasron

Sea Surface Circulation in the Straits of Malacca and the Andaman Sea Using Twenty-Three Years Satellite Altimetry Data 77
K. N. A. A. K. Mansor, M. F. Pa'suya, A. H. M. Din, M. A. Abbas,
M. A. C. Aziz and T. A. T. Ali

Accuracy Assessment of TanDEM-X DEM and Global Geopotential Models for Geoid Modeling in the Southern Region of Peninsular Malaysia 91
Muhammad Faiz Pa'suya, Ami Hassan Md Din, Zulkarnaini Mat Amin, Noradila Rusli, Amir Hamzah Othman, Mohamad Azril Che Aziz and Mohd Adhar Abd Samad

The Evaluation of Performance and Quality of Preschool Using Fuzzy Logic Approach 101
Nur Syuhada Muhammat Pazil, Norwaziah Mahmud, Siti Hafawati Jamaluddin, Umi Hanim Mazlan and Afiqah Abdul Rahman

Evaluating Fuzzy Time Series and Artificial Neural Network for Air Pollution Index Forecasting 113
Nordianah Jusoh and Wan Juliyana Wan Ibrahim

Preliminary Findings: Revising Developer Guideline Using Word Frequency for Identifying Apps Miscategorization 123
Azmi Aminordin, Mohd Faizal Abdollah, Robiah Yusof and Rabiah Ahmad

Creativity in Mathematical Thinking Through Constructivist Learning Approach for Architecture Students 133
Nor Syamimi Samsudin, Ismail Samsuddin and Ahmad Faisol Yusof

Forecasting Rainfall Based on Fuzzy Time Series Sliding Window Model 143
Siti Nor Fathihah Azahari, Rizauddin Saian and Mahmod Othman

e-Diet Meal Recommender System for Diabetic Patients 155
Mahfudzah Othman, Nurzaid Muhd Zain and Umi Kalsum Muhamad

Comparative Study of Fuzzy Time Series and Artificial Neural Network on Forecasting Rice Production 165
Norpah Mahat, Rohana Alias and Suhailah Muhamad Idris

Scholarship Eligibility and Selection: A Fuzzy Analytic Hierarchy Process Approach 175
Norwaziah Mahmud, Nur Syuhada Muhammat Pazil, Umi Hanim Mazlan, Siti Hafawati Jamaluddin and Nik Nurnadia Che Hasan

Assessment of Energy Efficiency Level on UiTMPP's *Dewan Besar* Building 185
Mohamad Hilmi Akmal Zakaria, Mohamad Adha Mohamad Idin, Muhammad Firdaus Othman and Noorezal Atfyinna Mohd Napiah

Contents

Assessment of Energy Efficiency Level on Unit Kesihatan UiTMPP's Building 195
Muhammad Daniel Muhamad Nor, Mohamad Adha Mohamad Idin,
Muhammad Firdaus Othman and Noorezal Atfyinna Mohd Napiah

Assessment of Energy Efficiency Level on UiTMPP's Baiduri College Building 203
Amirul Ashraf Samad, Mohamad Adha Mohamad Idin,
Muhammad Firdaus Othman and Noorezal Atfyinna Mohd Napiah

A Compact SWB Antenna Using Parasitic Strip 215
Md. Moinul Islam, Md. Mehedi Hasan, Mohammad Rashed Iqbal Faruque
and Mohammad Tariqul Islam

A Multi-band Planar Double-Incidence Miniaturized Double-Negative Metamaterial 225
Mohammad Jakir Hossain, Mohammad Rashed Iqbal Faruque,
Md. Jubaer Alam and Mohammad Tariqul Islam

Depiction of a Combined Split P-Shaped Compact Metamaterial for Dual-Band Microwave Application 235
Md. Jubaer Alam, Mohammad Rashed Iqbal Faruque,
Mohammad Jakir Hossain, Mohammad Tariqul Isla
and Sabirin Abdullah

Photovoltaic/Thermal Water Cooling: A Review of Experimental Design with Electrical Efficiency 245
Zahratul Laily Edaris, Mohd Sazli Saad,
Mohammad Faridun Naim Tajuddin and Sofian Yusoff

Success Factor for Site Management in Industrialized Building System (IBS) Construction 255
Nor Izzati Muhammad Azmin, U. Kassim and Mohd Faiz
Mohammad Zaki

Street Network Analysis by SPOT Imagery for Urban Morphology Study. Case Study: Melaka 263
Marina Mohd Nor, Norzailawati Mohd Noor and Sadayuki Shimoda

Controlling Leapfrog Sprawl with Remote Sensing and GIS Application for Sustainable Urban Planning 273
Nur Aulia Rosni, Norzailawati Mohd Noor, Alias Abdullah
and Isoda Setsuko

Constructing and Modeling 3D GIS Model in City Engine for Traditional Malay City 285
Ahmad Afiq Aiman Abdullah, Norzailawati Mohd Noor
and Alias Abdullah

3D Data Fusion Using Unmanned Aerial Vehicle (UAV) Photogrammetry and Terrestrial Laser Scanner (TLS) 295
Mohamad Aizat Asyraff Mohamad Azmi, Mohd Azwan Abbas, Khairulazhar Zainuddin, Mohamad Asrul Mustafar, Mohd Zainee Zainal, Zulkepli Majid, Khairulnizam M. Idris, Mohd Farid Mohd Ariff, Lau Chong Luh and Anuar Aspuri

Spatial Mapping and Analysis of Carbon Dioxide Emissions from Electricity in UiTM Perlis for Assessment of Low Carbon Campus ... 307
Noradila Rusli, Nurhani Nadirah Hamzah, Muhammad Faiz Pa'suya and Suhaila Hashim

Estimation of *Acacia mangium* Aboveground Biomass and Wood Volume Through Landsat 8 317
Aqilah Nabihah Anuar, Ismail Jusoh and Affendi Suhaili

Screening of Transovarial Dengue Virus (DENV) Transmission in Field-Collected *Aedes albopictus* from Dengue Active Transmission Areas in Shah Alam, Selangor, Malaysia 327
H. Mayamin, A. Nurul Adilah, R. Nurul Hidayah, I. Nurul-Ain, M. Mohd Fahmi, Tengku Shahrul Anuar, C. D. Nazri, I. Rodziah, H. A. Abu and S. N. Camalxaman

Changes in Total Phenolics, β-Carotene, Antioxidant Properties and Antinutrients Content of Banana (*Musa Cavendishii L. Var. Montel*) Peel at Different Maturity Stages 333
Aishah Bujang, Siti Sarah Jamil and Nazrahyatul Hidayah Jemari

Exploring Attitude Toward Online Video Marketing in Malaysia 341
Akeem Olowolayemo, Norsaremah Saleh, Nafisat Toyin Adewale and Fatemah Shatar

Inventory Policies in Blood Bank Management 351
Farah Hani Jamaludin, Jasmani Bidin, Noorsham Mansor, Sharifah Fhahriyah Syed Abas and Zurina Kasim

Comparing Methods for Lee–Carter Parameter's Estimation for Predicting Hospital Admission Rates 361
Siti Zulaikha Zulkarnain Yap, Siti Meriam Zahari, Zuraidah Derasit and S. Sarifah Radiah Shariff

Data Pre-Processing Using SMOTE Technique for Gender Classification with Imbalance Hu's Moments Features 373
Ahmad Haadzal Kamarulzalis, Muhamad Hasbullah Mohd Razali and Balkiah Moktar

Optimizing Efficiency of Electric Train Service (ETS) Ticket Pricing 381
Noraini Noordin and Nur Syamila Mohd Ali Amran

Contents

Fuzzy Rules Enhancement of CHEF to Extend Network Lifetime 393
Muhammad A'rif Shah Alias, Habibollah Haron and Teresa Riesgo

Energy Consumption of Wireless Sensor Based on IoT in a Parking Space With and Without Fuzzy Rules 403
Muhamad Yazid Che Abdullah, Habibollah Haron and Teresa Riesgo

Forecasting River Water Level Using Artificial Neural Network 413
Norwaziah Mahmud, Nur Syuhada Muhammat Pazil, Izleen Ibrahim, Umi Hanim Mazlan, Siti Hafawati Jamaluddin and Nurul Zulaikha Othman

Optimization of Melt Filling Distribution for Multiple Gate System ... 421
Saiful Bahri Mohd Yasin, Sharifah Nafisah Syed Ismail, Zuliahani Ahmad, Noor Faezah Mohd Sani and Siti Nur Asiah Mahamood

Improvement on Ride Comfort of Quarter-Car Active Suspension System Using Linear Quadratic Regulator..................... 431
Sharifah Munawwarah Syed Mohd Putra, Fitri Yakub, Mohamed Sukri Mat Ali, Noor Fawazi Mohd Noor Rudin, Zainudin A. Rasid, Aminudin Abu and Mohd Zamzuri Ab Rashid

Evaluation of Protease Enzyme on *Pomacea canaliculata* Eggs 443
Noor Hasyierah Mohd Salleh and Nurhadijah Zainalabidin

A New Analytical Technique for Solving Nonlinear Non-smooth Oscillators Based on the Rational Harmonic Balance Method 453
Md. Alal Hosen, M. S. H. Chowdhury, M. Y. Ali and A. F. Ismail

CDIO Implementation in Separation Processes Course for Chemical Engineering................................. 465
Arbanah Muhammad, Salmi Nur Ain Sanusi, Siti Hajar Anaziah Muhamad, Sharifah Iziuna Sayed Jamaludin, Meor Muhammad Hafiz Shah Buddin, Mohd Zaki Sukor, Sitinoor Adeib Idris, Nur Shahidah Ab Aziz and Muhammad Imran Ismail

Usewear Experiment to Determine Suitability of Rock Material from the Crater of Meteorite Impacts as a Prehistoric Stone Tool.... 475
Siti Khairani Abd Jalil, Jeffrey Abdullah and Mokhtar Saidin

Left-Handed Network-Shaped Metamaterial for Visible Frequency ... 485
Md. Mehedi Hasan, Mohammad Rashed Iqbal Faruque and Mohammad Tariqul Islam

Volume Change Behaviour of Clay by Incorporating Shear Strength: A Review 495
Juhaizad Ahmad, Mohd Ikmal Fazlan Rosli, Abdul Samad Abdul Rahman, Syahrul Fithri Senin and Mohd Jamaludin Md Noor

Mechanical Properties and X-Ray Diffraction of Oil Palm Empty Fruit Bunch All-Nanocellulose Composite Films 505
Nur Liyana Izyan Zailuddin, Azlin Fazlina Osman, Salmah Husseinsyah, Zailuddin Ariffin and Faridah Hanum Badrun

Efficiency Cooling Channel at Core Side Incorporating with Baffle and Bubbler System 515
Saiful Bahri Mohd Yasin, Noor Faezah Mohd Sani, Salwa Adnan, Zahidahthorwazunah Zulkifli, Zuliahani Ahmad
and Sharifah Nafisah Syed Ismail

Potential Antibacterial Activity of Essential Oil of *Citrus hystrix* and *Chromolaena odorata* Leaves 523
Hamidah Jaafar Sidek and Fatin Fathihah Abdullah

Characterisation and Application of Diatomite in Water Treatment ... 533
Komathy Selva Raj, Megat Johari Megat Mohd Noor, Masafumi Goto and Pramila Tamunaidu

Reactive Red 4 Dye-Sensitized Immobilized TiO$_2$ for Degradation of Methylene Blue Dye 541
W. I. Nawawi and F. Bakar

The Influence of Water Quality Index (WQI) Assessment Towards Water Sports Activity at Taman Tasik Titiwangsa, Kuala Lumpur, Malaysia 551
Nor Hanisah Mohd Hashim, Balqis Dayana Badarodin
and Wan Hazwatiamani Wan Ismail

Phytochemical Screening, Antioxidant and Enzyme Inhibition Activity of *Phoenix dactylifera* Ajwa Cultivar 561
Muhamad Nabil Md Nor, Nur Syafiqah Rahim, Sarina Mohamad, Saiyidah Nafisah Hashim, Zainab Razali and Noor Amira Muhammad

Determination of Two Different Aeration Time on Food Waste Composting 571
Khairul Bariyah Binti Abd Hamid, Mohd Armi Abu Samah, Mohd Huzairi Mohd Zainudin and Kamaruzzaman Yunus

The Synergistic Antibacterial Effect of *Azadirachta indica* Leaves Extract and *Aloe barbadensis* Gel Against Bacteria Associated with Skin Infection 581
Hamidah Jaafar Sidek, Mohamad Azhar Azman
and Muhamad Shafizul Md Sharudin

Prediction of Dengue Outbreak in Selangor Using Fuzzy Logic 593
Mohd Fazril Izhar Mohd Idris, Amjad Abdullah
and Shukor Sanim Mohd Fauzi

Comparison of Characterization and Osteoblast Formation Between Human Dental Pulp Stem Cells (hDPSC) and Stem Cells from Deciduous Teeth (SHED) 605
Farinawati Yazid, Nur Atmaliya Luchman, Rohaya Megat Abdul Wahab and Shahrul Hisham Zainal Ariffin

Detection of Nicotine in Nicotine-Free E-Cigarette Refill Liquid Using GC–MS 615
Reena Abd Rashid, Asmira Nabilla Adnan, Sohehah Maasom and Gillian Taylor

Minimizing Warehouse Operation Cost 625
Tracy Adeline Ajol, Shirley Sinatra Gran and Awang Nasrizal Awang Ali

Bioaugmentation of Oil Sludge Using Locally Isolated Hydrocarbon Microbial in Single and Mixed Cultures Assisted by Aerated Static Pile (ASP): A Laboratory Scale 635
Nur Zaida Zahari and Mohd Tuah Piakong

Modeling Relationship Between Cocoa Beans Commodity Export Volatility and Stock Market Index (KLCI) 649
Siti Nor Nadrah Muhamad, Izleen Ibrahim, Nordianah Jusoh@Hussain, Siti Hannariah Mansor and Wan Juliyana Wan Ibrahim

Factors Influencing Internationalization of Malaysian Construction Firms 659
Nor Fazilah Omar, Che Maznah Mat Isa and Ruslan Affendy Arshad

The Optimization of Crop Production: A Case Study at the Farming Unit of UiTM Perlis 673
Nuridawati Baharom and Nurul Amalina Abu Bakar

Preliminary Study on Food Preferences of Common Palm Civet (*Paradoxurus hermaphroditus*) in Captivity 687
Zakirah Zaki, Tun Firdaus Azis, Syafiqah Rahim and Sarina Mohamad

The Influence of Root Zone Temperature Manipulation on Strawberry Yields in the Tropics 695
Mohd Ashraf Zainol Abidin, Desa Ahmad, Ahmad Syafik Suraidi and Josephine Tening Pahang

Detection of Class 1 Integron and Antibiotic Resistance Genes in *Aeromonas Hydrophila* Isolated from Freshwater Fish 705
Hamdan Ruhil Hayati, Mohd Daud Hassan, Ong Bee Lee, Hamid Nur Hidayahanum, A. Mohamed Nora Faten, Manaf Sharifah Raina, Tan Li Peng and Nik Mohd Fauzi Nik Nur Fazlina

Relationship of Pre-competition Anxiety and Cortisol Response in Individual and Team Sport Athletes 719
Jamilah Ahmad Radzi, Sarina Md Yusuf, Nurul Hidayah Amir and Siti Hannariah Mansor

Relationship Between Muscle Architecture and Jumping Abilities Among Recreationally Active Men 729
Ali Md Nadzalan, Nur Ikhwan Mohamad, Jeffrey Low Fook Lee and Chamnan Chinnasee

Metabolic Cost of Continuous Body Weight Circuit Training with Aerobic-Based Exercise Interval for Muscle Strength and Endurance on Young Healthy Adults 737
Nur Ikhwan Mohamad, Raiza Sham Hamezah and Ali Md Nadzalan

Effects of Body Weight Interval Training on Fitness Components of Primary Schoolgirls 747
Nurul Afiqah Bakar, Erie Zuraidee Zulkifli and Mohd Nidzam Jawis

Effect of Annealing Temperature on Structural Properties of $Ba_{0.9}Er_{0.1}TiO_3$ Thin Films 757
Zeen Vee Ooi and Ala'Eddin Ahmad Saif

The Effect of Zinc Addition on the Characteristics of Sn–2.0Ag–0.7Cu Lead-Free Solders 767
Ramani Mayappan and Amirah Salleh

Influence of Grain Size on the Isothermal Oxidation of Fe–40Ni–24Cr Alloy 777
Noraziana Parimin, Esah Hamzah and Astuty Amrin

Performance Measurement of Small and Medium Enterprises (SMEs) in Malaysia: Implications for Enhanced Competitive Advantage Toward the Implementation of TQM 787
Wan Ahmad Yusmawiza Wan Yusoff, Naim Ben Ali and M. Boujelbene

The Effect of Ethanol on Drying Defects of Oil Palm Trunk (OPT) ... 799
Suhaida Azura, Ahmad Fauzi Awang, Shaikh Abdul Karim Yamani Zakaria, Junaiza Ahmad Zaki and Izzah Azimah Noh

The Effect of Fly Ash and Bottom Ash Pile in Problematic Soil Due to Liquefaction 807
Mohd Ikmal Fazlan Rozli, Juhaizad Ahmad, Mohd Asha'Ari Masrom, Syahrul Fithri Senin and Abdul Samad Abdul Rahman

A Method for the Full Automation of Euler Deconvolution for the Interpretation of Magnetic Data 817
Nuraddeen Usman, Khiruddin Abdullah and Mohd Nawawi

Enhancing Ride Comfort of Quarter Car Semi-active Suspension System Through State-Feedback Controller 827
Muhamad Amin Zul Ifkar Mohd Fauzi, Fitri Yakub,
Sheikh Ahmad Zaki Shaikh Salim, Hafizal Yahaya, Pauziah Muhamad,
Zainudin A. Rasid, Hoong Thiam Toh and Mohamad Sofian Abu Talip

Heavy Metals Concentration in Water Convolvulus (*Ipomoea aquatica* and *Ipomoea reptans*) and Potential Health Risk 839
Siti Nuur Ruuhana Saidin, Farah Ayuni Shafie, Siti Rohana Mohd Yatim and Rodziah Ismail

Analyzing Throughput in a Smartphone-Based Grid Computing 847
Alif Faisal Ibrahim, Muhammad Amir Alias and Syafnidar Abdul Halim

UAV/Drone Zoning in Urban Planning: Review on Legals and Privacy 855
Norzailawati Mohd Noor, Intan Zulaikha Mastor and Alias Abdullah

Indigenous and Produce Vegetable Consumption in Selangor, Malaysia 863
Nur Filzah Aliah and Emmy Hainida Khairul Ikram

3D-MSWT: An Alternative Tool in Developing Students' Understanding in Landform Interpretation Course 875
Ruwaidah Borhan, Azran Mansor and Nur Hanim Ilias

Analysis of Weld Line Movement for Hot Runner System 883
Saiful Bahri Mohd Yasin, Zuliahani Ahmad, Sharifah Nafisah Syed Ismail, Noor Faezah Mohd Sani and Zafryll Amir Zulkifly

Ethnobotanical Study on Plant Materials Used in Malay Traditional Post-partum Bath *(Mandi Serom)* Among Malay Midwives in Kedah 891
Nur Illani Abdul Razak, Rashidi Othman and Josephine Tening Pahang

Meiofaunal Responses to Azoic Sediment in a Sandbar-Regulated Estuary in the East Coast of Peninsular Malaysia 899
Rohayu Ramli, Zaleha Kassim and Muhammad Akmal Roslani

Coastal Deposit Characteristic Influenced by Terrestrial Organic Matter and Its Sedimentary Structure at Jangkang Beach, Bengkalis District, Riau Province—Indonesia 909
Yuniarti Yuskar, Tiggi Choanji, Dewandra Bagus EP, Adi Suryadi and Rani A. Ramsof

Influence of Grain Size on Isothermal Oxidation of Fe–33Ni–19Cr Alloy 917
Zahraa Zulnuraini, Noraziana Parimin and Izzat Mohd Noor

Differentiation of Displacement Factor for Stiff and Soft Clay in Additional Modulus of Subgrade Reaction of Nailed-Slab Pavement System 927
Anas Puri

Decision Support System of Vegetable Crop Stipulation in Lowland Plains 935
Ause Labellapansa, Ana Yulianti and Arif Zalbiahdi

Geology and Geochemistry Analysis for Ki Index Calculation of Dompak Island Granite Bauxites to Determine the Economical Mineral 947
Catur Cahyaningsih, Arrachim Maulana Putera, Gayuh Pramukti and Mohammad Murtaza Sherzoy

Accuracy of Algorithm C4.5 to Study Data Mining Against Selection of Contraception 955
Des Suryani, Ause Labellapansa and Eka Marsela

Home Monitoring System Based on Cloud Computing Technology and Object Sensor 963
Evizal Abdul Kadir, Apri Siswanto and Ari Yulian

Geological Mapping of Silica Sand Distribution on the Muda Island and Ketam Island, Estuary of Kampar River, Indonesia 973
Husnul Kausarian, Tiggi Choanji, Detri Karya, Evizal Abdul Kadir and Adi Suryadi

Expert System Diagnoses on Degenerative Diseases Using Bayes Theorem Method 983
Ause Labellapansa, Ana Yulianti and Islahudin

The Analysis Factor to Determine Modern Store Location in Pekanbaru City 997
Puji Astuti, Febby Asteriani, Yoghi Kurniawan and Idham Nugraha

Landslide Hazard Map Using Aster GDEM 30m and GIS Intersect Method in Tanjung Alai, XIII Koto Kampar Sub-District, Riau, Indonesia 1009
Tiggi Choanji, Idham Nugraha, Muhammad Sofwan and Yuniarti Yuskar

Landscape Design for Residential Area

Mohd Ramzi Mohd Hussain, Izawati Tukiman and Ismawi Hj Zen

Abstract Landscape design for residential area should be favorable for the ambience of ecological environment and for the enhancement of both the local climate of a residential and the environmental quality of life. The paper provides information that will enable people toward a better understanding of landscape design for residential. A good-quality residential area can be highlighted as an important issue in creating a sustainable living environment. It begins by describing landscape focuses on landscape design in residential development. It also reviews the issues related to the residential landscape design and residential landscape for green neighborhood in order to obtain a clearer and better picture of the basic needs of landscape for residential developments.

Keywords Landscape design · Residential area · Ecological environment
Local climate · Green neighborhood

1 Introduction

In theory, landscape design acts as a problem-solving process through enriching the qualities of nature with human landscape preferences. Landscape design is 'an art and science of organizing and enriching outdoor space through the placement of plants and structures in agreeable and useful relationships with natural environment' (Hussain et al. 2014a, b, c). According to Williams and Tilt (2006), landscape design is the art of developing property for its greatest use and satisfaction. They add that effective landscape design is also a form of science because it involves

---

M. R. M. Hussain (✉) · I. Tukiman · I. H. Zen
Department of Landscape Architecture, Kulliyyah of Architecture and Environmental Design (KAED), International Islamic University Malaysia (IIUM), Jln Gombak, 53100 Kuala Lumpur, Malaysia
e-mail: ramzi@iium.edu.my

understanding the environment around our home and selecting plants that perform well in that environment. This view is also supported by VanDerZanden and Rodie (2008) who state that landscape design blends arts with environmental, physical, and biological of science which mainly focus on outdoor space. They also further explain that well-defined landscape space can enhance the quality of living areas which meet people's preferences. In either case, a well-conceived landscape design, properly installed and well maintained, adds value to property and enhances the quality of life. Thus, landscape design can create a sense of place to the area.

2 Residential Landscape Design

Besides that, according to VanDerZanden and Rodie (2008), landscape design is not only limited to plants material only. It also focuses on the hardscape that complement the plants in order to create a successful design. A well-defined landscape space can create a quality of environment and provide a conducive living space in residential. This landscape design might be able to become a factor influencing the price and value of the property. In relating to landscape design and housing property, Ekin (as cited in Smith et al. 2008) suggests that housing is an activity that is deeply connected to sustainable environment. The connections are as follows:

 i. Housing is a basic human need and its quality, price, and availability are crucially important to quality of life.
 ii. The location, planning, layout, and landscape design of house make an important contribution to community spirit.

Thus, housing can be seen as the central element which can link together economic development, environment, and social welfare in achieving a sustainable environment and society (Fig. 1). According to Gause et al. (2007), landscape design can be used as a tool to make a community more sustainable and contribute

Fig. 1 Importance of housing in a sustainable society (Smith et al. 2008)

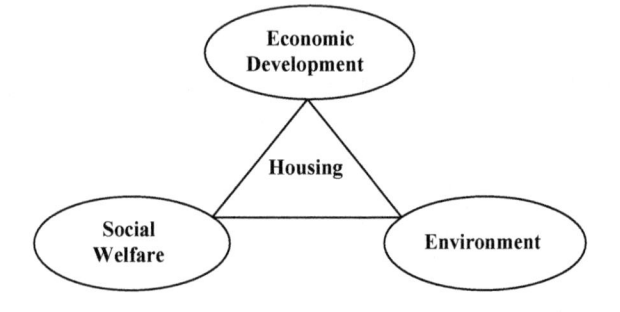

to a balanced environment. Natural and constructed landscape design can modify the residential community to become a better living environment and also increase the value of the housing property.

Landscape design is the art of developing property for its greatest use and enjoyment (Williams and Tilt 2006). They verify that effective landscape design is also a science because it involves understanding the environment around our home and selecting plants that perform well in that environment. There are four ways in which the landscape is valuable: aesthetically, economically, functionally, and environmentally (Helfand et al. 2006; Williams and Tilt 2006):

1. Aesthetic Value

An attractive landscape is aesthetically valuable because it adds beauty or is pleasing to our senses. The visual beauty of our home and property can be enhanced through creative landscaping while undesirable features can be downplayed. The sounds that a landscape offers, like a breeze rustling the leaves in the trees or the sounds of birds or of water splashing in a fountain, enhance the aesthetic qualities of your home environment. The aroma of flowers or the smell of a freshly mowed lawn and even the taste of fruit from plants that we might have in the landscape are soothing. The sense of touch also can be an aesthetically valuable feature of the landscape. Consider lying on the lawn in the shade of a stately oak, the feel of the cool grass on your back on a hot summer day.

2. Economic Value

The well-done landscape adds economic value to our home and property. The value of our home can be increased by as much as 6–15% as a result of a good landscape. However, the landscape is not the most valuable feature of your property; the house is. The function of the landscape is to enhance the beauty and therefore the economic value of our house. Thoughtful landscaping can also reduce energy bills by buffering seasonal temperatures. In addition, trees and shrubs can be used to reduce wind speed, making our outdoor living area more comfortable.

3. Functional Value

Landscape design offers a special functional value, too. Well-placed trees, shrubs, turf, and construction features increase your use of the property. A little shade in the right place, a little sun in another, a place for the kids to play, a private patio, pool, or deck all add to the enjoyment of being outside. Landscape design helps us solve landscape problems and cut down on maintenance. For example, ground cover used on a steep hill in the yard can help us avoid lawn maintenance headaches and, on a very steep slope, groundcover may be essential to prevent erosion.

4. Environmental Value

The landscape not only has functional worth, but it can enhance the environment. Through careful landscaping, temperatures can be buffered in the summer and winter. Glare and wind can be reduced, and water can be used more efficiently. In addition, plants in the landscape help clean the air of dust and some pollutants. Landscape design also provides a habitat for all kinds of wildlife (Helfand et al. 2006; Williams and Tilt 2006).

There are many attributes (e.g., vegetation and associated plants, spatial configuration of landscape elements, the topography and bodies of water among others) that determine the quality of a visual landscape (Bae 2011; Zheng et al. 2011). Additionally, the role of each attribute is dependent on the context and its interaction with the other attributes (Zheng et al. 2011). One of the important features of landscape in the context of the residential landscape is wilderness versus neatness. According to Zheng et al. (2011), most people agree that they prefer a neat environment than wilderness. Neatness is one of the factors for an attractive landscape for residential areas. However, according to Bae (2011), formulating and maintaining planting/vegetation is not easy because it requires consideration of not only the usual factors for general open space design such as ground-level ecological capacity and residents' preference but also plants' hydrophilic aspects. Different types of plants with varied heliophilicities should be cultivated in areas with various environments. Suitable plants are arranged in conformable areas based on plant characteristics (Guo et al. 2010).

The plants would not harm human beings and can be cultivated in multiple types of plants to form multi-level plant communities comprising of trees, bushes, and grasses based on local climate and natural distribution of plants and there should be at least three trees in every 100 m^2 of green land (Guo et al. 2010). The standards specify that the plant configuration should show local characteristics, the abundance of local plants resources, and plant landscapes with distinguishing features. Multi-level greening through a combination of trees, bushes, and grasses should be adopted to form a greening system with rich levels and favorable ecological benefits.

3 Issues Related to the Residential Landscape Design

Residential plays important roles in enhancing people's quality of life. This is because people's preferences when they buying/rental the residence and conducting their daily activities are closely related to the quality of life. However, there are some issues related to the residential landscape design that can hinder people from enjoying the facilities provided in residential areas. Abdul Mohit and Hassan Elsawahli (2010) stated that safeties become the major issues that influence people's decision. This is because crimes in residential area become common issues that happen nowadays. These have increased the fear within the community.

Other than that, issue on the age categories also affect people preferences in residential landscape design. This is due to urbanization causes stress in the community due to many factors. Regan and Horn (2005) state that the age factor also has become important factor that influence people preferences. They were conducted the study based on four group which is primarily students; aged between 18 and 25 years old, people who settling down in terms of career and family life; aged between 26 and 45 years old, people who starting to look ahead; aged between 46 and 60 years old, and people who already retire and close to that stage; aged 60 and above. The result shows that each of categories demonstrates different preferences as well as moods to the surrounding areas.

Apart from that, Ahmad (2003) mention that resident satisfaction is important since it can affect the social integration. Among the reasons are house compound; in terms of physical structures, lack of maintenance on public facilities, and poor social and physical environment which adversely affect the achievement of greater social integration. He also states that resident satisfaction can promote the process of socializing among people.

4 Residential Landscape for Green Neighborhood

The development of green neighborhood is related to the landscape design in order to generate a favorable living environment especially in town and urban areas. Based on the objective of the National Landscape Policy, the role of landscape design is important in providing a comfortable and quality housing area for each residential area. This initiative can be achieved through the idea of green neighborhood that can offer a neighborhood community with several green initiatives such as green corridor, green technology, and product that can create a sustainable neighborhood environment (Figs. 2 and 3). This initiative of green neighborhood can be applied in macro- and microscales of residential areas.

The overall planning of a residential area can be determined based on the pattern of the street design. There are five types of street patterns that are generally used, namely the grid, oblong grid type 1, oblong grid type 2, loops, and cul-de-sac patterns. A basic type of residential layout that has been identified in most areas in the Klang Valley area is oblong grid 2 pattern. Figure 4 shows an illustration of a typical pattern of residential development based on street pattern design.

Figure 4 shows that a residential area with oblong grid 2 has more buildable area than other types of residential patterns. This is because the area that can be used for the development of building and plinth area is bigger and leads to more of plinth area for landscape design purposes. Besides that, the minimum requirement is also influenced by the size of the plinth area of the house development. Plinth area can be identified as the built-up covered area measured at the floor level of the basement of the storey. It is stated in the Federal Territory Act 1982 (2006) that a management of plinth area allows more planning of public green area and children's play to be developed in residential areas. Figure 5 shows example of a plinth area for one

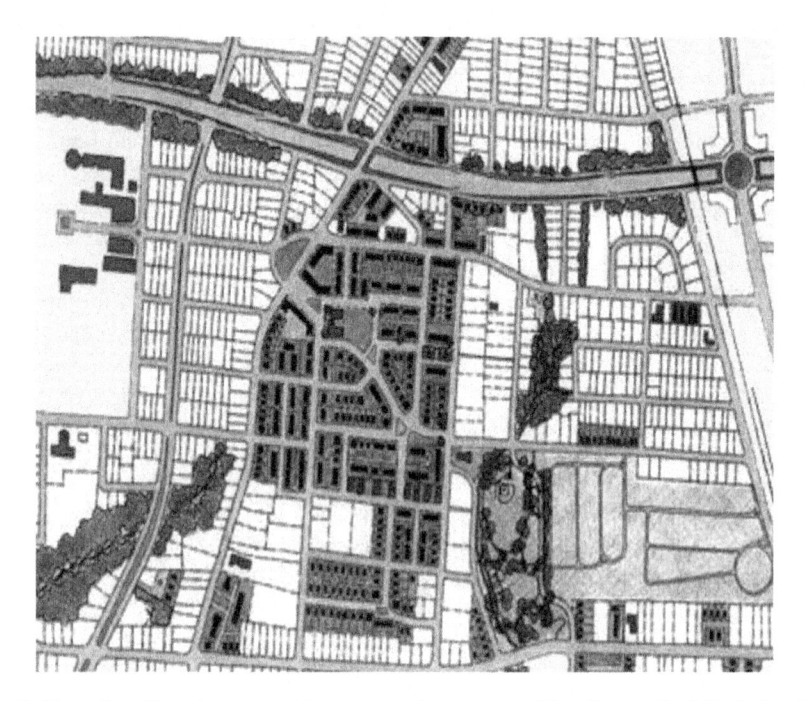

Fig. 2 Example of housing area with a systematic green corridor. *Source* Draf Garis Panduan Perancangan Perumahan (2011, p. 20)

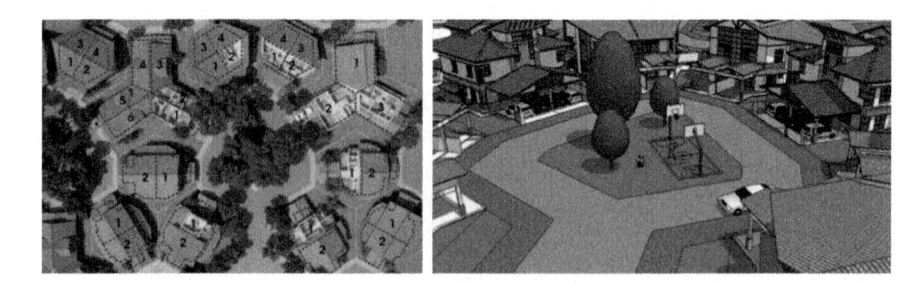

Fig. 3 A conducive arrangement of house layout and the cul-de-sac give an easy accessibility for users. Cul-de-sacs can also act as a place for interaction and recreation for residential communities. *Source* Draf Garis Panduan Perancangan Perumahan (2011, p. 17)

single building that consists of a front yard, a side yard, and a rear yard that can be used for landscape design purposes.

It is also stated in the Federal Territory Act 1982 (2006) that in order to identify the size of the plinth area for a housing area, two basic information need to be

	Square grid	Oblong grid	Oblong grid 2	Loops	Culs-de-sac
Percentage of area for streets	36.0%	35.0%	31.4%	27.4%	23.7%
Percentage of buildable area	64.0%	65.0%	68.6%	72.6%	76.3%

Fig. 4 A comparison of area for residential areas based on type of street pattern. *Source* http://www.cmhc-schl.gc.ca/publications/en/rh-pr/tech/socio75.html. Retrieved on 13th July 2012

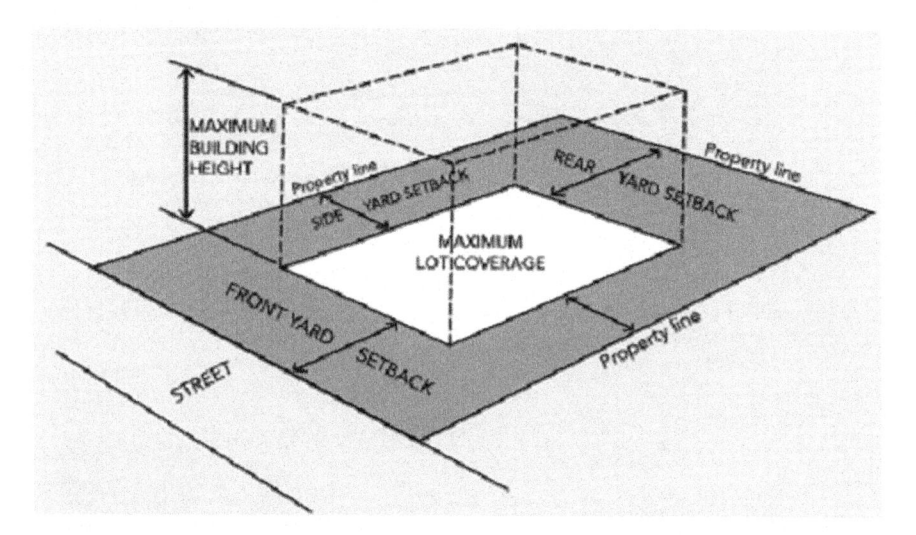

Fig. 5 Small building plinth area over an area of the site provides harmony with availability of more space on the ground floor in a development. *Source* Garis Panduan Pengiraan Nisbah Plot Dan Kawasan Plinth (2010, p. 3)

identified which are size for overall site/house area and size of built-up area. The size of plinth area is measured by Eq. 1.

$$\text{Plinth area} = \frac{\text{Size of built-up area}}{\text{Size of overall site/house area}} \times 100 \qquad (1)$$

The built-up area can be understood as an area surrounded by a building area that considers the perimeter of the external wall of the building. Structures that are not considered as plinth area are recreation facilities such as pergola, gazebo/*wakaf*, post guard, porch area, balcony, roof and fountain or pond (Fig. 6).

The minimum requirement of plinth area is different based on the type of house development area. As shown in Fig. 7, the bungalow type of residential area has a

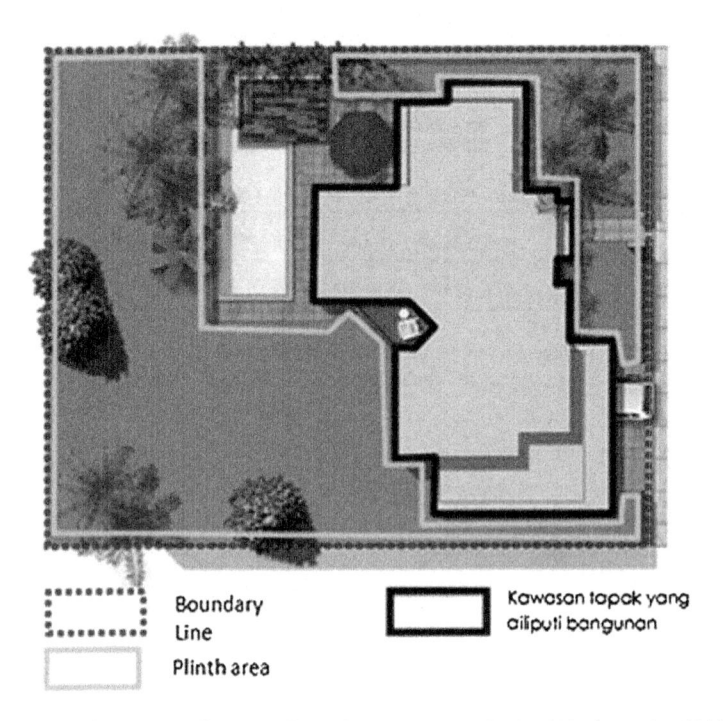

Fig. 6 Example of plinth area for type of housing area. *Source* Federal Territory Act 1982 (2006)

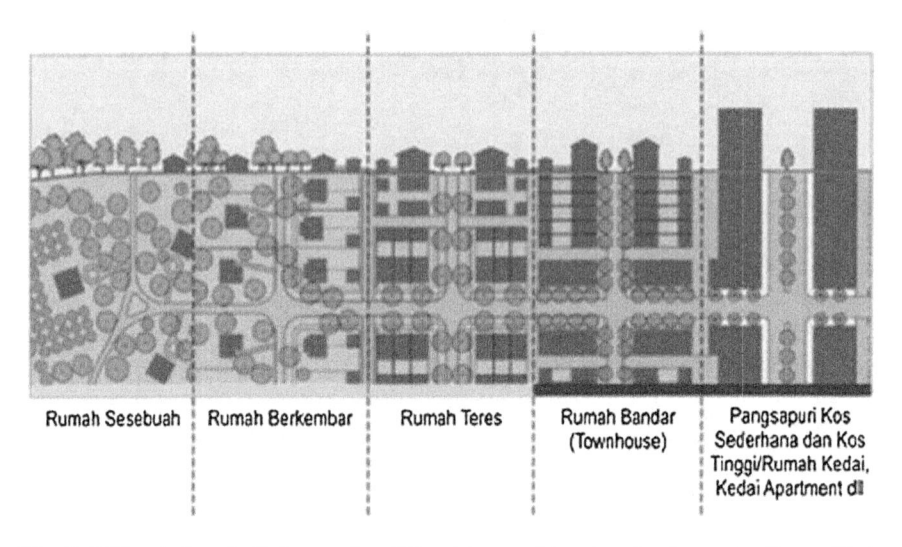

Fig. 7 Different size of plinth area for different types of housing developments. *Source* Garis Panduan Perancangan Kejiranan Hijau (2012, p. 28)

bigger plinth area compared to other types of residential houses. It is determined through the green area provided for landscape design in the house compound area.

5 Conclusion

The understanding of landscape design for residential is importance in creating a sustainable development for the residential area. The review on the issues and the concept of green neighborhood is essential to identify a basic needs and demands of landscape design in residential areas. Furthermore, the basic need of residential properties such as plants, open spaces, or recreation areas can become benchmarks for residential areas in applying more other landscape design methods in residential area. It is proven that proper landscape planning and design with sustainable concept and approach help to create conducive and responsive environment of the residential properties. People now look toward better environment as a package for their better living environment. However, limited green spaces within the proximity of residential properties are not supporting the landscape space and are not conducive for living space. It is stated in the Guidelines for Housing Design and Landscaping (2006), the overall value of residential properties needs to be enhanced by skillful landscape design and standard garden maintenance. Furthermore, the initiative to create a better living environment for residential communities has been supported by Local Agenda 21. Local Agenda 21 (LA21) is a program where the public and private sectors, the District Councils, the Municipal Councils, the City Councils, and the City Hall, cooperate in planning and managing the environment to achieve sustainable development. Besides that, through the Local Agenda 21, local communities can be involved to identify and examine the issues of sustainable development. Sustainable development is development that balances economic development with the requirements of social and environmental. Therefore, the importance of landscape design in creating a sustainable development has now become an issue that should be made aware of to all people and organizations. Moreover, the element and character of landscape design need to be enhanced in most of residential developments in order to support the needs of neighborhood communities. The basic need of landscape design in residential areas can become a benchmark in creating a sustainable residential environment in the future.

References

Abdul Mohit, M., & Hassan Elsawahli, H. M. (2010). Crime and housing in Malaysia: Case study of Taman Melati terrace housing in Kuala Lumpur. *Asian Journal of Environment-Behaviour Studies, 1*(3), 25–36.

Ahmad, H. H. (2003). Residential satisfaction and social integration in public low cost housing in Malaysia. *Pertanika Journal of Social Sciences and Humanities, 11*(1), 1–10.

Bae, H. (2011). Urban stream restoration in Korea: Design consideration and residents' willingness to pay. *Urban Forestry & Urban Greening, 10*, 119–126 (Elsevier).
Draf Garis Panduan Perancangan Perumahan. (2011). Jabatan Perancang Bandar dan Desa.
Federal Territory Act 1982. (2006). The commissioner of law revision, Malaysia.
Gause, J., Franko, R. A., Heid, J. J., Kellenberg, S., Kingsbury, J., McMahon, E. T., et al. (2007). *Developing sustainable planned communities*. Washington D.C: Urban Land Institute.
Garis Panduan Pengiraan Nisbah Plot dan Kawasan Plinth. (2010). Jabatan Perancang Bandar dan Desa.
Garis Panduan Perancangan Kejiranan Hijau. (2012). Jabatan Perancang Bandar dan Desa.
Guideline for Housing Design and Landscaping. (2006). Waterways. Live on the water, p. 14.
Guo H. F., Ge, J., Yue, M., Zhou, X., & Jin, W. (2010). Landscape design method for a green community based introduction to green wall technology, benefits & design on green building design theory. *Journal of Zhejiang University-Science (Applied Physical & Engineering), 11*(9), 691–700 (China).
Helfand, G. E., Park, J. S., Nassauer, J. I., & Kosek, S. (2006). The economics of native plants in residential landscape designs. *Landscape and Urban Planning, 78*, 229–240 (Elsevier).
Hussain, M. R. M., Nizarudin, N., & Tukiman, I. (2014a). Landscape design as part of green and sustainable building design. *Advanced Materials Research, 935*, 277–280.
Hussain, M. R. M., Tukiman, I., & Zen, I. H. (2014b). Landscape design as added value in Klang Valley. *Advanced Materials Research, 935*, 281–284.
Hussain, M. R. M., Tukiman, I., Zen, I. H., & Shahli, F. (2014c). The impact of landscape design on house prices and values in residential development in urban areas. *APCBEE Procedia, 10*, 311–315.
Regan, C. L., & Horn, S. A. (2005). To nature or not to nature: Associations between environmental preferences, mood states and demographic factors. *Journal of Environmental Psychology, 25*, 57–66.
Smith, C., Clayden, A., & Dunnett, N. (2008). *Residential landscape sustainability; A checklist tool*. Blackwell Publishing.
The Green Building Evaluation Standards. (2006). National Standard of the People's Republic of China
VanDerZanden. A., M., & Rodie, S., N. (2008). *Landscape design; Theory and application*. Canada: Thomson Delmar Learning.
Williams, J. D., & Tilt, K. (2006). *Residential landscape design*. Alabama Cooperative Extension System. ANR-813. www.aces.edu.
Zheng, B., Zhang, Y., & Chen, J. (2011). *Preference to home landscape: Wildness or neatness? Landscape and Urban Planning, 99*, 1–8 (Elsevier).

A Review on Tropical Green Roof Maintenance Practice of High-Rise Residential Buildings

Shafikah Saharuddin, Natasha Khalil and Alia Abdullah Saleh

Abstract The Malaysia Plan is slowly moving towards the green growth for sustainability and resilience to develop a green environment and culture. Few initiatives have been developed, and one of them is with the development of Green Building Index (GBI) as Malaysia's green building rating system where the green roof is also the important criteria needed to achieve green status. Hence, the green roof has received significant attention in Malaysia over the years especially in the environmental performance of the system. However, the maintenance management of green roofs is often being neglected and does not being looked in. There are also lack of scientific research regarding the maintenance management practice of green roof in this tropical climate. This research is carried out to assess the green roof maintenance practice in high-rise residential buildings and to analyse the guidelines and codes being applied in other developed countries. The research has been conducted through an extensive reading on the literature from journals and articles. From the literature reviews, there are few successful green roof projects in high-rise residential buildings located in urban cities of Kuala Lumpur, and the current maintenance practice of green roofs is also being analysed. Generally, it can be

---

S. Saharuddin (✉)
Centre of Postgraduate Studies, Faculty of Architecture, Planning and Surveying,
Universiti Teknologi MARA, Perak Branch, Seri Iskandar Campus,
32610 Bandar Baru Seri Iskandar, Perak, Malaysia
e-mail: shafikahsaharuddin@gmail.com

N. Khalil
Department of Quantity Surveying, Faculty of Architecture, Planning
and Surveying, Universiti Teknologi MARA, Perak Branch, Seri Iskandar
Campus, 32610 Bandar Baru Seri Iskandar, Perak, Malaysia

A. A. Saleh
Department of Building Surveying, Faculty of Architecture, Planning
and Surveying, Universiti Teknologi MARA, Perak Branch, Seri Iskandar
Campus, 32610 Bandar Baru Seri Iskandar, Perak, Malaysia

concluded that maintenance works in high-rise residential buildings are being out-tasked to the related teams in order to maintain the green roof as well as the building in a proper way. A better maintenance management practice will inspire greener roof implementation in this country in future.

Keywords Green roof · Maintenance practice · High-rise residential buildings Tropical climate

1 Introduction

Towards achieving an advanced Malaysian economy by 2020, the country has come out with a momentous milestone of the Eleventh Malaysia Plan 2016–2020. The Eleventh Plan has developed six strategic thrusts and six game changers, where one of them is focusing on pursuing green growth for sustainability and resilience. This is where the country will embark on green growth and it will be a way of Malaysians' life. Not only the public and government sectors who are trying to take the initiatives in developing the green environment and culture, but also the private companies and NGOs. For instance, Kementerian Tenaga, Teknologi Hijau dan Air (KeTTHA) prime objective is to enhance the green technology and environment (KeTTHA). Meanwhile, the Institute of Architect Malaysia/Pertubuhan Arkitek Malaysia (PAM) and the Association of Consulting Engineers Malaysia (ACEM) have developed the Green Building Index (GBI) (Ting 2012) as a rating tool for evaluating and valuing the green status of a building. Malaysia Green Technology Corporation (GreenTechMalaysia) also has been established to enact the legal mechanisms to regulate and enforce green technology, and to define the role of every government agency involved in the implementation of green technology in the country (Rahman et al. 2013).

Other than that, there are also lot of exhibitions and conferences on green technology that have been held and organized in Malaysia itself, for example the region's largest green technology exhibition and conference, International Greentech and Eco Products Exhibition and Conference Malaysia (IGEM), organized by Ministry of Energy, Green Technology and Water (KeTTHA) every year since 2010. This shows that Malaysia has noticed and is moving towards environment sustainability living with the prominent need of energy efficiency, water efficiency, sustainable and planning management, material and resources, indoor environmental quality as well as new innovation. Although the GBI rating tool was just being introduced, it can act as a benchmark in order for the building to achieve green status. One of the criteria of the GBI rating assessment is the installation of green roof (Fauzi et al. 2013). Hence, it may lead to the emergence of green roof technology in Malaysia and help the country to implement as well as encourage its people to move towards the green revolution and technology of green roof system.

Even though green roof application in Malaysia is still at the preliminary stage, there are some successful green roof applications applied in few buildings in Malaysia (Ismail et al. 2012). Nevertheless, the application of green roofs in the country is still low but in urban cities, it has slowly becoming a trend in building industry to have such green roofs or rooftop gardens in a certain building due to the merits it offers. Based on the previous studies, among the benefits of green roof are on hydrological performance regarding run-off retention and storage, carbon sequestration, noise reduction, building and urban cooling, pollutant trap, food production, biodiversity and social benefits where it can be classified into physical impacts, ecological impacts and socio-economic impacts (Hoang and Fenner 2014). With the various benefits available, it pointed out that green roof should be promoted and used extensively over the country.

Nevertheless, there is an issue where the non-existence of framework or guidelines specify on the green roof maintenance in our country, Malaysia is not being observed and explored intensely where it is always been denied. Unlike other countries that are very advance in green roof technologies such as Germany, Canada, Japan and America, they already have their own guidelines for green roof system (Hui 2010). Meanwhile, in United Arab Emirates (UAE), Dubai also has already evolved a design guideline by Dubai Municipality with cooperation of Dubai Green Roof (Aziz and Ismail 2011). Moreover, a brief dialogue with Ishamuddin Mazlan, the Minister of Energy, Green Technology and Water (KeTTHA), stated that there are not yet been introduced the design guidelines for green roof systems in Malaysia and no explicit design standards and guideline stipulated to the Malaysian industry on green roof systems (Aziz and Ismail 2011; Chow and Abu Bakar 2016; Fauzi et al. 2013; Ismail et al. 2008a, b). Even though the Town and Country Planning Department (JPBD) has initiated various planning guidelines that included green roofs since 2004 (Zahir et al. 2014), the green roof system in Malaysia does not been looked in and explored deeply and the awareness on this matter is often been denied.

Even so, the use of green roofs varies according to the temperature and climate of certain places. The hot and humid weather in Malaysia will influence the types of vegetations used as well as its operation and maintenance systems. As compared to Singapore, Malaysia is said to be far behind in terms of policies, research done, technology and practice in this area; because with the same climate and weather, Malaysia is lacking of incentives, guidance as well as the government and green building standard organization supports. This shows that the lack of knowledge and research on maintenance system mainly for tropical climate has led to a zero interest in using the system among the building owners and developers. In managing green roof system, its maintenance can be said to be a crucial part of the process after the construction and installation stage. This paper discussed the practice of maintenance for green roof in high-rise residential buildings mainly in urban city of Kuala Lumpur, Malaysia.

2 Literature Review

2.1 Overview of Green Roof

2.1.1 Definition of Green Roof

Green roof is defined as a flat or sloping rooftop designed to help support the vegetation despite it is being designed to provide a fully functioning roof to a building (Ismail et al. 2008a, b; Kamarulzaman et al. 2014). However, Aziz and Ismail (2011) and McIntyre and Snodgrass (2010) stated that green roof is where it uses plants on the top layer of the roof or roof covered by vegetation in a layer of soil or growing medium to enhance roof performance as well as its appearance or both. Meanwhile, it is also clarified as a vegetated roof system where growing plants on the rooftops in order to replace the vegetated footprint that was decimated as the building were built (Chow and Abu Bakar 2016; Fauzi and Malek 2013; Getter and Rowe 2006). In short, green roof is basically a flat or sloping roof covered and layered with vegetation in order to maximize the use of green space especially in urban areas. It can be divided into two types which are extensive and intensive green roofs.

2.1.2 Types of Green Roof

Extensive green roof is thinner, simple and lighter kind of green roof where normally it only needs shallow substrate of a depth about 15 cm. Basically, this green roof is inaccessible to the public and requires a low cost. It is highly demand in Europe as proven by Philippi more than 80% of green roof is of the extensive type in Germany (McIntyre and Snodgrass 2010). Meanwhile, the intensive green roofs or also known as roof gardens are heavier, strong enough to support additional load, and require deeper substrate of a minimum depth of 15 cm. It is usually accessible to the public and is high in cost as it requires intensive maintenance.

2.2 High-Rise Residential Buildings

The scarcity of land in the urban areas such as Kuala Lumpur, Selangor, Penang and Johor Bahru has increase the demand for the development of houses. As for these high-density areas, the landed residential properties are very costly. Thus, individuals nowadays tend to have a higher demand on the non-landed residential properties or also refer to high-rise residential buildings rather than the landed residential properties. Hence, there are many developments of high-rise residential buildings in urban areas as compared to the development of landed residential buildings (Ta 2006; Wahab et al. 2015).

High-rise residential building is mainly a type of housing that has multi-dwelling units built on the same lot of land (Wahab et al. 2016). Meanwhile, Craighead (2009) defined high-rise residential building as multi-story structure of around seven stories or higher and contains separate residences where a person may live or regularly stay. However, Malaysia has no national building code or guidelines clarifying the minimum height or number of floors of high-rise buildings, but Mirrahimi et al. (2016) stated that the height of high-rise buildings in Malaysia ranges between 5 and 88 stories. In spite of that, Lau et al. (2016) described high-rise building in his study based on International Building Code IBC 2009 and National Fire Protection Association NFPA code, where both codes interpret high-rise buildings as buildings with a minimum height of 75 feet (22.9 m) above ground level which means that the minimum number of floors acceptable as high-rise in the study is seven. It may also be known as an apartment, a residence, a tenement or a condominium.

Basically, this type of building provides basic facilities, such as covered parking, garbage chute, elevators, 24-h security system, swimming pool, attractive landscape garden. Residents living in high-rise residential buildings are presumed to share the entire basic amenities provided by the management corporation, and in order to maintain the good condition of those facilities, they are expected to follow the rules and regulations stipulated (Aizuddin et al. 2010; Wahab et al. 2015).

2.3 Overview of Maintenance

2.3.1 Definition of Maintenance

As the buildings aged, they require maintenance. Nowadays, it is a truism that no building is maintenance-free. Even under the "Pekeliling Am Bil 1 Tahun 2003, Arahan Penyelenggaraan Bangunan Kerajaan Di Putrajaya dated February 11th 2003", the Prime Minister Department has given an instruction saying that the building maintenance must be efficiently and properly executed (Mohd-Noor et al. 2011).

Different people have different perceptions on the definition of maintenance. Maintenance is defined as the required processes and services undertaken to preserve, repair, protect and care for a building's fabric and services after the completion, repair, refurbishment or replacement to current standards in order to allow it to fulfil its intended functions throughout its entire lifespan without drastically upsetting its basic features and use (AbdulLateef 2010). However, Idrus et al. (2010) clarify maintenance as the continuous care and protection entailing minor repair works done to certain elements of a building so that it remains in good condition and at the same time extending the life of the element and entire building for as long as possible with significant administrative and managerial expertise. Meanwhile, maintenance can also be explained as a process where the structure and component of a building undergo reservation and restoration activity (Akasah et al. 2008). In a simple word, maintenance is the work that is done to maintain an asset such as roof, in order to preserve its continuity of use and function to a reasonable level of performance without any unforeseen renewal or major repair activities.

2.3.2 Building Maintenance Strategy

Generally, the owners or the developers are fully responsible for the maintenance as well as the safety factors regarding the green roof maintenance in private buildings such as residential buildings. They have to ensure that the green roof is being maintained to a set standard. Traditionally, building maintenance strategies can be divided into corrective, preventive and condition-based maintenance strategies.

Corrective maintenance is done when an element of the building breaks down and cannot be functioned anymore. Usually, the element of the building is used until it fails to perform its required function and then maintenance is applied. According to David and Arthur as cited in Horner et al. (2009), corrective maintenance tasks commonly take places in an ad hoc manner where it happens when necessary and not planned in advance in response to breakdowns or as requested by users. Corrective maintenance is sometimes referred to as failure-based or unplanned maintenance (Fawaz 2013). Meanwhile, preventive maintenance is carried out at a predetermined plan at regular, fixed intervals, which may minimize the probability of occurrence of failure and preventing sudden failure. It is also referred to as time-based maintenance, planned maintenance or cyclic maintenance (Fawaz 2013; Horner et al. 2009) as the maintenance cycle is planned according to needs. Kelly and Harris as cited in Fawaz (2013), Horner et al. (2009) clarify condition-based maintenance as "Maintenance carried out in response to a significant deterioration in a unit as indicated by a change in monitored parameter of the unit condition or performance". It is performed as there are changes in the condition or performance of an element from a normal to an abnormal condition. Before the maintenance is carried out, it is being monitored first in order to determine which part of the elements is about to deteriorate by using suitable condition monitoring tools and techniques.

Meanwhile, when maintaining the green roof, plants, waterproofing membranes, cost of the maintenance and inspection are the most important factors to be considered. The predictive maintenance in building with green roof involves the planning of inspections, both visual and monitoring. A visual inspection is the most common one where the maintenance actions are being identified in the survey. It includes the checking of any blockage of the drainage system or the control of infested vegetations. These measures are crucial to diagnose visible anomalies.

Meanwhile, monitoring inspections incorporate with the in situ campaigns to evaluate the in-service performance of each maintenance source element. Basically, they need complementary diagnosis techniques. For instance, the indirect information about any inner anomaly can be gained through the measurement of heat fluxes or superficial temperatures and moisture of the ceiling slab (Coelho et al. 2015). Finally, preventive maintenance actions will enhance the in-service performance of green roofs which include regular cleaning, repairs or replacements. And most importantly, all maintenance to be carried out at roof level ought to be in full compliance with the suitable health and safety regulations, and specifically dealing with those working at height.

2.3.3 Maintenance Procurement Strategy

The maintenance procurement strategy is divided into two, which is out-source or in-house maintenance. Maintenance of facilities and services can be done either in-house or out-source. It depends on the organization to choose which maintenance procurement strategy they want to use as there are no specific rules on this matter. Furthermore, it also has been agreed by Barret as mentioned in Kamarazaly that there are no hard and fast rules involving what should be remained in-house and what should be contracted out.

Outsourcing is a procedure where an organization employs a separate company, under a contract to carry out a certain work or task which includes the people and its management responsibility. Other than that, outsourcing is also referring to contracting out, which means the facilities management services that rely upon an organization is being placed into the hands of external service providers (Atkin 2003). It is also referring to where the certain activities or tasks are being out-tasked to the external suppliers or companies.

Meanwhile, in-house is clarified as an approach that continues to deal internally with product or services that demand skill and knowledge in order to cater for customer's needs. In-house is also defined as a "service that is provided by a dedicated resource directly employed by the client organization, where monitoring and control of performance are normally conducted under the terms of conventional employer/employee relationship, although internal service-level agreements may be employed as regulating mechanisms" (Barrett and Baldry 2003, p. 17). From a somewhat different perspective, Atkin (2003) described in-house as the retained personnel of the organization set up as a team, group or division and is under the direct control of senior management. In a better explanation, in-house procurement strategy is where an organization of the building itself carries out the services of the building. The services are directly employed by the client organizations, but the monitoring and control of performance are under the employer.

2.4 Guidelines and the Climate Conditions

Other cities such as Germany, Canada, Japan and America have already established their own standardized guideline (Aziz and Ismail 2011), for example the FLL guidelines, GRO Green Roof Code, ASTM International, Hong Kong Planning Standards and Guidelines, The International Green Roof Association (IGRA) and many more. The same goes to United Arab Emirates (UAE), Dubai; with the cooperation with Dubai Green Roof they have already managed to develop their own design guidelines on green roof which includes the planning, execution and maintenance of green roof as well as green wall. Thus, Malaysia should start moving towards a guideline that is essential for the installation of green roof particularly regarding to the maintenance works.

Many researches on green roof have been carried out for the past few years. Hence, they can act as references and a great source of information to the implementation of such guidelines in Malaysia. For instance, a study by Tolderlund has discovered green roof design, implementation and maintenance recommendations and requirements which can be applied to green roofs in the semi-arid and arid regions. Meanwhile, there is another researcher Velazquez explores on the design of green roofs in America with details on system components, maintenance, cost issues and the range possible applications (Aziz and Ismail 2011). On the other hand, Hui (2010) has developed technical guidelines for green roof systems in Hong Kong. All these guidelines are unsuitable and inappropriate to be applied in Malaysia as there is difference in types of environmental conditions between the countries, nevertheless they can be used as guidance in developing guidelines for green roofs in this country. At the present time, Malaysian industry still does not have a clear guideline on green roof systems.

The establishment of guidelines in adapting green roof system is usually being determined by using different criteria as each country has different characteristics especially in terms of climatic zones. An author concludes that the types of plant suitability must be assessed in terms of management and maintenance practices of different locations and climatic regions (Fauzi et al. 2013). In general, the climatic region of the earth consists of cold, temperate, dry and tropical climates. The tropical climate cities include Salvador (Brazil), Fort Lauderdale (USA), Pucallpa (Peru), Apia (Samoa), Bandar Seri Begawan (Brunei), Colombo (Sri Lanka), Kuala Lumpur (Malaysia), Medan (Indonesia) and Singapore (Ting 2012). Malaysia is also among the countries which are located in the tropical climate zone or also known as equatorial countries, where the green building benchmarks are a new phenomenon, but it is not in temperate countries. A tropical climate has no pronounced summer or winter; but it is hot and humid throughout the year with infrequent and heavy rainfall (Ting 2012). For this reason, the green roof in different climatic regions need different guidelines as the characteristics of each country is dissimilar. Hence, all these cities especially the Asian countries such as Singapore's guidelines, the Green Mark which is also the first green building rating system established for tropical climate, can be the best example for Malaysia in practicing green roofs in the country as they are in the same climate region as Malaysia.

3 Methodology

For this study, a comprehensive literature review was done on the green roof in high-rise residential buildings, and its maintenance towards the building and environment, technology and sustainability. The resources of the study are from leading journals and articles and also from conference proceedings. To ensure the most updated data on green roofs system, the articles and journals are mostly drawn from between the years 2006 and 2017. The papers selected and examined in order to identify the green roof projects for high-rise residential buildings and the current green roof maintenance practice in high-rise residential buildings in tropical climate. The maintenance practices of green roofs are then discussed based on the current scenario and the existing literature. Few of the existing guidelines on green roof system over the countries are also being referred to.

4 Analysis on Review and Discussion on Findings

4.1 Green Roof Projects in High-Rise Residential Buildings

Table 1 represents few of the green roofs that have been successfully installed in high-rise residential buildings, which are mostly located in urban cities such as Kuala Lumpur as stated by Ghazali et al. (2017) that Kuala Lumpur city area has the most high-rise buildings in Malaysia. Moreover, according to Rahman et al. (2013) based on his study, the most popular building category undertaken with green roofs projects in the country is residential with 46.7% compare to the other types of buildings. This is also being agreed by Zahir et al. (2014) that has concluded the recent trend of implementing green roofs and green facades in contemporary modern high-rise design in Kuala Lumpur. It is one of the efforts by the developers to enhance the environment as well as the quality of life.

4.2 Current Practices of Green Roof Maintenance

In the current practice of green roof in high-rise residential buildings in this country, most of the building owners would outsource or contracted out the maintenance

Table 1 List of green roof projects in high-rise residential buildings

Green roof project	Location	Type of green roof	Type of building	Level	Accessibility	Completion year
Kiara 9	Mont Kiara, Kuala Lumpur	Intensive	Condominium	Three and a half (consist of 16 gardens)	Private access	2011
Casa Desa Condominium	Taman Desa, Kuala Lumpur	Intensive	Condominium	Third	Private access	2008
The Saffron	Sentul East, Kuala Lumpur	Intensive	Condominium	Fourth	Private access	2008
Riana Green East	Wangsa Maju, Kuala Lumpur	Intensive	Condominium	Fourth	Private access	2009
The Tamarind	Sentul East, Kuala Lumpur	Intensive	Condominium	Fourth	Private access	2006
Setia Eco Villa	Shah Alam, Kuala Lumpur	Intensive	Bungalow	First	Private access	2007
Idaman Residence	Jalan P. Ramlee, Kuala Lumpur	Extensive	Condominium	Thirty-four	Private access	2008
Monte Bayu	Cheras, Kuala Lumpur	Intensive	Apartment	Seventh	Private access	…
Park Seven Condo	Persiaran KLCC, Kuala Lumpur	Extensive	Condominium	…	Private access	2008
Swiss Garden Residences	Jalan Pudu, Kuala Lumpur	Intensive	Apartment	Sixth	Private access	2011
Suasana Sentral Condominium	Sentral, Kuala Lumpur	Intensive	Condominium	Sixth	Private access	2002
Ritze Perdana 2	Petaling Jaya, Selangor	Intensive	Mixed-development (shops and condominium)	Sixth	Private access	2010
Perdana Exclusive	Petaling Jaya, Selangor	Intensive	Condominium	Fourth	Private access	…

(continued)

Table 1 (continued)

Green roof project	Location	Type of green roof	Type of building	Level	Accessibility	Completion year
Flora Damansara	Petaling Jaya, Selangor	Intensive	Condominium	Sixth	Private access	…
The Maple	Sentul West, Kuala Lumpur	Intensive	Condominium	Fifth	Private access	2006
Suasana Bangsar	Bangsar, Kuala Lumpur	Intensive	Condominium	…	Private access	2011
Sunway Vivaldi	Mont' Kiara, Kuala Lumpur	Intensive	Condominium	…	Private access	2011

works for green roof. This is because they need specialized organizations that are capable in handling out the works so that they will not damage the building and thus prolonging its lifespan. However, the supply of local materials, technology as well as the expertise related in the country is said to be very limited at the time being. As mentioned by Ismail et al. (2015) the current maintenance system of green roof in Malaysia is being carried out by building operators for the structural aspects but for the vegetation aspects, the operators would outsource it to the landscape consultants. It shows that the practice of maintenance for green roof in Malaysia is rather complex and is not standardized. Thus, to prevent the problems of leakages, mould growth, water ponding, mosquito breeding, present of unwanted animals, building defects that may shorten the lifespan of the roof and thus enforce more loads to the building structure and may ultimately lead to some other unwanted problems from occurring to the building, the managing agents have to do so despite the higher cost and unstandardized maintenance works. That is why it is very important to have local expertise in green roof systems that are specialized in installing, maintaining and problem solving (Ismail et al. 2012) in order to provide the industry with a more economical and affordable installation and maintenance cost of the green roof in the future.

4.3 Green Roof Guidelines, Manuals, Codes, and Standards in Developed Countries

Green roof application is a new method in building construction. Nevertheless, it has been widely used all over the world especially in the developed countries such as Germany, Japan and Singapore (Ismail et al. 2012) as well as in the UK, the USA, Canada and several European countries (Fauzi and Malek 2013). Japan, Singapore, Hong Kong and China are among few of the Asian countries which are currently active with the development of green roof in their countries. Malaysia, as one of the Asian country, has also begun to accept this system as the green roof may reduce the environmental problems such as global warming, urban heat island, as well as the energy consumption of a building. However, Malaysia's awareness on this matter and its implementation is still low because there are only a small number of buildings in Malaysia adopted green roof system despite the benefits it offers (Chow and Abu Bakar 2016; Zahir et al. 2014) as there is not yet any standardized guidelines being introduced. Hence, this shows that Malaysia is a little recede as compared to the other Asian countries such as Singapore, Japan, Hong Kong and United Arab Emirates regarding this system. Nevertheless, these may help Malaysia in developing guidelines, manuals, codes and standards for green roof implementation (Table 2).

Table 2 Legend for contents of guidelines

	Description		Description
a	Introduction to green roof	g	Installation techniques/executions/constructions
b	Ways to implement/planning	h	Maintenance
c	Policies and programs	i	Safety
d	Design consideration	j	Best practice
e	Specifications	k	Benefits
f	Selection of plants	l	Market green roof/cost

5 Conclusion

From the literature reviews, the practice of green roof is not being used extensively in this country even though the awareness on this matter has increased over the years. There are only few buildings that have been successfully installed with green roof, and the maintenance is often being neglected where sometimes it is being assigned to the unprofessional teams. Since maintenance plays a crucial role in sustaining green roofs for its intended function, a progressive effort should be introduced among the Malaysian researchers to conduct more research focuses on green maintenance management system in order to prolong the lifespan of a building with green roofs system in tropical climates. This is because different cities are located at a different zones and climates, thus the maintenance practice also differs. It is very crucial to improve the current maintenance practice of green roof in this country in order to achieve sustainability and enhance the performance of a healthy living ambience of the owners and residents of high-rise buildings installed with green roofs in the cities.

References

AbdulLateef, O. A. (2010). Quantitative analysis of criteria in university building maintenance in Malaysia. *Australasian Journal of Construction Economics and Building, 10*(3), 51–61.

Aizuddin, N., Noor, M., & Eves, C. (2010). Malaysia high-rise residential property management: 2004–2010 Trends & Scenario. In *17th Pacific Rim Real Estate Society Conference*, 2004–2010.

Akasah, Z. A., Shamsuddin, S. H., Rahman, I. A., & Alias, M. (2008). School building maintenance strategy: A new management approach (pp. 1–5).

Atkin, B. (2003). Contracting out or managing services in-house. *Nordic Journal of Surveying and Real Estate Research—Special Series, 1*, 18–33.

Aziz, H. A., & Ismail, Z. (2011). Design guideline for sustainable green roof system. In *2011 IEEE Symposium on Business, Engineering and Industrial Applications (ISBEIA), Langkawi, Malaysia Design* (pp. 198–203).

Barrett, P., & Baldry, D. (2003). *Facilities management: Towards best practice* (2nd ed.). Oxford: Blackwell Publishing Ltd.

Chow, M. F., & Abu Bakar, M. F. (2016). A review on the development and challenges of green roof systems in Malaysia. *International Journal of Civil, Environmental, Structural, Construction and Architectural Engineering, 10*(1), 16–20.

Coelho, A., Silva, C. M., & Flores-Colen, I. (2015). Green roofs in mediterranean areas e survey and maintenance planning. *Building and Environment, 94,* 131–143. https://doi.org/10.1016/j.buildenv.2015.07.029.

Craighead, G. (2009). High-rise security and fire life safety chapter 2: High-rise building definition, development, and use, August 2005.

Fauzi, M. A., & Malek, N. A. (2013). Green building assessment tools: Evaluating different tools for green roof system. *International Journal of Education and Research, 1*(11), 1–14.

Fauzi, M. A., Malek, N. A., & Othman, J. (2013). Evaluation of green roof system for green building projects in Malaysia. *International Journal of Environmental, Chemical, Ecological, Geological and Geophysical Engineering, 7*(2), 75–81.

Fawaz, B. (2013). Maintenance management of green buildings in Nairobi County, June.

Getter, K. L., & Rowe, D. B. (2006). The role of extensive green roofs in sustainable development. *HortScience, 41*(5), 1276–1285. https://doi.org/10.17776/csj.30292.

Ghazali, A., Salleh, E. I., Haw, L. C., Mat, S., & Sopian, K. (2017). Performance and financial evaluation of various photovoltaic vertical facades on high-rise building in Malaysia. *Energy and Buildings, 134,* 306–318.

Hoang, L., & Fenner, R. A. (2014). System interactions of green roofs in blue-green cities. In *13th International Conference on Urban Drainage, Sarawak, Malaysia,* September, 7–12, 2014.

Horner, R. M. W., El-Haram, M. A., & Munns, A. K. (2009). Building maintenance strategy: a new management. *Journal of Quality in Maintenance Engineering, 3*(4), 273–280. https://doi.org/10.1108/13552519710176881.

Hui, S. C. M. (2010). Development of technical guidelines for green roof systems in Hong Kong, November, 1–8.

Idrus, A., Khamidi, F., & Sodangi, M. (2010). Maintenance management framework for conservation of heritage buildings in Malaysia. *Modern Applied Science, 4*(11). https://doi.org/10.5539/mas.v4n11p66.

Ismail, A., Samad, M. H. A., & Rahman, A. M. A. (2008a). Using green roof concept as a passive design technology to minimise the impact of global warming. In *2nd International Conference on Built Environment in Developing Countries (ICBEDC 2008)* (pp. 588–598). Retrieved from http://eprints.usm.my/25136/1/BUILDING_ENGINEERING_AND_CONSTRUCTION_7.pdf.

Ismail, A., Samad, M. H. A., & Rahman, A. M. A. (2008b). Using green roof concept as a passive design technology to minimise the impact of global warming, (Icbedc), pp. 588–598.

Ismail, Z., Aziz, H. A., Nasir, N. M., Zafrullah, M., Taib, M., & Alam, S. (2012). Obstacles to adopt green roof in Malaysia. In *2012 IEEE Colloquium on Humanities, Science & Engineering Research (CHUSER 2012)* (pp. 357–361).

Kamarulzaman, N., Hashim, S. Z., Hashim, H., & Saleh, A. A. (2014). Green roof concepts as a passive cooling approach in tropical climate—An overview. In *E3S Web of Conferences,* (Vol. 3, p. 7). https://doi.org/10.1051/e3sconf/20140301028.

Lau, A. K. K., Salleh, E., Lim, C. H., & Sulaiman, M. Y. (2016). Potential of shading devices and glazing configurations on cooling energy savings for high-rise office buildings in hot-humid climates: The case of Malaysia. *International Journal of Sustainable Built Environment, 1,* 387–399.

McIntyre, L., & Snodgrass, E. (2010). *The green roof manual: A professional guide to design, installation, and maintenance.* Timber Press, Inc. Retrieved from https://books.google.com/books?hl=en&lr=&id=tJE6AwAAQBAJ&oi=fnd&pg=PA6&dq=+The+Green+Roof+Manual:+A+Professional+Guide+to+Design,+Installation,+and+Maintenance&ots=wLzEvn3_Lv&sig=P4IcMDCs5NM4MXJUCUFHDo_fosY.

Mirrahimi, S., Mohamed, M. F., Haw, L. C., Ibrahim, N. L. N., Yusoff, W. F. M., & Aflaki, A. (2016). The effect of building envelope on the thermal comfort and energy saving for high-rise buildings in hot–humid climate. *Renewable and Sustainable Energy Reviews, 53,* 1508–1519.

Mohd-noor, N., Hamid, M. Y., Abdul-ghani, A. A., & Haron, S. N. (2011). Building maintenance budget determination: An exploration study in the Malaysia government practice. *Procedia Engineering, 20,* 435–444. https://doi.org/10.1016/j.proeng.2011.11.186.

Rahman, S. R. A., Ahmad, H., & Rosley, M. S. F. (2013). Green roof: Its awareness among professionals and potential in Malaysian market. *Procedia-Social and Behavioral Sciences, 85,* 443–453. https://doi.org/10.1016/j.sbspro.2013.08.373.

Ta, T. L. (2006). Managing high-rise residential building in Malaysia: Where are we? In *2nd NAPREC Conference, INSPEN* (pp. 1–25).

Ting, K. H. (2012). Tropical green building rating systems. In *2012 IEEE Business, Engineering and Industrial Applications Colloquium (BEIAC)* (pp. 263–268).

Wahab, S. R. H. A., Ani, A. I. C., Sairi, A., Tawil, N. M., & Razak, M. Z. A. (2016). *A survey on classification of maintenance fund for high rise residential building in Klang Valley, 200010200.* https://doi.org/10.1063/1.4960843.

Wahab, S. R. H. A., Sairi, A., Ani, A. I. C., Tawil, N. M., & Johar, S. (2015). Building maintenance issues: A Malaysian scenario for high-rise residential buildings. *International Journal of Applied Engineering Research, 10*(6), 15759–15776.

Zahir, M. H. M., Raman, S. N., Mohamed, M. F., Jamiland, M., & Nopiah, Z. M. (2014). The perception of Malaysian architects towards the implementation of green roofs: A review of practices, methodologies and future research. In *E3S Web of Conferences* (Vol. 3, p. 1022). EDP Sciences. https://doi.org/10.1051/e3sconf/20140301022.

Cycling! A Way Forward

Emy Ezura A. Jalil, Lau Siong Heng, Tan Song Jun and Fong Sim Ling

Abstract In Malaysia, traffic congestion remains to be a major problem, especially in developing regions Penang and Kuala Lumpur, which results in massive delays, fuel wastage, monetary losses as well as environmental pollution. In order to deter these issues, a sustainable transport mode should be promoted which is a bicycle. The present study intends to look into the relationship between infrastructural factors and cycling intention. Availability, route planning and accessibility are the predictors' variables, and user intention is the criterion variable. The Ajzen's theory of planned behaviour (TPB) will be the underpinning theory in this research in order to understand the intention from the road users. The judgmental sampling method was used, and 400 usable data were collected from Penang population through a self-administered questionnaire. The relationship between availability of public bike (APB), route planning (RP) and accessibility of cycling-transit integration (ACT) is positively correlated with user's intention to cycle (CCB), and the strongest relationship was found between APB and CCB. Hence, the hypothesis testing proves that three variables are crucial in which findings show a significant relationship between each variable. The present study will be beneficial towards city planner in Malaysia by offering valuable information regarding the intention of residents in a congested area.

Keywords Urban transport · Infrastructures · Theory planned behaviour Sustainable transport

---

E. E. A. Jalil (✉) · L. S. Heng · T. S. Jun · F. S. Ling
School of Technology Management & Logistics, Universiti Utara Malaysia,
06010 Sintok, Kedah, Malaysia
e-mail: ezura@uum.edu.my

1 Introduction

Transportation brings significant effects towards our daily activities, despite; it is commonly being view negatively. Transportation is the largest end which contributes towards environment pollution such as global warming towards country (Maitra and Sadhukhan 2013). Generally, these emissions are generated from the automobiles on traffic, especially during congestion (Maitra and Sadhukhan 2013). There is a serious need to reduce the numbers of vehicles on traffic, for the purpose of congestion free and green environment. A general consensus indicating the 'right' urban design would most probably stimulate the use of public transport, resulting in a reduction of cars on traffic (de Souza et al. 2014). In the world where environmental sustainability, congestion and physical inactivity are significant concerns, cycling presents itself as a tenable approach to address these issues.

Investment has been made for the purpose of promoting cycling culture in Penang Island, Malaysia, such as the Bike on Fridays (BoF); however, the connection of bicycle network seems to be disjointed which becomes a deterrent on cycling intention (Buehler and Pucher 2012). Few notable studies have been conducted on cycling infrastructure and cycling intention (Ariffin 2016). Findings of these studies have reported that cycling infrastructures are crucial determinants towards cycling intention. Further, most of the research usually focuses on the relationship between behavioural aspect (attitudes, social norms, weather conditions) and cycling intention (Milković and Štambuk 2015). Despite considering both of these aspects, there is limited research conducted in Malaysia context.

Hence, this study intends to look into the relationship between the infrastructural factors and cycling intention in Malaysia. The present study contributes to the research on infrastructural factors and cycling intention in both knowledge and practical aspects. By studying the infrastructural factors in local context, the findings of this study will further reassure whether the infrastructural factors that established from past researches can be generalized in Malaysia. While majority of studies are conducted in Western context, little is known on the effects of infrastructural factors on the Eastern culture such as Malaysia. The results of this study will enable the city planner in Malaysia to identify the strength and weaknesses of the present infrastructures provided to the users and make further necessary improvement in order to increase the intention to cycle as well as towards a more sustainable transport system in Malaysia.

2 Literature Reviews

2.1 Intention

The concept of intention has been an important topic especially in philosophical discussion since 1950s (Schröder et al. 2014). Although there are numerous debates on what are intentions, however in a broadly speaking, it can be defined in some way in which the mental state is linked to phenomena such as decision, desire and belief (Kaplan et al. 2015). The intention can be considered as intentional and unintentional. Milakis (2015), intentional intention is referring to a person's intended to do somethings on rationally whereby unintentional intention are unaware. Theory planned behaviour (TPB) has been one the most successful theories that has been widely applied in studying human behaviour (Milakis 2015). The main concept of this theory is the intention that represents if a person would perform a particular behaviour, how many efforts he or she is willing to invest (Milković and Štambuk 2015). Researchers usually refer to these three factors: attitude, subjective norm and perceived behavioural control (PBC) when discussing intention (Ajzen 2011; Kaplan et al. 2015).

Among the three factors, the present research tends to focus more on PBC instead the other two factors, where the questionnaire questions are also designed based on PBC. PBC refers to the perceived ease or difficulty of performing specific behaviour, reflecting past experience, as well as anticipating an obstacle which an individual is expected to overcome in order to perform that behaviour (de Souza et al. 2014). Cycling intention has been intensively discussed in past studies, subject in the area of transportation, particularly for cycle use (Kaplan et al. 2015; Fernández-Heredia et al. 2016). According to Milković and Štambuk (2015), Ajzen's theory of planned behaviour (TPB) is the most favourable theories applied to predict the cycling intentions.

Further, past studies have shown that availability and accessibility of the infrastructures have a positive effect on users' intention (Hazen et al. 2015; Kabra et al. 2015). Meta-analysis of 185 studies also showed that components of TPB, particularly perceived behavioural control, show the strongest relationship with behavioural intention ($r = 0.43$) (Milković and Štambuk 2015).

2.2 Availability of Public Bike

According to Kaplan et al. (2015), an easy access to cycle and availability constraints is associated with cycling intentions. This statement is further supported by a research conducted in Klang Valley, Malaysia, where the availability of public transport system as well as non-motorized mode of transport around neighbourhood would increase the use of the services (Khoo and Ong 2015). Public bike is becoming more popular (Shaheen et al. 2014) and recently emerged in major cities over the world rapidly (Vogel et al. 2011). Bike share began in Europe since 1965 and now it is available in 50 countries with 712 cities, operating approximately 806,200 bicycles at 37,500 stations (Shaheen et al. 2014). Bicycle is being one of the sustainable transport options where it can reduce the reliability on motorized vehicle especially in short-distance travel. Buehler and Pucher (2012), express in their new book City Cycling that using public bike or bike share are able to overcome the challenges such as car dependence, population health issues, livability, traffic congestion and oil dependence as well as to addressing climate change. Further, according to Fishman et al. (2015) public bike or bicycle share provides support for multimodal transport connection and acting as 'last mile' that connecting to public transport.

Moreover, Vogel et al. (2011), claim that bike availability is a crucial factor in encouraging people for the use of public bike or bike sharing system. Bike availability is then further being discussed by Kabra et al. (2015), as the likelihood of finding a bike at the station or the ease to access to the bike. The availability of bike is closely related to user satisfaction, as it would affect the decision to cycle. The findings of the research have suggested that the proportion of actively commuting was decreased mainly because of the decreases of bike use over the time (Grize et al. 2010). Moreover, Kabra et al. (2015), have also expressed their bike availability that is closely to the concept of operation management's service level. Hence, the bike availability is the probability of successful bike rental (Vogel et al. 2011).

2.3 Route Planning for Urban Cycling

In additional, route planning seems to be common in logistics context which generally is defined as planning for the best route for cost-effective and efficient distribution in which route planning needs to be able to quickly respond to any of the events occurred (Murray 2016). Referring to Jiang (2014), route optimization is crucial which could significantly increase productivity and efficiency, at the same time reduce costs. Past studies on route planning were usually focusing on algorithms or mathematical calculation, for calculating ideal logistics routes or networks (Sayarshad et al. 2012). However, the present study aims to seek whether route planning is one of the criteria that affect cycling intention. Based on the study of Shin, a sample about 1000 cyclists has been conducted based on two cities in Netherlands and Sweden which are Groningen and Vaxjo, respectively, and had shown that among several criteria, distance is the most important criteria affecting cycling route option.

2.4 Accessibility to Public Transit

Past studies have also shown that built environment such as dedicated cycle path actually influences cycling rates. The environmental factors such as the density of bike lanes are positively correlated with bicycle commute in US study. In addition, findings on Australia respondents have also shown that the accessibility to bicycle path as well as the facilities along road were positively associated with cycling for transport (Litman 2015). Hence, a comprehensive system of separated bicycle lanes or paths, providing cyclist reserved right of way, would probably be the most visible commitment for people to cycle (Fishman et al. 2015). In the logistics industry, accessibility is an important location factor which translates to lower transportation costs and shorter time to markets (Maitra and Sadhukhan 2013). Litman (2015) indicated that if transportation is evaluated based on accessibility then it would need to take into consider about rideshare, roadway, public transit, as well as the improved walking and cycling conditions. According to Litman (2015), accessibility combines the simple concept of mobility with an understanding that travel is driven by a desire to reach destination.

Further, Litman (2015) stated that the development and evaluation of accessibility policy are significantly related to urban structure, transport system quality, individual characteristics and purchasing power. This shows that the functions of accessibility play an important role in distribution activities, ability and desire of users to overcome the spatial separation between activities (Pucher et al. 2010). Hence, accessibility is essentially crucial for economic development as it enables the movement of people and goods to support the functioning of the economy (Ford et al. 2015). Litman (2015) suggests that transit services mainly provide basic mobility for non-drivers. Rahul and Verma (2014) stated that cycling has the limit distance for the residents ride on it. In order to facilitate cycle–transit coordination, installing bicycle infrastructures at the front of transit buses or allowing bicycles to be brought on board rail cars might be the alternatives that enlarge the public transit catchment areas (Midgley 2011).

Based on Litman (2015), the industries near transit accessible locations may help to access gap that can be created when clusters are promoted without regional multimodal accessibility in mind. Mingardo (2013) suggests that the more the park and ride of bicycle provided, residents might be more likely to use bike. A great cycling system has to link directly to the hierarchy of the road network (Ribeiro et al. 2014). Ribeiro et al. (2014) also indicate that cycling network is referring to the compatibility between existing traffic characteristics and the implementation of a bicycle infrastructure. The strategies to promote cycling can be grouped into the categories of travel-related infrastructure, end-of-trip facilities, transit integration, promotional and other programs, bicycle access and regulations (Pucher et al. 2010). Cycling facilities must be available in order to assure any form of inter-modality.

3 Methodology and Result

3.1 Research Methodology

The current research is conducted based on non-contrived setting where events occur in nature and do not involve the manipulation of respondents' experiences. A correlation study is to determine the relationship between predictors' variables (availability, route planning and accessibility) and the criterion variable (intention to cycle). Quantitative approach is used to conduct the present study. The sample size in the present study is 400 Penang residents. A judgmental sampling method was applied in this study due to present study mainly focusing on Penang's residents.

3.2 Result and Discussion

Demographic analyses implied that 42.5% of them are male and 57.5% are female. The proportion of ownership of private vehicle shows 53.8% owned and used private vehicle; however, 46.3% road users have not used and owned a vehicle to commute. As well as, findings show that 38.8% of respondents own a bicycle and 61.3% does not. The construction of the questionnaire is adopted from past studies (Hazen et al. 2015). Respondents were asked to use a 7-point Likert-type scale to indicate the extent to which they will intend to cycle if the infrastructures are well provided. Response choice alternatives ranged from 1 (strongly disagree) to 7 (strongly agree).

The study used chi-square ($X2$) tests, and the findings show that there is no major difference between nominal data which is demographics. However, results shown in Table 1 suggest there is still a difference in perception between gender towards cycling intention ($X2$ (11, $N = 400$) = 21.24, $p < 0.05$) and the relation between occupation and the variables in which occupation and ACT were found significant ($X2$ (15, $N = 400$) = 35.18, $p < 0.05$). Further, the test shows a difference in perception on ownership of bicycle towards cycling intention ($X2$ (11, $N = 400$) = 22.29, $p < 0.05$).

The present research objective is to determine the relationship between availability (APB), route planning (RP), accessibility (ACT) and user's intention to cycle (CCB). The correlations between constructs and scale reliability values are presented in Table 2. Findings from Table 4 indicates that all the predictors variables are significantly correlated with cycling intention (availability $r = 0.345$, route planning $r = 0.340$, accessibility $r = 0.295$, all $p < 0.05$). The strongest correlation was found between availability (APB) and cycling intention (CCB).

The result from this correlation test preliminary supports the alternative hypothesis that the variables have a significant relationship with user's intention to cycle. Findings have shown consistent with the previous study indicating bike availability (Vogel et al. 2011; Schlote et al. 2013), route planning (Kaplan et al. 2015)

Table 1 Chi-square test for gender differences

Item	Value	Df	Asymp Sig (two-sided)
Gender*MeanCCB	21.23	11	0.03
Occupation*MeanACT	35.18	15	0.00
Ownership*MeanCCB	22.29	11	0.02

Table 2 Correlations

Item	Mean	SD	APB	RP	ACT	CCB
APB	5.893	0.620	1	0.734**	0.530**	0.345**
RP	6.075	0.658	0.734**	1	0.535**	0.340**
ACT	5.872	0.696	0.530**	0.535**	1	0.295**
CCB	6.114	0.751	0.345**	0.340**	0.295**	1

**Correlation is significant at the 0.01 level (two-tailed)

and accessibility (Cheng and Liu 2012) are significant predictors towards the use of public bike.

A multiple regression analysis was conducted to test the extent to which the infrastructural factors influence user's cycling intention. The findings of the regression analysis (Table 3) for the current study hypothesis were statistically significant, ($R2 = 0.146$, Adjusted $R2 = 0.140$, F (3396) = 22.608, $p < 0.05$) which means that 14% of the variance in user's cycling intention is explained by the predictors variables (APB, RP, ACT).

The hypothesis of current study predicted that availability (H1a), route planning (H1b) and accessibility (H1c) are related to cycling intention. Further, based on Table 4, results show that the regression model was statistically significant ($R2 = 0.146$ F (3396) = 22.608, $p < 0.05$). By referring to Table 5, availability (APB) was positively related to user's intention to cycle (CCB) ($\beta = 0.171$, $p < 0.05$) indicating where the higher the availability of public bike, the higher the user's intention to cycle (CCB). Route planning (RP) ($\beta = 0.147$, $p < 0.05$) was positively related to CCB, which suggest that the higher the level of efficiency of the route planning, the higher the user's intention to cycle.

Finally, accessibility (ACT) ($\beta = 0.126$, $p < 0.05$) was positively related to CCB, showing that the higher the level of accessibility, the higher the user's

Table 3 Model summary

Model	R	R square	Adjusted R square	Std. Error of the estimate
1	0.38	0.15	0.14	0.70

Dependent variable: MeanCCB
Predictors: (constant), MeanAPB, MeanRP, MeanACT

Table 4 ANOVA

Model	Sum of square	Df	Mean square	F	Sig.
1 regression	32.92	3	10.97	22.61	0.00
Residual	192.20	396	0.49		
Total	225.12	399			

Dependent variable: MeanCCB
Predictors: (constant), MeanAPB, MeanRP, MeanACT

Table 5 Coefficient

	Unstandardized coefficients		Standardized coefficients		
Model	B	Std. Error	Beta	T	Sig.
1 (constant)	3.08	0.37		8.29	0.00
MeanAPB	0.21	0.09	0.17	2.43	0.02
MeanRP	0.17	0.08	0.15	2.09	0.04
MeanACT	0.14	0.06	0.13	2.22	0.03

Dependent variable: MeanCCB

intention to cycle. Hence, the findings of present study suggest that users will more likely commute by bike, if there is a quality and comprehensive cycling infrastructures.

4 Conclusion

To conclude, the outcome of this study indicates that the infrastructural factors are positively associated with cycling intention. Further, most of the infrastructures are not yet widely implemented in Malaysia, where respondent's view and thought would be different. The present study has been focusing on a single geographical location which is in Penang; however, the results might not generalize well to other states in Malaysia or overseas countries. Besides, respondents may have a varying knowledge regarding the present bicycle infrastructures which may affect their response to the present study.

Acknowledgments This work is an outcome from student's final year project on urban sustainable management.

References

Ajzen, I. (2011). The theory of planned behaviour: Reactions and reflections. *Psychology & Health, 26*(9), 1113–1127. https://doi.org/10.1080/08870446.2011.613995.
Ariffin, N. F. M. (2016). Analyzing the factors impact on social-behavioral aspects of cyclists in UPM Serdang Campus toward promoting green transportation system. *Research Journal of Fisheries and Hydrobiology, 11*(3), 74–81.
Buehler, R., & Pucher, J. (2012). Cycling to work in 90 large American cities: New evidence on the role of bike paths and lanes. *Transportation, 39*(2), 409–432.
Cheng, Y. H., & Liu, K. C. (2012). Evaluating bicycle-transit users' perceptions of intermodal inconvenience. *Transportation Research Part A: Policy and Practice, 46*(10), 1690–1706. https://doi.org/10.1016/j.tra.2012.10.013.

de Souza, A. A., Sanches, S. P., & Ferreira, M. A. (2014). Influence of attitudes with respect to cycling on the perception of existing barriers for using this mode of transport for commuting. *Procedia-Social and Behavioral Sciences, 162,* 111–120.
Fernández-Heredia, Á., Jara-Díaz, S., & Monzón, A. (2016). Modelling bicycle use intention: The role of perceptions. *Transportation, 43*(1), 1–23. https://doi.org/10.1007/s11116-014-9559-9.
Fishman, E., Washington, S., Haworth, N., & Watson, A. (2015). Factors influencing bike share membership: An analysis of Melbourne and Brisbane. *Transportation Research Part A: Policy and Practice, 71,* 17–30.
Ford, A., Barr, S., Dawson, R., & James, P. (2015). Transport accessibility analysis using GIS: Assessing sustainable transport in London. *ISPRS International Journal of Geo-Information, 4*(1), 124–149. https://doi.org/10.3390/ijgi4010124.
Grize, L., Bringolf-Isler, B., Martin, E., & Braun-Fahrländer, C. (2010). Trend in active transportation to school among Swiss school children and its associated factors: Three cross-sectional surveys 1994, 2000 and 2005. *International Journal of Behavioral Nutrition and Physical Activity, 7*(1), 1.
Hazen, B. T., Overstreet, R. E., & Wang, Y. (2015). Predicting public bicycle adoption using the technology acceptance model. *Sustainability (Switzerland), 7*(11), 14558–14573. https://doi.org/10.3390/su71114558.
Jiang, T. (2014). *The ins and outs of route optimization.* Retrieved October 20, 2016 from https://www.geotab.com/blog/ins-outs-route-optimization/.
Kabra, A., Belavina, E., & Girotra, K. (2015). Bike-share systems: Accessibility and availability. Chicago Booth Research Paper, (15-04).
Kaplan, S., Manca, F., Nielsen, T. A. S., & Prato, C. G. (2015). Intentions to use bike-sharing for holiday cycling: An application of the theory of planned behavior. *Tourism Management, 47,* 34–46. https://doi.org/10.1016/j.tourman.2014.08.017.
Khoo, H. L., & Ong, G. P. (2015). Understanding sustainable transport acceptance behavior: A case study of Klang valley, Malaysia. *International Journal of Sustainable Transportation, 9*(3), 227–239.
Litman, T. (2015). Evaluating accessibility for transportation planning measuring people's ability to reach desired goods and activities, January 2015, 57.
Maitra, B., & Sadhukhan, S. (2013). Urban public transportation system in the context of climate change mitigation: Emerging issues and research needs in India. In *Mitigating climate change* (pp. 75–91). Berlin Heidelberg: Springer.
Midgley, P. (2011). *Bicycle-sharing schemes: Enhancing sustainable mobility in urban areas,* (pp. 1–12). United Nations, Department of Economic and Social Affairs.
Milakis, D. (2015). Will Greeks cycle? Exploring intention and attitudes in the case of the new bicycle network of Patras. *International Journal of Sustainable Transportation, 9*(5), 321–334.
Milković, M., & Štambuk, M. (2015). To bike or not to bike? Application of the theory of planned behavior in predicting bicycle commuting among students in Zagreb. *Psychological Topics, 24*(2), 187.
Mingardo, G. (2013). Transport and environmental effects of rail-based Park and Ride: Evidence from the Netherlands. *Journal of Transport Geography, 30,* 7–16. https://doi.org/10.1016/j.jtrangeo.2013.02.004.
Murray, M. (2016). *Route planning for logistics and distribution companies.* Retrieved September 22, 2016 from https://www.thebalance.com/route-planning-2221322
Pucher, J., Dill, J., & Handy, S. (2010). Infrastructure, programs, and policies to increase bicycling: An international review. *Preventive Medicine, 50,* S106–S125.
Rahul, T. M., & Verma, A. (2014). A study of acceptable trip distances using walking and cycling in Bangalore. *Journal of Transport Geography, 38,* 106–113. https://doi.org/10.1016/j.jtrangeo.2014.05.011.
Ribeiro, P., Neiva, C. L., & Lemos, P. (2014). Planning process of a cycling network: Case study of Ponte de Lima, Portugal. In *Recent Advances in Environmental Science and Biomedicine* (pp. 168–174).

Sayarshad, H., Tavassoli, S., & Zhao, F. (2012). A multi-periodic optimization formulation for bike planning and bike utilization. *Applied Mathematical Modelling, 36*(10), 4944–4951. https://doi.org/10.1016/j.apm.2011.12.032.

Schlote, A., Chen, B., Sinn, M., & Shorten, R. (2013). The effect of feedback in the assignment problem in shared bicycle systems. In *2013 International Conference on Connected Vehicles and Expo (ICCVE)*.

Schröder, T., Stewart, T. C., & Thagard, P. (2014). Intention, emotion, and action: A neural theory based on semantic pointers. *Cognitive Science, 38*(5), 851–880. https://doi.org/10.1111/cogs.12100.

Shaheen, S. A., Martin, E. W., Chan, N. D., Cohen, A. P., & Pogodzinski, M. (2014). North America: Public bikesharing in North America during a period of rapid expansion: Understanding business model, industry trends, and user impacts.

Vogel, P., Greiser, T., & Mattfeld D. C. (2011). *Understanding bike-sharing systems using data mining: Exploring activity patterns* (Vol. 20, pp. 514–523). Germany: University of Braunschweig.

A Review Factors Affecting Residential Property Price

Wan Nur Ayuni Wan Ab Rashid, Thuraiya Mohd and Lizawati Abdullah

Abstract Currently, there are limited studies conducted in industry about factors affecting residential price within Malaysia. Questions arise regarding issue in development with are (i) what are the attributes factors that affect green residential price in Malaysia? and (ii) how the relationship between this factors and property price. The purpose of this paper is to present the factors that affect property price. This papers use method of comprehensive search and review of extant literature across internal factors and external factors that affect the developed of residential property price was undertaken. It is also to understand the factors involved and its relevance to property market. The review permitted developing a comprehensive set of characteristics that are associated with the residential property. The findings contribute a conceptual framework of factors affected residential property price organized according to the following dimensions is presented and discussed. This paper advances theoretical knowledge in the residential property market domain by offering a conceptual framework of factors that affect growth of residential property.

Keywords Residential building · Conceptual framework · Residential market price

1 Introduction

Malaysia is one of the country this rapidly growth with green building development with the area of approximately 329,750 km^2 (127,317 square miles), has over the past years. Despite the increasing environment issue like other cities around the world, Malaysia changes the development climate into green development. The

---

W. N. A. W. A. Rashid (✉) · T. Mohd · L. Abdullah
Faculty of Architecture, Planning and Surveying, Universiti Teknologi MARA,
Seri Iskandar Campus, 32610 Seri Iskandar, Perak, Malaysia
e-mail: ayunirashid@yahoo.com

condition and uncontrolled environment issue in development, Malaysia became one of the country that practice green development.

Questions arise regarding issue in development with are (i) what are the attributes factors that affect green residential price in Malaysia? and how the relationship between this factors and property price. Thus, the conceptual research was conducted to analyze the influencing factors for housing price in Malaysia property market.

After finding the factors that are uniquely influencing the prices for housing products in Malaysia, reliability and factor analysis will be conducted to ensure the result and the importance rate for those factors.

This research is important for real estate development in Malaysia, as currently there are limited studies conducted in this industry within Malaysia. The present study is expected to produce significant findings for real estate industry in Malaysia and major cities all around Malaysia, as currently there are only a few studies discussing the relationship between green residential developments and housing price in Malaysia. This research can also be valuable for the real estate developers in the other metropolitans in Southeast Asia, as some of the behaviors in purchasing a residential product are similar in the cities such as Indonesia, Singapore, Bangkok, and Manila.

The residential property market in Malaysia is one of the key components for domestic economy. The housing development activities are usually used to encourage the economic activities. The Ministry of Finances, Valuation and Property Services Department segments the property market in Malaysia into six classifications, which are the residential, commercial, industrial, agricultural, development land, and others subsectors. This research is focused on the residential sector. The residential sector can be in general classified into houses, flats, apartments, and condominiums.

This research is intended to be the first to pioneer the study on factors of preferences on green residential products in Malaysia. The conceptual framework and methods can be replicated and implemented in other major cities all around Malaysia such as Selangor, Kuala Lumpur, Johor, and Pulau Pinang. At the end, a national database on factors of preferences on green residential products in Malaysia can be developed and maintained for the benefits of real estate developers, residential consumers, governments, and the next generation of researchers in this field.

2 Literature Review

2.1 Malaysia Real Estate Market in Green Building

As mentioned previously, the research on green residential industry is not easily found. One of the firth researches discussing green residential industry in Malaysia was conducted by Mohd Thas Thaker and Chandra Sakaran, (2016), Elias and Lin (2015), Ibrahim et al. (2014), and Darus and Hashim (2012). The recommendation on Mohd Thas Thaker and Chandra Sakaran (2016) paper suggested that due to the increase in demand for residential property and the scarcity of land for development spin landed residential properties in major urban areas in Malaysia such as Penang, Kuala Lumpur, Selangor and Johor Bahru has resulted in the rapid development of housing, especially high-rise residential schemes in these high-density areas. The urban crisis and the real estate development have been obstructed and negatively affected. This urban crisis resulted in the rising price of residential product that eventually affected the whole economic condition in Malaysia.

The price of the residential property in Malaysia experiencing grow in value over time considering the significant increase in price of housing products in just over the past five years, as demonstrated by price in many growing areas in major urban.

At the same time, the surrounding nature must be preserved from any damages caused by the pursuit of economic growth through the heavy development. The application of green development has been practiced in various countries such as UK, Australia, and America. Besides, Asian countries also go to the green such as China, Singapore, and Indonesia. Green development is not only important to the advanced countries but also important to the developing countries such as Malaysia and Thailand (Elias and Lin 2015).

Therefore, this situation has superiority to a growing awareness around the world especially Malaysia to change the current method of design and built into a more sustainable approach that fulfills current development needs and requirements without sacrificing the environment (Ibrahim et al. 2014).

In Malaysia, government concerns on the sustainable development which is in Tenth Malaysia Plan focus on the sustainable housing and tax incentives for buildings and designs that are environmental friendly. However, the reality is incentives that government provided are not really attract the stakeholder like Local Governments, developer and construction firm (Yee, Rahim, and Mohamed, n.d.). Thus, for the coming years, development might change the direction to green development, but at this moment, green movement is still at a slow stride.

2.2 Previous Research on Reviewing the Factors Affecting Property Price

There are several factors that affect property price that has been discussed from the previous research. Figure 1 summarizes the research finding of factors that influence the property price. The analysis shows factors that affect property price which is age. After that, it is followed by gross floor area, floor, transaction price, parking, swimming pool, and green labels/certificate. Another certificate is influencing the prices which are land area, floor, and type of property. Additional certificates for different literature are building coverage, distance to open space, distance to CBD, density, view, orientation, ease of accessibility, population density, security, and infrastructure development (Clausen and Hirth 2016; Fuerst and McAllister 2011; Högberg 2013; Jayantha and Man 2013; McCord et al. 2014; Oloke et al. 2013; Ponsignon et al. 2015).

Factors AffectProperty Price

—— total

[Radar chart showing factors: Structural (8), Transaction Price, Age (7), Density (7), Land Area, Gross floor area (sq ft) (8), Floor (3), Swimming pool (4), Parking (4), Building coverage, Type of property, Location, View, Orientation, Distance to open space, Distance to CBD (2), Environmental (3), Green labels/certifications (5), Neighborhood (3), Population density (4), Security (6), Development infrastructure (8)]

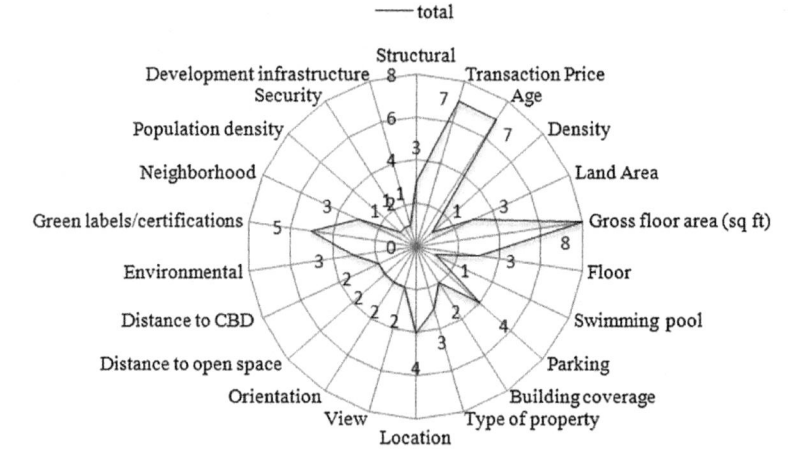

Fig. 1 Summarization of the factors that influence the property price

A brief explanation of the structural variables used in the conceptual models follows: Housing prices are defined as the real transacted price in a year. Housing density Clausen and Hirth (2016) reported is calculated by dividing the total number of housing units by site area. Gross floor area is the total buildup area per square feet. Number of bedroom is the number of bedroom in the building coverage. The average floor is average of floor levels of different buildings within the residential property. Age is the property age calculated by subtracting construction year. The variable of parking is the vector of the number of parking lots and swimming pool including as facilities variable. To ensure reliability of the findings, the sample size across all property types was investigated to confirm representation within the sales sample dataset.

The location factors is road access with Central Business District (CBD) (Abidoye and Chan 2016) include location near activity center, location near open space (Clausen and Hirth 2016; McCord et al. 2014), good security system (Oloke et al. 2013), ease of accessibility and good social communication (Rahadi et al. 2015) and location uniqueness (Jayantha and Man 2013) is one of the major price boosters for housing product sales. Unique location characteristics are the properties of the environment view itself and orientation of the building (Jayantha and Man 2013) such as facing west, facing east, facing north, or facing south that give an effect to the property price.

The environment variable referred to the green assessment tools consists of environmental elements such as electricity efficiency, water efficiency, indoor environmental quality, sustainable site planning and management, materials and resources, and innovation (Green Building Index 2011). The green certificate will make sure the building or developments practice the green concept.

Lastly, the rapid development at urban area can cause increasing population density. It is one of the variables that affect property price (Clausen and Hirth

2016). Besides, neighborhood security (Oloke et al. 2013) significantly influences property value in the city. Others, the variable that influences the market price is infrastructure development (Oloke et al. 2013) at that area such as road and drainages as well as good estate plan and quality designs.

2.3 Factors Affecting Residential Property Price

According to the review of the literature, this research can conclude that there are four (4) main factors that give an effect to the property price.

2.3.1 Structural

From reviewing the previous research, the structural refers to the characteristic of the property which are transaction price, age, area, gross floor area, density, type of property, parking, swimming pool, and number of bedroom. The transaction price is important to analyze the market at the area. The transaction price records ratio between green buildings and nongreen buildings (Jayantha and Man 2013). Another structural factor will affect the property price either positive or negative. Many previous researches did not look at the direct impact of housing density on housing prices. The result shows that household density has a negative effect on housing unit price (Clausen and Hirth 2016).

2.3.2 Location

The location environmental factors include distance to Central Business District (CBD) which are direct toll road access, location near family, location near working area, location near activity center, location near shopping center, location near education center, location near religious center, good security system, ease of accessibility, good social communication, distance to open space, and location uniqueness. The location uniqueness means the view of the building such as building view or open space view. The orientation of the building also will give an effect to the property price such as facing west, facing east, facing north, or facing south.

2.3.3 Environment

An environment factor is a green certificate which is green assessment tools. The element or criteria in green assessment tools including overall aspect in protecting environment which are the main criteria such as electricity efficiency, water efficiency, indoor environmental quality, sustainable site planning and management, materials and resources, and innovation (Green Building Index 2011).

2.3.4 Neighborhood

From the reviewing analysis, there are three influencing factors for housing products such as population density like planned community or gated concept, prestige or reputation, and security. Infrastructure development referred to the facilities that provided at the area. Utility is the ability of a product to satisfy a human want, need, or desire (Oloke et al. 2013).

3 Method

This research found four (4) main factors influencing green residential price. The method that will be used to fulfill the highlight of objective in this research is based on secondary sources which are a review and analysis data from the previous research. Secondary sources on the other hand referred to as the existing data from researchers who have studied the same topic or data from researchers who are conducting current research. Reviewing the secondary sources of the literature consists of published and unpublished data such as journals, conference, article, and thesis.

4 Findings from the Literature

4.1 Factors Influencing Property Price

The author created a synthesis from the literature reviews, discussing the influencing factors affecting housing price. The factors are chosen according to the literature review, and it will be set as an independent variable. The variable factors will analysis using SPSS to determine the percentage variable affect the property price in the further paper. At the preliminary study, all factors variable from the literature review will be included. The factors that are not significant will be dropped from this research.

4.2 Conceptualizing the framework

From the reviews on factors that affect housing price, it seems appropriate to outline four (4) important variables in modeling the housing price. The variables are structural, location, environment, and neighborhood preferences. The conceptualization of the framework will be based on the four (4) variables as mentioned earlier (structural, location, environment, and neighborhood preferences). Then, the construction of questionnaire survey will include the importance of structural, location, environment, and neighborhood (Clausen and Hirth 2016; Fuerst and McAllister

A Review Factors Affecting Residential Property Price 43

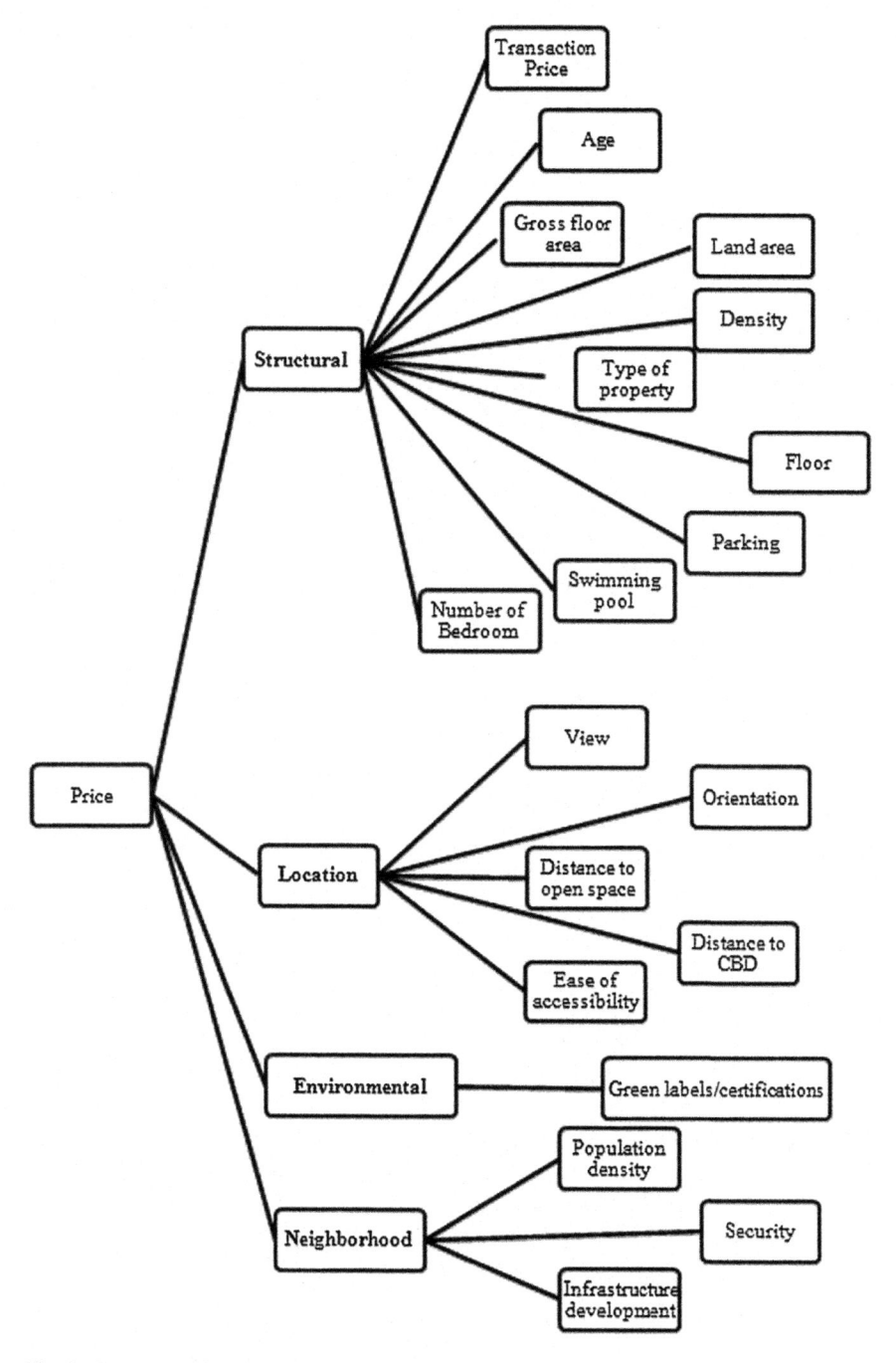

Fig. 2 Conceptual framework of factors

2011; Högberg 2013; Jayantha and Man 2013; McCord et al. 2014; Oloke et al. 2013; Ponsignon et al. 2015). Figure 2 shows the conceptual framework of factors affecting house price.

5 Summary

The main objective of this research was to investigate the factors that affect residential property price. The reviewing from previous research into deep some factors has been founded. The results show several main factors that affect the property price such as structural, location, environment, and neighborhood regarding the conceptual of framework. All the factors found in previous research are reported as elements that give an effect to the residential property price.

The findings of this research trigger some important implications. For valuer, the results of this research may shed light on the main assumption underlying the widespread implementation of the valuing residential property price. The factors analysis also will give awareness to the public to capitalize environmental consideration in investment decisions.

The factors in this paper will be used in the further papers. Besides that, one of the factors found is environment factor that gives an impact on the residential property price. In the further papers, we will discuss in detail about the environment factors that give effect to the residential property price.

References

Abidoye, R. B., & Chan, A. P. C. (2016). Critical determinants of residential property value: Professionals' perspective. *Journal of Facilities Management, 14*(3), 283–300. https://doi.org/10.1108/JFM-02-2016-0003.

Clausen, S., & Hirth, S. (2016). Measuring the value of intangibles. *Journal of Corporate Finance, 40,* 110–127. https://doi.org/10.1016/j.jcorpfin.2016.07.012.

Darus, A. Z. M., & Hashim, N. A. (2012). Sustainable building in Malaysia: The development of sustainable building rating system. *Sustainable Development-Education Business and Management-Architecture and Building Construction-Agriculture and Food Security,* (pp. 1–33; 342 p.) http://doi.org/10.5772/27624.

Elias, E. M., & Lin, C. K. (2015). The empirical study of green buildings (residential) implementation: Perspective of house developers. *Procedia Environmental Sciences, 28,* 708–716. https://doi.org/10.1016/j.proenv.2015.07.083.

Fuerst, F., & McAllister, P. (2011). Green noise or green value? Measuring the effects of environmental certification on office values. *Real Estate Economics, 39*(1), 45–69. https://doi.org/10.1111/j.1540-6229.2010.00286.x.

Green Building Index. (2011). Gbi assessment criteria contents, March, 0–57.

Högberg, L. (2013). The impact of energy performance on single-family home selling prices in Sweden. *Journal of European Real Estate Research, 6*(3), 242–261. http://doi.org/10.1108/JERER-09-2012-0024.

Ibrahim, F. A., Mohd Shafiei, M. W., Ismail, R., & Said, I. (2014). Green homes development: Factors affecting housing developers' readiness. *ARPN Journal of Engineering and Applied Sciences, 9*(6), 971–980. https://doi.org/10.5829/idosi.wasj.2013.28.03.13798.

Jayantha, W. M., & Man, W. S. (2013). Effect of green labelling on residential property price: A case study in Hong Kong. *Journal of Facilities Management, 11*(1), 31–51. https://doi.org/10.1108/14725961311301457.

McCord, J., McCord, M., McClusky, W., Davis, P. T., & McIlhatton, D. (2014). Effect of public green space on residential property values in Belfast metropolitan area. *Journal of Financial Management of Property and Construction, 19*(2), 117–137. https://doi.org/10.1108/JFMPC-04-2013-0008.

Mohd Thas Thaker, H., & Chandra Sakaran, K. (2016). Prioritisation of key attributes influencing the decision to purchase a residential property in Malaysia. *International Journal of Housing Markets and Analysis, 9*(4), 446–467. https://doi.org/10.1108/IJHMA-09-2015-0052.

Oloke, O. C., Simon, F. R., & Adesulu, A. F. (2013). An examination of the factors affecting residential property values in Magodo neighbourhood, Lagos state. *International Journal of Economy, Management and Social Sciences, 2*(8), 639–643.

Ponsignon, F., Klaus, P., Maull, R. S., Ponsignon, F., Klaus, P., & Maull, R. S. (2015). Article information. http://doi.org/10.1108/JEIM-07-2014-0077.

Rahadi, R. A., Wiryono, S. K., & Koesrindartoto, D. P. (2015). Factors influencing the price of housing in Indonesia. *International Journal of Housing Markets and Analysis, 8*(2), 169–188. http://doi.org/10.1108/JEIM-07-2014-0077.

Yee, C. S., Rahim, M. H. I. A., & Mohamed, S. H. (n.d.). An insight of sustainable housing in Malaysia. Department of Construction Management Faculty.

Tensile Properties and Thermal Characteristics of Linear Low-Density Polyethylene/Poly(Vinyl Alcohol) Blends Containing 3-(Trimethoxysilyl) Propyl Methacrylate

Razif Nordin, Hanafi Ismail, Zuliahani Ahmad and Ragunathan Santiago

Abstract Tensile properties and thermal characteristics of blends made from linear low-density polyethylene (LLDPE) and polyvinyl alcohol (PVA) have been investigated. 3-(Trimethoxysilyl)propyl methacrylate (Si A174) was used as a coupling agent. The blends were melted and mixed using a Haake Rheomix at 150 °C and 50 rpm for 10 min. Results show that the incorporation of Si A174 resulted in the increase of the tensile strength and Young's modulus but decrease in the elongation at break of the LLDPE/PVA blends. The crosslinked formation between LLDPE, PVA, and Si A174 was confirmed by decrease of melting temperature and degree of crystallinity as demonstrated by differential scanning calorimetry study.

Keywords Linear low-density polyethylene · Poly(vinyl alcohol) Blends · Silane coupling agent

---

R. Nordin (✉)
Department of Chemistry, Faculty Applied Sciences,
Universiti Teknologi MARA, 02600 Arau, Perlis, Malaysia
e-mail: razifmn@perlis.uitm.edu.my

H. Ismail
Polymer Division, School of Materials and Mineral Resources Engineering, Universiti Sains Malaysia, Engineering Campus, 14300 Nibong Tebal, Pulau Pinang, Malaysia

Z. Ahmad
Department of Chemistry, Faculty Applied Sciences, Universiti Teknologi MARA, 40450 Shah Alam, Selangor, Malaysia

R. Santiago
School of Bioproses Engineering, Universiti Malaysia Perlis,
Kompleks Pusat Pengajian Jejawi 3, 02600 Arau, Perlis, Malaysia

© Springer Nature Singapore Pte Ltd. 2018
R. Saian and M. A. Abbas (eds.), *Proceedings of the Second International Conference on the Future of ASEAN (ICoFA) 2017 – Volume 2*,
https://doi.org/10.1007/978-981-10-8471-3_5

1 Introduction

In situ silane crosslinking technology in polyethylene has offered several advantages over chemical treatment and radiation methods. This technique has been used extensively as the insulation compounds on the electrical cable system ranging from underground to Aerial Bundle Cable network (Celina and George 1995). The improvement in properties of silane crosslinked polyethylene included process operation and production, mechanical properties, degradation stability, and resistance to creep.

The treatment of filler or polymer by silane coupling agent has been reviewed by different authors (Bengtsson and Oksman 2006; Prachaywarakorn et al. 2008). It is observed that, in general, the silane crosslink polyethylene consists of at least two stages. The first stage is grafted silane via its vinyl groups on polyethylene through radical reaction. The second stage is silane crosslinking reaction of resultant copolymer via the exposure to hot water with the presence of catalyst through hydrolysis and condensation reaction.

In our previous work (Ismail et al. 2009, 2010), we have reported the tensile and thermal properties of linear low-density polyethylene/poly(vinyl alcohol) (LLDPE/PVA) blends. Results indicated that in situ incorporation of maleic acid in LLDPE/PVA blends showed increases in the tensile strength and Young's modulus but slightly decreases the elongation at break. In this work, 3-methacryloxypropyltrimethoxysilane was used to crosslink linear low-density polyethylene/poly(vinyl alcohol) blends while the polymers in the molten state. It is expected that formation of crosslinked network will be reflected on the tensile and thermal properties.

2 Materials and Methods

2.1 Materials

Poly(vinyl alcohol) with average molecular weight of 89,000–98,000 and 99+% hydrolyzed, 3-(trimethoxysilyl)propyl methacrylate (Si A174), and dicumyl peroxide (DCP) was obtained from Sigma-Aldrich (Malaysia) Sdn. Bhd. Linear low-density polyethylene (LLDPE) with melt mass flow rate (MFR) (190 °C/2.16 kg) 1.0 g/10 min and melting temperature of 123 °C was obtained from Polyethylene (Malaysia) Sdn. Bhd.

2.2 Mixing Procedure

The formulation of LLDPE/PVA blends with and without Si A174 is shown in Table 1. The preparation of blends was carried out in a Haake Rheomix Polydrive

Table 1 Formulation used in LLDPE/PVA blends

Blends (php/php)	Control blends					Blends with Si A174				
LLDPE	90	80	60	50	40	90	80	60	50	40
PVA	10	20	40	50	60	10	20	40	50	60
Si A174	–	–	–	–	–	3	3	3	3	3
DCP	–	–	–	–	–	0.1	0.1	0.1	0.1	0.1

Table 2 Mixing sequences of components in preparation of the blends

Control blends		Blends with Si A174
Time (min)	Operation	
0	Rotor started and loading of LLDPE	Rotor started and loading of LLDPE
2	PVA	PVA, Si A174, DCP
10	Dump	

at temperature 150 °C and rotor speed of 50 rpm for about 10 min. The blends were mixed according to mixing sequence as shown in Table 2.

Sample of blends was compression molded in an electrically heated hydraulic press. Hot-press procedures involved preheating at 150 °C for 4 min followed by compressing for 4 min at the same temperature and subsequent cooling under pressure for 3 min.

2.3 Measurement of Tensile Properties

Tensile test was carried out according to ASTM D 638 using Instron 3366. A crosshead speed of 50 mm/min was used, and the test was performed at 25 °C. Tensile strength, Young's modulus, and elongation at break were evaluated from the stress–strain data.

2.4 Differential Scanning Calorimetry

Differential scanning calorimetry (DSC) measurements of blends were carried out by using Perkin Elmer Pyris 1 DSC under nitrogen atmosphere. Each blend was first heated from 30 to 170 °C at heating rate 10 °C/min, annealed for 5 min at 170 °C, cooled from 170 to −50 °C at cooling rate 10 °C/min, and rescanned from −50 to 300 °C at heating rate 10 °C/min (second scan). The blend weight was 10 ± 0.5 mg. Thermal properties such as melting temperature (T_m), crystallization temperature (T_c), and heat of fusion (ΔH) were determined from DSC thermograms.

3 Results and Discussion

3.1 Tensile Properties

Figure 1a, b shows the effect of PVA content on the tensile properties of LLDPE/PVA blends with and without Si A174. It is clear that the tensile strength and elongation at break of LLDPE/PVA blends without Si A174 decrease steadily with increasing PVA content, whereas Young's modulus (Fig. 2) shows the opposite trend. This is due to the poor interfacial adhesion between two opposite polarity polymers and resulted in poor stress transfer between matrix and the dispersed phase. It is believed that PVA forms agglomerates due to strong intramolecular hydrogen bonds between the hydroxyl groups which resulted in poor dispersion and adhesion in the LLDPE particularly at higher PVA content (Ismail et al. 2009, 2010). But as Si A174 were added into LLDPE/PVA blends, the tensile strength and Young's modulus of the blends increase with increasing PVA content, while

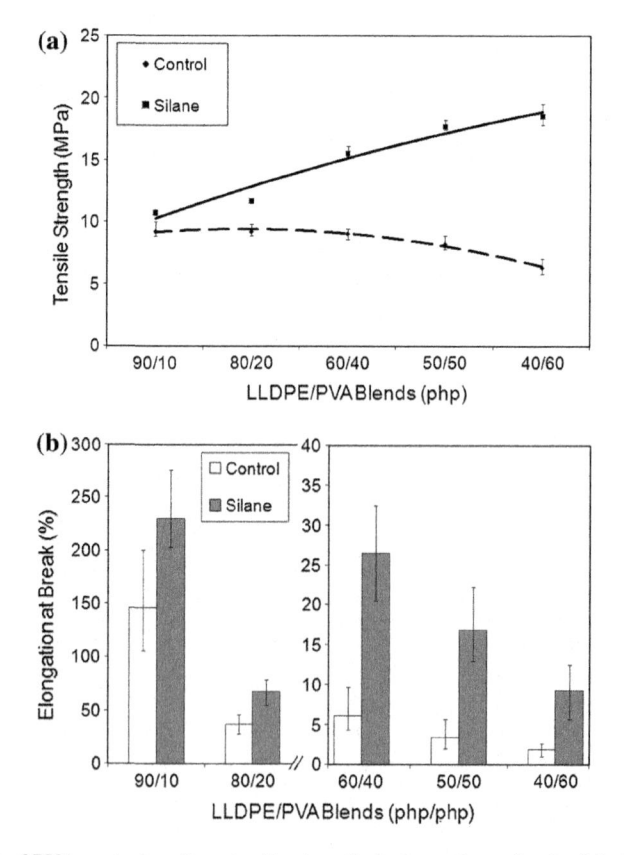

Fig. 1 Effect of PVA content on the **a** tensile strength, **b** elongation at break of the LLDPE/PVA blends with and without silane coupling agent (Si A174)

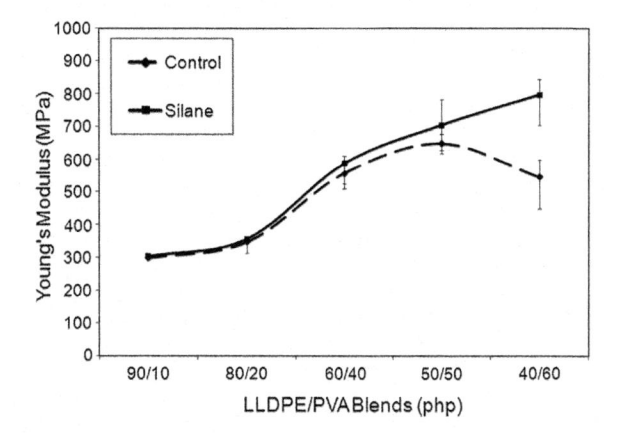

Fig. 2 Relationship of variation of PVA content on Young's modulus of the LLDPE/PVA blends with and without silane (Si A174)

the elongation at break shows the opposite trend. This is due to the ability of Si A174 to improve compatibility between the PVA and LLDPE through crosslink reaction. Figures 3 and 4 show the proposed reaction occurred between LLDPE and PVA with the presence of Si A174.

3.2 Thermal Properties

The crystallinity (X_c), melting endotherms, melt crystallization temperatures (T_c), and the melting peak temperatures of the blends from DSC analysis are summarized

Fig. 3 Proposed chemical reaction between LLDPE and silane initiated by dicumyl peroxide

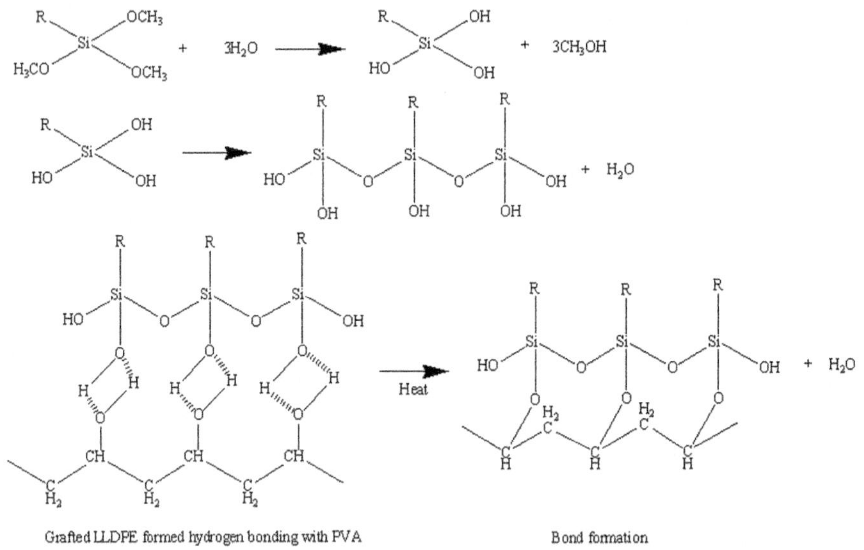

Fig. 4 Plausible chemical reaction between LLDPE, silane, and PVA

Table 3 Melting parameters of LLDPE and PVA in LLDPE/PVA blends with and without Si A174

LLDPE/PVA	ccc	LLDPE	ccc	ccc	PVA
(php/php)	ccc	Tm (°C)	Tc (°C)	Xc (%)	Tm (°C)
Without Si A174	100/0	121.35	108.56	33.7	–
ccc	60/40	123.34	115.57	12.5	226.74
ccc	50/50	123.75	116.47	7.4	227.73
ccc	40/60	123.45	116.79	4.9	227.3
Si A174	60/40	122.72	109.86	8.6	227.09
ccc	50/50	122.53	110.88	6.1	227.45
ccc	40/60	122.55	115.55	1.6	227.63

in Table 3. The lower melting temperature and degree of crystallinity in crosslinked LLDPE/PVA blends compared to blends without Si A174 may be related to the crosslinking network initiated during melt processing. The network structure makes the macromolecular LLDPE chain less flexible, lowering the T_c and making crystallization more difficult. Interestingly, the T_c of crosslinked LLDPE/PVA blends with ratios of 60/40 and 50/50 are close to that of the pure LLDPE. This result indicates that Si A174 enhances the compatibility between PVA and LLDPE.

4 Conclusions

The following conclusion can be drawn from this study:

i. The presence of Si A174 has increased the tensile strength, elongation at break, and Young's modulus of LLDPE/PVA blends.
ii. The compatibility of the LLDPE/PVA blends is enhanced by depression of melting temperature.

Acknowledgements One of the authors is thankful for the financial support from research grant of Universiti Teknologi MARA.

References

Bengtsson, M., & Oksman, K. (2006). Silane crosslinked wood plastic composites: Processing and properties. *Composites Science and Technology, 66,* 2177–2186.

Celina, M., & George, G. A. (1995). Characterization and degradation studies of peroxide and crosslinked polyethylene. *Polymer Degradation and Stability, 48,* 297–312.

Ismail, H., Ahmad, Z., Nordin, R., & Rashid, A. R. (2009). Processibility and miscibility studies of uncompatibilized linear low density polyethylene/poly(vinyl alcohol) blends. *Polymer Plastics Technology and Engineering, 48*(1–7), 2009.

Ismail, H., Ahmad, Z., Nordin, R., & Rashid, A. R. (2010). Processability and miscibility studies of linear low density polyethylene/poly(vinyl alcohol) blends: In situ compatibilized with maleic acid. *Iranian Polymer Journal, 19,* 297–308.

Prachaywarakorn, J., Khunsumled, S., Thongpin, C., Kositchaiyong, A., & Sombatsompop, N. (2008). Effects of silane and MAPE coupling agents on the properties and interfacial adhesion of wood filled PVC/LLDPE blend. *Journal of Applied Polymer Science, 108,* 3523–3530.

A Review Analysis of Accident Factor on Road Accident Cases Using Haddon Matrix Approach

Nur Fatma Fadilah Yaacob, Noradila Rusli and Sharifah Norashikin Bohari

Abstract The present paper provides a comprehensive review of past research that examined accident factor on the road accident. Articles and publications were selected for relevance and research strength through a comprehensive search of major databases such as Science Direct, UiTM EZAccess, and Transportation Research Information Services (TRIS). A total of 40 articles were analyzed on the accident factors and categorized using Haddon matrix method. There are three major factors that contribute to the road accident which are human behavior, road environment, and vehicle condition. The result of findings shows that human behavior contributes the highest factor on road accident by accounting for 41% and then followed by environment factor contributed 26% and vehicle factor contributed 33%.

Keywords Accident factors · Haddon matrix · Human behavior Road · Vehicle

1 Introduction

Road accident is an event that occurs on a road where one or more road users are injured or died, and at least one vehicle is involved. It is a rare, random, multi-factor event always preceded by a situation in which one or more road users have failed to cope with their environment (San 2013). Approximately, almost 23–34 million people injured in road accidents annually, an average of 1.24 million died, and it

---

N. F. F. Yaacob (✉) · N. Rusli · S. N. Bohari
Green Technology and Environment (GREENTech) Research Group,
Centre of Studies for Surveying Science and Geomatics, Faculty
of Architecture, Planning and Surveying, Universiti Teknologi MARA
(UiTM), Perlis Branch, Arau Campus, 022600 Arau, Perlis, Malaysia
e-mail: cikfatma@yahoo.com

© Springer Nature Singapore Pte Ltd. 2018
R. Saian and M. A. Abbas (eds.), *Proceedings of the Second International Conference on the Future of ASEAN (ICoFA) 2017 – Volume 2*,
https://doi.org/10.1007/978-981-10-8471-3_6

was nine ranked cause of death in the world, where the ranking is projected to rise to third by year 2020 (Manyara 2013). According to Liang, Malaysia recorded 489,606 road accidents and 6706 deaths caused by accidents for the year 2015. The number fatalities caused by road accidents in Malaysia have consistently been above 6000 cases since year 2010. Due to the increasing number of accident shows, it is critical issues that need to be considered in many aspects. The critical evaluation on causes and effect must be investigated. According to the Ahmed (2013), the matrix developed by Dr. William Haddon stated that there are three factors that contribute to the road accident including human factor, environmental factor, vehicles, and equipment factor. Human behavior consists of drug or alcohol influence, excess speed, driving attitude, use mobile phone, and others. Environment includes weather effect, road design, impact of wind speed, improper of road intersection, and so on. In other hand, vehicle consists of its design, vehicle type, traffic volume, and traffic composition on road. The Haddon matrix (Table 1) is an analytical tool to help in identifying these factors which related to road accident. Once, the factor was identified and analyzed, and then, the countermeasures can be developed as a guideline for the short-term and long-term period.

Previous research that conducted using Haddon matrix approach such as Masoumi et al. (2016) and Chen et al. (2016) analyzes the factor based on road accident data. Therefore, this research intends to implement Haddon matrix approach that focusing on road accident factors by reviewing 40 journal paper on related research around the world.

Table 1 Haddon matrix (Ahmed 2013)

			Factors	
Phase	Human	Vehicles and equipment	Environment	
Pre-crash	Crash prevention	Information	Roadworthiness	Road design
		Attitudes	Lighting	Road layout
		Impairment	Braking	Speed limits
		Police enforcement	Handling	Pedestrian facilities
			Speed management	
Crash	Injury prevention during the crash	Use of restraint	Occupant restraint	Crash-protective roadside objects
		Impairment	Other safety device	
			Crash-protective design	
Post-crash	Life sustaining	First aid skill	Ease of access	Rescue facilities
		Access to medics	Fire risk	Congestion

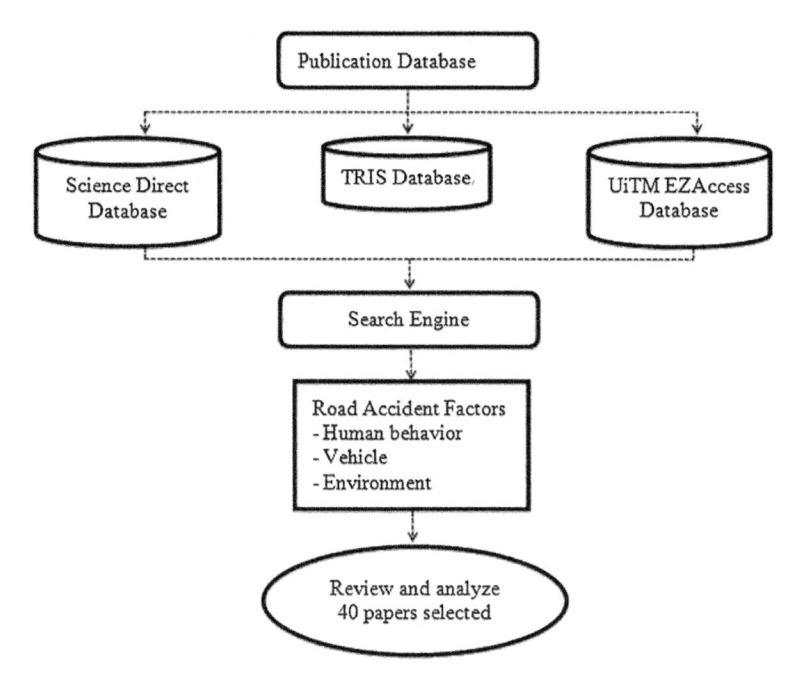

Fig. 1 Overall process of study

2 Methodology

Figure 1 describes method in reviewing road accident factors by previous researchers based on three categories such as human behavior, vehicle, and environment. Total 40 papers were reviewed by assessing Science Direct, UiTM EZAccess, and Transportation Research Information Services (TRIS) database. The papers selected worldwide such as study area in Thailand, UK, Europe, Kenya, Indonesia, Malaysia, and others. Then, the analysis was carried out after reviewing the factors.

3 Road Accident Factors

The road safety and accident-related studies have been discussed by many researchers for the last two decades. There were various analyses carried out such as the study on spatial-temporal pattern, emergency response, and hotspot area of road accident. Besides that, factor of road accident is also a popular analysis among researchers in worldwide such as Wang and Ohsawa (2016), Haadi (2014) and

Rastegarfar et al. (2013). Previous research offers valuable insight into the underlying road accident factor mechanism. However, with the increase of rapid urbanization and motorization which was changed to traffic composition in recent years, prior research findings should be updated on these factors. Hence, the present paper provides a review of road accident factors by previous researchers in their study in Table 2. There are 20 papers published by researchers in Malaysia and 20 papers from researchers in other countries such as Thailand, Europe, Indonesia, Kenya, and others.

Based on Table 2, human behavior factor was stated by 26 researchers such as Hammoudi et al. (2014), Kim et al. (2016), and Nordin et al. (2015). Besides that, vehicle factor was stated by 16 researchers such as Zhang et al. (2014), Suraji and Tjahjono (2012), and Masuri et al. (2011). Meanwhile, 21 researchers believe that road accidents that occur are caused by environment (Sundara et al. 2016; Perrels et al. 2015; Jung et al. 2010).

4 Results and Findings

Based on 40 papers reviewed, human behavior leads the most factors contributed to the road accident. Second is the environment factor and followed by vehicle factor. Figure 2 shows the percentage of accident factors for each category.

There are 13 different journals or articles of 40 papers reviewed such as *Elsevier journal, medical journal, engineering journal, proceeding journal, and transportation journal* which are showed in Fig. 3.

The most publications which discuss road accident were Elsevier journal, medical journal, engineering journal, transportation journal, and proceedings. Others publications just publish one until five papers.

Human factor is variable of human characteristics such as level of experience, lack of attention while driving, and person fatigue level (Elslande et al. 2008). It can give negative impact to driver's driving performance. Besides that, excessive speed and rash driving can lead to road accident (Raut et al. 2016; Sultan et al. 2016; Manyara 2013). Involving in road accident due to the influence of alcohol or drug is higher shown by statistics (Hammoudi et al. 2014). This causes a serious problem especially to teenagers, and local authority should do something to solve this problem.

Meanwhile, vehicle factor includes vehicle design and vehicle maintenance (Bun 2012; Ponboon et al. 2010; Mako 2015). Even though many cars have seat belts and airbags to reduce accident, but if tires and brakes are not functioning very well, the accident will occur. Thus, road users need to alert about their vehicle condition. On other hand, lost in balance for two wheels vehicle also cause the accident (Masuri et al. 2011). The riders can fall easily. The small size of motorcycle may

Table 2 Road accident factors

Country	Author	Human behavior	Vehicle	Environment
Western Cape, South Africa	Vogel and Bester (2005)	– Negligence – Excess speed – Driving attitude	– Maintenance	– Daylight – Traffic hour – Pedestrian facilities
Europe	Elslande et al. (2008)	– Level of experience – Lack of attention – Fatigue level	x	x
Wisconsin, USA	Jung et al. (2010)	x	x	– Weather effect
UK	Wang (2010)	– Not wear seat belt – Drug or alcohol influence – Driving attitude	x	x
Thailand	Ponboon (2010)	– Driving attitude – Drug or alcohol influence	– Design	– Road design
Nigeria	Bun (2012)	– Excess speed – Driving attitude – Visual – Decision making ability	– Design – Maintenance	– Road design
Jordan	Obaidat and Ramadan (2012)	– Excess speed	x	– Road design – Traffic hour – Pedestrian facilities – Daylight
Indonesia	Suraji and Tjahjono (2012)	x	– Vehicle type	x
Asia & Pacific Area	Ahmed (2013)	x	x	– Road design
Kenya	Manyara (2013)	– Driving attitude – Excess speed – Drug or alcohol influence	x	x
Europe	Martensen and Dupont (2013)	x	– Vehicle type	– Road design – Daylight

(continued)

Table 2 (continued)

Country	Author	Human behavior	Vehicle	Environment
Ghana	Haadi (2014)	– Use mobile phone – Excess speed – Drug or alcohol influence – Fatigue level	x	x
United Arab Emirates	Hammoudi et al. (2014)	– Not wear seat belt – Drug or alcohol influence – Use mobile phone	x	x
Thailand	Ratanavaraha and Suangka (2014)	– Excess speed	x	x
China	Zhang et al. (2014)	x	– Vehicle type	x
Finland	Perrels et al. (2015)	x	x	– Weather effect
Europe	Mako (2015)	– Excess speed	– Vehicle type	– Road design – Pedestrian facilities
India	Mohanty and Gupta (2015)	– Demographic criteria – Drug or alcohol influence	x	– Road design – Traffic hour – Daylight – Weather effect
Korea	Kim et al. (2016)	– Excess speed – Drug or alcohol influence – Demographic criteria	x	x
Aurangabad	Raut et al. (2016)	– Excess speed – Driving attitude – Fatigue level	– Maintenance	– Improper location of signboard – Road design
Malaysia	Mustafa (2010)	– Excess speed	– Combination of traffic composition – Traffic volume	– Improper intersection design – Street lightning – Pedestrian facilities – Large number of traffic light in urban area

(continued)

Table 2 (continued)

Country	Author	Human behavior	Vehicle	Environment
Malaysia	Kee et al. (2010)	– Excess speed – Level of experience – Fatigue level	x	– Road design
Malaysia	Masuri et al. (2011)	x	– Vehicle type	x
Malaysia	Arun et al. (2012)	– Driving attitude	x	x
Malaysia	Rizati et al. (2013)	x	x	– Pedestrian facilities
Malaysia	San (2013)	– Excess speed	x	x
Malaysia	Oxley et al. (2013)	– Excess speed – Not wear helmet – Demographic criteria	– Vehicle type	– Road design
Malaysia	Ahmed et al. (2014)	– Excess speed	x	– Road design
Malaysia	Isa et al. (2014)	x	– Traffic volume	x
Malaysia	Manan (2014)	x	– Vehicle type	x
Malaysia	Ramli et al. (2014)	– Excess speed – Drug or alcohol influence – Driving attitude	– Vehicle type	– Road design
Malaysia	Manan and Varhelyi (2015)	x	– Vehicle type – Traffic volume	– Road design
Malaysia	Masuri et al. (2015)	– Driving attitude – Demographic criteria	x	x
Malaysia	Nordin et al. (2015)	– Excess speed – Driving attitude – Demographic criteria	x	x
Malaysia	Rusli et al. (2015)	x	x	– Road design
Malaysia	Sullman et al. (2015)	– Driving attitude – Demographic criteria	x	x
Malaysia	Shahid et al. (2015)	x	x	– Land use category

(continued)

Table 2 (continued)

Country	Author	Human behavior	Vehicle	Environment
Malaysia	Sukor et al. (2016)	– Driving attitude	– Vehicle type	x
Malaysia	Sultan et al. (2016)	– Excess speed – Drug or alcohol influence	x	x
Malaysia	Sundara et al. (2016)	x	x	– Impact of wind speed

Fig. 2 Percentage of accident factor

Percentage of Accident Factor

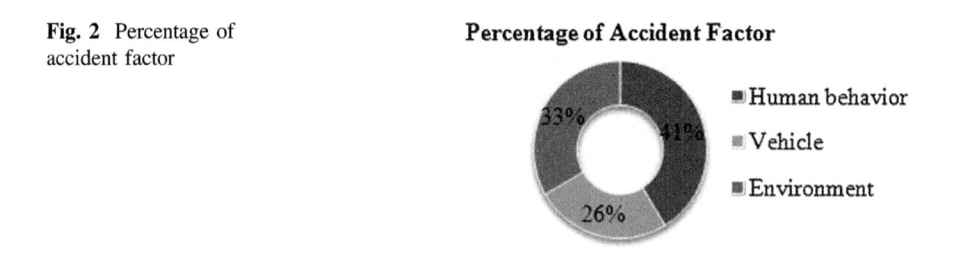

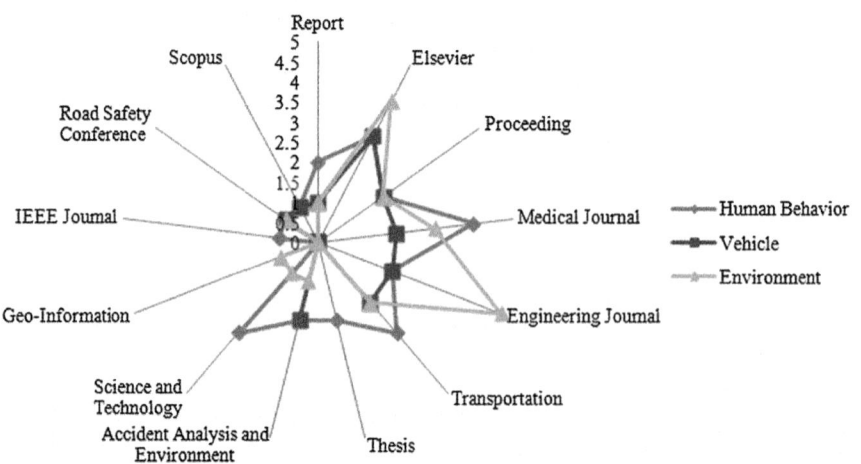

- Human behavior 41%
- Vehicle 26%
- Environment 33%

Fig. 3 Road accident factors by different journal/article

limit other user visibility especially when rider wears dark clothes during night. Besides that, it is limited choice to riders to protect them from injury.

Environmental factors include the common factors of climate and environment, lighting conditions of road, and time of accident occur (Mohanty and Gupta 2015).

Meanwhile, Rusli et al. (2015) revealed driving along mountainous road is dangerous than non-mountainous road and will lead to road accident. Through their research, they showed the road accident occurred, respectively, about 4.2 and 2.8 times on mountainous road than non-mountainous road.

5 Conclusion

In the nutshell, this paper presents a review about road accidents factor based on previous research using Haddon matrix approach. The most categories of road accident factors highlighted by previous research are human behavior, environment, and vehicle. Numerous approaches have been proposed and reported in various publications about road safety and emergency services. However, road accident cases are still increasing every year. Thus, research on road accident still requires to improve the aspect of road safety which is encouraged by Prime Minister of Malaysia.

Acknowledgements This research work is supported by Green Technology and Environment (GREENTech) Research Group, UiTM Perlis Branch.

References

Ahmed, A., Sadullah, A. F. M., & Yahya, A. S. (2014). Accident analysis using count data for unsignalized intersections in Malaysia. *Procedia Engineering, 77,* 45–52.

Ahmed, I. (2013). Road infrastructure and road safety. *Transport and Communications Bulletin for Asia and the Pacific, 83,* 19–25.

Arun, S., Sundaraj, K., & Murugappan, M. (2012). Driver inattention detection methods : A review. In *Conference on Sustainable Utilization and Development in Engineering and Technology* (pp. 1–6) October.

Bun, E. (2012). Road traffic accidents in Nigeria. *African Medical Journal, 3*(2), 34–36.

Chen, T., Zhang, C., & Xu, L. (2016). Factor analysis of fatal road traffic crashes with massive casualties in China. *Advances in Mechanical Engineering, 8*(4), 1–11.

Elslande, P. Van, Naing, C., & Engel, R. (2008). Analyzing human factors in road accidents: TRACE WP5 Summary Report Table of Contents.

Haadi, A.-R. (2014). Identification of factors that cause severity of road accidents in Ghana: A case study of the Northern Region. *International Journal of Applied Science and Technology, 4*(3), 242–249.

Hammoudi, A., Karani, G., & Littlewood, J. (2014). Road traffic accidents among drivers in Abu Dhabi, United Arab Emirates. *Journal of Traffic and Logistics Engineering, 2*(1), 7–12.

Isa, N., Yusoff, M., & Mohamed, A. (2014). A review on recent traffic congestion relief approaches. In *International Conference on Artificial Intelligence with Applications in Engineering and Technology* (pp. 121–126).

Jung, S., Qin, X., & Noyce, D. A. (2010). Rainfall effect on single-vehicle crash severities using polychotomous response models. *Accident Analysis and Prevention, 42*(1), 213–224.

Kee, S., Tamrin, S. B. M., & Goh, Y. M. (2010). Driving fatigue and performance among occupational drivers in simulated prolonged driving. *Global Journal of Health Science, 2*(1), 167–177.

Kim, H., Yoon, D., Shin, H., & Park, C. H. (2016). Driving characteristics analysis of young and middle-aged drivers. In *International Conference on Information and Communication Technology Convergence* (pp. 864–867).

Mako, E. (2015). Evaluation of human behaviour at pedestrian crossings. In *International Conference on Cognitive Infocommunications* (pp. 443–447).

Manan, M. M. A. (2014). Motorcycles entering from access points and merging with traffic on primary roads in Malaysia: Behavioral and road environment influence on the occurrence of traffic conflicts. *Accident Analysis and Prevention, 70,* 301–313.

Manan, M. M. A., & Varhelyi, A. (2015). Motorcyclists' road safety related behavior at access points on primary roads in Malaysia—A case study. *Safety Science, 77,* 80–94.

Manyara, C. G. (2013). Combating road traffic accidents in Kenya: A challenge for an emerging economy. In *KESSA Proceedings* (pp. 1–7).

Martensen, H., & Dupont, E. (2013). Comparing single vehicle and multivehicle fatal road crashes: A joint analysis of road conditions, time variables and driver characteristics. *Accident Analysis and Prevention, 60,* 466–471.

Masoumi, K., Forouzan, A., Barzegari, H., Asgari Darian, A., Rahim, F., Zohrevandi, B., et al. (2016). Effective factors in severity of traffic accident-related traumas; an epidemiologic study based on the Haddon matrix. *The Official Journal of Emergency Department, 4*(2), 78–82.

Masuri, M. G., Dahlan, A., Danis, A., & Isa, K. A. M. (2015). Public participation in shaping better road users in Malaysia. *Procedia-Social and Behavioral Sciences, 168,* 341–348.

Masuri, M. G., Isa, K. A. M., & Tahir, M. P. M. (2011). Children, youth and road environment. *Asian Journal of Environment-Behaviour Studies, 2*(6), 13–20.

Mohanty, M., & Gupta, A. (2015). Factors affecting road crash modeling. *Journal of Transport Literature, 9*(2), 15–19.

Nordin, F., Rahman, N. A., Rashdi, M. F., Yusoff, A., Rahman, R. A., Sulong, S., et al. (2015). Oral and maxillofacial trauma caused by road traffic accident in two university hospitals in Malaysia: A cross-sectional study. *Journal of Oral and Maxillofacial Surgery, Medicine, and Pathology, 27*(2), 166–171.

Obaidat, M. T., & Ramadan, T. M. (2012). Traffic accidents at hazardous locations of urban roads. *Jordan Journal of Civil Engineering, 6*(4), 436–447.

Oxley, J., Ravi, M. D., Yuen, J., & Hashim, H. H. (2013). Fatal motorcycle collisions in Malaysia, 2007–2011. In *16th Road Safety on Four Continents Conference* (pp. 1–12), May.

Perrels, A., Votsis, A., Nurmi, V., & Pilli-Sihvola, K. (2015). Weather conditions, weather information and car crashes. *ISPRS International Journal of Geo-Information, 4*(4), 2681–2703.

Ponboon, S., Boontob, N., Tanaboriboon, Y., Islam, M., & Kanitpong, K. (2010). Contributing factors of road crashes in Thailand: Evidences from the accident in-depth study. *Journal of the Eastern Asia Society for Transportation Studies, 8,* 1–13.

Ramli, R., Oxley, J., Noor, F. M., Abdullah, N. K., Mahmood, M. S., Tajuddin, A. K., et al. (2014). Fatal injuries among motorcyclists in Klang Valley, Malaysia. *Journal of Forensic and Legal Medicine, 26,* 39–45.

Rastegarfar, A., Zahertahan, A., Armannia, A., & Branch, D. (2013). Factors in road accident reduction strategies and the use of intelligent systems its a case study of the Dezful-Shushtar. In *Fifth International Conference on Geo-Information Technologies for Natural Disaster Management* (pp. 52–61).

Ratanavaraha, V., & Suangka, S. (2014). Impacts of accident severity factors and loss values of crashes on expressways in Thailand. *IATSSR, 37*(2), 130–136.

Raut, U. M., Nalawade, D. B., & Kale, K. V. (2016). Mapping and analysis of accident black spot in Aurangabad city using geographic information system. *International Journal of Advanced Research in Computer Science and Software Engineering, 6*(1), 511–518.

Rizati, H., Ishak, S. Z., & Endut, I. R. (2013). The utilization rates of pedestrian bridges in Malaysia. In *Business Engineering and Industrial Applications Colloquium* (pp. 646–650).

Rusli, R., Haque, M., King, M., & Shaw, W. (2015). A Comparison of road traffic crashes along mountainous and non-mountainous roads in Sabah, Malaysia. In *Proceeding of the 2015 Australasian Road Safety Conference* (pp. 14–16).

San, H. P. (2013). *Influence of speed and flow rate on road accidents: A case study along federal Route 50 (FT50)*. Degree of Master of Civil Engineering. Universiti Tun Hussein Onn Malaysia.

Shahid, S., Minhans, A., Puan, O. C., Hasan, S. A., & Ismail, T. (2015). Spatial and temporal pattern of road accidents and casualties in Peninsular Malaysia. *Jurnal Teknologi, 14*, 57–65.

Sukor, N. S. A., Tarigan, A. K. M., & Fujii, S. (2016). Analysis of correlations between psychological factors and self-reported behavior of motorcyclists in Malaysia, depending on self-reported usage of different types of motorcycle facility. Transportation Research Part F (pp. 1–15).

Sullman, M. J. M., Stephens, A. N., & Yong, M. (2015). Anger, aggression and road rage behaviour in Malaysian drivers. *Transportation Research Part F: Psychology and Behaviour, 29*, 70–82.

Sultan, Z., Ngadiman, N. I., Kadir, F. D. A., Roslan, N. F., & Moeinaddini, M. (2016). Factor analysis of motorcycle crashes in Malaysia. *Journal of Malaysian Institute of Planners*, 135–146.

Sundara, P., Tukimat, N. N. A. B., Ali, M. I. Bin, & Yusof, N. B. M. (2016). Effect of wind induced on road accident along east coast expressway (ECE). *International Journal of Applied Engineering Research, 11*(16), 8858–8862.

Suraji, A., & Tjahjono, N. (2012). A confirmatory factor analysis of accidents caused by the motorcycle aspect in urban area. *International Journal for Traffic and Transport Engineering, 2*, 60–69.

Vogel, L., & Bester, C. J. (2005). A relationship between accident types and causes. In *Southern African Transport Conference* (pp. 233–241), July.

Wang, J., & Ohsawa, Y. (2016). Evaluating model of traffic accident rate on urban data. *Proceedings of the Federated Conference on Computer Science, 8*, 181–186.

Zhang, G., Yau, K. K. W., & Zhang, X. (2014). Analyzing fault and severity in pedestrian-motor vehicle accidents in China. *Accident Analysis and Prevention, 73*, 141–150.

GIS Efficiency Analysis on Traffic Congestion for Emergency Responses in Alor Setar, Kedah

Ummu Syira-Ain Redzuwan, Sharifah Norashikin Bohari, Noradila Rusli and Nursyahani Nasron

Abstract Traffic congestion is a major problem in developing areas, where it disrupts emergency response and more severe that may cause delays and lead to loss of life and loss of valuable goods. This study carried out with the aim of applying geographic information system (GIS) to provide an effective and efficient database for emergency responses in Alor Setar. It examines a methodology where to identify the closest emergency responses and to solve the problem of emergency responses to each incident or vice versa by considering the historical traffic data year 2015 using shortest path and time slider analysis. The finding of this study is development of dynamic road network model for incorporating historical traffic data and performing the best way for a given time and date. Hence, the road network model is considered dynamic in the sense that cost attributes such as travel speed change with respect to time. This study is beneficial to emergency responses as they able to utilize the database to reduce congestion and contribute to efficient emergency responses management.

Keywords Emergency response · Level of service · Geographic information system · Traffic congestion

1 Introduction

The urbanization in developing countries indicates that more people live in cities than before. The trend of urbanization, population increase and the increase in number of registered vehicles brings pressure on traffic movements and makes living in urban area more challenging. According to Pan et al. (2012), the two

---

U. S.-A. Redzuwan (✉) · S. N. Bohari · N. Rusli · N. Nasron
Green Environment and Technology (GREENTech) Research Group, Centre of Study for Surveying Science and Geomatics, Faculty of Architecture, Planning and Surveying, Universiti Teknologi MARA (UiTM), Cawangan Perlis, 02600 Arau, Perlis, Malaysia
e-mail: ummusyira@gmail.com

(2) most important commodities of the twenty-first century are time and energy, but traffic congestion wasting both. In other words, traffic congestion is a major problem in urban areas and can disrupt emergency response. Reported by Traffic Enforcement Investigation Department (JSPT) Bukit Aman (2016), there were 489,606 road accidents cases that lead to 6193 fatalities in year 2015

All countries over the world had encountered traffic congestion problem, especially in the developing countries. Recently, traffic congestion occurred in Malaysia has become more problematic and widespread, thus affecting emergency response performance. As stated by Irdalynna, traffic congestion has been a never-ending problem in Malaysia and lead to the highest numbers of car owners globally with an estimated 997 cars per 1000 people. Moreover, by the area of 9425.00 km^2 and 12 districts, the state of Kedah also contributes traffic congestion towards Malaysia. Therefore, the traffic congestion is unavoidable but a different solution can be created to solve this problem (Rajesh and Benedict 2011).

In the case of emergency medical services, the survival rates for some types of medical problems like heart attack or bleeding resulting from accidents are reliant on the quick intervention of emergency medical team. Derekenaris et al. stated that no one survive cardiac arrest if the response time is more than eight (8) minutes, supporting this statement Kuehnert affirms that the cardiac arrest patient will have a high survival rate if life support cardiopulmonary resuscitation (CPR) is provided within four (4) minutes than if the patient gets the CPR after four minutes. Same goes for the prevention of an intending criminal or terrorist act as well as the quick evacuation of affected population in event of a disaster. Even in the fire service response being the focal point of interest time is a deciding factor to the successes of firefighting. It is also supported the study by Rajesh and Benedict (2011), and the ambulance routing and finding feasible path for safe transportation are the probable necessities to avoid traffic and path constraints. The ambulance service is the exemplary gift showered by the nature, to save life found handicapped in the present existing condition. This reflects the inability of the ambulance driver to reach the spot on-time.

1.1 Study Area

The study area is Alor Setar, Kedah, as shown in Fig. 1. It was chosen because of its strategic importance as the capital and administrative heart of Kedah. Alor Setar becomes an important city as it placed Central State Administration and Administrative Centre at Bandar Muadzam Shah and Kedah Royal City of Bandar Anak Bukit. In short, it is a developing city and accounts for urban current sprawl that impacts on congestion. It is a second largest city after Sungai Petani and administrated by the Local Authority (LA) of Alor Setar City Council.

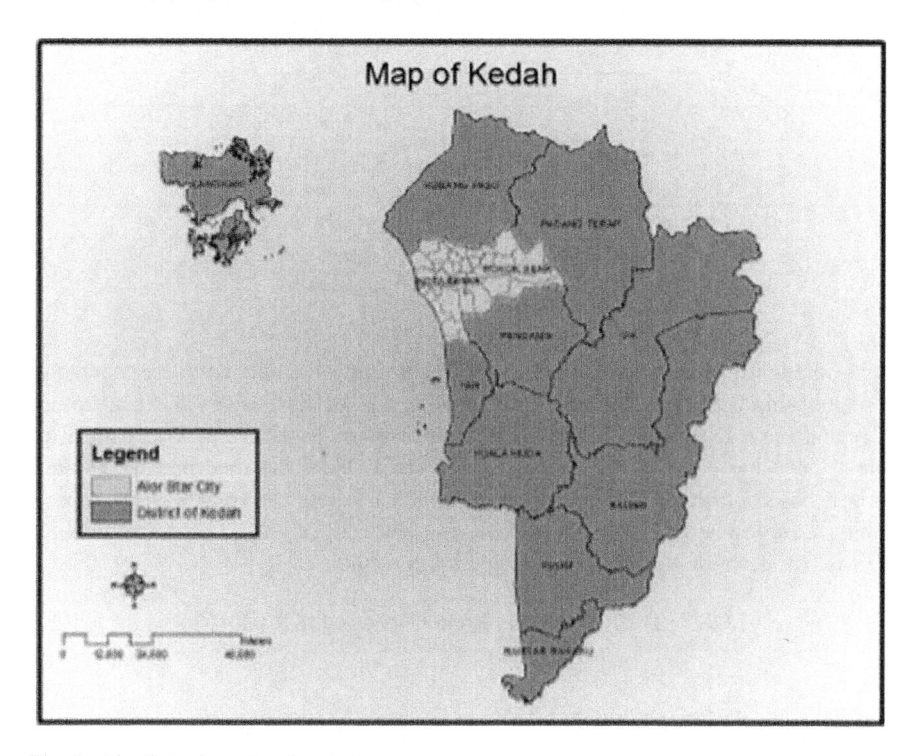

Fig. 1 Alor Setar boundary for study area

2 Data and Methodology

2.1 Dataset

There are four (4) main data collected which were base map of Kedah, streets of Kedah and traffic volume in Alor Setar year 2015 that obtained from Public Work Department (PWD) and also the location of the emergency responses units such as police station, fire station and hospital in the Alor Setar.

The traffic volume data was in the hardcopy format included the numbers of vehicles that pass at certain point at every 60 min, the types of vehicles considered were car, van, lorry, bus and motorcycle and the observation was made within 16 h in a day from 6 am to 9 pm. This data had been observed twice which is March and September every year.

2.2 Methodology

There are about three (3) main phases to complete this study which are data collection, data processing and data analysis using ArcGIS software. In processing

Table 1 Level of service categories

Type	Level	Remark
A	0–0.24	Free flow
B	0.25–0.39	Stable flow
C	0.40–0.69	Stable flow
D	0.70–0.89	Approach unstable flow
E	0.90–1.00	Unstable flow
F	Above 1.00	Congest flow

stages, the ID of street layer and table of traffic profile were utilized to create a table of relationship between street and traffic profile, and all these layers are then stored in geodatabase to generate a network dataset. The traffic data need to calculate to obtain the value of level of service (LOS). The LOS of the road had been categorized based on traffic flow on the road which is a letter from A to F, A representing free flow and F is the worst flow as shown in Table 1. The determination LOS is based on the traffic volume calculation formula.

$$\text{LOS} = \text{Traffic Volume}/\text{Road Capacity (JKR, 2015)} \qquad (1)$$

where

Road capacity for one line = 1200
Road capacity for two line = 3000

Then, from network dataset, time slider tool was activated and able to visualize travel speed with time. Besides, network analyst tool also activated and able to identify closest facility and next; find the best route to avoid congestion. Using historical traffic data, the travel speeds for every hour can be visualized. With the tool of time slider in ArcMap, the road network model is considered dynamic with the cost attributes of travel speed change with respect to time. With time slider, the date and time can be adjusted besides and can be set to current time.

3 Results and Analysis

3.1 Traffic Profile on Weekend and Weekday

In this study, traffic profile divided into two main areas which are rural and urban area during weekend and weekday. Jalan Titi Haji Idris/Datuk Kumbar is one of urban road areas as shown in Fig. 2 which has a peak period at 8.00 am with LOS value is 1.12 while at rural area, Jalan Pokok Sena also has morning peak period at 7.00 am with LOS value is 0.27. The morning peak period was very prominent during weekdays, and it took the lead in all weekdays either in city area or rural

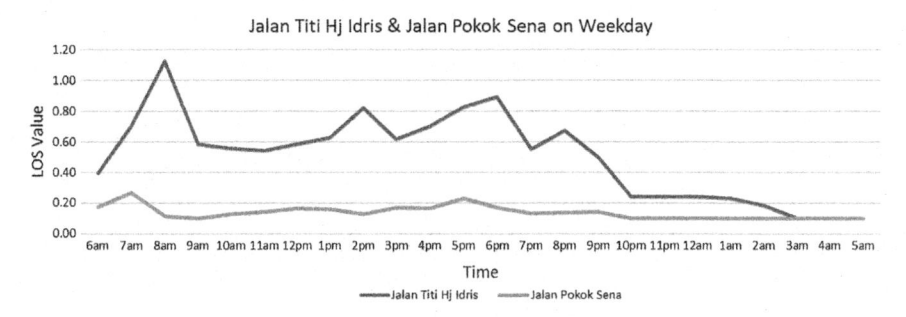

Fig. 2 Traffic profile of Jalan Titi Haji Idris/Datuk Kumbar (blue) and Jalan Pokok Sena (red) on weekday

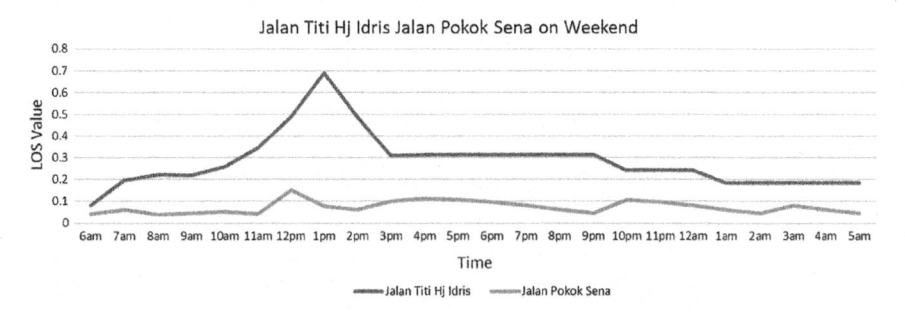

Fig. 3 Traffic profile of Jalan Titi Haji Idris/Datuk Kumbar (left) and Jalan Naka/Bukit Payung (right) on weekend

area. Generally, travelling in the morning during weekday is more difficult because many activities tend to limit the road capacity.

While on weekend, Jalan Titi Haji Idris/Datuk Kumbar at urban area has a peak period at 1.00 pm with LOS value is 0.69, while Jalan Pokok Sena in rural area has a peak period at 12.00 pm with LOS value is 0.15 as shown in Fig. 3, indicates that the roads were calm at morning and start congested on evening due to the day off and people start to do their activities on the evening.

Hence, the city had a high traffic volume and contributed to worst LOS and experienced heavy congested compared to rural area where the LOS at their peak period was still in free flow condition during both weekday and weekend.

3.2 Effect of LOS on Travel Speed on Weekday and Weekend

LOS may affect on travel speed that cause delay and time. In this study, the different colours have been used to represent different categories of travel speed classes on the road. The colour used was followed the standard; green indicates the road is free flow, yellow indicates the road is moderate or stable flow, orange indicates the road is slow or unstable flow and red indicates the road is stop and go or forced flow. Based on Fig. 4, the travel speed at city centre on weekday during 8.00 am (Jalan Titi Haji Idris/Datuk Kumbar) shows the highest morning peak period where the road section was heavily congested with stop and go condition of speed. While towards the city centre, it was observed that the other roads such as Lebuhraya Sultanah Bahiyah were in slow condition of speed and Jalan Langgar was in moderate condition of speed. This condition is also observed during weekend period in city area as shown in Fig. 5.

While in rural area in most condition during weekend and weekday, for example in Jalan Pokok Sena were in free flow condition. These types of results are expected on weekdays, as people have to go out to work and school. All the areas referred to city centre, located in Alor Setar, where there are many school, government and private organizations located in the same zone. The people in district of Pokok Sena

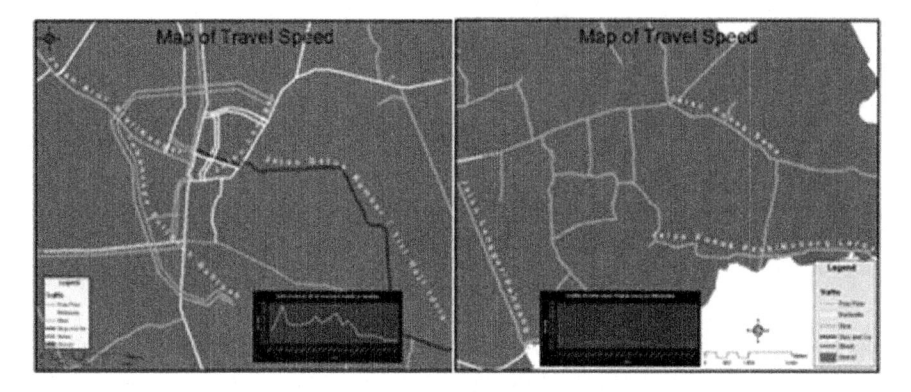

Fig. 4 Travel of speed at city centre (left) and rural area (right) on weekday at 8.00 am

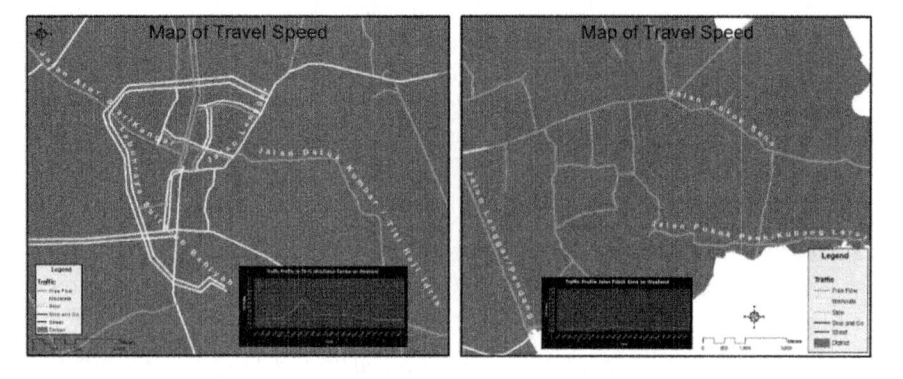

Fig. 5 Travel of speed at city centre (left) and rural area (right) on weekend at 8.00 am

also need to drive to city centre to carry out their activities and make it in free flow condition in all the time. While, on weekend as most of the people in Malaysia fulfil their weekend activity at evening, during morning period the road has not congested as compared with the evening. In short, the city areas attract more traffic than rural areas on weekdays and weekend.

3.3 Effect of Travel Speed on Emergency Responses

In this study, the closest facility was identified and based on the shortest path; it was utilized historical traffic data to model time-varying costs of travelling on the network. Time-varying or time-dependent travel time costs are used to find the best route from an origin to a destination or vice versa.

For example, incident point represents as red dot. In this case, three nearest facilities were selected and identified. Therefore, the three-suggested road provided to the incident point will be different in terms of distance. Then, the shortest distance was selected to go to the incident point. So, based on the shortest distance, Pasukan Bomba Sukarela, Kuala Kedah, is the closest facility to incident point as compared to other facilities as shown in Figs. 6, 7 and 8.

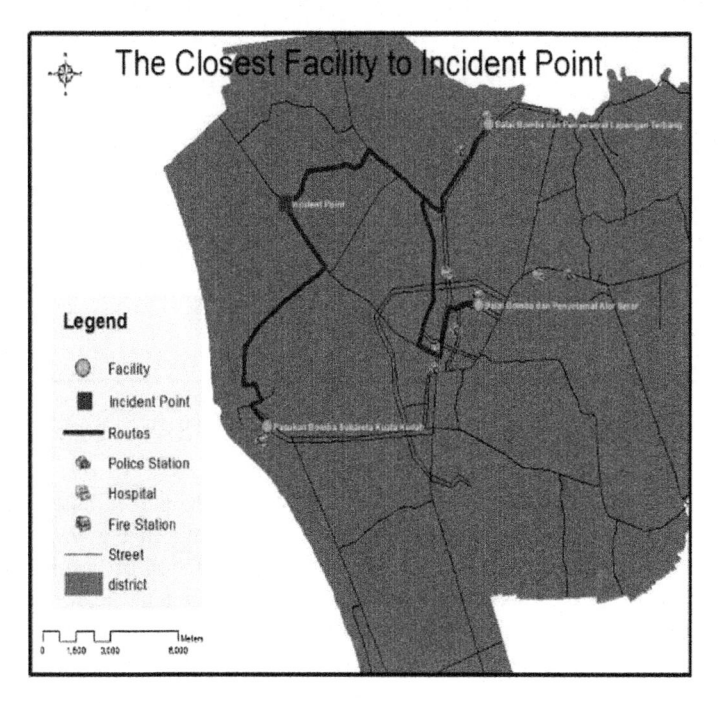

Fig. 6 Closest Facility distance

The Closest Facility	Distance
Pasukan Bomba Sukarela Kuala Kedah	7.3mi
Balai Bomba dan Penyelamat Alor Star	10.9mi
Balai Bomba dan Penyelamat Lapangan Terbang	7.8mi

Fig. 7 Closest facility to incident point

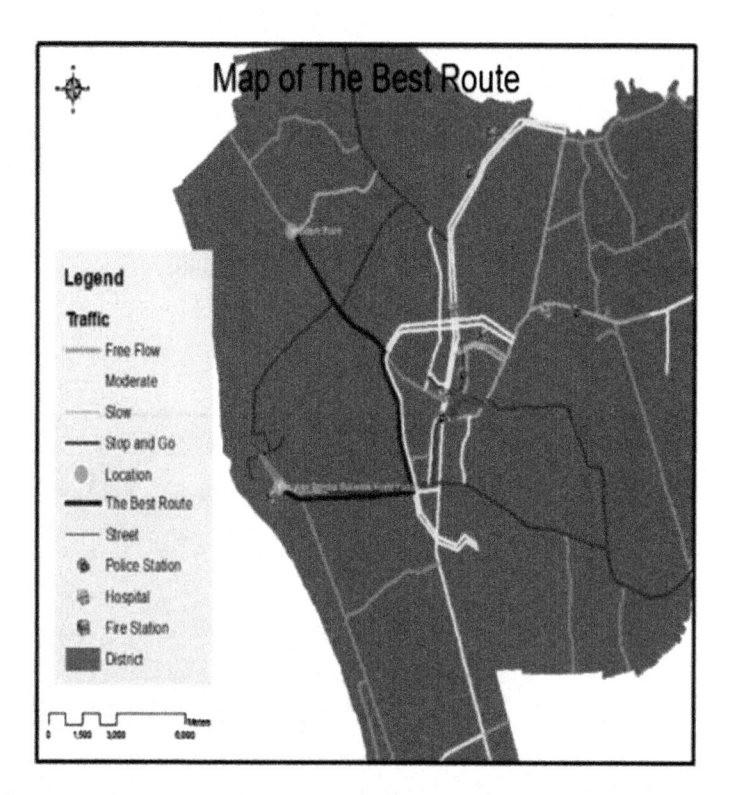

Fig. 8 Best route direction from fire station to incident point at 1.00 pm on weekday

After the identification of closet facilities, then it will be guided based on travel speed of the road. Visualize as purple line is the best route in Fig. 7, the map of travel speed shows during 1.00 pm. In the case of fire, station of Pasukan Bomba Sukarela, Kuala Kedah, needs to reach the incident point at Jalan Alor Setar/Kangar, and the fire truck should get through the Jalan Kubang Rotan/Seberang

```
Begin route Pasukan Bomba Sukarela Kuala Kedah - Incident Point
1: Start at Pasukan Bomba Sukarela Kuala Kedah
2: Go east on Jalan Alor Star /Kuala Kedah toward Jalan Kuala Kedah / Yan Kecil
Drive 3.5 mi - 11 min
3: Turn left on Lebuhraya Sultanah Bahiyah
Drive 3 mi - 4 min
4: Turn left on Jalan Alor Star/Kangar
Drive 3.5 mi - 13 min
5: Finish at Incident Point, on the left
Total time: 29 min
Total distance: 10 mi
Start time: 19-Jan-16 8:00 AM
Finish time: 19-Jan-16 8:29 AM
End of route Pasukan Bomba Sukarela Kuala Kedah - Incident Point
```

Fig. 9 Road direction to incident location

Kuala Kedah as the shortest way to incident point (Fig. 9). However, at 1.00 pm the route was heavily congested which at stop and go condition. Hence, the fire truck not able to go through the route as it will bring delay in travel time and effect in loss of life and properties.

The best direction provided by route solver where the fire truck needs to through Jalan Alor Setar/Kuala Kedah, Lebuhraya Sultanah Bahiyah and Jalan Alor Setar/Kangar to reach destination in fastest time. The direction may different by different time. Without consideration of road capacity, the fire truck will be stuck in congestion that causes delay.

3.4 Visualize the Travel Speed of Traffic in Real Time

Figure 10 was an example of real time of visualizing traffic using time slider in ArcMap database. On this database, emergency responses were able to query from unit station to the incident location or back from incident location to unit station in order to quickly proceed to the next emergency cases. This traffic congestion database enhances emergency responses.

4 Conclusion

In conclusion, GIS is a veritable tool for vital decision-making in traffic congestion for emergency responses provided on a well-designed database. Alor Setar being the economic and administrative heartbeat of Kedah requires a very effective and

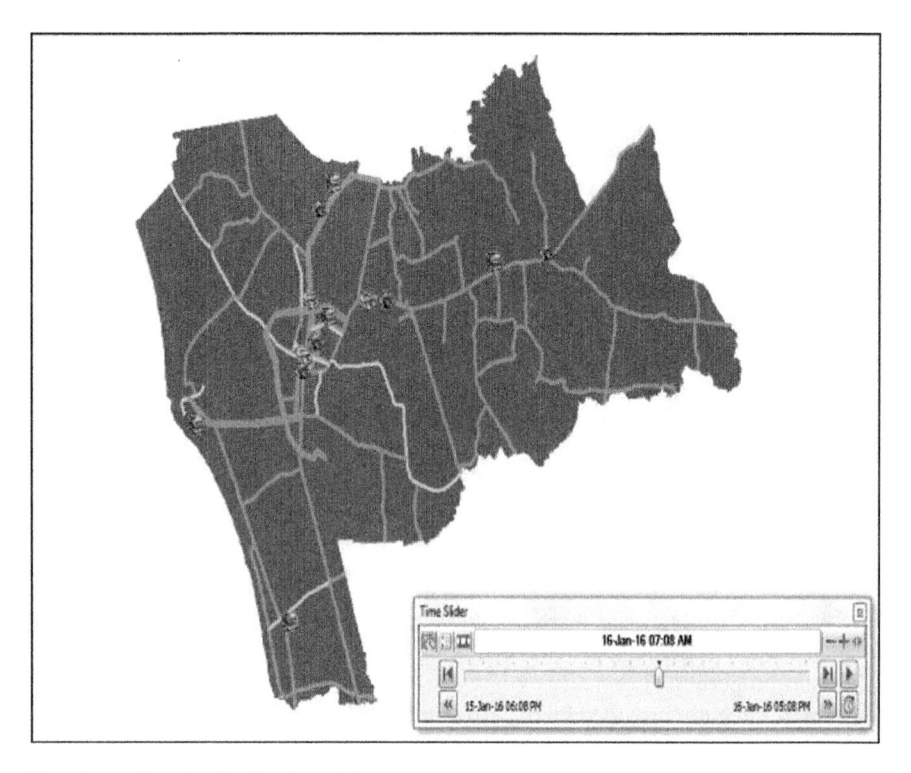

Fig. 10 Traffic visualization in real time

efficient fire emergency coverage service to cope with the growing population and urbanization. It was believed that the dynamic routing based on cost attributes derived from historical travel time data and applied to network edges could help response vehicles avoid congested areas and improve travel times.

Acknowledgements Authors also gratefully acknowledge the helpful comments and suggestions of the reviewers, which have improved the writing content.

References

Pan, B., Demiryurek, U., & Shahabi C. (2012). Utilizing realworld transportation data for accurate traffic prediction. In *2013 IEEE 13th International Conference on Data Mining*, December 2013.

Rajesh Kumar, V., & Benedict P. (2011). Development of route information system for ambulance services using GPS and GIS—A study on Thanjavur town. *International Journal of Geomatics and Geosciences, 2*(1), Integrated Publishing services.

Sea Surface Circulation in the Straits of Malacca and the Andaman Sea Using Twenty-Three Years Satellite Altimetry Data

K. N. A. A. K. Mansor, M. F. Pa'suya, A. H. M. Din, M. A. Abbas, M. A. C. Aziz and T. A. T. Ali

Abstract The ocean circulation in the Straits of Malacca is derived using the satellite altimeter data from January 1993 until 2015. The satellite altimeters are TOPEX, Jason-1, Jason-2, ERS-1, ERS-2, Envisat, SARAL, and Cryosat. The sea surface height derived from the satellite altimeter has been very useful in the study of the ocean circulation but still is not appropriate for the oceanographic application. This is because it is a superposition of geophysical effect such as the tidal effect. The tidal models are more suitable to be used in the open sea like the South China Sea. The tidal effect is rather complex to be determined, especially in a shallow water area like the area in the Strait of Malacca. In order to remove the tidal effect, the best ocean tide model needs to be examined to determine the ocean circulation in the Straits of Malacca. To verify the result, the sea level anomaly (SLA) data were compared to the tide gauge data and the pattern obtained are regularly the same. In order to check the ocean circulation using the altimetric data, the result was compared with the trajectories drifter from Marine Environmental Data Station (MEDS) of Canada. The trajectories of the drifter have confirmed that the current pattern around studied region during June 9, 1999 until July 9, 1999.

Keywords Ocean circulation · Satellite altimeter · Strait of Malacca
Tidal effect

K. N. A. A. K. Mansor (✉) · M. F. Pa'suya · M. A. Abbas · M. A. C. Aziz · T. A. T. Ali
Faculty of Architecture, Planning and Surveying, Universiti
Teknologi MARA, Shah Alam, Perlis, Malaysia
e-mail: kufeezakumansor@yahoo.com

A. H. M. Din
Faculty of Geoinformation and Real Estate, Universiti
Teknologi Malaysia, Johor Bharu, Johor, Malaysia

1 Introduction

In the last four decades, altimeters have been shown to be a powerful tool to measure the sea surface height with its high precision for oceanographic and geodetic application. In oceanographic application, the altimetry data are very useful for the study of ocean circulation and eddy variability due to the long-time series, good resolution and it covers large area. A number of satellite altimeters were launched, but the most important satellite for the study of ocean circulation are TOPEX/Poseidon and Jason-1/2 that had given the most precise altimetry data to date compared to other missions. Unfortunately, the use of altimetric products in the marginal or shallow water is still difficult and the challenge exists in the form of the satellite altimeter observation. According to Durand et al. (2008), there are two major problems when using the altimetry data in the coastal domain, which are the interference of radar echo with the surrounding land and that the standard processing procedure to derive the sea level from altimetry. This is because most of the current-processing procedure has been designed for the open sea. Besides, accurate tidal correction of the altimetry data has to be resolved before applying the altimetry data to the marginal seas (Morimote and Matsuda 2005). Therefore, much attention has been given to study the ocean circulation using altimetry data in the open seas compared to the marginal seas.

However, thanks to the improvement in spatial and vertical resolution brought by SARAL/Altika, it promises a lot for remote sensing of the coastal sea level and provides data products for monitoring of the main continental water levels (Verron et al. 2015). This mission is a Ka-band altimeter that has a narrower footprint than the standard C band altimeters such as TOPEX/Poseidon (T/P hereafter) and Jason mission. Therefore, this mission is expected to perform better in the marginal or coastal region compare to other missions.

2 The Straits of Malacca and Andaman Sea

The Straits of Malacca separates Peninsular Malaysia and Sumatra, forming a funnel-shaped waterway as it narrows to the south. This strait is one of the busiest ways of shipping routes where form the shortest route connecting both the Middle Eastern oil suppliers with the Far Eastern economic giants of China, Japan, and South Korea. In general, Straits of Malacca is a marginal sea with depth water changes slightly from approximately 30 m in the south to 200 m at the north (Rizal et al. 2012). Not much attention has been given to study the ocean circulation in this region compared to the South China Sea and Indian Ocean. According to Sakmani et al. (2013), the direction of ocean current in this strait is almost constant whereby flows northward from the South China Sea towards the Indian Ocean with current speed of approximately 2 m/s.

Andaman Sea is part of the Indian Ocean whereby the Andaman and Nicobar Islands separate the region from the Bay of Bengal. In general, the average depth of

Fig. 1 Topography of Andaman Sea and Malacca Strait, depths are given in meters

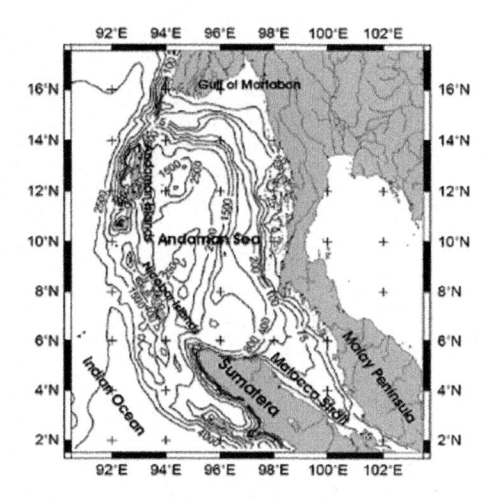

the Andaman Sea is about 1000 m. According to Rizal et al. (2012), the water depth in this region changes rapidly from over 3000–4000 m in the Indian Ocean to approximately 200 m in the area around the islands, returning to deeper than 2500 m in the centre of the Andaman Sea as shown in Fig. 1. The coral reefs and islands in this region are popular tourist destinations besides it is used for fishery and transportation. Like the Straits of Malacca, the ocean circulation study in this region has not been getting much attention. Thus, more and more observations in the Bay of Bengal have made prominent in order to understand the circulation in the region compared to the Andaman Sea of which has been left behind. Varkey et al. (1996) have pointed out that there is an anticlockwise gyre in the Andaman Sea during northeast monsoon (summer) and anticlockwise during southwest monsoon (winter) near the Andaman Island.

3 Satellite Altimeter

The basic principle of the satellite altimetry is based on the simple fact that a period of time is equivalent to a distance, whereby the distance between the satellite and the sea surface is measured from the round-trip travel time of microwave pulses emitted downward by the satellite radar, reflected back from the ocean, and received again on-board as shown in Fig. 2. Meanwhile, independent tracking systems are used to compute the satellite's three-dimensional position relative to an earth-fixed coordinate system. By combining these two measurements, profiles of the sea surface height, H, with respect to a reference ellipsoid is obtained (Din et al. 2012).

$$H = h - R' - \Sigma \Delta Rj \tag{1}$$

Fig. 2 Schematic view of satellite altimeter measurement. Adopted from Fu and Cazenave (2001)

However, the range measured by the satellite altimeters, R, contains a few biases, $\Sigma \Delta Rj$, and the method of yielding sea level profiles is far more complex in practice. Several factors have to be taken into account, such as atmospheric corrections, reference systems, precise orbit determination, varied satellite characteristics, instrument design, calibration, and validation (Naeije et al. 2008; Din et al. 2015). In the oceanographic application, accurate estimates of R and H (corrected from bias) are not sufficient to estimate the dynamic effects of geostrophic ocean currents because the sea surface height given by Eq. 1 is still superposition of a number of geophysical effects such as geoid undulation, tidal height variation and atmospheric pressure loading (Fu and Cazenave 2001). These effects need to be removed in order to get the sea surface's height above the ocean rest state that is the part of the sea surface height that arises from the ocean's circulation and is referred to as the dynamic ocean topography (hd). The dynamic ocean topography is thus estimated as

$$\text{hd} = H - R_{\text{obs}} - \Delta h_{\text{dry}} - \Delta h_{\text{wet}} - \Delta h_{\text{iono}} - \Delta h_{\text{ssb}} - h_{\text{geoid}} - h_{\text{tides}} - h_{\text{atm}} \qquad (2)$$

where

H Satellite altitude
R_{obs} $c\,t/2$ is the computed range from the travel time, observed by the on-board ultra-stable oscillator (USO), and c is the speed of the radar pulse neglecting refraction (approximate 3×10^8 m/s)
Δh_{dry} Dry tropospheric correction
Δh_{wet} Wet tropospheric correction

Δh_{iono} Ionospheric correction
Δh_{ssb} Sea state bias correction
H_{mss} Mean sea surface correction
h_{tides} Tides correction
h_{atm} Dynamic atmospheric correction.

In the present study, efforts have been made to present and describe the accuracy of altimetry data derived ocean circulation pattern in the shallow region that is the Strait of Malacca and the Andaman Sea using a 23 years multi-mission and long-term altimetry data set.

4 Data and Method

The sea level anomaly data from January 1993 to December 2015 were derived from a multi-mission satellite altimetry data from TOPEX, Jason-1, Jason-2, ERS-1, ERS-2, EnviSat, CryoSat-2, and SARAL/Altika using Radar Altimeter Database System (RADS). The altimetry sea level anomaly data extracted ranges between $2°N \leq$ Latitude $\leq 18°N$ and $91°E \leq$ Longitude $\leq 102°E$, covering the Malacca Straits and Andaman Sea. Figure 3 shows the overview of altimeter data processing flows in RADS.

The sea level data are corrected for orbital altitude; altimeter range corrected for the instrument, sea state bias, ionospheric delay, dry and wet tropospheric corrections, solid earth and ocean tides, ocean tide loading, pole tide electromagnetic bias, and inverse barometer corrections. The bias is reduced by applying specific models for each satellite altimeter mission in RADS as shown in Table 1. In this study recent, DTU13 MSS model released by the Denmark Technical University is used as a datum to derive the sea level anomaly. Crossover adjustments have been performed after applying altimeter corrections and removing bias to adjust the sea

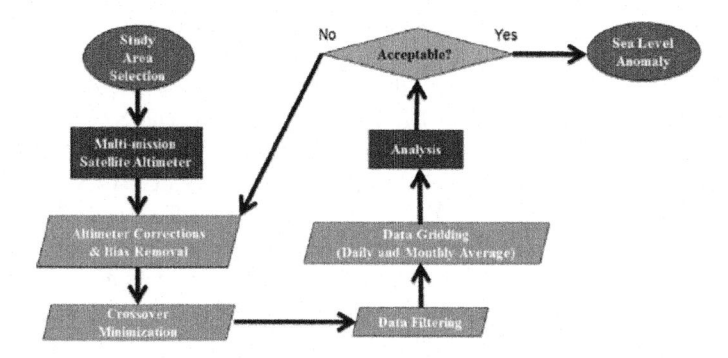

Fig. 3 Overview of altimeter data processing flows in RADS. Adapted from Din et al. (2014)

Table 1 Corrections and models applied for RADS altimeter processing

Correction/model	Editing (m) Min	Editing (m) Max	Description
Orbit/gravity field			All satellites: EIGEN GL04C ERS: DGM-E04/D-PAF
Dry troposphere	−2.4	−2.1	All satellites: atmospheric pressure grids (ECMWF)
Wet troposphere	−0.6	0.0	All satellites: radiometer measurement
Ionosphere	−0.4	0 04	All satellites: smoothed dual-frequency ERS: NIC09
Dynamic atmosphere	−1.0	1.0	All satellites: MOG2D
Ocean tide	−5.0	5.0	All satellites: GOT4.10
Load tide	−0.1	0.1	All satellites: GOT4.10
Solid earth tide	−1.0	1.0	Applied (elastic response to tidal potential)
Pole tide	−0.1	0.1	Applied (tide produced by polar wobble)
Sea state bias	−1.0	1.0	All satellites: CLS non-parametric ERS: BM3/BM4 parametric
Reference engineering flag	−1.0	1.0	DTU13 mean sea surface applied
Reference surface			Jason-1 Jason-2 TOPEX

surface heights (SSH) from different satellite missions to a "standard" surface. In this process, the orbit of the TOPEX-class satellites (T/P and Jason) held fixed and those of the ERS-class satellites adjusted concurrently.

As discussed before, shallow or marginal water area like the Straits of Malacca and the Andaman Sea gives a number of challenges to the users of satellite altimeter due to the tidal effect. In the altimeter data processing, global tide model usually used to remove the tidal effect but the global tide model does not have sufficient accuracy to remove the effect (Yanagi et al. 1997) and the effect larger and complex in coastal region (Andersen 1999). Thus, efforts have been made to evaluate the accuracy of three (3) global tide models that are FES2004, GOT4.8, and GOT4.10 for the study area. The Finite Element Solution (FES2004) includes nine short period waves such as Q1, O1, K1, P1, 2N2, N2, M2, K2, and S2. The Goddard ocean tide model GOT4.7 includes ten short period waves such as K1, O1, P1, O1, S1, K2, M2, N2, S2, and M4. The Goddard ocean tide model GOT4.8 differs from GOT4.7 only in its S2 component. The different between GOT4.8 and GOT4.10 are GOT4.8 data come from only TOPEX data but GOT4.10 only on Jason data. The GOT4.10 is now the default tide model in altimeter data processing in RADS. At the final stage processing, all the data were merged and gridded into $0.25° \times 0.25°$ latitude/longitude using Gaussian weighting function with sigma 1.5. The corrected and gridded SLA that has been processed from RADS is used to estimate the geostrophic current using the geostrophic approximation;

$$\text{ui}: \frac{-g}{f} \frac{(h_4-h_3)}{(y_2-y_1)} \quad \text{vi}: \frac{g}{f} \frac{(h_2-h_1)}{(x_1-x_3)} \tag{3}$$

$$x: R \times (\lambda) \times \pi/180° \times \cos(\phi)$$
$$y: R \times (\phi) \times \pi/180°$$
$$f: 2 \times \Omega \times \sin(\phi)$$

where ui and vi are zonal (east direction) and meridional (north direction) components of velocity, g is the gravitational acceleration, f is the Coriolis force with Ω being the earth's rotational rate (7.292115×10^{-5}), λ is the longitude, φ is the latitude, h is the absolute dynamic topography, x and y are the local east and north coordinate.

5 Result and Discussion

5.1 Estimation the Best Global Ocean Tide Model

In order to estimate the best global tide model to remove the tidal signal in altimetric data for study area, the sea level from the altimeter corrected using four global tide models is compared with the tidal data from the selected tide gauge stations around west coast of Peninsular Malaysia, Langkawi, Port Klang and Pulau Pinang, Kukup, and Tanjung Kling. The data are obtained from Permanent Service for Mean Sea Level Web site. The monthly averages of SLA are computed and interpolated into the tide gauge position using inverse distance weighting method to compute the monthly time series of SLA. The accuracy for each model is computed, and the result is shown in Table 2.

Table 2 shows the root mean square error (RMSE) for each model. For family of Finite Element Solution (FES), FES 2012 gets better accuracy with the range of 0.040–0.086 compare with FES2004 (0.056–0.094). For Goddard Ocean Tide (GOT), GOT 4.10 has a better accuracy compare to GOT 4.8 when it only has a

Table 2 Accuracy of ocean tide model

	FES2004	GOT4.8	GOT4.10	FES2012
Langkawi	0.062	0.049	0.049	0.049
Port Klang	0.088	0.056	0.058	0.058
Pulau Pinang	0.056	0.040	0.040	0.040
Kukup	0.094	0.087	0.086	0.086
Tanjung Kling	0.086	0.079	0.080	0.080

slight difference at tide gauge for Port Klang, Kukup, and Tanjung Kling. According to these tide gauges, GOT 4.10 is more dominant. GOT 4.10 come with the accuracy within the range of 0.040–0.086 but GOT 4.8 with the range of 0.040–0.087. Same as regression, GOT 4.10 and FES 2012 also share the same accuracy. But, note that, tide loading effects have not yet been computed for FES2012 and need to use load tide from GOT load tide models combined with ocean tide from FES2012. In addition, GOT 4.10 is a default tide model in RADS. The GOT 4.10 on the other hands includes an adjustment for the geocenter but not for GOT 4.8. The geocenter is described as an earth's motion in inertia space and serves as the orbital center for all earth satellites. With this, GOT 4.10 has been chosen as the best global ocean tide model to be used in this research.

Based on the best global ocean tide model (GOT4.10), the pattern of the sea level trends has been plotted. Figure 4 shows the pattern of the sea level trends between the tide gauge and the altimeter. The pattern is relatively similar to each other. The west coast Peninsular Malaysia seems like noisy and inconsistence. This part became noisy and inconsistence because the area is exposed to the narrow and shallow water.

5.2 Validation of the Surface Current Pattern with the GPS-Tracked Drifter

In order to evaluate the estimated sea surface circulation pattern from the altimetric data, the results are compared with the selected Argos-tracked Drifting Buoys. The drifter data obtained from Marine Environmental Data Station (MEDS) in Canada (http://www.meds-sdmm.dfo-mpo.gc.ca/meds/Database). Figure 5a, b shows the trajectories of drifter in the Strait of Malacca and geostrophic current pattern derived from altimeter data. From Fig. 5a, one can see the drifter 1162798 release at 8.078°N 95.032°E on June 9, 1999 from Andaman Sea moved all the way southward before turn eastward at around ~ 5.5°N, 97°E toward Peninsular Malaysia. However, the drifter turns cyclonically northward before changes direction to southward until end at 4.307°N 99.6°E on July 9, 1999. Probably, it is indicating the existence of anticyclonic circulation in the Straits of Malacca. In general, the pattern of geostrophic current is derived in this study almost consistent with the trajectories of drifter.

5.3 Seasonal Ocean Surface Circulation

The seasonal surface current is computed from a long-term averaged SLA (1993-2015) and plotted based on the monsoon season, and the results have been compared with the results from the hydrographic observation by Wyrtki. Figure 6a,

Fig. 4 Sea level trends at selected tide gauge

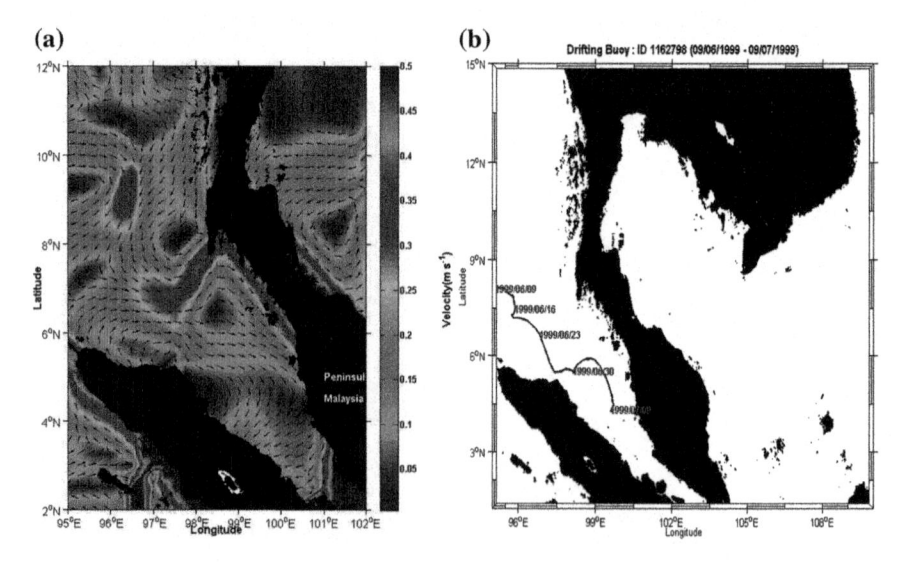

Fig. 5 Geostrophic current derived from altimetric data (**a**) and trajectories of drifter (**b**)

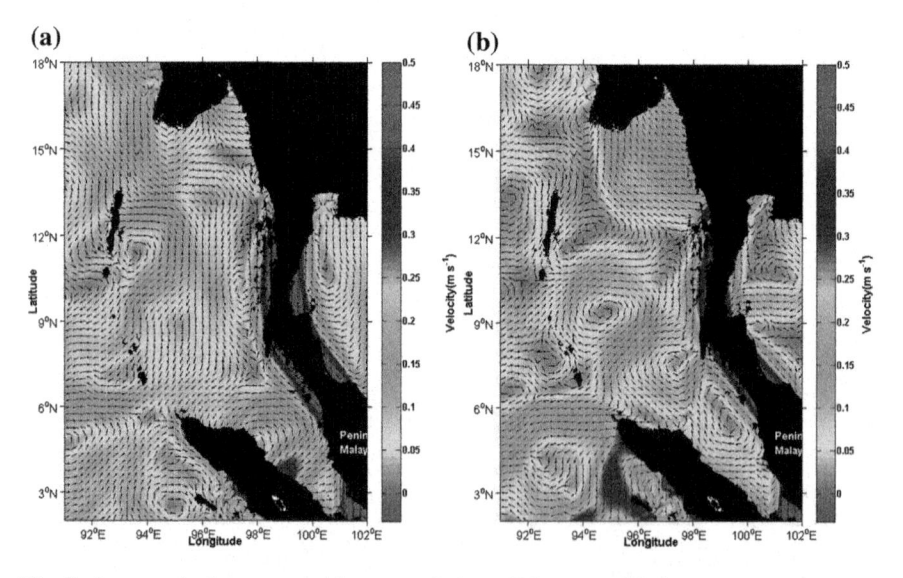

Fig. 6 Average absolute geostrophic current during **a** February and **b** August

b shows the composite plots of average absolute geostrophic current for February and August which represents the Northeast Monsoon (NE) and Southwest Monsoon (SW), respectively, superimposed on the geostrophic current velocity. Meanwhile, Fig. 7a, b shows the results from hydrographic observation by Wyrtki (1961). In general, one can see the surface current flow from the Bay of Bengal and entering

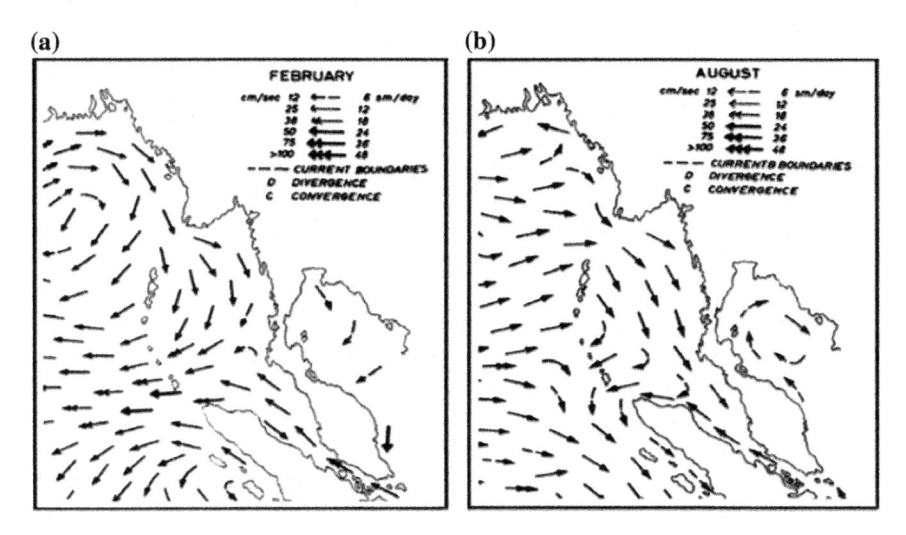

Fig. 7 Surface currents during **a** February and **b** August, according to Wyrtki (1961)

the Andaman Sea at the northern part of the region and flow southward during February. At around $\sim 13°N–15°N$, part of the current flows eastward towards the coastal area, before continuing south. The water leaves the Andaman Sea in the wide area between south of the Andaman Islands and Sumatra to the Bay of Bengal. An anticyclone has detected near the Andaman Island centered at ($\sim 12°N\ 94°E$) and a cyclonic eddy formed near the coastal region centered at ($\sim 10°N\ 98°E$). In general, the circulation patterns during February is almost consistent to that of Wyrtki (1961) except the existence of eddies in the Andaman Sea during February, which is not reported by Wyrtki (1961). However, the existence of anticyclonic eddy near the Andaman Island is almost consistent with the result by Varkey et al. (1996), whereby they found that there is one clockwise gyre form in the centre of the Andaman Sea.

In the Straits of Malacca, the surface current flows northeastward from the middle part of Sumatra's east coast towards the Peninsular Malaysia before turn cyclonically eastward and flow westward towards Sumatra. Apart from that, part of the current also flows northward along the west coast of Peninsular Malaysia. In general, the circulation patterns are not similar to that of Wyrtki (1961). They found that the surface flow is directed north-westward towards the Andaman Sea.

During August, the surface circulation in the Andaman Sea is rather complex than during February. At the northern part of the Andaman Sea, the surface water from Bay of Bengal flow enters the Andaman Sea before converging with the westward current from coastal region and flow southward along the Andaman Island. Around $9°N–11°N$, a second surface water enters the Andaman Sea from Bay of Bengal at the southern part of Andaman Island and flow northeastward before converging with westward current from coastal region. Part of the current flows cyclonically eastward and forms a clockwise gyre centered at ($\sim 9°N\ 95°E$).

Interestingly, the anticyclonic eddy exists near the east coast of Andaman Island has changed to a cyclonic eddy during this season. The ocean surface circulation pattern derived from altimeter data is not similar to what was reported by Wyrtki (1961). They only reported that the water moves southward toward Straits of Malacca before leaving the Andaman Sea in the wide area between south of the Andaman Islands and Sumatra to the Indian Ocean. The existence of eddies in the Andaman Sea during August also not report by the Wyrtki (1961). However, the cyclonic eddy near the Andaman Island almost consistent what reported by Varkey et al. (1996).

In the Straits of Malacca, a clockwise gyre was clearly seen and had been formed in the middle part of the strait. At the northern part of the strait, the surface water from the Andaman Sea enters the Straits of Malacca before form a clockwise gyre centered at ($\sim 5°N$ 99°E). A weak cyclonic eddy also has been detected at the northern part of the Malacca Straits centered at ($\sim 8°N$ 98°E). In general, the ocean circulation pattern in the Straits of Malacca during August is totally different from what has been reported by Wyrtki (1961).

6 Conclusion

The present study aims to study the ocean circulation in the Straits of Malacca and the Andaman Sea using the altimeter data from January 1993 until February 2016 (23 years). Basically, the sea surface geostrophic current derived using satellite altimeter is the alternative approach to understand the ocean circulation. The geostrophic current considers only one factor which is the pressure gradient compared to other research used through the combination such as from the wind and tidal factors in investigating the surface current. Actually, the altimetric data has a long-term observation data, and also, high-resolution data is enough to represent the ocean surface current in this study area, but a more detailed study needs to be done to confirm the presence of the eddy surround. Based on this research, the surface current pattern derived from the altimeter data is consistent with the trajectories of the drifter provider by Marine Environmental Data Station (MEDS) of Canada. The existence of eddies near the Andaman Island during February and August is almost consistent with the result by Varkey et al. (1996) but was not reported by the Wyrtki (1961). Interestingly, this study also reveals the existence of a clockwise gyre in the middle part of the Straits of Malacca during August and is not reported by any previous study. The existence of the gyre is almost consistent with the trajectories of drifter 1162798.

Acknowledgments The authors would like to thank the TU Delft, NOAA, Altimetrics LLC for providing altimetry and Marine Environmental Data Station (MEDS) of Canada for drifter data. Special thanks to Universiti Teknologi MARA, and Ministry of Higher Education Malaysia for funding this project under the Research Acculturation Grant Scheme (RAGS) Fund; Vote Number RAGS/1/2014/STWN04/UITM.

References

Andersen, O. B. (1999). Shallow water tidal determination from altimetry—The M 4 constituent. *Bollettino Di Geofisica Teorica Ed Applicata, 40*(3–4), 427–437.

Din, A. H. M., Omar, K. M., Naeije, M., & Ses, S. (2012). Long-term sea level change in the Malaysian seas from multi-mission altimetry data. *International Journal of Physical Sciences, 7*(10), 1694–1712, March 2, 2012. https://doi.org/10.5897/ijps11.1596.

Din, A. H. M., Reba, M. N. M., Omar, K. M., Pa'suya, M. F., & Ses, S. (2015). Sea level rise quantification using multi-mission satellite altimeter over Malaysian Seas. In *The 36th Asian Conference on Remote Sensing (ACRS 2015)*. October 19–23, 2015. Manila, Philippines. Scopus—Proceeding.

Din, A. H. M., Ses, S., Omar, K. M., Yaakob, O., Naeije, M., & Pa'suya, M. F. (2014). Derivation of sea level anomaly based on the best range and geophysical corrections for Malaysian seas using radar altimeter database system (RADS). *Jurnal Teknologi (Sciences And Engineering), 71*(4), 83–91.

Durand, F., Shankar, D., Birol, F., & Shenoi, S. S. C. (2008). Estimating boundary currents from satellite altimetry: A case study for the east coast of India. *Journal of Oceanography, 64* (6), 831–845.

Fu, L. L., & Cazenave, A. (2001). *Altimetry and earth science, a handbook of techniques and applications Vol. 69 of international geophysics series*. London: Academic Press.

Morimote, A., & Matsuda, T. (2005). The application of altimetry data to the Asian Marginal Seas. In *Proceedings of IEEE International IGARSS 2005* (Vol. 8, pp. 5432–5435).

Naeije, M., Scharroo, R., Doornbos, E., & Schrama, E. (2008). Global altimetry sea level service: GLASS. NIVR/SRON GO project: GO 52320 DEO.

Rizal S., Damm P., Wahid M. A., Sundermann J., Ilhamsyah Y., & Iskandar T. (2012). General circulation in the Malacca Strait and Andaman Sea: A numerical model study. *American Journal of Environmental Sciences, 8*(5), 479–488.

Sakmani, A. S, Lam, W. H., Hashim, R., Chong, H. Y. (2013). Site selection for tidal turbine installation in the strait of Malacca. Renewable and Sustainable Energy Reviews 2013, in press.

Varkey, M. J., Murty, V. S. N., & Suryanarayana A. (1996). Physical oceanography of the Bay of Bengal and Andaman Sea. *Oceanography Marine Biology: An Annual Review, 34*, 1–70.

Verron, J., Sengenes, P., Lambin, J., Noubel, J., Steunou, N., & Guillot, A. et al. (2015). The SARAL/AltiKa altimetry satellite mission. *Marine Geodesy, 38*(sup 1), 2–21. https://doi.org/10.1080/01490419.2014.1000471.

Yanagi, T., Morimoto A., & Ichikawa, K. (1997) Co-tidal and co-range charts for the East China Sea and the yellow Sea derived from satellite altimetry data. *Journal of Oceanography, 53*, 303–310.

Accuracy Assessment of TanDEM-X DEM and Global Geopotential Models for Geoid Modeling in the Southern Region of Peninsular Malaysia

Muhammad Faiz Pa'suya, Ami Hassan Md Din,
Zulkarnaini Mat Amin, Noradila Rusli, Amir Hamzah Othman,
Mohamad Azril Che Aziz and Mohd Adhar Abd Samad

Abstract In modeling of geoid model, global digital elevation models (GDEMs) and global geopotential models (GGMs) involve in most part of the geoid computation process. Any errors in GDEMs and GGMs will introduce errors directly in geoid computation. Therefore, this study aims to evaluate the six recent GGMs and new digital elevation model from TanDEM-X, as well as the previously available GDEMs, SRTM GDEMs, over the southern region of Peninsular Malaysia. The evaluation of GDEMs has been performed with the use of high-precision Global Navigation Satellite System (GNSS) and EGM96 as vertical reference consisting of 277 stations. Meanwhile, the evaluation of GGMs is carried out using sixty-two (62) collocated GPS/leveling benchmarks (BMs). Based on the statistical analysis, it is shown that the improvement of DEM from TanDEM-X data is compared to the previously available DEMs, SRTM GDEMs. DEM from TanDEM-X of 30-m arc resolution is much better than TanDEM-X of 12-m arc resolution, as well as SRTM 30m and 90m. Comparison of GGMs with GNSS leveling shows that geoid height from GOCO05c fits well with the local geoid model.

Keywords Geoid · GDEMs · SRTM · TanDEM-X · GGMs

M. F. Pa'suya (✉) · N. Rusli · M. A. C. Aziz · M. A. A. Samad
Green Environment and Technology (GREENTech) Research Group, Centre of Study for Surveying Science and Geomatics, Faculty of Architecture Planning and Surveying, Universiti Teknologi MARA, Perlis Branch Arau Campus, 02600 Arau, Perlis, Malaysia
e-mail: faiz524@perlis.uitm.edu.my

A. H. Md Din · Z. M. Amin · A. H. Othman
Faculty of Geoinformation and Real Estate, Universiti Teknologi Malaysia,
81310 Johor Bahru, Johor, Malaysia

1 Introduction

Since the surveying and mapping activities have been revolutionized by the use of Global Navigation Satellite System (GNSS), vertical reference frame has become very imperative. The ability of GNSS to define position with the accuracy up to millimeter (mm) is undeniable, but the ellipsoidal height provided by GNSS cannot be satisfied in surveying and mapping as they have no physical meaning. The height must be transformed to orthometric heights (H), which are referred to as geoid (Erol and Celik 2004) by combining geoidal heights (N) and ellipsoidal heights (h). Precise knowledge of the geoidal height can lead to the estimation of orthometric height. Thus, determining a precise geoid model has become a major task in geodesy today. It has been one of the main research areas in science of geodesy to achieve the goal for a "1-cm geoid" accuracy and to be able to determine the orthometric height using the GNSS without leveling.

In the modeling of a geoid model, global digital elevation models (GDEMs) and global geopotential models (GGMs) have become imperatively crucial in providing information and involvement in most part of the geoid computation process. GDEMs are very important in the computation of Bouguer gravity correction, combined topographic correction, and the downward continuation effect. Meanwhile, GGMs play a major role to overcome sparse and inaccuracy of the terrestrial gravity observation issue that usually occurs and will host error in the local gravimetric geoid model. Thus, the best way is to use the GGMs and combine it with the terrestrial gravity data to compute a precise regional gravimetric geoid (Kiahmehr 2006; Triarahmadhana and Heliani 2014). Nowadays, a number of the most-recent GGMs and GDEMs have been obtained and freely released to the users for scientific studies. However, the accuracy of GGMs and GDEMs should be analyzed in order to choose the optimum GGMs and GDEMs for the geoid modeling. According to Ssengendo (2015), any error in the GDEMs will affect the accuracy of the geoid model. Thus, the goal of this study is to evaluate the accuracy of several GDEMs and GGMs using GNSS and leveling data in the southern part of Peninsular Malaysia as shown in Fig. 1.

2 GDEMs and GGMs Data

2.1 Global Digital Elevation Models (GDEMs)

In this study, two (2) GDEMs with different resolutions are evaluated and those are Shuttle Radar Topography Mission (SRTM) GDEM and the newly released GDEM from Tandem-X mission. SRTM GDEM is the most well-known GDEM and has been applied in most scientific studies. Many studies have evaluated the accuracy of SRTM GDEM using various reference datasets such as Khalid et al. (2016),

Fig. 1 Distribution of 277 points for evaluation of GDEMs superimposed on SRTM GDEMs

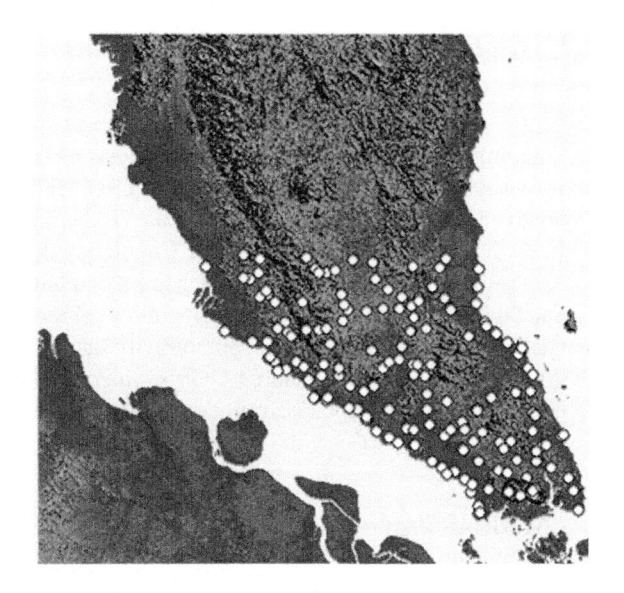

Rodriguez et al. (2016), Mukherjee et al. (2013), Kiahmehr (2006), and others. SRTM GDEMs provide the most complete highest resolution DEM of the world at resolution levels of 1 arc-sec (∼30 m) and 3 arc-sec (∼90 m). The SRTM DEMs 1 arc-sec (∼30 m) and 3 arc-sec (∼90 m) are downloaded via the csi.cgiar Web site (http://srtm.csi.cgiar.org/) and the USGS Web site (http://earthexplorer.usgs.gov/). Since 2010, the German Aerospace Center (DLR) has been operating Germany's first two formations flying Synthetic Aperture Radar (SAR) satellites, TerraSAR-X and TanDEM-X, to generate the new global DEM which is the generation of a worldwide, consistent, current, and high-precision DEM, with a spatial resolution of 0.4 arc-seconds (Wecklich et al. 2015). Various studies have evaluated the accuracy of Tandem-X DEM (Becek et al. 2016; Erasmi et al. 2014; Wecklich et al. 2015) and have found that the DEM derived from TanDEM-X is the most accurate global DEM to date in terms of its vertical accuracy. For this study, TanDEM-X data with two (2) resolutions, 30 and 12 m, from DLR's TerraSAR-X/TanDEM-X satellite which provided by the German Aerospace Centre were used. Table 1 shows the characteristics of the TanDEM-X and SRTM GDEMs.

Table 1 Description of SRTM and TanDEM-X GDEMs

No.	Model	Spatial resolution	Horizontal datum	Vertical datum
1	SRTM	30, 90 m	WGS84	EGM96
2	TanDEM-X	12, 30 m	WGS84	WGS84 reference ellipsoid

2.2 Global Geopotential Models (GGMs)

Currently, GGMs can be classified into three groups. The first group is satellite-only, which is derived from the analysis of the orbits of artificial satellites such as advent of satellite gravity mission CHAMPS (Challenging Mini-satellite Payload), GRACE (Gravity and Climate Experiment recovery), and GOCE (Gravity field and steady-state Ocean Circulation Explorer). The second group combines the gravity field models which are a combination of gravity sources such as satellite data, land and ship-track gravity observations, and marine gravity anomalies, and more recently airborne gravity data to yield high-degree combined GGMs. The third group is tailored GGMs which is the satellite-only, or combined GGMs is adjusted using unused higher resolution gravity data.

3 Methodology

In order to evaluate the accuracy of TanDEM-X and SRTM GDEMs, each model was evaluated using 277 GNSS observations. The GNSS data were collected by the Department of Survey and Mapping Malaysia around the study area using the Real-Time Kinematic (RTK)-GNSS and Static-GNSS methods. The data are provided for this study through a special request made by the researcher, and the distribution of points is shown in Fig. 1. Since the vertical datum for the SRTM is EGM96, the ellipsoidal height derived from GNSS observation, h, is transformed to orthometric height using EGM96 (H_{EGM96}) using the following equation:

$$H_{\text{Reference}} = h - N_{\text{EGM96}} \qquad (1)$$

Kriging approach from SURFER software (http://www.goldensoftware.com/) is applied to interpolate the GDEMs height, and the comparison is traditionally based on the following model of the difference between the two datasets:

$$\Delta H = H_{\text{Reference}} - H_{\text{GDEMs}} \qquad (2)$$

where

H orthometric height,
h WGS84 ellipsoid height.

Meanwhile, the height DEM derived from TanDEM-X was directly compared with the ellipsoidal height derived from GNSS observation, h. The vertical accuracy was computed by the mean error (ME) and vertical root mean square error (RMSE) by following Eqs. (3) and (4) where n is the total number of points.

$$ME = |\Delta H|/n \qquad (3)$$

$$RMSE = \text{sqrt}\left(\sum \Delta H^2\right)/n \qquad (4)$$

In this study, six (6) GGMs model of which are recently published in the year 2016 are evaluated using the 62 GNSS/leveling points and the distribution of the points is shown in Fig. 2. The GGMs is computed from the International Centre for Global Earth Models (ICGEM) Web site (http://icgem.gfz-potsdam.de/tom_longtime), and the characteristic of utilized GGMs is shown in Table 2.

Fig. 2 Distribution of 62 GNSS leveling points for evaluation of GGMs

Table 2 Description of GGMs evaluated in the study

Model	Year	Degree	Data	Reference
HUST-Grace2016s	2016	160	S(Grace)	Zhou et al. (2017)
ITU_GRACE16	2016	180	S(Grace)	Akyilmaz et al. (2016b)
ITU_GGC16	2016	280	S(Grace, Goce)	Akyilmaz et al. (2016a)
EIGEN-6S4v2	2016	300	S(Goce, Grace, Lageos)	Förste et al. (2016)
GOCO05c	2016	720	S, G, A (see model)	Pail et al. (2016)
GGM05C	2016	360	S(Grace, Goce), G, A	Ries et al. (2016)

Source http://icgem.gfz-potsdam.de/tom_longtime

The GGMs are evaluated by comparing the GGM-derived geoid height (N_{GGMs}) with GNSS/leveling-derived geoid height (N_{GNSS}) based on the model where h is WGS84 ellipsoid height:

$$N_{\text{GNSS}} = h - H_{\text{levelling}} \tag{5}$$

$$\Delta N = N_{\text{GNSS}} - N_{\text{GGMs}} \tag{6}$$

The vertical accuracy is computed by mean error (ME) and vertical root mean square error (RMSE) by following Eqs. (7) and (8) where n is total numbers of points.

$$\text{ME} = |\Delta N|/n \tag{7}$$

$$\text{RMSE} = \text{sqrt}\left(\sum \Delta N^2\right)/n \tag{8}$$

In order to minimize biases due to reference ellipsoid, all the GGMs were computed from the ICGEM Web site that will be referred to Tide-Free system and WGS84 as the normal field.

4 Result and Analysis

4.1 GDEMs Accuracy Assessments

The statistical summary of the GDEMs is shown in Fig. 4. As expected, the comparison between SRTM 30 m and SRTM 90 m shows SRTM 30 m performs well compared to SRTM 90 m GDEM for the southern region of Peninsular Malaysia. The minimum and maximum values between the height derived from GNSS/EGM96 and SRTM 30 GDEM were 0.002 and 19.834 m, respectively. The MRE and RMSE value for this GDEM was calculated as 3.969 and ±5.068 m, respectively. Meanwhile, statistical analysis for TanDEM-X DEM highlights that the accuracy of TanDEM-X 30 m is much better than TanDEM-X 12 m where the MRE and RMSE values are 2.376 and ±3.813 m, respectively. In general, TanDEM-X 30 m performs well compared to others GDEMs (see Fig. 3).

Figure 4 shows the graphic plots and the summary of linear correlation of the GNSS-derived elevations against the SRTM and TanDEM-X GDEMs. Based on the results, it indicates that the two GDEMs have showed strong positive correlations with the elevations derived from GNSS. Overall, the correlation value R^2 is about ~0.97 and TanDEM-X 30 m shows the much strong correlation which is 0.9750 as shown in Fig. 4.

Accuracy Assessment of TanDEM-X DEM ...

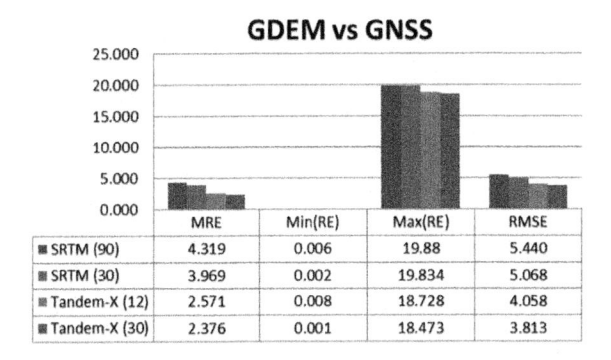

GDEM vs GNSS

	MRE	Min(RE)	Max(RE)	RMSE
SRTM (90)	4.319	0.006	19.88	5.440
SRTM (30)	3.969	0.002	19.834	5.068
Tandem-X (12)	2.571	0.008	18.728	4.058
Tandem-X (30)	2.376	0.001	18.473	3.813

Fig. 3 Statistics of the differences between heights derived from GDEM and GNSS

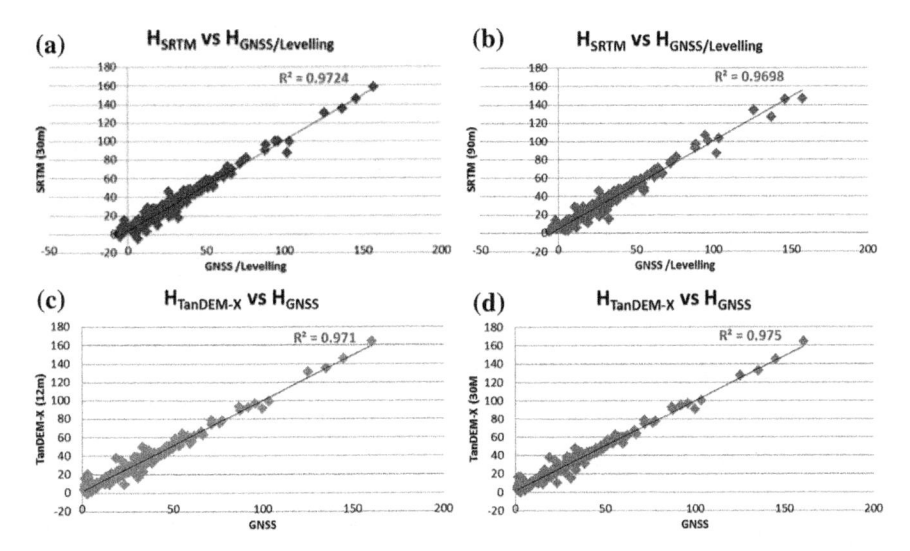

Fig. 4 Linear correlation, R^2 between GNSS field survey and GDEMs. **a** SRTM_30m, **b** SRTM_90m, **c** TanDEM-X_12m, **d** TanDEM-X_30m

4.2 GGMs Accuracy Assessments

Figure 5 represents the comparison of geoid height derived from GNSS leveling and GGMs. It can clearly be seen that almost all of the GGMs geoid height fit very well with the GNSS leveling derived from geoid height except the ITU_GRACE16, of which has shown a big difference. The basic statistic of geoid heights residuals is shown in Fig. 6. The absolute comparison with GNSS leveling data has shown that GOCO05c model gives better representation of the gravitational field in the southern region of Peninsular Malaysia, followed by GGM0GC and the *EIGEN-6S4v2*, where

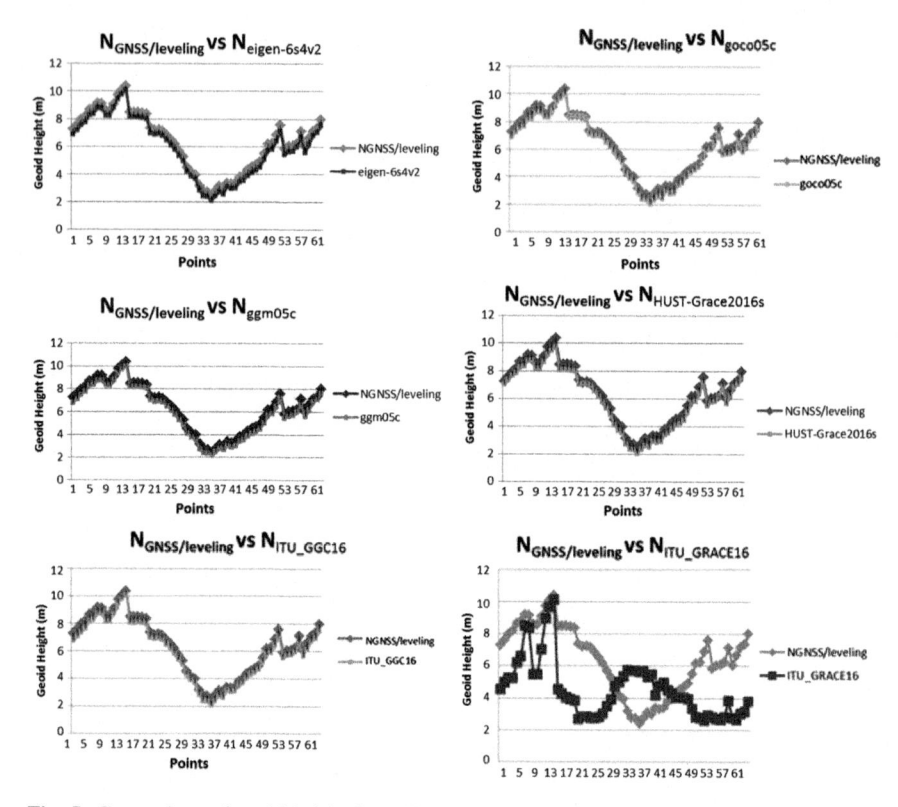

Fig. 5 Comparison of geoid height from GNSS leveling and GGMs

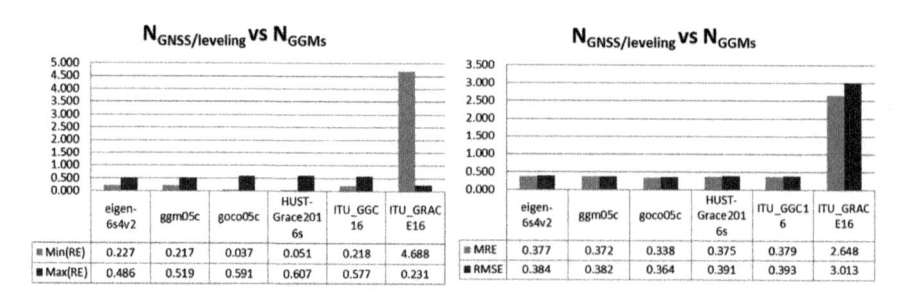

Fig. 6 Statistics of the differences between geoid heights derived from GNSS/leveling and GGMs (unit meter)

the RMSE are ±0.364, ±0.382, and ±0.384 m, respectively. Based on the results, it is indicated that the combination of GGM from the incorporation of gravity and altimetry datasets models presents a better fit to the investigated region.

5 Conclusion

In this study, the global digital elevation models and the global geopotential model were investigated using the GNSS observation. Statistical computation of GDEMs has shown that the DEM from TanDEM-X presents better result than the SRTM GDEMs. Surprisingly, TanDEM-X of spatial resolution 30 m is much better than TanDEM-X of spatial resolution 12 m where the statistical computation shows that the minimum residual, maximum residual, mean residual error, and RMSE are 0.001, 18.473, 2.376, and 3.813 m, respectively. Meanwhile, the comparison of six GGMs has shown that GOCO05c is the highest accuracy where its gravitational field fits well with the local gravitational field and the lowest accuracy derived using the ITU_GRACE16. The comparison of the GOCO05c GGMs with the GNSS/leveling has shown that the accuracy of the gravitational field from the model in terms of their RMSE and mean residual error is ±0.364 and 0.338 m, respectively.

Acknowledgements Thanks to the U.S. Department of the Interior U.S. Geological Survey for providing the GDEMs data for study region. Also special thanks to the Department Survey and Mapping Malaysia (DSMM) for providing GNSS Leveling data in the study region. The authors also would like to thank German Aerospace Center for providing DEM data from DLR's TerraSAR-X/TanDEM-X satellite.

References

Akyilmaz O., Ustun A., Aydin C., Arslan N., Doganalp S., & Guney C. et al. (2016a). *The combined global gravity field model including GRACE & GOCE data up to degree and order 280*. http://doi.org/10.5880/icgem.2016.005.

Akyilmaz O., Ustun A., Aydin C., Arslan N., Doganalp S., & Guney C. et al. (2016b). *The global gravity field model including GRACE data up to degree and order 180 of ITU and other collaborating institutions*. http://doi.org/10.5880/icgem.2016.006.

Becek, K., Wolfgang, K., & Kutoğlu, Ş. H. (2016). Evaluation of vertical accuracy of the WorldDEM™ using the runway method. *Remote Sensing, 8*(11), 934. https://doi.org/10.3390/rs8110934.

Erasmi, S., Rosenbauer, R., Buchbach, R., Busche, T., & Rutishauser, S. (2014). Evaluating the quality and accuracy of TanDEM-X digital elevation models at archaeological sites in the Cilician Plain, Turkey. *Remote Sensing, 6*, 9475–9493. https://doi.org/10.3390/rs6109475.

Erol, B., & Çelik, R. N. (2004). Precise local geoid determination to make GPS technique more effective in practical applications of geodesy. FIG Working Week 2004, 22–27 May, Athens, Greece.

Förste C., & Bruinsma S. L. (2016). *A time-variable satellite-only gravity field model to d/o 300 based on LAGEOS, GRACE and GOCE data from the collaboration of GFZ Potsdam and GRGS Toulouse*. http://doi.org/10.5880/icgem.2016.004.

Khalid, N. F., Din, A. H. M., Omar, K. M., Khanan, M. F. A., Omar, A. H., & Hamida, A. I. A. et al. (2016). Open-source digital elevation model (dems) evaluation with GPS and lidar data. In *The International Archives of the Photogrammetry, Remote Sensing and Spatial Information Sciences, Volume XLII-4/W1. International Conference on Geomatic and Geospatial Technology (GGT)* 2016, October 3–5, 2016, Kuala Lumpur, Malaysia.

Kiahmehr, R. (2006). A hybrid precise gravimetric geoid model for Iran based on recent GRACE and SRTM data and the least squares modification of Stokes's formula. Available at: http://www.geophysics.ut.ac.ir/JournalData/1385A/Kiamehr.pdf.

Mukherjee, S., Joshi, P. K., Ghosh, A., Garg, R. D., & Mukhopadhyay, A. (2013). Evalution of vertical accuracy of open source digital elevation model (DEM). *International Journal of Applied Earth Observation and Geoinformation, 21,* 205–217.

Pail, R., Gruber, T., Fecher, T., & The GOCO Project Team. (2016). *The combined gravity model GOCO05c.* http://doi.org/10.5880/icgem.2016.003.

Ries J., Bettadpur S., Eanes R., Kang Z., Ko, U., & McCullough, C. et al. (2016). *The combined gravity model GGM05C.* http://dx.doi.org/10.5880/icgem.2016.002.

Rodriguez, E., Morris, C. S., & Belz, J. E. (2016). A global assessment of the SRTM performance. *Photogrammetric Engineering & Remote Sensing, 72*(3), 249–260.

Ssengendo, R. (2015). *A height datum for Uganda based on a gravimetric quasigeoid model and GNSS/levelling.* A Ph.D. Thesis from Royal Institute of Technology, Stolkhom Sweeden.

Triarahmadhana, B., & Heliani, L. S. (2014). Evaluation of GOCE's global geopotential model to the accuration of local geoid (Case Study : Java Island, Indonesia). FIG Congress 2014 (pp. 1–13).

Wecklich, C., Gonzalez, C., Bräutigam, B., Jasiewicz, J., Zwolinski, Z., & Mitasova, H. et al. (2015). Height accuracy for the first part of the global TanDEM-X DEM data. In *Geomorphometry for Geosciences* (pp. 5–8), Poznan, Poland: Poznan University.

Zhou, H., Luo, Z., Zhou, Z., Zhong, B., & Hsu, H. (2017). HUST-Grace2016s: A new GRACE static gravity field model derived from a modified dynamic approach over a 13-year observation period. *Advances in Space Research.* https://doi.org/10.1016/j.asr.2017.04.026.

The Evaluation of Performance and Quality of Preschool Using Fuzzy Logic Approach

Nur Syuhada Muhammat Pazil, Norwaziah Mahmud,
Siti Hafawati Jamaluddin, Umi Hanim Mazlan and Afiqah Abdul Rahman

Abstract Malaysia is currently transforming its education system to achieve world-class status. High-quality preschool education improves children's health and promotes their development and learning. Educators have a role in promoting the quality of early education and childcare by examining the issues and challenges involved. The performance of children includes four major factors which are based on the physical aspect, socio-emotional, spiritual, and intellectual of preschool children in their learning process. This paper presents a methodology to improve these four factors by analyzing the performance of individual students. This study focuses on 17 preschools in Johor which consist of five selected schools in the year 2015. For this purpose, the whole data from the four aspects are divided into various ranges. This is done by fuzzy logic system. It has been shown that the performance of the five selected schools is successful in their process of learning.

Keywords Fuzzy logic · Performance · Preschool · Quality

1 Introduction

Preschool education plays an important role in developing high-quality human resource. A focus on holistic children development is needed to prepare them with the ability to compete and to have the survival skills to meet global changes. In order to reach the quality of preschool students before entering Standard One, assessment and evaluation of the development on preschool children are important because they help for the progress of children and to identify the overall variety of students' potential.

N. S. Muhammat Pazil (✉) · N. Mahmud · S. H. Jamaluddin · U. H. Mazlan
A. Abdul Rahman
Universiti Teknologi MARA, Cawangan, Shah Alam, Perlis, Malaysia
e-mail: syuhada467@perlis.uitm.edu.my

Most preschool students faced difficulties when they reached Standard One. As a result, students are unable to score in academic since they are facing difficulties either they are weak students or those who are already skilled. Since there is no streaming process carried out by the school administrator for year one students, this leads to many problems later on. For example, some of the students have difficulties to catch up with the learning process in class since they need more attention from the teachers. Preschool teachers need to give their attention toward the students who want to be noticed and record the students' progress in detail.

There are three main objectives of this study. The first objective is to determine the potential of preschool students' strengths and weaknesses in learning process from time to time. Secondly, it is to help teachers to identify the effective teaching approach through assessment and evaluation, and the last objective is to identify the preschool quality performance before Standard One by using the application of fuzzy logic.

This study will focus on preschool students in Johor which consist of five selected schools in the year 2015. The students will be evaluated based on the physical aspect, socio-emotional, spiritual, and intellectual in their learning process. The purpose of this study is to help the school administrator especially during the intake of Standard One students. It is also to measure the quality of preschool students' education before entering Standard One by observing students based on several criteria. Fuzzy logic can give many benefits to the administrator in order to carry out the streaming process for Standard One students.

2 Literature Review

2.1 Preschool Education

Preschool education is one of the most important education systems in order to transform the education and achieve world-class status. There are several aspects of preschool learning process such as socio-emotional, intellectual, physical, and spiritual that should be observed in preschool education before they enter year one.

Cress et al. (2015) stated that the Preschool Behavioral and Emotional Rating Scales (PreBERS) is a normative assessment of emotional and behavioral strengths in preschool children. The PreBERS has well-established reliability and validity for typically developing children as well. Besides that, the assessment of social, emotional, and behavioral functioning is particularly important at the preschool level because of the number of children impacted, the persistence, and the stability of behavioral problems. Thus, the implications suggest that the PreBERS items are reliable scores that can be used to identify behavioral strengths in preschool children.

2.1.1 Socio-emotional Aspect

According to Adela et al. (2011), normally children who lack in social and emotional competencies will face behavioral problems in developing their emotion. Usually, educators need to have skills and knowledge about all the programs that will be implemented in kindergartens which make them easier to monitor and evaluate the preschool children. Meanwhile, Barblet and Maloney (2010) found that children who develop with positive social, emotional growth, and well-being become an important part of positive mental health, early school success, and integration of development domain which affect on language, communication skill, early literacy, and numeracy. For that reason, the children's social and emotional growth in early childhood program depends on how well the teachers know the children and how skilled they are at gaining meaningful information about children's social.

2.1.2 Intellectual Aspect

In intellectual aspect, there are understanding personalities in preschool children with intellectual disability (ID) and parent-oriented approach affected by child (Bostrom et al. 2011). Meanwhile, Simatwa (2010) stated that instructional management focuses on planning, execution, and evaluation of learning experiences. Therefore, for teachers in pre-secondary schools, they need to have a good understanding of process of cognitive development in children.

2.1.3 Physical Aspect

A study conducted by Andrea and Holden (2015) stated that the selected preschool physical education programs lacked the necessary resources for effective physical education activities. They found that these resources are needed for children with disabilities to develop appropriately throughout their lifespan. The physical education curriculum for pre-kindergarten students incorporates movement as a primary factor for students to acquire the necessary skills to function at an appropriate level.

2.1.4 Spiritual Aspect

Spiritual aspect can be applied by integrating new educational approaches into preschool education. Therefore, one of the approaches that should be implemented is social entrepreneurship education (Sarikaya and Coskun, 2015). By providing social entrepreneurship education, it is possible to develop children's abilities and enable them to produce innovative solutions to social problems. Moreover, social entrepreneurship education supports children's self-sufficiency, creativity, empathy, entrepreneurship skills, and rational thinking.

2.2 Fuzzy Logic

The application of fuzzy logic techniques is important to evaluate the students' performance. This technique is usually used in new performance evaluation and is compared to the classical evaluating method. By using fuzzy logic, this method can provide several preferences which are flexible and constant through the mathematical calculations (Gokmen et al. 2010). This study is significant in order to identify students' performance through classical assessment method by using fuzzy logic model. Hence, not only it is suitable for laboratory application but it is also used in evaluation of theoretical lesson.

Fuzzy logic is also used to evaluate students' knowledge and skills (Voskoglou 2013). The main goal to develop fuzzy model is to assess student group's knowledge and skills which include analogical reasoning abilities and their understanding of problem-solving skills. Most technique skills are evaluated through students' performance. As a result, the implementation of fuzzy model can provide an effective technique which is characterized by a degree of vagueness and uncertainty.

3 Methodology

This section elaborates on the research framework that was used throughout the project as shown in Fig. 1. All the gathered information during the data acquisition phase was analyzed to obtain the required data for the project. The analyzed data then were fed into fuzzy rule-based system in order to evaluate the performance and quality of the preschools.

3.1 Data Acquisition

For this study, 17 preschools from each of the five schools in Johor were selected. The five schools are SK. Bandar Easter, SK. Seri Setia Jaya, SK. Bukit Mahkota, SK. Sedili Kecil, and SK. Tunjuk Laut. There are four main aspects of the

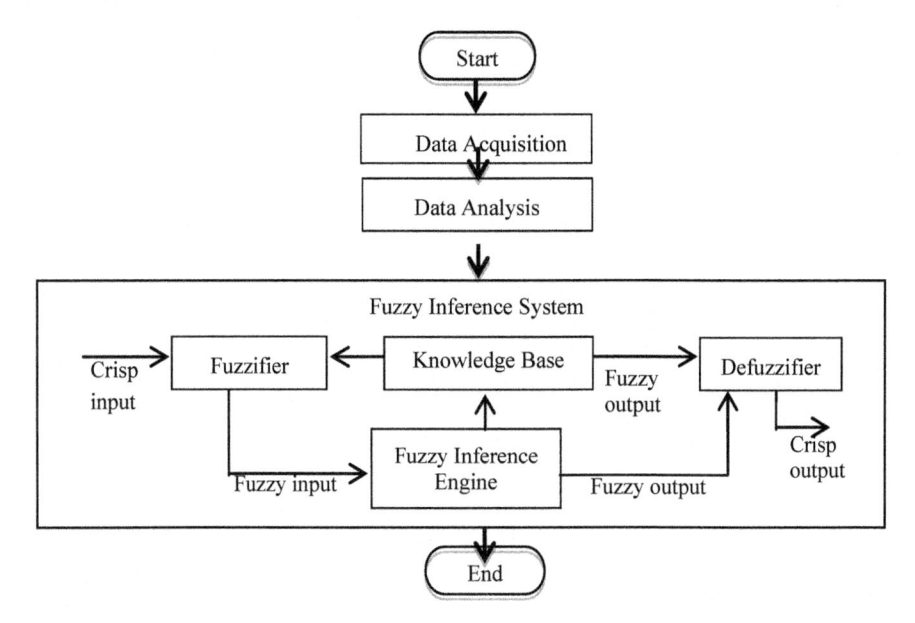

Fig. 1 Research framework

preschool performance which were considered during the data collection. The most dependable aspects are physical, socio-emotional, spiritual, and intellectual. All the aspects are necessary because if children have same factors which are good, but other factors are not, they will not fall into the category of very good performance. The teachers were asked to evaluate the students using the questionnaire provided by the State Education Department.

3.2 Data Analysis

The data analysis phase involves three activities; identifying non-numeric linguistic variables, determining maximum, minimum, and average values, and generating membership function.

3.2.1 Identifying Non-numeric Linguistic Variables

In fuzzy logic applications, the non-numeric linguistic variables are often used to facilitate the expression of rules and facts. In this study, Very_Low, Low, High, and Very_High are identified as linguistic variables for all aspects.

3.2.2 Determining the Maximum, Minimum and Average Values

In order to create a membership function, the maximum, minimum, and average values are calculated based on Eqs. 1–3.

Let U be the universal set $U = \{X_1, X_2, X_3, \ldots, X_n\}$. A fuzzy set A of the universal set U can be defined by a membership function as follows:

$$A = \{(x, \mu_A(x) | x \in U)\}, \qquad (1)$$

where $\mu_A(x) \in [0, 1]$ is the membership function of fuzzy set A, U is the universal set, x is an element in U, and A is a fuzzy subset in x.

Let A and B be two fuzzy sets defined in the universal set U. The union between the fuzzy sets A and B can be defined as:

$$\mu_{AB}(x) = \max(\mu_A(x), \mu_B(x)), \qquad (2)$$

where μ_A and μ_B are the membership functions of the fuzzy sets A and B, respectively; $\mu_A : U[0, 1]$ and $\mu_B : U[0, 1]$.

Let A and B be two fuzzy sets defined in the universal set U. The intersection between the fuzzy sets A and B can be defined as follows:

$$\mu_{AB}(x) = \min(\mu_A(x), \mu_B(x)), \qquad (3)$$

where μ_A and μ_B are the membership functions of the fuzzy sets A and B, respectively; $\mu_A : U[0, 1]$ and $\mu_B : U[0, 1]$.

3.2.3 Generating the Membership Function

The triangular membership function is used for converting the crisp set into fuzzy set. A triangular membership function is specified by three parameters (a, b, c) as shown in Fig. 2.

With the help of membership function, the input and output parameters are divided into different ranges. The membership function of a fuzzy set is a generalization of the indicator function in classical sets. In fuzzy logic, it represents the

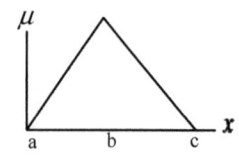

Fig. 2 Triangular membership function

The Evaluation of Performance and Quality ...

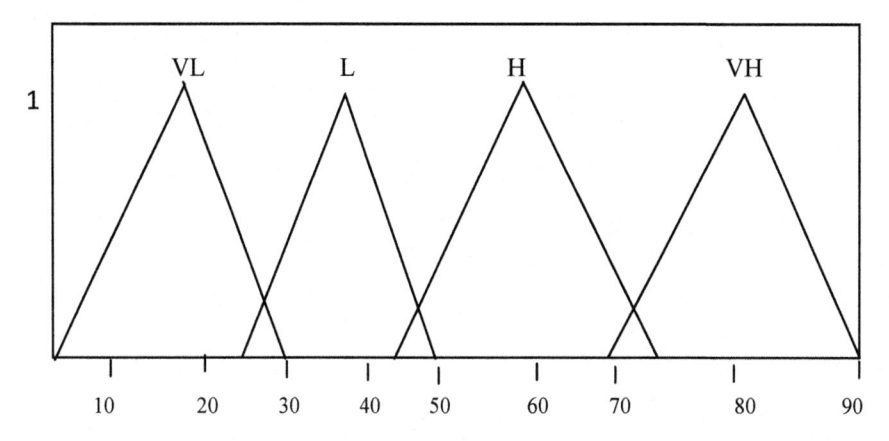

Fig. 3 Input membership function of physical, socio-emotional, spiritual, and intellectual aspect

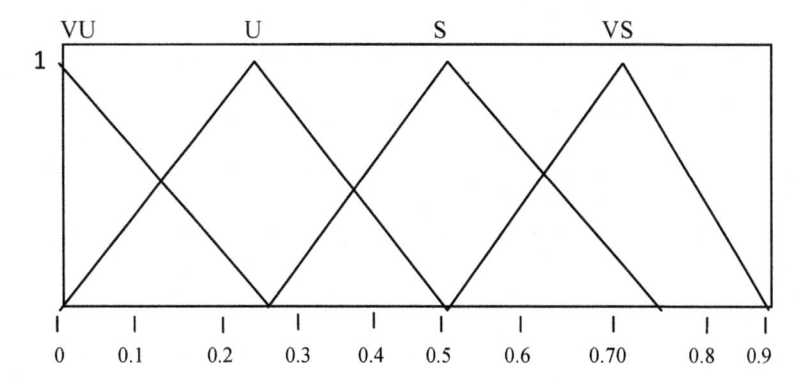

Fig. 4 Output membership function of physical, socio-emotional, spiritual, and intellectual aspect

degree of truth as an extension of valuation. The degree of truth is often confused with probabilities although they are conceptually distinct because fuzzy truth represents membership in vaguely defined sets. Therefore, the ranges of physical aspect, socio-emotional, spiritual, and intellectual performance were translated in a form of percentage. Figures 3 and 4 indicate the membership function of input and output, respectively.

3.3 Fuzzy Inference System

The fuzzy inference system involves three steps, namely fuzzification, fuzzy inference engine, and defuzzification. The Mamdani approach is applied in this study because it has a simple structure of min-max operation that is commonly used in an application.

3.3.1 Fuzzification

Fuzzification is the method to change a real scalar value into a fuzzy value. The crisp values are used to determine the membership values. This procedure generated the fuzzification input as shown in Table 1.

3.3.2 Fuzzy Inference Engine

Fuzzy rules are a pool of linguistic statements that describe how the fuzzy inference system should make a decision about categorizing an input or controlling an output. To formulate the conditional statements that comprise of fuzzy logic, IF-THEN rule statements are used. In the simplest form, the fuzzy IF-THEN rule can be written as "IF <antecedent>, THEN <consequent>".

The premise or antecedent part is the input variable, while the conclusion part is the output variable. Since the Mamdani approach is applied in this study, the rule consequence is defined by "IF x is A and y is B, THEN x is C."

The rule inference of fuzzy logic was determined based on the value of membership function. The range value of each aspect can be defined by using this rules inference. For each category, a total of 256 rule statements have been produced by fuzzy sets to classify the preschool before they enter year one.

3.3.3 Defuzzification

To solve a decision problem, the output must be in crisp value. Therefore, the fuzzy set must be transformed into a single numerical value, which is called crisp set. This crisp number is attained by a process called defuzzification. In order to identify the center of the area of defuzzification, the most popular defuzzification method, namely centroid method, is used. Table 2 indicates the fuzzy set of output.

Table 1 Fuzzy set of input variables

Linguistic expression	Symbol	Interval
Very low	VL	(0, 15, 30)
Low	L	(25, 38, 50)
High	G	(45, 60, 75)
Very high	VG	(70, 85, 100)

Table 2 Fuzzy set of output variables

Linguistic expression	Symbol	Interval
Very unsuccessful	VU	(0, 0, 0.25)
Unsuccessful	U	(0, 0.25, 0.50)
Successful	S	(0.25, 0.50, 0.75)
Very successful	VS	(0.50, 0.75, 1.0)

4 Results

This study focuses on preschool students in Johor from five selected schools in the year 2015. The schools are SK. Bandar Easter, SK. Seri Setia Jaya, SK. Bukit Mahkota, SK. Sedili Kecil, and SK. Tunjuk Laut. This study aims to identify the best school that performs well in preschool performance and has quality preschool students before going to Standard One in Johor. Table 3 shows the performance achieved by 17 preschool students from the five selected schools in Johor as mentioned earlier.

Table 3 Preschool performance

No.	SK. Bandar Easter	SK. Seri Setia Jaya	SK. Bukit Mahkota	SK. Sedili Kecil	SK. Tunjuk Laut
1	0.75	0.75	0.75	0.65	0.75
2	0.75	0.75	0.75	0.75	0.75
3	0.75	0.75	0.75	0.75	0.75
4	0.75	0.75	0.75	0.69	0.75
5	0.75	0.75	0.75	0.75	0.69
6	0.75	0.75	0.75	0.69	0.75
7	0.75	0.75	0.75	0.75	0.75
8	0.75	0.75	0.75	0.75	0.69
9	0.75	0.75	0.75	0.75	0.75
10	0.75	0.75	0.75	0.75	0.75
11	0.75	0.75	0.75	0.65	0.69
12	0.75	0.75	0.75	0.75	0.69
13	0.75	0.75	0.75	0.75	0.75
14	0.75	0.75	0.75	0.75	0.75
15	0.75	0.75	0.75	0.75	0.75
16	0.75	0.75	0.75	0.65	0.69
17	0.75	0.75	0.75	0.75	0.68

For each preschool student, the performance of four categories was fuzzified by the means of triangular membership function with different input values for every category. Active membership functions were calculated according to rule table, using the Mamdani Fuzzy Decision Techniques. The output which is the performance value was calculated and then defuzzified by calculating the center (centroid) of resulting geometrical shape. The average result of performance value for five selected schools in Johor is 0.75 which means successful in their process of learning.

5 Conclusion

In order to achieve three main objectives, this must be classified based on four aspects which are physical aspect, socio-emotional aspect, spiritual aspect, and intellectual aspect. At the application stage, course converters can edit the rules and membership functions are used for all students taking some lessons. This study identified that every five selected preschools in Johor performed successfully in their learning process and it also important for the preschool student to perform well before entering Standard One. Furthermore, it also gives advantages for teachers to provide a range of remedial and enrichment activities in order to help children who are identified as facing difficulties when teaching process is carried out.

References

Adela, M., Mihaela, S., Adriana, T. E., & Monica, F. (2011). Evaluation of a program for developing socio-emotional competencies in preschool children. *Journal of Social and Behavioral Sciences, 30,* 2161–2164.
Andrea, W. S., & Holden, G. (2015). The strengths and weaknesses of physical education programs in selected preschools in Central North Carolina. *Journal of Research Initiatives, 1* (3), 1–10.
Barblett, L., & Maloney, C. (2010). Complexities of assessing social and emotional competence and wellbeing in young children. *Australasian Journal of Early Childhood, 35*(2), 13–18.
Bostrom, P. K., Broberg, M., & Bodin, L. (2011). Child's positive and negative impacts on parents-a person-oriented approach to understanding temperament in preschool children with intellectual disabilities. *Research in Developmental Disabilities, 32,* 1860–1871.
Cress, C., Lambert, M. C., & Epstein, M. H. (2015). Factor analysis of the preschool behavioral and emotional rating scale for children in head start programs. *Journal of Psychoeducational Assessment, 2,* 1–14.
Gokmen, G., Akincib, T. C., Tekta, M., Onatc, N., Kocyigita, G., & Tekta, N. (2010). Evaluation of student performance in laboratory applications using fuzzy logic. *Journal of Procedia Social and Behavioral Sciences, 2,* 902–909.

Sarikaya, M., & Coskun, E. (2015). A new approach in preschool education: Social entrepreneurship education. *Journal of Procedia-Social and Behavioral Sciences, 195,* 888–894.

Simatwa, E. M. W. (2010). Piaget's theory of intellectual development and it implication for instructional management at pre-secondary school level. *Educational Research and Reviews, 5*(7), 366–371.

Voskoglou, M. G. (2013). Fuzzy logic as a tool for assessing students' knowledge and skills. *Journal of education sciences, 3,* 208–221.

Evaluating Fuzzy Time Series and Artificial Neural Network for Air Pollution Index Forecasting

Nordianah Jusoh and Wan Juliyana Wan Ibrahim

Abstract Forecasting of air pollutant levels is very important in environmental science research today. The rise of air pollution in both developed and developing countries has attracted much global attention. In view of current world environment quality, this pollution may affect health, ecosystem, forest species, and agriculture. In the recent years, many researchers have studied several methods and models to predict Air Pollution Index (API). Therefore, this paper presents methods of artificial neural network (ANN) and fuzzy time series to forecast the API values in Port Klang, Malaysia. This research also employs popular measure errors which are mean squared error (MSE), mean absolute percentage error (MAPE), and root mean squared error (RMSE) to identify the best model with the smallest error. The result shows that ANN gives the smallest forecasting error to forecast API compared to FTS. Hence, it is proven that the ANN method can be applied successfully as tools for decision making and problem-solving in forecasting research.

Keywords Air pollutant · Fuzzy time series · Artificial neural network Alyuda NeuroIntelligence

1 Introduction

Good air quality is very important to all species as it is a basic necessity of life. Every day, the average person breathes about 14,000 L of air. The existence of pollutants in the air and its implication can harmfully affect people's health. Due to

---

N. Jusoh (✉) · W. J. W. Ibrahim
Faculty of Computer and Mathematical Sciences,
Universiti Teknologi MARA, Shah Alam, Malaysia
e-mail: dianah642@melaka.uitm.edu.my

human activities, industrialization, and natural factors, there is an increase in air pollution rate, which is not safe for breathing. It has become a major environmental issue in both urban and rural areas for such a long time.

Haze is a common problem in Southeast Asia region. Slash and burn practices in Indonesia to clear land for oil palm plantation dating back to as early as 1991 have been the major causal factor to smog occurrences in many regions in Malaysia. Greenpeace reported that the health of millions living in Sumatera has been at risk due to haze phenomenon.

Since 1989, Air Pollution Index (API) has been used as an indicator of air quality in Malaysia. The API value is based on five criteria of air pollutants: ozone (O_3), carbon monoxide (CO), nitrogen dioxide (NO_2), sulfur dioxide (SO_2), and particulate matter with a diameter of less than 10 μm (PM_{10}) (Azid et al. 2013).

Forecasting of air pollutant is very important for planning, proper actions, and controlling strategies. Fuzzy time series (FTS) and artificial neural network (ANN) have been applied for forecasting purposes. This study aims to forecast API in Port Klang for the year 2017 using FTS and ANN. This study aims to determine the best model between FTS and ANN at forecasting the API in Port Klang based on the smallest error measure by using mean squared error (MSE), mean absolute percentage error (MAPE), and root mean squared error (RMSE).

2 Literature Review

Fuzzy time series, artificial neural network, Box–Jenkins, and Kriging model have been identified as effective measures in many research areas related to forecasting including air pollution cases. In recent years, FTS and ANN methods have attracted much attention. In particular, time series models have utilized fuzzy theory to solve forecasting problems such as air quality forecasting, financial forecasting, temperature forecasting, and rainfall forecasting, some of which are displayed in Table 1.

From Table 1, this study has compared the effectiveness of FTS and ANN at forecasting API measures at Port Klang for the year 2017.

3 Methodology

This section describes the use of FTS and ANN to forecast API in Port Klang. In particular, all simulations for ANN have been done through Alyuda NeuroIntelligence.

Table 1 Previous researches involving FTS and ANN

Author (year)	Objective of research	Method	Findings
Elena et al. (2012)	To forecast arrivals of tourists to Bali	Applied different models of FTS and seasonal autoregressive integrated moving average (SARIMA)	Chen's FTS is more accurate than SARIMA and other FTS models
Lee et al. (2013)	To construct and develop an accurate forecasting model to predict the monthly API To evaluate ability of models to monitor air quality status	Applied FTS, ANN, and autoregressive integrated moving average (ARIMA)	ANN gives better performance comparable to FTS and ARIMA models
Azid et al. (2013)	To predict API at seven selected Malaysian air monitoring stations	Applied feed-forward ANN and principal component analysis (PCA)	ANN method shows better performance
Kaushik and Singh (2013)	To forecast production of sugar in India To compare neural network method with different FTS models	Applied ANN and different FTS models (Chen, Huarng, Chen higher order and Singh)	ANN is more accurate and gives better result than FTS
Kottur and Mantha (2015)	To predict the levels of air pollutants in Mumbai and Navi Mumbai	Applied an integrated model combining ANN and Kriging techniques	Kriging is able to predict pollutant values for unknown locations, but ANN gives better R value
Kar (n.d.)	To predict stock market indices	Applied ANN	ANN is a more effective forecasting method when applied with ANN and backpropagation algorithm

3.1 Fuzzy Time Series

Fuzzy time series has been widely used in the field of air pollution cases. This analysis method was first defined by Song and Chissom (1993). Figure 1 displays the seven-step procedure of this method.

Using these seven steps, the development of fuzzy time series forecasting models can be undertaken as follows:

i. Define the universe of discourse U to accommodate the time series data in which complete production history can be covered.
ii. Partition the universe of discourse into seven equivalent length intervals.

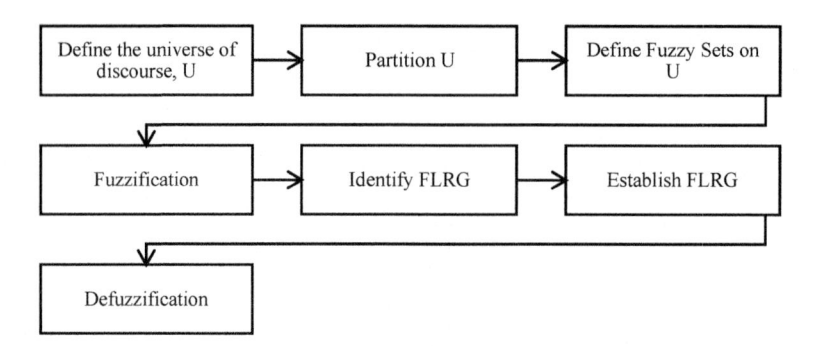

```
┌─────────────────────┐    ┌─────────────┐    ┌──────────────────────┐
│ Define the universe │───▶│ Partition U │───▶│ Define Fuzzy Sets on │
│   of discourse, U   │    │             │    │          U           │
└─────────────────────┘    └─────────────┘    └──────────────────────┘
           │                                              │
           ▼                                              ▼
┌─────────────────────┐    ┌─────────────┐    ┌──────────────────────┐
│    Fuzzification    │◀───│Identify FLRG│◀───│    Establish FLRG    │
└─────────────────────┘    └─────────────┘    └──────────────────────┘
           │
           ▼
┌─────────────────────┐
│   Defuzzification   │
└─────────────────────┘
```

Fig. 1 Procedures for FTS model

iii. Describe the fuzzy set by defining the membership functions in respective partitions.
iv. Fuzzification is the process of identifying associations between historical values in the data set and fuzzy sets. Each historical value is fuzzified according to its highest degree of membership. Fuzzification is done for further calculation in the next step.
v. Fuzzy relationships are identified from the fuzzified historical data. Suppose $F(t-1)$ as A_i and A_j a fuzzy logical relationship can be expressed as $A_i \rightarrow A_l$, where A_i and A_j are called the left-hand side and right-hand side of the fuzzy logical relationship correspondingly.
vi. Develop the fuzzy logical relationships into the fuzzy logical relationship based on the similar fuzzy number on the left-hand side of fuzzy logical relationships. If the transition occurs to the similar fuzzy set, make a distinct logical relationship group.
vii. The forecasted value at t, $F(t)$, is demonstrated by three of heuristic rule. Assuming the fuzzified API of $F(t-1)$ is A_j, the forecasted output of $F(t)$ is determined using the following rules:

Rule 1: If there exists a one-to-one relationship group of A_j, say $A_j \rightarrow A_k$, and the highest degree of belongingness of A_k occurs at interval u_k, then the forecasted output of $F(t)$ equals the midpoint of u_k.
Rule 2: If A_j is empty set, i.e., $A_j \rightarrow \phi$, and the interval where A_j has the highest degree of belongingness is u_j, then the forecasted output equals the midpoint of u_j.
Rule 3: If there exists a one-to-many relationship group of A_j, say $A_j \rightarrow A_1, A_2, \ldots, A_n$, and the highest degree of belongingness occurs at set

$u_1, u_2, \ldots, u_n$, then the forecasted output is completed as the average of the midpoints $m_1, m_2, \ldots, m_n$ of $u_1, u_2, \ldots, u_n$. This equation can be expressed as

$$\frac{m_1 + m_2 + \cdots + m_n}{n} \qquad (1)$$

3.2 Artificial Neural Network

ANN is an information processing unit analog to the neuron network in biological system (Silverman and Dracup 2000). ANN is widely known as a method to provide better forecasting. This model consists of three categories of neuron layers: input layer, hidden layer, and output layer. In this study, ANN model was manipulated using Alyuda NeuroIntelligence software and the special ANN software has proven its usefulness and is easy to use. In addition, this study has selected the Levenberg–Marquardt with the rest functions set containing default values as the training algorithm. By using Levenberg–Marquardt algorithm, the output values and error correction rate can be determined to maximize the convergence.

Many algorithms have been developed to be used with ANN method like the well-known backpropagation algorithms (BPAs) or the most effective Levenberg–Marquardt algorithms (LMAs). It is developed to be used with ANN that is based on minimizing a criterion (which is most frequently based on the error between the desired and the obtained output). Most of them are based on derivative calculations of the error as a mean to minimize it. LMA was chosen because of its strength and faster convergence. LMA curve-fitting method is actually a combination of two minimization methods which are gradient descent method (GDM) and the Gauss–Newton method (GNM). In the GDM, the sum of the squared errors is reduced by updating the parameters in the steepest-descent direction, while for GNM, the sum of the squared errors is reduced by assuming the least square function is locally quadratic and by finding the minimum of the quadratic. LMA acts more like a GDM, when the parameters are far from their optimal value, and acts more similar to the GNM when the parameters are close to their optimal value.

Several steps conducted in this software are described in Fig. 2.

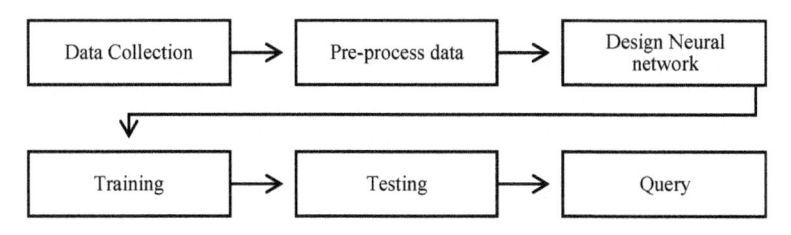

Fig. 2 Procedures for ANN model

Referring to Fig. 2, after collecting data, data is pre-processed to normalize the raw data in order to guarantee the accuracy of the data. The analyzed data were also pre-processed. The data of input layer were reflected to [−1,1], and the data of output layer were reflected to [0,1]. In designing the neural network, the number of layers and hidden neurons is chosen. Furthermore, it is possible to change the activation and error functions. However, there are no theoretical criteria to choose the best number of hidden neurons. The simulation training of Alyuda NeuroIntelligence software has 7 options of algorithm. After multiple simulation training, this study selected the Levenberg–Marquardt algorithm with rest functions set with the default values. After the training simulation, the training errors of the training and validation sets and some further statistics can be seen. In testing the performance stage, the difference between actual and forecasted value is compared. The accuracy can be determined from the correlation graphs. If the difference between training results was too large, procedures 1–4 should be repeated for training. Hence, the results can be seen in the query stage.

3.3 Accuracy Measurement

For identification of models performance of FTS and ANN, statistical criteria chosen are mean squared error (MSE), mean absolute percentage error (MAPE), and root mean squared error (RMSE), as given by the following equations:

$$\text{MSE} = \frac{\sum_{t=1}^{n}(Y_t - \hat{Y}_t)^2}{n} \qquad (2)$$

$$\text{MAPE} = \frac{\sum_{t=1}^{n}\left|\frac{Y_t - \hat{Y}_t}{Y_t}\right|}{n} \times 100\% \qquad (3)$$

$$\text{RMSE} = \sqrt{\frac{\sum_{t=1}^{n}(Y_t - \hat{Y}_t)^2}{n}} \qquad (4)$$

where

Y_t The actual value at time t,
$\hat{Y}_t$ The fitted value at time t,
n The number of observations.

4 Results and Discussion

ANN was manipulated using Alyuda NeuroIntelligence software, while FTS was calculated manually. Forecast API values for ANN and FTS after running Alyuda NeuroIntelligence software and calculated manually are given in Table 2.

The forecasted result for FTS is the same since it can only forecast based on previous month (previous fuzzified enrollment). Since December 2016 forecasted value is 63.7, then the value for the year 2017 will be 63.7.

In this study, MSE, MAPE, and RMSE were used to evaluate the forecasting accuracy based on the smallest value of error measures. Results obtained after running Alyuda NeuroIntelligence software and calculated manually are given in Table 3 containing summary value of error measures. It shows that RMSE for ANN model has the smallest error which is 9.6772. The comparisons between actual and forecasted values for ANN are shown in Fig. 3. In 2017, there is a slight increase and decrease in the forecasted result. The average of forecasted API value is between 50.22 and 68.31.

Table 2 Forecast results for FTS and ANN model

Method	FTS	ANN
Month	Year	
	2017	2017
Jan	63.7	52.12738
Feb	63.7	55.291017
Mar	63.7	56.531003
Apr	63.7	56.937603
May	63.7	57.046688
June	63.7	57.066513
July	63.7	57.06558
Aug	63.7	57.062406
Sept	63.7	57.060424
Oct	63.7	57.059521
Nov	63.7	57.059178
Dec	63.7	57.059068

Table 3 Summary of error measures for ANN and FTS

	MSE	MAPE	RMSE
ANN	93.6487	10.9075	9.6772
FTS	127.9710	15.6851	11.3124

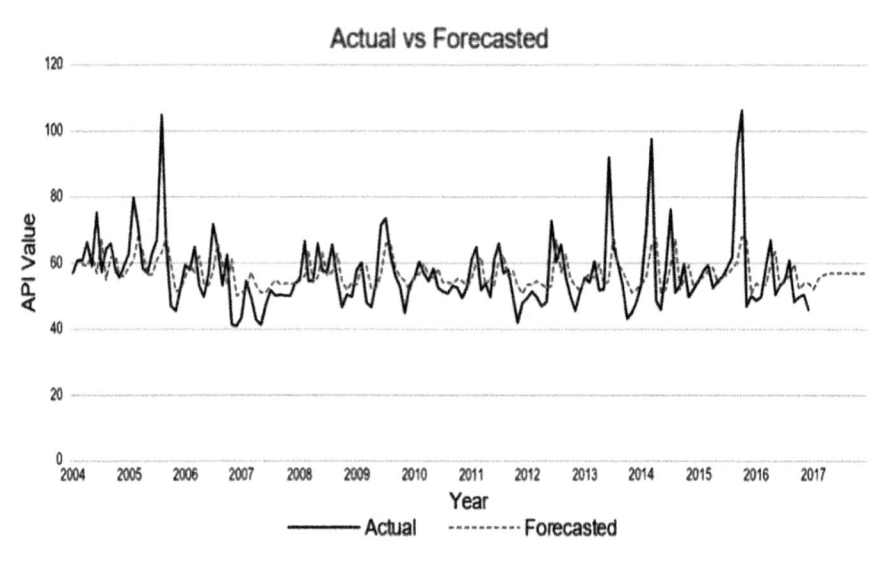

Fig. 3 Actual and forecast API values for ANN models

5 Conclusion

The FTS and ANN model in this study has predicted API at Port Klang for 2017 based on historical data from the Department of Environment. Using the results from this study, further prediction of future API may take place. This study has used MSE, MAPE, and RMSE to identify the best model between FTS and ANN based on smallest error measure. The findings in this study have shown preference of the ANN method over FTS in forecasting API for 2017. RMSE of ANN shows the smallest error which is 9.6772. Clearly, ANN is useful in decision-making processes for air quality control and management. Manual calculations of the fuzzy time series may have errors; thus, future works may be effective in improving the accuracy of forecasting. It is recommended that neural network be combined with other models such as Particle Swarm Optimization, Ant Colony Optimization, Genetic Algorithm, and Backpropagation Neural Network in order to improve forecasting accuracy.

Acknowledgements This work is partially supported by Universiti Teknologi MARA (UiTM). The researchers would like to thank the Department of Environment Malaysia, UiTM, for providing data, permission to utilize data, advice, guidance, and support for this study.

References

Azid, A., Juahir, H., Latif, M. T., Zain, S. M., & Osman, M. R. (2013). Feed-forward artificial neural network model for air pollutant index prediction in the southern region of Peninsular Malaysia. *Journal of Environmental Protection, 4,* 1–10. https://doi.org/10.4236/jep.2013.412a001.

Elena, M., Lee, M. H., Suhartono, S., Hossein, J. S., Haizum, N., & Bazilah, N. A. (2012). Fuzzy time series and Sarima model for forecasting tourist arrivals to Bali. *Jurnal Teknologi (Sciences and Engineering), 57,* 69–81.

Kar, A. (n.d.). *Stock prediction using artificial neural network.* Department of Computer Science and Engineering, IIT Kanpur. http://www.eecs.berkeley.edu/~akar/IITK_website/EE671/report_stock.pdf.

Kaushik, A., & Singh, A. K. (2013). Long term forecasting with fuzzy time series and neural network: A comparative study using sugar production data. *International Journal of Computer Trends and Technology (IJCTT), 4,* 2299–2305.

Kottur, S. V., & Mantha, S. S. (2015). An integrated model using artificial neural network (ANN) and Kriging for forecasting air pollution using meteorology data. *International Journal of Advanced Research in Computer and Communication Engineering, 4,* 149–152.

Lee, M. H., Rahman, N. H. A., Latif, M. T., & Suhartono, S. (2013). Forecasting of air pollution index with artificial neural network. *Jurnal Teknologi (Science and Technology), 63,* 59–64.

Silverman, D., & Dracup, J. A. (2000). Artificial neural networks and long-range precipitation in California. *Journal of Applied Meteorology, 31,* 57–66. https://doi.org/10.1175/1520-0450(2000)039<0057:annalr>2.0.co;2.

Song, Q., & Chissom, B. S. (1993). Fuzzy time series and its models, fuzzy sets and systems. *International Journal of Forecasting, 54,* 269–277.

Preliminary Findings: Revising Developer Guideline Using Word Frequency for Identifying Apps Miscategorization

Azmi Aminordin, Mohd Faizal Abdollah, Robiah Yusof and Rabiah Ahmad

Abstract Number of application in Google Play Store is increasing at a rapid rate. It is currently holding more than 3 million apps. With a large number of application files and information, locating the right apps into the right category can be quite challenging. We observed more than one thousand apps to prove that there are miscategorizations of apps in Android official marketplace. We revise the subject inside the guideline provided by developer console and kept the sub-category remain. In order to have more specific subjects for each sub-category, we revise the standard guideline by their description using word frequency. Top five ranked in terms of weighting without duplication from existing guideline was chosen as new subject in particular sub-categories. Furthermore, we remove redundant subject across sub-category. Finally, we calculate the miscategorization apps based on new guideline. The result shows that "Lifestyle" sub-category contributes to large number of misplaced apps. The result shows only 61% apps are correctly inserted into their category for all 1105 collected data. Having fine category will be a feeder to the further research related to malware detection.

Keywords Android · Miscategorization · Word frequency · Guideline

1 Introduction

Mobile devices, especially smartphones, made a great contribution towards environment of fast and massive information sharing. The popularity and demand of smartphones are increased rapidly with the different types of Operating System

A. Aminordin (✉) · M. F. Abdollah · R. Yusof · R. Ahmad
Universiti Teknikal Malaysia Melaka, Durian Tunggal, Melaka, Malaysia
e-mail: azmi1107@melaka.uitm.edu.my

© Springer Nature Singapore Pte Ltd. 2018
R. Saian and M. A. Abbas (eds.), *Proceedings of the Second International Conference on the Future of ASEAN (ICoFA) 2017 – Volume 2*,
https://doi.org/10.1007/978-981-10-8471-3_12

(OS) also by the increasing of functionality of the mobile (Zhou et al. 2012). To date, most of the mobile devices are supported by four types of popular OS, namely Android by Google, iOS from Apple, Windows Phone and Blackberry. Among them, Android dominated the world mobile OS market since 2012 until present. In second quarter of 2015, Android conquered with an 82.8% market share (International Data Corporation 2015). The popularity of Android OS brought to positive and negative impact. According to Symantec, 97% of malware application targets Android. In order to mitigate the threats on Android, several static features were analysed and one of them is category inside the application metadata.

Although Google Play Store provides category for applications, many developers still published in wrong category and bring challenge to app downloaders to find useful applications.

In an Android environment, in order to publish apps, the developers are freely choosing the category by their own. Wrongly inserted apps by the developers sometime happen. Even though Google had provided a guideline and standard for developers to insert the apps into their repository, the miscategorizations of apps are still happening. Moreover, this situation will affect the number of downloads an app will receive ultimately affecting the earning of application publishers. Thus, the coarse applications category presented on Google Play Store must be turned into fine-grained category before moving to further processes.

In this paper, we revised subjects in Android apps sub-category as the guideline to developer. We then present findings to show the existence of miscategorization on Android application in Google Play Store also the process of revising the guideline.

The remainder of the paper is organized as follows. We present the related works in Sect. 2. Section 3 presents a methodology. In Sect. 4, findings are explained. Last section explained regarding conclusion and the future works.

2 Related Works

Categorization or classification is an important phase as to facilitate further operations such as maintenance (McMillan et al. 2011; Panichella et al. 2015), searching (Yang and Tu 2012) and malware detection (Qadir et al. 2014). There are numerous solutions to categorize software or application. These include using topic selection, sentiment analysis and text categorization in software or application categorization (Maalej and Nabil 2015; Panichella et al. 2015).

Traditionally, there are two approaches to categorizing information, manual and automatic where each with its own advantages and drawbacks. Manual categorization is not just the Yahoo model, where editors put each Web page into one and only category. This approach return highly effective because it reflects the subtlety of human judgement and expert experience (Kamman 2001). Behind it, the main drawback of this technique is it takes a lot of expert attention to configure, populate and maintain.

The comparison between manual and Lucene's kNN categorization done by Beyer and Pinzger (2014) to classify Android app development issues on Stake Overflow. Lucene's kNN with 10 k-folds achieved more significant value of precision.

Struggling from effort of categorization manually has changed the acceptance of automation towards software application in repositories increasingly (Di Lucca et al. 2002; Kawaguchi et al. 2003). Although there are many research to overcome automatic software categorization issue, the approach cannot be directly applied to Android applications because Android applications are compiled and spread in the native APK. This APK needs the parsed process which will bring to complicate the automation stage (Yang and Tu 2012).

The increasing number of document availability in digital form brought to automated categorization (Sebastiani 2002). As more data become available, more ambitious problem can be tackled. As a result, machine learning is widely used in computer science and other fields (Domingos 2012).

Works done by Jia and Mu (2010) achieved good classification accuracy when applying Support Vector Machine (SVM) for text mining large-scale systems on Grid-based Open DSS. Work done by Maalej and Nabil (2015) shows that machine learning with multiple binary classifications outperforms single multi-class classifiers.

Many other text classification used machine learning approach done by several researchers such as Kawaguchi et al. (2003, 2006), Tian et al. (2009) Qiang (2010), Yang and Tu (2012). Those text classifications were applied using either Latent Dirichlet Allocation (LDA) or Latent Semantic Indexing (LSI).

The categorization also can be a feeder to malware detection as mentioned by Jagtap and Lomte (2015), Alatwi et al. (2016). Mapping the correct sensitive API (Yuksel et al. 2014) and other static information (Feizollah et al. 2015) with category will enhance the detection of malware in an Android mobile environment.

In this research, we analyse official Android apps category. However, we focused on the apps title (subject) and description based on the guideline given. For the categorization of the apps, we applied word frequency instead of automated approaches, such as LDA. Furthermore, we create new subject into the guideline template to make it easier for the developer to make a choice.

Table 1 Number of collected apps

Sub-category name	# of apps
Books & Reference	314
Education	421
Lifestyle	370

3 Methodology

This section explains on how the data were collected and the technique on how to revise the standard guideline.

3.1 Data Collection

One thousand and five free apps were collected from Android official market store. For this preliminary study, we only cover three sub-categories from the "APPS" main category as shown in Table 1. Apart from that, we also download the apk file for our future use. We use two types of online apk downloader, namely apkleecher[1] and evozi[2] in order to download an Android apk file.

3.2 Standard Guideline

In this research, we focused on top free apps within above-mentioned sub-categories. Apps Id and apps title of all collected apps are copied and located into spreadsheet file. The apps descriptions are also copied and located into separate text file.

We only calculate word frequency based on apps title and the description. Existing guidelines are too simple and confusing (George 2016). Therefore, the subject given was redundant with other category and unclear. Standard guideline used was taken from Developer Console Help[3] as shown in Table 2.

Table 2 Apps example by sub-category

Sub-category	Subjects/examples
Books & Reference	Book readers, reference books, textbooks, dictionaries, thesaurus, wikis
Education	Exam preparations, study aids, vocabulary, educational games, language learning
Lifestyle	Style guides, wedding and party planning, how-to guides

---

[1]http://apkleecher.net/.
[2]https://apps.evozi.com/apk-downloader.
[3]https://support.google.com/googleplay/android-developer/answer/113475?

3.3 Word Frequency

Word frequency is where a language's words grouped by frequency of occurrence within some given text corpus. It can be managed either by levels or as a ranked list. Every word in corpus will be separated; then, the same word will be grouped together. Finally, the system will calculate the number of word in every group before displaying the result. While calculating all words frequency, we remove *stop word*, which is commonly used word (such as "the", "as", "a").

3.4 Proposed Guideline

One of the main components that affect the importance of a term in a document is term frequency (TF) (Gaigole et al. 2013). In this study, we observe collected applications by the apps title and apps descriptions. Some apps can be easily classified by the apps title. For example, "English Dictionary" can be easily spotted into "Book & Reference" sub-category. On the other hand, the clue of an application is only shown or described in their description through its functions. Therefore, we firstly examine the apps title than followed by description. After that, the keywords related to the app and category were recorded. Figure 1 shows the

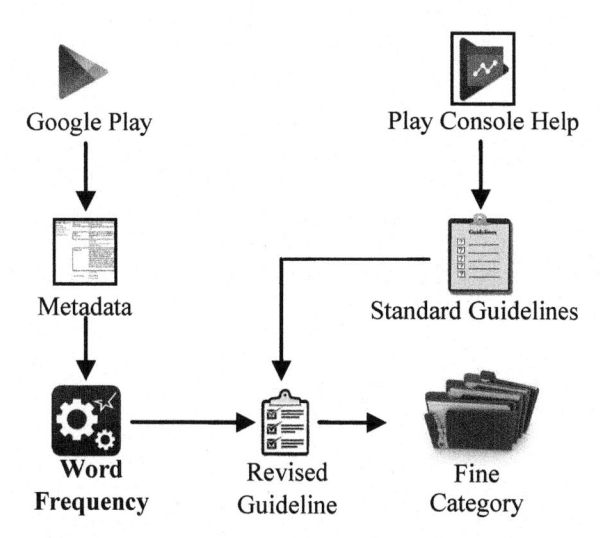

Fig. 1 Workflow of Android apps categorization

workflow for classification use in this research. Two main processes for revising guideline are as follows:

i. *Remove duplicate subjects*—We remove the "educational games" under Education sub-category because the educational games had their own sub-category (GAMES → Education).
ii. *Add new subjects*—We added few more subjects in the table according to the term frequency of word located into same category. For example, the frequencies of apps title "Guide" appeared in "Book & Reference" are more than other sub-categories; thus, this subject will be added into the guideline.

Moreover, we combined similar subjects using subject heading as introduced by Library of Congress Subject Heading (LCSH) (Library 2014). For example, we group (Quran, Vedas, Tripitaka, Bible etc.) becoming *Holy Scripture*. Lastly, we add "Holy Scripture" subject into "Book & Reference."

4 Result

Table 3 shows the new subjects in every sub-category. All sub-categories have additional subject as the evolution of apps. The fine category is obtained by firstly matching the app title with the new guideline. If the title does not match with any subject, we proceed to look into the apps description. This technique usually applies for library cataloguing and classification (Sandbergja 2015).

4.1 Precision

For this study, we only calculate the precision (positive predictive value). True positive (TP) is where the numbers of correct applications are identified stored into right sub-category, while false positive (FP) is where the applications were wrongly inserted to the category. Results are shown in Table 4.

Table 3 New subjects by sub-category

Sub-category	Subjects
Books & Reference	Guide (Kiblah, pray), holy scripture (Quran, Bible, Hadith, Vedas), zikr, commentary, pray (Doa's)
Education	Classroom tools, learning course, online learning, course tutorials, school (preschool, college, university)
Lifestyle	Horoscope, zodiac, lottery (lucky draw, 4D, Toto, Damacai), make-up (hair, eyebrow, tattoo), date and time (prayer, alarm, calendar)

Table 4 Result by sub-category

Sub-category name	TP	FP	PPV (%)
Books & Reference	283	31	90.13
Education	229	192	54.39
Lifestyle	157	213	42.43

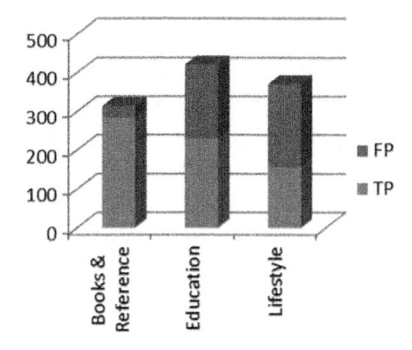

Fig. 2 Result of TP and FP by sub-category

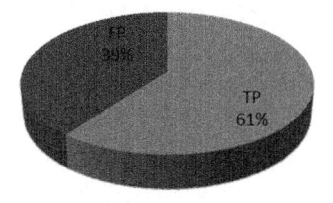

Fig. 3 TP and FP for all sub-categories

Among these three sub-categories, "Lifestyle" shows the highest value of false positive where more than 50% collected apps were wrongly published by the developers. "Book & Reference" gain the highest with 90.13% of precision. Figure 2 shows the chart for each category TP and FP.

Figure 3 shows overall TP and FP from three categories. For all collected apps, it only achieved 60.54% of TP.

5 Conclusion

The result shows that miscategorization occurred in Google Play Store with 39% precision of false positive. Lack of developer concern and no automate checking system available on Google Market brought to this situation. However, manual classification method can contribute to several issues such as inconsistent classification (Goren-Bar et al. 2000), expensive and time consuming (Malik and Bhardwaj 2011).

As mentioned earlier, this preliminary study is to revise the subject in sub-category to facilitate developers and to prove that miscategorization happens in Android official market. For the future, we planned to automate the categorization system by applying machine learning algorithm. Static code features such as permissions in AndroidManifest.xml and API calls in classes.dex will be taken into consideration on our next research. Preprocessing rules such as stop word removal, tokenizing, and stemming for document classification and text analysis will involve before weighting the word by its frequency in the corpus. More sub-categories and applications will be involved in our future works. Lastly, we planned to detect malware based on the newly refined category.

References

Alatwi, H. A., Oh, T., & Fokoue, E. (2016). Android malware detection using category-based machine learning classifiers. In *Proceedings of the 17th Annual Conference on Information Technology Education* (pp 54–59).

Beyer, S., & Pinzger, M. (2014). A manual categorization of android app development issues on stack overflow. In *2014 IEEE International Conference on Software Maintenance and Evolution* (pp. 531–535).

Di Lucca, G. A., Di Penta, M., & Gradara, S. (2002). An approach to classify software maintenance requests. In *International Conference on Software Maintenance* (pp. 93–102).

Domingos, P. (2012). A few useful things to know about machine learning. *Communications of the ACM, 55,* 78–87. https://doi.org/10.1145/2347736.2347755.

Feizollah, A., Anuar, N. B., Salleh, R., & Wahab, A. W. A. (2015). A review on feature selection in mobile malware detection. *Digit Investig, 13,* 22–37. https://doi.org/10.1016/j.diin.2015.02.001.

Gaigole, P. C., Patil, L. H., & Chaudhari, P. M. (2013). Preprocessing techniques in text categorization. In *National Conference on Innovative Paradigms in Engineering & Technology (NCIPET-2013)* (pp. 1–3).

George T (2016) Which category of Play Store should my utility app be in? In AndroidCentral. https://forums.androidcentral.com/ask-question/706378-category-play-store-should-my-utility-app.html.

Goren-Bar, D., Kuflik, T., & Lev, D. (2000). Supervised learning for automatic classification of documents using self-organizing maps. In *Proceedings of the First DELOS Network of Excellence Workshop on "Information Seeking, Searching and Querying in Digital Libraries"*, Zurich.

International Data Corporation (2015) *Smartphone OS Market Share, 2015 Q2.* http://www.idc.com/prodserv/smartphone-os-market-share.jsp.

Jagtap, A. H., & Lomte, A. C. (2015). Android app categorization using Naïve Bayes classifier. *International Journal of Computers and Applications, 122,* 26–29.

Jia, Z., & Mu, J. (2010). Web text categorization for large-scale corpus. *In ICCASM 2010–2010 International Conference on Computer Application and System Modeling, Proceedings* (pp. 188–191).

Kamman, J. (2001). *A people-smart approach to categorization.* Glob. Wisdom Inc (pp. 1–8).

Kawaguchi, S., Garg, P. K., Matsushita, M., & Inoue, K. (2003). On automatic categorization of open source software. In *3rd Workshop on Open Source Software Engineering* (pp. 79–83).

Kawaguchi, S., Garg, P. K., Matsushita, M., & Inoue, K. (2006). MUDABlue: An automatic categorization system for open source repositories. *Journal of Systems and Software, 79,* 939–953. https://doi.org/10.1016/j.jss.2005.06.044.

Library EER. (2014). Library of congress subject headings. In University of Alaska. https://library.uaf.edu/ls101-lc-subject. Accessed March 23, 2017.

Maalej, W., & Nabil, H. (2015). Bug report, feature request, or simply praise? On automatically classifying app reviews. In *2015 IEEE 23rd International Requirements Engineering Conference (RE)* (pp. 116–125).

Malik, H. H., & Bhardwaj, V. S. (2011). Automatic training data cleaning for text classification. In *2011 IEEE 11th International Conference on Data Mining Workshops* (pp. 442–449).

McMillan, C., Linares-Vásquez, M., Poshyvanyk, D., & Grechanik, M. (2011). Categorizing software applications for maintenance. In *IEEE International Conference on Software Maintenance*, ICSM (pp. 343–352).

Panichella, S., Di Sorbo, A., & Guzman, E. et al. (2015). How can i improve my app? Classifying user reviews for software maintenance and evolution. In *2015 IEEE 31st International Conference on Software Maintenance and Evolution, ICSME 2015*—Proceedings (pp. 281–290).

Qadir, M. Z., Jilani, A. N., & Sheikh, H. U. (2014). Automatic feature extraction, categorization and detection of malicious code in Android applications. *International Journal of Information and Network Security, 3,* 12–17.

Qiang, G. (2010). An effective algorithm for improving the performance of Naive Bayes for text classification. In *2nd International Conference on Computer Research and Development, ICCRD 2010* (pp. 699–701).

Sandbergja (2015) Cataloging and classification. In WIKIBOOKS. https://en.wikibooks.org/wiki/Cataloging_and_Classification.

Sebastiani, F. (2002). Machine learning in automated text categorization. *ACM Computing Surveys, 34,* 1–47.

Tian, K., Revelle, M., & Poshyvanyk, D. (2009). Using latent dirichlet allocation for automatic categorization of software. In *2009 6th IEEE International Working Conference on Mining Software Repositories* (pp. 163–166).

Yang, C.-Z., & Tu, M.-H. (2012). LACTA: An enhanced automatic software categorization on the native code of android applications. *Lecture Notes in Computer Science, 1,* 769–773.

Yuksel, A. S, Zaim, A. H., & Aydin, M. A. (2014) A comprehensive analysis of android security and proposed solutions. *International Journal of Computer Network and Information Security, 6,* 9–20. https://doi.org/10.5815/ijcnis.2014.12.02.

Zhou, Y., Wang, Z., Zhou, W., & Jiang, X. (2012). Hey, you, get off of my market: Detecting malicious apps in official and alternative android markets. In *Proceedings of the 19th Annual Network and Distributed System Security Symposium* (pp. 5–8).

Creativity in Mathematical Thinking Through Constructivist Learning Approach for Architecture Students

Nor Syamimi Samsudin, Ismail Samsuddin and Ahmad Faisol Yusof

Abstract Theoretically, improving student's performance in mathematics is challenging for today education. In the context of typical classrooms that adopt conventional teaching method, students are usually taught using structured rules based on the given academic syllabus. However, teaching architecture students needs a different approach. This is because architecture students learn by understanding the application into practice rather than by only solving the principle problem. In this research, a teaching experimented was conducted on 26 groups of architecture students. The experiment was designed based on the constructivist learning approach to study the mathematical creativity of the students. The findings indicated that their mathematical creativity has improved when they are learning about structure in architecture.

Keywords Constructivist learning approach · Creativity in mathematical thinking Experimental study for architecture students

1 Introduction

In Malaysia, most educational practices of teaching mathematics are still referring to the traditional method of learning instructions and curricula (Zanzali 2000). Basically, they are based on the transmission, or absorption, view of teaching, and structured learning by using structured teaching and learning method. In this perspective, students are required to memorize mathematical structures which restricted their understanding on the application of the real site, especially for architecture students. According to Nayak (2007) mention its application in the real

---

N. S. Samsudin (✉) · I. Samsuddin · A. F. Yusof
Faculty of Architecture, Planning and Surveying, Universiti Teknologi MARA,
Campus Seri Iskandar, 32610 Shah Alam, Perak, Malaysia
e-mail: norsya992@perak.uitm.edu.my

© Springer Nature Singapore Pte Ltd. 2018
R. Saian and M. A. Abbas (eds.), *Proceedings of the Second International Conference on the Future of ASEAN (ICoFA) 2017 – Volume 2*,
https://doi.org/10.1007/978-981-10-8471-3_13

site and situation, teaching practices will be more effective if students are given the possibility to explore and emphasize on their understanding into practicality. Nadjafikhah et al. (2012) agreed with Laycock (1970) that creativity in mathematics can be achieved by analyzing a given problem from different perspectives, such as seeing patterns, looking for differences and similarities, generating multiple ideas, and choosing a proper method to deal with unfamiliar mathematical situations. As a conclusion toward engaging mathematics and student's creativity, Inan (2013) claims that visual materials play a role in expanding exploration and storing organizing in the long-term memories (as cited in Erkan 2006). This theory shows that information is stored in long-term memory both in visually and verbally. In relation to above statement, it can be simply understood mathematical thinking can be injected through creative problem solving, divergent and convergent thinking by engaging students in lesson class.

1.1 Problem Statement

Mathematics has the ability to form a negative feeling of frightening, confusing, and demotivating learners all around the world. Considering above mention, the experimental study has been done and they carried out with two reasons why students having difficulties in this subjects. As continuing reading Inan (2013), the reason was students are lacking motivation and conceptual abstractness (as cited in Durmus 2004). It could be stated that this situation can be improved through teaching and constructivist learning approach which provide students with the opportunity to expand their knowledge and deliver new information through practice (Inan 2013). According to Idris and Nor (2010), mathematics can foster creativity through a dynamic mental process including divergent thinking. Therefore, the real challenge is to provide an environment of practice and stimulates creativity through problem solving and divergent thinking especially in the classroom.

1.2 Purpose of Study

The purpose of study is to investigate how creativity can be generated through mathematical thinking among architecture students. This study will involve a group of architecture students which involve traditional instruction and the influence of constructivist learning approach in the experimental group. As referring to the above purpose, the instructional material used in study (model making) to be engage in challenging problems and experience the aspect of creative problem solving. The experimental study was constructed through the material evaluation (on '*remember me*' test) developed by Inan (2013).

1.3 Objective of the Study

The main objective of this research is to study on the effects of constructivist learning approach toward mathematical creativity for architecture students.

In order to examine the effect of constructivist learning approach toward high achievement in mathematics hypothesis (on *'remember me'* test), several stages are developed based on the study by Ervynck (1991). Those stages are as follows:

- Stage 1: *A preliminary technical stage.*
  Students understand the concept of loading on structure by experiencing the process of model making.
- Stage 2: *Algorithmic activity stage.*
  Students are capable to apply mathematical techniques which involve mathematical operations, to calculate, manipulate, and solve.
- Stage 3: *The creative (conceptual, constructive) activity.*
  Students analyze theory into non-algorithmic decision (info graphics diagram) and solve the calculation.

2 Literature Review

Jonassen (1992) explained that constructivism is concerned with how we construct knowledge from our experiences, mental structures, and beliefs that are used to interpret objects and events. Wilson (1995) also defines a constructivist learning environment as a place where learners may work together and support each other as they use a variety of tools and information resources in their pursuit of learning goals and problem-solving activities. Therefore, it can be argued that constructivism generates student's divergent thinking of authentic learning as long as the tasks closely replicate the real activity.

In addition to the above statement, Jarmon et al. (2008) reviewed the literature on the use of 3D virtual worlds for teaching and learning and supporting this statement, citing a great deal of research; they found that such activities have the potential to increase student motivation, collaboration, discovery, social interaction, creativity, and address different learning styles.

However, Kim (2005) states that there are three fundamental differences between constructivist and traditional approach. Firstly, learning involves an active constructive process rather than the process of knowledge accession. Secondly, instead of delivering the information to learners, constructivist approach evolves the learner's process of thinking through creative problem solving. Lastly, constructivist approach is a learning–teaching concept rather than a teaching–learning concept.

In other words, constructivist approach is toward student-centered learning. Therefore, students will be able to develop their creativity through critical thinking and problem-based learning by creating ideas and communicate with their colleagues mathematically.

3 Methodology

The research will be conducted by using purposive sampling method, Torrance Test of Creative Thinking. The participants consist of 52 semester four architecture students of University Technology Mara Campus Seri Iskandar, Perak. Students are assigned to form a group of two for the task. The study took one month to conduct, and it focused on one topic only from the syllabus.

While preparing the experimental group, the table of study plan has been prepared based on syllabus content. According to this table, the researcher has divided four weeks of lesson unit as shown in Table 1. First week will be an introduction to structural component, and theories follow by introduction to forces. The Structural Assessment Test (SAT) was developed by researcher to justify on the mathematical creativity development analysis. In order to collect the relevant data for the study, the researcher prepares and uses the (i) Torrance Test of Creative Thinking (TTCT) and (ii) Pilot Test in order to get some student's feedback about the experimental test.

3.1 Data Collection Tools

As the data collection tool, there are two stages of experimental constructivist learning approach. The researcher developed instructional material based on three

Table 1 Teaching plan by following weeks prepared by the researcher

Syllabus content	Week in 2017 (duration: 1 month)
Introduction: Introduction to structural analysis, theories, and components in building structure Units, symbol, and definitions	1
Forces: Loading on structures Finding resultant forces	2
Forces: Forces in equilibrium Moment of force	3
Structural assessment test (SAT)	4

stages of development of mathematical creativity hypothesis. Instruction consisted of introduction of new material (model making), the formulation a problem through creating formula, and followed by diagrammatic solving. Students are required to design a structure in order to achieve balance. In other words, the principle of moment will be applied so that there is no obvious movement of force (turning points) to achieve equilibrium of a system.

Students are given a series of time in order to complete all the development stage during Structural Assessment Test (SAT). Depending on design of the structure model, students are given to solve the structure formulation and answer with the cooperative group member. '*Remember me*' test has been applied accordingly by repeating same stages of development but with different groups of students. Considering above mention, students are required to change their module among other groups and repeat the same stages. The degree of difficulty of task may be different for each group but it still maintained same principle of previous knowledge. This task encourages student interest and creativity toward constructivist leaning model.

As mentioned earlier, constructivist approach involved five steps: (1) inviting ideas; (2) exploring; (3) proposing; (4) explanation and solution; and (5) taking action (Yager, 2000). Throughout the test, measuring instrument of mathematical achievement of student has been developed by researcher and transferred into tables. At the end of the experimental test, the students were asked to do a Pilot Test regarding their own feelings and thought about the Structure Achievement Test (SAT) as part of their learning approach.

3.2 Experimental Design and Procedure

The present study was conducted according to non-equivalent pretest/posttest design as shown in Table 2.

At the beginning, the researcher evaluated Structure Achievement Test (SAT) pretest to ensure whether the groups achieved on the hypothesis stage one. At this point student achievement are considered through making stability of structure model and use their creativity in terms of design and material as shown in Fig. 1. In order to strengthen the task given, each group is required to identify 1: 100 scales of model, module length, and weight. Subsequently, the model needs to

Table 2 Experimental structure

Assigned group	Stages involvement	Treatment
Pretest (original module)	S1, S2, S3	Learning constructivist framework
Posttest (change module)	S1, S2, S3	Learning constructivist framework

S1 A preliminary technical stage (p. 42)
S2 Algorithmic activity stage (p. 43)
S3 The creative (conceptual, constructive) activity (p. 43)

Fig. 1 Developing creativity through involvement of model making

be reassembled. Students are given a series of time to build and to present their concept idea and understanding of the application of the structure model based on previous lesson.

Once they have achieved the results in stage one, the students are required to proceed to next stage of formulation concept and solve the calculation. At the end of the experiment, the same Structure Achievement Test (SAT) is repeated among other groups and Pilot Test to complete the experimental analysis.

4 Results and Discussion

Findings and discussions of this study will focus on student's achievement on creativity in mathematical thinking through constructivist approach and their learning success in understanding on the mathematical application.

4.1 Evaluate Structural Assessment Test Pre-/Post-improvement

Pre/post from the Torrance Test of Creative Thinking result can be seen in Fig. 2. Analysis shows that there is an increment in student's divergent thinking and problem solving on posttest. The researcher noticed that student's performance on SAT definitely changes during the posttest. This was indicated by the student's increased level of using their own creativity in order to solve the mathematical tricks. The level of difficulties is slightly higher during posttest for each stage as shown in Fig. 2. As being mentioned on previous three stages of hypothesis, most of the students complied with above statement and managed to complete the task within the time given.

Creativity in Mathematical Thinking ...   139

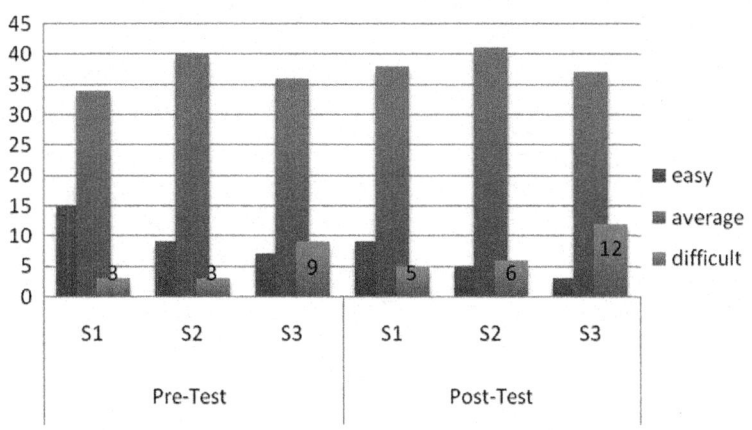

Comparison of Torrance Pre-Test and Post-Test Improvement based on the hypothesis stages (N =52)

Fig. 2 SAT pre-/post-improvement result

4.2 Review on Students Views Based on Torrance Creativity Test (SAT)

In these sections, students have been rated based on three category sections of divergent thinking which lead to creativity development in mathematical principle. This consists of self-rated creativity, originality, and extending or breaking boundaries as shown in Figs. 3, 4 and 5. Based on the data collection, the researcher concludes the constructivist approached successfully improved on student's creativity and understanding in mathematics application.

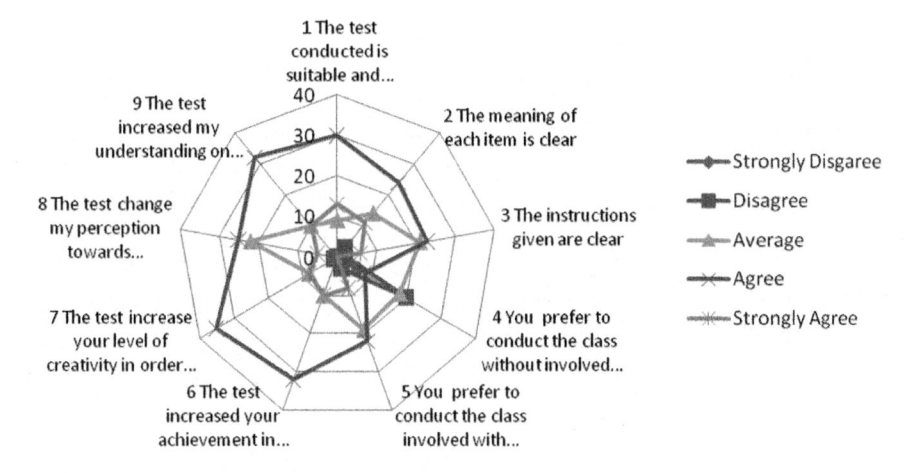

Fig. 3 Torrance assessment review on self-rated creativity

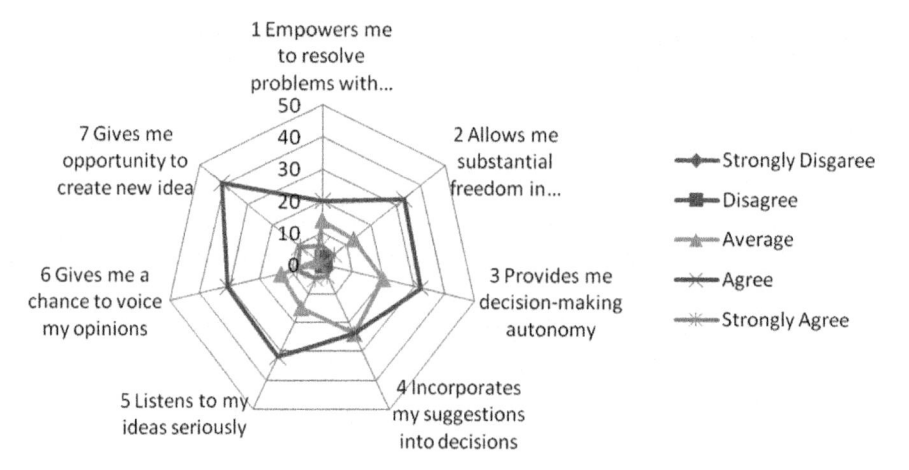

Fig. 4 Torrance assessment review on originality

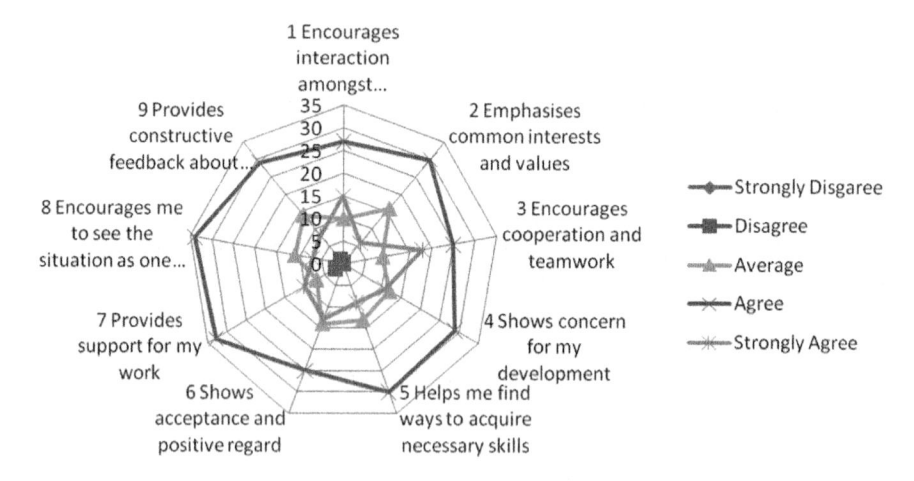

Fig. 5 Torrance assessment review on extending or breaking boundaries

In addition, for this module each Web shows positive self-development as in Figs. 3, 4 and 5. Surprisingly, Fig. 3 at point number 4 shows that majority of students disagree if we conducted the class without involvement of the activity. These findings show that students would prefer thought in constructivist learning environment which remarkably strengthens their understanding and application abilities in mathematical knowledge.

5 Conclusion

Fostering mathematical thinking through creativity in learning environment can be effective by implementing constructivist approach. Developing creative thinking for architecture students basically is not only by teaching numbers and formulas in mathematical problems. The students also need to look at mathematics from both the divergent and convergent thinking perspective. This will allow them to be more creative in solving architectural problems related to structure.

Nowadays, it is a common issue among the teachers and educators to see many students who fail in mathematics. Unfortunately, many educators are focusing to improve subject content rather than the instructional practice to overcome the failure problems. This study shows that the role of instructor is also essential to improve student's creativity through mathematics thinking.

As mentioned previously, the outcomes of the main objective of this research have shown that application of constructivist approach in learning can improve student's mathematical creativity. Students become more active, enjoyable, and participative while doing the task given to them. Constructivist learning, the emphasis is on learning and on the student-centric the learning environment. Students become active participants in their own learning processes including problem solving, critical thinking, communication, collaboration, and self-management.

Therefore, this study is to provide the practical evidences of students learning in constructivist approach which have significant impacts on student's creativity and achievement in mathematics. Those impacts can be seen from the evaluation based on their understanding and applicability on the integrations of their previous learning concept to developed knowledge. In addition, adapting constructivist learning approach in this study also enhances student's soft skills ability such as sharing opinions, learning from peers and communication ability.

The main limitations of this study are the numbers of architecture participants and the size of the task given. Perhaps, those numbers and size can be increased to get more substantial data for a more profound findings and insights into creativity and learning.

Acknowledgements A special gratitude to all architecture students of University of Technology MARA, semester 04 for this constitution of the research. This work is partially supported by Research Management Institute (RMI) of University of Technology MARA. The authors also gratefully acknowledge the helpful comments and suggestions of the reviewers, which have improved the presentation.

References

Ervynck, G. (1991). Mathematical creativity. In *Advanced mathematical thinking* (pp. 42–53).
Idris, N., & Nor, N. M. (2010). Mathematical creativity: Usage of technology. *Procedia-Social and Behavioral Sciences, 2*(2), 1963–1967.
Inan, C. (2013). Influence of the constructivist learning approach on students' levels of learning trigonometry and on their attitudes towards mathematics. *Hacettepe Üniversitesi Eğitim Fakültesi Dergisi, 28*(3).
Jarmon, L., Traphagan, T., & Mayrath, M. C. (2008). Understanding project-based learning in Second Life with the pedagogy, training and assessment trio. *Education Media International, 45*(3), 157–176.

Jonassen, D. H. (1992). Evaluating constructivistic learning. In T. M. Duffy & D. H. Jonassen (Eds.) *Constructivism and the technology of instruction: A Conversation* (pp. 137–148). Hillsdale, NJ: Lawrence Erlbaum Associates.

Kim, J. S. (2005). The effects of a constructivist teaching approach on student academic achievement, self-concept, and learning strategies. *Asia Pacific Education Review, 6*(1), 7–19.

Laycock, M. (1970). Creative mathematics at Nueva. *Arithmetic Teacher, 17,* 325–328.

Nadjafikhah, M., Yaftian, N., & Bakhshalizadeh, S. (2012). Mathematical creativity: Some definitions and characteristics. *Procedia-Social and Behavioral Sciences, 31,* 285–291.

Nayak, D. K. (2007). A study on effect of constructivist pedagogy on students' achievement in mathematics at elementary level. Noida: National Institute of Open Schooling, MHRD.

Wilson, B. G. (1995). Metaphors for instruction: Why we talk about learning environments. *Educational Technology, 35*(5), 25–30.

Yager, R. E. (2000). The constructivist learning model. *Science Teacher, 67*(1), 44–45.

Zanzali, N. A. A. (2000). *Designing the mathematics curriculum in Malaysia: Making mathematics more meaningful.* Skudai: Universiti Teknologi Malaysia.

Forecasting Rainfall Based on Fuzzy Time Series Sliding Window Model

Siti Nor Fathihah Azahari, Rizauddin Saian and Mahmod Othman

Abstract Many researchers are deploying fuzzy time series model for forecasting purposed including rainfall forecasting. However, the limited division size of intervals, u_i to obtain subintervals in that model, influences the accuracy of rainfall forecasting. Hence, in this study the model is enhanced and combined with sliding window algorithm to propose a forecasting model of fuzzy time series sliding window. The proposed model is enhanced in division of intervals, u_i, where several types of division groups of intervals are applied to obtain the better subintervals of model which can increase the accuracy of forecasting result. Another enhancement is introduced which identifies temporal prediction values that will produce the final result of forecasting. The function of subintervals is used to categorize temporal prediction values to observe the trend of rainfall forecasting based on rules of forecasting. In order to validate the proposed model, an error measurement of root-mean-squared error (RMSE) is deployed. The RMSE values between several different divisions of intervals in proposed model and previous study of sliding window algorithm are compared. The result shows that the proposed model, with division of interval groups of 5, 4, 3, 2, and remain unchanged (RU), is better among them. The proposed model is suggested to be tested with other types of data for forecasting.

Keywords Fuzzy time series · Rainfall · Sliding window algorithm

---

S. N. F. Azahari (✉) · R. Saian
Universiti Teknologi MARA, 02600 Arau, Perlis, Malaysia
e-mail: fathihah08@gmail.com

R. Saian
e-mail: rizauddin@perlis.uitm.edu.my

M. Othman
Universiti Teknologi Petronas, 32610, Seri Iskandar, Perak, Malaysia
e-mail: mahmod.othman@utp.edu.my

© Springer Nature Singapore Pte Ltd. 2018
R. Saian and M. A. Abbas (eds.), *Proceedings of the Second International Conference on the Future of ASEAN (ICoFA) 2017 – Volume 2*,
https://doi.org/10.1007/978-981-10-8471-3_14

1 Introduction

Time series forecasting is used to forecast the future event based on previous data. Rainfall is also one of the events that is forecasted among researches. Many conventional forecasting models have been performed, for example fuzzy concept (Dani and Sharma 2013; Xiong et al. 2001) artificial neural network (El-Shafie et al. 2011; Hung et al. 2008), applications of satellite, and radar precipitation products (Abera et al. 2016; Moreno et al. 2012).

In this study, a rainfall forecasting model, fuzzy time series sliding window (FTSSW), is introduced. This model is a combination between fuzzy time series (FTS) model and sliding window algorithm (SWA) (Kapoor and Bedi 2013). The problem statement of FTS is related to the division of interval, u_i, in forming subintervals. The size of division forms a limited subinterval. Hence, this study is aiming to enhance several steps in FTS and SWA and then combine the model together to increase the accuracy of forecasting model. Two types of historical rainfall data from Kampung (Kg.) Behor Lateh which is one of the districts in Perlis are collected from Department of Irrigation and Drainage (DID). Rainfall distribution of tested data is forecasted using FTSSW model and validated using root-mean-squared error (RMSE). The RMSE value is compared between previous SWA and several division group intervals to choose the better model between them. Therefore, the result shows that the FTSSW model, with division group intervals of 5, 4, 3, 2, and RU, is the better model than others.

2 Related Work

Fuzzy time series is introduced by Song and Chissom. The model is proposed to forecast data based on linguistic or numerical of historical data. It has been used in various forecasting studies such as tourism (Tsaur and Kuo 2011), pollution concentration (Domańska and Wojtylak 2012), and also student enrollment (Chen and Hsu 2004; Chen et al. 2009). In fact, many researchers had enhanced fuzzy time series model in various ways to obtain the best forecasting result in their studies.

For example, Chen in his study has reduced the computation overhead of time-variant and time-invariant models by simplified model including simple arithmetic operation (Chen 1996). In 2001 and 2006, the FTS model is improved by Hurang and Yu (2006) to explore ways to determine the useful lengths of intervals in fuzzy time series. The use of ratios is suggested apart from equal lengths of intervals that more properly represent the intervals among observations. Ratio-based lengths of intervals are actually proposed to reduce error of forecasting values. Erol (2012) proposed a new method of fuzzy time series that employs membership functions of fuzzy sets. The new method determines elements of fuzzy relation matrix based on genetic algorithms. The proposed method uses first-order fuzzy time series forecasting model, and it is applied to the several data sets.

Therefore, most of the improvement that was performed by researchers had satisfied result of forecasting.

There is another forecasting model, sliding window algorithm (SWA), which is adopted in this study by combining it into FTS and produces fuzzy time series sliding window (FTSSW) model. Sliding window is introduced by Datar et al. (2002) which is a popular model for processing infinite data streams in small space when the goal is to compute properties of data that has arrived in the last window of time. Sliding window is setting for data streams, which aggregates and statistic are computed over a "sliding window" of the N most of recent items in the data stream. A number of interesting results presented on estimating functions over a sliding window for a single stream are obtained. A basic problem that they considered determines the number of 1's in a sliding window which is known as the Basic Counting problem. Sliding window algorithm is widely used in various temporal applications like weather forecasting (BenYahmed et al. 2015; Kapoor and Bedi 2013), and database system (Arasu and Manku 2004) and medical (d'Arcy et al. 2002).

2.1 Basic Theory of Fuzzy Time Series

If U is the universe of discourse, $U = \{u_1, u_2, ..., u_n\}$, and if X is a fuzzy set in the universe of discourse, U is defined as follows:

$$X = f_x(u_1)/u_{1=} + f_x(u_2)/u_{2=} + \cdots + f_x(u_n)/u_n \tag{1}$$

where f_x is the membership function of X, $f_x: U \to [0,1]$, $f_x(u_1)$ means the grade of membership of u_i in the fuzzy set X, $f_x(u_i) \in [0,1]$, and $1 \leq i \leq n$ symbol "/" splits the elements from the grades of membership in the universe of discourse U. Meanwhile, "+" indicates "union."

If $W_a(t)$ $(t = ..., 0, 1, 2, ...)$ is the universe of discourse and is the subset of R_a, let fuzzy set $x_i(t)$ $(i = 1, 2, ...)$ be described in $W_a(t)$. Let $X_a(t)$ be a collection of $x_i(t)$ $(i = 1, 2, ...)$. Then, $X_a(t)$ is called a fuzzy time series of $W_a(t)$ $(t = ..., 0, 1, 2, ...)$. Then, X_a is caused by $X_a(t-1)$, indicated by $X_a(t-1) \to X_a$, where this relationship can be represented by $X_a = X_a(t-1) \circ R_a(t, t-1)$, where the symbol "∘" is a composition operator symbol of the Max–Min; $R_a(t, t-1)$ is a fuzzy relation between $X_a(t)$ and $X_a(t-1)$ and is known as the first-order model of $X_a(t)$.

Meanwhile, let $X_a(t)$ be a fuzzy time series and also $R_a(t, t-1)$ be a first-order model of $X_a(t)$. If $R_a(t, t-1) = R_a(t-1, t-2)$ for any time t, then $X_a(t)$ is known as a time-variant fuzzy time series. If $R_a(t, t-1)$ is dependent on time t, in addition $X_a(t)$ is known as a time-variant fuzzy time series. Then, the stepwise procedures define the fuzzy time series of Song and Chissom model.

Step 1: The universe of discourse U is defined from the historical data. The smallest data value is assumed as $D_{\min}$, and the largest data value is assumed as $D_{\max}$.

Hence, the universe $U = [D_{min}, D_{max}]$. Make sure that the D_{min} and D_{max} are in the whole number.

Step 2: The universe of discourse U is divided into equal length of interval; $u_1, u_2, ..., u_n$. Assign each of the intervals with fuzzy set $A_1, A_2, ..., A_n$.

Step 3: Each of the fuzzy set A_i from step 2 is defined where if the fuzzy set is $A_1, A_2, ..., A_n$, then the fuzzy set $A_i, \forall i = 1, 2, ..., n$ can be described as $Ai = f_{Ai}(u_1)/u_1 + f_{Ai}(u_2)/u_2 + \cdots + f_{Ai}(u_n)/u_n$.

Step 4: The historical data is fuzzified.

Step 5: A suitable parameter v, where $v > 1$, $R^v(t, t-1)$ and predict the data as follow:

$$X_a(t) = X_a(t-1)R^v(t, t-1) \qquad (2)$$

where $X_a(t)$ indicates the forecast fuzzy data of year t, $X_a(t-1)$ indicates the fuzzified data of year $t-1$, and

$$R^v(t, t-1) = X_a(t-1) \times X(t-1) \cup X(t-2) \times X(t-3) \cup \cdots \cup X_a(t-w+1) \qquad (3)$$

where w is denoted as "model basis" denoting the number of years before t, $F(t)$ is the forecasting fuzzy of year t, "$\times$" is the symbol of Cartesian product operator, and T is the transpose operator.

Step 6: The forecasted output is determined. If the time series data $A(t), \forall t \in [1, n]$, $i = 1, 2, ..., n$ then the forecast of $F_a(k + 1)$ is obtained as:

$$F_a(k+1) = Ai \circ R, \forall t \in [1, n], i = 1, 2, ..., n \qquad (4)$$

Step 7: The forecasted output is interpreted. Each of the forecasting value is defuzzified based on the centroid method.

3 Methodology

The proposed model of this study, fuzzy time series sliding window (FTSSW), is introduced where Fig. 1 shows the overall 16 steps.

3.1 Numerical Example

Based on Fig. 1, the first step is defined the Universe U of rainfall data of Kg. Behor Lateh is [0, 20] and divided into several interval, u_i that produce $u_1 = [0, 4.0]$, $u_2 = [4.0, 8.0]$, $u_3 = [8.0, 12.0]$, $u_4 = [12.0, 16.0]$, and $u_5 = [16.0, 20.0]$. Then, the rainfall data are classified based on suitable range in interval, u_i. Table 1 shows the classification of rainfall data into suitable interval, u_i.

Based on Table 1, the classification of rainfall data is aggregated and sorted from the highest frequency (first) interval to the lowest frequency (nth + 1) interval. Then, each of the intervals is divided into certain division of subinterval, Sj, that are shown in third to ninth row of Table 2. Table 2 shows one of the examples of re-division of intervals which is division based on 5, 4, 3, 2, and remain unchanged.

Then, construct 12 × 1 matrix size of current year (CY) and 24 × 1 matrix size of previous year (PY). The examples to construct the matrix size are shown in Tables 3 and 4. Then, identify 13 sliding windows of size 12 × 1 from matrix PY, compute Euclidean distance, Edi, of sliding window, and define the mean of Edi. Table 6 shows a set of sliding window that computes Edi and mean of Edi.

From Table 5, sliding window, W_2, is selected as minimum mean of Edi based on Eq. 5 that computes Euclidean distance, Edi for each sliding window and Eq. 6 which calculates mean of Edi. Then, rename the minimum mean of Edi as mean variation previous (VP). Meanwhile, the variation of current year CY is also computed and renamed as mean variation current (VC). Hence, according to Eqs. 8 and 10, the VP is −0.28 and VC is −0.02.

Next step, the predicted variation "V" of Kg. Behor Lateh is identified based on Eq. 11 that equals to 0.15. Then, the predicted variation "V" is total to the average of monthly, AM, to obtain temporal prediction (TP) value of each month. The TP values are classified into suitable rule of forecasting in Step 15. Finally, defuzzify TP value to obtain forecasting rainfall (FR) values. Table 6 shows the classification of TP value into rule of forecasting to obtain FR.

4 Result and Discussion

A rainfall forecasting model, FTSSW, is introduced in this study to forecast rainfall distribution in Kg. Behor Lateh rainfall gauge station, Perlis, Malaysia. RMSE is used in this study to validate the models. Based on Lazim (2007), the smallest value of RMSE shows the best algorithm because of less error between forecasted rainfall and actual data. Firstly, the model is validated on FTSSW model with different divisions of intervals that were computed in Table 1. The different divisions which are producing different subintervals are aiming to choose the best subintervals that satisfy accurate result. Then, the chosen subintervals are proposed in FTSSW model to compare with previous model which is SWA (Kapoor and Bedi 2013). The equation of RMSE is as follows:

$$\text{RMSE} = \sqrt{\frac{1}{n}\sum_{i=1}^{n}\left|\frac{(\text{Actual rainfall}_i - \text{Forecast rainfall}_i)^2}{\text{Number of sample}}\right|} \qquad (15)$$

Step 1: Define the Universe $U = [AR_{min}, AR_{max}]$.
Step 2: Categorize each data based on interval. Universe U is divided into seven equidistant intervals of $u_1, u_2, ..., u_7$.
Step 3: Identify the frequency of data. Sort the highest to the lowest frequency.
Step 4: Re-divide interval to produce sub-interval.
Step 5: Construct matrix size 12x1 for Current Year (CY) and 24x1 for Previous Year (PY).
Step 6: Identify 13 sliding window from matrix PY.
Step 7: Compute Euclidean distance, *Edi* of sliding window. Find mean of *Edi*;

$$Edi = \sqrt{(x_i - y_i)^2} \qquad (5)$$

$$\text{Mean of } Edi = \sum \frac{Ed_i}{n} \qquad (6)$$

Step 8: Select the minimum mean value of Euclidean distance, *Edi* from sliding window.
Step 9: Compute variation for min. Euclidean distance, *Edi* and rename as Mean Variation Previous (VP).

$$\text{Find variation; } V = AR_t - AR_{t-1} \qquad (7)$$

$$\text{Mean of Variation} = \frac{\sum \text{variation of each month}}{n} \qquad (8)$$

Step 10: Compute variation of CY and rename as Mean Variation Current (VC).

$$\text{Find variation; } V = AR_t - AR_{t-1} \qquad (9)$$

$$\text{Mean of Variation} = \frac{\sum \text{variation of each month}}{n} \qquad (10)$$

Step 11: Determine the Predicted Variation "V".

$$\text{Predicted Variation "V"} = \frac{\text{mean VC} + \text{mean VP}}{2} \qquad (11)$$

Step 12: Define average monthly (AM) of rainfall 1990 to 2013.
Step 13: Define temporal prediction (TP).
$$TP = AM + \text{Predicted Variation "V"}. \qquad (12)$$
Step 14: Determine sub interval, *Sj*, for each temporal prediction, *TP*.
Step 15: Identify the rule of forecasting for temporal prediction, *TP*.

If $(TP_t - TP_{t-1}) - (TP_{t-1} - TP_{t-2}) = +ve$, $\qquad (13)$
then TP is classify in Rule 2.
If $(TP_t - TP_{t-1}) - (TP_{t-1} - TP_{t-2}) = -ve$, $\qquad (14)$
then TP is classify in Rule 3.

(Note: TP_t is temporal prediction of current month, TP_{t-1} is one month before current month, TP_{t-2} is two month before current month).
Step 16: Defuzzify TP to obtain forecasting rainfall distribution 2014 based on forecasting rule.

Fig. 1 Fuzzy time series sliding window model

Table 1 Classification of rainfall data

Month	u_1	u_2	u_3	u_4	u_5
Jan.	23	1	–	–	–
Feb.	20	4	–	–	–
Mar.	13	10	–	1	–
Apr.	8	12	3	1	–
May	9	11	4	–	–
June	8	11	5	–	–
July	6	8	9	–	1
Aug.	4	10	7	3	–
Sept.	2	9	8	3	2
Oct.	3	10	6	5	–
Nov.	6	9	7	–	2
Dec.	20	3			1
Total	122	98	49	13	6
Sort	1st	2nd	3rd	4th	5th
Divide	4	3	2	RU	RU
Divide	5	4	3	2	RU
Divide	6	5	4	3	2
Divide	7	6	5	4	3
Divide	8	7	6	5	4
Divide	9	8	7	6	5
Divide	10	9	8	7	6

Note RU is remain unchanged

Table 2 Re-division interval to obtain subinterval

Divide by						
5	4	3				
u_i	Sj	u_i	Sj	u_i	Sj	
$u_{1,1}$	$S_1 = [0, 0.8]$	$u_{2,1}$	$S_6 = [4.0, 5.0]$	$u_{3,1}$	$S_{10} = [8.0, 9.33]$	
$u_{1,2}$	$S_2 = [0.8, 1.6]$	$u_{2,2}$	$S_7 = [5.0, 6.0]$	$u_{3,2}$	$S_{11} = [9.33, 10.66]$	
$u_{1,3}$	$S_3 = [1.6, 2.4]$	$u_{2,3}$	$S_8 = [6.0, 7.0]$	$u_{3,3}$	$S_{12} = [10.66, 12]$	
$u_{1,4}$	$S_4 = [2.4, 3.2]$	$u_{2,4}$	$S_9 = [7.0, 8.0]$			
$u_{1,5}$	$S_5 = [3.2, 4.0]$					
2	RU					
u_i	Sj	u_i	Sj			
$u_{4,1}$	$S_{13} = [12.0, 14.0]$	u_5	$S_{15} = [16.0, 20.0]$			
$u_{4,2}$	$S_{14} = [14.0, 16.0]$					

Table 3 Matrix size of 12 × 1 of CY

Current year (CY)	
Month/year	Rainfall
Jan-13	0.6
.	.
.	.
.	.
Dec-13	0.6

Table 4 Matrix size of 24 × 1 of PY

Previous year (PY)	
Month/year	Rainfall
Jan-12	3.4
.	.
.	.
.	.
Dec-11	3.3

Table 5 A Set of Sliding Window that Compute Ed_i and Mean of Ed_i

W_1		W_2				W_{13}	
R	Ed_1	R	Ed_2			R	Ed_{13}
1.7	1.14	3.5	2.93			1.2	0.56
.	.	.	.			.	.
.	.	.	.	.	.	.	.
.	.	.	.			.	.
6.5	5.17	1.2	0.14			2.4	1.12
Mean	3.22	Mean	3.55			Mean	2.82

Table 6 Classification of TP value into rule of forecasting to obtain FR

Month	Sj	TP	Diff. between month	Rule	FR
Jan.	2	1.26	–	–	1.2
Feb.	3	2.09	0.84	1	2
Mar.	6	4.22	1.3	2	4.25
Apr.	7	5.67	−0.7	3	5.5
May	7	5.57	−1.5	3	5.5
June	7	5.9	0.4	2	5.5
July	9	7.39	1.2	2	7.25
Aug.	9	7.83	−1	3	7.5
Sept.	10	9.31	1	2	8.67
Oct.	10	8.03	−2.8	3	8.67
Nov.	9	7.41	0.7	2	7.5
Dec.	5	3.36	−3.4	3	3.6

Fig. 2 Result of RMSE between several divisions and SWA

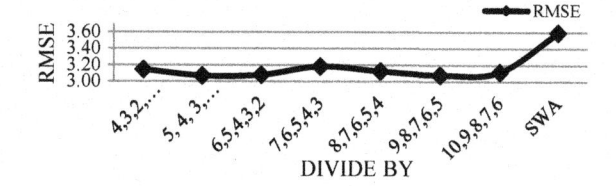

Figure 2 shows the result of RMSE between different divisions of intervals and SWA.

Based on the figure, the line graph shows that RMSE for FTSSW model that is divided by 5, 4, 3, 2, and remain unchanged is 3.07. The RMSE value is the lowest among other division of intervals including SWA. The lowest error measurement of RMSE is favorable for an efficient model. Therefore, the value of RMSE of FTSSW model with division of intervals of 5,4,3,2, and remain unchanged shows the better model as compared to others which also tells the satisfaction accuracy of forecasted rainfall values.

5 Conclusion

In this study, a model of rainfall forecasting, fuzzy time series sliding window (FTSSW), is proposed. Rainfall data from Kg. Behor Lateh which is one of the rainfall gauge stations in Perlis, Malaysia, are taken from Department of Irrigation and Drainage (DID) and tested to the proposed model. The forecasting is carried out based on sixteen steps in Fig. 1. Several types of division of intervals, u_i, are executed to identify the best subintervals that can increase the accuracy of rainfall forecasting. The previous forecasting model by Kapoor and Bedi (2013), sliding window algorithm (SWA), is also tested to identify the best model among them. Model validation is executed to the proposed model and SWA by using root-mean-squared error (RMSE). According to the result in Fig. 2, the proposed model with division of intervals 5, 4, 3, 2, and remain unchanged (RU) shows the lowest value of RMSE as compared to other division of intervals and SWA. Hence, the proposed model with division of intervals 5, 4, 3, 2, and RU is the better forecasting model among them. Therefore, the application of this forecast model should be able to provide more accurate predictions on rainfall distribution in order to avert any disaster that is caused by rainfall. For extended work, this study can be proposed to other forecasting problems.

Acknowledgements The study was funded by the "Long Term Research Grant (LRGS) (UUM/ RIMPC/P-30)." A huge thanks go to the Faculty of Computer and Mathematical Sciences, Universiti Teknologi MARA and Malaysia's Drainage and Irrigation of Department for providing the laboratory facilities, the information of rainfall distribution data to complete this study.

References

Abera, W., Formetta, G., Brocca, L., & Rigon, R. (2016). Water budget modelling of the Upper Blue Nile basin using the JGrass-NewAge model system and satellite data. *Atmospheric Research, 178–179*, 471–483.

Arasu, A., & Manku, G. S. (2004). Approximate counts and quantiles over sliding windows. *Paper presented at the Proceedings of the twenty-third ACM SIGMOD-SIGACT-SIGART Symposium on Principles of Database Systems*, Paris, France.

BenYahmed, Y., Abu Bakar, A., Hamdan, A. R., Ahmed, A., & Syed Abdullah, S. M. (2015). Adaptive sliding window algorithm for weather data segmentation. *Journal of Theoretical and Applied Information Technology, 80*(2), 322–333.

Chen, S. M. (1996). Forecasting enrollments based on fuzzy time series. *Fuzzy Sets and Systems, 81*(3), 311–319.

Chen, S. M., & Hsu, C. C. (2004). A new method to forecast enrollments using fuzzy time series. *International Journal of Applied Science and Engineering, 2*(3), 234–244.

Chen, S. M., Wang, N. Y., & Pan, J. S. (2009). Forecasting enrollments using automatic clustering techniques and fuzzy logical relationships. *Expert Systems with Applications, 36*(8), 11070–11076.

d'Arcy, J., Collins, D., Rowland, I., Padhani, A., & Leach, M. (2002). Applications of sliding window reconstruction with cartesian sampling for dynamic contrast enhanced MRI. *NMR in Biomedicine, 15*(2), 174–183.

Dani, S., & Sharma, S. (2013). Forecasting rainfall of a region by using fuzzy time series. *Asian Journal of Mathematics and Applications, 2013*.

Datar, M., Gionis, A., Indyk, P., & Motwani, R. (2002). Maintaining stream statistics over sliding windows. *SIAM Journal on Computing, 31*(6), 1794.

Domańska, D., & Wojtylak, M. (2012). Application of fuzzy time series models for forecasting pollution concentrations. *Expert Systems with Applications, 39*(9), 7673–7679.

El-Shafie, A. H., El-Shafie, A., El Mazoghi, H. G., Shehata, A., & Taha, M. R. (2011). Artificial neural network technique for rainfall forecasting applied to Alexandria, Egypt. *International Journal of Physical Sciences, 6*(6), 1306–1316.

Erol, E. (2012). A new time-invariant fuzzy time series forecasting method based on genetic algorithm. *Advances in Fuzzy Systems, 2012*, 6.

Huarng, K., & Yu, T. H.-K. (2006). Ratio-based lengths of intervals to improve fuzzy time series forecasting. *IEEE Transactions on Systems, Man, and Cybernetics, Part B (Cybernetics), 36*(2), 328–340.

Hung, N. Q., Babel, M. S., Weesakul, S., & Tripathi, N. K. (2008). An artificial neural network model for rainfall forecasting in Bangkok, Thailand. *Hydrology and Earth System Sciences Discussions, 5*(1), 183–218.

Kapoor, P., & Bedi, S. S. (2013). Weather forecasting using sliding window algorithm. *ISRN Signal Processing, 2013*, 5.

Lazim, M. A. (2007). *Introductory business forecasting: A practical approach*: Pusat Penerbitan Universiti, Universiti Teknologi MARA.

Moreno, H. A., Vivoni, E. R., & Gochis, D. J. (2012). Utility of quantitative precipitation estimates for high resolution hydrologic forecasts in mountain watersheds of the Colorado Front Range. *Journal of Hydrology, 438*, 66–83.

Tsaur, R. C., & Kuo, T. C. (2011). The adaptive fuzzy time series model with an application to Taiwan's tourism demand. *Expert Systems with Applications, 38*(8), 9164.

Xiong, L., Shamseldin, A. Y., & O'Connor, K. M. (2001). A non-linear combination of the forecasts of rainfall-runoff models by the first-order Takagi-Sugeno fuzzy system. *Journal of Hydrology, 245*(1–4), 196–217.

e-Diet Meal Recommender System for Diabetic Patients

Mahfudzah Othman, Nurzaid Muhd Zain and Umi Kalsum Muhamad

Abstract This paper discusses the development of a recommender system specifically built for diabetic patients. With the constraints faced by the health officers to monitor patients' diabetes care effectively and efficiently, this study is aimed to provide a simple meal recommender system for diabetic patients to increase their awareness in self-managing their daily food consumption and monitoring their own diabetes care. The recommender engine was developed using the neighborhood-based collaborative filtering technique, or more specifically the user-based collaborative filtering algorithm. The algorithm will adapt the Pearson correlation coefficient to find similarities among patients and make meal recommendations based on the patients' ratings. In order for the system to provide the first meal recommendations, patients need to submit their current blood glucose readings and body mass index (BMI) beforehand and then later need to rate the recommended meals. Based on the ratings, a list of detailed recommended meals will be generated by the recommender engine ranging from the most recommended meals to the least recommended ones. From here, patients will be able to review the recommended meals according to their diabetic conditions and hopefully becoming increasingly aware of their own daily dietary intake and diabetes care.

Keywords Diabetes · Recommender system · Collaborative filtering

1 Introduction

According to the National Health and Morbidity Survey (NHMS) done in 2015 by the Malaysia Ministry of Health (MOH), there was a general increasing trend in prevalence of diabetes mellitus among Malaysians aged 18 years old and above

M. Othman (✉) · N. M. Zain · U. K. Muhamad
Universiti Teknologi MARA, Perlis Branch, 02600 Arau, Perlis, Malaysia
e-mail: fudzah@perlis.uitm.edu.my

where the prevalence was slightly higher in the urban areas and much higher among the female respondents (Institute for Public Health 2015).

Based on the study done by Zanariah et al., the increase in the prevalence of type 2 diabetes has contributed significantly to the increment in the prevalence of overweight and obesity among Malaysians. Abdominal obesity caused by poor diet management and lack of physical activities is one of the main factors that contribute to the high increment of diabetic patients apart from other factors such as ethnicity, genetic inheritance, and family health backgrounds (Mohamud et al. 2012; Fuziah et al. 2012). As an intervention to control the disease, most of the public healthcare facilities in Malaysia are currently providing diabetes-monitoring programs through consultations with dietitians, diabetes nurse educators, and pharmacists. In recent years, diabetes mellitus treatment adherence clinics have also been established in all major public hospitals in Malaysia, which are coordinated and operated by pharmacists in order to increase awareness and deliver diabetes education to the diabetic patients.

Nonetheless, with the increasing number of patients being referred to the healthcare providers, it is impossible for the dieticians to monitor the dietary intakes of every patient effectively and efficiently. Poor adherence to dietary recommendations is likely to be high among the patients (Tan et al. 2011). Further investigations also claimed that the lack of healthcare officers trained in diabetes education often leads to lack of supervision, guidance, and support for diabetes self-management. Consequently, it will affect patients' conditions and will lead to more complications and comorbidities.

Therefore, the main objective of this study is to develop an electronic diet recommender system for diabetic patients in order to assist them in self-managing their diabetes disease. This system is aimed to provide personalized recommended meals on daily basis for diabetic patients. To achieve these goals, a collaborative filtering technique was used in order to provide the recommendations by adapting the user-based collaborative filtering algorithm with Pearson correlation coefficient similarity measure to calculate similarities among users and recommends meals based on the users' ratings. A list of recommended meals will be displayed for the patients to review and make choices for their dietary regiments. By doing this, the patients will be increasingly aware of their diabetic conditions and could monitor their own daily food consumption.

2 Background of Study

2.1 Web-Based System for Diet Management

Web-based diet management system is very useful for the patients that have limited time or constraints to go and see their personal doctors. Because of the low achievements among diabetic patients that follow the dietary regiments given by the dieticians and lack of involvement in self-management programs, Web-based system can be seen as an intervention to bridge the gap between the diabetes care and self-management (Tan et al. 2011; Yu et al. 2014).

Among the benefits of using the Web-based system as a platform for self-monitoring and managing diabetes care are: It may be considered as cost-effective, more convenient, and easier to use and learn from anywhere without patients having to present physically at the healthcare facilities (Viral and Satish 2015). Patients also found it easier to understand diabetes and its complications, thus becoming increasingly aware to personalize and self-manage diabetes via the Web-based system (Viral and Satish 2015).

2.2 Personal Recommender System

Recommender systems are tools that are responsible for interacting with complex and large data spaces. They provide some personalized view of such prioritizing, space items so that it will be more interesting to the user (Burke et al. 2011). A recommendation system is important because it helps users to upload and share information with other people. Park et al. (2011) defined recommender system as the supporting system that helps users to search products, services, or information by analyzing and aggregating suggestions from users. This is achieved through reviews from various user attributes and authorities, aggregates, and directs to appropriate recipients. Often, the primary transformation is in the aggregation and its ability to make good matches between the recommenders and those seeking recommendations (Park et al. 2011).

In addition, personal recommender system can be divided into four techniques that can be used to make recommendations to users in the Web-centric domain. The techniques are collaborative filtering, content-based, knowledge-based, and hybrid-based recommender system. According to Liu et al. (2009), collaborative filtering system means that a user will be recommended to the objects that are similar to the previous users' similarities and preferences. Meanwhile, the content-based recommender system is the type of recommender system that will give the prediction of the scores from the object content information (Liu et al. 2009). On the other hand, knowledge-based recommender system will make predictions based on user-specified requirements rather than the history of the user (Aggarwal 2016). Thus, the hybrid recommender system is actually a recommender engine that combines all the aforementioned recommender systems described before (Liu et al. 2009).

2.2.1 Collaborative Filtering Technique

In this study, the collaborative filtering technique was used to develop the recommender engine. According to Liu et al. (2009), collaborative filtering technique can be divided into two categories, which are the memory-based and model-based collaborative filtering system. The former is also known as neighborhood-based collaborative filtering and consists of two different algorithms, which are the user-based collaborative filtering algorithm and item-based collaborative filtering algorithm (Aggarwal 2016). For the purpose of this study, the user-based collaborative filtering algorithm was selected where the ratings in the system are predicted using the ratings of neighboring users, thus finding the similarities among the users. In addition to find the similarity, the Pearson correlation coefficient was adapted where it will calculate the users' similarities and make recommendations based on the user ratings (Aggarwal 2016).

2.3 Pearson Correlation Coefficient

One measure that captures the similarity between the rating vectors of two users is the Pearson correlation coefficient (Aggarwal 2016). The Pearson correlation coefficient is used to calculate the correlation between sets of data, whereby the data will be measured as how far they are related. The formula for Pearson correlation coefficient used in this study is as depicted in the equation below.

$$r = \frac{n(\sum xy) - (\sum x)(\sum y)}{\sqrt{\left[n\sum x^2\right] - (\sum x)^2\right]\left[n\sum y^2 - (\sum y)^2\right]}} \quad (1)$$

The result for Pearson correlation will be between −1 and 1 where if the value of "r" is very close to zero, then the variation of the data points around the line of best fit is greater (Andale 2012). To summarize, if the result is +1, it indicates a very strong positive correlation. Meanwhile, −1 indicates a strong negative correlation and a zero shows no correlation.

In addition, the results can be transformed into graph forms as depicted in Fig. 1. Based on Fig. 1, for the positive correlation, it shows that as x-value increases, the y-value will get larger and vice versa for the negative correlation.

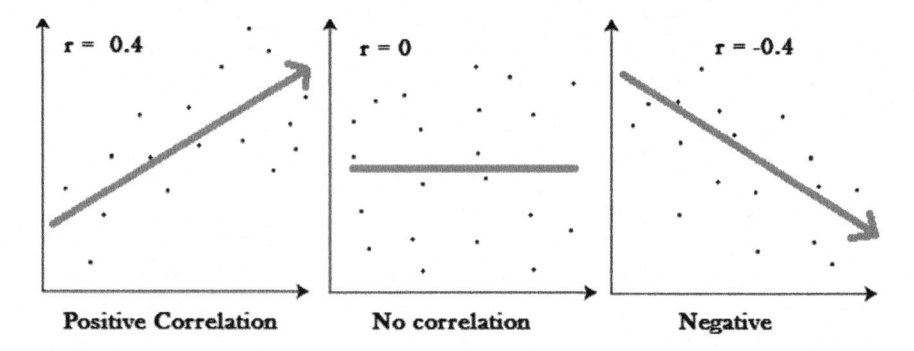

Fig. 1 Pearson correlation categories depicted in graphs (Andale 2012)

3 Materials and Methods

The overall development of the e-Diet Meal Recommender System prototype involved a broader process as depicted in the system development life cycle (SDLC) waterfall model in Fig. 2. The overall process involved five phases, which are information gathering and analysis phase, design phase, construction phase, testing and documentation phase. Nevertheless, the discussion in this paper is more focused on the construction of the recommender engine, which only involved the first three phases.

Briefly, the first phase involved information gathering activities and investigations among diabetic patients in two public medical centers in Kangar and Arau, Perlis, which are known as Klinik Kesihatan Arau and Klinik Kesihatan Kangar. Within a week in the beginning of the study, questionnaires were distributed to 60 respondents, which were also the diabetic patients that underwent medical treatments at the respective medical centers. The aim of the investigation was to further understand the backgrounds of the diabetic patients and how they self-managed the disease. Interviews were also conducted with the dietician and health officers in both healthcare facilities to acquire information about the consultation sessions with patients and how the dietary intakes were recommended. From the investigation, it has revealed that about 50% of the diabetic patients were Malays aged between 40 and 60 years old.

Furthermore, the consultations normally done by individual basis as recommended by the physicians. During the consultation session, patients were encouraged to do self-monitoring and manage their daily food intakes to control their blood glucose levels. The most common and easiest way to monitor the blood glucose level is by using the portable glucometer that provides patients with daily readings. Other than that, patients were also advised to monitor their weight gain or body mass index (BMI) to ensure that their diabetic conditions are in control and

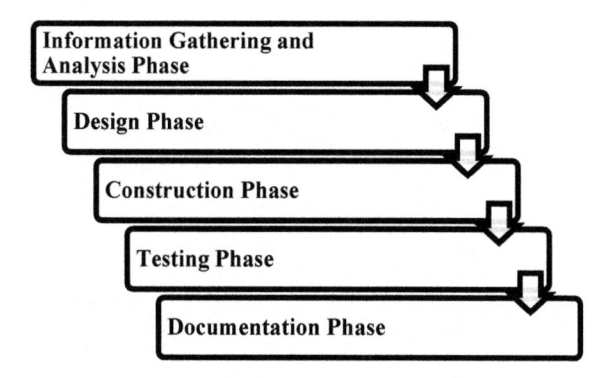

Fig. 2 System development life cycle (SDLC) waterfall model

manageable as many studies have showed strong correlation between overweight or obese patients and diabetes disease.

The second phase, which is the design phase, involved two main steps, which are designing the Web-based system and designing the algorithm for the recommender engine as shown in Fig. 3. In designing the Web-based system, we have used several tools to design the interface and database such as storyboard and data flow and process flow diagrams. Nonetheless, because the main focus of the

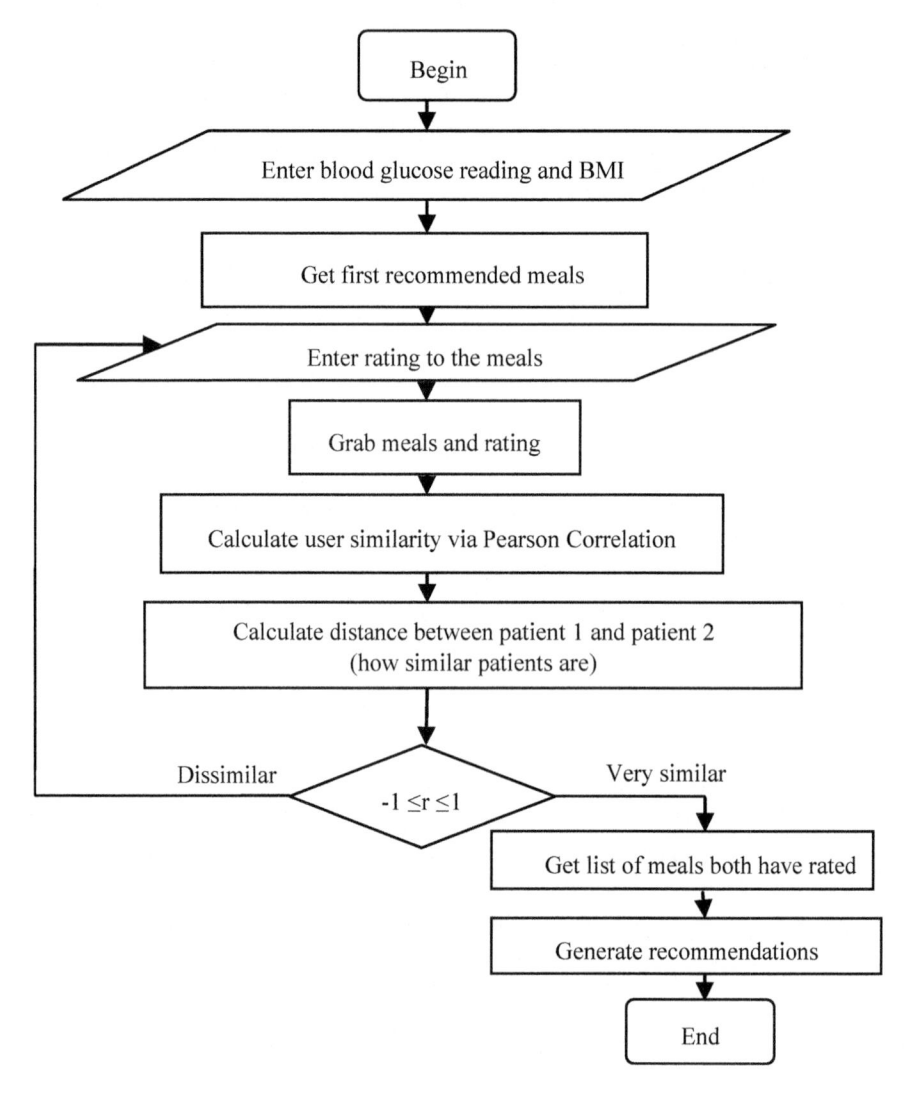

Fig. 3 User-based collaborative recommender algorithm

discussion in this paper is the recommender engine, the design of the storyboard and data and process flows will not be discussed in detail in this paper.

Based on the investigation in the first phase, we have selected two variables that will determine the patients' first meal recommendations, which are the patients' blood glucose readings and BMIs because of the simple, economical, and non-invasive procedures. Therefore, as displayed in Fig. 3, patients need to enter their current blood glucose readings together with their BMIs beforehand in order to get their first recommended meals of the day.

Later, patients will need to rate the recommended meals where the recommender engine will grab the ratings, calculate the similarities, and generate a list of more detailed recommended meals that are suitable for their dietary regimes. For the purpose of this study, the Web-based application tool known as Xampp, which includes MySQL and PHP, was used to develop the system. The algorithm designed from the previous phase was translated into meaningful codes called as routines in the Xampp for the recommender engine to work successfully in the Web-centric paradigm. Example of the routine constructed in the Xampp is as depicted in Fig. 4. A detailed process on how the recommender engine works will be discussed in the next section.

```
// Phase 1: grab ratings
$patientRatings = array();     // [userId][talkId] = rating
$mealsTitles = array();        // we'll store these for later


// Phase 2: Calculate user similarity (via Pearson correlation)
$pearson = array();

$sql5 = mysql_query("Select * from mrfdp.uniqDiet ORDER BY user_ic ASC");
$j = 0;

while ($row = mysql_fetch_array($sql5))
{
    $res_return[] = $row;
}
//echo "</br>";
//echo "</br>";
//echo "</br>";
//print_r($res_return);
//echo "</br>";
//echo "</br>";
//echo "</br>";
$patientRatings = $res_return;
//var_dump($patientRatings);
$patients = array_keys($patientRatings);
```

Fig. 4 Coded routines for meal recommender engines in Xampp platform

4 e-Diet: Meal Recommender System for Diabetic Patients

Because most of the patients involved in this study were Malays aged between 40 and 60 years old, simple Malay language was used to provide instructions in the recommender system. There were few functions created to support the recommendation, where the main function is where the patients need to submit their current blood glucose readings and BMIs to get their first daily recommended meals. The patients can review the recommended meals and then later rate the meal as shown in Fig. 5.

The scale used to rate the meals was the Likert scale where scale 1 represents the least recommended meal, while scale 5 represents the most recommended meal as displayed in Fig. 6. This process eventually will involve the routine in the

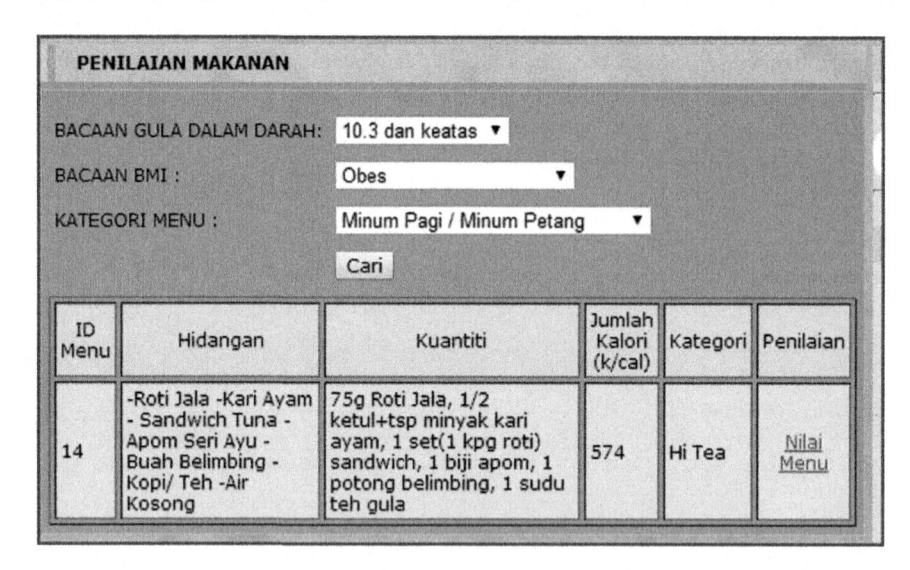

Fig. 5 Recommended menu based on patient's blood glucose reading and BMI

Fig. 6 Patients need to rate the meal recommended to them

e-Diet Meal Recommender System for Diabetic Patients

MAKANAN CADANGAN

ID	Nama Hidangan	Kuantiti	Jumlah Kalori (k/cal)	Kategori	Rating
1	Kuih Apam kukus, Roti jala gulung, Tembikai merah kopi/ teh O	- 2 ketul kecil kuih apam - 1 keping roti jala - 1 potong buah * gula satu sudu teh (minuman)	200	Minum Pagi / Minum Petang	5
3	Nasi tomato, ayam masak merah, acar timun/nenas, dalca, betik. Air kosong	- 1 cawan nasi - 2 ketul ayam - 1/2 cawan acar timun/nenas - 1/2 cawan dalca - 1 potong betik	594	Makan Tengahari/ Makan Malam	4
14	-Roti Jala -Kari Ayam - Sandwich Tuna - Apom Seri Ayu -Buah Belimbing -Kopi/ Teh -Air Kosong	75g Roti Jala, 1/2 ketul+tsp minyak kari ayam, 1 set(1 kpg roti) sandwich, 1 biji apom, 1 potong belimbing, 1 sudu teh gula	574	Hi Tea	3
16	-Nasi Lemak (ikan bilis goreng, telur rebus, timun, kacan tanah, kuah sambal) -Oren -Kopi/ Teh O	Nasi Lemak (1/2 cawan nasi, 1 sudu makan ikan bilis, 1/4 biji telur, 3 keping timun, 1 sudu makan kacang, 2 sudu makan kuah sambal), 1/4 biji oren, gula 1 sudu makan teh	330	Sarapan Pagi	3
4	Nasi putih, ikan kembung kukus, air asam, sambal tauhu, kangkung goreng, buah belimbing. Air kosong	- 1 cawan nasi - 1 ekor ikan - 1/2 sambal tauhu - 1/2 cawan kangkung - 1 potong buah	371	Makan Tengahari/ Makan Malam	2
20	Karipap, Kuih lapis, Tembikai merah, Kopi/ Teh O	1 keping karipap, 1 keping kuih lapis, 1 potong tembikai, gula 1 sudu teh	310	Minum Pagi / Minum Petang	1

Fig. 7 List of recommended meals derived from the recommender engine

recommender engine to grab the meals id, together with the patients' ratings, and will start to calculate the similarity to recommend a more detailed list of meals to the patients.

Each patient needs to rate their first meals of the day to provide the recommender engine a sufficient dataset to make other meal recommendations that are more suitable for their diabetic conditions. Example of the list of recommended meals derived from the recommender engine is as depicted in Fig. 7.

5 Conclusion

The prototype of the e-Diet Meal Recommender System has not yet been tested with the diabetic patients and dietician in the healthcare centers. We will conduct the usability and user acceptance tests once the prototype system undergoes few modifications such as including audio instructions to assist the elderly patients and video presentations to make it more interactive. Future recommendation also involves the development of a mobile application to make the recommender system accessible via mobile technologies.

References

Aggarwal, C. C. (2016). *Recommender systems: The textbook* (1st ed.) Springer.
Andale (2012) *Pearson correlation: Definition and easy steps for use*. Retrieved from Statistics How To. http://www.statisticshowto.com/what-isthe-pearson-correlation-coefficient/. Accessed 9 May 2017.
Burke, R., Felfering, A., & Goker, M. H. (2011). Recommender system: An overview. *AI Magazine, 32*(3), 13–18.
Fuziah, M. Z., Hong, J. Y. H., Wu, L. L. (Eds.). (2012). *2nd report of diabetes in children and adolescents registry (DiCARE) 2006–2008*. Kuala Lumpur, Malaysia: Ministry of Health.
Institute for Public Health. (2015). *National Health and Morbidity Survey 2015 (NHMS 2015). Vol. II: Non-communicable diseases, risk factors and other health problems*. Kuala Lumpur, Malaysia: National Institute of Health, Ministry of Health.
Liu, J. G., Chen, M. Z., Chen, J., Deng, F., Zhang, H. T., Zhang, Z. K. et al. (2009). Recent advances in personal recommender systems. *International Journal of Information and Systems Sciences*, 230–247.
Mohamud, W. N., Ismail, A. A., Sharifuddin, A. et al. (2012). Prevalence of metabolic syndrome and its risk factors in adult Malaysians: Results of a nationwide survey. *Diabetes Research and Clinical Practice, 96*, 91–97.
Park, D. H., Choi, I. Y., Kim, H. K., & Kim, J. K. (2011). A review and classification of recommender system research. In *2011 International Conference* (pp. V1-290–V1-294). Singapore: Social Science and Humanity.
Tan, S. L., Juliana, S., & Sakinah, H. (2011). Dietary compliance and its association with glycemic control among poorly controlled type 2 diabetic outpatients in Hospital Universiti Sains Malaysia. *Malays J Nutr, 17*(3), 287–299.
Viral, N. S., & Satish, K. G. (2015). Managing diabetes in the digital age. *Clinical Diabetes and Endocrinology, 2015*(1), 16. https://doi.org/10.1186/s40842-015-0016-2.
Yu, C. H., Parsons, J. A., Hall, S., Newton, D. et al. (2014). User-centered design of a web-based self-management site for individuals with type 2 diabetes—Providing a sense of control and community. *BMC Medical Informatics and Decision Making*. https://doi.org/10.1186/1472-6947-14-60.

Comparative Study of Fuzzy Time Series and Artificial Neural Network on Forecasting Rice Production

Norpah Mahat, Rohana Alias and Suhailah Muhamad Idris

Abstract Rice is crucial in Malaysian diet, and Malaysia only produces 4.3 t/Ha (Ministry of Agriculture and Agro-based Industry) of what it needs to support itself. The increasing population of Malaysia requires further increase in rice production for consumption of the country. In recent years, the Malaysian government has been trying to encourage rice production by giving subsidies to the farmers. Accurate forecasting of rice production can provide useful information for the government, planners, decision- and policy makers. The purpose of this paper is to compare between two forecasting methods for rice production estimates in Kedah, Malaysia. The two methods considered and applied on the 35 yearly rice production data are Modified Approach Fuzzy Time Series and Artificial Neural Network. Alyuda NeuroIntelligence software is used for the Artificial Neural Network forecasting. The best model can be determined based on minimum value of mean square error (MSE), root-mean-squared error (RMSE) and mean absolute per cent error (MAPE). The findings of this study indicate that Artificial Neural Network is the best model to be used to forecast rice production in Kedah, Malaysia.

Keywords Fuzzy forecasting · Artificial Neural Network · Rice production

1 Introduction

Rice is a staple food for Malaysians, and many eat it practically every day. Due to its importance, Malaysian paddy and rice industry is seriously emphasized by the government. In fact, as an effort to strengthen the industry, Malaysian government has privatized North Pacific Research Board (NPRB) in 1996 and its name changed

---

N. Mahat (✉) · R. Alias · S. Muhamad Idris
Faculty of Computer and Mathematical Sciences, Universiti Teknologi MARA Cawangan Perlis, 02600 Arau, Perlis, Malaysia
e-mail: norpah020@perlis.uitm.edu.my

© Springer Nature Singapore Pte Ltd. 2018
R. Saian and M. A. Abbas (eds.), *Proceedings of the Second International Conference on the Future of ASEAN (ICoFA) 2017 – Volume 2*,
https://doi.org/10.1007/978-981-10-8471-3_16

to Padiberas Nasional Berhad (BERNAS) (Fahmi et al. 2013). The government has also allocated billions of *Ringgit* in order to increase rice production. This includes research and development, credit facilities, subsidized retail price, guaranteed minimum price, extension support, fertilizer subsidies and irrigation investment.

There are many challenges faced by the farming community in Malaysia which includes insect and pest problems, global warming and the outmigration of youth. All these challenges affect the level of rice productivity. This pattern of rice production needs to be studied to avoid reduction of rice production and to ensure that the supply of rice can meet the increasing demand of rice. Thus, accurate forecasting of rice production is essential in providing necessary information for planning, decision- and policy makers.

This study focuses on the production of rice in Kedah, the highest producer of rice in East Malaysia. The yearly rice production from 1980 to 2014 is obtained from the official portal of Department of Agriculture, Malaysia. The main objective of this study is to compare and identify the best forecasting technique of rice production between two forecasting techniques: Fuzzy Time Series and Artificial Neural Network. The study further considers only Conjugate Gradient Descent algorithm for the forecast of rice production by Artificial Neural Network.

2 Methodology

This study will compare and analyse the best method to forecast rice production in Kedah, Malaysia. The methods involved are Fuzzy Time Series and Artificial Neural Network. Fuzzy Time Series followed the eight steps as described in Sect. 2.1. The Artificial Neural Network focused only on the Conjugate Gradient Descent method, which includes data analysis, data pre-processing, design, training, testing and validation. Statistical tests and data were analysed using Microsoft Excel and Alyuda NeuroIntelligence software. The data for the period 1980–2014 are used for estimation purposes and are obtained from the official portal of Department of Agriculture, Malaysia. Limitations of the study include the availability of the data is only up to 2014, two methods are considered, and other factors affecting rice production in Kedah are not considered.

2.1 Fuzzy Time Series—Proposed Method

There are many previous techniques to forecast crop production per year. However, this paper proposed using Fuzzy Time Series to forecast rice production. Yadav et al. (2013) explained that there are eight steps in Fuzzy Time Series method to obtain the forecasted value. The following steps were implemented to forecast rice production in Kedah:

Step 1: The historical data on production of rice in Kedah, Malaysia were collected from 1980 to 2014.

Step 2: The maximum and minimum values of rice production are identified as 1357492 and 562823, respectively. Let $D_1 = 823$ and $D_2 = 2508$. Then, the universe of discourse is defined us as $U = [D_{min} - D_1, D_{max} + D_2]$, $U = [562000, 1360000]$.

Step 3: U is partitioned into seven intervals with equal length of 114000: U_1 to U_7. The intervals are: $U_1 = [562000, 676000]$, $U_2 = [676000, 790000]$, $U_3 = [790000, 904000]$, $U_4 = [904000, 1018000]$, $U_5 = [1018000, 1132000]$, $U_6 = [1132000, 1246000]$ and $U_7 = [1246000, 1360000]$. The frequencies of the rice production data in intervals U_1, U_2 U_3, U_4, U_5, U_6 and U_7 are 1, 6, 4, 15, 3, 5 and 1, respectively.

Based on this distribution, each of the intervals was further divided to establish new sub-intervals of different lengths: $v_1 = [562000, 676000)$ with the length of 114000; $v_2 = [676000, 695000)$, ..., $v_7 = [771000, 790000)$ with the length of 19000; $v_8 = [790000, 818500)$, ..., $v_{11} = [875500, 904000)$ with the length of 28500; $v_{12} = [904000, 911600)$, ..., $v_{26} = [1010400, 1018000)$ with the length of 7600; $v_{27} = [1018000, 1056000)$, ..., $v_{29} = [1094000, 1132000)$ with the length of 38000; $v_{30} = [1132000, 1154800)$, ..., $v_{34} = [1223200, 1246000)$ with the length of 22800; and $v_{35} = [1246000, 1360000)$ with the length of 114000.

Step 4: The fuzzy trapezoidal number is defined by $A_1 = [448000, 562000, 676000, 695000]$, $A_2 = [562000, 676000, 695000, 714000]$, ..., $A_{34} = [1200400, 1223200, 1246000, 1360000]$, $A_{35} = [1223200, 1246000, 1360000, 1474000]$.

Step 5: The fuzzified historical data are classified into the corresponding fuzzy numbers. If the value of rice production data is in the range of v_j, then it belongs to fuzzy number A_j. For example, the actual rice production in 1980 is 839715 and is assigned to the fuzzified number A_9.

Step 6: For all fuzzified data, the fuzzy logical relationship is established. $A_9 \rightarrow A_9$ and $A_3 \rightarrow A_4$ are examples of the fuzzy logical relationship used.

Step 7: The fuzzy logical relationships are arranged into fuzzy logical relationship groups based on the same fuzzy number on the left of the fuzzy logical relationships. For example, fuzzy set $A_9 \rightarrow A_9$ has one-to-one relationship and fuzzy sets $A_4 \rightarrow A_3, A_6$ have one-to-many relationship.

Step 8: The forecasted rice production is calculated for different fuzzy logical relationship groups. This study used only Rule 2 and Rule 3. Using Rule 2, the fuzzified rice production for year 1984 is A_1, and paralleling to fuzzy logical relationship group, $A_1 \to A_9$ are used to forecast for the year 1985. This is done by repeating Step 4 to obtain the fuzzy trapezoidal number for A_9 where it is found that A_9 = [790000, 818500, 847000, 875500].

From A_9, $t_2 = 818500$, $t_1 = 818500 - 790000 = 28500$, $t_3 = 847000$ and $t_4 = 875500 - 847000 = 28500$.
Then, $t_4 - t_1 = \frac{28500 - 28500}{4}$ and $t_4 + t_1 = \frac{28500 - 28500}{2} = 28500$.

$$NSTFN(A_9) = \left[\left(t_2 + \frac{t_4 - t_1}{4} \right) - \frac{t_4 + t_1}{2}, t_2 + \left(\frac{t_4 - t_1}{4} \right), t_3 \right.$$
$$\left. + \left(\frac{t_4 - t_1}{4} \right), \left(t_3 + \frac{t_4 + t_1}{2} \right) + \frac{t_4 + t_1}{2} \right]$$
$$= [818500 - 28500, 818500$$
$$+ 0, 847000 + 0, 847000 + 28500]$$
$$= [790000, 818500, 847000, 875500]$$

According to Rule 2, the value of Fv_t for the year 1985 is defined as $R[NSTFN(A_9)] = \frac{790000 + 818500 + 847000 + 875500}{4}$, where $Fv_t = 832750$. The fuzzified rice production by using Rule 3 is $A_9 \to A_4, A_6, A_9, A_{12}$. As A_{12} = [875500, 904000, 911600, 919600], $t_2 = 904000$, $t_1 = 904000 - 875500 = 28500$, $t_3 = 911600$, $t_4 = 919200 - 911600 = 7600$, $t_4 - t_1 = \frac{7600 - 28500}{4} = -5225$ and $t_4 + t_1 = \frac{28500 + 7600}{2} = 18050$. Then, the calculated value of NSTFN(A_{12}) is given by

$$NSTFN(A_{12}) = \left[\left(t_2 + \frac{t_4 - t_1}{4} \right) - \frac{t_4 + t_1}{2}, t_2 + \left(\frac{t_4 - t_1}{4} \right), t_3 \right.$$
$$\left. + \left(\frac{t_4 - t_1}{4} \right), \left(t_3 + \frac{t_4 + t_1}{2} \right) + \frac{t_4 + t_1}{2} \right]$$
$$= [898775 - 18050, 904000 + (-5225), 911600$$
$$+ (-5225), 906375 + 180050]$$
$$= [880725, 898775, 906375, 924425].$$

Similar process is used to find NSTFN(A_4), NSTFN(A_6) and NSTFN(A_9). The results are NSTFN(A_4) = [695000, 714000, 733000, 752000], NSTFN(A_6) = [733000, 752000, 771000, 790000] and NSTFN(A_9) = [790000, 818500, 847000, 875500]. According to Rule 3, the forecasted value for 1985 is given by

$$Fv_t = R \left[\frac{NSTFN(A_{k1}) + NSTFN(A_{k2}) + \ldots + NSTFN(A_{kp})}{p} \right]$$

Therefore,

$$Fv_t = R \begin{bmatrix} \left(\dfrac{695000 + 733000 + 790000 + 880725}{4}\right), \\ \left(\dfrac{714000 + 752000 + 818500 + 898775}{4}\right), \\ \left(\dfrac{733000 + 771000 + 847000 + 906375}{4}\right), \\ \left(\dfrac{752000 + 790000 + 875500 + 924395}{4}\right). \end{bmatrix}$$

$Fv_t = R[\,774681.25, \quad 795818.75, \quad 814343, \quad 835481.25\,] = 805081.25$

The forecasted values for all other fuzzy logical relationship groups are found using similar process of Rule 2 and Rule 3.

2.2 Artificial Neural Network

Artificial Neural Network is computational model encouraged by the central nervous systems of animals (particularly the brain) that are capable of machine learning and pattern recognition in computer science and related fields. They are usually presented as systems of interconnected 'neurons' that can calculate values of input through the network by providing information (Dahikar and Rode 2014). Co (2007) compared the performance of Artificial Neural Network with traditional methods such as exponential smoothing and ARIMA model to forecast rice exports from Thailand, and it was shown that Artificial Neural Network is better than traditional methods.

This study applied Artificial Neural Network to forecast rice production in Kedah. There are many algorithms in training Artificial Neural Network. This study focused on Conjugate Gradient Descent. The data of rice crop production were implemented in the Alyuda NeuroIntelligence software following the six steps as shown in Fig. 1 (Lan et al. 2012).

2.2.1 Conjugate Gradient Descent

Conjugate Gradient (CG) algorithm is an important optimization algorithm. Shen et al. (2015) stated that CG algorithm is used to speed up the learning process. During CG training, back propagation of error derivatives was employed to fine-tune the weights for optimal reconstruction. It is shown that the optimized two-phase training procedure enables fast convergence. Besides that, CG methods

Fig. 1 Flow chart steps in Alyuda NeuroIntelligence software

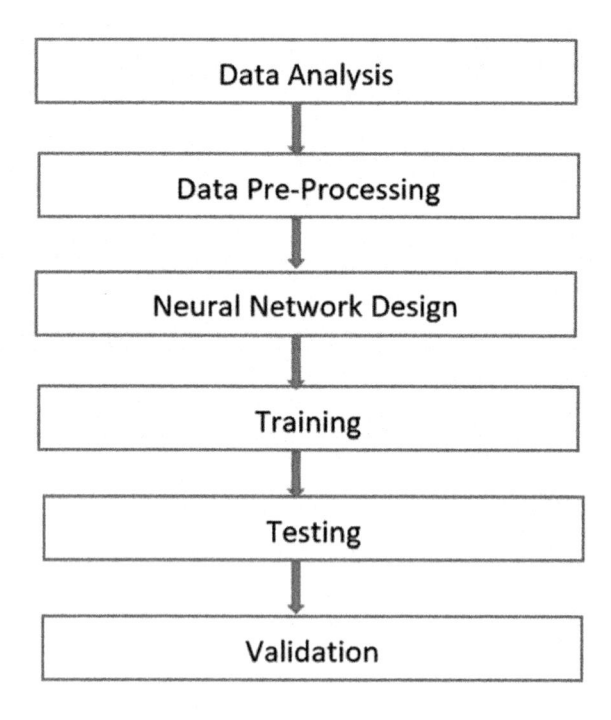

Data Analysis → Data Pre-Processing → Neural Network Design → Training → Testing → Validation

are probably the most famous iterative methods for efficiently training neural networks in scientific and engineering computation. They are characterized by the simplicity of their iteration, numerical efficiency and their low memory requirements (Livieris and Pintelas 2013).

The CG training process can be realized by minimizing the mean square error (MSE) function defined by

$$\text{MSE} = E(e^\text{T} e) = E\left((o_d - o)^\text{T}(o_d - o)\right).$$

where $E(.)$ is the mathematical expectation function and $o_d - o$ is the error of network output. The general purpose of the CG training is to search for an optimal set of connection weights in the manner that errors defined by the MSE can be minimized. According to Livieris and Pintelas (2013), these methods generate a sequence of weights $\{w_k\}$ using the iterative formula:

$$w_{k+1} = w_k + n_k d_k, k = 0, 1, \ldots,$$

where k is called epoch, $w_0 \in \Re^n$ is a given initial point, $n_k > 0$ is the learning rate, and d_k is a descent search direction defined by

$$d_k = \begin{cases} -g_0, & \text{if } k = 0; \\ -g_k + \beta_k d_{k-1}, & \text{otherwise,} \end{cases}$$

where β_k is a scalar and $g_k = \nabla E(w_k)$.

In the beginning of CG method, the initial search direction d_0 is calculated by

$$d_0 = -\nabla \text{MSE}(w_0) = -g_0.$$

Each direction d_{k+1} is chosen to be a linear combination of the steepest descent direction $-g_{k+1}$ and the previous direction d_k. The formula is given by

$$d_{k+1} = -g_{k+1} + \beta_k d_k,$$

where the scalar d_k and d_{k+1} must fulfil the conjugacy property. β_k is the CG updated parameter, and there are different CG methods for different choices of β_k. One choice is given by $\beta_k = \frac{\|g_{k+1}\|^2}{\|g_k\|^2}$.

3 Results and Discussion

The fitted values of rice production for 1980–2014 using Fuzzy Time Series are illustrated in Fig. 2. It represents the actual rice production and the forecasted value. The graph indicates that the forecasted values closely imitate the fluctuations of the actual rice production.

The result of neural network forecasting was imported from the Alyuda NeuroIntelligence software as shown in Fig. 3. It shows the comparison between target and output values for rice production forecasting using the Conjugate Gradient Method. Table 1 shows the best network architecture [2-5-1] which has the lowest value of training, testing and validation errors.

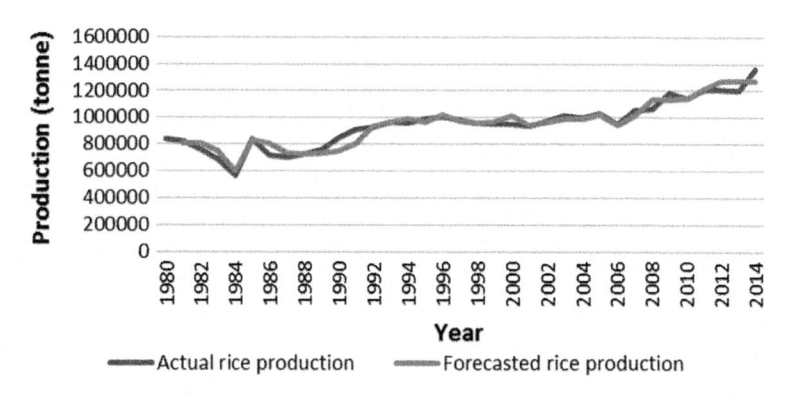

Fig. 2 Rice production fitted using Fuzzy Time Series

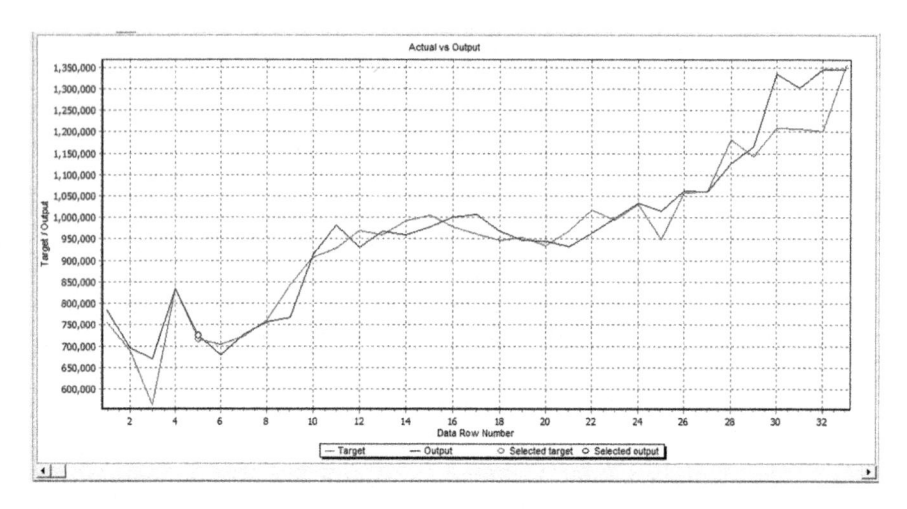

Fig. 3 Comparison between output and target values of rice production using Conjugate Gradient method (Alyuda NeuroIntelligence software)

Table 1 Network architecture for rice production

Architecture	Weights	Training error	Validation error	Testing error
[2-5-1]	21	51066.54	48694.83	46487.25

Table 2 Summary statistics

	Target	Output	AE	ARE
Mean	954703.69	967285.70	35053.39	0.036
Standard deviation	172592.83	184984.21	37529.29	0.041
Min	562823	670817.33	555.12	0.000582
Max	1357492	1344770.26	144084.26	0.19188

Correlation: 0.963584
R-squared: 0.922932

The statistical summaries are shown in Table 2. The correlation coefficient value of 0.96 indicates that there is a strong positive relationship between the target and output values of rice production. In addition to that, the value of 0.9229 for the coefficient of determination, r^2, indicates that 92.29% of the variation in the output can be explained by the model and 7.71% by random errors.

The two forecasting techniques, Fuzzy Time Series and Artificial Neural Network, were evaluated using within-sample estimate data procedure. Table 3 validates the results of the evaluation where the values of the MSE, RMSE and MAPE are calculated. The performance of the two techniques is compared using the sizes of the error measures. The best technique to forecast rice production is the Artificial Neural Network since all error measures calculated for this model are

Table 3 Summary of error measures

Method	MSE	RMSE	MAPE
Fuzzy Time Series	2050471374	45282.13	3.57
Artificial Neural Network	1366573771	36967.20	3.18

the smallest compared to Fuzzy Time Series. The values of MSE, RMSE and MAPE are 1366573771, 36967.20 and 3.18, respectively.

4 Conclusion

In contrast to the Fuzzy Time Series Technique, the Artificial Neural Network is a more complex method and requires more effort to train the network repeatedly to find the best model. Furthermore, the Alyuda NeuroIntelligence used different sets of data each time when running the network to avoid memorization.

This study compared the two different techniques, which are Fuzzy Time Series and Artificial Neural Network in forecasting rice production in Kedah. Artificial Neural Network has been proven to be the best method for forecasting rice production in Kedah based on the criterion of minimum error measures. In addition to that, the Artificial Neural Network Technique also indicated a strong positive relationship between target and output values and is well fitted.

References

Co, H. C. (2007). Forecasting Thailand's rice export: Statistical techniques vs artificial neural networks. *Computer and Industrial Engineering, 53*, 610–627.

Dahikar, S. S., & Rode, S. V. (2014). Agricultural crop yield prediction using artificial neural network approach. *International Journal of Innovative Research in Electrical, Electronic, Instrumentation and Control Engineering, 2*(1), 683–686.

Fahmi, Z., Samah, B. A., & Abdullah, H. (2013). Paddy industry and paddy farmers well-being: A success recipe for agriculture industry in malaysia. *Asian Social Science, 9*(3), 17–181.

Lan, T., Chen, P., Lan, C., & Chuang, K. (2012). A study of using back-propagation neural network for the sales forecasting of the thin film sputtering process material. *International Journal of Advancements in Computing Technology, 4*(8), 118–127.

Livieris, I. E., & Pintelas, P. (2013). A new conjugate gradient algorithm for training neural networks based on a modified secant equation. *Applied Mathematics and Computation, 221*, 491–502.

Official Portal of Ministry of Agriculture and Agro-based Industry. www.moa.gov.my.

Shen, F., Chao, J., & Zhao, J. (2015). Neurocomputing forecasting exchange rate using deep belief networks and conjugate gradient method. *Neurocomputing, 167*, 243–253.

Yadav, V. K., et al. (2013). A comparative study of neural-network & fuzzy time series forecasting techniques-Case study: Marine fish production forecasting. *Indian Journal of Geo-marine Sciences, 42*(6), 707–716.

Scholarship Eligibility and Selection: A Fuzzy Analytic Hierarchy Process Approach

Norwaziah Mahmud, Nur Syuhada Muhammat Pazil, Umi Hanim Mazlan, Siti Hafawati Jamaluddin and Nik Nurnadia Che Hasan

Abstract A scholarship is an award of financial aid for a student to further his or her education. This research focuses on Federal Scholarship awarded by the Malaysian Education Ministry. It demonstrates the scholarship selection with cases in Sekolah Menengah Kebangsaan To' Uban, Pasir Mas, Kelantan, in 2015. The scholarship was given to form one to form five students and was awarded to those who performed in their academics and co-curricular activities but only for students whose parents with low income. The selection of eligible students in receiving the scholarship has become more challenging throughout the years due to competitiveness and an increasing number of applicants. In order to overcome this problem, Fuzzy Analytic Hierarchy Process (AHP) method was applied in this research. The objective is to identify the most important criteria and sub-criteria for the selection of students and aims to select the students who fulfil the criteria and sub-criteria. The result shows that the most important criteria are academic performance (0.712) followed by parents' income (0.220) and co-curriculum (0.068), respectively. Moreover, for the sub-criteria of academic performance, the most important sub-criteria is "High" (0.760). The criteria for the parents' income is left out since the sub-criteria exists is only one. Lastly, for the criteria co-curriculum the most important sub-criteria is "Sports" (0.741). The final results show that Alternative 9 (0.056) is the best choice in selecting the student.

Keywords Fuzzy AHP · Scholarship · Selection

N. Mahmud (✉) · N. S. Muhammat Pazil · U. H. Mazlan · S. H. Jamaluddin
N. N. Che Hasan
Universiti Teknologi MARA Cawangan Perlis, Kampus Arau, Arau, Malaysia
e-mail: norwaziah@perlis.uitm.edu.my

1 Introduction

Scholarships are financial aids given to students where they do not have to repay the money awarded to them. They are given to students who are eligible and meet the criteria given by the funder to receive the scholarship. There are a few advantages of scholarships for students. One of the advantages of scholarship for students is to help them in their expenses throughout their academic years in the institution. Students who are excellent in academic performance but do not have enough money to continue their education can apply for a scholarship provided by the Malaysian Education Ministry. Students who need scholarships are mostly from poor family background. They need scholarships not only to pay their tuition fees, but also to buy books and various other related expenses. In addition, scholarships will encourage students to be successful in their studies. In this country, there are many scholarships offered by the government or private sectors. This scholarship is not only offered for higher education students but it is also provided to primary and secondary school students. This study focuses on the selection for 'Biasiswa Kecil Persekutuan' (BKP) which is offered by the Ministry of Education for secondary school students. There are three main criteria that will be taken into consideration which are academic performance, parents' income and students' involvement in extra-curricular activities. Selecting students who are qualified to receive the Federal Scholarship awarded by the Malaysian Education Ministry is a challenging decision-making process. Therefore, in order to solve this problem, multiple selection criteria have to be measured concurrently. This study aims to develop a selection model using Fuzzy Analytic Hierarchy Process (FAHP) method to select eligible students for this scholarship. The data were collected from Sekolah Menengah Kebangsaan To' Uban, Pasir Mas, Kelantan, in 2015 which involved 30 applicants. The data contain academic performances which are Ujian Penilaian Sekolah Rendah (UPSR) results, parents' income and the involvement in co-curricular activities.

2 Related Works

2.1 Fuzzy Analytic Hierarchy Process (FAHP)

Fuzzy AHP was proposed by various authors (Van Laarhoven and Pedrycz 1983; Buckley 1985; Chang 1996). However, the most commonly used is the FAHP which was extensively analysed by Chang (1996). The FAHP variants require an additional defuzzification procedure to convert fuzzy weights to crisp weights and fuzzy preference programming technique to derive the crisp weights from fuzzy pairwise comparison judgment matrix. There are many applications of FAHP such as in personnel selection in human resource management, supplier selection, teachers' ranking. For example, in personnel selection, AHP method is suggested to solve academic staff selection based on triangular fuzzy numbers (Rouyendegh and Erkan 2012). There are four parts involved in this selection which are identifying the problem, constructing an FAHP and carrying out the calculation, explaining the proposed model in the real world; the last part is conclusion, and future study areas will be discussed. Based on FAHP, the candidate with the highest score was chosen for the job selection. Besides that, Fuzzy AHP technique is used to evaluate teachers' ranking (Hota et al. 2013). There are three criteria involved in this study, namely communication, knowledge and interaction. The hierarchy is built to evaluate teachers' ranking. Make a pairwise comparison in between alternative to alternative and for each criterion. The weights are obtained to decide the ranking of teachers. In this study, Fuzzy AHP method is applied in order to rank the teachers.

2.2 Past Research on Selecting Students for Scholarship

Saptarini and Prihatini (2015) had used the data of 25 students to construct an experiment on decision support system for scholarship in Bali State Polytechnic (BSP) which is one vocational education institution located in Jimbaran, Bali. In this case, the three criteria were measured. The Analytical Hierarchy Process (AHP) and the Technique for Order of Preference by Similarity to Ideal Solution (TOPSIS) were used. AHP was used to give weight to each criterion based on its priority; meanwhile, the method of TOPSIS was used to rank the students based on its values of each criterion. In order to get the weight of each criterion, the pairwise matrix was compared to others in AHP. The fuzzification was done for every student, and the result was used to calculate the closeness value using TOPSIS. The first priority to get the scholarship was based on TOPSIS method which was measured based on a student who had the highest closeness value.

3 Research Methodology

3.1 Data Collection

For this study, the data were collected from 30 form one students in Sekolah Menengah Kebangsaan To' Uban, Pasir Mas, Kelantan. All 30 students applied for the scholarship. In order to identify the relevant factors, an interview with the teachers of the school was conducted. As a result, three criteria were identified, namely academic performance (C_1), parents' income (C_2) and co-curricular activities (C_3). Each criterion later on was characterized by its sub-criteria such as "High" (C_{11}), "Moderate" (C_{12}) and "Low" (C_{13}) for academic performance and "Sports" (C_{31}), "Club" (C_{32}) and "Uniform" (C_{33}) for co-curricular activities, while for parents' income, there was only one sub-criteria which is "Poor" since the candidates come from families with low-income background.

3.2 Construct FAHP Model

The FAHP is adopted to model the student selection. This selection process involves six steps which are listed as follow:

Step 1: Selection of an Expert Group for Decision-Making
The teachers are chosen as expert decision-makers who are responsible for the selection of students.

Step 2: Calculation of Fuzzy Triangular Number
At this phase, the triangular fuzzy number will be set up. A pairwise comparison between the criteria and sub-criteria will be done by the experts to determine the relative score. As a replacement of crisp value, the fuzzy AHP is a range of values to fit in the decision-makers' uncertainty. The pairwise comparison matrix is given by the following matrix (1):

$$C_{ij} = \begin{array}{c} c_1 \\ c_2 \\ \vdots \\ c_k \end{array} \begin{bmatrix} c_1 & c_2 & \cdots & c_k \\ f_{11} & f_{12} & \cdots & f_{1k} \\ f_{21} & f_{22} & \cdots & f_{2k} \\ \vdots & \vdots & \cdots & \vdots \\ f_{k1} & f_{k2} & \cdots & f_{kk} \end{bmatrix} \quad (1)$$

for $i = 1, 2 \ldots, k$ and $j = 1, 2 \ldots, k$, where $f_{11}, f_{12}, \ldots f_{1k}$ is a triangular fuzzy number. The fuzzy conversion scale has been employed by Chang (1996), fuzzy prioritization approach. The decision-maker will be asked to compare the two criteria by referring to the linguistic variable as shown in Table 1.

Step 3: Calculation of Geometric of Fuzzy Comparison Value
The geometric mean of fuzzy comparison values of each criterion will be calculated using the Buckley (1985) approach (2–5), and Ayhan (2013).

$$g_i = \left(\prod_{j=1}^{c_k} f_{ij} \right)^{\frac{1}{k}}, \quad i = c_1, c_2, \ldots, c_k \quad (2)$$

Table 1 Pairwise comparison matrix for criteria and sub-criteria

Triangular fuzzy number	Linguistic variable
(1, 1, 1)	Equally Important (EI)
(2, 3, 4)	Weakly More Important (WI)
(4, 5, 6)	Strongly More Important (SI)
(6, 7, 8)	Very Strongly Important (VSI)
(9, 9, 9)	Absolutely More Important (AI)

where
$$l_i = (l_{i1} \otimes l_{i2} \otimes l_{ik})^{\frac{1}{k}}, \quad i = 1, 2, \ldots, k \tag{3}$$

$$m_i = (m_{i1} \otimes m_{i2} \otimes m_{ik})^{\frac{1}{k}}, \quad i = 1, 2, \ldots, k \tag{4}$$

$$u_i = (u_{i1} \otimes u_{i2} \otimes u_{ik})^{\frac{1}{k}}, \quad i = 1, 2, \ldots, k \tag{5}$$

Step 4: Calculate Fuzzy Weights

To calculate the fuzzy weights, the vector summation of each geometric mean will be calculated. Then, the reciprocal of summation vector will be computed and the fuzzy triangular number will be replaced in ascending order. Later, to obtain the fuzzy weight (w_i) of criterion, each geometric mean must be multiplied with the following reverse vector:

$$w_i = \left[\frac{l_i}{\sum_{i=1}^{k} u_i}, \frac{m_i}{\sum_{i=1}^{k} m_i}, \frac{u_i}{\sum_{i=1}^{k} l_i} \right] \tag{6}$$

Step 5: Defuzzification

In this defuzzification step, the fuzzy weights need to be defuzzified by centre of area method because the weights are still in fuzzy triangular number. This method was proposed by Chou and Chang (2008) through the Eq. (7) below:

$$d_i = \frac{sw_i + tw_i + uw_i}{3} \tag{7}$$

Step 6: Normalization

The final stage is to normalize the defuzzification value using the following equation:

$$N_i = \frac{d_i}{\sum_{i=1}^{k} d_i} \tag{8}$$

3.3 Determination of Fuzzy Priorities for Each Alternative

In order to determine the fuzzy priorities for each alternative, all the six steps need to be repeated. The only difference is the symbols used for each step in criteria and alternative. The pairwise comparison for step 1 is given by the matrix below:

Table 2 TFN for the linguistic variables for the priorities alternative

Importance intensity	Triangular fuzzy number
Very good	(3, 5, 5)
Good	(1, 3, 5)
Moderate	(1, 1, 1)
Poor	(1/5, 1/3, 1)
Very poor	(1/5, 1/5, 1/3)

$$C_{mn_j} = \begin{array}{c} \\ A_1 \\ A_2 \\ \vdots \\ A_l \end{array} \begin{array}{c} A_1 \quad A_2 \quad \cdots \quad A_k \\ \begin{bmatrix} f_{11} & f_{12} & \cdots & f_{1l} \\ f_{21} & f_{22} & \cdots & f_{2l} \\ \vdots & \vdots & \cdots & \vdots \\ f_{l1} & f_{l2} & \cdots & f_{ll} \end{bmatrix} \end{array} \quad (9)$$

for $m = 1, 2, \ldots, l$, $n = 1, 2, \ldots, l$, and $j = 1, 2, \ldots, l$, where f_{11} is a triangular fuzzy number. For each alternative which is candidate of scholarship selection, the global weight will be determined and the fuzzy priorities will be calculated according to the linguistic variable that was proposed by Chan et al. (2000) as illustrated in Table 2.

The geometric mean of each alternative (step 3) is denoted as p_{m_j} while the fuzzy weight of each alternative (step 4) is denoted as V_{m_j}; the defuzzified weight of each alternative (step 5) is denoted as h_{m_j}, and the normalized defuzzified weight of each alternative with respect to the sub-criteria (step 6) is denoted as Z_{m_j}, for $j = 1, 2, \ldots, l$ and $m = 1, 2, \ldots, l$.

3.4 Calculation of Alternatives' Score

The score for each alternative can be calculated using the following equation:

$$\text{Score}_l = \sum_{j=1}^{l} N_i \times Z_{l_j} \quad \begin{array}{l} m = 1, 2, \ldots, l \\ j = 1, 2, \ldots, k \end{array} \quad (10)$$

3.5 Selection of Alternatives

Alternatives must be selected based on the score. Based on the FAHP, the candidate with the highest score is the most eligible to be chosen as a candidate to receive the scholarship.

4 Results and Discussions

Table 3 shows the pairwise comparison matrix and normalized weights for each criterion in the selection of students. Based on the table, the most important criterion in the selection of students is academic performance with normalized weight, 0.712, followed by parents' income and co-curriculum with 0.220 and 0.068, respectively. This indicates that the students, who perform in academic performance, will be the first one to receive the scholarship.

Based on Tables 4 and 5 above, the most important sub-criteria for academic performance is "High" with normalized weight, 0.712, followed by "Moderate" and "Low". However, for co-curriculum, the most important sub-criteria is "Sports" with normalized weight, 0.741, followed by "Club" and "Uniform". Since there is only one sub-criteria exists for parents' income criteria, the normalized weight for parents' income is 1.

In FAHP approach, the alternative with the highest score is more preferred to receive a scholarship compared to the lower score. Thus, based on Table 6, Alternative 9 (0.056) is the most preferred to be selected as a scholarship recipient.

Table 3 Pairwise comparison matrix and normalized weights of each criterion

y	C_1	C_2	C_3	Normalized weight
C_1	(1, 1, 1)	(4, 5, 6)	(6, 7, 8)	0.712
C_2	(1/6, 1/5, 1/4)	(1, 1, 1)	(4, 5, 6)	0.220
C_3	(1/8, 1/7, 1/6)	(1/6, 1/5, 1/4)	(1, 1, 1)	0.068

Table 4 Pairwise comparison matrix and the normalized weights of academic performance sub-criteria

Sub-criteria	C_{11}	C_{12}	C_{13}	Normalized weight
C_{11}	(1, 1, 1)	(6, 7, 8)	(9, 9, 9)	0.760
C_{12}	(1/8, 1/7, 1/6)	(1, 1, 1)	(6, 7, 8)	0.0192
C_{13}	(1/9, 1/9, 1/9)	(1/8, 1/7, 1/6)	(1, 1, 1)	0.048

Table 5 Pairwise comparison matrix and the normalized weights of co-curriculum sub-criteria

Sub-criteria	C_{31}	C_{32}	C_{33}	Normalized weight
C_{31}	(1, 1, 1)	(6, 7, 8)	(6, 7, 8)	0.741
C_{32}	(1/8, 1/7, 1/6)	(1, 1, 1)	(6, 7, 8)	0.203
C_{33}	(1/8, 1/7, 1/6)	(1/8, 1/7, 1/6)	(1, 1, 1)	0.056

Table 6 Final score for selection of students

Ranking	Alternatives	Criteria			
		Academic performance	Parents' income	Co-curriculum	Score
1	A9	0.046	0.097	0.035	0.056
2	A28	0.048	0.064	0.034	0.051
3	A1	0.045	0.033	0.075	0.044
4	A14	0.044	0.031	0.063	0.043
5	A18	0.047	0.031	0.035	0.042

5 Conclusion

In conclusion, the most important criteria to select students who are entitled to receive the scholarship are academic performance followed by parents' income and co-curricular activities. The final score shows that Alternative 9 has the highest score due to appropriate choice followed by Alternative 28, Alternative 1, Alternative 14, Alternative 18 and other remaining alternatives. The candidates with the highest score are the most suitable for the selection in order to receive the scholarship. As for future study, it is suggested that other multi-criteria approaches may be applied to solve the problem related to students' scholarship selection such as Analytic Network Process (ANP), Fuzzy Multi Attribute Making Decision (MADM) with Technique for Order Preference by Similarity to Ideal Solution (TOPSIS) and Fuzzy Analytic Network Process (FANP). The result can be compared in the process of students' scholarship selection. The comparison of various methods in the selection of scholarship recipients may help in finding out the accuracy, appropriateness, suitability and efficiency.

References

Ayhan, M. B. (2013). A fuzzy AHP approach for supplier selection problem: A case study in a gearmotor company. *International Journal of Managing Value and Supply Chains, 4*(3), 11–23.

Buckley, J. J. (1985). Fuzzy hierarchical analysis. *Fuzzy Sets and Systems, 17*(3), 233–247.

Chan, F. T. S., Chan, M. H., & Tang, N. K. H. (2000). Evaluation methodologies for technology selection. *Journal of Materials Processing Technology, 107*(1), 330–337.

Chang, D. Y. (1996). Applications of the extent analysis method on fuzzy AHP. *European Journal of Operational Research, 95*(3), 649–655.

Chou, S. W., & Chang, Y. C. (2008). The implementation factors that influence the ERP (enterprise resource planning) benefits. *Decision Support Systems, 46*(1), 149–157.

Hota, H. S., Pavani, S., & Gangadhar, P. V. (2013). Evaluating teachers ranking using fuzzy ahp technique. *International Journal of Soft Computing and Engineering, 2*(6), 485–488.

Rouyendegh, B. D., & Erkan, T. E. (2012). Selection of academic staff using the fuzzy analytic hierarchy process (FAHP): A pilot study. *International Journal of Advances in Sciences and Technology, 19*(4), 923–929.

Saptarini, N. G., & Prihatini, P. M. (2015). Decision support system for scholarship in Bali state polytechnic using AHP and topsis. *International Conferences on Information Technology and Business, 4*(5), 38–46.

Van Laarhoven, P. J. M., & Pedrycz, W. (1983). A fuzzy extension of Saaty's priority theory. *Fuzzy Sets and Systems, 11*(1–3), 229–241.

Assessment of Energy Efficiency Level on UiTMPP's *Dewan Besar* Building

Mohamad Hilmi Akmal Zakaria, Mohamad Adha Mohamad Idin, Muhammad Firdaus Othman and Noorezal Atfyinna Mohd Napiah

Abstract The Malaysian Ministry of Education (MOE) and The Malaysian Ministry of Higher Education (MOHE) have urged staffs and students to save energy in education's building. In line with the government's initiative to reduce energy usage in government buildings, this study will focus on energy efficiency level at Dewan Besar, Universiti Teknologi MARA, Pulau Pinang (UiTMPP). This study presents energy efficiency level and provides the solution to reduce electricity cost at UiTMPP's Dewan Besar. Building energy index (BEI) is a tool to calculate the energy performance of a building. Gross floor area and power usage at UiTMPP's Dewan Besar are required to calculate BEI. This (BEI) will be compared with Malaysia Standard MS1525 requirement that it should be below than 220 kWh/m^2/year. The BEI of UiTMPP's Dewan Besar is 156.2 kWh/m^2/year. Several suggestions have been given in reducing electricity cost. It estimated that it reduces 5% electricity at UiTMPP's Dewan Besar.

Keywords Building energy index · Energy efficiency index · Energy level

1 Introduction

Increasing number of people in the world can cause increased demand for electricity. The increased demand for electricity leads to the depletion of energy resources of the non-renewable sources (Noranai and Kammalluden 2012). Electric energy is one of the most expensive forms of energy used in buildings. Thus, reduction in electric energy consumption may save more money (Kamaruzzaman et al. 2009).

The purpose of this study is to identify the energy performance of a building and gives a suggestion to reduce the building's energy consumption. The study was

---

M. H. A. Zakaria (✉) · M. A. Mohamad Idin · M. F. Othman · N. A. Mohd Napiah
Universiti Teknologi MARA Pulau Pinang, Permatang Pauh, Malaysia
e-mail: emy_akmal93@yahoo.com

© Springer Nature Singapore Pte Ltd. 2018
R. Saian and M. A. Abbas (eds.), *Proceedings of the Second International Conference on the Future of ASEAN (ICoFA) 2017 – Volume 2*,
https://doi.org/10.1007/978-981-10-8471-3_18

conducted at Dewan Besar in Universiti Teknologi MARA, Pulau Pinang (UiTMPP). The building's energy performance will be calculated using building energy index (BEI). BEI of UiTMPP's Dewan Besar will show energy performance of that building and use this index to identify any wastage energy. By using this data, a suggestion will be given to UiTMPP's Facility Management Section for reducing and minimizing the electric wasted at UiTMPP's Dewan Besar. This suggestion will reduce energy usage; thus, it will reduce the financial cost of electric bill at UiTMPP's Dewan Besar.

1.1 Building Energy Index

BEI is the green building index tool. It is introduced by Persatuan Arkitek Malaysia based on the references to other rating systems in the world. BEI or also known as energy efficiency index (EEI) is introduced to calculate the energy usage of buildings (Najihah et al. 2014). BEI is usually expressed as kWh/m^2/year which shows amount of total energy usage for one year in a building by kilowatts hours which is divided by the gross floor area of building in square meters (Moghimi et al. 2011). Malaysia Standard MS1525 is being a reference and a guideline for energy efficiency. According to MS1525 standard, the recommended BEI in Malaysia is 220 kWh/m^2/year. Based on the research, most of the building in Malaysia does not meet this standard. Majority of the building users or building owners do not care whether their building achieves or not this requirement (Noranai and Kammalluden 2012).

1.2 Green Building Index

The government in Malaysia has set up a system which is known as the green building index (GBI). This GBI system was created to be environmental-friendly throughout the life cycle of the building construction. In the green building rating systems, there are six elements that are involved, which are energy efficiency, indoor environmental quality, materials and resources, sustainable site planning and management, water efficiency, and innovation. Four classifications are given for the green building certifications to show the gradient of compliance with the requirements, namely platinum, gold, silver, and certified certification (Department of Standards Malaysia 2007). This GBI is used to encourage the existing buildings to reduce non-renewable energy sources, energy usage, and pollution while maintaining the comfort, safety, and health of occupants (Chua and Oh 2011). Table 1 shows a building energy rating.

Table 1 Building index rating

Building index rating	BEI kWh/m^2/year
Average Malaysia building	250
Meets MS1525	200–220
GBI certified	150–180
GBI silver	120–150
GBI gold	100–120
GBI platinum	<100

2 Methodology

There are three stages to be gone through:

(1) Selected Buildings

In this study, UiTMPP's Dewan Besar has been selected. This building is in University Technology Mara Pulau Pinang (UiTMPP). The building was located at Permatang Pauh, Pulau Pinang, Malaysia. This building operates if UiTM has an event such as for the examination hall, convocation, and many more. Front view of UiTMPP's Dewan Besar was shown in Fig. 1.

(2) Collect and Calculate Data

Energy usage has been measured at Substation UiTMPP's Dewan Besar on November 18, 2016, until December 6, 2016, using three-phase quality recorder (Fluke 1750). Besides that, the total gross area of UiTMPP's Dewan Besar is 3540 m^2. Data of the energy usage and total gross floor area at that building need to be used in calculating BEI. BEI is calculated simply by dividing the total energy of the building (kWh/year) with its total gross floor area of the building (m^2). Figure 2 shows the formula to calculate BEI:

Fig. 1 Front view of UiTMPP's Dewan Besar

$$\textit{Dewan Besar}\ \text{UiTMPP Building Energy Index} = \frac{\text{Total Energy Usage}}{\text{Total Gross FloorArea}}$$

Fig. 2 Calculation of building energy index

(3) Data Analysis

BEI at UiTMPP's Dewan Besar that has been calculated will be analyzed. This BEI will be analyzed either this index is achieved or not compared to the recommend index by Malaysia Standard MS1525 requirement. This BEI should be below than 220 kWh/m^2/year. The potential energy saving for the buildings also will be suggested.

3 Result

3.1 Electricity Bill of UiTMPP

Tables 2 and 3 show the electric bills for UiTMPP in year 2014 and 2016. Both of these electricity bills will be compared either the trend is decreasing or increasing.

Table 2 is the electric bill record from January until December in 2014. According to the summary, UiTMPP total annual electric bill in 2014 was 17,904,697 kWh, 47,081 kW and total cost was RM7,198,788.

Table 3 shows the electric bill record from January until December in 2016. According to the table, total electricity consumption in 2016 was 16,407,599 kWh,

Table 2 Total cost, energy usage, and energy demand for each month in 2014

Month	kWh	kW	Cost (RM)
January	1,679,306	4141	573,756.60
February	1,644,488	3875	656,219.75
March	1,551,534	4126	632,150.05
April	1,809,071	4239	721,235.40
May	1,461,123	4078	600,644.95
June	1,301,375	3562	553,031.65
July	1,645,429	4309	668,558.35
August	1,396,991	4132	580,736.65
September	1,416,532	3759	576,924.10
October	1,485,227	3930	604,589.25
November	1,312,899	3510	535,437.25
December	1,200,722	3420	495,504.00

Table 3 Total cost, energy usage, and energy demand for each month in 2016

Month	kWh	kW	Cost (RM)
January	1,376,800	3872	564,700.00
February	1,406,696	3873	590,464.56
March	1,190,318	3950	519,379.35
April	1,579,931	3976	652,207.49
May	1,400,391	4107	595,193.55
June	1,305,271	4012	560,163.64
July	1,346,866	4003	573,998.82
August	1,241,178	3823	532,892.59
September	1,559,496	4050	647,452.08
October	1,511,788	3991	629,549.54
November	1,260,508	3453	528,588.05
December	1,228,356	3878	530,160.24

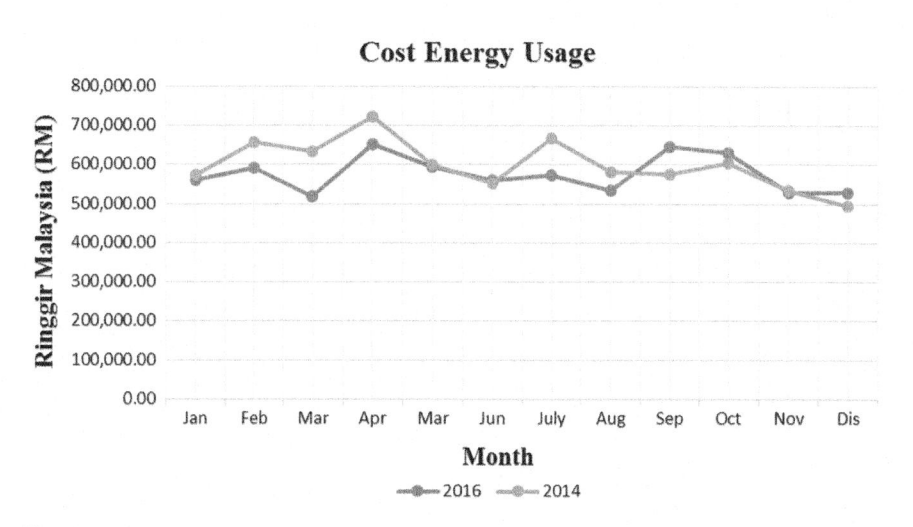

Fig. 3 Electricity cost for each month in 2014 and 2016

46,988 kW and total cost was RM6,924,749. Figure 3 exhibits the electricity cost for Dewan Besar in 2014 and 2016.

The twelve months electricity consumptions were analyzed, and the cost energy usage is plotted. Almost a constant cost of electricity usage over the year can be observed for each year that average is range between RM700,000 until RM500,000. The highest cost electricity was in April for both years that was RM721,235 and RM652,207. The lowest cost electricity was in March of year 2016 that was RM519,379 and in December of 2014 that was RM495,504. According to the summary, UiTMPP total cost annual electric bill was RM7,198,788 for 2014 and RM6,924,749 for 2016. It had shown a decrease of total electricity cost.

3.2 Energy Usage at UiTMPP's Dewan Besar

Energy usage has been measured at Substation UiTMPP's Dewan Besar on November 18, 2016. The energy usage has been measured for two weeks until December 6, 2016.

From Table 4, the total power usage and power demand for 18 days are 27642.52 kWh, and the maximum power demand on 18 days is 344.75. It can be approximated that for a month and a year, the power usage is 46070.87 kWh and 552850.44 kWh. So it can be concluded that electricity usage at Dewan Besar is just 3.8% from the overall of electricity at UiTMPP.

3.3 BEI of UiTMPP's Dewan Besar

Total Energy usage for 2016 = 552, 840 kWh

Total *Dewan Besar* UiTMPP Gross Floor Area = 3540 m^2.

BEI at UiTMPP's Dewan Besar = Total Energy Usage/Total Gross FloorArea
= 156.2 kWh/m^2/year

Table 4 Events at UiTMPP's Dewan Besar

Date	Events	Power demand (kW)	Power usage (kWh)
18/11/2016	Preparation for convocation	243.38	2703.95
19/11/2016	Preparation for convocation	277.38	4519.34
20/11/2016	Pre-convocation	323.97	4790.47
21/11/2016	Convocation	344.75	4502.25
22/11/2016	–	29.66	549.86
23/11/2016	–	26.30	275.87
24/11/2016	Preparation for TYT	284.10	3010.91
25/11/2016	Preparation for TYT	292.74	3066.73
26/11/2016	TYT	331.34	2585.9
27/11/2016	–	11.31	210.5
28/11/2016	–	26.90	245.87
29/11/2016	–	12.54	107.39
30/11/2016	–	10.18	102.1
01/11/2016	–	6.26	118.95
02/11/2016	–	16.18	131.75
03/11/2016	–	22.65	170.71
04/11/2016	–	4.47	99.88
05/11/2016	Meeting diploma student with PA	271.34	450.09

BEI of UiTMPP's Dewan Besar is 156.2 kWh/m^2/year. The best BEI practice and which is recommended by Malaysian Standard is 200–220 kWh/m^2/year. Therefore, BEI of UiTMPP's Dewan Besar is lower compared to recommended value. Based on the building index rating that has been created by Malaysia Government, this UiTMPP's Dewan Besar has a BEI falling within a range of 150–180 kWh/m^2/year. It means that UiTMPP's Dewan Besar has a potential to achieve green building certified based on the building index rating.

By the way, BEI at Dewan Besar not really shows the actual building performance index because power usage of this UiTMPP's Dewan Besar is not fully utilized in whole year. Estimated the result might high from current result because the data of power usage at Dewan Besar were measured during earlier semester, but Dewan Besar is fully utilizes at the end of semester where this building is used for an examination hall and it will use a lot of energy usage.

University building operation are not practice like shopping mall building, general office building or factory building where there are operation almost a whole year. University building is less used during semester break where student is not in the university campus.

3.4 Suggestion to Improve Energy Consumption

Usually, UiTM's Dewan Besar is operating from 9 a.m. to 5 p.m. It means the average hour that used in a day is 8 h. In addition, the average of UiTM's Dewan Besar used in a month is 15 times. Table 5 shows the energy usage at UiTMPP's Dewan Besar before the suggestion. Table 6 shows the total cost in a month and year before the suggestion.

First, change the fluorescent tube light with the LED type (lamp without ballast). The fluorescents, despite being rated at 36 W, while the LED tube light rated at 16 W. Next, change the compact fluorescent bulb (18 W) with the LED bulb light

Table 5 Energy usage at UiTMPP's Dewan Besar before the suggestion

Type	Quantity	Power (W)	Total power (kW)
Fluorescent tube lamp	98	36	3.528
Spotlight	42	2500	105
High bay light	9	240	2.160
Compact fluorescent light	85	18	1.53
Air conditioner (6 hp)	4	4250	17
Air conditioner (3 hp)	2	2450	4.9
Air conditioner (1 hp)	3	950	2.85
HVAC (hall)	1	220 k	220
Speaker (PA system)	1	2000	2
Other equipment	–	–	–
Total			358.497

Table 6 Total cost UiTMPP's Dewan Besar before the suggestion

Total power (month) = Total power × 8 h × 15 times
Total power (month) = 358.497 kW × 8 × 15 = 43,076 kWh (month)
Total cost (month) = Total power (month) × RM0.365
Total cost (month) = 43,076 kWh × RM0.365 = RM15,723

Table 7 Energy usage at UiTMPP's Dewan Besar after the suggestion

Type	Quantity	Power (W)	Total power (kW)
LED tube light	98	16	1.568
Spotlight	42	2500	105
High bay light	9	240	2.16
LED bulb light	85	9.5	0.808
Air conditioner (6 hp)	–	–	–
Air conditioner (3 hp)	2	2450	4.9
Air conditioner (1 hp)	3	950	2.85
HVAC (hall)	1	220 k	220
Speaker (PA system)	1	2000	2
Other equipment	–	–	–
Total			339,286

Total Power (month) = Total Power x 8 hour x 15 times

Total Power (month) = 339,286kW x 8 x 15 = 40,714 kWh (month)

Total cost (month) = Total Power (month) x RM0.365

Total cost (month) = 40,714 kWh x RM0.365 = RM14,860

Fig. 4 Total cost UiTMPP's Dewan Besar after the suggestion

(9.5 W). The light quality provides the same amount of light as the old fluorescents. Besides that, LEDs use less power (Watts) per unit of light generated (lumens). Thus, LEDs help to lowers electric bills. Lastly, four split type air conditioners (6 hp) should not be installed or operated since the air condition with the chiller has been installed in the hall (Table 7 and Fig. 4).

The difference of total cost for a month before and after the suggestion is RM862.30. Estimated that it can reduce 5% of electricity at UiTMPP's Dewan Besar. Even though the difference between before and after the suggestion is not much, at least, it can reduce energy that has been wasted. Another suggestion in reducing energy usage is building owner, and building user also needs to be introduced about energy awareness program. This energy awareness program is to increase the occupants' awareness of how energy is consumed and then let them know how to save energy in their building.

4 Conclusion

For the conclusion, this study aims to determine energy efficiency level on UiTMPP's Dewan Besar. BEI of UiTMPP's Dewan Besar is 156.2 kWh/m^2/year. BEI of UiTMPP's Dewan Besar is lower compared to Malaysian Standard that it should be below than 220 kWh/m^2/year. In addition, several suggestions have been given in reducing electricity cost. Based on the suggestion, 5% of electricity UiTMPP's Dewan Besar can be reduced. Besides that, there are several factors need to be considered for reducing energy consumption in university campuses, such as educate users, create awareness, monitoring, and others. This program can be created convenient with the government's initiative to reduce energy usage in government buildings.

Acknowledgements The author would like to thank the research supervisor, En Mohamad Adha bin Mohamad Idin and Muhammad Firdaus Othman, for his guidance throughout the development of the research. Furthermore, the authors are thankful to Universiti Teknologi MARA for providing the financial support.

References

Chua, S. C., & Oh, T. H. (2011). Green progress and prospect in Malaysia. *Journal of Renewable and Sustainable Energy Reviews, 15*(6), 2850–2861 [e-journal]. Available through: Universiti Tunku Abdul Rahman Library website http://www.sciencedirect.com/science/article/pii/S1364032111001080. Accessed January 21, 2015.

Department of Standards Malaysia. (2007). *Malaysian standard: Code of practice on energy efficiency and use of renewable energy for non-residential buildings, 1.*

Kamaruzzaman, S. N., Ali, A. S., Abdul-samad, Z., & Zawawi, E. M. A. (2009). Energy performance of electrical support facilities: The case of adaptive re-used historical buildings in Malaysia. *International Journal of Physical Sciences, 4*(12), 752–757.

Moghimi, S., Mat, S., Lim, C. H., Zaharim, A, & Sopian, K. (2011). Building Energy Index (BEI) in large scale hospital : Case study of Malaysia. *Biomedicine,* 167–170.

Najihah, N., Bakar, A., Hassan, M. Y., Abdullah, H., Rahman, H. A., Hussin, F., et al. (2014). *Identification building energy saving using energy efficiency index* (pp. 366–370).

Noranai, Z., & Kammalluden, M. N. (2012). Study of Building Energy Index in Universiti Tun Hussei Onn Malaysia. *Academic Journal of Science, 1*(2), 429–433. Retrieved from http://universitypublications.net/ajs/0102/html/TRN168.xml.

Assessment of Energy Efficiency Level on Unit Kesihatan UiTMPP's Building

Muhammad Daniel Muhamad Nor, Mohamad Adha Mohamad Idin, Muhammad Firdaus Othman and Noorezal Atfyinna Mohd Napiah

Abstract This project presents the study of energy efficiency level on a selected building. Currently, UiTM Penang suffers from a high cost of electricity. The intention of this study is to collect data based on the previous electric bill and in field data acquisition. Thus, the data will be analyzed to compute the level of energy efficiency. This project is a part of an initiative to reduce energy wastage in Universiti Teknologi Mara (UiTMPP) building. This project will introduce the building energy index (BEI) in order to interpret the building energy efficiency level. Apart from that, data acquisition was held at Unit Kesihatan UiTMPP's Main Distribution Board (MSB) room. The data acquisition was done by using Fluke 1750 Power Recorder. In addition, the project also can identify any wastage of energy and what aspect can be improved in the building. This project found that the energy usage monthly of this building is 4% of total energy consumed by the whole area of UiTM Penang.

Keywords Building energy efficiency · Energy consumption · Energy efficiency

1 Introduction

In this century, the increase of population cannot be controlled easily. Thus, the number of building growing rapidly causes the energy demand upsurge intensively. Building sector is major energy consumers where it consumes 48% of the total electrical energy (Noranai and Kammalluden 2012). According to guideline in Malaysian Standard MS1525, the energy index in Malaysia should be near 135 $kWh/m^2/year$ (Tang and Chin 2013). Nevertheless, practically, very small number of building meets this standard. Apart from that, the implementation of energy efficiency is needed in order to achieve the standard.

M. D. Muhamad Nor (✉) · M. A. Mohamad Idin · M. F. Othman
N. A. Mohd Napiah
Universiti Teknologi MARA Pulau Pinang, Permatang Pauh, Malaysia
e-mail: dane_daniel03@yahoo.com

© Springer Nature Singapore Pte Ltd. 2018
R. Saian and M. A. Abbas (eds.), *Proceedings of the Second International Conference on the Future of ASEAN (ICoFA) 2017 – Volume 2*,
https://doi.org/10.1007/978-981-10-8471-3_19

The energy efficiency can be determined by energy efficiency index (EEI) or building energy index (BEI) (Bakar 2014). These indexes act as indicator to monitor the performance of energy consumption in building. The index is determined by the energy consumption per hour divided by gross area of the building. BEI is usually expressed as kWh/year/m^2. If the BEI reading is high, it means the energy consumption is not efficient.

$$BEI = \frac{\text{Energy Consumption (kWh)}}{\text{Area}(m^2)} \quad (1)$$

Energy efficiency can be achieved if the energy usage is in the optimum state. When the energy-efficient concept is applied, there are several advantages that everyone can be appreciated such as the power plant will reduce energy cost as well as CO_2 emission. Besides that, when the building achieves energy efficiency, it will increase the asset value while strengthening the organization image (Bakar 2014).

Substantial financial savings can result from the application of high-grade energy saving systems, especially when considering that high-grade energy replacement costing higher about 3–10 times as much per kWh, as the equivalent unit of low-grade energy appliances (Kamaruzzaman et al. 2009). Furthermore, high-grade electrical accounts for only 40% of energy consumption, but these high-grade energy uses account for 66% of energy expenditure (Kamaruzzaman et al. 2009).

2 Methodology

2.1 Case Study

This study was carried out at Unit Kesihatan building in Universiti Teknologi Mara (UiTM). The building is single stories building with total floor area of 541 m^2. The main activities in the building are treatment, consulting, and curing. There are eight medical officers working in the building and variable number of patient every day. The building consists of four main facilities which are the treatment room, sick bay, waiting room, and pharmacy. A site visit was conducted to record the available or used electrical appliances in the building.

2.2 Energy Consumption

Table 1 shows the load of energy consumption for the building. The main contribution for the total energy consumption is consumed by air conditioner which is 16.343 kW. Lighting becomes second highest energy consumption then followed by other loads.

Assessment of Energy Efficiency Level on Unit Kesihatan ...

Table 1 Power consumption of electrical appliances in Unit Kesihatan building

Type of appliances	Rated power (W)	Quantity	Power consumption (kW)
Air conditioner	4540 (6 HP)	2	9.08
	1798 (2.5 HP)	2	3.60
	1429 (2 HP)	1	1.43
	746 (1 HP)	3	2.40
Lighting	18	24	0.43
	36	76	2.74
Fridge	1200	1	1.20
Stand fan	75	3	0.23
PC/Laptop	90	3	0.27
Water heater	1800	1	1.80
Total power consumption	23.18		

2.3 Operating Schedule Analysis

The operating schedule of Unit Kesihatan is apparently quite fixed. Table 2 shows the premise operating timetable through a week. From Monday to Thursday, it opens from 8.00 a.m. to 5 p.m. and the premise closes at 1.00 p.m. to 2.00 p.m. for a break. While on Friday, this facility opens at 8.00 a.m. to 5.00 p.m. also but closed temporarily at 12.15 p.m. and will open back at 2.45 p.m. However, on weekend and public holiday, this Unit Kesihatan is closed entire day. The only energy consumed just for security light at corridor.

2.4 Proposed Energy Efficiency Index

This project proposed a building energy index logarithm based on total gross floor area, as shown in Eq. 2.

$$\text{BEI} = \frac{\text{Total Energy Consumption kWh/year}}{\text{Total gross floor area m}^2} \qquad (2)$$

Table 2 Operating timetable of Unit Kesihatan UiTMPP

Day	Working time
Monday–Thursday	8.00 a.m.–1.00 p.m.
	2.00 p.m.–5.00 p.m.
Friday	8.00 a.m.–12.15 p.m.
	2.45 p.m.–5.00 p.m.
Saturday and Sunday	Close
Public holiday	Close

The total energy consumption is defined as total energy of electricity consumed by the building per annum. For the total gross floor area is defined as total area in the building included the area occupied or unoccupied measured in meter square, m^2. The unit of BEI remains expressed in $kWh/m^2/year$. The relationship between total gross area and the total energy consumption can be determined by using the index above.

3 Result and Discussion

3.1 Electric Bill of Universiti Teknologi MARA (UiTM) Penang Branch

Based on graph in Fig. 1, in December 2014 shows that the lowest cost of the year. This is because in that month, majority diploma student was in semester break. A part from that, only office and degree student occupy in UiTMPP. While in April 2014, the usage seems to be the highest among the month. This could be the usage of Hall for examination are hardly use. The chiller was constantly worked to chill the overall hall area. The average cost for this year is RM650,679.50. This price can be reduce by reducing the energy consumption by applying the energy efficiency concept.

Based on the graph in Fig. 2, in January 2016 shows that the lowest consumption of the year. This is because degree students were just started their semester break so the usage of energy is low. Then, diploma student just started their new semester in February which makes cost of electric bill to increase rapidly. In April 2016, the cost seems to be the higher of the year. This situation may cause by the usage of facilities were fully occupied by users. The average cost monthly is RM552,886 which is more than half million. This number needs to be reduced by applying energy efficiency concept. In order to get bill for Unit Kesihatan building

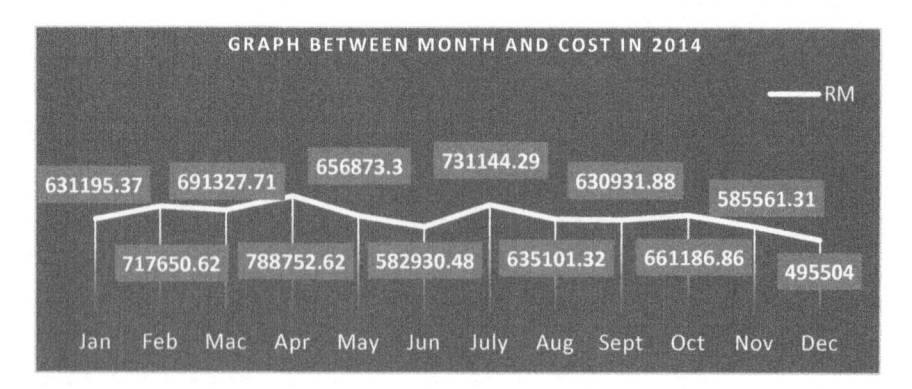

Fig. 1 A graph of monthly electric cost of UiTMPP's in 2014

Assessment of Energy Efficiency Level on Unit Kesihatan … 199

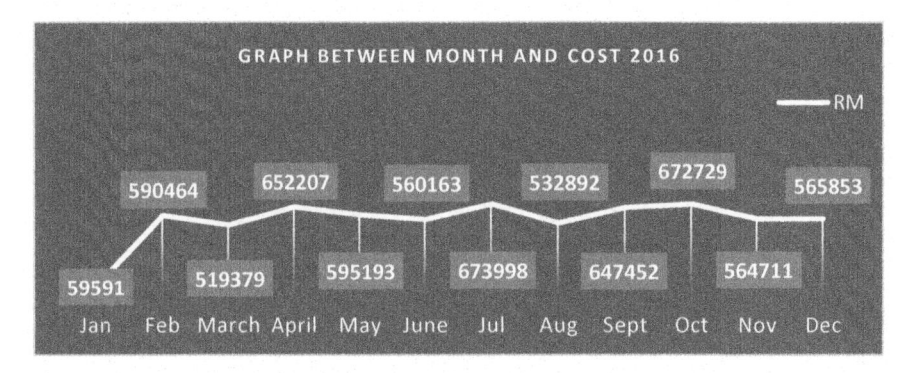

Fig. 2 A graph of monthly electric cost of UiTMPP's in 2016

only, the manual calculation is required because the billing meter is not installed by service provider (TNB).

3.2 Power Monitoring

A power monitoring was conducted by using Fluke 1750 Power Recorder in order to record the power consumed by the building. This experiment was carried out for two weeks with approximately 12 days of power recording. The necessity of this experiment is because the building was not being equipped with reading meter for the electric usage.

From the Fig. 3, the pattern is nearly constant and predictable. The measurement was taken at 11.22 a.m. on November 7, 2016, and ended at 11.24 a.m. on November 18, 2016. The figure shows clearly that the minimum energy consumed is on weekend and during night. This is because only the security peripheral and a refrigerator are turn on. The data then has been collected and is given in Table 3.

Based on the table above, the total cost for approximately 12 days of power recorded is RM9586.20. In average, the building consumed about RM23,965.50 for one month. That is 4% of total UiTMPP's electrical bill which is RM564,711.32 for November 2016. This calculation considers the tariff with two different rates for power (kW) multiple by 30.3 and energy (kWh) multiple by 0.365, respectively.

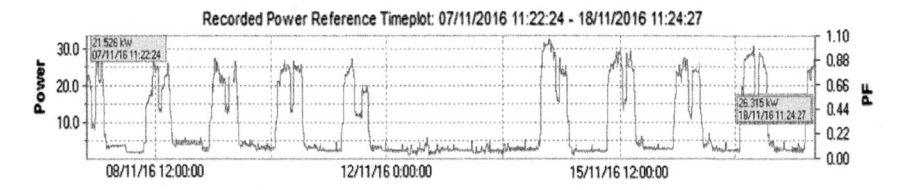

Fig. 3 Monitoring data taken for 2 weeks

Table 3 Power and energy usage recorded

Date	Power (kW)	Energy (kWh)
7/11/16	26.09	141.61
8/11/16	29.32	231.93
9/11/16	26.05	239.48
10/11/16	25.62	264.6
11/11/16	25.88	231.86
12/11/16	4.59	61.9
13/11/16	4.24	63.68
14/11/16	32.34	294.6
15/11/16	28.97	283.77
16/11/16	27.07	263.48
17/11/16	29.37	280.6
18/11/16	27.18	104.36
Total	286.72	2461.87
Total cost (RM)	8687.62	898.58

3.3 Building Energy Index (BEI) Calculation

From the experiment, the data has been start taken at 11.22 a.m. and ended at 11.24 a.m. next week. Hence, the incomplete duration of data is not included which is November 7, 2016, and November 18, 2016 data.

$$\text{Energy usage in } 10 \text{ days} = 2215.9 \text{ kWh}$$

$$\text{In average, } 1 \text{ day usage} = 221.59 \text{ kWh}$$

$$30 \text{ day usage} = 6647.7 \text{ kWh}$$

$$1 \text{ Year usage} = 79,772.4 \text{ kWh}$$

$$\text{Gross floor area} = 541 \text{ m}^2$$

$$\text{BEI} = \frac{79,772.4 \text{ kWh/year}}{541 \text{ m}^2}$$
$$= 147 \text{ kWh/year/m}^2$$

4 Recommendation to Improve Energy Efficiency

This recommendation can be applied to reduce energy consumption resulting in increase of the energy efficiency level. Therefore, replacing the lightings from fluorescent to LED type is the best chance in term of upgrading current appliances.

Table 4 Comparison between LED tube light and fluorescents tube light

Type of load	T8 LED tube (600 mm)	T8 fluorescents tube (600 mm)	T8 LED tube (1200 mm)	T8 fluorescents tube (1200 mm)
Properties				
Power (W)	10	18	18	36
Lumens	800	800	1600	1600
Kilowatts electricity used at 30 unit per year (kWh)	864	1555.2	1555.2	3110.4
Annual operating cost for 30 unit per year	RM315.36	RM567.65	RM567.65	RM1135.30
Life span (h)	>30,000	>15,000	>30,000	>15,000

The comparison between fluorescent light and light-emitting diode (LED) light is given in Table 4.

Apart from that, user can manage to reduce energy consumption by resetting the air conditioner to 24 °C. This will help the compressor to run at half power and save energy consumption at same time occupants still comfort. By doing this to all unit air conditioner, building can save 30% energy consumption yearly. To be exact, the saving is 9509.79 kWh/year from 31,699.2 kWh/year. The building will save about RM 3471 per year.

5 Conclusion

In conclusion, the Unit Kesihatan of UiTMPP is not energy efficient enough. This is because BEI of the building exceed Malaysian standard which is 145 kWh/m^2/year. The Malaysian standard for commercial building is at 135 kWh/m^2/year. The lower the index, the higher the level of efficiency. In addition, this study found that the total energy consumed by the building is 4% of the total energy cost of UiTMPP. The facility department should take initiative to enhance the energy efficiency level. Authorities should take part to achieve an energy-efficient building thus reducing the operating cost. The retrofitting concept is the best way and low cost to improve the energy efficiency level on the building.

Acknowledgements The authors gratefully acknowledge the helpful comments and suggestions of the reviewers, which have improved the presentation. Furthermore, the authors are thankful to Universiti Teknologi MARA for providing the grant and supports.

References

Bakar, N. N. A. (2014). Identification building energy saving using Energy Efficiency Index approach. In *2014 IEEE International Conference on Power and Energy (PECon)* (Vol. 965, pp. 366–370).

Kamaruzzaman, S., Ali, A. S., Abdul-Samad, Z., & Zawawi, E. M. A. (2009). Energy performance of electrical support facilities: The case of adaptive re-used historical buildings in Malaysia. *International Journal of Physical Sciences, 965,* 752–757.

Noranai, Z., & Kammalluden, M. N. (2013). Study of Building Energy Index in University Tun Hussein Onn Malaysia. *Academic Journal of Science, 965,* 429–433.

Tang, C., & Chin, N. (2013). Development of JKR/BSEEP technical passive design guidelines for Malaysian Building Industry. *Building Sector Energy Efficiency Project/UNDP-JKR Malaysia, 965,* 252–255.

Assessment of Energy Efficiency Level on UiTMPP's Baiduri College Building

Amirul Ashraf Samad, Mohamad Adha Mohamad Idin, Muhammad Firdaus Othman and Noorezal Atfyinna Mohd Napiah

Abstract Nowadays, renewable energy and energy efficiency become the most significant issues that needed to be considered when it comes to the sustainability of energy management. This project is developed to present the energy efficiency level of UiTMPP's Baiduri College building. The energy efficiency is used to refer energy usage without wasting the energy. This project is carried out due to the unavailability of electrical meter for each building. The main objective of this project is to reduce the electric bill and energy consumption at UiTMPP's Baiduri College. The method used for this project is by monitoring, collecting, and analyzing electric energy usage by using three-phase quality recorder (Fluke 1750). Building energy index (BEI) is used as a rating point to indicate the building energy performances. This technique is mostly used to calculate the energy efficiency in the building. This BEI also can be expressed as $kWh/m^2/year$ which shows that the total of energy used for one year in building in kilowatt hours (kWh) is divided by the gross area of building in m^2. The result for this project has been determined by calculating the electricity bill and the power usage for the building. A total of BEI for this building is 925.27 $kWh/m^2/year$. This BEI will be compared with Malaysia Standard MS1525 requirement that it should be lower than 220 $kWh/m^2/year$. There are a few suggestions on how to reduce and save the energy usage based on collected results.

Keywords Building energy index · Energy efficiency index · Energy level

---

A. A. Samad (✉) · M. A. Mohamad Idin · M. F. Othman · N. A. Mohd Napiah
Universiti Teknologi MARA Pulau Pinang, Permatang Pauh, Malaysia
e-mail: army_rulekiba93@yahoo.com

1 Introduction

In this era, the numbers of world population are increasing, causing huge demand of building and electric energy. For the increasing population, the efficient usage of the energy is one being studied and implemented to reduce the electric energy demand. Besides that, many buildings in Malaysia almost do not meet recommended standard building index. This building was not aware whether their building achieved or not this requirement (Bakar 2014). Hence, it will focus on to identify building energy index (BEI) and calculate the BEI in college. This project is focusing on power consumption at UiTMPP's Baiduri College. The energy used at the Baiduri College needs to calculate manually based on the monitoring electricity usage at that building. The main objective of this project is to calculate the building index and to give suggestion on how to reduce the electric energy consumption. The suggestions have been made to reduce and improve the energy efficiency at the Baiduri College's building.

1.1 Energy Efficiency Index

Energy efficiency index (EEI) is used to maintain more efficiency reducing the energy consumption and electric usage (Zain et al. 2013). Furthermore, energy efficiency also has to determine any wastage of energy, and there needs to improve that design (Bakar 2014). Based on the research, energy requirement will be extremely increased in Malaysia (Moghimi et al. 2011) toward year 2020. For the building sector is the mostly electric energy consumer about 48% of the total electricity energy (Bakar 2014). From the analysis of the energy efficiency that a several suggestion can be make such as usage of energy efficient lighting and appliances, shutting down non-essential equipment at time and improve control and monitoring of electrical equipment that will reducing electricity energy and can save more money (Bakar 2014). Hence, this method can reduce electric bill as well. Energy efficiency index (EEI) is a tool that is used to track the performance of energy consumption. The measurement of EEI can be defined based on the energy equipment used in the building and a factor that is related to the energy using component of the organization. Examples of factor that is related to energy use are:

- Weight of product produced,
- Number of items produced,
- Weight of raw material used,
- Period of production,
- Period of plant usage,
- Floor area of building.

In addition, the EEI can be defined in terms of a factor that is related to the energy component such as the weight of product produced, number of items produced, weight of row material used, period of production, floor area building, and period of plant usage. The several advantages when the energy efficiency can be reduced are (Kamaruzzaman et al. 2009).

- Reduce energy cost,
- Improve building operating performance,
- Improve occupant.

1.2 Electricity Energy Consumption

Generally, energy consumption is commonly used to determine the building energy performance against its building size for the purpose of comparison (Moghimi et al. 2011). Energy consumption has to be analyzed relative to its building size in order to determine whether the building overconsumed or underconsumed energy. In addition, the increase in number of commercial buildings has some impact to the nation development and the energy demand. To calculate the energy consumption, it normally used the BEI formula.

Based on the several researches, the building description and their characteristic are one of the reasons for the change of power. The most electricity consumption in Malaysia are industry 53.2%, public lighting 0.7%, commercial 28.7%, mining 0.3%, residential 17.0%, and export 0.1%[3] (Fig. 1).

1.3 Building Energy Index (BEI)

BEI is a normal method used to calculate the BEI (Noranai et al. 2012). This method also can be explaining the electrical system from the sole energy-consuming device. BEI also can be expressed as kWh/m^2/year which shows that the total of energy used for one year in building in kilowatt hours is divided by the gross area of building in m^2 (Moghimi et al. 2011). Before using this method,

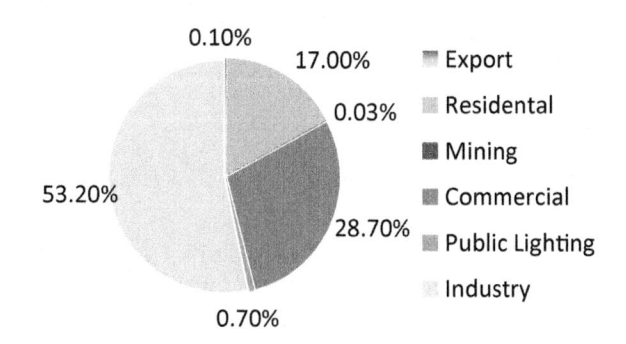

Fig. 1 Chart of electricity consumption

the total energy consumption energy must be calculated first. The BEI calculation using Eq. 1 is:

$$\text{Baiduri College BEI} = \frac{\text{Total Energy Usage}}{\text{Total Gross Floor Area}} \quad (1)$$

This BEI also has several concept of energy efficiency index such as to organize an effective energy efficiency scheme for the future and understanding on their building energy utilization (Kamaruzzaman et al. 2009). This concept also has to calculate and define the ratio of the energy input per factor related to the energy using component (Kamaruzzaman et al. 2009). According to MS1525 standard, recommended BEI in Malaysia is 135 kWh/m^2/year, even though almost nowadays the building in Malaysia does not meet this standard.

From the some previous research, that was found the BEI for the engineering building in UiTM Shah Alam is 149 kWh/m^2/year (Kamaruzzaman et al. 2009). That amount of energy can be reduce about 20% with the several experiment and design the building to reduce the electric energy usage. Using this method also that can identify any wasted of energy particularly and the scope for improvement as well (Bakar 2014).

2 Methodology

The aim will be focused on the study of BEI in UiTMPP's Baiduri College. Energy performance index (EUI) or sometimes known as BEI is calculated simply by dividing the total annual energy consumption of the building (kWh/year) with its total occupied floor area of the building (m^2). Using this method, result of UiTMPP's Baiduri College BEI will be calculated and discussed.

Several data have been collected including monthly electrical bill, gross floor area, as well as calculating the building index. Based on the collected data, the electric energy use in that college building can be determined. There-phase quality recorder (Fluke 1750) is used in order to record the energy used. Besides, energy consumption also can be determined.

Then, the studies will be focused to analysis on the BEI to monitor the electrical energy that has been used and wasted. The calculated BEI will be compared to the Malaysia Standard MS1525.

2.1 System Operation

Mostly, the loads used at the Baiduri's Colleges are ceiling fan and fluorescent lamp. Data acquisition was recorded within two weeks. The data obtained in kWh and hence the detailed energy consumption for that particular building were calculated. In additional, to be more precise, the manually calculated data also have been provided based on the load used. From the observation, this project has to reduce the electrical energy wasted. As the expected result, 10% or more energy reduction can be achieved (Noranai et al. 2012).

2.2 Data Analysis of Electrical Bill

Table 1 shows the summarization of the electrical bill recorded from January to December 2014. The total cost energy used was RM7,198,788. From the table also shows the energy (kWh) used almost 18 million. The highest power consumed was 4309 kW contributed by July. In the other hand, December was the lowest power consumed which is 3420 kW.

Table 2 shows the summarization of the electrical bill recorded from January to December 2016. The total cost energy used was RM6,924,749.91. The highest power consumed was 4107 kW contributed by May. In the other hand, November was the lowest power consumed by 3453 kW.

Figure 2 shows that the maximum electrical bill for each month in the years 2014 and 2016. From Fig. 2, total cost in year 2014 is much higher compared to year 2016. The total cost in year 2016 was RM6,924,749.91, while the total cost in year 2014 was RM7,198,788. However, the graph shows April contributed higher electric bill for both years. This is because students in diploma and degree levels both stay in the university. Besides that, other factors contributed to the energy consumption due to laboratory and lecture rooms that run non-stop, starting from 8 a.m. to 10 p.m. Among others factor, air conditioners are the main contribution toward high energy consumption. Air conditioner is installed everywhere around the university except college building.

Table 1 Total cost, energy usage, and energy demand for each month in 2014

Month	Power (kW)	Energy (kWh)	Cost (RM)
January	4141	1,679,306	573,756.60
February	3875	1,644,488	656,219.75
March	4126	1,551,534	632,150.05
April	4239	1,809,071	721,235.40
May	4078	1,461,123	600,644.95
June	3562	1,301,375	553,031.65
July	4309	1,645,429	668,558.35
August	4132	1,396,991	580,736.65
September	3759	1,416,532	576,924.10
October	3930	1,485,227	604,589.25
November	3510	1,312,899	535,437.25
December	3420	1,200,722	495,504.00
Total	47,081	17,904,697	7,198,788

Table 2 Total cost, energy usage, and energy demand for each month in 2016

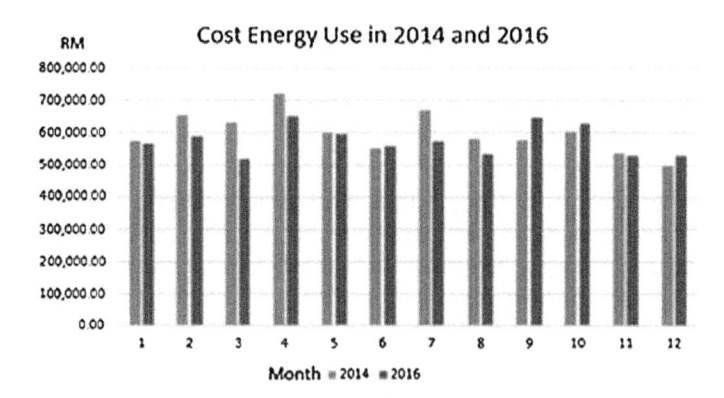

Month	Power (kW)	Energy (kWh)	Cost (KM)
January	3872	1,376,800	564,700.00
February	3873	1,406,696	590,464.56
March	3950	1,190,318	519,379.35
April	3976	1,579,931	652,207.49
May	4107	1,400,391	595,193.55
June	4012	1,305,271	560,163 64
July	4003	1,346,866	573,998.82
August	3823	1,241,178	532,892.59
September	4050	1,559,496	647,452.08
October	3991	1,511,788	629,549.52
November	3453	1,260,508	528,588.07
December	3878	1,228,356	530,160.24
Total	46,988	16,407,599	6,924,749.91

Fig. 2 Graph of total electric bill in January–December 2014 and 2016

3 Result and Discussion

Based on the data collected, several results had been concluded. Firstly, analysis of energy consumption that shows the electrical bill based on monthly basis in university was not consistent. The differences in electric bill depend on the intensity of electrical usage during the semester break, semester term, holiday, and others. In addition, benchmarking calculation of electricity cost in 2014 and 2016 has been made. From this analysis, the BEI reading is high that means the energy consumption used in the college's building is also high based on Eq. 1.

The overall load factor is also the main target in this project which is below than 0.5 load factor that means the distribution is poor. The energy wasted in the college building that can measure by comparing between BEI measurement and based line energy consumption usage. From the general observation, maybe the energy can be reduced around 5–10% by creating awareness from all student in the college.

3.1 Monitoring Result Taking from College Building Using Three-Phase Quality Recorder (Fluke 1750)

Based on Fig. 3, the result shows one-day result after taping using Fluke 1750 at the college building. From the result, the waveform shows the power consumption is not consistent. This happens because the power consumption depends on the consumer used at that particular time. In addition, high power consumption was mostly at the nighttime. This situation occurs because most of the students study and do their assignments by using lamp and fan individually. This situation caused an increase in electrical power consumption used at that time. Apart from that, the result also shows the different energy consumption usage between the weekdays and the weekends. At the weekend, the energy consumption is lower than the energy used at the weekdays. This happens because at the weekend most of the student leaves college building for short vacation. Based on the result, the average power factor is almost 0.9 so that the power factor is also in good condition. From the research, it is mentioned that the power factor below than 0.5 means that the distribution is poor.

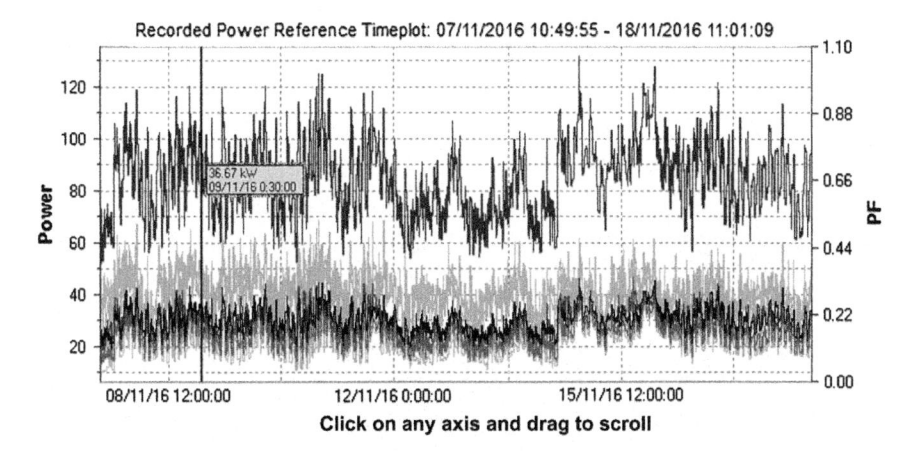

Fig. 3 Result on November 9, 2016, at 12.30 a.m.

3.2 Data Equipment at UiTMPP's Baiduri College

Figure 4 shows manually data collection at the UiTMPP's Baiduri College. This data collection have been taking manually because in the UiTMPP that only one meter for the entire UiTMPP but do not have a specific meter for each building. From that problem, the energy used at the Baiduri College needs to be calculated manually based on the monitoring electricity usage at that building. Table below also shows that mostly equipment that is used in the Baiduri College is fluorescent lamp, fan, and aircon. From the observation, in this building normally the facilities in UiTMPP use the same type of brand for the fluorescent lamp. The brand that is use for the fluorescent lamp is Phillips lamp that consists 32 W power for each lamp. Besides that, the usage of fluorescent lamp in television room is greater than fluorescent lamp used in study room (Table 3).

3.3 Manual Data Collection at Baiduri College's

Table below shows the data calculation at the Baiduri College building. From that table, the total energy used in one month, total energy used in one years, total BEI, and total cost for one month have been calculated. In general, BEI is expressed as kWh/m^2/year which shows the amount of total energy usage for one year in kilo-watts hours which is divided by the gross floor area of building in square meters. Based on that table, the BEI for Baiduri College building is 925.97 kWh/m^2/year that means this BEI for Baiduri College is not too good. The best BEI practice

LOCATION	LAMP (PHILLIPS)	FAN CEILING (DC12V48D2)	FAN WALL (KDK M40MS-GY)	AIRCON (ALC20B AFCA)	LUX
TELEVISION ROOM	120 (32 watt)	28 (40 watt)	-	-	219-290
SIDE WALK	16 (32 watt)	-	-	-	208-215
STUDY ROOM	52 (32watt)	12 (40watt)	-	4 (5hp)	668-709
CORRIDOR	9 (32 watt)	-	-	-	199-209
ROOM	7 (32 watt)	2 (40watt)	1 (20watt)	2 (1hp)	215-253
CORRIDOR	12 (32 watt)	2 (40 watt)	-	-	177-201
STUDENT ROOM (each)	2 (32 watt) 6 (16 watt)	1 (40 watt)	1 (20watt)	-	215-285
TOILET	18 (32 watt)	-	-	-	65-102

*note: Ground Floor / 1st Floor / 2nd – 9th Floor

Fig. 4 Data equipment at Baiduri College

Table 3 The power demand and power usage based on daily tapping

Date	Day	Power demand (kW)	Power usage (kWh)
8/11/2016	Tuesday	109.01	1120.09
9/11/2016	Wednesday	112.43	2103.63
10/11/2016	Thursday	114.46	2116.16
11/11/2016	Friday	110.58	2119.81
12/11/2016	Saturday	96.42	2109.98
13/11/2016	Sunday	97.54	1858.44
14/11/2016	Monday	119.78	1845.43
15/11/2016	Tuesday	71.29	2058.55
16/11/2016	Wednesday	122.16	2327.07
17/11/2016	Thursday	112.11	2252.80
18/11/2016	Friday	106.34	2082.33

recommended by Malaysian Standard is 200–220 kWh/m^2/year. The result not really shows the actual building performance index because power usage of UiTMPP's Baiduri College is not fully utilized in whole year. Estimated the result might higher from current result because Baiduri College were fully utilized during earlier semester toward the end of semester. This is because energy in this building run continuously usage of electricity. Besides that, the cost for one month in this building is RM5420.51. This value is quite high for this category of building, and most of the students are not aware of the energy-saving method. The cost for the building might be higher if the total hour of used energy is more than calculated hour.

3.4 Suggestion to Improve Energy Consumption

As all know, most of the people still are not aware of the important of saving energy. Some of them need to pay high amount of electricity bill for every month because of lack of knowledge about the significance of energy saving. There are a few steps for the community to follow the guidelines in order to achieve better energy consumption. These are some suggestion that can be implemented by the community:

- Change the fluorescent lamp with the LED type (lamp without Ballast),
- Make sure all the electric equipment is approved by SIRIM and has the SIRIM sticker,
- Set the air conditioner at 24 °C,
- Buy electrical appliances that have inverter.

In addition, several factors need to be considered to improve BEI in Baiduri College such to educate users, to create awareness, monitoring, and others.

- Turn off lights when they are not in use. Occupancy sensors and timers can help, but a less expensive alternative would be to educate and motivate employees to turn off lights at the end of the day. Some equipment cannot be turned off entirely, but turning it down to minimum levels where possible can save energy
- Make sure that your HVAC system is regularly cleaned and serviced and can help to prevent costly heating and cooling bills.

4 Conclusion

The result of the study has answered the research question that the BEI based on the benchmarking of the energy consumption in building is relevant but not too good. As a benchmark that should know the recommended BEI should not be more than 149 kWh/m^2/year. Based on data collection from Baiduri College shows the power consumption for this building is high. Due to this problem, this will affect the total cost for UiTMPP electric bill. Besides that, this project was done to find out the energy consumption by miscellaneous equipment installed in the building. This is comply with the second sub objective is to study the effect of energy efficiency based on calculating the energy consumption at UiTMPP's Baiduri College. For the future work, the campus needs to launch awareness campaign to the benchmark about the important of the saving energy in daily life. The poster or reminder sticker that show the energy-saving step need to be placed everywhere during the campaign. Furthermore, improving the energy efficiency that have several strategy can be take such as the MCB room can be control and manage by the management system by supplying energy based on the request only. Besides that, the facilities also need to change the fluorescent lamp with the LED type because the lamp without ballast that can save more energy and reduce cost. By applying this strategy the reducing energy cost can be achieve.

Acknowledgements The authors wish to thank the research supervisor, En Mohamad Adha bin Mohamad Idin, for his sound advice, great guidance, and his enormous patience throughout the development of the research. Furthermore, the authors are thankful to Universiti Teknologi MARA for providing the grant and support. Lastly, the author also would like to offer their deepest thankful and blessings to family and friends, for the unfailing support and encouragement throughout this research.

References

Bakar, N. N. A., et al. (2014). Identification building energy saving using Energy Efficiency Index approach. In *2014 IEEE International Conference on Power and Energy (PECon)* (pp. 366–370), Kuching.

Kamaruzzaman, S. N., Ali, A. S., Abdul-samad, Z., & Zawawi, E. M. A. (2009). Energy performance of electrical support facilities: The case of adaptive re-used historical buildings in Malaysia. *International Journal of Physical Sciences, 4*(12), 752–757.

Moghimi, S., Mat, S., Lim, C. H., Zaharim, A., & Sopian, K. (2011). Building energy index (BEI) in large scale hospital: Case study of Malaysia. In *Recent Researches in Geography, Geology, Energy, Environment and Biomedicine—Proceedings of the 4th WSEAS International Conference on EMESEG'11, 2nd International Conference on WORLD-GEO'11, 5th International Conference on EDEB'11* (pp. 167–170).

Noranai, Z., & Kammalluden, M. N. (2012). Study of building energy index in Universiti Tun Hussein Onn Malaysia. *Academic Journal of Science, 1*(2), 429–433.

Zain, Z. M., Aziz, M. B. A., Kassim, A. H., Hadi, R. A., Ismail, I., & Baki, S. R. M. S. (2013). Energy efficiency benchmarking in UiTM engineering complex Shah Alam. In *2013 IEEE Conference on Systems, Process & Control (ICSPC)* (pp. 252–255), Kuala Lumpur.

A Compact SWB Antenna Using Parasitic Strip

Md. Moinul Islam, Md. Mehedi Hasan,
Mohammad Rashed Iqbal Faruque and Mohammad Tariqul Islam

Abstract A compact super-wideband (SWB) antenna has been presented based on parasitic strip in this paper. The radiating patch of the antenna is constructed by the circular disc fed with a microstrip line and parasitic strip on the ground plane. This antenna has been fabricated on 1.60-mm-thick epoxy resin fibre material where the total dimension of the antenna is 25 mm × 33 mm. High-frequency structural simulator has been adopted for performing all the simulations and analysis. The 17.10 GHz (from 2.90 to 20 GHz) impedance bandwidth is achieved in the measurements with the stable gain (3.2–6.22 dBi) through the proposed antenna.

Keywords Microstrip antenna · Parasitic strip · Super-wideband

1 Introduction

Due to the rapid growth in modern communication technology, the demand for antennas with wide bandwidth, miniaturized dimensions, high gain and high data rate transfer has significantly increased. The ultra-wideband technology (UWB) fulfilled these requirements in the short-range communication frequency spectrum. Super-wideband technology is defined as the technology having large

---

Md. Moinul Islam (✉)
Department of Software Engineering, Daffodil International University,
Dhaka 1207, Bangladesh
e-mail: mmoiislam@yahoo.com

Md. Mehedi Hasan · M. R. I. Faruque
Space Science Centre (ANGKASA), Institute of Climate Change, Universiti
Kebangsaan Malaysia, 43600 Bangi, Selangor Darul Ehsan, Malaysia

M. T. Islam
Department of Electrical, Electronics and Systems Engineering, Universiti
Kebangsaan Malaysia, 43600 Bangi, Selangor Darul Ehsan, Malaysia

bandwidth and high data rates. For covering both short- and long-range areas for communication, the bandwidth ratio should be at least 10:1. Antenna structures with a bandwidth ratio greater than or equal to 10:1 are termed as super-wideband antenna structures. The Federal Communications Commission has approved 3.1-10.6 GHz frequency range as a standard for short-distance unlicensed wireless communication (Islam et al. 2015a, b, 2016). UWB antennas play an important role in wireless communications due to having attractive properties such as low cost, miniaturization, low interferences, high transmission rate. It is not so easy to design a miniaturized antenna for operating UWB frequencies and also still a challenging now. Different types of method have been existing to design the UWB antennas. These methods include creating various types of slots on the radiating patch/ground plane, etching of split-ring resonators and use of metamaterial structure as radiating patch/ground plane/parasitic elements on the antenna structure (Hasan et al. 2016, 2017a, b, c). Besides, to achieve wideband performance, multiple antenna elements with dimensions corresponding to different frequencies are required. The antenna element size should be greater than or equal to one-quarter of the wavelength for efficient radiation. Few UWB antennas are reviewed in the study. In Barbarino and Consoli (2010), a circular-shaped asymmetrical dipole antenna was presented. This reported antenna attained an operating frequency band covering from 0.79 to 17.46 GHz including antenna size of 90×135 mm^2. A monopole antenna was proposed for SWB applications (Chen et al. 2011). This antenna covers the UWB demands, including gain and bandwidth. At lower frequencies (2–2.5 GHz), the input impedance is not matched properly and the dimension is too large, that is 35×77 mm^2. A SWB antenna was proposed with printed patch and tapered feed region in Manohar et al. (2014). The input impedance being mismatched at 18–19 GHz frequencies creates variant group delay, and the antenna structure is too large, 40×30 mm^2. A wideband antenna was reported for multiband applications (Liu et al. 2010). This antenna acquired wide bandwidth covering from 1:08 to 27:4 GHz, with a dimension of 124×120 mm^2. A monopole antenna of the compact disc was designed for future UWB applications where operating bandwidth is from 3:50 to 31:90 GHz with a dimension of 35×30 mm^2 (Srifi et al. 2011). Based on an iterative octagon, a fractal antenna covering super-wideband was mentioned (Azari 2011). This reported antenna attained an operating frequency band covering from 10 to 50 GHz including antenna size of 60×60 mm^2, which is not appropriate for C, S, UWB, L applications. Circular–hexagonal fractal super-wideband antenna was shown that covers bandwidth from 2.18 to 44.50 GHz (Dorostkar et al. 2013). A circular metallic patch has been used to construct the reported antenna with a transmission line. Super-wideband has been attained with the antenna size 31×45 mm^2. A circular patch antenna for SWB applications had been presented, which had an impedance bandwidth of 26.0 GHz (from 2.40 to 28.40 GHz) with a small dimension of 30×40 mm^2 (Mishra and Sahu 2016). A metamaterial structure-based Sierpinski fractal pedal antenna was introduced for super-high frequency (SHF), which covers the C, X, K and Ka bands. Moreover, the SHF antenna radiated at 10, 21.70, 24.50, 25.80 and 29.60 GHz, and the overall dimension of SHF antenna was 35 × 52 × 1.57 mm^3 (Khanum and Amit 2016). A CPW-fed octagonal super-wideband fractal antenna was presented that comprised four iterations of an octagonal slot-loaded octagonal radiating patch, coplanar waveguide feed line and modified ground plane loaded with a pair of rectangular notches. An impedance bandwidth of 3.80–68.0 GHz (179%) means 17.89:1 ratio bandwidth was achieved (Singhal and Singh 2016).

In this paper, a disc-shaped monopole antenna with a structure of the parasitic element accomplishes with a compact super-wideband profile physically to obtain nearly omnidirectional radiation characteristics, high gain, reasonable current distribution, and time domain performance. The mentioned SWB antenna is made of circular radiating patch and the ground plane containing the parasitic element on the upper portion, generating a super-wide bandwidth from 2.90 to more than 20.0 GHz. The parasitic element structure is inserted on the upper portion of the ground plane to generate super-frequency band for SWB applications. By virtue of significant selection of the parasitic element structure, it is observed that the reported antenna can obtain the operating SWB frequency band. The paper is oriented in this manner; design of the proposed antenna with the schematic top and bottom views as well as 3D view is in Sect. 2; besides, the parametric analysis by varying the effective parameters has been shown in this same section, results and discussion analysis are in Sects. 3 and 4 concludes this paper.

2 Antenna Design Architecture

Figure 1 displays the antenna structures with detailed configurations such as 3D view, front view and back view. This antenna has been fabricated on 1.60-mm-thick epoxy resin fibre material with a microstrip line feeding. The designed antenna consists of circular disc radiating patch, microstrip feed and parasitic strip on the ground plane. The parasitic element is implanted on the upper portion of the partial ground plane to create a super-frequency band and resonance frequencies. The parasitic element is made of four embedded rectangular strips, and the position of the parasitic element plays an important role to create more the capacitance effect in the antenna. If the capacitance of the antenna is raised, then the resonances are shifted towards the higher frequency bands and the gaps or slots between the parasitic elements make more resonance points in the frequency bands. In addition, this parasitic strip-inspired ground maintained the impedance bandwidth by creating additional surfaces of current paths in the designed antenna. The patch is connected to the feed line. A 50 Ω SMA connector is connected with the end of the feeding strip and the ground plane. Finite element method-based EM simulator HFSS is used for designing, simulation, and investigation purpose. The optimized values of the proposed SWB antenna are listed in Table 1.

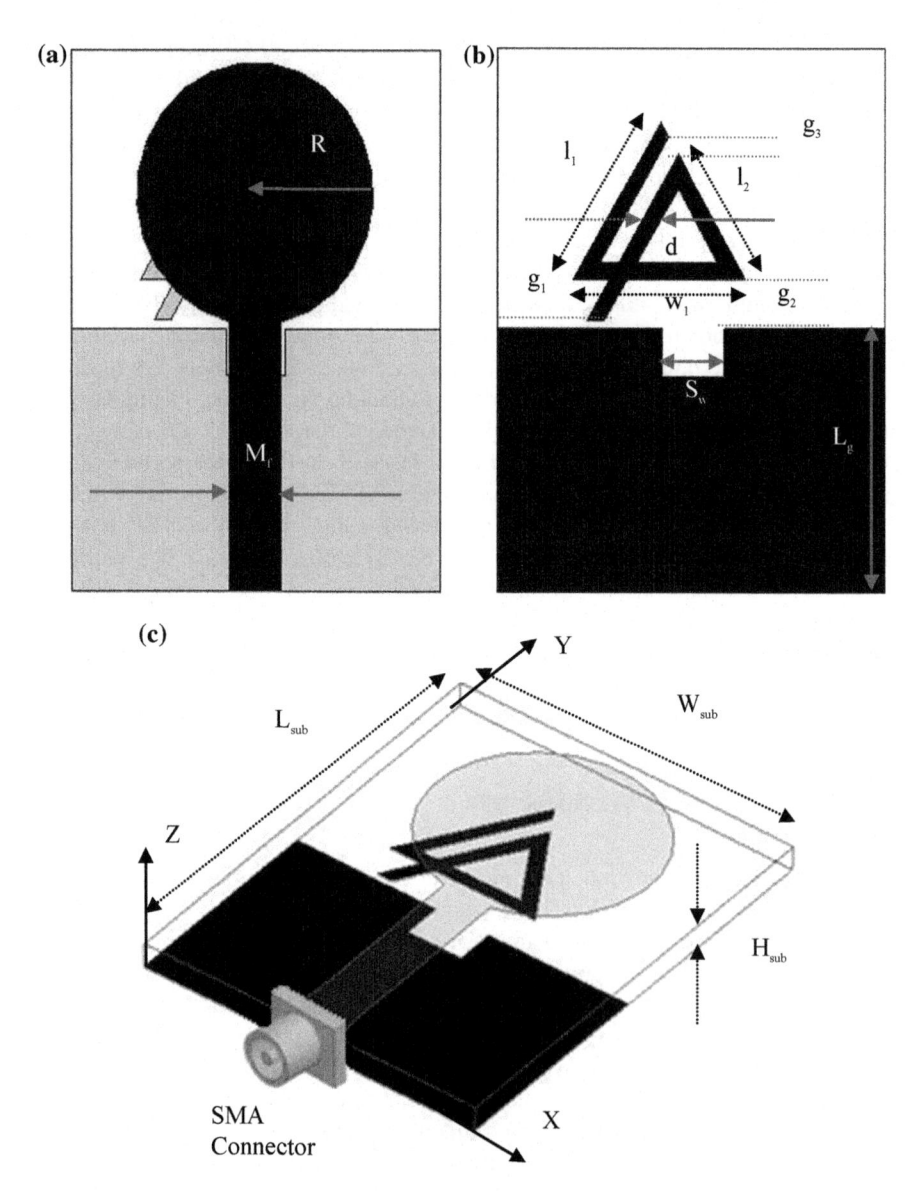

Fig. 1 Proposed antenna structure of **a** top view, **b** bottom view, **c** 3D view

The parametric analysis by varying the effective parameters has been performed. Figure 2 shows the effects of reflection coefficient on the thickness variation of parasitic stub, d. It can be realized from Fig. 2 that the proposed antenna with $d = 1.23$-mm-thick (optimum value) parasitic stub has covered the SWB band in terms of reflection coefficient. Figure 3 shows the effect of reflection coefficient on

Table 1 Optimized dimensions of the SWB antenna

Parameters	L_{sub}	W_{sub}	H_{sub}	l_1	l_2	g_1	g_2
Dimensions (mm)	25.0	33.0	1.60	11.0	8.70	0.50	3.0
Parameters	g_3	d	w_1	R	M_f	L_g	S_w
Dimensions (mm)	1.10	1.23	11.05	8.0	3.50	16.0	4.0

Fig. 2 Reflection coefficient on the thickness variation of parasitic stub, d

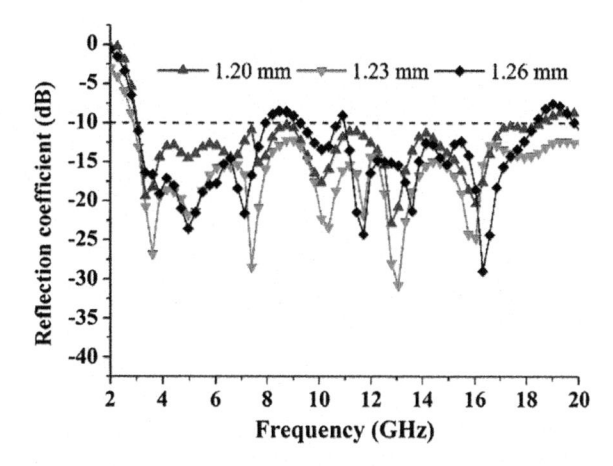

Fig. 3 Reflection coefficient on the width variation of microstrip feed, M_f

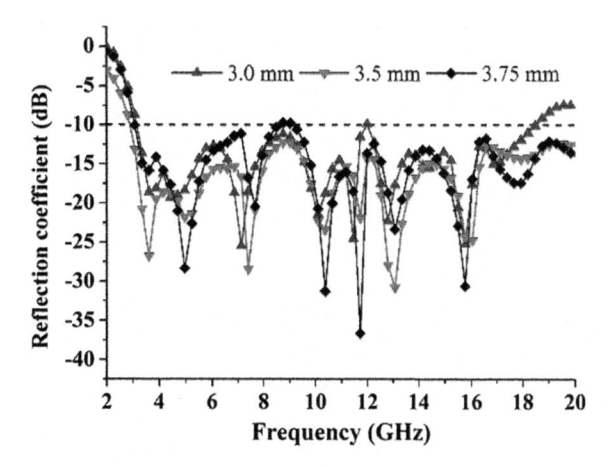

the width variation of microstrip feed, M_f. It is observed that the proposed antenna with M_f = 3.5-mm-thick (optimum value) feed width has covered the operating SWB band. Figure 4 exhibits the reflection coefficient on the radius variation of circular patch, R. It is shown in Fig. 4 that the patch radius = 8 mm provides the optimal reflection coefficient.

Fig. 4 Reflection coefficient on the radius variation of circular patch, *R*

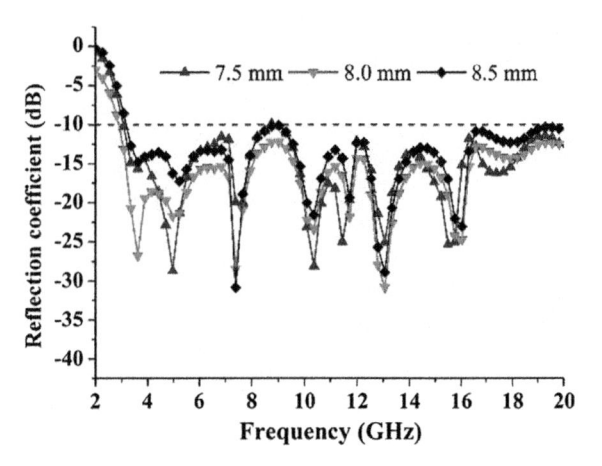

3 Results and Discussion

The performance characteristics of the proposed SWB antenna are explained, studied and optimized using HFSS. The simulated and measured reflection coefficient of the proposed SWB antenna is shown in Fig. 5. The measured results demonstrate that the designed line-fed antenna indicates a broadband impedance-matched properties and a coverage frequency area (2.90—more than 20 GHz). The little discordance between the measurement and simulation results is owing to fabrication tolerance, extended ground effect and the effect of the improper soldering of the SMA connector. However, the measured results are also almost coinciding with the expected results, so these results are adoptable. From Fig. 6, it is observed that the parasitic element on the upper portion of the ground plays an important role to create resonances and achieve super-frequency bands. The parasitic stub has a major effect at frequencies 3.50, 4.25 GHz on the ground plane. The resonances depend on the permittivity of the substrate material. If the

Fig. 5 Simulated and measured reflection coefficient of the designed SWB antenna

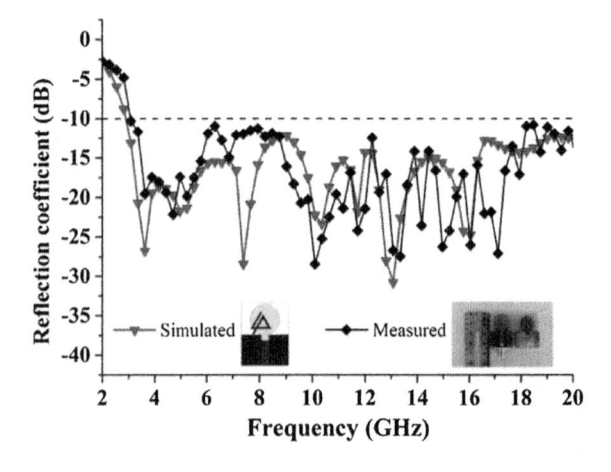

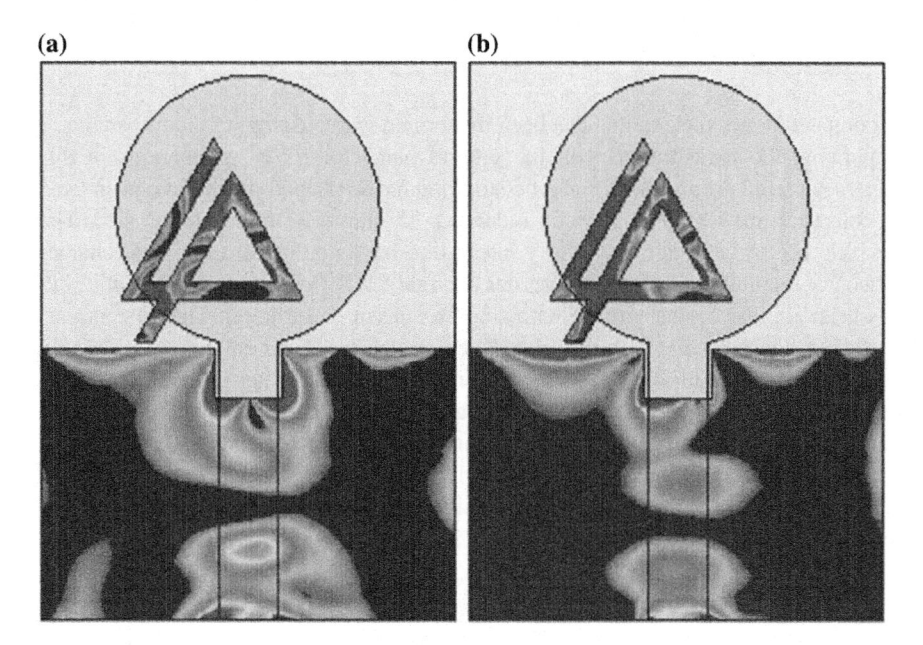

Fig. 6 Surface current distributions on the ground plane in frequencies **a** 3.50 GHz, **b** 4.25 GHz

permittivity is increased, then the resonance peaks are shifted towards the lower frequency. The shift of the resonance frequency can be explained by the overall capacitance changes of the resonator. Increasing the dielectric constant causes an increase for each capacitance values between the ground plane and resonator, where the capacitance has a major impact on the resonance frequencies. Further, the dependency of this parasitic stub has confirmed the properties of the designed SWB antenna. It can be observed clearly that the average gain of the SWB antenna is 3.78 dBi, where the maximum peak gain is 6.22 dBi, which is accepted for SWB applications from Fig. 7.

Fig. 7 Designed antenna with measured gain

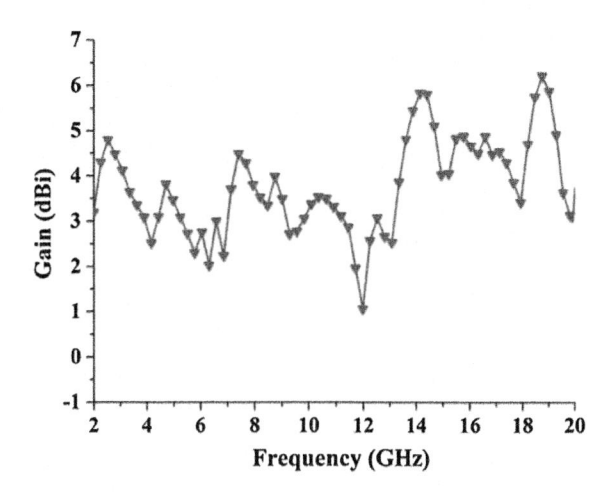

4 Conclusion

A compact monopole antenna has been mentioned with a design evolution analysis, implanting the parasitic strip on the ground plane for SWB applications in this paper. A partial ground with parasitic strip and a disc-shaped patch have been used to construct this designed antenna including 25 mm × 33 mm antenna size. The simple, ease of fabrication and easy integration have confirmed the planar characteristic of the designed antenna. In order to attain SWB frequency, the parasitic stub has been applied on the ground with smooth current distribution. The antenna has been matched properly with the impedance, which leads the antenna to obtain an impedance bandwidth of 17.10 GHz. Moreover, due to having sharp surface current, radiation characteristics, feeding characteristics, large bandwidth and better gain, the designed super-wideband antenna will be a promising one for various potential applications such as mobile application, wireless access system, satellite application, defence system, wideband high definition television, small aperture terminal/satellite news gathering.

References

Azari, A. (2011). A new super wideband fractal microstrip antenna. *IEEE Transactions on Antennas and Propagation, 59*, 1724–1727.

Barbarino, S., & Consoli, F. (2010). Study on super-wideband planar asymmetrical dipole antennas of circular shape. *IEEE Transactions on Antennas and Propagation, 58*, 4074–4078.

Chen, K. R., Sim, C., & Row, J. S. (2011). A compact monopole antenna for super wideband applications. *IEEE Antennas and Wireless Propagation Letters, 10*, 488–491.

Dorostkar, M. A., Islam, M. T., & Azim, R. (2013). Design of a novel super wide band circular-hexagonal fractal antenna. *Progress in Electromagnetics Research, 139*, 229–245.

Hasan, M. M., Faruque, M. R. I., & Islam, M. T. (2017a). Inverse E-shape chiral metamaterial for long distance telecommunication. *Microwave and Optical Technology Letters, 59*, 1772–1776.

Hasan, M. M., Faruque, M. R. I., & Islam, M. T. (2017b). Multiband left handed biaxial meta-atom at microwave frequency. *Materials Research Express, 4*, 035015.

Hasan, M. M., Faruque, M. R. I., & Islam, M. T. (2017c). A single layer negative index meta atom at microwave frequencies. *Microwave and Optical Technology Letters, 59*, 1450–1454.

Hasan, M. M., Faruque, M. R. I., Islam, S. S., & Islam, M. T. (2016). A new compact double-negative miniaturized metamaterial for wideband operation. *Materials, 9*(10), 830.

Islam, M. M., Islam, M. T., Faruque, M. R. I., Misran, N., Samsuzzaman, M., Hossain, M. I., et al. (2016). A compact disc-shaped super wideband patch antenna with a structure of parasitic element. *International Journal of Applied Electromagnetics and Mechanics, 50*, 11–28.

Islam, M. M., Islam, M. T., Samsuzzaman, M., & Faruque, M. R. I. (2015a). Compact metamaterial antenna for UWB applications. *Electronics Letters, 51*, 1222–1224.

Islam, M. T., Islam, M. M., Samsuzzaman, M., Faruque, M. R. I., & Misran, N. (2015b). A negative index metamaterial-inspired UWB antenna with an integration of complementary SRR and CLS unit cells for microwave imaging sensor applications. *Sensors, 15*, 11601–11627.

Khanum, T. F., & Amit, S. (2016). A compact wideband sierpinski aantenna loaded with metamaterial. In *International Conference on Electrical, Electronics, and Optimization Techniques* (pp. 3–5), India.

Liu, J., Esselle, K. P., & Zhong, S. S. (2010). A printed extremely wideband antenna for multi-band wireless systems. In *IEEE Antennas and Propagation Society International Symposium (APSURSI)* (pp. 11–17), Canada.

Manohar, M., Kshetrimayum, R. S., & Gogoi, A. K. (2014). Printed monopole antenna with tapered feed line, feed region and patch for super wideband applications. *IET Microwaves, Antennas and Propagation, 8,* 39–45.

Mishra, G., & Sahu, S. (2016). Compact circular patch antenna for SWB applications. In *International Conference on Communication and Signal Processing* (pp. 6–8), India.

Singhal, S., & Singh, S. A. (2016). CPW-fed octagonal super-wideband fractal antenna with defected ground structure. *IET Microwaves, Antennas and Propagation, 11,* 370–377.

Srifi, M. N., Podilchak, S. K., Essaaidi, M., & Antar, Y. M. (2011). Compact disc monopole antennas for current and future ultra-wideband (UWB) applications. *IEEE Transactions on Antennas and Propagation, 59,* 4470–4480.

A Multi-band Planar Double-Incidence Miniaturized Double-Negative Metamaterial

Mohammad Jakir Hossain, Mohammad Rashed Iqbal Faruque, Md. Jubaer Alam and Mohammad Tariqul Islam

Abstract In this research, a new planar double-negative (DNG) metamaterial for multi-band operation has been designed and simulated that can work in the microwave region. The proposed metamaterial structure offers DNG properties for normal and parallel incidence means two axes (z and x) electromagnetic (EM) wave propagation through the metamaterial structure. For along the z-axis EM wave propagation, metamaterial exhibits DNG characteristics in the C band. Similarly, for along the x-axis EM wave propagation, it displays DNG characteristics in the L band. Commercially available computer simulation technology electromagnetic simulator was adopted to observe the design of the metamaterial. The metamaterial has exhibited multi-band retaliation in coincidence with the DNG characteristics and miniaturization factor over the certain frequency bands in the microwave spectra.

Keywords Double-negative metamaterials · Double-incidence · Miniaturization Multi-band

1 Introduction

Prevalent research in the electromagnetic platform has been controlled recently with the target of building novel non-natural metamaterial structure or double-negative (DNG) materials at microwave bands. Governing the electromagnetic properties of materials, going beyond the border that is achievable with naturally prevailing substances, has become an authenticity with the advent of metamaterials (Veselago

---

M. J. Hossain (✉) · M. R. Iqbal Faruque · Md. J. Alam
Space Science Centre (ANGKASA), Institute of Climate Change,
Universiti Kebangsaan Malaysia, 43600 Bangi, Selangor, Malaysia
e-mail: jakir@siswa.ukm.edu.my

M. T. Islam
Department of Electrical, Electronics and Systems Engineering,
Universiti Kebangsaan Malaysia, 43600 Bangi, Selangor, Malaysia

© Springer Nature Singapore Pte Ltd. 2018
R. Saian and M. A. Abbas (eds.), *Proceedings of the Second International Conference on the Future of ASEAN (ICoFA) 2017 – Volume 2*,
https://doi.org/10.1007/978-981-10-8471-3_22

1968; Pendry et al. 1999). DNGs are engineered metamaterials with effective negative permittivity ($\varepsilon < 0$) and permeability ($\mu < 0$) simultaneously, which have exotic properties in terms of well-known phenomena like Snell's law, Cherenkov radiation, and Doppler Effect. (Veselago 1968; Shelby et al. 2001). Because of these exotic properties of the metamaterial, it can be used many vital applications such as electromagnetic absorption reduction, invisibility cloaking, absorber design. (Bowen et al. 2016; Kwak et al. 2017; Ni et al. 2015). In order to specific applications, many structures of metamaterials have been suggested in the literatures. On the other hand, metamaterial with double-incidence DNG properties is infrequently originated. A mono-axial-nested U-shape metamaterial has been utilized for C band and X band operations by Turkmen et al. (2012). They utilized multirings for finding multi-band application, but their metamaterial does not display DNG property. "Z" shaped metamaterial was proposed by Dhouibi et al. (2012) which was got working as a C band only with single negative characteristic for single-axis EM wave propagation. Moreover, recently, a structure of metamaterial was proposed in Islam et al. (2016) for S-band operations; on the contrary, it was presented ε-negative property for the EM wave propagation along x-axis. Likewise, there are some DNG metamaterials found in the literature, namely Karamanos et al. (2012) that suggested a dual-frequency mono-axial DNG metamaterial in C band; however, they utilized two different unit cell structures on the opposite sides of the substrate. A multi-band negative refractive index metamaterial was suggested in Islam et al. (2016), in contrast, the negative refractive index property of the EM wave propagation along z-axis only. In addition, Hossain et al. (2017a, b) claimed a DNG multi-band metamaterial; on the other hand, they used EM wave propagation along z-axis. The miniaturization of the metamaterial is directed by improved effective medium ratio (EMR).

In this research, a new double Π-U-shaped metamaterial has been designed that exhibits DNG properties for the EM wave propagation along z- and x-axis in the microwave spectra. The proposed metamaterial structure demonstrates DNG properties in L, and C band applications with better EMR property of EM wave propagation along x-, and z-axis, respectively. The construction of the metamaterial size is being miniaturized and follows the superior EMR. To determine the scattering parameters, namely the reflection coefficient (S_{11}) and transmission coefficient (S_{21}), the commercially available CST electromagnetic simulator 2015 was used.

2 Methodology

2.1 Design of Double-Negative Metamaterial

The DNG metamaterial structure consists of two layers. The front side and back side of a dielectric slab are a planar double Π-U-shaped copper metal structure and no copper, respectively. A combination of double Π- and double U-shaped split ring resonators with square split ring resonator was utilized to achieve unconventional characteristics of metamaterials that were usually not found in nature. The proposed metamaterial structure and design parameters are shown in Fig. 1a. The dimension of the substrate is $10.5 \times 11 \times 1.6$ mm³ where FR4 lossy material is used as substrate material. All elements of the resonators are made of copper with conductivity of 5.8×10^7 S/m, rho of 8930 kg/m³, Young's modulus of 120 KN/mm², and the thickness of copper resonators are 0.035 mm that is printed on a substrate with standard permittivity, $\varepsilon = 4.3$, El. tand, $\delta = 0.025$ as well as permeability, $\mu = 1$. The parameters of the structure are $L = 10$ mm, $W = 10.6$ mm, $W_1 = 0.4$ mm, $W_2 = 0.65$ mm, $W_3 = 0.7$ mm, $W_4 = 0.6$ mm, $W_5 = 0.7$ mm, $W_6 = 0.9$ mm, $W_7 = 0.6$ mm, $W_8 = 0.7$ mm, $L_1 = 6.7$ mm, $g = 0.3$ mm, and $s = 0.4$ mm.

In this paper, the finite integration method-based CST simulator is adopted to examine this design structure. To examine the double-incidence operation of the metamaterial, primarily, the z-axis EM wave propagation is performed, and then the x-axis EM wave propagation has been executed. The electric field and magnetic field have been polarized along the x-axis and the y-axis, respectively, whereas z-axis has been utilized for electromagnetic wave traveling. The boundary conditions of perfect magnetic conductor (PEC) and the perfect electric conductor (PMC) are utilized along the x-axis and y-axis, individually, and two waveguide ports are placed on the positive and negative z-axis. The schematic diagram of the proposed design is illustrated in Fig. 1, and a simulation setup is shown in Fig. 2a. To determine the transmission coefficient and the reflection coefficient in simulation, a frequency domain solver is utilized and 1001 frequency samples have been reserved. The frequency range between 1 and 15 GHz was used to simulate the design of metamaterial.

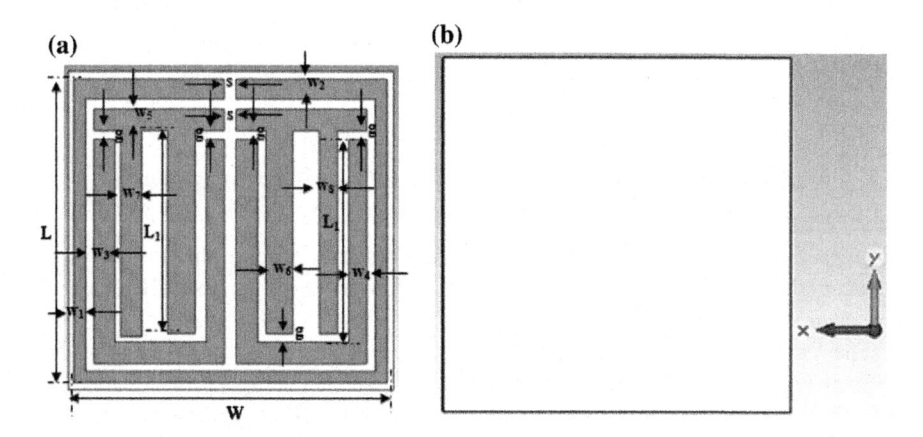

Fig. 1 Proposed metamaterial structure **a** front side **b** back side

(a) (b)

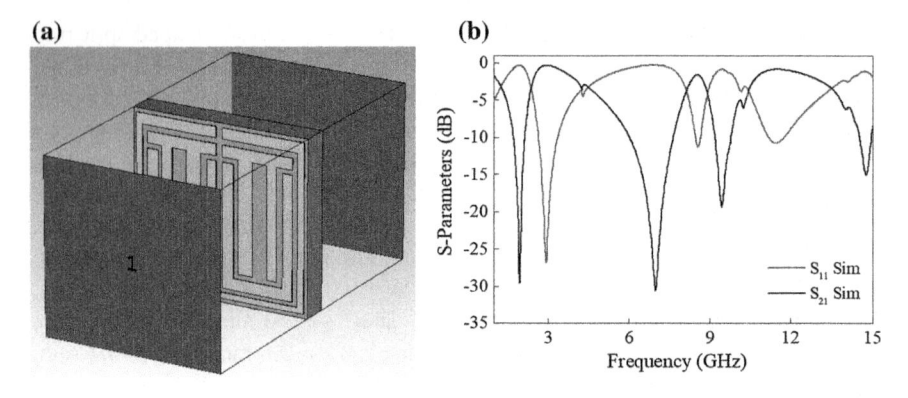

Fig. 2 a Simulation setup b S-parameters along z-axis propagation of the proposed structure

2.2 Effective Scattering and Medium Parameters Calculation

The Nicolson-Rose-Weir (NRW) method is utilized to determine the medium parameters like effective permeability (μ) and permittivity (ε) from simulated scattering parameters such as transmission coefficient (S_{21}) and reflection coefficient (S_{11}). The direct refractive index method is applied to calculate the effective refractive index (n) from the simulated complex S-parameters (Hossain et al. 2017a, b).

3 Results and Discussion

The real and imaginary both values of the permittivity, permeability, and refractive index are justified to characterize the proposed metamaterial. In this paper, unit cell with different types of EM wave propagation such as along z-axis and x-axis have been analyzed.

3.1 Normal Incidence of the EM Wave (z-axis Propagation)

Propagation along z-axis analysis has been performed by placing the metamaterial between the two waveguide ports at the z-axis to confirm z-axis EM wave propagation.

The PEC and PMC boundary conditions were used in the x-axis and y-axis, respectively. The simulation result of double Π-U-shaped compact metamaterial is offered. The simulated reflection coefficient (S_{11}) and transmission coefficient (S_{21}) of the unit cells are demonstrated in Fig. 2. Figure 2b illustrates the mathematical values of the four frequency ranges of resonance frequencies such as 1.73–2.13,

6.03–7.64, 9.19–9.76, and 14.52–14.94 GHz that designates L, C, X and Ku bands applications.

Figure 3 exhibits the electric and magnetic field distributions of the proposed metamaterial structure. Electric field and magnetic field perform simultaneously, but orthogonally. In addition, the electric field is very coherent with the magnetic field, that is why, at the positions where the electric field is low; the magnetic field is high and vice versa at the resonance frequency 6.88 GHz where it disclosures double-negative properties.

Figure 4a, b reveals the retrieval effective permeability, $\mu = \mu' + i\mu''$, and effective permittivity, $\varepsilon = \varepsilon' + i\varepsilon''$, respectively. Figure 4a displays the real values of the relative negative permittivity 1.69–2.50, 4.07–6.89, 8.45–9.08, 10.34–12.47, and 14.52–15.00 GHz for the propagation direction along the z-axis of structure of metamaterial. The electric and magnetic dipole moment can keep route with the applied electromagnetic field at lower frequency.

In contrast, the dipole moment cannot content with the applied field. As a consequence, the fluctuation of scattering parameters has been occurred at higher frequencies. Figure 4b interprets the real values of the relative negative permeability 6.80–13.87 GHz, for the propagation direction along the z-axis of design structure of metamaterial. The negative properties of the permittivity and permeability have changed a little bit owing to the polarization effect on the interior construction of metamaterial structures. Anisotropic materials contain unequal lattice structure that manipulates the applied electromagnetic field due to the polarization of dipole moment inside materials. Hence, the change of design structure also changes the properties of the effective permittivity and permeability of the metamaterials. Figure 4c exhibits $n = n' + in''$ that indicates the retrieval effective refractive index. The magnitude of the real values of the relative negative refractive index 6.82–7.19, 8.39–9.76, and 10.02–13.95 GHz of the design structure of metamaterial is shown in Fig. 4c.

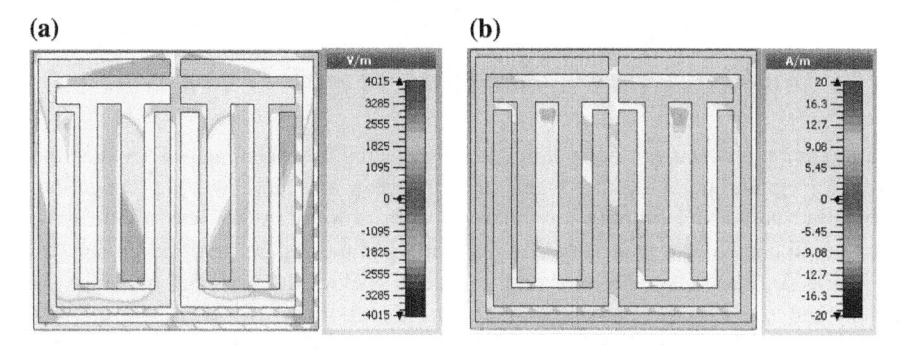

Fig. 3 a Electric field b magnetic field along z-axis propagation at resonance frequency of 6.88 GHz of the proposed structure

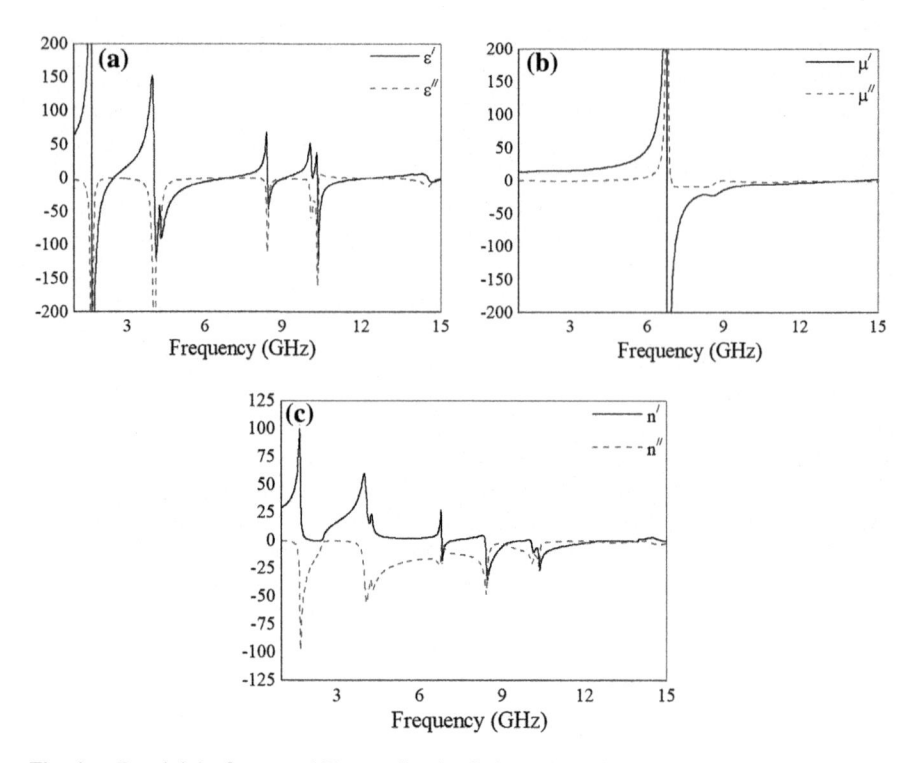

Fig. 4 a Permittivity b permeability c refractive index values along

The curves of the effective refractive index become negative when the curves of the permittivity and permeability are negative, simultaneously. The design structure of metamaterial has shown DNG properties at 6.88 GHz because the permittivity, permeability and refractive index were negative at that point simultaneously.

3.2 Parallel Incidence of the EM Wave (x-axis Propagation)

Furthermore, analysis has been completed by placing the metamaterial between the two waveguide ports at the x-axis to approve x-axis EM wave propagation. The PEC and PMC boundary conditions were utilized in the rest of the axis. The amplitudes of the real values of scattering parameters are revealed in Fig. 5b. Figure 5 depicts the simulation setup, reflection coefficient (S_{11}), and transmission coefficient (S_{21}) of the metamaterial for the direction of EM wave propagation along the x-axis. Figure 5b demonstrates the mathematical values of the peak resonances at the frequencies such as 1.40 GHz at −35.84 dB, 5.38 GHz at −21.11 dB, 9.29 GHz at −43.42 dB, and 12.72 GHz at −13.50 dB that designates L, C, X, and Ku bands applications.

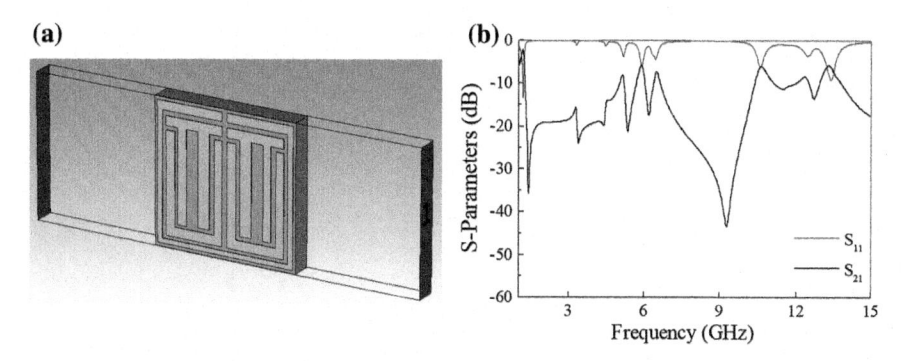

Fig. 5 a Simulation setup **b** S-parameters along *x*-axis propagation of the proposed structure

The electric and magnetic field distributions of the proposed metamaterial structure are shown in Fig. 6. The electric field distribution and magnetic field distribution at 1.34 GHz where it exposes double-negative properties are shown in Fig. 6. The electric field distribution and magnetic field distribution have changed owing to the previously referred reason. The real curves of retrieval effective permittivity of the metamaterial for the propagation direction of the *x*-axis are depicted in Fig. 7a. From Fig. 7a, a large range of frequency from 1.22 to 7.38 GHz of negative magnitude has been obtained. Moreover, it also reveals a negative value from 1.22 to 7.38 GHz, and 12.98 to 13.38 GHz, in the microwave spectra. Similarly, from Fig. 7b, it has been observed that a negative peak of the effective permeability is 1.18–1.34, 5.75–5.93, 7.72–10.59, and 10.62–14.69 GHz. Hence, it seems that the metamaterial exhibits negative amplitude for both permittivity and permeability at the resonance of 1.34 GHz in the microwave region. Generally, the properties of permittivity and permeability are affected by the polarization because

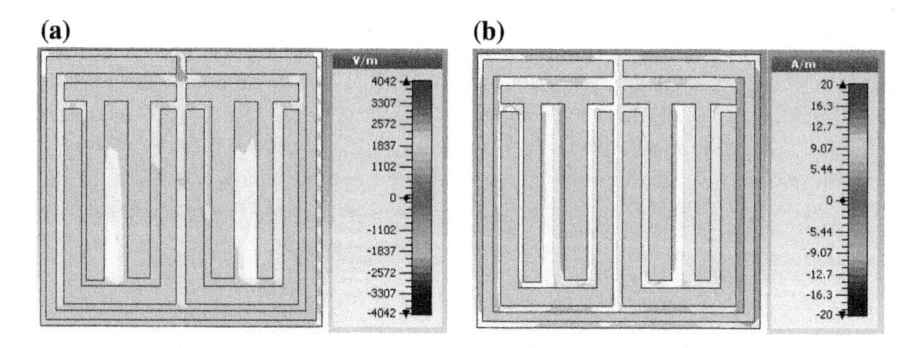

Fig. 6 a Electric field **b** magnetic field along *x*-axis propagation at resonance frequency of 1.34 GHz of the proposed structure

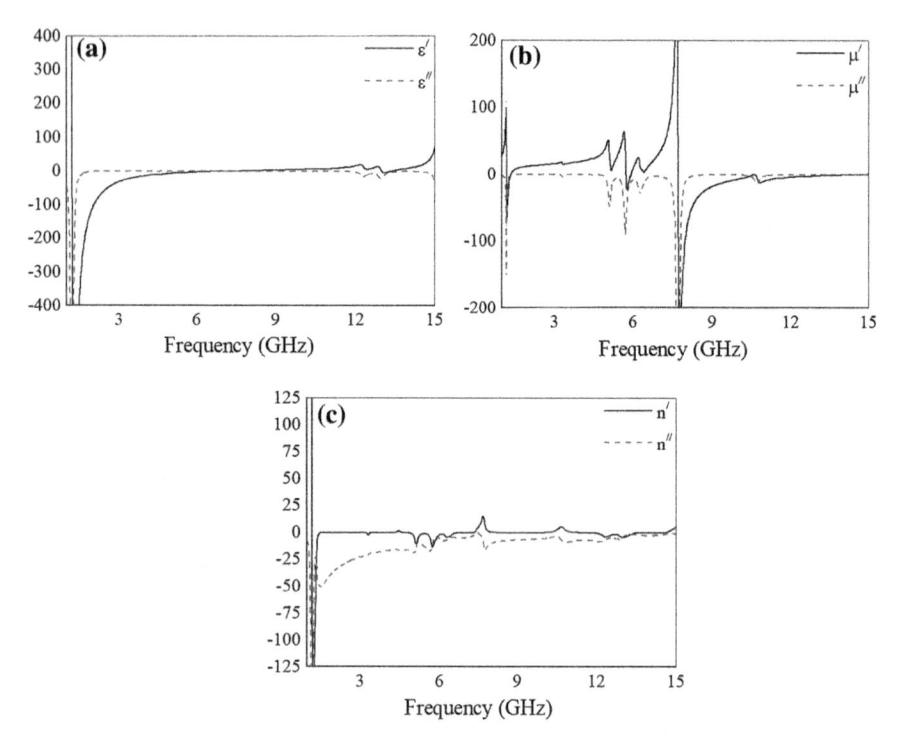

Fig. 7 a Permittivity **b** permeability **c** refractive index values along x-axis propagation of metamaterial

of the internal structure of the metamaterial. When EM waves pass in anisotropic materials, which contain unequal lattice axes that is affected by the polarization inside the material. Therefore, the refractive index curve of the material is also affected by the polarization like as permittivity and permeability. The real amplitude of refractive index is shown in Fig. 7c for x-axis EM wave propagation, where peak resonances are observed in the L, S, C and X band with a value of refractive index n = −38.30, n = −0.68, n = −1.23 and n = −0.025, respectively, at the frequencies of 1.34, 3.38, 5.38, and 11.69 GHz of the microwave region.

Table 1 demonstrates the comparisons of the covered bands and EMR of the proposed design with EM wave propagation for different axes. The proposed design

Table 1 Comparison of EMR with EM wave propagation for different axes

Direction of EM wave propagation	First resonance frequency (GHz)	Covered band	Effective medium ratio (λ/a)
Along z-axis	1.94	L, C, X and Ku band	14.06
Along x-axis	1.38	L, S, C and X band	19.62

illustrates different EMRs for different directions of EM wave propagation. EM wave propagation along the *x*-axis shows the most compactness compare to the *z*-axis propagation.

4 Conclusion

A new design of the compact DNG metamaterial structure was proposed for multi-band applications in this research. This design has exhibited higher EMR with DNG characteristics along the parallel and normal incidence of EM wave propagation. The finite integration-based CST tool was applied to determine the metamaterials properties. A comparative analysis like EMR was also carried out consistent with EM wave propagation along the parallel and normal incidence of EM. The metamaterial structure was compact in dimension and follows improved EMR which were more suitable in microwave regime. Moreover, the proposed metamaterials were applicable for military telemetry, GPS, GSM, long-distance radio telecommunications, and satellite communications.

Acknowledgements This work was supported by the Ministry of Education (MOE) under Fundamental Research Grant Scheme Top Down, Code: FRGS TOP DOWN/2014/TK03/UKM/ 01/1 and Research Universiti Grant, Dana Impak Perdana Code: DIP-2015-014.

References

Bowen, P. T., Baron, A., & Smith, D. R. (2016). Theory of patch-antenna metamaterial perfect absorbers. *Physical Review A, 93,* 063849.
Dhouibi, A., Burokur, S. N., De Lustrac, A., & Priou, A. (2012). Study and analysis of an electric z-shaped metamaterial. *Advanced Electromagnetics, 1,* 64–70.
Hossain, M. J., Faruque, M. R. I., & Islam, M. T. (2017a). Design and analysis of a new composite double negative metamaterial for multi-band communication. *Current Applied Physics, 17,* 931–939.
Hossain, M. J., Faruque, M. R. I., & Islam, M. T. (2017b). An effective medium ratio following miniaturized concentric meta-atom for S- and C-band Applications. *Microwave Opt Technol Lett, 59,* 1233–1240.
Islam, S. S., Faruque, M. R. I., Hossain, M. J., & Islam, M. T. (2016). A new wideband negative-refractive-index metamaterial. *Materiali in Tehnologije, 50,* 873–877.
Karamanos, T. D., Dimitriadis, A. I., & Kantartzis, N. V. (2012). Compact double-negative metamaterials based on electric and magnetic resonators. *IEEE Antennas and Wireless Propagation Letters, 11,* 480–483.
Kwak, S., Sim, D.-U., Kwon, J. H., & Yoon, Y. J. (2017). Design of PIFA with metamaterials for body-SAR reduction in wearable applications. *IEEE Transactions on Electromagnetic Compatibility, 59,* 297–300.
Ni, X., Wong, Z. J., Mrejen, M., Wang, Y., & Zhang, X. (2015). An ultrathin invisibility skin cloak for visible light. *Science, 349,* 1310–1314.
Pendry, J. B., Holden, A. J., Robbins, D., & Stewart, W. (1999). Magnetism from conductors and enhanced nonlinear phenomena. *IEEE Trans Microwave Theory Tech, 47,* 2075–2084.

Shelby, R. A., Smith, D. R., & Schultz, S. (2001). Experimental verification of a negative index of refraction. *Science, 292,* 77–79.

Turkmen, O., Ekmekci, E., & Turhan-Sayan, G. (2012). Nested u-ring resonators: A novel multi-band metamaterial design in microwave region. *IET Microwaves, Antennas and Propagation, 6,* 1102–1108.

Veselago, V. G. (1968). The electrodynamics of substances with simultaneously negative values of ε and μ. *Sov Phys, 10,* 509–514.

Depiction of a Combined Split P-Shaped Compact Metamaterial for Dual-Band Microwave Application

Md. Jubaer Alam, Mohammad Rashed Iqbal Faruque,
Mohammad Jakir Hossain, Mohammad Tariqul Isla
and Sabirin Abdullah

Abstract The paper imparts the architecture of a compact and combined split P-shaped metamaterial unit cell that is felicitous for dual band of microwave frequency, and it shows negative permeability and permittivity at those frequencies. Two split P-shaped resonators are connected to each other by a metal ink to increase the electrical length, and two modified split-ring resonators (SRRs) reside inside each of it. A correlation was made on the performance after the analysis of unit cell and 1 × 1 array structure. A great transmission coefficient of almost 13 GHz with a 500 MHz band gap at the middle is demonstrated for all of these configurations. The resonator covers L, C, X and Ku band separately with double-negative phenomena at X and Ku band. To justify the performance of the proposed resonator, an analogy is conferred. Due to its compact and auspicious design, double-negative characteristics and the proposed metamaterial have potential to be used for dual-band application.

Keywords Metamaterial · Double negative · Dual band

1 Introduction

Metamaterials are the special type of materials that are usually not available in nature. They are actually engineered materials; they need embedding periodic unit cell for their formation to create naturally unavailable electromagnetic properties. The size of their periodically organized metallic components is smaller than the wavelength. Moreover, these materials have the power to control the electromag-

---

Md. J. Alam (✉) · M. R. I. Faruque · M. J. Hossain · S. Abdullah
Space Science Centre (ANGKASA), Institute of Climate Change, Universiti Kebangsaan Malaysia, 43600 Bangi, Selangor Darul Ehsan, Malaysia
e-mail: jubaer.alam@iubat.edu

M. T. Isla
Department of Electric, Electronics and Systems Engineering,
Universiti Kebangsaan Malaysia, 43600 Bangi, Malaysia

netic wave beams to show their unorthodox characteristics than conventional. These unusual features of the metamaterials totally depend on the geometry of the atomic construction. It has been started from the year 1968; observed unique properties of materials having negative permittivity (ε) and permeability (μ). But it was not appreciated until 2000 when Smith et al. fortunately validated a new unreal with these unconventional properties (both permittivity and permeability were negative) is called left-handed metamaterial. In case of negativity, it has been categorized as single-negative (either permittivity is negative or permeability is negative), double-negative (both permittivity and permeability are negative). There is also a term called near-zero refractive index metamaterial (NZRI) where the permittivity and permeability of a material become approximately to zero of a particular range of frequency. Having these captivating electromagnetic phenomena, necessary applications, like SAR reduction (Faruque et al. 2012; Islam et al. 2015a, b), super lenses, antenna design (Khan et al. 2014), filters (Singh et al. 2013), and invisibility cloaking (Islam et al. 2015a, b). Currently, multi-band metamaterial absorbers have become an auspicious application in explosives detection, bolometers, and thermal detectors. Moreover, a very few studies have been made on designing this type of materials (Shen et al. 2011). Different alphabetic shapes have become popular for particular operations; like, Benosman et al. (2012) introduced a double S-shaped metamaterial that showed negative values of η from 15.67 to 17.43 GHz. Mallik et al. proposed various U-shaped rectangular array structures left-handed aspect at approximately 5, 6, and 11 GHz. A V-shaped metamaterial was presented by Ekmekci et al., the architecture showed double-negative characteristic. Zhou and Yang (2015) designed an S-shaped 15 × 15 mm^2 chiral metamaterial for X and Ku band application. Though the EMR was not higher than four. For the purpose of application on S and C bands, Hossain et al. (2015) design G-shaped DNG for different unit cells and array sizes.

A metamaterial unit cell of combined split P-shape has been proposed in this paper which has two SRR in the center of each of the P-shaped resonators. The structure covers multiple bands (L, C, X, and Ku) of frequencies for the transmission coefficient (Islam et al. 2014). And for effective parameters, it covers the X and Ku bands with double-negative characteristic. With the help of commercially available Computer Simulation Technique (CST) Microwave Studio, an observation is made on the reflection (S_{11}) and transmission coefficient (S_{21}) of the unit cell. The results of S_{11} and S_{21} are applied to obtain ε and μ for the unit cell and single-array structure.

2 Methodology

The diagram of the prospective combined P-shaped unit cell composition is itemized in Fig. 1. The structure contains two SRRs of copper inside each of the P-shaped resonators. Each unit cell comprises with 10 mm in length and 10 mm in width. All elements have the thickness of 0.35 mm. Each P-shaped split resonator

Depiction of a Combined Split P-Shaped Compact Metamaterial … 237

Fig. 1 Geometry of metamaterial unit cell

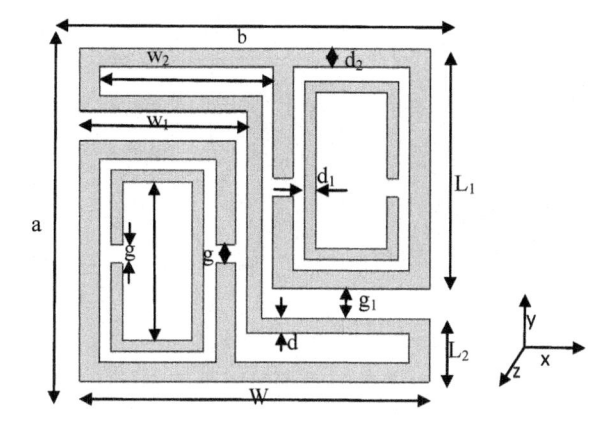

Table 1 Parameters of the unit cell

Parameter	Optimum value (mm)	Parameter	Optimum value (mm)
a	10	b	10
W	9	g_1	0.8
g	0.4	d	0.4
w_1	4.3	w_2	4.5
d_1	0.3	d_2	0.5
L_1	6.5	L_2	1.7

has the width of 0.5 mm with a same split gap. The outer length of the resonator is 9 mm where the distance between each of the resonators and the connected line is 0.5 mm. The entire patch (made of copper) is developed on a substrate called FR-4. It has a dielectric constant of $\varepsilon_r = 4.3$, a dielectric loss tangent of $\tan\delta_\varepsilon = 0.025$. Sides of the substrate are $a = b = 10$ mm, and the thickness is $t = 1.6$ mm. Designed parameters of the proposed metamaterial are enlisted in Table 1.

CST Microwave Studio is used to get the result of S_{11} and S_{21} with the help of hundred frequency samples. Two waveguide ports are used to propagate the electromagnetic waves to excite the configuration on two opposite directions of z-axis. PEC and PMC were used along the vertical direction of x- and y-axis, respectively. And for the free-space simulation purposes, a frequency domain solver was utilized. Moreover, for the analysis purpose of these configurations, a tetrahedral mesh was used with a flexible mesh. The normalized impedance was 50 Ω, and the system was performed from 1 to 18 GHz.

To differentiate the effective permittivity (ε_r) and permeability (μ_r) with S_{11} and S_{21}, the NRI method is applied. To such a degree the ε_r and μ_r can be determined by

(a) **(b)**

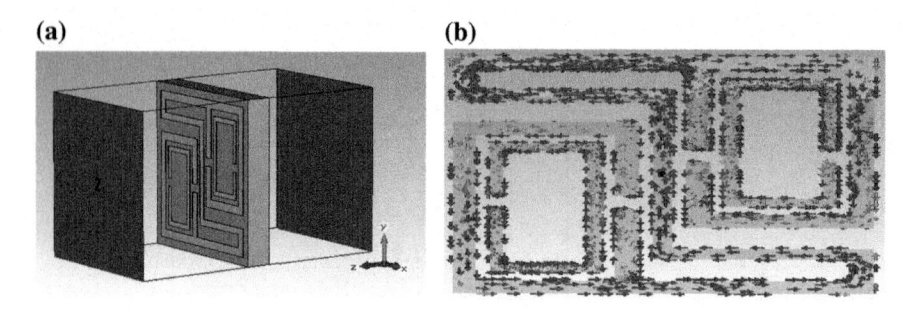

Fig. 2 a Simulation setup of the proposed unit cell; **b** current distribution of the unit cell at 1.63 GHz

$$\varepsilon_r = \frac{c}{j\pi fd} \times \frac{(1-V_1)}{(1+V_1)} \qquad (1)$$

$$\mu_r = \frac{c}{j\pi fd} \times \frac{(1-V_2)}{(1+V_2)} \qquad (2)$$

The effective refractive index (η_r) can also be calculated from S_{21} and S_{11} (Islam et al. 2014):

$$(\eta_r) = \frac{c}{j\pi fd} \times \sqrt{\frac{(S_{22}-1)^2 - S_{11}^2}{(S_{22}+1)^2 - S_{11}^2}} \qquad (3)$$

By settling the perspective unit cell in between, the waveguides as per Fig. 2a to determine the scattering parameters accurately of the combined P-shaped split metamaterial.

3 Results and Discussions

There are plenty of ways to find out the effective parameters of a unit cell like NRW strategy, DRI. The main concern of this paper is to highlight the electromagnetic properties using the real values of ε, μ, and η using S_{11} and S_{21}.

3.1 Analysis of Unit Cell

As the unit cell is built on Fr-4 substrate with an area of 100 mm^2, it has been consistent within a frequency ranging from 1 to 18 GHz. The simulation was performed by CST MWS to get the result of the transmission coefficient (S_{21}). The transmission coefficient exhibits a wide band with a coverage of L, C, X and

Ku band. first resonance is found in the L band at frequency 1.63 GHz. Then a wide band from 4.68 to 17.18 GHz with a little band gap of 500 MHz.

However, the optimized resonance frequency is 8.3 GHz. Figure 2b shows the current distribution of the unit cell at 1.63 GHz. Due to the antithetical geometry of the structure, a reverse current flow is noticed in the metal fillet of the configuration. Moreover, opposite current flows through the inner and outer surfaces of the resonator create the stop band at this frequency (Hasan et al. 2016). Figure 3a shows magnitude of the transmission coefficient (S_{21}).

By using S_{21} and S_{11} parameters, the effective parameters, i.e., effective permeability and effective permittivity can be obtained (Luukkonen et al. 2011). Figure 3b, c shows the result of effective permittivity and effective permeability, respectively.

Figure 3b shows negative permittivity at resonating points. It shows negativity at 1.6–1.75, 3.9–7.5, 8.3–9.5, 11.88–13.46, and 15.53–17.85 GHz. Figure 3c shows the negative permeability at 7.75–12, 12.15–15.53, and 16.98–17.81 GHz. At lower frequencies, the current flow matches with the applied field. But in case of higher frequencies, it is not possible for the current to cope up with the applied field when the permeability becomes negative. In the gap, there is a charge produced of a SRR is regulated to a fluctuating magnetic field. At low frequency, the current

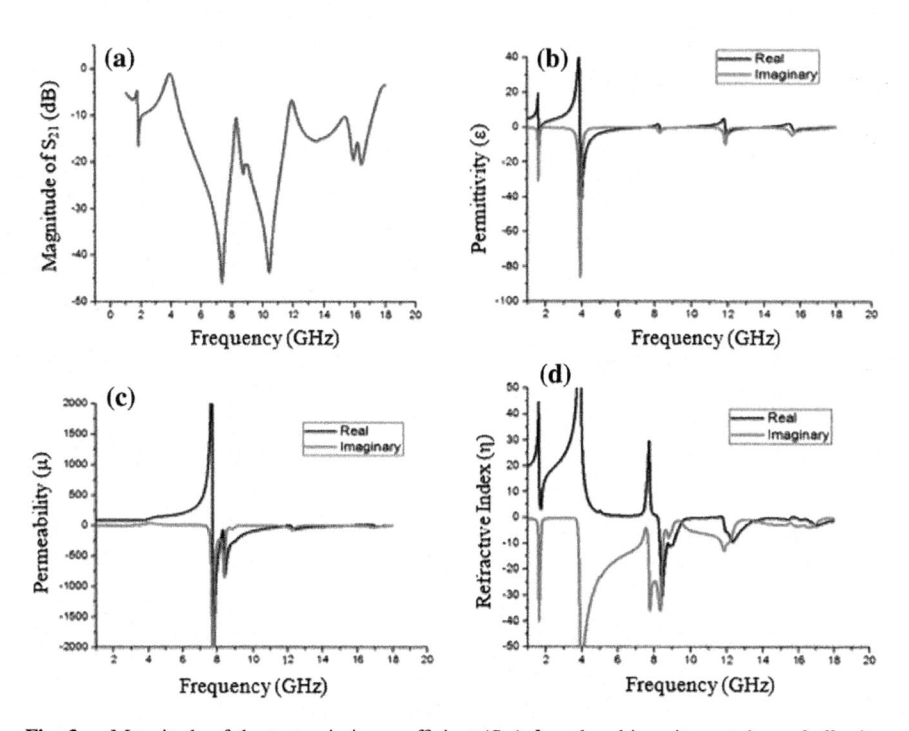

Fig. 3 a Magnitude of the transmission coefficient (S_{21}); **b** real and imaginary values of effective permittivity (ε) versus frequency; **c** real and imaginary values of effective permeability (μ) versus frequency; **d** real and imaginary values of refractive index (η) versus frequency

Table 2 Frequency range of effective parameters

Effective parameters	Frequency range (GHz)	Covered bands	Values (dB) at 8.3 and 17 GHz
Permittivity (ε)	1.6–1.75, 3.9–7.5, 8.3–9.5, 11.88–13.46, 15.53–17.85	L, C, X and Ku	−1.06 & −0.57
Permeability (μ)	7.75–12, 12.15–15.53, 16.98–17.81	X and Ku	−80.56 & −3.45
Refractive index (η)	8.3–11.37, 11.8–15.55, 15.82–18	X and Ku	−9.82 & −3.11

remains in phase with the applied field, but it fails to remain in phase in higher frequencies, and as a result, negative permeability produces.

In Fig. 3d, real and imaginary parts of η are plotted as a function of frequency. The curve shows negativity at 8.3–11.37, 11.8–15.55, and 15.82–18 GHz. Table 2 shows the frequency range of refractive indices with effective parameters of the unit cell at different resonating frequency bands. The refractive index shows negativity when the permittivity and permeability become negative. Here η shows certain negativity at different bands of frequencies. Hence, the designed unit cell has significant portions where all the three effective parameters become negative. Therefore, this unit cell can be claimed as double-negative metamaterial as it has negative peaks at 8.3 and 17 GHz in all the three effective parameters which is shown in Table 2 with bandwidths.

3.2 (1 × 1) Array Analysis

Figure 4 represents the array formation of the unit structure which is placed vertically on the basic unit structure on the same Fr-4 substrate. The array structure is analyzed within the frequency range of 1–18 GHz. Both the units are placed 0.5 mm apart from each other on the substrate, and the similar approach was used to assess the attainment of the array.

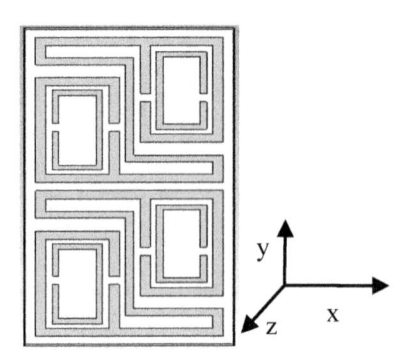

Fig. 4 Array formation of the unit structure

Figure 5a shows the transmission coefficient of the 1 × 1 array. It is apparent that the resonances of the frequencies are found at the same points as the unit cell, but having greater negative magnitudes. The S_{21} improves a bit as there is no band bap in between 4.68 and 17.18 GHz.

Figure 5b, c shows the real value of the ε and μ as a function of frequency of the 1 × 1 array. From Fig. 5b, c, it is observed that the negative values for the unit cell and the array are almost in similar positions. The differences among them are the amplitudes or magnitudes. The negative magnitude decreases in case of permittivity. But in case of permeability, the negative magnitude increases at resonating points. Figure 5d shows the real values of the refractive index, and only the negative quatities are counted for this parameter.

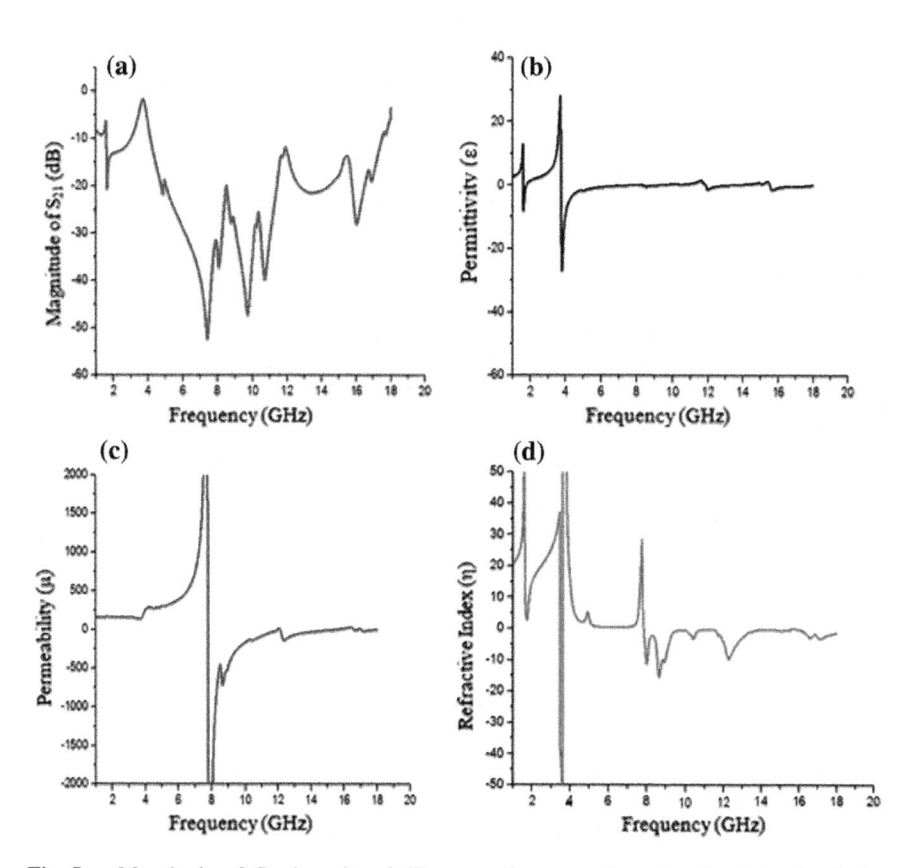

Fig. 5 **a** Magnitude of S_{21} in units of dB versus frequency, **b** real and values of effective permittivity (ε) versus frequency; **c** real values of effective permeability (μ) versus frequency; **d** real and values of refractive index (η) versus frequency for the 1 × 1 array of the unit structure

Table 3 Frequency range of effective parameters for 1 × 1 array structure

Effective parameters	Frequency range (GHz)	Covered bands	Values at 8.3 and 17 GHz
Permittivity (ε)	1.6–1.765, 3.75–7.5, 8.3–9.25, 11.86–13.34, 15.51–17.95	L, C, X and Ku	−0.11 & −0.347
Permeability (μ)	7.76–11.95, 12.13–15.15, 16.58–16.75, 16.98–17.9	X and Ku	−793.16 & −0.20
Refractive index (η)	3.46–3.58, 7.83–15.56, 15.7–18	X and Ku	−2.22 & −2.83

There is a dissimilarity in the array parameter than the unit cell. A small spike of negative refractive index is found at 3.5 GHz which was not present in the structure. All the effective parameters of this 1 × 1 array are summarized in Table 3. Still the array shows double-negative characteristics at 8.3 and 17 GHz.

4 Comparative Analyses of the Configurations

In this paper, total observation is made on S-parameters, effective permittivity, effective permeability, and refractive index. All the results have shown unique but not contradictory information throughout the methodology. Based on the comparison of unit cell and 1 × 1 array structure, it is found that the metamaterial shows double negativity at X and Ku bands. It has covered 8.29–9.5 GHz (bandwidth of 1.21 GHz), 12.15–13.46 (bandwidth of 1.31 GHz), and 16.98–17.81 (bandwidth of 0.83 GHz) in basic unit structure. Among these set of results, 8.31 and 17 GHz are the two frequencies where the double-negative character is found for both the configurations. Table 4 shows the covered area and relative bandwidths by refractive index of different configurations for double-negative characteristic. It is evident that all the configurations show similar double-negative characteristic on the respective frequency range.

Table 4 Covered area and relative bandwidths by refractive index of different configurations for double-negative characteristic

Structure	Frequency range with bandwidth (GHz)	Covered bands	Type of metamaterial
Unit cell	8.29 to 9.5 (BW of 1.21), 12.15 to 13.46 (BW of 1.31), 16.98 to 17.81 (BW of 0.83)	X and Ku	DNG
1 × 1 Array	8.29 to 9.24 (BW of 0.95), 12.13 to 13.34 (BW of 1.21), 16.58 to 16.75 (BW of 0.17), 16.98 to 17.9 (BW of 0.92)	X and Ku	DNG

5 Conclusion

In this paper, a compact metamaterial has been introduced and analyzed on transmission coefficient, effective permittivity, effective permeability, and refractive index. Firstly, the unit cell is proven to be a metamaterial. Then, the analysis and the comparison are made on unit cell and 1 × 1 array structure. The transmission coefficient covered L, C, X and Ku bands for all the configurations. Negative effective permittivity and permeability are also found for all the structures. However, unit cell and the array structure have shown good commitment to the effective parameters. Even the negative values of each of the effective parameters are found on the X and Ku bands at 8.31 and 17 GHz with a bandwidth of more than 1.31 GHz. It certainly represents the dual-band double-negative characteristic of the proposed compact design. Thus, these structures are valid for the application of dual bands. These can also be a promising choice for double negativity. These structures can be auspicious alternatives to new metamaterials, especially in utilizations where metamaterials are the only requirement.

Acknowledgements This work was supported by the Ministry of Education (MOE) under Fundamental Research Grant Scheme Top Down, Code: FRGS TOP DOWN/2014/TK03/UKM/01/1 and Research University Grant, Dana Impak Perdana Code: DIP-2015-014.

References

Benosman, H., & Hacene, N. B. (2012). Design and simulation of double "S" shaped metamaterial. *International Journal of Computer Science, 9*, 534–537.

Faruque, M. R. I., Islam, M. T., & Misran, N. (2012). Design analysis of new metamaterial for EM absorption reduction. *Progress in Electromagnetics Research, 124*, 119–135.

Hasan, M., Rashed, M., Faruque, I., & Islam, S. S. (2016). A new compact double-negative miniaturized metamaterial for wideband operation. *Materials, 9*, 830. https://doi.org/10.3390/ma9100830.

Hossain, M. I., Faruque, M. R. I., Islam, M. T., & Ullah, M. H. (2015). A new wide-band double-negative metamaterial for C- and S-B and applications. *Materials, 8*, 57–71.

Islam, S. S., Faruque, M. R. I., & Islam, M. T. (2015a). a near zero refractive index metamaterial for electromagnetic invisibility cloaking operation. *Materials, 8*, 4790–4804.

Islam, M. M., Islam, M. T., Samsuzzaman, M., & Faruque, M. R. I. (2015b). Compact metamaterial antenna for UWB applications. *Electronics Letters, 51*, 1222–1224.

Islam, S. S., Rashed, M., Faruque, I., & Islam, M. T. (2014). The design and analysis of a novel split-H-shaped metamaterial for multi-band microwave applications. *Materials, 7*, 4994–5011.

Khan, O. M., Islam, Z. U., Islam, Q. U., Bhatti, F. A. (2014) Multiband high-gain printed Yagi array using square spiral ring metamaterial structures for S-Band applications. *IEEE Antennas and Wireless Propagation Letters, 13*.

Luukkonen, O., Maslovski, S. I., & Tretyakov, S. A. (2011). A stepwise Nicolson–Ross–Weir-based material parameter extraction method. *IEEE Antennas and Wireless Propagation Letters, 10*, 3588–3596.

Shen, X., Cui, T. J., Zhao, J., Ma, H. F., Jiang, W. X., & Li, H. (2011). Polarization-independent wide-angle triple-band metamaterial absorber. *Optics Express, 19,* 9401–9407.
Singh, R., Al-Naib, I., Cao, W., Rockstuhl, C., Koch, M., & Zhang, W. (2013). The fano resonance in symmetry broken terahertz metamaterials. *IEEE Transactions on Terahertz Science and Technology, 3,* 1–7.
Zhou, Z., & Yang, H. (2015). Triple-band asymmetric transmission of linear polarization with deformed S-shape bilayer chiral metamaterial. *Applied Physics A, 119,* 115–119.

Photovoltaic/Thermal Water Cooling: A Review of Experimental Design with Electrical Efficiency

Zahratul Laily Edaris, Mohd Sazli Saad, Mohammad Faridun Naim Tajuddin and Sofian Yusoff

Abstract Photovoltaic (PV) cells can produce electrical energy from irradiance; however besides than electrical energy transformation, the remainder energy can also be transformed into thermal energy which increases the temperature of PV cell. The high operating temperature in PV cell will affect the electrical power output and PV cell lifespan. Numbers of researchers had designed different Photovoltaic/Thermal (PV/T) collector for cooling with better electrical and thermal efficiency. This paper presents a review of literature available that cover different configuration and design of water Photovoltaic/Thermal (PV/T) and their performances in terms of electrical efficiency. The review covers detailed previous researchers work of experimental design and the evaluation of the electrical efficiency. It can be concluded that, the performance of PV/T system depends on the factor of different cooling methods of passive and active, type and size of the cell, and climatic factors such as solar radiation and ambient temperature.

Keywords Photovoltaic cooling · Passive cooling · Active cooling Water cooling · Electrical efficiency

---

Z. L. Edaris (✉) · M. F. N. Tajuddin
School of Electrical System Engineering, University Malaysia Perlis, Perlis, Malaysia
e-mail: zahratul@studentmail.unimap.edu.my

M. S. Saad
School of Manufacturing Engineering, University Malaysia Perlis, Perlis, Malaysia

S. Yusoff
Mechanical Engineering Department, Polytechnic Sultan Abdul Halim
Mu'adzam Shah, Jitra, Malaysia

1 Introduction

Solar energy is one of the most abundant resources besides wind energy and biomass energy to be the next reliable energy source after the disappearance of fossil fuels. Solar energy is able to produce electricity in two ways, electrical energy and thermal energy (Grubisšić-Čabo et al. 2016; Jakhar et al. 2016). Photovoltaic (PV) cells build of semiconductor material produce electrical energy from direct sunlight or irradiance. The transformation of energy from solar irradiance to electrical energy takes place without any moving parts or environmental releases through the process (Shukla et al. 2017). The overall efficiency of solar irradiance converted to electrical energy ranges from 4 to 20% depending upon the type of the solar cells used and climate conditions (Gedik 2016). Hence, the big part of the energy was extracted from heat where the temperature parameter rise in PV affects the overall electrical efficiency of PV cells produced. The operating temperature a PV cell has upon its electrical efficiency is well studied and documented by Audwinto et al. (2015); Gedik (2016). It has been shown that as the solar radiation increased, temperature of PV module will also be increased which directly affects the photovoltaic module electrical efficiency to decrease. Hence, an effective way to increase efficiency of a PV module is to reduce the operating temperature of PV module. This may be performed by using cooling techniques of the PV module during the process. The techniques of cooling such as air cooling, water cooling, thermoelectric cooling, use of heat pipe and phase change materials have already been proposed previously by many researchers in order to remove heat. Each of the techniques produced different volume of electrical and thermal efficiency as has been reviewed by Moradi et al. (2013). From the studies, Moradi et al. had concluded that water cooling technique was found to be better than air cooling primarily from thermal point of view because of water's high heat capacity and conductivity. Moharram et al. (2013) also concluded from the literature survey that the water cooling was the promising medium in order to reduce heat than using air. Due to the better efficiency of electrical and thermal energy produced by water, a great number of researchers proposed different research designs of water cooling technique. Thus, the objective of this paper is to identify the research design and electrical efficiency of various existing project researches for the water cooling techniques in order to determine the appropriate and suitable research design for better electrical efficiency. Cooling methods in PV can be distributed into active and passive cooling. Active cooling involves the use of pump or other external power input to generate the cooling PV process in contrast to passive cooling which uses natural conduction or convention to enable the heat extraction. The next topic compiles recent works and achievements on passive and active PV water cooling focusing on the experimental design and electrical efficiency.

2 Passive Water Cooling Method

Literature studies carried out by Grubisšić-Čabo et al. (2016) concluded that most of the water passive cooling techniques applied the immersion of PV into the water. The immersion cooling of PV has been made with front and back cooling and thus resulted in higher electrical efficient because of the higher thermal capacity of water. Rosa-Clot and Rosa-Clot (2008) submerged the panel of monocrystalline silicon at different depth of water in a pool. The data analysis shows that the electrical efficiency of panel was increased at the depth of 5 cm and relatively increased to 15% at the depth of 40 cm of water. El-Seesy et al. (2012) designed collector tubes of metallic absorber plate attached to PV. A tank of 80 L is build of thermosyphon system to circulate water from storage tank to valve as a heat exchanger. The efficiency of PV with collector was increased about 2%. Chandrasekar et al. (2013) developed a simple passive cooling system with cotton wick structures for standalone flat PV modules. The thermal and electrical performances of flat PV module with cooling system consisting of cotton wick structures in combination with water, Al_2O_3/water nanofluid, and CuO/water nanofluid were investigated experimentally. The cotton wick structures were wrapped spirally at the back of the module. The efficiency was increased to 10.4%. Abdulgafar et al. (2007) investigated passive water cooling of submerging photovoltaic into different depths of distilled water. Results prove the increase of water depth for PV had increased the efficiency. At 6 cm water depth of the panel, the efficiency has increased to 11%. Gakkhar et al. (2016) cooled a PV using liquid cooling on both sides of the panel. A novel design for the liquid immersion cooling system of Concentrated Photovoltaic (CPV) along with Parabolic Trough Collector (PTC) is presented. The Heat Transfer Fluid (HTF) flows through inside as well as outside of absorber tube, thus cooling the CPV cells from both sides for better performance. Results show that fluid flow within the annulus is better where heat can be carried away easily from both sides of the cells. The operating efficiency of CPV cells then was increased. Shenyi and Chenguang (2014) designed a passive cooling method which utilized rainwater as cooling media and a gas expansion device to distribute the rainwater. The gas was thermally expanded from receiving solar radiation as such the amount of water it pushes to flow over the PV cells is proportional to the solar radiation received. The passive cooling system reduced the temperature of the cells and increased electrical efficiency of the PV panel by 8.3%. Mehrotra et al. studied the immersion technique of passive water cooling. The PV was submerged in water and was monitored in real climatic conditions with PV surface temperature from 31–39 °C. The electrical efficiency has increased about 17.8% at water depth 1 cm.

3 Active Water Cooling Method

Studies of literature survey regarding PV cooling methodologies have been carried out by Shukla et al. (2017). Fraisse et al. (2007) discussed cooling of PV by designing a PV/T collector combined with a solar water heating collector and PV cells. The combination of Direct Solar Floor was selected in the study due to its low operating temperature level. A numerical study using software TRNSYS. It is based on connecting elementary modules called TYPE, which are either components of the studied system such as storage tanks or particular functions such as the weather data reader that allows connecting the selected area weather conditions to the system. In this study, it has been shown that PV and thermal field have a very strong relationship to yield a unique industrial element that is incorporated with the thermal absorber and the PV cells. Dubey and Tiwari (2009) evaluated the performance of partially or fully covered flat plate water collectors connected in series using theoretical modeling. A DC motor was used to circulate the water in forced mode. Detailed analysis of energy, exergy, electrical energy, and cost analysis has been presented by varying the number of collectors by considering four weather conditions for five different cities. It can be concluded that fully covered collectors are beneficial when the primary requirement is electrical energy yield. Zhu and Si. (2012) identified the experimental study on water-cooled PV module and air-cooled PV module. It can be concluded that water-cooled PV module produced lower temperature and higher electrical power than air-cooled PV module while the PV/T collector using amorphous silicon PV cells can reach high electrical efficiency. Mishra and Tiwari (2012) designed a Hybrid Photovoltaic/Thermal (HPV–T) Solar Water Heater with the study of thermal energy, exergy, and electrical energy analysis. Four different weather conditions for cities in India have been carried out, and resulted fully PV/T is better for thermal energy and partially PV/T is better for electrical energy efficiency. Moharram et al. (2013) designed a PV power plant to study the effect of cooling by water on the performance of PV panels. The cooling system had the element of water pump to suck the water from the middle of the water tank then pass it through the water filter, and it is sprayed over the PV modules for cooling. Water is sprayed using water nozzles, which are installed at the upper side of the modules. Results show that the cooling system which operated for 5 min had decreased the solar cells temperature by 10 °C and increased the solar cell efficiency by 12.5%. Pierrick et al. (2015) had developed a dynamic 3D numerical model for high efficiency glass-covered water-based PV–T prototype. The energy balance of conductive and convective heat transfer equations from the fluid flow interpretation was considered in the model developed. The model was centered on energy balance for each constituent control volume. The model was found to be accurate to 1–2 °C for a stable flow rate and responsive to immediate changes in flow rate. PV/T water cooling design involved different configuration such as the sheet and tube collectors, channel collectors, water free-flow collectors and two absorber collectors. Ozgoren et al. (2013), Rawat et al. (2014), and Yazdanpanahi et al. (2015) designed tube collectors and absorber mounted at the backside and rear of the PV panel for cooling system. The electrical efficiency for the tube collector configuration was increased to 9, 11.5–13.6, and 7.57%, respectively. Ozgoren et al. (2013) agreed that the PV temperature difference will decrease with the increase value in mass flow rate of the water. It has been proven through the experiment carried out. Abdolzadeh and Ameri (2009), Krauter (2004) designed the PV/T water cooling by spraying water on top of PV cell. Through the experiment performed and analyzed (Abdolzadeh and Ameri 2009; Krauter 2004), the configuration of water spraying on top of PV was able to decrease the cell's temperature and cell's reflection. The water spray improved the optical performance by 1.8%. Besides that, Dorobanțu et al. (2013), Hosseini et al. (2011) designed the PV/T water cooling by film water free flowing on top of PV module. The advantage of the thin water or film running on top of the PV is better electrical efficiency can be obtained due to the decrement of reflection loss. Irwan et al. (2015) designed the PV/T water cooling by water flow at front of the PV using solar simulator. The solar simulator was fabricated because of the experiment can be carried out any chosen time, continued for 24 h a day and in indoor test. Nižetić et al. (2016) performed and analyzed PV water cooling on front and backside of PV. The circumstances of simultaneous front and backside PV panel cooling resulted in maximal relative increase in panel power output.

4 Results

Summaries of previous researchers reported for passive and active water cooling were critically summarized in Tables 1 and 3. The configuration of PV immersion resulted in higher electrical efficiency compared to thermosyphon and capillary effect in passive water cooling configuration. The configuration of absorber and tube collector, free-flow water on top of PV, and water spraying on both sides of PV produced higher electrical efficiency for active cooling. Table 2 shows the collective literature for water cooling provided by Shukla et al. (2017). Shukla et al. listed literature works for active water and air cooling in PV. It was reported that air and water cooling of active cooling produced higher electrical efficiency. However, study carried out by Moharram et al. (2013) resulted in the active water cooling able to remove heat until to 10 °C with 12.5% electrical efficiency. Tables 1 and 3 summarized that the active water cooling was able to remove heat and reduce the temperature from 18.7 to 23 °C. In addition to that, the configuration of free-flow water spraying on top of PV and water spraying on both sides of PV resulted in a better electrical efficiency from 9 to 22%. These techniques are proven to be promising designs in contributing higher electrical efficiency. However, factors such as size of cells and climatic factors did affect the performance of PV/T systems.

Table 1 Selected parameters that different researchers reported for passive water cooling

Author, Year	Configuration	Aperture area (m^2)	Temp. w/out collector (°C)	Temp. with collector (°C)	Solar irradiation (W/m^2)	Electrical efficiency
Rosa-Clot and Rosa-Clot (2008)	Immersion	–	–	–	–	15%
El-Seesy et al. (2012)	Thermosyphon effect	0.2600	57.50	47.60	845	2%
Chandrasekar et al. (2013)	Capillary effect	0.3600	59.00	54.00	–	10.4%
Alami (2014)	Evaporative	–	85.00	45.00	905	16.7%
Han et al. (2013)	Immersion	0.0019	–	–	30000	15.2%
Abdulgafar et al. (2007)	Immersion	0.0015	36.00	30.00	700	11%
Gakhkar et al. (2016)	Liquid Immersion of CPV	–	39.86	35.85	–	Increase with temperature reduction
Mehrotra et al. (2014)	Liquid immersion	0.033	34.80	30.80	1170	17.8% at depth = 1 cm

Table 2 Selected parameters for water cooling that different researchers reported (Shukla et al. 2017)

Author, Year	Configuration	PV type	Technique (Experimental/simulation)	Electrical efficiency	Thermal efficiency	PV cell temperature	Conclusion
Fraiss et al. (2007)	Water circulation	PV/T with a glazed cover	Numerical modeling	6.8%	–	35 °C	–
Dubey and Tiwari (2009)	Forced circulation of water and air	Thermal collector partially covered with PV	Theoretical	8.5–10.5%	38%	40–90 °C	–
Zhu and Si (2012)	Forced circulation	Si PV cell with air and water	Experimental	–	37.5%–62.1%	33.1 °C	–
Mishra and Tiwari (2012)	Forced circulation of water and air	PV–T with fully and partially covered	Experimental	12%	–	–	–
Moharram et al. (2013)	Forced circulation of water	Monocrystalline silicon	Experimental	12.5%	–	45 °C	Reduced by 10 °C
Haurant et al. (2015)	A glass-covered water-based PV–T prototype	–	Numerical 3D dynamic model	8–11%	–	–	–

Table 3 Selected parameters that different researchers reported for active PV/T water cooling

Author, Year	Configuration	PV type	Technique (Experimental/simulation)	Electrical efficiency	Thermal efficiency	PV cell temperature	Conclusion
Bahaidarah et al. (2013)	Hybrid water cooling with tube collector	Monocrystalline	Experimental	9%	–	30.5 °C	
Ozgoren et al. (2013)	Absorber and tube collector	Monocrystalline	Experimental	11.5–13.6%	51%	30–45 °C	
Rawat et al. (2014)	Tube collector	Polycrystalline	Experimental	7.57%	45.5%	50–70 °C	
Yazdanpanahi et al. (2015)	Tube collector	Monocrystalline	Experimental and simulation	–	–	–	Exergy efficiency 13.95%
Abdolzadeh and Ameri (2009)	Free-flow water on top of PV (spraying)	Polycrystalline	Experimental	13.5%	–	30–38 °C	Reduce up to 23 °C
Krauter (2004)	Free-flow water on top of PV (spraying)	Monocrystalline and Polycrystalline	Experimental	–	–	–	Reduce up to 22 °C
Dorobantu et al. (2013)	Free-flow water of film on top	Monocrystalline	Experimental	9.5%	–	38.5–41.5 °C	
Hosseini et al. (2011)	Free-flow water of film on top	–	Experimental	Relative difference of 33.3%	–	–	Reduce up to 18.7 °C
Irwan et al. (2015)	Free-flow water of film on top	Monocrystalline	Experimental	9–22%	–	–	Reduce from 5 to 23 °C
Nižetić et al. (2016)	Water spraying on both sides of PV	Monocrystalline	Experimental	14.4%	–	30–38 °C	Reduce up to 22 °C

5 Conclusion

Cooling configuration for PV/T water has been detailed and analyzed. It is shown that active cooling had a higher efficiency compared to passive cooling. The configuration of free flow water either by spraying or thin film of water running at front PV prove to results large gap of temperature reduction up to 23 °C. It was shown that free-flow water configuration was the best solution due to the simultaneous result of producing greater electrical efficiency and able to act as self-cleaning element for the PV module especially in hot climate. Hence, this analysis can be used in the future for the implementation of PV/T cooling system.

References

Abdolzadeh, M., & Ameri, M. (2009). Improving the effectiveness of a photovoltaic water pumping system by spraying water over the front of photovoltaic cells. *Renewable Energy, 34* (1), 91–96. https://doi.org/10.1016/j.renene.2008.03.024.

Abdulgafar, S. A., Omar, O. S., & Yousif, K. M. (2007). Improving the efficiency of polycrystalline solar panel via water immersion method. *International Journal of Innovative Research in Science, Engineering and Technology (An ISO Certified Organization), 3297*(1).

Audwinto, I. A., Leong, C. S., Sopian, K., & Zaidi, S. H. (2015). Temperature dependences on various types of Photovoltaic (PV) panel. *IOP Conference Series: Materials Science and Engineering, 88,* 12066. https://doi.org/10.1088/1757-899X/88/1/012066.

Chandrasekar, M., Suresh, S., Senthilkumar, T., & Ganesh Karthikeyan, M. (2013). Passive cooling of standalone flat PV module with cotton wick structures. *Energy Conversion and Management, 71,* 43–50. https://doi.org/10.1016/j.enconman.2013.03.012.

Dorobanțu, L., Popescu, M. O., Popescu, C. L., & Crăciunescu, A. (2013). Experimental assessment of PV panels front water cooling. *Strategy, 1*(11), 1–4.

Dubey, S., & Tiwari, G. N. (2009). Analysis of PV/T flat plate water collectors connected in series. *Solar Energy, 83*(9), 1485–1498. https://doi.org/10.1016/j.solener.2009.04.002.

El-Seesy, I. E., Khalil, T., & Ahmed, M. H. (2012). Experimental investigations and developing of photovoltaic/thermal system. *World Applied Sciences Journal, 19*(9), 1342–1347. (ISSN 1818-4952). https://doi.org/10.5829/idosi.wasj.2012.19.09.2794.

Fraisse, G., Ménézo, C., & Johannes, K. (2007). Energy performance of water hybrid PV/T collectors applied to combisystems of Direct Solar Floor type. *Solar Energy, 81*(11), 1426–1438. https://doi.org/10.1016/j.solener.2006.11.017.

Gakkhar, N., Soni, M. S., & Jakhar, S. (2016). Analysis of water cooling of CPV cells mounted on absorber tube of a parabolic trough collector. *Energy Procedia, 90,* 78–88. https://doi.org/10.1016/j.egypro.2016.11.172.

Gedik, E. (2016). Experimental investigation of module temperature effect on photovoltaic panels efficiency. *Journal of Polytechnic, 19*(4), 569–576.

Grubišić-Čabo, F., Nizetić, S., & Marco, T. G. (2016). Photovoltaic panels: A review of the cooling techniques: EBSCOhost. *Transactions of FAMENA, 40*(1), 63–74.

Hosseini, R., Hosseini, N., & Khorasanizadeh, H. (2011). An experimental study of combining a photovoltaic system with a heating system. In *World Renewable Energy Congress* (pp. 2993–3000).

Irwan, Y. M., Leow, W. Z., Irwanto, M., Fareq, M., Amelia, A. R., Gomesh, N., et al. (2015). Analysis air cooling mechanism for photovoltaic panel by solar simulator. *International Journal of Electrical and Computer Engineering, 5*(4), 636–643.

Jakhar, S., Soni, M. S., & Gakkhar, N. (2016). Historical and recent development of concentrating photovoltaic cooling technologies. *Renewable and Sustainable Energy Reviews*. https://doi.org/10.1016/j.rser.2016.01.083.

Krauter, S. (2004). Increased electrical yield via water flow over the front of photovoltaic panels. *Solar Energy Materials and Solar Cells, 82*(1), 131–137. https://doi.org/10.1016/j.solmat.2004.01.011.

Mishra, R. K., & Tiwari, A. (2012). Study of Hybrid Photovoltaic Thermal (HPVT) solar water heater at constant collection temperature for Indian climatic conditions. *Ashdin Publishing Journal of Fundamentals of Renewable Energy and Applications, 2*. https://doi.org/10.4303/jfrea/r120310.

Moharram, K. A., Abd-Elhady, M. S., Kandil, H. A., & El-Sherif, H. (2013). Enhancing the performance of photovoltaic panels by water cooling. *Ain Shams Engineering Journal, 4*(4), 869–877.doi: https://doi.org/10.1016/j.asej.2013.03.005.

Moradi, K., Ali Ebadian, M., & Lin, C.-X. (2013). A review of PV/T technologies: Effects of control parameters. *International Journal of Heat and Mass Transfer, 64*, 483–500. https://doi.org/10.1016/j.ijheatmasstransfer.2013.04.044.

Nižetić, S., Čoko, D., Yadav, A., & Grubišić-Čabo, F. (2016). Water spray cooling technique applied on a photovoltaic panel: The performance response. *Energy Conversion and Management, 108*, 287–296. https://doi.org/10.1016/j.enconman.2015.10.079.

Ozgoren, M., Aksoy, M. H., Bakir, C., & Dogan, S. (2013). Experimental performance investigation of Photovoltaic/Thermal (PV-T) system. In *EPJ Web of Conferences, 1106*(45). https://doi.org/10.1051/epjconf201/34501106.

Pierrick, H., Christophe, M., Leon, G., & Patrick, D. (2015). Dynamic numerical model of a high efficiency PV-T collector integrated into a domestic hot water system. *Solar Energy, 111*, 68–81. https://doi.org/10.1016/j.solener.2014.10.031.

Rawat, P., Debbarma, M., Mehrotra, S., Sudhakar, K., & Sahu, P. K. (2014). Performance evaluation of solar photovoltaic/thermal hybrid water collector. In *Impending Power Demand and Innovative Energy Paths* (pp. 268–275). https://doi.org/10.1016/s0038-092x(00)00153-5.

Rosa-Clot, M., & Rosa-Clot, P. (2008). Submerged photovoltaic solar panel: SP2.

Shenyi, W., & Chenguang, X. (2014). Passive cooling technology for photovoltaic panels for domestic houses (pp. 1–9). https://doi.org/10.1093/ijlct/ctu013.

Shukla, A., Kant, K., Sharma, A., & Biwole, P. H. (2017). Cooling methodologies of photovoltaic module for enhancing electrical efficiency: A review. *Solar Energy Materials and Solar Cells, 160*, 275–286. https://doi.org/10.1016/j.solmat.2016.10.047.

Yazdanpanahi, J., Sarhaddi, F., & Mahdavi Adeli, M. (2015). Experimental investigation of exergy efficiency of a solar photovoltaic thermal (PVT) water collector based on exergy losses. *Solar Energy, 118*, 197–208. https://doi.org/10.1016/j.solener.2015.04.038.

Zhu, Q., & Si, L. (2012). Electrical outputs and thermal outputs of water/air cooled amorphous-silicon photovoltaic modules. *Advances in Biomedical Engineering, 8*, 83.

Success Factor for Site Management in Industrialized Building System (IBS) Construction

Nor Izzati Muhammad Azmin, U. Kassim and Mohd Faiz Mohammad Zaki

Abstract The Industrialized Building System (IBS) is the modern method construction which enables on-site prefabricated or precast building components manufactured at factories for building materials and housing products. Site management in IBS construction is crucial when dealing with the material and equipment at the workplace which can help the planning process of the project. The objectives of this research study are to discover the reason for using IBS construction in view of site management, to investigate the factors to improve towards site management in IBS construction (management, workplace, work team, safety, and material and equipment), and to produce a recommendation to improve the factors towards site management for IBS construction. The collected data has been analysed by using the Statistical Package Social Science (SPSS) version 21. This quantitative method research has analysed with frequency distribution method from SPSS v21 software. All three objectives have been achieved by finding the most important factor to improve site management which is the management factor with the highest mean 5.82 followed by the work team factor with 5.70 mean. The material and equipment factor is the lowest mean from the analysis data with 5.56 and safety factor with 5.67 higher than the workplace factor which is 5.64 mean. Thus, it clearly shows that the management factor is the most influenced factor that needs to improve in site management. In future, this research is suggested by doing a factor analysis in order to get the accurate result for the most influenced factor to improve site management.

Keywords Industrialized building system (IBS) · Success factor
Site management · Construction · Manufacture

N. I. Muhammad Azmin (✉) · U. Kassim · M. F. Mohammad Zaki
School of Environmental Engineering, Universiti Malaysia Perlis,
Kompleks Pengajian Jejawi 3, 02600 Arau, Perlis, Malaysia
e-mail: izzati.azmin94@gmail.com

1 Introduction

The Industrialized Building System (IBS) is the modern method construction which enables on-site prefabricated or precast building components manufactured at factories for building materials and housing products. This new construction method required knowledge and skilled worker to ensure the quality product produced. Compared to the conventional method regarding the safety, IBS system has implemented IBS Roadmap 2003–2010 with a target of zero number of accident and fatalities. Apart from this, it can reduce the labour intensity and construction standardization, which enables the cost saving and quality improvement that also offers minimal wastage, less site materials, cleaner and neat environment, controlled quality, and lower total construction cost.

This research focuses on the Johor area since Johor is one of the states in Malaysia that has a rapid development apart from the capital of Malaysia, Kuala Lumpur. Furthermore, Johor is also an interesting place to do this research because it is near to the Singapore which is also one of the success countries that has implemented IBS in their construction industry. Johor also indirectly uses the innovative ways by getting the components of IBS from Singapore. This international collaboration is a good thing to discover in terms of material and equipment factor. This research study covers 9 out of 10 districts in Johor. The total number of the IBS companies that involved in this research is 36 companies in Johor.

2 Issues in Managing IBS

IBS projects that have a poor management are frequently prompted to numerous troubles, which wind-up to extend delays, unsuitable qualities, and higher cost (Haas and Fagerlund 2002; Kamar et al. 2012; Rahman and Omar 2006). The accomplishment of IBS execution in the Malaysian development industry is exclusively relying upon the temporary workers who deal with the procedures required in the IBS lifecycle. Separated from that, the dedication of contractual worker in dealing with the project is critical to accomplish most extreme well-being of that project (Ismail et al. 2013). The contractor ought to be able and experienced in dealing with the construction activities (Lou and Kamar 2012). Likewise, the contractor ought to think about every one of the issues in their administration home so as to make progress in the usage of IBS projects (Table 1).

From the literature, 28 issues have been recognized that generally emerge in overseeing projects of IBS construction. Among of these issues are high beginning cost, gigantic volume of work to equal the initial investment, absence of gear and apparatus, absence of testing facility and IBS components, necessity of aptitude work, challenges to apply changes, deficient preparing nearby levels, hard to draw in new specialists join the workforce and retrain them with new IBS abilities and building absconds.

Table 1 Table for determining sample size for finite population (Krejcie and Morgan 1970)

N	S	N	S	N	J
10	10	220	140	1200	291
15	14	230	144	1300	297
20	19	240	148	1400	302
25	24	250	152	1500	306
30	28	260	155	1600	310
35	32	270	159	1700	313
40	36	280	162	1800	317
45	40	290	165	1900	320
50	44	300	169	2000	322
55	48	320	175	2200	327
60	52	340	181	2400	331
65	56	360	186	2600	335
70	59	380	191	2800	338
75	63	400	196	3000	341
80	66	420	201	3500	346
85	70	440	205	4000	351
90	73	460	210	4500	354
95	76	480	214	5000	357
100	80	500	217	6000	361
110	86	550	226	7000	364
120	92	600	234	8000	367
130	97	650	242	9000	368
140	103	700	248	10,000	370
150	108	750	254	15,000	375
160	113	800	260	20,000	377
170	118	850	265	30,000	379
180	123	900	269	40,000	380
190	127	950	274	50,000	381
200	132	1000	278	75,000	382
210	136	1100	285	1,000,000	384

3 Methodology

In research methodology, the method in the research objectives can be questioned. The method that has been chosen is the quantitative method to achieve the objectives of the research, and the random sampling method is used to get the data collection process. In Fig. 1, the general flow chart of the research procedure is as shown below. The collected data will be analysed by using the Statistical Package Social Science (SPSS).

Fig. 1 Flow chart of research methodology

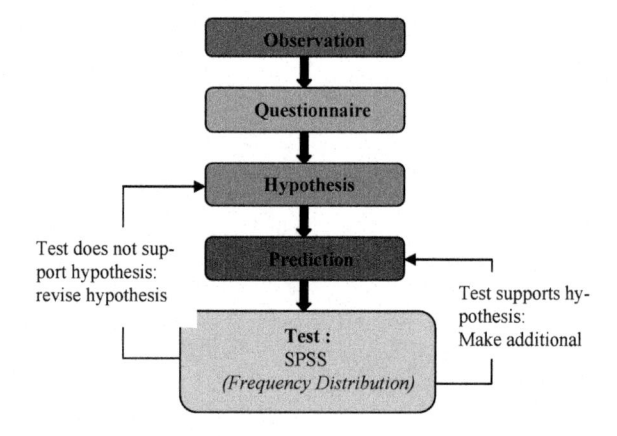

3.1 Sampling of Population

This research will be conducted in the state of Johor. The state of Johor in the southern region of Malaysia has been chosen as a research place because the state is one of the most rapid developments of IBS in Malaysia with the international export and import with Singapore in the construction using IBS. The research was focused on the manufacturer in all categories of IBS. The total population for IBS construction in Johor that registered under IBS Centre is not a big number since it is still a new practice system in Malaysia.

3.2 Sample of Respondent

There will be selected all thirty-six (36) manufacturer with the actual sample size that need based on table (Krejcie and Morgan 1970) is thirty-two (32) sample for thirty-five (35) population. The questionnaire will be sent out to the respondent according to the list of manufacturer that has been registered under IBS Centre (Fig. 2).

Fig. 2 Total mean of factors to improve site management

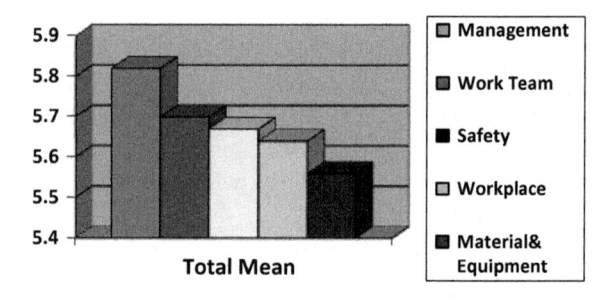

4 Result and Discussion

A total set of 35 questionnaires was analysed and the results were divided into seven parts which are demographic profile, success factor of management, success factor of workplace, success factor of work team, success factor of safety, success factor of material and equipment, and recommendation part (action and education). The total set of 35 of questionnaires responded was returned, and it represents a response rate of 100%.

4.1 Reason for Using IBS

In this section, it consists of five components of the reasons using IBS in construction industry which are fast completion, minimizing risk lost, less workers, quality construction, and better than conventional. The descriptive analysis used the average value of each component in determining the reason of using IBS among the contractors.

4.1.1 Summary of the Reason for Using IBS

Table 2 shows the summarization statistics of the reason for using IBS in Malaysia. From the mean analysis, the reason for less workers needed during handling IBS projects is the most higher reason with mean 6.20, and it is been categorized as mostly good. However, from the overall summary, result with total mean is 5.84, and the other reasons are also in mostly good categories. So, it can be concluded that IBS is a better method to use in construction industry nowadays.

4.2 Factor to Improve Site Management

This section consists of seven dimensions of the important factors which are almost never important, usually never important, rarely important, occasionally important, often important, usually always important, and almost always important in site

Table 2 Summary of reason for using IBS

Components	Mean	Category
Less workers	6.20	Mostly good
Fast completion	6.14	Mostly good
Better than conventional	5.74	Mostly good
Minimizing risk lost	5.69	Mostly good
Quality construction	5.43	Good
Overall	5.84	Mostly good

management of IBS. Descriptive analysis used the average value of each component in determining the important level among the respondents.

From the frequency distribution data, the highest mean goes to the management factor which is 5.82 followed by the work team factor with 5.70 mean. The material and equipment factor is the lowest mean from the analysis data with 5.56 and safety factor with 5.67 higher than the workplace factor which is 5.64 mean. However, all the factors are still in the same range. As expected, the management factor becomes the highest since management includes various elements in construction industry such as organizing the project, which requires a special knowledge, and also a supporting discipline which is very important for site management.

From the previous journal, the results are slightly different where the top 3 of the higher data for site management goes to management, work team, and workplace. Whereas, in this research, it shows that management, work team, and safety are the top 3 factors that should be considered in site management. However, the overall results can be concluded that these four factors, management, work team, workplace, and safety, are important to ensure that site management achieves the best management in construction which indirectly can reduce the weaknesses in site management such as delay components, and projects can be avoided.

4.3 Recommendation

Based on the feedback from the respondents, the topic points out several recommendations and suggestions about the site management of IBS in Johor which are:

i. The education syllabus in higher learning institutes especially for the subject IBS must stress the importance of flexibility as to create perfect and effective management in IBS construction.
ii. For the development and smooth-running of the site management in IBS process in the future, the site management act needs to be implemented as an important score in IBS construction.
iii. Those who are going to join with IBS in construction industry need to have very high skills in managing IBS product. For the fresh graduates in order to have very good skills, the companies have to train their employees so that they know how to handle all the problems and issues that critically happen in the real world of construction industry.

5 Conclusion

The purpose of this research is to investigate the success factors that can be improved in site management of IBS construction. The conclusions based on the objective of this research have been summarized as there are five components of the reasons for using IBS in construction industry. The results find that the reason for the construction industry to use IBS because of the less worker and fast completion of project. Next, the most important factor to improve the site management of IBS in construction industry is the management factor where management covers various components in construction industry. All the components in industry such as technology, innovative products, and modern methods that have been invented in current situation for construction purpose need to have very effective management skills.

This research has successfully discovered that the reasons for construction nowadays have implement IBS and management is the important factor to improve site management. The results show about 83% respondents agree that management is the most important factor which is satisfying the finding that can be a main role of the success factor for site management in IBS construction.

Acknowledgements The authors also gratefully acknowledge the helpful comments and suggestions of the reviewers, which have improved the presentation.

References

Haas, C. T., & Fagerlund, W. R. (2002). *Preliminary research on prefabrication, pre-assembly, modularization and off-site fabrication in construction.* The Construction Industry Institute, The University of Texas at Austin.

Ismail, F., Einur, H., Baharuddin, A., & Marhani, M. A. (2013). Factors towards site management improvement for Industrialised Building System (IBS) construction. *Procedia—Social and Behavioral Sciences, 85,* 43–50.

Kamar, K. A. M., Kamar, M., Abd, Z., Maria, H., Mohd, Z., Ahmad, Z., et al. (2012). Drivers and barriers of industrialised building system (IBS) roadmaps in Malaysia. *Malaysian Construction Research Journal, 9,* 1–8.

Krejcie, R. V., & Morgan, D. W. (1970). Determining sample size for research activities Robert. *Educational and Psychological Measurement, 38*(1), 607–610.

Rahman, A. B. A., & Omar, W. (2006). Issues and challenges in the implementation of industrialised building systems in malaysia. In *Proceedings of the 6th Asia-Pacific structural Engineering and Construction Conference (Apsec 2006)* (pp. 5–6), Kuala Lumpur. Malaysia.

Street Network Analysis by SPOT Imagery for Urban Morphology Study. Case Study: Melaka

Marina Mohd Nor, Norzailawati Mohd Noor and Sadayuki Shimoda

Abstract This paper is based on the study of morphological changes in Melaka, Malaysia. In order to identify the morphology of this city, the street network has been studied to determine the evolutionary of urban form and structure. The objective of this paper is to examine the movement and direction of the street pattern in influenced the morphology of the city from the year 1993 to 2016. Three series of satellite images on the years 1993, 2005 and 2015 from SPOT satellite have been used in detecting the development of street network pattern aided by remote sensing and GIS software. By extracting the street in 20 years of development, a comparison of the street pattern will be taken into consideration to examine the direction of the expansion of Melaka city. The finding shows that the street expansion grows fast as the more settlements were built along this process, and the expansion was merely influenced by the location of the site which near to Malacca Straits and as a port for trade sector. Furthermore, the movement of streets is expanding inwards to the inner city and along the shoreline. Finally, this study shows a street network as one of the principle elements in the urbanization process, and it provides an understanding key on how cities are formed and developed in order to achieve a resilient city.

Keywords Urban morphology · Street extraction · SPOT · Remote sensing Urban planning

---

M. Mohd Nor (✉) · N. Mohd Noor
Department of Urban and Regional Planning, Urban and Regional Planning Department, Kulliyah Architecture and Environmental Design, International Islamic University of Malaysia, 50728 Kuala Lumpur, Malaysia
e-mail: marina.mn@gmx.com

S. Shimoda
Department of Architecture and Civil Engineering, Kumamoto National College of Technology, Yatsushiro City, Kumamoto, Japan

1 Introduction

Monitoring the changes in urban area or can be called as urban morphology has become a very interesting topic to be tackled by a researcher in order to analyse how the city exists. With the advancement of technology, urban morphology can be traced more easily with the technology of remote sensing. Urban morphology is an important aspect of the planning system, which contributes to the existence of the city, history of the city, how the city is developed, planning system of the city and so forth, which contribute to the sustainable of the city. There are many aspects of analysing the morphology of the city such as buildings, land use, urban design, street network, urban landscape or others.

In this paper, we will analyse the street network of Melaka city as part of the criteria in the evolutionary process of development. Thus, to examine the pattern and development of the street network, road extraction method will be used to extract the road pixels in the satellite images. Road extraction approach becomes one of the significant areas to explore and heated among the researchers. There are a quite number of ways in terms of strategies, type and resolution, processing and so forth to extract road from high-resolution images. For this study, IMAGINE Objective tools will be used to classify and extract features in order to provide accurate geospatial content.

2 Street Network as Elements in Urban Morphology

Street network analysis can help to identify the transformation of the urban form development and the evolutionary of urban form and structure which can determine the morphology of the city (Cheng 2011). Urban morphology occurred based on certain characteristic such as the configuration of urban fabrics, natural and man-made structures, street network/layout, architectural complexity, open space and other physical element (Moudon 1997; Li and Yeh 2004; Sharifah et al. 2013; Paul 2008). Therefore, the street network can represent as an indicator to control the development activity and provide opportunities and constraints for city-building processes, such as land subdivision, infrastructure development or building construction.

Street network provides ease of movement, accessibility, reachability and safety within the area. In urban planning and development, street or road is one of the most significant criteria in forming a city. Moreover, it creates a better connection between the existing settlement and the new development area.

Furthermore, the growth of the city can be determined by the effective and systematic of the street layout as one of the urban forms in urban planning. Hence, the street system in the city is important in creating a sustainable and resilient city in future and influences the development of urban planning which represents a system of streets or roads and one of the tools in the urbanization process (Barthelemy 2015).

Cities are constantly developing even though with or without planning. Indeed, the human population is getting higher every year, and it would lead to the migration of people from rural areas to city areas and increase the urban growth of the particular area. The urban form such as street networks is expanding and permanent physical aspect in the cities which are unalterable and have a stronger influence on the town planning. It gives influence on its surrounding, upon the safety, comfortable to the people and the efficiency of government and public services in shaping the urban development. Besides that, the street networks directly determine people how shall they travel, live, work and play.

In addition, the morphology of cities should be understood not only in economic, historical and physical terms, but also in cultural heritage terms. Therefore, we suggest that the preservation, redevelopment, urban policy and future expansion of cities should incorporate analyses of the cities' historical and contemporary morphological development. Streets are one of the important elements in the urban form that lead to the urban history and morphology studies which give impact in many aspects such as social and economic life (Shpuza 2014). The study of streets can provide the information on the urban growth of the city and enlighten the evolution of the urban form in the particular area.

Indeed, streets have greater structural important, which have a tendency to attract people to have more experience and making them more prone to imageability. Moreover, streets can provide unique features to the user due to the pattern of the streets such as narrowness and width of the street, and the activity along the streets (Omer and Jiang 2010; Lynch 1960). Therefore, streets' network not only pointed on the traffic load, but also it has significant value to the society which can perform livable street as public spaces and provide safety to the user as well.

3 Study Area

This study was conducted in Bandar Melaka, Melaka, Malaysia. Melaka is situated in the southern region of the Malay Peninsula, next to the Straits of Malacca. This historical city centre has been listed as a UNESCO World Heritage Site since 7 July 2008.

The character of the Historic City of Melaka is strengthened by the unique townscape qualities of the streets and the buildings that shaped the quality of space created by these streets. Different with other towns, Melaka townscape is quite distinctive in character because of its sense of enclosure and mixture of houses, shops and places of worship (Fig. 1).

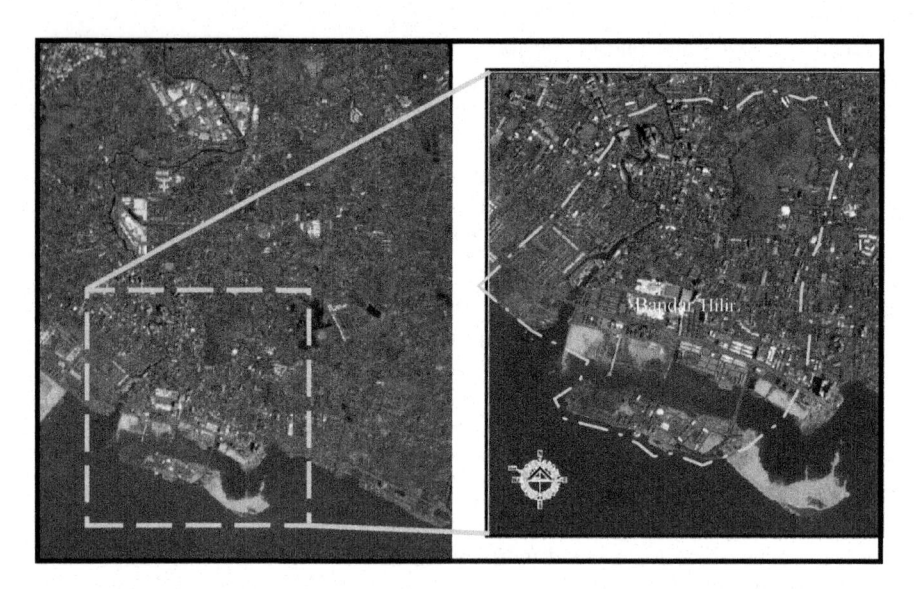

Fig. 1 Study area, Melaka

4 Materials and Method

4.1 Data

Three different SPOT satellite images (nominally 1993, 2005 and 2015) are used to analyse the morphology of street network in the study area (Table 1).

4.2 Methods

All the digital form of data is readily used in data processing in digital image processing. ERDAS IMAGINE 2014 is the main software that used to analyse the morphology of street network by extracting the road. The formation of street network or road in the study area will be monitored based on the years 1993, 2005 and 2015. Hence, the expansion of road network will determine the development of study area within 20 years.

Table 1 Materials used in this study

Data types	Year	Spatial resolution (m)	Provider
SPOT-2 XS	1993	20	Malaysian Remote Sensing Agency (MRSA)
SPOT 5	2005	5	
SPOT 6	2015	1	

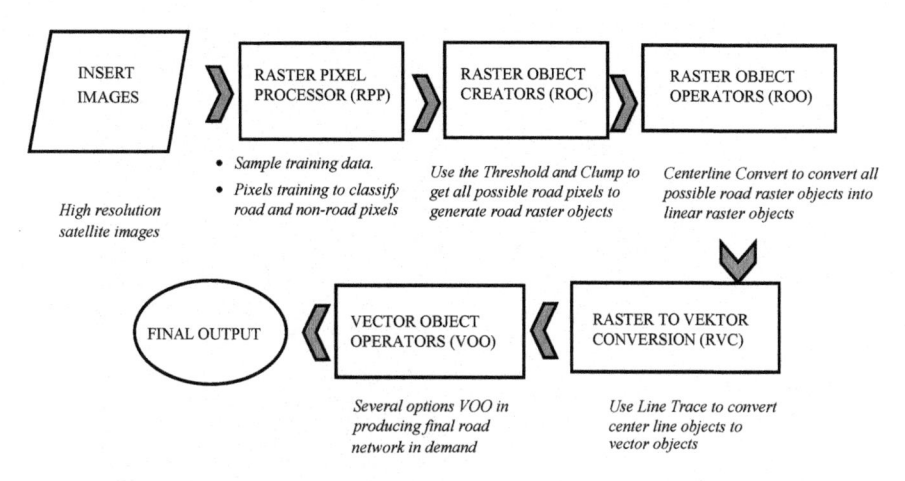

Fig. 2 Flow of methodology adopted in this study

The method consists of a sequence of operators and processes. This method successfully enables to use to extract any information such as residential rooftops, commercial and industrial buildings, road, tree crowns and so forth. Thus, for this study, road extraction using IMAGINE Objective will be a great help to separate the roads from the background. Below are steps to extract road by using IMAGINE Objective (Fig. 2).

5 Results and Discussion

5.1 Road Extraction

The satellite images will be trained to classify the pixels and objects or both in order to segregate pixels between the road and non-road pixels. This training road pixel is important to classify the pixels and turn non-related pixels to the background, which it will help to see clearly the road pixel (Fig. 3). This step will generate and indicate the probability of being a road pixel (Table 2).

From the classification result, Raster Object Creators (ROC) process node will generate all possible road pixels to road raster objects with the use of threshold and clump operator. Then, further reduction in the number of non-road rasters must be applied in Raster Object Operator (ROO). Several options on the operator can be used such as size filter which will remove tiny objects and for road extraction, centreline convert operator will convert all possible road raster objects into linear raster objects, which all have single pixel width (Fig. 4).

Fig. 3 Before and after training and classify the pixels of road and non-road pixels during raster pixel processor stage. The non-pixels road will turn out to be background in the sampling images

Table 2 Changes of street network from 1993 to 2015

Year	Percentage (%)
1993	14
2005	30
2015	55

Line trace operator will convert centreline objects to vector objects in Raster-to-Vector Conversion (RVC) process node. Finally, Vector Object Operators (VOOs) can be used to finalizing the road network demand based on the series of operators that need to use such as Line Link, Smooth, Line Snap and Ribbon Remove. The series of VOOs can be adjusted by changing the parameters or rearrange the process nodes. Below is the result of road extraction for the satellite images that have been processed.

The result shows the extraction of the street network that varies from the years 1993, 2005 and 2015. The percentage of street network area is increasing, and formula for this is

$$\frac{\text{Area(Raster attributes)} \div \text{Total Area}}{100} \quad (1)$$

But, continuous processing is a need in order to have a good result due to training sample during RPP Process which is variable during pixel extraction.

Fig. 4 Road extraction results in years 1993, 2005 and 2015

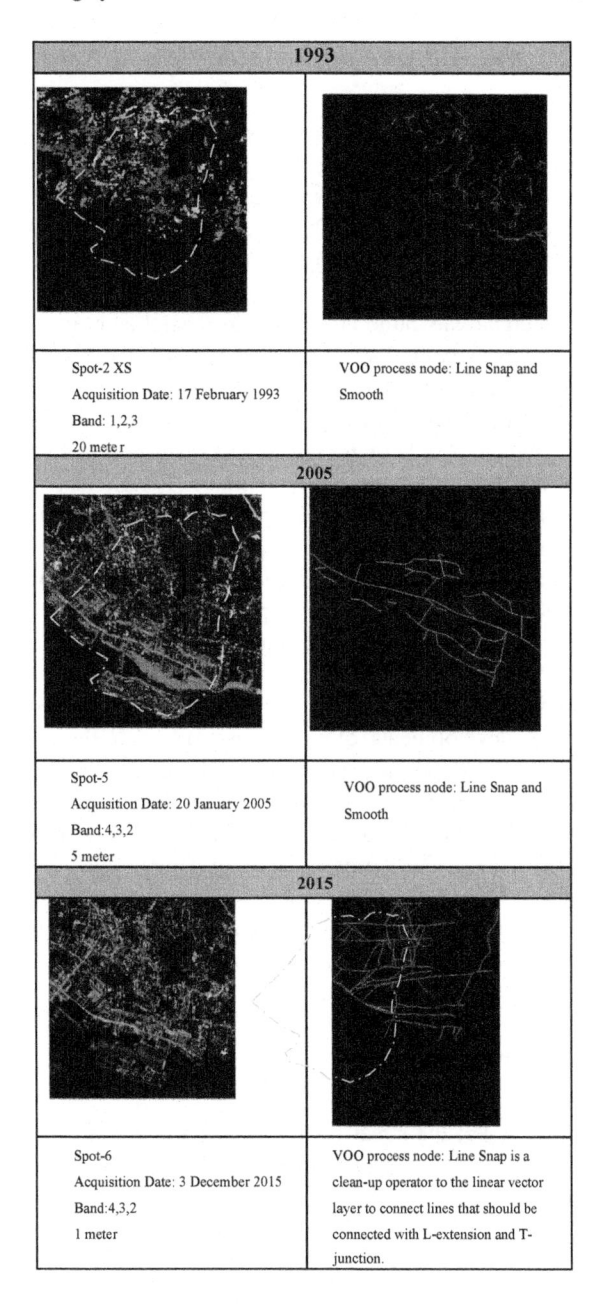

1993	
Spot-2 XS Acquisition Date: 17 February 1993 Band: 1,2,3 20 meter	VOO process node: Line Snap and Smooth
2005	
Spot-5 Acquisition Date: 20 January 2005 Band:4,3,2 5 meter	VOO process node: Line Snap and Smooth
2015	
Spot-6 Acquisition Date: 3 December 2015 Band:4,3,2 1 meter	VOO process node: Line Snap is a clean-up operator to the linear vector layer to connect lines that should be connected with L-extension and T-junction.

5.2 The Morphology of Street/Road Network

Based on the result, the road or street network showed some changes between the years 1993, 2005 and 2015. The complexity of road network develops from the inner of the city to the coastal area. In the year 1993, the development of the city is not crowded as the year 2005 which the new development was trigger over the Malacca Strait. The opening of Pulau Selat as new development contributes a new road network that connects between Bandar Hilir to Pulau Selat. Thus, the new commercial was build and generates social activity within this street.

The expansion of street network towards the Malacca Strait becomes wider year by year due to the development of this city. The street during the year 1993–2005 was tremendously developed as the transportation facilities become improve and the needs of people to use the street to transport from one location to other location. During that time, the introductions of vehicular transportation become overwhelmed and occur high demand on the transportation needs.

Because of the needs of transportation in high demand, the opening of new roads is needed. This necessity is important in the process of city's development to cater the needs and population of the people. It is important to the urban planning process in making the city more systematic and well organized to transport people from one particular destination to others. Most of the road within this city is one-way route which to control the traffic congestion, especially during peak hour.

The developments of the streets occur at the centre of the city, and urban commercial streets are more livable due to much social activity and public spaces offer to the people. As a Heritage City that recognized by UNESCO, this vibrant city attracts most of local and tourist to come and visit this city, thus it does affect the liveliness of the streets. The city centre, Bandar Hilir, has variety numbers of heritage buildings, and no new development is allowed in this area due to the dense and saturated area. Thus, beautification of this area is needed to make sure the place was user-friendly and make the spaces more livable.

6 Conclusion

In this paper, the street network is an important aspect in determining the evolutionary process of the city. From the existence of this city, the development of street network becomes more important not only to cater the transportation itself but to make the city more livable with attractive street design and the enhancement of street network system. The evolutionary of the street network during British administration until now makes Melaka city street system improve, but at the same time, control development is taken care in order to avoid any harm to the heritage value. With the advancement of remote sensing technology, street network analysis can be done with the method of road extraction. While doing this research, there is some area that can be improved such as the resolution of images which are suitable to conduct the extraction features such as a road. High-resolution satellite images are a must in order to extract features such as road, building, roof and henceforth. Indeed, this method can analyse high-resolution imagery but still maintain large-scale mapping and geospatial database.

Acknowledgements The authors greatly acknowledge the Malaysian Remote Sensing Agency and Town and Urban Planning Department for providing invaluable respective data used in this study. Authors sincerely thank all referees for their suggestions to improve the manuscript.

References

Barthelemy, M. (2015). *From paths to blocks: New measures for street patterns*. Environment and Planning B: Planning and Design.
Cheng, J. (2011). Exploring urban morphology using multi-temporal urban growth data: A case study of Wuhan, China. *Asian Geographer, 28*(2), 85–103.
Li, X., & Yeh, A. G.-O. (2004). Analyzing spatial restructuring of land use patterns in a fast growing region using remote sensing and GIS. *Landscape and Urban Planning, 69*(4), 335–354.
Lynch, K. (1960). *The Image of the city*. MIT Press.
Moudon, A. V. (1997). Urban morphology as an emerging interdisciplinary field. *International Seminar on Urban Form, 1*, 3–10.
Omer, I., & Jiang, B. (2010). *Geospatial analysis and modelling of urban structure and dynamic*. Berlin: Springer.
Paul, S., & Congress, J. (2008). *QUT digital repository: Presenting the past: The impact of urban morphology in shaping the form of the city*.
Sharifah, K. S. O., Mohamad, N. H. N., & Abdullah, S. M. S. (2013). The influence of urban landscape morphology on the temperature distribution of hot-humid urban centre. *Procedia—Social and Behavioral Sciences, 85*, 356–367.
Shpuza, E. (2014). Allometry in the syntax of street networks: Evolution of Adriatic and Ionian coastal cities 1800–2010. *Environment and Planning B: Planning and Design, 41*(3), 450–471.

Controlling Leapfrog Sprawl with Remote Sensing and GIS Application for Sustainable Urban Planning

Nur Aulia Rosni, Norzailawati Mohd Noor, Alias Abdullah and Isoda Setsuko

Abstract This paper aims to measure and characterise urban sprawl development in Kuala Lumpur city using leapfrog geospatial indices. The researcher utilised remote sensing satellite data such as Landsat and Spot images for two different times. The remote sensing application was subsequently integrated with GIS database to detect changes and analyse the pattern of growth for urban areas in Kuala Lumpur. From the finding of land use change detection, the leapfrog sprawl index was calculated by using geospatial indices formula. The results proved that Kuala Lumpur is the most highly developed city in Malaysia with new development is leapfrogging towards the periphery of the city and infill development pattern seemingly increase vastly filling up the leapt areas cause by leapfrog sprawl. Improper planning will create this type of urban sprawl that is predicted to expand beyond the border towards other states locating adjacent to Kuala Lumpur which is now called Greater Kuala Lumpur. The current scenario has become an absolute threat towards Malaysia planning goal to achieve sustainable urban planning development.

Keywords Urban planning · Urban sprawl · Change detection
Remote sensing GIS · Leapfrog geospatial indices

N. A. Rosni (✉) · N. Mohd Noor · A. Abdullah
Department of Urban and Regional Planning, Kulliyah of Architecture
and Environmental Design, International Islamic University Malaysia,
Jalan Gombak, 53100 Kuala Lumpur, Malaysia
e-mail: nuraulia9@gmail.com

I. Setsuko
Architectural and Civil Design Engineering Department, Kumamoto
National College of Technology (KOSEN), Yatsushiro, Japan

1 Introduction

A city is a highly complex, socio-economic and spatial entity with a hierarchical order that also serves as a hub for human activities (Ji et al. 2006; Rajeshwari 2006). Apostolos (2007) suggests that the issues of emerging towards the sustainable and resilient city are highly related to the understanding of urban development phenomena. Urban development which includes the process of urban growth and urbanisation is a universal phenomenon taking place in the world. This process drives the changes in land use patterns and to what extent the changes occurred determined the process of urban sprawl (Ramachandra et al. 2014; Sudhira et al. 2004). On the other hand, urban sprawl is a critical issue in today's world and has become one of the most important matters in the urban area around the world since the twenty-first century whose importance is still growing today (Altieri et al. 2014; Arribas-Bel et al. 2011). While being a manifestation of development, it is known for its negative environmental and social impacts (Crawford 2007; Feng et al. 2015). Urban sprawl refers to the outgrowth of urban areas caused by uncontrolled, uncoordinated, and unplanned growth. The inability to visualise such growth during planning, policies and decision-making process has resulted in the sprawl that is both unsustainable and inefficient. Moreover, the unavailability of the precise definition prevents researcher to determine urban sprawl in the correct manner (Bhatta et al. 2010; Wilson et al. 2003). In short, sprawl is widely discussed but poorly understood. Thus, in order to measure and characterised urban sprawl development in Kuala Lumpur by using leapfrog geospatial indices with GIS and remote sensing application, urban sprawl is defined as inefficient and dysfunctional urban development pattern that significantly damages the utilisation of land in the landscape context which leads to the degradation of environment, economic and social well-being of a community.

2 Theoretical Review

Discontinuity of land use is one of the most cited dimensions of urban sprawl. Ibrahim and Sarvestani (2009) defined urban sprawl as a discontinuous development while Barnes et al. (2001) and Coisnon et al. (2014) identified leapfrog sprawl as a discontinuous pattern of development especially in urban area. According to them, the discontinuous and leapfrogging development is often fragmented, is widely separated and has blurred boundaries (Galster et al. 2001; Hamidi and Ewing 2014; Yue et al. 2016). Torrents and Alberti (2000) added that the scattering characteristics of sprawl manifest themselves in a variety of semblances, for example fragmentation, leapfrogging, discontinuous development, dispersal, and piecemeal development. According to Barnes et al. (2001), leapfrog sprawl is caused by different factors such as the physical geography which make continuous development prohibitively expensive to be executed or other factors such as different land use policies between political jurisdictions. Nevertheless, Yue et al. (2016) and Galster et al. (2001) asserted that natural features like water bodies, preserved wetlands, forests, public reservations and facilities are not viewed as interruptions of continuous development. Vyn (2012) argues that while the objectives of policies in an area are achieved, an unintended consequence of leapfrog effect might occur in another zone. In the beginning, leapfrog development occurred due to lack of government ability to adequately control sequencing of development growth (Hasse and Lathrop 2003). Crawford (2007) has explicitly regarded leapfrog sprawl as residential growth occurring at a significant distance from existing housing area. In this research, the leapfrog sprawl is defined as any new residential development that took place outside the 1500 m radius from the central business district (CBD).

3 Study Area

The study area is Kuala Lumpur, Malaysia, which covers an administrative area spread over 24,221.05 ha. The majority of the land cover and land use patterns consist of built-up areas (residential, commercial, industrial, institution, recreational area, road, infrastructure and utilities) and unbuilt areas (agriculture, forest, bare land), open space (recreational area) and water bodies. For development purposes, Kuala Lumpur is divided into six strategic planning zones (SPZ) by the Kuala Lumpur City Hall, namely (1) Sentul-Menjalara, (2) Wangsa Maju-Maluri, (3) City Centre, (4) Bandar Tun Razak-Sungai Besi, (5) Bukit Jalil-Seputeh and (6) Damansara-Penchala. The population of Kuala Lumpur was 1.6 million in the year 2005 and is expected to increase to be approximately 2.2 million in the year 2020 (Kuala Lumpur City Plan, 2020). Kuala Lumpur also served as the capital and premier city which is the primary location for business and trade. Kuala Lumpur accommodates the regional headquarters of national and multinational companies, international and local commercial and financial services, specific high-end retail services, high technology manufacturing activities, education training services and national cultural institutions. The future growth of the conurbation is expected to be even more challenging as cities in Selangor expands, and the need to manage regional balance and environmental sensitivity must remain the main agenda for the conurbation. To achieve better future urban development and infrastructure planning, it is crucial for the Kuala Lumpur City Hall to recognise sprawl phenomenon happening in Kuala Lumpur including on how to measure and determine urban sprawl factors and patterns to control its growth.

3.1 Materials

This research mainly depended on the data obtained from Malaysian Remote Sensing Agency (MRSA), Department of Survey and Mapping Malaysia (JUPEM), Federal Department of Town and Country Planning (FDTCP) and the local authority (Kuala Lumpur City Hall). The primary data sources include satellite data of multiband Landsat 5 and Landsat 8 data from the years 2005 and 2015 as well as panchromatic and multispectral images of Spot 5 from the year 2005 and Spot 6 from the year 2014. The GIS database consists of Kuala Lumpur land use, zoning and road network data ranging from the year 2005 to 2014. On the other hand, the secondary data consisted of published journal, articles, book and legal documents from government organisations. The software is used to process and analyse the raw data, and to generate the results included ERDAS Imagine 2014, ArcGIS 10.2, E-Cognition and MapInfo Professional 12.0.

3.2 Method

There are many approaches to measuring leapfrog development sprawl, but the most commonly used method is through a land use change model. The land use change patterns could be linked to changes in farmland values in areas where leapfrog development may be occurring (Vyn 2012). In this research, the first step of land use change detection includes data preparation and preprocessing. The image to map procedure applied to satellite images. The geo-referencing techniques include radiometric and geometric correction using ground control points (GCPs) defined from the hybrid classification methods (integration of unsupervised and supervised classification) as well as from Kuala Lumpur map. The images were set to a particular map projection (Universal Transverse Mercator WGS 84). Maximum likelihood classification is used because it assumes that the statistics for each class in each band are usually distributed and calculates the probability that a given pixel belongs to a particular class. The accuracy assessment of the satellite data achieves overall accuracy above 85% in the processing stage. The change detection technique is used to extract the changes of residential land use from 2005 to 2015 (10 years gap) for the purpose of analysing the leapfrog sprawl growth pattern in Kuala Lumpur.

The leapfrog sprawl in Kuala Lumpur was measured by the distance of newly built residential areas from the CBD in each SPZ. The grade range that is considered as smart growth, natural development and leapfrog sprawl are specified in Table 1. The parameters presented are based on the study conducted by Hasse and Kornbluh (2004) where they grade urban sprawl occurrence based on accessibility. The area that is easily accessible by walking (which means the area are automatically approved as ideally accessible by another mode of mobility, i.e. bicycle and automobile) being considered as convenient thus smart growth. Based on the study

Table 1 Leapfrog development sprawl indicators

Grade	Distance from central business area	Parameters	Annotation
A	0–500 m	Walking smart growth	Smart growth
B	501–1500 m	Bicycle smart growth	Common development
C	>1500	Suburban sprawl	Leapfrog sprawl

Sources Modified from Hasse and Kornbluh (2004)

by Inani et al. (2012), the maximum distance of walkability for Malaysia's urban areas is 600 m and the most comfortable walking distance is 400 m (5 min) for all age groups with consideration of several factors such as pedestrian age and health limits, safety, walkway facilities and environmental factors such as climate. Since the walking distance in the indicator is 500 m which is still within the maximum and comfortable walking distance in Malaysia, the researcher decided to use the same length generated by Hasse and Kornbluh (2004). The second grade is the area that is moderately accessible by walking but easily accessible by bicycle and automobile is considered as bicycle smart growth in their research work or common development in this research. Lastly, the area that is poorly accessible by walking and cycling but only ideally accessible by automobile is considered as sprawling. In this research context, the accessibility is discussed between the newly developed residential area and the CBDs. This measurement method is expected to solve the problem of sprawl due to different boundary governed by various state and local authority as currently facing many cities in Malaysia including Kuala Lumpur.

The newly developed residential areas for leapfrog sprawl measurement adopted the mapping unit by Irwin et al. (2006) indicating minimum mapping unit for newly developed residential area is 10 acres in urban area and 40 acres for all other classes. This mapping unit is used to exclude the infill, redevelopment or any other development types that hinder the accuracy of leapfrog sprawl measurement. In this research, the newly developed residential area is any land area within Kuala Lumpur that was converted and currently functions as residential use in the year 2015, which is derived from the land use and land cover change detection using both remote sensing and GIS applications.

Later, the selected residential development in 2015 with an area of 10 acres is buffered using 400 m radius, creating patches of the new residential area. Using the GIS spatial analyst tool, these residential patches are later converted into grid layers of 400 m cells resolutions and being overlaid with the layer of residential patches. The grid cells outside the buffered radius were eliminated as well as cells that are being covered by only ¼ (25%) and less of the buffered radius. Next, the CBD for each strategic planning zone as identified in Kuala Lumpur City Plan 2020 was buffered using distance as determined by three different grades as shown in Table 1. The next step includes overlay analysis between layers of new residential grid cells with the buffered CBD grade distance. The query and geographical tools in the GIS software were used to calculate the percentage of grid cells according to the parameters.

4 Analysis and Findings

The results and findings of leapfrog development sprawl measurement were indicated by the number of grid cells within the radius of buffered CBDs in each strategic planning zone. Patches of new residential growth that occur within the buffered radius of Grade C from existing CBD are considered as leapfrog sprawl, whereas residential development occurred within the buffered radius of Grade A is considered as smart growth. Strategic planning zone in Kuala Lumpur with patches of new growth with high leapfrog percentage is considered as sprawling, whereas low ratios are considered as smart growth. The value of leapfrog development sprawl (LF_{Gx}) was calculated by using the formula modified from Hasse, 2004 as in Eq. 1:

$$(LF_{Gx}) = \frac{\sum GC_{Gx}}{\sum GC_{Unit}} \times 100 \qquad (1)$$

where

(LF_{Gx}) = Leapfrog development sprawl according to strategic planning zone
(GC_{Gx}) = Number of grid cell according to CBD grade distance buffered area
      = (G_a) 0–500 m from central business district
      = (G_b) 501–1500 m from central business district
      = (G_C) > 1500 m from central business district
(GC_{Unit}) = Number of grid cell units

The leapfrog development sprawl measurement method was conducted for all SPZ in Kuala Lumpur. The results and findings were presented in percentages value as shown in Table 2 in three main categories, namely Grade A (smart growth), Grade B (common development) and lastly Grade C.

The findings identified that, of all new residential development in Kuala Lumpur, three out of six SPZ scored as common development (Grade B), followed by the remaining three SPZ that scored leapfrog sprawl (Grade C). Sentul-Menjalara scores the highest percentage of common development with (15.63%) followed by Damansara-Penchala strategic zone (13.67%) and lastly the city centre with only 0.78%. The strategic zone that is currently facing the leapfrog sprawl problem includes Bandar Tun Razak-Sg.Besi (13.28%), Bukit Jalil-Seputeh (9.77%) and Wangsa Maju-Maluri (7.81%). Apparently, the percentage of leapfrog development sprawl of Wangsa Maju-Maluri SPZ is same with the leapfrog percentage of Damansara-Penchala SPZ. Although only half of the Kuala Lumpur SPZ is regarded as leapfrog sprawl, the remaining SPZ did not score smart growth (Grade A) either. As previously stated in this research, SPZ that scored that common development indicator is prone to sprawling as this area depended highly on the successful implementation of good urban planning and management by the authority. As for now, the total findings from research measurement prove that Kuala Lumpur is facing a major problem of leapfrog development sprawl with

Table 2 Leapfrog development sprawl land area by SPZ

Kuala Lumpur strategic zone	Grade A (0–500 m) (%)	Grade B (501–1500 m) (%)	Grade C (1501–3000 m) (%)	Total (%)	Annotations
Sentul-Menjalara	7.03	15.63	7.42	30.08	Common development
Wangsa Maju-Maluri	1.56	3.13	7.81	12.50	Leapfrog sprawl
City Centre	0.00	0.78	0.00	0.78	Common development
Bandar Tun Razak-Sg.Besi	3.52	3.91	13.28	20.70	Leapfrog sprawl
Bukit Jalil-Seputeh	0.00	0.00	9.77	9.77	Leapfrog sprawl
Damansara-Penchala	4.69	13.67	7.81	26.17	Common development
Total	16.80	37.11	46.09	100.0	Leapfrog sprawl

dominant percentage scores of 46.09% over smart growth that only scores 16.80% and common development 37.11%.

5 Discussion

Leapfrog sprawl has brought significant negative effects on not only on the agriculture land in rural area, but also on the environment and city life as well. This is because leapfrog urban sprawl has led to huge loss of high-quality fertile lands in the suburbs and has encroached upon limited open space, such as forests, grassland and water area. Moreover, the leapfrog sprawl caused the growing traffic burden by increasing the distance between the newly developed land and city centres where job opportunities concentrate upon (Feng 2008). According to Osman et al. (2006), this form of development is the most costly with respect to providing urban services such as water and sewerage.

At first glance, the leapfrog sprawl phenomenon is likely to occur in Kuala Lumpur when the Kuala Lumpur City Hall (2011) clearly stated that the land use pattern had been the result of past practices and development trends, which were based on single land use zoning. Thus, the past practice of urban planning in Kuala Lumpur has characterised the land use phenomena either as monocentric versus polycentric forms, centralised versus decentralised patterns and continuous versus discontinuous developments. As a result, Kuala Lumpur is experiencing population loss to surrounding the areas of the city due to decentralisation of economic development to other more industrialised zones and the availability of relatively cheaper housing development in other parts of Klang Valley. Moreover, the

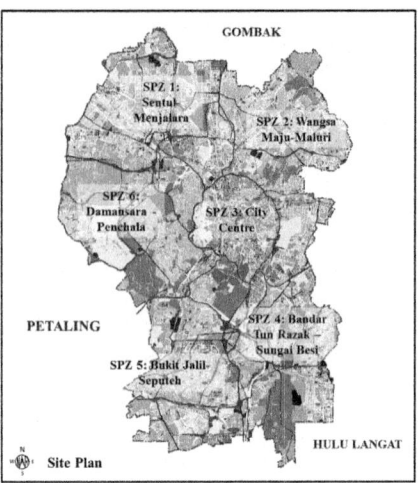

Fig. 1 Kuala Lumpur study area

establishment of Putrajaya as the national federal administrative centre in 1999 has caused the population to shift out from Kuala Lumpur where most of the administrative office previously stood (Kuala Lumpur City Plan, 2020) (Fig. 1).

For this research, the land use geospatial indices model of leapfrog development sprawl identifies one of the urban development categories, namely the outlying urban growth (leapfrog development) from infill and common land use expansion. The infill growth is characterised by a newly developed residential area in the year 2015 with an area less than 10 acres. It can also be defined as the residential development of a small tract of land mostly surrounded by a built-up land cover, while the common expansion development types are newly developed residential areas within the appropriate specified distance from the CBDs. Lastly, the outlying growth or leapfrog development is characterised by a change from non-residential to residential development occurring beyond the specified distance from the existing CBDs and likely approaching the urban fringe (Fig. 2).

The city of Kuala Lumpur shares boundaries with several districts that fall under different municipalities or local planning authorities within the state of Selangor, which is also regarded as Greater Kuala Lumpur. Based on Fig. 2, the pattern of leapfrog development growth in Kuala Lumpur tends to move towards the fringe of Wangsa Maju-Maluri SPZ that adjoined with Gombak district, Bandar Tun Razak-Sg.Besi SPZ with Hulu Langat district and Bukit Jalil-Seputeh SPZ with Petaling district. The dynamic urban development of Kuala Lumpur has great impacts towards these surrounding areas, and this situation led to many unwanted issues towards the various municipalities governing the areas as well as the Kuala Lumpur City Hall.

In the same manner, the new development within the strategic planning zone of Kuala Lumpur is also highly influenced by the development pattern of its

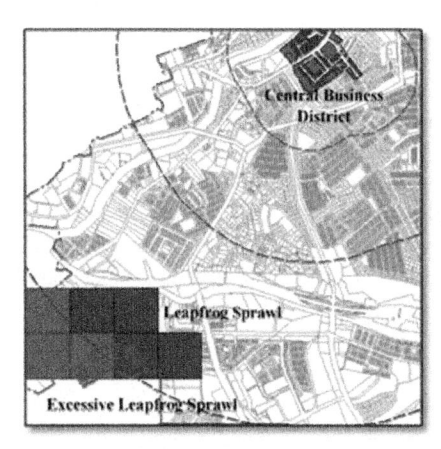

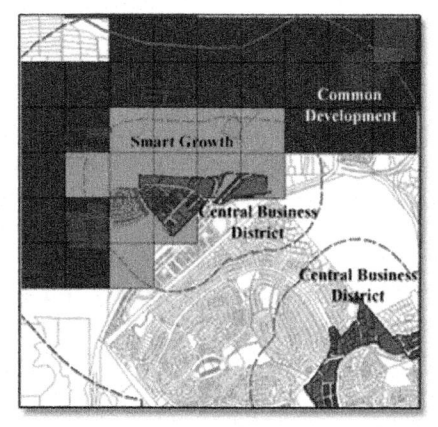

Fig. 2 Leapfrog development sprawl pattern (left) vs. smart growth (right)

surrounding districts. Undeniably, both change detection analysis and leapfrog analysis conducted in this research proved that the urban development in Kuala Lumpur is moving towards its fringe. Sentul-Menjalara SPZ, City centre SPZ and Damansara-Penchala SPZ are currently showing moderate leapfrog sprawl tendency. However, this situation might turn to be leapfrogging in the future if measures to control its growth are not effectively executed.

Historically, the Malaysian land use planning system lacks ability to monitor the sequencing of new development growth adequately either caused by natural topography factor, land ownership, the price of real estate, accessibility, infrastructure or policies by the government. This disorganised pattern of development has a domino effect on property functionality. The current trend in Kuala Lumpur is more development demand of residential land use at the fringe area due to lower property price, and there are a lot of access and modes of mobility towards the centre. Moreover, the Greater Kuala Lumpur area plays a significant role in accommodating this type of sprawl. It produces an increasingly fragmented land use pattern that in turn leads to elevated transportation requirements and other infrastructure and utilities. Moreover, Holcombe et al. (1999) revealed that leapfrog development also nurtures the strip sprawl growth. The vacant land in between employment centre and residential zones attracted some developer to use it for commercial activities and attract the commuting population to stop thus contributing to strip sprawl. Fortunately for Kuala Lumpur, the leapfrog development sprawl problem does not contribute to strip sprawl yet, but without any efforts to kerb its growth, the leapfrog sprawl might causing not only strip sprawl in the future but other types of land use geospatial indices sprawl as well.

6 Conclusion

The complex nature of land use pattern in urban sprawl requires measures to employ some multiple geospatial indices. The paper examines one of the five land use geospatial indices (LUGI) of urban sprawl, namely leapfrog development using remote sensing imagery data and GIS approach. The other LUGI includes (1) segregated land use, (2) development planning consistency, (3) urban density and (4) strip sprawl. The application of technology in city management is necessary since cities were growing rapidly in most developing countries. Leapfrog development index provides a significant method for identifying, comparing and contrasting sprawl development in a more detailed manner for further investigation of the underlying process at play. As urban patterns for given region change with time, which are reflected in changing sprawl index value and its technological tools, they may offer insights into the long-term trends, underlying process and likely the consequences of spreading development compared to the smart growth analysis.

Acknowledgements The authors are grateful especially to the International Islamic University Malaysia, anonymous reviewers and other contributors in improving this paper. The research is especially indebted to the Malaysia Remote Sensing Agency (MRSA) for the digital land use satellite imagery database. The authors would like to mention International Institute for Applied System Analysis (IIASA) and Academy of Sciences Malaysia (ASM) for the opportunity to join their program thus provide knowledge and valuable experience contributing to this research.

References

Altieri, L., Cocchi, D., Pezzi, G., Scott, E. M., & Ventrucci, M. (2014). Urban sprawl scatterplots for urban morphological zones data. *Ecological Indicators, 36,* 315–323. https://doi.org/10.1016/j.ecolind.2013.07.011.

Apostolos, L. (2007). Fractal analysis of the urbanisation at the outskirts of the city: Models, measurement and explanation. *Cyber, 39,* 1–15.

Arribas-Bel, D., Nijkamp, P., & Scholten, H. (2011). Multidimensional urban sprawl in Europe: A self-organizing map approach. *Computers, Environment and Urban Systems, 35*(4), 263–275. https://doi.org/10.1016/j.compenvurbsys.2010.10.002.

Barnes, K. B., Morgan, J. M., Roberge, M. C., & Lowe, S. (2001). *Sprawl development: Its patterns, consequences, and measurement.* Baltimore, Maryland.

Bhatta, B., Saraswati, S., & Bandyopadhyay, D. (2010). Urban sprawl measurement from remote sensing data. *Applied Geography, 30*(4), 731–740. https://doi.org/10.1016/j.apgeog.2010.02.002.

Coisnon, T., Oueslati, W., & Salani, J. (2014). Urban sprawl occurrence under spatially varying agricultural amenities. *Regional Science and Urban Economics, 44*(1), 38–49. https://doi.org/10.1016/j.regsciurbeco.2013.11.001.

Crawford, T. W. (2007). Where does the coast sprawl the most? Trajectories of residential development and sprawl in coastal North Carolina, 1971–2000. *Landscape and Urban Planning, 83*(4), 294–307. https://doi.org/10.1016/j.landurbplan.2007.05.004.

Feng, L. (2008). Applying remote sensing and GIS on monitoring and measuring urban sprawl. A case study of China. *International Journal of Applied Earth Observation and Geoinformation, 10*(1), 47–56.

Feng, L., Du, P., Zhu, L., Luo, J., & Adaku, E. (2015). Investigating sprawl along China's urban fringe from a spatio-temporal perspective. *Applied Spatial Analysis and Policy*, (September). http://doi.org/10.1007/s12061-015-9149-z.

Galster, G., Hanson, R., Ratcliffe, M. R., Wolman, H., Coleman, S., & Freihage, J. (2001). Wrestling sprawl to the ground: Defining and measuring an elusive concept. *Housing Policy Debate, 12*(4), 681–717. https://doi.org/10.1080/10511482.2001.9521426.

Hamidi, S., & Ewing, R. (2014). A longitudinal study of changes in urban sprawl between 2000 and 2010 in the United States. *Landscape and Urban Planning, 128,* 72–82. https://doi.org/10.1016/j.landurbplan.2014.04.021.

Hasse, J. E., & Lathrop, R. G. (2003). Land resource impact indicators of urban sprawl. *Applied Geography, 23*(2–3), 159–175. https://doi.org/10.1016/j.apgeog.2003.08.002.

Hasse, J., & Kornbluh, A. (2004). Measuring accessibility as a spatial indicator sprawl. *Middle States Geographer, 37,* 108–115.

Holcombe, R. G., Bast, J. L., Brown, M., Shaw, J. S., Benjamin, D. K., & Avenue, S. (1999). *Urban sprawl: Pro and con* (Vol. 17). Bozeman, MT.

Ibrahim, A. L., & Sarvestani, M. S. (2009). Sensing and GIS—Case study Shiraz City, Iran. *Urban Remote Sensing Joint Event.*

Inani, D., Abdul, H., Zamreen, M., & Amin, M. (2012). Comparing the walking behaviour between urban and rural residents. *Procedia—Social and Behavioural Sciences, 68,* 406–416. https://doi.org/10.1016/j.sbspro.2012.12.237.

Irwin, E. G., Bockstael, N. E., & Cho, H. J. (2006). Measuring and modelling urban sprawl: Data, scale and spatial dependencies. In *Urban Economics Sessions, 53rd Annual North American Regional Science Association Meetings of the Regional Science Association International.* Toronto, Canada.

Ji, W., Ma, J., Twibell, R. W., & Underhill, K. (2006). Characterising urban sprawl using multi-stage remote sensing images and landscape metrics. *Computers, Environment and Urban Systems, 30*(6), 861–879. https://doi.org/10.1016/j.compenvurbsys.2005.09.002.

Kuala Lumpur City Hall. (2011). *Draft Kuala Lumpur City Plan 2020.* Percetakan Nasional Malaysia Berhad.

Osman, N. H., Wahab, H. A., Mahmud, M. R., Mohd Sopian, N. F., Wan Mustafa, W. M. R., Yip, K. C., et al. (2006). *Remote sensing and geographical information system for urban sprawl analysis. Remote sensing training camp.* Johor Bahru, Malaysia.

Rajeshwari, (2006). Management of the urban environment using remote sensing and geographical information systems. *Journal of Human Ecology, 20*(4), 269–277.

Ramachandra, T. V., Bharath, a. H, & Sowmyashree, M. V. (2014). Monitoring urbanisation and its implications in a mega city from space: Spatiotemporal patterns and its indicators. *Journal of Environmental Management.* https://doi.org/10.1016/j.jenvman.2014.02.015.

Sudhira, H. S., Ramachandra, T. V., & Jagadish, K. S. (2004). Urban sprawl: metrics, dynamics and modelling using GIS. *International Journal of Applied Earth Observation and Geoinformation, 5*(1), 29–39. https://doi.org/10.1016/j.jag.2003.08.002.

Torrents, P. M., & Alberti, M. (2000). Measuring sprawl (No. 27). In *Association of Collegiate Schools in Planning Conference.* London.

Vyn, R. J. (2012). Examining for Evidence of the Leapfrog effect in the context of strict agricultural zoning. *Land Economics, 88*(August), 457–477.

Wilson, E. H., Hurd, J. D., Civco, D. L., Prisloe, M. P., & Arnold, C. (2003). Development of a geospatial model to quantify, describe and map urban growth. *Remote Sensing of Environment, 86*(3), 275–285. https://doi.org/10.1016/S0034-4257(03)00074-9.

Yue, W., Zhang, L., & Liu, Y. (2016). Measuring sprawl in large Chinese cities along the Yangtze River via combined single and multidimensional metrics. *Habitat International, 57,* 43–52. https://doi.org/10.1016/j.habitatint.2016.06.009.

Constructing and Modeling 3D GIS Model in City Engine for Traditional Malay City

Ahmad Afiq Aiman Abdullah, Norzailawati Mohd Noor and Alias Abdullah

Abstract Malay urbanization is one of the oldest urbanization in the human history. Traditional Malay also refers to the Malay races which full of custom, heritage, and also full with decency values while Traditional Malay City refers to the Malay City which establish before the British colonial era. Traditional Malay City have their own way of socioeconomic aspect without adapting the Western influence. The formation of the Malay settlement and Malay Traditional Cities has their own uniqueness and specialty in terms of social life and integration between the land uses and building. Rapid development and uses of geographic information system (GIS) as a tool in urban planning nowadays have pushed the GIS technique further in visualization and presentation of urban planning data. Therefore, the main objective of this paper is to study on the formation and constructing the existing Traditional Malay city.

Keywords Malay city · 3D GIS modeling · CityEngine and urban planning

1 Introduction

The paper was studied the form of urban form in Traditional Malay and extend to modeling a 3D of Malay city through the use of CityEngine software. It integrates the 2D data GIS data with the 3D GIS analysis. For 3D GIS, researchers explore the CityEngine software which is one of the ESRI families in order to construct basic 3D models for the sample city of Traditional Malay City. The objective of the paper is to construct the 3D model based on 3D GIS application for the Traditional Malay City which in this paper, Kota Bharu was selected as the case study.

A. A. A. Abdullah (✉) · N. Mohd Noor · A. Abdullah
Department of Urban and Regional Planning, Kuliyyah of Architecture
and Environmental Design, International Islamic University Malaysia,
Selangor, Malaysia
e-mail: irme.aiman@gmail.com

© Springer Nature Singapore Pte Ltd. 2018
R. Saian and M. A. Abbas (eds.), *Proceedings of the Second International Conference on the Future of ASEAN (ICoFA) 2017 – Volume 2*,
https://doi.org/10.1007/978-981-10-8471-3_28

Urban forms are based on result of the urban growth that started based on migration from rural area to a city and planned settlement. The importance of urban morphology as a field of academic research and practice began mainly in the second half of the twentieth century, with the development of an urban analysis not limited to the physical elements, but also to their context and the social and cultural aspects that characterize the urban space and its connections (Fumega et al. 2014). There are a few different meanings that express the urban form, for instances, Moudon (1997) indicates that urban form is defined by three fundamental physical elements—buildings and their related open spaces, plots or lots, and streets. Levy shows common elements on the research on urban form analysis: plot, street, constructed space, and the open space. While Lynch indicates that urban form encompasses the spatial disposition of people developing their activities, the resulting spatial movements, products and information, and the physical characteristics that modify significantly the space for those actions.

2 Traditional Malay Urban Form

Traditional Malay refers to the period of the Malay city or Malay settlement before the colonial period, especially by British. There are a lot of Malay Kingdom, especially in the Southeast Asia region. The old Malay Kingdom such as Malacca, Acheh, Palembang were among establish and famous Malay Kingdom during the fourteenth Century (Harun and Raj 2012). Besides, Kelantan also is recorded as one of the Malay states with the earliest Malay civilization and one of the states that free from the British colonization until years 1900. With the uniqueness based on the location and history of the civilization and war between the Kings, Kelantan until nowadays still preserves most of the historical building and sources that can be used as the reference for the paper. In Malaysia, old traditional city have a lot of treasure and unique heritages and historical element in terms of their history of settlement, architecture, identity of the city, culture, and also the urban morphology. Old city of the day still carry past memories that tell their own tales of history they have seen.

2.1 3D GIS Modeling

3D modeling was one of the planning tools that gives new dimension in development planning and decision-making process. Usually, 3D model is used as one of the presentation methods to give clear view on the new development, but in the paper, 3D model will focus on the 3D GIS application where to construct the 3D GIS model for the old traditional Malay City. The main objective the use of 3D GIS based for the 3D model is to look into the CityEngine software, and to see if it was a compatible and suitable to use as a new planning tool, especially in Malaysia context. Even though the paper is on constructing a city, but only a certain portion of the city area will be used as the case study in the paper.

2.1.1 3D GIS Software: CityEngine

CityEngine one of the ESRI's family whose among the big players in GIS industry. CityEngine was established well since its commercial release in 2008 and this software has been used extensively to create the highly detailed 3D. One of the specialties of CityEngine software is it has the option to import data either from architectural or geographic dataset such as, CAD and ArcGIS and its possibilities of modeling real landscape are quite promising. By using CityEngine, user can choose the existing data on the street pattern either radial, organic, and raster, or user also can determine by himself on the type of road either based on existing road or new one. After growing the streets patterns, CityEngine will create the lots between the roads and divided into polygon shapes. The user can then assign a CGA (computer generated architecture) rule file to each of these polygons shapes which defines the shape and split grammar. The user can also use height maps and obstacle maps to limit the city; the use of population data is also acceptable to influence street growth patterns (Maren et al. 2012).

2.1.2 How CityEngine Works

CityEngine basically works based on the script as the main of its operation. The script will determine the rule and the parameter for the polygon or shape in 3D model. This rule can be user-defined rule or can use from the sample rule provided by the ESRI. There are a few sources of data that can be used to get the based map for sit which in this paper Kota Bharu as the model. Script in CityEngine plays the vital role as the scripts will define the rule and parameter sets by the user. Script will determine the process in the CityEngine such as extruding into 3D model, roof creation, coloring, setback, and some more. In the paper, researchers explore on a few script to extrude the block as the massing and put basic color on the 3D massing.

3 Methodology

3.1 Materials and Methods

Based on the qualitative type of research, this paper will strengthen on the constructing the 3D model for Kota Bharu City in order to analyze and determine the pattern and formation of the Traditional Malay City. Methodology is divided into two parts in which the first part is to determine the study area for the Traditional Malay City and categorized by the element of the traditional urban component. Second part of the paper is to construct the existing area of the traditional model based on the existing data. In this paper, researchers explore on the open street map (OSM) which is one of the open sources that shows the current street and building on map as the main source of 2D GIS data. The OSM data can be exported from www.Openstreetmap.org which user only need to search for the site location and edit and updated directly from the OSM for the existing new building and land use data. Using 3D GIS software, CityEngine, the model will extrude into 3D massing block and will differentiate between the main land use categories such as historical building, commercial building, residential area, and government agencies building.

3.1.1 Case Study: Kota Bharu

Kota Bharu, which is located in state of Kelantan, becomes the case study for the research. Due to the unique history, and also one of the state that free from the British colonization until 1900 was one of the best example for the paper. Kota Bharu significantly means as "New Fort" which resembles the establishment of new city during 1845. Traditional Malay city make the fort boundaries to resemble their main city boundaries which contain the administrative center for the city. Istana Balai Besar which is the symbol of proud for Kelantanese was one of the early palaces of Sultan Kelantan during eighteenth century. Sultan Muhammad II built the Istana Balai Besar as a new palace and as the administrative center in Kota Bharu. From that point, the Kota Bharu city being planned by the local planner and architect during that time and growth as one establish cities until British came to Kelantan during nineteenth century.

3.1.2 Methods

In summary, there are three stages of methodology in this paper: first part is the data collection stages, second part is data processing stages, and the last part is the output and the discussion.

Data Collection

There are two types of data adopted in conducting this research: primary data and secondary data. Secondary data are taken from (i) Majlis Perbandaran Kota Bharu Bri, (ii) JUPEM, and (iii) Majlis Agama dan Istiadat Kelantan MAIK. Type of the secondary data includes the 2D spatial data, old maps, reports, and also land title. A few trip to Kota Bharu has been done in order to collect for the primary data for this paper where the primary data consist of 2D images of the existing land use and building, and input from the interview with the representative such as, State

Muzium Officers, Historian, and a few NGOs where active in historical data research and collection. In order to determine the original area for the Traditional Malay City in Kota Bharu, the historical data need to dig in so that, the segregation between the building existed before British colonization and after colonization period. This is important so that identifying the traditional city pattern can be analyzed and differentiated between city development done by local or ruler and British government. In this paper, the construction of 3D model will focus on the current land use in order to see the height and the location of the current building, so that the based GIS data that used are imported from OSM street data.

Data Processing

During data collection stages, there are bulk of data and historical record being obtained for the paper. All the data need to be sort out accordingly, especially where involved the timeline record. For example, land survey from JUPEM, where each of the plan varies based on the year the plan, was drawn. Using the overlay technique, the land plan is sorted out and tallied with the old map of Kelantan in year 1910. Figure 1 shows, map for Kota Bharu during 1910 where the main and earliest development for the Kota Bharu city. The radius of 500 m from the Istana Balai Besar has been selected as the main area for the study, where it represents the closest area from the administration center and the other city element. Besides, this area has been identified as the area that consists of the most historical building and the settlement area which already exist from 1845. Based on the primary and secondary data, the area of 500 m radius from Istana Balai Besar was demarcated using the GIS data where OSM data and MapInfo data were used in the paper.

In order to get clear and updated spatial data, the OSM data need to refine by drawing the polygon for the existing building and aligned with the primary and secondary data. The updated polygon of the existing building which updated in the OSM data will automatically created as the polygon when exported into the CityEngine software. The polygon will categorize and grouped into land use class order and the type of the building use. In order to differentiate between the historical building and new building, all the polygon which represents the building from 1945 till 1910 was grouped as "Historical Building" group. Figure 2 shows the different group of the updated polygon according to the building and land use class order.

CityEngine software was script-based software where, all the command and 3D construction based on script and rule format including. In order to construct basic 3D block massing, the new script known as CGA Rule need to be set up. Under the left console, there are space that where user will enter the script to define the characteristic of the 3D models. The script need to specify the attribute where meet the processing the polygon in CityEngine. In order to construct basic 3D model based on the height of the building, the attribute which represents by "attr" will define the command for the process. The other important script also is the start rule. Building footprint is calculated the area of the polygon shape in the CityEngine which in this paper, area used was generated directly from the OSM data. To

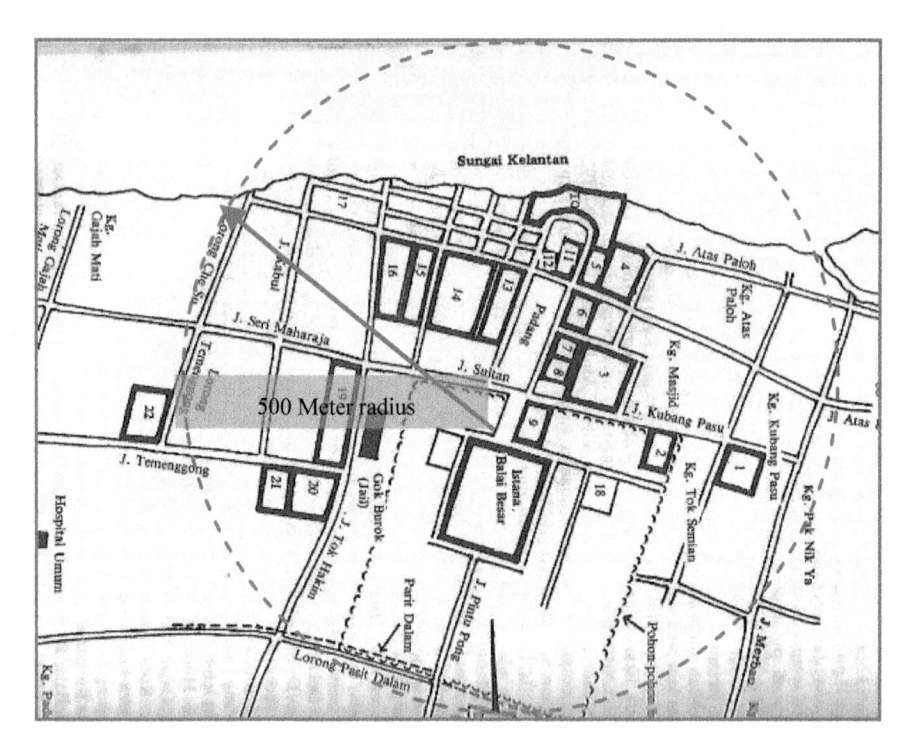

Fig. 1 Kota Bharu map in 1910

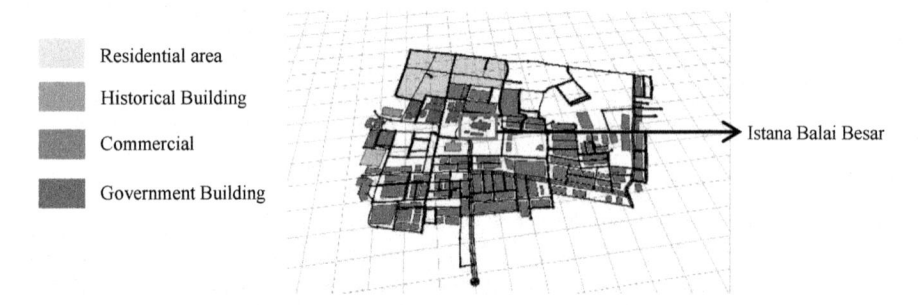

Fig. 2 Classification of polygon in CityEngine

differentiate between buildings heights, each of the new shape layer was assigned by different rule where the height is based on the existing building height range. Figure 3 shows the basic rule script used for the basic 3D model for Kota Bharu where extrude command defined the height of the building and wall, and color used to differentiate between the different land use and building classes.

Fig. 3 Example script for 3D construct

```
@Range (3,30)
attr Height=3
attr RoofType="Flat"

@StartRule
BuildingFootprint -->
    extrude(3)
    comp(f) {top:Roof|side:Wall}

Wall-->
    color(0,1,0)
Roof -->
    color (1,0,1)
```

4 Result and Discussion

The main result in this paper is to come out with basic 3D model for the existing land use and building of Traditional Malay City and to look on the differentiation between the land use pattern between traditional period and after colonization period. As Fig. 4 shows the 3D massing block for the Istana Balai Besar building footprint and its surrounding which generated based on different height. Besides height, the color of the building also was assigned to show the difference between the historical building and after colonization building in Kota Bharu. Also using script rule, the color attribute still needs to be defined in the rule. In addition, flat roof was assigned into the 3D massing model for better visualization and further analysis of the models. Combination and integration between primary and secondary data, the 3D block massing in Fig. 5 shows the different between the building height and the land use pattern for Kota Bharu. The historical building represents the early building as the city establishment in 1845 still exists as a proof of the originality of the Traditional Malay City in Kelantan. The land use pattern can be analyzed based on the timeline and date of the building was built. Most of the commercial building which represent by yellow block were built during the colonization period while the settlement area and the historical building were originally from the early era of Malay City where establish by Sultan Muhammad II in 1845. From the 3D models, the location of the historical building which function as the administrative center during the early period and the residential area located within 500 meters radius, and the land use pattern for the traditional Malay City more toward organic pattern on the land use, street and block alignment. Exporting and overlaying the 3D City model into google earth as shown in Fig. 5 can open for future research or others analysis on the city formation and can relate with planning policy and guidelines, future development planning, preservation, tourism planning or disaster management can be explore more since the 3D model in CityEngine are

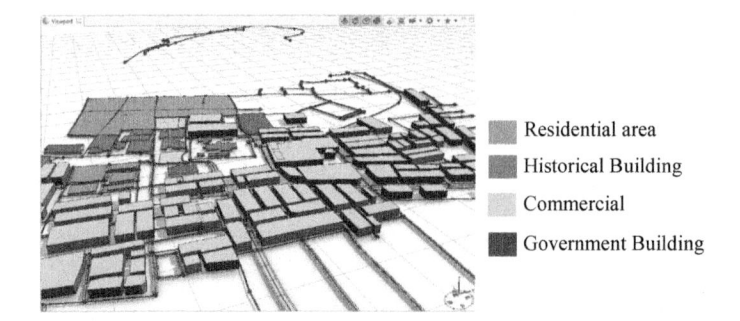

Fig. 4 Kota Bharu 3D massing in CityEngine

Residential area
Historical Building
Commercial
Government Building

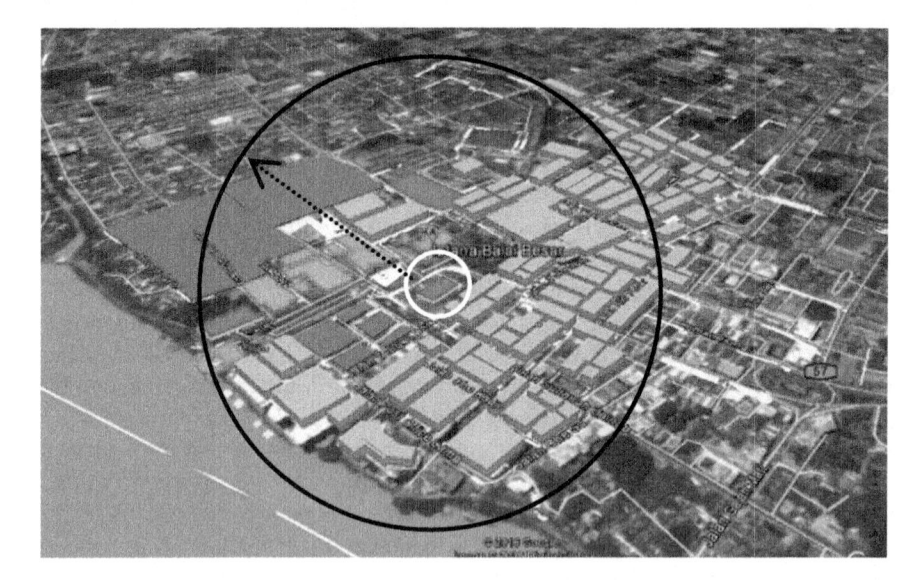

Fig. 5 3D GIS overlay

based on rule and parameter set by the user, and this nature of CityEngine can give more efficient and fast result of analysis.

5 Conclusion

CityEngine is one of the latest 3D GIS software that really can help in development planning and decision-making process. This software can act as the 3D GIS software that is powerful in creating realistic 3D models. This software also well suited in generating 3D GIS model based from the 2D spatial data such as ESRI, Google earth, OSM, and also AutoCAD product. Besides only for visualization, City Engine also can act as the parameter for new development based on the existing policy and parameters, especially by local authority. In this paper, research only focuses on the constructing based on the existing data, for the Traditional Malay City building, in future more research can be done, especially on much details element and component in Traditional Malay city.

Acknowledgements The authors greatly acknowledge the International Islamic University of Malaysia for a research grant on Trans-Disciplinary Research Grant Scheme (TRGS) (TRGS16-03-002-00002), Ministry of Higher Education Malaysia for providing invaluable respective data used in this study. Authors sincerely thank all referees for their suggestions to improve the manuscript.

References

Arran Rothwell-Eyre. (2014). *Introduction to 3D modelling VERSION 8.X with additions by Dave Green & Ciaran Abrams.*
Fumega, J., Niza, S., Ferrão, P. (2014). *Identification of urban typologies through the use of urban form metrics for urban energy and climate change analysis.*
Harun, N., & Raj, J. (2012). The morphological history of The Malaysian urban form. *IPEDR, V48* (24), 111–116. https://doi.org/10.7763/IPEDR.
Maren, G. V., Shephard, N., Schubiger, S. (2012). *Developing with Esri Cityengine*, viewed September 29 at http://Proceedings.Esri.Com/Library/Userconf/Devsummit12/Papers/Developing_With_ESri_Cityengine.Pdf.
Moudon, A. V. (1997). Urban morphology as an emerging interdisciplinary field. *Urban Morphology, 1*(1), 3.

3D Data Fusion Using Unmanned Aerial Vehicle (UAV) Photogrammetry and Terrestrial Laser Scanner (TLS)

Mohamad Aizat Asyraff Mohamad Azmi, Mohd Azwan Abbas, Khairulazhar Zainuddin, Mohamad Asrul Mustafar, Mohd Zainee Zainal, Zulkepli Majid, Khairulnizam M. Idris, Mohd Farid Mohd Ariff, Lau Chong Luh and Anuar Aspuri

Abstract Recognizing the various advantages offered by 3D new metric survey technologies for 3D modelling reconstruction, this study presents a method or procedure of complete 3D data acquisition, using different sensors, and their possible fusion. Besides, this involved a discussion of the capability of UAV photogrammetry in order to collect data for the rooftop of the building with the low cost and combine the data with terrestrial measurement technique. Meanwhile, the aim of this study is to identify the minimum network configuration as to combine TLS and UAV photogrammetry data fusion. Hence, some procedure has been done in the methodology by registering TLS and UAV point clouds by employing multi-network configuration. Furthermore, the data fusions' results were evaluated on the registration error of UAV towards TLS coordinate system via the various number of network configuration samples. Based on the experimental results, the sample of network configuration for the three registration points results in a good condition due to the well-distributed registration points and less errors of both point clouds' registration. As conclusion, a good quality of data fusion between TLS and UAV photogrammetry is determined by a good selection number of the registration points via several samples of network configuration.

Keywords Terrestrial laser scanner · UAV · Photogrammetry · Building 3D data · Point clouds · Registration · Network configuration · Data fusion

---

M. A. A. Mohamad Azmi (✉) · M. A. Abbas · K. Zainuddin · M. A. Mustafar · M. Z. Zainal
Faculty of Architecture, Planning and Surveying, Universiti Teknologi MARA, Perlis, Malaysia
e-mail: aizatazmi92@yahoo.com

Z. Majid · K. M. Idris · M. F. Mohd Ariff · L. C. Luh · A. Aspuri
Faculty of Geoinformation and Real Estate, Universiti Teknologi Malaysia, Skudai, Johor, Malaysia

© Springer Nature Singapore Pte Ltd. 2018
R. Saian and M. A. Abbas (eds.), *Proceedings of the Second International Conference on the Future of ASEAN (ICoFA) 2017 – Volume 2*,
https://doi.org/10.1007/978-981-10-8471-3_29

1 Introduction

The integration of data and knowledge from several sources is known as data fusion. Data fusion is the process of integration of multiple data and knowledge representing the same real-world object into a consistent, accurate and useful representation (Buckley et al. 2013). Besides, data fusion processes are often categorized as low, intermediate or high, depending on the processing stage at which fusion takes place. The expectation result is that fused data is more informative and similar than the original inputs (Guarnieri et al. 2006).

TLS technology is based on Light Detection and Ranging (LiDAR) and is referred to as ground-based LiDAR. Other than that, laser pulses which are emitted by the scanner and observables include the range and intensity of pulse returns reflected by the surface or object being scanned is an active imaging system (Pfeifer and Briese 2007). Moreover, the generation of high-resolution 3D maps and images of surfaces and objects over scales of metres to kilometres with centimetre to sub-centimetre precision is the main capability of TLS (Buckley et al. 2013).

Unmanned aerial vehicle (UAV) LiDAR and photogrammetry imaging applications are increasing rapidly. This is not surprising for the surveyors as using GPS-enabled UAV for aerial surveying is very cost-effective compared to hiring an aircraft with photogrammetry equipment (Corrigan 2016). Nowadays, UAV platform is a valuable source of data for inspection, surveillance, mapping and 3D modelling issues. Moreover, new applications in the short and close range domain are introduced which makes the UAV photogrammetry a low-cost alternative to the classical manned aerial photogrammetry (Remondino et al. 2012).

Unfortunately, there is a limitation of both UAV photogrammetry and TLS. UAV photogrammetry just focuses on the rooftop through the cost-effective budget instead of the implementation of UAV LiDAR which is not sufficient or appropriate with the small projects (Phoenix 2014). Meanwhile, dense point clouds acquired by the laser scanner technique allowed to develop a detailed 3D data on the building facades but with a lack of information on the roofs and ledges (Bastonero et al. 2014).

In addition, the different platforms of data sources, where TLS produces 3D data and while UAV photogrammetry produces 2D aerial photograph, were also taken into account as factors that influenced on this integration of both data. Besides, the integration of multisensor, multiresolution and multiplatform data may result in terms of improved information extraction as well as to increase the reliability of the information (Ehlers et al. 2010). Thus, the fusion of UAV photogrammetry and TLS data will yield a complete 3D data acquisition. However, there are some issues worth to be studied. As example, the network configuration in order to perform processing.

2 Methodology

This research was conducted based on the four major phases that have been categorized as illustrated in Fig. 1.

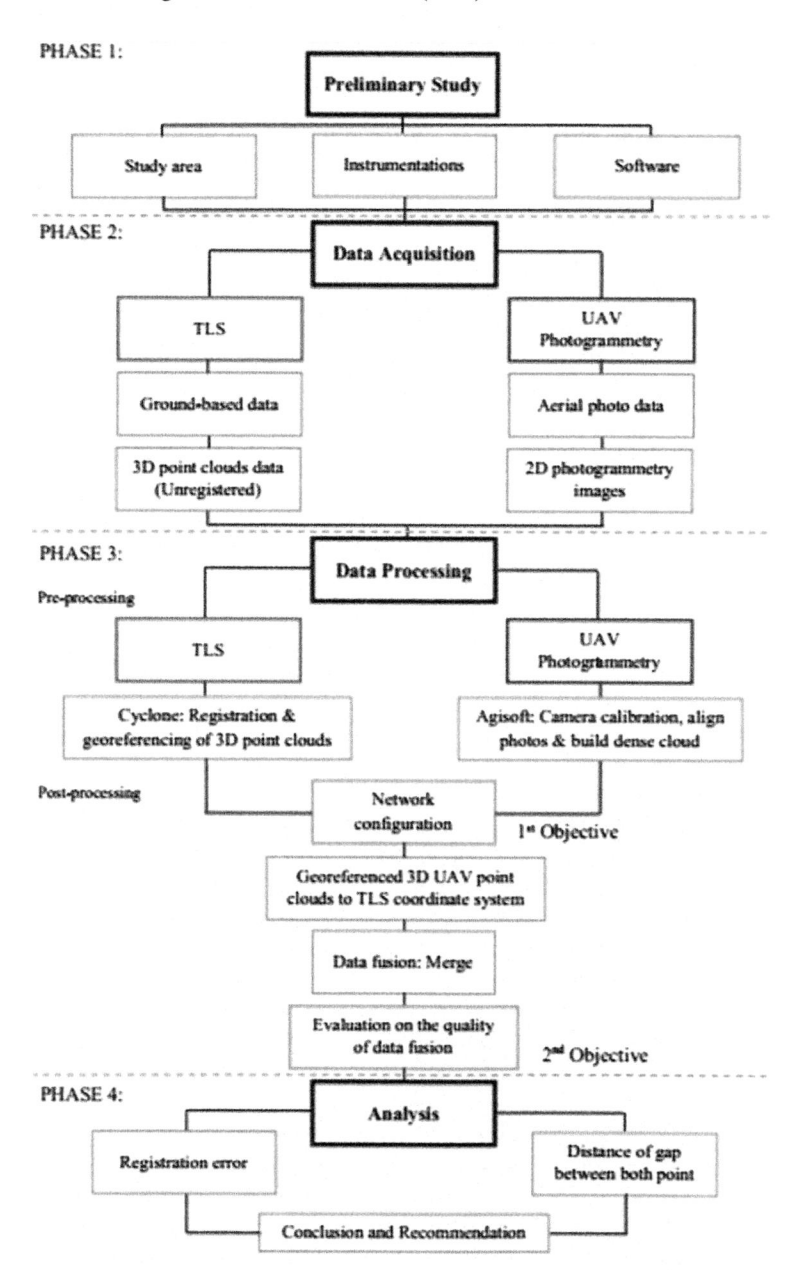

3D Data Fusion Using Unmanned Aerial Vehicle (UAV) ...

Fig. 1 Research methodology workflow

2.1 Data Acquisition

Laser scanning survey has been done by using a Leica ScanStation C10 together with black/white and sphere targets. Besides that, terrestrial scanning is done for ground-based data collection which focuses on the facade of the sports complex building. The achieved data was 3D dense point clouds (unregistered). Figure 2 shows the TLS data collection.

Seven scanstations were placed for this TLS scanning procedure. Moreover, the resolution of scanning used for this research is medium resolution of building facade with high resolution of registration targets. The registration targets are placed as well-distributed, and it will be used for the registration of all seven scanstations.

UAV photogrammetry survey has been done by using DJI Phantom 2 Standard drone with FC200 Aptina camera by focal length of 5 mm with a fisheye lens. Besides, this UAV photogrammetry is also done for the purpose of aerial photograph data collection and focusing the feature on top of the building (rooftop).

Fig. 2 TLS data collection

Fig. 3 UAV photogrammetry data collection

The achieved data was 2D photogrammetry images. Figure 3 shows the UAV photogrammetry data collection.

In this UAV photogrammetry survey work, the area involved is 14,843 m square. Furthermore, the total number of images captured is ninety-eight (98) aerial photographs. Thus, the planning of the aviation can be conclude by the number of the flight line which is fourteen (14) lines, and each of the flight line consists of seven (7) photographs. Besides, the altitude of the drone from the ground is 50 m.

2.2 Data Processing

The data which has been collected from TLS survey was then processed by using Leica Geosystems HDS Cyclone. The data was named as SUG653.imp. All these scan data were then undergo the registration process. Then, the registration process of scanstation begins by identifying the types of target, marking and labelling all the registration targets. Besides, just two targets have been used in this research, sphere and black/white. After completing the process of marking and labelling the registration targets, all the scanstations were then registered together. The registration error is then controlled by below than 10 mm.

The aerial photograph data which has been collected from UAV photogrammetry survey was then processed by using Agisoft PhotoScan Professional. The total images obtained from the aerial photograph are 98. All these photographs will undergo the process of camera calibration, align photograph and build dense cloud. The main step is to do the camera calibration. The purpose of this camera calibration is to specify the camera parameter in order to make a correction for those photographs. This parameter will be used for photograph align.

While carrying out photograph alignment, PhotoScan estimates both internal and external camera orientation parameters, including nonlinear radial distortions. For the estimation to be successful, it is crucial to apply the estimation procedure separately to photographs taken with different cameras. By referring to Agisoft PhotoScan (2012), once photographs have been loaded in the program, PhotoScan will automatically divide them into calibration groups according to the image resolution and Exchangeable image file format (EXIF) metadata like camera type and focal length.

If extra-wide lenses were used to get the source data, standard PhotoScan camera model will not allow to estimate camera parameters successfully. Fisheye camera-type setting will initialize implementation of a different camera model to fit ultra-wide lens distortions (Agisoft PhotoScan, 2012). Next step is to align the photographs. At this stage, PhotoScan finds the camera position and orientation for each photograph and builds a sparse point cloud model. In this align photograph, the accuracy that has been set is medium. Medium setting causes image downscaling by factor of 4 (two times by each side). This setting is limited due to time constraints. If high accuracy is selected, a time that will be taken to process this data is for a week. In fact, this is the duration of time by using high-performance personal computer (PC).

After completing the aligning photograph process, next it goes to building dense cloud process. At the stage of dense point cloud generation reconstruction, PhotoScan calculates depth maps for every image. Due to some factors, like poor texture of some elements of the scene, noisy or badly focused images, there can be some outliers among the points. To sort out the outliers, PhotoScan has several built-in filtering algorithms that answer the challenges of different projects. The setting for the depth filtering that has been chosen is Aggressive. This setting can be explained by Agisoft PhotoScan (2012) where if the area to be reconstructed does not contain meaningful small details, then it is reasonable to choose Aggressive depth filtering mode to sort out most of the outliers. The final output of this UAV photogrammetry processing was then converted into LASer (.las) format to enable it to be imported into cyclone software for the purpose of data fusion

2.3 Network Configuration and Data Fusion

There are five (5) samples of network configuration that have been created in order to get the best result of minimum number of registration point needed for the fusion of data. It started with the 20 registration points. Then, step by step it reduced to 15, 10, 5, until minimum 3 registration points.

This process is done in cyclone software. It starts with a sample of 20 registration points. First, the twenty registration points were marked and labelled on the natural feature in TLS point clouds data by well-distribute. Then, the marked natural feature points are extracted out and exported into survey (.svy) format named as TLS3.svy. Meanwhile, the same steps are performed on 3D point clouds of UAV photogrammetry with the same points of natural features except the export data and named as UAV3. In addition, the TLS3.svy data is imported in Sukan_Aizat cyclone database and becomes TLS3 scanstation. Next, the new registration project was created for the purpose of this 20 registration points of network configuration.

This is the stage where the UAV photogrammetry data is georeferenced to TLS coordinate system. In this registration part, the TLS3 data and UAV3 data were then added and the TLS3 data is set as Home Scanworld. Furthermore, the constraints were added into the registration by Auto-Add Constraints (Target ID only). When completing all the steps above, both data were then registered. On the other hands, both data which have been successfully registered were then merged together with TLS point clouds data. This process is called data fusion. All of the steps above were then repeated for another samples of 15, 10, 5 and 3 registration points.

3 Results and Discussion

3.1 TLS Scanstation Registration Result

There were seven scanstations being placed to cover the whole area of the sports complex building, so that it can produce a finest and dense point clouds observation. Furthermore, the total number of scanstation needed to be suitable with the area in order to avoid any miss point clouds which are obstruct by the near object. This scanstation was located surrounding the sports complex building and was completely focused on the sports complex facade. Figure 4 shows the Scanworld after registration process is completed.

The black area on top of the building is where the TLS cannot reach. So, it will need a help from the other sources of data to perform a complete shape or 3D point clouds of the building. There were 19 constraint points that were used for registration purpose, 14 sphere targets and 5 black/white (vertex) targets.

Meanwhile, Fig. 5 shows the sample of registration error for seven scanstations to form a complete 3D point clouds data of the sports complex building facade. The error shown on the red box is where the report produced. Besides, the minimum registration error produce was 0.001 m, while the maximum registration error produce from this registration process is 0.010 m.

3.2 UAV Photogrammetry Result

In this study, Agisoft PhotoScan Professional software has been used in order to create and form a dense point clouds data from 2D images. Besides that, the

Fig. 4 A completed registration of seven TLS scanstations

Fig. 5 Sample of registration error for seven scanstations

observation has been made by using DJI Phantom 2 Standard with FC200 Aptina camera and focal length of 5 mm. Figure 6 shows the aerial photographs processed results in good point clouds on the rooftop but lacks at the facades. The edge roof of the building is quite well because these photographs are processed by using medium accuracy of align photographs and medium resolution of quality for the

Fig. 6 3D point clouds data processed from the UAV photogrammetry

building of dense point clouds. Besides, a total of 12,104,506 point clouds are generated by this medium quality. All these photographs were processed by using a software, so it is not surprising with the quality of the point clouds generate because of the computer vision and some algorithm involved in order to transform from 2D photographs into 3D dense point clouds.

3.3 Data Fusion Result

Based on the results in Table 1, it shows five samples of final products' fusion data. The results were displayed as two perspectives. The first is expressed visually (quality), and the second is distance of gap (quantity) between both TLS and UAV photogrammetry point clouds.

From the visual results, the two blue points there show the end of the roof edge, one for TLS and another one is for UAV photogrammetry. Meanwhile, the point clouds of TLS are orange coloured, while UAV photogrammetry is blue coloured. In addition, these results show the quality of integrating data and it has been amplified by the distance of gap between both point clouds.

Theoretically, the more distribution of registration points, the more accurate results will be obtained. But it happened vice versa. In Fig. 7, the graph shows less registration points which may result in good merged data with distance gap between both point clouds. By using only three (3) registration points, the distance gap

Table 1 Data fusion result via samples of registration

No. of registration points	Visual result (roof edge)	Distance of point clouds gap (m)
20		0.508
15		0.373
10		0.196
5		0.212
3		0.093

Fig. 7 Graph of relationship between number of registration points and gap of both point clouds data

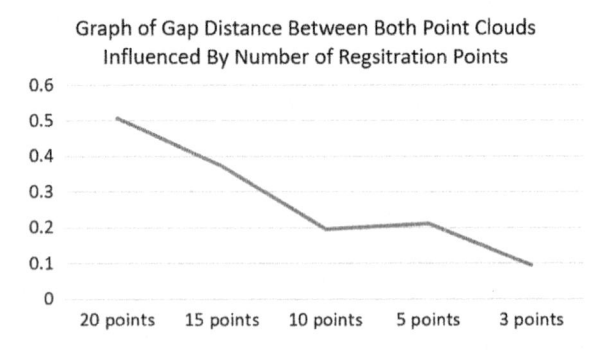

between both point clouds was result in 0.093 m. It was very tiny gap for the different of combination data from two different platforms. However, at the stage of ten (10) registration points, the distance gap decreased to 0.196 m. Unfortunately, the distance increased again at stage of 15 and 20 registration points, while the visual quality of the fusion point clouds is increased far from each other.

4 Conclusion

The combination of TLS and UAS in this project is a logical consequence, because they complement each other in terms of overcoming problems of the other method. Only their combination allows a dense and accurate 3D model of the entire building and is favourable from an economic point of view. In the photogrammetry, each object point must be recognized exactly in multiple images. In contrast, the TLS as the angle- and distance-based measurement method depends on the reflection properties of the measurement spots which are affected by the surface properties and the angle of incidence. For the UAV photogrammetry 3D model, the problem areas are at the lower building areas, start from the ground to the eaves height of the aisles, while the TLS in contrast is more accurate, but has data gaps in the higher areas. The accuracy of comparisons against the TLS reference areas shows that for continuous surfaces, high accuracies were achieved with a standard deviation in the range a few cm and with low noise.

Acknowledgements Faculty of Architecture, Planning and Surveying, Universiti Teknologi MARA (UiTM), Perlis, Malaysia. Special thanks to Sr Dr. Mohd Azwan Abbas and Sr Khairulazhar Zainuddin. Grateful appreciation towards beloved parents. Sincere gratitude to friends and person who directly or indirectly involved in this study, and also thanks to Universiti Teknologi Malaysia (UTM), Johor Bahru, Malaysia, for providing equipment in data collection, thus enabling this study to be carried out.

References

Bastonero, P., Donadio, E., Chiabrando, F., & Spanò, A. (2014). Fusion of 3D models derived from TLS and image-based techniques for CH enhanced documentation. *ISPRS Annals of Photogrammetry, Remote Sensing and Spatial Information Sciences, II-5*(June), 73–80.

Buckley, S. J., Kurz, T. H., Howell, J. A., & Schneider, D. (2013). Terrestrial lidar and hyperspectral data fusion products for geological outcrop analysis. *Computers & Geosciences, 54*, 249–258.

Corrigan, F. (2016, September 17). 10 Top Lidar Sensors for UAVs and So Many Great Uses. *DroneZon.* Retrieved from https://www.dronezon.com.

Ehlers, M., Klonus, S., Johan Åstrand, P., & Rosso, P. (2010). Multi-sensor image fusion for pansharpening in remote sensing. *International Journal of Image and Data Fusion, 1*(1), 25–45.

Guarnieri, A, Remondino, F., & Vettore, A. (2006). Digital photogrammetry and TLS data fusion applied to cultural heritage 3D modeling. *36*(5), 1–6.

Pfeifer, N., & Briese, C. (2007). Geometrical aspects of airborne laser scanning and terrestrial laser scanning. *International Archives of Photogrammetry and Remote Sensing, 36*(3), 311–319.

Phoenix. (2014, October 21). Phoenix aerial introduces low cost UAV LiDAR system. *sUAS News.* Retrieved from https://www.suasnews.com.

Remondino, F., Barazzetti, L., Nex, F., Scaioni, M., & Sarazzi, D. (2012). UAV photogrammetry for mapping and 3D modelling—Current status and future perspectives. *ISPRS—International Archives of the Photogrammetry, Remote Sensing and Spatial Information Sciences, XXXVIII-1* (September), 25–31.

Spatial Mapping and Analysis of Carbon Dioxide Emissions from Electricity in UiTM Perlis for Assessment of Low Carbon Campus

Noradila Rusli, Nurhani Nadirah Hamzah, Muhammad Faiz Pa'suya and Suhaila Hashim

Abstract The increase of energy usage is the cause of greenhouse gas emission, especially carbon dioxide. Furthermore, one of the factors of high carbon emission is electricity, which is one of the energy sources needed for campus activities. The aim of this study is to analyse the amount of carbon dioxide emissions produced by the source of electricity in UiTM Perlis, for the assessment of low carbon dioxide compliance in campus. There are three types of data utilized, which are total electric consumptions in UiTM Perlis from 2013 to 2015, the building plan of UiTM Perlis, the base map of UiTM Perlis and the spatial analysis of (GIS) geographical information system. The assessment of low carbon in campus is based on the calculation of the amount of carbon dioxide emission, and it is then mapped based on five building categories. The total consumption of electrical energy by the buildings is used to determine the amount of carbon emission using the formula for carbon dioxide emission. Carbon dioxide emissions per unit square meter (ktCO_2)/m^2 in the three years in UiTM Perlis are 58.34, 56.24 and 55.31, respectively. Based on the comparison of these results with carbon dioxide emission guideline per unit square meter (ktCO_2)/m^2 which is 56.5 ktCO_2, it can be seen that UiTM Perlis complies with the guideline for year 2014 and 2015.

Keywords Carbon dioxide emissions · Low carbon campus · Electricity

1 Introduction

Carbon dioxide is one of the greenhouse gases, which results to global warming and gives negative impacts to the environment. Based on various research conducted over the past few decades, it is shown that the increasing atmospheric concentration

---

N. Rusli (✉) · N. N. Hamzah · M. F. Pa'suya · S. Hashim
Green Environment and Technology (GREENTech) Research Group, Center of Study for Surveying Science and Geomatic, Faculty of Architecture Planning & Surveying, Universiti Teknologi MARA, Perlis, Arau Campus, 02600 Arau, Perlis, Malaysia
e-mail: radhiutm@gmail.com

of greenhouse gases, such as carbon dioxide (CO_2), and other gas emissions has a negative impact on the environment (Mahlia 2002). The low carbon dioxide emission in campus is one of the steps towards the reduction of carbon emission and energy sustainability in the campus. Additionally, electricity is one of the factors of the highest carbon dioxide emission in campus.

The issue of carbon emission is more prominent, especially in universities with large population and large spatial size. Electricity is the main factor of carbon emission in universities. This is proven by the research of "Carbon footprint analysis of student behavior for a sustain-able university campus in China" by Li et al. (2015), which claimed that one of the sectors that can provide solutions to reduce carbon emission is the higher education sector, which is unexpectedly growing in China. Electricity is used for most daily activities and services in campus, either by the staff of the campus or students. Based on the assessment of the amount of carbon emission in Universiti Teknologi Malaysia (UTM) Skudai and University College Sedaya International (UCSI), it is proven that electricity is the main factor of the highest carbon dioxide emission in campus, followed by transportation (Adeyemi and Ho 2015; Hooi et al. 2011).

Worldwide, many researches on the amount of carbon dioxide emissions from energy use in universities were conducted, such as by Riddell et al. (2009). These researches also performed an assessment of carbon dioxide emissions from electricity, which is known as (HVAC) high-voltage alternating current, and the use of hot water in US university. Meanwhile, Ho et al. (2014) conducted a research using multi-objective programming model for energy conservation and renewable energy structure of low carbon campus. The objective of the research was to find the model appropriate for energy conservation and renewable energy, based on the sources of energy which cause the release of carbon in the campus university offers a great potential for global sustainability and the assessment of carbon emission. This is one of the steps for energy sustainability in campus. The location of electrical energy is the factor of high carbon emission. The quantity and source of it can be identified through the assessment of the amount of carbon emission. The assessment involves determining the quantity of carbon dioxide in university produced by the sources of energy consumption, which can help provide options on how to achieve low carbon dioxide compliance in campus. Hence, this study will estimate the amount of carbon dioxide emission caused by electricity and perform a spatial analysis and mapping on it.

2 Data and Methodology

2.1 Study Area

Universiti Teknologi MARA (UiTM) Perlis is located at coordination 6° 26′ 59″ N and 100° 16′ 48″ E with an area around 135.57 ha. The climate condition in Arau,

Perlis, is categorized as tropical, with short dry season. Furthermore, the total population of UiTM Perlis consists of 7225 full-time students and 779 numbers of staff. Most of the buildings in UiTM Perlis are high rise, such as academic buildings, administration and facility buildings, library, resident's hostel buildings and ICT buildings. These are the examples of the buildings in UiTM Perlis which require electricity. However, the building with the highest requirement of electricity is the hostel building, as electricity is consumed for almost the whole day (Fig. 1).

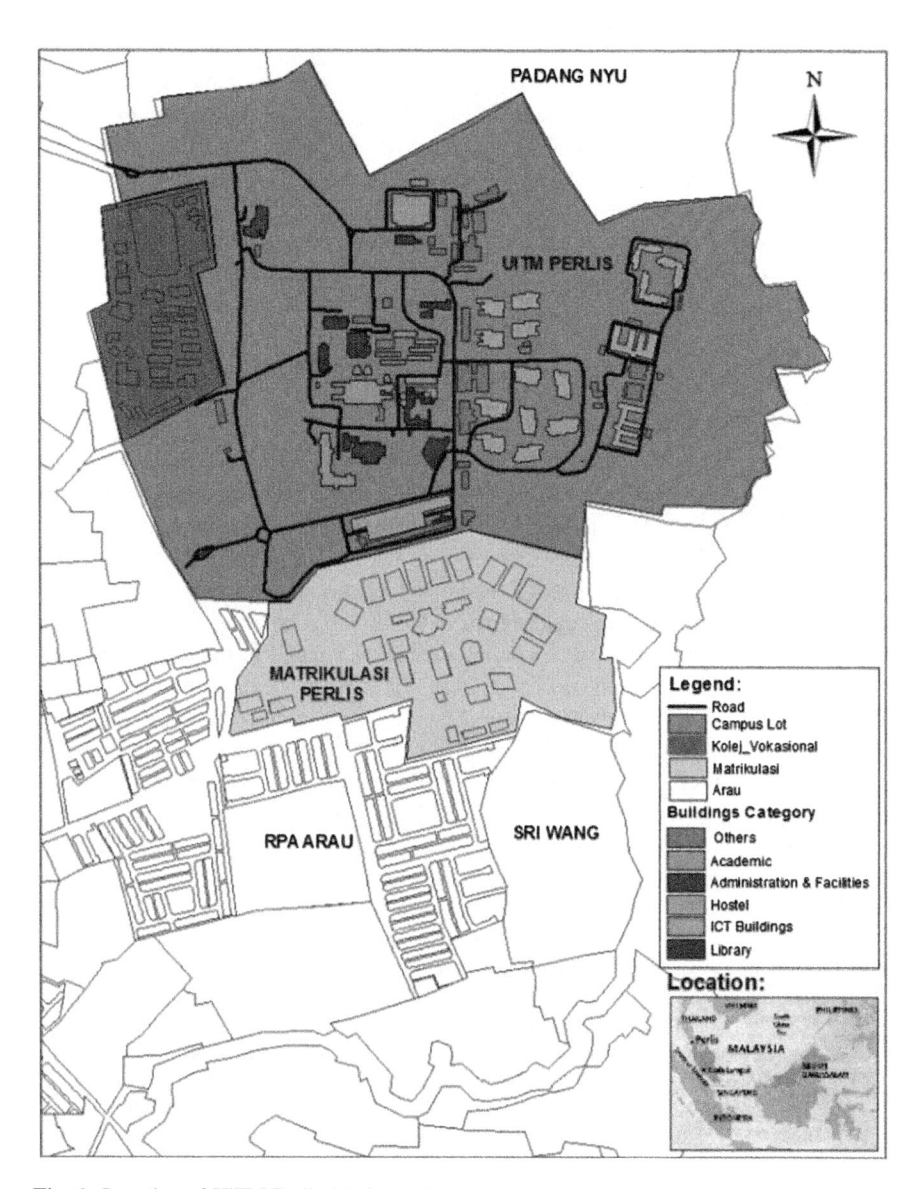

Fig. 1 Location of UiTM Perlis (study area)

2.2 Data Collection

There were three types of data which were required for this study. The first data were the total electric consumptions in UiTM Perlis. The data of electric consumption consisted of the amount of the consumption from year 2013 to 2015. The source of the data was obtained from the facility unit of UiTM Perlis, which is responsible for storing all the records of the campus's electric consumption. The data consisted of the amount of electric consumption for every month in each year, which covers the whole amount of electric consumption in UiTM Perlis.

The second and third data focused on the building plan of UiTM Perlis and the base map of UiTM Perlis. Both data were collected from Faculty of Architecture, Planning and Surveying (FSPU) UiTM Perlis. The building plan encompassed all buildings in UiTM Perlis. The plan showed the dimension of the building, including room measurements. Meanwhile, the base map showed the topographical area of UiTM Perlis and the position of each building. Additionally, the base map was used in this study to map and analyse the amount of carbon dioxide emissions in UiTM Perlis.

2.3 Data Processing

The data processing of this study consisted of five stages. The first stage involved the calculation of electrical energy consumption for each category of building. The total building area was calculated according to the buildings plan of UiTM Perlis and measured in square meter (m^2). Then, the electricity consumption for each space (m^2) of the building was determined using the total building area, which had been calculated. The electricity consumption was calculated in kilowatt-hours per m^2 (kWh/m^2), and it was determined from the electricity bills. Moreover, the electricity consumption for each room in the buildings was calculated, based on the amount of the electricity consumption for each m^2, which would be used to determine the total electricity consumption for each building. The second stage of data processing involved classifying the buildings into five categories, and calculating the electricity consumption based on the operational periods for each category. The building categories are specified in Table 1.

Table 1 Buildings categories for calculating carbon dioxide emission

Building categories	Total hours (h)	Operational period
Hostels	13	6.00 p.m.–7.00 a.m.
Academic buildings	10	8.00 a.m.– 6.00 p.m.
Administration and facilities buildings	10	8.00 a.m.– 6.00 p.m.
ICT buildings	10	8.00 a.m.–6.00 p.m.
Library	14	8.00 a.m.–10.00 p.m.

Then, the third stage involved the calculation of the quantity of electric consumption in each building/m^2. The total electric consumption in each building was calculated using Eq. (1).

Total usage of energy:

$$\text{E-electric} = \text{E-hostel} + \text{E-academic} + \text{E-administration} + \text{E-ict} + \text{E-library} \quad (1)$$

Specifically:
Measurement unit of E-electric is (kWh)
E-hostel is the energy used in hostel
E-academic is the energy used in Academic building
E-administration is the energy used in Administration and facilities buildings
E-ict is the energy used in ICT buildings
E-library is the energy used in Library

After that, the fourth stage of data processing involved the calculation of carbon dioxide emission quantity, which was released by each building/m^2. By referring to the calculated total electricity consumption, the quantity of carbon dioxide emission could be defined. The quantity of carbon dioxide emission was calculated in metric tons of carbon dioxide (tCO$_2$). The calculation utilizing equation is shown in (2).

$$CO_2 = AME \times EEF \quad (2)$$

Specifically:
AME: Average Monthly Electricity used (kWh)
EEF: Electricity Emission Factor (CO$_2$e/kWh)
(The average EEF of West Malaysia is 0.585 CO$_2$e/MWh)
MWh: Megawatt-hours
Source (Hooi et al. 2011).

3 Result and Discussion

This study produced the result of the analysis of carbon dioxide emission level in UiTM Perlis, whether it was high or low. All five building categories had different operational period per day and different total operation day per month. The operational period and total operational day were according to the major and the average duration of electricity use in the buildings, as displayed in Table 2.

The campus hostels consisted of 15 buildings with the total area of 67050.10 m^2. Based on the total area of the each building and the building category, the building

Table 2 Buildings categories and operational period

No	Category	Operational hours/day	Operational day/month	Operational hours/month	% of coverage area
1	Hostels	13	30	390	63
2	Academic buildings	10	22	220	20
3	Administration and facilities	10	22	220	10
4	ICT buildings	10	22	220	5
5	Library	14	30	420	2
			Total	1470	100

Table 3 Quantity of carbon dioxide emissions in UiTM Perlis

	Category					Carbon dioxide
Year	Hostels	Academic	Admin.	ICT	Library	emission (ktCO$_2$)
2013	3,911,910.29	1,386,371.94	611,337.61	98,296.44	282,098.19	6,290,014.48
2014	3,771,993.53	1,336,785.76	589,471.98	94,780.69	272,008.43	6,065,040.38
2015	3,708,377.04	1,314,240.23	579,530.25	93,182.17	267,420.87	5,962,750.56

with the largest percentage area was the hostel, with 63% coverage, 20% coverage of academic buildings category, 10% coverage of administration and facilities category, 5% coverage of ICT buildings and 2% coverage area of library. This is the smallest coverage area for a building category. The total electricity consumption for year 2013, 2014 and 2015 was decreasing, given the values were 10752161.50, 10367590.40 and 10192736.00 kWh, respectively. Based on the calculation in Table 3, hostels emitted highest carbon dioxide, compared to other four categories.

The emitted carbon dioxide was then spatially mapped in 3D, and the results are displayed in Fig. 2. Overall, the highest amount of carbon dioxide (marked as red) was emitted in hostel buildings, and the lowest (green) amount of carbon dioxide was emitted in library building, from year 2013 to 2015.

3.1 Assessment of Low Carbon Dioxide Compliance in UiTM Campus

Table 4 shows the result of carbon dioxide emission per unit square meter (ktCO$_2$)/m^2. It is based on the total area of the building category and the total quantity of carbon dioxide emission from year 2013 to 2015. The value of carbon dioxide emission

Spatial Mapping and Analysis of Carbon Dioxide Emissions ... 313

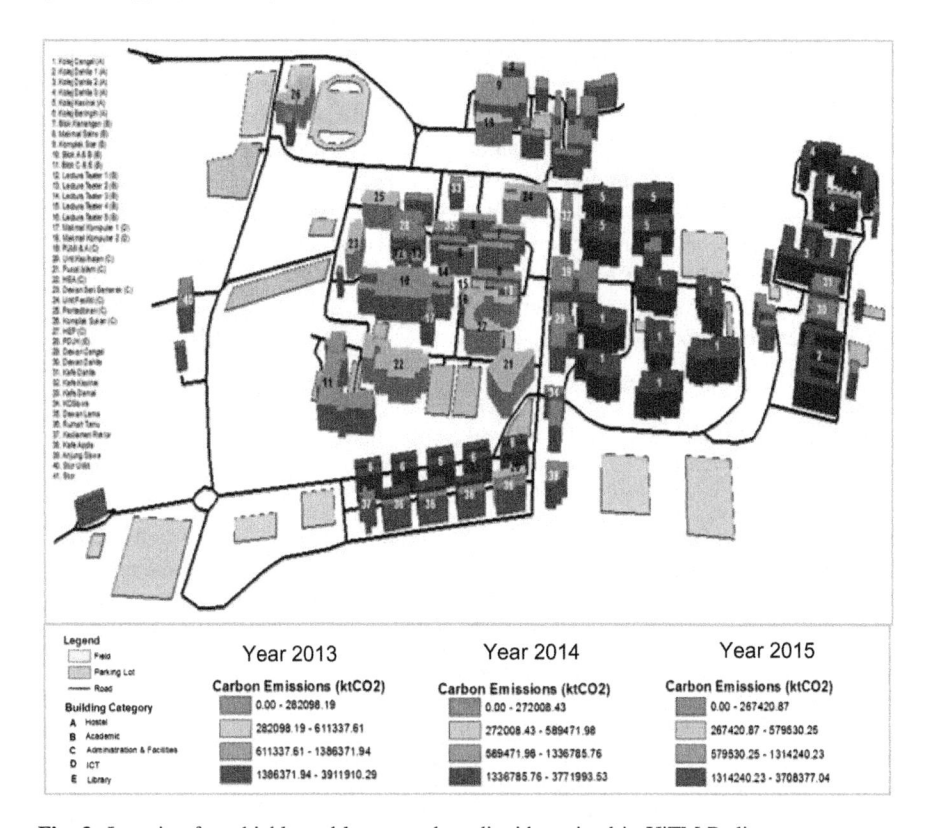

Fig. 2 Location from highly and lowest carbon dioxide emitted in UiTM Perlis

Table 4 Carbon dioxide emissions per unit square meter

Year	Total area (m^2)	Electrical Consumption (kWh)	Carbon dioxide emission (ktCO$_2$)	Carbon dioxide emission per unit square meter (ktCO$_2$)/m^2
2013	107810.78	10752161.50	6290014.478	58.34
2014		10367590.40	6065040.384	56.26
2015		10192736.00	5962750.56	55.31

provided by the previous research was 56.5 ktCO$_2$/m^2 (Adeyemi and Ho 2015). This research was used as a benchmark due to the study area for the study was conducted in Universiti Teknologi Malaysia (UTM)'s campus. This is the only current research involving carbon emission in university's campus in Malaysia. Therefore, the quantity of carbon dioxide emission should be lower or the same with the given value, in order to classify UiTM Perlis as a low carbon dioxide emission campus. As referred to the quantity of carbon dioxide emitted in 2013, the quantity of emission per unit square meter was 58.34 ktCO$_2$/m^2. This amount exceeded the quantity of emission provided in the previous research. Meanwhile, in the year 2014, the quantity of carbon dioxide emission per unit square meter was 56.26 $ktCO_2/m^2$. On the other hand, the quantity was 55.31 $ktCO_2/m^2$ in 2015. From here, it can be seen that the values of carbon dioxide emission per unit square meter in the year 2014 and 2015 were compliant with the low carbon dioxide emission in previous research in UTM.

The Save Energy Campaign is proven successful, as recognized by UiTM Perlis. This can be seen from the reduction of electricity consumption, which leads to the decrease of carbon dioxide emission. This is a positive step towards promoting a sustainable green campus, and the enhancement of available methods for the estimation of carbon dioxide emission.

4 Conclusion

The increase of electricity consumption results to the increase of carbon dioxide emission. This study's assessment of carbon dioxide emission in UiTM Perlis was able to identify the quantity of carbon dioxide emission, which previously had never been estimated in the campus. This allows the students and staff of the campus to realize that electricity, especially the one used in hostels, contributes to high carbon dioxide emission in the campus. Furthermore, this assessment can help in providing the suitable method in achieving the goals of energy sustainability in a campus. Moreover, this study has managed to assist a process of formulating the suitable solution in achieving the goals of energy sustainability. The formulated solution will be essential for the university to decrease the amount of greenhouse gases emission. Besides that, this process can also be made as a reference or model for other institutions to replicate a similar method. This method will be useful to asses low carbon dioxide compliance in campus. Last but not least, this study has supported Malaysia's objective of reducing the quantity of carbon emissions at 40% of reduction rate, in the year 2020 (Siong et al. 2013).

References

Adeyemi, I., & Ho, C. S. (June 4, 2015). Realizing low carbon emission in the university campus towards energy sustainability. *Open Journal of Energy Efficiency*, 15–27.

Ho, Y. F., Chang, C. C., Wei, C. C., & Wang, H. L. (2014). Multi-objective programming model for energy conservation and renewable energy structure of a low carbon campus. *Energy and Buildings, 80,* 461–468.

Hooi, K. K., Hassan, P., & Jami, N. A. (2011). An Assessment of Carbon Footprint at UCSI University and Proposed Green Campus Initiative Framework. *Sustainable Education, 12,* 342–347.

Li, X., Tan, H., & Rackes, A. (2015). Carbon footprint analysis of student behavior for a sustainable university campus in China. *Journal of Cleaner Production, 106,* 97–108. https://doi.org/10.1016/j.jclepro.2014.11.084.

Mahlia, T. M. I. (2002). Emissions from electricity generation in Malaysia. *Renewable Energy, 27* (2), 293–300.
Riddell, W., Bhatia, K. K., Parisi, M., Foote, J., & Iii, J. I. (2009). Assessing carbon dioxide emissions from energy use at a university. *International Journal of Sustainability in Higher Education, 10,* 266–278.
Siong, C., Matsuoka, Y., Simson, J., & Gomi, K. (2013). The case of Iskandar Malaysia development corridor. *Low Carbon Urban Development Strategy in Malaysia, 37,* 43–51.

Estimation of *Acacia mangium* Aboveground Biomass and Wood Volume Through Landsat 8

Aqilah Nabihah Anuar, Ismail Jusoh and Affendi Suhaili

Abstract Aboveground biomass (AGB) and wood volume are two useful parameters in showing the important role of forest in carbon cycling and practicing sustainable forest management. However, monitoring these parameters through conventional method such as destructive sampling has proven to be laborious, cost-ineffective, and time-consuming especially on a large forested area. A more logical approach with acceptable accuracy is to use satellite imagery data such as Landsat 8 to estimate AGB and wood volume of planted forest. The objectives of this study were to identify which spectral bands in Landsat 8 and vegetation indices that most correlated to AGB and wood volume of *Acacia mangium* plantation. Correlation and simple linear regression analyses were performed to determine the relationships between bands reflectance, vegetation indices, AGB, and wood volume. Results showed that reflectance of band 2 and band 5 is correlated to both AGB and wood volume. Using vegetation indices, correlation between Landsat bands reflectance and studied parameters improved significantly. Normalized difference vegetation index (NDVI) and modified vegetation index (ND52) from band 2 and band 5 showed significantly negative correlations with AGB; $r = -0.73$ and $r = -0.76$, respectively. Wood volume was also correlated with NDVI ($r = -0.75$) and ND52 ($r = -0.77$). The results suggest that AGB and wood volume of *A. mangium* plantation can be possibly estimated using NDVI and ND52 at an acceptable level of accuracy.

Keywords Aboveground biomass · Wood volume · Landsat 8
Bands reflectance · Vegetation indices

A. N. Anuar (✉) · I. Jusoh
Department of Plant Environment and Ecology, Faculty of Resource Science and Technology, Universiti Malaysia Sarawak (UNIMAS), 94300 Kota Samarahan, Malaysia
e-mail: aqilahanuar5093@gmail.com

A. Suhaili
Forest Operation Branch, Forest Department Sarawak, Wisma Sumber Alam, Petra Jaya, 93660 Kuching, Sarawak, Malaysia
e-mail: affendi@sarawak.gov.my

1 Introduction

Aboveground biomass and wood volume help us determine how much worth the forest values are, both from ecological and from economical aspects (Kumar et al. 2015). In many developing countries, forest is a valuable asset for its timber and non-timber products. Reforestation and afforestation are conducted throughout many countries including Thailand, China, Indonesia, and Malaysia for this general purpose (Suratman 2003; Iglesias 2007; Liu et al. 2014). Conducting a large-scale field sampling on the exact amount of biomass and wood volume would require a tremendous effort and time. Although the accuracy of the data is no doubt, in order to cover a large forested area, this has become less cost-efficient and laborious, thus making it less efficient in management level.

Remote sensing is the science combining with art in obtaining information of an object without having any direct contact with the collecting device known as sensor (Lillesand et al. 2014). Remotely sensed data collect the variation of the electromagnetic energy (EM) after interacting with the object. From a fixed distribution of the incident radiation (for all passive sensor is the illumination from the Sun), the properties of surface object it interacted give out a unique spectral signature of the specific targets. Remote sensing instruments help collecting the information in the form of digital data. As the sensor does not come in contact with targets, the digital data collected can cover a wide-ranging area, providing the information in a synoptic view and timely manner (Lu et al. 2004). Hence, resolving the use of Landsat 8 satellite imagery is a considerably reasonable approach, but poor understanding on the relationship between the spectral reflectance with wood volume and AGB, plus with the lack of skills and expertise in the application, contributes to the finding of this study.

Vegetation indices were used to distinguish vegetation canopy from other land covers, along with interaction of every bands reflectance: band 2 (blue), band 3 (green), band 4 (red), band 5 (near-infrared), band 6 (shortwave infrared 1), and band 7 (short-wave infrared 2) interaction with the wood volume and AGB of *A. mangium* plantation in Sarawak, Malaysia. Bands reflectance interacts differently based on the target's property. In vegetation, the radiation from electromagnetic energy interacts with the leaves, down to its cellular level. Chlorophyll *a*, *b*, β-carotene and other components in leaf pigment absorb blue and red parts in EM spectrum while reflecting the green, making vegetation seen as green in our naked eyes. Near-infrared (NIR) reacts differently against vegetation depending on the amount of water present within the spongy mesophyll tissue and cavities within the leaf. In NIR region, radiation is higher in green and the radiation differs between species (Gausman et al. 1969). *A. mangium* interactions with individual reflectance, visible light region (blue, green, red), NIR, and shortwave infrared (SWIR) were studied in this paper.

Spectral vegetation indices (VIs) are mathematical transformation of the bands reflectance. They are developed through linear combination to improve the sensitivity of the individual bands in characterizing the vegetation canopies (Basso et al. 2004). Normalized difference vegetation index (NDVI) is a classic index, having long been established to analyze biophysical parameters. The index makes use of the differences between the NIR and red reflectance over the sum of NIR and red reflectance, creating a ratio that normalized the external noise interference (Basso et al. 2004). The range of the value is between −1 to +1. The higher the index value indicates the radiation on the NIR reflected is greater than the radiation detected on the red region. More green vegetation is spotted as the ratio grew closer to +1 (Weier and Herring 2000). Vegetation index is a form of tool in image processing, and variation occurs as to how they interact between trees species (Wulder 1998). This paper applied the concept of VIs in illustrating the dependency between spatial data to ABG and wood volume by identifying the most correlated spectral bands with AGB and volume, analyzed the current established VIs with AGB and volume, and then modified some of the well-established vegetation indices to find the best-fit model in predicting the overall AGB and wood volume of *A. mangium* trees in the study area.

2 Methodology

2.1 Study Area

The study site is an *A. mangium* plantation area located in Bintulu, Sarawak, Malaysia. Areas with stand ages of 8 years old and above were selected for this study. The plantation is owned by Daiken Sarawak Sdn. Bhd., a Malaysian-Japanese joint venture company.

2.2 Image Preprocessing

Landsat 8 OLI satellite image data, with the path/row of 119/58, were downloaded from the USGS database. Data with less than 20% cloud cover and ranging from March 2016 until September 2016, were downloaded. The Landsat bands went through atmospheric correction using dark object subtraction (DOS) method (Lu et al. 2002). The software used is Quantum GIS 2.8.2. Topographic correction was done using digital elevation model, solar angles (elevation and azimuth), and radiance (Jiang et al. 2011).

2.3 Field Sampling

Twenty-five sample plots of 30 m × 30 m were established in the study area. The parameters measured in all plots were diameter at breast height (DBH), total height, and merchantable height, conducted between March 2016 and September 2016. Estimation of AGB and wood volume was based on previously established equations by Adam (2015).

$$\text{Above ground biomass} = 0.1173\, x^{2.454} \quad (1)$$

$$\text{Wood volume} = 0.00008869\, x^{2.72} \quad (2)$$

where x is the diameter at breast height (DBH) of individual trees (cm)

2.4 Image Processing

Six individual bands reflectance. blue, green, red, NIR, SWIR 1, and SWIR 2, were studied. Correlation between these bands and AGB and wood volume were determined.

2.5 Determination of Vegetation Indices

Apart from bands reflectance. vegetation indices were also determined. Determination of normalized difference vegetation index (NDVI) of the reflectance was carried out to distinguish the vegetation from bare soil. Equation 3 was introduced by Rouse et al. (1973):

$$\text{NDVI} = \frac{(\text{NIR} - \text{Red})}{(\text{NIR} + \text{Red})} \quad (3)$$

Other vegetation indices such as atmospheric and soil vegetation index (ASVI), soil-adjusted vegetation index (SAVI), and modified soil-adjusted vegetation index (MSAVI), were also determined:

$$\text{ASVI} = \frac{(\text{NIR} - 2\text{Red} + \text{Blue})}{(\text{NIR} + 2\text{Red} + \text{Blue})} \quad (4)$$

$$\text{SAVI} = (2\text{NIR} + 1) - \frac{\sqrt{(2\text{NIR} + 1)^2 - 8(\text{NIR} - 2\text{Red} + \text{Blue})}}{2} \quad (5)$$

$$\text{MSAVI} = (2\text{NIR} + 1) - \frac{\sqrt{(2\text{NIR} + 1)^2 - 8(\text{NIR} - 2\text{Red})}}{2} \quad (6)$$

A new index ND52, modified from NDVI, was introduced to determine its relationship with AGB and wood volume. This index takes into consideration the Sun azimuth angle and topography of the study area:

$$\text{ND52} = \frac{(\text{NIR} - \text{Blue})}{(\text{NIR} + \text{Blue})} \quad (7)$$

2.6 Data Analysis

Correlation and simple linear regression analyses were carried out to determine the relationships between bands reflectance and vegetation indices, and AGB and wood volume.

3 Results and Discussion

3.1 Image Analysis

Landsat 8 OLI provides good set of quality data with its spatial resolution of 30 m, and the temporal resolution is within 16-day interval. For the past three decades, Landsat images have been contributing to studies based on forest aboveground biomass because of its freely available database, and the data provided are in medium spatial resolution. Thus, it is a reliable source for collecting the multi-spectral data (Zhao et al. 2016). Dark object subtraction (DOS) is an empirical atmospheric correction technique that removed the darkest pixel value on the data, resulted from atmospheric scattering (Exelis Visual Information Solution 2017). The spectral rationing performed through NDVI helps to highlight the subtle variation in the spectral responses from the various targets and thus improve the classification performed afterward. As shown in Fig. 1, the comparison between the original image taken from DigitalGlobe and digitized NDVI after image processing enhanced the vegetation crown cover of *A. mangium* plantation. Higher intensity of the crown cover from vegetation is able to be distinguished from area with least vegetation. NDVI of 0.5 and above was represented in the darker shade of green, while the index between 0.4999 and 0.4 showed moderate vegetation covers. A negative value of the index indicates water, and area with least vegetation gives

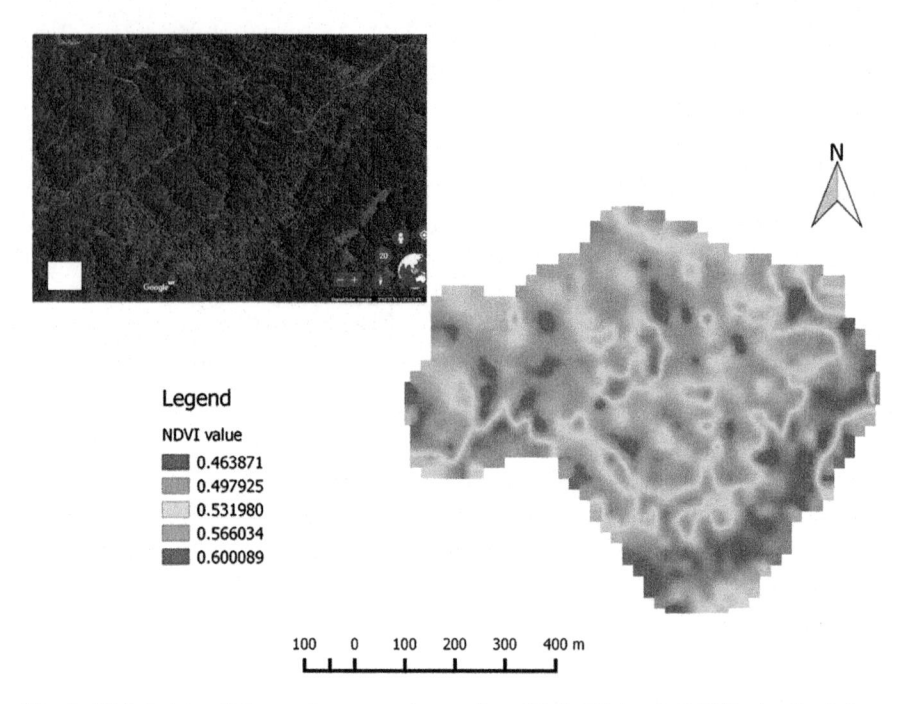

Fig. 1 Digital data of the study area, **a** image from DigitalGlobe, **b** NDVI visualized from Landsat 8 reflectance

the value closest to 0, while 0.2–0.4 for shrubs and grassland (Weier and Herring 2000). The closer the NDVI value to 1, the higher the possibility of having dense green leaves or vegetation covers.

Correlation coefficient on the bands reflectance and vegetation indices against AGB and volume of *A. mangium* trees within the study plots is summarized in Table 1. The data showed that vegetation indices significantly improved the

Table 1 Correlation coefficent between AGB, wood volume, bands reflectance, and vegetation indices

Reflectance	Biomass	Volume
Band 2 (blue visible light)	−0.51	−0.53
Band 4 (red visible light)	−0.24	−0.28
Band 5 (near-infrared)	0.54	0.52
Vegetation indices		
NDVI	−0.75[a]	−0.75[a]
ND52	−0.73[a]	−0.76[a]
ASVI	−0.63	−0.62
SAVI	−0.69[a]	−0.68[a]
MSAVI	−0.63	−0.62

[a]Statistically significant at 5% level

sensitivity of reflectance toward the studied biophysical parameters. Aboveground biomass and wood volume of acacia trees responded to reflectance in band 2 and band 5; however, their correlation coefficients were moderately negatively and positively correlated, respectively. After implementing vegetation indices, the correlations were significantly improved. Aboveground biomass and wood volume were negatively correlated with NDVI, ND52, and SAVI. The correlations of reflectance and vegetation indices vary as they depend on the characteristic of the study area (Lu et al. 2004).

Regression analysis (Fig. 2) shows significant relationship between vegetation index, ND52 with AGB and wood volume. Both relationships showed that as ND52 increases AGB and wood volume decrease. Regression analysis conducted on ND52 discovered that more than 50% aboveground biomass and wood volume can be explained by ND52. Equations in Fig. 2 can be used to estimate AGB and wood volume of *A. mangium* plantation using ND52 as predictor variable. Thus, making the study outcome relevant in estimating the amount of current aboveground biomass and wood volume present using concurrently updated Landsat 8 data.

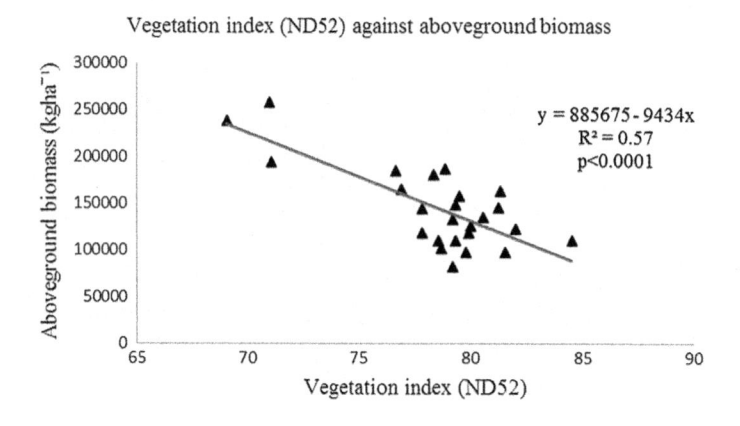

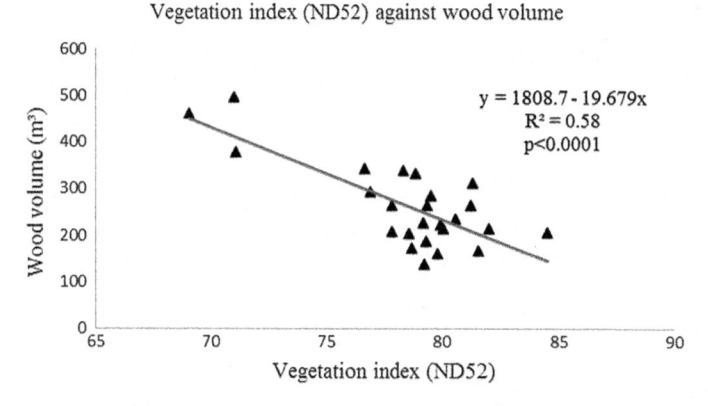

Fig. 2 Relationship between ND52, AGB, and wood volume

4 Conclusion

Conducting a forest inventory on *A. mangium* plantation in Sarawak using Landsat 8 is possible and allowed us to open up a whole new perspective in sustainable forest management. By applying the concept of vegetation indices, mapping the distribution of biomass and wood volume has becoming a lot easier and less time-consuming. The findings of this study showed that AGB and wood volume of *A. mangium* are correlated with band 2 and band 5 of Landsat 8. Integrating mathematical transformation using NDVI and modifying its concept to produce new index (ND52) improved estimation on the AGB and wood volume of *A. mangium* plantation.

Acknowledgements The authors would like to thank Daiken Sarawak Sdn. Bhd. for allowing their plantation area as the study site. This study was funded by FRGS/STWN02(02)/1142/2014 (09) grant.

References

Adam, N. S. (2015). *Carbon storage and sequestration potential of second generation Acacia mangium and acacia hybrid*. Unpublished master's thesis, Universiti Malaysia Sarawak, Kota Samarahan, Sarawak, Malaysia.

Basso, B., Cammarano, D., & Vita, P. D. (2004). Remotely sensed vegetation indices: Theory and applications for crop management. *Rivista Italiana di Agrometereologia, 1,* 36–53.

Exelis Visual Information Solution. (2017). *SPEAR Atmospheric correction (Using ENVI) Harris Geospatial Docs Centre*. https://www.harrisgeospatial.com/docs/spearatmosp hericcorrection.html. Accessed February 2, 2017.

Gausman, H. W., Allen, W. A., Myer, V. I., & Cardenas, R. (1969). Reflectance and internal structure of cotton leaves, *Gossypium hirsutum* L. *Agronomy Journal, 61,* 374–376.

Iglesias, C. O. (2007). *Determination of carbon sequestration and storage capacity of Eucalyptus plantation in Sra Kaew Province, Thailand using remote sensing*. MSc Thesis, Mahidol University, Thailand.

Jiang, K., Zhao, Y. & Geng, X. (2011). A simple topographic correction method based on smoothed terrain. *International Symposium on Image and Data Fusion*, Tengchong, Yunnan, pp. 1–4, https://doi.org/10.1109/isidf.2011.6024286.

Kumar, L., Sinha, P., Taylor, S., & Alqurashi, A. F. (2015). Review of the use of the remote sensing for biomass estimation to support renewable energy generation. *Journal of Applied Remote Sensing, 9*, https://doi.org/10.1117/1.irs.9.097696.

Lillesand, T., Kiefer, R. W., & Chipman, J. (2014). *Remote sensing and image interpretation* (7th ed.). United States: Wiley.

Liu, L., Peng, D., & Wang, Z. (2014). Improving artificial forest biomass estimates using afforestation age information from time series Landsat stacks. *Environmental Monitoring Assessment, 186,* 7293–7306.

Lu, D., Mausel, P., Brondizio, E., & Moran, E. (2002). Assessment of atmospheric correction methods for Landsat TM data application to Amazon basin LBA research. *International Remote Sensing, 23,* 2651–2671.

Lu, D., Mausel, P., Brondizio, E., & Maron, E. (2004). Relationship between forest stand parameters and Landsat TM spectral responses in the Brazilian Amazon Basin. *Forest Ecology and Management, 198,* 147–167.

Rouse, J. W., Haas, R. H., Schell, J. A., & Deering, D. W. (1973). Monitoring vegetation systems in the Great Plains with ERTS. *Third ERTS Symposium, 1,* 48–62.

Suratman, M. N. (2003). Applicability of Landsat TM Data for Inventorying and Monitoring of Rubber (Hevea brasiliensis) Plantation in Selangor, Malaysa: Linkages to Policies. PhD Thesis. The University of British Columbia.

Weier, J. & Herring, D (2000). *Measuring vegetation (NDVI & EVI).* https://earthobservatory.nasa.gov/Features/MeasuringVegetation/. Accessed April 4, 2017.

Wulder, M. (1998). Optical remote-sensing techniques for the assessment of forest inventory and biophysical parameters. *Progress in Physical Geography, 22,* 449–476.

Zhao, P., Lu, D., Wang, G., Wu, C., Huang, Y., & Yu, S. (2016). Examining spectral reflectance saturation in Landsat imagery and corresponding solutions to improve forest aboveground biomass estimation. *Remote Sensing, 8,* 469. https://doi.org/10.3390/rs8060469.

Screening of Transovarial Dengue Virus (DENV) Transmission in Field-Collected *Aedes albopictus* from Dengue Active Transmission Areas in Shah Alam, Selangor, Malaysia

H. Mayamin, A. Nurul Adilah, R. Nurul Hidayah, I. Nurul-Ain, M. Mohd Fahmi, Tengku Shahrul Anuar, C. D. Nazri, I. Rodziah, H. A. Abu and S. N. Camalxaman

Abstract Reported incidences of dengue outbreaks have increased rapidly over the years in Malaysia and such rise poses questions on the possibility of transovarial infection during the events. This study was conducted to screen the natural transovarial DENV transmission in *Aedes albopictus* population during outbreak seasons. Samples were collected using ovitraps from six dengue hotspot areas in Shah Alam, Selangor. Field-collected eggs were reared to adulthood in controlled environments under insectary conditions and morphologically identified as *A. albopictus*. Screening of transovarial transmission of dengue virus (DENV) was conducted using a commercially available multiplex qRT–PCR. None of the 120 female adult *A. albopictus* mosquitoes examined was tested positive for DENV RNA. Our findings indicate that transovarial transmission may not play a significant role in the epidemiology of dengue infection in Malaysia.

Keywords *Aedes albopictus* · Dengue · DENV · qRT–PCR · Transovarial transmission

---

H. Mayamin (✉) · A. N. Adilah · R. N. Hidayah · I. Nurul-Ain
M. M. Fahmi · T. S. Anuar
Department of Medical Laboratory Technology, Faculty of Health Sciences, Universiti Teknologi MARA, 42300 Puncak Alam, Selangor, Malaysia
e-mail: sitinazrina@salam.uitm.edu.my

C. D. Nazri · I. Rodziah
Department of Environmental Health and Safety, Faculty of Health Sciences, Universiti Teknologi MARA, 42300 Puncak Alam, Selangor, Malaysia

H. A. Abu · S. N. Camalxaman
School of Biological Sciences, Universiti Sains Malaysia, 11800 Gelugor, Pulau Pinang, Malaysia

© Springer Nature Singapore Pte Ltd. 2018
R. Saian and M. A. Abbas (eds.), *Proceedings of the Second International Conference on the Future of ASEAN (ICoFA) 2017 – Volume 2*,
https://doi.org/10.1007/978-981-10-8471-3_32

1 Introduction

Dengue fever (DF), an important arthropod-borne viral infection has emerged as a global problem in recent decades, afflicting as many as 390 million individuals per year (Bhatt et al. 2013). Symptoms of DF are broad ranging; from asymptomatic and mild flu-like to a more severe and fatal dengue hemorrhagic fever/dengue shock syndrome (DHF/DSS) (Martina et al. 2009; Rohani et al. 2014). The causative agent is the dengue virus (DENV), which can be transmitted by *Aedes aegypti* and *A. albopictus* (Service 2012), particularly in tropical and subtropical nations. Both vectors are capable of carrying and transmitting all four strains of the DENV, namely DENV 1–4. However, being the secondary vector, *A. albopictus* has received less limelight compared to its primary counterpart despite its ability to adapt well to new environments and become the dominant species found in all types of breeding sites (Rohani et al. 2014; Dom et al. 2016).

In Malaysia, the Ministry of Health (MOH) has reported a steady increase in dengue cases, with the state of Selangor being the highest contributors of the cases nationwide based on data retrieved between the years 2014–2016. Prevention of dengue largely depends on the control of its vector. Nevertheless, dengue-related cases are still on the rise despite numerous vector control strategies such as the use of chemical insecticides, biological control agents and social intervention efforts including the Communication for Behavioral Impact (COMBI) initiatives by local authorities.

The epidemiology of DENV transmission and its persistence in nature can be attributed to its transovarial transmission (TOT) as demonstrated in previous findings (Rosen et al. 1983; Lee and Rohani 2005; Rohani et al. 2007; Buckner 2013). The study on natural TOT is an integral part of vector surveillance which can be used to help predict future endemic events (Angel et al. 2016). Hence, this study was conducted to screen the natural TOT of DENV by *A. albopictus* at several dengue hotspot areas in Selangor, Malaysia, using a real-time reverse transcriptase–polymerase chain reaction (qRT–PCR) technique.

2 Materials and Methods

2.1 Sample Sites and Collection Method

The sampling sites in this study were dengue hotspot areas located in Shah Alam, Selangor, selected due to its high population density as documented by Dom et al. (2016). Mosquito eggs were collected from their natural habitats in six different localities in Shah Alam municipality areas, namely Subang Bestari U5, Subang Murni U5, Pangsapuri Damai U4, Desa Alam U12, Teres D Seksyen 7 and Pangsapuri Perdana S13 from January 2015 to March 2015 (Fig. 1). During this

(a)

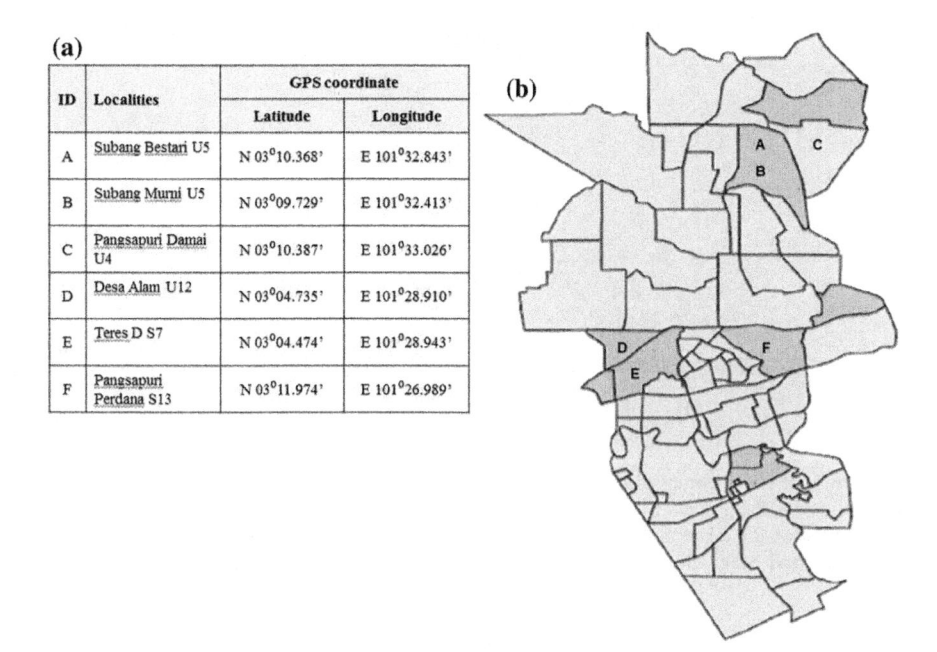

ID	Localities	GPS coordinate	
		Latitude	Longitude
A	Subang Bestari U5	N 03°10.368'	E 101°32.843'
B	Subang Murni U5	N 03°09.729'	E 101°32.413'
C	Pangsapuri Damai U4	N 03°10.387'	E 101°33.026'
D	Desa Alam U12	N 03°04.735'	E 101°28.910'
E	Teres D S7	N 03°04.474'	E 101°28.943'
F	Pangsapuri Perdana S13	N 03°11.974'	E 101°26.989'

Fig. 1 Characteristic of the study sites. **a** Geographical description of the study sites. **b** Map of Shah Alam showing the location of study sites symbolized as A, B, C, D, E, F

period, the average maximum temperature, average rainfall amount and relative humidity are between 34 and 36 °C, 52–150 mm and 73–78%, respectively.

Eggs were collected from ovitraps, transported to the laboratory within a week of sampling, and reared under standard insectary conditions. The temperature, relative humidity (RH) and photoperiod conditions were maintained at 28 ± 2 °C, 70 ± 10% and 12:12 photoperiod, respectively. Adult mosquitoes were supplied with 10% (w/v) of sucrose ad libitum and morphologically identified based on pictorial keys (Rueda 2004). Identified adult *A. albopictus* were segregated based on localities and stored in pools of 20 mosquitoes at −80 °C until further analysis.

2.2 Viral RNA Extraction

Viral RNA was extracted using QIAamp® Viral RNA Mini Kit (QIAGEN, Germany) according to the manufacturer's instructions with slight modifications as described by Thongrungkiat et al. (2010). Detection of DENV in mosquito pools was performed using a one-step *ab*TES™ DEN qPCR I Kit (AITbiotech, Singapore) alongside a commercially available positive control. Furthermore, the negative control was prepared using a pool of 20 mosquitoes from a laboratory strain of *A. albopictus* (F1611).

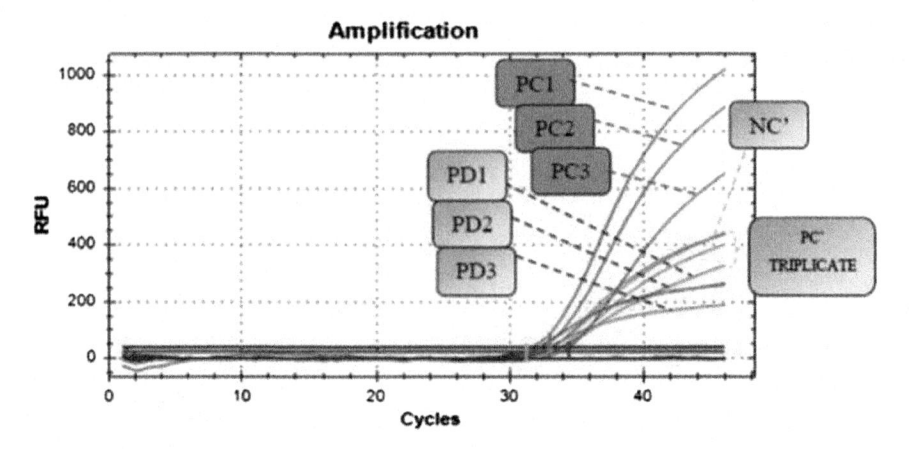

Fig. 2 Amplification curves of positive control (PC) by FAM fluorophore (blue peak) were seen. Meanwhile, for Pangsapuri Damai (PD) and negative control (NC), no amplification was seen. Internal control for all samples (PC, PD, and NC) is all positive indicated by the Texas Red fluorophore (red peak). Similar findings were also observed for all samples tested (Color figure online)

3 Results

From a total of 120 *A. albopictus* samples screened, none were positive for DENV. All of the different controls used in the assay yielded their intended result (Fig. 2), thus confirming the validity of the essay.

Based on the findings obtained, no transovarial transmission occurred at any of the localities sampled, even though the areas have been classified as dengue active transmission areas.

4 Discussion

In this study, one-step qRT–PCR was used to screen for DENV due to its specificity and sensitivity in detecting very low yields of viral load. The test was performed in triplicates in the presence of controls in order to control the validity of the method applied and to ensure the reproducibility of results. Findings from this study revealed that no DENV was detected in all *A. albopictus* samples. Since no DENV detection obtained, thus no absolute quantification methods were subsequently pursued. Our results imply that transovarial transmission of DENV did not occur in the pooled mosquitoes, the numbers of which is sufficient to represent the population of the study areas.

The negative findings also imply that no maintenance of DENV occurred among *A. albopictus* populations in Subang Bestari U5, Subang Murni U5, Pangsapuri

Damai U4, Desa Alam U12, Teres D Seksyen 7 and Pangsapuri Perdana S13, all of which has been classified as endemic areas by local authorities. This finding is parallel with a study reported by Hutamai et al. (2007) that also failed to detect transovarial transmission by adults *A. albopictus* in re-epidemic areas. The previous study postulated that the negative findings observed may be due to the mode of transmission in nature (if any) is very low in the vector populations.

This situation can be related to the present study, which targeted the urbanized areas where the human populations are higher. Ovitraps were placed outdoors representing the *A. albopictus*. Nevertheless, in the abundance of blood meal sources, transovarial transmission may not be feasible. This is supported by Martins et al. (2012) that documented transovarial transmission of DENV in females *Aedes aegypti* deprived of blood feeding, while on the other hand, producing a negative finding in blood-fed adult females *Aedes* mosquitoes. This fact strengthens the notion that the vector only used transovarial transmission as a mechanism in order to maintain the viruses in an environment which lack blood meal sources.

However, the contradict finding in this study with most of the similar study done previously in Malaysia may also be due to some limitations. Different stage of mosquito life cycle used as samples could also reflect the result obtained. Most of the previous study use immature stages of mosquito; larvae, as samples (Lee and Rohani 2005; Rohani et al. 2014), unlike this study which focuses more on the adult mosquito, the infectious stage. Despite this, this study could be improved by using larger sample size. The smaller sample size used in the current study is due to the low total amount of collected eggs during the sampling period. In addition, the sampling was done during the hot and dry season which could be one of the reasons for population reduction of mosquitoes (Rohani et al. 2014). This small-scale study was also limited due to time constraints.

5 Conclusion

In conclusion, this finding suggested that the transovarial transmission of DENV may be very low (if any) in the population and this mode of transmission may not play an important role in maintaining the DENV in the vector populations during an epidemic period. However, further investigation on improving the time of sample collection during the inter-epidemic period and frequent duration of sample collection is recommended in the future study besides isolating the serotypes of DENV circulating in the vector populations surrounding re-epidemic areas.

Acknowledgements The thank the Ministry of Education, Malaysia, for providing financial support under the Fundamental Research Grant Scheme (FRGS/1/2014/SKK10/UITM/03/1) and the General Director of Shah Alam Municipal Council for granting permission to collect samples in specified localities. In addition, the authors also thank the Faculty of Health Sciences, Universiti Teknologi MARA for providing research facilities, and Universiti Sains Malaysia for providing the laboratory strain.

References

Angel, A., Angel, B., & Joshi, V. (2016). Rare occurrence of natural transovarial transmission of dengue virus and elimination of infected foci as a possible intervention method. *Acta Tropica, 155,* 20–24.

Bhatt, S., Gething, P. W., Brady, O. J., Messina, J. P., Farlow, A. W., Moyes, C. L., … Hay, S. I. (2013). The global distribution and burden of dengue. *Nature, 496,* 504–507.

Buckner, E. A., Alto, B. W., & Lounibos, L. P. (2013). Vertical transmission of Key West dengue-1 virus by *Aedes aegypti* and *Aedes albopictus* (Diptera: Culicidae) mosquitoes from Florida. *Journal of Medical Entomology, 50,* 1291–1297.

Dom, N. C., Madzlan, M. F., Yusoff, S. N. N., Ahmad, A. H., Ismail, R., & Camalxaman, S. N. (2016). Profile distribution of juvenile *Aedes* species in an urban area of Malaysia. *Transactions of the Royal Society of Tropical Medicine and Hygiene, 110,* 237–245.

Hutamai, S., Suwonkerd, W., Suwannchote, N., Somboon P., Prapanthadara, L., Borne, V., … Mai, C. (2007). A survey of dengue viral infection in *Aedes aegypti* and *Aedes albopictus* from re-epidemic areas in the north of Thailand using nucleic acid sequenced based amplification assay. *Southeast Asian J Trop Med Public Health, 38,* 448–454.

Lee, H. L., & Rohani, A. (2005). Transovarial transmission of dengue virus in *Aedes aegypti* and *Aedes albopictus* in relation to dengue outbreak in an urban area in Malaysia. *Dengue Bulletin, 29,* 106–111.

Martina, B. E. E., Koraka, P., & Osterhaus, A. D. M. E. (2009). Dengue virus pathogenesis: An integrated view. *Clinical Microbiology Reviews, 22,* 564–581.

Martins, V. E. P., Alencar, C. H., Kamimura, M. T., de Carvalho Araújo, F. M., de Simone, S. G., Dutra, R. F., et al. (2012). Occurrence of natural vertical transmission of dengue-2 and dengue-3 viruses in *Aedes aegypti* and *Aedes albopictus* in Fortaleza, Ceará, Brazil. *PLoS ONE, 7,* 1–9.

Rohani, A., Azahary, A. R. A., Malinda, M., Zurainee, M. N., & Rozilawati, H. (2014). Eco-virological survey of Aedes mosquito larvae in selected dengue outbreak areas in Malaysia. *Journal of Vector Borne Diseases, 51,* 327–332.

Rohani, A., Zamree, I., Lee, H. L., Mustafakamal, I., Norjaiza, M. J., & Kamilan, D. (2007). Detection of transovarial dengue virus from field-caught *Aedes aegypti* and *Aedes albopictus* larvae using C6/36 cell culture and reverse transcriptase-polymerase chain reaction (RT-PCR) techniques. *Dengue Bulletin, 31,* 47–57.

Rosen, L., Shroyer, D. A., Tesh, R. B., Freier, J. E., & Lien, J. C. (1983). Transovarial transmission of dengue viruses by mosquitoes: *Aedes albopictus* and *Aedes aegypti*. *Journal of Vector Borne Diseases, 32,* 1108–1119.

Rueda, L. M. (2004). Pictorial keys for the identification of mosquitoes (Diptera: Culicidae) associated with dengue virus transmission. Zootaxa 589.

Service, M. (2012). *Medical entomology for students* (5th ed., p. 303). United States of America: Cambridge Universities Press.

Thongrungkiat, S., Maneekan, P., & Wasinpiyamongkol, L. (2010). Prospective field study of transovarial dengue-virus transmission by two different forms of *Aedes aegypti* in an urban area of Bangkok, Thailand. *Journal of Vector Ecology, 36,* 147–152.

Changes in Total Phenolics, β-Carotene, Antioxidant Properties and Antinutrients Content of Banana (*Musa Cavendishii L. Var. Montel*) Peel at Different Maturity Stages

Aishah Bujang, Siti Sarah Jamil and Nazrahyatul Hidayah Jemari

Abstract Banana peel has been reported to contain both nutrients and antinutrients contents which have good and bad health effects, respectively. The objective of this study was to assess the changes in total phenolics, β-carotene, antioxidant activities and selected antinutrients at different maturity indexes of *Musa cavendishii* banana peel. In general, the results showed significant changes ($p < 0.05$) in all components measured with the changes in maturity of the peel. The highest amount of total phenolics was obtained at index 4 and lowest at index 1. Peel at index 7 showed the strongest free radical scavenging and lowest is at index 4, while index 4 exhibited the highest reducing activity and lowest was at index 1. For β-carotene, index 7 showed highest concentration and lowest at index 1. For antinutrients, saponin and phytate content were found to be the highest at index 7. Both antinutrients were found to increase significantly with the increase of maturity index. In contrast, tannic acid was found to decrease significantly with the increase of maturity index. Highest tannic acid was at index 1 and lowest was at index 7. The information obtained in this study will be useful for the application of banana peel as functional ingredient in food products.

Keywords Antioxidant · Total phenolics · Saponin · Tannin · Phytate
Banana peel

A. Bujang (✉)
Malaysia Institute of Transport (MITRANS), Universiti
Teknologi MARA, Shah Alam, 40450 Selangor, Malaysia
e-mail: aisah012@salam.uitm.edu.my

A. Bujang · S. S. Jamil · N. H. Jemari
Faculty of Applied Sciences, Universiti Teknologi MARA,
Shah Alam, 40450 Selangor, Malaysia

1 Introduction

Banana is one of the most widely grown tropical fruits, cultivated all over the world. Edible bananas are derived from Autralimusa and Emusa series, which have different origins from the same genus. Three most common species of Musa are *M. cavendishii, M. paradisiaca and M. sapientum*. The plant is known to have originated from the Indian subcontinent, East Asia (Malaysia and Japan) and Africa (Mohapatra et al. 2010). In Malaysia, banana is one of the premier and popular fruit and is widely cultivated.

Banana peels were reported to be rich in nutrients such as dietary fibre (50% dry basis), proteins (7% dry basis), essential amino acids, polyunsaturated fatty acids and potassium (Emaga et al. 2008). About 40% of the total weight of fresh banana is contributed by its peels, and the amount of fruit waste from the peel is expected to increase which will cause environmental fouling. On the contrary, dried banana peels have been reported to be used in making banana charcoal which is an alternative source of cooking fuel while the wet peels have been used to create liniment for reducing arthritis, aches and pains (Saheed et al. 2012). Antinutrients such as tannin, oxalate and saponin are among secondary metabolites found in plants. Their presence in food causes interference to the absorption of nutrients. For example, tannin interferes with absorption of iron and other minerals and has the capability to bind and precipitate proteins that leads to decreasing in digestibility (Ekop 2007).

Very limited studies were conducted on the *Musa cavendishii* planted in Malaysia. Thus, this study will further provide information on the nutritional values and antinutrients content in banana peel of Cavendish variety. The amount of these components at different maturity stages will give knowledge on the development of each chemical components during ripening stages.

The aim of this study was to determine total phenolic content, antioxidant activities, β-carotene, saponin, tannin and phytate content of *Musa cavendishii* banana peel at three different maturity indexes.

2 Materials & Method

2.1 Materials

Banana *Musa cavendishii L. Var. Montel* of three different maturity indexes (Table 1) were purchased at wet market, Section 6, Shah Alam. Analytical grade chemicals used in this study were obtained from Chemo Lab Company, Petaling Jaya, Selangor.

Table 1 Maturity index of M. cavendishii banana

Ripening index	Description of ripening stage
1	Dark green overall. Immature fruit to be harvested
4	More yellow than green. Nearly ripe fruit
7	Amber. Yellow overall with a brown speckles. Overripe fruit

2.2 Preparation of Sample

Banana peels were dried in cabinet dryer at 50 ± 2 °C for 12 h and then grounded. The powdered peel obtained was stored at −18 °C in amber bottle prior to analysis.

2.3 Determination of Total Phenolic, Antioxidant Activities and β-Carotene Content

Total phenolic content (TPC) was measured according to the method describe by Katalinic et al. (2006). Absorbance was measured at 760 nm using spectrophotometer (UVA-160921, Helios-α, England). Gallic acid was used as standard and results obtained were expressed as mg gallic acid equivalent (GAE)/100 g. Free radical scavenging activity was determined using DPPH method with absorbance measured at 517 nm. Reducing capability was determined using ferric reducing power (FRAP) assay according to the method describe by Pulido et al. (2000). Absorbance was measured using at 539 nm. Beta-carotene content was analysed following standard procedure of AOAC.

2.4 Determination of Antinutrients (Tannin, Phytate and Saponin) Content

Saponin content was measured gravimetrically and calculated in percentage according to the method described by Donatus. Tannin content was determined spectrometrically following the method by Okwu and Friday (2008) with absorbance measured at 605 nm within 10 min. Results were expressed as tannic acid equivalent (mg TAE/100 g). Phytate content was measured according to the method described by Reddy et al. (1982).

2.5 Statistical Analysis

All data were expressed as mean ± standard deviation of triplicate measurements. The data were analysed using one-way ANOVA using SPSS ver. 20.0, and Duncan's new multiple range test was used to determine significant difference at 5% probability level ($p < 0.05$).

3 Results and Discussion

3.1 Changes in Total Phenolics and Antioxidant Activities of *M. Cavendishii L. Var. Montel* Peel

It is known that chemical changes occur during ripening of fruits which directly affects the nutrient composition as well as edibility of the fruits. Table 2 shows the total phenolics content, antioxidant properties and β-carotene content of Cavendish banana peel.

The results showed that total phenolics were significantly highest at index 4 with an increase of 35.4% compared to index 1. The amount was then reduced as the banana fruit ripens with 5.0% decrease at index 7 compared to index 4. These showed that presence of phenolic compounds varies at different ripening stages of banana. Similar finding was observed by Christelle et al. (2013) that studied the composition of bioactive compounds in Cavendish banana variety. They stated that differences in total phenolics were due to the developmental stages as part of the physical ripening process. The decrease in total phenolics may be due to the oxidation of polyphenols by polyphenol oxidase into other quinones compound during maturity stages (Amiot et al. 1995; Kulkami and Aradhya 2005; Shwartz et al. 2009). Also, at a certain stage of ripening, stoppage of new biosynthesis of

Table 2 Health beneficial components of Cavendish banana peel at three different maturity stages

Parameters	Index 1	Index 4	Index 7
Total phenolics (mg GAE/100 g)	68.90 ± 2.13[c]	93.28 ± 0.85[a]	88.62 ± 1.29[b]
Free radical scavenging (%)	34.25 ± 1.55[b]	26.25 ± 1.18[c]	42.18 ± 2.01[a]
Ferric reducing antioxidant (mg/100 g)	36.86 ± 1.01[c]	55.71 ± 0.64[a]	42.57 ± 1.75[b]
β-carotene (μg/L)	1.05 ± 0.01[c]	1.22 ± 0.02[b]	1.50 ± 0.02[a]

Different superscript alphabet of the same row showed significant difference at $p < 0.05$

polyphenols during fruit maturation will cause the reduction in total phenolics as observed at higher maturity index (Kulkami and Aradhya 2005).

For antioxidant activities, the results for % free radical scavenging showed that highest activity was at index 7 followed by index 1 and the lowest is at index 4. The scavenging effects of banana peel showed significant decrease by 23.4% from index 1 to index 4. However, as the maturity stage increase to index 7, the scavenging activity increases by 60.7% while the results obtained for ferric reducing antioxidant power (FRAP) showed similar trend with total phenolics. Highest reducing activity was observed at index 4 with an increase of 51.1% from index 1 and then reduced by 23.6% at index 7. This indicates that reducing activities were closely related to the content of phenolic compounds present in the peel.

In previous study conducted by Gil et al. (2000) and Fischer et al. (2011) on relationship between phenolic composition and antioxidant activity in pomegranate, they stated that reduction in antioxidant activities during pomegranate fruit development is associated with apparent decrease in quantity of polyphenols in the fruit. As the fruit maturation progress develop, the polyphenols are converted into o-diphenol and then into o-quinones. In this study, the increase and decrease of % free radical scavenging and total phenolics showed an opposite trend. According to Hassanien (2008), the attribution of different composition of phenolic compounds as well as composition of other non-phenolic antioxidant presence in the fruit contributes to the differences although in some study, antioxidant activity was found to positively correlate with total phenolics.

Significant increase was observed for β-carotene content with the increase of maturity index of banana. An increase of 16.2% of β-carotene content was obtained between index 1 to index 4 and further increase of 22.9% at index 7. Carotenoids provide an attractive fruit colour during ripening due to chlorophyll degradation thus contributes to changes in colour as maturation progress (Ding et al. 2007). Studies on bananas grown in Africa and South America showed similar findings on the relationship between yellow-to-orange flesh colouration and higher carotenoid content (Amorim et al. 2009; Newilah et al. 2008). This means that the yellow colour of banana observed upon maturation is due to the increasing concentration of carotenoid content.

3.2 Changes in Antinutrients Content of *M. Cavendishii* *L. Var. Montel Peel*

Table 3 shows the results obtained for antinutrient component in banana peel of Cavendish variety.

Table 3 Antinutrient composition in banana peel at three different maturity stages

Parameters	Index 1	Index 4	Index 7
Saponin (%)	6.53 ± 0.09^c	10.21 ± 0.15^b	12.53 ± 0.16^a
Tannic acid (mg TAE/100 g)	103.32 ± 5.79^a	82.58 ± 3.77^b	68.58 ± 2.04^c
Phytate (mg/g)	1.46 ± 0.05^c	1.59 ± 0.15^b	1.83 ± 0.06^a

Different superscript alphabet of the same row showed significant difference at $p < 0.05$

The changes in saponin and phytate content at different maturity indexes of banana showed a similar trend. The amount increases with the increase in maturity index. Saponin content increases by 56.3% from index 1–4 and by 91.8% to index 7, while phytate content increases by 8.9% from index 1–4 and subsequent increased by 26.2% as the maturity reaches index 7.

A study conducted by Ahwange (2008) showed that the amount of saponin in *Musa sapientum* banana peel grown in Nigeria was 24 ± 0.27%. This is about half of the amount of saponin content at index 7 obtained in banana of *M. cavendishii L. var. Montel* in this study. Ahwange (2008) also reported a lower amount of phytate content with only 0.28 mg/g in *M. sapentium* peel as compared to the amount obtained in this study. The difference is probably due to the different variety of banana used and also differences in cultivation practices and soil composition.

In contrast to saponin and phytate, tannin was found to decrease with the increase of ripening stages. Tannin reported as tannic acid in this study was reduced by 20.1% from index 1–4 and further decrease by 33.6% to index 7. Green banana peels provided high level of tannin content and then decreased during the ripening process. According to Emaga et al. (2008), during ripening, tannin migrates into the pulp or was degraded by polyphenol oxidases and peroxidases. Also, the decrease in amount of tannin could be explained by the polymerisation process of proanthocyanidins occurring during fruit development (Lu and Foo 2000).

4 Conclusion

Changes in chemical components including antioxidant activities occur as the fruit ripen. In this study, highest amount of phenolic compounds and reducing capacity of banana peel were observed at index 4, while the lowest was observed at index 1. However, β-carotene was found to increase with the increase in maturity stages. Similarly, antinutrients namely saponin and phytate were also increased with the increase in maturity. In contrast, tannin decreases with the increase in maturity. The amount of nutrients and antinutrients present at different maturity stages is a useful data in application of waste materials such as banana peel as functional ingredients.

Acknowledgements The author wishes to thank Faculty of Applied Sciences for technical and laboratory facilities provided. The authors also gratefully acknowledge the helpful comments and suggestions of the reviewers, which have improved the presentation.

References

Amiot, J. M., Tacchini, M., Aubert, S. Y., & Oleszek, W. (1995). Influence of cultivar, maturity stage and storage conditions on phenolic composition and enzymatic browning of pear fruit. *Journal of Agriculture and Food Chemistry, 43,* 1132–1137.
Amorim, E. P., Vilarinhos, A. D., Cohen, K. O., Amorim, V. B. O., dos-Santos, S. J. A., Silva, S. O., Pestana, K. N., dosSantos VJ, ..., dos Reis, R. V. (2009). Genetic diversity of carotenoid-rich bananas evaluated by Diversity Arrays Technology (DArT). *Genetic Molecular Biology, 32*(1), 96–103.
Anhwange, B. (2008). Chemical composition of *Musa sapientum* (Banana) peels. *Journal of Food Technology, 6*(6), 263–266.
Christelle, B. B., Oliver, H., Didier, M. M., & Dominique, Patrick P. (2013). Effect of physiological harvest stages on the composition of bioactive compounds in cavendish bananas. *Journal of Zhejiang University Science B., 14*(4), 270–278.
Ding, P., Ahmad, S. H., Abd Razak, A. R., Saari, N., & Mohamed, M. T. M. (2007). Plastid ultrastructure, chlorophyll contents, and color expression during ripening of Cavendish banana (*Musa acuminate* 'William') at 18 and 27 °C. *N. Z. J. Crop Horticulture Science, 35,* 201–210.
Ekop, A. S. (2007). Determination of chemical composition of *Gnetum africanum* (Afang) seeds. *Pakistan Journal of Nutrition, 6*(1), 40–43.
Emaga, T., Robert, C., Ronkart, S. N., Wathelet, B., & Paquot, M. (2008). Dietary fibre components and pectin chemical features of peels during ripening in banana and plantain varieties. *Bioresource Technology, 99*(10), 4346–4354.
Fischer, U. A., Carle, R., & Kammerer, D. R. (2011). Identification and quantification of phenolic compounds from pomegranate (*Punica granatum* L.) peel, mesocarp, aril and differently produced juices by HPLC-DAD–ESI/MSn. *Food Chemistry, 127,* 807–821.
Gil, M. I., Tomas-Berberan, F. A., Hess-pierce, B., Holcroft, D. M., & Kader, A. A. (2000). Antioxidant activity of pomegranate juice and its relationship with phenolic composition and processing. *Journal of Agriculture Food Chemistry, 48,* 4581–4589.
Hassanien, M. A. R. (2008). Total antioxidant potential of juices, beverages, and hot drink consumed in Egypt screened by DPPH in vitro assay. *59*(3), 254–259.
Katalinic, V., Milos, M., Kulisic, T., & Jukic, M. (2006). Screening 70 medicinal plant extracts for antioxidant capacity and total phenols. *Food Chemistry, 94,* 550–557.
Kulkami, A. P., & Aradhya, S. M. (2005). Chemical changes and antioxidant activity in pomegranate arils during fruit development. *Food Chemistry, 93,* 319–324.
Lu, Y., & Foo, Y. (2000). Antioxidant and radical scavenging activities of polyphenols from apple pomace. *Food Chemistry, 68,* 81–85.
Mohapatra, D., Mishra, S., & Sutar, N. (2010). Banana and its by-product utilisation: An overview. *Journal of Scientific & Industrial Research, 69,* 323–329.
Newilah, G. N., Lusty, C., Van den Bergh, I., Akyeampong, E., Davey, M., & Tomekpe, K. (2008). Evaluating bananas and plantains grown in Cameroon as a potential sources of carotenoids. *Food, 2*(2), 135–138.
Okwu, D. E., & Friday, I. F. (2008). Phytochemical composition and biological activities of *Uvaria chamae* and *Clerodendoron splendens*. *E-Journal of Chemistry, 6*(2), 553–560.
Pulido, R., Bravo, L., & Saura-Calixto, F. (2000). Antioxidant activity of dietary polyphenols as determined by a modified ferric reducing/antioxidant power assay. *Journal Agricultural Food Chemistry, 48,* 3396–3402.

Reddy, N. R., Sathe, S. K., & Salunkhe, D. K. (1982). Phytates in legumes and cereals. *Advanced in Food Resources, 28,* 1–92.

Saheed, O. K., Jamal, P., & Alam, Z. (2012). Optimization of Bio-Protein Production from Banana Peels by Sequential Solid State Bioconversion. In *International Annual Symposium on Sustainability Science and Management,* (July), 982–989.

Shwartz, E., Glazer, I., Bar-Ya'akov, I., Matityahu, I., Bar-Ilan, I., Holland, D., & Amir, R. (2009). Changes in chemical constituents during the maturation and ripening of two commercially important pomegranate accessions. *Food Chemistry, 115,* 965–973.

Exploring Attitude Toward Online Video Marketing in Malaysia

Akeem Olowolayemo, Norsaremah Saleh, Nafisat Toyin Adewale and Fatemah Shatar

Abstract Online video marketing is still in its nascent stage, specifically in Malaysia. This form of marketing has not been entirely capitalized to its fullest potentials by businesses, especially among small and medium businesses (SMEs). The various available platforms such as YouTube, Wistia, Vidyard, and Vimeo for video marketing are usually taken over by mainly the big players in the industry. As a result of marketing costs and multipurpose use of the platforms, the SMEs are left out owing to their lesser capacity to spend on marketing and advertising of their products and services. To solve this problem confronted by SMEs, a dedicated online video marketing platform is proposed. The platform is intended to allow businesses to upload video contents to market and promote their products and services. In order to predict the success of implementation of this platform prior to its adoption, a technology acceptance model (TAM) has been adopted to evaluate the intention to use by SMEs. This paper discusses the TAM model with a view to assisting the inventor of the platform gauge the behavioral intention to use the dedicated online video marketing platforms by SMEs.

Keywords Online video marketing · Dedicated platforms · SMEs
TAM

1 Introduction

As the paradigm in business marketing shifts from the traditional medium, such as the printed media and even television, the surge in online content consumption has increased tremendously over these past years and people are now more engaged in the online space than ever before. For instance, a recent report by Kemp (2016)

A. Olowolayemo (✉)
University Malaysia Sarawak, Kota Samarahan, Malaysia
e-mail: oakeem@unimas.my

N. Saleh · N. T. Adewale · F. Shatar
International Islamic University Malaysia, Kuala Lumpur, Malaysia

gave very interesting statistics regarding the research conducted on digital usage in 2016. It indicated that the number of active users of the Internet in Malaysia to date is at 20.62 million out of a population of about 30.54 million, of which 77% use the Internet daily. Also, the average amount of time spent on the Internet media either by using a PC or a tablet is above 4 h a day. This is more than twice the average time spent watching TV, roughly 2 h a day, and more than the mean time spent using mobile 3½ hours a day. On e-commerce, the report also found that about 59% of the Internet users had searched online for products/services to buy, while 45% visited an online retail store, 44% made online purchase through PC/laptop, and 31% made online purchase via mobile devices (Kemp 2016).

These statistics show an immense increase in the penetration of Internet usage and indicated that more and more people are engaged in the online space and are turning toward the Internet to seek information about products and services as well as purchasing them online. This is especially crucial as consumer attention is becoming more and more challenging to achieve and there is a growing concern among companies and advertising agents on getting the attention of consumers (Teixeira 2014). One of the areas that are growing at an amazing pace is online video advertising. For instance, YouTube and others have several ads formats such as display ads; placed on the right of the feature video, and on the upper part of the video suggestion list, or the overlay ads; semi-transparent overlay ads that are imposed on about 20% portion of the feature video. There are also the skippable and non-skippable ads which are ads where the users could skip after a set time, typically, 5 s and those that cannot be skipped, respectively.

However, as laudable as the idea of a dedicated platform is, there is the need to investigate the behavioral intention to use for effective implementation of the DOVM platform for advertisement among SMEs as well as big players in the industry. Thus, the main aim of this research is to develop a theoretical framework based on the technology acceptance model (TAM) (Wei et al. 2015) to assess the readiness of the SMEs in Malaysia toward online video marketing. Therefore, this study aims to address the research questions are set out thus: (1) What is the level of awareness regarding online video advertisements among SMEs in Malaysia? (2) What is the likely level of acceptance of online video marketing based on the level of awareness? (3) Is there any significant difference in awareness as well as acceptance among SMEs in Malaysia?

2 Related Work

There has not been noticeable work on video marketing specifically in Malaysia. Most of the related previous work either focused online marketing generally such as social networks or they sought to compare between the classical advertising media such as TV, newspapers, and online advertising (Draganska et al. 2014; Sall 2009).

There has not been definite focus on online video marketing. Due to the relevance and pervasiveness of online video marketing on different platforms such as YouTube, Vimeo, Dailymotion, and VidYard, it is imperative that an effort is made to study the impact as well as acceptance of online video marketing may have on advertising. Again, other forms of advertising such as TV or newspapers are highly costly for small and medium enterprises, and probably less engaging as the online video media such as YouTube videos. Various work had tried to compare the effect of print, TV, and online advertising generally. Notable previous work includes relative comparison between different forms of advertisement, namely effects of online, print, and TV advertisements (Draganska et al. 2014; Sall 2009). Findings from these works demonstrated that online advertisements generally are more useful for brand identification, though TV advertisements are associated with highest advertisement appreciation among the various mediums of advertisement. The main focus of the research was on consumer engagement and advertising experiences.

With respect to advertising in Malaysia, specifically digital advertising, most resources found were findings from online advertising agents. For instance, Warc.com (2016) reported that digital ads spend in Malaysia is estimated to have reached MYR 726 Millions in 2015, placing Internet as the third-largest ad channel in Malaysia. Furthermore, GDP growth slows to sub-5%, and with marketing budget under pressure, it is only natural that marketers will turn to the 'lower cost', more traceable medium of digital, said Sandeep Mark Joseph, head of strategy and digital at Zenith Malaysia. More specific targeting via interests and habits, location-based advertising, and e-commerce linkages are some clear trends to watch for 2016. Moreover, some interesting stats have proven the compelling nature of video contents as a powerful marketing tool. It is reported in (Insivia 2016) that often after watching a video, 64% of viewers are more likely to buy a product online according to ComScore, 90% enjoyment of video ads increase purchase intent by 97% and brand association by 139% according to Unruly, and Cisco now predicts that 80% of all Internet traffic will be streaming video content by 2019 (Insivia 2016). What can be gleaned from these impressive statistics is that video has become the main medium for seeking information, aiding e-commerce to thrive and whetting the shopping appetite of the emerging Internet users. Contrast to reading and looking at images of product/services, video tells a story and passes indelible message directly to viewers, consequently, leaving a lasting impression or impact beyond words on the users.

In a study of Malaysian consumers in Wei et al. (2015), a great number of companies spent huge amounts of money on online advertising. Now, online advertising is becoming the driving force in many advertising initiatives and efforts (Armstrong et al. 2014). Malaysia's digital consumer spending and presence in the advertising space is small and in its infancy. For example, Internet advertising accounts for 5% of total advertising expenditure in 2014 and is only expected to increase to 8% in 2019 (Menezes and Graham 2015). Much of Malaysia's digital potential remains untapped despite the growth in Internet access and mobile Internet penetration. To harness this segment and compete effectively in the digital space, entertainment and media companies must get connected to consumers. It is now more prevalent to hear people talk about something they viewed online rather than what they watched on TV.

3 Methodology

The approach adopted in this study is a quantitative approach while an online survey has been utilized for data collection. Online surveys have several advantages when compared with the traditional paper-based surveys. Notably, it is known to shorten data collection time as well as the cost to be incurred during data collection due to the fact that it is not constrained to a specific geographic area (Wright 2006). The online survey was thought to be appropriate tool for use to reach wider respondents comprising different categories of respondents. The online survey is designed with the view to examining the relationship between variables proposed in the research model.

3.1 Model & Hypothesis

This study adopted the technology acceptance model (TAM), with the inclusion of an external variable (Job relevance) based on previous research (Venkatesh et al. 2003). Although quite a number of theories/models of measuring technology acceptance exist (e.g., Theory of Planned behavior, Diffusion of Innovation (Everett 1995), the Unified Theory of Acceptance and Use of Technology—UTAUT (Venkatesh et al. 2003), the TAM model is the most preferred as most of these frameworks focus on technical factors (Al-busaidi and Al-shihi 2010) and for measuring behavioral intention, especially in the field of information system and is known to have a high validity (Alharbi and Drew 2014).Thus, the research model is depicted in Fig. 1, followed by the study's hypotheses.

Based on the TAM model, it is expected that the behavioral intention to use pitch video marketing positively is influenced by attitude toward its usage. Meanwhile, attitude toward usage of pitch video marketing is also expected to positively influence by its perceived ease of use and perceived usefulness. In addition, "Job relevance" is expected to positively predict perceived ease of use and the perceived usefulness of the pitch video marketing. Sequel to the research model depicted in Fig. 1, the following hypotheses were tested in the study:

(1) *Perceived ease of use positively affects the perceived usefulness of online video marketing.*

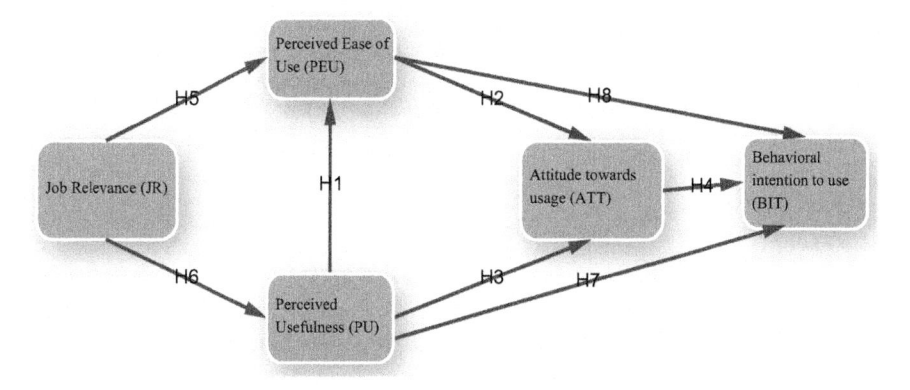

Fig. 1 Research model

(2) *Perceived ease of use positively determines attitude towards using online video marketing.*
(3) *Perceived usefulness positively affects attitude towards using online video marketing.*
(4) *Attitude towards usefulness of online video marketing positively determines the behavioral intention to use online video marketing.*
(5) *Job relevance positively affects perceived ease of use of online video marketing.*
(6) *Job relevance positively affects perceived usefulness of online video marketing.*
(7) *Perceived usefulness positively determines the behavioral intention to use online video marketing.*
(8) *Perceived ease of use positively determines the behavioral intention to use online video marketing.*

3.2 Instrumentation & Questionnaire

The questionnaire adopted for the study was evaluated for content validity using two approaches. Firstly, the questionnaire was adapted from the original measurement scales previously designed and used in TAM (Davis 1989) as well as other literatures (Shroff et al. 2011; Wu et al. 2011). Secondly, some modifications were made to the existing TAM questionnaires while also changing some few wording. Finally, the questionnaire was validated to fit within the context of online video marketing usage. Previous study has confirmed that pre-tests are essential to ensure that wording problems do not significantly influence accuracy (Sekaran and Bougie 2013).

The research instrument consists of two main sections. The first section incorporates a nominal scale to identify respondents' demographic information. The second section uses 7-point Likert response scale where 7: Strongly disagree, 6: Moderately disagree, 5: Slightly disagree, 4: Neutral, 3: Slightly agree, 2: Moderately agree, and 1: Strongly agree. This section includes TAM constructs.

4 Results

This section presents the results from the study. A total of 400 respondents were sampled out of which only 360 responses were usable. The rest were either uncompleted or improperly completed. The response rate is therefore around 90%. The result is presented in the next frames. The discussion starts with the demographic distribution of the respondents. This was followed with the test of reliability of the instrument, and subsequently tests of the hypotheses put forward in the preceding section. The discussion of the entire results was presented at the end of the section.

4.1 Demographics

The demographics of the respondents is such that the ratio of male to female respondents is about one to two, while majority of the respondents are below 30 years of age, and majorly single undergraduate with area of specialization distributed across all domains. Majority have consistent access to the Internet via different devices especially mobile phones.

4.2 Instrument Reliability

In order to ensure the internal consistency of the variables under study, the reliability test was done using Cronbach's alpha. According to Hair (2010), reliability is usually assessed using the Cronbach's alpha. Reliability assessment is important in order to ensure internal consistency among variables (Alharbi & Drew 2014). A threshold of 0.7 is required (Hair 2010). The result of the reliability test which was performed using the SPSS software presented in Table 1.

Table 1 Instrument Reliability

Variables	No of items	Cronbach's alpha
PEU	7	0.963
PU	6	0.975
ATT	5	0.958
BIT	3	0.933
JR	2	0.944

The result of the reliability test shows that all variables are above the 0.7 threshold, ranging between 0.933 and 0.975. This implies that the instrument is reliable.

4.3 Hypothesis Testing

In order to achieve the objectives of the study, correlation analysis using the SPSS software was done so as to assess the relationship between the variables of concern. As such, the decision on whether the hypotheses were supported or otherwise was based on the outcome of the correlation analysis. The result of the correlation analysis is presented in Table 2.

From the result of the correlation analysis presented in Table 2, there is a statistically significant positive relationship between the pairs of variables considered in the study, with r-values ranging between **0.739** and **0.867**, p-value of 0.000, and significance level of 0.01. Hence, all the study hypotheses were supported. It was found that a statistically significant relationship exists between Perceived Ease of Use (PEU) and Perceived usefulness (PU), Perceived Ease of Use (PEU) and Attitude (ATT), Perceived Usefulness (PU) and Attitude (ATT), Attitude (ATT) and Behavioral Intention (BIT). Similarly, a direct positive relationship was found between Job Relevance (JR) and Perceived Ease of Use (PEU) as well as between Job Relevance (JR) and Perceived Usefulness PU. Finally, a direct positive relationship was found between Perceived Usefulness and the Behavioral intention to use the Pitch online video marketing.

Table 2 Correlation analysis based on study hypotheses

Hypothesis		Correlation coefficient (r-value)	Decision
1	PEU-PU	0.857**	Supported
2	PEU-ATT	0.739**	Supported
3	PU-ATT	0.766**	Supported
4	ATT-BIT	0.867**	Supported
5	JR-PEU	0.713**	Supported
6	JR-PU	0.759**	Supported
7	PU-BIT	0.756**	Supported

$P = 0.000$, **significant at 0.01 level (2-tailed)

4.4 Discussion

Apparently, findings from this study indicate that the technology acceptance model (TAM) is applicable in measuring the behavioral intention to use the online video marketing platform. Since the acceptance of a new information system is predicted by behavioral intention and attitude toward usage and posited by TAM, this study findings also suggest that SMEs have the behavioral intention to use the dedicated online video marketing platform. In addition, the two internal beliefs (Perceived Usefulness and Perceived Ease of Use) have also positively predicted the attitude toward the use of the dedicated online video marketing platform, which is in accordance with the TAM model. The result of the correlation between perceived usefulness and the behavioral intention to use the online video marketing platform also indicates that there exists a positive relationship between the two factors. Hence, SMEs are positive about using the online video marketing as they believe that this will increase their performance. Results also show that SMEs perceive the online video marketing to be relevant to their job. According to (Venkatesh and Davis 2000), job relevance is "an individual's perception regarding the degree to which the target system is applicable to his or her job" (p. 191). Therefore, both perceived usefulness and perceived ease of use of the online video marketing platform are positively determined by job relevance among the SMEs.

From all indications, the findings from this study further confirm the TAM model in predicting the behavioral intention to use a new system such as the pitch online video marketing platform. All the hypotheses tested regarding the relationship between the pairs of factors were statistically significant and highly correlated. Hence, all the study hypotheses were supported. This is an indication that SMEs are positive about the behavioral intention to use the online video marketing platform. The support for the technology acceptance model as found in this study is not peculiar, as most studies which adopted the model found similar results even though the participants were diverse. For instance, studies conducted among academics (Afshari et al. 2009; Albalawi 2007; Alharbi and Drew 2014) also show similar results. In relation to online video sharing, a study conducted by (Yang et al. 2010) also shows that positive relationships exist among the TAM constructs perceived usefulness, perceived ease of use, and attitude toward use and the behavioral intention to use YouTube to share video among online users.

5 Conclusion

The study is focused on predicting the behavioral intention to use a new online video marketing platform specifically for SMEs. The work considered several variables such as the Perceived Ease of Use (PEU), Perceived usefulness (PU), Attitude (ATT), Behavioral Intention (BIT), Job Relevance (JR), and Perceived

Ease of Use (PEU). The study demonstrated that all hypotheses postulated are supported. This shows that there is the high likelihood that online video marketing would be a welcome idea among the potentially increasing SMEs populations.

Future work would include usability and usage evaluation of the online video marketing.

Acknowledgements The author would like to acknowledge the faculty of cognitive sciences and human development, Univerisiti Malaysia Sarawak as well as the faculty of information and communication technology, International Islamic University Malaysia. The authors also gratefully acknowledge the helpful comments and suggestions of the reviewers, which have improved the presentation.

References

Afshari, M., Bakar, K. A., Luan, W. S., Samah, B. A., & Fooi, F. S. (2009). Factors affecting teachers' use of information and communication technology. *Online Submission, 2*(1), 77–104.

Al-busaidi, K. A., & Al-shihi, H. (2010). Instructors â€™ acceptance of learning management systems: A theoretical framework.

Albalawi, M. S. (2007). *Critical factors related to the implementation of web-based instruction by higher-education faculty at three universities in the Kingdom of Saudi Arabia*. The University of West Florida.

Alharbi, S., & Drew, S. (2014). Using the technology acceptance model in understanding academics' Behavioural intention to use learning management systems. *International Journal of Advanced Computer Science and Applications (IJACSA), 5*(1), 143–155. Retrieved from http://ijacsa.thesai.org/.

Armstrong, G., Adam, S., Denize, S., & Kotler, P. (2014). *Principles of marketing*. Australia: Pearson.

Davis, F. D. (1989). Perceived Usefulness, Perceived Ease of Use, and User Acceptance of Information Technology. *MIS Quarterly, 13*(3), 319–340. https://doi.org/10.2307/249008.

Draganska, M., Hartmann, W. R., & Stanglein, G. (2014). Internet versus television advertising: A brand-building comparison. *Journal of Marketing Research, 51*(5), 578–590. https://doi.org/10.1509/jmr.13.0124.

Everett, R. M. (1995). *Diffusion of Innovations* (3rd ed.).

Hair, J. F. (2010). *Multivariate data analysis: A global perspective*. Upper Saddle River, N.J., [etc.]: Pearson.

Insivia. (2016). 50 Must-know stats about video marketing 2016. Retrieved from http://www.insivia.com/50-must-know-stats-about-video-marketing-2016/.

Kemp, S. (2016). *Digital in 2016*. Www.Wearesocial.Com. Retrieved from http://wearesocial.com/sg/special-reports/digital-2016%5Cnhttp://www.slideshare.net/wearesocialsg/digital-in-2016.

Menezes, I., & Graham, M. (2015). Malaysia entertainment and media outlook 2015–2019, 20. Retrieved from https://www.pwc.com/my/en/assets/publications/entertainment-media-outlook-2015.pdf.

Sall, M. (2009). Brand identification and advertisement appreciation of online, print, and TV advertisements. *Wharton Research Scholars, 28*.

Sekaran, U., & Bougie, R. (2013). *Research methods for business: A skill-building approach* (6th ed.). Wiley.

Shroff, R. H., Deneen, C. C., & NG, M. W. E. (2011). Analysis of the technology acceptance model in examining students' behavioural intention to use an e-portfolio system. *Australasian Journal of Educational Technology, 27*, 600–618.

Teixeira, T. S. (2014). *The Rising Cost of Consumer Attention: Why You Should Care, and What You Can Do about It. HBS Working Paper.* Retrieved from http://www.hbs.edu/faculty/Pages/item.aspx?num=46132.

Venkatesh, V., & Davis, F. D. (2000). A theoretical extension of the technology acceptance model: Four longitudinal field studies. *Management Science, 46*(2), 186–204. https://doi.org/10.1287/mnsc.46.2.186.11926.

Venkatesh, V., Morris, M. G., Davis, G. B., & Davis, F. D. (2003). User acceptance of information technology: Toward a unified view. *MIS Quarterly, 27*(3), 425–478. Retrieved from http://www.jstor.org/stable/30036540.

Warc.com. (2016). Malaysia sees digital adspend hike. Retrieved January 19, 2017, from https://www.warc.com/Content/News/N36461_Malaysia_sees_digital_adspend_hike.content?.

Wei, K. K., Jerome, T., & Shan, L. W. (2015). Online advertising: A study of malaysian consumers. *International Journal of Business and Information, 5*(2) (2010). Retrieved from http://ijbi.org/ijbi/article/view/51.

Wright, K. B. (2006). Researching internet-based populations: advantages and disadvantages of online survey research, Online questionnaire authoring software packages, and web survey services. *Journal of Computer-Mediated Communication, 10*(3), 00–00. https://doi.org/10.1111/j.1083-6101.2005.tb00259.x.

Wu, I.-L., Li, J.-Y., & Fu, C.-Y. (2011). The adoption of mobile healthcare by hospital's professionals: An integrative perspective. *Decision Support Systems, 51*(3), 587–596. https://doi.org/10.1016/j.dss.2011.03.003.

Yang, C., Hsu, Y.-C., & Tan, S. (2010). Predicting the determinants of users' intentions for using YouTube to share video: moderating gender effects. *Cyberpsychology, Behavior and Social Networking, 13*(2), 141–152.

Inventory Policies in Blood Bank Management

Farah Hani Jamaludin, Jasmani Bidin, Noorsham Mansor, Sharifah Fhahriyah Syed Abas and Zurina Kasim

Abstract Blood bank operation is a facility which involves the process of collecting, testing, processing and storing blood for later use in blood transfusion. Blood bank management faces a crucial problem due to shortage of blood supply that sometimes does not meet the demand. Blood inventory is difficult to organize and control especially in festive seasons when demand is high due to road accidents and casualties. Blood bank management needs to balance the demand and supply to avoid shortage, and since blood is perishable, the management also needs to avoid wastage. To solve this issue, an effective inventory method should be implemented. This study compares two policies: the continuous review policy and the periodic review policy. The continuous review policy observes the monitory level more accurately and on time, while the periodic policy applies only periodically inventory control. This study looks at the relationships between the service level, safety stock and inventory cost that are based on three case studies, which are variable demand and constant lead time, variable lead time and constant demand and variable demand and variable lead time. The result shows that the continuous review policy should be implemented in blood bank management due to lower inventory cost and being more systematic with higher efficiency.

Keywords Blood bank management · Inventory method · Continuous review policy · Periodic review policy

1 Introduction

Blood is a need for everyone. Health care operations whose job is to supply sufficient blood to the people who need it have to manage the job properly. The key component of health care system is the blood service operations or more well

F. H. Jamaludin (✉) · J. Bidin · N. Mansor · S. F. Syed Abas · Z. Kasim
Faculty of Computer & Mathematical Sciences, Universiti Teknologi MARA Cawangan Perlis (Kampus Arau), 02600 Arau, Perlis, Malaysia
e-mail: farahhanijamaludin@gmail.com

© Springer Nature Singapore Pte Ltd. 2018
R. Saian and M. A. Abbas (eds.), *Proceedings of the Second International Conference on the Future of ASEAN (ICoFA) 2017 – Volume 2*,
https://doi.org/10.1007/978-981-10-8471-3_35

known as the blood bank. Blood bank management involves the process of collecting, testing, processing, transporting and storing of blood for later use by applying certain procedures. In order to have enough and healthy blood supply, hospitals have to practise good management of blood bank. Blood bank supply is important at all time to ensure effective health care operations and in saving people's lives. Many blood banks are facing serious problems in maintaining sufficient blood supply and to ensure that demands are met. At the same time, since blood plasma is perishable, the management needs to ensure that wastage of blood does not occur. There is a need to manage properly available blood of each group and to always maintain a good stock. Inventory management encompasses all activities associated with ordering, storing, handling and issuing of blood. An effective management system should be implemented in the blood bank system in order to have a minimal inventory level.

The purpose of this study is to determine the best inventory policy review that should be practised in blood bank management, and the relationship between service level, safety stock and inventory cost. Periodic policy review and continuous policy review have been analysed and compared to obtain the best inventory policies to be applied in blood bank operations.

2 Literature Review

Blood banks are responsible for preparing sufficient blood supplies, but at certain times they have to face a crucial shortage of supply. Cimaroli et al. (2012) agreed that the balance between the supply and demand is a big question, and factors affecting this include an aging population and increasing immigrant communities with lower donation rates. Ma et al. (2013) stated that an increasing demand for blood has elevated day by day, and this becomes an issue for the health care society in requesting for more blood donors. Although there is initiative to create awareness towards blood donation, people are still hesitating in donating their blood. Balkees and Mohammad (2014) declared that the range of socio-demographic, organizational and physiological factors affects the people's willingness to donate blood.

According to Blake and Hardy (2014), in making the decision to order blood supply, one does not only need to consider the stock amount in hand, but also the lifespan of the perishable item. Thus, an efficient inventory method needs to be implemented to ensure effective management of blood supply. Duan, Li, Tien and Huo (2012) studied the inventory models with backlogging and without backlogging. Backlogging is defined as reserved or unfulfilled orders on a given day. When a product is highly perishable, the sellers prefer backlog demand in order. In sales, when a product is highly perishable, the sellers prefer backlog demand in order. Stanger et al. (2012) stated that blood management is very costly due to major components that make up the inventory management and blood distribution.

According to them, inventory transparencies, procedure of simple inventory, freshness focus, ordering patterns and stock levels, training and human resources are the key points in having a good blood inventory management.

Haijema (2012) applied a model of stochastic dynamic programming that optimizes the ordering and disposal decision on the perishable products, while Gunpinar and Centeno (2015) applied a model of stochastic integer programming in order to reduce wastage and shortage of blood products in hospitals. In order to have an efficient blood resource management, optimization models were developed for hospitals. The stochastic and the deterministic models were considered which involved demand for two types of patients, the cross-match transfusion ratio (CT ratio) and uncertain demand rates. Results showed that the method helped in reducing wastages, shortages of blood products and the total cost involved.

3 Methodology

The statistics of blood demand was obtained from the National Blood Centre in Kuala Lumpur. The data collected was for weekly demand. Information on the value of lead time, monitoring cost, holding cost and set up cost was also given. The weekly blood demand in the year 2014 was analysed. The constant variables for this study were set up cost, holding cost and monitoring cost.

Two significant systems of inventory control policies were used. The first was the continuous inventory system, also known as the fixed-order-quantity inventory. The second was the periodic inventory system, also known as fixed-time-period inventory. The continuous inventory system involves continuous updates, where at the time when the level of inventory decreases to a certain level, the replenishment is made based on the previous fixed quantity. The periodic inventory system involves periodic updates, where an order is placed in a variable amount after specific regular intervals. The comparison between the two policies is shown in Table 1.

Table 1 Comparison of continuous review policy and periodic review policy

	Continuous review policy	Periodic review policy
Order time	Variable order intervals	Fixed order interval
Order quantity	Fixed order quantity	Variable order quantity
Ease of management	Review with high frequency	Review with low frequency
Safety stock	Smaller	Larger
Advantages	Accurate inventory counting	Lower investment for counting
Disadvantages	High investment on counting inventory level	Inaccurate inventory counting

Source Ma et al. (2013)

Three cases were considered in this study. In the first case, the demand was assumed as a random variable while the lead time as a constant. In the second case, the lead time was a random variable while the demand was a constant. In the third case, both demand and the lead time were assumed as random variables.

All the cases were analysed with different service levels within each of the inventory control systems. The service level was assumed in the data. This study was developed using the service level of 90% in both inventory policies, denoting that there was no shortage of supply in 90% of the cases. Both policies were tested using four different service levels as shown in Table 2.

The values of Z were determined by using the method of interpolation inventory.

The basic formula used in all three cases were to calculate the mean demand d, demand deviation σ_d, lead time ℓ and lead time deviation σ_ℓ. The mean, probability and the standard deviation of the demand and lead time were calculated by using Microsoft Excel.

$$d = \sum_{i=1}^{53} (\text{weekly demand})_i \times (\text{pobability})_i$$
$$d = (2905 \times 0.017) + (3167 \times 0.018) + \cdots + (3186 \times 0.018)$$
$$= 3347 \, \text{unit/week} \tag{1}$$

$$\sigma_d = \sqrt{0.017 \times (2905 - 3347)^2 + \cdots + 0.018 \times (3186 - 3347)^2} = 326. \tag{2}$$

$$\ell = (1 \times 0.1) + (2 \times 0.2) + (3 \times 0.3) + (4 \times 0.4) = 3 \, \text{weeks}. \tag{3}$$

$$\sigma_\ell = \sqrt{0.1 \times (1-3)^2 + 0.2 \times (2-3)^2 + \cdots + 0.4 \times (4-3)^2} = 1. \tag{4}$$

Equation 1 represents the mean demand, Eq. 2 represents the value of demand deviation, Eq. 3 represents the value of mean lead time and Eq. 4 indicates the value of lead time deviation. The mean demand was 3347 units per week, the demand deviation was 326, while the value for the mean lead time was 3 weeks with the lead time deviation of 1.

The total inventory cost which included annual holding cost, set up cost and monitoring cost was also calculated. The monitoring cost (mc) for the continuous review policy and the periodic review policy were RM 1840 and RM 368,

Table 2 Z value and service level

Z	Service Level (%)
1.28	90
1.65	95
2.33	99
3.08	99.90

Inventory Policies in Blood Bank Management

Table 3 Inventory model

Continuous review policy		Periodic review policy
	Economic order quantity $Q = \sqrt{\frac{2ad}{h}}$	Time order interval $T = \sqrt{\frac{2a}{dh}}$
Variable demand and constant lead time	Safety stock, $SS = Z \times \sigma_d \times \sqrt{\ell}$ Reorder point, $ROP = d \times \ell + SS$	Safety stock, $SS = Z \times \sigma_d \times \sqrt{T+\ell}$
Variable lead time and constant demand	Safety stock, $SS = Z \times d \times \sigma_\ell$ Reorder point, $ROP = d \times \ell + SS$	Safety stock, $SS = Z \times \sigma_d \times \sqrt{\ell}$
Variable demand and variable lead time	Safety stock, $SS = Z \times \sqrt{\ell \times \sigma_d^2 \times d^2 \times \sigma_\ell^2}$ Reorder Point, $ROP = d \times \ell + SS$	Safety stock, $SS = Z \times \sqrt{\ell \times \sigma_d^2 \times d^2 \times \sigma_\ell^2} \times \sqrt{\ell + T}$
	Total inventory cost $TIC = [h \times SS] + [a \times 52] + [h \times (Q/2)] + mc$	

respectively. The holding cost (h) and set up cost (a) were constant at RM 4 and RM 368, respectively. All the formulas that were being used throughout the study in blood bank management were represented in Table 3.
where,

$$TIC = \text{Total Inventory Cost}$$
$$Q = \text{quantity per order}$$
$$SS = \text{safety stock}$$
$$h = \text{holding cost}$$
$$a = \text{set up cost}$$
$$mc = \text{monitoring cost}$$

4 Findings

The findings are summarized in the following Table 4 which shows that total inventory cost and safety stock for both inventory policies in all the three cases using service level of 90, 95, 99 and 99.9%. The safety inventory stock shows all positive values under both policies. Under the four service levels, the total inventory cost for the continuous review policy, under the uncertain demand and constant lead time condition, is larger than the total inventory cost under the periodic review policy. This is because management for uncertain demand needs more facilities and

Table 4 Risk analysis for two inventory policies

Inventory management policy		Continuous review policy				Periodic review policy			
Service level (%)		90	95	99	99.90	90	95	99	99.90
Variable demand and constant	Safety stock	723	923	1316	1740	746	962	1358	1796
	Inventory cost	25,439	26,275	27,812	29,507	24,059	24,922	26,508	28,257
Lead time variable lead time and constant	Safety stock	3012	5523	7799	10,309	7663	9879	13,950	18,440
	Inventory cost	34,595	44,636	53,740	63,781	51,727	60,588	76,873	94,834
Demand variable demand and variable lead time	Safety stock	4344	5600	7908	10,454	7771	10,017	14,145	18,699
	Inventory cost	39,923	44,946	54,178	64,360	52,158	61,143	77,656	95,869

labour. However, the total costs for the continuous review policy under the other two cases are lower than the cost under the periodic review policy. This happens because the periodic review policy requires more labour to check the inventory level periodically, especially in the case of uncertain lead time and demand.

The highlighted figures in Table 4 are the total inventory cost and safety inventory stock for all three cases using the lowest service level at 90%. The total inventory cost is at its lowest under the periodic review policy in the case where variable demand and constant lead time are applied. But when the two other cases are applied, the continuous review policy costs less when compared to periodic review policy. However, in all cases, the safety inventory stock is larger under the continuous review policy.

The results also show that in all cases, as the service level increases, both the safety stock and total inventory cost increase. The main thing that can be seen is that the safety stock only relies on service level, while the inventory cost depends on both service level and the conditions of demand and lead time. The observation shows that more safety stock is available as the blood bank updates its inventory, and thus the demand of blood can be fulfilled.

Hence, the results are analysed and represented in bar chart. Figures 1, 2 and 3 below show the comparison between the total inventory cost of the continuous review policy and the periodic review policy based on the three different cases.

Figure 1 shows the total inventory cost for the case of variable demand and constant lead time. The total inventory cost for the continuous review policy is higher than the total inventory cost for the periodic review policy. The factor of uncertain demand influences the total inventory cost.

Figure 2 shows the total inventory cost for the case of variable lead time and constant demand. The total inventory cost for the continuous review policy is lower than the inventory cost for the periodic review policy. It can be concluded that the factor of lead time influences the total inventory cost.

Inventory Policies in Blood Bank Management

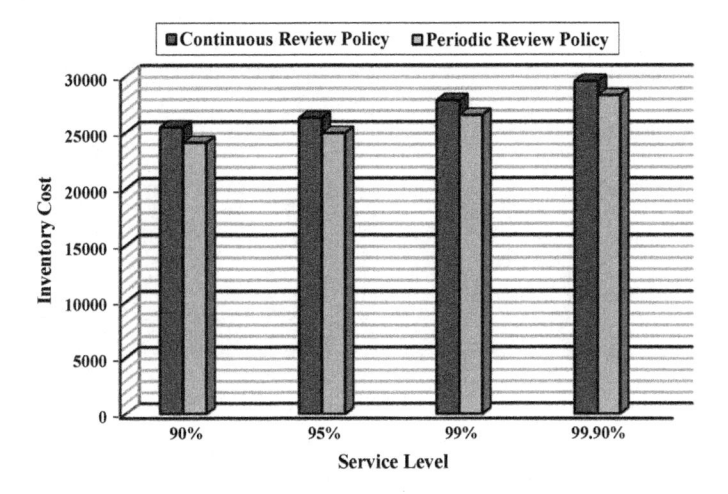

Fig. 1 Total inventory cost for variable demand and constant lead time

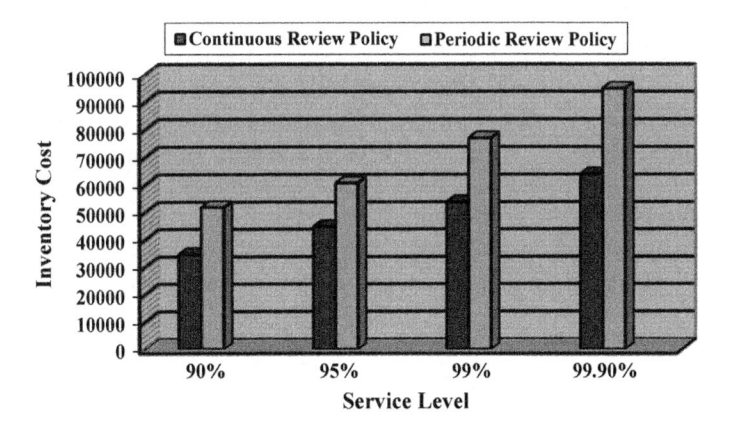

Fig. 2 Total inventory cost for variable lead time and constant demand

Figure 3 shows the total inventory cost for the case of variable demand and variable lead time. The total inventory cost for the continuous review policy is less than the total inventory cost for periodic review policy. The total inventory cost for continuous review policy is lower in two cases out of three cases. So, the continuous review policy should be implemented in the health care management for blood bank inventory.

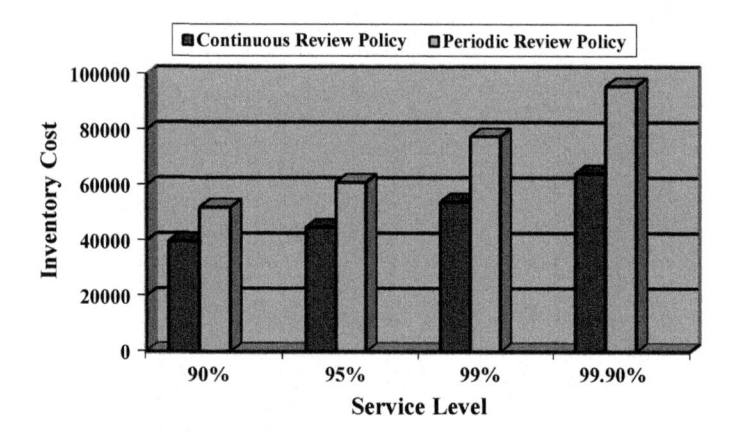

Fig. 3 Total inventory cost for variable demand and variable lead time

5 Conclusion and Recommendation

Results show that the periodic review policy has a smaller total inventory cost than the continuous review policy in the case of uncertain demand and constant lead time. Meanwhile, in the cases of uncertain lead time and constant demand, and uncertain demand and uncertain lead time, the total inventory cost for the continuous review policy is lower. Generally, it can be concluded that the continuous review policy is more economical and it also provides an on-time inventory control method.

A manageable website or inventory system should be designed in order to have efficient and systematic operations with minimal costs. The system should also be able to monitor the blood supply and demand for each blood type.

Acknowledgements We would like to express our gratefulness to the reviewers for their insightful and valuable comments and criticism to improve this paper.

References

Balkees, H. A., & Mohammad, Y. N. S. (2014). Investigating knowledge and attitudes blood donors and barriers concerning blood donation in Jordan. *Procedia-Social Science & Medicine Journal, 96*, 86–94.

Blake, J. T., & Hardy, M. (2014). A generic modelling framework to evaluate network blood management policies: The Canadian Blood Services experience. *Operational Research for Health Care Journal, 3*(3), 116–128.

Cimaroli, K., Paez, A., Newbold, K. B., & Heddle, N. M. (2012). Individual and contextual determinants of blood donation frequency with a focus on clinic accessability: A case study of Toronto Canada. *Health and Place Journal, 18*(2), 424–433.

Duan, Y., Li, G., Tien, J. M., & Huo, J. (2012). Inventory models for perishable items with inventory level demand rate. *Applied Mathematical Modelling, 36*(10), 5015–5028.

Gunpinar, S., & Centeno, G. (2015). Stochastic Integer Programming Models for reducing wastages and shortages of blood product at hospitals. *Computers and Operations Research Journal, 54,* 129–141.

Haijema, R. (2012). Optimal ordering, issuance and disposal policies for inventory management of perishable products. *Production Economics, 157,* 158–169.

Ma, J., Lei, T., & Okudan, G. E. (2013). EOQ-based inventories control policies for perishable items: The case of blood plasma inventory management. In A. Krishnamurthy, W. K. V. Chan (Eds.), *Proceedings of the 2013 industrial and systems engineering research conference,* pp. 1–10. Pennsylvania USA: University Park.

Stanger, S. H. W., Yates, N., Wilding, R., & Cotton, S. (2012). Blood inventory Management: Hospital Besut Best Practice. *Transfusion Medicine Reviews, 26,* 153–163.

Comparing Methods for Lee–Carter Parameter's Estimation for Predicting Hospital Admission Rates

Siti Zulaikha Zulkarnain Yap, Siti Meriam Zahari, Zuraidah Derasit and S. Sarifah Radiah Shariff

Abstract The government hospital in Malaysia is prominent for the low cost of health care and medical treatment, and it had been reported that the hospital admission is increasing annually. This has led to the widespread problem of overcrowding. In order to assist the government to plan and manage demands for health services and health care needs, the prediction for future admission is outlined. The model is fitted to the matrix of admission rates, spanning the period of historical data from 2001 to 2011 for estimation. In particular, there are three types of estimation approaches used to generate the parameters of the Lee-Carter model which are singular value decomposition (SVD), iterative Newton–Raphson method (NR) and Poisson maximum likelihood estimation (PMLE). The resulting estimation generated by these three estimation approaches is then being compared and evaluated by two types of error measures which are mean squared error (MSE) and mean absolute percentage error (MAPE). The smallest values of MSE and MAPE indicate the better performance of the estimated parameters. The best estimation approach is subsequently used to forecast the admission rates of HRPZ II, Kota Bharu, for ten years ahead; using the out samples data from period 2012–2015. At the end of the study, it can be generally concluded that the occurrence of admission to HRPZ II, Kota Bharu, for both genders in each of the all broad age groups will slowly decrease along with the increment of years (2016–2025).

Keywords Lee-Carter model · Singular value decomposition · Iterative Newton–Raphson method · Poisson maximum likelihood estimation

S. Z. Z. Yap · S. M. Zahari (✉) · Z. Derasit · S. S. R. Shariff
Centre for Statistics and Decision Science, Faculty of Computer & Mathematical Sciences, Universiti Teknologi MARA, 40450 Shah Alam, Selangor, Malaysia
e-mail: mariam@tmsk.uitm.edu.my

1 Introduction

Hospital is a core component of a health care system as it plays an essential role in ensuring the high-quality service provided to the community and global development. In accordance with the progress of the country towards a developed nation, which is to achieve the Vision of 2020, the number of hospitals in Malaysia has shown a significant increase, either in the private or public sector (Utusan Online, May 22, 2015). This in line with the government's desire to press forward the medical field in the country and thus match with the other private hospitals in the country; meanwhile the equipment and facilities provided are equivalent to the international hospitals (Sadiq 2003).

Forecasting is an important support to many areas of health care field, particularly in hospital management, which typically comprise of managing bed and patient flow by minimizing patient waiting list, determining the health services demand of hospital unit and planning on surgery scheduling or nursing care slot as well as accessing the information on future daily admission. There are many studies attempting to predict the future trends of patient volumes using different forecasting approaches based on given time series data (Finks et al. 2011). In this study, the admission rates of HRPZ II, Kota Bharu, will be adapted and forecasted using the most influential approaches to the stochastic modelling which is Lee-Carter model (Lee and Carter 1992). The stochastic modelling is referred as a form of model that comprises one or more random variables that purposely to estimate how probable consequences are within a forecast to predict circumstances for different situations.

Some studies may employ time series approaches, such as in the study by Kim et al. (2014) that carried out a comparative study of different time series forecasting methods in order to predict the patient volumes in hospital medicine. The univariate methods under study include exponential smoothing, autoregressive integrated moving average (ARIMA), seasonal ARIMA and generalized autoregressive conditional heteroscedasticity (GARCH) methods. Time series approaches offer reasonable estimate of future admissions. The accuracy depends on the availability of historical time series data. However, the predictability of the model may not be reliable over long periods of time.

1.1 Forecasting Model: The Lee-Carter Model

In order to forecast the mortality, there are a lot of approaches have been established using stochastic model a few decades ago. This study adapts the Lee-Carter method (Lee and Carter 1992) to forecast the hospital admission rates which originally established to forecast mortality rates, and it has very few applications with other variables. The Lee-Carter model is widely used because of its simplicity, precision and robustness in the context of age-specific log mortality rates have linear trends

(Booth et al. 2006). This method comprises of age and time factors. The method depicts the logarithm of a time series of age-specific death rates by means of the sum of an age-specific parameter, that is, independent of time as well as the product of a time-varying index of general level of mortality, and an age-specific parameter that constitutes how quickly or gradually mortality at each age fluctuates when the index of general level of mortality changes. The most important feature in this model is time-varying index of level of mortality, k_t, where the index of level of mortality illustrates the trend of death from time to time. The extrapolation of estimated index obtained is then modelled and forecasted through the selection of appropriate time series model. The estimation of the general pattern of mortality by age, a_x, is simply the average value of $\ln(m_{x,t})$ over time where $\ln(m_{x,t})$ is the logarithm of death rates at age x in year t. Meanwhile, the estimation of relative speed of change at each age, b_x, and index of the level of mortality at time t, k_t are estimated using the least square solution.

2 Methodology

In this study, the Lee-Carter model is adapted to predict the hospital admission rates in Kota Bharu. The process of modelling the admission rates is similarly as mortality rates, that is, based on the logarithm of matrix of age-specific admission rates, m_{xt}, which hold the linear relationship with only two main factors related to age (x) and year (t). According to Jakstas (2014), the equation that describes the modelling of admission rates is expressed as:

$$\ln(m_{x,t}) = a_x + b_x k_t + \varepsilon_{x,t} \qquad (1)$$

The parameters are interpreted as follows:

$\ln(m_{x,t})$ is logarithm of admission rate at age x in year t. It can be calculated as $\hat{m}_{x,t} = \frac{y_{x,t}}{e_{x,t}}$
where $y_{x,t}$ is the number of hospital admissions at age x and year t, and $e_{x,t}$ is referred as exposure for number of population at age x and year t.

a_x represents average pattern of hospital admissions by age across years and is calculated as the average of logarithm of the admission rates, $\ln(m_{x,t})$, over time T, such that
$\hat{\alpha}_x = \sum_{t=t_1}^{t_1+T-1} \ln(m_{x,t})/T$
where t_1 is the starting year and T is the number of years in the data.

b_x measures relative rates of change of hospital admissions at each age.

k_t measures a time trend of hospital admission index at year t.

$\varepsilon_{x,t}$ the error term at age x and year t, and it should be white noise with 0 mean and relatively small variance (Lee 2000).

2.1 Estimation Approaches

In this study, to estimate the system of Lee-Carter model equation, there are three estimation approaches applied in this study in order to generate the estimations for these three sets of Lee-Carter parameters, namely a_x, b_x and k_t. There are known as singular value decomposition (SVD), iterative Newton–Raphson method (NR) and Poisson maximum likelihood estimation (PMLE).

Singular Value Decomposition Estimation
The steps of estimation using SVD are as follows:

Step 1: The parameter of a_x is computed as the average over time of the logarithm of the admission rates to minimize the sum of squares of residuals

$$\hat{a}_x = \frac{1}{n} \sum_{t=t_1}^{t_n} \ln(m_{x,t})$$

Step 2: Create a matrix Z as such $Z_{x,t} = \ln(m_{x,t}) - \hat{a}_x = b_x k_t$ to estimate b_x and k_t and apply the SVD to produce the matrices $PdQ' = SVD(Z_{x,t}) = d_1 P_{z1} Q_{t1} + \cdots + d_X P_{xX} Q_{tX}$ where d is the singular value, P is the age component and Q is the time component. The first term gives the estimates $\hat{b}_x = P_{x1}$ and $\hat{k}_t = d_1 Q_{t1}$.

Iterative Newton–Raphson Method Estimation
Instead of resorting for singular value decomposition, an iterative Newton–Raphson method also can be used to obtain the parameter estimates of a_x, b_x and k_t. The objective function is given as:

$$F(a,b,k) = \sum_{x=x_1}^{x_m} \sum_{t=t_1}^{t_n} \left(\ln(m_{x,t}) - \hat{a}_x - \hat{b}_x \hat{k}_t \right)^2$$

The idea of Newton–Raphson method is to update the parameters using a univariate Newton–Raphson scheme. Starting from initial guess value of $x^{(0)}$ as a root of function $f(x)$, the $(k+1)^{\text{th}}$ iterative provide $x^{(k+1)}$ from $x^{(k)}$ by

$$x^{(k+1)} = x^{(k)} - \frac{f(x^{(k)})}{f'(x^{(k)})}$$

The updating scheme of parameter estimates of Lee-Carter model based on a Newton–Raphson algorithm is as follows:

Step 1: The starting values for all parameters are being chosen, such as

$$\hat{a}_x^{(0)} = 0, \hat{b}_x^{(0)} = 1, \hat{k}_t^{(0)} = 0$$

Step 2: Update parameter $\hat{a}_x$

$$\hat{a}_x^{(k+1)} = \hat{a}_x^{(k)} + \frac{\sum_{t=t_1}^{t_n}\left[\ln(m_{x,t}) - \hat{a}_x^{(k)} - \hat{b}_x^{(k)}\hat{k}_t^{(k)}\right]}{t_n - t_1 + 1}$$

$$\hat{a}_x^{(k+1)} = \hat{a}_x^{(k+1)} + \hat{b}_x^{(k)} E_t\left(\hat{b}_x^{(k)}\right)$$

Step 3: Update parameter $\hat{k}_t$

$$\hat{k}_x^{(k+1)} = \hat{k}_x^{(k)} + \frac{\sum_{x=x_1}^{x_n} \hat{b}_x^{(k)}\left(\ln(m_{x,t})\right) - \hat{a}_x^{(k+1)} - \hat{b}_x^{(k)}\hat{k}_t^{(k)}\right]}{\sum_{t=t_1}^{t_n}\left(\hat{b}_x^{(k)}\right)^2}$$

$$\hat{k}_x^{(k+1)} = \left(\hat{k}_x^{(k+1)} - E_t\left(\hat{k}_x^{(k+1)}\right)\right)\sum_x \hat{b}_x^{(k)}$$

Step 4: Update parameter $\hat{b}_x$

$$\hat{b}_x^{(k+1)} = \hat{b}_x^{(k)} + \frac{\sum_{t=t_1}^{t_n} \hat{k}_t^{(k)}\left(\ln(m_{x,t}) - \hat{a}_x^{(k+1)} - \hat{b}_x^{(k)}\hat{k}_t^{(k+1)}\right)}{\sum_{t=t_1}^{t_n}(\hat{k}_t^{(k+1)})^2}$$

$$\hat{b}_x^{(k+1)} = \frac{\hat{b}_x^{(k+1)}}{\sum_x \hat{b}_x^{(k+1)}}$$

The sequences of 2–4 are repeated until convergence. The procedure will stop when there is a very small increase in objective function which is smaller than the default value of 10^{-6}.

Poisson Maximum Likelihood Estimation

In order to estimate the parameter in the Lee-Carter model, Wilmoth (1993) and Alho (2000) proposed using the estimation of maximum likelihood method. This approach assumes that the number of admissions at age x and year t follows Poisson distribution. Suppose $D_{x,t}$ is the number of admissions in year t of patients from age x, and $N_{x,t}$ is the total population at age x and year t. Hence, the observed admission rate presented is:

$$D_{x,t} \sim \text{Poisson}\ (N_{x,t}, m_{x,t}) \text{ where } \ln(m_{x,t}) = a_x + b_x k_t$$

By maximizing the full log-likelihood function, the coefficients a_x, b_x and k_x are estimated (Wilmoth 1993) and can be represented as follows:

$$F(a,b,k) = \sum_{x=x_1}^{x_m} \sum_{t=t_1}^{t_n} \left[(D_{x,t} \cdot (a_x + b_x k_t) - N_{x,t} \cdot \exp(a_x + b_x k_t) \right]$$

To estimate the log-linear models with bilinear terms, the iterative method is proposed by Goodman (1979). In the iteration step $v+1$, a single set of parameters is updated to fix the other parameters at their current estimates. The updating scheme is as follows:

Step 1: The starting values for all parameters are being chosen, such as

$$\hat{a}_x^{(0)} = 0, \hat{b}_x^{(0)} = 1, \hat{k}_t^{(0)} = 0$$

Step 2: Update parameter $\hat{a}_x$

$$\hat{a}_x^{(k+1)} = \hat{a}_x^{(k)} + \frac{\sum_{t=t_1}^{t_n} \left[D_{x,t} - N_{x,t} \cdot \exp\left(\hat{a}_x^{(k)} + \hat{b}_x^{(k)} \hat{k}_x^{(k)} \right) \right]}{\sum_{t=t_1}^{t_n} \left[N_{x,t} \cdot \exp\left(\hat{a}_x^{(k)} + \hat{b}_x^{(k)} \hat{k}_x^{(k)} \right) \right]}$$

$$\hat{a}_x^{(k+1)} = \hat{a}_x^{(k+1)} + \hat{b}_x^{(k)} E_t(\hat{k}_x^{(k)})$$

Step 3: Update parameter $\hat{k}_t$

$$\hat{k}_x^{(k+1)} = \hat{k}_x^{(k)} + \frac{\sum_{x=x_1}^{x_n} \left[D_{x,t} - N_{x,t} \cdot \exp\left(\hat{a}_x^{(k+1)} + \hat{b}_x^{(k)} \hat{k}_x^{(k)} \right) \right] \hat{b}_x^{(k)}}{\sum_{t=t_1}^{t_n} \left[N_{x,t} \cdot \exp\left(\hat{a}_x^{(k)} + \hat{b}_x^{(k)} \hat{k}_x^{(k)} \right) \right] \left(\hat{b}_x^{(k)} \right)^2}$$

$$\hat{k}_x^{(k+1)} = \left(\hat{k}_x^{(k+1)} - E_t\left(\hat{k}_x^{(k+1)} \right) \right) \sum_x \hat{b}_x^{(k)}$$

Step 4: Update parameter $\hat{b}_x$

$$\hat{b}_x^{(k+1)} = \hat{b}_x^{(k)} + \frac{\sum_{t=t_1}^{t_n} \left[D_{x,t} - N_{x,t} \cdot \exp\left(\hat{a}_x^{(k+1)} + \hat{b}_x^{(k)} \hat{k}_x^{(k+1)} \right) \right] \hat{k}_t^{(k+1)}}{\sum_{t=t_1}^{t_n} \left[N_{x,t} \cdot \exp\left(\hat{a}_x^{(k+1)} + \hat{b}_x^{(k)} \hat{k}_x^{(k+1)} \right) \right] \left(\hat{k}_t^{(k+1)} \right)^2}$$

$$\hat{b}_x^{(k+1)} = \frac{\hat{b}_x^{(k+1)}}{\sum_x \hat{b}_x^{(k+1)}}$$

The sequences of 2–4 are repeated until convergence. The procedure will stop when there is a very small increase in log-likelihood function which is smaller than the default value 10^{-6}.

2.2 Model Validation

In general, model validation is describing the process of evaluating the performance of models to find the best one to forecast the time series data. There are basically three stages involved in validating the model.

The first stage is called initial data preparation. During the first stage, the data series will be divided into two parts. The first part known as within samples or fitting parts, that is, used to estimate the model forecasting performance. Meanwhile, the second part is to evaluate the model, called as out samples, evaluation or holdout part. In this study, the fitting part encompasses 99 of observations between January 2001 and December 2011, whereas the remaining 36 observations of January 2012 up to December 2015 denoted as evaluation part of model forecasting performance.

In the second stage, the within-sample statistics is used to estimate the model. The three parameters of Lee-Carter model in this study are estimated using different estimation approaches, executing by statistical package of R Development Software. The best estimation approach is selected based on the outcomes of comparing their error measures performances. For this purpose, MSE and MAPE are used with the hope that the final outcome of the competition is as conclusive as possible. For example, a model with the estimation approach that has at least two out of three error measures calculated with the smallest values it is then judged as the winner.

3 Results

The descriptive statistics are part of the statistical study tools, techniques or procedures used to depict or describe a collection of data or observations that have been made by the creation of graphs, charts or diagrams.

In this study, the main source of the time series data used for the study is the annual hospital admission data. It is obtained from medical records unit of Hospital Raja Perempuan Zainab II (HRPZ II), Kota Bharu. Apart from that, this study also used the data regarding population distribution in Kelantan, and it is obtained from the website of Department of Statistics Malaysia (DOSM), Kota Bharu. The requested data sets are both comprised of main variables in yearly basis with specifies on gender (male and female) and age groups, covering the period of 2001 until 2015.

3.1 Data Description

The information provided by HRPZ II, Kota Bharu, and DOSM are standardized and includes the following types of variables in the data (Table 1).

Table 1 Description of data

Variables	Description	Measure	Measurement unit
Year	Selected years are from 2001 until 2015	Numeric	Year
Age group	The age group of male and female are 0–4, 5–19, 20–24, 25–34, 35–44, 45–54, 55–64, 65–74, 75–84, 75+	Categorical	Year
Total admission	The number of male and female patients being admitted to HRPZ II that specifies on age groups	Numeric	Number of male and female distribution
Population size	The number of male and female population in Kelantan that specifies on age groups	Numeric	Number of male and female distribution

Table 2 Age group based on DOSM and HRPZ II data

Variable	Sources		The chosen range
	DOSM	HRPZ II	
Age group	0–4, 5–9, 10–14, 15–19, 20–24, 25–29, 30–34, 35–39, 40–44, 45–49, 50–54, 55–59, 60–64, 65–69, 70–74, 75–79, 80–84, 85+	0–4, 5–19, 20–24, 25–34, 35–44, 45–54, 55–64, 65–74, 75–84, 75+	0–4, 5–19, 20–24, 25–34, 35–44, 45–54, 55–64, 65–74, 75–84, 75+

There exists a problem of inconsistent data sets since the age groups of data set provided by DOSM are not matched with the one given by HRPZ II, Kota Bharu. Thus, in order to solve this problem, the age groups of both data are synchronized to be similar in range as shown in the Table 2, for instance, the values under age group of 55–59 and 60–64 from DOSM data will be added up in order to produce the new range of age group, 55–64. The purpose is to make the data consistent in such that it matched with the age group of 55–64 from HRPZ II, Kota Bharu, data.

3.2 Fitting the Lee-Carter Model

After preliminary computations of applying the Lee-Carter model to the admission data of HRPZ II, Kota Bharu, the estimations for three sets of Lee–carter parameters, namely as a_x, b_x and k_t are generated with different estimation approaches singular value decomposition (SVD), iterative Newton–Raphson method (NR) and Poisson maximum likelihood estimation (PMLE). The Lee-Carter model is applied separately to male and female data, by considering the time horizon from 2001 to 2011 as estimation part and maximum age is 80. Note that regarding the chosen value of maximum age in the study; it had been supported by the National Statistics Reports published on 16 February 2016. The report stated that, in general, the average time of an individual is expected to live considering on the birth year, current age and other demographic factors are around 78.8 years. Thus, this supported the relevance of using the value of 80 to be the optimum age for this study.

The actual values of estimated parameters are summarized in Tables 3, 4 and 5. The properties of model parameters estimation provided by these three different estimation approaches are evaluated and compared by two different error measures in order to measure the performance for forecasting purposes.

Table 3 Values of SVD, NR and PMLE of a_x and b_x in the Lee-Carter model for male

	SVD		NR		PMLE	
Age group	Male		Male		Male	
	a_x	b_x	a_x	b_x	a_x	b_x
0–4	−3.2311	−0.0135	−3.2311	−0.0135	−3.1508	0.0821
5–19	−4.2911	0.0727	−4.2911	0.0727	−4.251	0.0706
20–24	−3.6646	0.7282	−3.6646	0.728	−3.764	0.4027
25–34	−3.6119	0.1117	−3.6119	0.1117	−3.5742	0.0604
35–44	−3.6177	0.0634	−3.6177	0.0634	−3.5709	0.0757
45–54	−3.3114	0.0289	−3.3114	0.0289	−3.2601	0.075
55–64	−2.7714	0.0408	−2.7714	0.0408	−2.7187	0.0803
65–74	−2.2064	−0.0539	−2.2064	−0.0539	−2.1109	0.079
75+	−2.2643	0.0219	−2.2643	0.022	−2.1822	0.0744

Table 4 Values of SVD, NR and PMLE of a_x and b_x in the Lee-Carter model for female

	SVD		NR		PMLE	
Age group	Female		Female		Female	
	a_x	b_x	a_x	b_x	a_x	b_x
0–4	−3.4727	−1.6205	−3.4727	−1.6155	−3.4697	−0.8871
5–19	−4.4572	0.1512	−4.4572	0.1507	−4.456	0.0814
20–24	−2.6886	1.2068	−2.6886	1.203	−2.686	0.6434
25–34	−2.1132	1.4909	−2.1132	1.4862	−2.1129	0.7904
35–44	−2.746	1.8829	−2.746	1.8773	−2.7444	1.0238
45–54	−3.5182	0.2431	−3.5182	0.2427	−3.5145	0.1934
55–64	−3.15	0.1924	−3.15	0.1923	−3.1455	0.1773
65–74	−2.8013	−1.3458	−2.8013	−1.3407	−2.7827	−0.5267
75+	−2.9449	−1.2009	−2.9449	−1.196	−2.9198	−0.4958

Table 5 Values of SVD, NR and PMLE of in the Lee-Carter model

Year	SVD Male	SVD Female	NR Male	NR Female	PMLE Male	PMLE Female
	k_t					
2001	2.7031	0.1342	3.1676	0.1252	5.2842	0.2408
2002	−6.2773	0.0975	−0.1420	0.0890	−1.5151	0.1827
2003	−6.0250	0.0767	−0.1726	0.0738	−1.3053	0.1434
2004	−3.5191	0.0364	−0.2961	0.0215	−1.1743	0.0642
2005	−6.0673	−0.0014	−0.4726	0.0036	−1.4405	−0.0021
2006	1.6952	−0.0035	−0.3590	−0.0910	0.0144	−0.0976
2007	−0.5187	−0.0435	−0.4114	−0.0502	−0.8287	−0.0920
2008	−1.1541	−0.0413	−0.4074	−0.0467	−0.9031	−0.0933
2009	4.3929	−0.0696	0.7132	−0.0537	3.5540	−0.1228
2010	−4.0142	−0.1078	−0.6235	−0.0379	−1.9517	−0.1403
2011	2.2835	−0.0606	−0.9961	−0.0334	0.2661	−0.0829

3.3 The Accuracy of Estimation Approaches

Several error measures had been considered in order to examine the accuracy of estimation approaches for both genders including mean square error (MSE) and mean absolute percentage error (MAPE). The better estimation approach should be come up with smaller values of MSE and MAPE.

From the evidence available (Table 6), the MSE shows that the estimation method of NR is the clear winner since it produces the smallest value of MSE compared to the other two estimation approaches which are 0.0003475 and 0.0000174, respectively, for both genders. On MAPE, the smallest value of MAPE is recorded from the estimation approaches of NR with 12.9905690 and PMLE with 5.0651302 for male and female, respectively.

Above all the reasons mentioned, it can be concluded that the estimation approach of iterative Newton–Raphson method turned out to be the best estimation approaches for males group as it provides the smallest values of all error measures MSE and MAPE. Thus, it can be concluded that Newton–Raphson method will be used in forecasting the admission rates for males to HRPZ II, Kota Bharu.

In the meantime, as for females group, the smaller values of MAPE are according to the estimation approach of Poisson maximum likelihood despite of lowest MSE is notified from iterative Newton–Raphson method. Note that the value

Table 6 Different error measures by different estimation approaches and gender

	SVD MSE	SVD MAPE	NR MSE	NR MAPE	PMLE MSE	PMLE MAPE
Male	0.004375	53.86621	0.000348	12.99057	0.000465	20.35971
Female	2.52E−05	5.450342	1.74E−05	5.338162	1.86E−05	5.06513

of MSE of iterative Newton–Raphson method and Poisson maximum likelihood estimation differs by only the small changes of value (0.0000012). Therefore, since the majority of smallest error measures are devoted from Poisson maximum likelihood estimation, the method is then will be used to forecast the admission rates for females to HRPZ II, Kota Bharu.

4 Conclusion

The issue on overcrowding has become widespread problem in the hospitals across the developed countries due to the increasing number of patient admissions annually. Thus, there is a significant need to forecast the hospital admission rates. As for this study, the Lee-Carter method is used to predict accurately the future admission rates based on historical time series data of admission obtained from HRPZ II, Kota Bharu, and Kelantan population distribution data from DOSM, Kota Bharu.

The singular value decomposition (SVD) along with the other two alternatives estimation approaches which are iterative Newton–Raphson (NR) and Poisson maximum likelihood estimation (PMLE) are used to generate the parameters of the Lee-Carter model. Based on the results of MSE and MAPE, they generally show that the iterative NR method is the best estimation approaches for males, and the PMLE is the estimation approaches for females. Thus, it can be concluded that based on the admission data of HRPZ II, Kota Bharu, the iterative NR method is appropriate for the estimation of the parameters of the Lee-Carter model and it will be used in forecasting the hospital admission rates for males. Analogously, the PMLE is appropriate for the estimation of the parameters of the Lee-Carter model, and it will be used in forecasting the hospital admission rates for females.

There is still need a research to carry out in some areas due to the limitation of the research or to improve the quality of the research. Despite treating the Lee-Carter model as a benchmark and the fact that its application is mostly well-known in dealing with mortality data, the Lee-Carter method adapted in this hospital admission rates study might be compared with other extension of Lee-Carter or other stochastic mortality forecasting models in the future.

Acknowledgements This study was made possible by the continuous support from Universiti Teknologi MARA Grant No. 600-IRMI/MyRA 5/3/LESTARI (0135/2016).

References

Alho, J. M. (2000). "The Lee-Carter Method for Forecasting Mortality, with Various Extensions and Applications". Ronald Lee, January 2000. *North American Actuarial Journal, 4*(1), 91–93.
Booth, H., Hyndman, R., Tickle, L., & De Jong, P. (2006). Lee-Carter mortality forecasting: A multi-country comparison of variants and extensions.

Caj hospital kerajaan hanya lima hingga 10 peratus termasuk rawatan untuk pesakit kronik. (2016, March 20). Retrieved from Utusan Online: http://www.utusan.com.my/berita/nasional/caj-hospital-kerajaan-hanya-lima-hingga-10-peratus-1.202499.

Finks, J. F., Osborne, N. H., & Birkmeyer, J. D. (2011). Trends in hospital volume and operative mortality for high-risk surgery. *New England Journal of Medicine, 364*(22), 2128–2137.

Goodman, L. A. (1979). On the estimation of parameters in latent structure analysis. *Psychometrika, 44*(1), 123–128.

Jakstas, G. (2014). Quantifying Longevity Risk.

Kim, K., Lee, C., O'Leary, K., Rosenauer, S., & Mehrotra, S. (2014). *Predicting patient volumes in hospital medicine: A comparative study of different time series forecasting methods*. Tech. rep.: Northwestern University.

Lee, R. (2000). The Lee-Carter method for forecasting mortality, with various extensions and applications. *North American Actuarial Journal, 4*(1), 80–91.

Lee, R. D., & Carter, L. (1992). Modeling and Forecasting the Time Series of US Mortalityl. *Journal of the American Statistical Association, 87*(659), 71.

Sadiq Sohail, M. (2003). Service quality in hospitals: more favourable than you might think. *Managing Service Quality: An International Journal, 13*(3), 197–206.

Wilmoth, J. R. (1993). *Computational methods for fitting and extrapolating the Lee-Carter model of mortality change*. Technical report, Department of Demography, University of California, Berkeley.

Data Pre-Processing Using SMOTE Technique for Gender Classification with Imbalance Hu's Moments Features

Ahmad Haadzal Kamarulzalis, Muhamad Hasbullah Mohd Razali and Balkiah Moktar

Abstract Imbalance data is common in real-world applications like text categorization, face recognition for gender classification, medical diagnosis, fraud detection, oil-spills detection of satellite images. Most of the algorithms in machine learning are focusing on classification of majority class while ignoring or misclassifying minority sample. The minority samples are those that rarely occur but very important. It is commonly agreed that standard classifiers such as neural networks, support vector machines, and C4.5 are heavily biased in recognizing mostly the majority class since they are built to achieve overall accuracy to which the minority class contributes very little. In this study, we demonstrate how the synthetic minority over-sampling technique (SMOTE) can significantly improve the imbalance problem in gender classification from the data-level perspective. Hu's moment of the face images was generated as the numerical descriptors with different imbalance ratio and classified using a supervised decision tree (J48) algorithm. The results show that prior to preprocessing the data with SMOTE, the minority group was severely misclassified as the majority group. Our claims are confirmed through the application of SMOTE in reducing the imbalance effects before inducing the decision tree.

Keywords Imbalanced data · SMOTE · Hu's moments · J48 decision tree

1 Introduction

Most classifiers work well when the class distribution in the response variable of the dataset is well balanced. Problems arise when the dataset is imbalanced. When there is a class that has a small number of dataset, while another class has the majority, the imbalance problem occurs. The accuracy of the classifier will show outstanding

A. H. Kamarulzalis (✉) · M. H. Mohd Razali · B. Moktar
Faculty of Computer and Mathematical Sciences, Universiti
Teknologi MARA Cawangan Perlis, Arau, Malaysia
e-mail: haadzal9301@gmail.com

performance, but it is meaningless since the majority class has influenced the minority class.

There are three main objectives in this study. First is to generate the Hu's moments of face images. Second is to predict the gender with decision tree (J48) classification technique on the generated numerical features. Last is to compare the performance of the classifier with different imbalance ratio with and without SMOTE preprocessing technique.

2 Literature Review

2.1 Hu's Moments Invariants for Classification

Hu (1962) describes invariants moments that designed in order to achieve maximum utility and flexibility. Hu's tries to come out with two categories to achieve maximum utility. According to Palaniappan et al. (2000), moment can handle a type of symmetric and noise problem regarding the image.

2.2 Image Classification

Khan et al. (2013) conducted a study to verify that gender classification can be done by using a decision tree (J48) approach. Phillips (1998) conducted a study to view how to do facial recognition using support vector machine (SVM) combined with principal component analysis (PCA) and discovered SVM is useful in classification regarding two class problems. Maturana et al. (2010) studied the issues of how face recognition for the binary pattern can be done. Decision trees and structure local binary patterns (LBPs) has been used, and it results low in discriminative features. Moallem and Mousavi (2013) used fuzzy inference system (FIS) to classify gender. Zernik memories, local binary pattern (LBP) to calculate pixels in the neighborhood, and active appearance model (AAM) are being used as extraction mechanism to run this study. Jia and Martinez (2009) used support vector machine (SVM) and partial support vector machine (PSVM).

Study on face recognition across facial by using binary decision trees by Riaz et al. (2008) processes the image using AAM, active shape model (ASM), and principal components analysis (PCA) to fulfill the requirement in face recognition procedures. Based on the findings, a binary decision tree is efficient to classify the image. Baumann et al. (2013) used random forest that introduced in 2001 by Leo Breiman in order to achieve a robust classification and to improve detection sensitivity of the data. The finding shows the random forest is useful for multi-class detection.

2.3 Synthetic Minority Over-Sampling Technique (SMOTE) for Imbalance Data

Chawla et al. (2002) introduced synthetic minority over-sampling technique (SMOTE) by creating synthetic samples from the minority class instead of creating copies. The algorithm selects two or more instances using a distance measure and instance one attribute at a time by a random amount within the difference to the neighboring instances. Furthermore, SMOTE can improve the accuracy of a minority class besides act as a new approach for over-sampling by increasing the region of minority values.

3 Methodology

3.1 Data Collection

The image was captured using a digital camera from 5 male and 5 female student Bachelor of Science (Hons.) Management Mathematics (CS248). Each student should provide five different expressions, and finally, there are 50 images which were captured in total.

3.2 Generating the Hu's Moments

An image moment is a gross characteristic of the contour computed by generating (integrate or sum) over all of the pixels of the contour. The moment ("order moment") of order $(p+q)$ for 2D continuous function $f(x,y)$ is characterized as

$$M_{pq} = \int_{-\infty}^{\infty} \int_{-\infty}^{\infty} x^p y^q f(x,y) \mathrm{d}x \mathrm{d}y. \tag{1}$$

Equation 1 represents the function that has been used before adjusting to gray scale without changing the quality of the images. The function $f(x,y)$ is also uniquely determined by the moment of sequence $\{M_{pq}\}$. The invariant features can be achieved using central moment which are defined as follows

$$\mu_{pq} = \int_{-\infty}^{\infty} \int_{-\infty}^{\infty} (x - \bar{x}^p)(y - \bar{y})^q f(x,y) \mathrm{d}x \mathrm{d}y \tag{2}$$

For $p, q = 0, 1, 2, \ldots$ where

$$\bar{x} = \frac{M_{10}}{M_{00}} \quad \bar{y} = \frac{M_{01}}{M_{00}}.$$

The pixel points $(\bar{x}, \bar{y})$ are the centroid of the images $f(x, y)$. The centroid moment μ_{pq} is equivalent to M_{pq} whose center has been shifted to the centroid of the image. Scale moment can be generated by normalization. The normalization central moment can be define as follows

$$\eta_{pq} = \frac{\mu_{pq}}{\mu_{00}^{\gamma}}, \gamma = (p+q=2)/2, \text{ where } p+q = 2, 3, \ldots \tag{3}$$

Based on normalized central moments, Hu introduced seven moment invariants that we call it as Hu's moment, which are the linear combinations of the central moments

$$\phi_1 = \eta_{20} + \eta_{02} \tag{4}$$

$$\phi_2 = (\eta_{20} - \eta_{02})^2 + 4\eta_{11}^2 \tag{5}$$

$$\phi_3 = (\eta_{30} - 3\eta_{12})^2 + (3\eta_{21} - \eta_{03})^2 \tag{6}$$

$$\phi_4 = (\eta_{30} + \eta_{12})^2 + (\eta_{21} + \eta_{03})^2 \tag{7}$$

$$\begin{aligned}\phi_5 = (\eta_{30} - 3\eta_{12})(\eta_{30} + \eta_{12})\left[(\eta_{30} + \eta_{12})^2 - 3(\eta_{21} + \eta_{03})^2\right] \\ (3\eta_{21} - \eta_{03})(\eta_{21} + \eta_{03})\left[3(\eta_{30} + \eta_{12})^2 - (\eta_{21} + \eta_{03})^2\right]\end{aligned} \tag{8}$$

$$\phi_6 = (\eta_{20} - \eta_{02})\left[(\eta_{30} + \eta_{12})^2 - (\eta_{21} + \eta_{03})^2 + 4\eta_{11}(\eta_{30} + \eta_{12})(\eta_{21} + \eta_{03})\right] \tag{9}$$

$$\begin{aligned}\phi_7 = (3\eta_{21} - \eta_{03})(\eta_{30} + \eta_{12})\left[(\eta_{30} + \eta_{12})^2 - 3(\eta_{21} + \eta_{03})^2\right] \\ + (\eta_{30} - 3\eta_{12}), (\eta_{21} + \eta_{03})\left[3(\eta_{30} + \eta_{12})^2 - (\eta_{21} + \eta_{03})^2\right]\end{aligned} \tag{10}$$

Hu's moment equation can find the value of image corresponding to the different expression, and this logically will classify an expression according to the class (gender). The idea in Hu's invariant moments is that by combining the different normalized central moments, it is possible to create invariant functions representing different aspects of the image in a way that is invariant to scale, rotation, and reflection. The moments were generated with the Python-OpenCV library.

3.3 Smote

Firstly, for each minority class sample, gets its *k*-nearest neighbors from other minority class sample among the *k*-neighbors. Finally, generates the synthetic sample x_{new} by interpolating between x and $\tilde{x}$ as follows

1.
$$x_{\text{new}} = x + r \text{ and } (0, 1)(\tilde{x} - x) \tag{11}$$

3.3.1 Image Classification for Balanced and Imbalance Data Using Decision Tree (J48) Algorithm

Numerical features extracted using Hu's moments then were applied to compare their classification performance using WEKA software. Two approaches were tested which were called as balanced versus imbalanced dataset. The performance was compared using cross-validation to see the prediction accuracy. The framework of the study summarized in Fig. 1.

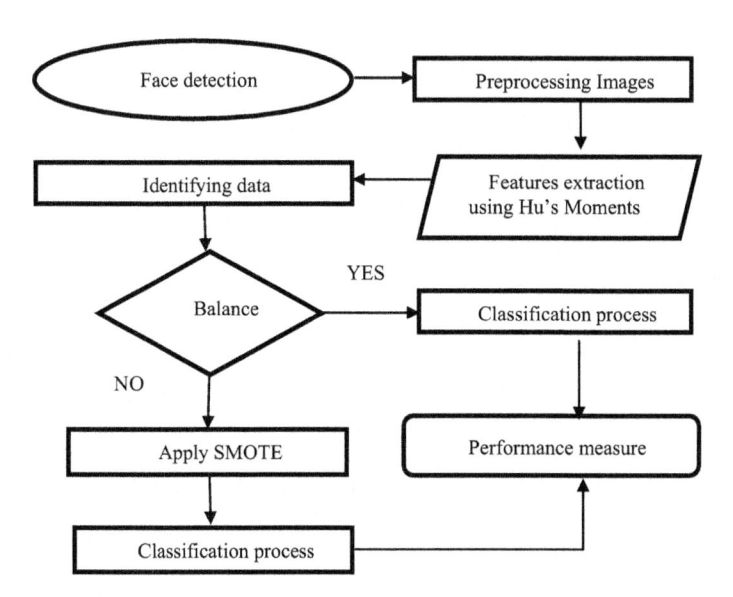

Fig. 1 Theoritical framework of the study

Table 1 Classification result for balance data (25:25)

	Balanced	
Gender	Male	Female
Male	25	0
Female	0	25
Sensitivity	100%	
Specificity	100%	
Accuracy/validity test	100%	
Error rate	0%	

Table 2 Classification result for imbalanced

	Without Smote (2:25)		Smote (4:25 and 8:25)			
Gender	Male	Female	Male	Female	Male	Female
Male	0	2	4	0	8	0
Female	0	25	0	25	0	25
Sensitivity (%)	0		100		100	
Specificity (%)	92.5926		100		100	
Accuracy/validity test (%)	92.5926		100		100	
Error rate (%)	7.4074		0		0	

4 Results

Table 1 shows the classification of balanced dataset which was correctly classified according to the gender.

Table 2 shows the output of imbalance image expression classification. From the output, only 92.5926% were correctly classified, and two males (rare but important class) have been misclassified as female when we put the imbalanced situation. The result after incorporating SMOTE with ratio 4:25 and 8:25 shows that the decision tree (J48) classifier can correctly classify the gender perfectly. In such a small-scale study, apparently, the first replication of the minority class (4:25) has sufficiently shifted the decision region boundary for the minority class.

5 Conclusion

In this study, we have presented the data preprocessing technique using SMOTE prior to inducing the decision tree J48. Through an empirical investigation, we have demonstrated the potential of using the numerical image moment features for gender identification purposes. This paper is believed to contribute in terms of providing an insight into how statistical pattern recognition, computer vision, and machine learning can improve the existing procedure of image identification. For future work, the extension would be to generate different face image descriptors considering the challenge to capture the distinctive patterns among genders. Furthermore, more data could be collected in order to evaluate the robustness of the classifiers over large scale of data with different imbalance ratios.

References

Baumann, F., Ehlers, A., Vogt, K., & Rosenhahn, B. (2013). Cascaded random forest for fast object detection. In *Image Analysis,* (pp. 131–142). Berlin Heidelberg: Springer.

Chawla, N. V., Bowyer, K. W., Hall, L. O., & Kegelmeyer, W. P. (2002). SMOTE: Synthetic minority over-sampling technique. *Journal of Artificial Intelligence Research, 321–357.*

Hu, M. K. (1962). Visual pattern recognition by moment invariants. *Information Theory, IRE Transactions, 8*(2), 179–187.

Jia, H., & Martinez, A. M. (2009, June). Support vector machines in face recognition with occlusions. In I*EEE conference on computer vision and pattern recognition, 2009.* CVPR 2009. (pp. 136–141).

Khan, M. N. A., Qureshi, S. A., & Riaz, N. (2013). Gender classification with decision trees. *International Journal of Signal Processing, Image Processing and Pattern Recognition, 6,* 165–176.

Maturana, D., Mery, D., & Soto, A. (2010). Face recognition with decision tree-based local binary patterns. In *Computer Vision–ACCV 2010,* (pp. 618–629). Berlin Heidelberg: Springer.

Moallem, P., & Mousavi, B. S. (2013). Gender classification by fuzzy inference system. *International Journal of Advanced Robotic Systems,* 10.

Morgan, R. E., & Mason, B. J. (2014). Crimes against the elderly, 2003–2013. Special Report (NCJ 248339). Washington, DC: United States Department of Justice, Office of Justice Programs, Bureau of Justice Statistics.

Palaniappan, R., Raveendran, P., &Omatu, S. (2000). Improved moment invariants for invariant image representation. In *Invariants for pattern recognition and classification'* (pp. 167–187). Singapore: World Scientific Publishing Co.

Phillips, P. J. (1998). Support vector machines applied to face recognition(Vol. 285). US Department of Commerce, Technology Administration, National Institute of Standards and Technology.

Riaz, Z., Mayer, C., Wimmer, M., & Radig, B. (2008). Model based face recognition across facial expressions. *Journal of Information and Communication Technology, 2.*

Optimizing Efficiency of Electric Train Service (ETS) Ticket Pricing

Noraini Noordin and Nur Syamila Mohd Ali Amran

Abstract Ticket prices for the Electric Train Service (ETS) provided by Keretapi Tanah Melayu (KTMB) have been observed to increase without due consideration to passengers. Some KTMB services are double the costs of other public transport. These fares have also not been priced according to distance travelled. In an attempt to determine the minimum price for the ETS ticket according to type of ticket, this study has investigated the relationship between price and type of passenger, distance travelled and type of seat. Estimation of distances between stations in the data collected from KTMB was done using Google Maps. Other procedures included (i) adapting categorization procedures used by Mijares et al. (2013) for data development, (ii) adapting Monte Carlo Simulation (MCS) optimization procedures by Cerrato and Cheung (2006) and (iii) using variance reduction technique (VRT) to calculate pricing efficiency. Random numbers were generated for the MCS based on total ticket price and total distance. Findings indicate that MCS was able to optimize all silver ticket prices, but was only able to minimize gold and platinum ticket price for stations with distances of less than 40 km. More importantly, efficiency scales less than 1 were obtained for all types of tickets (adult and children); thus, pricing efficiency was achieved for this research. Future works in this area may use other optimization procedures to minimize tickets for distances greater than 40 km between stations.

Keywords Monte Carlo simulation · Pricing efficiency · Random number Variance reduction technique

---

N. Noordin (✉) · N. S. Mohd Ali Amran
Faculty of Computer and Mathematical Sciences,
Universiti Teknologi MARA, Shah Alam, Malaysia
e-mail: noraininoordin@perlis.uitm.edu.my

1 Introduction

Rising operation cost for all public transportations has affected transportation fares. In particular, Electric Train Service (ETS) fares provided by Keretapi Tanah Melayu (KTMB) have been observed to increase without due consideration to passengers. Most frequent complaint from KTMB passengers is the expensive price of ETS and KTMB ticket (Sunif 2011). The high ETS fares have discouraged the public from using the service regularly (Khalid et al. 2014).

There is not much difference in the quality of the seats for ETS platinum and gold classes, but the platinum class ticket has been priced higher than the gold class ticket (KTMB 2016). Higher ticket price during peak seasons may not be suitable for the less affluent regular passengers (Department for Transport 2013). However, peak season ETS ticket price is almost twice that of off-peak tickets. Currently, ETS ticket price is double the cost of tickets for other public transports.

Qiao et al. (2013) identified several factors affecting ticket price: transfer times, number of passenger, disabilities, passenger category—student, military or old people, and demand for seats (gold, silver or platinum seat). Ticket price will normally differ between types of transportation. All fares for the same distance travelled by KTMB are significantly different from those offered by other public transportations; thus, ETS fares have not been priced per distance travelled. In this study, the relationship between price and passenger type, distance travelled and seat type is being investigated and modelled using MCS in an attempt to determine the minimum price for the ETS ticket according to ticket type.

2 Literature Review

ETS ticket pricing by KTMB has significantly affected passengers. This section will discuss this issue according to the following perspectives: development of data set for ETS fare, optimization of price using MCS and pricing efficiency.

2.1 Data Set Development for ETS Fare

Different categorizing procedures were carried out during sampling in research works ticket pricing. Lantseva et al. (2015) compared and analysed flight fares for two large Russian airlines based on different travel classes in airline services and the seasons (peak or off-peak) when the tickets were bought. Winston and Maheshri (2007) calculated the short run total cost by summing up operating cost and depreciation of rail transit capital using data on number of stations served, distance between two links, shortest route between two stations and total land area served. In a study concerning rail fare policy and passenger overload, data was categorized according to fare level, fare structures, ticket types, distance, zone and time of day for analysis (Mijares et al. 2013; Mijares et al. 2016a, b). Variables in data development by Li et al. (2012) included rail line length, number and the location of stations, the fares and the headway of trains. Data set for the current study involved actual data like rail line length, number and location of station, fares and headway of trains. In particular, categorization of data in the data development process will adapt the method used by Mijares et al. (2013).

2.2 Optimization of Price Using Monte Carlo Simulation (MCS)

Simulation is a reality learning technique by experimenting with a model that represents the system in order to estimate the effects of various actions and decisions. MCS also known as static simulation is mostly used to model the probability of different outcomes in process that is difficult to predict due to intervention of random number. By using finite sample performance of some recent Monte Carlo estimator under different market scenarios, Cerrato and Cheung (2006) found that least square method MCS can find the accuracy and efficiency of an estimator and was remarkable for Asian Bermudan option price. In trying to understand and implement the American option pricing using least squares MCS to improve computational cost and accuracy of it, Coskan (2008) concluded that it did not improve American spread option price but only increased the computational cost. Moreno and Navas (2003) used least squares MCS in simple regression to analyse the impact of different basic function on American option prices. Wang and Wang (2011) applied MCS to pricing the barrier option in three steps, namely (i) simulate and sample pats of underlying asset price over time interval, (ii) calculate payoff of option and (iii) average the discounted payoff of the option of the path. The current study has adapted the MCS as applied by Cerrato and Cheung (2006) because the study has categorized the price according to different ticket types.

2.3 Efficiency of Pricing

Speed and efficiency of a simulation can be improved using techniques like variance reduction technique (VRT) (Goddard Consulting 2017). It can increase the precision of estimation obtained for a given number of iterations. All random variable outputs from simulation are associated with limit precisions of simulation results. In particular, VRT is used to get a statistically efficient simulation and to obtain smaller confidence intervals for output of random variable of interest. In the current study, efficiency of pricing was determined using VRT as used by Wang and Wang (2011). Their research concerned the reduction of error when simulation was continuously increasing. They also compared price value with different initial price for barrier level among different type of ticket classes.

3 Methodology

Excel spreadsheet has been used to organize data collected from KTMB according to distance from SP to PB. Actual distances between stations were calculated based on distances in Google Maps. Data was also categorized according to type of ticket. A random number was calculated using this data. The random number was used to generate the new ticket price according to type of ticket.

3.1 Optimizing ETS Fare According to Ticket Type Using MCS

Cerrato and Cheung (2006) stated that MCS method is used when unattainable analytical solution occurs and failure domain cannot be expressed or approximated by analytical form. Mathematical formulation of MCS is relatively simple. MCS is capable of handling practically all possible cases regardless of its complexity. The objective function of MCS can be stated as below:

$$N_H = \sum N_0 \qquad (1)$$

$$D = \sum D_0 \qquad (2)$$

$$I(x_j) = \frac{N_H}{D} \qquad (3)$$

$$\min P_t \cong D * I(x_j) \qquad (4)$$

where

D_0 Distance between stations
$I(x_j)$ Random number
D Total distance between station
N_H Total older price
P_t New price
N_0 Older price.

3.2 Generating a Random Number

In the research by Cerrato and Cheung (2006), random number was calculated using square-integrable random variable. In this study, random numbers were generated using Excel by getting the average between total distance and total ticket price for adult and children, as follows:

$$\text{Random number} = \frac{\text{total ticket price}}{\text{total distance}} \tag{5}$$

Equation 5 is used to get the random number for adult ticket and child ticket separately. This random number is then used to find a new adult ticket price and a new child ticket price according to type of ticket.

3.3 Determining Price Efficiency Using VRT

Price efficiency in this study was determined using the VRT. The purpose of using VRT was to improve efficiency and accuracy of MCS method. The efficiency scale must be less than 1. The formula for VRT is given below:

$$Z_j = P_t - N_0 \tag{6}$$

$$Z(n) = \sum_{j=1,2,\ldots n} Z_j \tag{7}$$

$$\text{Var}[Z(n)] = \frac{\text{Var}(Z_j)}{n} \tag{8}$$

where

Z_j Difference between old price and new price
$Z(n)$ Total of difference between old price and new price
P_t New price
N_0 Older price
n Total number of stations.

For this study, the efficiency value for pricing according to type of ticket and type of customer is calculated as follows:

$$\text{Efficiency} = \frac{1}{n} \text{variance}[\text{sum}(\text{newprice} - \text{oldprice})] \tag{9}$$

where

$$n = \text{total number of stations}$$

4 Results and Discussion

Data collected from KTMB was categorized into three types of ticket (gold, platinum and silver). This new data set was then used to get the optimal price for ETS ticket. Table 1 displays data developed data set for ETS Gold Fares in Excel.

Table 1 Data of ticket price and distance between stations for ETS gold type from SP to PB

ETS gold fare (RM)				
Station (start)	Station (end)	Price (adult)	Price (children)	Distance (km)
SP	Gr	11.00	10.00	21.40
SP	Kb	14.00	11.00	38.60
SP	AS	16.00	12.00	56.40
SP	AB	18.00	13.00	64.50
SP	Kd	19.00	14.00	87.25
SP	Ar	20.00	14.00	95.10
SP	BK	21.00	15.00	105.50
SP	PB	23.00	16.00	124.10
Gr	Kb	11.00	10.00	17.20
Gr	AS	13.00	11.00	35.00
Gr	AB	16.00	12.00	43.10
Gr	Kd	16.00	12.00	65.85
Gr	Ar	17.00	13.00	73.70
Gr	BK	18.00	13.00	84.10
Gr	PB	21.00	15.00	102.70
Kb	AS	11.00	10.00	17.80
Kb	AB	13.00	11.00	25.90
Kb	Kd	14.00	11.00	48.65
Kb	Ar	14.00	11.00	56.50
Kb	BK	16.00	12.00	66.90
Kb	PB	18.00	13.00	85.50
AS	AB	11.00	10.00	8.10
AS	Kd	12.00	10.00	30.85
AS	Ar	12.00	10.00	38.70
AS	BK	14.00	11.00	49.10
AS	PB	16.00	12.00	67.70
AB	Kd	9.00	9.00	22.75
AB	Ar	10.00	9.00	30.60
AB	BK	11.00	10.00	41.00
AB	PB	14.00	11.00	59.60
Kd	Ar	9.00	9.00	7.85
Kd	BK	11.00	10.00	18.25
Kd	PB	13.00	11.00	36.85
Ar	BK	10.00	9.00	10.40
Ar	PB	12.00	10.00	29.00
BK	PB	11.00	10.00	18.60

Abbreviations *AB* Anak Bukit, *Ar* Arau, *AS* Alor Setar, *BK* Bukit Ketri
Gu Gurun, *Kb* Kobah, *Kd* Kodiang; *PB* Padang Besar
Sample Calculation Distance (Ar–PB) = Distance (Ar–BK) + Distance (BK–PB) = (10.40 + 8.60) km

Optimizing Efficiency of Electric Train Service (ETS) … 387

In this study, MCS has categorized ticket price according to type of ticket and distance between stations. Table 2 shows simulation results for price of gold ticket. Comparing columns 3 and 4 to columns 8 and 9, only tickets with ETS distance of less than 40 km were lowered except Gr-AB and AB-BK routes (highlighted in blue). Yellow and green boxes denote prices that were not minimized. The same results were true for silver class tickets. Interestingly, the entire new price was lower for all platinum tickets type as shown in Table 3.

Table 4 shows the percentage and probability of the change in price. As can be seen, low percentage change in the price occurred for routes SP-AS, Kb-Kd, AS-BK, AB-BK for adult ticket and child ticket.

Table 5 displays the efficiency value for pricing according to type of ticket and type of customer. As can be seen, all efficiency values for all types of ticket (adult and children) are below than 1. Therefore, the price efficiency has been achieved for all types of ticket.

Table 2 New ETS gold ticket price

Station (start)	Station (end)	Price (adult)	Price (children)	Distance (km)	Price per km (adult)	Price per km (child)	New price (adult)	New price (child)
SP	Gr	11.00	10.00	21.40	0.29	0.23	6.17	4.92
SP	Kb	14.00	11.00	38.60	0.29	0.23	11.14	8.87
SP	AS	16.00	12.00	56.40	0.29	0.23	16.27	12.95
SP	AB	18.00	13.00	64.50	0.29	0.23	18.61	14.81
SP	Kd	19.00	14.00	87.25	0.29	0.23	25.17	20.04
SP	Ar	20.00	14.00	95.10	0.29	0.23	27.44	21.84
SP	BK	21.00	15.00	105.50	0.29	0.23	30.44	24.23
SP	PB	23.00	16.00	124.10	0.29	0.23	35.80	28.50
Gr	Kb	11.00	10.00	17.20	0.29	0.23	4.96	3.95
Gr	AS	13.00	11.00	35.00	0.29	0.23	10.10	8.04
Gr	AB	16.00	12.00	43.10	0.29	0.23	12.43	9.90
Gr	Kd	16.00	12.00	65.85	0.29	0.23	19.00	15.12
Gr	Ar	17.00	13.00	73.70	0.29	0.23	21.26	16.93
Gr	BK	18.00	13.00	84.10	0.29	0.23	24.26	19.32
Gr	PB	21.00	15.00	102.70	0.29	0.23	29.63	23.59
Kb	AS	11.00	10.00	17.80	0.29	0.23	5.14	4.09
Kb	AB	13.00	11.00	25.90	0.29	0.23	7.47	5.95
Kb	Kd	14.00	11.00	48.65	0.29	0.23	14.04	11.17
Kb	Ar	14.00	11.00	56.50	0.29	0.23	16.30	12.98
Kb	BK	16.00	12.00	66.90	0.29	0.23	19.30	15.37
Kb	PB	18.00	13.00	85.50	0.29	0.23	24.67	19.64
AS	AB	11.00	10.00	8.10	0.29	0.23	2.34	1.86
AS	Kd	12.00	10.00	30.85	0.29	0.23	8.90	7.09
AS	Ar	12.00	10.00	38.70	0.29	0.23	11.16	8.89
AS	BK	14.00	11.00	49.10	0.29	0.23	14.17	11.28

(continued)

Table 2 (continued)

Station (start)	Station (end)	Price (adult)	Price (children)	Distance (km)	Price per km (adult)	Price per km (child)	New price (adult)	New price (child)
AS	PB	16.00	12.00	67.70	0.29	0.23	19.53	15.55
AB	Kd	9.00	9.00	22.75	0.29	0.23	6.56	5.23
AB	Ar	10.00	9.00	30.60	0.29	0.23	8.83	7.03
AB	BK	11.00	10.00	41.00	0.29	0.23	11.83	9.42
AB	PB	14.00	11.00	59.60	0.29	0.23	17.19	13.69
Kd	Ar	9.00	9.00	7.85	0.29	0.23	2.26	1.80
Kd	BK	11.00	10.00	18.25	0.29	0.23	5.27	4.19
Kd	PB	13.00	11.00	36.85	0.29	0.23	10.63	8.46
Ar	BK	10.00	9.00	10.40	0.29	0.23	3.00	2.39
Ar	PB	12.00	10.00	29.00	0.29	0.23	8.37	6.66
BK	PB	11.00	10.00	18.60	0.29	0.23	5.37	4.27
		515.00	410.00	1785.10				

Abbreviations *AB* Anak Bukit, *Ar* Arau, *AS* Alor Setar, *BK* Bukit Ketri, *Gr* Gurun, *Kb* Kobah, *Kd* Kodiang; *PB* Padang Besar
Formulas: Price per km (adult) = total price (adult)/total distance; Price per km (child) = total price (child)/total distance
Total distance = sum of all distances
New price (adult/child) = price per km (adult/child) × distance

Table 3 New ETS platinum ticket price

ETS Platinum fare (RM)								
Station (start)	Station (end)	Price (adult)	Price (children)	Distance (km)	Price per km (adult)	Price per km (child)	New price (adult)	New price (child)
SP	Gr	–	–	21.40	0.11	0.08	–	–
SP	Kb	–	–	38.60	0.11	0.08	–	–
SP	AS	19.00	14.00	56.40	0.11	0.08	6.03	4.27
SP	AB	22.00	15.00	64.50	0.11	0.08	6.90	4.88
SP	Kd	–	–	87.25	0.11	0.08	–	–
SP	Ar	34.00	16.00	95.10	0.11	0.08	10.18	7.19
SP	BK	–	–	105.50	0.11	0.08	–	–
SP	PB	29.00	19.00	124.10	0.11	0.08	13.28	9.39
Gr	Kb	–	–	17.20	0.11	0.08	–	–
Gr	AS	–	–	35.00	0.11	0.08	–	–
Gr	AB	–	–	43.10	0.11	0.08	–	–
Gr	Kd	–	–	65.85	0.11	0.08	–	–
Gr	Ar	–	–	73.70	0.11	0.08	–	–
Gr	BK	–	–	84.10	0.11	0.08	–	–
Gr	PB	–	–	102.70	0.11	0.08	–	–
Kb	AS	–	–	17.80	0.11	0.08	–	–
Kb	AB	–	–	25.90	0.11	0.08	–	–
Kb	Kd	–	–	48.65	0.11	0.08	–	–

(continued)

Table 3 (continued)

		ETS Platinum fare (RM)						
Kb	Ar	–	–	56.50	0.11	0.08	–	–
Kb	BK	–	–	66.90	0.11	0.08	–	–
Kb	PB	–	–	85.50	0.11	0.08	–	–
AS	AB	12.00	10.00	8.10	0.11	0.08	0.87	0.61
AS	Kd	–	–	30.85	0.11	0.08	–	–
AS	Ar	14.00	11.00	38.70	0.11	0.08	4.14	2.93
AS	BK	–	–	49.10	0.11	0.08	–	–
AS	PB	19.00	14.00	67.70	0.11	0.08	7.24	5.12
AB	Kd	–	–	22.75	0.11	0.08	–	–
AB	Ar	12.00	10.00	30.60	0.11	0.08	3.27	2.31
AB	BK	–	–	41.00	0.11	0.08	–	–
AB	PB	19.00	16.00	59.60	0.11	0.08	6.38	4.51
Kd	Ar	–	–	7.85	0.11	0.08	–	–
Kd	BK	–	–	18.25	0.11	0.08	–	–
Kd	PB	–	–	36.85	0.11	0.08	–	–
Ar	BK	–	–	10.40	0.11	0.08	–	–
Ar	PB	11.00	10.00	29.00	0.11	0.08	3.10	2.19
BK	PB	–	–	18.60	0.11	0.08	–	–
		RM191.00	RM135.00	1785.10				

Abbreviations *AB* Anak Bukit, *Ar* Arau, *AS* Alor Setar, *BK* Bukit Ketri, *Gr* Gurun, *Kb* Kobah, *Kd* Kodiang; *PB* Padang Besar

Formulas: Price per km (adult) = total price (adult)/total distance; Price per km (child) = total price (child)/total distance

Total distance = sum of all distances

New price (adult/child) = price per km (adult/child) × distance

Table 4 Probability and percentage of change in price for adult and child ticket

Station (start)	Station (end)	Z_j (adult)	Z_j (child)	Probability of change in price (adult)	Probability of change in price (child)	% of change in price (adult)	% of change in price (child)
SP	Gr	4.83	5.08	0.44	0.51	43.87	50.85
SP	Kb	2.86	2.13	0.20	0.19	20.46	19.40
SP	AS	0.27	0.95	0.02	0.08	1.70	7.95
SP	AB	0.61	1.81	0.03	0.14	3.38	13.96
SP	Kd	6.17	6.04	0.32	0.43	32.48	43.14
SP	Ar	7.44	7.84	0.37	0.56	37.18	56.02
SP	BK	9.44	9.23	0.45	0.62	44.94	61.54
SP	PB	12.80	12.50	0.56	0.78	55.66	78.14
Gr	Kb	6.04	6.05	0.55	0.60	54.89	60.50
Gr	AS	2.90	2.96	0.22	0.27	22.33	26.92
Gr	AB	3.57	2.10	0.22	0.18	22.29	17.51

(continued)

Table 4 (continued)

Station (start)	Station (end)	Z_j (adult)	Z_j (child)	Probability of change in price (adult)	Probability of change in price (child)	% of change in price (adult)	% of change in price (child)
Gr	Kd	3.00	3.12	0.19	0.26	18.74	26.04
Gr	Ar	4.26	3.93	0.25	0.30	25.07	30.21
Gr	BK	6.26	6.32	0.35	0.49	34.79	48.58
Gr	PB	8.63	8.59	0.41	0.57	41.09	57.25
Kb	AS	5.86	5.91	0.53	0.59	53.32	59.12
Kb	AB	5.53	5.05	0.43	0.46	42.52	45.92
Kb	Kd	0.04	0.17	0.00	0.02	0.25	1.58
Kb	Ar	2.30	1.98	0.16	0.18	16.43	17.97
Kb	BK	3.30	3.37	0.21	0.28	20.63	28.05
Kb	PB	6.67	6.64	0.37	0.51	37.04	51.06
AS	AB	8.66	8.14	0.79	0.81	78.76	81.40
AS	Kd	3.10	2.91	0.26	0.29	25.83	29.14
AS	Ar	0.84	1.11	0.07	0.11	6.96	11.11
AS	BK	0.17	0.28	0.01	0.03	1.18	2.52
AS	PB	3.53	3.55	0.22	0.30	22.07	29.58
AB	Kd	2.44	3.77	0.27	0.42	27.07	41.94
AB	Ar	1.17	1.97	0.12	0.22	11.72	21.91
AB	BK	0.83	0.58	0.08	0.06	7.53	5.83
AB	PB	3.19	2.69	0.23	0.24	22.82	24.44
Kd	Ar	6.74	7.20	0.75	0.80	74.84	79.97
Kd	BK	5.73	5.81	0.52	0.58	52.14	58.08
Kd	PB	2.37	2.54	0.18	0.23	18.22	23.06
Ar	BK	7.00	6.61	0.70	0.73	70.00	73.46
Ar	PB	3.63	3.34	0.30	0.33	30.28	33.39
BK	PB	5.63	5.73	0.51	0.57	51.22	57.28

Formulas: Difference = |new price–old price|
Probability of price change = difference between new and old price/old price
% of price change = probability × 100

Table 5 Efficiency value according to type of ticket and type of customer

Ticket type	Passenger type	Efficiency	Passenger type	Efficiency
Gold	Adult	0.35	Children	0.35
Platinum	Adult	0.32	Children	0.27
Silver	Adult	0.35	Children	0.34

5 Conclusion

MCS method in this study was able to minimize all platinum type ticket price but was only able to minimize gold and silver ticket price for distances of less than 40 km on the SP-PB route, except Gr-AB and AB-BK links. Above all, efficiency scale (<1) was achieved for all ticket types. Future works at minimizing ticket price may consider using other alternative methods for distances over 40 km, and the scope may include more stations.

Acknowledgements The authors wish to express their gratitude to KTMB for allowing access to their data which was needed in this study.

References

Cerrato, M., & Cheung, K. K. (2006). Valuing American style options by least squares methods. London Metropolitan University.
Coskan, C. (2008). Pricing American @@lation. Bogaziçi University.
Department for Transport. (2013). *Rail fares and ticketing: Next steps*. London: Great Minster House.
Goddard Consulting. (2017). Monte Carlo option pricing variance reduction. Retrieved May 13 2017, from http://www.goddardconsulting.ca/option-pricing-monte-carlo-vr.html.
Khalid, et al. (2014). User perceptions of rail public transport services in Kuala Lumpur, Malaysia: KTM Komuter. *Procedia—Social and Behavioral Sciences, 153*, 566–573.
KTMB. (2016). Mengenai ETS. from http://www.ktmb.com.my/ktmb/index.php?r=portal/index.
Lantseva, et al. (2015). data-driven modeling of airlines pricing. *Procedia Computer Science, 66*, 267–276.
Li, et al. (2012). Modeling the effects of integrated rail and property development on the design of rail line services in a linear monocentric city. *Transportation Research Part B: Methodological, 46*(6), 710–728.
Mijares, et al. (2013). Equity analysis of urban rail fare policy and passenger overload delay: An international comparison and the case of metro ManilaMRT-3. *Journal of Eastern Asia Society for Transportation Studies, 10*.
Mijares, et al. (2016a). An analysis of metro manila mrt-3 passenger's perception of their commuting experience and its effects using structural equation modelling (SEM) (6th ed.). The Transportation Research Board.
Mijares, et al. (2016b). Passenger satisfaction and mental adaptation under adverse conditions: Case study in Manila. *Journal of Public Transportation, 19*(4), 144–160.
Moreno, M., & Navas, J. F., (2003). On the robustness of least-squares Monte Carlo (LSM) for pricing American derivatives. *Review of Derivatives Research, 6*, 107–128.
Qiao, et al. (2013). Passenger route choice model and algorithm in the urban rail transit network. *Journal of Industrial Engineering and Management, 6*(1).
Sunif. (2011). Perkhidmatan Pengangkutan Awam di Malaysia. (Bachelor Degree), Universiti Teknologi Malaysia.
Wang, B., & Wang, L. (2011). Pricing barrier options using Monte Carlo methods. Department of Mathematics, Uppsala University.
Winston, C., & Maheshri, V. (2007). On the social desirability of urban rail transit systems. *Journal of Urban Economics, 62*, 362–382.

Fuzzy Rules Enhancement of CHEF to Extend Network Lifetime

Muhammad A'rif Shah Alias, Habibollah Haron and Teresa Riesgo

Abstract Current implementation to extend the network lifetime is by using clustering algorithms such as LEACH and CHEF. LEACH is based on the probability model and it can cause big overhead, while CHEF is an extension of LEACH and one of the protocols that implement fuzzy logic in reducing the overhead by collecting and calculating information. CHEF uses two descriptors which are local distance and energy. In this paper, an enhanced fuzzy rule based on the existing CHEF fuzzy rule which is called iCHEF is proposed that add a third descriptor called BS distance. Three performance measurements namely number of node, number of round, and first node die are used to evaluate the performance of the proposed iCHEF. The result shows that the proposed iCHEF successfully improves the network lifetime two times compared to CHEF in terms of independent network.

Keywords Wireless sensor network · Fuzzy logic · CHEF · Network lifetime

1 Introduction

Wireless sensor network (WSN) is a network that consists of many nodes that have limited energy, computation, and memory that deployed in the area to monitor, collect information from the environment, and then report to base station (BS). WSNs are used for a variety of purposes such as object tracking, monitoring the environment, and military device in surveillance (Cerpa et al. 2001; Shen et al.

M. A. S. Alias · H. Haron (✉)
Department of Computer Science, Faculty of Computing,
Universiti Teknologi Malaysia, UTM Skudai, Johor, Malaysia
e-mail: habib@utm.my

T. Riesgo
Centre of Electronic Industrial, Universidad Politecnica de Madrid,
C/Jose Gutierrez Abascal 2, 28006 Madrid, Spain

2001). Normally, it will be distributed in the area that is far away and dangerous such as detection of chemical or biological agent threat (Chong and Kumar 2003).

Since it is impracticable to replace the battery after it has been deployed, it is important to take energy as the main factor to be considered. Clustering is one of the approaches that can be used to extend the network lifetime by dividing the node into the cluster wherein one cluster contains several nodes and one cluster head (CH). The CH acts as the leader where it will collect all the information from the nodes in its cluster and compress the data to report to the BS. By using the approach, the overheads of the all node report to the BS can be reduced.

The clustering can extend the network lifetime, but it has some problems where the energy consumption is not evenly distributed on the CH. The solution to this matter is LEACH protocol by Heinzelman et al. (2000) which localized the cluster where each node elects itself as a CH randomly based on the probability model. There are some cases that cause inefficient CH to be elected and some CH may have been located in the illogical place as CH, and this can reduce the energy efficiency [5]. CHEF protocol by Kim et al. (2008) implemented the fuzzy logic (FL) to overcome the previous problem by considering the remaining energy and the local distance of the CH to the node within its radius, but it does not consider the distance of the CH to the BS. To maximize the network lifetime, this paper enhances the fuzzy rule proposed by Kim et al. (2008) in the CHEF protocol with the addition of the descriptor which is the distance of the CH to BS to extend the network lifetime. Three performance measurements used are number of alive node (NN), a number of rounds (NR), and first node die (FND). NN indicates the number of alive nodes at round equal to 500, NR indicates the number of rounds when node equal to 0, while FND indicates the number of rounds when the first node die.

The remaining of this paper is structured as follows. Section 2 summarizes the related work. In Sect. 3, the proposed enhanced CHEF mechanism is described. The proposed fuzzy model is presented in Sect. 4. Section 5 shows the result and analysis, while Section 6 concludes this paper.

2 Related Work

There are several types of protocol which are network flow, location-based, data-centric, and hierarchical. This section only focuses on the hierarchical protocol where the clusters are formed with one CH lead one cluster. The CH is responsible for collecting information from nodes within its cluster and compresses the information to be transmitted to the BS. Two existing hierarchical protocols and their weaknesses are described below.

2.1 Low Energy Adaptive Clustering Hierarchy (LEACH)

LEACH by Heinzelman et al. (2000) is one of the famous clustering algorithms where the node elects itself as CH based on the probability model. CH consumes a lot of energy; therefore to prevent this matter it will reset the CH each round and the previous CH cannot be the CH for the next round until the countdown pass. A node can be CH if its random value which is generated by itself is lower than the threshold value $T(n)$. CH advertised the message to the other nodes around its cluster. The other nodes that receive the message must calculate the distance between CH and itself and then send the message back to join the nearest CH. Equation 1 defines the $T(n)$ where P is the requested ratio of CH, r is the current round, G is the set of nodes that were not elected as CH in last $1/P$ rounds.

$$T(n) = \begin{cases} \frac{P}{1-P \times \left(r \bmod \frac{1}{P}\right)}, & \text{if } n \in G \\ 0, & \text{otherwise} \end{cases} \quad (1)$$

2.2 Cluster Head Election Fuzzy (CHEF)

In CHEF by Kim et al. (2008), FL if-then rules have been implemented to compute Chance where Chance is used to elect the CH. It uses two descriptors which are remaining energy and the local distance or proximity distance to be the key descriptor in computing the Chance. This approach allows the node with the highest energy and most proximity distance to be elected as CH. CHEF has shown that it has 22.7% more efficiency than LEACH.

2.3 Weakness of LEACH and CHEF

In Heinzelman et al. (2000) and Kim et al. (2008), there are some critical points that need consideration. For example, in LEACH the elected cluster heads may be very close to each other. This is because LEACH depends on only the probability model. LEACH does not consider the local information such as energy remains of each node. The nodes with small energy remains can be elected as cluster heads. This can shorten the network lifetime. While in CHEF and LEACH, the cluster head may be on the edge of the network. This is because it did not consider the distance of the CH to BS. In this case, the transmission energy of the CH is increasing over the area. Therefore, the current fuzzy rule of CHEF should be enhanced and additional descriptor should be added to improve the problem that is faced in CHEF.

3 The Proposed Enhanced CHEF Mechanism (ICHEF)

This section consists of description of four stages on how the proposed iCHEF is developed. First, a simulation model is adopted from the CHEF network topology proposed by Kim et al. (2008). Next, an improved of CHEF which is called

Fig. 1 Network design and topology

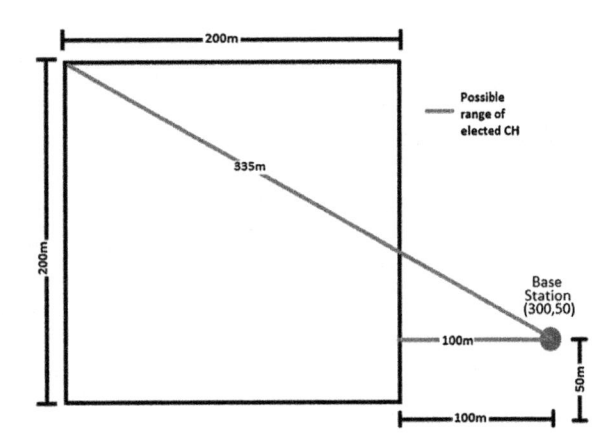

CHEF$_{TB}$ is developed. Third, the proposed iCHEF mechanism is developed. Fourth, an enhanced fuzzy if-then rule is added into the iCHEF mechanism.

3.1 Network Topology

The network topology or design that is used in CHEF by Kim et al. (2008) is adopted in this project where the area of the WSN is 200 time 200, and its BS is located in position (300, 50). The network design can be seen in Fig. 1. From Fig. 1, it shows that CH can be elected in the range of 100–335 m distance to the BS.

3.2 The CHEF Simulation Implementation (CHEF$_{TB}$)

To develop a new enhanced CHEF, a simulation representing the current CHEF by Kim et al. (2008) must be developed. In developing the simulation for CHEF, it applies all the implementation that had been implemented in Kim et al. (2008) such as the clustering algorithm, all the given formula, the membership function in the fuzzy if-then rule, and all the descriptor parameter as experimental setup.

3.3 The ICHEF Clustering Algorithm

Similar to LEACH, CHEF, and CHEF$_{TB}$, the cluster will be configured in each round and then reset before another round. The proposed clustering algorithm is shown in Fig. 2. In every round, the evaluation of Chance is computed and it will be sorted in descending order. The random value generated is between 0 and 1, and if it is below the P_{opt}, the node is elected as CH and then CH_Message is

Fig. 2 iCHEF clustering algorithm

> *The Proposed Clustering Algorithm*
> 1. /* for every round*/
> 2. compute the three descriptor from local information.;
> 3. evaluate the Chance using if-then rule ;
> 4. sort the Chance in descending order;
> 5.   if rand() < $P_{opt\ then}$
> 6.   myCH = N;
> 7.   Adv CH_Message
> 8.   else
> 9.   On receiving CH_Message
> 10.   Select the closest CH;
> 11.   Send ClusterJoin_Message to closest CH;
> 12. end if

advertised. The normal nodes that receive the CH_Message will send a CluterJoin_Message to the closest CH.

3.4 The Enhanced if-then Fuzzy Rule

In this paper beside energy and local distance, a new descriptor called BS distance has been added in the fuzzy if-then rule. The descriptors are defined as:

- Energy: the remaining energy of the nodes.
- Local distance: total distance of the node to CH in the radius.
- BS distance[new]: distance of each node to the BS.

The mechanism uses the fuzzy if-then rule to compute Chance where the bigger Chance means the node has higher possibility to become CH. To compute Chance, the fuzzy computes from three descriptors as mentioned above. The energy and local distance are adopted from the CHEF proposed by Kim et al. (2008). The BS distance[new] is taken from percentage of the distance of CH to the BS which is calculated using Eq. 2.

$$\text{BS Distance} = \frac{\sqrt{(n.x - \text{BS}.x)^2 + (n.y - \text{BS}.y)^2}}{335} \times 100\% \qquad (2)$$

In Eq. 2, *n.x* and *n.y* indicate the position of node at *x*- and *y*-axis while for *BS.x* and *BS.y* indicate for BS location that is (300, 50). The equation is divided into 335 because it is the maximum BS distance for the network topology. The range for the BS distance uses the percentage of 0–50%, 50–65%, 65–75%, 75–85%, and 85–100% which is later shown in fuzzy inference system.

Table 1 is a basis to derive 45 rules because of permutation of three fuzzy descriptors with repetition of 3 × 3 × 5 = 45. There are 45 rules to be implemented in the proposed iCHEF that is based on these three fuzzy descriptors shown

Table 1 Fuzzy descriptor list

Fuzzy descriptor		
Energy	Local distance	BS distance[new]
Low	Far	Too far
Medium	Medium	Far
High	Close	Medium
		Close
		Too close

in Table 1. The energy has three levels which are low, medium, and high, while for the local distance it has far, medium, and close. For the addition descriptor of fuzzy compared to CHEF, BS distance contains five levels which are too far, far, medium, close, and too close.

3.5 The Proposed ICHEF Mechanism

In this section, the proposed iCHEF that uses fuzzy if-then rule to maximize the network lifetime is explained. iCHEF is based on CHEF by Kim et al. (2008), but it has added descriptor and is different in the steps of clustering algorithm.

3.5.1 Clustering Algorithm

The proposed clustering algorithm is different with the CHEF by Kim et al. (2008) where it computes all the node in P_{opt} first and then evaluate it Chance using fuzzy if-then rule. As shown in Fig. 2, firstly the Chance is calculated and then is sorted in descending order. Then all nodes undergo the comparison with P_{opt}. Node that is less than P_{opt} becomes CH which is elected based on the highest value in the order.

3.5.2 Fuzzy Inference System

The process of fuzzification is shown in Fig. 3 which is known as the fuzzy inference system, while Fig. 4 shows the fuzzy set of three descriptors. The characteristic of iCHEF can be summarized as energy, local distance, and BS distance that are considered in electing the CH. There will be no CH that will be elected at the end of the edge.

Fuzzification module is needed to use system input into fuzzy sets. System input is calculated before the evaluation as shown in Line 2, Fig. 2. For the knowledge base, it stores the if-then rule which is 45 rules. In inference engine or where the

Fig. 3 Fuzzy inference system

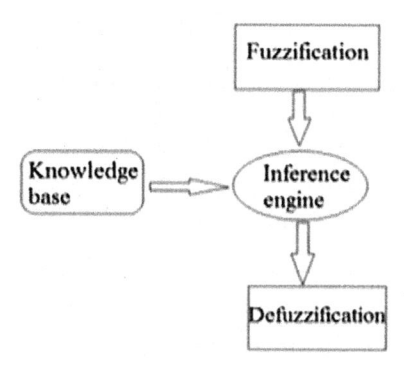

evaluation occurs as shown in Line 3, Fig. 2, it uses the input from the fuzzification from Fig. 4 and if-then rule from the knowledge base to simulate fuzzy reasoning in producing fuzzy inference. In defuzzification module, it translates fuzzy set that produces from inference engine into a value which is Chance. Next, Chance values are sorted as shown in Line 4, Fig. 2, in descending order. Lastly, the election of CH occurs which is similar to Heinzelman et al. (2000) and Kim et al. (2008).

4 Result and Analysis

This section shows comparison result of four protocols namely LEACH, CHEF, $CHEF_{TB}$, and iCHEF based on three performance measurements that are NR, NN, and FND. The result of LEACH and CHEF is taken by Kim et al. (2008), while the result of $CHEF_{TB}$ and iCHEF is based on the experiment conducted. Table 2 shows the performance measurement value between three performance measurements of four protocols that is based on the graph from Kim et al. (2008), Figs. 5 and 6. The first two column of Table 2 is taken from previous work by Kim et al. (2008), while the last two column is the result taken from the experiment conducted.

In general, $CHEF_{TB}$ result has big difference compared to CHEF (Kim et al. 2008). For NN, the value is near the CHEF result, but the NR and FND have big difference. In NR, $CHEF_{TB}$ positive ranges about 250 rounds until all node died, while in FND $CHEF_{TB}$ negative ranges in 232.23 rounds. For FND, value of $CHEF_{TB}$ is faster to go but for NR it can last longer than CHEF. These are not very good since the result supposed to be not too much different in three performance measurement. This shows that it still has difference in clustering algorithm or fuzzy if-then rule configuration. Therefore, $CHEF_{TB}$ cannot be properly used as the simulation for CHEF by Kim et al. (2008).

For NR is positively enhanced compared to the other three protocols. This shows that iCHEF WSN can last two times longer than CHEF. For NN, it also shows that

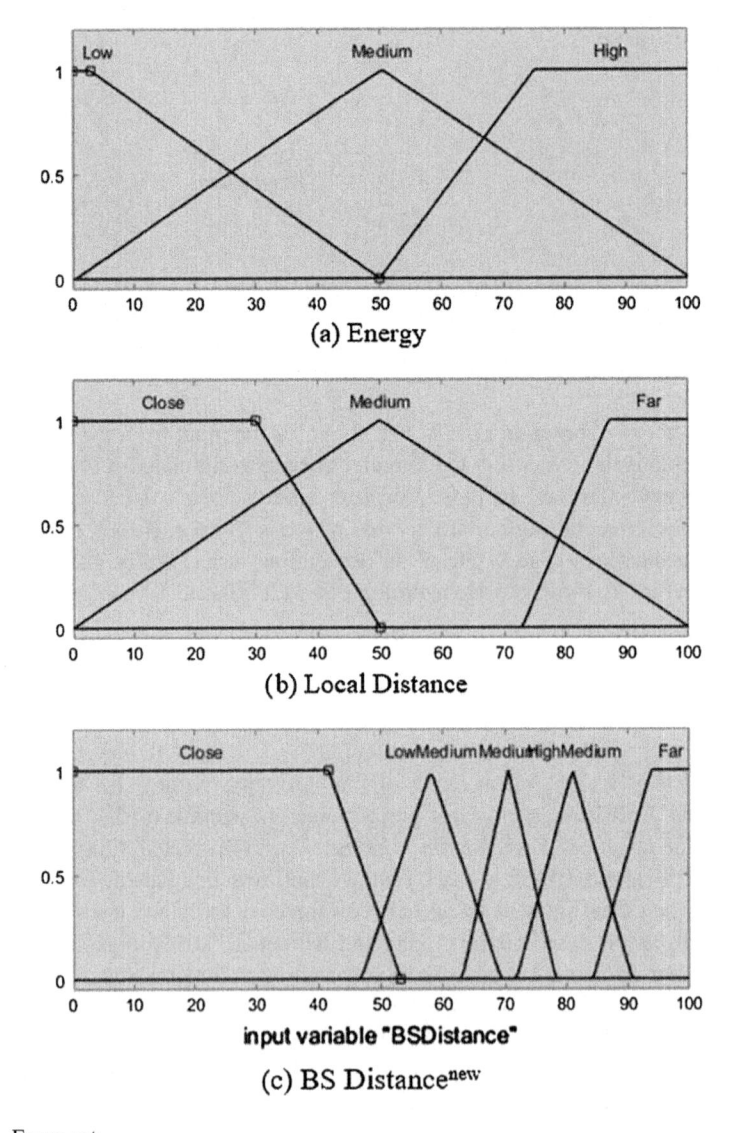

Fig. 4 Fuzzy set

Table 2 Performance measurement

Performance	Type of protocol			
	LEACH	CHEF	$CHEF_{TB}$	iCHEF
NR	470	450	725	867
NN	0	0	27	300
FND	307.41	397.69	165.46	151.35

Fuzzy Rules Enhancement of CHEF to Extend Network Lifetime   401

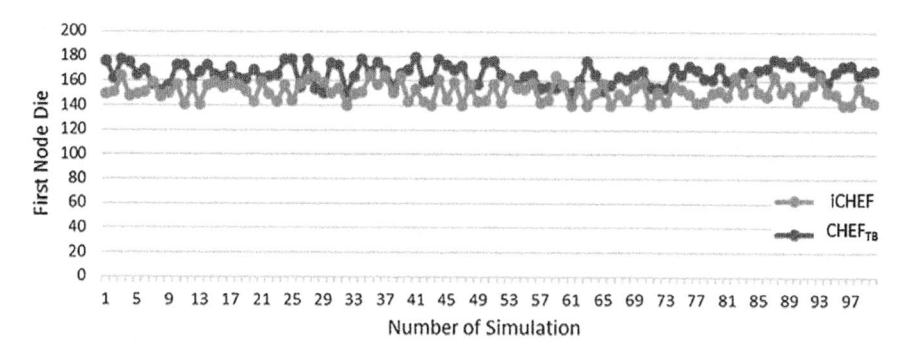

Fig. 5 Number of first node die for 100 simulation

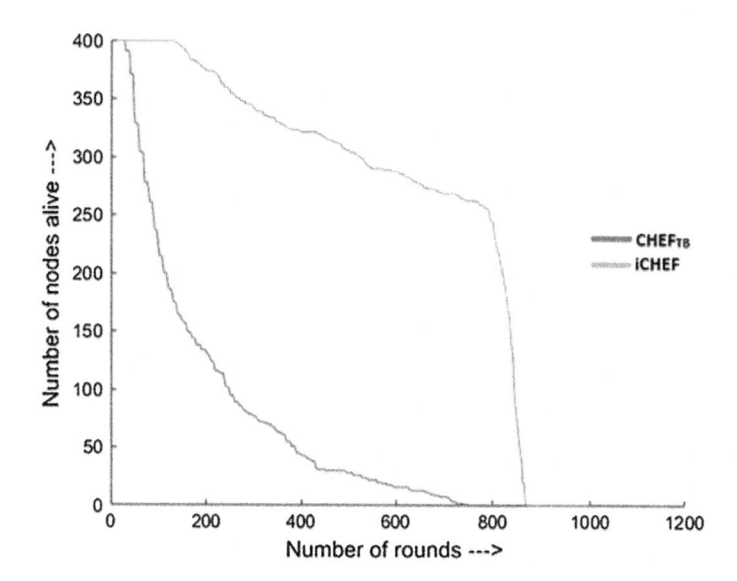

Fig. 6 Number of nodes alive for each round

at round 500 iCHEF still has 300 nodes that alive while LEACH and CHEF already zero nodes and CHEF$_{TB}$ is 27 nodes. For FND, iCHEF shows problem where it has the earliest round of the first node to die at averagely round 151.35. While for CHEF$_{TB}$ is 165.46, CHEF is 397.69 and LEACH 307.41. This shows that iCHEF not suitable for WSN that dependent to all of its nodes because iCHEF result for FND gives the negative impact on the network lifetime. It concludes that iCHEF is suitable for the network that needs to work for a long time and did not have problem with the lack of some functionalities because it shows that it can work two times longer than CHEF and LEACH.

5 Conclusion and Future Work

In this paper, an enhanced fuzzy rule for CHEF is proposed and is called iCHEF. The enhancement has successfully improved the network lifetime compared to the proposed CHEF algorithm proposed by Kim et al. (2008) in terms of the independent network. For future work, the proposed iCHEF can be extended and improved by implementing fuzzy-based master cluster head election (F-MCHEL) protocol proposed by Sharma and Kumar (2012) to find the optimal fuzzy set.

Acknowledgements The authors would like to show gratitude to Mohd Ridhwan Hilmi Mohd Adnan for assistance with the understanding of FL implementation which provide insight and expertise that greatly assisted this project. This work is also supported by Malaysian Ministry of Higher Education (MOHE) Fundamental Research Grant Scheme with UTM reference number R.J 130000.7828.4F860.

References

Cerpa, A., Elson, J., Estrin, D., Girod, L., Hamilton, M., & Zhao, J. (2001). Habitat monitoring: Application driver for wireless communications technology. *ACM SIGCOMM Computer Communication Review, 31*, 20–41.
Chong, C. Y., & Kumar, S. P. (2003). Sensor networks: Evolution, opportunities, and challenges. *Proceedings of the IEEE, 91*(8), 1247–1256.
Heinzelman, W. R., Chandrakasan, A., & Balakrishnan, H. (2000). Energy Efficient Communication Protocol for Wireless Microsensor Networks. In *Proceedings of the 33rd hawaii international conference on system sciences*, 2:10.
Kim, J. M., Park, S. H., Han, Y. J., & Chung, T. M. (2008). CHEF: Cluster head election mechanism using fuzzy logic in wireless sensor networks. In *International conference on advance communication technologies*, 654–659.
Sharma, T., & Kumar, B. (2012). F-MCHEL: Fuzzy based master cluster head election Leach protocol in wireless sensor network. *International Journal Computing Science Telecommunication, 3*(10), 8–13.
Shen, C., Srisathapornphat, C., & Jaikaeo, C. (2001). Sensor information networking architecture and applications. *IEEE Personal Communications, 8*(4), 52–59.

Energy Consumption of Wireless Sensor Based on IoT in a Parking Space With and Without Fuzzy Rules

Muhamad Yazid Che Abdullah, Habibollah Haron and Teresa Riesgo

Abstract Wireless sensor and Internet of Things (IoT) are among important technologies of recent times. Energy consumption of wireless sensor is an important consideration especially when designing and implementation of any network design such as in a parking space. Fuzzy logic is one of the artificial intelligence techniques that represent fuzzy rule that can be combined with IoT functions in order to calculate the energy consumption. To calculate the energy consumption, formulas available in the Arduino function and formulas from established works of previous researchers are used. The objective in this paper is to design a parking space with four parking lots, to develop an IoT prototype system representing the parking space, and to define the non-fuzzy rule and fuzzy rule to control the components in the parking space. The IoT prototype system for the parking space is developed using C programming language in the Arduino hardware and software. In the programming, declaration and new definition of function and rules are created. The energy consumption, kilowatt per hour (kWh), is measured using INA219 sensor that monitors the voltage, current, power, and energy consumed. By applying the fuzzy rule compared to non-fuzzy rule in parking space, the result shows that the energy consumption has been reduced by 92.6%. To conclude, this project can contribute to optimize the energy consumption by applying fuzzy logic for a parking system.

Keywords Fuzzy rule · Sensor · Energy consumption · Arduino

M. Y. Che Abdullah (✉) · H. Haron
Faculty of Computing, Universiti Teknologi Malaysia,
UTM Skudai, 81310 Johor Bahru, Malaysia
e-mail: habib@utm.my

T. Riesgo
Centro de Electrónica Industrial, Universidad Politécnica de Madrid,
C/Jose Gutierrez Abascal 2, 28006 Madrid, Spain

© Springer Nature Singapore Pte Ltd. 2018
R. Saian and M. A. Abbas (eds.), *Proceedings of the Second International Conference on the Future of ASEAN (ICoFA) 2017 – Volume 2*,
https://doi.org/10.1007/978-981-10-8471-3_40

1 Introduction

Energy consumption of wireless sensor is an important consideration in designing and implementation of any network design. Wireless sensors are distributed autonomous sensor to monitor physical or environmental conditions such as sound, pressure, and temperature. In this project, parking space that is self-created has been chosen as case study. In order to evaluate energy consumption, functions and rules are created to generate the assigned parking lot.

The rules to evaluate energy consumption are based on Boolean Logic that is created without and with fuzzy rule. Also, it can be used to derive conclusions as outputs from user inputs and the fuzzy reasoning process. The input will be number of cars enter and park into the parking space. For the function, every single sensor used in this paper calculates its voltage, current, energy consumed, and operation time remaining

To represent the function and rules, an Arduino software IDE embedded with fuzzy logic library is used. As a result, the energy consumption can be evaluated by comparing the result of the energy consumption used for non-fuzzy and with fuzzy rule. The analysis is then performed for both of the results.

Paper organization consists of seven sections. The sections are introduction, literature reviews, experimental setup, methodology, design and implementation, result and analysis, and conclusion and future works.

2 Literature Reviews

2.1 Related Work

Due to the explosive growth of automobiles, parking space especially near to public area gradually becomes one of the most annoying things to car owners. In most cases, most of the users find the indoor parking spaces nearby are always full, and they have to drive around to look for any available parking space on the street. With the continuous growth of automobiles, the situation becomes worse and worse. So, the demand for street parking guidance service is expected to grow rapidly in the near future. Wireless sensors have lots of potential toward providing an ideal solution for street parking service, such as their low power, small size, and low cost (Baronti et al. 2007; Fraifer and Fernström 2016; Sachdeva et al. 2016).

2.2 Energy Consumption

The energy consumption in wireless sensor is determined by three main components that are sensing, processing, and transmission. Sensing energy consumption for sensor node is determined by the specific characteristics of the sensor, and its value is determined based on the device datasheet. In parking space, the wireless devices which are battery-powered are used as the sensors to track the number of incoming car and out coming car in parking space. To calculate the energy consumption, the type of sensor used and energy efficient method can produce a better result depending on the situation that is described. In previous works, there are three equations, Eqs. 1–3, that are used to calculate the energy consumption (Zhang et al. 2013).

$$\text{Current}, I = \text{Voltage}, V / \text{Resistance}, R \qquad (1)$$

$$\text{Power}, P = \text{Voltage}, V \times \text{Current}, I \qquad (2)$$

$$\text{Energy}, E = \text{Power}, P \times \text{time}, t \qquad (3)$$

2.3 Light-Emitting Diode (LED)

A light-emitting diode (LED) is a two-lead semiconductor light source. It emits light when activated based on Houghton Mifflin Company (2005). In the proposed parking system, LED is used at entrance, exit gate, and all parking lots. Green LED (parking lamp) is used in parking lots to determine the availability of the parking lot, while white LED (indicator lamp and light source) is used as the source of light at the gates and the parking lot.

2.4 Light Dependent Resistor (LDR)

Based on Robert (2005), LDR is a type of sensor that works on the principle of photoconductivity. Photoconductivity is an optical phenomenon in which the materials conductivity is increased when light is absorbed by the material. LDR decreases when light falls on them and increases if it is kept in dark. The advantage of LDR is that it is cheap, smaller in size, consumed very small power and voltage for its operation. The disadvantage of this sensor is that it is highly inaccurate and the response time is quite slow which is tens or hundreds of milliseconds.

2.5 INA219 Sensor

Based on Ada (2016), INA219 sensor is one of the best sensors in terms of energy measurement and monitoring the problem. It is a small chip and very smart because it can handle high side current which is great for tracking battery life. This chip is used to measure the energy consumed by LEDs in kilowatt per hour (kWh).

2.6 Servomotor and Counter

There are two servomotors located at the entrance and exit gate. The function of them is to open and close the gate when the car enters or leaves the parking space. The counter is used to calculate the number of car enters and leaves in the parking space.

2.7 Fuzzy Rule

Fuzzy set theory can deal with the vagueness and uncertainty residing in the knowledge acquired by human beings or implicated in the numerical data. Fuzzy rule-based expert system contains fuzzy rule in its knowledge base and derives conclusions as outputs from the user inputs and the fuzzy reasoning process (Zadeh 1965).

In this project, fuzzy rule is designed to differentiate the process in the parking system. This fuzzy rule creates inputs and outputs depending on the situation given. In terms of parameter of energy consumption, this rule is compared to non-fuzzy rule for the final result of the output.

3 Experimental Setup

A parking layout is designed based on the list of sensors that are used. The arrangement of the sensors and how the sensors connect with each other in circuit of the parking design are shown in Fig. 1.

Figure 1 shows the layout for the overall parking design. There are four parking lots available in the parking system. Each parking lot contains two types of LED which are green LED and white LED. Green LED is used as the parking lamp as the

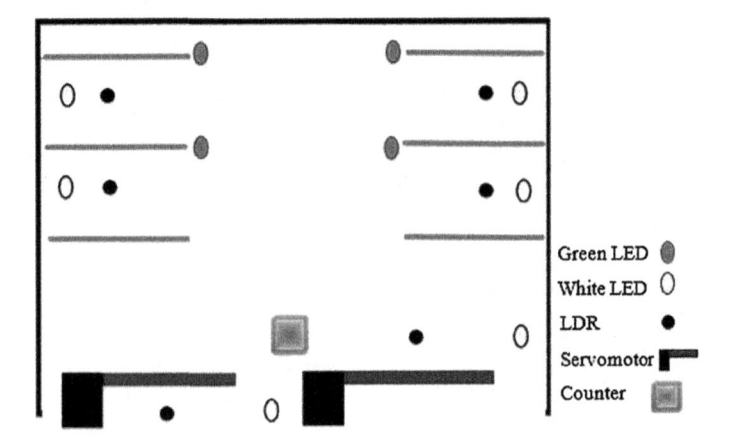

Fig. 1 Layout of parking system

indication to tell the other users that the parking is occupied, while white LED acts as the indicator lamp as the guidance where the user or driver parks the car that is assigned to the parking lot. Also, there are four LDRs installed at each of the parking lot. The LDRs are the input that reacts with the light intensity that changes the output of indicator and parking lamp.

At the entrance and exit gate, there are two LDRs, two servomotors, and one counter near the gates. The LDRs are used to operate the servomotor at both of the gates. If the car blocked the LDR sensor, it gives the signal to Arduino and allows the current to flow from the Arduino to the servomotor. Hence, the servomotor is operated except if the parking is full. The counter is used to count the number of the car entering and leaving the parking space.

4 Methodology

There are five steps used to create the parking system with the completed functions and rules to produce the results.

Identify Input and Output: Once the parking layout has been completely designed, identify all the input and the output based on the parking design itself. The input that is used in this paper is LDRs, and the output is the display, servomotor, and LEDs.

Define Boolean Logic of the Input and Output: Create the table to define Boolean Logic to identify the relationship between the input and output in the circuit. With Boolean Logic, every single sensor can be programmable using this logic. There will be two types of Boolean Logic that will be defined which is non-fuzzy and with fuzzy rule.

Data Collection and Evaluation of Energy Consumption: There are three pre-defined datasets of ten cars in parking system per hour for each dataset with random time duration in the parking lot. For example, Table 1 gives the dataset of energy consumed by parking lamp for non-fuzzy.

Table 1 Dataset of energy consumed by parking lamp for non-fuzzy

No. of car	Arrival time	Departure time	Duration (min)	Parking lot
1	9.00	9.15	15	P1
2	9.10	9.28	18	P4
3	9.20	9.48	28	P3
4	9.25	9.39	14	P2
5	9.30	9.38	8	P1
6	9.35	10.00	25	P4
7	9.40	10.00	20	P2
8	9.45	10.00	15	P1
9	9.50	10.00	10	P3
10	9.55	Null	Null	Full

Result for Non-Fuzzy and Fuzzy Rule: For the fuzzy rule, the programming method used is based on Boolean Logic that is related with fuzzy. The difference is that it consists of more than one declaration of variables which is the input to control whether one or more outputs.

Compare Both Results and Make Analysis: After the result without and with fuzzy rule has been achieved, the comparison will be made on both of them. The analysis will be conducted to compare the results which is the energy consumed by the Arduino boards (sensors) without and with the fuzzy rule.

5 Design and Implementation

5.1 Parking System Operation

Each component gives the response to certain inputs by giving its outputs, and the outputs will be other input sources of other component. These inputs and outputs give a circulation to the system to be more responsive and do the job consequently and stable.

5.2 Design of Prototype

The parking space is designed using the LDR sensors, two types of LEDs which are parking lamp (green LED) and indicator lamp (white LED), servomotors for the gate, and a counter to count the number of car enters and leaves the parking space. Figure 2 depicts example of the operation of the parking system for non-fuzzy.

5.3 Input, Output, and Counter

When designing the parking system, it is essential to identify the input and output that are placed in the parking space. Input is one of the important variables that is measured to determine what the output is. There are six inputs that have been identified when conducting this experiment. These inputs are represented by the sensor called LDRs. In the parking space, they are placed at the entrance gate, exit gate, and at the parking lot. For the output, there are two servomotors, four green LEDs (parking lamp), and four white LEDs (indicator lamp). For each parking lot, one green LED and one white LED are placed. The counter which is a seven-segment display counts the number of car enters and leaves the parking space.

Fig. 2 Operation of the parking system for non-fuzzy

5.4 Boolean Logic for Non-Fuzzy

For non-fuzzy rule, there is no rule applied in this parking system. The user can leisurely enter the parking space and park the car anywhere that the user wants to. If the parking is full, the user cannot enter into the parking space. As for the parking lamp and indicator lamp, all the indicator lamps are on when the parking lot is empty and once the user parked the car in the parking lot, indicator lamp turns off and parking lamp turns on.

5.5 Boolean Logic for Fuzzy Rule

In fuzzy rule, queue method is applied in the parking system which is based on priority of the parking lot. The parking lots are labelled based on their priority as P1, P2, P3, and P4. The priority is based on the first parking lot and so on. When a user enters the parking space, only P1 turns on the indicator lamp. Once the user parked in P1, indicator lamp turns off and parking lamp turns on. Another example is when P1 and P4 are occupied, the next car enters the parking space and there are two parking lots available which are P2 and P3. P2 is chosen based on the priority.

6 Result and Analysis

To measure the energy consumption, INA219 sensor is needed because it can monitor and measure the energy consumed. The results are based on energy consumed by the indicator lamps for all the parking lots. The result of energy consumed is shown in Figs. 3, 4, and 5 as the first, second, and third dataset, respectively.

As in Figs. 3, 4, and 5, the difference of energy consumption for indicator lamps are in a huge gap to each other. This is how the fuzzy rule plays an important role because it made a big change to the indicator lamps operation. Fuzzy rule contributed a lot especially to the management to maintain all the sensors in the parking

Fig. 3 Result for the first dataset

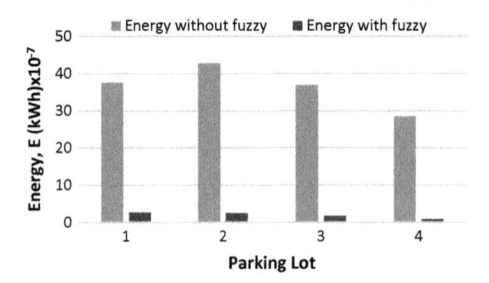

Fig. 4 Result for the second dataset

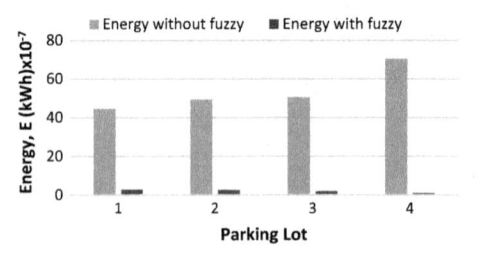

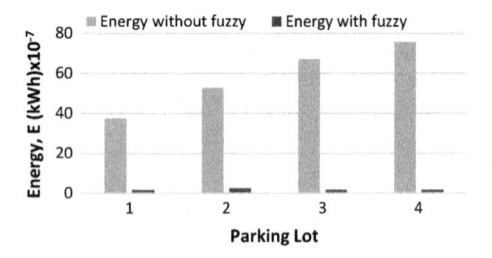

Fig. 5 Result for the third dataset

Energy Consumption of Wireless Sensor Based on IoT ...

Table 2 Total value of energy consumption for non-fuzzy and fuzzy in three datasets

No. of datasets	Energy consumption for non-fuzzy rule (kWh) $\times 10^{-7}$	Energy consumption for fuzzy rule (kWh) $\times 10^{-7}$
1	145.8	7.6
2	214.8	7.6
3	232.9	7.6

lot which can save a lot of money, time, and energy. For the calculation, Table 2 refers to the total value of energy consumption for non-fuzzy and the total result of energy consumption for fuzzy rule in three datasets.

$$\text{Average energy for non} - \text{fuzzy} = (145.8 + 214.8 + 232.9)/3 = 197.83 \text{ kWh}$$

$$\text{Average energy for fuzzy} = (7.6 + 7.6 + 7.6)/3 = 7.6 \text{ kWh}$$

$$\text{Total Energy used} = 197.83 + 7.6 = 205.43 \text{ kWh}$$

$$\text{Energy reduction} = 197.83 - 7.6 = 190.23 \text{ kWh}$$

$$\text{Percentage, } \% \text{ for energy reduction} = 190.23/205.43 = 92.6\%$$

The average value of energy for non-fuzzy and fuzzy rule in all three datasets is 197.83 and 7.6 kWh. In terms of energy optimization, the percentage by using fuzzy rule to non-fuzzy is 92.6% which means that fuzzy rule saves about 92.6% energy compared to non-fuzzy.

7 Conclusion and Future Works

To conclude, this paper has presented a fuzzy rule for a simple parking system integrated with IoT components. Results show that with the help of fuzzy rule, it creates more organized ways to design a control system of a parking system. The layout of the parking system and its IoT components and fuzzy rule can be applied to optimize energy consumption. The energy saved by 92.6% for a simple layout shows that it can save more energy if applied to bigger parking system.

In future, adding more advanced sensors that are able to detect the incoming target in more efficient ways should be considered so that the sensors will be able to collect the data more accurate and systematic.

Acknowledgements This work is also supported by Malaysian Ministry of Higher Education (MOHE) Fundamental Research Grant Scheme with UTM reference number R. J130000.7828.4F860.

References

Ada L. (2016), Adafruit INA219 Current Sensor Breakout, Adafruit Indutries.

Baronti, P., Pillai, P., Chook, V. W. C., Chessa, S., Gotta, A., & Hu, Y. F. (2007). Wireless sensor networks: A survey on the State of the Art and the 802.15.4 and ZigBee standards. *Computer Communications, 30*(7), 1655–1695.

Fraifer, M., & Fernström, M. (2016). Smart car parking system prototype utilizing CCTV nodes: A proof of concept prototype of a novel approach towards IoT-concept based smart parking. In *IEEE 3rd world forum on internet of things (WF-IoT)*, 12–14 Dec, pp. 649–654.

Robert, D. (2005). Photoresistors (Photoconductive Cells). *Electronic Devices: System and Applications, 1,* 497–498.

Sachdeva, E., Porwal, P., Vidyulatha, N., Shrestha, R. (2016). Design of low power VLSI-architecture and ASIC implementation of fuzzy logic based automatic car-parking system. In *IEEE annual india conference* (INDICON), pp. 1–6.

Zadeh, L. A. (1965). Fuzzy sets. *Information and Control, 8*(3), 338–353.

Zhang, Z., Li, X., Yuan, H., Yu, F. (2013). A street parking system using wireless sensor networks. *International Journal of Distributed Sensor Networks, 9*(6).

Forecasting River Water Level Using Artificial Neural Network

Norwaziah Mahmud, Nur Syuhada Muhammat Pazil, Izleen Ibrahim,
Umi Hanim Mazlan, Siti Hafawati Jamaluddin
and Nurul Zulaikha Othman

Abstract Flood is a natural phenomenon that occurs without warning. The occurrence of flood can cause many damages in terms of property, money, and even death. In order to prevent more loss, an early warning system of the increment of river water level needs to be forecasted in a more accurate and effective way to predict the best result. In this study, Artificial Neural Network (ANN) is implemented in order to forecast the river water level in Pelarit River, Perlis. The objectives of this study are to determine the efficiency of ANN in forecasting the river water level and to forecast the water level of Pelarit River for one week in advance. This study uses Multilayer Feed Forward Artificial Neural Network with two input neurons, seven hidden neurons, and one output neuron in order to forecast the water level. Three different types of algorithms, namely quick propagation, conjugate gradient descent, and Levenberg–Marquardt algorithms, are compared to get the most significant result in forecasting one week of river water level. Among all the algorithms used, conjugate gradient descent is proved to be the most reliable value to forecast the Pelarit River in terms of the lowest value of Root-Mean-Square Error and the highest value of correlation.

Keywords Flood · River water level · ANN · Forecasting

1 Introduction

Flood is natural disaster that usually happens in many countries around the world including Malaysia. In Malaysia, out of 189 river basins, 89 are located in Peninsular Malaysia, 78 in Sabah, and 22 in Sarawak with the main channel flows directly to the South China Sea. The types of flood in Malaysia can be categorized into flash flood and monsoon flood. In December 2005 and April 2006, the worst

N. Mahmud (✉) · N. S. Muhammat Pazil · I. Ibrahim · U. H. Mazlan
S. H. Jamaluddin · N. Z. Othman
Universiti Teknologi MARA Cawangan Perlis, Kampus Arau, Perlis, Malaysia
e-mail: norwaziah@perlis.uitm.edu.my

flood occurred in Perlis after 30 years where two-thirds of Perlis were affected (Lawal et al. 2014). In addition, flood that occurred in Perlis in the year 2010 had caused a damage worth RM 34,747 224 (D/iya et al. 2014).

Every river in Perlis has different water level depending on their own depths taken from the sea level. The function of water level system is to produce a sign of immediate action that needs to be taken by authorized organizations when heavy rainfalls occurred in certain areas. It is important to analyze the accuracy of the river water level in order to take an immediate action to prevent more loss and damage. Therefore, the main purpose of this study is to apply a forecasting technique to predict the river water level in Pelarit River at Kaki Bukit in Perlis.

2 Artificial Neural Network (ANN) Modelling

Artificial Neural Network (ANN) is created based on the description of behavior on human's perceptions into a simplified computational system or mathematical model. ANN is commonly used in forecasting field as it is suitable to be implemented for problems that have complete data and observation but the solutions require specific information that is difficult to be identified (Suliman et al. 2013). It is also a good model in forecasting because of its ability in solving dynamic nonlinear time series problems (Bustami et al. 2007).

Basically, this model includes three types of layer, namely input layer, hidden layer, and output layer, in which all the layers are interconnected among the simple elements called artificial neurons (Alvisi et al. 2006). The input layer consists of number of neurons or nodes where all the neurons will work together in a parallel form to generate information for the final output layer (Kia et al. 2012). The parameters in input layer depend on the received data obtained from different sources, or it may come from the investigators' or researchers' opinion. The input layer may also contain the influences of the phenomenon that are being investigated (Sztobryn 2013). For the hidden layer, the number of neurons is processed as a trial and error procedure in order to search for the lowest number of neurons without disturbing the model efficiency. Each of the hidden neurons will respond according to the neuron connection via input layer. All the elements inside the hidden layer will transform into the nonlinear transfer function (Kia et al. 2012).

3 Methodology

3.1 Data Acquisition

This study will be using the river water level of Pelarit River that was recorded daily from January 1, 2012, to December 31, 2014. The recorded data were obtained from the National Hydrological Network Management System (SPRHiN). The water level system is categorized into four types, namely normal level, alert level, warning level, and danger level. According to the Department of Irrigation and Drainage of Perlis, for Pelarit River, the normal level is 35.60 m, whereas the maximum water level which is also the danger level is 39.00 m. The water level reaches the alert level when the water level rises to 38.60 and 38.72 m is the warning level.

3.2 Data Preprocessing

Data preprocessing is implemented when there exists a complex data such as non-linearity and non-stationary data which cause difficulty in forecasting the case study using the chosen models. In ANN, preprocessing is a modification of the data before it can be used in the neural network, and it is a data transformation to develop a suitable neural network. Furthermore, data preprocessing also improves the quality of data because it involves filtering the outliers and approximating the missing values. The activities involved in data preprocessing are data cleaning, data reduction, outlier, and data normalization.

3.2.1 Data Cleaning

In order to obtain unbiased result, missing value must be overcome first. To solve the problem of missing value in the data, this study will be using average method. The average of the daily river data is calculated separately for every month from 2012 to 2014 using Waikato Environment for Knowledge Analysis (WEKA) to replace the missing value.

3.2.2 Data Reduction

The total of 1096 of daily water level data will be reduced to the average of 156 weeks in order to forecast the average of one week of river water level. The following formula will be applied for the data reduction:

$$\text{Weekly Data}_k = \sum_{n=0}^{6} \text{Raw Water Level Day}_{n+1}/7, \qquad (1)$$

where $k = 1, 2, 3, \ldots, 156$ and $n = 0, 1, 2, \ldots, 1095$.

3.2.3 Outlier

After the data reduction, the presence of the outliers has been detected at week 144 and week 155 where the values of the average water level are the highest. For this study, the numeric outlier is substituted by median since it will not be affecting the forecasting of one week of river water level.

3.2.4 Data Normalization

Data normalization is used to scale the data into (0, 1) and (−1, 1). The input of the average water level data at time t and the lag average water level at time $t-1$ will be normalized using the formula in Eq. (2) (Gazzaz et al. 2012), while the output of the next water level will be using logistic sigmoid activation function as in Eq. (3).

$$X_S = [(b-a)*(X_0 - X_{min})/(X_{max} - X_{min})] + a, \qquad (2)$$

where X_s and X_0 express the normalized and raw observations of parameter X, respectively, while a and b represent the lower and upper limits of the normalization range. X_{min} and X_{max} are minimum and maximum values of parameter X.

$$y = \frac{1}{(1+e^{-kx})}, \qquad (3)$$

where y is the sigmoid value, k is the sigmoid steepness coefficient, and x represents the data value, namely the total number of the input and weight values.

3.3 Network Design

Five possible network architectures are verified using the Artificial Neuro-Inteligence software. Later on, the architecture design is chosen based on the highest number of correlation coefficient, r, where the closer the value of r to 1, the better the network is. Figure 1 shows the best architecture design for this study that consists of two neurons of input layer, seven neurons in hidden layer, and one neuron for output layer, (2–7–1).

3.4 Training an ANN Using Algorithms

This study chooses three different algorithms, namely quick propagation, conjugate gradient descent, and Levenberg–Marquardt, in order to identify the best algorithms that can produce better results. For this study, the number of iteration that represents

Fig. 1 Best architecture design for Pelarit River

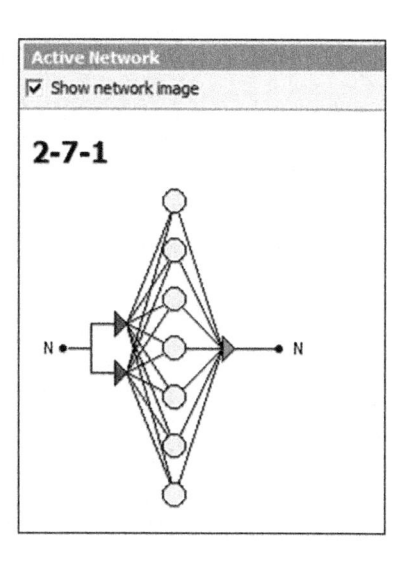

a complete presentation of the training set to the network training is 500. Furthermore, the value of 0.001 is chosen as the mean square error (MSE) as the stopping condition for the over-fitting and minimum error.

3.5 Testing the Algorithms

After the training process is completed, the forecasted values and the actual values are verified. Based on the acquired results, it shows that the conjugate gradient descent algorithm is the most efficient algorithm in forecasting the water level for Pelarit River.

3.6 Performance Evaluation

The performance of the actual water level and the corresponding neural network is evaluated by using different measurements such as Root-Mean-Square Error, RMSE (Eq. 4), and Nash–Sutcliffe, NS (Eq. 5). RMSE describes the average magnitude of error of the observed and predicted values, while NS coefficient of efficiency is widely used to describe the forecasting accuracy (Sulaiman et al. 2011).

$$\text{NS} = 1 - \frac{\sum_{i=1}^{N}(O_i - F_i)^2}{\sum_{i=1}^{N}(O_i - \bar{O})^2}, \tag{4}$$

$$\text{RMSE} = \sqrt{\frac{\sum_{i=0}^{N}(O_i - F_i)^2}{N}}, \tag{5}$$

where O_i is represented as an actual value, F_i is represented as a predicted value, $\bar{O}$ is mean of the actual value, and N is the number of data being evaluated. Error is calculated from the differences between the actual and predicted values. The formula is given as the following:

$$\text{Error} = (O_i - F_i), \tag{6}$$

where O_i is an actual value and F_i is a predicted value. The values obtained indicate the accuracy of the forecasting results for the developed model.

4 Results and Discussions

4.1 Analysis of Training Algorithms

Based on the training process from quick propagation algorithm, conjugate gradient descent algorithm, and Levenberg–Marquardt algorithm as shown in Table 1, the best result for the training process in this study is conjugate gradient descent

Table 1 Results of the training process for quick propagation, conjugate gradient descent, and Levenberg–Marquardt algorithms

Algorithm						
Parameter	Quick propagation		Conjugate gradient descent		Levenberg–Marquardt	
	Training	Validation	Training	Validation	Training	Validation
Absolute error	0.1040	0.1131	0.0944	0.1522	0.1686	0.1503
Network error	0.0085	0	0.0077	0	0.0152	0
Error improvement	0.000006		0.000003		0.000321	
Correlation coefficient, r	0.8482	0.8307	0.8487	0.8463	0.8408	0.8412
R-squared	0.5994	0.6430	0.5791	0.6782	0.2651	0.3397

algorithm with the lowest value of absolute error and the highest value of correlation that indicate that there is a strong positive relationship between the actual values and the network outputs. Apart from that, conjugate gradient descent also has the smallest network error compared to others.

Based on the result, the value of network error of conjugate gradient descent algorithm in the training set is 0.0077 which is higher than the network error in the validation set, 0, and this proves that the overtraining is being controlled during the iterations. Overtraining usually occurs when the value of network error in validation set is increasing, while the value of the network error in training set is decreasing. Moreover, the lowest value of the error improvement for conjugate gradient descent algorithm also indicates the ability of the neural network to improve the forecast value.

4.2 Analysis of Performance Algorithm

The performance of the actual river water level and the forecasted is evaluated by using the Root-Mean-Square Error (RMSE) and Nash–Sutcliffe (NS) as shown in Table 2. Based on the table, it shows that the value of NS for each of the algorithms reached the optimal value. However, RMSE of conjugate gradient descent algorithm has the minimum value compared to others.

4.3 Analysis of River Water Level Forecasting

Table 3 shows the forecast value of water level for Pelarit River in week 157 for conjugate gradient descent algorithm. Based on the result, the average one week ahead shows a decreasing pattern from the previous week.

Table 2 RMSE and NS for quick propagation, conjugate gradient descent, and Levenberg–Marquardt algorithms

	Quick Propagation	Conjugate gradient descent	Levenberg–Marquardt
RMSE	0.1862	0.1860	0.2098
NS	1.000	1.000	1.000

Table 3 Forecast value for the average of one week ahead

Week	Average water level (m)
156	36.13
157	35.98

5 Conclusion

In conclusion, the conjugate gradient descent algorithm is proved to be the most reliable value to forecast the water level of Pelarit River. The forecast value for week 157 is 35.98 m which is lower than 36.13 m in week 156. Therefore, the forecast value indicates that the water level is at a normal level. It is hoped that this study can be applied to forecast river water level in various areas that have risk of flood. In future, it is also recommended to consider other variables such as weather, river water level from the nearest river, and time taken in collecting the data.

References

Alvisi, S., Mascellani, G., Franchini, M., & Bárdossy, A. (2006). Water level forecasting through fuzzy logic and neural network approaches. *Hydrology and Earth System Sciences, 10*(1), 1–17.

Bustami, R., Bessaih, N., Bong, C., & Suhaili, S. (2007). Artificial neural network for precipitation and water level predictions of Bedup River. *IAENG International Journal of Computer Science, 34*(2), 228–233.

D/iya, S. G., Gasim, M. B., Toriman, M. E., & Abdullah, M. G. (2014). Floods in Malaysia historical: Reviews, causes, effects and mitigations approach. *International Journal of Interdisciplinary Research and Innovations, 2*(4), 59–65.

Gazzaz, N. M., Yusoff, M. K., Aris, A. Z., Juahir, H., & Ramli, M. F. (2012). Artificial neural network modelling of the water quality index for Kinta River (Malaysia) using water quality variables as predictors. *Marine Pollutin Bulletin, 64*(11), 2409–2420.

Kia, M. B., Pirasteh, S., Pradhan, B., Mahmud, A. R., Sulaiman, W. N. A., & Moradi, A. (2012). An Artificial Neural Network model for flood simulation using GIS: Johor River Basin, Malaysia. *Environmental Earth Sciences, 67*, 251–264.

Lawal, D. U., Matori, A. N., Yusuf, K. W., Hashim, A. M., & Balogun, A. L. (2014). Analysis of the flood extent extraction model and the natural flood influencing factors: A GIS-based and remote sensing analysis. *IOP Conference Series: Earth and Environmental Sciences, 18*, 1–6.

Sulaiman, M., El-Shafie, A., Karim, O., & Basri, H. (2011). Improved water level forecasting performance by using optimal steepness coefficients in an artificial neural network. *Water Resources Management, 25*, 2525–2541.

Suliman, A., Nazri, N., Othman, M., Malek, M. A., & Ku-Mahamud, K. R. (2013). Artificial neural network and support vector machine in flood forecasting: A review. In *Proceeding of the 4th international conference on computing and informatics*, pp. 237–332.

Sztobryn, M. (2013). Application of Artificial Neural Network into the water level modelling and forecast. *The International Journal on Marine Navigation and Safety of Sea Transportation, 7*(2), 219–223.

Optimization of Melt Filling Distribution for Multiple Gate System

Saiful Bahri Mohd Yasin, Sharifah Nafisah Syed Ismail,
Zuliahani Ahmad, Noor Faezah Mohd Sani
and Siti Nur Asiah Mahamood

Abstract In this study, five virtual trials have been conducted on a single cavity of hard disk frame incorporated with a multiple gates (known as spoke/spider gate) under varies of runner diameter modification. Grow from and weld line results were recorded in relation to a balanced of melt filling distribution. In the grow from analysis, the difference percentage between maximum and minimum of melt filling distribution for the 1st, 2nd, 3rd, 4th, and 5th trials was found to be 60, 38.69, 28, 22.30, and 7.35%, respectively. In the weld line analysis, the length of weld lines for the 1st trial to 5th trial was changed from 17.5 to 8.5 cm, respectively, where a reduction of weld line length reduced to 51.43%. Then, both results show that a balanced of grow from leads to a balanced of melt filling distribution and a reduction in the existing of weld line length on the product surface.

Keywords Injection molding · Feeding system · Runner diameter
Grow from · Flow front · Weld line · Overpacking · Underpacking
Flashing · Sink mark

1 Introduction

Injection molding is one of the most widely used manufacturing processes in producing thermoplastic parts. It consists of filling, packing, and cooling stages. In filling stage, in order to yield a high-quality product, a molten polymer must fill the

---

S. B. Mohd Yasin (✉) · S. N. Syed Ismail · Z. Ahmad · S. N. A. Mahamood
Polymer Technology Department, Faculty of Applied Sciences,
Universiti Teknologi MARA Cawangan Arau Perlis, Arau, Perlis, Malaysia
e-mail: sbmyasin@gmail.com

N. F. M. Sani
Faculty of Applied Sciences, Universiti Teknologi MARA, Perak Branch,
Tapah Campus, Tapah Road, 35400 Shah Alam, Perak, Malaysia
e-mail: faezah125@uitm.edu.my

cavity with minimizing weld lines, avoid air entrapments and other defects (Huszar et al. 2016). Nowadays, it is necessary to carry out a simulation process in reducing molding defects and as a continuous process to achieve dimensional stability and high mechanical properties for a molded product.

In the injection molding process, some problems occurred due to the product that made from polymeric materials, for instance, molding underpacking, overpacking, and prone to surface defects that can avoided only if material flow smoothly (Sabrina et al. 2015) and evenly. Thus, the simulation processes can be an early step before starting manufacturing process to ensure the molded product fully optimized, such as to run flow analysis in order to imitate the actual molding process. Flow analysis calculates the flow front advancement that grows through a cavity space to form a molded product, in which the melt comes from single location or multiple injection locations. It also provides the ability to simulate a uniform distribution of the melt filling inside the cavity space in order to reduce molding underpacking and overpacking issues. As a consequence, the molding defects such as flashing, dimensional instability, and sink mark can be avoided.

Furthermore, the common defect in injection molding is the presence of weld lines. Weld lines can compromise the appearance and strength of the plastic parts. Flow analysis can be used to predict weld line, temperature distribution (to determine visibility of weld line), and meeting angle prediction (to distinguish weld line and melt line). In eliminating weld line, the meeting angle can enlarge by optimizing wall thickness, change gate, and runner design, reduce the temperature distribution, and then re-validate the result by using flow analysis (Han and Im 1997). However, in multiple gate system, the weld line cannot be avoided and acts as a potential defect.

Thus, this analysis emphasizes on the balancing of material distribution in the multiple gate system by modifying the runner diameter size. In addition, multiple gates will create a lot of weld line due to the flow front comes many gates location. This research is also carried to identify the weld line distance before and after balancing the melt filling. CAE analysis technique is used as a part of a scientific approach to get an optimum parameter setting as well as an understanding in balancing the flow of polymeric materials.

2 Methodology

2.1 Material Description

Polypropylene CM 10902 was selected to be used in this simulation process. The injection pressure for this material was set at 180 MPa. The processing parameter of melt temperature was set at 220.00 °C (with range of minimum 180 °C and maximum 260 °C). The mold temperature was set at 50.00 °C (with range of minimum 20 °C and maximum 80 °C). The mechanical property of the material for elastic modulus first principle direction is 1574.77 MPa and the second principle direction is 1530.86 MPa. The Poisson's ratio (v12) is found to be 0.358, and the Poisson's ratio (v23) is 0.44. The shear modulus of polypropylene is 523.9 MPa.

2.2 Product Design and Modeling

In this study, a plastic hard disk frame was designed using CAD software as illustrated in Fig. 1. The dimension of hard disk frame was 141 mm (length) 125 mm (width) × 3 mm (thickness). The suitability of the product design is to provide a square hollow space at the center of product and significant to create a multiple point gate. This gate type is also known as spoke or spider gate, where its disadvantage is possibility of the weld lines.

Upon completion design process, the design was imported as STereoLithography (STL) model that can be used in the computer-aided engineering (CAE) software, for instance, Moldflow Advisor. Based on Fig. 2, there are

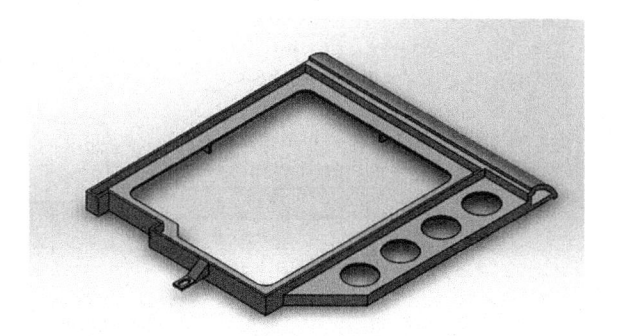

Fig. 1 Modeling of hard disk drive

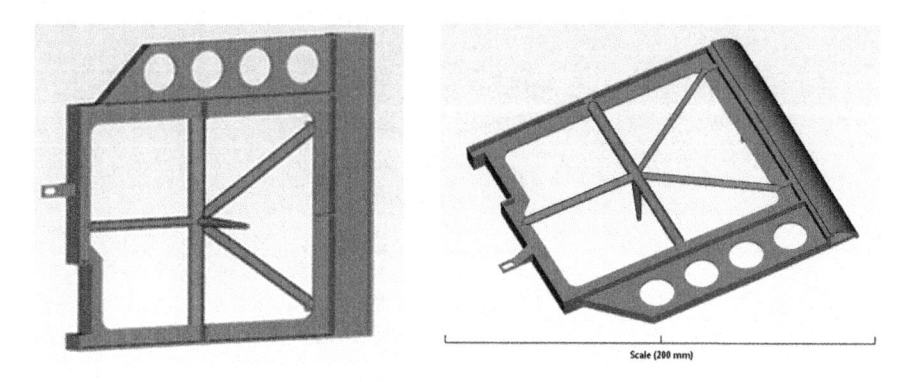

Fig. 2 Modeling of hard disk drive: front view (left) and bottom view (right)

five injection locations as a part of feeding system that are present inside of the .STL model, where it is capable to fill a large product and increase the filling time. The feeding system contains a sprue gate at the center of the model, connected with five different diameters of the runner and gates was set at the end of runner to allow the melt fills inside the cavity space. There are two main considerations in designing feeding system: (1) runner system designed with varied diameter to allow the modification in obtaining a balance of the melt filling and (2) the gates designed with a fixed design to ensure no variation in the gate sealing time and reduce the warpage issue.

Upon completion of product modeling with the feeding system, the filling analysis was conducted in order to determine grow from analysis for each gates and presence of weld lines. Research layout was simplified as sequential processes of flow analysis as shown in Fig. 3. The analysis was repeated by adjusting the size of runners until a balanced of the melt filling obtained. The grow from (visualized in color form) and weld line results were taken to make the comparison for each gate. The areas of grow from were calculated in order to determine the distribution of

Fig. 3 Flowchart in grow from analysis

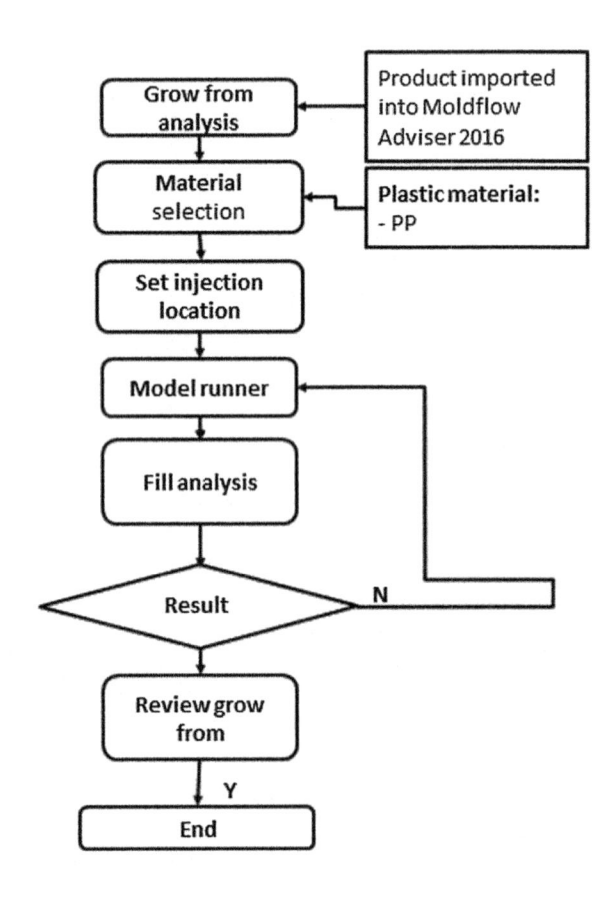

materials fill in cavity for each gates, while the weld lines occurred were measured in order to determine the length of each weld line.

3 Results and Discussion

A series of analyses were conducted to demonstrate the distribution of materials fill in cavity to form grow from and weld lines as illustrated in Fig. 4. The analyses were conducted by modifying the runner diameter size until obtaining a balanced

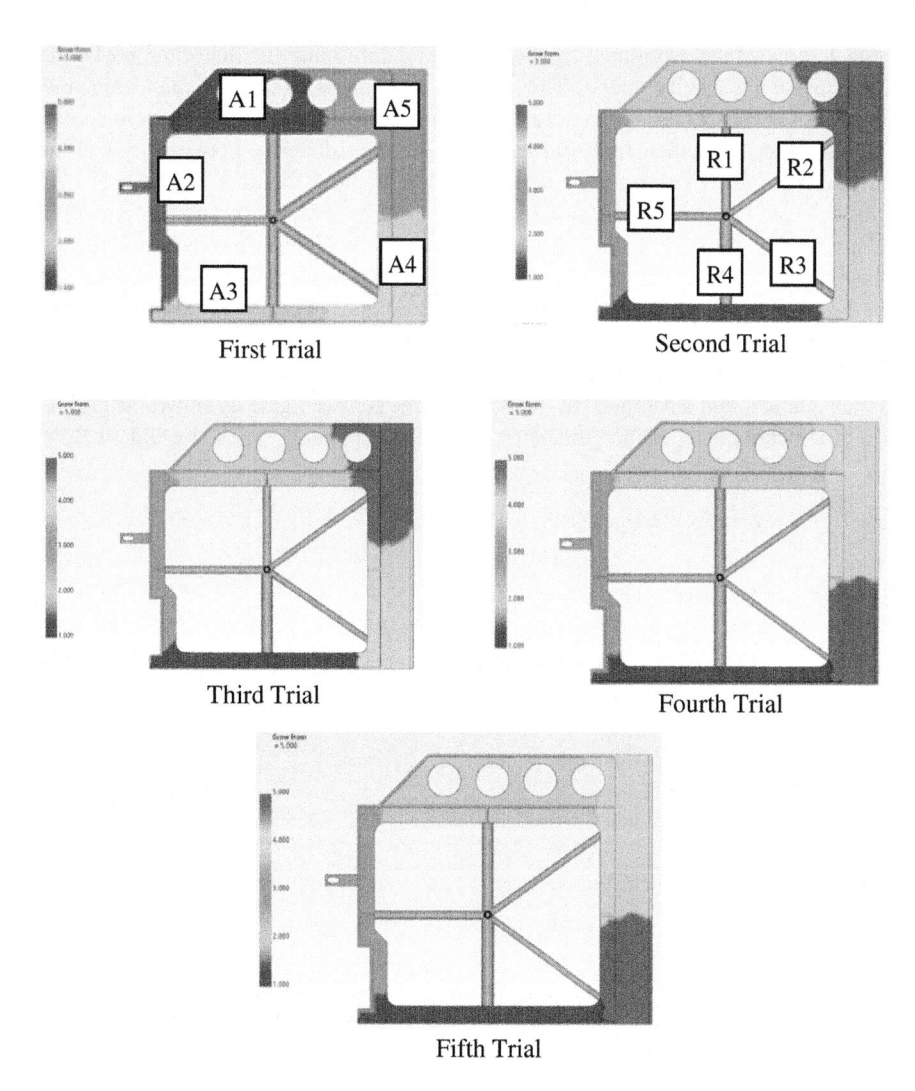

Fig. 4 Grow from result

grow from as shown in fifth trial. In addition, the modeling is labeled with A1, A2, A3, A4, and A5 as shown in first trial to indicate the area for each grow from that comes from different gate locations. These indications are used to make comparison of the melt filling distribution for each region, while R1, R2, R3, R4, and R5 indicate the runner diameter sizes that change upon the runner modification in order to obtain a balanced grow from.

3.1 Effect of Runner Sizes on the Grows from

Table 1 shows grow from area for each gate in determining the distribution of melt filling inside the cavity space. The area of each grow from was calculated and compared for each trial. Upon completion of fifth trial, the area of grow from shows a significance better than first trial by obtaining the difference between maximum and minimum grow from area, where the difference was found to be 7.4 cm^2 (1st trial), 5.3 cm^2 (2nd trial), 4.2 cm^2 (3rd trial), 3.3 cm^2 (4th trial), and 1.0 cm^2 (5th trial). Clearly, the finding indicates that the melt distribution has become almost balanced upon the runner diameter modification. Thus, the defects such as parting line flashing, sink mark, and others can be avoided due a reduction of the grow from differences.

Table 2 shows the runner diameter modification for the grow from optimization. R1, R2 R3, R4, and R5 represent the runner sizes for each gate as shown in Fig. 4. The modifications of runner diameter were done manually with the aid of flow

Table 1 Grow from area for each gate

Trial		1st trial	2nd trial	3rd trial	4th trial	5th trial
Grow from (cm^2)	A1	15.4	16.8	14.3	12.3	12.8
	A2	10.0	10.3	10.8	11.5	13.1
	A3	8.0	11.5	11.5	13.0	13.6
	A4	12.0	13.0	14.0	13.8	12.6
	A5	20.0	13.8	15.0	14.8	13.2
Distribution of the material		Poor	Poor	Poor	Moderate	Good

Table 2 Runner diameter modification for grow from optimization

Runner size and location		1st trial	2nd trial	3rd trial	4th trial	5th trial
Runner size (mm)	R1	6.5	4.5	4.5	4.5	5.5
	R2	5.0	3.2	3.2	3.2	3.2
	R3	5.0	3.5	4.0	3.0	3.0
	R4	6.0	6.0	6.0	6.0	6.0
	R5	4.0	4.0	4.0	4.0	4.0

Optimization of Melt Filling Distribution for Multiple Gate ... 427

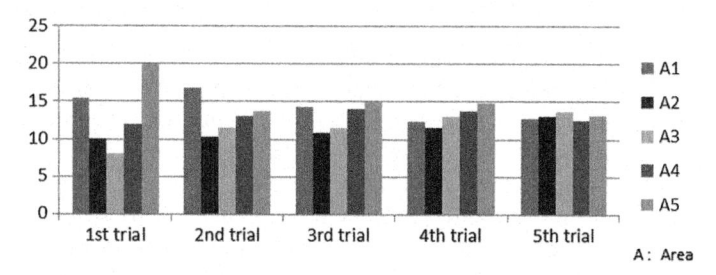

Fig. 5 Distribution of the melt filling in each grow from area

analysis. The series of analyses were carried out until 5th trial, where it is found to give almost balanced for each grow from.

The results of grow from optimization in determining a balanced melt filling of hard disk drive were tabulated in a bar chart as illustrated in Fig. 5. Based on the bar chart, the optimized grow from area for each gate upon completion of the runner modification is the 5th trial with respect to other trials. The optimization of grow from provides a balanced filling behavior of the thermoplastic material in the mold cavity during filling phase. In order to achieve the balanced filling, each gate must provide an equal amount of material into the cavity. Thus, the distribution of material from each gate must be nearly balanced as shown in 5th trial.

Feeding system for the 1st trial was designed and simulated with manually setting up the runner diameter size. As a result, each grow from areas are found to be unbalanced; this could lead to defects in the actual molding processing, for instance, overpacking and underpacking issues. The unbalanced grow from areas are related to the distribution of material filled in cavity. The unbalanced filled within the product indicates that the material is poorly distributed from each gate in the multiple gate system. Therefore, some modification of runner diameter was carried out in order to balance the flow of material within the cavity for the next trial. As a consequence, the grow from area of each gate for 2nd, 3rd, and 4th trials was improved due to influence of changes in runner diameter size. Thus, the distribution of the material filled inside the cavity was also improved, where the difference percentage between maximum and minimum of melt distribution for the 1st, 2nd, 3rd, and 4th was found to be 60, 38.69, 28, and 22.30%, respectively. Upon reaching the 5th trial, the grow from areas are almost balanced, where the difference percentage between maximum and minimum of melt distribution was measured to be 7.35%.

3.2 Effect of Weld Lines on the Grow from Optimization

One of significant effects of the grow from optimization is the reduction of weld line length. Based on the graph as shown in Fig. 6, it shows a decreasing trend of weld line length for each trial. In comparison, the weld lines of the first trial to fifth trial were changed from 17.5 to 8.5 cm, respectively. It shows a reduction of weld line length reduced to 51.43%. This occurs due to the adjustment of the runner diameter sizes. Weld lines are unavoidable in spoke or spider gates. In fact, they offer the areas with local weakness in strength due to lack of polymer chain entanglement at the merging flow fronts (Cho et al. 1997). However, the effect of weld line can be minimized by considering the balanced of grow from inside the cavity space. Based on this study, the flow pattern changed upon modifying the runner diameter as illustrated in Fig. 7. Therefore, the finding suggests that the feeding system can be re-designed to influence the position of weld lines in a balanced filling and in avoiding the defect such as flashing, sink mark, and others.

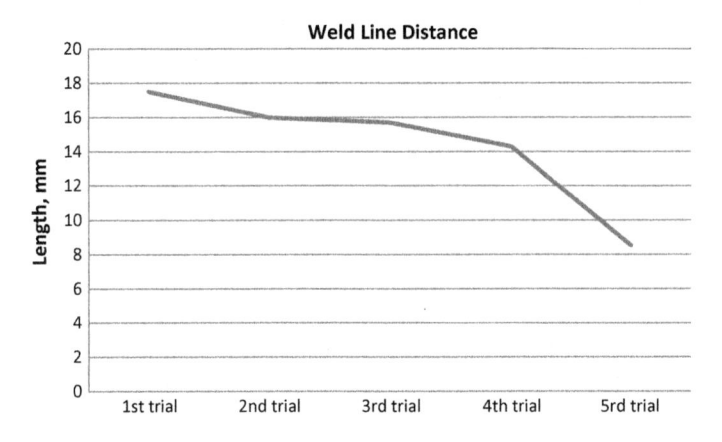

Fig. 6 Length of weld line presence in each of the trials

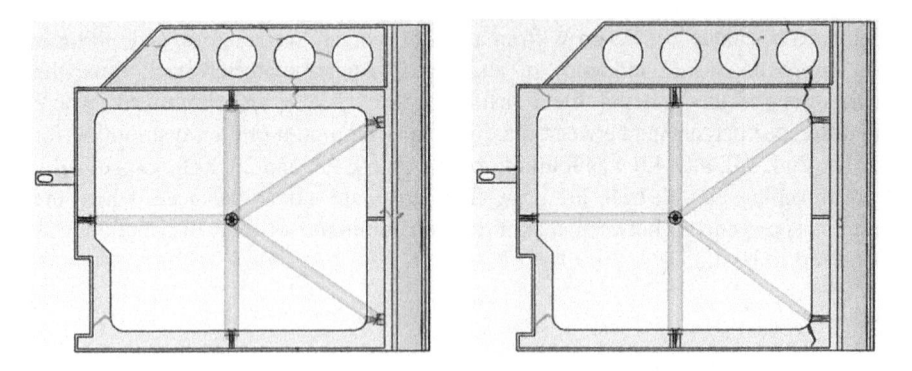

Fig. 7 Comparison weld line on the product surface: 5th trial (left) and 1st trial (right)

4 Conclusion

In this study, five virtual trials of the grow from were successfully conducted in order to determine the runner diameter size for a balanced material distribution in multi-gated system (spoke/spider gate) and reduce the weld line length. The minimum and maximum of grow from for the 1st, 2nd, 3rd, 4th, and 5th trials were found to be 60, 38.69, 28, 22.30, and 7.35%, respectively. For the last trial (5th trial), it gave almost balanced of the grow from and leads to a uniform of melt filling distribution. Furthermore, in the weld line calculation, it shows a reduction in weld line length for each trial. For the 1st trial, the length of weld line was measured to be 17.5 cm; meanwhile, the length of 5th trial was found to be 8.5 cm, and thus, the weld line reduction was estimated to reduce approximately 51.43%.

In future work, it is recommended to precede this simulation to the actual molding process and compare the actual processing result with respect to simulation result in order to obtain the coefficient of error that produced by simulation process. Otherwise, it is also recommended that to study on how the weld line length could affect to the product stiffness, where it could be done using finite element analysis (FEA). This is because the current computer-aided engineering (CAE) software cannot simulate and integrate the presence of weld line on the product surface with the possible load to be applied. Thus, it is quite difficult to determine product stiffness with effect of weld line unless to do so in real product testing which might be costly.

References

Cho, K., Ahn, S., Park, J., Park, C. E., & An, J. H. (1997). Evaluation of the weld-line strength of thermoplastics by compact tension test. *Polymer Engineering & Science, 37*(7), 1217–1225.

Han, K., & Im, Y, (1997). Compressible flow analysis of filling and post-filling in injection molding with phase-change effect, *38*(l), 179–190.

Huszar, M., Belblidia, F., Alston, S., Wlodarski, P., Arnold, C., Bould, D., et al. (2016). The influence of flow and thermal properties on injection pressure and cooling time prediction. *Journal of Applied Mathematical Modelling, 40,* 7001–7011. https://doi.org/10.1016/j.apm.2016.03.002.

Sabrina, M., Adriano, F. S., Jackson, M., & Ihar, Y. (2015). Design of conformal cooling for plastic injection moulding by heat transfer simulation. *Polímeros*. https://doi.org/10.1590/0104-1428.2047.

Improvement on Ride Comfort of Quarter-Car Active Suspension System Using Linear Quadratic Regulator

Sharifah Munawwarah Syed Mohd Putra, Fitri Yakub,
Mohamed Sukri Mat Ali, Noor Fawazi Mohd Noor Rudin, Zainudin
A. Rasid, Aminudin Abu and Mohd Zamzuri Ab Rashid

Abstract The goal of this paper is to investigate the performance of an active suspension system via linear quadratic regulator (LQR) control and proportional–derivative–integral (PID) control. This project presents the mathematical models of the two degrees of freedom of a quarter-car active suspension system. This project introduces the design of a controller performance used for an active suspension system. The equations of motion of the quarter-car active suspension system model are developed. In the passive suspension system, there are huge oscillations or vibrations that occur in the suspension system. This phenomenon will lead to uncomfortable ride among the passengers or the driver. Besides, it takes longer time to reduce the vibration. Therefore, a good controller design must be able to reduce the vibration and produce a fast settling time. This project is focused on designing a controller for active suspension system by using MATLAB and Simulink software for both PID and LQR controllers to enhance the performance of ride comfort. This research also aims to study the effect of disturbances such as road bump and holes to the response time of the vibration of the vehicle. The result shows that the response of LQR control gives the best output performances in minimizing the vibration and gives faster settling time than that of the PID control.

Keywords Ride comfort · Active suspension · Quarter-car model
LQR controller · PID controller

S. M. Syed Mohd Putra · F. Yakub (✉) · M. S. Mat Ali · N. F. M. N. Rudin ·
Z. A. Rasid · A. Abu
Malaysia-Japan International Institute of Technology, Universiti Teknologi
Malaysia, Jalan Sultan Yahya Petra, 54100 Kuala Lumpur, Malaysia
e-mail: mfitri.kl@utm.my; irtif81@gmail.com

M. Z. A. Rashid
Faculty of Electrical Engineering, Universiti Teknikal Malaysia Melaka,
Hang Tuah Jaya, 76100 Durian Tunggal Melaka, Malaysia

1 Introduction

A car suspension system is the mechanism that splits the car body from the tire of the car. The suspension system is an important aspect of car design because it affects both the ride comfort and safety of the passengers. There are two main element that any automotive suspension systems need which are to provide a comfortable ride and good handling when random disturbances from road unevenness upon the running vehicle. The random disturbances are bump and holes. Passenger comfort can be known the reduction of car body acceleration or as car body vertical displacement minimization, while good handling can be characterized as reduction of wheel acceleration. Besides, the role of car suspension is also to provide excellent road holding for a multiple of road conditions (Ghosh and Nagraj 2004; Coelho and Pereira 1986).

Suspension system that uses combined sets of springs and dampers is passive suspension system to reduce the vibration which is broadly used on current vehicle suspension system. The springs are very flexible elements that have the ability to store the energy applied in the form of loads and deflection. Besides, when the spring is bended or compressed to a shorter length, it should be able to absorb the energy (Rao 2000).

Three types of suspension systems exist in automotive suspension which are passive suspension system, active suspension system, and semi-active suspension system. Passive suspension system is a conventional suspension system that includes a spring fixed parallel with a damper, located between unsprung mass and sprung mass (Ghazaly and Moazz 2014; Yagiz and Hacioglu 2008). The spring and the damper have constant characteristic. Therefore, this passive suspension does not involve mechanism for feedback control. Meanwhile, the semi-active suspension system has the same elements of conventional system, but the damper has two or more selectable damping rates and needs a high force at low velocities and a low force at high velocities, and is able to move rapidly between the two. An active suspension has an actuator that allows the improvement to the passenger comfort when this element is placed in parallel to the damper and the spring between the car body and the wheel (Alvarez-Sanchez 2013).

The main goal of suspension system is to produce shock absorption. Design engineers take into account the impact absorption in automobile as an important aspect. The main causes of vibration of vehicle body through the suspension system and wheel are the bump and holes (Aly 2012; Sun 2002). After a road shock is given, a vibration oscillation will be subjected to the wheels that will be transferred to their axles. On the other hand, the passenger comfort gets affected by overshoot and settling time of vehicle that experiences vibration. As such, it is necessary to use suspension system of a better quality (Dukkapati 2006).

Many existing control methods that have been explored by researchers are linear quadratic Gaussian, fuzzy logic, adaptive sliding mode control, linear quadratic regulator (LQR), and proportional–integral–derivative (PID) control methods. The most broadly used controller is the PID controller. This is because PID does not require complex learning mechanism and it has the ability of tuning the gains such as in the case of fuzzy logic where it involves advanced self-learning optimal intelligent. A quarter-car model has been presented by Agharkakli and Sabet (2012). This study proposed an enhancing performance of a quarter-car active suspension system via PID controller and LQR controller. The aim of the controller's design is to minimize vibration and damping in an active suspension system model. The controller must be able to minimize car body displacement due to road disturbances. Hence, the ride comfort can be achieved.

This paper is structured as follows: Section 2 describes the mathematical model of quarter-car active suspension system. The MATLAB and Simulink software is used to run the simulations and study the response of the system with and without controller. Meanwhile in Sect. 3, suitable controller called LQR is designed in order to reduce the vibration that occurs in the passive suspension system based on the previous simulation that has been conducted. The results are discussed in Sect. 4 by comparing between the LQR and PID controllers. Lastly, Sect. 5 describes the conclusion and the possible future works.

2 System Modeling of Quarter Car

In order to design a controller, we need to understand every subsystem involves in quarter-car active suspension system for comfort motion. Hence, the controller can be designed to meet the criteria needs. The aim of mathematical modeling is to get a state-space representation of the quarter-car model. The type of suspension involved in this study is linear suspension system.

Figure 1 shows the two-degree-of-freedom system model of quarter car. The following are variables in the quarter-car model: m_1 (sprung mass), m_2 (unsprung mass), spring constant of sprung mass denoted as k_1, while spring constant of unsprung mass denoted as k_2, control force u, and damping constant of sprung mass b_1.

All the contact forces acting on the body of the vehicle and the wheels are described using free-body diagrams. By drawing free-body diagram, all the equations of motions can be obtained. Then, all the equations will be transferred to the MATLAB and Simulink software.

The equation from the free-body diagram shown in Fig. 2 can be denoted as follows:

For m_1, $F = ma$

$$m_1 \ddot{x}_1 = -k_1(x_1 - x_2) - b_1(\dot{x}_1 - \dot{x}_2) + u \qquad (1)$$

For m_2, $F = ma$

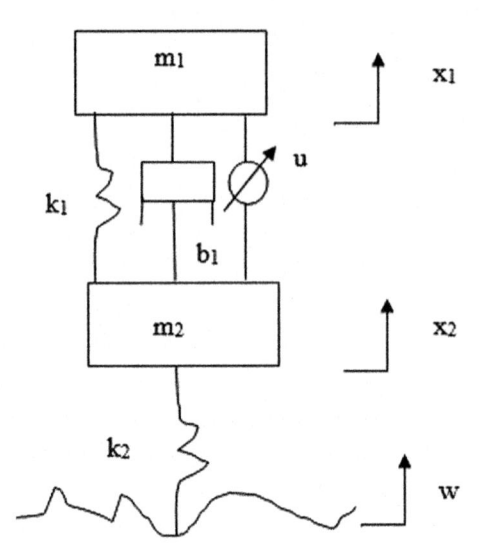

Fig. 1 Quarter-car model (Dukkapati et al. 2012)

Fig. 2 Free-body diagram **a** sprung mass and **b** unsprung mass

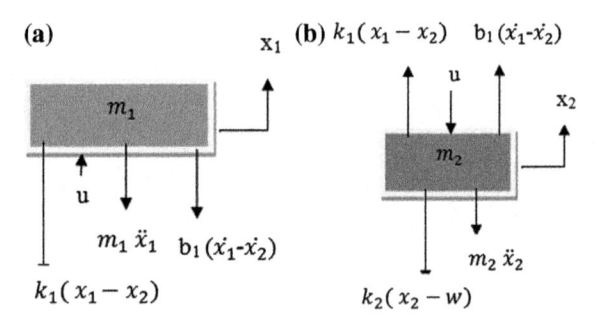

$$m_2\ddot{x}_2 = -k_1(x_1 - x_2) - b_1(\dot{x}_1 - \dot{x}_2) + k_2(x_2 - w) - u \quad (2)$$

In the state-space equation, the state variables are created using

$$\dot{X}(t) = Ax(t) + Bu(t) + f(t) \quad (3)$$

where

$$\begin{aligned}
\dot{X}_1 &= \dot{x}_1 - \dot{x}_2 \sim x_2 - x_4 \text{ (suspension deflection)} \\
\dot{X}_2 &= \dot{x}_1 \text{ (car body velocity)} \\
\dot{X}_3 &= \dot{x}_1 \text{ (car body displacement)} \\
\dot{X}_4 &= \dot{x}_2 - w \text{ (wheel deflection)}
\end{aligned} \quad (4)$$

Improvement on Ride Comfort of Quarter-Car Active Suspension ...   435

Table 1 Model parameters for a sedan car (Agharkakli et al. 2012)

Symbol	Description	Value
m_1 (kg)	Sprung mass	290
m_2 (kg)	Unsprung mass	59
k_1 (n/m)	Spring constant of suspension system	16,812
k_2 (n/m)	Spring constant of wheel and tire	190,000
b_1 (n/m/s)	Damping constant of suspension system	1000

Rewriting Eq. 3 into the matrix form produces

$$\begin{bmatrix} \dot{X}_1 \\ \dot{X}_2 \\ \dot{X}_3 \\ \dot{X}_4 \end{bmatrix} = \begin{bmatrix} 0 & 1 & 0 & -1 \\ \frac{-k_1}{m_1} & \frac{-b_1}{m_1} & 0 & \frac{b_1}{m_1} \\ 0 & 0 & 0 & 1 \\ \frac{-k_1}{m_2} & \frac{b_1}{m_2} & \frac{-k_2}{m_2} & \frac{-b_1}{m_2} \end{bmatrix} \begin{bmatrix} x_1 \\ x_2 \\ x_3 \\ x_4 \end{bmatrix} + \begin{bmatrix} 0 \\ \frac{1}{m_1} \\ 0 \\ \frac{1}{m_2} \end{bmatrix} u + \begin{bmatrix} 1 & 0 & 0 & 0 \\ 0 & 1 & 0 & 0 \\ 0 & 0 & 1 & 0 \\ 0 & 0 & 0 & 1 \end{bmatrix} \dot{w} \quad (5)$$

Table 1.

3 Controller Design

In this section, the controller is divided into two parts which are PID controller and LQR controller. Figure 3 demonstrates the Simulink block diagram of quarter-car system.

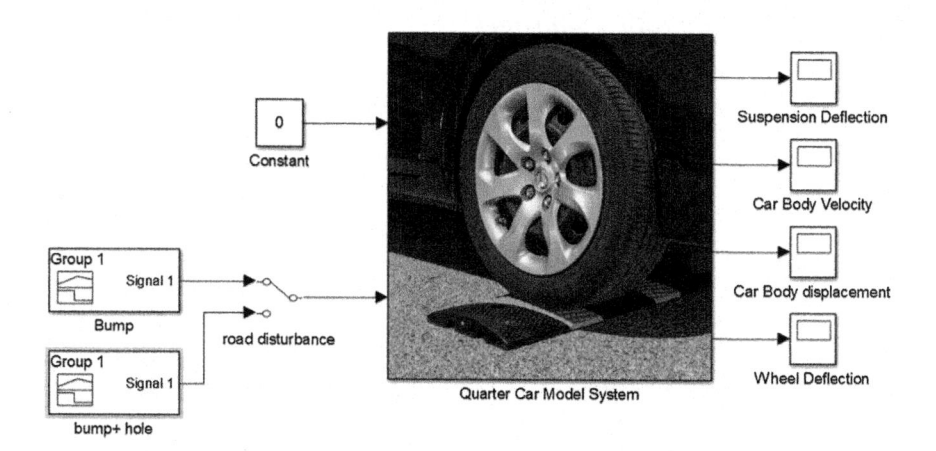

Fig. 3 Simulink block diagram of quarter-car system

3.1 PID Controller

In order to design a controller, we must understand its control structure which consists of plant, controller, and reference. In other words, whatever external disturbance is coming, the body should remain flat on the road. To minimize the vibration in a suspension system, the PID equation is used as stated below and shown in Fig. 4. The PID controller is the most used feedback control design. PID is a short form for proportional–integral–derivative, showing the three terms operating on the error signal to produce a control signal. PID controller is the most well-established class of control systems. However, they cannot be used in several more complicated cases such as in the multiple input and multiple output systems. In order to minimize the vibration in a suspension system, the PID equation is used as stated below (Yakub et al. 2016b):

$$u(t) = K_p\, e(t) + K_i \int_0^t e(\tau)d\tau + K_d\, \mathrm{d}e(t)/\mathrm{d}t \qquad (6)$$

From Eq. 6, K_p is denoted as the proportional which is nonnegative variable, K_i is the integral coefficient, K_d is the derivative coefficient, and the error signal is represented by e. In this study, the trial and error method is chosen to see the characteristic function of proportional, integral, and derivative. Then, several tunings and simulations are conducted in MATLAB and Simulink software until the good response is obtained.

3.2 Linear Quadratic Controller

Linear quadratic regulator is one of the popular control approaches that have been broadly used by many researchers in suspension system control. The implementation of LQR in this linear system is to improve the ride comfort for a quarter-car

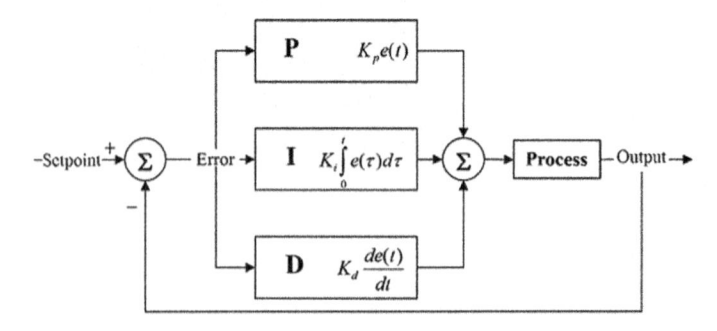

Fig. 4 PID controller system

model (Yakub et al. 2016a). Consider the state variable feedback regulator for the system as follows:

$$u(t) = Kx(t) \qquad (7)$$

where K is the constant state feedback gain matrix ($m \times n$) that gives the closed-loop state equation with the desired performance characteristics.

Figure 5 illustrates the Simulink block diagram of LQR controller design. The enhancement technique comprises of deciding the control input u, which reduces the performance index J. The performance index J is denoted as controller input limitation. The optimal controller of a given framework is characterized as controller is defined as controller design, which minimizes the following performance index.

$$J = \frac{1}{2} \int_0^t (x^T Q x + u^T R u) dt \qquad (8)$$

The matrix K is represented by

$$K = R^{-1} B^T P \qquad (9)$$

The P matrix must fulfill the reduced-matrix Riccati equation which is given as follows:

$$P + PA - PBR^{-1}B^T P + Q = 0 \qquad (10)$$

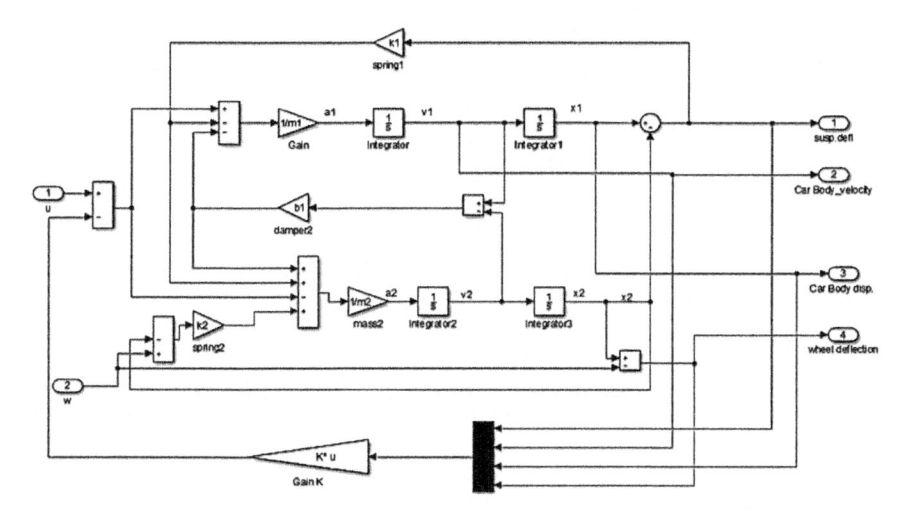

Fig. 5 Simulink block diagram of LQR controller design

Then, the feedback regulator u is

$$u(t) = -(R^{-1}B^T P)x(t) \tag{11}$$

$$u(t) = -Kx(t) \tag{12}$$

The determination of Q and R decides the optimality in the optimal control law. These matrices depend on the designers' choices. The preferred method in order to decide these matrices is trial and error simulation. In other words, the Q and R matrices are chosen to be diagonal. A large matrix R is needed for a small input. The corresponding diagonal element should be large in order a state to be small in magnitude. To keep J small, the Q matrix must be larger and the state $x(t)$ must be smaller. Selecting the larger R, the control input $u(t)$ must be smaller in order to obtain the small value of J. Different values of Q and R will exhibit the difference in response.

4 Results and Discussion

The simulation of the results is done by using MATLAB and Simulink software. The parameters that will be observed are suspension deflection, car body displacement, and the wheel deflection. The main goal of this research is to achieve the shortest time for the vibration to settle down and to minimize the vibration effect by comparing these three parameters.

There are two types of road disturbances that are assumed in this paper where the two bumps are declared as the disturbance input with the same height (10 cm) but the difference in bump width. We want to see the behavior of the system when the two different input disturbances are used for passive, PID, and LQR controllers. The simulation results are simulated via MATLAB and Simulink software. The results from Figs. 6, 7, and 8 show the comparison of suspension between passive, PID, and LQR controllers for two types of bumps with different criteria.

$$Q = \begin{bmatrix} 1000 & 0 & 0 & 0 \\ 0 & 500 & 0 & 0 \\ 0 & 0 & 1 & 0 \\ 0 & 0 & 0 & 10 \end{bmatrix} \text{ and } R = 0.001$$

Therefore, the value of gain K is $K = [29.7\ 3624.6\ 24.9\ 244.8]$. The time is set for 10 s to show the time taken when the vehicle reaches the stability state. From the graph, the vibration of suspension deflection, car body displacement, and wheel deflection of the vehicle needs 5 s to be stable after hitting the 10 cm of road bump with difference in width.

The suspension deflection represents the spring behavior when the car hits the road bump. As shown in Fig. 6, when the wheel is at the bump surface, the

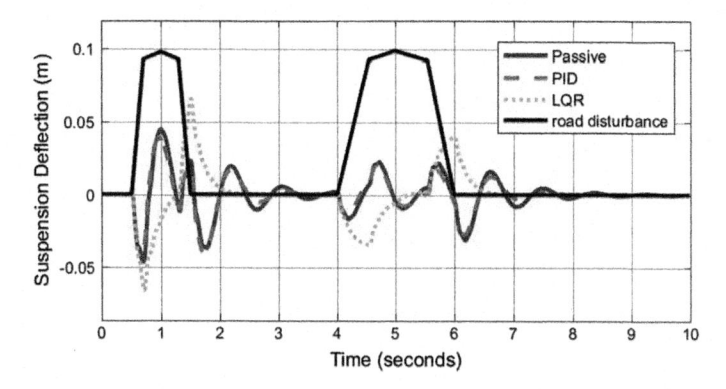

Fig. 6 Suspension deflection

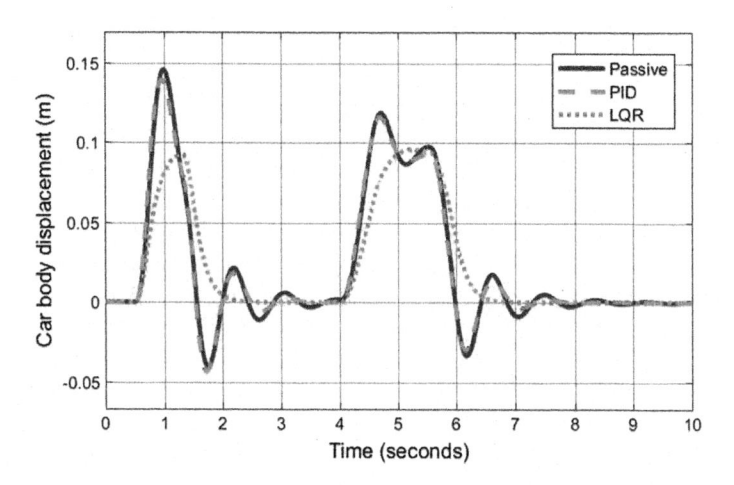

Fig. 7 Car body displacement

condition of the spring is being compressed and bended into a shorter length. After hitting the road bump, the spring will extend back until it reaches the stable state. Figures 6, 7, and 8 demonstrate that the amplitude and the vibration have been minimized using the PID by 2 s. However, the LQR control shows the best output performances as it can minimize the vibration and has the faster settling time of about 0.5 s to reach the stable steady state compared to passive and PID control for all the three parameters.

In Fig. 8, the simulation shows the results of wheel deflection. This characteristic is also considered in designing the controller of car suspension system. This criterion is also taken into account in order to increase the ride quality and observe the diagram of tire dynamic compression. We can see that after implementing the LQR control, the amplitude is slightly higher but the settling time is optimized. This part expresses that the tire is in contact with the road. Using PID control is the best

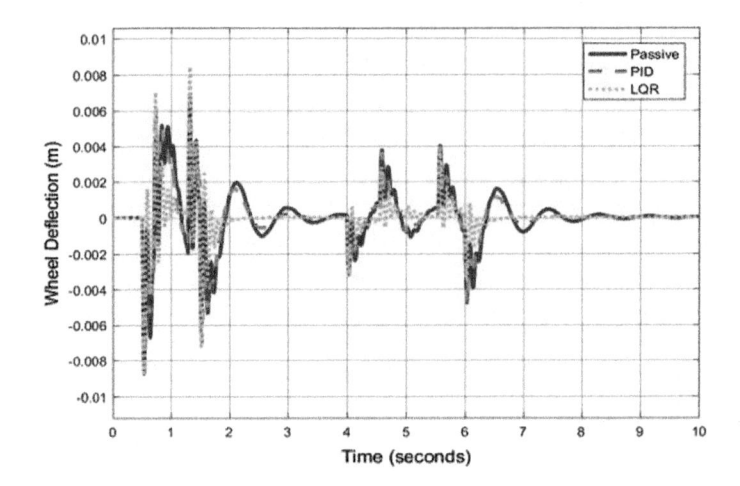

Fig. 8 Wheel deflection

approach as it minimizes the vibration but it takes approximate time of about 2 s to settle down compared to the LQR, which is 0.5 s. Using LQR controller is the best selection as it increases the ride quality of the suspension system.

By comparing the performances of the quarter car with and without controller, it is clearly shown that different road disturbances will exhibit different responses. The output performances with the two bumps happen to have difference in amplitude. As for example, from Fig. 6 at the height of 10 cm and 1 cm width of bump length gives the highest amplitude compared to the second bump at the width of 2 cm. Meanwhile in Fig. 7, the amplitude is slightly higher from the reference input of 10 cm where it reaches 15 cm. This shows that at this point, the passengers or the driver experiences the uncomfortable ride where the vibration is higher after the car hits the first and the second road bumps. The car body displacement has been indicated as ride comfort. The amplitude of car body displacement determined using the LQR technique is improved compared to the passive suspension system. Besides, the settling time also gives the faster response and reduces the vibration of the vehicle displacement by 0.5 s.

5 Conclusion

The vibration of suspension deflection, wheel deflection, and car body displacement is minimized. The time taken to settle down the vibration has also been optimized by using the suitable controller which is the LQR controller. However, there is some problem when using and implementing the LQR in the system. For most of the parameters measured, LQR control cannot reduce amplitude compared to passive suspension. However, LQR technique can provide faster settling time compared to passive suspension system.

In this paper, the control of suspension only used the quarter-car model. The output performances will produce different responses when different values of matrices Q and R, and parameters of K_p, K_i, and K_d were used. These parameters should undergo several tunings to get better responses through simulations. Another concern is about the different vehicle parameters can also give different outputs either for passive or for active suspension systems. In this case study, we used the vehicle parameter of sedan car, and with different parameters of heavy vehicle, the result might be changed. Finally, using the different type of controllers such as the intelligent based approach are recommend to get more robust results than the LQR approach.

Acknowledgements The authors would like to thank Dr Shamsul Sarip and Dr Hatta Ariff for their comments and advice. This work was financially supported through research grant Universiti Teknologi Malaysia under Vot number 11H67.

References

Agharkakli, A., Sabet, G. S., & Barouz, A. (2012). Simulation and analysis of passive and active suspension system using quarter car model for different road profile. *International Journal of Engineering Trends and Technology, 3*, 636–644.
Alvarez-Sánchez, E. (2013). A quarter-car suspension system: Car body mass estimator and sliding mode control. *Procedia Technology 7*, 208–214.
Aly, A. A. (2012). Car suspension control systems: basic principles. *International Journal of Control, Automation and Systems, 1*(1) 41–46.
Coelho, H., & Pereira, L. M. (1986). Automated reasoning in geometry theorem proving with Prolog. *Journal of Automated Reasoning, 2*(3), 329–390.
Dukkapati, R. V. (2006). *Analysis and Design of control Systems using MATLAB*. Connecticut, USA: New Age International Publishers.
Ghazaly, N. M., & Moazz, A. O. (2014). The future development and analysis of vehicle active suspension system. *IOSR Journal of Mechanical and Civil Engineering, 11*(5) 19–25.
Ghosh, M. K., & Nagraj, A. (2004). Turbulence flow in bearings. *Proceedings of the Institution of Mechanical Engineers, 218*(1), 61–64.
Rao, P. N. (2000). *Manufacturing technology foundry, forming and welding* (2nd ed., pp. 53–68). Singapore: McGraw Hill.
Sun, L. (2002). Optimum design of 'road friendly' vehicle suspension systems subjected to rough pavement surfaces. *Applied Mathematical Modelling, 26*, 635–652.
Yagiz, N., & Hacioglu, Y. (2008). Back stepping control of a vehicle with active suspension system. *Control Engineering Practice, 16*, 1457–1467.
Yakub, F., Lee, S., & Mori, Y. (2016a). Comparative study of MPC and LQC with disturbance rejection control for heavy vehicle rollover prevention in an inclement environment. *Journal of Mechanical Science and Technology, 30*(8), 3835–3845.
Yakub, F., Muhamad, P., Thiam, T. H., Fawazi, N., Sarip, S., Ali, M. S. M,, & Zaki, S. A. (2016b) Enhancing vehicle ride comfort through intelligent based control. In *IEEE international conference on automatic control &i ntelligent systems*, 22 October, Shah Alam, Malaysia, pp. 66–71.

Evaluation of Protease Enzyme on *Pomacea canaliculata* Eggs

Noor Hasyierah Mohd Salleh and Nurhadijah Zainalabidin

Abstract Pomacea canaliculata (PC) is one of the most destructive pests as it causes severe damage to the paddy field. Most of the approaches are targeting its flesh rather than its egg; however in the same time, it has high fecundity and hatchability rate. Thus, effective approach is necessary to control the hatchability, which in turn affects its life cycle. Protease has been shown to have the capability to suppress the hatchability of Pomacea canaliculata eggs (PCE). The objective of this study is to further support the findings, where it discusses the chemical assessment of protease-treated PCE. The assessment focuses on the effect of protease treatment on the PCE in terms of the following factors, namely non-hatchability, cuticle protein reduction, gas exchange and water loss activity. Non-hatchability of 42 and 91% after treating with 0.75 and 5 U/ml protease, respectively, was recorded. With these findings, it is safe to consider that protease applicable to be used as biocontrol against PCE. The non-hatchability of protease-treated eggs occurred as a result of cuticle disruption as supported by the present chemical studies.

Keywords Protease · *Pomacea canaliculata* eggs · Cuticle · Non-hatchability

1 Introduction

Pomacea canaliculata, generally known as the golden apple snail, is classified as one of the most destructive pests as it completely destroys the young tender stems and leaves of paddy (Wu et al. 2005). The damage caused by the snail is much more severe as compared to other pests such as mice and insects. For example, in Perlis, one of the rice producing regions in Malaysia, 68 and 11% of the plantation area

---

N. H. M. Salleh (✉)
Department of Chemical Engineering Technology, Faculty of Engineering Technology, UniMAP, 02000 Sg Chucuh, Padang Besar, Perlis, Malaysia
e-mail: hasyierah@unimap.edu.my

N. Zainalabidin
School of Bioprocess Engineering, UniMAP, 02600 Arau, Perlis, Malaysia

© Springer Nature Singapore Pte Ltd. 2018
R. Saian and M. A. Abbas (eds.), *Proceedings of the Second International Conference on the Future of ASEAN (ICoFA) 2017 – Volume 2*,
https://doi.org/10.1007/978-981-10-8471-3_44

were affected by snails and mice, respectively, in 2013 (Annual report 2013). The extensive damage and the losses caused by the snail have prompted researchers, farmers and various agro-related institutions to seek for appropriate methods to manage the snail invasion.

So far, several methods have been introduced and reviewed (Noorhasyierah et al. 2012). Most of these methods target the flesh of the snail, and only a few deal with snail eggs. However, while reducing the snail population is important, it is also necessary to destroy snail eggs for the following reasons. Firstly, targeting snail alone does not necessarily disrupt the eggs from hatching because the eggs are usually deposited above the water level at a distance from the snail. Secondly, targeting the snail alone might not be effective because the snail prefers to hide under the soil; hence, it is un-reachable by pesticides. Additionally, eggs are fragile, immobile and locate above on the water level. Due to these reasons, it is deemed necessary to find effective agents to suppress the hatchability of the snail eggs. These agents should preferably give minimal effect towards non-targeting organisms and should be affordable by farmers.

Previous reports concerning *P. canaliculata* eggs' suppression include the application of fungus (Maketon et al. 2009; Chobchuenchorm and Bhumiratana 2003; Nurhadijah et al. 2014), water submersion (Wang et al. 2012; Horn et al. 2008) and wax (Wu et al. 2005). Recently, protease enzyme has been shown to be effective agents in suppressing *P. canaliculata* eggs' hatchability (Noorhasyierah et al. 2013, 2014). However, despite its effectiveness, the actual mechanism remains to be investigated.

Therefore, the objective of the present paper is to discuss supporting evidences of protease action in suppressing *P. canaliculata* eggs' hatchability in terms of protease effect on cuticle thinning, gas exchange and egg water loss together with morphological changes of the egg cuticle. This finding might enhance understanding on the protease action in suppressing *P. canaliculata* eggs' hatchability.

2 Materials and Method

2.1 Collection of *P. canaliculata* Eggs

Fresh *P. canaliculata* eggs were randomly collected from several paddy fields in Perlis, Malaysia. The paddy field with high density of the eggs was chosen as it reflects the density of the snails. The fresh 1-day-old eggs were identified by the milky pink colour, soft-fragile condition and mucus covered. About 50 clutches of eggs were collected without damaging their structure. The eggs were treated according to the previous work (Meyer-Willere and Santos-Soto 2006) but with a slight modification. Basically, the eggs were transferred to an open breeding chamber (62 cm × 40 cm × 46 cm) which was filled with tap water to provide a humid condition.

2.2 Un-Hatchability Studies

The 3-day-old eggs in the breeding chamber were carefully weighted to 0.4 g/clutch. The un-hatchability studies were carried out by applying several concentrations of protease solution on the eggs. The protease solutions (0–10 U/ml) (Sigma-Aldrich, USA) and ethylenediaminetetraacetic acid (EDTA) (Hmbg, Germany) in 0.02–0.1 M were prepared accordingly. EDTA-treated eggs were served as positive control. For each treatment, 1 ml of the solution was pipetted onto 0.4 g/clutch of 3-day-old eggs in a plastic container in a stepwise manner to ensure all the surface of the eggs was covered by the enzyme solution. The protease- and EDTA-treated eggs were incubated in the breeding chamber for 14 days, after which they were observed for un-hatchability. The un-hatchability is defined as the non-emergence of juveniles to crack and move out from the eggshell. Each treatment was prepared in five (5) replicates. The untreated eggs were served as negative control. The un-hatchability against protease concentration was statistically analysed by ANOVA while lethal concentration for LC_{50} and LC_{95} was calculated using goal seek analysis in Excel.

2.3 Cuticle Protein Measurement

Commercial protease from *Aspergillus oryzae* (Sigma-Aldrich, USA) was prepared in several concentrations ranging from 0 to 10 U/ml by diluting in 0.02-M phosphate buffer, pH7 (Noorhasyierah et al. 2013). Five (5) g of 3-day-old eggs were placed in a net basket (5 cm × 3 cm) and were dipped in 100 ml of protease solutions for 30 min under mild stirring. The experiment was performed in such a way so that the whole eggs were in contact with the protease solutions and no crack. The eggs were then rinsed several times with distiled water and further treated with 0.5% SDS for 10 min to release the remaining protein (Martel et al. 2012). The protein dissolved in SDS solutions was then assayed using Lowry method. All experiments were performed in triplicates. The protein standard curve was developed using BSA as an internal standard.

2.4 Gas Exchange Studies

The same clutch of eggs undergo three treatments which were water, protease 5 and 10 U/ml accordingly. Twenty grams (20 g) of 3-day-old eggs was immersed in 100 ml of water for 30 min before placing the eggs in an airtight ventilation chamber. The ventilation chamber (5 cm × 3 cm × 7.5 cm) with probes of oxygen and carbon dioxide was used (Vernier Software Technologies, UK) as was suggested by Delpech (2006) with some modifications. The probes were connected to the data Logger Pro software version 3.8.6.1 (Vernier Software Technologies,

UK). The gas exchange for water-treated eggs was recorded for 15 min. After that, the water-treated eggs were then immersed in 100 ml of 5 U/ml of protease solution for 30 min and recorded the gas exchange in a similar manner with water-treated eggs before continuing with 10 U/ml protease. The data collection was repeated for five times each. The water-treated eggs were served as control.

2.5 Egg Water Loss Studies

The 3-day-old eggs were divided into 0.5 g/clutch each before placing them in the weighted container. The eggs were then treated with 1 ml of water, protease 5 and 10 U/ml, respectively. After 10 min of the treatment, the samples were weighted again and labelled as day 1 treatment before placing in the desiccator with 75% relative humidity (RH). The RH of 75% was prepared from saturated Na_2CO_3 (Merck, Malaysia) (Peebles et al. 1987). The desiccator was then placed in 30 °C. The samples' weight was taken daily until day 8 and was further kept in the desiccator until it hatches (approximately 12 days). The hatching eggs were counted and recorded. Each treatment was prepared in five replicates. The egg water loss was calculated by dividing the daily egg's weight with the initial egg's weight. The water-treated eggs were served as control.

3 Results

3.1 Effect of Protease Treatment on Un-Hatchability Studies

The influence of protease treatment on the hatchability of 3-day-old eggs was studied as presented in Fig. 1. EDTA treatment represents as a positive control against protease. It shows that the un-hatchability of eggs increases proportionally with protease concentration. The lowest concentration of protease, 0.75 U/ml, suppresses 71% of *P. canaliculata* eggs' hatchability as compared to untreated eggs which have low un-hatchability, 14%. The un-hatchability of protease-treated egg gets saturated after treatment with 7.5 and 10 U/ml, respectively, due to substrate saturation. Therefore, it suggests that at this concentration, the cuticle protein has been digested by protease and hence lost its capability to protect the embryo. The similar hatchability pattern was observed when EDTA was used in place of protease to digest the protein within the cuticle (Fig. 2). Based on the p value <0.05 and $F > F_{crit}$ as shown in Table 1, the protease gives significant effect to the un-hatchability. The lethal concentration for LC_{50} and LC_{95} is 06.07 and 3.798 U/ml protease, respectively.

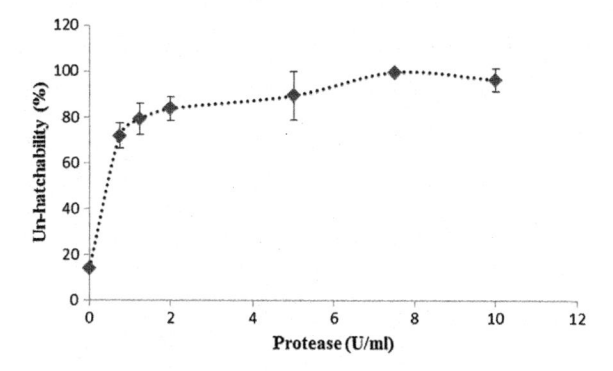

Fig. 1 Effect of protease on the un-hatchability of *P. canaliculata* eggs

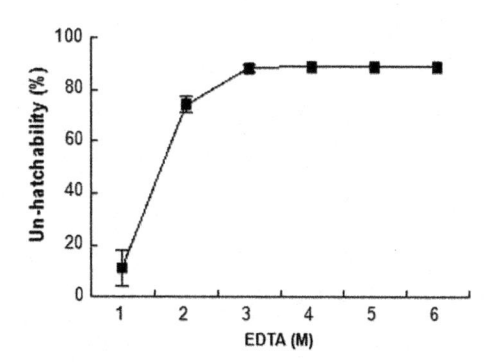

Fig. 2 Effect of EDTA on the un-hatchability of *P. canaliculata* eggs

Table 1 ANOVA of protease-treated *P. canaliculata* eggs

Factor	F	*p* value	F_{crit}
Protease concentration	44.78	<0.001	4.084

3.2 Effect of Protease on the Cuticle Protein

Upon application of protease, the cuticle protein of *P. canaliculata* eggs was investigated (Fig. 3). It shows that the protein decreases proportionately with the protease concentration.

3.3 Effect of Protease on the Gas Exchange

Upon protease treatment, the effect on the gas exchange was studied. The rate of O_2 uptake and the rate of CO_2 production are shown in terms of slope as in Fig. 4.

Fig. 3 Protease effect on the cuticle protein of *P. canaliculata* eggs

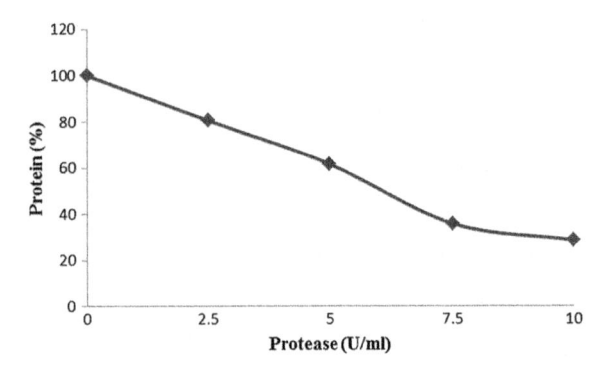

Fig. 4 Effect of protease on O_2 consumption and CO_2 production rate of *P. canaliculata* eggs

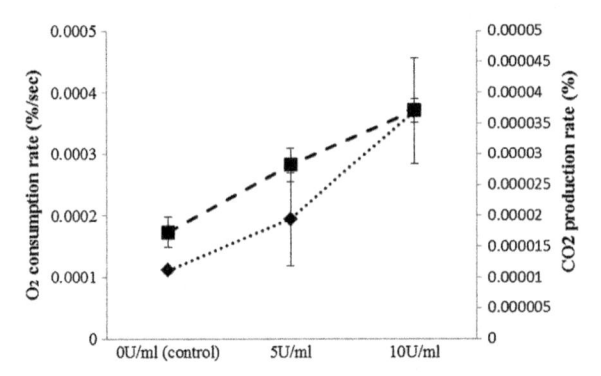

From the graph, it shows that the protease increases the rate of O_2 consumption and CO_2 production of *P. canaliculata* eggs upon treated with 5–10 U/ml protease, respectively, as compared to control (water treated). In addition, the changes in the gas exchange of protease-treated *P. canaliculata* eggs are also significantly relates to the un-hatchability as presented in Fig. 1.

3.4 Effect of Protease on the Egg Water Loss

Effect of protease on the egg water loss is also investigated. It shows that the water loss rate of *P. canaliculata* eggs under treatment with 5, 10 U/ml and 0.1 M EDTA increases 1.6, 9.5 and 21%, respectively, as compared to water treatment (Fig. 5). However, for the control (water treated) *P. canaliculata* eggs, high density of cuticle covers the eggs and blocks the pores, thus creating a barrier to water loss and consequently decelerating the reduction of egg weight. Extra losses of water vapour consequently affect the un-hatchability, as presented in Fig. 1.

Fig. 5 Effect of protease on the water loss rate of *P. canaliculata* eggs

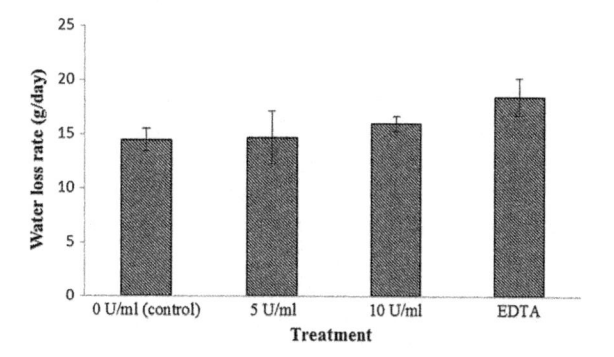

4 Discussion

P. canaliculata eggs, as with many other eggs, are covered by a white thin layer of cuticle which plays an important part in the hatching process to ensure that the embryo can develop well and hatch successfully. However, after protease application, 71% of *P. canaliculata* eggs un-hatch (Fig. 1). It reveals the effectiveness of protease in suppressing the *P. canaliculata* eggs' hatchability and hence has high potentiality to be used as a safe molluscicide against *P. canaliculata* eggs. A similar pattern was also observable through the EDTA studies. When the cuticle structure was altered by using EDTA instead of protease, the hatchability of the eggs decreased at a similar pattern to the protease treatment (Fig. 2). Thus, it can be deduced that the reduction in egg hatchability is related to the disruption of the cuticle due to the action of protease. In this case, the protease digests the protein within the cuticle, so that the higher the protease, the lower the remaining cuticle protein and the more susceptible the eggs to the attack from the surrounding environment which in turn lowering the hatchability.

The un-hatchability of *P. canaliculata* eggs caused by protease is supported by the cuticle protein studies (Fig. 3). Protease is well known as a digestive enzyme that is capable to break down the peptide bond. From the graph, it explains the role of protease in digesting the cuticle protein where higher protease treatment consequently reduces the remaining cuticle protein. This condition consequently affects the hatchability. This observation is in line with previous studies on a plant-parasitic nematode eggs' cuticle (Huang et al. 2003). The authors showed the role of protease, chitinase and collagenases in hydrolysing the nematode eggs. Additionally, in another works conducted by Khan et al. (2004), they found that protease excreted from *Paecilomyces lilacinus* hydrolysed the nematode eggs (Khan et al. 2004).

Besides cuticle protein changes, the un-hatchability of the eggs was also supported by the gas exchange and water loss studies as shown in Figs. 4 and 5, respectively. It describes the action of protease in disrupting the cuticle layer, thus reducing the barrier made from the cuticle and exposing the pores blocked by cuticle before the treatment. This eventually accelerates the gas exchange process and water loss rate. The gas exchange and water loss processes are important processes that related between each other since the process through the same pores that blocked by the thick cuticle. It was reported that thick cuticle acts as a barrier for water vapour and gas diffusion (Deeming 2011). These processes play an important role in embryo development where embryo takes oxygen and releases off carbon dioxide and water vapour. Generally, the cuticle controls the gas exchange and water vapour diffusion. Therefore, disruption of the cuticle protein affects the gas exchange and water loss rate and in turn affects the embryo development

5 Conclusion

In the light of the above discussion, it is safe to consider that the protease is potentially applicable to be used as a biopesticide to disrupt the *P. canaliculata* eggs. The application offers some benefits. Firstly, the protease is only active for a limited period, thus causing less damage to the surrounding environment. Secondly, due to the fact that protease actions are known to be specific, it might prove useful for specific target such as *P. canaliculata* eggs in this case. This research provides supporting evidence in terms of reduction of cuticle protein in suppressing *P. canaliculata* eggs' hatchability. The application of protease offers easy, safe and effective alternative of molluscicide in *P. canaliculata* eggs' management.

Acknowledgements The researchers would like to thank the Malaysia Ministry of Science, Technology and Innovation (MOSTI) under Science Fund Grant (9005-00056) for financial support, the Perlis Department of Agriculture for the supporting snail invasion data and the School of Bioprocess Engineering, University Malaysia Perlis for their facilities and continuing support.

References

Annual report. (2013). Department of Agriculture, Perlis, Malaysia.
Chobchuenchorm, W., & Bhumiratana, A. (2003). Isolation and characterization of pathogens attacking *Pomacea canaliculata*. *World Journal of Microbiology & Biotechnology, 19,* 903–906.
Deeming, D. C. (2011). A review of the relationship between eggshell colour and water vapour conductance. *Avian Biology Research., 4,* 224–230.
Delpech, R. (2006). Making the invisible visible: monitoring levels of gaseous carbon dioxide in the field and classroom. *School Science Review, 87,* 320.
Horn, K. C., Johnson, S. D., Boles, K. M., Moore, A., Gilman, E., & Gabler, C. A. (2008). Factors affecting hatching success of golden apple snail eggs: Effect of water immersion and cannibalism. *Wetlands, 28,* 544–549.
Huang, H. C., Liao, S. C., Chang, F. R., Kuo, Y. H., & Wu, Y. C. (2003). Molluscicidal saponins from *Sapindus mukorossi*, inhibitory agents of golden apple snails, *Pomacea canaliculata*. *Journal Agricultural Food Chemistry, 51,* 4916–4919.
Khan, A. K., Williams, L., & Nevalainen, H. K. M. (2004). Effects of *Paecilomyces lilacinus* protease and chitinase on the eggshell structure and hatching of *Meloidogyne javanica* juveniles. *Biological Control, 31,* 346–352.

Martel, M. R., Du, J., & Hincke, M. T. (2012). Proteomic analysis provides new insight into the chicken eggshell cuticle. *Journal of Proteomics, 75,* 2697–2706.

Meyer-Willere, A. O., & Santos-Soto, A. (2006). Temperature and light intensity affecting egg production and growth performance of the Apple snail *Pomacea patula* (Baker 1922). *Advances en Investigacion Agropecuaria, 10,* 41–58.

Maketon, M., Suttichart, K., & Domhom, J. (2009). Effective control of invasive apple snail (*Pomacea canaliculata* Lamarck) using *Paecilomyces lilacinus* (Thom) Samson. *Malacologica, 51,* 181–190.

Noorhasyierah, M. S., Dachyar, A., Mohamed Zulkali, M. D., Nilawati, P., & Rohaina, N. (2012). Management of Siput Gondang Emas in Northen of Malaysia: Mini review. *APCBEE Procedia, 2,* 129–134.

Noorhasyierah, M. S., Dachyar, A., & Mohamed Zulkali, M. D. (2013). Preparation and thermal protease stability of biopesticide to control the *Pomacea canaliculata*'s eggs. *Journal of Life Sciences and Technologies, 1,* 123–126.

Noorhasyierah, M. S., Dachyar, A., Mohamed Zulkali, M. D., & Ahmad Z. A. (2014). Protease-based biopesticide for disruption of *Pomacea canaliculata* eggs, presented in: International Conference of Engineering and Management, organized by University Malaysia Perlis, Malaysia, 25–26 January 2014.

Nurhadijah, Z. A., Noorhasyierah, M. S., Dachyar, A., & Mohamed Zulkali, M. D. (2014). Application of *Bacillus thuringiensis* subs *kurstaki* as biological agents for disrupting Golden Apple Snail (*Pomacea canaliculata*) eggs and juveniles, presented in International Conference of Engineering and Management, organized by University Malaysia Perlis, Malaysia, 25–26 January 2014.

Peebles, E. D., Brake, J., & Gildersleeve, R. P. (1987). Effect of eggshell cuticle removal and incubation humidity on embryonic development & hatchability of broilers. *Poultry Science, 66,* 834–840.

Wang, Z., Tan, J., Tan, L., Liu, J., & Zhong, L. (2012). Control the egg hatching process of *Pomacea canaliculata* (Lamarck) by water spraying and submersion. *Acta Ecologica Sinica, 32,* 184–188.

Wu, D. C., Yu, J. Z., Chen, B. H., Lin, C. Y., & Ko, W. H. (2005). Inhibition of egg hatching with apple wax solvent as a novel method for controlling golden apple snail (*Pomacea canaliculata*). *Crop Protection, 24,* 483–486.

A New Analytical Technique for Solving Nonlinear Non-smooth Oscillators Based on the Rational Harmonic Balance Method

Md. Alal Hosen, M. S. H. Chowdhury, M. Y. Ali and A. F. Ismail

Abstract In the present paper, a new analytical technique based on the rational harmonic balance method (RHBM) has been introduced to determine approximate periodic solutions for the nonlinear non-smooth oscillator. A frequency–amplitude relationship has also been obtained by a novel analytical way. The standard rational harmonic balance method (SRHBM) cannot be used directly; it is possible if we rewrite the nonlinear differential equations (NDEs). To overcome this previously stated issue, we offered a modified rational harmonic balance method (MRHBM). It is noticed that a MRHBM works very well for the whole range of initial amplitudes and the excellent agreement of the approximate frequencies as well as the corresponding periodic solutions with its exact ones. The method is basically illustrated by the nonlinear non-smooth oscillators, but it is additionally useful for other nonlinear oscillatory problems with mixed parity arising in recent development of nonlinear sciences and engineering.

Keywords Analytical technique · Approximate angular frequencies
Nonlinear non-smooth oscillator · Power series solution · Rational harmonic balance method

---

Md. A. Hosen
Department of Mathematics, Rajshahi University of Engineering
and Technology (RUET), Rajshahi 6204, Bangladesh

M. S. H. Chowdhury (✉)
Department of Science in Engineering, Faculty of Engineering,
International Islamic University Malaysia, Jalan Gombak, 53100
Kuala Lumpur, Malaysia
e-mail: sazzadbd@iium.edu.my

Md. A. Hosen · M. Y. Ali
Department of Manufacturing and Material Engineering, Faculty of Engineering,
International Islamic University Malaysia, Jalan Gombak, 53100
Kuala Lumpur, Malaysia

A. F. Ismail
Department of Mechanical Engineering, Faculty of Engineering, International Islamic
University Malaysia, Jalan Gombak, 53100 Kuala Lumpur, Malaysia

1 Introduction

Along with the rapid progress of nonlinear sciences, an intensifying interest among scientists and researchers has been emerged in the field of nonlinear oscillating systems with the nonlinear non-smooth oscillators because this issue is very applicable in dynamics of structures which is stated in Chopra 1995. Nowadays, obtaining exact solutions of the nonlinear oscillatory problems is one of the biggest challenges. In general, it is often more difficult to obtain an analytic approximation than a numerical one. A few nonlinear systems can be solved explicitly, and numerical methods, especially the most popular Runge-Kutta fourth-order method, are frequently used to calculate approximate solutions. However, numerical schemes do not always give accurate results, especially the class of stiff differential equations, chaotic differential equation, which present a more serious challenge to numerical analysis. And also, the frequency-amplitude relationship cannot be obtained. The popular method for solving nonlinear differential equations (NDEs) associated with oscillatory systems is perturbation method (Nayfeh 1973; Azad et al. 2012) which is the most versatile tools available in nonlinear analysis of engineering problems, and they are constantly being developed and applied to even more complex problems. However, for the strongly nonlinear regime perturbation method cannot yield desired results.

As a result, due to conquering these weak points, in recent past, numerous researchers have devoted their time and effort to find potent approaches for investigating the nonlinear phenomena. As the earliest effort, they developed a large variety of approximate methods commonly used for strongly nonlinear oscillators including homotopy perturbation method (Belendez 2009; Ozis and Akci 2011), modified He's homotopy perturbation method (Belendez et al. 2007), He's modified Lindsted-Poincare method (Ozis and Yildirim 2007), max–min approach method (Ganji and Azimi 2012), global residue harmonic balance method (Peijun 2015), energy balance method (Hosen 2016, 2017), He's energy balance method (Askari et al. 2014), rational energy balance method (Daeichin et al. 2013), iteration method (Ikramul et al. 2013; Mickens 2006), harmonic balance method (Mickens 2010; Hosen et al. 2012; Cveticanin 2009; Lim et al. 2005; Gottlieb 2003), and so on. However, the results obtained by most of the mentioned methods only first-order approximation has been considered which leads insufficient accuracy. Furthermore, the solution procedures are tremendously difficult task and cumbersome, especially for obtaining higher-order approximation. In this situation, we will see that the rational harmonic balance method (RHBM) considered in this paper can be applied to nonlinear non-smooth oscillator. The RHBM discussed by Mickens and Semwogerere (1996), for instance, has rarely been applied to the determination of periodic solutions of the nonlinear problems. In fact, to the best of our knowledge, recently Belendez et al. (2008) and Yamgoue et al. (2010) used it to solve a simple-term oscillator equation of plasma physics in a completely analytic fashion. Generally, a set of complicated nonlinear algebraic equations are found when RHBM is applied. Sometimes analytical solutions of these algebraic equations fail, especially for large amplitude. In the present study, this limitation is removed. The nonlinear algebraic equations have been approximated using power series solution (a new small parameter). Consider the interesting issue that the proposed technique provides accurate results and it is more convenient and efficient for solving more complex nonlinear problems.

2 Solution Procedure by the Standard Rational Harmonic Balance Method

Consider a general second-order nonlinear differential equation with mixed parity which is of the following form as

$$\ddot{x} = -\varepsilon x^{\frac{1}{2n+1}} \text{ and the initial condition } x(0) = a_0, \dot{x}(0) = 0, \tag{1}$$

where $x^{\frac{1}{2n+1}}$, $n = 1, 2, 3, \cdots$ is a fractional-order nonlinear function and ε is a constant.

The nth-order periodic solution of Eq. 1 can be considered as

$$x(t) = \frac{A_1 \cos \varphi + A_3 \cos 3\varphi + A_5 \cos 5\varphi + \cdots}{1 + u \cos 2\varphi + v \cos 4\varphi + w \cos 6\varphi + \cdots}. \tag{2}$$

where $\varphi = \omega t$ and A_1, A_3, A_5, u, v, w are unknown constants. The solution of Eq. 2 does not satisfied of Eq. 1 directly; it is possible if we rewrite the Eq. 1. Then applying Eq. 2 into the rewritten Eq. 1, it can be transformed into

$$\begin{aligned}A_1^{2n+1} \omega^{4n+2}[(1 + u^2 + \cdots) \cos(\omega t) + (1 - u + \cdots) \cos 3\varphi + \cdots] \\ = -\varepsilon[F_1(A_1, u, \cdots) \cos \varphi + F_3(A_3, u, \cdots) \cos 3\varphi + \cdots]\end{aligned} \tag{3}$$

By comparing the coefficients of equal harmonic terms of Eq. 3, one could obtain as

$$\begin{aligned} A_1^{2n+1} \omega^{4n+2}(1 + u^2 + \cdots) &= -\varepsilon F_1(A_1, u, \cdots) \\ A_1^{2n+1} \omega^{4n+2}(1 - u + \cdots) &= -\varepsilon F_3(A_3, u, \cdots), \\ A_1^{2n+1} \omega^{4n+2}(1 - v + \cdots) &= -\varepsilon [F_5(A_5, u, \cdots)] \end{aligned} \tag{4}$$

With help of the first equation, ω^{4n+2} is eliminated from all the remaining equations of Eq. 4. Thus, second and third equations of Eq. 4 can be expressed as

$$u = G_1(\varepsilon, a_0, u, v, \cdots,), \quad v = G_2(\varepsilon, a_0, u, v, \cdots,), \cdots, \tag{5}$$

where $G_1, G_2, \cdots$ exclude, respectively, the linear terms of $u, v, \cdots$.

Whatever the values of ε and a_0, there exists a parameter $\lambda_0(\varepsilon, a_0) \ll 1$, such that $u, v, \cdots$ are expandable in following series

$$u = U_1\lambda_0 + U_2\lambda_0^2 + \cdots, \quad v = V_1\lambda_0 + V_2\lambda_0^2 + \cdots, \quad \cdots \tag{6}$$

where $U_1, U_2, \cdots, V_1, V_2, \cdots$ are constants.

Finally, substituting the values of $u, v, \cdots$ from Eq. 6 into the first equation of Eq. 4, the unknown angular frequency ω is determined. This completes the determination of all related functions for the proposed periodic solution as given in Eq. 2.

3 Solution Procedure by the Modified Rational Harmonic Balance Method

Here, solution Eq. 2 is applied into Eq. 1 directly, if we expand the fractional nonlinear terms $x^{\frac{1}{2n+1}}$ in a Fourier series as

$$x^{\frac{1}{2n+1}} = \sum_{n=0}^{\infty} b_{2n+1} f(x) = b_1 \cos(\omega t) + b_3 \cos(3\omega t) + \cdots. \tag{7}$$

where $b_1, b_3, \cdots$ will be calculated by using the following integration

$$b_{2n+1} = \frac{4}{\pi} \int_0^{\pi/2} x^{\frac{1}{2n+1}} \cos[(2n+1)\varphi] d\varphi; \quad n = 0, 1, 2, 3, \cdots \tag{8}$$

where $\varphi = \omega t$.

Substituting Eqs. 2, 7–8 into Eq. 1 and then Eq. 1 can be transformed into an algebraic identity as

$$\begin{aligned} A_1^{2n+1} \omega^{4n+2} &[(1 + u^2 + \cdots) \cos(\omega t) + (1 - u \cdots) \cos 3\varphi + \cdots] \\ &= -\varepsilon [F_1(A_1, u, \cdots) \cos \varphi + F_3(A_3, u, \cdots) \cos 3\varphi + \cdots] \end{aligned} \tag{9}$$

By comparing the coefficients of equal harmonics of Eq. 9, the following non-linear algebraic equations can be found as

$$\begin{aligned} A_1^{2n+1} \omega^{4n+2}(1 + u^2 + \cdots) &= -\varepsilon F_1 \\ A_1^{2n+1} \omega^{4n+2}(1 - u + \cdots) &= -\varepsilon F_3(A_3, u, \cdots), \\ A_1^{2n+1} \omega^{4n+2}(1 - v + \cdots) &= -\varepsilon F_5(A_5, u, \cdots) \end{aligned} \tag{10}$$

With help of the first equation, ω^{4n+2} is eliminated from all the remaining equations of Eq. 10. Thus, second and third equations of Eq. 10 can be expressed into the following form as

$$u = G_1(\varepsilon, a_0, u, v, \cdots,), \quad v = G_2(\varepsilon, a_0, u, v, \cdots,), \quad \cdots \quad (11)$$

where $G_1, G_2, \cdots$ exclude, respectively, the linear terms of $u, v, \cdots$.

Whatever the values of ε and a_0, there exists a parameter $\lambda_0(\varepsilon, a_0) \ll 1$, such that $u, v, \cdots$ are expandable in following series as

$$u = U_1 \lambda_0 + U_2 \lambda_0^2 + \cdots, \quad v = V_1 \lambda_0 + V_2 \lambda_0^2 + \cdots, \quad \cdots \quad (12)$$

where $U_1, U_2, \cdots, V_1, V_2, \cdots$ are constants.

Finally, substituting the values of $u, v, \cdots$ from Eq. 12 into the first equation of Eq. 10, the unknown angular frequency ω is determined. This completes the determination of all related functions for the proposed periodic solution as given in Eq. 2.

4 Application of the Standard Rational Harmonic Balance Method (SRHBM)

Consider $n = \varepsilon = 1$ into Eq. 1, the nonlinear non-smooth oscillator (Belendez 2009; Ozis and Yildirim 2007; Mickens 2006, 2010) can be written as

$$\ddot{x} + x^{1/3} = 0, \quad x(0) = a_0, \quad \dot{x}(0) = 0. \quad (13)$$

This is a conservative system, and the solution to Eq. 13 is periodic. We observe that in Eq. 13 direct application of SRHBM does not work. To apply the SRHBM, we rewrite the Eq. 13 as

$$\ddot{x}^3 + x = 0. \quad (14)$$

Now, the solution Eq. 2 can be expressed by Eq. 13. From Eq. 2, the second-order approximation solution of Eq. 14 can be supposed as

$$x(t) = \frac{A_1 \cos \varphi}{1 + u \cos 2\varphi} = \frac{A_1 \cos(\omega t)}{1 + u \cos(2\omega t)}. \quad (15)$$

Now using Eq. 15 in the Eq. 14 and then setting the coefficients of $\cos(\omega t)$ and $\cos(3\omega t)$ equal to zero, the following nonlinear algebraic equations can be obtained as

$$A_1^2 [\omega^6 (3/4 + 9u^2/16 - 6u^3 - 699u^4/32 + \cdots)] = 1 + 4u + 14u^2 + 21u^3 + \cdots, \quad (16)$$

$$A_1^2[\omega^6(1/4 - 15u/4 - 141u^2/16 + 91u^3/4 + \cdots)] = 4u + 7u^2 + 21u^3 + \cdots. \tag{17}$$

After simplification, Eq. 16 can be written as

$$\omega^6 = \frac{(1 + 4u + 14u^2 + 21u^3 + 105u^4/4 + 35u^5/2 + 35u^6/4 + \cdots)}{A_1^2(3/4 + 9u^2/16 - 6u^3 - 699u^4/32 + 105u^5 - 30645u^6/256)} \tag{18}$$

By elimination of ω^6 from Eq. 17 with the help of Eq. 18, the equation of u can be written as

$$u = \lambda_0 \left(1 - \frac{409u^2}{4} - 311u^3 - \frac{637u^4}{8} + \frac{17577u^5}{16} + \frac{119113u^6}{64} + \frac{3179u^7}{8} + \cdots \right), \tag{19}$$

where $\lambda_0 = \frac{1}{23}$.

The power series solution of Eq. 19 can be derived in terms of λ_0 as

$$u = \lambda_0 - \frac{409}{4}\lambda_0^3 - 311\lambda_0^4 + \frac{41661}{2}\lambda_0^5 + \frac{2561557}{16}\lambda_0^6 - \frac{40034213}{8}\lambda_0^7 + \cdots \tag{20}$$

Substituting the value of u from Eq. 20 into Eq. 18 and using $A_1 = a_0(1 + u)$, the approximate angular frequency can be determined as

$$\omega(a_0) = \sqrt[6]{\frac{(1 + 4u + 14u^2 + 21u^3 + 105u^4/4 + 35u^5/2 + 35u^6/4 + \cdots)}{A_1^2(3/4 + 9u^2/16 - 6u^3 - 699u^4/32 + 105u^5 - 30645u^6/256)}}$$
$$= \frac{1.063575}{a_0^{1/3}} \tag{21}$$

Thus, the approximation solution of Eq. 13 is $x(t) = \frac{A_1 \cos(\omega t)}{1 + u \cos(2\omega t)}$ where u and ω are, respectively, given by Eqs. 20-21.

5 Application of the Modified Rational Harmonic Balance Method (MRHBM)

We can apply the MRHBM directly in Eq. 13. The second term of Eq. 13 i.e. $x^{1/3}$ can be expanded in a Fourier series as

$$x^{1/3} = \sum_{n=0}^{\infty} b_{2n+1} x^{1/3} = b_1 \cos(\omega t) + b_3 \cos(3\omega t) + \cdots. \tag{22}$$

Herein $b_1, b_3, \cdots$ are calculated by the following integration as

$$b_{2n+1} = \frac{4}{\pi} \int_0^{\pi/2} x^{1/3} \cos[(2n+1)\varphi] d\varphi, \qquad (23)$$

setting $\varphi = \omega t$.

Now substituting Eq. 15, 22–23 into the Eq. 13 and then equating the coefficients of $\cos(\omega t)$ and $\cos(3\omega t)$, the following nonlinear algebraic equations are obtained as

$$-(1+u-11u^2/2)A_1\omega^2 + \frac{A_1^{1/3}(17820-2376u+1881u^2-938u^3)I_0}{5940\sqrt{\pi}} = 0, \quad (24)$$

$$(3u-9u^2/4)A_1\omega^2 - \frac{A_1^{1/3}(3564+3267u-531u^2+1211u^3)I_0}{5940\sqrt{\pi}} = 0, \quad (25)$$

where $b_1, b_3, \cdots$ are determined as

$$b_1 = \frac{A_1^{1/3}(17820-2376u+1881u^2-938u^3)I_0}{5940\sqrt{\pi}}, \qquad (26)$$

$$b_3 = -\frac{A_1^{1/3}(3564+3267u-531u^2+1211u^3)I_0}{5940\sqrt{\pi}}, \text{ where } I_0 = \frac{Gamma[\frac{7}{6}]}{Gamma[\frac{2}{3}]} \quad (27)$$

and so on.

After disentanglement, Eq. 24 can be written into another form as

$$\omega^2 = \frac{A_1^{1/3}(17820-2376u+1881u^2-938u^3)I_0}{5940(1+u-11u^2/2)A_1\sqrt{\pi}} \qquad (28)$$

By omitting ω^2 from Eq. 25 with the help of Eq. 28 and then some modification, one could obtain the following nonlinear algebraic equation of u as

$$u = \lambda_0\left(1 + \frac{3373\,u^2}{396} - \frac{56555\,u^3}{7128} + \frac{44711\,u^4}{14256} - \frac{8771\,u^5}{3564}\right), \text{ where } \lambda_0 = \frac{12}{157} \quad (29)$$

The power series solution of Eq. 29 in terms of λ_0 is

$$u = \lambda_0 + \frac{3373\lambda_0^3}{396} - \frac{56555\lambda_0^4}{7128} + \frac{7748693\lambda_0^5}{52272} - \frac{960746707\lambda_0^6}{2822688} + \cdots . \qquad (30)$$

Now substituting the value of u from Eq. 30 into Eq. 28 and using $A_1 = a_0(1+u)$, the approximate angular frequency can be obtained as

$$\omega(a_0) = \sqrt{\frac{A_1^{1/3}(17820 - 2376u + 1881u^2 - 938u^3)I_0}{5940(1+u-11u^2/2)A_1\sqrt{\pi}}} \qquad (31)$$
$$= \frac{1.077845}{a_0^{1/3}}$$

Therefore, the modified approximate solution of Eq. 13 is $x(t) = \frac{A_1 \cos(\omega t)}{1+u\cos(2\omega t)}$ where u and ω are, respectively, given by Eqs. (30)–(31).

6 Results and Discussions

The approximate angular frequencies have been obtained by standard rational harmonic balance method and modified harmonic balance method for the nonlinear non-smooth oscillators. For this nonlinear problem, the exact value of the frequency is

$$\omega_{ex}(a_0) = \frac{1.070451}{a_0^{1/3}},$$

which is stated in (Gottlieb 2003). The approximated angular frequencies have been plotted in Figs. 1 and 2. It is highly remarkable that the approximated results show a good agreement with the corresponding exact frequency. Moreover, the solution

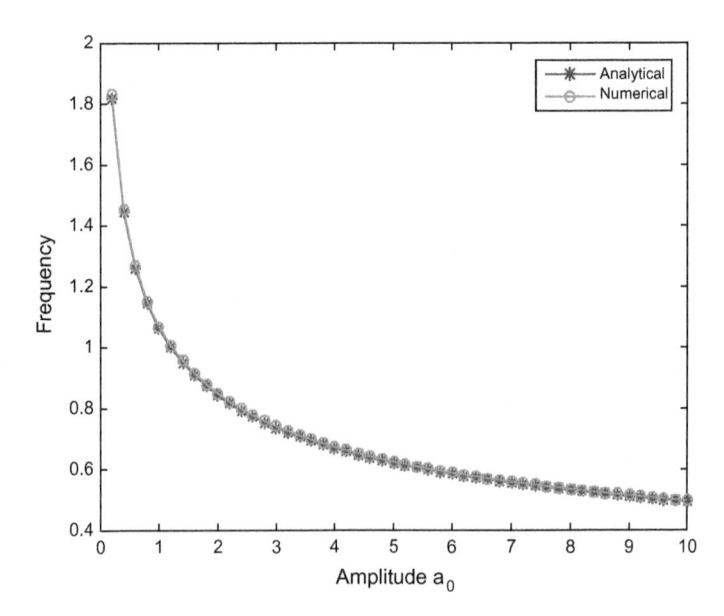

Fig. 1 Employed standard rational harmonic balance method

A New Analytical Technique for Solving Nonlinear Non-smooth … 461

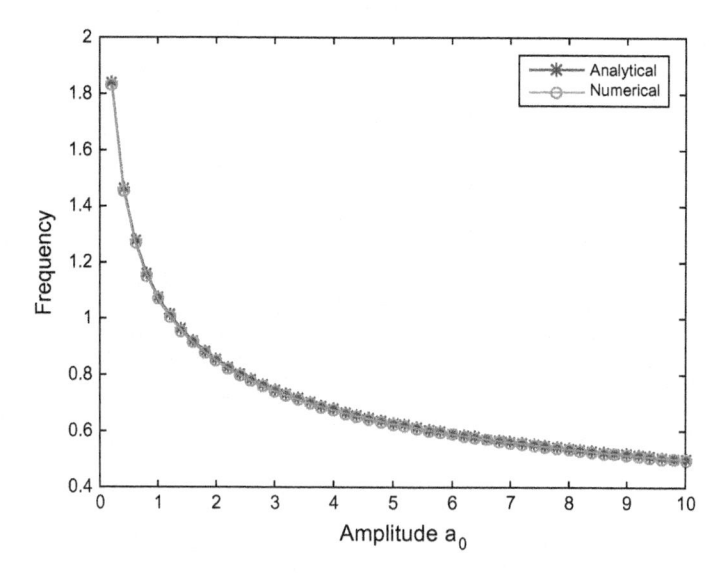

Fig. 2 Modified harmonic balance method

procedure of the proposed method is simple, straightforward, and quite easy. The advantages of this method include its analytical simplicity and computational efficiency, and the ability to objectively find better results.

7 Conclusion

A new analytical technique based on the rational harmonic balance method (RHBM) has been investigated to obtain approximate angular frequencies for the nonlinear non-smooth oscillators. The approximated angular frequencies give almost similar as compared to its exact ones. Moreover, in comparison with previously published methods the determination procedure of approximate solutions is straightforward and simple. The high accuracy and validity of the approximate frequencies assured about the results and reveal this method can be used easily for nonlinear non-smooth oscillators. To entirety up, we can say that the technique offered in this study for solving nonlinear non-smooth oscillators can be considered as powerful, an efficient alternative of the previously existing methods.

Acknowledgements The authors would like to acknowledge the financial supports received from the International Islamic University Malaysia, Ministry of Higher Education, Malaysia, through the research grant FRGS-14-143-0384.

References

Askari, H., Saadatnia, Z., Esmailzadeh, E., et al. (2014). Multi-frequency excitation of stiffened triangular plates for large amplitude oscillations. *Journal of Sound and Vibration, 333,* 5817–5835.

Azad, A. K., Hosen, M. A., & Rahman, M. S. (2012). A perturbation technique to compute initial amplitude and phase for the Krylov-Bogoliubov-Mitropolskii method. *Tamkang Journal of Mathematics, 43*(4), 563–575.

Belendez, A. (2009). Homotopy perturbation method for a conservative $x^{1/3}$ force nonlinear oscillator. *Computers & Mathematics with Applications, 58*(11–12), 2267–2273.

Belendez, A., Gimeno, E., Fernandez, E., et al. (2008). Accurate approximate solution to nonlinear oscillators in which the restoring force is inversely proportional to the dependent variable. *Physica Sripta, 77,* 065004.

Belendez, A., Pascual, C., Gallego, S., et al. (2007). Application of a modified He's homotopy perturbation method to obtain higher-order approximation of an $x^{1/3}$ force nonlinear oscillator. *Physics Letters A, 371,* 421–426.

Chopra, Ak. (1995). *Dynamic of structures, theory and application to earthquake engineering.* New Jersey: Prentice-Hall.

Cveticanin, L. (2009). Oscillator with fractional order restoring force. *Journal of Sound and Vibration, 320,* 1064–1077.

Daeichin, M., Ahmadpoor, M. A., Askari, H., et al. (2013). Rational energy balance method to nonlinear oscillators with cubic term. *Asian-European j. math., 6*(2), 1350019.

Ganji, D. D., & Azimi, M. (2012). Application of max min approach and amplitude frequency formulation to nonlinear oscillation systems. *UPB Scientific Bulletin, Series A: Applied Mathematics and Physics, 74*(3), 131–140.

Gottlieb, H. P. W. (2003). Frequencies of oscillators with fractional-power non-linearities. *Journal of Sound and Vibration, 261,* 557–566.

Hosen, M. A., Chowdhury, M. S. H., Ali, M. Y., et al. (2016). A new analytical approximation technique for highly nonlinear oscillations based on the energy balance method. *Results in Physics, 6,* 496–504.

Hosen, M. A., Chowdhury, M. S. H., Ali, M. Y., et al. (2017). An analytical approximation technique for the Duffing oscillator based on the energy balance method. *Italian Journal of Pure and Applied Mathematics, 37,* 455–466.

Hosen, M. A., Rahman, M. S., & Alam, M. S. (2012). An analytical technique for solving a class of strongly nonlinear conservative systems. *Applied Mathematics and Computation, 218,* 5474–5486.

Ikramul, B. M., Alam, M. S., & Rahman, M. M. (2013). Modified solutions of some oscillators by iteration procedure. *Journal of the Egyptian Mathematical Society, 21,* 142–147.

Lim, C. W., Lai, S. K., & Wu, B. S. (2005). Accurate higher-order analytical approximate solutions to large-amplitude oscillating systems with general non-rational restoring force. *Nonlinear Dynamics, 42,* 267–281.

Mickens, R. E. (2006) Iteration method solutions for conservative and limit-cycle $x^{1/3}$ force oscillators. *Journal of Sound and Vibration, 292,* 964–968.

Mickens, R. E. (2010). *Truly nonlinear oscillations.* Singapore: World Scientific Publishing Co., Pte. Ltd.

Mickens, R. E., & Semwogerere, D. (1996). Fourier analysis of rational harmonic balance approximation for periodic solutions. *Journal of Sound and Vibration, 195,* 528–530.

Nayfeh, A. H. (1973). *Perturbation Methods.* New York: J. Wiley.

Ozis, T., & Akci, C. (2011). Periodic solutions for certain non-smooth oscillators by iteration homotopy perturbation method combined with modified Lindstedt-Poincare technique. *Meccanica, 46,* 341–347.

Ozis, T., & Yildirim, A. (2007). Determination of periodic solution for a $u^{1/3}$ force by He's modified Lindstedt-Poincare method. *Journal of Sound and Vibration, 301,* 415–419.

Peijun, J. (2015). Global residue harmonic balance method for Helmholtz-Duffing oscillator. *Applied Mathematical Modelling, 39*(8), 2172–2179.

Yamgoue, S. B., Bogning, J. R., & Jiotsa, A. K. (2010). Rational harmonic balance-based approximate solutions to nonlinear single-degree-of-freedom oscillator equations. *Physica Scripta, 81,* 035003.

CDIO Implementation in Separation Processes Course for Chemical Engineering

Arbanah Muhammad, Salmi Nur Ain Sanusi,
Siti Hajar Anaziah Muhamad, Sharifah Iziuna Sayed Jamaludin,
Meor Muhammad Hafiz Shah Buddin, Mohd Zaki Sukor,
Sitinoor Adeib Idris, Nur Shahidah Ab Aziz
and Muhammad Imran Ismail

Abstract This study introduces the concept of Conceive, Design, Implementation, and Operate (CDIO) as an effective teaching method in engineering education to improve student's knowledge as well as to enhance their soft skills while learning. CDIO concept was incorporated in a course named separation processes which was offered to students undertaking Diploma of Chemical Engineering in UiTM Pasir Gudang. Specifically, the concept was applied through their liquid–liquid extraction (LLE) laboratory work where students were given specific tasks that comply with CDIO standards. The effectiveness of CDIO implementation was evaluated through survey questions. The data available suggested that students had a fun learning experience as compared to the conventional method of teaching though most of them agreed that this course is the most difficult course for their level of study. Moreover, they gained greater depth of knowledge as they were exposed to the real chemical engineer roles in industry through this activity. Based on the outcome, CDIO concept was proven to be successfully implemented in separation process; hence, it is highly recommended that this approach to be enforced in other courses in chemical engineering program.

Keywords Separation processes · Chemical engineering · Project-based learning
CDIO

A. Muhammad (✉) · S. N. A. Sanusi · S. H. A. Muhamad · S. I. Sayed Jamaludin
M. M. H. Shah Buddin · M. Z. Sukor · M. I. Ismail
Faculty of Chemical Engineering, Universiti Teknologi MARA,
Cawangan Johor, Kampus Pasir Gudang, Jalan Purnama, Bandar
Seri Alam, Masai, 81750 Johor Bahru, Johor, Malaysia
e-mail: arbanah7188@johor.uitm.edu.my

S. A. Idris · N. S. Ab Aziz
Faculty of Chemical Engineering, Universiti Teknologi MARA,
Shah Alam 40450, Selangor, Malaysia

1 Introduction

The chalk and talk method which was widely used as conventional way of teaching seems to show a declining trend in terms of its effectiveness to attract students' attention in classrooms. Students nowadays have better access to reliable and attractive information as they are well equipped with electronic gadgets together with good internet connection. As a consequent, they have less interest in learning through the conventional method due to its unattractiveness and time-consuming process. This phenomenon resulted in failure of knowledge transfer process as students unable to fully understand the lecture. In fact, students are unable to redeliver and apply the fundamental knowledge in some circumstances, and it is highly likely to happen among engineering students. They also lack in good communication and technical report writing skills where both skills are of utmost important for engineers. To curb this issue, activities in classroom, especially for engineering courses should be conducted frequently and interestingly while the focus should be student centered rather than lecturer centered.

Apart from fundamental knowledge in their respective field, the Ministry of Higher Education, Malaysia, also pointed out the importance of having good soft skills among graduates. Graduates especially from engineering background must be able to absorb knowledge as well as able to translate the knowledge gained into design and work. The introduction of Conceive, Design, Implementation, and Operate (CDIO) concept in engineering education is an improvised teaching and learning method in classrooms through relevant activities. This process allows students to implement their knowledge besides improving their skills to serve the ever-increasing needs of professional engineers. Furthermore, the Institute of Chemical Engineering (IChemE) emphasized that the chemical engineering education should stimulate and develop student's talents with a good chemical engineering program. Therefore, Massachusetts Institute of Technology (MIT) and few universities in Sweden have outlined CDIO framework, consisting 4 parts syllabus, 13 skill sets, and 12 standards.

This well-structured framework focuses on student-centered learning by designing an outcome-based curriculum that utilizes active learning to promote skills in conceiving, designing, implementing, and operating a product or systems using a life-cycle approach together with soft skills. The skills developed along the process include teamwork, communication, various thinking skills, and others. As mentioned by Cheah (2009), teaching through CDIO approach is contextual that should reflect real-world environment. This learning concept of CDIO was nearly similar to problem-based learning (PBL) which stimulates problem-solving thinking among students. However, to the best of our knowledge reports on the application of CDIO concept in students' laboratory work scarce.

The key objective of chemical engineering education is to synthesize subjects into designing system, component, process, or experiments (Cheah 2008). The introduction of CDIO model in Chemical Engineering can meet the increasing demand among employers for graduate who possess not only technical skills, but wide range of soft skills or higher order skills including communication, presentation, teamwork, and leadership (Bryne 2006). The present study was conducted to evaluate the effectiveness of CDIO model in students' laboratory work of chemical engineering education. The laboratory experiment used in this study was liquid–liquid extraction (LLE), known to be a component in separation processes course's syllabus for Diploma in Chemical Engineering program in UiTM Pasir Gudang.

2 Materials and Methods

2.1 CDIO Activity in Liquid–Liquid Extraction (LLE) Experiment

CDIO concept was incorporated in liquid–liquid extraction (LLE) experiment, which was a component in separation processes course's syllabus for academic session December 2016–April 2017. This course was taken by 191 students during their third year of study in Diploma in Chemical Engineering. Instructions and guidelines for LLE experiment were designed based on CDIO framework and the experiment was conducted at a pilot plant in Fakulti Kejuruteraan Kimia, UiTM Pasir Gudang.

The total number of students took part in this study is presented in Table 1. However, based on the table, the total number of responses for each survey varies and only valid responses will be counted throughout this study. Considering CDIO syllabus, the selection of topics must comprise personal and professional skills, attributes and interpersonal skills, as suggested by Karpe et al. (2011). The topic and skills obtainable is presented in Table 2.

Table 1 Number of students enrolled and valid response

Subject	Number of responses
Number of students enroll in this course	191
Number of students conducted LLE laboratory (CDIO)	171
Number of responses for post-experiment	91
Number of valid responses for post-experiment	87

Table 2 Topics as suggested by Karpe et al. (2011)

Skills	Topic
Personal and professional skills and attributes	Engineering reasoning and problem-solving, experimentation and knowledge discovery, system thinking, personal skills and attributes, professional skills, and attributes
Interpersonal skills	Teamwork, communications

2.2 Group Distribution

Students, as respondents to the survey were divided into groups of 12 where they were required to work as a team. The groups were divided according to their academic performance in the previous academic session where each group consists of weak, average, and good students. To enhance their interpersonal skills, each group has appointed one member among themselves to serve different roles; manager, safety officer, process engineer, and senior technician. These roles were chosen due to the importance of its existence in a chemical processing plant. As their first assignment, they were required to select a leader to allow the experimental works to be carried out efficiently and effectively. Good interpersonal skills must be shown by the manager as the person must delegate tasks among the team members. Theoretically, the characters of the respondents are highly dependent on their specific roles. The manager must be able to identify special characters of each team member to properly delegate tasks among the team members as suggested in Table 3.

Once this process completed, they were given a situation as the instruction in order to conduct the LLE experiment as follows:

"You have just joined Maju Jaya Ltd; a company that purifies acetone from acetone–water mixture using toluene as solvent. For the first task, you and your team is required to operate a liquid–liquid Extraction (LLE) column with a recovery unit. Upon completion of this task, you need to prepare a report. Refer to the theory given as guidance".

Each member in a group is responsible to take on the assigned role where they need to understand and must deliver the given tasks. The laboratory session is allowed to be conducted once they are well prepared to run the experiment with

Table 3 Roles and its specific tasks

Roles	Tasks
Managers	1. Distribute roles to each team members 2. To ensure the smoothness of the overall process 3. Responsible for submission of the final memo at a given time to the CEO (Lecturers)
Safety officer	1. To ensure all necessary personal protective equipment (PPE) to be worn by every member 2. To ensure chemicals to be handled properly 3. To prepare and implement site safety and health plan 4. Aware of active and developing situations at all time
Process engineer	1. Thoroughly understand and analyze critically the overall process 2. To prepare process flow diagram (PFD) and explain the process to each member 3. In occurrence of any problems, please solve them accordingly
Senior technician	1. Understand the explanation provided by the process engineer 2. To operate the unit accordingly 3. In occurrence of any problems, please provide assistance to solve them

minimal supervision. Students are expected to be able to apply and relate fundamental knowledge of chemical engineering that they have gained throughout their two years of study.

Once they completed the experiment, they are required to present their work to the CEO, which played by the lecturer via an oral presentation. Questions will be asked according to their roles and some questions were randomly given to any members of the group to verify their involvement and understanding in the experimental work. As a knowledgeable person, the CEO will pointed out their wrongdoings during the experiment (if any), and they need to defend themselves. The process was purposely carried out under a stressful condition to test the students' limit to work under pressure. They were assessed through this process and marks were given based on the oral presentation. Also, respondents will be evaluated on scale 1 to 10 based on these criteria; work distribution, technical report, safety performance, clean-up, as well as pre-experiment survey/post-experiment survey. Online pre- and post-laboratory experiments surveys were carried out at the beginning and end of the course to evaluate the effectiveness of CDIO model implementation in LLE laboratory work.

3 Results and Discussion

Implementation of CDIO concept was carried out via one course at a time as it is highly unlikely to change overall program structure at once without knowing the effectiveness this new teaching concept. Hence, this study applies CDIO concept specifically in separation processes as it incorporates multiple fundamental chemical engineering knowledge in this course. The course supports the CDIO elements as it consists of a laboratory work involved for liquid–liquid Extraction (LLE) topic where students able to experience real working environment just like engineers do. To be a proficient engineer, students need to equip themselves with professional skills such as critical thinking, problem-solving and interpersonal skills. These are conventional skills heavily gravitated toward content and knowledge acquisition (Karpe et al. 2011).

In this study, students are required to accomplish few major tasks; perform pre-laboratory experiments, perform experimental works and oral presentation. These tasks demanded their higher order thinking skills, interpersonal skills, knowledge understanding and learning quality. After the laboratory session, students produced a detail technical report. Evidently, it was found that chemical engineering students have poor technical report writing skills. Since this course was taken by students who were in their third year of study, they should have an excellent fundamental knowledge in chemical engineering in order to conduct the CDIO incorporated LLE experiment. Besides separation processes, there are four other main courses taken by them during the current academic session, including heat transfer and equipment, separation processes, plant safety and occupational health, process control and instrumentation. Students have shown impressive presentation skills during their oral presentation and able to answer questions by linking the fundamental knowledge with their hands-on experience.

Next, students answered an online post-experiment survey and the data collected is presented in Table 4 and Fig. 1. Questions were divided into two sections, namely individual assessments and course assessments. Questions numbered 1, 2, 7, and 9 are individual assessments while questions numbered 3, 4, 5, 6, and 10 assessed the course. The respondents need to answer based on the following scale; 1 (strongly disagree), 2 (disagree), 3 (mixed feeling), 4 (agree), and 5 (strongly agreed).

Based on Fig. 1, Question 3 showed the highest mean value (4.05 ± 0.85) as compared to the others. Majority of the students expected that this course is the most difficult course in Chemical Engineering program. Though students agreed that this course is difficult, but a high mean value for Question 2 (4.03 ± 0.81) reveals that respondents acknowledged the importance of this course in preparing themselves to be a chemical engineer. This is due to the fact that respondents could obtain greater depth of knowledge through CDIO activities as well as able to experience a real working environment. Along this process, students able to self-gain the knowledge as they applied the concept of implementation and operate.

The lowest mean value (3.60 ± 0.83) was found in Question 4 which concerns the student's ability to define and comprehend the concept of mass transfer. This is

Table 4 Survey questions for CDIO implementation in separation processes

Question	Mean	Individual/course assessment
Your interest in this course	3.82 ± 0.81	Individual
Expected course relevancy in preparing you to be an engineer	4.03 ± 0.81	Individual
Expected difficulty level of this course	4.05 ± 0.85	Course
Ability to define and explain the concept of mass transfer	3.60 ± 0.83	Course
Ability to identify and differentiate between extract and raffinate phases as well as between solute, carrier, and solvent	3.76 ± 0.86	Course
Level of understanding of fundamental concepts for the respective separation processes	3.63 ± 0.82	Course
Fun learning experience	3.94 ± 0.97	Individual
Do you think this learning method helped you to develop new skills that you did not acquire through the conventional learning method?	3.87 ± 0.82	Course
Do you think you can gain more/deeper knowledge through this learning experience?	3.99 ± 0.84	Individual
Will you recommend this learning method to other students?	3.91 ± 0.9	Course

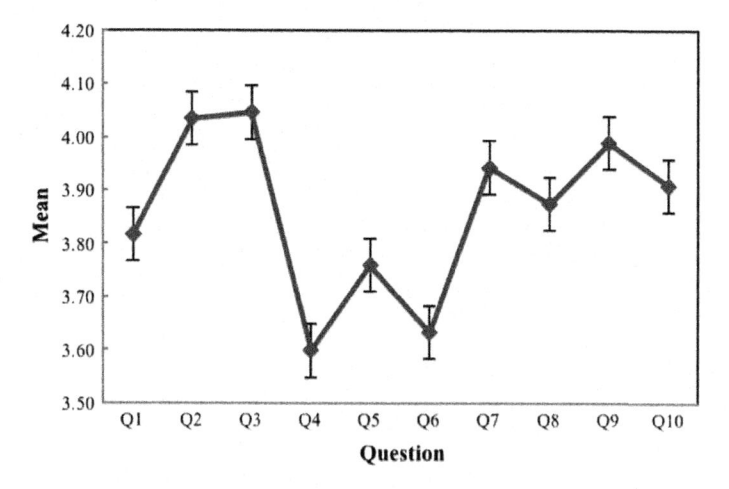

Fig. 1 Mean results for question 1–10

the fundamental concept in separation processes. It can be seen that students were unable to relate the basic knowledge of mass transfer in LLE operation process at satisfactory level. In other words, students tend to merely memorize the meaning of mass transfer but have the difficulty to relate this concept while conducting the experimental works. Nevertheless, they able to describe basic terms relating to LLE operation such as extract, raffinate, solute, solvent, and carrier. This situation was proven via Question 5 as the recorded mean value was slightly higher (3.76 ± 0.86) as compared to Question 4. However, declining mean value (3.63 ± 0.82) was obtained in Question 6 where it tested their level of understanding on fundamental concepts. It is highly possible that low mean values were recorded for these questions due to the application of conventional teaching method when delivering the fundamental concepts in classrooms. As CDIO concept was solely used in their experimental works, the results between these two teaching methods differ significantly.

The CDIO activities in this course had proven that students can develop new skills which are almost impossible to be obtained through the conventional learning method. This phenomena was proven by the high mean value (3.87 ± 0.82) for Question 8. Besides, students recognized that the implementation of CDIO concept in their course has yielded a fun learning experience for them, as asked in Question 7.

On the other hand, Question 9 showed a high mean value (3.99 ± 0.84). This question inquired their knowledge level after conducting the experiment. This value affirmed that CDIO implementation in this LLE experiment has a great potential in increasing students' level of knowledge. Self-direct learning concept was applied in this study, as suggested by Savin-Baden's (2003). This concept was adopted as the students were given a scenario as stated in Sect. 2.2. Through the instructions, the self-direct learning questions were triggered and this factor has contributed to the recorded results. Students realized that they have obtained greater depth of knowledge as they able to link the knowledge they gained classrooms with their experience either via hands-on as well as via visual. Additionally, it is also claimed by Robison and Aronica (2009) that a good working environment able to help students in self-direct learning. Via this method, students able to utilize the knowledge they gained in classrooms to complete the task by working in a harmonious group.

Lastly, most of the students agreed that the CDIO implementation should be recommended to other courses (Question 10), at mean value 3.91 ± 0.9).

4 Conclusion

The work of utilizing CDIO concept in Diploma in Chemical Engineering is a breakthrough and more activities need to be developed to improve student's motivation and results in the course taken. The integrated CDIO framework in chemical engineering needs to be expanded to meet future demand. It can be concluded that the introduction of CDIO in separation process able to improve student's interest and performances as well as their level of knowledge. Activities introduced in this course were designed to meet CDIO approach where students can prepare themselves to be chemical engineers as they acted as one. Based on 10 questions given, the implementation of CDIO in separation processes was assessed. Even though students found out that this subject is the most difficult subject in chemical engineering, the implementation of CDIO has successfully created fun learning experience and at the same time their knowledge in Chemical Engineering widens. The implementation of CDIO should be further implemented on other courses for the benefit of the students.

Acknowledgements This work is partially supported by Academic and Research Assimilation (ARAS) project no 600-IRMI/DANA 5/3/ARAS (0065/2016). Authors also gratefully acknowledge the help from students and staff of Faculty of Chemical Engineering, UiTM Johor, Pasir Gudang Campus who involved directly and indirectly in the survey.

References

Bryane, E. P. (2006). The role of specialization in the chemical engineering curriculum, education for Chemical Engineers. Tranns I Chem E, Part D 3–15. https://doi.org/10.1205/ece05008.

Cheah, S. M. (2008). Revamping the diploma in chemical engineering curriculum: Issues and challenges. 2nd International Symposium on Advances in Technology Education, At Kumamoto, JAPAN.

Cheah, S. M. (2009). Using CDIO to revamp the chemical engineering curriculum, Singapore Polythecnic. Proceedings of the 6th International CDIO Conference, École Polytechnique, Montréal, June 15–18, 2010."

Karpe, R. J., Maynard, N., Tade, M. O., & Atweh, B. (2011). Taking CDIO into chemical engineering classroom: Alingning curriculum, pedogogy, assessment. Proceedings of the 7th International CDIO Conferences. Technical University of Denmark, Copenhagen, Jun 20–23.
Robinson, K., & Aronica, L. (2009). The element: How finding your passion changes everything, Allen Lane, imprint of Penguin Books, Victoria, Australia.
Savin-Bade, M. (2003). Facilitating problem-based learning: Illuminating perspectives, Society for Research into Higher Education, Open University Press, Maindenhead, UK.

Usewear Experiment to Determine Suitability of Rock Material from the Crater of Meteorite Impacts as a Prehistoric Stone Tool

Siti Khairani Abd Jalil, Jeffrey Abdullah and Mokhtar Saidin

Abstract Bukit Bunuh has been identified as a crater result from the impact of meteorite. The impacted area has a variation type of rocks. However, through excavations, it was found that there are three types of rocks in the area which have been used for producing stone tools, for instance, suevite stone (impact rocks), quartzite and metasediment rock such as cherty. The significant difference between the three types of stones is their texture. Suevite rocks have a heterogeneous texture, while quartzite and cherty have homogenous and fine grains. However, quartzite rocks differ from cherty rocks in terms of the grain size where quartzite stones are coarser compared to cherty rocks. Normally, a homogenous material and fine rocks are suitable to be used as a tool. Since suevite rock is also used in this experiment, therefore, the analysis is done to investigate the differences and compatibility of the textures for these two types of rocks, heterogeneous and homogenous. The study also aims to determine the effect of attributes on the different materials. For this purpose, usewear experimental was used. Observation was done on the attributes results from the experiments which are roundness, fraction, polish distribution, type of polish distribution and linear feature. The result shows that suevite rock suitable to be used as stone tools because its edge is wear-resistant and harder but difficult to obtain sharp edges. For cherty rocks, it is easy to flake and wear but it has sharper edge compared to suevite rock. This means that the impacted rocks are suitable to be used as stone tools but will produce different usewear attributes according to the rock's texture.

Keywords Attributes · Usewear · Texture · Suevite · Cherty

S. K. Abd Jalil (✉) · J. Abdullah · M. Saidin
Centre for Global Archaeological Research, Universiti Sains Malaysia,
Gelugor, Malaysia
e-mail: sitikhairanijalil@gmail.com

1 Introduction

Lenggong Valley is located in Hulu Perak, Malaysia, has been known as the capital of prehistoric people (Zuraina 2003). This area was gazetted as a world heritage site under UNESCO (Department of National Heritage 2011). Many Palaeolithic sites have been studied in this area including Kota Tampan 1987 (Zuraina 1996), Kota Tampan 2005 (Hamid 2007), Kampung Temelong (Mokhtar 1997), Lawin (Mokhtar 1998), Gua Gunung Runtuh (Zuraina 1996) and Bukit Jawa (Zuraina 1996). One of the most important sites in this valley is Bukit Bunuh. Bukit Bunuh is one of the places which was impacted by meteorite around 1.83 million years ago by refering to the chronometric dates (Mokhtar 2010: 82–83). According to the research done by Mokhtar (2006, 2010), Nur Asikin (2013) and Nor Khairunisa (2013), the affected area shows evidence of Palaeolithic era which made the area as a complex site. The impact of these meteorites had led to the existence of different types of rocks, such as suevite, cherty, metasediment, quartz, quartzite (Mokhtar 2006; Nur Asikin 2013). As a result from the discovery and excavation on the open site in the meteorite impacted area, there are two types of impact rocks which were selected to be used as a tool by Palaeolithic society. The selected rocks are suevite and cherty.

Both types of this impact rocks are different in physical aspects and their formations. However, both of these rocks were used by Palaeolithic society (Mokhtar 2006; Nur Asikin 2013). Cherty is metamorphic rock which consists of two hornfels rocks. It is most likely originated from sediment mudstone (Nur Asikin 2013). While suevite rock is an affected breccia stone which consists of different type of rocks (Nur Asikin 2013). Therefore, the usewear experiment was designed by using two types of rocks from Bukit Bunuh which are suevite and cherty. These two rocks have been chosen based on their properties which are cherty has finest texture and suevite has coarsest texture (Nur Asikin 2013).

2 Objective

The objective of this study is to investigate the differences between the usewear effect on suevite and cherty rocks based on selected attributes. Besides, this research also to study the development of attributes on these two different types of rocks.

3 Material

For this purpose, two types of materials are used, namely suevite and cherty rocks (Fig. 1). This rock is found in the central part of the impacted meteorite crater (Jinmin et al. 2014). In this crater area, there are many rocks which were produced such as suevite, quartzite, cherty, argite and quartz. However, only the suevite and

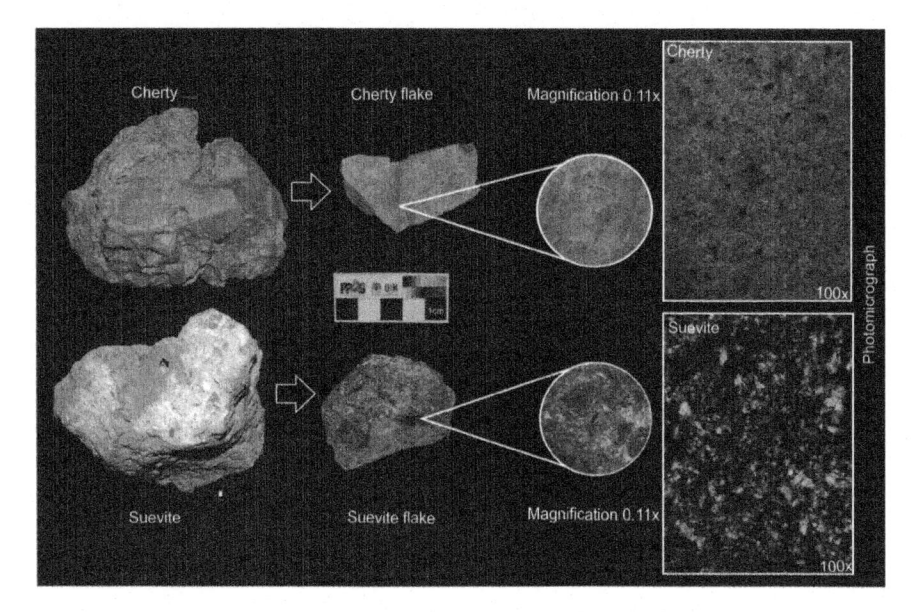

Fig. 1 Hand specimen of suevite and cherty showing the difference texture under stereomicroscope and polarizing petrographic microscope

cherty rocks have been used as a material to create stone tools by prehistoric societies (Nor Khairunisa 2013; Nur Asikin 2013).

Impact rock such as suevite is a heterogeneous rock. It consists of various types of minerals with various sizes. It is said to be hard because the molten fluid filling the spaces between the minerals. This melt liquid is formed due to high pressure resulting from meteorite impact. Therefore, according to Nur Asikin (2013), suevite rock is suitable as a tool due to its hardness. Only homogeneous, fine grain and delicate rocks are said to be suitable for stone tools (Sharon 2008). This is because the fine-textured rock when broken will form a conchoidal. To prove the suitability of these stones as a tool, the usewear experiment was conducted.

4 Methodology

For these purposes, the usewear experimental method has been carried out. Usewear experiments were done by selecting the working method. For this experiment, sawing method has been applied by using rattan as a working medium. This experiment requires an observation on specific attributes which are roundness, fraction, polish distribution, type of polish development and linear. These attributes are observed before flake stone is used and subsequently every 20 min for one hour. Comparison of these attributes between two types of stones is conducted to determine the changes in attributes over time (Fig. 2).

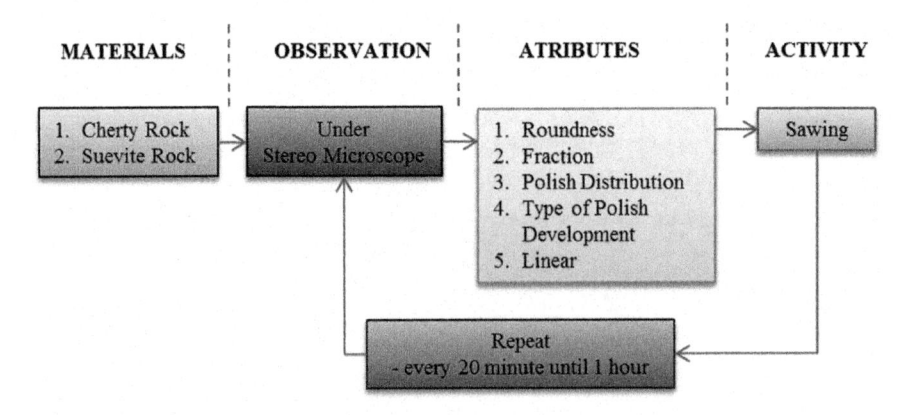

Fig. 2 Flowchart of usewear experiment on suevite and cherty flake

5 Result and Discussion

These experiments focused on the changes of usewear attributes on the edge of the flake. The data obtained from these experiments will be used to compare the development on the edge of fine and rough flake tools.

Fine rocks usually produce fractures in conchoidal shaped as shown in Fig. 3. They are different with coarse rocks, where they will produce snap fractures. This is because the effect from the usage of flake on minute 20–60 which caused the edge of cherty flake became blunt and worn.

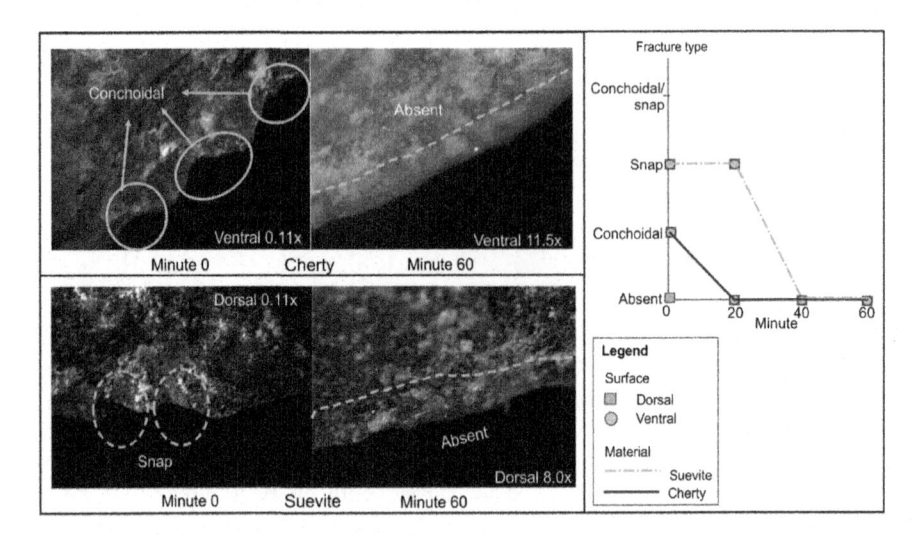

Fig. 3 Fracture type on cherty and suevite flake

Meanwhile suevite flake rocks produced snap fractures on minute 0. These snap fractures were formed during the process of producing flake. Besides, the fractures formed because the suevite is a coarse rock. On minute 20, the sawing activities produced snap fracture on the edge (Fig. 3). It was formed due to friction between the tools and working medium which in this case is rattan. The snap fracture which only happens on the thin edges will show horizontal movements as it often happens when slicing or sawing activity was done by using untrimmed flake.

This is because while sawing activities in progress, the lateral movement of the tools will flexure the edge of tools and produced snap fracture (Grace 2012; Hamon and Plisson 2008; Keeley and Newcomer 1977; Keeley 1974). Meanwhile, on minute 40, the fractures become increasingly blunt and wear out due to the constant use on the edge of flakes. Therefore, there is no fracture on minute 60 and the edge of the tools become falter and blunt.

Suevite flake produces roundness attributes relatively slower compared to cherty flake. This is because suevite flake has hard and rough edges. Therefore, this condition makes it difficult to form a smooth and roundness surface. Because of that, the level of roundness is very low on minute 40 and started to reach high roundness level on minute 60 (Fig. 4).

On the other hand, for cherty flake, the roundness formation is high on minute 40 and minute 60 because the surface characteristic is fine and makes it easy for edge of the flake stones to become smooth and round. According to the observation by using microscope, the edge of cherty flake on minute 60 (Fig. 4) produced roundness and shiny edge as the same as roundness effect which was defined by Sussman (1985, 1988) and Keeley (1974, 1980). According to research done by Keeley (1980), roundness is a round effect on the shiny edge which look like a melting snow.

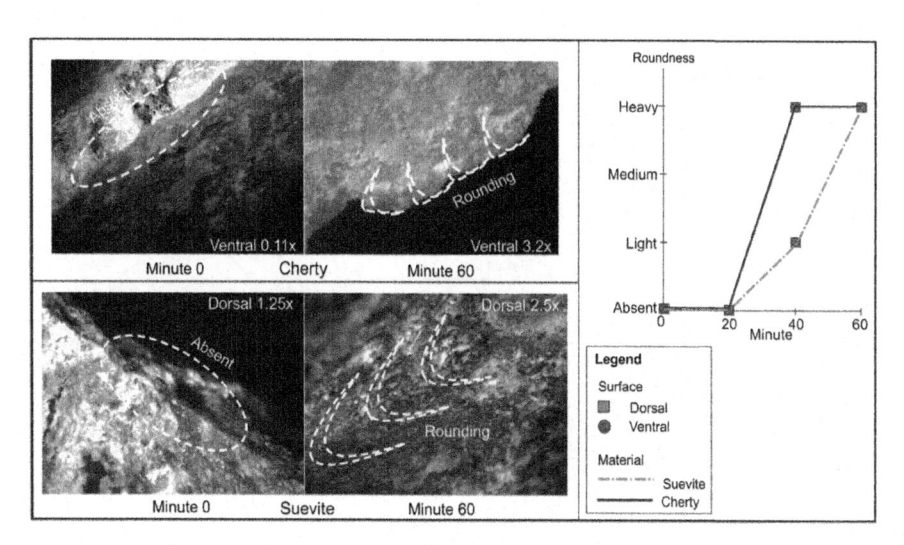

Fig. 4 Roundness of cherty and suevite flake

The polished distribution consists of two main features which are continuing and intermittent. For suevite flake rocks, the polished distribution is formed more continually compared to cherty flake, which formed more continuous on the edge of flake stones. The suevite flake formed the continuous polished distribution attributes on minute 60 but only on dorsal side of flake. Whereas, ventral side of suevite flake on minute 60 is still in intermittent distribution (Fig. 5). The intermittent polished distribution formed when the convex and uneven surface of the edge rummage with working medium.

Since cherty flake has smooth and flat surface, the continuous polished distribution was formed on the edge of the flake stones. Flat surface will facilitate the dissemination and it will produce a constant distribution (Barton et al. 1998; Grace et al. 1985; Grace 1990; Olle and Verges 2014; Sussman 1985). Therefore, the edge of cherty flake formed continuous polished distribution compared to suevite flake which formed intermittent polished distribution on the edge of flake (Fig. 5).

According to Keeley (1980), the type of polished distribution which was formed depends on the shape of topography surface on the edge of flake. Therefore, cherty flake formed the polished on the edge only and then evenly distributed on minute 20 until 60, meanwhile, suevite flake is formed isolated polished. This is because cherty flake has smooth and flat surface compared to suevite flake where it has coarse and uneven surface. Flat and smooth surface facilitates the formation of the polished on the edge of flake rocks.

Suevite flake rocks formed isolated polished on the ventral side of flake rocks. It takes a long time to become evenly distributed to dorsal side which is on minute 60. It is because the surface of the dorsal and ventral side on the edge of the flake rocks has coarse and uneven topography. Therefore, polished distribution only formed on

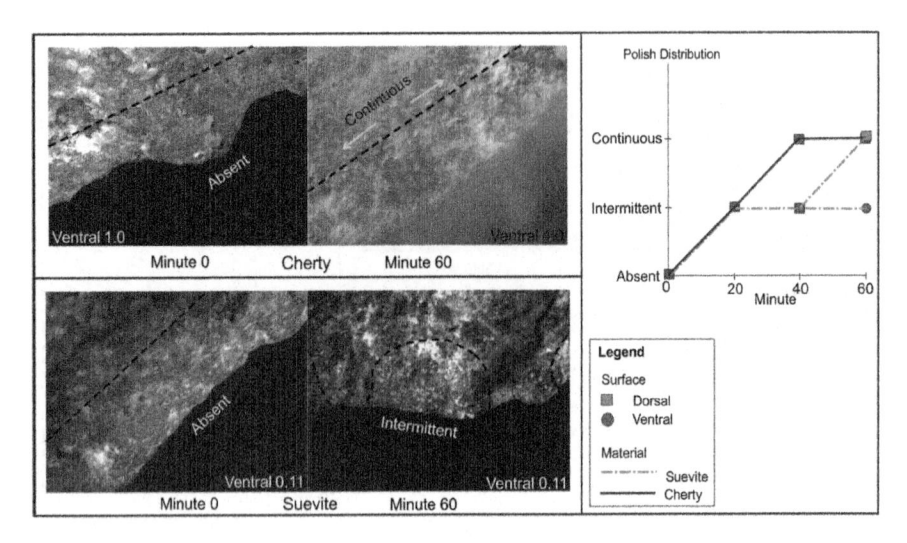

Fig. 5 Polished distribution on cherty and suevite flake

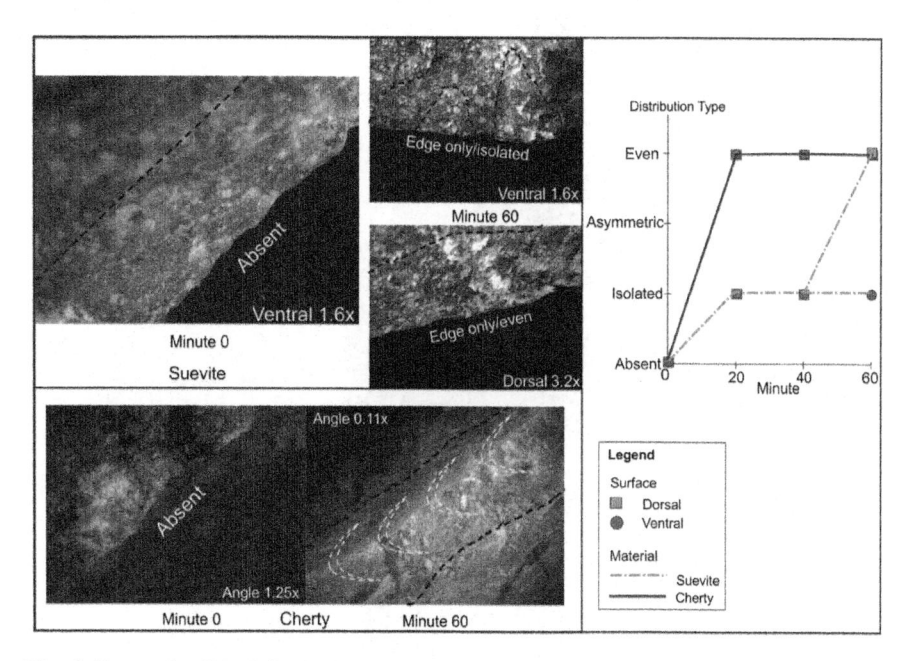

Fig. 6 Types of polished distribution on suevite and cherty

dorsal side of the flake stones because this surface almost flat compared to surface of ventral side (Fig. 6).

Linear feature is divided into four types, which are linear parallel, vertical, angular and circle. This type of linear identified by polished lines resulting from a scratch on the edge of the flake (Lerner 2007). According to Semenov (1964) and Kay (1996, 1998), the linear polished is one of the most important attributes to identify the function of the tools. Linear polished which is formed at the edge of the cherty and suevite flake is parallel because these two tools carry out the same work which is sawing. The difference between these two flake is the development of linear strip on the edge of the flake tools (Fig. 7).

Cherty flake has produced linear strip in its early usage which is on minute 20 until 60. This proves that the fine-grained flake rocks easily shown an effect on the edge of flake. Besides that, with fine and homogeneous characteristic plus the flat surface of rocks, it promotes the development of polished on the edge of the flake. Continuous polished distribution will form linear strip on the edge of it.

The linear development of suevite and cherty flake is different. According to the observation, on minute 40, the linear strip started to appear on the dorsal side of flake and obviously seen on minute 60 (Fig. 7). Meanwhile, there is no linear strip observed on ventral side after it was used on minute 20, 40 and 60. This is because the surface of the dorsal side is nearly flat which results in the development of polished in minutes 60 then leads to development of linear strip on the edge of the flake. If the development of polished with gaps, the linear line will not form on the

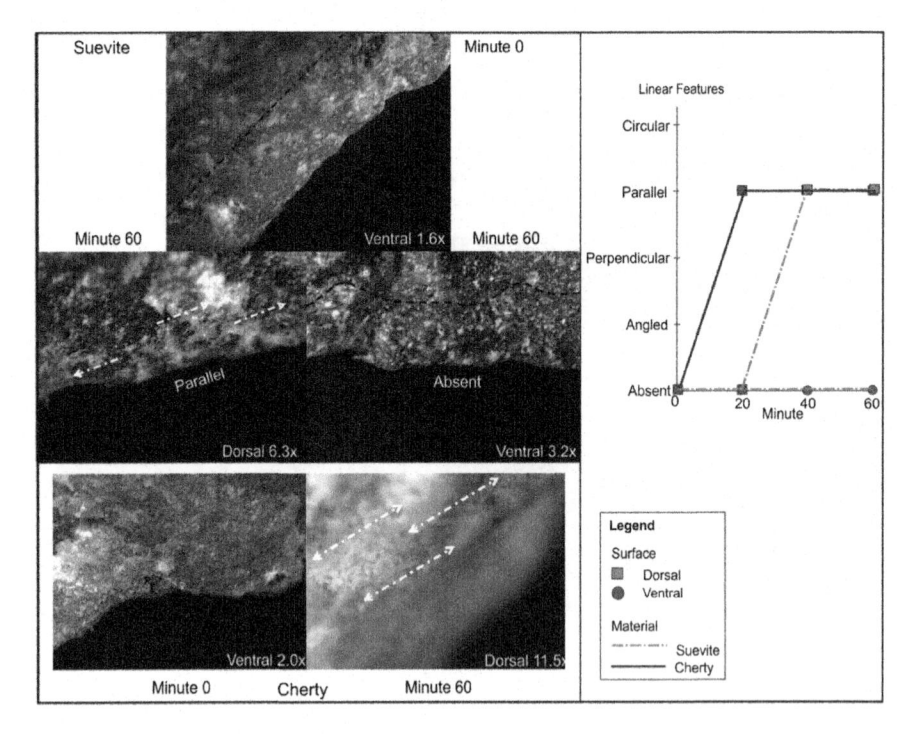

Fig. 7 Linear features on suevite and cherty flake

edge of flake. This is because it is only formed when there is continuous polished on the edge of the flake. Therefore, there is no linear strip on the ventral side of it.

6 Conclusion

Result from the experiment shows that different rocks material will produce different usewear attributes. The fine and homogeneous textures rocks such as cherty have homogenous textured surface and it eases the development of attributes on the edge of flake. Whereas, coarse and heterogeneous textures rocks such as suevite is difficult to form attributes because of coarse texture on the surface. Furthermore, the edge of suevite flake stones takes longer time to wear out, therefore, the usewear attributes also taking longer time to develop compare to cherty flakes. This is because the smooth and homogenous textured surface of cherty flake make it easier for polished development. Besides of the time taken, the other factor should be considered for both materials to have similar attributes is working medium. Soft medium needs longer time to form similar attribute compared to the harder material. Therefore, the harder working medium, the sooner they have similar attributes.

Therefore, it is concluded that the size of grain influence the development of attributes. Besides that, the size of grain also affects the sharpness and durability of the edge of the tools. Suevite rocks are suitable to be used as a tool because it has very hard edge and difficult to wear out. However, this type of rock difficult to form or flake because of its characteristic. Moreover, cherty rocks also suitable as a tool. Although this type of stones easy to wear out, but its edge is sharp and easy to flake to be used as a tool because of its fine grain and also conchoidal.

Acknowledgements We would like to express our gratitude to the Vice-Chancellor Prof. Datuk Dr. Asma Ismail and former Vice-Chancellor, Prof. Dato' Dr. Omar Osman, who has lent support to this archaeological project. This research also would not be possible without funding from Project Litik Technology in Palaeolithic at Malaysia Grants (1001/PARKEO/870013), Archaeological Research Malaysia and Global Grants (1001/PARKEO/270015), Sabah North Archaeological Research: Exploration of early human evidence Projects Grant (203/PARKEO/6730139) and Project Evidence Of Migration Routes During Palaeolithic Era In Eastern Sabah, Sabah (1001/PARKEO/8016020). Our thanks to the staff of Centre for Global Archaeological Research, USM, for their patience in providing guidance and assistance to our friends as well as everyone who was involved directly and indirectly throughout the execution of this study.

References

Barton, H., Torrence, R., & Fullagar, R. (1998). Clues to stone tool function re-examined: Comparing starch grain frequencies on used and unused obsidian artifacts. *Journal of Archaeological Science, 25*(12), 1231–1238.

Department of National Heritage. (2011). Archaeological heritage of the Lenggong Valley: Nomination dossier for inscription on the UNESCO world heritage list (Vol. 1). Malaysia: Department of National Heritage, Ministry of Information Communications & Culture, p. 418.

Grace, R., Graham, I. D. G., & Newcomer, M. H. (1985). The quantification of micro-wear polishes. *World Archaeology, 17*(1), 112–120.

Grace, R. (1990). The limitations and applications of use-wear analysis. The interpretive possibilities of Micro-wear analysis. *Uppsala: AUN, 14*, 9–14.

Grace, R. (2012). *Interpreting the function of stone tools*. : Ikarus Books.

Hamid Mohd Isa. (2007). *70,000 Tahun Dahulu) di Kota Tampan, Lenggong, Perak*. Master's Thesis. Universiti Sains Malaysia, Penang. (Unpublished).

Hamon, C., & Plisson, H. (2008). Which analytic framework for the functional analysis of grinding stones? The blind test contribution. In Which analytic framework for the functional analysis of grinding stones? The blind test contribution (pp. p-29). *BAR International Series,* Archaeopress, Oxford.

Jinmin, M. Rosli Saad, Mokhtar Saidin dan Nur Azwin Ismail, (2014). The Bukit Bunuh possible meteorite impact, Malaysia: Final stage results of impact crater from 2-D electrical resistivity tomography survey. *The Electronic Journal of Geotechnical Engineering, 19,* Bundle F.: 1499–1504.

Kay, M. (1996). Micro-wear analysis of some clovis and experimental chipped stone tools. In G. H. Odell, (Eds.), *Stone tools: Theoretical insights into human prehistory* (pp. 315–344). Plenum Press, New York.

Kay, M. (1998). Scratching the surface: Stone artifact micro-wear evaluation. In M. B. Collins, (Eds.), *Wilson-Leonard: An 11,000-year archaeology record of hunter-gatherers in central texas*. Volume III: Artifacts and special artifacts studies (pp. 743–794). Studies in Archaeology 31, Texas Archaeological Research Laboratory, University of Texas at Austin.

Keeley, L. H., & Newcomer, M. (1977). Micro-wear Analysis of experimental flint tools: A test case. *Journal of Archaeological Science, 4,* 29–62.

Keeley, L. H. (1974). Technique and methodology in micro-wear studies. A critical review. *World Archaeology, 5,* 323–336.

Keeley, L. H. (1980). *Experimental determination of stone tool uses.* Chicago: University of Chicago Press.

Lerner, H. J. (2007). Digital image analysis and use-wear accrual as a function of raw material: an example from northwestern New Mexico. *Lithic technology,* 51–67.

Mokhtar Saidin. (1997). Kajian perbandingan tapak paleolitik Kampung Temelong dengan Kota Tampan dan sumbangannya terhadap zaman Pleistosen Akhir di Asia Tenggara. *Malaysia Museum Journal*: 32.

Mokhtar Saidin. (1998). *Kebudayaan Paleolitik di Malaysia - Sumbangan Tapak Lawin, Perak dan Tingkayu, Sabah.* Ph.D. dissertation. Universiti Sains Malaysia: Penang (Unpublished).

Mokhtar Saidin. (2006). Bukit Bunuh, Lenggong, Malaysia: New evidence of late Pleistocene culture in Malaysia and Southeast Asia. In A. B. Elisabeth, C. G. Ian, C. Vincent (Eds.), *Uncovering Southeast Asia's Past: Selected Papers from the 10th International Conference of the European Association of Southeast Asian Archaeologists* (pp. 60–64). Pigott Blackwell Publishing Ltd.

Saidin, M. (2010). Out of Malaysia: Putting Malaysia on the map of human development. In D. A. Razak, (Ed.), *Transforming higher education for a sustainable tomorrow: 2009 laying the foundation.* Pulau Pinang: Universiti Sains Malaysia.

Talib, N. K. (2013). *Ekskavasi Tapak Bukit Bunuh, Lenggong, Perak: Sumbangan kepada Pemahaman Kebudayaan Paleolitik.* Master's Thesis. Universiti Sains Malaysia, Penang. (Unpublished).

Nur Asikin Binti Rashidi. (2013). *Pemilihan Jenis Batuan Oleh Masyarakat Prasejarah di Kawasan Impak Meteorit Bukit Bunuh, Lenggong, Perak dan Sumbangannya kepada Teknologi Paleolitik.* Master's Thesis. Universiti Sains Malaysia, Penang. (Unpublished).

Ollé, A., & Vergès, J. M. (2014). The use of sequential experiments and SEM in documenting stone tool micro-wear. *Journal of Archaeological Science, 48,* 60–72.

Semenov, S. A. (1964). *Prehistoric Technology,* translated by M.W. Thompson, Cory, Adams and Mackay, London, 211 p.

Sharon, G. (2008). The impact of raw material on Acheulian large flake production. *Journal of Archaeological Science, 35,* 1329–1344.

Sussman, C. (1985). Microwear on quartz: fact or fiction? *World Archaeology, 17*(1), 101–111.

Sussman, C. (1988). *A Microscopic analysis of use-wear and Polish Formation on Experimental Quartz Tools.* BAR International Series 398. Oxford: BAR.

Majid, Z. (1996). *Prasejarah Malaysia: Sudahkah Zaman Gelap Menjadi Cerah?* Universiti Sains Malaysia.

Majid, Z. (2003). *Archaeology in Malaysia.* Pulau Pinang: Pusat Penyelidikan Arkeologi Global.

Left-Handed Network-Shaped Metamaterial for Visible Frequency

Md. Mehedi Hasan, Mohammad Rashed Iqbal Faruque and Mohammad Tariqul Islam

Abstract In this paper, a network-shaped left-handed metamaterial for visible frequency applications has been presented. The proposed metamaterial is designed on the epoxy resin composite with woven glass fibre by a complex structure is driven by the average connectivity of the metallic network. Finite integration technique-based electromagnetic simulator Computer Simulation Technology Microwave Studio has been utilized to design, simulation and purpose of the proposed design. The designed structures exhibit resonance at 104.95, 140.61, 164.2 and 195.19 THz as well as the left-handed characteristics at 182 THz. Finally, the structure is also analysed by rotating 22.5°, 45°, 67.5° and 90°, respectively, in the xy-plane for observing the rotation effects on the result of the reflection (S_{11}) and transmission (S_{21}) coefficient.

Keywords Left handed · Network shape · Terahertz · Visible frequency

1 Introduction

Metamaterials are artificially engineered planar materials, which have exotic optical properties that are unattainable in natural materials and designed to control electromagnetic fields. The massive exploitation of metamaterials caused a leap in controlling light at the nanoscale, bringing forward a manifold of entirely new functionalities. Traditionally, the design of metamaterial has been inspired by the

---

Md. M. Hasan (✉) · M. R. I. Faruque
Space Science Centre (ANGKASA), Institute of Climate Change, Universiti Kebangsaan Malaysia, 43600 Bangi, Selangor Darul Ehsan, Malaysia
e-mail: mehedi20.kuet@gmail.com

M. T. Islam
Dept of Electric, Electronics and Systems Engineering, Universiti Kebangsaan Malaysia, 43600 Bangi, Selangor Darul Ehsan, Malaysia

© Springer Nature Singapore Pte Ltd. 2018
R. Saian and M. A. Abbas (eds.), *Proceedings of the Second International Conference on the Future of ASEAN (ICoFA) 2017 – Volume 2*,
https://doi.org/10.1007/978-981-10-8471-3_48

assembly of periodically spaced building blocks, which control their optical properties. Moreover, metamaterials have been utilized to facilitate a wide range of applications including satellite applications, microwave filter applications, optics lenses, electromagnetic cloaks, electromagnetic absorption (Wang and Wang 2016), hyper-spectral imaging, medical imaging, SAR reductions, energy harvesting, high-power applications (Bossard et al. 2016). In 2000, D. R. Smith et al. exhibited a material that shows the negative permittivity and permeability at a same time with some infrequent properties (2000). In 2003, Ziolkowski developed a metamaterial by capacitor loaded strips and split-ring resonators, which exhibited negative permittivity and permeability both at the X band frequencies (2003). In 2016, Hasan et al. proposed a compact metamaterial that was applicable for C and X band operations (2016). In 2015, Galinski et al. suggested new complex structures that were driven by the average connectivity of the metallic network to focus on the use of metallic nanoscale networks, sustaining plasmonic resonances. At variance with classical metasurface, which were mostly lattice-based structures, the key element of network architecture was the local connectivity between different elements (2016). In 2017, Hasan et al. presented a chiral metamaterial based on the inverse E-shape combined with the outer ring resonator printed on Rogers RT 5880 material. The proposed structure shows resonance at C band and 5.14 GHz bandwidth from 4.0 to 9.14 GHz. The effective medium ratio of the designed metamaterial was 6.83 and left-handed characteristics in 7.61 GHz (2017b). In 2016, a new metamaterial-based absorber was proposed for solar cell applications that show an outstanding single band with 99.7% absorption in the visible frequency, whereas the resonance was at 614.4 THz with an excellent 15.5% absorption bandwidth. Besides, the simulation results at different incident angles and different polarization confirm the quality by showing how insensitive it was to both the defined incident angles and different polarization angles of electromagnetic waves (Rufangura and Sabah 2016). In 2017, Hasan et al. projected a negative index meta-atom, resonance at C, X and Ku bands with wide negative refractive index bandwidth from 7.0 to 12.81 GHz. Negative index characteristics, bandwidth, compactness of the proposed structure are analysed at z-axis by propagating the electromagnetic waves. In addition, the meta-atom presented negative index characteristics at X band and Ku band (Hasan et al. 2017a). In 2016, a negative index metamaterial inspired antenna was presented for mobile communications. The presented antenna was covered most of the mobile bands like GSM, Bluetooth and WLAN. In addition, the antenna prepared by a semi-circular patch, a 50 Ω micro-strip feed line with metamaterial ground plane. The measured bandwidth is 1.34 GHz (from 1.66 to 3.0 GHz) and 1.44 GHz (from 4.4 to 5.84 GHz) for GSM, Bluetooth and WLAN. However, 72.11 and 75.53% of specific absorption rate were reduced by the designed antenna, respectively, at 1.8 and 2.4 GHz (Alam et al. 2016). A planar Inverted-F antenna with an artificial magnetic conductor structure for body specific absorption rate reduction in a wideband CDMA-band. The results show the proposed structure works in the WCDMA-band and provided 43.3% reduction of SAR in 2017 (Kwak et al. 2017). In 2016, a wideband and low-loss high-temperature superconducting band-pass filter developed by composite right- or left-handed stepped impedance resonator is introduced by Liu et al. The structure was designed in an entirely printed circuit technology based on the CRLH theory and the resonance at 2.6 GHz for LTE2600 application (2016). A chiral selective plasmonic absorber designed by ŋ-shaped resonators in the visible frequency was suggested by Tang et al. in 2017. The metasurface structure enabled chiral selective absorption bands and both simulated and measured exhibited 80% exceeding absorptions (Tang et al. 2017). In 2017, a metasurface was designed for multiband coherent perfect absorption in the mid-infrared region. The operating wavelengths and bandwidth of the coherent perfect absorption could be precisely tailored through the designed structure (Xiao et al. 2017). In 2015, a perfect absorber for 99.7% absorption over a large area without any lithographic processing, with spectral tunability from the visible-to-near-infrared spectrum, was invented by Akselrod et al. The absorbers were based on the colloidally synthesized silver nanocubes situated over a metal film, separated by the well-controlled nanoscale spacer (Akselrod et al. 2015).

In this study, a network-shaped metamaterial at visible frequency (from 100 to 200 THz) has been proposed, which exhibits resonance at 104.95, 140.61, 164.2 and 195.19 THz. The metamaterial shows left-handed characteristics at 182 THz, whereas the permittivity, permeability and refractive index are, respectively, −1.75, −0.87 and −1.31. However, the designed 3D structure also investigated by rotating 22.5°, 45°, 67.5° and 90° in the xy-plane for observing the rotation effects on the result of the reflection (S_{11}) and transmission (S_{21}) coefficient. The paper is adorned in this fashion; methodology explained elaborately with the schematic view, simulated diagram, retrieval methods of effective medium parameters and equivalent circuit model of the proposed metamaterial in the Sect. 2. Results analysis is shown in Sect. 3, whereas the effect of wave propagation along the x-axis, y-axis and z-axis effects on the scattering parameters is also explained. Finally, Sect. 4 accomplishes this paper.

2 Methodology

The proposed network metamaterial which is developed by a complex structure is driven by the average connectivity of the metallic network. Epoxy resin fibre is used as a substrate material in which dielectric constant and loss tangent are, respectively, 4.5 and 0.002. The thickness of the substrate material is considered as 0.1 μm. The total dimension of the designed metamaterial structure is 5.1 × 5 μm^2, whereas the small single unit cell is 1 × 0.85 μm^2. The schematic view, top view

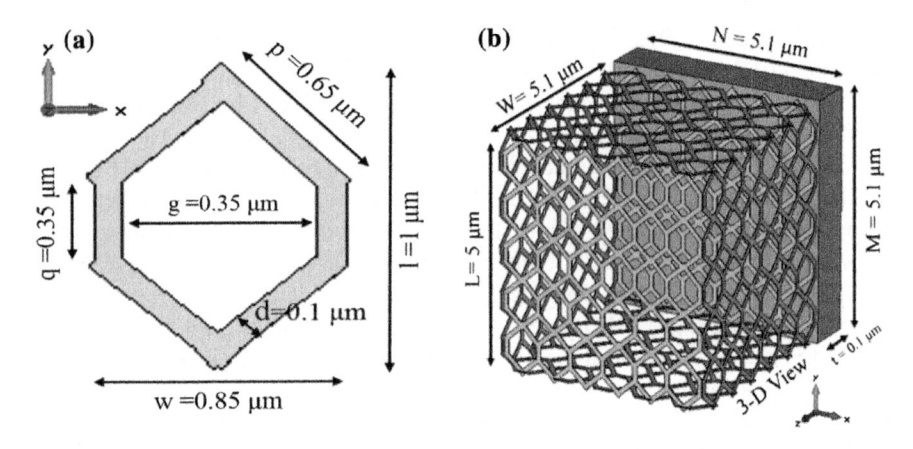

Fig. 1 Schematic geometry of the **a** single unit cell and the **b** 3D view of the proposed network metamaterial

Table 1 Design specification of the proposed network structure

Parameters	L	W	p	q	M	N
Dimensions (µm)	5	5.1	0.65	0.35	5.1	5.1
Parameters	l	w	d	g	t	h
Dimensions (µm)	1	0.85	0.1	0.35	0.1	0.017

and the 3D view of the proposed metamaterial structures have been shown in Fig. 1a–c (Table 1).

Finite integration technique-based Computer Simulation Technology has been applied for the simulation of the mentioned metamaterial, which helps to compute the parameters with the complex scattering constitutive parameters. For the simulation purpose, the proposed network structure has been placed between the waveguides. The electromagnetic waves are propagating along the z-axis, whereas the x-axis and y-axis are, respectively, considered as a perfect electric and magnetic boundary. Frequency domain solver with standardized impedance 50 Ω which has been set for simulation from 0 to 100 THz is shown in Fig. 2a. Moreover, the retrieving procedure of the effective parameters is given as follows,

$$X = \left(\frac{\mu_r}{\varepsilon_r}\right)^{1/2} = \left(\frac{1+\Gamma_{12}}{1-\Gamma_{12}}\right) \quad (1)$$

$$\text{Reflection coefficient, } \Gamma_{12} = \left\{\frac{1-(S_{21}^2 - S_{11}^2)}{2S_{11}}\right\} \pm \left[\left\{\frac{1-(S_{21}^2 - S_{11}^2)}{2S_{11}}\right\}^2 - 1\right]^{1/2} \quad (2)$$

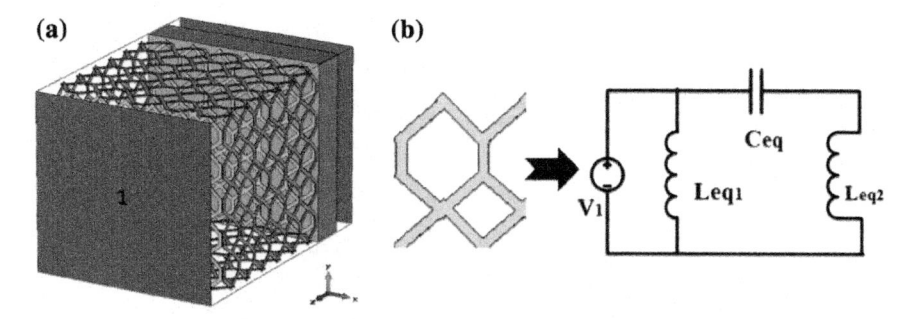

Fig. 2 **a** Simulation view with boundary condition in the CST-MWS and **b** equivalent lumped circuit of the designed network metamaterial

$$Z = \left\{ \frac{(S_{11}+S_{21}) - \Gamma_{12}}{1 - (S_{11}+S_{21})\Gamma_{12}} \right\} \quad (3)$$

Effective Permittivity, $\varepsilon_{\text{eff}} \approx (Z/X)$ \quad (4)

Effective Permeability, $\mu_{\text{eff}} \approx (X \times Z)$ \quad (5)

Effective Refractive index, $n_{\text{eff}} \approx (\varepsilon_{\text{eff}} \mu_{\text{eff}})^{1/2} \approx j\frac{c_o}{d\omega}[\ln|Z| + j \arg(Z)]$ \quad (6)

Equivalent circuit of the proposed network metamaterial structure unit cell is shown in Fig. 2b, where C_{eq}, L_{eq}, V_1 are represented as, respectively, capacitance, inductance and external source of the lumped LC-circuit model. Moreover, metal strips formed inductive effects, whereas the gaps are accountable for capacitive effect. Besides, there is a parasitic coupling effect for the mutual inductance and capacitance (Su et al. 2015; Hasan et al. 2017c).

3 Results Analysis

Figure 3a shows opposite current flowing which is seen in the two opposite side arrow of the network metamaterial. This flow of opposite current results a sharp transmittance at that frequency. The arrows are showing the direction of the currents, and colour express the intensity and flowing opposite directions, as well as nullify each other at a certain frequency. However, the electric field density at 77.8 THz is exhibited in Fig. 3b.

In Fig. 4a, the transmission (S_{21}) and reflection (S_{11}) coefficient for the proposed metamaterial have been shown. It demonstrates that the extent of transmission

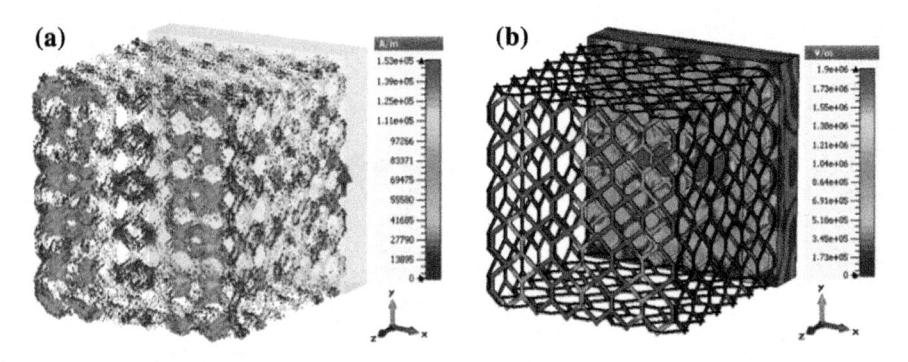

Fig. 3 **a** Surface current distribution and **b** electric field pattern, in 104.95 THz

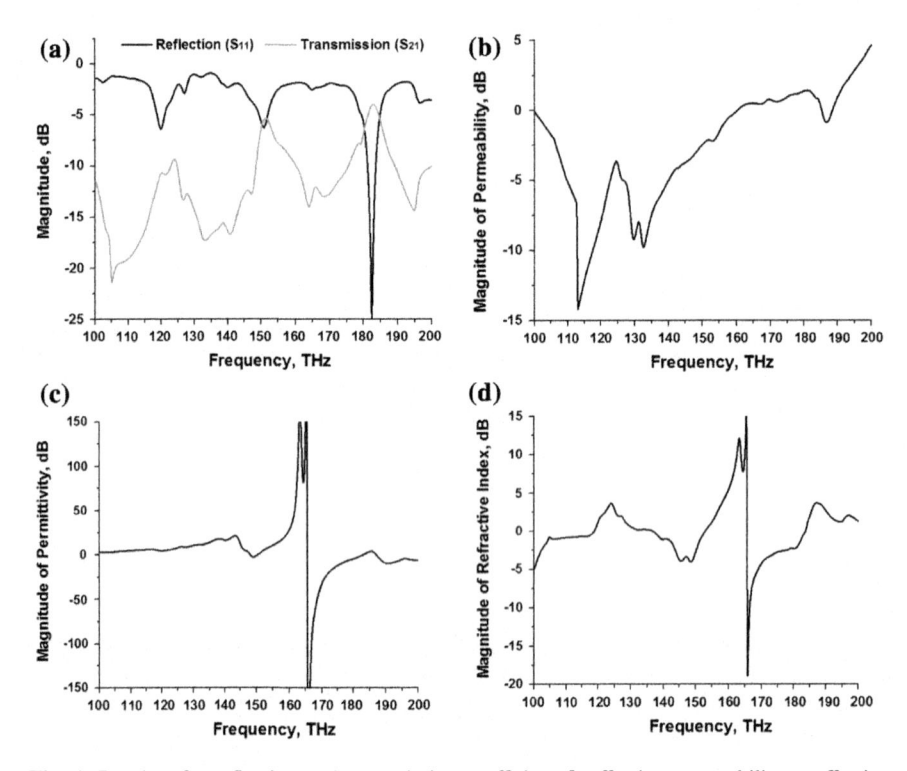

Fig. 4 Results of **a** reflection and transmission coefficient, **b** effective permeability, **c** effective permittivity and **d** effective refractive index of the proposed metamaterial

Table 2 Value of effective medium parameters for the left-handed characteristics

Resonance frequency	Permeability (μ)	Permittivity (ε)	Refractive index (η)
182 THz	−0.87	−1.75	−1.31

Table 3 Performance of the proposed network metamaterial at different rotation angle

Rotation Angle	Resonance of the reflection coefficient (S_{11}) (THz)	Value of the reflection coefficient	Resonance of the transmission coefficient (S_{21}) (THz)	Value of the transmission coefficient
22.5°	129.1	−19.54	130.98	−20.52
	148.27	−25.31	176.40	−19.46
	170.85	−28.69	195.68	−22.10
45°	103.27	−42.10	123.62	−26.40
	116.82	−26.10	160.44	−22.21
	130.35	−25.60	167.95	−23.10
67.5°	121.32	−39.35	108.72	−42.96
	151.43	−25.66	134.43	−25.94
90°	187.40	−37.28	175.91	−13.66

coefficient displays a resonance at the frequency of 104.95, 140.61, 164.2 and 195.19 THz. The figured permeability (μ) in addition with permittivity (ε) against frequency is depicted in Fig. 4b and c separately. In Fig. 4c, the real magnitude of the permittivity displays negative value from the frequency of 165.5–183 THz and 187.4–200 THz that covers nearly 17.5 and 12.6 THz bandwidth. Figure 4d depicts the refractive index of the material, where the curve displays the negative peak from the frequency of 100–118.6 THz, 136.9–152.84 THz and 165.77–183.68 THz that covers, respectively, around 18.6, 15.94 and 17.91 THz bandwidth. Moreover, at 182 THz frequency zones of refractive index curve and the permeability and permittivity curves also display negative peak. That is why the material can be characterized as left-handed metamaterial in these regions (Table 2).

Further analysis was done with the network metamaterial by rotating the structure at 22.5°, 45°, 67.5° and 90° in xy-plane, where the electromagnetic waves flow through the metamaterial along the z-axis. The results of the reflection (S_{11}) and transmission (S_{21}) coefficient of the 22.5°, 45°, 67.5° and 90° angular rotations are shown in Table 3. However, from Fig. 5a to d, simulated configuration at 22.5°, 45°, 67.5° and 90° rotations is shown and results of the reflection (S_{11}) and transmission (S_{21}) coefficient at 22.5°, 45°, 67.5° and 90° are presented, respectively, from Fig. 5e to h.

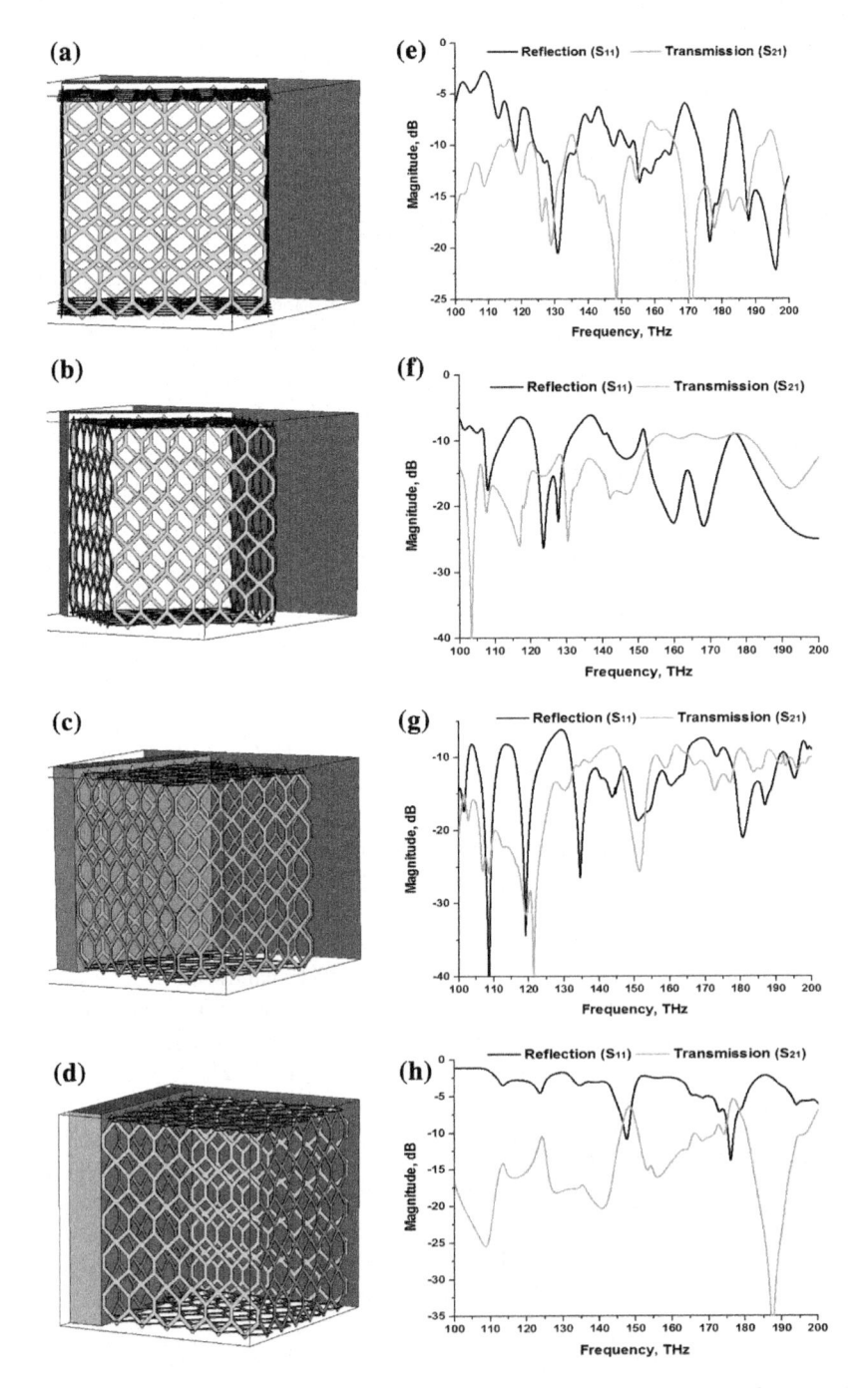

Fig. 5 Simulated configuration at **a** 22.5° rotation, **b** 45° rotation, **c** 67.5° rotation, **d** 90° rotation and simulated results of scattering parameters at **e** 22.5° rotation, **f** 45° rotation, **g** 67.5° rotation, **h** 90° rotation

4 Conclusion

A new network-shaped left-handed metamaterial is presented for visible range frequency applications. The metamaterial exhibits left-handed characteristics for z-axis wave propagation at 182 THz. The designed structure investigated by rotating 22.5°, 45°, 67.5° and 90° in the xy-plane for observing the rotation effects on the result of the reflection (S_{11}) and transmission (S_{21}) coefficient. However, the proposed structure can be evolved from the inductance–capacitance resonator and could find wide applications in absorption filter, hot-electron collection devices, optical applications, etc.

Acknowledgements This work was supported by the Research Universiti Grant, Geran Universiti Penyelidikan (GUP), and code: 2016-029.

References

Akselrod, G. M., Huang, J., Hoang, T. B., Bowen, P. T., Su, L., Smith, D. R., et al. (2015). Large-area metasurface perfect absorbers from visible to near infrared. *Advanced Materials, 27,* 7897.

Alam, T., Faruque, M. R. I., & Islam, M. T. (2016). Specific absorption rate analysis of broadband mobile antenna with negative index metamaterial. *Applied Physics A, 122,* 1–6.

Bossard, J. A., Scarborough, C. P., Wu, Q., Campbell, S. D., Werner, D. H., Werner, P. L., et al. (2016). Mitigating field enhancement in meta-surfaces and metamaterials for high-power microwave applications. *IEEE Transactions on Antennas and Propagation, 64,* 5309–5319.

Galinski, H., Fratalocchi, & A., Capasso, F. (2016). Network metamaterials: An alternative platform for optical materials. Meta Conference. Spain, pp. 529–530.

Hasan, M. M., Faruque, M. R. I., & Islam, M. T. (2017a). A single layer negative index meta atom at microwave frequencies. *Microwave and Optical Technology Letters, 59,* 1450–1454.

Hasan, M. M., Faruque, M. R. I., & Islam, M. T. (2017b). Inverse E-shape chiral metamaterial for long distance telecommunication. *Microwave and Optical Technology Letters, 59,* 1772–1776.

Hasan, M. M., Faruque, M. R. I., & Islam, M. T. (2017c). Left-handed metamaterial using Z-shaped SRR for multiband application by azimuthal angular rotations. *Materials Research Express, 4,* 4.

Hasan, M. M., Faruque, M. R. I., Islam, S. S., & Islam, M. T. (2016). A new compact double-negative miniaturized metamaterial for wideband operation. *Materials, 9*(10), 830.

Kwak, S. I., Sim, D. U., Kwon, J. H., & Yoon, Y. J. (2017). Design of PIFA with metamaterials for body-SAR reduction in wearable applications. *IEEE Transactions on Electromagnetic Compatibility, 59,* 297–300.

Liu, H., Wen, P., Jiang, H., & He, Y. (2016). Wideband and low-Loss high-temperature superconducting band pass filter based on metamaterial stepped-impedance resonator. *IEEE Transactions on Applied Superconductivity, 26,* 1500404.

Rufangura, P., & Sabah, C. (2016). Polarisation insensitive tunable metamaterial perfect absorber for solar cells applications. *IET Opto electron, 10,* 211–216.

Smith, D. R., Padilla, W. J., Vier, D. C., Nemat- Nasser, S. C., & Schultz, S. (2000). Composite medium with simultaneously negative permeability and permittivity. *Physical Review Letters, 84,* 4184–4187.

Su, L., Naqu, J., Contreras, J. M., & Martín, F. (2015). Modelling metamaterial transmission lines loaded with pairs of coupled split-ring resonators. *IEEE Antennas and Wireless Propagation Letters, 14,* 68–71.

Tang, B., Li, Z., Palacios, E., Liu, Z., Butun, S., & Aydin, K. (2017). Chiral- selective plasmonic meta-surface absorbers operating at visible frequencies. *IEEE Photonics Technology Letters.* https://doi.org/10.1109/LPT.2016.2647262.

Wang, B. X., & Wang, G. Z. (2016). Quad-band terahertz absorber based on a simple design of metamaterial resonator. *IEEE Photonics Journal, 8,* 5502408.

Xiao, D., Tao, K., Wang, Q., Ai, Y., & Ouyang, Z. (2017). Meta-surface for multi wavelength coherent perfect absorption. *IEEE Photonics Journal, 9,* 6800108.

Ziolkowski, R. W. (2003). Design, fabrication, and testing of double negative metamaterials. *IEEE Transactions on Antennas and Propagation, 51,* 1516–1529.

Volume Change Behaviour of Clay by Incorporating Shear Strength: A Review

Juhaizad Ahmad, Mohd Ikmal Fazlan Rosli, Abdul Samad Abdul Rahman, Syahrul Fithri Senin and Mohd Jamaludin Md Noor

Abstract Soft clay settlement is well known as a major problem in the civil engineering structures such as highways, buildings and bridges. The problem arises due to the current settlement analysis which is derived from the effective stress concept. It considers the volume change increase with the increase in the effective stress, whereas the settlement problem is not always happened in this fashion. In fact, the inundation problem that causes the sudden settlement during the effective stress decrease really put a tremendous blow to this concept. It is proved that the empirical effective stress concept that is widely used by the engineer is not the appropriate analysis. In fact, this is one of the main factors that cause many settlement problems. The objective of this paper is to review several settlement models available in the literature and to distinguish the applicability of the model to predict soil settlement. It also reviewed the volume change behaviour by incorporating shear strength concept that is more appropriate to predict the settlement. Hopefully, this paper will help the new researcher to uncover the unique behaviour of clay settlement and at the same time will reduce the settlement problem cases.

Keywords Volume change · Settlement · Anisotropic · Mobilised shear stress Effective stress

---

J. Ahmad (✉) · M. I. F. Rosli · S. F. Senin
Faculty of Civil Engineering, Universiti Teknologi MARA,
13500 Permatang Pauh, Penang, Malaysia
e-mail: jetz77@gmail.com

A. S. Abdul Rahman · M. J. Md Noor
Faculty of Civil Engineering, Universiti Teknologi MARA,
40450 Shah Alam, Selangor, Malaysia

1 Introduction

Soft clay can be found in many parts of Malaysia, as shown in Fig. 1. It is classified as soft marine clay. This is a major problem faced by the geotechnical engineers. The high compressibility and low in strength have caused this soil to be very problematic. This soil is located along the east and west coast of Peninsula Malaysia. The rapid development around this area makes it inevitable. Therefore, the geotechnical engineers always find a way to work on this site although the risk of differential settlement is awaiting if the foundation is not strong enough to resist the applied loads.

2 Volume Change of Clay

Volume change of clay refers to the complex interactions between the soil fabric and the voids. It also depends on various factors such as soil mineralogy, fluid composition, void ratio and stress level (Di Maio et al. 2004). The influence of volume change on clay behaviour is widely acknowledged since the nineteenth

Fig. 1 Soft soil sites in Peninsula Malaysia (Mineral and Geoscience Department, Malaysia 2016)

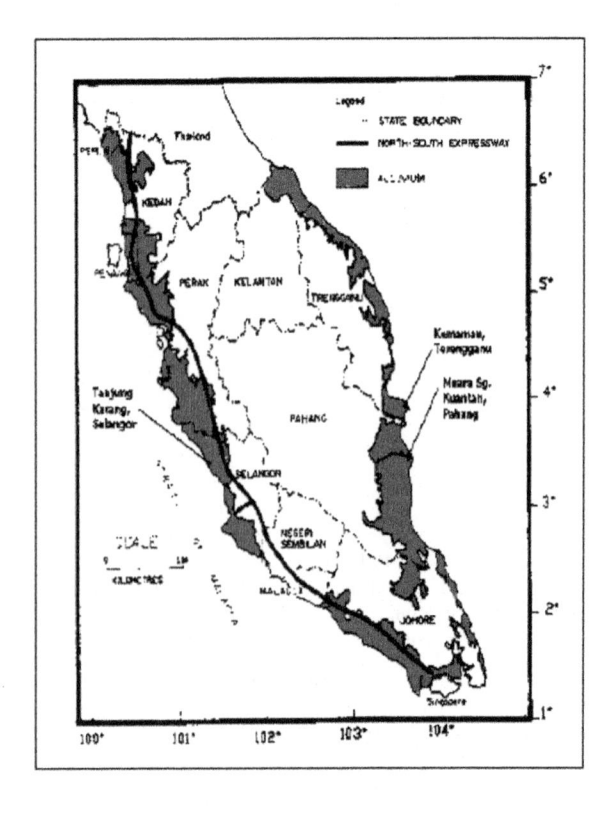

century. Then, many reports that addressed the importance of this basic soil mechanics fundamental have been published. But, it is still a room for debates to improve the current understanding on the volume change behaviour of clay (Nyugen 2006). This is especially on the tropical marine clay that exhibits many problems for the construction.

The volume change behaviour of clay is well known with the spring analogy concept. It is being used to explain the basic fundamental regarding on the consolidation process of clay and how water flow out from the soil body when compressed. The change in the void ratio is monitored with the increase in effective stress. While the researches are obsessed with this fundamental, not many realised that there is another important parameter missed out by the researcher. It is the strength development of the soil after compression (Md. Noor and Anderson 2015; Md. Noor and Abd-Rahman 2014; Saffari et al. 2016). It is common for all types of soils to settle until it reaches the zero-void ratio. The phenomenon is always viewed in terms of void ratio and stress only, whereas the soil particles attained its stability throughout the settlement process.

One-dimensional compression test is the only test that has been used to predict the consolidation settlement for clay. This method is widely used because of simplicity and the ability to capture many parameters such as coefficient of compressibility, pre-consolidation pressure and many others. Later, this method is used to predict settlement by using many equations by various researches (Janbu et al. 1956; Terzaghi 1943; Steinbrenner 1934; Terzaghi and Peck 1967). Sadly, none of them were formulated per the actual behaviour of clay. Clay is anisotropic in nature; therefore, this parameter must be considered. Moreover, the analysis still considers the settlement as the function of effective stress which not able to explain the inundation and wetting collapse phenomenon (wetting collapse is defined as sudden decrease in volume of soil, without any increment on effective stress on the soil due to moisture).

Later, the soil model development had reached its ultimatum with the creation of critical state soil model. The modified Cam clay model is considered as the best model to represent the actual clay behaviour. Cam clay models of critical state soil mechanics are widely used in many geotechnical applications involving numerical predictions of stability and deformation behaviour of compressible soil materials such as soft clays (Atkinson 1981; Roscoe and Burland 1968; Schofield and Wroth 1968). The general features of these models include pressure sensitivity, hardening response with plastic volumetric compaction, softening response with plastic dilation and coupled volumetric and deviatoric plastic deformations, which are essential to model the prototype granular material behaviour realistically (Borja and Tamagnini 1998).

Modified Cam clay model is widely used to predict the behaviour of the stress–strain curve (Munda et al. 2014; Rao et al. 2017; Hao and Yao 2013; Gu et al. 2014; Wen et al. 2013). However, there are many shortcomings in this model. The most obvious weakness in this model is it relates volume change as the main factor for settlement problem. But the main factor for settlement problem is the mobilised shear strength envelope. It is also modelled to predict the behaviour of natural soil, but again it is not perfectly replicate the behaviour of natural soil. The elliptical yield locus does not have the same shape as obtained from experimental test for many natural soils.

The concept of modified Cam clay model is given by Powrie (2004). Firstly, the change in net mean stress, p' produces no distortion. For example, shear strain, deq. Secondly, the change in the distortional deviator stress, q produces no change in volume. Thirdly, yield loci expand at constant shape but have different size. The expand of yield loci indicates the plastic deformation. Fourthly, the expansion is associated with the hardening of the soil and is related to the normal compression of the soil. Fifthly, the size of the elliptical yield locus is controlled by tip stress, po'. Lastly, the expansion of the yield locus is related to plastic strains (shear + volumetric).

In the past years, many development works have been made to improve the performance of modified Cam clay model. One of it is done by Mita et al. (2004). The MCC model has been proven to satisfactorily predict the behaviour of normally to lightly overconsolidated (OC) clays that lie in the subcritical region. But, its prediction for the stress–strain behaviour of heavily OC clays in the supercritical region is not satisfactory, as the adopted yield curves tend to overestimate peak strength in this region. Hence, Mita et al. (2004) have come out with 3D Hvorslev–MCC model which can coop with the problem. It is a good improvement for the critical state. However, critical state is still lack of charisma because it still cannot solve inundation settlement problems.

3 Anisotropic Behaviour of Clay

Terzaghi model is widely used in the industry although this model does not consider the anisotropic behaviour of a real clay soil. Although there are a lot of researches that proved that clay is anisotropic in suffusion and hydrocompression processes. Voottipruex et al. (2014) have used different horizontal and vertical coefficient of permeability in vertical drain equation. Also, Mesri and Funk (2016) used the same concept in Kansai Airport site. Therefore, the Terzaghi equation should be understood as the basic fundamental model only and not being used to predict the clay behaviour for construction site. The process of volume change is very complex. Therefore, a comprehensive study must be conducted to come out with a correct model.

4 Recent Settlement Prediction Models

Settlement of clay is commonly encountered in many parts of the world. It is inevitable phenomenon that causes a lot of problems and incurs high cost for repairing works. Therefore, geotechnical engineers continue to research on this unique behaviour. They have produced various clay settlement prediction models. The first consolidation settlement model was coined out by Terzaghi (1943). The theory based on elastic theory which is unreal for clay. According to Seawsirikul et al. (2015), the calculated and recorded data ratio at site showed that is in the range of 0.44–1. It is proved that the elastic theory is underestimated the settlement although at some points the prediction is within acceptable value. This is supported by the work by Taylor (1948) and Davis and Raymond (1965) that Terzaghi method showed similar than those predicted for normally consolidated clays. On the other hand, Oda et al. (2015) predict the consolidation settlement using Artificial Neural Network (ANN). It is claimed that this model can give prediction up to 1.1 times than the measured settlement. Besides that, Park et al. (2015) used Genetic Algorithm (GA) to calculated settlement. The results from GA were compared with the widely used graphical methods (Asoka and hyperbolic methods). The advantages of using GA are it can accommodate for multi-layered soil and modified surcharging plans. It was concluded that GA can give the best prediction compared to the two graphical methods with less than 200 mm margin. Bari and Shahin (2014) used Simplified Probabilistic Method to predict the inherent properties of coefficient of consolidation because spatial variability of horizontal coefficient of consolidation is commonly ignored. But, it is very significant in the consolidation model. All these models were based on empirical laboratory or site records. It is hard to find a semi-empirical model that can capture a unique relationship that exists on many types of clay that can be found in this world. Detailed study needs to be conducted to look at this matter seriously to provide a reliable clay settlement prediction model.

5 Settlement Prediction According to Rotational Mobilised Shear Strength Envelope

The volume change behaviour of clay that incorporated shear strength is given by Md. Noor and Anderson. Rotational mobilised shear strength envelope is a model intended to predict the behaviour of saturated and unsaturated soil. The conventional shear strength equations are not suitable to be applied in unsaturated soil as the important parameter such as suction is not incorporated. Hence, rotational mobilised shear strength envelope is the solution for it. Moreover, the rotational mobilised shear strength envelope brings a new dimension in soil mechanics because there are no other models which can solve engineering problem such as inundation settlement and shallow slope failure. Therefore, it is simply more elegant than other soil mechanics model such as modified Cam clay model.

Before moving to the prediction of soil behaviour according to this model, it is essential to understand the actual fundamental related to strength and volume change behaviour of soil. The critical state model considered the critical state strength governs the settlement, whereas it is not the critical state strength governs the settlement of soil. Actually, the settlement of soil is governing by mobilised shear strength. The mobilised shear strength envelope plays an important role in defining yield surface in the model. The maximum failure shear strength is the curve which resulted from failure points of various stress states in stress–strain curve. The intermediate points on the stress–strain curve defining the mobilised shear strength envelope of the soil. Soil memorised maximum mobilised shear strength. Hence, during unloading, the stress will decrease and after reloading, the stress will follow the elastic path before the past maximum applied load is exceeded, then the stress will follow back the plastic deformation path. The unloading point and reloading point are the yield point for the soil, but it is not necessary means the failure point if the stress will continue to increase after reloading. Figure 2 shows the behaviour of soil during loading and unloading of specimen (Md Noor 2006).

The theory behind the success of this model to predict the behaviour of soil is the unique relationship between effective mobilised shear strength and axial strain. Figure 3 shows the relationship of effective mobilised minimum friction angle and axial strain. It can be used to predict the deviator stress of soil to predict the stress–strain curve of soil (Md Noor 2006).

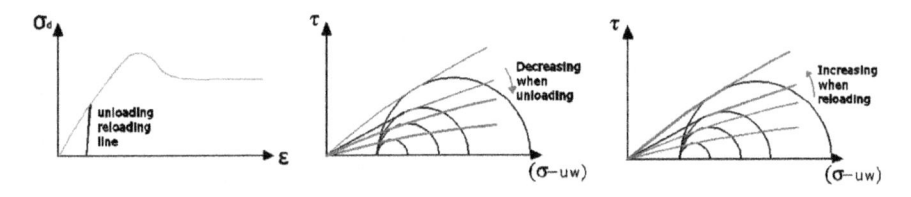

Fig. 2 Behaviour of soil during loading and unloading of specimen (Md Noor 2006)

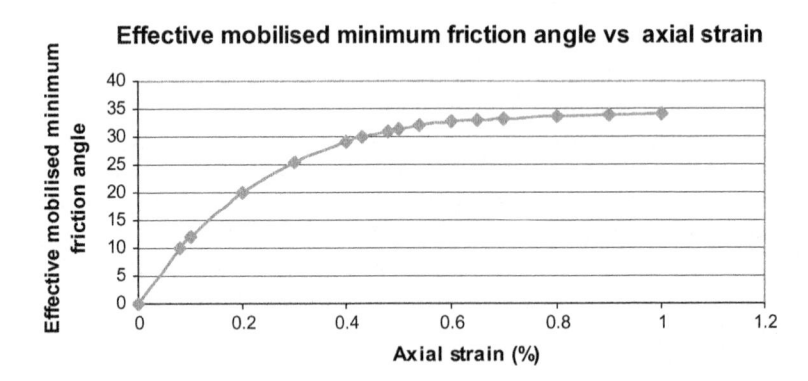

Fig. 3 Relationship of effective mobilised minimum friction angle and axial strain (Md Noor 2006)

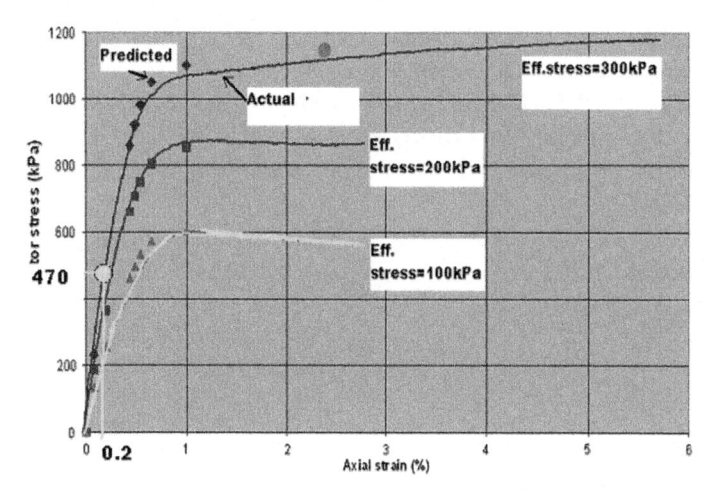

Fig. 4 Predicted stress–strain curve over actual stress–strain curve (Md Noor 2006)

From various stress states plotted in the Mohr circle graph, it produces mobilised shear strength angles for each corresponding axial strain in the stress–strain curve. Consider a point in the stress–strain curve, let say for confining pressure equals to 300 kPa and the corresponding axial strain is 0.2. By plotting Mohr circle for that stress, the effective mobilised minimum friction angle can be obtained. For axial strain 0.2, the effective mobilised minimum friction angle is 20°. The value of minor (σ_1') and major effective stress (σ_3') can be identified which are 300 and 770 kPa, respectively. Hence, the deviator stress can be determined. The minimum mobilised friction angle is deduced from various stress states from corresponding axial strain. Hence, the deviator stress can be calculated because the minimum mobilised friction angle is based on major and minor stress in Mohr circle. The deviator stress is calculated as $\Delta\sigma_d = \sigma'_1 - \sigma_3 + u$, which equals to 470 kPa. Then, the deviator stress versus axial strain graph can be plotted. It produced predicted stress–strain curve. The predicted graph shows a good agreement between predicted and actual test results. Figure 4 shows the predicted stress–strain curve over actual stress–strain curve (Md Noor 2006).

6 Conclusion

Various models for settlement prediction were reviewed. The conclusions can be drawn from the review are:

1. The settlement models can be classified as empirical or semi-empirical. The empirical models are widely used. It is originated from the one-dimensional compression test results. Also, the settlement models were derived from site

measurement and modelled using numerical or graphical methods. These models are based on effective stress concept.
2. The rotational mobilised shear strength envelope is considered as the best model to predict the settlement for many types of soils including clay. It is also capable of explaining the inundation problem. This is not observed for the other models stated above.

Acknowledgements This work is partially supported by Universiti Teknologi MARA. The authors also gratefully acknowledge the helpful comments and suggestions of the reviewers, which have improved the presentation.

References

Atkinson, J. H. (1981). *Foundations and slopes: An introduction to applications of critical state soil mechanics*. New York: Halsted Press.
Bari, M. W., & Shahin, M. A. (2014). Probabilistic design of ground improvement by vertical drains for soil of spatially variable coefficient of consolidation. *Geotextiles and Geomembranes, 42*(2014), 1–14.
Borja, R. I, & Tamagnini, C. (1998). Cam-Clay plasticity Part III: Extension of the infinitesimal model to include finite strains. *Computer Methods in Applied Mechanics and Engineering, 73*.
Davis, E. H., & Raymond, G. P. (1965). A non-linear theory on consolidation. *Geotechniques, 15*(2).
Department of Geoscience Malaysia. (2016). Malaysian soft clay soils distribution map. Available from: https://www.researchgate.net/figure/279202501_fig1_Figure-1-Malaysian-Soft-Clay-Soils-Distribution-Map-5. Accessed 9 September, 2016.
Di Maio, C., Santoli, L., & Sciavone, P. (2004). Volume change behavior of clay: The influence of mineral composition, pore fluid composition and stress state. *Mechanics of Minerals, 36*, 435–451.
Gu, X., Zhou, T., & Cheng, S. (2014). The soft soil foundation consolidation numerical simulation based on the model of modified cam-clay. *Applied Mechanics and Materials, 580–583*, 3223–3226.
Hou, W., & Yao, Y. (2013). Comparison between feature of modified Cam-Clay model and UH model. In Q. Yang, J. M. Zhang, H. Zheng, & Y. Yao (Eds.), *Constitutive Modeling of Geomaterials. Springer Series in Geomechanics and Geoengineering.* Berlin, Heidelberg: Springer.
Janbu, N., Bjerrum, L., & Kjaernsli, B. (1956). Veiledring ved llosning av fundamnentering-soppgaver. Norwegian Geotechnical Institute Publication No 16. Oslo.
Md Noor, M. J. (2006) Shear strength and volume change behaviour of unsaturated soils. PhD Thesis. Univeristy of Sheffield, U.K.
Md. Noor, M. J., & Abd-Rahman, A. S., (2014). Role of soaking in rainfall-induced slope failure according to non-linear failure envelope. In Proceedings of World of Land Slide Forum 2014, International Consortium on Landslides (ICL) China Geological Survey (CGS). Beijing, China.
Md. Noor, M. J., & Anderson, W. F. (2015). Concept of effective stress and shear strength interaction in rotational multiple yield surface framework and volume change behaviour of Banting clay. In Proceedings of the 11th International Conference on Applied and Theoretical Mechanics (MECHANICS '15). Kuala Lumpur. April 2015.
Mesri, G., & Funk, J. R. (2016). Closure to settlement of the Kansai international airport islands. *Journal of Geotechnical and Geoenvironmental Engineering, 142*(6).

Mita, K. A., Dasari, G. R., & Lo, K. W. (2004). Performance of a three-dimensional Hvorslev-Modified Cam clay model for overconsolidated clay. *International Journal of Geomechanics, 4,* 304.

Munda, J., Pradhan, P. K., & Nayak, A. K. (2014). A review on the performance of Modified Cam Clay model for fine grained soil. *Journal of Civil Engineering and Environmental Technology., 1*(5), 65–71.

Nguyen, A. M. (2006). An investigation of the anisotropic stress-strain-strength characteristics of an Eocene clay. PhD Thesis. Imperial College London.

Oda, K., Yokota, K., & Bu, L. D. (2015). Stochastic estimation of consolidation settlement of soft clay layer with artificial neural network. In The 15th Asian Regional Conference on Soil Mechanics and Geotechnical Engineering. http://doi.org/10.3208/jgssp.JPN041.

Park, H. I., Kima, K. S., & Kimb, H. Y. (2015). Field performance of a genetic algorithmin the settlement prediction of a thick soft clay deposit in the southern part of the Korean peninsula. *Engineering Geology, 196*(2015), 150–157.

Powrie, W. (2004). Soil mechanics concepts and applications (2nd ed., Vol. 86). Spon Press.

Rao, P. P., Chen, Q., Nimbalkar, S., & Cui, J. (2017). Elastoplastic solution for spherical cavity expansion in modified Cam-Clay soil under drained condition. *International Journal of Geomechanics ASCE.*

Roscoe, K. H., & Burland, J. H. (1968). On the generalized stress-strain behaviour of 'wet' clay. In J. Heyman & F. A. Leckie (Eds.), *Engineering plasticity* (pp. 535–609). Cambridge: Cambridge Univ. Press.

Saffari, P., Md. Noor, M. J., & Ashaari, Y. (2016). Stress-strain response of Malaysian granitic residual soil grade V, according to Rotational Multiple Yield Surface Framework. Web of Conference, E-UNSAT 2016.

Schofield, A., & Wroth, P. (1968). *Critical state soil mechanics.* New York: McGraw-Hill.

Seawsirikul, S., Chantawarangul, K., & Vardhanabhuti, B. (2015). Evaluation of differential settlement along bridge approach structure on soft Bangkok clay. IOP Press. https://doi.org/10.3233/978-1-61499-580-7-614.

Steinbrenner, W. (1934). Tafeln zur Setzungsberechnung. *Die Strasse, 1,* 121–124.

Taylor, D. W. (1948). *Fundamentals of soil mechanics.* New York: John Wiley.

Terzaghi, K. (1943). *Theoretical of soil mechanics.* New York: John Wiley and Sons.

Terzaghi, K., & Peck, R. B. (1967). *Soil mechanics in engineering practice.* New York: John Wiley.

Voottipruex, P., Bergado, D. T., Lam, L. G., & Hino, T. (2014). Back-analyses of flow parameters of PVD improved soft Bangkok clay with and without vacuum preloading from settlement data and numerical simulations. *Geotextiles and Geomembranes, 42*(2014), 457–467.

Wen, Y., Yang, G., Zhong, Z., Fu, X., & Zhang, Y. (2013). A similar Cam-clay model for sand based on the generalized potential theory. In Second International Conference on Geotechnical and Earthquake Engineering IACGE 2013. ASCE.

Mechanical Properties and X-Ray Diffraction of Oil Palm Empty Fruit Bunch All-Nanocellulose Composite Films

Nur Liyana Izyan Zailuddin, Azlin Fazlina Osman, Salmah Husseinsyah, Zailuddin Ariffin and Faridah Hanum Badrun

Abstract The all-nanocellulose composite films were prepared using cellulose solvent system N-dimethylacetamide/lithium chloride (DMAc/LiCl). The process includes partial dissolution of oil palm empty fruit bunch (OPEFB) and microcrystalline cellulose (MCC) in the N-dimethylacetamide/lithium chloride (DMAc/LiCl) solution followed by the regeneration process. The regeneration process also includes the removal of the N-dimethylacetamide/lithium chloride (DMAc/LiCl) solvent by using distilled water and drying of the films. The all-nanocellulose composite films with OPEFB contents in the range of 1–4 wt% were prepared and analyzed for their mechanical properties and crystallinity. The all-nanocellulose composite film with 1 wt% of OPEFB content showed the best tensile strength and modulus of elasticity with the value of 7.95 and 179.77 MPa, respectively. This could be due to a good dispersion of the cellulose particles in the films. However, the elongation at break for the composite film with 1 wt% of OPEFB content showed lower value than the ones contained higher percentage of OPEFB contents. As the content of OPEFB increases the tensile strength decreases especially at 4 wt% of OPEFB content of the composite film. The X-ray diffraction (XRD) analysis suggests that the all-nanocellulose composite film with 1 wt% of OPEFB content has higher crystallinity compared to the all-nanocellulose composite film with 4 wt% of OPEFB content. This could explain why this particular all-nanocellulose-based composite system performed greater tensile strength and modulus at low OPEFB content (1 wt%).

Keywords Oil palm empty fruit bunch · All-cellulose composite N-dimethylacetamide/lithium chloride · Microcrystalline cellulose

---

N. L. I. Zailuddin (✉) · A. F. Osman · S. Husseinsyah
School of Materials Engineering,
Universiti Malaysia Perlis, 02600 Arau, Perlis, Malaysia
e-mail: azlin@unimap.edu.my

Z. Ariffin · F. H. Badrun
Faculty of Applied Sciences, Universiti Teknologi Mara (UiTM),
02600 Arau, Perlis, Malaysia

© Springer Nature Singapore Pte Ltd. 2018
R. Saian and M. A. Abbas (eds.), *Proceedings of the Second International Conference on the Future of ASEAN (ICoFA) 2017 – Volume 2*,
https://doi.org/10.1007/978-981-10-8471-3_50

1 Introduction

It has been known that polymers in the form of natural resources have been the focus around the globe due to their beneficial properties that are able to be an alternative to the traditional petrochemical (Guansen et al. 2012; Pang et al. 2013). In the composite world especially for biocomposite study, instead of using the conventional petroleum-based matrices, natural polymers such as palm oil resins or starch are used in the formation of the composite. The purpose of using these renewable resources is due to their biodegradable properties and is less harmful to the environment (Huber et al. 2012).

Palm oil plantation is one of the vast agricultural industries in Malaysia, and the increase in the production of palm oil generates lignocelluloses residues in the form of oil palm empty fruit bunch (OPEFB), oil palm fiber, and other biomass wastes (Ariffin et al. 2008; Sahari and Sapuan 2011; Sulaiman et al. 2011; Chang 2014). Idle wastes like OPEFB have the potential to be used in fields such as a carbon source for bioconversion and biomass feedstock (Ariffin et al. 2008; Ghaderi et al. 2014; Shariff et al. 2014). The composition of OPEFB can be categorized into cellulose (50%) which is the highest components followed by lignin (25%) and hemicelluloses (25%) (Kavitha et al. 2013). OPEFB cellulose fiber can be used as natural resources in many fields especially in the composite area. In addition, the OPEFB cellulose can also be further studied in the production of biocomposite films (Ghaderi et al. 2014).

A newly developed composite based on natural resources known as all-cellulose composites (ACCs) is a composite where the cellulose fibers are impregnated with a cellulose matrix which leads to excellent mechanical properties (Soykeabkaew et al. 2009a, b). This type of composite focuses on phasing out the chemical incompatibilities that exists between fiber reinforcement and matrix components by using a single material that can function as both components (Huber et al. 2012).

In recent times, a new method of producing ACCs has been proposed which is the surface selective dissolution of cellulose fibers using suitable cellulose solvents whereby the cellulose surface is partially dissolved using cellulose solvents. Then, it will be regenerated into the matrix phase surrounding the non-dissolved fiber which acts as the reinforcement. This does not only simplify the composite's preparation process but can somewhat improve the fiber/matrix interface which can lead to excellent mechanical properties similar to those prepared by a traditional impregnation method (Soykeabkaew et al. 2009a, b; Huber et al. 2012). Potential or possible application of all-cellulose composite is in biomedical-related field. For example, good mechanical properties of all-cellulose composite might be used as substitution of bone or cartilage material (Huber et al. 2012). The uses of regenerated cellulose fibers can be seen in the production of films, membranes, and sponges (Klemm et al. 2005).

Based on previous studies, microcrystalline cellulose (MCC) is used in the production of composite films. This MCC material is usually purchased commercially. A research by Govindan et al. (2014) on preparation and characterization of regenerated cellulose using ionic liquid showed that the dissolution of microcrystalline cellulose (MCC) using DMAc/LiCl can improve the tensile strength and modulus of elasticity of the biocomposite films. Zailuddin et al. (2017) proved that the treated oil palm empty fruit bunch regenerated cellulose biocomposite films with butyl methacrylate that have improved tensile strength, modulus of elasticity, thermal stability, and crystallinity index as compared to the untreated ones.

Considering that more researches are now focusing on implementing environmental friendly resources, most current study is to incorporate cellulose nanoparticles (CN) to produce all-cellulose nanocomposite films. CN is known as a particle type that consists of at least one dimension in the nanoscale (Moon et al. 2011). The incorporation of nanoscale cellulose in the biocomposite system can improve some properties. This is due to the nanoscale cellulose characteristics of being lightweight, high aspect ratio for good reinforcement, biodegradable, and renewable (Siro and Plackett 2010).

In the development of the all-nanocomposite films, the processing technique of cellulose is thought to be crucial as the natural polymer does not melt or dissolve in conventional solvent due to their inter- and intra-hydrogen bond (Zadagen et al. 2010; Han et al. 2013). It is known that cellulose can be difficult to dissolve. Cellulose can dissolve and swell in several solvents, for example N-dimethylacetamide/lithium chloride (DMAc/LiCl) (Zhang et al. 2012a, b; Zailuddin and Husseinsyah 2016), N-methylmorpholine-N-oxide (NMMO) (Biganska and Navard 2002; Zhao et al. 2007), 1-butyl-3-methylimidazolium chloride (Mahmodian et al. 2012a, b; Soheilmoghaddam and Wahit 2013; Soheilmoghaddam et al. 2013a, b), 1,3-demethyl (2-limidazolimide/Lithium chloride (DMI/LiCl) (Tamai et al. 2004), and 1-butyl-3-methylimidazolium acetate ([Bmim]Ac) (Liu et al. 2011).

The use of dimethylacetamide/lithium chloride (DMAc/LiCl) solvent is common among scientist and researches (Dupont 2003). An activation step is important and is included in the DMAc system to shorten the time of cellulose dissolution that can reach up to a few months (Huber et al. 2012). A research by Ishii et al. (2008) showed that the DMAc-treated celluloses dissolved more rapidly as compared to the dissolution of the acetone-treated celluloses and the untreated one. Therefore, in this study, DMAc/LiCl solvent was used to prepare the all-nanocellulose composite films using OPEFB. In addition, the use of OPEFB cellulose in the form of nanoparticles can explore more advantages of OPEFB waste. Thus, the nanotechnology can be applied to further develop OPEFB for wider applications.

2 Materials and Methodology

2.1 *Materials*

Oil palm empty fruit bunch (OPEFB) was obtained from Malaysian Palm Oil Board (MPOB) Bangi, Selangor. The OPEFB cellulose fiber was ground using ball mill into powder form. The particle size between 35.9 and 50.90 nm was observed through FESEM analysis. Microcrystalline cellulose (MCC) with particle size of

Table 1 Formulation of OPEFB all-nanocellulose composite films

Materials	OPEFB all-nanocellulose composite films
OPEFB cellulose (wt%)	1, 2, 3, 4
Microcrystalline cellulose (MCC) (wt%)	3

50 μm was purchased from Aldrich. N,N-Dimethylacetamide (DMAc) was supplied by Merck, and lithium chloride (LiCl) was purchased from Across, Belgium.

2.2 Preparation of OPEFB All-Nanocellulose Composite Films

The raw OPEFB has undergone acid hydrolysis pretreatment according to Salehudin et al. (2014) and Zailuddin (2014) methods to extract the cellulose component. The OPEFB cellulose and MCC were activated for 5 h in distilled water at room temperature for swelling purpose. The swollen OPEFB cellulose and MCC were then dehydrated in acetone for 5 h and DMAc for 4 h. The activated OPEFB cellulose and MCC were dissolved in DMAc for 10 min. LiCl at 8 wt% was added to the solution and was continuously stirred for another 20 min. The regenerated cellulose solution was then poured onto a glass plate and was left overnight for the films to develop. The next day, distilled water was used to wash and remove the residual DMAc/LiCl solution from the films. The films were then dried at room temperature for 24 h. Table 1 shows the formulation of OPEFB all-nanocellulose composite films.

3 Characterizations

The tensile test was operated using Instron Universal Testing Machine, model 5590, according to ASTM D 882. The tensile strength, elongation at break, and modulus of elasticity were recorded from the software. The cross-head speed of testing was at 10 mm/min.

X-ray diffraction was conducted using X-ray diffractometer model. Bruker DS advance diffractometer with Cu-Kd ($\lambda = 1.5418$ Å) was used as the X-ray source.

4 Results and Discussions

4.1 Mechanical Properties

Figure 1 shows the effect of OPEFB contents on the tensile strength of all-nanocellulose composite films. The results show that OPEFB contents at 1 wt%

have the highest tensile strength compared to OPEFB contents at 2, 3, and 4 wt%. The high strength of tensile at 1 wt% of OPEFB content is probably due to good dispersion and good interaction between the OPEFB cellulose matrix and OPEFB cellulose fiber in the all-nanocellulose composite films. A research by Zhang et al. (2012a, b) on regenerated cellulose/graphene nanocomposite films prepared in DMAC/LiCl solution stated that the mechanical property of nanocomposite films improved with the addition of the graphene which is due to the strong interaction that exist between graphene and the polymer matrix. A study by Soheilmoghaddam et al. (2013a, b) on regenerated cellulose/halloysite nanotube nanocomposite films prepared with an ionic liquid showed that the tensile strength at 6 wt% increased possibly due to good dispersion of the halloysite nanotube inside the regenerated cellulose films. The study also mentioned that when the contents of halloysite nanotube were further increased beyond 6 wt%, the tensile strength decreased (Soheilmoghaddam et al. 2013a, b). A study by Zailuddin and Husseinsyah (2016) on oil palm empty fruit bunch regenerated cellulose biocomposite films which suggests that good tensile strength can be achieved when OPEFB content of 2 wt% was employed because well-dispersed OPEFB particles were obtained in the solvent. On the other hand, an increase in OPEFB content led to the decrement of tensile strength especially at 4 wt% of OPEFB contents. Similarly, the results of this study also showed the increase of OPEFB content causes the tensile strength to decrease probably due to the aggregation of the cellulose particles.

Figure 2 illustrates the effect of OPEFB contents on the elongation at break of all-nanocellulose composite films. The elongation at break of all-nanocellulose composite films increases from 1 to 4 wt% of OPEFB contents. The decrement of elongation at break at 1 wt% of OPEFB contents may be due to the more orderly structure of the chains that restrain the mobility of the films. This can be related to

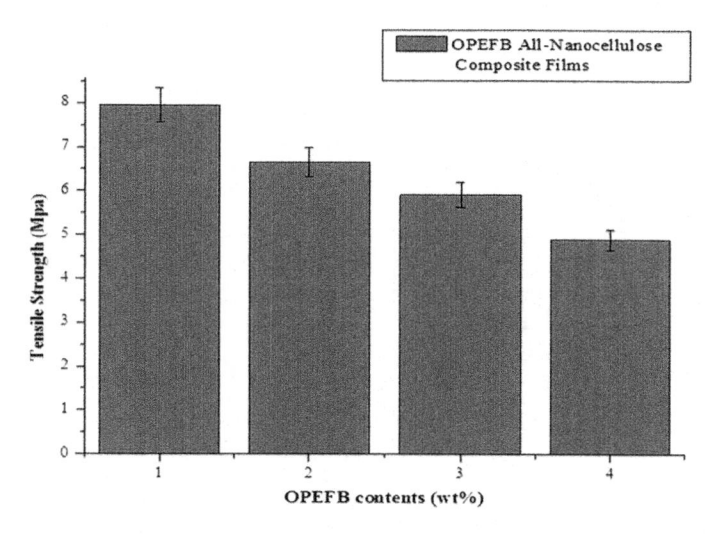

Fig. 1 Effect of OPEFB contents on the tensile strength of all-nanocellulose composite films

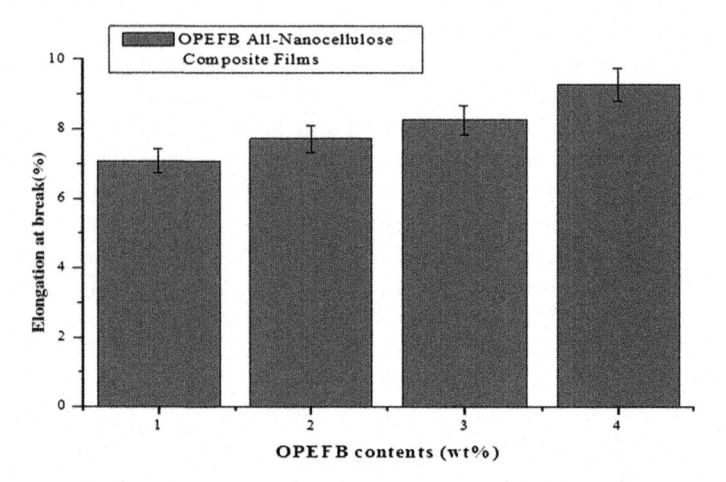

Fig. 2 Effect of OPEFB contents on the elongation at break of all-nanocellulose composite films

Zailuddin and Husseinsyah (2016) where the regenerated cellulose biocomposite films at 2 wt% of OPEFB contents showed more rigidity compared to other OPEFB contents thus decreased the mobility of the chains in films.

Figure 3 represents the effect of OPEFB contents on the modulus of elasticity of all-nanocellulose composite films. From Fig. 3, it can be seen that the modulus of elasticity of all-nanocellulose composite films decreased from 1 to 4 wt% of OPEFB contents. The reason for the decrement could probably be due to the increase of the OPEFB contents which can cause the aggregation of the particles. The modulus of elasticity is lower at 3 and 4 wt% of OPEFB contents is probably due the particles that were not dispersed well in the solvent system which was reported by Zailuddin

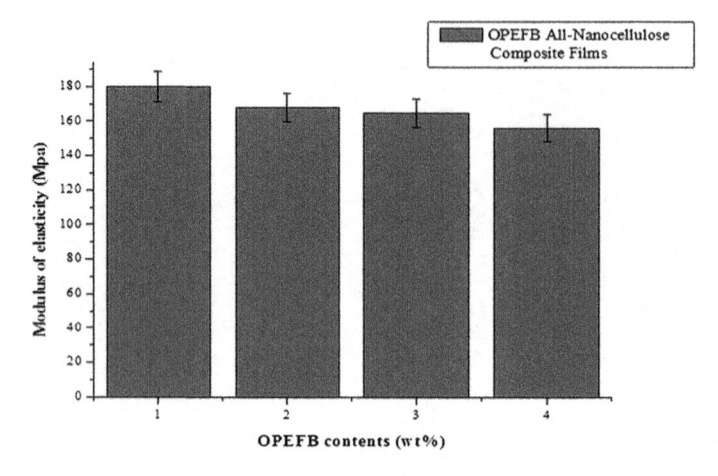

Fig. 3 Effect of OPEFB contents on the modulus of elasticity of all-nanocellulose composite films

and Husseinsyah (2016). Also in Zailuddin and Husseinsyah (2016) it is mentioned that when the OPEFB contents increased, the tensile strength decreased due to the aggregation of OPEFB particles.

4.2 X-Ray Diffraction (XRD)

Figure 4 illustrates the XRD curves of all-nanocellulose composite films. From Fig. 4, the curve for OPEFB 1 wt% has more intense peak compared to OPEFB at 4 wt% suggesting high crystallinity. The decrease in crystallinity for OPEFB at 4 wt% is probably due to the high contents of OPEFB that can lead to agglomeration of the cellulose particles in the films. Pullawan (2012) reported that the decrease in the crystallinity of the all-cellulose nanocomposites at a concentration of 20 v/v% of cellulose nanowhiskers (CNWs) volume fraction is due to the increase in the cellulose nanowhiskers content which could bring about the whiskers aggregation. Mahmoudian et al. (2012) stated that in regenerated cellulose/montmorillonite 8 wt% (RC/MMT-8) nanocomposite films, with high content of MMT had lead to agglomeration in the regenerated cellulose matrix. The findings by Zailuddin et al. (2017) evidenced that the peak of XRD curves for untreated OPEFB RC biocomposite films at 2 wt% is higher than 4 wt% of untreated OPEFB RC biocomposite films. Another study by Soheilmoghaddam et al. (2014) showed that increase of the amount of sepiolite leads to the reduction of crystallinity in the nanocomposite films.

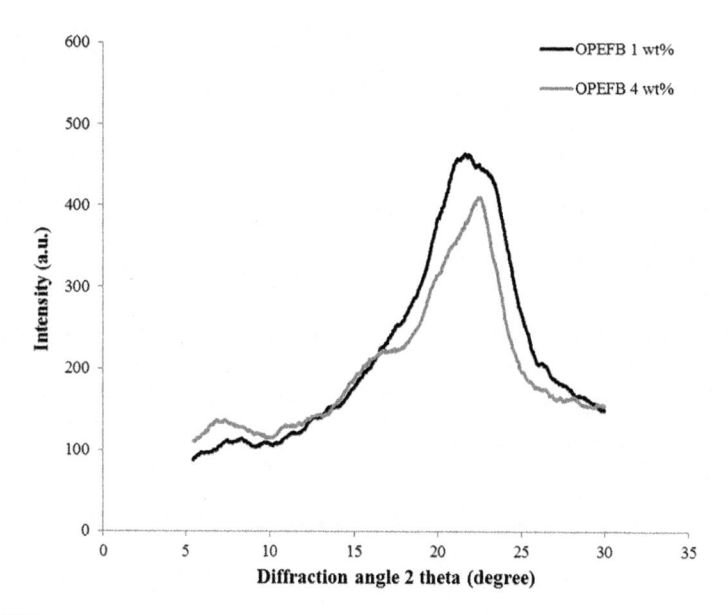

Fig. 4 XRD curves of all-nanocellulose composite films with 1 and 4 wt% of OPEFB contents

5 Conclusion

All-nanocellulose composite films were produced from the dissolution of OPEFB with DMAc/LiCl. The tensile strength and modulus of elasticity values of the composite film with 1 wt% of OPEFB content are higher than other composite films with higher content of OPEFB. These could be due to the greater dispersion of cellulose particles in the composite and higher degree of crystallinity of the material when low content of OPEFB was used.

References

Journal articles

Ariffin, H., Hassan, M. A., Umi Kalsom, M. S., Abdullah, N., & Shirai, Y. (2008). Effect of physical, chemical and thermal pretreatments on the enzymatic hydrolysis of oil palm empty fruit bunch (OPEFB). *Journal of tropical agriculture and food science, 36*(2), 1–10.

Biganska, O., & Navard, P. (2002). Crsytallization of cellulose/N-methilmorpholine-N-Oxide solution. *Polymer, 43*, 6139–6145.

Chang, S. H. (2014). An overview of empty fruit bunch from oil palm as feedstock for bio-oil production. *Biomass and Bioenergy, 62*, 174–181.

Dupont, A. L. (2003). Cellulose in Lithium Chloride/N-N-Dimethylacetamide, optimization of a dissolution method using paper substrates and stability of the solutions. *Polymer, 44*(15), 4117–4126.

Ghaderi, M., Mousavi, M., & Yousefi, Labbafi M. (2014). All-cellulose nanocomposite film made from bagasse cellulose nanofibers for food packaging application. *Carbohydrate Polymers, 104*, 59–65.

Govindan, V., Hussiensyah, S., Leng, T. P., & Amri, F. (2014). Preparation and characterization of regenerated cellulose using ionic liquid. *Advances in Environmental Biology, 8*(8), 2620–2625.

Guansen, J., Weifeng, H., Lin, L., Xiao, W., Fengjian, P., & Yumei, Z. (2012). Structure and properties of regenerated cellulose fibers from different technology processes. *Carbohydrate Polymers, 87*, 2012.

Han, J., Zhou, C., Frech, A. D., Han, G., & Wu, Q. (2013). Characterization of cellulose II nanoparticles regenerated from 1-butyl-3-methylimidazolium chloride. *Carbohydrate Polymers, 94*, 773–778.

Huber, T., Mussig, J., Curnow, O., Pang, S., Bickerton, S., & Staiger, M. P. (2012). A Critical review of all-cellulose composites. *Journal Materials Science, 47*, 1171–1186.

Ishii, D., Tatsumi, D., & Matsumoto, T. (2008). Effect of solvent exchange on the supramolecular structure, the molecular mobility and the dissolution behavior of cellulose in LiCl/DMAc. *Carbohydrate Research, 343*, 919–928.

Kavitha, B., Jothimani, P., & Rajannan, G. (2013). Empty fruit bunch—A potential organic manure for agriculture. *International Journal of Science, Environment and Technology, 2*(5), 930–937.

Klemm, D., Heublein, B., Fink, H.-P., & Bohn, A. (2005). Cellulose: Fascinating biopolymer and sustainable raw material. *Angewandte Chemie Int Ed, 44*, 3358–3393.

Liu, Z., Wang, H., Li, Z., Lu, X., Zhang, X., Zhang, S., et al. (2011). Characterization of the regenerated cellulose films in ionic liquid and rheological properties of the structures. *Materials Chemistry and Physics, 128,* 220–227.

Mahmodian, S., Wahit, M. U., Imran, M., Ismail, A. F., & Balahrishnan, H. (2012a). A Facile Approached to prepare regenerated cellulose/grapheme nanoparticles nanocomposite using room temperature ionic liquid. *Journal of Nanoscience and Nanotechnology, 12,* 5233–5239.

Mahmodian, S., Wahit, M. U., Ismail, A. F., & Yussuf, A. A. (2012b). Preparation of regenerated cellulose/montmorillonite nanocomposite films via ionic liquids. *Carbohydrate Polymers, 88,* 1251–1257.

Moon, R. J., Martini, A., Nairn, J., Simonsen, J., & Youngblood, J. (2011). Cellulose nanomaterials review: Structure, properties and nanocomposites. *Chemical Society Reviews, 40,* 3941–3994.

Pang, J., Liu, X., Zhang, X., Wu, Y., & Sun, R. (2013). Fabrication of cellulose film with enhanced mechanical properties in ionic liquid 1-Allyl-3-methylimidaxolium chloride (AmimCl). *Materials, 6,* 1270–1284.

Sahari, J., & Sapuan, S. M. (2011). Natural fibre reinforced biodegradable polymer composites. *Advanced Materials Science, 30,* 166–174.

Salehudin, M. H., Salleh, E., Mamat, S. N. H., & Muhamad, I. I. (2014). Starch based active packaging film reinforced with empty fruit bunch (EFB) cellulose nanofiber. *Procedia Chemistry, 9,* 23–33.

Shariff, A., Aziz, N. S. M., & Abdullah, N. (2014). Slow pyrolysis of oil palm empty fruit bunches for biochar production and characterisation. *Journal of Physical Science, 25*(2), 97–112.

Siro, I., & Plackett, D. (2010). Microfibrillated cellulose and new nanocomposite materials: A review. *Cellulose, 17,* 459–494.

Soheilmoghaddam, M., & Wahit, M. U. (2013). Development of regenerated cellulose/halloysitenanotube bionanocomposite films with ionic liquid. *International Journal of Biological Macromolecules, 58,* 133–139.

Soheilmoghaddam, M., Wahit, M. U., Mahmoudian, S., & Hamid, N. A. (2013a). Regenerated cellulose/halloysite nanotube nanocomposite films prepared with an ionic liquid. *Materials Chemistry and Physics, 141,* 936–943.

Soheilmoghaddam, M., Wahit, M. U., & Akos, N. I. (2013b). Regenerated cellulose/epoxidized natural rubber blend film. *Materials Letters, 111,* 221–224.

Soheilmoghaddam, M., Wahit, M. U., Yussuf, A. A., Al-Saleh, M. A., & Whye, W. T. (2014). Characterization of bio regenerated cellulose/sepiolite nanocomposite films prepared via ionic liquid. *Polym Test, 33,* 121–130.

Soykeabkaew, N., Nishino, T., & Peijs, T. (2009a). all-cellulose composites of regenerated cellulose fibres by surface selective dissolution. *Composites: Part A, 40,* 321–328.

Soykeabkaew, N., Sian, C., Gea, S., Nishino, T., & Peijis, T. (2009b). All-cellulose nanocompositesby surface selective dissolution of bacterial cellulose. *Cellulose, 16,* 435–444.

Sulaiman, F., Abdullah, N., Gerhauser, H., & Shariff, A. (2011). An outlook of Malaysian energy, oil palm industry and its utilization of wastes as useful resources. *Biomass and Bioenergy, 35*(9), 3775–3786.

Tamai, N., Tatsumi, D., & Matsumoto, T. (2004). Rheological properties and molecular structure of tunicate cellulose in LiCl/1,3-dimethyl-2-imidazolidinone. *Biomacromolecules, 5*(2), 422–432.

Zailuddin, N. L. I., & Husseinsyah, S. (2016). Tensile properties and morphology of oil palm empty fruit bunch regenerated cellulose biocomposite films. *Procedia Chemistry, 19,* 366–372.

Zailuddin, N. L. I., Husseinsyah, S., Hahary, F. N., & Ismail, H. (2017). Characterization and properties of treated oil palm empty fruit bunch regenerated cellulose biocomposite films with butyl methacrylate using ionic liquid. *Polymer - Plastics Technology and Engineering, 56*(2), 109–116.

Zadagen, S., Hossanimalippier, M., Ghassari, H., Rezak, H. R., & Naimi-Jamal, M. R. (2010). Synthesis of cellulose-monohydroxyapatite composite in 1-n-butyl-3 methylimidazolium chloride. *Ceramic International, 36,* 73–778.

Zhang, X., Feng, J., Liu, X., & Zhu, J. (2012a). Preparation and characterization of regeneratedcellulose/poly (vinylidene fluoride) (PVDF) blend films. *Carbohydrate Polymers, 89*, 67–71.

Zhang, X., Liu, X., Zheng, W., & Zhu, J. (2012b). Regenerated cellulose/graphene nanocomposite films prepared in DMAC/LiCl solution. *Carbohydrate Polymers, 88*, 26–30.

Zhao, H., Kwak, J. H., Wang, Y., Franz, J. A., White, J. M., Holladay, J. E. (2007). Interactions between cellulose and N-methylmorpholine-N-oxide. Carbohydr Polym 67:97–103.

Dissertations

Pullawan, T. (2012). Interfacial micromechanics of all-cellulose nanocomposites using Raman Spectroscopy, The University of Manchester.

Zailuddin, N. L. I. (2014). Tensile and thermal properties of oil palm empty fruit bunch renerated cellulose biocomposite films using ionic liquid. University Malaysia Perlis.

Efficiency Cooling Channel at Core Side Incorporating with Baffle and Bubbler System

Saiful Bahri Mohd Yasin, Noor Faezah Mohd Sani, Salwa Adnan, Zahidahthorwazunah Zulkifli, Zuliahani Ahmad and Sharifah Nafisah Syed Ismail

Abstract In injection molding process, cooling system is considered as the most significant phase among filling and ejection phase as it affects both productivity and part quality. These can be achieved by uniform distribution of temperature on molded parts from the core and cavity of mold. In this study, three different designs of cooling channel were developed to be employed in hollow plastic part (plastic vase). Computer-aided engineering (CAE) software was used to assess the efficient of the three different cooling systems which are conventional, baffle, and bubbler cooling channel. The results from the simulated part revealed that the baffle and bubbler cooling system supplied more uniform temperature distribution as compared to conventional system. Thus, it leads to the reduction in percentage of shrinkage that causes to warpage to the part. The range of cooling temperature for both systems is 85–98 °C and 114–193 °C for the latter.

Keywords CAE · Cooling system · Warpage · Hollow part · Baffle Bubbler

---

S. B. M. Yasin (✉) · S. Adnan · Z. Zulkifli · Z. Ahmad · S. N. S. Ismail
Polymer Technology Department, Faculty of Applied Sciences, Universiti Teknologi MARA Cawangan Perlis, Kampus Arau, 02600 Arau, Perlis, Malaysia
e-mail: sbmyasin@gmail.com

S. Adnan
e-mail: aizawa1794@yahoo.com

Z. Zulkifli
e-mail: zahidahthorwazunah@gmail.com

Z. Ahmad
e-mail: zuliahani@perlis.uitm.edu.my

S. N. S. Ismail
e-mail: sh.nafisah86@yahoo.com

N. F. M. Sani
Faculty of Applied Sciences, Universiti Teknologi MARA, Perak Branch, Tapah Campus, 35400 Tapah Road, Perak, Malaysia
e-mail: faezah125@uitm.edu.my

© Springer Nature Singapore Pte Ltd. 2018
R. Saian and M. A. Abbas (eds.), *Proceedings of the Second International Conference on the Future of ASEAN (ICoFA) 2017 – Volume 2*,
https://doi.org/10.1007/978-981-10-8471-3_51

1 Introduction

Injection molding process is a major process in producing plastic part. The molten plastic injects into the cavity to form a desired shape of plastic part. The main phase in injection molding process is filling, cooling, and ejection process. The cooling process plays a main role in the molding process where it takes up 70–80% of injection molding cycle. A mold is cooled by designed cooling channel around the cavity and core insert and using water or oil as the cooling media. The position and design of cooling channels are significant elements in the design of the mold (Pirnc et al. 2014). A suitable design of cooling system is required to ensure optimum heat transfer process between plastic molten and the mold plate. Uniform cooling temperature improves the quality and maintains dimensional of part by reducing the difference of residual stress (Patil and Surange 2016), volumetric shrinkage, and warpage. Practically, the cooling system usually uses water or replaced with oil when the temperature of mold set to be higher than 90 °C (Paclt 2011).

Conventional cooling system consists of straight drilled holes in the mold. It has a disadvantage when used for hollow part such plastic vase, pipe, and others as it resulted in inadequate cooling process to the internal surface of the part. Thus, conventional cooling system gave uneven temperature distribution and the part may have a warpage issue. In avoiding this issue, two types of cooling channel at core side are suggested: baffle and bubbler system. In this system, baffle is inserted in a drilled blind hole and is used as a stopper in providing the coolant flow up one side and down the other side as shown in Fig. 1a. Baffles provide maximum cross section for the coolant in eliminating the disadvantage of baffle system. In this system, bubbler is inserted by fitting a tube in the drilled blind hole to form an

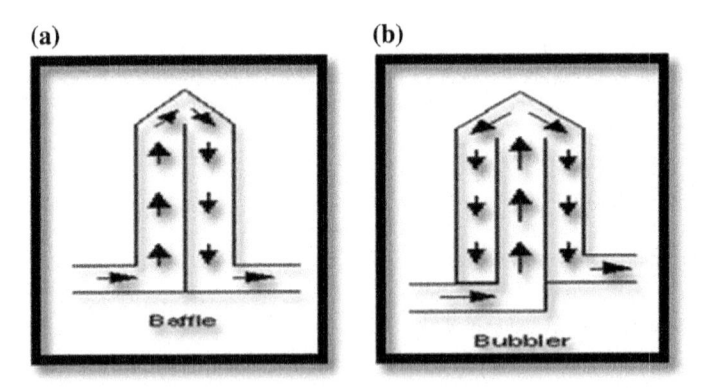

Fig. 1 Baffle and bubbler cooling system

annular channel as shown in Fig. 1b. The coolant flow up through the inner tube and flows down outside of the tube. Thus, bubbler is found the most effective cooling system for the hollow molded part.

Hence, this study is introduced to analyze the efficiency of the baffle or bubbler cooling system in order to approach a favorable cooling distribution to entire of hollow part with respect to the conventional cooling system. Thus, defects such as different residual stress, uneven volumetric shrinkge, and warpage can be reduced.

2 Design of Part and Cooling Systems

The steps of cooling phase analysis are shown in Fig. 2. In this study, plastic vase with hollow section was imported to perform CAE analysis. Length, width, and thickness of the modeling were 203 mm × 229.4 mm × 2 mm as illustrated in Fig. 3. The polypropylene (PP) was selected as the material, where the initial parameter setting such as mold temperature and melt temperature was set to be 50 and 220 °C, respectively. The location of gate was set at the center of the base plastic vase and used sprue gate. Based on the modeling, it required to insert a baffle

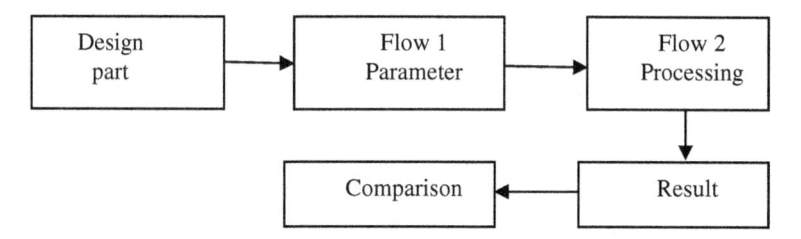

Fig. 2 Cooling analysis flow chart

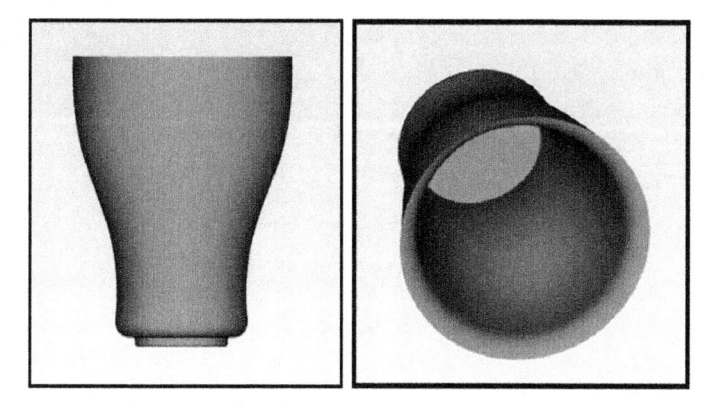

Fig. 3 Plastic vase model with hollow section

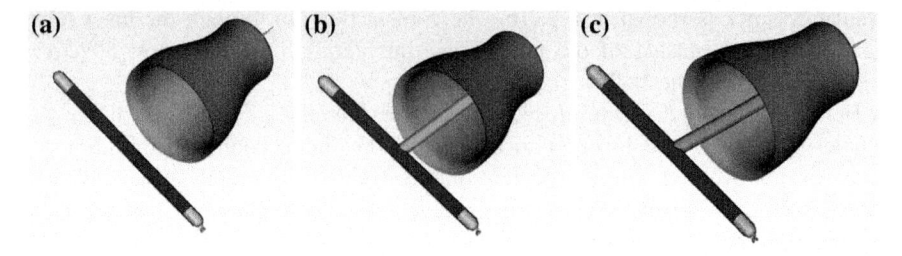

Fig. 4 Type of cooling system. **a** Conventional, **b** Baffle, and **c** Bubbler

or bubbler system into hollow section in order to provide cooling channel in the core side, thus results an uniform cooling temperature distribution as shown in Fig. 4. Cooling channel was designed with the diameter 24 mm.

3 Plastic Simulation Study

Molding window analysis was used to analyze favourable parameter setting such as melt temperature, mold temperature, and injection time as shown in Fig. 5. The melt temperature is found to be 221.0 °C, while mold temperature is found to be 63.64 °C.

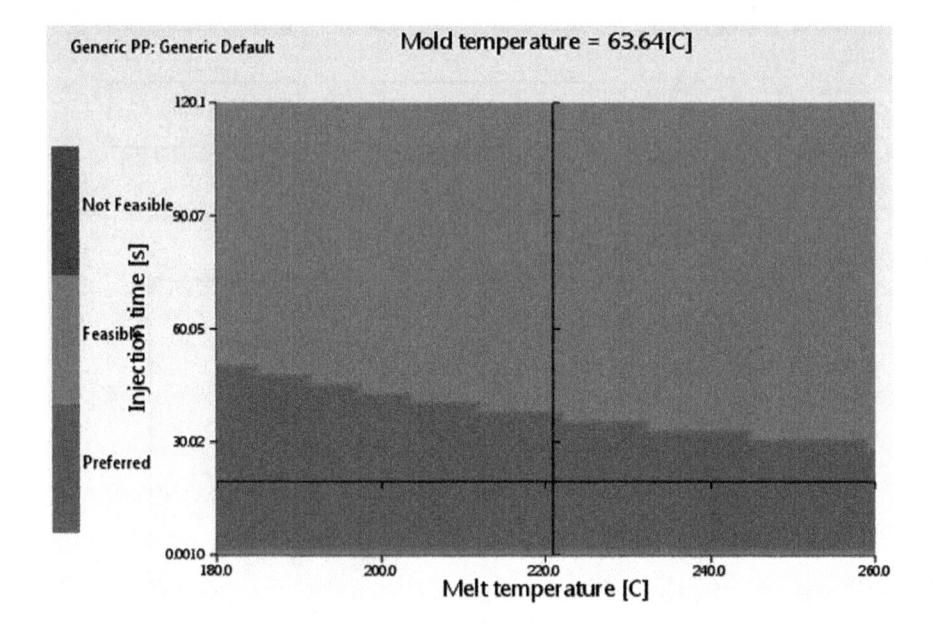

Fig. 5 Molding window

4 Results and Discussion

Figure 6 shows the temperature comparisons of three different cooling systems. In this analysis, temperature distribution was displayed by numerical values and different colors in cross-sectional area of the vase model. Red and blue shade indicates high temperature and low temperature, respectively. The part was divided into three stages to represent the distance from cooling channel to end of baffle and bubbler system.

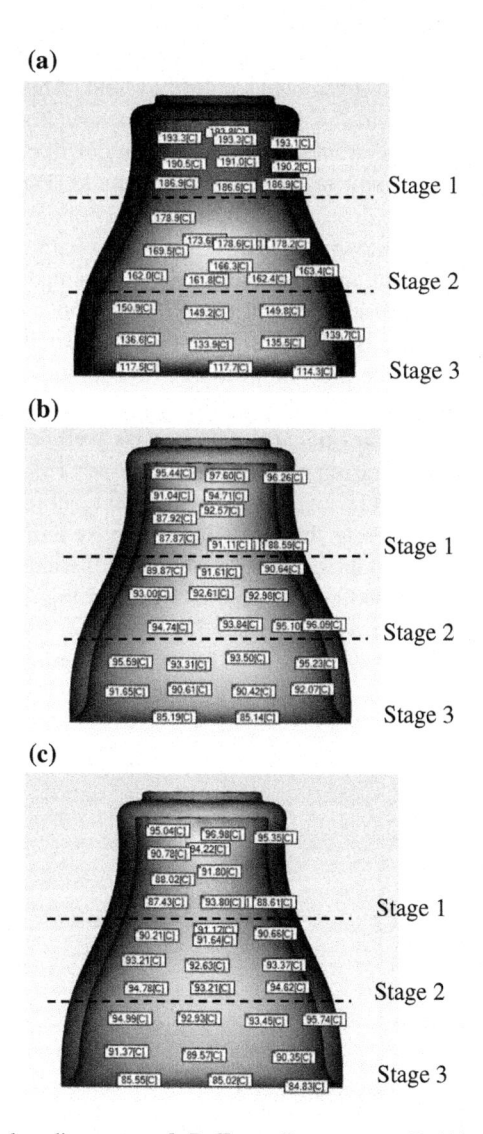

Fig. 6 a Conventional cooling system, b Baffle cooling system, c Bubbler cooling system

Figure 6a displays a wide range of colors which depicts an inconsistency of the temperature distribution inside the hollow part when conventional cooling system is employed. The minimum temperature of the part during cooling process is 114.3 °C, while maximum temperature is 193.3 °C and the average temperature is 154.25 °C.

Figure 6b gives the schematic structure of temperature uniformity for baffle cooling system. Minimum and maximum temperature was measured at 85.14 and 97.60 °C, respectively, with the average, 91.37 °C

The results for bubbler cooling system are represented in Fig. 6c, where the minimum is 84.33 °C, maximum is 96.93 °C, and the average is 90.97 °C. The range of temperature in the hollow vase model for baffle and bubbler cooling system is smaller than in conventional cooling system. This clarify, baffle and bubbler cooling system provided a better removal of heat from the mold to cool down the hollow section. The results attributed to contour-like channel that able to absorb heat away from the molten plastic inside the mold (Dimla et al. 2005).

The stages of temperature distribution for the three cooling systems represent in Fig. 7. It can be seen that hollow vase temperature using baffle and bubbler system is more uniform than conventional cooling system. The temperature different from maximum to minimum temperature is 79, 12.46, and 12.60 °C for conventional, baffle, and bubbler cooling system, respectively. This indicates, conventional cooling system needs other cooling system support to cool the molten polymer from the inside of hollow part. Thus, it can be concluded that baffle and bubbler cooling system is able to cool all zones of part uniformly as well as more efficiently by supplied uniform temperature even on complex shape part (Mohamed et al. 2013).

One of most significant affecting on the temperature difference is volumetric shrinkage, where it is occurred as the molded part to have a different shrinkage on the part surface during the cooling process. Effect of different cooling channel on volumetric shrinkage in molded hollow vase is shown in Fig. 8. Injection molding machine which employed conventional cooling system revealed to have higher percentage (9.52%) of shrinkage due to different temperature during the cooling

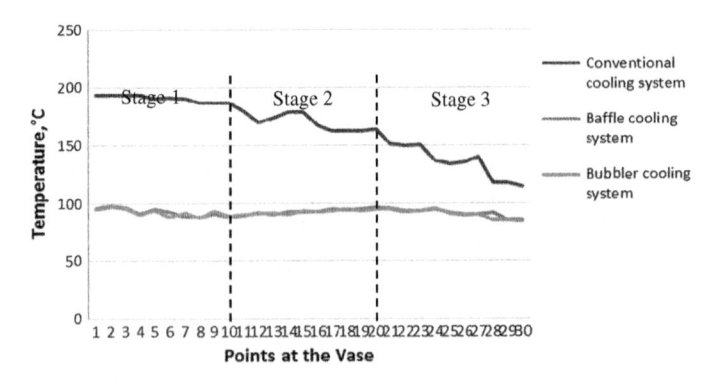

Fig. 7 Temperature Distribution for the three cooling system

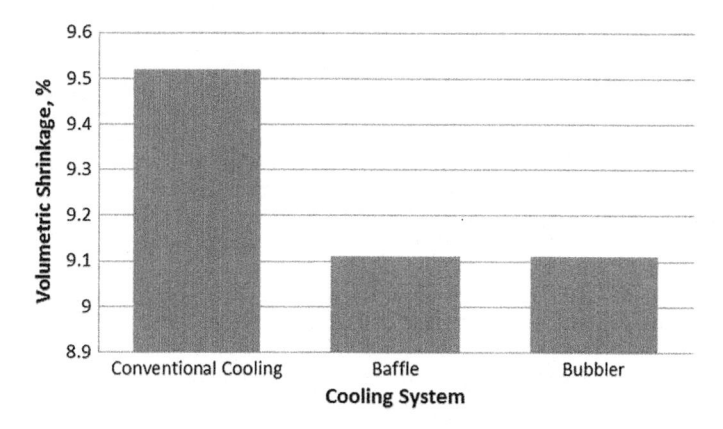

Fig. 8 Percentage of the average volumetric shrinkage

cycle, whereas the percentage of average volumetric shrinkage recorded for baffle and bubbler cooling system drop to 9.11% for both. The reduced temperature can be explained due to uniform distribution of heat. By decrease, the percentage of shrinkage in hollow vase cooled by baffle and bubbler system, post-injection warpage of part also can be reduced (Dimla et al. 2005).

5 Conclusion

In conclusion, baffle and bubbler system was successfully analyzed with the difference temperature distribution recorded at 12.77 and 12.55%, and thus it is recommended as the most suitable cooling system for plastic vase compared to conventional cooling system. It gives better cooling part temperature and less volumetric shrinkage. With the baffle and bubbler installation, it provides efficient cooling process for the hollow vase model because they are able to distribute uniform temperature to all zones of part. Then, it showed that it is possible to improve the productivity and part quality by predicting the best type of cooling system.

It is recommended to test the actual process for pilot test and comparing the result with the simulation data. Otherwise, it is proposed to analyze the cooling system with series of the baffle or bubbler system that may require large part/core insert in order to find uniform cooling temperature and produce lowest possible temperature.

References

Dimla, D. E., Camilotto, M., & Miani, F. (2005). Design and optimisation of conformal channels in injection molding tools. *Journal of Materials Processing Technology, 164–165,* 1294–1300.

Mohamed, O. A., Masood, S. H., & Saifullah, A. (2013). A simulation study of conformal cooling channels in plastic injection molding. *International Journal of Engineering Research, 2*(5), 344–348.

Paclt, R. (2011). Cooling/heating of the injection mold. *Journal for Technology of Plasticity, 36*(2), 147–154.

Patil, R., & Surange, J. (2016). A review on cooling system design for performance enhancement of injection molding machine. In International Conference on Global Trends in Engineering, Technology and Management.

Pirnc, N., Schmidi, F., Mongeau, M., Bugarin, F., & Chinesta, F. (2014). Optimization of 3D cooling channels in injection molding using DRbem and model reduction.

Potential Antibacterial Activity of Essential Oil of *Citrus hystrix* and *Chromolaena odorata* Leaves

Hamidah Jaafar Sidek and Fatin Fathihah Abdullah

Abstract Both *Chromolaena odorata* and *Citrus hystrix* leaves are known to have various biological activities. This study was designed to determine the antibacterial potential of essential oil of both plants. The essential oil was tested against the selected Gram-positive (*Enterococcus faecalis* and *Staphylococcus aureus*) bacteria and Gram-negative (*Pseudomonas aeruginosa* and *Escherichia coli*) bacteria strains by using Kirby Bauer Disc Diffusion method. *C. hystrix* essential oil exhibited the highest inhibitory effect against *E. faecalis* where the highest inhibition zone was 16.33 ± 0.33 mm, followed by *E. coli*, with 15.4 ± 0.33 mm inhibition zone compared to the other bacteria. The result showed that both the gastrointestinal bacteria which were *E. faecalis* and *E. coli* were susceptible to the essential oil. The essential oil of *C. odorata* showed a potential antibacterial activity against *E. faecalis* and *S. aureus* with 17.3 ± 0.33 mm and 17.0 ± 0.58 mm of inhibition zones, respectively. In addition, a combination effect study was done to examine the possible synergistic antibacterial activity from bioactive compounds of both essential oil. The inhibitory effect of combined essential oil against *E. coli* showed the highest inhibition zone which was at 19.0 ± 0.58 mm, while the inhibition zones of bacteria growth for *E. faecalis* and *S. aureus* were at 17.7 ± 0.33 mm and 18.0 ± 0.58 mm, respectively. The results from this study showed that essential oil from both plants separately and in the combined state has the potential to be significant antibacterial agents in the pharmaceutical industry.

Keywords Antibacterial activity *Citrus hystrix* · *Chromolaena odorata* Combination effect study · Essential oil

---

H. J. Sidek (✉) · F. F. Abdullah
Faculty of Applied Sciences, Universiti Teknologi MARA, 02600 Arau, Perlis, Malaysia
e-mail: hamidahjs@perlis.uitm.edu.my

© Springer Nature Singapore Pte Ltd. 2018
R. Saian and M. A. Abbas (eds.), *Proceedings of the Second International Conference on the Future of ASEAN (ICoFA) 2017 – Volume 2*,
https://doi.org/10.1007/978-981-10-8471-3_52

1 Introduction

Modern medicine become household choice because of their proven effectiveness to disclose and treat a mass number of different types of medical conditions, especially the ones initiated and triggered by bacteria, viruses and other sorts of infectious agents (Ventola 2015). However, in recent years, modern medicine is desperately short of new treatment due to the growing of drug resistance bacteria and disease which is mainly caused by the misuse of medications like antibiotics and other lifesaver medicines (Carteret 2011). Therefore, scientists are modernising traditional medicine by making them available in the mainstream by incorporating the knowledge, theories and beliefs into modern healthcare to ensure they meet safety and efficacy standards, and thus, they may provide new and lucrative drugs resources (Shetty and Rinaldi 2010).

In Southeast Asia, especially Malaysia, a lot of medicinal herbs has been used traditionally but never been properly commercialized due to their lack of in vitro, in vivo and clinical testing. Kaffir Lime or scientifically known as *Citrus hystrix* is one of the famous herbs that has been used in many cuisine in Malaysia, Thailand and Indonesia (Hassan et al. 2016). Although Kaffir Lime is mainly used in cuisine, several scientifically testing has been done that showed a great potential to be commercialized as another medicinal product as it exhibited great potential antibacterial and antioxidant properties (Kooltheat et al. 2016; Ng et al. 2011; Ratseewo et al. 2016). *Chromolaena odorata* is a scientific name for Siam Weed or also known as 'Daun Kapal Terbang' in Malay, is a plant that has been widely used traditionally in many South East countries to stop bleeding and help in wound healing process (Atindehou et al. 2013; Vaisakh and Pandey 2012). In previous studies, Siam Weed also showed its potential in antimicrobial activity (Atindehou et al. 2013; Javed et al. 2013; Nurul Huda et al. 2004).

The individual effects of both plant extracts of Kaffir Lime and Siam Weed have been extensively reported on their antimicrobial and antioxidant properties (Mbajiuka 2015; Ratseewo et al. 2016). However, their combination effects on the properties or bioactivities seem to be lacking. Therefore, the present study is done to investigate the combination effect of the essential oil of both plants whether it gives out synergistic or additional value on the antibacterial testing.

2 Materials and Methods

2.1 Plant Materials

Both Kaffir Lime (*C. hystrix*) and Siam Weed (*C. odorata*) leaves were collected in Chuping, Perlis, Malaysia. Kaffir Lime leaves were dried under shade for one day and then were placed in a hot-air oven with temperature at 50 °C (Tomar et al. 2013). Siam Weed leaves were collected washed, dried under sunlight and then placed in a hot-air oven with the temperature controlled at 60 °C (Hanphakphoom et al. 2016). After sufficiently dried, the plants dried leaves were ground into powder by using a blender. Then, the leaves powder of both plants were placed separately in airtight container and labelled to their respective plant. The extraction yield was calculated using Eq. 1.

$$\text{Yield (\%)} = \frac{\text{Weight of essential oil (EO) recovered}}{\text{Weight of dried leaves}} \times 100\% \qquad (1)$$

2.2 Hydrodistillation

The extraction of essential oil was carried out by using hydrodistillation apparatus. The extraction yield was obtained after 4 h. The distillate obtained was then subjected to separation process where there were two layers formed where the top layer corresponded to oil while the bottom layer was water. The distillate sample was transferred into separating funnel. Dicholoromethane was added into the funnel to form immiscible solvent where the oil and dichloromethane solvent at the bottom layer and water at the upper layer. The oil at lower layer was taken and the liquid or water at the upper layer was discarded. The evaporating process was continued to dry up remaining water contained in oil layer by using sodium anhydrous acid powder. The sodium anhydrous acid and dichloromethane acted as catalyst to accelerate the evaporating process using Vacuum Rotary Evaporator (Rotavap). The essential oil was labelled and stored at 4 °C in the dark (Kasuan et al. 2013). Finally, the extracted oil sample was subjected for antibacterial study and combine effect study.

2.3 Microorganisms

The antibacterial activity was evaluated using Gram-negative bacteria: *P. aeruginosa, Escherichia coli* and Gram-positive bacteria: *E. faecalis* and *S. aureus*. All bacteria strains were provided from microbiology laboratory UiTM Perlis Branch. Bacterial strain was maintained by subculture in the nutrient broth favourable to their growth within 18–24 h in the incubator at 37 °C. Then, the turbidity of bacteria was checked with 0.5 McFarland standard. If the bacteria culture was cloudier than 0.5 McFarland, the saline water was added in it in order to dilute the suspension to appropriate density. Meanwhile, if the bacteria culture was clear compared to standard, the bacteria culture was left to incubate more.

2.4 Antibiotics

The antibiotics standard used for antibacterial testing in this study was Gentamicin (120 µg).

2.5 Antibacterial Susceptibility Test (Kirby Bauer Method)

The ability of the essential oil to inhibit the growth of the isolated strain was determined in vitro using disc diffusion method. The Mueller-Hinton (MH) agar was poured (estimated 20–25 ml), in sterile Petri dishes (90 mm diameter). Basically, the essential oil was impregnated on blank disc or paper disc and left covered for about 5 min. The paper discs (6 mm diameter) that were impregnated with 15 µl of each pure essential oil and antibiotic were then placed on the inoculated agar surface. Petri dishes were allowed to stand for 15 min at room temperature before incubation at 37 °C for 24 h. The effect of essential oil was reflected by the appearance of a clear zone surrounding the disc which indicated the absence of growth of bacteria around that area. The diameter of inhibition zone was measured in mm. The larger the diameter of the area the more susceptible is the strain towards the effect of the essential oil. The diameter of inhibition zone was then compared to the table on the bacteria response range with the specific antibiotic.

2.6 Combination Effect Study

To evaluate the synergistic or combination effect, the essential oil of both plants was combined together with 1:1 ratio and then was impregnated on the paper disc and then placed on the inoculated agar surface.

3 Result and Discussion

3.1 Percentage Yield of Essential Oil (EO)

The extraction yields of both samples were indicated in Table 1. After drying and grinding process, the weight of Kaffir Lime dried leaves was 250 g, while Siam Weed leaves were 300 g. After evaporation process by using rotary evaporator, percentage of Kaffir Lime oil and Siam Weed was 3.487 and 1.971%, respectively. For Kaffir Lime, its percentage yield was slightly higher from previous study that used the same method (Chanthaphon et al. 2008). In contrast, the percentage yield of Siam Weed essential oil were 1.971%. It is worth to note that the percentage

Table 1 Percentage yield of essential oil

Essential oil	Percentage yield (%)
Kaffir Lime	3.487
Siam Weed	1.971

yield of this study was slightly higher compared to two other studies (Oladoye et al. 2014; Owolabi et al. 2010).

Both of plants' essential oil percentage yields were higher compared to previous study. There were many considerable conditions which can affect the percentage yield of plants. In the present study, the harvest time of these leaves may contribute to their higher percentage yield. For Kaffir Lime, the leaves were harvested in the morning, where the weather condition was sunny and grown in moist and well-drained soil where these conditions were ideal conditions for its optimal growth. As the Kaffir Lime leaves were harvested in the morning, the attribution of sunlight was low and moist condition increased the chances of higher percentage yield of Kaffir Lime extraction. Meanwhile, Siam Weed leaves were harvested in the late evening. Similarly, Siam Weed's percentage yield also showed higher percentage yield when compared to previous study. Interestingly, a study by Oladoye et al. (2014) stated that morning harvest of Siam Weed leaves has higher percentage yield and compounds compared to afternoon harvest of Siam Weed leaves. In the present study, interestingly, the Siam Weed leaves were harvested in late evening, where similar conditions of light attribution may affect the percentage yield of these leaves. In comparison of both essential oil, the extraction yield of Kaffir Lime essential oil was higher compared to Siam Weed essential oil as can be seen in Table 1.

3.2 Antibacterial Activity of Essential Oil

The essential oil was tested for antibacterial activity against Gram-positive bacteria which were *S. aureus* and *E. faecalis*, and also Gram-negative bacteria which were *P. aeruginosa* and *E. coli*. Based on Table 2, Kaffir Lime oil was the most effective against *E. faecalis* where it showed the highest inhibition zone of 16.33 ± 0.33 mm and effective against *E. coli*, with 15.4 ± 0.33 mm inhibition zone compared to other bacteria. For both *S. aureus* and *P. aeruginosa*, essential oil of Kaffir Lime only inhibits the bacteria growth at 14.0 ± 0.58 and 13.3 ± 0.89 mm, respectively, which was in the bacterial response of the intermediate range. From the result based on the zone of inhibition of the tested bacteria, we can conclude that the essential oil of Kaffir Lime was effective in inhibiting the growth of the gastrointestinal tracts bacteria.

Siam Weed essential oil showed strong antibacterial properties against *E. faecalis* and *S. aureus* with 17.3 ± 0.33 mm and 17.0 ± 0.58 mm diameter of inhibition zones, respectively. In addition, *P. aeruginosa* also showed a susceptible

Table 2 Zone of inhibition of different kinds of bacteria towards the essential oil

Bacteria	Kaffir Lime	*Br	Siam Weed	*Br	Positive Control	*Br
Staphylococcus aureus	14.0 ± 0.58	I	17.0 ± 0.58	S	20.0 ± 0.47	S
Pseudomonas aeruginosa	13.3 ± 0.89	I	15.0 ± 0.58	S	23.0 ± 0.58	S
Escherichia coli	15.4 ± 0.33	S	13.4 ± 0.33	I	19.0 ± 0.00	S
Enterococcus faecalis	16.33 ± 0.33	S	17.3 ± 0.33	S	19.0 ± 0.58	S

Data were expressed as means of triplicate zones inhibition ± Standard Error of Mean (SEM) = S. D/n
S.D Standard Deviation, *n* Number of test plates (3)
*(*BR* Bacteria Response, *S* Susceptible, *I* Intermediate)

response towards the inhibitory effect of the essential oil towards the growth of bacteria with inhibition zone of 15 ± 0.58 mm. In contrast, the zone inhibition for *E. coli* by Siam Weed essential oil was 13.4 ± 0.33 mm and showed intermediate response. From the finding, we can conclude that Siam Weed oil has a better inhibitory effect towards the growth of Gram-positive bacteria as the diameter of inhibition zones was higher compared to the zone inhibition of Gram-negative bacteria.

Based on the results in Table 2 type of bacteria, it can be concluded that antibacterial properties in Kaffir Lime essential oil was more effective against gastrointestinal bacteria, whether Gram-positive or Gram-negative bacteria. The finding on antibacterial activity of this study of Kaffir Lime essential oil was slightly different from past study that has been done (Ng et al. 2011; Srisukh et al. 2012), as their finding showed that Kaffir Lime oil was more effective against *S. aureus* and other Gram-positive bacteria.

Siam Weed essential oil showed a more effective result against Gram-positive bacteria. It is worth to note that Siam Weed essential oil showed a better inhibitory effect compared to Kaffir Lime essential oil in this antibacterial study. A better extracting method and identification of active compounds though several purification methods should be examined in future studies for a better result of both essential oil as shown in Fig. 1.

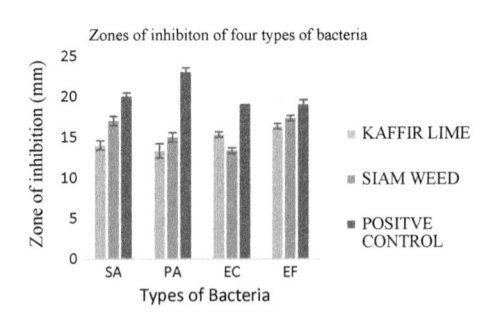

Fig. 1 Comparison between the inhibitory effect of Kaffir Lime and Siam Weed essential oil with positive control towards the four types of bacteria

3.3 Combination Effect Study

The evaluation of the combination effect of essential oil was done by combining both of essential oil in 1:1 ratio. The combine essential oil then was tested for its inhibitory effect on different species of bacteria which were *S. aureus, E. faecalis, P. aeruginosa* and *E. coli*. The combination effect study was done purposely to highlight feasible potential of the essential oil's combination as an antibacterial agent.

A combination effect study was done to examine the possible synergistic antibacterial activity from bioactive compounds of Kaffir Lime and Siam Weed essential oil which both of them previously have exhibited sufficient results individually in disc diffusion method. In previous study, they have shown promising antibacterial activity against a number of microbes when administered separately in vitro (Hanphakphoom et al. 2016). But their antibacterial combination effect study seems to be scarce and limited. From the finding recorded in Table 3 showed that combined essential oil was most effective against *E. coli* in contrast with *P. aeruginosa* where the finding showed the combined essential inhibited the lowest data against this bacteria. Combined essential oil showed the highest inhibitory effect against *E. coli* where the inhibition zone was at 19.0 ± 0.58 mm. For both *E. faecalis* and *S. aureus,* combined essential oil inhibits the bacteria growth at 17.7 ± 0.33 mm and 18.0 ± 0.58 mm, respectively, which was in the bacterial response of the susceptible range. Out of the four bacteria, *P. aeruginosa* showed the lowest inhibition response where the combined essential oil only inhibit bacteria at 15.7 ± 0.33 mm which was also in the bacterial response of the susceptible range.

In comparison with the previous foregoing finding in Sect. 3.2, individual result showed the antibacterial activity of Kaffir Lime essential oil was highest against *E. faecalis,* while the antibacterial activity for Siam Weed essential oil was highest against both *E. faecalis* and *S. aureus* where the inhibition zones were at 16.33 ± 0.33 mm, 17.3 ± 0.33 mm and 17.0 ± 0.58 mm, respectively. Overall, based on Table 3, the inhibition effect of combined essential oil proved

Table 3 Inhibitory effect of the combination of essential oil from two plant samples on different species of bacteria

Bacteria	Kaffir Lime + Siam Weed essential oil	*BR	Positive control	*BR
Staphylococcus aureus	18.0 ± 0.58	S	18.3 ± 0.67	S
Pseudomonas aeruginosa	15.7 ± 0.33	S	17.0 ± 0.58	S
Escherichia coli	19.0 ± 0.58	S	15.3 ± 0.33	S
Enterococcus faecalis	17.7 ± 0.33	S	16.0 ± 0.58	S

Data were expressed as means of triplicate zones inhibition ± Standard Error of Mean (SEM) = S. D/n, *S.D* Standard Deviation, *n* Number of test plates (3)
*(*BR* Bacteria Response, *S* Susceptible, *I* Intermediate)

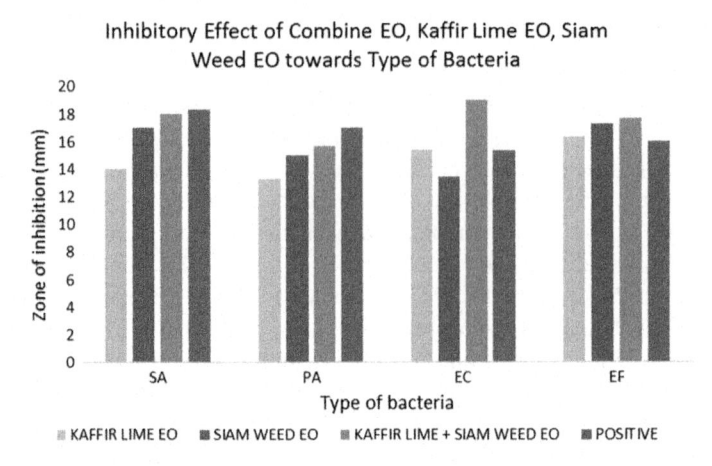

Fig. 2 Comparison of inhibitory effect between combined essential oil, Kaffir Lime essential oil, Siam Weed essential oil and positive control against four types of bacteria

more effective and showed stronger antibacterial activity compared to inhibition effect individually. Remarkably, the combined essential oil inhibits the growth of *E. coli* was better than positive control used in this study which was gentamicin where it only inhibited the bacteria growth at 15.3 ± 0.33 mm. Specifically, this result indicated that at certain point, combined essential oil showed a significant result as a better antibacterial agent against the growth of *E. coli* compared to positive control, gentamicin. The comparison of inhibitory effect against bacteria between each essential oils either individually or combined and also positive control was illustrated in Fig. 2. The result of individual essential oil was taken from Table 3.

4 Conclusion

Both *C. hystrix* and *C. odorata* essential oil separately have shown good antibacterial properties against several types of Gram-positive or Gram-negative bacteria. Based on the result of this study, both of the essential oil can be considered as potential antibacterial agents against different types of bacteria. The combination of both of the essential oil greatly helped to increase their potential as antibacterial agent against Gram-positive bacteria especially towards *E. faecalis*. Further studies and more experimental works are needed to be done in the extraction, purification and identification of active compounds as both of the essential oil may provide new compounds that have inhibitory effect against several types of bacteria that are pathogenic to human.

References

Atindehou, M., Lagnika, L., Guérold, B., Marc Strub, J., Zhao, M., Van Dorsselaer, A., ... Metz-Boutigue, M.-H. (2013). Isolation and identification of two antibacterial agents from *Chromolaena odorata* L. Active against four diarrheal strains. *Advances in Microbiology, 3*(1), 115–121. https://doi.org/10.4236/aim.2013.31018.

Carteret, M. (2011). Traditional Asian health beliefs & Healing practices—Dimensions of culture. Retrieved December 15, 2016, from http://www.dimensionsofculture.com/2010/10/traditional-asian-health-beliefs-healing-practices/.

Chanthaphon, S., Chanthachum, S., & Hongpattarakere, T. (2008). Antimicrobial activities of essential oils and crude extracts from tropical Citrus spp. against food-related microorganisms. *Journal of Science and Technology, 30*(April), 125–131.

Hanphakphoom, S., Thophon, S., Waranusantigul, P., Kangwanrangsan, N., & Krajangsang, S. (2016). Antimicrobial activity of Chromolaena odorata extracts against bacterial human skin infections. *Modern Applied Science, 10*(2), 159–171. https://doi.org/10.5539/mas.v10n2p159.

Hassan, M. A., Othman, S. N. A. M., Nahar, L., Basar, N., Jamil, S., & Sarker, S. D. (2016). Essential oils from the Malaysia Citrus (Rutaceae) medicinal plants. *Medicines, 3*(13), 1–11. https://doi.org/10.3390/medicines3020013.

Javed, S., Ahmad, R., Shahzad, K., Nawaz, S., Saeed, S., & Saleem, Y. (2013). Chemical constituents, antimicrobial and antioxidant activity of essential oil of Citrus limetta var. Mitha (sweet lime) peel in Pakistan. *Journal of Microbiology Research, 7*(24), 3071–3077. https://doi.org/10.5897/AJMR12.1254.

Kasuan, N., Muhammad, Z., Yusoff, Z., Hezri, M., Rahiman, F., Taib, M. N., & Haiyee, Z. A. (2013). Extraction of Citrus hystrix D.C. (Kaffir Lime) essentail oil using automated steam distillation process: Analysis of volatile compounds. *Malaysian Journal of Analytical Sciences, 17*(3), 359–369.

Kooltheat, N., Kamuthachad, L., Anthapanya, M., Samakchan, N., Sranujit, R. P., Potup, P., ... Usuwanthim, K. (2016). Kaffir Lime leaves extract inhibits biofilm formation by Streptococcus mutans. *Nutrition, 32*(4), 486–490. https://doi.org/10.1016/j.nut.2015.10.010.

Mbajiuka, C. S. (2015). Antimicrobila effects of the leaf extracts of Chromolaena Ordorata (Siam Weeds) on some human pathogens. *Journal of Pharmaceutical, 4*(8), 209–220. Retrieved from www.wjpr.net.

Ng, D. S. H., Rose, L. C., Suhaimi, H., Mohamad, H., Rozaini, M. Z. H., & Taib, M. (2011). Preliminary evaluation on the antibacterial activities of citrus hystrix oil emulsions stabilized by tween 80 and span 80. *International Journal of Pharmacy and Pharmaceutical Sciences, 3*(SUPPL. 2), 209–211.

Nurul Huda, A. K., Mamat, A. S., Effendy, A. W. M., Hussin, Z. M., & Hasan Sayed, M. Z. (2004). The antimicrobial effect of Chromolaena odorata extract on gram-positive bacteria.pdf. In 11th International Conference of the Association of Institutions for Tropical Veterinary Medicine and 16th Veterinary Association Malaysia Congress (pp. 342–343). Petaling Jaya.

Oladoye, S. O., Bello, I. A., & Abdul-hammed, M. (2014). Effect of harvest time on essential oil composition of Chromolaena odorota (L.) leaves from Nigeria. *Journal of Natural Sciences Research, 4*(10), 18–23. Retrieved from www.iiste.org.

Owolabi, M. S., Ogundajo, A., Yusuf, K. O., Lajide, L., Villanueva, H. E., Tuten, J. A., et al. (2010). Chemical composition and bioactivity of the essential oil of Chromolaena odorata from Nigeria. *Records of Natural Products, 4*(1), 72–78.

Ratseewo, J., Tangkhawanit, E., Meeso, N., Kaewseejan, N., & Siriamornpun, S. (2016). Changes in antioxidant properties and volatile compounds of Kaffir Lime leaf as affected by cooking processes. *International Food Research Journal, 23*(1), 188–196.

Shetty, P., & Rinaldi, A. (2010). Integrating modern and traditional medicine: Facts and figures. Retrieved December 15, 2016, from http://www.scidev.net/global/indigenous/feature/integrating-modern-and-traditional-medicine-facts-and-figures.html.

Srisukh, V., Bunyapraphatsara, N., Puttipipatkhachorn, S., Tungrugsasut, W., Pongpan, A., Bunsiriluk, S., et al. (2012). Fresh produce antibacterial rinse from Kaffir Lime oil. *Journal of Pharmaceutical Sciences, 39*(2), 15–27.

Tomar, A., Chandeshwar, I., & Pradesh, U. (2013). Pharmacological importance of citrus fruits. *International Journal of Pharmaceutical Sciences and Research, 4*(1), 156–160.

Vaisakh, M. N., & Pandey, A. (2012). The invasive weed with healing properties: A review on Chromolaena odorata. *International Journal of Pharmaceutical Sciences, 3*(1), 80–83.

Ventola, C. L. (2015). The antibiotic resistance crisis: Part 1: Causes and threats. *P & T: A Peer-Reviewed Journal for Formulary Management, 40*(4), 277–83. Retrieved from http://www.ncbi.nlm.nih.gov/pubmed/25859123.

Characterisation and Application of Diatomite in Water Treatment

Komathy Selva Raj, Megat Johari Megat Mohd Noor, Masafumi Goto and Pramila Tamunaidu

Abstract Conventional drinking water treatments are often inapplicable in developing countries, due to its high cost, lack of unfitting infrastructures, and availability and side effects of chemicals. Diatomite is a sedimentary rock containing diatoms with fine frustules surrounding it. It is used in various industrial applications due to its unique physical and chemical properties. This study investigates the properties and effectiveness of refined diatomite obtained from China in water treatment. The refined diatomite has enhanced adsorption and purity. Diatomite was observed under the Field Emission Scanning Electron Microscope coupled with Energy-Dispersive X-ray (FESEM–EDX) to study on its characterization. It was then tested as a coagulant to treat a different range of artificial turbid water (kaolin suspension) from low to high turbidity. The jar test was conducted to determine the optimum dosage of diatomite for turbidity removal.

Keywords Characterization · Diatomite · Water treatment

1 Introduction

Water treatment is essential to meet increasing demand for clean water worldwide and declining availability of water resources. Statistically, about 663 million of people are suffering from a shortage of safe drinking water worldwide (Water.org 2017). Although people gain access to water in many forms such as rain, river, and lakes, yet, the basic human need for clean water could not be fulfilled. This could be because of the rapid urbanization and industrialization that produce more contaminants thus elevating the cost of water treatment process. Despite various types of water treatments available, a need for a cost-effective, safe, specific and efficacious wastewater treatment has driven the interest of scientists into treating water

K. S. Raj · M. J. M. M. Noor · M. Goto · P. Tamunaidu (✉)
Malaysia-Japan International Institute of Technology,
Universiti Teknologi Malaysia(UTM), Kuala Lumpur, Malaysia
e-mail: pramila@utm.my

using natural coagulants and diatomite had the potential as a coagulant candidate in this study. Though the development of water treatment using coagulants has been in the discussion a few years ago, the evolution of natural coagulant usage has been stagnant or progressing slowly. In this study, diatomite which is a non-metallic material formed by deposition of fossilized fragmentary remains of diatoms in nature such as seas and lakes (Alyosef et al. 2014; Yuan et al. 2015; Wu et al. 2005) is tested for its efficiency as a coagulant in synthetic wastewater treatment.

The fragmentary remains of diatomite influence its shapes and sizes. It mainly composed of amorphous silica and small constituent of other materials such as carbonates and organic matter (Gomez et al. 2014; Du et al. 2014; Gao et al. 2005). The major factor that contributes to its high absorption capacity in water treatment is the unique integration of both physical and chemical properties such as high porosity about 80–90% voids, chemical inertness, low thermal conductivity and combination of small pore sizes, and large specific surface area (Mendonça et al. 2011; Zhao et al. 2014; Bakr 2010). In addition to that variety of potential diatomite products and applications signify its importance in water purification such as filtering mediums and adsorbents, decontamination of sewage water, and decolorization of textile dyes in textile industries in many countries (Hao 2000).

2 Methods

The material used in this study was diatomite sourced from Teng Chong, a province in China and transferred to the laboratory. The samples were air dried and powdered before storing it in a desiccator to prevent moisture prior to use. A commercially available kaolin was purchased from R&M Chemicals to generate a synthetic wastewater to evaluate its performance.

Three analyses such as Field Emission Scanning Electron Microscopy (FESEM) coupled with Energy-Dispersive X-ray (EDX) spectroscopy (FEI Nova NanoSEM 230, USA) and Brunner–Emmett–Teller (BET) (BELSORP-mini II, Japan) were done. A conventional jar test (PB 900, Phipps and Bird, USA) was used to evaluate optimum dosage required for the addition of coagulant in the sample water as shown in Fig. 1. A stock turbidity solution was prepared by adding 10 g of kaolin clay mixed rapidly with 1000 mL of deionised water for 30 min. The solution was then allowed to settle for 24 h, before extracting the supernatant which is the stock turbidity solution using a syringe. About 1000 mL of water sample was measured and added into each of the jars for mixing. The speed of rapid mixing was set at 125 rpm for the duration of four minutes while the speed of slow mixing at 30 rpm for thirty minutes. To determine the turbidity of the initial water sample, the sample was taken at the depth of 3 cm from water surface using a pipette and the sample collected was analyzed using turbidimeter (2100Q, Hach, USA) to determine residual turbidity (Mohammad et al. 2013). Different levels of turbidity were generated which were 50 NTU (low turbidity), 100 NTU (medium turbidity), and 250 NTU (high turbidity).

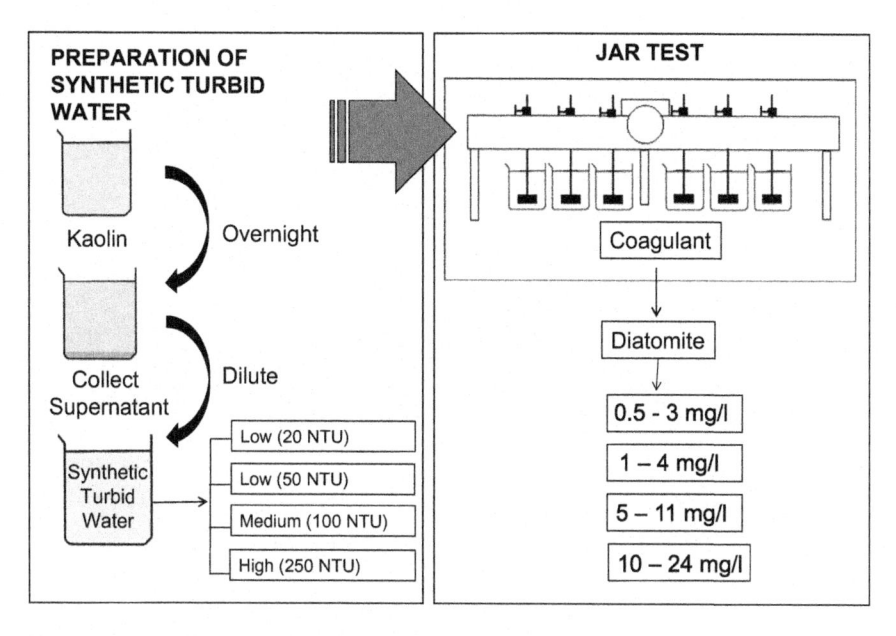

Fig. 1 Methodology of turbid water preparation and jar test

3 Results and Discussion

Figure 2 shows morphological features of diatomite analyzed using FESEM. The results revealed that the diatomite samples were mainly comprised of fossilized algae species with well-distributed macropore and micropore. Diatomite frustules are made up of silica found in the water forming an outer layer and categorized under two different types, namely centric (discoid) and pennate (elongated to filiform) (Khraisheh et al 2004). The images in Fig. 2 conform various types and shapes of algae species present in the sample used in this study.

Interestingly, each shape and type contributes to individual species of diatoms. The species variation depends on geographical locations as such that species found in Egypt has species like *Melosora granulate, Stephanodiscus aegyptiacus*, and *Cyclostephanos dubius* (Fatah et al. 2015). This is because of the quantity and types of minerals found in earth composition that affect the shapes and structures of diatomite.

The compositional elements were determined using EDX. The EDX spectra are shown in Fig. 3, with two major peaks present which are alumina and silica. According to the elemental periodic table, the Kα energy value for alumina and silica is 1.486, 1.740, respectively (EDAX 2016). Therefore, the combination of silica and alumina may provide support and stability to diatomite. This is because silica contributes to its unique structure and the network of pores in which creates a multi-functional material. Porosity is important for coagulation process to capture

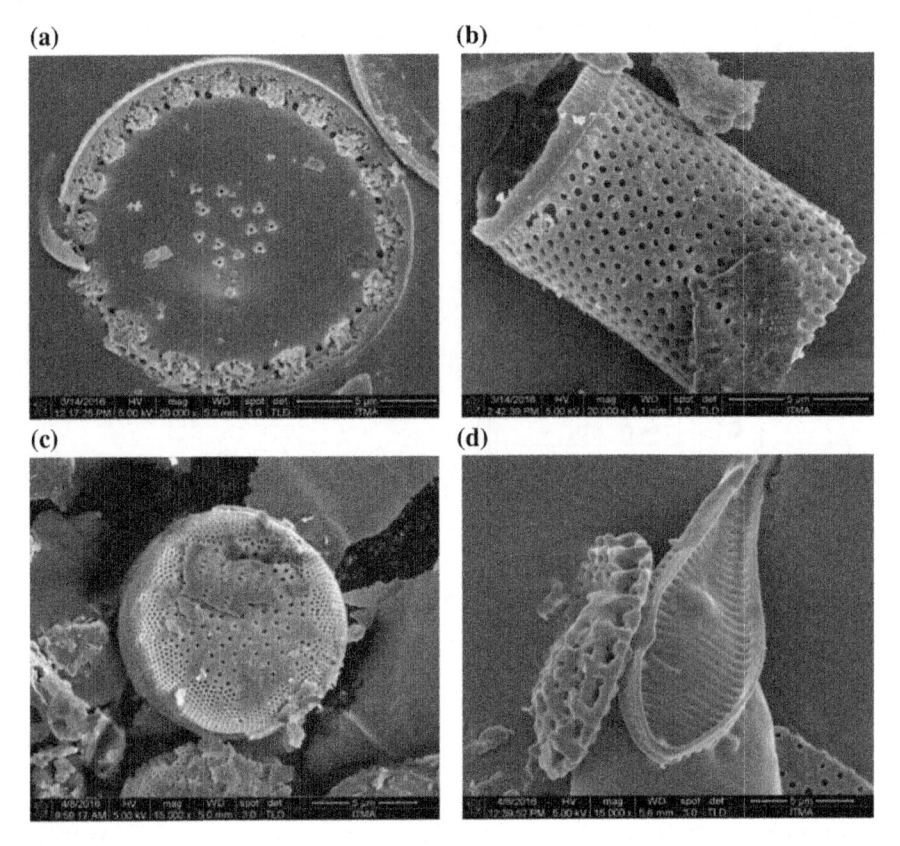

Fig. 2 FESEM micrograph of diatomite at magnification of **a**, **b** 20,000 × magnification, **c**, **d** 15,000 × magnification

Fig. 3 EDX spectra of refined diatomite

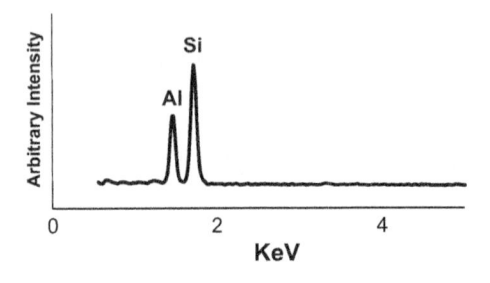

particles in wastewater. Similarly, alumina has properties of high strength and hardness to control the stability.

Physico-chemical characterization of diatomite is shown in Table 1. Based on the results, diatomite is an acidic solid material. The bulk density measured is about 0.645 g/ml shows its high porosity. The bulk density is an essential parameter in regard to its application as a coagulant. The total pore volume and average pore diameter of 0.051 cm^{3}g^{-1} and 56.16 nm, respectively, increase its capacity as a coagulant. The carbonate content of 0.15 indicates the presence of low amount impurities. Its high water holding capacity and compressibility indicated that diatomite can be also used as supporting material. These characteristics make it a coagulant with the second highest bulk-specific surface area (60 mil m^2/m^3), only next to activated carbon (with 100 mil m^2/m^3) and at least 50,000 times more than commercial biofilm carriers. The comparison of diatomite to other commercial carriers can be seen in Table 2.

The diatomite was then tested for its effectiveness as a coagulant to treat a different range of artificial turbid water (kaolin suspension) at 20, 50,100, and 250 NTU turbidity. The jar test was conducted to determine the optimum dosage of diatomite for turbidity removal. At low turbidities of 20 NTU, a low amount of diatomite dosage of 2.5 mg/L could almost reach the minimum allowable standard of 5 NTU. Similarly, turbidity reduction of 83.46% with 3.5 mg/l at 50 NTU (low), and at 100 NTU (medium), the turbidity reduction of 90.70% was achieved with 10.0 mg/L dosage of diatomite. At a high turbidity of 250 NTU, the turbidity reduction observed was 96.42%. Based on the results obtained, it shows that diatomite works highly efficient for turbidity more than 100 NTU (reduction efficiency above 90%) as compared to low turbid water of 50 NTU Fig. 4.

Table 1 Physical and chemical properties of diatomite

Properties	Refined diatomite
Texture	Powder
pH (10% slurry)	4.09
Moisture (%)	1.08
Bulk density (g/ml)	0.645
Total pore volume (cm^3 g^{-1})	0.0514
Average pore diameter (nm)	56.115
Carbonate content	0.15
Water content	12.37
Water holding capacity (HC)	14.11
Compressibility index	12.903

Table 2 Comparison of diatomite with typical biofilm carriers (Metcalf & Eddy/Acecom 2013)

Type of biofilm carrier	Specific gravity	Example of nominal dimensions, mm	Example bulk-specific surface area, m^2/m^3
Diatomite	2.2–2.3	0.01–0.20	60,000,000
Sponge	0.95	15 × 15 × 12 depth	850
Plastic wheel (KI)	0.96–0.98	7 × 10 dia	500
Plastic wheel (K3)	0.96–0.98	4 × 25 dia	800
Plastic wheel (K5)	0.96–0.98	9 × 25 dia	500
Biochip (P)	0.96–1.02	3 × 45 dia	900
Biochip (M)	0.96–1.02	2 × 48 dia	1200
Plastic square	0.96	15 × 15 × 10 depth	680
Roper	NA	45 dia. rope	2.85 m^2/m

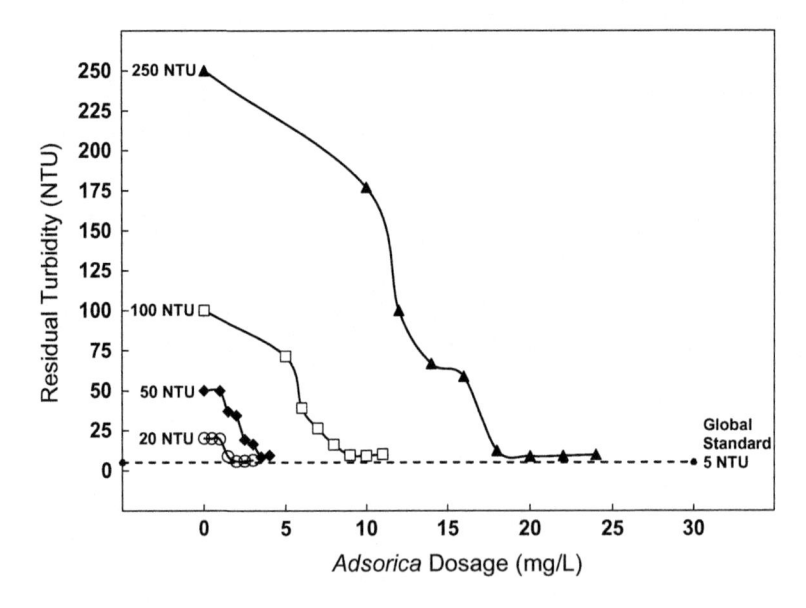

Fig. 4 Turbidity removal efficiency at initial turbidity of 250 NTU

4 Conclusion

These physical and chemical characteristics such as porosity, high strength structure, and high water holding capacity clearly show that diatomite has high potential value and versatility as a promising candidate for coagulant applications in water treatment. The difference in structure and network of pores also contributes to its unique applications. As a coagulant, diatomite works efficiently at high turbidity more than 100 NTU (reduction efficiency above 90%) as compared to low turbid water.

Acknowledgements This work is supported by Ministry of Higher Education (MOHE) and Universiti Teknologi Malaysia (UTM) for providing Flagship grant (Vote No. 03G24), Malaysia–Japan International Institute of Technology (MJIIT) and Felda Water Sdn Bhd for the provision of research support and financial assistance to the researchers. Support from all sources is greatly appreciated.

References

Alyosef, et al. (2014). Effect of acid treatment on the chemical composition and the structure of Egyptian diatomite. *International Journal of Mineral Processing, 132,* 17–25.
Bakr, H. E. G. M. M. (2010). Diatomite: Its characterization, modifications and applications. *Asian Journal of Materials Science, 3,* 121–136.
Du, et al. (2014). MnO_2 nanowires in situ grown on diatomite: Highly efficient absorbents for the removal of Cr (VI) and As (V). *Microporous and Mesoporous Materials, 200,* 27–34.
EDAX INC, (2016). EDAX's Interactive Periodic Table of Elements. Edax.com. Retrieved July 23, 2016, from http://www.edax.com/periodic-table/index.aspx.
Fatah, H. M., & El, Abd. (2015). Pleistocene Stephanodiscaceae diatoms from deposits in El Fayoum depression., Egypt. *Vegetos—An International Journal of Plant Research, 28*(2), 15.
Gao, B., Jiang, P., An, F., Zhao, S., & Ge, Z. (2005). Studies on the surface modification of diatomite with polyethyleneimine and trapping effect of the modified diatomite for phenol. *Applied Surface Science, 250,* 273–279.
Gomez, J., Gil, M. L. A., de la Rosa-Fox, N., & Alguacil, M. (2014). Formation of siliceous sediments in brandy after diatomite filtration. *Food Chemistry, 170,* 84–89.
Hao, O. J., Kim, H., & Chiang, P. (2000). Decolorization of wastewater. *Critical Reviews in Environmental Science and Technology, 30*(4), 449–505.
Khraisheh, M. A. M., Al-degs, Y. S., & Mcminn, W. A. M. (2004). Remediation of wastewater containing heavy metals using raw and modified diatomite. *Chemical Engineering Journal, 99,* 177–184.
Mendonça, E., Picado, A., Cunha, M. A, & Catarino, J. (2011). Environmental management in practice, InTech. http://www.intechopen.com/books/environmental-management-in-practice/environmental-management-ofwastewater-treatment-plants-the-added-value-of-the-ecotoxicologicalappr Accessed July 5, 2017.
Mohammad, T. A., Mohamed, E. H., Megat Mohd Noor, M. J., & Ghazali, A. H. (2013). Coagulation activity of spray dried salt extracted Moringa oleifera. *Desalination and Water Treatment, 51* (7-9), 1941–1946.
Water.org. (2017). The water crisis. https://water.org/our-impact/water-crisis/. Accessed July 23, 2017.
Wu, J., Yang, Y.S., & Lin, J. (2005). Advanced tertiary treatment of municipal wastewater using raw and modified diatomite. *Journal of Hazardous Materials,* 196–203.
Metcalf & Eddy/Acecom. (2013) Wastewater engineering: Treatment and resource recovery (5th ed., Vol 2, p. 999). McGraw - Hill.
Yuan, et al. (2015). Novel hierarchically porous nanocomposites of diatomite-based ceramic monoliths coated with silicalite-1 nanoparticles for benzene adsorption. *Microporous and Mesoporous Materials, 206,* 184–193.
Zhao, S., Huang, G., Fu, H., & Wang, Y. (2014) Enhanced coagulation/flocculation by combining diatomite with synthetic polymers for oily wastewater treatment. *Separation Science and Technology, 49,* 999–1007.

Reactive Red 4 Dye-Sensitized Immobilized TiO$_2$ for Degradation of Methylene Blue Dye

W. I. Nawawi and F. Bakar

Abstract The purpose of this study is to enhance photocatalytic degradation of cationic methylene blue (MB) dye. A 0.3 g of TiO$_2$ and polymer binder was coated onto a clean glass plate using brush technique to develop an optimum immobilized TiO$_2$ system. A comparison study between immobilized TiO$_2$/PEG (Im/TiO$_2$/PEG) system with and without RR4 sensitizer was carried out under 55-W fluorescent lamp. The photocatalytic degradation of MB was significantly enhanced for RR4 dye-sensitized Im/TiO$_2$/PEG with first-order rate constant was *ca.* 0.0962 min^{-1} under 55-W fluorescent lamp. Same observation was observed under visible light irradiation, whereby those RR4-sensitized immobilized photocatalysts system were successfully improved the photocatalytic activity as compared with immobilized photocatalysts without RR4 as sensitizer. The photocatalytic enhancement under Im/TiO$_2$/PEG/RR4 are due to the ability of RR4 dye to become electron (e$^-$) donor for conduction band (CB) of TiO$_2$, thus making TiO$_2$ CB riches with electron, eventually this e$^-$ is used to remove MB dye by producing hydroxyl radical. This could be done due to the lower band gap of RR4 compared to the TiO$_2$ itself. Therefore, low energy was needed by electron from valence band of RR4 dye to excite its conduction band. In addition, the LUMO energy for RR4 is higher than TiO$_2$, and this allows electron transferring from the sensitizer to the semiconductor. The value of band gap was measured by using UV-Vis DRS, and the stability of RR4 in the formulation was detected using FTIR spectroscopy by observing the related bond present in the sensitized immobilized.

Keywords Titanium dioxide · Decolorization · Dye sensitizer PEG Methylene blue

---

W. I. Nawawi (✉)
Faculty of Applied Sciences, Universiti Teknologi MARA, 02600 Arau, Perlis, Malaysia
e-mail: wi_nawawi@perlis.uitm.edu.my

F. Bakar
Faculty of Applied Sciences, Universiti Teknologi MARA, 40450 Shah Alam, Selangor, Malaysia
e-mail: faezah_bakar@ymail.com

1 Introduction

Globalization influences the level of pollution in today's worlds. The bad impact is it creates more pollution, especially into the water system. This is as results from the increasing demand of textile industries from all over the world which lead to the released of various types of dyes into the water streams and causes deleterious effect on the well-being of mankind. Textile dyes are potentially toxic because of their low removal rate and if untreated would cause long-term health concerns (Chatterjee et al. 2008). It is estimated that approximately 15–20% of the synthetic textile dyes used are lost in wastewater streams during manufacturing or processing operations. Therefore, textile dyes and other commercial colourants had emerged as a focus of environmental remediation efforts. Among dyestuffs, reactive azo dyes constitute a significant portion of dye pollutants and probably have the least desirable consequences to the surrounding ecosystem (Chatterjee et al. 2008). For that reason, water pollution issue has easily become the centre point of research efforts in today's scientific world. It can be seen that pollutant which is emitted from various sources poses severe ecological problem as biodegradable of these pollutants is very slow and conventional method used are mostly not environmentally compatible and costly. In addition, some pollutants found in water system are recalcitrant to some applied method (Bergendahl and O'Shaughnessy 2004).

Therefore, this issue has grabbed attention of many researchers to develop a method in order to reduce or overcome this pollution. These factors contribute to the development of many new methods in physical, chemical and biological field. Titanium dioxide TiO_2 appeared to be the most attractive means for water purification due to the facts that they are inexpensive, non-toxic, having high surface area and having tunable properties which can be modified by size reducing, doping and sensitizer (Chatterjee and Dasgupta 2005). However, the band gap of TiO_2 is so wide which is 3.2 eV, which limits its photocatalytic activity under visible light region (Wang et al. 2008). This only allows TiO_2 to absorb the ultraviolet light (<387 nm) which only occupies a small fraction (3–5%) of the solar photons (Li et al. 2008). As a result, this will limit the photocatalytic activity of TiO_2 over its wide use.

Therefore, modification of TiO_2 photocatalyst for degradation of pollutant is demanding area of research among scientist, especially in visible light region (Chatterjee et al. 2008). Among these modification methods, dye sensitization has been able to improve the photocatalytic activity of TiO_2 under visible light (Li et al. 2012). Many efforts have been done by focusing on a novel method to modify TiO_2, where photosensitization is an important way to excite TiO_2 to the wavelength of the visible light (Pei and Luan 2012).

Dye-sensitized photocatalytic process appeared to have several advantages over direct photocatalysis. This is because, it extends the range of excitation energies of the semiconductor TiO_2 into visible light region, thus making a more complete use of solar energy and could promote removal of coloured pollutants more effectively (Chatterjee et al. 2008). There are many types of effective sensitizers that have been studied before such as inorganic sensitizers, organic dye and coordination metal complexes. Under appropriate condition, these sensitizers can be adsorbed onto semiconductor by several means which are electrostatic, hydrophobic, or chemical interaction, upon excitation, injects electron into conduction band of semiconductor (Pei and Luan 2012). In recent years, plenty of organic dyes has got attention and has been tested as photosensitizer such as eosin Y, riboflavin, rose bengal, cyanine, cresyl violet and merocyanine. In order to be functioned effectively, the band gap of the sensitizer must be near the appropriate value for the optimum utilization of solar energy (Pei and Luan 2012).

2 Experimental

2.1 Materials

TiO_2 Degussa P25 powder was used as the starting material in the preparation of immobilized TiO_2. Methylene blue dye (chemical formula: $C_{16}H_{18}ClN_3S \cdot xH_2O$, molecular weight: 319.86 g mol^{-1}, λ max: 661 nm). Reactive red (RR4) dye or commonly known as Cibacron Brilliant Red (Colour Index Number: 18105, chemical formula: $C_{32}H_{23}ClN_8Na_4O_{14}S_4$, molecular weight: 995.23 g mol^{-1}, λ max: 517 nm). Ultra-pure water was used to prepare all solutions in this work.

2.2 Preparation of TiO_2 Immobilized System

TiO_2 solution was prepared by adding 6.5 g of titanium dioxide (TiO_2) Degussa P25 powder (20% rutile, 80% anatase) powders into 50 mL distilled water. The solution was then added with 1 mL of PEG (mw = 6000) polymer in a reagent bottle, followed by sonicated under ultrasonic vibrator for 30 min or until homogenized. The white solution of TiO_2 was then coated using brush-coating method onto a clean glass plate with dimension of 50 mm × 10 mm × 80 mm (Length × Wide × Height), which has been taped before with double-sided adhesive tape (DSAT). The formulation was then dried by using 850 W hot blower with temperature about 120 °C until dry, forming Im/TiO_2/PEG, and coating process was repeated until 0.3 g loading of Im/TiO_2/PEG was achieved. Im/TiO_2/PEG samples were undergoing washing process by exposing immobilized sample with 55-W of fluorescent lamp under aerated condition for 30 min.

2.3 Preparation of RR4-Sensitized TiO_2 Immobilized System

The formulation was prepared by adding 6.5 g of titanium dioxide (TiO_2) Degussa P25 powder (20% rutile, 80% anatase) powders and 1 mL of PEG into 50 mL

RR4, forming slightly pink solution. The solution was then coated onto a clean glass plate which has been taped with double-sided adhesive tape (DSAT). The coating was done by using brush-coating technique until 0.3 g loading was achieved, forming Im/TiO$_2$/PEG/RR4 and then was dried by using 850 W hot blower with temperature about 120 °C until dry. The samples were undergoing washing process by exposing them to 55-W of fluorescent lamp under aerated condition for 30 min.

2.4 Photodecolorization of Methylene Blue Dye

Those immobilized sample was tested for its photocatalytic activity by applying the plate in a glass cell containing methylene blue (MB) dye with a chemical formula: $C_{12}H_{15}O_6$. The process was carried out by immersing immobilized plate into a glass cell filled with 20 mL of 12 mg L^{-1} of MB dye and irradiated with 55-W fluorescent lamp, model Ecotone with visible light intensity measured for about 461 and 6.7 Wm^{-2} of UV light detected as UV leakage and named as normal light. An aquarium pump model NS 7200 is used as an aeration source to ensure enough oxygen supply into the system. The dimension of glass cell used in this study is 50 mm × 10 mm × 80 mm (Length × Wide × Height). The decolourization degree of MB dye was determined by 3 mL aliquot of treated MB dye was taken from glass cell at 15 min time interval until complete decolourization is achieved in one hour. The colour reduction value of MB was determined by measuring the absorbance using HACH DR 1900 spectrophotometer with 661 nm wavelength. The experimental procedures were repeated by applying the same steps for different parameters which are catalysts loading, sources of light, pH and aeration flow rate. The set-up of the experiment was shown in Fig. 1.

2.5 Recyclability Study

The recyclability study was carried out to observe the effect of TiO$_2$ immobilized RR4 sensitizer towards the photodecolorization activity of tested dye. The experiment was conducted initially through photodecolorization method of methylene blue (MB) dye. Immobilized TiO$_2$/PEG/RR4 was then undergone washing process using distilled water and irradiated for 30 min. Both photodecolorization and washing processes for immobilized TiO$_2$/PEG/RR4 sample were then repeated until 8th cycles. Photodecolorization percentage of MB dye in every cycle was recorded at every 15 min interval until MB has become colourless.

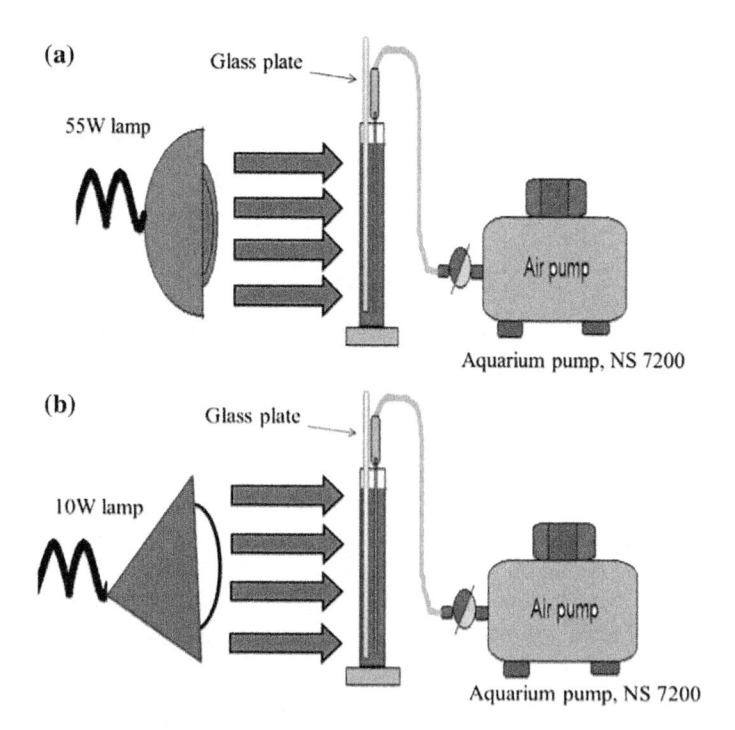

Fig. 1 Set-up of experiment for decolorization of MB dye under **a** normal light **b** visible light

2.6 Characterization Test for Im/TiO$_2$/PEG/RR4

X-ray diffraction (XRD) spectra were obtained using a Rigakuminiflex II, X-ray diffractometer. Structural information of the films was obtained in the range of 2θ angles from 3 to 80° with a step size increment of 1.00 s/step. FTIR spectra of powder samples were recorded on Perkin Elmer Spectrum Version equipped with an attenuated total reflectance device (Frontier) with a diamond crystal. Spectra were collected in frequency range of 600–4000 cm^{-1} with 4 scans and spectral resolution 4 cm^{-1}. The surface area of the immobilized TiO$_2$ film powders were measured by nitrogen adsorption using the BET equation at 77 K (Micrometrics ASAO 2020 M+C, America). A UV–vis spectrophotometer UV-25, Perkin Elmer was used to obtain the UV–vis reflectance spectrum of the powder sample.

3 Result and Discussion

This study was conducted to improve the performance of immobilized TiO$_2$ in degrading organic pollutant, where methylene blue (MB) dye was used as model of pollutant. The modification technique was done by applying reactive red 4 (RR4)

Table 1 Rate of decolorization in presence of light sources

Light sources	RR4 (mgL^{-1})	Pseudo-first-order (min^{-1})
Fluorescent lamp	200	0.0465
Fluorescent lamp	300	0.0962
Fluorescent lamp	400	0.0882

Table 2 Rate of decolorization in the absence of light sources

Light sources	RR4 (mgL^{-1})	Pseudo-first-order (min^{-1})
Absence	200	0.0067
Absence	300	0.0229
Absence	400	0.0135

dye which acts as sensitizer to the immobilized TiO$_2$. Table 1 shows the experimental result and pseudo-first-order rate constant of immobilized TiO$_2$ with and without the presence of RR4 sensitizer and control sample in degrading MB dye. As shown in Table 1, the addition of RR4 dye into the formulation has made a significant difference in degradation of MB dye. The results show an increase in pseudo-first-order rate constant reading, thus improving the performance of immobilized TiO$_2$. The addition of RR4 from to 300 mgL^{-1} has increased the decolorization rate of MB dye. However, the reading drops when the concentration of RR4 become 400 mgL^{-1}. The increasing rate of decolorization might be due to the simultaneous occurrence of both photocatalysis and adsorption process (Table 2).

In this study, photocatalysis process is more dominant compared to adsorption. This is because, without irradiation from any light source, the rate of decolorization significantly decreases. In addition, the adsorption process could also lead to the increase in MB dye concentration on the TiO$_2$ interface where charge transfer process occurs due to the oxidation of chemisorbed dyes. In fact, when 400 mgL^{-1} of RR4 is used as sensitizer, the reading drops because too much RR4 will cover the surface of TiO$_2$. It is important to avoid excessive use of sensitizer to determine the optimum efficiency of both photocatalysis and adsorption processes. For that reason, it is obviously shown that when the concentration of RR4 dyes increases to the maximum, the efficiency of photodecolorization of MB dye decreases where the optimum concentration of RR4 dye in this study is 300 mgL^{-1}. Therefore, from the results showed in Table 3, it is obviously shown that the presence of RR4 in the formulation makes a significant difference in the efficiency of immobilized TiO$_2$ towards decolorization of MB dye.

Table 3 Optimum rate of decolorization by using 300 mgL^{-1} of RR4 dye

Light sources	Presence of RR4 dye (min^{-1})	Absence of RR4 dye (min^{-1})
Normal light	0.0962	0.0571
Visible light	0.0846	0.0418

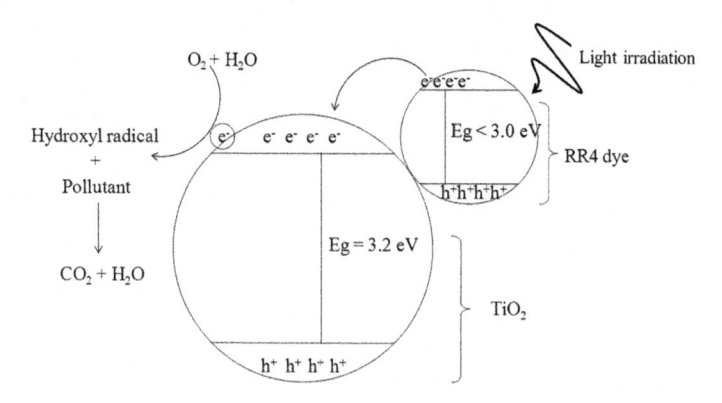

Fig. 2 Mechanism of dye-sensitized immobilized TiO$_2$

The proposed mechanism between immobilized TiO$_2$ and RR4 dye is shown in Fig. 2. Dye sensitizer is able to increase the range of wavelength response absorb by the TiO$_2$ surface which is a crucial aspect for photocatalyts to operate under visible light (Gupta and Tripathi 2011). Nishikiori et al. stated that increasing in crystallinity of TiO$_2$ nanoparticles will increase the surface quality of the nanoparticles, which then improve the anchoring geometry of the dye on particles surface and lead to electron injection. When dye is adsorbed into the surface of TiO$_2$ under light irradiation, the surface-adsorbed dye will be excited. During the excited state, the ejected electron from the dye sensitizer particles is transferred to the conduction band of TiO$_2$ semiconductor. This electron is then taken up by oxygen molecule present in the system and form superoxide radical anion and hydroxyl radical, which attacks the dye contaminant repeatedly to converts it into non-toxic harmless end products (Gupta and Tripathi 2011).

3.1 XRD Analysis

The crystal phases of TiO$_2$ were analysed by means of X-ray diffraction as shown in Fig. 3. As being shown, the peaks at 25.3° and 27.4° for pristine TiO$_2$ are the characteristic reflection of anatase and rutile, respectively (Zhou et al. 2012). All peaks for the TiO$_2$/PEG/RR4 samples also show the appearance of anatase and rutile indicating that no phases transformation occur since all the processes for sensitization were prepared under low temperature. Therefore, it did not destroy the characteristic structure of TiO$_2$. According to Akpan et al., calcination can influence the photocatalytic activity of TiO$_2$ depending on the method of preparation of the semiconductor. The efficiency of TiO$_2$ increased rapidly and was optimum at 500 °C as the calcination temperature increased 300–500 °C. However, further increase

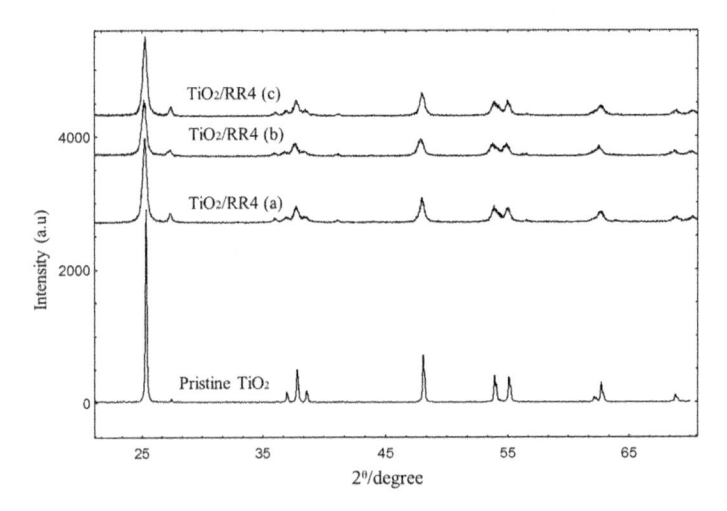

Fig. 3 XRD spectrum for photocatalysts

in temperature from 500 to 700 °C lowered the performance of the TiO$_2$. This is because temperature beyond 500 °C would promote the phase transformation from anatase to rutile, which has little photocatalytic activity. By referring to Fig. 3, pristine TiO$_2$ shows the highest crystallinity compared to TiO$_2$/PEG/RR4 samples, due to the presence of PEG binder and RR4 dye in the formulation which has increased the porosity of TiO$_2$.

3.2 FTIR Analysis

Fourier transform infrared spectroscopy is a powerful tool used to determine the difference in vibrational energy of each molecule present in the sample. The FTIR characterization shows there were peaks which appeared at 3000–3500 and 1600 cm^{-1} wave numbers. The broad peak at 3000–3500 cm^{-1} indicates that the presence of O–H stretching of the hydroxyl functional group and 1600 cm^{-1} wavenumber for the presence of chemisorbed water (H–O–H) in the sample (Tristantini and Mustikasari 2011). After PEG and RR4 being immobilized and sensitized on the TiO$_2$, the basic characteristic peaks of TiO$_2$ also changed. As shown in Fig. 3, the hydroxyl groups are stronger and broader in pristine TiO$_2$ compared than in immobilized sensitized TiO$_2$, indicating that the hydroxyl group decrease with the addition of PEG and RR4. According to Cho and Choi 2001 when TiO$_2$ was buried in the polymer matrix (PEG), the surface OH$^-$ groups will gradually deplete with irradiation and become broader compared to pristine TiO$_2$ (Cho and Choi 2001).

4 Conclusion

The photocatalytic degradation of MB was significantly enhanced for RR4 dye-sensitized Im/TiO$_2$/PEG with first-order rate constant was *ca.* 0.0962 min^{-1} under 55-W fluorescent lamp. Same observation was observed under visible light irradiation, whereby those RR4-sensitized immobilized photocatalysts system were successfully improved the photocatalytic activity as compared with immobilized photocatalysts without RR4 as sensitizer. The photocatalytic enhancement under Im/TiO$_2$/PEG/RR4 is due to the ability of RR4 dye to become electron (e$^-$) donor for conduction band (CB) of TiO$_2$, thus making TiO$_2$ CB riches with electron; eventually, this e$^-$ is used to remove MB dye by producing hydroxyl radical. This could be done due to the lower band gap of RR4 compared to the TiO$_2$ itself. Therefore, low energy was needed by electron from valence band of RR4 dye to excite its conduction band. In addition, the LUMO energy for RR4 is higher than TiO$_2$, and this allows electron transferring from the sensitizer to the semiconductor. The value of band gap was measured by using UV-Vis DRS and the stability of RR4 in the formulation was detected using FTIR spectroscopy by observing the related bond present in the sensitized immobilized.

Acknowledgements We would like to thanks the Malaysian Ministry of Education (KPM) for providing generous financial support under RAGS grants: 600-RMI/RAGS 5/3 (35/2014) in conducting this study and Universiti Teknologi MARA (UiTM) for providing all the needed facilities.

References

Bergendahl, J., & O'Shaughnessy, J. (2004). Applications of advanced oxidation for wastewater treatment. *A Focus on Water Management, 2,* 3–7. Retrieved from http://www.wpi.edu/Images/CMS/NEABC/wastewatersummary.pdf.

Chatterjee, D., & Dasgupta, S. (2005). *Visible light induced photocatalytic degradation of organic pollutants, 6,* 186–205. https://doi.org/10.1016/j.jphotochemrev.2005.09.001.

Chatterjee, D., Rupini, V., Sikdar, A., Joshi, P., Misra, R., & Rao, N. N. (2008). Kinetics of the decoloration of reactive dyes over visible light-irradiated TiO$_2$ semiconductor photocatalyst, *156,* 435–441. http://doi.org/10.1016/j.jhazmat.2007.12.038.

Cho, S., & Choi, W. (2001). Solid-phase photocatalytic degradation of PVC–TiO$_2$ polymer composites. *Journal of Photochemistry and photobiology A: Chemistry, 143*(2–3), 221–228. https://doi.org/10.1016/S1010-6030(01)00499-3.

Gupta, S. M., & Tripathi, M. (2011). A review of TiO$_2$ nanoparticles. *Chinese Science Bulletin, 56* (16), 1639–1657. https://doi.org/10.1007/s11434-011-4476-1.

Li, X., Shi, L., Wang, D., Luo, Q., & An, J. (2012). Visible light photocatalytic activity of TiO$_2$/heat-treated PVC film, (February), 1187–1193. http://doi.org/10.1002/jctb.3747.

Li, X., Wang, D., Cheng, G., & Luo, Q. (2008). Preparation of polyaniline-modified TiO$_2$ nanoparticles and their photocatalytic activity under visible light illumination, *81,* 267–273. http://doi.org/10.1016/j.apcatb.2007.12.022.

Pei, D., & Luan, J. (2012). Development of visible light-responsive sensitized photocatalysts. http://doi.org/10.1155/2012/262831.

Tristantini, D., & Mustikasari, R. (2011). Modification of TiO_2 Nanoparticle with PEG and SiO_2 For Anti-fogging and Self-cleaning Application. *International Journal of Engineering*, (April).

Wang, D., Wang, Y., Li, X., Luo, Q., An, J., & Yue, J. (2008). Sunlight photocatalytic activity of polypyrrole – TiO_2 nanocomposites prepared by " in situ" method, 9, 1162–1166. http://doi.org/10.1016/j.catcom.2007.10.027.

Zhou, X., Ji, H., & Huang, X. (2012). Photocatalytic degradation of methyl Orange over Metalloporphyrins Supported on TiO_2 Degussa P25, 1149–1158. http://doi.org/10.3390/molecules17021149.

The Influence of Water Quality Index (WQI) Assessment Towards Water Sports Activity at Taman Tasik Titiwangsa, Kuala Lumpur, Malaysia

Nor Hanisah Mohd Hashim, Balqis Dayana Badarodin
and Wan Hazwatiamani Wan Ismail

Abstract Taman Tasik Titiwangsa lakes are popular for water sport activity. Thus, it is important to keep the health of the lakes at an acceptable level for the optimum usage of water sport activity. The water quality index is used to assess the water quality condition of the lakes. Water quality index (WQI) and interim national water quality standards (INWQS) for Malaysia are used to monitor the health of the lakes. A total of two sample stations were collected at Lake 1 and Lake 2 of Taman Tasik Titiwangsa. Six selected parameters (biological oxygen demand, chemical oxygen demand, dissolved oxygen, pH, suspended solid and ammoniacal nitrogen) were used to calculate the water quality index. From the analysis, it showed that both Lake 1 and Lake 2 are Class II condition and visitors can use it for water sport activity. Several recommendations are noted to improve the WQI value for the use of Taman Tasik Titiwangsa visitors especially for water sport activity.

Keywords Water quality index (WQI) · Taman tasik titiwangsa
Water sport activity · Mining area

N. H. M. Hashim (✉)
Faculty of Architecture, Planning and Surveying, Universiti Teknologi MARA, Centre of Studies for Parks and Amenity Management, 40450 Shah Alam Selangor, Malaysia
e-mail: norhanisah@salam.uitm.edu.my

B. D. Badarodin
School of Social, Development and Environmental Studies, Faculty of Social Science and Humanities, Universiti Kebangsaan Malaysia UKM, Bangi, Selangor, Malaysia
e-mail: balqisdayana@gmail.com

W. H. W. Ismail
Faculty of Architecture, Planning and Surveying, Universiti Teknologi MARA, Centre of Postgraduate Studies, 40450 Shah Alam, Selangor, Malaysia
e-mail: amanie_wan91@yahoo.com

© Springer Nature Singapore Pte Ltd. 2018
R. Saian and M. A. Abbas (eds.), *Proceedings of the Second International Conference on the Future of ASEAN (ICoFA) 2017 – Volume 2*,
https://doi.org/10.1007/978-981-10-8471-3_55

1 Introduction

Historically, public open spaces within the crowded urban areas have been considered as an important asset for citizen (Rung 2005). Ayeghi (2014) stated that parks and urban greens have been valued as physical settings because they fulfil many leisure, recreation and social needs of urban residents involving the park user's satisfaction. Taman Tasik Titiwangsa was an example of Public Park that received a lot of visitors. The beautiful green landscape of Taman Tasik Titiwangsa with its tranquillity and the calmness of the lake attracted many visitors to visit Taman Tasik Titiwangsa. Apart from the beautiful landscape of Taman Tasik Titiwangsa, water sport activity such as boat paddling, kayaking and water ball is among the popular sport activities offered at the park (Sharifudin 2014). Even though most of the visitors purposely visited Taman Tasik Titiwangsa for the water sport activity, the quality of the water lake must be taken into consideration at all times.

The importance of water for recreation is becoming apparent, especially in countries, where open water surfaces are limited and where standards of living are high.

Besides, according to Ling, water quality of river and lakes is one of the most common issues in Malaysia. It is advised that we should protect the lakes and make sure it is in a good and safe condition.

In a study by Rung (2005), the characteristics of urban park have been classified into six categories such as facilities and amenities; condition of the parks likes maintenance, accessibility, the availability and proximity, the aesthetics value that has the attractiveness and appeal. Other than that, safety value such as the personal security and fear, and policies in the management and budget. Thus, the objectives of this study are:

1. To examine water quality index (WQI) that affects water recreational activities.
2. To determine water sport user's satisfaction on types of facilities provided by the management.

2 Literature Review

2.1 Water Quality Monitoring

Many chemical, physical and biological parameters are considered the determining factors of water quality in the aquatic system (Sargaonkar and Deshpande 2003). Therefore, parameters or variables such as dissolved oxygen (DO), pH, total dissolved solids, faecal coliform and chemical oxygen demand (COD) are measured,

Table 1 Water quality monitoring

Development component	Indicated
pH	The contamination and acidification
Chemical oxygen demand (COD)	Measure the requirement of oxygen of water sample which is oxidised by a strong chemical oxidant
Biochemical oxygen demand (BOD)	The amounts of organic pollutants in water
Ammoniacal nitrogen (AN)	The nutrient status, organic enrichment and health of the water body
Suspended soil (SS)	The small solid particles which remain in suspension in water as a colloid due to the motion of water
Dissolved oxygen (DO)	Measures the amount of oxygen dissolved or carried in the water

and the results are used in the assessment and classification of water quality (Boyacioglu 2007). Parameters chosen for water quality index (WQI) are dissolved oxygen (DO), biochemical oxygen demand (BOD), chemical oxygen demand (COD), ammoniacal nitrogen (AN), suspended solid (SS), and pH. Subindices are used in the calculations of the selected parameters namely SIBOD, SIDO, SIAN, SISS, SIPH and SICOD (Table 1).

$$\text{DOE} - \text{WQI} = 0.22 \times \text{SI DO} + 0.19 \times \text{SI BOD} \\ + 0.16 \times \text{SI COD} + 0.15 \times \text{SI AN} \\ + 0.16 \times \text{SI SS} + 0.12 \times \text{SI pH} \quad (1)$$

Water quality index (WQI) is a tool that is used to evaluate the quality of water. The variability in these water quality parameters is usually due to anthropogenic and other natural factors (Simeonov et al. 2002).

2.2 Study Area

Taman Tasik Titiwangsa was chosen as the study area because it was the top three Best Public Park in Kuala Lumpur beside Lake Garden and KLCC Park (Attraction Malaysia 2015). To date, no study has been done on water quality index at Taman Tasik Titiwangsa. Originally, Taman Tasik Titiwangsa was an abandon tin mining area and over the time the area was inhabited by weeds, fishes, earthworms, snakes and other living organisms (Malek 2004). For this study, samples were collected at Lake 1, a formerly mining area and Lake 2, a man-made lake (Fig. 1).

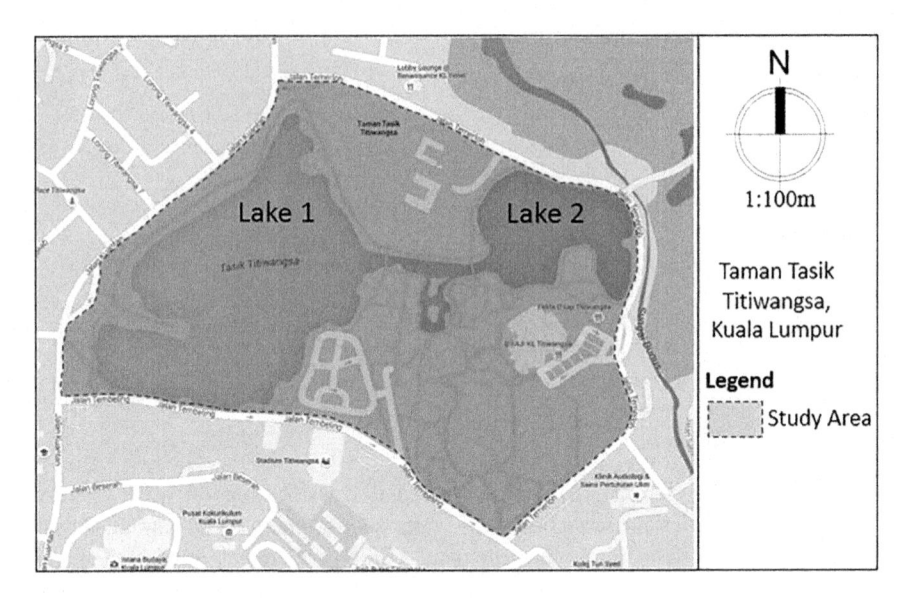

Fig. 1 Location of Taman Tasik Titiwangsa

3 Methodology

3.1 Research Design and Data Collection

a. **Laboratory test**

In situ data collection using multiparameter YSI Probe that measured dissolved oxygen, salinity, conductivity, pH, total dissolved solids (TDS), temperature, resistivity, specific conductance, ORR, ammonium, chloride and nitrate is executed at Lake 1 and Lake 2. Water samples from sampling locations were collected to determine the biochemical oxygen demand (BOD), chemical oxygen demand (COD), suspended solids (SS) and ammoniacal nitrogen (AN) results for water quality index calculations.

b. **Questionnaire survey**

In order to determine, water sport users' satisfaction on types of facilities provided by the management, questionnaire survey were distributed to the respondents or water sport users. The questionnaire was distributed during weekends from 8 a.m. to 12 noon and from 5 p.m. to 7 p.m. where many people came for recreation and water sport activity. The questionnaires were outlined based on certain criteria that emphasis on the demographic profile (gender, age, race and occupation) of the water sport users, the satisfaction level of the water sport users' activity, facilities provided at Taman Tasik Titiwangsa, and maintenance works executed by the management.

Table 2 DOE water quality index classification

Parameters	Unit	Class				
		I	II	III	IV	V
Ammoniacal nitrogen	mg/l	<0.1	0.1–0.3	0.3–0.9	0.3–0.9	>2.7
Biochemical oxygen demand	mg/l	<1	1–3	3–6	6–12	>12
Chemical oxygen demand	mg/l	<10	10–25	25–50	50–100	>100
Dissolved oxygen	mg/l	>7	5–7	3–5	1–3	<1
pH	–	>7	6–7	5–6	<5	>5
Total suspended solid	mg/l	<25	25–50	50–150	150–300	>300
WQI		>92.7	76.5–92.5	51.9–76.5	51.9–76.5	51.9–76.5

3.2 Data Analysis

a. **Laboratory analysis**

Data acquired from the laboratory study were classified by calculating individual parameter quality index or grouped to calculate WQI which referring to dissolved oxygen, biochemical oxygen demand, chemical oxygen demand, ammoniacal nitrogen (AN), suspended solid and pH (Table 2).

$$\text{DOE} - \text{WQI} = 0.22 \times \text{SI DO} + 0.19 \times \text{SI BOD} + 0.16 \times \text{SI COD} + 0.15 \times \text{SI AN} + 0.16 \times \text{SI SS} + 0.12 \times \text{SI pH}$$

b. **Survey analysis**

Data analysis for the questionnaire was done by using the common statistical software SPSS (Statistical Package for Social Science) version 19. Chi-square test was executed to see the association between demographic profile of the respondents and satisfaction of the facilities provided at Taman Tasik Titiwangsa. All questions were individually analysed, taking into considerations all the available factors and supported with descriptive and inferential analysis.

4 Findings

4.1 Water Quality Index (WQI) Result

$$\text{DOE} - \text{WQI} = 0.22 \times \text{SI DO} + 0.19 \times \text{SI BOD} + 0.16 \times \text{SI COD} + 0.15 \times \text{SI AN} + 0.16 \times \text{SI SS} + 0.12 \times \text{SI pH}$$

Table 3 Results for water quality index (WQI) parameters at Lake 1 and Lake 2

Parameters	Lake 1	Lake 2
pH	8.87	8.04
BOD	6.5 mg/l	11 mg/l
COD	14 mg/l	7 mg/l
AN	0.29 mg/l NH3-N	0.12 mg/l NH3-N
SS	35	60
DO (%)	77.2	360.8

Table 4 DOE water quality classification based on water quality index

Water quality index	Status		
	Clean	Slightly polluted	Polluted
Water Quality Index (WQI)	81–100	60–80	0–59

Table 5 Water classes and uses

Class	Uses
Class I	Conservation of natural environment
	Water Supply I—practically no treatment necessary
	Fishery I—very sensitive aquatic species
Class IIA	Water Supply II-conventional treatment required,
	Fishery II—sensitive aquatic species
Class IIB	Recreational use with body contact
Class III	Water supply III—extensive treatment required
	Fishery III—common, of economic value and tolerant species; livestock drinking
Class IV	Irrigation
Class V	None of the above

Based on the WQI formula, the WQI for Lake 1 is 77.63 and Lake 2 is 82.28 where it shows that Lake 2 is in clean status while Lake 1 in slightly polluted status (Table 4) (Tables 3 and 5).

4.2 Questionnaire Survey Results

Based on the study, it shows that male respondents are more interested to participate in the survey (54.7%) as compared to female respondents (45.3%). According to Breen et al. (2017), men are more preferred to do active activities such as water sports activity compared to women that are more likely to do passive activities such as reading and visiting art galleries. Adolescents (age group between 13 and 19 years old) are among the majority who came for the water sport activity (34.2%) whereas stated by Erikson that the teens are keen to cope with social and academic demands as compared to older age groups. Another age group that came for the activity was those in the age between 20 and 40 years old (44.4%). According to Erikson, this group is particular about having relationship or friendship with others. Teenagers and youngsters come to Taman Tasik Titiwangsa and enjoyed boat paddling and water ball activity with their partners, families or their friends. Most of the respondents are Malays (64.7%) followed by Chinese (21.9%), Indian (13.1) and others (0.3%). Majority of the respondents were unemployed, students and housewives with 44.2%, where they have ample leisure time to visit Taman Tasik Titiwangsa, Kuala Lumpur. Secondly, respondents worked with the private sector (35.3%) where they came for water sport activity at Taman Tasik Titiwangsa, Kuala Lumpur. Besides that, respondent from government sector is 16% and others (4.6%).

Table 6 shows the significant associations between the demographic profiles of the respondents and types of water sport activity, facilities and maintenance works provided by the management at Taman Tasik Titiwangsa. From Table 6, there are

Table 6 Summary of significant associations between gender, age, race and occupation of the respondents with types of water sport activity, facilities and maintenance works

Variables	Asymp. Sig			
	Gender	Age	Race	Occupation
A. *Types of water sport activity*				
Water ball	0.039*	0.290	0.932	0.920
Kayaking	0.105	0.003*	0.666	0.000*
Boat Paddling	0.525	0.019*	0.621	0.338
B. *Facilities*				
Fees counter	0.623	0.149	0.965	0.846
Safety equipment	0.660	0.116	0.968	0.691
Watchmen	0.007*	0.018*	0.938	0.959
Boats	0.201	0.257	0.839	0.556
Paddle	0.533	0.703	0.411	0.601
Information board	0.238	0.080	0.731	0.150
C. *Maintenance work*				
The maintenance work done by the management team is good	0.233	0.046*	0.644	0.552
The service given by the management team is good	0.24	0.001*	0.267	0.225
It is easy to communicate with the staff	0.21	0.000*	0.459	0.061
Security system	0.547	0.001*	0.658	0.091
Landscape	0.329	0.168	0.124	0.466
Cleanliness	0.222	0.649	0.217	0.096
Smell	0.579	0.320	0.240	0.037*
Aesthetic value	0.434	0.000*	0.061	0.087
Water sport activity equipment	0.146	0.002*	0.619	0.470

*denotes significant value less than $p < 0.05$

significant values between gender water ball activity ($p = 0.039$) and watchmen ($p = 0.007$). Male respondents feel dissatisfied with water ball activity because they keen to play in larger group and in an adventure way. Male respondents also showed their dissatisfaction with the watchmen because the watchmen are gender biased. Respondents' age shows more significant associations compared to gender. The significant associations are shown from kayaking ($p = 0.003$) and boat paddling activity ($p = 0.019$) while from facility there is association with watchmen ($p = 0.018$). Other than that significant associations are noted for 'the maintenance work done by the management team is good' ($p = 0.046$), 'the service given by the management team is good' ($p = 0.001$), 'It is easy to communicate with the staff' ($p = 0.000$), 'security system' ($p = 0.001$), 'aesthetic value' ($p = 0.000$) and 'water sport activity equipment' ($p = 0.002$). Respondents with different age background show disparities of satisfactions of the maintenance works. Younger groups (13–19 years old, 20–40 years old and 41–64 years old) are more concerned with the maintenance work by the management compared to elder group. No significant association is recorded for race. For occupation, the significant associations only noted for kayaking activity ($p = 0.000$) and maintenance work for odour or smell at the area ($p = 0.037$). These significant associations refer to the respondents who work in the government sector and also from the other group who are dissatisfied with the condition of the smelling area at the Lake.

5 Conclusions

In conclusion, proper management should be taken to improve the facilities and maintenance of the recreation water sport facility at Taman Tasik Titiwangsa. Intensive research on water quality index (WQI) should be done regularly to update the status of the level of safety for water recreation activity.

Lake 1 where the venue for the water recreational activity took place showed a WQI value of 77.63 which is a slightly polluted status. This is because the lake has been used for various water sport activities. However, water user's sport activity may have direct contact with both Lake 1 and Lake 2 as water in the lakes is safe.

References

Attraction Malaysia. (2015). Taman Tasik Titiwangsa. Retrieved http://kuala-lumpur.attractionsinmalaysia.com/Titwangsa-Lake-Gardens.php.

Ayeghi, A. N. U. (2014). The Impacts of Physical Features on User Attachment to Kuala Lumpur City Centre (KLCC) Park, Malaysia. Malaysian Journal of Society and Space, 44-54.

Breen, B., Curtis, J and Hynes, S. (2017). Recreational use of Public Waterways and the impact of water quality. ESRI Working Paper No. 552.

Boyacioglu, H. (2007). Surface water quality assessment using factor analysis. *Water SA, 32*(3), 389–393.

Malek, N. A. (2004). Assessment of satisfactions, preferences, need and use patterns in quality neighbourhood park. Malaysia.

Rung, A. L. (2005). The significance of parks to physical activity and public health. *American Journal of Preventive Medicine,* 159–168.

Sargaonkar, A., & Deshpande, V. (2003). Development of an overall index of pollution for surface water based on a general classification scheme in Indian context. *Environmental Monitoring and Assessment, 89*(1), 43–67.

Sharifudin, M. Z. (2014, July 26). Water sport activity at Taman Tasik Titwangsa. Retrieved http://ww1.utusan.com.my/utusan/Keluarga/20140726/ke_02/Riadah-di-Taman-Tasik-Titiwangsa.

Simeonov, V., Einax, J. W., Stanimirova, I., Kraft, J. (2002). Environmetric modeling and interpretation of river water monitoring data. Anal. Bioanal. Chem. 374(5), 898-905.

Phytochemical Screening, Antioxidant and Enzyme Inhibition Activity of *Phoenix dactylifera* Ajwa Cultivar

Muhamad Nabil Md Nor, Nur Syafiqah Rahim, Sarina Mohamad, Saiyidah Nafisah Hashim, Zainab Razali and Noor Amira Muhammad

Abstract Ajwa date (*Phoenix dactylifera*) is one of a special variety of Saudi Arabian dates with many health benefits such as nutrient and fiber. Despite extensive studies on the pharmacological properties of *P. dactylifera* and its constituents, the study on its potential as acetylcholinesterase inhibitor has not been reported so far; therefore, Ajwa cv. was selected as sample in this study. Water extract of Ajwa cv. fruit flesh was analyzed for their antioxidant activity through phytochemical screening and DPPH assay. The sample also tested on inhibition activity toward acetylcholinesterase (AChE) enzyme using thin-layer chromatography plate. The extracts show the presence of phenols, alkaloids, flavonoids, tannins, carbohydrates which are good antioxidant sources, of the extracts exhibited a concentration dependence pattern across the range tested. 100 mg/ml of Ajwa dates showed the highest antioxidant activity with 75.47% of inhibition. For acetylcholinesterase inhibition test, the findings showed no positive sign of enzyme inhibition found as no white spot discovered on the TLC plate. It could be done on seed itself to prove that the seed plays the main role as the acetylcholinesterase inhibitor.

Keywords *Phoenix dactylifera* · Phytochemical screening · Antioxidant activity Alzheimer disease · Acetylcholinesterase inhibitor

---

M. N. M. Nor (✉) · N. S. Rahim · S. Mohamad · Z. Razali
Faculty of Applied Sciences, Universiti Teknologi MARA,
Perlis Campus, 02600 Arau, Perlis, Malaysia
e-mail: muhamadnabil28@gmail.com

S. N. Hashim · N. A. Muhammad
Faculty of Applied Sciences, Universiti Teknologi MARA, Perak Branch,
Tapah Campus, 35400 Tapah Road, Perak, Malaysia

1 Introduction

Phoenix dactylifera is very commonly consumed in many parts of the world and is a vital component of the diet in most of the Arabian countries. There are different cultivars (cv.) of *P. dactylifera* that exhibit different characteristics and benefits. Study done by Al-Yahya et al. (2015) found that Ajwa cv. is a special variety of Saudi Arabian dates. This is because the Ajwa cv. rich with nutrients, fibers, and bioactive molecules such as anthocyanins, ferulic acid, protocatechuic acid, and caffeic acid. In the present study, we, therefore, examined whether diet rich in date palm fruits could improve memory. According to Subash et al. (2015), date palm fruits may represent protective strategies to minimize the risk of developing Alzheimer's disease (AD).

Zehra et al. (2015) describe that antioxidants have always helped in preventing the damage done to cells by free radicals that are released during normal metabolic process of oxidation. These free radicals include reactive oxygen free radical species (ROS), reactive hydroxyl radicals (OH$^-$), superoxide anion radical (O_2^-), hydrogen peroxides (H_2o), and peroxyl (ROO$^-$). It is well known that *P. dactylifera* is a rich source for natural phytochemical antioxidants including vitamins (ascorbic acid, Vitamin A, and a-tocopherols), carotenoids, and phenolic compounds (Allaith 2008). According to Tang et al. (2013), *P. dactylifera* contains antioxidant that may provide essential nutrients and potential health benefits to consumer. The paper also reveals that palm date extract can inhibit protein oxidation as well as neutralize superoxide and hydroxyl radicals. However, there is still lack of antioxidant activity study on the flesh of Ajwa cultivar.

Acetylcholinesterase enzyme (AChE) functions to terminate the neurotransmission cholinergic synapses by splitting the neurotransmitter acetylcholine (Tripathi and Srivasta 2008). The principal role of acetylcholinesterase is the termination of nerve impulse transmission at the cholinergic synapses by rapid hydrolysis of acetylcholine (Mukherjee et al. 2007). Acetylcholinesterase affects the cholinergic system because it regulates the level of acetylcholine and terminates nerve impulses by catalyzing the hydrolysis of acetylcholine. Its inhibition causes death. Irreversible inhibitors have been developed as insecticides such as organophosphates and carbamates (Njoroge et al. 2016).

An acetylcholinesterase inhibitor is a chemical or a drug that inhibits the acetylcholinesterase enzyme from breaking down acetylcholine, thereby increasing both the level and duration of action of the neurotransmitter acetylcholine. The drugs are commonly used for Alzheimer's disease (AD) therapy by counteracting the acetylcholine deficit and enhance the acetylcholine level in brain (Heinrich and Teoh 2004). Acetylcholine is involved in synapses for signal transfer. After being delivered in the synapses, acetylcholine is hydrolyzed into choline and acetyl group in a reaction catalyzed by the enzyme acetylcholinesterase (Voet and Voet 1995). Some of the drugs approved for therapeutic use show hepatotoxicity (Knapp et al. 1994), and consequently, there has been a continuous search for new drugs especially from plant source. There has been a lot of research on the biological effect of plants traditionally used either in infusions or in traditional remedies as acetylcholinesterase inhibitors in vitro and also as memory enhancers in vivo (Perry et al. 2000; Heinrich and Teoh 2004).

Test on acetylcholinesterase inhibition activity is commonly done by using thin-layer chromatography (TLC) technique as it is very suitable to be used in synthetic chemistry for identifying compounds. It also able to determine the purity of the compounds and following the progress of a reaction. Currently, enzyme inhibition assays such as inhibition of xanthine oxidase acetylcholinesterase (AChE) and tyrosinase on TLC have greatly expanded the use of TLC bioautography as a screening method (Gu et al. 2015).

This study is a first study done to discover the antioxidant activity of Ajwa cv. fruit flesh as previous studies only focused on the mixture of palm date fruit. The finding of this study will be very beneficial to identify the potential of Ajwa water extract as acetylcholinesterase inhibitor and as an alternative source of drugs in treating AD.

2 Materials and Methods

2.1 Preparation of Date (Phoenix dactylifera) Extracts

Extraction was performed according to Maged and Abbas (2013) with slight modification. 500 g of Ajwa fruit flesh was separated from the seed. The seedless dates were dried for 12 days in the oven at 50 °C to eliminate the water content. 200 g of the dried dates was then ground into powder. Apart from it, 20 g of the sample powder were dissolved in 250 ml of water with continuous shaking for 24 h at room temperature. The Ajwa water extract was then further filtered using Whatman Paper No. 1. The filtrate was then put into the deep freeze in inclined position for 18 h at −80 °C followed by drying process for 72 h. The crystalline powder obtained from this extraction was kept in refrigerator at 4 °C for further use.

2.2 Phytochemical Screening on Ajwa Fruit Flesh

Detection of phenols was performed according to Al-Dawah and Ibrahim (2013) with slight alteration. 50 mg of Ajwa powder was dissolved with 5 ml of distilled water in a test tube. Few drops of neutral ferric (III) chloride were dropped into the test tube and the color change was recorded. The appearance of dark green color indicates the presence of phenols.

The presence of alkaloid was tested by dissolving 0.5 g of the powder extract into 5 ml of 1% hydrochloric acid in steam bath. Six drops of Dragendorff's reagent was added into the solution. The formation of precipitate, brown indicates the presence of alkaloids in Ajwa dates.

Next, Ajwa date extract powder weighed 0.5 g was dissolved in distilled water, and to these, 10 drops of dilute hydrochloric acid followed by a small piece of magnesium were added. The presence of green blue and violet coloration indicated the presence of flavonoids (Kodangala et al. 2010).

Tannin detection was performed according to Kodangala et al. (2010) with modification. 1–2 ml of Ajwa date extract was added with 3 ml of 5% w/v ferric (III) chloride solution. The appearance of blue–black color indicates the presence of tannins.

Lastly, detection of carbohydrates was done according to Bhandary et al. (2012) with modification. 3 ml of Ajwa date extract was mixed with few drops of Benedict's reagent (alkaline solution containing cupric citrate complex) and boiled in water bath. The formation of reddish brown precipitate showed the presence of carbohydrate.

2.3 Antioxidant Activity (DPPH Assay)

The antioxidant of Ajwa water extract was evaluated by free radical scavenging capacity through diphenylpicrylhydrazyl (DPPH) assay according to Kchaou et al. (2013). 500 µl of sample solution was dissolved in ethanol and mixed with 500 µl of 0.5 mM DPPH in ethanol. Then, the mixture was shaken vigorously and was incubated in the dark at room temperature for 30 min. By using spectrophotometer, ethanol was used as blank and the DPPH solution as control. The absorbance was measured at 517 nm. The antiradical activity had expressed as percentage of inhibition of the sample as calculated using Eq. 1.

$$Q\% = 100 \times (A_0 - A_c)/A \tag{1}$$

2.4 Acetylcholinesterase Test Method

Acetylcholinesterase test was prepared according to Yang et al. (2009) with slight modification. 500 mL of tris-hydrochloric acid buffer (0.05 mol/L, pH = 7.8) was dissolved in acetylcholinesterase. 500 mg BSA was added in the solution to stabilize the enzyme during bioassay. 150 mg of 1-naphthyl acetate was dissolved in 40 ml of ethanol and diluted with 60 ml distilled water. 50 mg of fast blue B salt was dissolved in 100 ml of distilled water. Sample was applied to silica gel TLC plate and migrated with solvent. The plate was dried for 60 s using hairdryer. Enzyme solution and 1-naphthyl acetate were sprayed to the TLC plate and the plate blew with hairdryer. TLC plate was put in container containing little humidity at 37 °C for 20 min. This process enabled the enzyme to react with 1-naphthyl acetate. Lastly, fast blue B salt was sprayed onto the TLC plate to observe the presence of white spot.

3 Results and Discussion

3.1 Phytochemical Screening

Phytochemical analysis of Ajwa water extract was done in order to identify the presence of bioactive compounds such as flavonoids, phenols, tannins, and saponins. Table 1 showed the results of bioactive compounds present in the sample. This is supported by Nasir et al. (2015) which revealed that dates were rich in bioactive compounds including phenols, sterols, carotenoids, anthocyanins, procyanins, flavonoids, tannins, carotinoids, alkaloids, and polyphenols which are good sources of antioxidant.

The potential health benefits of Ajwa dates have been partially attributed to their polyphenol contents, especially flavonoids that have much attention from the literature over the past decade for its biological effects. The flavonoids and phenolic acids are known to possess antioxidant activities due to the presence of hydroxyl groups in their structures, and their contribution to defense system against the oxidative damage due to endogenous free radicals is extremely important (Saggu et al. 2014). Phenolic compounds or polyphenols and alkaloids are secondary plant metabolites that are ubiquitously present in plants and their products. Most of them have been proven to have high levels of antioxidant activities (Razali et al. 2008). Due to their redox properties, these compounds such as flavonoids, tannins, and alkaloids contribute to the overall antioxidant activities. The tabulated data was supported with findings done by Saleh et al. (2011). Palm date fruit might be a good source of these active components and has a potent ability to suppress free radicals.

3.2 Antioxidant Activity (DPPH Assay)

Date fruit consists of high antioxidant activity as it is rich in phenolic compounds and flavonoid constituents with free radical scavenging. Findings showed that an increase of Ajwa dates extract concentration caused a significant increase in the concentration of DPPH inhibition. There were no statistically significant differences for all concentration as the percentage of inhibition is lower than p-value = 0.05

Table 1 Pythochemical screening of Ajwa dates' compound

Bioactive compounds	Crude extract of Ajwa
Phenols	+
Alkaloids	+
Flavonoids	+
Tannins	+
Carbohydrates	+

+ Indicates the positive result
− Indicates the negative result

Table 2 Comparison of percentage of inhibition of Ajwa dates and ascorbic acid

Concentration of Ajwa dates extraction (mg/ml)	% of Inhibition	
	Ajwa dates	Ascorbic acid
0.01	5.71 ± 1.54	85.61 ± 4.36
0.10	9.67 ± 2.68	90.11 ± 0.84
1	27.74 ± 8.33	91.49 ± 0.57
10	63.24 ± 6.97	93.82 ± 2.77
100	75.47 ± 0.84	96.07 ± 1.88

The values were expressed in mean ± SD

even though the concentration of the Ajwa dates extract increased by ten times. The highest percentage of inhibition is 75.47% at the concentration of 100 μg/ml while the lowest percentage of inhibition is 5.71% at 0.01 μg/ml (Table 2).

3.3 Acetylcholinesterase Test

It was found that plants with AChE inhibitory and antioxidant activity may help in preventing or alleviating patients suffering from AD (Ferreira et al. 2006). In this present study, Ajwa fruit flesh extract was tested on its inhibitory activity toward AChE. After several tests, the test found that there is no white spot appear on TLC plate. The plate that sprayed with naphthol and fast blue B should be the positive control as to indicate the negative result. The white spot is referring the inhibition action to deactivate the AChE enzyme from catalyzed the hydrolyzing reaction of acetylcholine into acetate and choline (Fig. 1). It was found that plants with AChE inhibitory and antioxidant activity may help in preventing or alleviating patients suffering from AD (Ferreira et al. 2006).

Fig. 1 No AChE inhibition of Ajwa cultivar on TLC plate

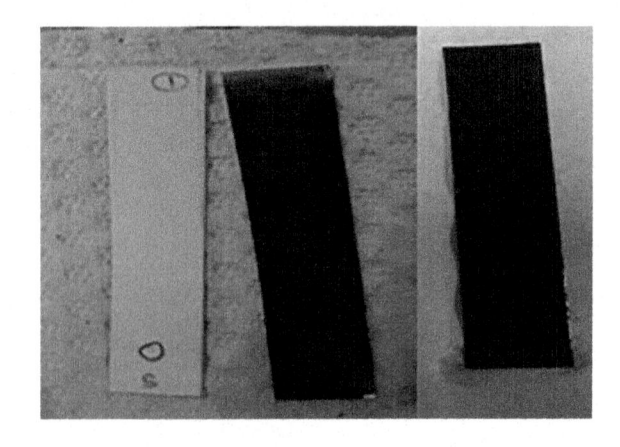

From previous study, Sekeroglu et al. (2012) mentioned that inhibitory activities of Ajwa date seed extract and flesh mixture had the highest inhibitory activity against the enzyme acetylcholinesterase (AChE) and butyrylcholinesterase (BChE). However in this experiment, different finding as Ajwa cv. flesh had not brought positive inhibitory action. In comparing with results by Sekeroglu et al. (2012), the good inhibitory activity of Ajwa seed and flesh mixture is mighly contribute by the seed.

4 Conclusion and Recommendation

It can be concluded that Ajwa flesh water extract contains several bioactive compounds such as phenols, alkaloids, flavonoids, tannins, and carbohydrates. Antioxidant result showed by highest percentage of inhibition; 75.47% at the concentration of 100 µg/ml. The findings based on DPPH assay, the sample also consist with antioxidant properties. Lastly, the inhibition effect of Ajwa flesh water extract toward acetylcholinesterase enzyme is giving negative result as no white spot existed on the TLC plate. For recommendations, high-performance liquid chromatography (HPLC) or gas chromatography (GC) should be applied next time to identify the composition of the Ajwa date sample. For AChE inhibitory activity, acetylcholinesterase test should be done on seed itself to prove that the seed plays the main role as the acetylcholinesterase inhibitor.

Acknowledgements We would like to thank Universiti Teknologi MARA Perlis for providing the laboratory facilities to conduct the research project.

References

Allaith, A. A. A. (2008). Antioxidant activity of Bahraini date palm (*Phoenix dactylifera L.*) fruit of various cultivars. *International Journal of Food Science & Technology, 43*(6), 1033–1040.

Al-Dawah, N. K., & Ibrahim, S. L. (2013). Phytochemical characteristics of Date Palm (*Phoenix dactylifera* L.) leaves extract. *Kufa Journal For Veterinary Medical Sciences, 4*(1).

Al-Yahya, M., Raish, M., AlSaid, M. S., Ahmad, A., Mothana, R. A., Al-Sohaibani, M., ... Rafatullah, S. (2015). 'Ajwa'dates (*Phoenix dactylifera L.*) extract ameliorates isoproterenol-induced cardiomyopathy through downregulation of oxidative, inflammatory and apoptotic molecules in rodent model. *Phytomedicine, 2*(1), 222.

Bhandary, S. K., Kumari, S. N., Bhat, V. S., Sharmila, K. P., & Bekal, M. P. (2012). Preliminary phytochemical screening of various extracts of *Punica granatum* peel, whole fruit and seeds. *Journal of Health Science, 2*(4), 35–38.

Ferreira, A., Proença, C., Serralheiro, M. L. M., & Araujo, M. E. M. (2006). The in vitro screening for acetylcholinesterase inhibition and antioxidant activity of medicinal plants from Portugal. *Journal of Ethnopharmacology, 108*(1), 31–37.

Gu, L. H., Liao, L. P., Hu, H. J., Bligh, S. A., Wang, C. H., Chou, G. X., et al. (2015). A thin-layer chromatography-bioautographic method for detecting dipeptidyl peptidase IV inhibitors in plants. *Journal of Chromatography A, 1411,* 116–122.

Heinrich, M., & Teoh, H. L. (2004). Galanthamine from snowdrop—the development of a modern drug against Alzheimer's disease from local Caucasian knowledge. *Journal of Ethnopharmacology, 92*(2), 147–162.

Kchaou, W., Abbès, F., Blecker, C., Attia, H., & Besbes, S. (2013). Effects of extraction solvents on phenolic contents and antioxidant activities of Tunisian date varieties (*Phoenix dactylifera* L.). *Industrial Crops and Products, 45,* 262–269.

Knapp, M. J., Knopman, D. S., Solomon, P. R., Pendlebury, W. W., Davis, C. S., Gracon, S. I., et al. (1994). A 30-week randomized controlled trial of high-dose tacrine in patients with Alzheimer's disease. *JAMA, 271*(13), 985–991.

Kodangala, C., Saha, S., & Kodangala, P. (2010). Phytochemical studies of aerial parts of the plant *Leucas lavandulaefolia. Scholars Research Library Der Pharma Chemica, 2*(5), 434–437.

Maged, N. Q. A., & Abbas, N. A. (2013). Antibacterial activity of *Phoenix dactylifera* L. leaf extracts against several isolates of bacteria. *Kufa Journal for Veterinary Medical Sciences, 4* (2).

Mukherjee, P. K., Kumar, V., & Houghton, P. J. (2007). Screening of Indian medicinal plants for acetylcholinesterase inhibitory activity. *Phytotherapy Research, 21*(12), 1142–1145.

Nasir, M. U., Hussain, S., Jabbar, S., Rashid, F., Khalid, N., & Mehmood, A. (2015). A review on the nutritional content, functional properties and medicinal potential of dates. *Science Lecture, 3,* 17–22.

Njoroge, A. W., Ngugi, M. P., Aliyu, U., Matheri, F., Gitahi, M. S., Mwangi, M. B., … Ngure, G. M. (2016). In Vitro Anti-Acetylcholinesterase activity of dichloromethane leaf extracts of *Carphalea glaucescens* in Chilo partellus Larvae.

Perry, N. S., Houghton, P. J., Theobald, A., Jenner, P., & Perry, E. K. (2000). In-vitro inhibition of human erythrocyte acetylcholinesterase by *Salvia lavandulaefolia* essential oil and constituent terpenes. *Journal of Pharmacy and Pharmacology, 52*(7), 895–902.

Razali, N., Razab, R., Junit, S. M., & Aziz, A. A. (2008). Radical scavenging and reducing properties of extracts of cashew shoots (*Anacardium occidentale*). *Food Chemistry, 111*(1), 38–44.

Saleh, E. A., Tawfik, M. S., & Abu-Tarboush, H. M. (2011). Phenolic contents and antioxidant activity of various date palm (*Phoenix dactylifera* L.) fruits from Saudi Arabia. *Food and Nutrition Sciences, 2011.*

Saggu, S., Sakeran, M. I., Zidan, N., Tousson, E., Mohan, A., & Rehman, H. (2014). Ameliorating effect of chicory (*Chichorium intybus* L.) fruit extract against 4-tert-octylphenol induced liver injury and oxidative stress in male rats. *Food and Chemical Toxicology, 72,* 138–146.

Şekeroğlu, Z. A., & Şekeroğlu, V. (2012). Effects of viscum album L. extract and quercetin on methotrexate-induced cyto-genotoxicity in mouse bone-marrow cells. *Mutation Research/ Genetic Toxicology and Environmental Mutagenesis, 746*(1), 56–59.

Subash, S., Essa, M. M., Braidy, N., Awlad-Thani, K., Vaishnav, R., Al-Adawi, S., … Guillemin, G. J. (2015). Diet rich in date palm fruits improves memory, learning and reduces beta amyloid in transgenic mouse model of Alzheimer's disease. *Journal of Ayurveda and integrative medicine, 6*(2), 111.

Tang, Z. X., Shi, L. E., & Aleid, S. M. (2013). Date fruit: Chemical composition, nutritional and medicinal values, products. *Journal of the Science of Food and Agriculture, 93*(10), 2351–2361.

Tripathi, A., & Srivastava, U. C. (2008). Acetylcholinesterase: a versatile enzyme of nervous system. *Annals of Neurosciences, 15*(4).

Voet, D., & Voet, J. G. (1995). Serine proteases biochemistry (2nd ed., p. 390). USA: John Wiley and Sons.

Yang, Z., Zhang, X., Duan, D., Song, Z., Yang, M., & Li, S. (2009). Modified TLC bioautographic method for screening acetylcholinesterase inhibitors from plant extracts. *Journal of Separation Science, 32*(18), 3257–3259.

Zehra, S., Saeed, A., & Fatima, S. (2015). Antioxidant and antibacterial studies of *Phoenix dactylifera* and its varieties. *International Journal of Applied Microbiology and Biotechnology Research, 3,* 81–88.

Determination of Two Different Aeration Time on Food Waste Composting

Khairul Bariyah Binti Abd Hamid, Mohd Armi Abu Samah, Mohd Huzairi Mohd Zainudin and Kamaruzzaman Yunus

Abstract Increasing volume of organic waste especially food waste due to increased population can affect human and also the environment if it is not managed properly. One of the methods to treat food waste is composting. However, there are many problems regarding composting, for example, time taken to complete the composting process. Thus, the objective for this study is to compare two aeration times used for composting. Methodology for this study started with collecting the food waste, drying and grinding process before starting the composting, and finishing the process in 3 days. Results show that from the two different aeration times, it shows that pH of Compost A (at 34th hour) become neutral faster than Compost B (at 55th hour). There

---

K. B. B. A. Hamid (✉)
Department of Biotechnology, Kulliyyah of Science, International Islamic University Malaysia, Jalan Sultan Ahmad Shah, Bandar Indera Mahkota, 25200 Kuantan, Pahang, Malaysia
e-mail: khairulbariyahabdhamid@gmail.com

M. A. A. Samah
Department of Chemistry, Kulliyyah of Science, International Islamic University Malaysia, Jalan Sultan Ahmad Shah, Bandar Indera Mahkota, 25200 Kuantan, Pahang, Malaysia
e-mail: marmi@iium.edu.my

M. H. M. Zainudin
Laboratory of Sustainable Animal Production and Biodiversity, Institute of Tropical Agriculture and Food Security, Universiti Putra Malaysia UPM, 43400 Serdang, Selangor, Malaysia
e-mail: mohdhuzairi@upm.edu.my

K. Yunus
Department of Marine Science, Kulliyyah of Science, International Islamic University Malaysia, Jalan Sultan Ahmad Shah, Bandar Indera Mahkota, 25200 Kuantan, Pahang, Malaysia
e-mail: kama@iium.edu.my

is no significant different for compost temperature between Compost A and Compost B (minimum and maximum). Besides that, total weight reduction for Compost B (45%) is higher compared with Compost A (31%). Even though Compost B is slower to reach pH neutral, it is faster in reduction of total weight. Thus, it can solve the problem especially in increasing volume of food waste.

Keywords Comparison · Aeration · Composting · Food waste

1 Introduction

Domination of the food waste contributes major environmental problems. Each year, high generation of food waste resulted in greater amount of municipal solid waste. Without proper management, all of the waste will end up at the landfill. Therefore, the lifespan of landfill become shorter and more space needed to dispose the waste. The untreated waste will affect three major aspects which are environment, economy, and social problems.

Currently, in Malaysia, total population is about 30 million (World Bank 2016) and more than 70% of population living in the urban areas. Thus, more volume of municipal solid waste was generated (Hamid 2012). From the total 38 thousand tonnes of solid waste generated per day, 15 thousand tonnes are food waste only (Corp 2014). Supposedly, 33 thousand tonnes of solid waste daily should be achieved in year 2020 but it already exceeded in the year 2016. This shows that the problem of solid waste is critical.

There is a lot of research regarding solid waste. Most of the research shows that food waste dominated solid waste stream. However, without proper treatment, all the waste ends up at the landfill. Hamid et al. (2012) stated that at the moment, unfortunately, government yet not seriously seek solution to solve this problem. They just aim to solve basic matter only regarding management of municipal solid waste. Food waste in Malaysia was not treated separately from municipal solid waste due to limited treatment and poor management.

According to Agumuthu et al. (2009), various methods to treat solid waste were proposed but the achieved target was always low. In year 2020, government planned to reduce landfilling about 50% by increasing composting rate to 8% (Johari et al. 2014). Rawat et al. (2013) stated that among various treatments to manage organic waste, decaying it using biological process is more suitable.

Historically, composting is the best way to treat the organic wastes. Compost is a humus-like substance produced from the conversion of solid organic material under the controlled biological decomposition. Composting is one of the low-cost biological decomposition processes where it is circuited by microbial activity but with consideration of the physical–chemical parameters which include temperature, aeration, moisture content, C/N ratio, and pH (Fathi et al. 2014). Furthermore, the process of composting is known to be an environmental-friendly method that does not release any hazardous chemical which can affect human health and known to be beneficial ingredients that are essential for organism growth such as in plants.

Composting is an alternative solid waste management system (SWM); it can be used for the recycling of organic matters into useful products. According to Shilev et al. (2006), although composting has its own benefits, production of practical common organic compost still required a lot of time. In addition, the scarcity of information in regard to the solid waste compositions was the common problems in Malaysia. This resulted in most researchers and government officials using or presenting outdated data in the estimation of future trends (Johari et al. 2012). However, the optimization of formula and composition in food waste was convenient to face the obstacle in performing quality of composting and encountered the slow process of composting (Ishii and Takii 2003). Thus, the objective for this study is to compare two aeration times used during composting process.

2 Methodology

Sample collection (food waste) was done at chosen restaurant. About 40 kg of food waste will be dried and grinded. Before starting the composting process, the food waste was divided into 2 bins which contain 20 kg each. Bin A was aerated for 5 min per 20 min, and Bin B was aerated for 15 min per 60 min. Each parameter was recorded every 1 h except for moisture content which was recorded for every 24 h. Figure 1 shows the summary of research flowchart.

Food waste collection was collected after the restaurant was closed. Next, the food waste was transferred to the composting site where the drying and grinding were done. The grinding process was done using heavy duty commercial blender.

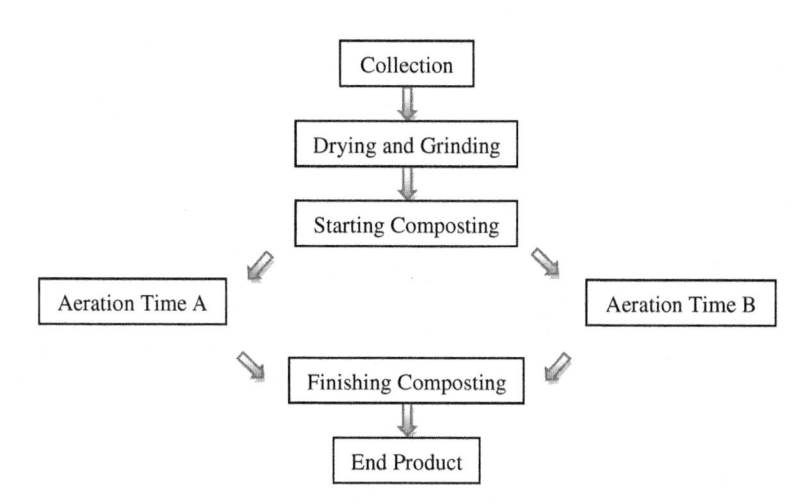

Fig. 1 Summary of research flowchart

During composting process, both of the bins were aerated continuously for 3 days. Compost temperature data was taken using soil thermometer while compost pH data was taken using pH meter. Data for moisture content was obtained using conventional method.

3 Result and Discussion

Total data collected during composting process is 73 for each parameter in 3 days. Table 1 shows all data of the parameter recorded during composting process. In Compost A (5 min per 20-min aeration), the minimum and maximum pH recorded is 5.47 and 6.37, respectively. In Compost B (15 min per 60-min aeration), the minimum and maximum pH recorded is 5.62 and 6.35, respectively. Compost A starting to maintain pH 6 at 34th hour while Compost B starting to maintain at pH 6 at 55th hour. We can conclude that by using different time aeration, the compost becomes more neutral faster. Compost temperature recorded for both compost is same for minimum and maximum which is 24 and 37 °C, respectively. The compost bin was located at the same place and same time.

Moisture content is the measure of the quantity of water present in compost product. Overly dry compost (<35% moisture) can be dusty and hard to handle, while very wet compost (55–60%) becomes clumpy and difficult to uniformly apply. The ideal moisture content needed in compost is 35–55%. In this research, the moisture content of the compost was measured every 24 h. As shown in Table 2, the starting value of moisture content in the compost before process was already below 35% which is 5.91% for Compost A and 7.63% for Compost B. This happens because of the drying process before starting composting process has decreased water content in the compost.

Table 3 shows total weight reduction of food waste after completing composting. With aeration time B, the weight reduction is higher than using aeration time A. Compost B was given longer aeration time compared to Compost A. Water content in the compost also plays important role to reduce the total weight.

Figure 2 shows the graph of compost pH versus time. Compost A becomes neutral (acidic to neutral) faster than Compost B. After that, pH for Compost A is maintained and recorded slightly higher than Compost B. It shows that different aeration time can affect the pH of the compost.

Figure 3 shows the graph of compost temperature versus time. There is no significant difference between the two. This may be happened because the compost was located with the same condition during the composting process. The high temperature was recorded during day while the low temperature was recorded during night.

Table 1 Compost pH and Compost temperature recorded during composting process

Time (hours)	C. pH A	C. pH B	C. Temp A	C. Temp B
0	6.01	6.01	31	31
1	5.92	5.91	32	32
2	5.94	5.91	32	34
3	5.47	5.67	37	34
4	5.79	5.66	34	33
5	5.73	5.7	31	31
6	5.79	5.64	28	29
7	5.82	5.73	27	27
8	5.84	5.73	25	26
9	5.88	5.76	25	26
10	5.85	5.81	25	25
11	5.92	5.77	25	25
12	5.98	5.73	25	25
13	5.87	5.76	25	25
14	6.02	5.77	25	25
15	6.15	5.62	25	25
16	5.96	5.75	25	25
17	5.90	5.66	25	25
18	6.02	5.71	24	24
19	6.06	5.80	24	24
20	6.14	5.74	24	24
21	5.98	5.74	25	25
22	6.09	5.79	27	28
23	6.04	5.76	30	31
24	5.94	5.77	32	32
25	5.86	5.77	31	33
26	5.88	5.80	32	31
27	5.90	5.87	34	32
28	5.82	5.85	34	32
29	6.01	5.88	30	30
30	5.93	5.86	28	28
31	5.94	5.93	27	27
32	5.88	5.77	26	27
33	5.93	5.84	26	27
34	6.00	5.82	26	27
35	6.02	5.80	26	26
36	6.05	5.80	26	26
37	6.04	5.69	26	26

(continued)

Table 1 (continued)

Time (hours)	C. pH A	C. pH B	C. Temp A	C. Temp B
38	6.09	5.80	26	27
39	6.09	5.79	25	26
40	6.07	5.95	25	26
41	6.11	5.97	25	25
42	6.16	5.99	25	25
43	6.21	6.01	24	24
44	6.15	5.98	25	25
45	6.14	6.04	26	26
46	6.13	6.02	28	28
47	6.21	5.96	32	31
48	6.13	5.89	35	33
49	6.03	5.94	35	37
50	6.12	5.92	33	36
51	6.08	5.94	37	36
52	6.08	5.94	36	33
53	6.07	6.03	29	32
54	6.12	5.96	29	29
55	6.14	6.03	27	28
56	6.19	6.00	27	27
57	6.19	6.02	26	27
58	6.21	6.05	26	27
59	6.26	6.03	26	26
60	6.23	6.01	26	26
61	6.28	6.05	26	26
62	6.22	6.06	26	26
63	6.27	6.03	25	25
64	6.35	6.10	25	25
65	6.30	6.10	24	25
66	6.30	6.13	25	25
67	6.32	6.26	24	24
68	6.35	6.35	24	24
69	6.29	6.08	24	24
70	6.36	6.12	27	25
71	6.37	6.11	29	30
72	6.35	6.15	32	33

Table 2 Moisture content recorded during composting process

Time (Hours)	Compost A			Compost B		
	Before	After	%	Before	After	%
0	60.4462	56.8728	5.91	59.7009	55.1482	7.63
24	61.3549	58.0938	5.32	60.6157	56.5385	6.73
48	58.5370	55.7457	4.77	58.6963	55.6331	5.22
72	59.7447	57.8714	3.14	61.3366	58.5990	4.46
Total loss			2.77 (46.87%)			3.17 (41.55%)

Table 3 Total weight reduction of food waste before and after composting

Time (Hours)	Compost A	Compost B
0	20 kg	20 kg
72	13.8 kg	11.0 kg
Loss	6.2 kg	9.0 kg
Reduction percentage	31%	45%

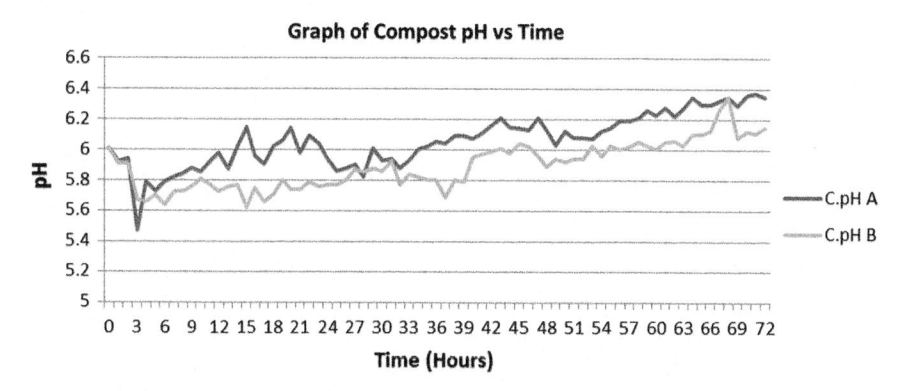

Fig. 2 Graph of compost pH versus time

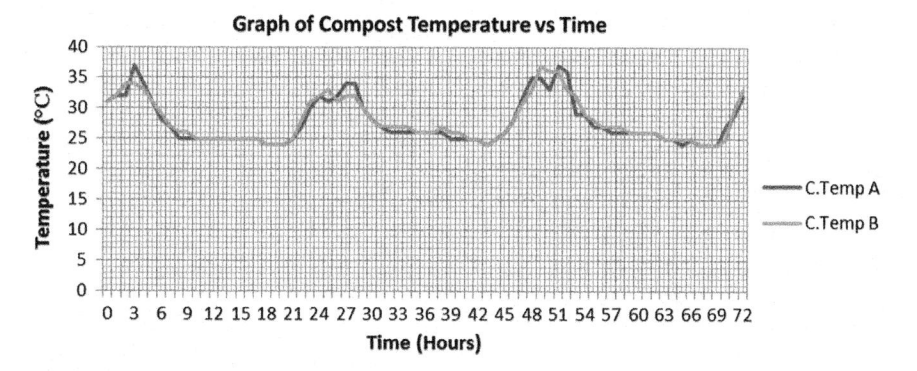

Fig. 3 Graph of compost temperature versus time

4 Conclusion

From the two different aeration times, we can conclude that pH of Compost A (at 34th hour) become neutral faster than Compost B (at 55th hour). For temperature, there is no significant different between Compost A and Compost B (minimum and maximum). Mass reduction for Compost B (45%) is higher compared with Compost A (31%). The most important part is that the waste management options to treat food waste using composting can be designed. Although this is undoubtedly the most difficult, the Malaysian government and non-government organizations should consider prompt implementation due to its current and long-term benefits. Hopefully, with this research outcome or results, the food waste problem can be solved. Thus, it improves and provides better environment for future generation.

Acknowledgements This work is partially supported by Research Initiatives Research Grant (RIGS), International Islamic University Malaysia. Alhamdulillah, praise to Allah. For the excellent guidance and patient assistance, I truly thank Dr. Mohd Armi Abu Samah. Also thanks to IRS for providing me with workplace and facility. Lastly, thanks for people who are helping directly and indirectly.

References

Agumuthu, P., Hamid, F. S., & Khidzir, K. (2009). Evolution of solid waste management in Malaysia: Impacts and implications of the solid waste bill. *Journal of Material Cycles Waste Management,* 96–103.

Corp, S. W. (2014). Pelan Strategik 2014–2020. Perbadanan Pengurusan Sisa Pepejal Dan Pembersihan Awam. Ministry Of Urban Planning, Housing And Local Government.

Fathi, H., Zangane, A., Fathi, H., & Moradi, H. (2014). Municipal solid waste characterization and it is assessment for potential compost production: A case study in Zanjan City, Iran. *American Journal of Agriculture and Forestry, 2*(2), 39–44.

Hamid, F. S. (2012). Trends in sustainable landfilling in Malaysia, a developing country. *Waste Management and Research,* 1–8.

Hamid, A. A., Ahmad, A., Ibrahim, M. A., & Rahman, N. N. N. A. (2012). Food waste management in Malaysia-Current situation and future management options. *Journal of Industrial Research and Technology, 2*(1), 36–39.

Ishii, K., & Takii, S. (2003). Comparison of microbial communities in four different composting processes as evaluated by denaturing gradient gel electrophoresis analysis. *Journal of Applied Microbiology, 95*(1), 109–119.

Johari, A., Ahmed, S. I., Hashim, H., Alkali, H., & Ramli, M. (2012). Economic and environmental benefits of landfill gas from municipal solid waste in Malaysia. *Renewable and Sustainable Energy Reviews, 16*(5), 2907–2912.

Johari, A., Alkali, H., Hashim, H., Ahmed, S. I., & Mat, R. (2014). Municipal solid waste management and potential revenue from recycling in Malaysia. *Modern Applied Science, 8*(4), 37–49.

Rawat, M., Ramanathan, A. L., & Kuriakose, T. (2013). Characteristics of Municipal Solid Waste Compost (MSWC) from selected Indian cities – a case study for its sustainable utilisation. *Journal of Environmental Protection, 4,* 163–171.

Shilev, S., Naydenov, M., Vancheva, V., & Aladjadjiyan, A. (2006). Composting of food and agricultural wastes. In V. Oreopoulu (Ed.), *Utilization of by-product and treatment of waste in food industry* (pp. 283–301). Springer.

World Bank. (2016). Population, total. http://data.worldbank.org/indicator/SP.POP.TOTL. Accessed May 23, 2017.

The Synergistic Antibacterial Effect of *Azadirachta indica* Leaves Extract and *Aloe barbadensis* Gel Against Bacteria Associated with Skin Infection

Hamidah Jaafar Sidek, Mohamad Azhar Azman and Muhamad Shafizul Md Sharudin

Abstract Combinations of two agents which create inhibitory effects that are greater than the individual effects produce a positive interaction known as synergism. This method can help in developing agents for antibacterial activity in order to treat bacterial infection. This study was done to assess the possible synergistic antibacterial effect of the combination between *Azadirachta indica* leaves extracts and *Aloe barbadensis* gel against five bacteria commonly associated with skin infection. Synergistic antibacterial activities from the interaction of both plants against *Escherichia coli*, *Staphylococcus aureus*, *Pseudomonas aeruginosa*, *Klebsiella aerogenes* and *Streptococcus faecalis* were measured by using Kirby–Bauer disc diffusion assay. *A. indica* ethanolic extract, *A. barbadensis* gel, combination of *A. indica* ethanolic extract and *A. barbadensis* gel, and commercialize antibiotics were tested on the five bacteria. The data were subjected to statistical analysis of one-way ANOVA and Tukey's post hoc tests ($\alpha = 0.05$). The results of this study showed significant inhibition of the bacteria from the synergistic effect of *A. indica* and *A. barbadensis* compared to the commercial antibiotic. The significant results may contribute to the development of stronger antibacterial agent in skin infection treatment in the pharmaceutical industry.

Keywords *Azadirachta indica* · *Aloe barbadensis* · Synergistic antibacterial activity · Kirby–Bauer disc diffusion assay versus skin infection bacteria

1 Introduction

Many types of plants have been used in treating diseases since the early ages, for example in the Ayurvedic medicines. Each part of different plants has been collected and studied to show the antibacterial effect, and the number of plants already

H. J. Sidek (✉) · M. A. Azman · M. S. M. Sharudin
Faculty of Applied Sciences, Universiti Teknologi MARA,
02600 Arau, Perlis, Malaysia
e-mail: hamidahjs@perlis.uitm.edu.my

introduced to the world is estimated to be about 250,000–500,000 species (Das et al. 2014). The amount of plants that are used for medicinal purposes are less than the amount of plants that are used as food by human and animals (Das et al. 2014).

As the production of medicine is increasing and developing rapidly, the emergence of disease has also grown. One of the common diseases that is associated with human is skin disease. Skin diseases are mostly caused by antibacterial infection. For example, cellulitis is caused by β-haemolytic *Streptococci* and *Staphylococcus aureus* (Gunderson 2011). Skin infections such as impetigo, folliculitis, furunculosis, cellulitis, abscesses are caused by methicillin-resistant *S. aureus* (Ullah et al. 2016). Thus, many types of research have been done in producing antibacterial agent that can kill bacteria that cause these skin diseases.

Synergism is one of the best methods where the highly potential antibacterial agents are combined to create new and strong antibacterial activities that produce an effect greater than the sum of their individual effects. The combined effect of bacteriocin and extracts of neem leaf is being tested at present that will assist to make a formulation for microbial infection on skin (Das et al. 2014).

In this study, two types of plants, *Azadirachta indica* and *Aloe barbadensis*, were tested to evaluate the synergistic antibacterial activity potential in order to overcome skin infection caused by bacteria. *A. indica* or commonly known as neem is the native tree from India and naturalized in most of tropical and subtropical countries such as Malaysia, Indonesia and Thailand (Juss et al. 2013). Neem tree can be found in at least 30 countries in Asia, Africa, Australia as well as Central and South Americas (Patel et al. 2016). Parts of the neem tree have been used as part of traditional medicine in various locations around the world, and the antimicrobial properties of their extract and compounds have been studied widely in pharmacological aspects (Quelemes et al. 2015). One of the compounds in the neem is *Azadirachtin*, which consists of antiviral, antifungal, antibacterial and insecticidal properties (Kashyap 2014). Extracts of the neem leaf have been found to possess immunomodulatory, anti-inflammatory, and anticarcinogenic properties (Elumalai et al. 2012). Based on recent study, neem has been the object of extensive phytochemical studies, due to its strong biological effect including antibacterial activity (Sujarwo et al. 2016). The International Scientific Community has included *A. indica* as the top ten of lists of plants to be studied and used for sustainable development of the planet and the health of living beings (Kashyap 2014). Thus, *A. barbadensis* or commonly known as *Aloe vera* is one of the ancient medicinal ailments for human being. *Aloe vera* has been used in folk medicine for over 2000 years and has remained an important component in the traditional medicine of many contemporary cultures, such as China, India, the West Indies and Japan (Radha and Laxmipriya 2015). *Aloe vera* is stem less or sometimes may be a very short-stemmed succulent plant growing up to 60–100 cm tall and has thick, fleshy green leaves with some varieties showing white flecks on the upper and lower stem surfaces (Irshad et al. 2011). It is a perennial succulent xerophyte, which develops water storage tissue in the leaves to allow it to survive in dry areas of low or erratic rainfall, and the innermost part of the leaf has a clear, soft, moist and slippery tissue that consists of large thin-walled parenchyma cells in which water is held in the form of vicious mucilage (Nejatzadeh-barandozi 2013). In recent studies, *Aloe vera* is recommended for treating all types of skin diseases (Bhat et al. 2014).

The main point of this study is to introduce new potential antibacterial source to treat skin infection by synergistic effect of two different plants without chemical interference. The usage of herbs in treating skin infection has been practised a long time ago, but based on past studies there is no specific documentation on the traditional treatment methods to cure skin diseases (Bhat et al. 2014). Nowadays, advancement of medicinal studies has increased the chances to reduce skin infection. This is only possible if various and repetitive research and development is being done.

2 Materials and Methods

2.1 Collection of Plant Materials

The leaves of *A. indica* were collected in the month of June 2016, and the leaves of *A. barbadensis* were collected in the month of August 2016 from the tree growing wildly in Bintong, Perlis, Malaysia.

2.2 Extraction of Ethanolic Compound from Azadirachta Indica *Leaves*

The ethanolic compounds were extracted according to the method used by (Abdussalam 2011). Initially, the fresh leaves were allowed to dry under shade for 14 days and ground into powder using a grinder. The powdered material was weighed using electronic weighing balance, and drying of the leaves was continued until a constant weight was obtained. An amount of two hundred and fifty grams of the powder was placed in a container and was defatted using petroleum ether, following which it was subjected to maceration using 300 ml of 95% (v/v) ethanol in order to obtain the ethanolic extract of the plant. The mixture was stirred up and kept for 24 h. The mixture was filtered, and another 300 ml of the ethanol was added to the residue and kept for another 24 h before filtration. This procedure was repeated 3 times, and the combined filtrate was subjected to rotary evaporator to obtain the crude extract.

2.3 Extraction of Aloe barbadensis *Gel*

The outermost part of *A. barbadensis* leaf was peeled off, and the inner part of the leaf was left in the form of vicious mucilage. The gel from the skinless leaf was stripped out, and about 5 ml of gel was obtained from single 25 cm length of *A. barbadensis* leaf.

2.4 Test Organisms

The micro-organisms used were *Escherichia coli* (ATCC 11303), *S. aureus* (ATCC 25923), *Pseudomonas aeruginosa* (ATCC 10145), *Klebsiella aerogenes* (ATCC 15380) and *Streptococcus faecalis* (ATCC 29212) and were obtained from Microbiology Laboratory 5 UiTM Perlis, Arau, Malaysia, to represent skin infection bacteria.

2.5 Sterilization of the Equipment and Disinfection

All the equipment was disinfected with cotton wool soaked in 70% ethanol so as to maintain sterility throughout the process. Wire loop, conical flask and beaker were sterilized by hot air oven at 160 °C for 45 min, whereas moisture-insensitive materials were sterilized by autoclaving at 121 °C for 15 min (Abdussalam 2011).

2.6 Preparation of Media

The Mueller–Hilton agar (MHA) consisted of (gm/litre) agar 17.0 g, beef extract 2.0 g, starch 1.50 g and acid hydrolysate of casein 17.50 g. An amount of 35 g of Mueller–Hilton agar was weighed and dissolved in 1000 ml of distilled water and adjusted to pH of 7.4 ± 0.2 at 25 °C. This was sterilized by autoclaving at 121 °C for 15 min at 15 psi pressure and was used for Kirby–Bauer disc diffusion tests.

2.7 Antibacterial Activity Assay of the Plant Extracts

Kirby–Bauer disc diffusion assay was carried out to get the zone of inhibition that showed the antimicrobial activity of *A. indica* leaves extract, *A. barbadensis* gel, and the combination *A. indica* leaves extract and *A. barbadensis* gel towards the five microbes: *E. coli* (ATCC 11303), *S. aureus* (ATCC 25923), *P. aeruginosa*

(ATCC 10145), *K. aerogenes* (ATCC 15380) and *S. faecalis* (ATCC 29212). All the bacteria were subcultured into McCartney bottle that contains nutrient broth before being spread on the Mueller–Hilton agar. First, the agar was removed from the refrigerator, placed in incubator and let it to come down to the room temperature. In the meantime, the laboratory bench was first disinfected with 70% ethanol and the Bunsen burner was lighted up to keep up with a sterile environment. Once the agar was warmed up, the agar was removed from the incubator for culturing. Forty-five blank paper discs were used in the antibacterial assay. Fifteen blank paper discs were soaked in crude extract of *A. indica* leaves, next fifteen blank paper discs were soaked in *A. barbadensis* gel, and another fifteen blank paper discs were soaked in the mixture of crude extract of *A. indica* leaves and *A. barbadensis* gel. Six discs of 10 µg ampicillin and nine disc of 20 µg gentamicin were used as the positive control, and five blank paper discs were soaked in ethanol as negative control. About 0.2 ml of the bacterium broth culture was transferred onto the Mueller–Hilton agar medium, aseptically. Then, the broth culture of *E. coli, S. aureus* and *P. aeruginosa, K. aerogenes* and *S. faecalis* was spread with L-shaped glass spreader. For precaution, the L-shaped glass spreader was dipped into ethanol first before being used in the next spreading. Then, in order to determine their inhibitory antibacterial effect, the soaked paper discs and positive control and negative control discs were placed onto the Mueller–Hilton agar surface. Each disc was slightly pressed down into the agar medium to ensure the complete contact with the Mueller–Hilton agar. All plates were inverted and incubated at 37 °C for 24 h. The diameter of the clear zones of inhibition of the test organisms in response to the crude leaves extract of *A. indica*, *A. barbadensis* gel, mixture of the crude leaves extract of *A. indica* and *A. barbadensis* gel, gentamicin, ampicillin, and ethanol was measured in millimetres.

2.8 Statistical Analysis

Data were statistically analysed using IBM SPSS statistics version 23. A one-way analysis of variance (ANOVA) followed by Tukey's post hoc test was applied for analysis of data with the level of significance set at $p < 0.05$.

3 Results and Discussion

3.1 Results

These experiments using Kirby–Bauer disc diffusion method have shown significant results where the zone of inhibition of the five tested bacteria for the mixture of the crude leaves extract of *A. indica* and *A. barbadensis* gel was obviously larger than the zone of inhibition by the commercialize antibiotic, the crude leaves extract of *A. indica* and *A. barbadensis* gel that were tested individually. All the five bacteria tested were susceptible towards the commercial antibiotic, the crude leaves extract of *A. indica* and the mixture of the crude leaves extract of *A. indica* and *A. barbadensis* gel (Table 1 and Fig. 1). The comparison of the antibacterial combination effect for the mixture of the crude leaves extract of *A. indica* and *A. barbadensis* gel was clarified in Fig. 1.

Table 1 Inhibitory effect of antibacterial sample on five tested bacteria

Bacteria	Zone of inhibition (mm)					
	Antibacterial sample					
	Neem extract	Aloe vera gel	Aloe vera gel and Neem extract	20 µg gentamicin	10 µg ampicillin	70% ethanol
Escherichia coli	192.82	132.16	322.16	192.54	Not tested	0
Pseudomonas aeruginosa	164.32	9,331.69	272.16	20,33.51	Not tested	0
Staphylococcus aureus	18.7	10,331.25	220.82	18,31.74	Not tested	0
Klebsiella aerogenes	18,662.36	100.82	220.82	Not tested	240	0
Streptococcus faecalis	19,670.47	10,330.47	22,332.06	Not tested	18	0

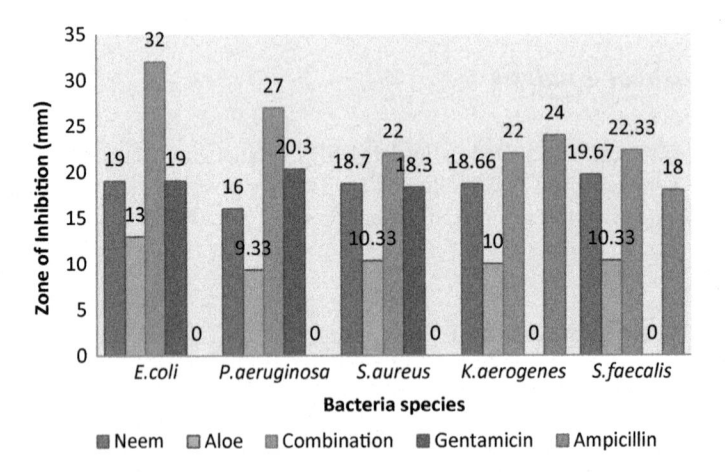

Fig. 1 Comparison of zone of inhibition of five skin infection bacteria from the inhibitory effect of neem extract, aloe vera gel, combination of neem extract and aloe vera gel, and commercialized antibiotics

3.2 Discussion

Plants are the larger source of potentially useful compound for the development of new antibacterial agent which can be tested in vitro with an antibacterial activity assay. The idea of finding synergistic antibacterial activity effect of two different plants is initiated by the need of a new type of antibacterial agent that can overcome skin infection caused by bacteria. Most of the commercial antibiotic is already ineffective against the bacteria, and there are chemical influence from the current antibiotic agent (Cock 2015). The increasing of frequency in use of antibiotics for treatment of humans and animals has developed the antibiotic resistance and multidrug resistance micro-organisms (Prasannabalaji et al. 2012). In this research, *A. indica* and *A. barbadensis* were chosen in determining the synergistic antibacterial activity against five skin infection bacteria. The results of this experiment are not compared with any other research since there is lack of research focusing on combining these two plants. Both of these plants are well known in treating skin infection. From the recent study, *Aloe vera* has been recommended for treating various kinds of skin disease (Bhat et al. 2014). The synergistic effects of both plants were measured by using disc diffusion assay on five different types of bacteria that are related to skin infection. The antibacterial activities were measured by the diameter of zone of inhibition and the larger the diameters of the zone of inhibition represent the stronger antibacterial activity. The result showed that the combination of *A. indica* leaves extract and *A. barbadensis* gel produced larger zone of inhibition on the tested bacteria compared to other individual extracts (Table 1). The largest average zone of inhibition for the combination of *A. indica* leaves extract and *A. barbadensis* gel was about 32 mm on *E. coli* which was susceptible, and the antibacterial activity on other bacteria has shown positive results. Commercial antibiotic has been used to compare with the samples which are gentamicin and ampicillin. Specific antibiotic was placed on different bacteria where the gentamicin is used on *E. coli*, *P. aeruginosa* and *S. aureus*, while ampicillin was used on *Klebsiella aerogene* and *S. faecalis*. Different antibiotics are designated to different bacteria in order to get optimum zone of inhibition since in past study has shown the effect of these antibiotics on the bacteria. There were large differences in the diameter of zone of inhibition for the combination of *A. indica* leaves extract and *A. barbadensis* gel compared to *A. indica* leaves extract, *A. barbadensis* gel and the Commercial antibiotic tested individually towards the five pathogenic bacteria (Fig. 1). The results of the synergistic antibacterial activity can be seen more after being analysed using ANOVA which showed that the significant value is positive since the value did not exceed $\alpha = 0.05$ and it was proceeded with Tukey's post hoc test that produced same results as shown in Table 2. The influence of diffusion of the bioactive compound from the extract into the media could be responsible for the results. The results of this finding are aligned with several

Table 2 Multiple comparison

Dependent Variable	(I) Antibacterial agent	(J) Antibacterial agent	Mean difference (I-J)	Std. error	Sig.	95% confidence interval	
						Lower bound	Upper bound
Escherichia coli	Aloe vera	Neem	-6	2.4037	0.103	-13.3752	1.3752
		Aloe neem	-19.00000*	2.4037	0.001	-26.3752	-11.6248
	Neem	Aloe vera	6	2.4037	0.103	-1.3752	13.3752
		Aloe neem	-13.00000*	2.4037	0.004	-20.3752	-5.6248
	Aloe neem	Aloe vera	19.00000*	2.4037	0.001	11.6248	26.3752
		Neem	13.00000*	2.4037	0.004	5.6248	20.3752
P. aeruginosa	Aloe vera	Neem	-6.66667	2.95647	0.14	-15.7379	2.4046
		Aloe neem	-17.66667*	2.95647	0.002	-26.7379	-8.5954
	Neem	Aloe vera	6.66667	2.95647	0.14	-2.4046	15.7379
		Aloe neem	-11.00000*	2.95647	0.023	-20.0713	-1.9287
	Aloe neem	Aloe vera	17.66667*	2.95647	0.002	8.5954	26.7379
		Neem	11.00000*	2.95647	0.023	1.9287	20.0713
Staphylococcus aureus	Aloe vera	Neem	-8.33333*	2.09054	0.017	-14.7477	-1.919
		Aloe neem	-11.66667*	2.09054	0.003	-18.081	-5.2523
	Neem	Aloe vera	8.33333*	2.09054	0.017	1.919	14.7477
		Aloe neem	-3.33333	2.09054	0.318	-9.7477	3.081
	Aloe neem	Aloe vera	11.66667*	2.09054	0.003	5.2523	18.081
		Neem	3.33333	2.09054	0.318	-3.081	9.7477
K. aerogene	Aloe vera	Neem	-8.66667*	1.51535	0.003	-13.3162	-4.0171
		Aloe neem	-12.00000*	1.51535	0.001	-16.6495	-7.3505
	Neem	Aloe vera	8.66667*	1.51535	0.003	4.0171	13.3162
		Aloe neem	-3.33333	1.51535	0.15	-7.9829	1.3162
	Aloe neem	Aloe vera	12.00000*	1.51535	0.001	7.3505	16.6495

(continued)

Table 2 (continued)

Dependent Variable	(I) Antibacterial agent	(J) Antibacterial agent	Mean difference (I–J)	Std. error	Sig.	95% confidence interval	
						Lower bound	Upper bound
S. faecalis	Aloe vera	Neem	3.33333	1.51535	0.15	−1.3162	7.9829
		Neem	−9.33333*	1.24722	0.001	−13.1601	−5.5065
		Aloe neem	−12.00000*	1.24722	0	−15.8268	−8.1732
	Neem	Aloe vera	9.33333*	1.24722	0.001	5.5065	13.1601
		Aloe neem	−2.66667	1.24722	0.162	−6.4935	1.1601
	Aloe neem	Aloe vera	12.00000*	1.24722	0	8.1732	15.8268
		Neem	2.66667	1.24722	0.162	−1.1601	6.4935

literature (Abdussalam 2011; Khan et al. 2010; Reynolds and Dweck 1999) which found that the plants possess significant antimicrobial activities against several pathogens. From the results, the finding of the synergistic antibacterial activity of *A. indica* and *A. barbadensis* looks promising to become potential antibacterial agent in treating skin infection caused by this pathogen.

4 Conclusion

From this study, it can be concluded that there is high potential of synergistic antibacterial activity from the combination of *A. indica* leaves extract and *A. barbadensis* gel against *E. coli, S. aureus, P. aeruginosa, K. aerogene* and *S. faecalis*. The tested bacteria *E. coli* were more susceptible to the combination of *A. indica* leaves extract and *A. barbadensis* gel. Furthermore, the inhibitory effect of the synergistic plants extract outperformed the inhibitory effect of the positive control of the 20 μg gentamicin disc, a commercialized antibiotic on *E. coli*. Thus, the use of synergistic antibacterial activity from the combination of *A. indica* leaves extract and *A. barbadensis* gel seems promising to be a potential antibacterial agent in treating skin infection caused by these bacteria. As this study was preliminary, further study is needed to be done by including minimum inhibitory concentration and using other parts of these plants and tested on clinical isolates of bacteria and resistant strains.

Acknowledgements We would like to thank Universiti Teknologi MARA Perlis Branch for providing the laboratory facilities to conduct the research project.

References

Abdussalam, B. (2011). Antibacterial and phytochemical screening of the ethanolic leaf extract of *Azadirachta indica* (neem) (Meliaceae) (Vol. 3, pp. 194–199). Maiduguri, Nigeria: Department of Pharmacology and Toxicology, Faculty of Pharmacy, University of Maiduguri, Department of Paediatrics, University of Maiduguri Teaching Hospital.

Bhat, P., Hegde, G. R., Hegde, G., & Mulgund, G. S. (2014). Ethnomedicinal plants to cure skin diseases—An account of the traditional knowledge in the coastal parts of Central Western Ghats. *Journal of Ethnopharmacology, 151*(1), 493–502. https://doi.org/10.1016/j.jep.2013.10.062.

Cock, I. E. (2015). Antimicrobial activity of *Aloe barbadensis* miller leaf gel components. *The Internet Journal of Microbiology, 4*.

Das, S., Chatterjee, S., & Mandal, N. C. (2014). Original research article enhanced antibacterial potential of ethanolic extracts of neem leaf (*Azadiracta indica* A. Juss). *upon combination with bacteriocin. International Journal of Current Microbiology* and *Applied Sciences, 3*(9), 617–621.

Elumalai, P., Gunadharini, D. N., Senthilkumar, K., Banudevi, S., Arunkumar, R., Benson, C. S., ... Arunakaran, J. (2012). Ethanolic neem (*Azadirachta indica* A. Juss) leaf extract induces apoptosis and inhibits the IGF signaling pathway in breast cancer cell lines. *Biomedicine & Preventive Nutrition, 2*(1), 59–68. https://doi.org/10.1016/j.bionut.2011.12.008.

Gunderson, C. G. (2011). Cellulitis: Definition, etiology, and clinical features. *AJM, 124*(12), 1113–1122. https://doi.org/10.1016/j.amjmed.2011.06.028.

Irshad, S., Butt, M., & Younus, H. (2011). In-Vitro antibacterial activity of *Aloe barbadensis* Miller (Aloe Vera). *International Research Journal of Pharmaceuticals, 1*(2), 59–64.

Juss, A., El-hawary, S. S., El-tantawy, M. E., Rabeh, M. A., & Badr, W. K. (2013). ScienceDirect DNA fingerprinting and botanical study of *Azadirachta indica*. *Beni-Suef University Journal of Basic and Applied Sciences, 2*(1), 1–13. https://doi.org/10.1016/j.bjbas.2013.09.001.

Kashyap, P. (2014). Azadirachta indica : A Plant With versatile potential. *Journal of Pharmaceutical Sciences, 4*(2) https://doi.org/10.5530/rjps.2014.2.2.

Khan, I., Srikakolupu, S. R., Darsipudi, S., & Gotteti, S. D. (2010). Phytochemical studies and screening of leaf extracts of Azadirachta indica for its anti-microbial activity against dental pathogens. *Archives of Applied Science Research, 2*(2), 246–250.

Nejatzadeh-barandozi, F. (2013). Antibacterial activities and antioxidant capacity of Aloe vera. *Organic and Medical Chemistry Letters, 3*(5), 1–8.

Patel, S. M., Venkata, K. C. N., Bhattacharyya, P., Sethi, G., & Bishayee, A. (2016). Potential of neem (*Azadirachta indica* L.) for prevention and treatment of oncologic diseases. *Seminars in Cancer Biology*. https://doi.org/10.1016/j.semcancer.2016.03.002.

Prasannabalaji, N., Muralitharan, G., Sivanandan, R. N., Kumaran, S., & Pugazhvendan, S. R. (2012). Antibacterial activities of some Indian traditional plant extracts. *Asian Pacific Journal of Tropical Disease, 2*, S291–S295. https://doi.org/10.1016/S2222-1808(12)60168-6.

Quelemes, P. V, Perfeito, M. L. G., Guimarães, M. A., Raimunda, C., Lima, D. F., Nascimento, C., ... Leite, S. A. (2015). Effect of neem (*Azadirachta indica* A. Juss) leaf extract on resistant *Staphylococcus aureus* bio fi lm formation and Schistosoma mansoni worms. *Journal of Ethnopharmacology, 175*, 287–294. https://doi.org/10.1016/j.jep.2015.09.026.

Radha, M. H., & Laxmipriya, N. P. (2015). *Journal of traditional and complementary medicine evaluation of biological properties and clinical effectiveness of Aloe vera: A systematic review, 5*, 21–26. https://doi.org/10.1016/j.jtcme.2014.10.006.

Reynolds, T., & Dweck, A. C. (1999). Aloe vera leaf gel : A review update. *Journal of Ethnopharmacology, 68*, 3–37.

Sujarwo, W., Keim, A. P., Caneva, G., Toniolo, C., & Nicoletti, M. (2016). Ethnobotanical uses of neem (*Azadirachta indica* A. Juss.; Meliaceae) leaves in Bali (Indonesia) and the Indian subcontinent in relation with historical background and phytochemical properties. *Journal of Ethnopharmacology, 189*, 186–193. https://doi.org/10.1016/j.jep.2016.05.014.

Ullah, N., Parveen, A., Bano, R., Zulfiqar, I., Maryam, M., Jabeen, S., ... Ahmad, S. (2016). Asian pacific journal of tropical disease. *Asian Pacific Journal of Tropical Disease, 6*(8), 660–667. https://doi.org/10.1016/S2222-1808(16)61106-4.

Prediction of Dengue Outbreak in Selangor Using Fuzzy Logic

Mohd Fazril Izhar Mohd Idris, Amjad Abdullah and Shukor Sanim Mohd Fauzi

Abstract This research encapsulates a case study based on fuzzy logic approach for predicting dengue outbreak in Selangor from 2010 to 2013 by looking at certain variables such as total rainfall and total rainy days obtained from Malaysian Meteorological Department and Ministry of Health. The prediction is made to determine the conditions contributing to this outbreak. The objective of this research is to evaluate the performance of the fuzzy logic prediction and to predict the dengue outbreak in Selangor using fuzzy logic. The data is gathered in order to achieve a pattern, and correspondingly the value between actual and predicted is analysed. Fuzzy logic is then evaluated by looking at the mean square error (MSE). Other than that the value for prediction is attained after the actual value is normalized using min–max normalization. The analysis of the result concludes that fuzzy logic is able to provide the value of prediction that can be compared with other methods. By using this fuzzy logic model, the authorities are able to use this model to predict the dengue outbreak at certain regions and consequently preventive measures can be taken as precautions.

Keywords Fuzzy logic · Prediction · Dengue outbreak · Mean square error

---

M. F. I. M. Idris (✉) · A. Abdullah · S. S. M. Fauzi
Faculty of Computer and Mathematical Sciences, Universiti Teknologi MARA, Shah Alam, Malaysia
e-mail: fazrilizhar@perlis.uitm.edu.my

A. Abdullah
e-mail: amjad6834@gmail.com

S. S. M. Fauzi
e-mail: shukorsanim@perlis.uitm.edu.my

1 Introduction

Dengue is a viral infection transmitted by the bite of affected female Aedes mosquitoes. Dengue virus consists of four distinct serotypes which are dengue 1, dengue 2, dengue 3 and dengue 4. Besides, there are many types of dengue conditions including mild dengue fever, Dengue Haemorrhagic Fever, Dengue Shock Syndrome and alike. This virus infection affects all level of age such as babies, young children and adults. The latest issue in Malaysia on the outbreak of dengue cases was found in Selangor. Statistically, Selangor recorded 75% rise in dengue cases and this accentuates the awareness of Selangor residents for potential risk of an outbreak (The Star Online 2015). Previous research showed that fuzzy logic is a proper method to diagnose (Salman et al. 2014; Faisal et al. 2012) and predict the dengue outbreak (Pham et al. 2016) besides by taking early precautions to reduce mortality of people especially in remote and rural areas due to the lack of medical experts.

Fuzzy logic is the method that uses linguistic variables. This refers to the method of utilizing either numerical or sentences in describing the variables. This method also utilizes the rules and classifiers for more accurate and detailed results. Fuzzy logic concepts satisfy the range of interval between 0 and 1. Fuzzy logic is an easy yet powerful troubleshooting technique with vast capabilities. It is currently used in business, system controls, electronics and traffic engineering (Murtha 2005). This method can be used to generate solutions to problem based on unclear, vague, qualitative, incomplete or inaccurate information (Jimoh et al. 2013).

Therefore, in this research, fuzzy logic model is proposed in order to predict the dengue outbreak in Selangor involving the parameters such as the total rainfall, total rainy days and number of dengue cases in addition to looking at the extent of the model's performance to make predictions.

2 Literature Review

Fuzzy logic has been applied in a wide range of studies. According to Buczak et al. (2012), the objective of the research is to obtain relationships between experimental, climatic and sociopolitical data from Peru. Fuzzy rule association is utilized, and the best rules will be selected, and a classifier is formed to estimate future dengue incidence whether it gives high or low linguistic variables. The three different fuzzy association rule models are built in predicting dengue incidence three and four weeks earlier. The third estimation involved a four-week period which is four to seven weeks from time of estimation. A positive estimation value, a negative estimation value, a sensitivity and a specificity are produced when previous unused test data for the period 4–7 weeks from time of prediction is used. The method is general and could be extended for use in any geographical region and has the potential to be extended to other environmentally influenced infections.

Furthermore, there has a research from Devi and Rani (2014) where the objective of their research is for prediction. Associative classification (AC) is chosen to be a suitable technique which has two data mining tasks which are classification and rule mining. Classification is used to determine the class labels while association rule is used to describe relationship between elements in a transactional database. Fuzzy weighted associate classifier is used in this study to predict mosquito-borne disease incidence. Therefore, associative classifier is having better accuracy as compared to traditional classifiers.

Sharma et al. (2013) proposed the design of the decision support system for mosquito-borne disease diagnosis. This is due to the lack of medical experts in rural and remote areas. By using MATLAB's GUI feature with the implementation of fuzzy, the proposed system is designed and developed. Due to non-availability of pathological and imaging-based medical diagnosis tool in remote areas, the patient's life is at risk of danger as it may lead to death due to improper diagnosis and treatment of diseases. Therefore, this system is beneficial in the diagnosis of disease and early detection of disease in which results in saving patients' life.

While Sapre and Bhatye (2016) introduced a simple and effective methodology for medical diagnosis based on fuzzy logic. Diagnosis of dengue and swine flu is considered as a vehicle to illustrate the concept while the developed methodology is suitable for application in a much wider range of diseases. A set of features F is defined relevant to the set of diseases featured D. The input case to be diagnosed is described by assigning a fuzzy value to each feature of the set F. Each disease of the set D is specified by its profile in the form of a fuzzy table obtained by consulting an expert physician. Two concepts have been used which are occurrence level and confirmation level. Fuzzy inference is applied to obtain a decision fuzzy set for each considered disease, and crisp decision values are obtained to state at which level the disease is present in the patient. Consequently, computer program prototype is developed and used to diagnose several typical inputs of case studies where the obtained results for all cases were highly satisfactory.

Meanwhile, Salman et al. (2014) proposed that mortality caused by Dengue Haemorrhagic Fever (DHF) remains increasing in Indonesia particularly in Jakarta. Diagnosis of the dengue shall be made as early as possible so that first aid can be delivered in the hope of decreasing death risk. The study was conducted by developing expert system based on computational intelligence method. During the first year, the study utilized the Fuzzy Inference System (FIS) method to diagnose Dengue Haemorrhagic Fever particularly in mobile device consists of smartphone. Expert system applications particularly use fuzzy system in mobile device were applied as they are useful in making early diagnosis of Dengue Haemorrhagic Fever that produces outcome faster than laboratory test. The evaluation of this application is conducted by performing accuracy test before and after validation using data of patient who have the Dengue Haemorrhagic Fever. This expert system application is easy, convenient and practical in which capable of conducting early diagnosis of Dengue Haemorrhagic Fever to avoid mortality in the first stage.

3 Methodology

3.1 Fuzzy Logic Research Framework

Based on Fig. 1, there are four main activities for Fuzzy Inference System (FIS) which are:

1. Fuzzification transforms the feature value of the effort drivers into proper linguistic fuzzy information.
2. Fuzzy rule base stores the knowledge and rules for deriving the outputs. These rules are expressed in the IF–THEN format.
3. Fuzzy inference engine takes the human feelings, thoughts and logical inference into account in order to obtain a reasonable result from an already-known fact and relevant fuzzy rules. There are basically two kinds of inference operators: minimization (min) and product (prod).
4. Defuzzification is responsible for transforming the fuzzy results from the fuzzy system into crisp values.

3.1.1 Fuzzification

Testing samples or crisp values are transformed into fuzzy input values that are called fuzzification. In this step, the range for linguistic variables such as 'very low', 'low', 'moderate', 'high' and 'very high' are assumed to get the better result. The variables involved for input data are total rainfall and total rainy days per week while for the output data is dengue case per week. The range for total rainy days is set up as the lowest is 0 day and the highest is 7 days. Figure 2 shows the membership function of total rainy days per week obtained from MATLAB.

Figure 3 illustrates the membership function obtained from MATLAB of input data for total rainfall per week. The range for total rainfall is set up as the highest value is 36.3 and the lowest value is −9.3. From the range, linguistic variables are set up such as 'very low', 'low', 'moderate', 'high' and 'very high'.

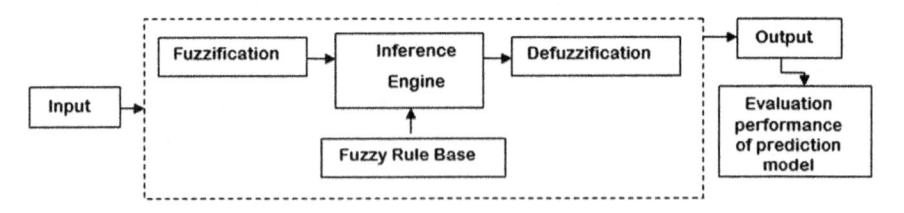

Fig. 1 Fuzzy inference system (FIS)

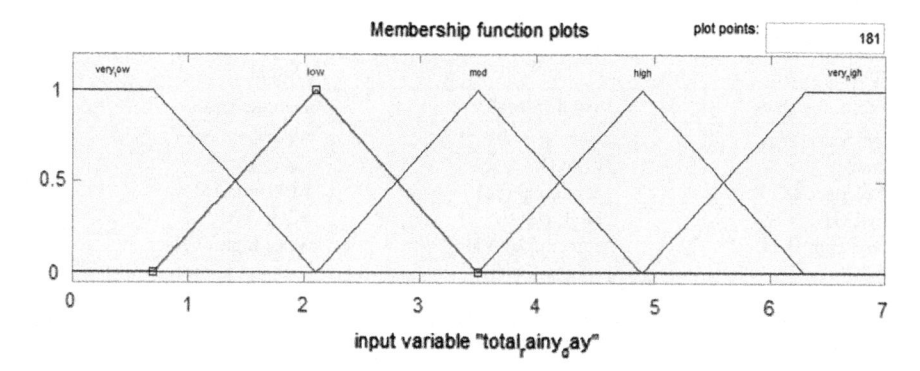

Prediction of Dengue Outbreak in Selangor Using Fuzzy Logic 597

Fig. 2 Membership function for total rainy days per week

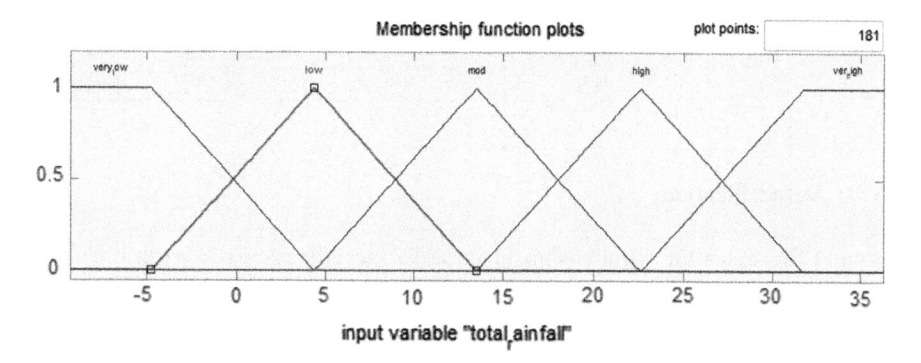

Fig. 3 Membership function for total rainfall per week

3.1.2 Fuzzy Rule-Based and Inference Engine

Inference engine is not complete if it is not combining with the rule-based technique in which this technique is used to obtain a reasonable result from relevant fuzzy rules. To complete the inference engine, the fuzzy rule-based needs to be performed to continue the defuzzification process. Table 1 shows the example of rule-based technique in table form by each parameter for input and output, while Table 2 shows the relationship between total rainy days, total rainfall and number of dengue cases.

Rule-based technique is being describe all the inputs with a parameter in which the input on this research have two and the parameter is divided by five phases. Thus, 25 rules have been obtained.

Table 1 Input and output of fuzzy-rule based

Input		Output
Total rainy days	Total rainfall	Dengue cases
Very low (VL)	Very low (VL)	Very low (VL)
Low (L)	Low (L)	Low (L)
Moderate (M)	Moderate (M)	Moderate (M)
High (H)	High (H)	High (H)
Very high (VH)	Very high (VH)	Very high (VH)

Table 2 Relationship between rainy days, rainfall and dengue cases

	Rainy days				
Rainfall	VL	L	M	H	VH
VL	VL	VL	L	L	M
L	VL	L	L	M	H
M	L	L	M	H	H
H	L	M	H	H	VH
VH	M	H	H	VH	VH

3.1.3 Defuzzification

Figure 4 illustrates the membership function for dengue cases per week obtained from MATLAB, and the range is set up as 1408 cases for the highest value and 70 cases for the lowest. Then, output value data is normalized by using the Eq. (1). From the range, the linguistic variables are assumed to be 'very low', 'low', 'moderate', 'high' and 'very high'.

$$\text{min--max normalization} = \left(\frac{\text{actual data} - \text{minimum data}}{\text{maximum data} - \text{minimun data}} \right) \quad (1)$$

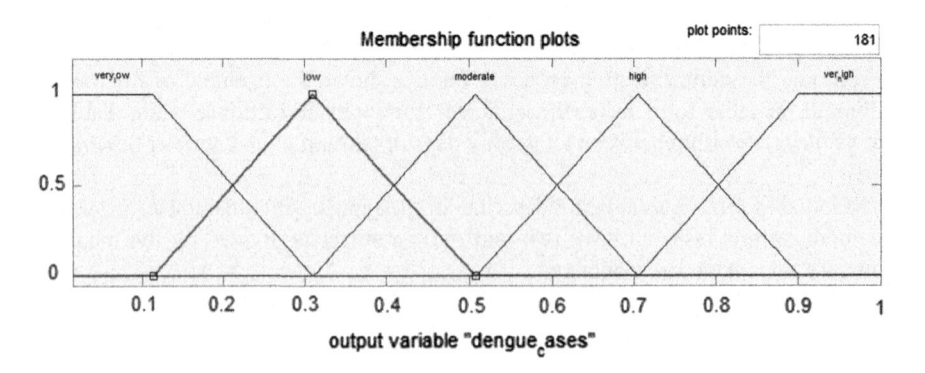

Fig. 4 Membership function for dengue cases per week

An example of the normalization for actual value of dengue cases is

$$\text{min–max normalization} = \left(\frac{583 - 70}{1408 - 70}\right) = 0.383.$$

3.1.4 Output

The dengue cases are obtained after the input data (total rainy days per week and total rainfall per week) are processing through Mamdani IF–THEN rules. Figure 5 shows one of the examples after the inputs data are running in the MATLAB and the output data which is dengue cases is produced. The example shows that if total rainy day = 5 and total rainfall per weeks = −1, then the predicted value for dengue cases obtained is 0.417.

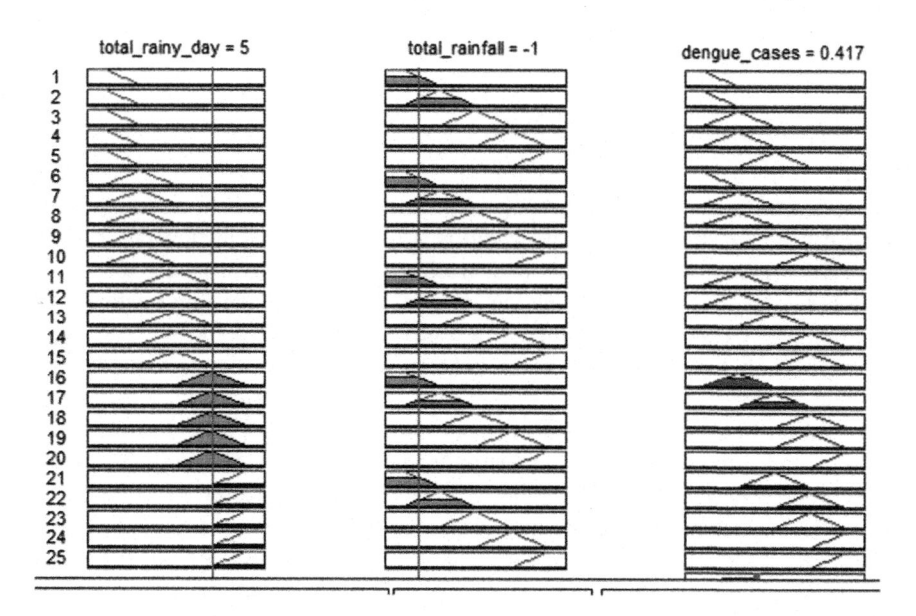

Fig. 5 Output of dengue cases

3.2 Evaluation the Performance

3.2.1 Mean Square Error (MSE)

MSE can be calculated using the Eq. (2):

$$\text{MSE} = \frac{\sum_{t}^{n} e_t^2}{n} \qquad (2)$$

where $e_t = y_t - \hat{y}t$; yt = Actual value in time t; $\hat{y}_t$ = Predicted value in time t; and n = Total of observations.

MSE is used in this study to determine whether the performance of the model has a good quality in prediction or not. The MSE value closer to zero indicates that the model is good.

4 Results and Discussions

4.1 Evaluation the Performance

According to the mean square error (MSE), the prediction result when using fuzzy logic can be categorized as small which is 0.199. Higher prediction rate can be achieved when using this method. Therefore, this method appears to be an appropriate method in predicting the dengue outbreak in Selangor where the parameters provide higher influence towards the dengue outbreak. The high accuracy predictions are essential in helping the authorities to predict dengue outbreak in Selangor.

Based on Fig. 6, the predicted value for number of dengue cases is somewhat stable as compared to the actual number of dengue cases. However, starting at week 32 of year 2013, the graph shows the error between actual and predicted value of dengue cases is smaller because the value of actual and predicted is closed to each other. This means that the closer the value to each other, the best MSE will be obtained.

4.2 Prediction of Dengue Outbreak

The predicted value of dengue cases for 2017 in Fig. 7 can be calculated using min–max normalization formula and is equal to (0.445 × 1338) + 70 = 665. The authorities can use this result to figure out the dengue outbreak at certain area. Hence, they can take precautionary measures with the possibility of the number of

Prediction of Dengue Outbreak in Selangor Using Fuzzy Logic

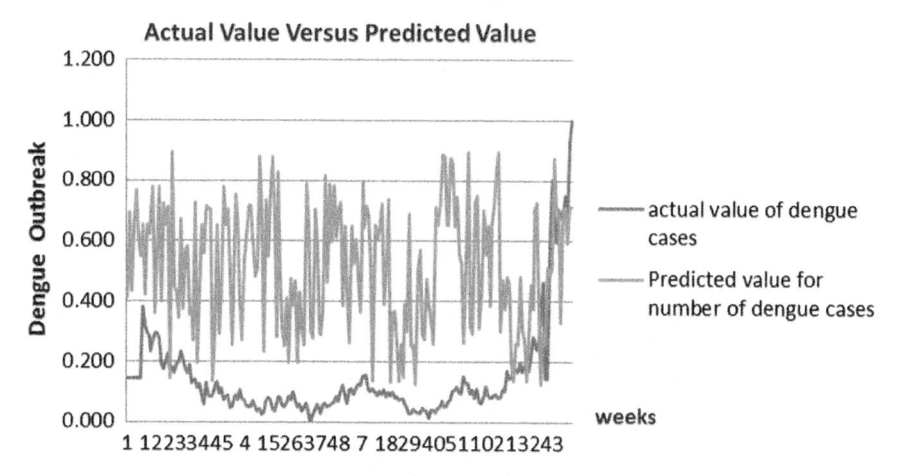

Fig. 6 Actual value versus predicted value

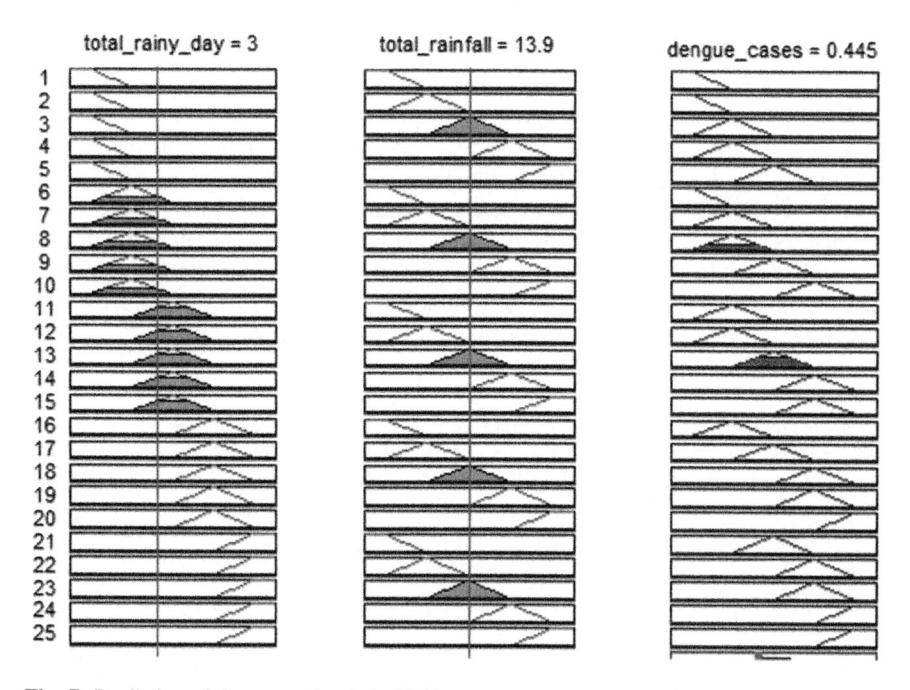

Fig. 7 Prediction of dengue outbreak in 2017

dengue cases predicted. In addition, the Malaysian Health Ministry can take early action in terms of preparation for treating dengue patients and preventive measures from continuing to spread.

5 Conclusion

In this research, the process of developing the fuzzy logic model for predicting the dengue outbreak based on the data obtained from Malaysian Meteorological Department and Ministry of Health in Selangor was investigated. This research uses fuzzy logic method where the analysis of the result concludes that the fuzzy logic approach is able to minimize the error based on MSE between the actual and predicted value. In this study, the results for testing samples are well match with the real data at the latest of the week for the year 2013. The fuzzy logic is working almost perfectly if it is used for a long-term prediction.

According to the results, it is proven that by using fuzzy logic, the dengue outbreak in Selangor can be predicted well. By using this fuzzy logic model, some interested organization and authorities may adopt the model to figure out which areas in Selangor tend to have higher dengue outbreak based on certain parameters.

For further study, several recommendations for improvement need to be address. In order to identify whether the fuzzy logic model can be applied in future, the number of input data and membership functions should be added and the rules to be constructed should be referred to the experts to obtain more accurate rules. In the current paper, the parameters measured were rather limited to the total rainfall and total rainy days. It is highly recommended for future studies to address representative numbers of parameters such as humidity, temperature, wind velocity and many more. Besides, this fuzzy model should be tested in other states to investigate the efficiency of the model.

Acknowledgements This research is partially supported by Academic Affairs Department of Universiti Teknologi MARA, Perlis. The authors also gratefully acknowledge the helpful comments and suggestions of the reviewers, which have improved the presentation.

References

Buczak, A. L., Koshute, P. T., Babin, S. M., Feighner, B. H., & Lewis, S. H. (2012). A data-driven epidemiological prediction method for dengue outbreaks using local and remote sensing data. *BMC Medical Informatics and Decision Making, 12*(124), 1–20.

Devi, M. K., & Rani, M. U. (2014). Mosquito borne disease incidence prediction system using fuzzy weighted associative classification. *International Journal of Computer Applications, 91*(13), 15–21.

Faisal, T., Taib, M. N., & Ibrahim, F. (2012). Adaptive neuro-fuzzy inference system for diagnosis risk in dengue patients. *Expert Systems with Applications, 39*(4), 4483–4495.

Jimoh, R. G., Olagunju, M., Folorunso, I., & Asiribo, M. (2013). Modeling rainfall prediction using fuzzy logic. *International Journal of Innovative Research in Computer and Communication Engineering*, 929–936.

Murtha, J. (2005). *Application of fuzzy logic in operational meteorology*. Carlifornia: Addison Wesley, Longman Inc.

Nation. (2015, February 5). Selangor records 75% rise in dengue deaths. The Star Online. Retrieved from http://www.thestar.com.my/news/nation/2015/02/05/dengue-17pc-rise-in-deaths-in-selangor/.

Pham, D. N., Nellis, S., Sadanand, A. A., Jamil, J. B., Khoo, J. J., Aziz, T., & ... Sattar, A. (2016). A literature review of methods for dengue outbreak prediction. *The Eighth International Conference on Information, Process, and Knowledge Management*, 7–13.

Salman, A., Lina, Y., & Simon, C. (2014). Computational intelligence method for early diagnosis dengue haemorrhagic fever using fuzzy on mobile device. *EPJ Web of Conferences, 68*, 1–6.

Sapre, R. G., & Bhatye, A. P. (2016). E-learning and teacher preparation in science and mathematics. *International Journal for Mathematic, 2*(2), 1–8.

Sharma, P., Singh, D. B. V., Bandil, M. K., & Mishra, N. (2013). Decision support system for malaria and dengue disease diagnosis (DSSMD). *International Journal of Information and Computation Technology, 3*(7), 633–640.

Comparison of Characterization and Osteoblast Formation Between Human Dental Pulp Stem Cells (hDPSC) and Stem Cells from Deciduous Teeth (SHED)

Farinawati Yazid, Nur Atmaliya Luchman, Rohaya Megat Abdul Wahab and Shahrul Hisham Zainal Ariffin

Abstract Dental pulp stem cells from permanent (hDPSC) and deciduous teeth (SHED) are adult mesenchymal stem cells that are less invasive and easily available for cellular therapy. The characterization and osteogenic differentiation of hDPSC and SHED are important in manipulating these cells for regenerative medicine and dentistry. Objectives: This study is to determine the characterization and osteogenic potential of adult stem cells of permanent and deciduous dental pulp. Methods: Dental pulp was extracted from permanent and deciduous teeth. Both cells were treated with enzymatic digestion and cultured until passage 3. Morphology of the cells was recorded with cellB software. The proliferation rate of hDPSC and SHED was assessed by 3-(4,5-dimethylthiazol-2-yl)-2,5-diphenyltetrazolium (MTT) assay. Meanwhile, the osteoblast differentiation potential of hDPSC and SHED was determined by biochemical analysis. Results: Both types of cells exhibited fibroblast-like morphology at passage 3. The proliferation rate of SHED was significantly higher than hDPSC ($p < 0.05$). This study also showed that the osteoblast differentiation using alkaline phosphatase (ALP) assay was significantly higher in hDPSC compared to SHED. Conclusions: The characterization of hDPSC and

F. Yazid (✉) · N. A. Luchman · R. M. A. Wahab
Faculty of Dentistry, Universiti Kebangsaan Malaysia,
Jalan Raja Muda Abdul Aziz, 50300 Kuala Lumpur, Malaysia
e-mail: drfarinawati@ukm.edu.my

N. A. Luchman
e-mail: atmaliyaluchman@yahoo.com

R. M. A. Wahab
e-mail: rohaya@medic.ukm.my

S. H. Z. Ariffin
School of Bioscience and Biotechnology, Faculty of Science and Technology,
Universiti Kebangsaan Malaysia, 43600 Bangi, Selangor, Malaysia
e-mail: shahroy8@gmail.com

SHED exhibited fibroblast-like morphology at passage 3. Both cells capable of proliferating and differentiating into osteoblast with SHED have higher proliferating rate meanwhile hPDSC demonstrates better osteogenic potential. However, further studies need to be done to evaluate the quality of bone regeneration in 3-dimensional culture using scaffold for in vitro and in vivo studies between hDPSC and SHED.

Keywords Dental pulp · Mesenchymal stem cells · Fibroblast-liked Osteoblast differentiation · Proliferation

1 Introduction

Mesenchymal stem cells (MSCs) are capable of self-renewal and multilineage differentiation with the advantages of no ethical controversies compared to embryonic stem cells. MSCs are defined as multipotent cells that adhere to plastic, show a spindle-shaped or fibroblast-like morphology, express a particular set of surface antigens, and differentiate into adipocytes, chondrocytes, and osteocytes in vitro (Dominici et al. 2006). Bone marrow stem cells (BMSCs) and adipose stem cells (ASCs) are the most readily sources of MSCs and have relatively abundant progenitors. However, BMSCs showed a higher degree of commitment to differentiate into chondrogenic and osteogenic lineages which make it more superior MSCs than ACSs (Li and Ikehara 2013; Szpalski et al. 2012). Even though BMSCs are well documented, but due to donor site morbidity, difficulty in obtaining a sufficient number of cells and loss of phenotypic behavior during culturing, alternative sources of stem cells from other organ and tissue in the body, including MSCs from oral tissues such as dental pulp stem cells need to be investigated.

Dental pulp stem cells demonstrate the characteristic of BMSCs by exhibiting similar surface markers and matrix proteins associated with the formation of mineralized tissue such as alkaline phosphates, osteocalcin, and osteopontin (Ranganathan and Lakshminarayanan 2012). Unlike bone, dental tissues will not undergo continuous remodeling; thus, they could be more committed in their differentiation potential in comparison with BMSCs (Huang et al. 2006; Huang et al. 2009). This alternative sources of stem cells from dental pulp should be further investigated in terms of their isolation and characterization based on the properties found on MSCs isolated from the bone marrow.

The un-mineralized soft tissue residing in the inner structure of teeth known as dental pulp is a noninvasive and promising source of adult MSCs. Usually, exfoliated teeth or teeth extracted due to clinical reason such as suggested for orthodontics are uselessly discarded with little or no trauma. Furthermore, in the year of 2000, Gronthos et al. successfully isolated stem cells from the human dental pulp (hDPSC) derived from impacted third molars and demonstrate that these isolated cells show potency as clonogenic, highly proliferative, and capable of regenerating a new tissue which is the properties that effectively define them as stem cells. Three years later, Miura et al. (2003) reported the potential of stem cells from human exfoliated deciduous teeth (SHED) that is capable to regenerate the adequate amount of bone when transplanted to a mouse. Together, these studies indicate that dental pulp from permanent and exfoliated teeth was the best candidates to be an alternative source of adult multipotent MSCs.

Establish an understanding of characterization and osteogenic potential of dental stem cells are significance especially for future utilization for in vivo and clinical application in regenerative medicine which involve cellular approach of stem cells. Therefore in this study, we extracted dental pulp from two different sources mainly permanent and exfoliated deciduous teeth with the aim to determine the difference in terms of their morphology, proliferation, and osteoblast differentiation potential.

2 Methodology

2.1 Permanent and Deciduous Tooth Sample Collection

This study was approved by Research Ethics Committee of Universiti Kebangsaan, Malaysia (Ethical approval number, UKM PPI/111/8/JEP-2016-524) for using human dental pulp sample. Clinically healthy human permanent and deciduous teeth were obtained with the subject's consent (adults aged 18–30 years old for the permanent tooth) or subject's parent's consent (parents of a child aged 3–12 years old for the deciduous tooth) who meets all the criteria necessary to be sampled for this study. Extraction was performed by dental officers or specialist at Faculty of Dentistry, Universiti Kebangsaan, Malaysia, Kuala Lumpur and Universiti Kebangsaan Malaysia Medical Centre, Cheras, Kuala Lumpur.

2.2 Isolation and Culture of hDPSC and SHED

Dental pulp tissue was cut into small fragments with a sterile blade and digested in 3 mg/mL collagenase 1A (Sigma, USA) in Knockout Dulbecco Modified Eagle Medium (KO-DMEM) at 37 °C for 40 min with every 15 min vortex to accelerate the digestion process. After 40 min, the enzyme action was neutralized with the addition of 10% (v/v) fetal bovine serum (FBS) (Gibco, Grand Island, NY, USA). Cells obtained were seeded in complete medium consist of KO-DMEM, 10% FBS and 1% (v/v) penicillin-streptomycin (HI Media, Mumbai, India) and 0.01x Glutamax (Gibco, Grand Island, NY, USA) in the T75 cm^2 culture flask and incubated at 37 °C, 95% (v/v) humidity and 5% (v/v) CO_2. Complete medium was changed after culturing for 24 h to remove non-adherent cells followed by every three days until cells reached 80–90% confluency.

2.3 Proliferation Assay of hDPSC and SHED

The proliferation ability of hDPSC and SHED was assessed using 3-(4,5-dimethylthiazol-2-yl)-2,5-diphenyltetrazolium (MTT) assay. Confluent cells were detached from the culture flask using 0.25% (v/v) Trypsin-EDTA and counted using hemocytometer. Cells were seeded at 1×10^4 cell/cm^2 for 9 days analysis with three-day interval (Day 0, 3, 6, and 9). MTT solution and complete media (ratio 1: 9) were added to each well-containing cells and incubated at 37 °C for 4 h. Formazan crystal formed as a result of cell metabolism was diluted in dimethyl sulfoxide (DMSO) with glycine buffer at pH 7.4. The absorbance reading at 570 nm with a reference wavelength of 655 nm was taken using ELISA microplate reader (Bio-Rad, USA). Absorbance values were plotted against day of analysis.

2.4 Osteoblast Differentiation

A total of 5000 cells/cm^2 hDPSC and SHED were cultured in 96 well plates and incubated for 24 h at 37 °C, 95% humidity and 5% CO_2. After 24 h of incubation, complete medium was removed and replaced with osteoblast differentiation medium, which consists of complete medium supplemented with 50 mg/mL ascorbic acid and 10 mM (w/v) β-glycerophosphate. Also, a complete medium without differentiation factors will be used as a negative control. The medium will be changed every three days.

2.5 Alkaline Phosphatase (ALP) Assay of Differentiated Cells

Differentiated hDPSC and SHED were analyzed every three days during the differentiation period and stored at −20 °C. Differentiated cells were harvested using a lysis buffer containing 0.1% (v/v) Triton X-100 (Sigma, USA) in cold Tris-buffered saline (TBS). ALP assay was performed by incubating the lysed cells with 0.1 M sodium bicarbonate–sodium carbonate buffer (pH 10.0) (MERCK, Germany) containing 0.1% (v/v) Triton X-100, 2 mM (w/v) magnesium sulfate (Sigma, USA), and 6 mM (w/v) p-nitrophenyl phosphate (pNPP) (Sigma, USA) and incubated for 30 min at 37 °C. The reaction was stopped by the addition of 1 M sodium hydroxide, and the absorbance measurement was taken at a wavelength of 405 nm with a spectrophotometer. Cellular ALP activity was determined by plotting a graph of specific activity against days of analysis.

2.6 Statistical Analysis

Results obtained are shown as mean ± standard deviation from experiments conducted in triplicate. Statistical significance was analyzed using Student's t-test. P values less than 0.05 were considered to be significant.

3 Results

3.1 Morphological Characterization of the Cells

Morphological characterization of hDPSC and SHED was analyzed using a cellB software. Figure 1 represents both cells at the various stages of confluency. The isolated hDPSC and SHED were observed to able to form a colony and exhibit fibroblast-like and epithelial-like morphology at the beginning of culture (Fig. 1a, d). After 4–5 weeks after isolation (initial plating), the cells reach about 80–90% confluency which makes it ready to be trypsinized at 0.25% concentration and transfer into a new culture flask (Fig. 1c, e). Both types of cells appear to be fibroblast-like morphology toward their confluency.

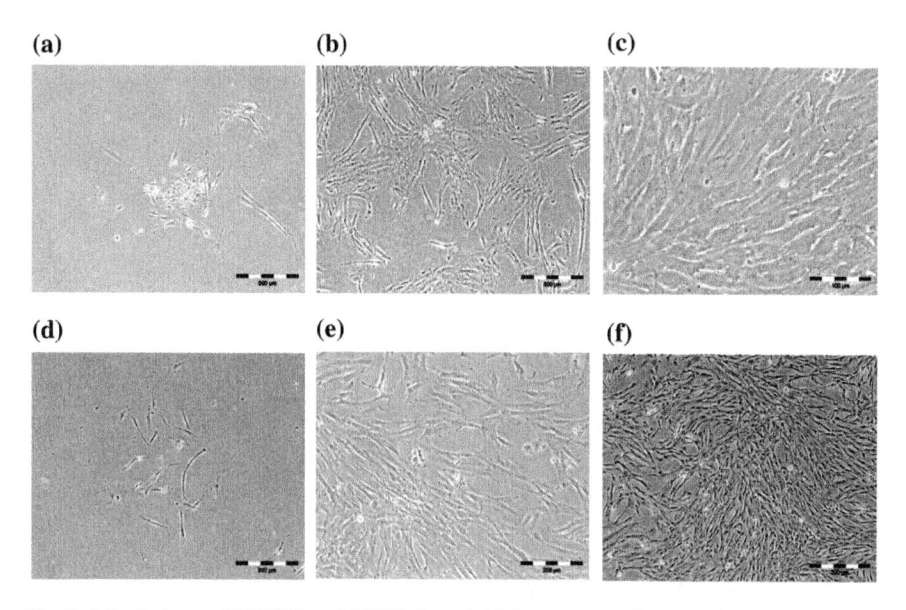

Fig. 1 Morphology of hDPSC and SHED from initial passage until confluency (Magnification 100 and 200 μm). **a** The formation of colonies hDPSC in the first week after isolation. **b** hDPSC subconfluent on day 16 of culture. **c** hDPSC confluent after 5 weeks of isolation. **d** SHED colony formation in the first week after isolation. **e** SHED subconfluent after 16 days of culture. **f** Confluent SHED after 4 weeks of isolation. A heterogeneous population of cells appears during the initial plating, but both types of cells exhibit fibroblast-like morphology when approaching confluency

3.2 Proliferation Ability of hDPSC and SHED

Proliferation analysis was assessed by MTT assay for 9 days of culture at passage 3. Statistical analysis was conducted by Student's t-test for comparison of proliferation from day 0 to 9 for hDPSC and SHED ($p < 0.05$, $n = 3$). The analysis result revealed that isolated hDPSC and SHED have a higher proliferative capability. However, the proliferation ability of SHED was significantly higher in comparison with hDPSC with more significant on day 3 and 6 (Fig. 2).

3.3 Osteoblast Differentiation of hDPSC and SHED

Alkaline phosphatase (ALP) assay was performed to evaluate the osteoblast differentiation of hDPSC and SHED. Comparison of ALP activity between the cells with the addition of osteoblast differentiation factor at day 0 (as control) and differentiated hDPSC and SHED was performed. ALP activity, which represents osteoblast marker, shows statistically significant difference on day 3, 6, and 9 in favor of differentiated cells ($p < 0.05$, $n = 3$) (Fig. 3). Therefore within 9 days of osteoblast differentiation, both types of cells able to differentiate into osteoblast with the potential of hDPSC to differentiate more prominent compared to SHED.

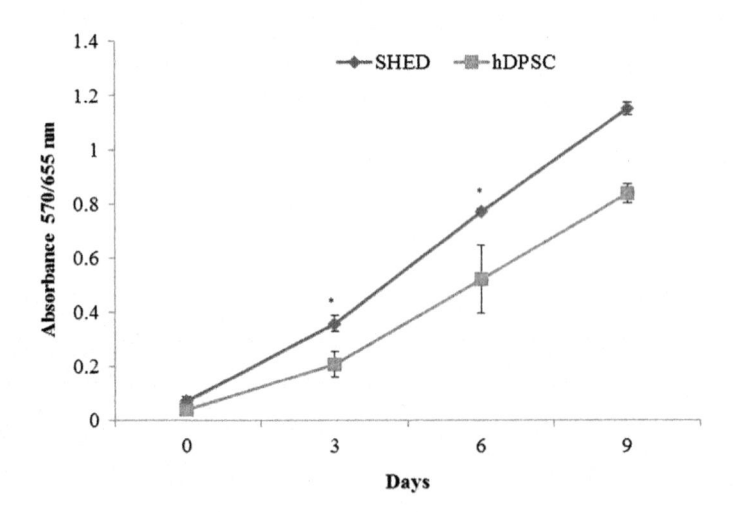

Fig. 2 Analysis of proliferation of hDPSC and SHED using an MTT assay. Both cells were cultured in complete medium for 9 days, and MTT assay was performed every three days interval. Absorbance at 570 nm was counted and plotted against days of analysis. Data are expressed as the mean ± standard deviation ($p < 0.05$, $n = 3$)

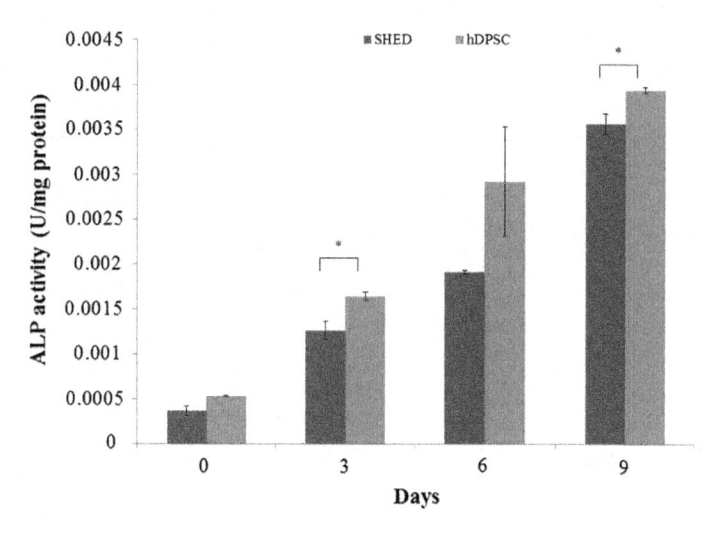

Fig. 3 Alkaline Phosphatase (ALP) profile of hDPSC and SHED in osteoblast differentiation medium. Comparison of data between differentiated hDPSC and SHED showed significant (*) on day 3 and 9. Statistical analysis was conducted using Student's t-test for comparison of ALP activity of hDPSC and SHED during days of analysis ($p < 0.05$, $n = 3$)

4 Discussion

The availability of resources is one of the main criteria for the selection of an alternative source of adult stem cells. Permanent and deciduous teeth are both accessible sources of adult mesenchymal stem cells. Extraction of permanent and deciduous teeth for obtaining hDPSC and SHED has an advantage over other adult stem cells as it is easy to access and can be obtained during adult life; meanwhile, exfoliated deciduous teeth can be secured at a very young age and the cells obtained can be stored for future use. Both types of teeth can be extracted with minimizing trauma experience to the donor. Most research on isolation of dental pulp stem cells has been carried out using different enzymatic treatments on the pulp which function to break down the extracellular matrix of tissue in a short period, thus obtaining single cell suspensions. de Souza et al. (2015) highlight the involvement of different enzyme for pulp digestion methods such as collagenase, dispase, trypsin, or combination of it. Based on this study, we found that enzymatic digestion of the pulp using collagenase 1A results in the release of a heterogeneous population of cells. Within one week after isolation, hDPSC and SHED capable of forming colonies that contained fibroblast-like and endothelial-like cells. This result is consistent with those of Huang et al. (2006) which stated that the time required for the hDPSC to form colonies might vary from one to two weeks. After the first week of culture, both types of cells become elongated indicating the start of fibroblast-like shape formation, while the cells were completely fibroblast-like shaped at the second week. Even though permanent teeth are significantly different from deciduous teeth in terms of developmental processes, tissue structure, and function, both type of cells isolated from these teeth in this study capable to adhere to plastic and exhibit fibroblast-like morphology during macroscopic examinations which is the main characteristic of mesenchymal stem cells during in vitro culture (Dominici et al. 2006; Miura et al. 2003; Mushegyan et al. 2014).

In this study, we compared proliferation ability between hDPSC and SHED within 9 days of analysis using 3-(4,5-dimethylthiazol-2-yl)-2,5-diphenyltetrazolium (MTT) assay. Those two dental stem cells are capable of forming colonies with extensive proliferation ability. However, cells isolated from deciduous teeth showed significantly higher proliferative capability compared to cells from permanent teeth when cultured in complete medium. Previous studies by Miura et al. (2003) and Tatullo et al. (2015) demonstrate that the proliferation of SHED is significantly high than that of hDPSC. Our results showed that SHED possessed similar cell proliferation ability as demonstrate by Miura et al. (2003) and Tatullo et al. (2015).

Osteoblast differentiation potential of hDPSC and SHED was evaluated by alkaline phosphatase (ALP) assay. ALP serves as a predictive parameter for a bone formation which functions in the mineralization of hard tissue as it provides free phosphate for the formation of hydroxyapatite crystals and hydrolysis pyrophosphate, an inhibitor of bone matrix formation (Štefková et al. 2015). Štefková et al. (2015) also emphasize the association of ALP activity with its expression and regulation especially mediated by the actual microenvironment rather than by some individual signaling pathways. Thus, ALP activity is regulated mainly through the developmental status of cells. Therefore, as stated above, ALP expression is a generally suitable marker of osteoblast differentiation processes both in vitro and in vivo for particular cell types. In the present study, we isolated hDPSC and SHED and demonstrated that they both had the ability to form colonies with fibroblast-like morphology, high proliferative ability and capable of differentiating into osteoblast. Analysis for ALP activity for both types of cells showed statistically significantly different on day 3, 6, and 9 between cells under osteoblast induction and the control group (Day 0). Moreover, when hDPSC and SHED were induced to differentiate into osteoblast using differentiation factors that consist of ascorbic acid and β-glycerophossphate, hDPSC is able to perform better than SHED with significant ALP activity at day 3 and 9. Within 9 days of osteoblast differentiation, both types of cells able to express ALP with hDPSC showed a higher potential of osteoblast differentiation. However, for a better ALP expression osteoblast differentiation needs to be done with a longer duration instead of 9 days of stimulation. This is because studies by Nourbakhsh et al. (2008), Raouf and Seth (2002), and Thomas et al. (2002) suggest that ALP activity for dental stem cells mostly occurring between 11 and 25 day; meanwhile, SHED ALP activity may increase after three weeks of osteoblast stimulation. Thus, longer duration of osteoblast differentiation may result in better ALP expression of hDPSC and SHED.

5 Conclusion

The characterization of hDPSC and SHED exhibited fibroblast-like morphology at passage 3. Both cells have osteogenic potential with SHED has higher proliferation rate; meanwhile, hPDSC shows better osteogenic potential ($p < 0.05$, $n = 3$). However, further studies need to be done to evaluate the quality of bone regeneration in 3-dimensional culture using scaffold for in vitro and in vivo studies between hDPSC and SHED.

Acknowledgements This study was supported by grants from Malaysia Ministry of Higher Education (MOHE) (FRGS/1/2015/SG05/UKM/02/2) and Universiti Kebangsaan Malaysia under Young Researcher Incentive Grant (GGPM-2015-006).

References

de Souza, L. M., Bittar, J. D., da Silva, I. C. R., de Toledo, O. A., de Macedo, Brígido M., & Fonseca, M. J. P. (2015). Comparative isolation protocols and characterization of stem cells from human primary and permanent teeth pulp. *Brazilian Journal of Oral Sciences, 9*(4), 427–433.

Dominici, M., Le, Blanc K., Mueller, I., Slaper-Cortenbach, I., Marini, F., Krause, D., et al. (2006). Minimal criteria for defining multipotent mesenchymal stromal cells. The International Society for cellular therapy position statement. *Cytotherapy, 8*(4), 315–317.

Gronthos, S., Mankani, M., Brahim, J., Robey, P. G., & Shi, S. (2000). Postnatal human dental pulp stem cells (DPSCs) in vitro and in vivo. *Proceedings of the National Academy of Sciences, 97*(25), 13625–13630.

Huang, G.-J., Gronthos, S., & Shi, S. (2009). Mesenchymal stem cells derived from dental tissues vs. those from other sources: Their biology and role in regenerative medicine. *Journal of Dental Research, 88*(9), 792–806.

Huang, G. T.-J., Sonoyama, W., Chen, J., & Park, S. H. (2006). In vitro characterization of human dental pulp cells: Various isolation methods and culturing environments. *Cell and Tissue Research, 324*(2), 225–236.

Li, M., & Ikehara, S. (2013). Bone-marrow-derived mesenchymal stem cells for organ repair. *Stem Cells International 2013*.

Miura, M., Gronthos, S., Zhao, M., Lu, B., Fisher, L. W., Robey, P. G., et al. (2003). SHED: Stem cells from human exfoliated deciduous teeth. *Proceedings of the National Academy of Sciences, 100*(10), 5807–5812.

Mushegyan, V., Horst, O., & Klein, O. D. (2014). Adult stem cells in teeth. In *Adult stem cells* (pp. 199–216). Springer.

Nourbakhsh, N., Talebi, A., Mousavi, B., Nadali, F., Torabinejad, M., Karbalaie, K., et al. (2008). Isolation of mesenchymal stem cells from dental pulp of exfoliated human deciduous teeth. *Cell J, 10*(2), 101–108.

Ranganathan, K., & Lakshminarayanan, V. (2012). Stem cells of the dental pulp. *Indian Journal of Dental Research, 23*(4), 558.

Raouf, A., & Seth, A. (2002). Discovery of osteoblast-associated genes using cDNA microarrays. *Bone, 30*(3), 463–471.

Štefková, K., Procházková, J., & Pacherník, J. (2015). Alkaline phosphatase in stem cells. *Stem Cells International 2015*.

Szpalski, C., Barbaro, M., Sagebin, F., & Warren, S. M. (2012). Bone tissue engineering: current strategies and techniques—Part II: Cell types. *Tissue Engineering Part B: Reviews, 18*(4), 258–269.

Tatullo, M., Marrelli, M., Shakesheff, K. M., & White, L. J. (2015). Dental pulp stem cells: Function, isolation and applications in regenerative medicine. *Journal of Tissue Engineering and Regenerative Medicine, 9*(11), 1205–1216.

Thomas, C. H., Collier, J. H., Sfeir, C. S., & Healy, K. E. (2002). Engineering gene expression and protein synthesis by modulation of nuclear shape. *Proceedings of the National Academy of Sciences, 99*(4), 1972–1977.

Detection of Nicotine in Nicotine-Free E-Cigarette Refill Liquid Using GC–MS

Reena Abd Rashid, Asmira Nabilla Adnan, Sohehah Maasom and Gillian Taylor

Abstract An electronic cigarette (e-cigarette) is a device which contains a battery, cartridge and heating elements that vaporizes liquid for inhalation. The popularity of the e-cigarette has been on the rise since 2001 in Malaysia, and it has become an alternative to conventional tobacco smoking. As per today, neither the FDA nor any international and/or local organizations have set regulations for refill liquid's compound limit, including nicotine. Refill liquid available in the local market, Malaysia, has 3–24 mg nicotine concentration. However, sellers and users have the ability to alter the nicotine concentration in the refill liquid. Quantification of nicotine in electronic cigarette refill liquid is necessary. This study aims to determine the concentration of nicotine in nicotine-free electronic cigarette refill liquid from the local market in Malaysia using GC–MS. Seven samples of nicotine-free refill liquid from different brands but with same flavour were purchased in the Selangor area, Malaysia. Sample was diluted with dichloromethane and analyzed using GC–MS. Nicotine concentration standards were plotted in calibration curves range 5–100 ppm. The results conclude to found traces of nicotine in two out of seven samples with nicotine concentration value 16.38 ppm and 9.63 ppm. T test was conducted on the concentration value with significant difference with $p < 0.5$.

Keywords Nicotine · E-cigarette · GC–MS

R. A. Rashid (✉)
Faculty of Applied Sciences, Universiti Teknologi MARA,
40450 Shah Alam, Malaysia
e-mail: reena1572@salam.uitm.edu.my

A. N. Adnan · S. Maasom
Chemistry (Forensic Analysis) Programme, Faculty of Applied Sciences,
Universiti Teknologi MARA, 40450 Shah Alam, Malaysia

G. Taylor
School of Science & Engineering, Teesside University, Middlebrough, UK

© Springer Nature Singapore Pte Ltd. 2018
R. Saian and M. A. Abbas (eds.), *Proceedings of the Second International Conference on the Future of ASEAN (ICoFA) 2017 – Volume 2*,
https://doi.org/10.1007/978-981-10-8471-3_61

1 Introduction

An electronic cigarette is a battery-powered device that is designed to transfer the content of the refill liquid in aerosol form to the respiratory tract of its users (Bansal and Kim 2016). Since the past decade, the number of e-cigarette users has increased compared to tobacco cigarette users globally. In 2015, it was reported approximately 3.2% of Malaysia populations (31.2 million) are regular user of e-cigarette where one-tenth of the users were high school students (Ganasegeran and Rashid 2016). Studies have shown that adolescent users select e-cigarettes over conventional cigarettes believing that it is safer than tobacco products because the refill liquid is water based and helps to reduce craving for nicotine (Gorukanti et al. 2017). Furthermore, many users have seen it as an alternative for quitting smoking tobacco but it is debated as to the role of e-cigarettes as an effective cessation tool efficacy (Schaller et al. 2013). Contradictory population-based studies on e-cigarettes indicate that users most commonly use it concurrently with conventional tobacco cigarettes (dual use) even after 1-year period, not ceasing the usage of tobacco cigarette (Etter and Bullen 2011; Goniewicz et al. 2013).

E-cigarette refill liquid contains propylene glycol, glycerine, nicotine and a flavouring reagent that creates a visible aroma of vapour when it is heated by the heating coil in the device (Goniewicz et al. 2015). The refill liquid available in the market specifically in Malaysia today are produced domestically and not manufactured by certified entity. The manufacturers of the raw materials and product are largely from China, Europe and USA. Some local seller will purchase the raw materials and form the final product in their shop without allocating a standard quality control procedures on the manufacturing products. The manufacturing label on the product content has also been reported to differ significantly from the actual content due to lack of standard and regulation by the Malaysia Ministry of Health (MOH).

Users can select nicotine content in the refill liquid upon purchased without going through any certified health organization for approval. Currently, there are no regulations set by the FDA and Pharmaceutical Services Division, Ministry of Health Malaysia on the production of e-cigarette and the refill liquid content validity. E-cigarette manufacturers can even advertise products without contravening rules of the government tobacco advertising restriction. Refill liquid can be flavour-less or have a fruity and sweet essence with nicotine range between high (36 mg/mL nicotine) to medium (16 mg/mL nicotine) to none (0 mg/mL nicotine). In Malaysia, the maximum permitted levels of tar and nicotine set by the National Pharmaceutical Regulatory Agency are 20 mg/g and 1.5 mg/g, respectively, for each stick of cigarette. There were no permitted levels disclosed for nicotine in e-cigarette refill liquid content. Refill liquid that are labelled as nicotine-free may contain traces of nicotine from nicotine contamination from the poor handling and storage of the materials and finished products (Goniewicz et al. 2015).

Nicotine (3-(1-methyl-2-pyrrolidinyl) pyridine) is an alkaloid that can be found a leaves of *Nicotiana tabacum* plants (Hossain and Salehuddin 2013). Proof of nicotine as a potential health hazards has been documented throughout the years where the effects extend from latent psychotic behaviour in adolescent to a risk factor of attention deficit hyperactivity disorder (ADHD) syndrome in prenatal exposure (Elgayar et al. 2016; Zhu et al. 2017). Presence of inorganic compounds and flavour chemicals in the refill liquid and tobacco cigarette boosts nicotine's pH and in return increases nicotine adsorption level (Stanfill et al. 2009). Nicotine can be readily absorbed through all routes of exposure including on skin which illustrates nicotine-patch application using transdermal penetration mechanism (Maina et al. 2017).

Majority of studies have focused on the determination of contaminants, toxicants and trace metals content and its effect in the biological system (Bahl et al. 2012; Behar et al. 2014; Bansal and Kim 2016; Beauval et al. 2016; Noble et al. 2017). Fewer studies have focused on the determination of nicotine content in the refill liquid. Past studies have analyzed nicotine using GC–FID, GC–MS and HPLC MS/MS on both biological and chemical samples. (Magni et al. 2016; El Golli et al. 2016). GC–FID was the first method to be verified as a robust method for nicotine analytical testing on tobacco and smokeless tobacco sample set by Centers for Disease Control in 1999. Over time, sophisticated flavours and complex chemical compound have emerged in the tobacco sample and these compounds interfere with detection of nicotine using GC–FID. The complex compounds and flavours can co-elute with nicotine and nicotine constituents that cause problems for the accurate determination of nicotine content (Burdock 2005). On the other hand, direct-injection GC–MS has shown to produce high sensitivity for nicotine chromatography peak confirmation in refill liquid, in both raw and certain highly flavoured tobacco products. GC–MS running time recorded varies from 6 to 20 min per samples with most studies shows elution time at 6–7.4 min compared to GC–FID which maximum time observed thus far is 27 min per sample (Hong et al. 2017; Magni et al. 2016; Hossain and Salehuddin 2013; Stanfill et al. 2009).

The objective of this study is to quantify the amount of nicotine in nicotine-free refill liquid brands sold in Selangor, Malaysia, using GC–MS. The findings of this study would assist to verify the concentration of nicotine in electronic cigarette refill liquid as stated on the manufacturer labelling.

2 Methodology

2.1 Sample Preparation for GC Analysis

Seven nicotine-free refill liquid of different brands (different manufacturer) with the same mango flavour was purchased from local retail stores in different parts of Selangor, Malaysia. Sample was acquired in September 2016 and was kept at room temperature pending analysis.

The following procedure was adopted from Pagano et al. (2015) with slight modification. Samples were prepared by adding 15 μL of refill liquid directly into GC vial with 1500 μL of dichloromethane. Sigma Aldrich analytical standard grade dichloromethane (GC purity ≥ 99.9%) was purchased from Merck (Darmstadt, Germany). All vials were sonicated for 20 min and then vortexed for 10 s at 3000 rpm to ensure mixing. Dichloromethane was used instead of using methanol for GC analysis sample preparation to reduce solvents effect. A total of three replicates were prepared for each brand.

2.2 Nicotine Standard for Preparation for GC Analysis

Liquid nicotine standard (purity ≥ 99.9%) was purchased from Merck (Darmstadt, Germany) with pH 10.2. Nicotine standards were prepared in 10 mL volumetric flask with dichloromethane. The standards were prepared in 5–100 ppm range to obtained calibration curve.

2.3 GC–MS

Gas chromatography (Agilent Technologies 6890 N) was used to analyze nicotine concentration of refill liquid and nicotine standards. 1 μL of each standard and sample was injected using an auto-sampler in triplicate and passed through a 5% phenyl 95% dimethylpolysiloxane capillary column (30 m × 250 μm × 0.25 μm HP5-MS capillary column)

The instrument was set up using helium gas and sample injection temperature 230 °C at split mode (1:50). The oven temperature upon injection was 80 °C and then increased at 25 °C/min to the final temperature of 245 °C. The final temperature was held for 4.5 min. Total running time for each sample was approximately 11 min.

The ion source for MS was kept at 180 °C, while the transfer line between GC and MS was held at 280 °C. The MS operated in total ion monitoring mode.

2.4 Quantitative Analysis

Internal standard was not used in this study. Quantitative analysis is achieved through this method. All samples were spiked with 60 ppm standard nicotine and run through GC–MS using the same conditions. The RF value and peak area obtained are used to compute nicotine concentration in % recovery.

2.5 LOD and LOQ

LOD and LOQ were calculated to ensure method reliability on the analytical procedures used under the defined conditions. Both LOD and LOQ values were obtained from the nicotine concentrations calibration curve and standard deviations, whose response provided a signal-to-noise (S/N) ratio of 3 and 10, respectively, as determined from the least abundant qualifier ion.

3 Results and Discussion

3.1 Standard of Nicotine

The presence of nicotine in refill liquid was determined by using GC–MS. Nicotine standard was detected at 4.294 min using GC–MS. A calibration curve (Fig. 1) was plotted from series of nicotine standards chromatogram range from 5 to 100 ppm with linearity value, $R^2 = 0.9964$. The standard was diluted using only dichloromethane. Dichloromethane boiling point is 39.6 °C and nicotine is at 247 °C, thus no overlapping with nicotine peaks. The series was injected using an auto-sampler of GC–MS.

3.2 Method Validation

Method validation is necessary to ensure that reliable analytical procedures are used under the defined conditions. To validate the GCMS method, limit of detection (LOD) and limit of quantification (LOQ) were calculated. The slope of the

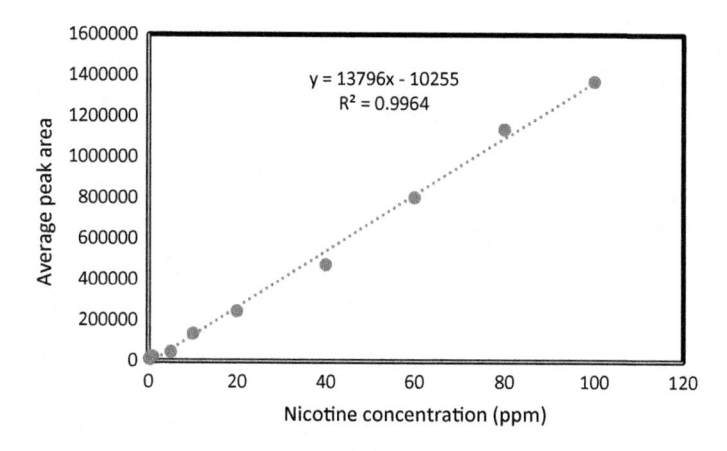

Fig. 1 Calibration curve of nicotine (0.3–100 ppm)

calibration curve was determined. The LOD of standard nicotine was found to be 3.48 ppm and LOQ, 11.6 ppm, respectively. The method was shown to be linear in the range of 5–100 ppm. Total ion monitoring modes of MS operation was used to quantify nicotine.

3.3 Nicotine Concentration in Sample

Nicotine peak appeared as 3-(1-methylpyrrolidin-2-yl) pyridine by MS with 4.201 min mean retention time. Samples labelled 1, 2, 3, 5 and 7 show no peaks of 3-(1-methylpyrrolidin-2-yl) pyridine detected. The determined relative concentration in Sample 4 is 16.38 ppm ± 1.27 ppm and in Sample 6, the value is 9.63 ppm ± 1.2 ppm (Fig. 2).

From this study, five out of the seven samples were accurately labelled as nicotine-free product. Only two out of seven samples contain trace of nicotine. All seven samples used similar flavouring which is mango. The effect of flavourings on nicotine detection are not studied and/or considered as a factor in this study. The significance differences between seven samples were calculated using t test and obtain p value < 0.5.

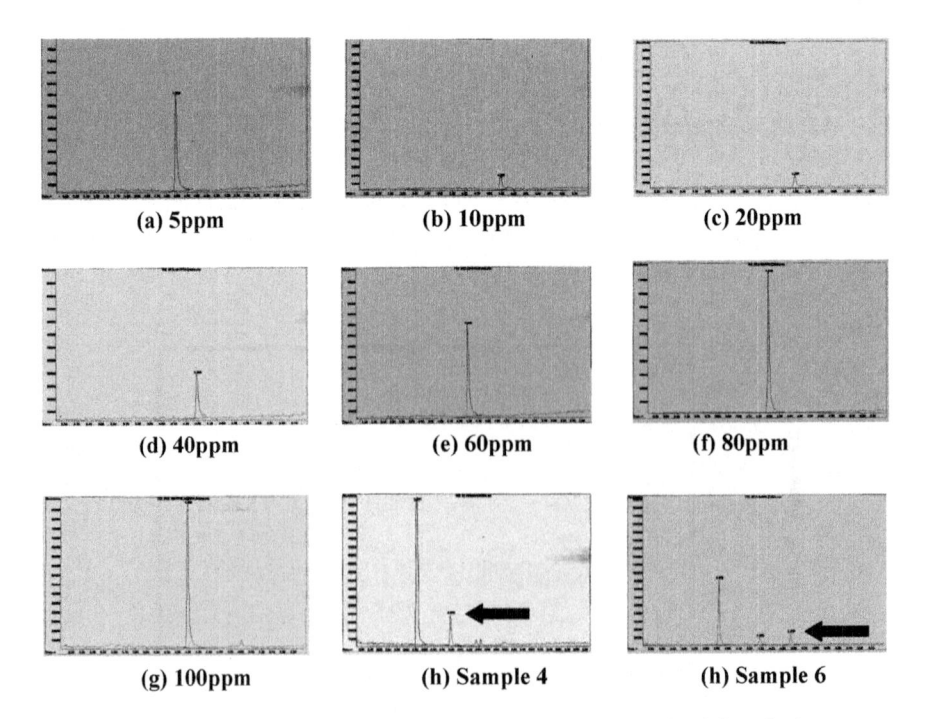

Fig. 2 GC–MS nicotine chromatogram of standard samples, Sample 4 and Sample 6

Nicotine concentration detected in all the sample was below the permitted levels set by National Pharmaceutical Regulatory Agency (Malaysia) which are 0.01638 mg/g (16.38 ppm) and 0.00963 mg/g (9.63 ppm). The permitted levels are 1.5 mg/g of nicotine content per cigarette. Nicotine can be potentially lethal at 30–60 mg or 0.8–1.0 mg/kg of adult body weight though it is rare to be intoxicated with high level of nicotine unless ingested (Morean et al. 2016). Both samples that contain traces of nicotine was a finish product purchased from an authorized seller in a licensed shop whereas other five samples were purchase from third-party seller that prepare the refill liquid manually upon purchase by adding the propylene glycol, glycerine, flavourings and nicotine level desired by buyer.

Presence of nicotine in two out of seven samples can be caused by various source of nicotine contamination. There is a possibility of contamination from the utensils or equipment used to load the refill liquid in the bottle or possibility of using reused bottle case that still have traces of nicotine (Copeland et al. 2017; Stanfill et al. 2009). Other suggested due to transfer of nicotine on packaging material and nicotine contamination between raw materials (Goniewicz et al. 2015; Benowitz and Goniewicz 2013). Regardless, these still show that nicotine can be present in nicotine-free refill liquid and the manufacturing process does not meet its standards. And currently there is no official governing body responsible to set the guidelines and become the regulators in this matter.

3.4 Quantitative Analysis

Samples without nicotine detection (Sample 1, 2, 3, 5 and 7) were spiked with 60 ppm nicotine and the percentage of recovery was calculated in Table 1. Samples were run through GC–MS using the same conditions to confirm that the sample does not contain any nicotine. Peak of nicotine appears at Sample 1 is 4.296 min, Sample 2 is 4.298 min, Sample 3 is 4.297 min, Sample 5 is 4.297 min and Sample 7 is 4.298 min.

Four out of five samples have less than 50% of nicotine percentage recovery. Loss of nicotine analyte can be due to the degradation of nicotine to nitrosamines metabolites such as cotinine after dilution and due to sample exposure to light for more than 1–2 h. (Bhalala 2003; AHFS 2009). It is unlikely for nicotine to be

Table 1 Nicotine recovery from samples

Sample	Average peak area	Concentration (ppm)	% recovery	% RSD
1	142560	28.51	47.52	0.69
2	138251	27.65	46.08	0.03
3	275240	55.05	91.75	0.29
5	112582	22.51	37.52	0.66
7	142084	28.42	47.36	0.06

Table 2 Retention time of standard and samples

Sample	Retention time (min)	Peak area (Hz * s)	Response factor (RF) (ppm/Hz * s)	Calculated nicotine concentration (ppm)
Nicotine standard (60 ppm)	4.293	286701	0.0002	57.34

volatile during the study as the boiling point of nicotine is at 244 °C and the temperature GC use in this study was gradually increased from 80 to 245 °C (Lide 2007).

From the data calculation, the response factor (RF) of standard nicotine was 0.0002 ppm/Hz * s. Using RF, the amount of nicotine present can be determined using the equation, RF multiply with the peak area of nicotine in sample (Table 2).

4 Conclusion and Recommendation

Nicotine can be detected and quantified in the refill liquid using gas chromatography mass spectrometry (GC–MS). The value detected in the sample was too low to cause adverse health effect if consumed. The health effects might be significant if consumed in redundant.

For recommendation, more sample brands should be used to represent equal distribution of the data of the refill liquid in Selangor, Malaysia. Other recommendation is the sampling should include refill liquid's flavour variety as a factor to observe whether a different type of flavour contributes to any significance difference for nicotine analysis.

Acknowledgements We would like to extend our appreciation to the laboratory staff of Forensic Science Analytical lab in Faculty of Applied Sciences, UiTM Shah Alam, for the assistance provided. We also would like to express our gratitude to the reviewers for their helpful comments and to Mr. M.A. Jamal for his wise insights.

References

American Society of Health System Pharmacists. (2009). *AHFS drug information 2009*. Bethesda. http://toxnet.nlm.nih.gov/cgibin/sis/search/. Accessed December 20, 2016.

Bahl, V., Lin, S., Xu, N., Davis, B., Wang, Y., & Talbot, P. (2012). Comparison of electron cigarette refill fluid cytotoxicity using embryonic and adult models. *Reproductive Toxicology, 34*(4), 529–537.

Bansal, V., & Kim, K. (2016). Review on quantitation methods for hazardous pollutants released by e-cigarette (EC) smoking. *TrAC Trends in Analytical Chemistry, 78*, 120–133.

Beauval, N., Howsam, M., Antherieu, S., Allorge, D., Soyez, M., Garçon, G., et al. (2016). Trace elements in e-liquids—Development and validation of an ICP-MS method for the analysis of electronic cigarette refills. *Regulatory Toxicology and Pharmacology, 79,* 144–148.

Behar, R. Z., Davis, B., Wang, Y., Bahl, V., Lin, S., & Talbot, P. (2014). Identification of toxicants in cinnamon-flavored electronic cigarette refill fluids. *Toxicology in Vitro, 28*(2), 198–208.

Bhalala, Oneil. (2003). Detection of cotinine in blood plasma by HPLC MS/MS. *MIT Undergraduate Research Journal, 8,* 45–50.

Burdock, G. A. (2005). *Fenaroli's handbook of flavor ingredients* (3rd ed., pp. 63, 88, 95, 113, 136, 196–197, 263, 274, 287, 561). Boca Raton, Florida: CRC Press.

Copeland, A., Peltier, M., & Waldo, K. (2017). Perceived risk and benefits of e-cigarette use among college students. *Addictive Behaviors, 71,* 31–37.

Elgayar, S., Hussein, O., Abdel-Hafez, A., & Thabet, H. (2016). Nicotine impact on the structure of adult male guinea pig auditory cortex. *Experimental and Toxicologic Pathology, 68*(2–3), 167–179.

El Golli, N., Dkhili, H., Dallagi, Y., Rahali, D., Lasram, M., Bini-Dhouib, I., & Asmi, M. A. (2016). Comparison between electronic cigarette refill liquid and nicotine on metabolic parameters in rats. *LifeSci, 146,* 131–138. https://doi.org/10.1016/j.lfs.2015.12.049.

Etter, J. F., & Bullen, C. (2011). Electronic cigarette: Users profile, utilization, satisfaction and perceived efficacy. *Addiction, 106,* 2017–2028.

Ganasegeran, K., & Rashid, A. (2016). Clearing the clouds—Malaysia's vape epidemic. *The Lancet Respiratory Medicine, 4*(11), 854–856.

Goniewicz, M. L., Lingas, E. O., & Hajek, P. (2013). Patterns of electronic cigarette use and user beliefs about their safety and benefits: An internet survey. *Drug and Alcohol Review, 32,* 133–140.

Goniewicz, M., Gupta, R., Lee, Y., Reinhardt, S., Kim, S., Kim, B., et al. (2015). Nicotine levels in electronic cigarette refill solutions: A comparative analysis of products from the US, Korea, and Poland. *International Journal of Drug Policy, 26*(6), 583–588.

Gorukanti, A., Delucchi, K., Ling, P., Fisher-Travis, R., & Halpern-Felsher, B. (2017). Adolescents' attitudes towards e-cigarette ingredients, safety, addictive properties, social norms, and regulation. *Preventive Medicine, 94,* 65–71.

Hossain, A. M., & Salehuddin, S. M. (2013). Analytical determination of nicotine in tobacco leaves by gas chromatography–mass spectrometry. *Arabian Journal of Chemistry, 6*(3), 275–278.

Hong, Y., Kim, K., Sang, B., & Kim, H. (2017). Simple quantification method for N-nitrosamines in atmospheric particulates based on facile pretreatment and GC-MS/MS Environmental Pollution (in Press) (Corrected proof).

Lide, D. R. (2007). *CRC handbook of chemistry and physics* (88th ed., pp. 3–386). Boca Raton: CRC Press, Taylor & Francis.

Magni, P., Pazzi, M., Vincenti, M., Alladio, E., Brandimarte, M., & Dadour, I. (2016). Development and validation of a GC–MS method for nicotine detection in Calliphora vomitoria (L.) (Diptera: Calliphoridae). *Forensic Science International, 261,* 53–60.

Maina, G., Castagnoli, C., Ghione, G., Passini, V., Adami, G., Filon, F. L., et al. (2017). Skin contamination as pathway for nicotine intoxication in vapers. *Toxicology in Vitro, 41,* 102–105.

Morean, M. E., Kong, G., Cavallo, D., Camenga, D. R., & Krishna-Sarin, S. (2016). Nicotine concentration of e-cigarettes used by adolescents. *Drugs and Alcohol Dependance, 167,* 224–227.

Noble, M., Longstreet, B., Hendrickson, R. G., & Gerona, R. (2017). Unintentional pediatric ingestion of electronic cigarette nicotine refill liquid necessitating intubation. *Annals of Emergency Medicine, 69*(1), 94–97.

Pagano, T., Bida, M. R., & Robinson, R. J. (2015). Laboratory activity for determination of nicotine in electronic cigarette liquid using gas chromatography-mass spectrometry. *Journal of Chemical Education, 3*(3), 37–43. https://doi.org/10.5923/j.jlce.20150303.01.

Schaller, K., Ruppert, Kahnert, S., Bethke, C., Nair, U., & Potschke-Langer, M. (2013). Electronic cigarettes: An overview. *Red Series Tobacco Prevention and Tobacco Control, 19*. Heidelberg, Germany: German Cancer Research Center (DKFZ).

Stanfill, S. B., Jia, L. T., Ashley, D. L., & Watson, C. H. (2009). Rapid and chemically-selective quantification of nicotine in smokeless tobacco products using gas chromatography/mass spectrometry. *Journal of Chromatographic Science, 47*.

Zhu, J., Fan, F., McCarthy, D. M., Zhang, L., Cannon, E. N., Spencer, T. J., et al. (2017). A prenatal nicotine exposure mouse model of methylphenidate responsive ADHD-associated cognitive phenotypes. *International Journal of Developmental Neuroscience, 58*, 26–34.

Minimizing Warehouse Operation Cost

Tracy Adeline Ajol, Shirley Sinatra Gran
and Awang Nasrizal Awang Ali

Abstract Storage location assignment problems (SLAPs) are one of the challenges in warehouse managerial decision making to find the most cost-effective strategy in order to reduce the warehouse operational cost. The decision made in relation to the SLAP will also affect not only the operation cost, but also to the warehouse performance. Hence, a suitable alternative storage location assignment policy is significant to be implemented in the warehouse. This study intends to minimize the operational cost which includes both storage space cost and handling cost. A mixed integer programming model class-based storage location assignment with combined closest open location storage policy is used in this study where the allocation of any incoming product based on the closest open location (COL) storage location assignment policy within each class has been considered. Then, the assignment of product to each class to find the minimum operational cost has been solved using Microsoft Excel Adds-In Solver. The model also considers the volume of the storage space used to ensure the utilization of the storage space is maximized. Three different policies which are the developed model, the classical class-based, and dedicated are compared to determine the most profitable policies in terms of storage space utilization (storage space cost) and also storage total travel distance (handling cost). It is found that implementation of the proposed model class-based storage location assignment with combined closest open location storage policy contributes to the lowest operational cost in warehouse operation compared to the other policies.

T. A. Ajol (✉) · S. S. Gran
Faculty of Computer and Mathematical Sciences, Universiti
Teknologi MARA, 96400 Mukah, Sarawak, Malaysia
e-mail: tracy@sarawak.uitm.edu.my

S. S. Gran
e-mail: shirley@sarawak.uitm.edu.my

A. N. A. Ali
Faculty of Civil Engineering, Universiti Teknologi MARA,
94300 Kota Samarahan, Sarawak, Malaysia
e-mail: awang295@sarawak.uitm.edu.my

© Springer Nature Singapore Pte Ltd. 2018
R. Saian and M. A. Abbas (eds.), *Proceedings of the Second International Conference on the Future of ASEAN (ICoFA) 2017 – Volume 2*,
https://doi.org/10.1007/978-981-10-8471-3_62

Keywords Storage location assignment problems (SLAPs) · Mixed integer programming (MIP) · Warehouse · Cost · Storage

1 Introduction

The storage location assignment problem (SLAP) is usually concerned with the allocation of incoming products to the locations in storage zones in order to reduce material handling cost and improve space utilization (Gu et al. 2007). With the implementation of storage location assignment policy, it will be able to minimize the handling cost while maximizing the space utilization in a warehouse effectively, efficiently.

Most of the previous study conducted only applied one storage policy in a warehouse, and it was found that class-based storage policy is the most effective approach involving SLAP (Muppani and Adil 2008b). They also only considered the floor space utilization. However, what if the warehouse combined class-based storage policy and closest open location policy within each class? Which policy is able to reduce the storage and handling cost? Will the combined storage policy be able to reduce the operational cost?

Therefore, this study will addresses the research questions on the policy on assignment of product to its class and formulate suitable mathematical model to represent the case study and thus solve the SLAP. The storage volume including the width, length, and height of the warehouse capacity is also considered, and it is expected that enhancing the utilization of storage space will improve the operation performance and increase the warehouse profit.

2 Literature Review

In business, profit can be increased through cost reduction and enhancing customer service level for gaining a competitive edge. The efficiency of warehouse operation management is challenging due to its complexity in operational decision problems. Therefore, it is very important to have an effective warehouse management to reduce the cost while increasing customer satisfaction.

The past studies related to SLAP has been reviewed. There are few aspects that can be used to divide items into classes for class-based storage assignment policies which include the aspect of inventory turnover (Montulet et al. 1998). The study was mainly developed for the purpose of minimizing the peak load with single command cycles and branch and bound algorithm was used to solve the problem.

Altay and Erel (1998) classified the items into classes based on multiple attributes by comparing the effectiveness on classifying the items using analytical hierarchy process (AHP) and generic algorithm (GA). The result has shown that GA can classify items consistently based on decision maker's performances.

Then, an artificial neural network (ANN) has been studied and proven to be more efficient than the AHP model (Partovi and Anandarajan 2002). It was found that it is important to classify items into specific classes where it can increase the efficiency of assigning incoming products. In classed-based storage policies, items can be divided into classes based on few aspects such as system cycle times.

Muppani et al. (2006) implemented dynamic programming algorithm (DPA) and merge sort heuristic (MSH) where they considered the cost of storage space and handling cost. Classes are formed to further reduced the storage area needed. It was found that MSH is more time efficient compared to DPA in handling large problems. They further their study by using simulated annealing (SA) approach to form a well-organized storage classes, and it shows that SA approach produced a better solution than DPA where the class-based policy reduces the cost of storage space and order picking compared.

A nonlinear integer programming model has also been developed (Muppani and Adil 2008a). A branch and bound algorithm (BBA) was used and it shows that BBA is computationally efficient than the baseline DPA. Compared to dedicated policy, the model developed which considered storage space and handling cost when using class-based storage assignment policy resulted in cost reduction.

Pan et al. (2012) studied on improving the order picking efficiency with the consideration of multiple pickers in a warehouse. Heuristics approach was used to include the travel distance and waiting time to reduce the order fulfillment time. It has been proven to be able to determine the number of aisles needed that is balanced to reduce blocking time and travel distance based on workforce level.

A study for unit load automated storage and retrieval system has also been conducted (Gangliardi et al. 2012) where they compared the performance under several storage policies which include, random storage policy, class-based storage policy, and full turnover-based policy. It was found that full turnover storage policy does not necessarily outperform other storage policy.

Liu et al. (2014) and proposed genetic algorithm which resulted in minimizing the operational distance of warehouse, whereas Xie et al. (2014) used a genetic programming-based (GP) hyper-heuristic Approach and shows that the GP approach can achieve good optimization compared to traditional integer linear programming (ILP) approach.

In general, four common storage assignment policies used in warehousing are dedicated, random, closest open location, and also class-based storage location assignment policy. When compared, class-based storage location assignment has the advantages of lower order picking cost and storage space to solve the problem. This policy consists of determining number of classes, assignment of products to classes, and storage location for each class where the products will be assigned randomly to the storage location within each class. The most profitable areas are assigned to the most critical classes (Guerriero et al. 2012).

Tracy and Ainon (2014) have conducted a case study to minimize the handling cost. The study proposed the implementation of an efficient class-based storage location assignment policy combined with closest open location storage policy for arriving pallets system within each class with the objective to minimize warehouse handling cost through implementation of the formulated model.

Khalili and Lotfi (2015) determined the optimal capacity of private warehouse. However, the available space and budget to create a private warehouse in the model are limited. Due to the vagueness, some parameters were simulated using expert-based triangular fuzzy numbers. The resulted solution indicated that the proposed method, unlike the existing forecasting methods which are only suitable for some cases, may easily be extended for other different condition of production environments. Notably, there are some complex conditions in which the queuing models are not singly applicable and therefore suggested to use a combination of the queuing and simulation methods.

Qu and Wang (2016) studied on the sizing, pricing, and common replenishment in a headquarters-managed centralized distribution center and has formulated two decision models, namely the integrated model and the bilevel programming model. It was found that the optimal reorder lot, the reserved space, the group company's total cost, and the HQ-CDC's profit increase from the integrated model to the bilevel programming model. The improvement of the HQ-CDC's profit is from the increased subsidiaries' cost as well as the group company's total cost.

Zhang et al. (2016) developed a mixed integer linear programming model to formulate the integrated optimization problem with the objective of minimizing the total cost of production and warehouse operations. The problem with real data is a large-scale instance that is beyond the capability of optimization solvers. A novel Lagrangian relax-and-fix heuristic approach and its variants are proposed to solve the large-scale problem.

This study develops a mixed integer programming model where the variables consist of integer and binary number. The developed model incorporates class-based and closest open location policy which includes the storage space and handling cost in warehouse operation. Storage and retrieval of the products are performed in single command cycle.

3 Model Development

A new mixed integer programming model was developed, combining class-based and closest open location policy which includes the storage space and handling cost in warehouse operation. Instead of using 2D floor space utilization, this paper takes into consideration the storage space volume in a warehouse with single command cycle and the model will be solved using Microsoft Excel Adds-In Solver.

The following notations are used in this study:

$c, c\prime$	Index of classes
$n, n\prime$	Index of product
$l, l\prime$	Index of location
C	Total number of class
L	Total number of storage location
N	Total number of products
h	Floor level of a storage location
P_n	Popularity value of product n
a_l	Footprint area of storage location
d_l	Distance of storage location l from the input/output point
K_n	Total number of picks (pallet) for product n at location l
f_p	Footprint density, that is footprint area required to store a pallet of product n
r	Storage handling cost in RM per meter
f	Storage space cost in RM per square feet
Q_p^t	Storage quantity in unit loads (pallet) for product n at time, t
A_n	Number of storage locations allocated for each product
x_{nc}	$\begin{cases} 1 & \text{if product } n \text{ is assigned to class } c \\ 0 & \text{otherwise} \end{cases}$
y_{nlc}	$\begin{cases} 1 & \text{if product } n \text{ is assigned to storage location } l \text{ in class } c \\ 0 & \text{otherwise} \end{cases}$
z_{lc}	$\begin{cases} 1 & \text{if product } l \text{ is assigned to class } c \\ 0 & \text{otherwise} \end{cases}$

The objective function (Eq. 1) is to minimize the total storage space and handling cost. Equation 2 is a constraint to assign product n with higher popularity to the closest empty storage location l in class c. Class c is positioned near to the input or output point than class c'. Equation 3 indicates that each product needs to be assigned to only one class. Equation 4 is to ensure the total number of storage locations assigned to each class for each product is less than or equal to the total number of locations allocate for each product. Equation 5 ensures that there is a sufficient storage space to store the products in each planning period. Equation 6 is the binary restriction on decision variables.

Objective function:

$$\text{Minimize} \quad f \cdot \sum_{c=1}^{C} \sum_{l=1}^{L} (a_l \cdot z_{lc}) + 2r \sum_{c=1}^{C} \left[\sum_{l=1}^{L} \left(d_l \sum_{n=1}^{N} K_n y_{nlc} \right) \right] \quad (1)$$

Subject to

$$P_n \cdot y_{nlc} \geq P_{n'} \cdot y_{n'l'c'} \quad \forall p \neq p', l < l', c < c' \quad (2)$$

$$\sum_{c=1}^{C} x_{nlc} = 1 \quad \forall n, l \quad (3)$$

$$\sum_{l=1}^{L} x_{nlc} \leq A_n \quad \forall n \quad (4)$$

$$\sum_{n=1}^{N} \left(Q_p^t \cdot f_p \cdot y_{nlc} \right) \leq \sum_{l=1}^{L} (a_l \cdot h) \quad \forall c, t \quad (5)$$

$$y_{nlc} \in \{0, 1\} \quad \forall n, l, c \quad (6)$$

To start with, data on the warehouse such layout, travel distance measurement, storage capacity, and product information, including type of product, demand, and number of pick was collected. The distance of each storage location from input/output point is then measured. After that the space required for each product was determined and the product is arranged according to its popularity (high demand). Assignment of product to class to obtain the minimum operation cost was solved using Excel Adds-In Solver. The calculation is based on single command cycle. Finally, the result for the proposed model, class based and dedicated is compared to see the difference in term of storage space cost and handling cost. The figures show elaboration on the three different policies used in this study.

Figure 1 is a policy where each type of product is stored at a fixed location within the warehouse (1 product = 1 class). Product with higher demand or higher picking frequency will be allocated to the class which is near to the input or output point to obtain the minimum total travel distance. The incoming product will be stored randomly at any available location within its class. Therefore, travel distance

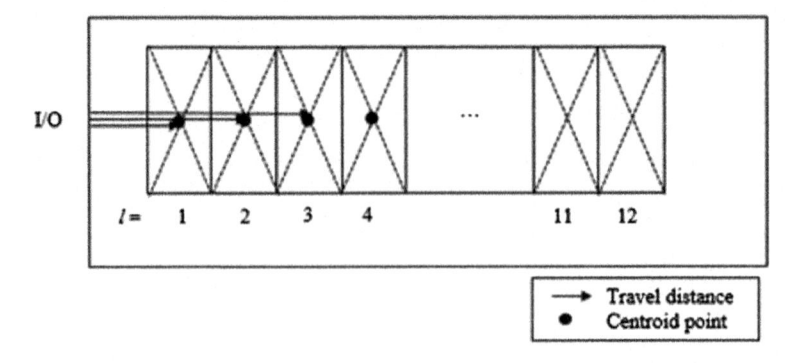

Fig. 1 Dedicated storage policy

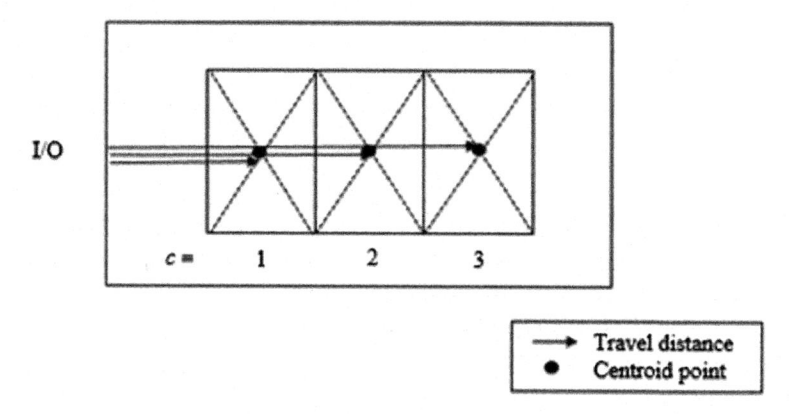

Fig. 2 Class-based storage policy

calculated is from the centroid point of each product storage area (class) to the input/output point. Since this study uses five products, there will be 5 class used to store the products. This will require more storage location to store the product.

Class-based model as shown in Fig. 2 applies the policy where each product will be assigned to one and only one class and each class can have more than one product. However, the incoming product will be stored randomly at any available location within its class. The distribution of the storage travel distance is spread evenly within its class as dedicated storage policy. Hence, the distance calculated is from the centroid point of each class to the input/output point.

The proposed model applies the class-based policy (each product assigned to one and only one class, each class can have more than one product) where an open location for incoming product within its class is selected based on the closest open location policy. The distance is calculated from each storage location to the input or output point as shown in Fig. 3.

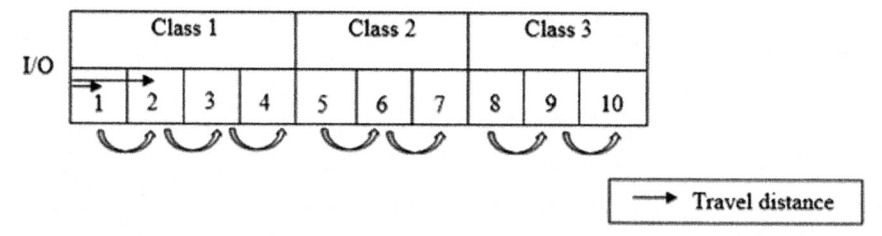

*If l is fully utilize, the incoming product will store to the next empty location, l', within each class.

Fig. 3 Proposed model (class-based + open location storage policy)

4 Results and Discussion

Assignment of product to class for obtaining the minimum handling cost was solved using Excel Adds-In Solver. The most popular product is stored at the nearest storage location to input/output point. This means that the product with higher frequency of pick belongs to the class which will be located nearer to the input/output point to reduce the traveling distance, and thus, reduce the handling cost, or otherwise. Meanwhile, some of the products will be stored within a class to reduce the storage space cost, thus decreasing the traveling distance to store and pick the products. The results obtained for the assignment of product to its particular class indicated that product, $p1$ is assigned to class, $c1$. Product, $p3$ is assigned to class, $c2$. As for the rest of the products, $p2$, $p4$, and $p5$ are all assigned to class, $c3$. Then, the results obtained will be used to compare the total operation cost which includes both handling cost and storage space cost when implementing the three different policies.

Table 1 shows that the proposed model for this study which implements the combined class-Based and combined closest open location policy contributes to the lowest total operational cost. This is because the implementation of class-based requires smaller storage space used and the assignment of incoming product within each class which is based on the nearest empty location within the class contributes to smallest travel distance, thus reducing the handling cost. It is also observed that the proposed model is able to reduce the overall operating cost at almost 8% compared to class-based and 5% when compared to dedicated model. The total handling cost for class-based model is higher compared to the proposed model. This happened since the incoming products were stored randomly, which means they were well-distributed (evenly) within the class. Therefore, it causes increase in the total travel distance from the storage location to the input/output point and thus increases the handling cost.

As for the total handling cost for dedicated model, it is lower compared to the class-based model but higher than the proposed model. Dedicated policy is where each product is stored only at specific location. It is the most common policy used in a warehouse since it is easier to be implemented. The incoming product is stored at a fixed area only even though there is an empty storage space nearer to the input/output point. Therefore, this policy will require huge storage space which leads to higher storage space cost. In this study, both proposed model and class-based model only

Table 1 Comparison cost for three policies

Policy model	Σ Handling + Storage cost (RM)
Proposed model (class-based + closest open location)	24,427
Class-based	26,541
Dedicated	25,817

used the minimum 10 storage location but the dedicated model requires up to 12 storage locations. Overall, the dedicated storage policy contributes to lower handling cost but higher storage space cost since the storage space was not fully utilized.

5 Conclusion

This research will be able to solve the storage location assignment problems by reducing the total operational cost through implementation of the proposed model of combined policy. The warehouse will be able to reduce the operational cost when both handling cost (total travel distance) and the storage space cost are reduced. Through computational experiments, the model demonstrated that it is able to minimize the total warehouse operation cost. Metaheuristic method is widely used to solve the storage location assignment problem in the past studies. These methods have been the preferred choice because it is suitable in solving large-scale problem and required reasonable computational times to solve the mathematical model for SLAP. However, with the development of computer software and the improvement in mathematical optimization and theory, it is now possible to solve SLAP using an exact method such as Excel Solver Adds and the result from this study will benefits the logistic and other related industry.

Acknowledgements This work is partially supported by Dana Kecemerlangan UiTM Sarawak [600-RMU/DANA 5/3 (12/2016)]. The authors also gratefully acknowledge the helpful comments and suggestions of the reviewers, which have improved the overall presentation.

References

Altay Guvenir, H., & Erel, E. (1998). Multicriteria inventory classification using a genetic algorithm. *European Journal of Operational Research, 105*(1), 29–37.
Gagliardi, J. P., Renaud, J., & Ruiz, A. (2012). On storage assignment policies for unit load automated storage and retrieval systems. *International Journal of Production Research, 50*(3), 879–892.
Gu, J., Goetschalckx, M., & McGinnis, L. F. (2007). Research on warehouse operation: A comprehensive review. *European Journal of Operational Research, 177*(1), 1–21.
Guerriero, F., Musmanno, R., Pisacane, O., & Rende, F. (2012). A mathematical model for the multi-levels product allocation problem in a warehouse with compatibility constraints. *Applied Mathematical Modelling*.
Khalili, S., & Lotfi, M. M. (2015). The optimal warehouse capacity: A queuing-based fuzzy programming approach. *Journal of Industrial and Systems Engineering*, 1–12.
Liu, S., Sun, J., & Wang, Q. (2014). *Optimization of storage performance for generic tiered warehouse by genetic algorithm*. Faculty of Information Sciences and Engineering.
Montulet, P., Langevin, A., & Riopel, D. (1998). Minimizing the peak load: An alternate objective for dedicated storage policies. *International Journal of Production Research, 36*(5), 1369–1385.

Muppani, V. R., & Adil, G. K. (2006). Formation of storage classes in the presence of space cost for warehouse planning. *International Journal of Services Operations and Informatics, 1*(3), 286–303.

Muppani, V. R., & Adil, G. K. (2008a). A branch and bound algorithm for class based storage location assignment. *European Journal of Operational Research, 189*(2), 492–507.

Muppani, V. R., & Adil, G. K. (2008b). Efficient formation of storage classes for warehouse storage location assignment: A simulated annealing approach. *Omega, 36*(4), 609–618.

Pan, J. C. H., Shih, P. H., & Wu, M. H. (2012). Storage assignment problem with travel distance and blocking considerations for a picker-to-part order picking system. *Computers & Industrial Engineering, 62*, 527–535.

Partovi, F. Y., & Anandarajan, M. (2002). Classifying inventory using an artificial neural network approach. *Computers & Industrial Engineering, 41*(4), 389–404.

Qu, T. Z., & Wang, H. X. (2016). Sizing, pricing and common replenishment in a headquarter-managed centralized distribution center. *Industrial Management & Data Systems*, 1–17.

Tracy & Ainon (2014). *Minimizing warehouse handling cost using combined class-based and closest open location storage policy (A case study at warehouse XYZ)* (Unpublished master dissertation). Universiti Teknologi Mara, Malaysia.

Xie, J., Mei, Y., Ernst, A. T., Li, X., & Song, A. (2014). A genetic programming-based hyper heuristic approach for storage location assignment problem. In: *Congress on Evolutionary Computation (CEG) 2014*.

Zhang, G., Nishi, T., Turner, S. D., Oga, K., & Li, X. (2016). An integrated strategy for a production planning. *Omega The International Journal of Management Science*, 1–29.

Bioaugmentation of Oil Sludge Using Locally Isolated Hydrocarbon Microbial in Single and Mixed Cultures Assisted by Aerated Static Pile (ASP): A Laboratory Scale

Nur Zaida Zahari and Mohd Tuah Piakong

Abstract Bioaugmentation is one of the best option treatment technologies in which hydrocarbon-degrading bacteria are added to contaminated soil to accelerate the degradation capacity at the site. In this study, three hydrocarbon-degrading strains (*Candida tropicalis*-RETL-Cr1, *Chromobacterium violaceum*-MAB-CR1, and *Pseudomonas aeruginosa*-BAS-CR1) were used to examine the efficiency of single strain and mixed cultures for remediation of oil sludge contaminated soil. Five different sets of treatment containing 10 kg of soil with 10% oil sludge were prepared as: Treatment A (contaminated soil with single strain *C. tropicalis*-RETL-Cr1), Treatment B (contaminated soil with single strain *C. violaceum*-MAB-CR1), Treatment C (contaminated soil with single strain *P. aeruginosa*-BAS-CR1), Treatment D (contaminated soil with mixed cultures, *C. tropicalis*-RETL-Cr1 + *C. violaceum*-MAB-CR1 + *P. aeruginosa*-BAS-CR1), and Treatment E [contaminated soil without bacterial inoculation (natural attenuation)]. The reduction of TPH in each bioreactor was observed during 60 days treatment periods. The physiochemical parameters such as pH, temperature, moisture content, and biological population in soil (CFUs/g) were investigated during the bioremediation process. The results obtained revealed bioaugmentation using mixed cultures was proven to be better approach to treat oil sludge with 88.2% degradation of TPH followed by Treatment C > Treatment B > Treatment A > Treatment E with 81.6%, 80.7%, 80%, and 57%, respectively. Hence, the results demonstrated that Treatment D with mixed culture was the most advantageous option for treatment of oil sludge as compared to single and natural attenuation treatment. This finding also demonstrated that all treatments except natural attenuation were able to degrade oil sludge up to 80% after 60 days of incubation period.

N. Z. Zahari (✉) · M. T. Piakong
Faculty of Science & Natural Resources,
Universiti Malaysia Sabah, UMS Road, 88400 Kota Kinabalu, Sabah, Malaysia
e-mail: zda.zarie@gmail.com

Keywords Bioaugmentation · Hydrocarbon-degrading bacteria
Single strain · Mixed cultures · Oil sludge

1 Introduction

Soil contaminated with oil sludge has become a major problem to the environment. The high production of oil and petroleum refining generates a huge amount of oil sludge, and this waste has to be treated and made harmless before its land disposal. Oily sludge is categorized as one of the hazardous wastes under the Environment Protection Act and Hazardous Waste Handling Rules. The main sources of oil sludge are from oil refinery operation and maintenance activity due to emulsion and wax treatment plant, crude oil storage tank, effluent treatment plant, free water knockout vessels, pigging operations, and sand from offshore platform. As this sludge contains a quantity of heavy oil, water, and sediments, the release of these wastes into the environment may pose severe impacts to the plants and animal ecosystems including human health (Imran et al. 2007; USEPA 1995). Severe diseases such as skin erythema, skin cancer, gastrointestinal cancer, and bladder cancer may occur when exposed to high level of oil sludge (ATSDR 2009). Soil contaminated with oil sludge will lose its fertility and have adverse impacts on seed germination thus resulting to the death of flora and fauna (Michelle and Peckol 2000). Hence, disposal of this oil sludge in an improper manner may cause a serious damage to our environment. A number of technologies are available for the treatment of oil sludge including incineration, stabilization, oxidation, and solidification. However, these methods are required to have a high capital and operational costs and not an environmental friendly solution (Hu and Cheng 2013). Bioaugmentation is one of the most viable options for remediating soil contaminated with organic and inorganic compounds considered detrimental to environment health. The addition of oil hydrocarbon-degrading bacteria into the contaminated soil can enhance the degradation capacity in the soil. As oil sludge components are made up of many different compounds, the degradation of these complex hydrocarbons usually requires the cooperation of more than one single species. In many cases, mixed cultures were more effective than single strain by the fact that intermediates of a catabolic pathway of one strain may be further degraded by other strains possessing suitable catabolic pathway (Heinaru et al. 2005). This fact can be attributed to the effects of synergistic interactions among the numbers of the cultures although the mechanism may be complex. Thus, this phenomenon strongly suggests that each strain has its role in hydrocarbon transformation process. Another critical factor in bioremediation of hydrocarbon is oxygen availability. In order to maintain aerobic biodegradation activities, many researchers only applied soil tilling and mechanical rotation without supplying adequate oxygen to the contaminated site (Ajoy Kumar et al. 2011; Sharma 2012; Vasudevan and Rajaram 2001). In this study, we are trying to assess the efficiency of bioaugmentation of oil sludge using locally isolated hydrocarbon-degrading microbial in single and mixed cultures assisted by continuous aerated static pile (ASP) in laboratory scale.

2 Materials and Methods

2.1 Bioreactor Design

A bioreactor made up of acrylic material with dimension of 60 cm × 40 cm 20 cm with 3 tubing for aeration at the side of the reactor connected to air pump (Model RESUN LP100 Low Noise Air Pump) (Vasudevan and Rajaram 2001) was prepared (Fig. 1). The reactor was divided into three parts where the base is filled with gravel-sized (1–1.5 cm) followed by sand and soil on the surface. The soils used in the study were mixed thoroughly before being put in the treatment plot.

2.2 Bacterial Strain

Bacterial strains used in this study were from Environmental Microbiology Laboratory, Faculty of Science & Natural Resources at Universiti Malaysia Sabah. These collections of hydrocarbon-degrading bacteria (Fig. 2) have been proven to degrade oil and phenol based on the previous research by Piakong et al. 2004, Nurulhuda and Piakong (2008).

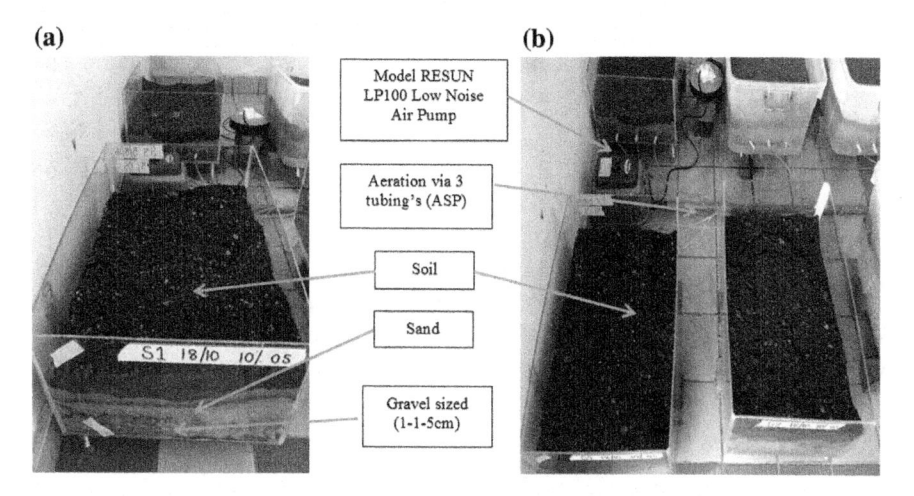

Fig. 1 Bioreactor setup for ASP bioaugmentation of oil sludge **a** side view and **b** top view

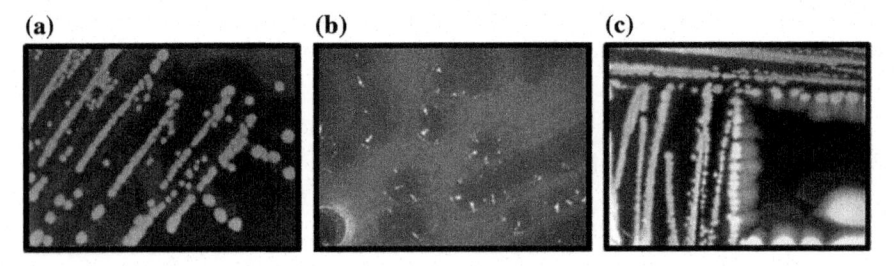

Fig. 2 Selected hydrocarbon-degrading bacteria used **a** *Candida tropicalis*-RETL-Cr1, **b** *Pseudomonas aeruginosa*-BAS-Cr1, **c** *Chromobacterium violaceum*-MAB-Cr1 supplied by Environmental Microbiology Laboratory, FSSA, UMS

2.3 Culture Medium

Ramsay broth was used in this study (Ramsay et al. 1983). Approximately, 2.0 gL^{-1} of NH_4NO_3, 0.5 gL^{-1} of KH_2PO_4, 1.0 gL^{-1} of $KHPO_4$, 0.1 gL^{-1} of KCl, 0.01 gL^{-1} of $CaCl_2 \cdot H_2O$, and 0.06 gL^{-1} of yeast extract were placed in the beaker and stirred using magnetic stirrer. The liquid medium was transferred to Schott bottle and autoclaved for 15 min at 112 °C. Glucose and $MgSO_4 \cdot 7H_2O$ were added to the medium after being autoclave.

2.4 Inoculum Preparation

Single colony and mixed cultures of bacteria isolated were inoculated into Ramsay broth at 30 °C for 24 h in an orbital shaker at 200 rpm. A total of 10% of the cultured bacteria with OD 0.5 and above at 600 nm were used as inoculum. The dense cultures (1×10^7) were harvested for further use.

2.5 Soil Preparation

Soil (10 kg) was sieved through 0.20 mm sieve size. The physical analysis of soil (pH, temperature, and moisture content) was determined according to the Alef and Nannipieri (1995). One liter of oil sludge (10% v/w) was sprinkled over the sieved soil and allowed to get adsorbed for 30 min for further bioremediation studies (Sharma and Rehman 2009).

2.6 Experimental Setup

The ability of bacteria isolate to remediate oil sludge contaminated soil was analyzed by carrying out the biodegradation experiment in bioreactor (60 cm × 40

Table 1 Treatment combinations

Treatment A:	Soil + oil sludge + single strains (C. tropicalis-RETL-Cr1)
Treatment B:	Soil + oil sludge + single strains (C. violaceum-MAB-Cr1)
Treatment C:	Soil + oil sludge + single strains (P. aeruginosa-BAS-Cr1)
Treatment D:	Soil + oil sludge + mixed cultures (C. tropicalis-RETL-Cr1 + C. violaceum-MAB-Cr1 + P. aeruginosa-BAS-Cr1)
Treatment E: (Natural attenuation)	Soil + oil sludge + without inoculation (control set)

cm × 20 cm). Five treatments (A, B, C, D, and E) were prepared in duplicate at open and air ventilated with the treatment combinations as in Table 1.

2.7 Technical Procedures Protocol

For each of the experimental unit, the soils were tillage every day and watering with sterile distilled water at an interval of 2 days to maintain the water holding capacity of soil. The bioreactors were supplied with 3 tubing for continuous aeration using air pump (Model RESUN LP100 Low Noise Air Pump). Inoculants (10% v/v) of each single and mixed culture were added to the plot containing 10 kg of soil contaminated with 10% (v/w) of oil sludge. The inoculation of single and mixed cultures into each plot was done for every 2 weeks during 60 days treatment period. Initial readings were taken immediately after the inoculation of the bacterial culture and further. The soil samples were drawn after every 7th day from the respective experimental unit and were analyzed up to 60 days of incubation period.

2.8 Physiochemical Analysis of Soil

2.8.1 pH

Analysis pH was carried out as according to ASTM (1995). Soil (20 g) was weighed and mixed in a beaker containing 40 ml of distilled water. Analysis was done by taking three readings for accuracy using the pH meter.

2.8.2 Temperature

The temperature of each plot was measured by using thermometer in three different places at each bioreactor treatment. The readings were recorded, and the averages were obtained.

2.8.3 Soil Moisture Content

The soil moisture was analyzed by using gravimetric method as described by Gardner (2001). An empty crucible was weighed and recorded as (M1). Then, 10 g of soil sample were grained with mortar and pestle and recorded as (M2). The soil samples were dried at 110 °C for 24 h in an oven. Let the soil being cooled in desiccator and weighed the crucible as (M3). The soil moisture content was measured by using Eq. 1:

$$\% \text{ Moisture content} = \frac{(M2 - M3)}{(M2 - M1)} \times 100 \quad (1)$$

M1   Weight of empty crucible (g)
M2   Wet of crucible + wet soil in (g)
M3   Weight of crucible + dry soil in (g)

2.8.4 Soil Microbial Population

The growths of microorganism in each plot soil were determined according to method APHA 9215. Analyses were carried out weekly in each plot study. Soil sample (10 g) was mixed with 100 ml sterile distilled water by diluting 1 ml of soil sample into 9 ml of sterile distilled water (dilution plate technique). The dilution series from 10^{-3} until 10^{-5} were chosen to calculate the biological population. Culture (0.1 ml) was spread into nutrient agar and incubated at 30 °C for 24 h. The total CFU was calculated based on the colony growth in the agar. The numbers of CFU were calculated by Eq. 2:

$$\text{CFU ml}^{-1} = \frac{\text{CFU per plate} \times \text{dilution factor}}{\text{volume of sample taken (ml)}} \quad (2)$$

2.9 Total Petroleum Hydrocarbon (TPH) Analysis

Total petroleum hydrocarbon was carried out based on gravimetric method (Soxhlet extraction) (USEPA 3540C) (USEPA 1996). Soil sample (20 g) was grained and placed in thimble and extracted with dichloromethane (DCM). Then the thimble was placed in Soxhlet extractor. Dichloromethane (175 ml) was added to round-bottomed flask (RBC). After that, the Soxhlet extraction was arranged and mixed with enough dichloromethane until it covers the thimble. The cooling temperature and mantel heater were set at 17 °C and 4 °C, respectively. The extraction process takes places for less than 24 h. Then the content was cooled, and let the DCM flow to the round-bottomed flask from the extractor. The total solvent was cleared completely with the vacuum evaporator at 40–50 °C. By using the rotary pump, DCM left the sample. The RBC together with the extract was cooled in desiccator after being dried in oven at 40 °C. The RBC was measured until constant weight is obtained. The percentage of total petroleum hydrocarbon (TPH) was calculated using the formula given below:

$$\% \, \text{TPH} = \frac{\text{Min extract weight in RBC}}{\text{Weight of sample}} \times 100$$

3 Results and Discussion

3.1 Oil Sludge Degradation

The variation of TPH in different treatment with single and mixed cultures versus time is shown in Fig. 3. On the contrary, a control set natural attenuation (NA) which consisted of a contaminated soil sample in the absence of active hydrocarbon-degrading bacteria but containing indigenous or native microorganisms soil was also analyzed.

Figure 3 shows the profile reduction of total petroleum hydrocarbon in oil sludge contaminated soil during 60 days of treatment. The initial TPH in the soil for all treatment sets was around 100 g/kg as determined by gravimetric analysis. The obtained results showed that the degradation of TPH in all bioreactor treatments was

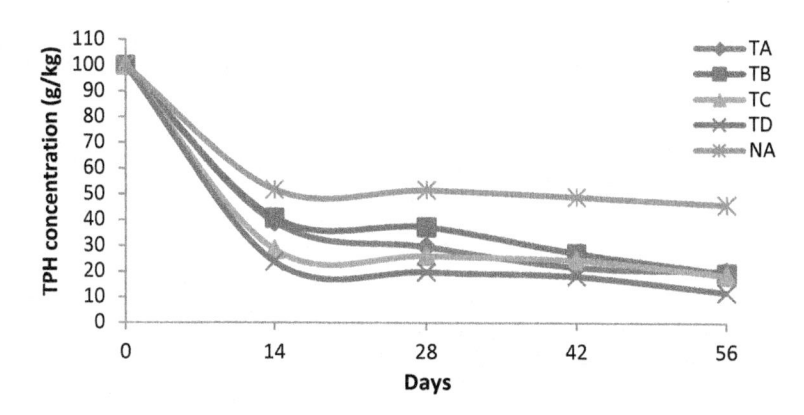

Fig. 3 Profile of TPH reduction in different treatment of oil sludge contaminated soil during 60 days of incubation period. Treatment A: soil + oil sludge + single strains (*C. tropicalis*-RETL-Cr1); Treatment B: soil + oil sludge + single strains (*C. violaceum*-MAB-Cr1); Treatment C: soil + oil sludge + single strains (*P. aeruginosa*-BAS-Cr1); Treatment D: soil + oil sludge + mixed cultures (*C. tropicalis*-RETL-Cr1 + *C. violaceum*-MAB-Cr1 + *P. aeruginosa*-BAS-Cr1); Treatment E: soil + oil sludge + without inoculation (natural attenuation control set)

the most rapid during the first 2 weeks of incubation (up to 76%) and the rate declined with the time. The effect of microbial preparation with single and mixed cultures on degradation of oil sludge had showed that the TPH decreased by 11–20 g/kg after 60 days of treatment. At the same time, activation of indigenous microorganisms (NA) decreased the TPH to 46 g/kg, respectively. The inoculation of the selected strains into the soil rapidly adapted to the environment efficiently degrading hydrocarbon compounds. This can be proven by the same trend profile in Fig. 3.1. The highest degradation (88.2%) was observed in Treatment D inoculated with mixed cultures (*Candida tropicalis*-RETL-Cr1 + *Chromobacterium violaceum*-MAB-Cr1 + *Pseudomonas aeruginosa*-BAS-Cr1) followed by Treatment C > B > A > E which are 81.6%, 80.7%, 80%, and 54%, respectively.

It is clear to note that the hydrocarbons were degraded mainly by the oil-degrading active bacteria as compared to indigenous microbes in soil. This study was found to be significant when compared to previous investigations. Anjana Sharma reported that biodegradation of diesel hydrocarbon in soil by bioaugmentation of *P. aeruginosa* in a bioreactor made up of cemented wall (12 cm × 17 cm) with 2.5 l diesel (10% v/v) without continuous oxygen supplied had showed that highest degradation with 66% during 60 days treatment period. Ouyang et al. (2005) performed a field scale bioremediation of oil sludge with four Sect. (2.5 m × 2.0 × 0.2 m) biopreparation under roof without a continuous supply of oxygen in the soil. The initial concentration of oil sludge was from 151 g/kg, and the experiment conducted lasted for 56 days under ambient temperature. He reported that the total hydrocarbon content (THC) was decreased by 53% (70 g/kg) in the soil augmented with two *Rhodococcus* oil-degrading strains. This study was also similar with Mishra et al. (2001) who conducted a field scale bioremediation study designed as A, B, and C with area 4000, 5900, and 100 m^2, respectively. The addition of three consortium oil-degrading bacteria namely *Acinobacter baumannii*, *Burkholderia cepacia*, and *Pseudomonas* had showed that the removal of TPH achieved 92% with the initial concentration 69.7 g/kg reduced to 5.53 g/kg within 360 days treatment. Thus, our finding showed that bioaugmentation with mixed cultures (*C. tropicalis*-RETL-Cr1 + *C. violaceum*-MAB-Cr1 + *P. aeruginosa*-BAS-Cr1) assisted by aerated static pile (ASP) was much better as compared to Ouyang et al. (2005) and Mishra et al. (2001) in terms of the duration (short time) needed to degrade the hydrocarbon in contaminated soil.

3.2 Environmental Parameters Monitoring During Bioaugmentation of Oil Sludge

The environmental parameters that have been monitored during 60 days treatments are shown in Fig. 4, 5 and 6. Figure 4 shows the profile of viable cell count at different treatments of oil sludge contaminated soil. The results showed that treatments A, B, C, and D with addition of single and mixed cultures showed slight increment of viable cell count from 6.3×10^6 CFUs/g soil to 4.0×10^7 CFUs/g of

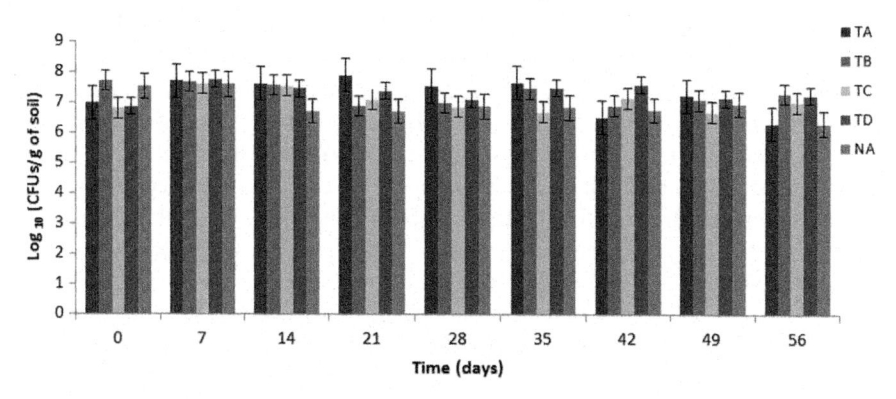

Fig. 4 Profile of viable cell count in different treatment of oil sludge contaminated soil during 60 days of incubation period

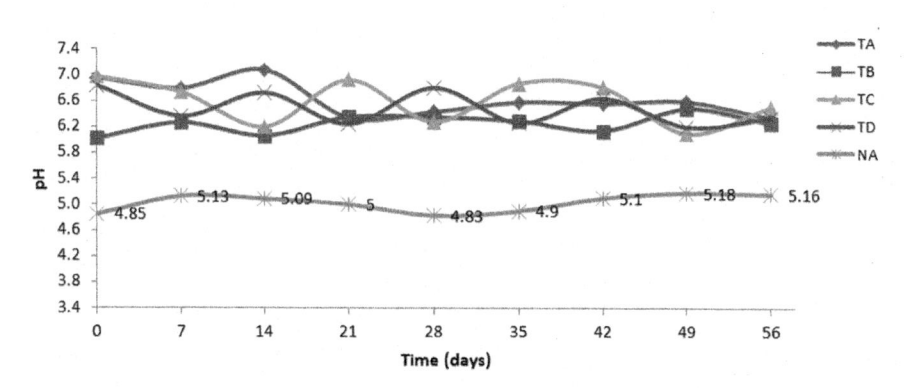

Fig. 5 Variation of pH during 60 days treatment of oil sludge contaminated soil

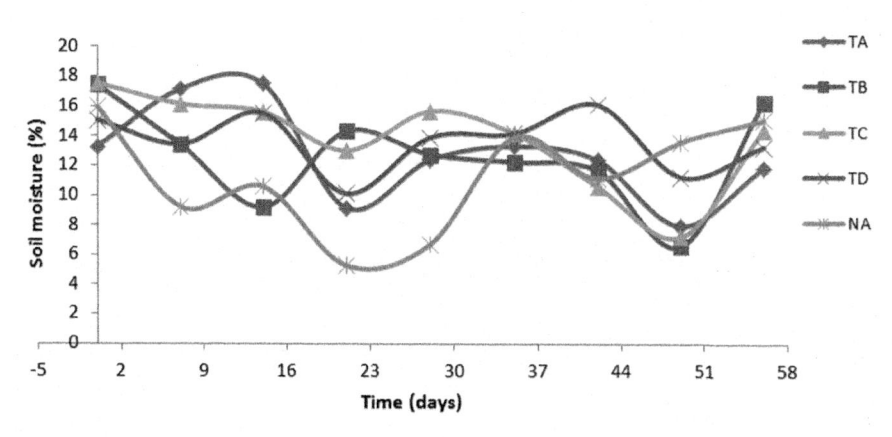

Fig. 6 Variation of soil moisture content during 60 days treatment of oil sludge contaminated soil

soil after 14 days of incubation. Then, the bacterial cell numbers slightly fluctuated until 60 days of incubation and remained at viable cell count of approximately 1.0×10^7 CFUs/g of soil. On the other hand, treatment natural attenuation (NA control set) which was not inoculated with any addition of bacterial inoculum or nutrient showed a viable cell count with only 3.3×10^6 CFUs/g of soil (indigenous microorganism) on the initial of experiment. The cell was declined along the incubation periods till 1.9×10^6 CFUs/g of soil, respectively. The cell reduction in the population density from day 0 to day 60 in control set can also be attributed to the toxic effect of oil sludge on the microorganisms present in the soil sample.

The results on viable cell count reveal that the single and mixed cultures inoculated in the treatment soil were capable of withstanding the toxicity of 10% oil sludge added. These strains successfully grew under the extreme condition exhibiting an increase in CFU for the first 14 weeks of study. This finding is similar to Kok Chang et al. (2011), who reported that the addition of bacteria inoculum into the soil showed increment of viable cell count up to 2.23×10^8 CFUs/g soil as compared to abiotic control (soil without inoculation) 1.55×10^5 CFUs/g soil, respectively. Farinazleen et al. (2004) also reported that the addition of consortium 2 (*Bacillus* sp. *Pseudomonas* sp. and *Micrococcus* sp.) in biodegradation of diesel oil had increased the microbial count from 5.13×10^7 CFUs/g soil to 7.76×10^8 CFUs/g soil after 15 days treatment. In all instances, our finding concluded that the cell counts in the inoculated soil with single and mixed cultures remained higher by 21-fold than in the soil without any microbial added.

The result of pH monitoring is shown in Fig. 5. The pH of the soil was ranged from 6.0 to 7.0 in all inoculated treatments (A, B, C, and D) which was within the optimum range for microbial activities (Vidali 2001). Then, this range continued to show slight increase and decrease during bioremediation process. The period of pH value fluctuation and increase in the inoculated bioreactor was synchronized with the time of the microbes' preparation addition proving that the preparation had an effect on decontamination of oil sludge. According to Eweis et al. (1998), biological activity in the soil is usually greatest within a pH range of 6–8, although some fungi gave optimal growth regions at pH levels of less than 5. Hence, it was expected that our strains *C. tropicalis*-RETL-Cr1, *C. violaceum*-MAB-Cr1, and *P. aeruginosa*-BAS-Cr1 may exhibit an optimal growth in the soil pH range of 6.0 to 7.0. For the natural attenuation treatment (control plot), the pH value was ranged from 4.8 to 5.1, respectively. Deviation of pH from the optimum range may result in reduced microbial population, because each species displays optimum growth at a particular pH.

The variations of soil moisture content for all treatments setup are shown in Fig. 6. The results showed that the percentage of soil moisture content for inoculated treatments with single and mixed cultures were ranged from 6.57 to 17.42%. Control plot (natural attenuation) showed the lowest soil moisture content with 5.28% during 21st day of incubation. Generally, the optimum activity for aerobic biodegradation occurs when the soil moisture is 25–28% of the field capacity. When moisture content is lower than 10% of the holding capacity, the bioactivity of

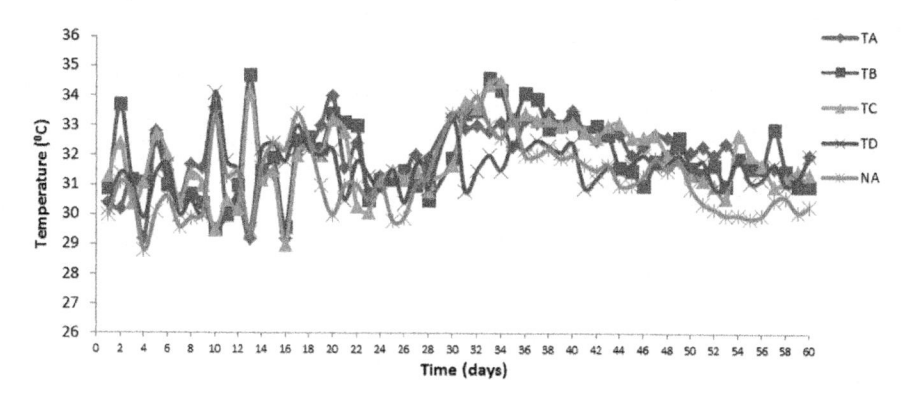

Fig. 7 Variation of temperature recorded during 60 days treatment of oil sludge contaminated soil

microorganisms will be slow thus inhibiting the degradation process (Testa and Winegardner 1991). The result indicated that at 25% moisture content TBP degradation was rapid, whereas in soil with only 10% moisture the degradation proceeded to a small extent.

The data on temperature monitoring for 60 days of incubation periods is shown in Fig. 7. The results obtained showed that temperature of all sets treatment is ranged from 28.8 to 34.7 °C. The highest temperature was observed in Treatment B at day 14 while the lowest temperature is shown in control plot (natural attenuation) during day 4 of incubation period. It was concluded that the range of temperature is dependent on the average daily air temperature. Hong et al. (2007), studying the effect of temperature on fenitrothion (nitrophenolic pesticide) degradation by inoculated *Burkholderia* sp. FDS-1, found that optimal parameters for bacteria activity were 30 °C whereas at 10 °C and 50 °C condition were unsuitable for pesticide detoxification. Malina et al. found a reduction in the rate of degradation of toluene and decane when soil temperature decreased from 20 to 10 °C. Vidali (2001) stated that the optimum temperature for oil degradation ranged from 20 to 30 °C and thus suggested that our data obtained is suitable for oil degradation microorganisms in all set treatments.

4 Conclusion and Recommendation

This study highlights the TPH degradation in oil sludge contaminated soil which was enhanced by bioaugmentation with strains *C. tropicalis*-RETL-Cr1, *C. violaceum*-MAB-Cr1, and *P. aeruginosa*-BAS-Cr1. It was observed that the inoculation of mixed cultures (*C. tropicalis*-RETL-Cr1 + *C. violaceum*-MAB-Cr1 + *P. aeruginosa*-BAS-Cr1) into oil sludge contaminated soil achieved better results in degrading oil sludge with 88.2% as compared to single strains. Conversely, the effect of natural attenuation was lower by 1.5-fold in degrading hydrocarbon confirming the requirement to feed the soil by addition of nutrients (e.g., NPK). It was noted that the assistance of aerated static pile (ASP) in this study can enhance the rate of oil degradation in soil. However, we recommend more frequency of soil watering in order to achieve the optimum condition of soil moisture needed for microbial activity.

Acknowledgements This study was financially supported by a grant (GUG0035-SG-P-1/2016) from the Center of Research & Innovation, Universiti Malaysia Sabah. We are indebted to Department of Environment (DOE), Malaysia, and Labuan Shipyard & Engineering Sdn. Bhd. for granting us permission to use and treat the oil sludge for this research.

References

American Society of Testing and Materials (ASTM). (1995). Annual book of ASTM standards, Designation D4972 - 95a: Standard test method for pH of soils.

Ajoy Kumar, M., Manab Sarma, P., Singh, B., Paul Jeyaseelan, C. (2011). Bioremediation: An environment friendly suitable biotechnological solution for remediation of petroleum hydrocarbon contaminated waste. *ARPN Journal of Science and Technology, 2*(1), 1–12.

Alef, K., & Nannipieri, P. (Eds.). (1995). *Methods in applied soil microbiology and biochemistry* (576 p). London: Academic Press.

ATSDR. (2009). Case studies in environmental medicine. toxicity of polycyclic aromatic hydrocarbons (PAHs). Atlanta: Agency for Toxic Substances and Disease Registry.

Eweis, J. B., Ergas, S. J., Chang, D. P. Y., & Schroeder, E. D. (1998). *Bioremediation principles* (p. 296). McGraw-Hill Companies, Inc.

Farinazleen, M. G., Rahman, R. N. Z. A., Salleh, A. B., & Basri, M. (2004). Biodegradation of hydrocarbons in soil by microbial consortium. *International Biodeterioration and Biodegradation, 54,* 61–67.

Gardner, C. M. K., Robinson, D. A., Blyth, K., & Cooper, J. D. (2001). Soil water content. In: K. A. Smith & C. E. Mullins (Eds.), *Soil & environmental analysis: Physical methods* (pp. 1–64). New York: Marcel Dekker.

Heinaru, E., Merimaa, M., Viggor, S., Lehiste, M., Leito, I., Truu, J., et al. (2005). Biodegradation efficiency of functionally important populations selected for bioaugmentation in phenol- and oil-polluted area. *FEMS Microbiology Ecology, 51,* 363–373.

Hong, Q., Zhang, Z., Hong, Y., & Li, S. (2007). A microcosm study on bioremediation of fenitrothion contaminated soil using Burkholderia sp. FDS-1. *International Biodeterioration Biodegradation, 59,* 55–61.

Hu, Y., & Cheng, H. (2013). Development and bottlenecks of renewable electricity generation in China: A critical review. *Environmental Science and Technology, 47*(7), 3044–3056.

Imran, M., Talpur, F. N., Jan, M. S., Khan, A., & Khan, I. (2007). Analysis of nutritional components of some wild edible plants. *Journal Chemical Society of Pakistan, 29*(5), 500–508.

Kok Chang, L., Ibrahim, D., & Omar, I. C. (2011). A laboratory scale bioremediation of Tapis crude oil contaminated soil by bioaugmentation of Acinobacter baumannii T30C. *African Journal of Microbiology Research, 5*(18), 2609–2615, September 16, 2011.

Mishra, S., Jyot, J., Kuhad, R. C., & Lal, B. (2001). Evaluation of inoculum addition to stimulate in situ bioremediation of oily-sludge-contaminated soil. *Applied and Environmental Microbiology, 67*(4), 1675–1681.

Michele, L. W., & Peckol, P. (2000). Effects of Bioremediation on Toxicity and Chemical Composition of No. 2 Fuel Oil: Growth Responses of the Brown Alga Fucusvesiculosus. *Marine Pollution Bulletin, 40*(2), 135–139.

Nurulhuda, Z., & Piakong, M. T. (2008). Isolation, characterization and screening of hydrocarbon-degrading bacteria from environmental samples for treatment of oil-sludge. In:

Proceeding International Conference on Environmental Research and Technology (pp. 521–525). Universiti Sains Malaysia.

Piakong, M. T., Jamaluddin, H., Hasila, S., Haron, R., Yahya, A., Md Salleh, M., & Noor Aini, A. R. (2004). Biodegradation of phenol by locally isolated strains from petrochemical wastewater treatment plants. *Water and Environmental Management Series*, 109–114.

Ramsay, B. A., Cooper, D. G., Margaritis, A., & Zajic, J. E. (1983). Rhodochorous bacteria: Biosurfactant production and demulsifying ability. *Microbial Enhanced Oil Recovery*, 61–65.

Sharma, S. (2012). Bioremediation: Features, strategies and application. *Journal of Pharmacy and Life Science, 2,* 201–213.

Sharma, A., & Rehman, M. B. (2009). Laboratory scale bioremediation of diesel hydrocarbon in soil by indigenous bacterial consortium. *Indian Journal of Experimental Biology, 47,* 766–769.

Testa, S. M., & Winegardner, D. L. (1991). *Restoration of petroleum contaminated aquifers.* London: Lewis Publishers.

United States Environmental Protection Agency (US EPA). (1996). *Method 3540C: Soxhlet extraction.* Washington: US EPA.

USEPA. (1995). Abstracts of remediation case studies. Office of Solid Waste and Emergency Response, US Environmental Protection Agency. Publication number EPA-542-R-95-001, Washington, DC.

Vasudevan, N., & Rajaram, P. (2001). Bioremediation of oil sludge contaminated soil. *Environment International, 26,* 409–411.

Vidali, M. (2001). Bioremediation: An overview. *Pure and Applied Chemistry, 73,* 1163–1172.

Ouyang, W., Liu, H., Murygina, V., Yu, Y., Xiu, Z., & Kalyuzhni, S. (2005). Comparison of bio-augmentation and composting for remediation of oily sludge: A field-scale study in China. *Process Biochemistry, 40*(2005), 3763–3768.

Modeling Relationship Between Cocoa Beans Commodity Export Volatility and Stock Market Index (KLCI)

Siti Nor Nadrah Muhamad, Izleen Ibrahim,
Nordianah Jusoh@Hussain, Siti Hannariah Mansor
and Wan Juliyana Wan Ibrahim

Abstract Volatility of stock market plays important roles in investment world. Since the export of cocoa has a great contribution in increasing Malaysia's revenue, it is important to know and understand the cocoa beans export volatility and the relationship with the stock market (KLCI). For that purpose, the well-known ARCH and GARCH family models were employed to investigate the relationship between volatility models and stock market (KLCI). These two models were also used to analyze empirically the volatility model of cocoa beans export. From the result, it showed that a unidirectional relation exists between stock market (KLCI) and cocoa bean volatility. The stock market (KLCI) had a causal effect on cocoa bean volatility in a short run.

Keywords Volatility · Stock market · ARCH · GARCH · Granger causality

1 Introduction

Malaysia is the fourth largest producer of cocoa and the tenth largest country that exports cocoa to selected countries all over the world. Malaysian Cocoa Board Web site reported that Malaysia exported 9.377 ton in 2004 and the amount went up to

---

S. N. N. Muhamad (✉) · I. Ibrahim · N. Jusoh@Hussain
S. H. Mansor · W. J. W. Ibrahim
Universiti Teknologi MARA, Cawangan Perlis, Arau, Perlis, Malaysia
e-mail: nadrahmuhamad@perlis.uitm.edu.my

I. Ibrahim
e-mail: izleen373@perlis.uitm.edu.my

N. Jusoh@Hussain
e-mail: dianah642@melaka.uitm.edu.my

S. H. Mansor
e-mail: sitihannariah@perlis.uitm.edu.my

W. J. W. Ibrahim
e-mail: wanjuliyana9@gmail.com

© Springer Nature Singapore Pte Ltd. 2018
R. Saian and M. A. Abbas (eds.), *Proceedings of the Second International Conference on the Future of ASEAN (ICoFA) 2017 – Volume 2*,
https://doi.org/10.1007/978-981-10-8471-3_64

17,927 ton in 2007. The production of cocoa beans is very volatile where it rapidly decreased to 7634 ton in 2008. Loss of production was due to weather conditions, spreads of diseases, and pests along with the competition from other crops such as palm oil particularly reversed the previous trend, when production grew at a rapid rate (Abdel Hameed et al. 2009). However, the production of cocoa beans increased dramatically in 2009 and steadily grew for six years until 2014. In 2016, the total export of 91,090 ton was recorded ("Export of Cocoa Beans and Cocoa Products," 2017). Malaysia cocoa export has a great contribution in generating and increasing not only the country's revenue but also the global cocoa industry. Domestic cocoa bean price is very volatile as it experiences an irregular change from time to time (Assis et al. 2010). The instability of cocoa prices creates major threat to producers, suppliers, consumers, and other parties involved in the cocoa market particularly in Malaysia. Usually, there is an inverse relationship between the world cocoa stock-to-grinding ratios and cocoa bean prices. Definitely, increasing prices are related to falling stocks-to-grindings ratio and vice versa. Therefore, it is important to know and understand the market volatility of cocoa export and the influence in the stock market (KLCI).

Volatility is a rate of measure of instability at which the price of a security increases or decreases for a given set of returns. In real financial markets, volatility is a measure of uncertainty or the risk of investment rate of return. It is measured by calculating the standard deviation or variance of the annualized returns over a period of time. It shows the range to which the price of a security may increase or decrease ("Volatility," 2017). A lower volatility means that the rate of returns or price is increasing steadily over a period of time. On the other hand, a higher volatility means that the investment rate of return can change dramatically over a short time period in either direction. There are many methods for measuring volatility. It can be done by using historical data or forecasting models based on the collected data. The familiar ways for modeling volatility are by using Autoregressive Conditional Heteroscedasticity (ARCH) and Generalized ARCH (GARCH) models.

2 Methodology

A common financial time series, the usual commodity spot price time series data always have properties like volatility clustering, excess kurtosis, and skewness. This study adopts the well-known Autoregressive Conditional Heteroscedasticity (ARCH) model and Generalized ARCH (GARCH) model to obtain the conditional volatility modeling of cocoa beans commodity exports in Malaysia.

2.1 Volatility Model

This study adopts the well-known ARCH model is the first model that provides a systematic framework for volatility introduced by Engle (2001) and developed by Bollerslve (1986) and Taylor as GARCH model in order to obtain the conditional volatility modeling of total export cocoa beans commodity in Malaysia and stock market Kuala Lumpur Composite Index (KLCI).

In financial time series, ARCH model is designed to capture volatility clustering and describes the conditional variance $(\sigma_t(\theta))$. An ARCH (p) model can be expressed as Eq. 1 represents the mean equation and $\varepsilon_t|I_{t-1}$ denotes the error term regarding the information in time $t - 1$. Where $I_{t-1} = \{y_{t-1}, x_{t-1}, y_{t-2}, x_{t-2}, \ldots\}$ and ε_t is normally distributed. Equation 2 shows the variance equation as h_t is the conditional variance of return series at time t. The shock ε_t of a return series is serially uncorrelated, but dependent of ε_t can be described by a simple function of its lagged values. Equation 3 shows conditional error distribution equation.

$$r_t = \mu + a_1 r_{t-1} + a_2 r_{t-2} + a_3 r_{t-3} + \ldots + a_p r_{t-p} + \varepsilon_t. \tag{1}$$

$$\varepsilon_t | I_{t-1} \sim N(0, h_t)$$

$$\sigma_t^2 = \alpha_0 + \alpha_1 \varepsilon_{t-1}^2 + \ldots + \alpha_p \varepsilon_{t-p}^2 \tag{2}$$

$$\varepsilon_t = \sigma_t z_t \tag{3}$$

where z_t is a sequence of independent and identically distributed random variable with mean 0 and variance 1, $\alpha_0 > 0$ and $\alpha_i \geq 0$ and $i = (1, 2, \ldots, p) > 0$.

Bollerslev (1986) introduced GARCH model. This model is the extended from ARCH model. The GARCH model is the most popular ARCH specification in empirical research, especially in modeling return series. GARCH model allows the conditional variance to be independent and linear combination of (q) lags of the squared residuals ε_t^2. The GARCH family models to be used in this study are the GARCH. The GARCH (p, q) conditional variance, where p is the order of GARCH terms while q is the order of ARCH process, is $\sigma_t^2 = \alpha_0 + \sum_{i=1}^{p} \alpha_i \varepsilon_{t-i}^2 + \sum_{j=1}^{q} \beta_j \sigma_{t-j}^2$ and $r_t = \mu_t + \varepsilon_t$ where $p \geq 0$, $q > 0$, $\alpha_0 > 0$, $\alpha_i \geq 0$, $\beta_j \geq 0$, $i = (1, 2, \ldots, q)$, $j = (1, 2, \ldots, p)$, and $(\alpha + \beta) < 1$ to ensure that the conditional variance is positive. Error (ε_t) is assumed normally distributed with mean zero and conditional variance σ_t^2.

2.2 Model Building

Fit the ARCH (p) for $p = 1, 2, \ldots, 9$. Then, GARCH (p, q) and EGARCH (p, q) model by selecting the lag of $(p = 1, q = 2)$ and $(p = 2, q = 1)$. Hence, select the adequate model based on the insignificance of the standardized residual $Q(12)$ and standardized square residual $Q^2(12)$. By comparing the models used the selection criteria by choosing the smallest AIC and SIC. For instance, the final model are usually has a greater and Log-likelihood (LL). If the selected model is inadequate among the models, then fit the AR (1)-GARCH (p, q), AR (1)-EGARCH (p, q) and AR (1),MA (1)–GARCH (p, q) and AR (1),MA (1)–EGARCH (p, q) model by selecting the lag of $(p = 1, q = 1)$, $(p = 1, q = 2)$, and $(p = 2, q = 1)$. Hence, choose among these models again.

2.3 Vector Autoregressive (VAR) and Vector Error Correction Model (VECM)

VAR model is to capture both dynamic and interdependent relationships between stock market (KLCI) and volatility model of cocoa beans commodity. Vector Autoregressive (VAR) method is suggested by Sims. In this stage, VAR has been the primary procedure to determine the relationship between stock market (KLCI) and total export of cocoa beans commodity. VAR model will be used when the variable in VAR model found to be co-integration.

3 Data Analysis and Result

The data consists of monthly series of stock market (KLCI) and monthly total export of commodity cocoa beans in millions of Malaysian ringgit. The analysis will cover a period from March 2004 to May 2015. The natural logarithm of the simple return called the "continuously compounded returns" or "log return" is employed to calculate the return of series as in Eq. 4.

$$r_t = \ln(1 + R_t) = \ln P_t - \ln P_{t-1} \qquad (4)$$

where r_t is the return at time t, ln is the natural logarithm, P_t is the current total monthly export, and P_{t-1} is previous total monthly export for agriculture commodity series. To further the analysis, the natural logarithm of monthly stock market index (KLCI) is employed in order to calculate the return of series as in Eq. 5;

$$s_t = \ln(S_t) \qquad (5)$$

where S_t is the monthly series of stock market index (KLCI) and s_t is monthly return of stock market index (KLCI).

The examinations of the time series plot in Fig. 1 are the total of monthly export (millions of Malaysian Ringgit) of cocoa beans in Malaysia that show the presence of trend component and non-stationary. The series present the growth and decline trend over the time period. Then, the data was transformed using natural logarithm of the simple return as Eq. 4. Figure 2 shows the monthly return series of monthly export (millions of Malaysian Ringgit) of cocoa beans in Malaysia.

Figure 2 shows the behavior of cocoa beans returns over the sample period. The monthly returns further suggest that volatilities of cocoa beans commodity are not constant throughout the study period. The plots appear to be stationary with most data locating around the mean of zero. The mean returns of all the series are constant, but the variances change over period of time. Volatility clustering is displayed in return plots where a condition of large changes tends to be followed by large changes while small changes tend to be followed by small changes (positive or negative changes).

ADF and PP tests were performed with the results as tabulated in Table 1. From the table, the null hypothesis of non-stationarity in the series is rejected.

Fig. 1 Monthly data plots export of cocoa beans in Malaysia from March 2004 to May 2015

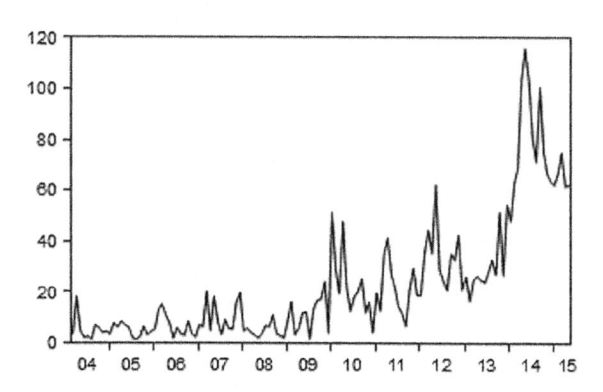

Fig. 2 Monthly return series plots export of cocoa beans in Malaysia from March 2004 to May 2015

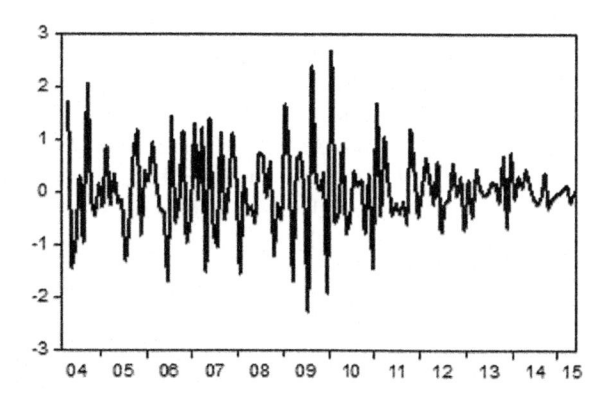

Table 1 Augmented Dickey-Fuller unit root (ADF) and Phillips-Perron (PP) test

	Level and intercept	
Test	ADF	PP
Commodity	Test statistic	Test statistic
Cocoa beans	−12.2292*	−41.6181*

*Significant at 1% significance level

Table 2 Descriptive statistics of monthly return for cocoa beans commodity

Descriptive statistics	Cocoa beans
Mean	0.0220
Median	−0.0419
Standard deviation	0.8226
Skewness	0.2649
Kurtosis	3.9607
Jarque-Bera normality test	6.7199 (0.0347*)
Sample Size	134

*Significant at 5% significance level. The value inside the bracket shows the probability value

Table 3 Result from ARCH LM test

	LM statistics	Probability
Cocoa beans	12.0959	0.0007*

*Significant at 5% significance level

The descriptive statistics of returns series for cocoa beans commodity is displayed in Table 2. The average price of total export for cocoa beans is 0.0220, and the skewness showed positively skewed which is skewed to the right. From the Jarque-Bera statistic, at 5% significance level, the null hypothesis of residuals following the normal distribution is rejected.

Heteroscedasticity test by using the Lagrange multiplier (LM) test is represented in Table 3. According to the results, the LM statistic for cocoa beans commodity is significant at 5% significance level. It concludes that cocoa beans commodity has the ARCH effect and suggesting the presence of ARCH in these models. Then, the analyses of ARCH (p) model for cocoa beans commodity were provided.

When the value of standardized and the square standardized residuals is found to be insignificant, the volatility equation is valid. The adequacy of the model is verified by comparing the lowest AIC and SIC as a best model to choose and describe the validity of volatility. Table 4 shows the adequate model which are defined as a best model to choose in order to describe the volatility model. AIC, SIC, log-likelihood values, and Ljung–Box statistic are utilized for choosing adequate models. GARCH (2, 1) is adequate to describe the conditional heteroscedasticity of the cocoa beans series. GARCH (2, 1) indicates that there are

Table 4 Results of fitted GARCH family models and diagnostic test

Commodity	Model	Log-likelihood	Box–Ljung	
			$Q(12)$	$Q^2(12)$
Cocoa beans	GARCH (2, 1)	−141.7848	24.5230 (0.0170)	14.4810 (0.2710)

The standardized residual $Q(12)$ and the squared standardized residuals, $Q^2(12)$ are distributed as $x^2(12)$ under the null hypothesis of no autocorrelation with the critical value of 26.217, 21.0261, and 18.5494 at 1, 5, and 10%, respectively. The value inside the bracket shows the probability value

Table 5 Estimate coefficient of volatility model for cocoa bean

Adequate model (volatility model)				
Cocoa bean				
Fitted model	GARCH (2, 1)			
Variance equation		Log-likelihood	−141.7848	
α_0	−0.005644 (0.0023)	AIC	2.1908	
α_1	0.232877 (0.1051)	SIC	2.2989	
α_2	−0.146387 (0.2944)	$Q(12)$	24.523 (0.017)	
β_1	0.911819 (0.0000)	$Q^2(12)$	14.481 (0.271)	

no serial correlations up to lag 12 describing the volatility series of cocoa beans. They indicate that the mean and variance equations are well fitted.

The adequacy of a fitted ARCH and GARCH model can be checked by examining the series. The Ljung–Box statistic $Q(12)$ and $Q^2(12)$ of standardized residual ($\hat{\varepsilon}_t$) and squared standardized residual ($\hat{\varepsilon}_t^2$) can be used to check the adequacy of the mean equation and test the validity of the volatility equation. When the value of standardized and the square standardized residuals is found to be insignificant, the volatility equation is valid. The adequacy of the model is verified by comparing the lowest AIC and SIC as a best model to choose and describe the validity of volatility. Table 5 shows the adequate model which are defined as a best model to choose in order to describe the volatility model. AIC, SIC, log-likelihood values, and Ljung–Box statistic are utilized for choosing adequate models. GARCH (2, 1) is adequate to describe the conditional heteroscedasticity of the cocoa beans series. GARCH (2, 1) indicates that there is no serial correlations up to lag 12 describing the volatility series of cocoa beans. They indicate that the mean and variance equations are well fitted.

4 Modeling Relationship Between Stock Market (KLCI) and Cocoa Beans

The co-integration requires the commodity need to be integrated at the same order; thus, cocoa stock market (KLCI) was integrated at the same order, I (1).

Table 6 Granger causality result of cocoa beans commodity export and stock market (KLCI)

Null hypothesis	Chi-square	P value	Result
Stock market does not Granger cause cocoa bean volatility	10.0965	0.0178**	KLCI → Cocoa Bean
Cocoa bean volatility does not Granger cause stock market	4.4959	0.2127	Cocoa Bean ↛ KLCI

**Significant at 5% significance level

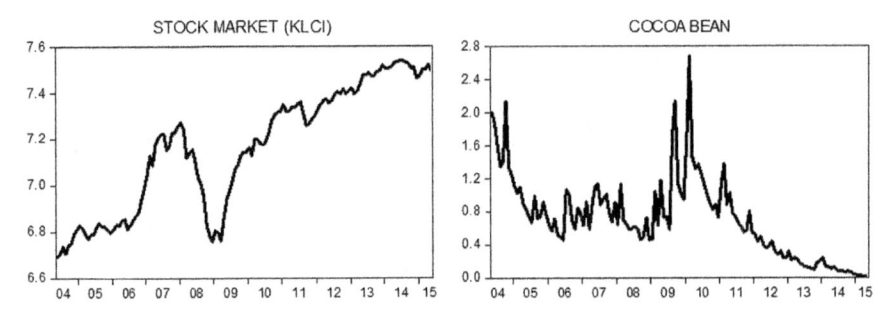

Fig. 3 Volatility plot of cocoa beans commodity export and stock market (KLCI)

Granger causality test is used for measuring the causality relation among variables. From Table 6, the result shows that a unidirectional relation exists between stock market (KLCI) and cocoa bean volatility. The null hypothesis is rejected at 5% significance level, saying that is stock market (KLCI) Granger-cause cocoa bean volatility. In a short run, stock market (KLCI) has a causal effect on cocoa bean volatility. Unidirectional relations exist between stock market (KLCI) and cocoa bean volatility. Accept the null hypothesis, saying that stock market does Granger-cause cocoa bean volatility.

The stock market (KLCI) shows the non-stationary series and the volatility clustering of cocoa bean shown in Fig. 3. The periods of low volatility are followed by periods of low volatility for a prolonged period; otherwise periods of high volatility are followed by periods of high volatility for a prolonged period.

5 Conclusion

This study examines the relationship between volatility model of cocoa bean commodities export in Malaysia and stock market Kuala Lumpur Composite Index (KLCI). The volatility of the series of cocoa bean commodity is estimated by using ARCH and GARCH models and found that GARCH (2, 1) is adequately fitted models for volatility. The best fit model was identified by comparing Akaike

Information Criterion (AIC) and Schwartz-Bayes Information Criterion (SIC), and the best fit model is the model that yields the least criterion value.

For instance, the relationship in short run is investigated by conducting Granger causality test. The stock market (KLCI) has a causal effect on cocoa bean volatility. The stock market (KLCI) has a causal effect on cocoa bean volatility and exists in unidirectional relationship.

References

Abdel, H., Awad, A., Hasanov, A., Idris, N., Abdullah, A. M., Mohamed Arshad, F., et al., (2009). *Supply and demand model for the Malaysian cocoa market*. Germany: University Library of Munich.

Assis, K., Amran, A., & Remali, Y. (2010). Forecasting cocoa bean prices using univariate time series models. *Researchers World, 1*(1), 71.

Bollerslev, T. (1986). Generalized autoregressive conditional heteroskedasticity. *Journal of Econometrics, 31*(1), 307–327.

Engle, R. (2001). GARCH 101: The use of ARCH/GARCH models in applied econometrics. *Journal of Economic Perspectives, 15*(4), 157–168.

Factors Influencing Internationalization of Malaysian Construction Firms

Nor Fazilah Omar, Che Maznah Mat Isa and Ruslan Affendy Arshad

Abstract Construction industry plays a vital role in the economic sector at both domestic and international levels. The international construction operations keep growing, and many construction firms are found continuously seeking opportunities in the new markets. However, in recent years, domestic prospects have not been good due to many unknown factors. In the last twenty years, Malaysian construction firms have delivered projects successfully in various regions around the world. The objective of this study is to determine the significant factors influencing internationalization of Malaysian construction firms. This study adopts a mixed method using questionnaire survey and semi-structured interviews. Thirty-five (35) out of 112 firms used as a sampling frame have participated with a response rate of 31%. The quantitative findings were further validated through in-depth interviews with six (6) firms currently venturing abroad. The findings indicate that to find opportunities abroad was the main predictor on Malaysian construction firms' internationalization. The firms chose to go abroad for a long-term opportunity for survival and longevity offered in international markets. Other important factors are because of the economic crisis, depreciation of ringgit, global drops in oil and gas prices and stagnant and saturated domestic market. The quantitative findings are supported by the interviews where one of the major factors on internationalization of Malaysian construction firms was due to limited projects offered in the domestic market. During the early stage of internationalization, firms will be exposed to various factors that impede their plans to expand. Thus, this study may

N. F. Omar (✉) · R. A. Arshad
Faculty of Architecture, Planning and Surveying, Universiti Teknologi MARA, Shah Alam, Selangor, Malaysia
e-mail: norfazilahomar@gmail.com

R. A. Arshad
e-mail: lanfnd@yahoo.com

C. M. M. Isa
Faculty of Civil Engineering, Universiti Teknologi MARA, Shah Alam, Selangor, Malaysia
e-mail: chemaznah_65@yahoo.com

assist promising local construction firms to understand important factors to internationalize in seizing the opportunities abroad.

Keywords Internationalization · Factors · Malaysian contractors

1 Introduction

For centuries, the construction industry has played a key role in the socioeconomic development, which ranges from providing infrastructure support to manufacturing products for other industries around the world. However, in recent years, the world economic downturn has impacted the Malaysian construction industry. It was recorded that since 1967, the Malaysian construction industry has suffered several economic downfalls and recovered (Abdullah et al. 2004). CIDB reported that Malaysia suffered economic recession during 1974–1978 due to the stock of building premises. Afterward, during 1984–1988, the recession was triggered by the US high-interest rate policy. In 1997, it was due to the Asian economic crisis and the latest was around mid-2007 until 2008 due to the global financial crisis where the rise of oil and diesel prices has dramatically influenced in the rise of material prices.

In the last twenty years, Malaysian construction firms have been constructing a lot of projects in regions that have never been thought possible by Malaysian professionals. According to the statistics provided by Construction Industry Development Board (CIDB), since 1982, Malaysian construction firms have done work valued at more than USD30 billion (CIDB 2015). Malaysian construction firms are known to have the competency and reliability to construct many diverse projects in more than 50 countries around the world including India, United Arab Emirates, and others involving diverse projects such as railways, highways, roads, water treatment systems, buildings, airports, bridges, power plants, and others. Despite the encouragement of Malaysian government through various plans, the number of projects secured by the construction firms and the number of construction firms operating in international markets have drastically reduced. The current study is in line with the three-step approach in the development of internationalization strategy by the Malaysian government (CITP: 2015–2020), namely to target potential markets, to prioritize the markets, and to define entry strategy such as entry mode. Thus, the main aim of this paper is to determine the factors influencing internationalization of construction firms.

2 Literature Review

This section outlines the Malaysian economic in relation to the construction industry and discusses on how Malaysian construction firms venture into international markets to grab the available opportunities.

2.1 Malaysian Economy and Construction Industry

Malaysian construction industry is currently not performing well due to the current world economic status (WEO 2015). It is shown that the Gross Domestic Product (GDP) and current account balance decreased from 2014 to 2015 and forecasted to decrease even further by 2016, resulting in the increasing of consumer price and unemployment for 2016. There are few external factors affecting Malaysia's economy such as the oil price plunge, China's economic slowdown, and the foreign capital outflow (Saleem 2015). Similar global economics scenario was experienced in 2007–2008 due to the rise of oil and diesel prices caused by supply disruptions and higher demand, and one of the consequences of the global financial crisis was the hike of material prices. In Bank Negara Malaysia Annual Report 2015, since September 2014, the continuous depreciation of the ringgit has increased concerns about the risk of higher inflation in Malaysia. The ringgit depreciation fosters the inflation in the country, and the weaker ringgit has not strengthened or improved import or exports as per what happened to Malaysia in 1997 due to the Asian Financial Crisis (AFC). The following section discusses the reasons companies venture abroad.

2.2 Reasons to Venture Abroad

Malaysian construction industry has largely been spurred by Malaysian government spending to develop the national infrastructure. However, there are not many mega projects in Malaysia going on. Even though construction industry contributed little to the national GDP, still it managed to indirectly increase GDP with other sectors such as educational institution, housing, commercial property, tourist attraction, and transportation infrastructure services such as airport, seaport, and road. Thus, the lack of mega projects is the general reason why some companies in Malaysia opt to go abroad.

Previous studies have shown that there were plenty of opportunities for Malaysian construction firms to work abroad (Mat Isa et al. 2006). Some of the reasons for the firms' foreign market expansion investigated by Abdul-Aziz were to solicit orders and exploit leverage to secure competitive advantages throughout the world to enhance their competitive standing. CIDB mentioned that the opening up of the Malaysian market due to globalization has resulted in increasing the competitive landscape, especially due to the increased amount of Free Trade Agreements (FTAs) entered into Malaysia over the last 10 years which reduces stagnancy. Several new FTAs have been agreed such as the Trans-Pacific Partnership Agreement (TPPA), putting the alert on local players to improvise and increase their competitive capabilities.

Thus, more import and export activities were anticipated and encouraged by Malaysian government due to liberalization through World Trade Organization (WTO) and ASEAN Free Trade Agreement (AFTA) (Che Senik et al. 2010). Several new FTAs have been agreed such as the Trans-Pacific Partnership Agreement (TPPA), putting the alert on local players to improvise and increase their competitive capabilities. As a result, international competitiveness was intensified.

Some of the push factors that initiated internationalization of Malaysian contractors are top management vision, long-term profitability, natural progress due to size, and specialist expertise technology, as well as the pull factors, such as stabilization of markets, globalization, and host government's incentives. In addition, Gunhan and Arditi recognized several reasons for construction firms' business expansion into the international markets, which includes stagnant and saturated domestic markets, spreading risks into new markets, competitive use of resources, and taking advantage of the opportunities offered by the global economy.

One of the main reasons that led many local firms to venture overseas is the stagnant domestic market. It was found that IJM and Eversendai decided to look for markets overseas mainly due to their failure to acquire local projects. Even though they have a huge reputation in Malaysia, this fact is very surprising. Nothing may trigger the internationalization of the IJM and Eversendai other than their failure to secure projects locally. According to CIDB, foreign players have been entering the local construction industry, as can be seen by the 10% increase in foreign construction firm market share from 2013 to 2014. Japan is the leading contributor of foreign with the biggest share in the foreign construction firm share market. They are slowly penetrating into several of projects, especially specialized and highly technical projects across Malaysia, indicating an increasing competition from foreign players within the industry.

Other reasons that have encouraged many firms to venture into foreign markets are to sustain the company's survival rate, to create a competitive advantage against other local competitors. Malaysia has been facing increased growth on the number of construction firms. The saturated Malaysian domestic market is also due to the high number of local contractors and also increased number of foreign contractors. Based on the statistic given by CIDB (2015), there are too many construction firms ranging from Grade 1, Grade 2, Grade 3, Grade 4, Grade 5, Grade 6, and Grade 7. There are more than 60,000 construction firms registered with CIDB. In addition, the presence of more foreign players has worsened this situation.

In addition, going abroad with the company's product and in its way of lengthening or renewing the product life cycle in other countries can avoid early market saturation in the home country. Thus, the firms can face the offset seasonal fluctuations and increase profits in general through exposure to a greater number of prospects in the overseas market.

Malaysian construction firms gain the competitive advantages by venturing abroad (Siong 2008). The firms' decisions to choose certain host countries to increase in global business activities are due to many available opportunities, for example in India and the Middle East. As stated in WEO (2015), if the economy is not performing well, it will impact the Malaysian construction industry.

In addition, going international can minimize the company's risk of losing market shares to foreign market. Awil and Abdul Aziz agreed that long-term profitability of the firm is connected with the survival of the firm. Thus, taking advantage of the global opportunities allows the Malaysian construction industry to lessen the effects of local market conditions and have greater control over its own development. The following section explains various opportunities available in international markets.

2.3 Opportunities Abroad

Many multinational corporations (MNEs) from developed countries have demonstrated their ability to internationalize in various industries including construction. There has been a growing interest by the US venture capital firms to turn to foreign countries worldwide in search of investment opportunities brought by globalization (Guler and Guillén 2009). At the same time, market liberalization which involves a freer and more liberalized trading environment has created more competitors around the world including the Malaysian home market. Majority of the Malaysian construction firms have firmed to revamp their aim by exploring international projects abroad for securing their businesses in a long run (Tan and Mohamed Ghazali 2011). The opportunities formed by overseas countries for infrastructure project have activated the process of internationalization of Malaysian construction firms.

India offers many opportunities in construction market due to the intensification of construction investments by their government, especially in the early years since the liberalization of its economy (Hamdan and Adnan 2008). One of the major factors helping the growth is India's ability to offer opportunities which will be sustained for the next 10 to 15 years. Other opportunities are offered from the Middle East regions, and decades after oil were discovered resulting in a rapid diversification of their economy as the construction sector boomed and foreigners are allowed to buy properties.

3 Methodology

This section describes the research design and methods used to analyze the data collected.

3.1 Research Design and Methods

This study adopts a quantitative and qualitative research methodology (mixed-methods research). The primary data were gathered through survey questionnaires and semi-structured interviews. A sampling frame is based on 112 construction firms under Grade 7 listed in the CIBD 2015/16 directory. In this study, the units of analysis are the Malaysian construction firms that engage in international business activities and have foreign market experience. The questionnaires were sent to all construction firms through mails, electronic mails, by hand, online survey and followed up with telephone surveys. The semi-structured interviews were conducted face-to-face and by telephone conversation with six (6) personnel of the selected firms. All 112 Malaysian construction firms are currently involved in international projects. However, only 35 responses were received giving a response rate of 31.25% which is acceptable, as most surveys carried out previously in Malaysia only resulted in a response rate of between 10 and 20% (Ramayah et al. 2010). Most of the questions adopted a five-point Likert scale to gather data for each construction of the research model. The respondents were requested to evaluate five (5) factors related to internationalization of construction firms based on the level of influence: 1 = Insignificant, 2 = Less significant, 3 = Quite Significant, 4 = Significant, and 5 = Very significant.

The second stage of data collection involved face-to-face interviews with the six (6) executives who gave their consent to be interviewed. This helped to validate the data collected from the questionnaires. Despite the rigor in data collection and validation, some might still argue that the findings must be treated with a certain degree of caution and limitation given the small sample population. The collected data were analyzed accordingly to answer the objectives by using descriptive, inferential, and content analysis as discussed in the following section.

3.2 Respondents' Profile

The following subsections explain the respondents' profile based on their designation and international experience.

Table 1 shows the respondents' background based on their designations. The respondents' designations are project directors (20%), project managers (42.9%), contract managers (14.3%), and other designation (project engineers) (22.9%).

Table 2 shows the respondents' number of years of international experience. The respondents also indicated their international experience; about 60% have experience between 1 and 5 years, 20% between 6 and 10 years, 11.4% between 11 and 15 years, and the rest (8.6%) has more than 15 years of experience. About 40% has greater international experience of more than 5 years. Hence, the profile findings indicate that the respondents fulfilled the required international construction background to participate in and provide reliable opinions for the survey.

Table 1 Respondents' designation

Position	Frequency	Percentage (%)
Project director	7	20.0
Project manager	15	42.9
Contract manager	5	14.3
Others	8	22.9
Total	35	100.0

Table 2 Respondents' international experience

Position	Frequency	Percentage (%)
1–5 years	21	60.0
6–10 years	7	20.0
11–15 years	4	11.4
>15 years	3	8.6
Total	35	100.0

The factors that influence internationalization of Malaysian construction firms are measured based the opinions of the thirty-five (35) respondents. The respondents gave their views based on the level of influence for each statement of the factors given a five-point Likert scale. To facilitate further analysis, the respondents' level of influence on each statement of factors related to internationalization of construction firms was analyzed using mean index, and standard deviation of each variable was calculated using descriptive analysis. This analysis consists of five questions that indicate the variables of factors namely: economic crisis; opportunities abroad; saturated local market; open market due to globalization and foreign players participation.

Table 3 shows the mean (M) ranking of factors influencing internationalization of Malaysian construction firms. The highest influential factor is associated with

Table 3 Mean, standard deviation, and ranking of factors (n = 35)

No.	Factors	Mean	Rank	Std. deviation
1	Economic crisis: suffers economic crisis, depreciation of ringgit, global drops in oil and gas prices and stagnant domestic market	4.09	1	0.951
2	Opportunities abroad: long-term opportunity for survival and longevity offered	3.80	2	0.719
3	Saturated local market: high number of construction firms (from Grade 1 to Grade 7) thus making the construction market become ultra-competitive and saturated	3.69	3	0.718
4	Open market due to globalization: Trans-Pacific Partnership Agreement (TPPA)	3.49	4	0.887
5	Foreign players participation: presence in Malaysian construction industry	3.49	5	0.781

"economic crisis" (M = 4.09, SD = 0.951), and the lowest influential factor is associated with "foreign players participation" (M = 3.49, SD = 0.781), while an average influential factor is associated with "saturated local market' (M = 3.69, SD = 0.718).

The highest influential factor explains that the main reason Malaysian construction firms to venture abroad was because of the economic crisis, depreciation of ringgit, global drops in oil and gas prices and stagnant domestic market. Economic downturn affects any country's GDP, and the local industry will be stagnant (WEO 2015). Therefore, many local construction firms seek opportunities abroad to minimize the effects of local market conditions and gain profits abroad. An earlier study by Abdul Rashid et al. (2006) also found that during the economic crisis of 1997, one of the construction firms was not able to proceed with the development projects as planned due to its financial problem. Thus, the firm has been aggressively seeking a joint venture partner to develop the property together and recover their investment as well as opportunity costs.

The least significant reason of Malaysian construction firm going abroad is because of foreign players' presence in Malaysia. This may be due to Malaysian government restrictions in open trade in protected industries. In many cases, it is imperative to have a local partner, usually a Bumiputra (ethnic Malay-owned) company, to effectively compete in the market as established under the Companies Act, 1965.

An average influential factor is the saturated local market. The saturated and competitive construction market in Malaysia has led many local firms to venture abroad as a means to effectively capitalize on special expertise (Mustaffa et al. 2012; Mat Isa et al. 2012). With insufficient local projects, bigger construction firms are starting to penetrate the international construction market. As a precaution, some Malaysian construction firms, they took step to go abroad to prevent early saturation of local market. Therefore, Malaysian construction firms must venture abroad to lessen the effects of local market conditions and have greater control over its own development (Mat Isa et al. 2012).

The following section presents the regression analysis carried out to determine the strongest predictor or factor influencing internationalization of Malaysian construction firms.

Table 4 shows the coefficients of predictor variables for factors influencing internationalization of Malaysian construction firms

By looking at the values under column labeled "Sig.", any factor corresponding to the value that is less than 0.05 will be the factor that contributes significantly to the predictive ability of the regression model. However, for this model, there is no statistical significance if the critical value is set to 0.05. Thus, when it was set to 0.10, the influential factor of "opportunity abroad" is found significant to be used in the regression model. The findings indicate that only one predictor, namely "opportunity abroad" factor, is statistically significant having Sig. values of 0.012 ($p < 0.10$). It is concluded that for the firms to internationalize, the respondents chose "opportunities abroad" as their most influential factors.

Table 4 Coefficients of predictor variables

Model predictors/ factors	Unstandardized coefficients		Standardized coefficients		
	B	Std. error	Beta		
(Constant)	1.291	0.965		1.338	0.191
Economic crisis	0.138	0.127	0.226	1.091	0.284
Saturated local market	0.081	0.138	0.100	0.586	0.562
Foreign players	0.066	0.157	0.089	0.423	0.675
Open market	0.032	0.114	0.050	0.284	0.778
Opportunities abroad	0.354	0.132	0.438	2.671	0.012

Internationalization offers opportunities to enhance profitability, improve capabilities and competitive advantages, to expand and establish international relationships (Mat Isa et al. 2014). There are opportunities offered by other countries such as in India and the Middle East. The growth of construction industry in India has boomed due to the intensification of construction investments by their government especially in the early years since the liberalization of its economy which offers long-term opportunities (Hamdan and Adnan 2008). Many contractors seek for these long-term opportunities in order to stay viable in the industry. Same goes to the Middle East where their economy rapidly diversified after the oil was discovered and they allowed foreigners to invest in their properties.

3.3 Validation of Empirical Findings Based on Semi-structured Interviews

The interview sessions were completed with six (6) respondents that have experience going abroad. The interview questions were designed to collect the crucial data and to validate and support the results obtained from the literature review and questionnaire surveys. The profiles of the interviewees are shown in Table 5. An average year of experience of the interviewees is about 6 years.

Table 6 shows the summary of transcript based on the interviews carried out . It can be summarized that the factors on why Malaysian construction firms ventured abroad are due to vast and ample opportunities to expand the business and also to sustain and survive due to domestic market are stagnant and slowing down. One of the thrusts in CITP is to boost growth of Malaysian construction industry abroad. This is one of the opportunities that Malaysian construction firms have to take advantage of since it becomes more competitive in order to survive in the construction industry.

Table 5 Profile of interviewee

Interviewee	Position	Qualification	Experience in Malaysia (years)	Experience abroad (years)
1	Operation quality manager	Master	16	4
2	QAQC manager	Degree	14	6
3	Senior contract manager	Degree	20	6
4	Project director	Degree	23	5
5	Project director	Professional	27	9
6	Planning manager	Degree	10	8

Table 6 Transcript summarization of factors

Interviewee	Feedbacks
1	New opportunities are vast and ample in the Middle East/African country/India as they are developing rapidly. Internationalization is also a way of expanding our business Government to government policy or joint venture arrangement with the local firm that has capability and strength at the host country helps to bring Malaysian construction firms
2	Potential business expansion and survival. While the local markets are slowing down and getting competitive, any business potential will be good for the firm to overcome the overhead cost of the firm operation
3	Construction firms are going internationally in order to learn the technology abroad which may not be available in Malaysia. Also, the contract values offered abroad are higher compared to local contract value Malaysian construction firms also have a chance to generate our technology, export expertise, bringing technology over to Malaysia or vice versa
4	Most Malaysian construction firms venture overseas project because they were invited by foreign firms. It is normally because of the recommendations by previous consultants due to a good track record Another reason is because of the demand in the international construction market
5	For greener pasture and business opportunity. Apart from planning to expand the firm and increase revenue, it is also a way of surviving in the construction industry Business offers, especially these few years where construction industry is slowing down in Malaysia Therefore, all of these construction firms must go abroad to look for business opportunity. It is about surviving
6	The major factor of Malaysian construction firms going abroad is due to limited big projects being offered in the local market Only a select few firms are big enough to take a major international job in terms of resources and financial It is good opportunity to enable them to do more complex projects in the future and to penetrate a new market that also garners good income

There is a vital need to improve the firm's capabilities and competitive advantages in order to achieve sustainable growth (Mat Isa et al. 2016). There is also demand in international market such as in the Middle East, African country, and India. Malaysian construction firm expands abroad to garner more profit to increase their revenue as the contract value offered abroad is higher than local contract values. Idris and Tey (2011) added that international joint venture will garner more profit, but the focal point is about to stay viable in the industry by penetrating the new market. Certain Malaysian construction firms have been invited and recommended by the previous consultant to perform project abroad due to their expertise and good track record. These are important measurements of the firm's capabilities; for example, WCT has been invited by their previous foreign consultants based on their experience and reputation in Sepang circuit project.

3.4 Consolidation of Findings Based on Quantitative and Qualitative Analysis

It can be summarized that the factors of internationalization of Malaysian construction firms ventured abroad are new opportunities to expand the business because the local market is slowing down and being more competitive in order to survive in the construction industry and gain more profit because the contract value offered abroad is higher than local contract values. Doing overseas projects enhances the firms' reputation and profile and increases the market share. Construction firms who are willing to venture abroad definitely will strengthen in terms of their capacities and capabilities (physically and financially) in order to compete and survive. Future direction for the study is to obtain information from the directors of the firms to reflect and strengthen the exact objectives of the study.

4 Conclusions

This paper investigates the factors of Malaysian construction firms' internationalization. The analysis revealed that the main influential factor lies in the economic woes and stagnant local market that have dampened the Malaysian local industries. A limited home market, with limited growth potential, may push firms to seek expansion abroad. Fierce competition in the home market is another factor pushing firms abroad, trying to generate more revenues and to lower costs, either through exploiting economies of scale or by lower production costs abroad. Malaysian construction firms internationalize for many reasons. Regression model further validates that the strongest predictor on internationalization of Malaysian construction firms is due to the abundance of opportunities abroad such as long-term opportunity for survival and longevity offered. Therefore, other construction firms can also internationalize to look for opportunities abroad in order to increase profitability, improve capabilities and competitive advantages, to expand and establish international relationships.

Acknowledgements This work is partially supported by Faculty of Architecture, Planning and Surveying, UiTM Shah Alam and Faculty of Civil Engineering, UiTM Shah Alam. The authors also gratefully acknowledge the helpful comments and suggestions of the reviewers, professionals, and managers from Malaysian construction firms which have improved the presentation.

References

Abdullah, F., Chiet, C. V., K. Anuar, K., & Shen, T. T. (2004). An overview on the growth and development of the malaysian construction industry. In *Workshop on Construction Contract Management 2004*. Universiti Teknologi Malaysia.

Abdul Rashid, A., Awil, A.-U., & Yi, H. S. (2006). Export capabilities of malaysian housing developers in the context of globalisation and liberalisation. In *Second NAPREC Conference, INSPENS*.

Bank Negara Malaysia Annual Report (2015).

Che Senik, Z., Mat Isa, R., Scott-Ladd, B., & Entrekin, L. (2010). Influential factors for SME internationalization: Evidence from Malaysia. *International Journal of Economics and Management, 4*(2), 285–304.

Construction Industry Transformation Plan (2015) CIDB Malaysia.

Guler, I., & Guillén, M. F. (2009). Institutions and the internationalization of US venture capital firms. *Journal of International Business Studies, 41*(2), 185–205. https://doi.org/10.1057/jibs.2009.35.

Hamdan, N. A., & Adnan, H. (2008). *An assessment of analysis on the penetration of Malaysian contractors into India*.

Idris, A., & Tey, L. S. (2011). Exploring the motives and determinants of innovation performance of Malaysian offshore international joint ventures. *Management Decision, 49*, 1623–1641.

Mat Isa, C. M., Adnan, H., & Endut, I. R. (2006). Malaysian contractors' opinions towards international market expansion. In *International Conference in the Built Environment in the 21st Century (ICiBE2006)* (Vol. 1, pp. 287–298).

Mat Isa, C. M., Mustafa, N. K., Mohd Saman, H., Mohd Nasir, S. R., & Che Ibrahim, C. K. (2012). Factors influencing Malaysian construction firms in venturing into international market. In *Innovation Management and Technology Research (ICIMTR), International Conference*.

Mat Isa, C. M., Mohd Saman, H., Preece, C. N., Jaafar, A., & Md Rani, N. I. (2014). Malaysian construction firms motives (Push and pull factors) for international market expansion. In *3rd International Conference on Technology Management, Business and Entrepreneurship*. June 23–24, 2014.

Mat Isa, C. M., Mohd Saman, H., Preece, C. N., & Che Ibrahim, C. K. (2016). Factors influencing Malaysian constructions firms' entry mode decisions into international markets. In *32nd Annual ARCOM Conference*, September 5–7, 2016 (Vol. 1, pp. 289–298). Manchester, UK: Association of Researchers in Construction Management.

Mustaffa, N. K., Adnan, H., & Zakaria, M. Z. (2012). Entry strategies for Malaysian construction related companies going abroad. *Australian Journal of Basic and Applied Sciences, 6*(6), 323–330.

Ramayah, T., Mohamad, O., & Jaafar, M. (2010). *Internationalization of Malaysian contractors*.

Saleem, S. (2015). *Malaysia's economic challenges: Implications of Ringgit's fall.*
Siong, J. C. (2008). The international expansion of Malaysian construction firms. *Master Dissertation.*
Tan, D. J. Z., & Mohamed Ghazali, F. E. (2011). Critical success factors for Malaysian contractors in international construction projects using analytical hierarchy process. In *EPPM, Singapore* (pp. 127–138).
World Economic Outlook. (2015). *Adjusting to lower commodity prices.*

The Optimization of Crop Production: A Case Study at the Farming Unit of UiTM Perlis

Nuridawati Baharom and Nurul Amalina Abu Bakar

Abstract Managing resources efficiently is essential in order to minimize the cost of operation of a company which produces products or services. In agriculture, farmers also need to allocate their resources such as land, labour hours, machinery and seedlings to avoid wastage. However, most farmers lack the knowledge on practical methods to allocate their resources. In this study, the linear programming approach was employed as a means to allocate the area of land for rubber, calamansi, *harumanis* and *kelapa matag*. This study also aimed to determine an optimal combination of crops among the selected crops that can maximize the annual revenue. The result was then compared with the traditional method. There were nine constraints of resources involved such as land, labour hours, machinery, pesticides, solid and liquid fertilizers, solid and liquid insecticides and seedlings. Based on the results, the best allocation was 11.34 and 11.51 acres of land for rubber and *harumanis*, respectively. However, calamansi and *kelapa matag* do not contribute high revenue and should be omitted from the system. The annual revenue obtained using the linear programming method also increased by 5% compared to the traditional method. Hence, the linear programming method proved to be the best method to allocate the limited resources in agriculture in order to maximize the revenue.

Keywords Linear programming · Resource allocation · Optimization

---

N. Baharom (✉) · N. A. A. Bakar
Faculty of Computer and Mathematical Sciences, Universiti Teknologi MARA Perlis, 02600 Arau, Malaysia
e-mail: nuridawati@perlis.uitm.edu.my

N. A. A. Bakar
e-mail: amalina_abubakar@ymail.com

1 Introduction

Thousands of years back in the pre-historic origin era, agriculture was one of the important activities to sustain and enhance human life. The development of the agricultural phase can be observed in the improvement of machinery used, for instance, from the use of baked clay sickle to the use of modern harvesting machinery. Apart from that, many years ago, agriculture was practised for personal or family consumption only. However, the revolution of agriculture has turned it into one of the vital economic contributions to a country. Since it contributes to the economic factor of a country, good planning of farm management is highly recommended.

What to produce, how to produce, how much to produce and how much to allocate limited resources are some of the common questions to the farmers. Farmers have used the traditional method in planning their farms as they may not know any jargon of the farm theory system. In the traditional method, farmers usually depend on their past experience and intuition as well as comparison with their neighbours' farms. Unfortunately, the traditional method eventually leads to wastage of limited resources as there is no appropriate system to determine the best amount of resources allocation. Hence, costs incurred by the farmers will increase.

In Malaysia, agricultural activities are under the supervision of the Ministry of Agriculture and agro-based industries. According to the Department of Statistics, Malaysia, the population of its citizens is projected to increase from 29,947,600 in 2013 to 32,441,200 in 2020. Therefore, the production of food as a basic need must be increased as population increases. The issue is how to increase food production with lower costs. Thus, farmers should have knowledge on allocation, distribution and utilization of resources. This is important in order to avoid wastage of resources as well as to reduce costs incurred. For instance, resources involved in agricultural activities are land, labour hours, weedicides, fertilizers, seedlings, insecticides and machinery.

Alongside with that, farmers have to allocate limited resources within optimal quantity in order to minimize costs as well as maximize output production such as revenue. Minimizing cost and maximizing revenue in farming activities are crucial in order to contribute a higher value to a country's economy. Apart from that, the optimal allocation of resources will help farmers to increase their wealth and improve their planting policy. Thus, this study was done to determine the best allocation for area of land such as rubber, calamansi, *harumanis* and *kelapa matag* that can maximize the annual revenue. This study also aimed to determine an optimal combination of crops among the selected crops. The result was then compared with the traditional method.

In real life, there are many methods that can be applied in the optimization of farm management. In particular, methods applied are linear programming (Majeke and Majeke 2010; Oni et al. 2013; Zira and Ghide 2013; Tanko et al. 2011; Abdus et al. 2010; Majeke et al. 2013; Ibrahim 2007), linear goal programming (Sen and Nandi 2012), mixed integer number programming (Issa et al. 2011) and fuzzy goal programming (Sharma et al. 2007). Generally, the linear programming method is widely used in solving the optimization of farm management.

2 Case Study Background

One of the goals in agriculture is to maximize the output of crops production in terms of revenue and profit. Resources for agriculture such as land area, fertilizers, labour, machinery, insecticides and weedicides must be allocated precisely in order to maximize the output of crop production in the Farming Unit of UiTM Perlis which has planted several combinations of crops such as rubber, calamansi, harumanis and coconut. However, currently, farmers are still using the traditional method in order to allocate areas of land for each crop since they may not know any practical method of resource allocation.

The use of the traditional method to allocate resources for each crop as well as to determine the crop combination is not practical as it leads to waste of resources and higher costs incurred. The resources are not fully used to maximize the revenue of crops production. Hence, the farming area needs a practical way, such as linear programming, to allocate the resources in order to ensure sustainable crop production as well as to prevent wastage.

2.1 Data Collection

In this study, data was collected at the Farming Unit of UiTM Perlis through an interview process. Data were based on input and output values such as the amount of resources required and revenue, respectively, on various crops such as rubber, calamansi, harumanis and kelapa matag in the year 2013. Next, the data for the crops investigated will be subjected to land areas, labour hours, machinery, weedicides, solid fertilizers, liquid fertilizers, solid insecticides, liquid insecticides and the number of seedlings. All the resource constraints involved are specifically based on crops that are being investigated. Land area will be measured in unit acres for each crop involved. Apart from that, labour will be measured in hours of work while machinery will be measured in units of machinery involved in crop production. Weedicides, liquid fertilizers and liquid insecticides are measured in units of litres while solid fertilizers and solid insecticides are measured based on their weight, which is kilogram.

2.2 Assumptions

In order to apply the method of linear programming in the maximization of the crop production revenue model, several properties of linear programming should be considered. Properties considered in this study are:

1. It assumes that all decision variables involved are nonnegative.
   In this study, areas of land allocated to each crop are in positive value.
2. It assumes that a condition of certainty does exist.
   All parameters in the model such as coefficient of constraints and objective function are constant as they do not change during the period of study.
3. It assumes that divisibility exists in the model.
   All coefficients of constraints and objective function can be subdivided into any fractional level. It considers that it may not be in a whole number only.
4. It assumes that all decision variables are independent.
   All decision variables in this study, which is land allocation for each crop, are assumed to be independent and there is no interaction with each other within their restricted limits which satisfy all the non-negativity constraints.
5. It assumes that linearity exists.
   The objective function and constraints in this study are expressed in linear equation and inequalities. All terms in objective function and constraints are of first degree only. Apart from that, all of the relationships in the models are based on the concept of linearity.
6. It assumes that proportionality does exist.
   Next, proportionality is applied for objective function and constraints. It indicates that each acre of land for each particular crop's activities is considered to receive the same management practice. For example, if one acre of *harumanis* needs 50.17 kg of solid fertilizer, then two acres of *harumanis* need 100.34 kg of solid fertilizer. Thus, the yield under the particular crop activity is constant.

2.3 Model Formulation

In this study, the linear programming model is adapted in the problem of optimization of crop production. The standard form of linear programming is written as in Eq. 1–3:

$$\text{Max or min } Z = \sum_{j=1}^{n} c_j x_j \qquad (1)$$

subject to

$$\sum_{j=1}^{n} a_{ij}x_j (\leq, \geq, =) b_i \qquad (2)$$

$$x_j \geq 0 \text{ for } i = 1,2,3,\ldots,m \text{ and } j = 1,2,3,\ldots,n \qquad (3)$$

The function Z is the overall value of objective function either to be maximized or minimized, x_j is the level of activity j (usually known as decision variable), c_j is an increase of each unit of level of activity jth that will result in an increase of Z value (usually known as coefficients of decision variables), b_i is the amount of resources i available to allocate in the activity, and a_{ij} is the amount of resources i is used by each unit of activity j (usually known as technical coefficients). The formulation for this study is as follows:

$$\text{Max } Z = c_1x_1 + c_2x_2 + c_3x_3 + c_4x_4 \qquad (4)$$

subject to

$$a_{11}x_1 + a_{12}x_2 + a_{13}x_3 + a_{14}x_4 \leq b_1 \text{. (Land constraint)} \qquad (5)$$

$$a_{21}x_1 + a_{22}x_2 + a_{23}x_3 + a_{24}x_4 \leq b_2 \quad \text{(Labour hour constraint)} \qquad (6)$$

$$a_{31}x_1 + a_{32}x_2 + a_{33}x_3 + a_{34}x_4 \leq b_3 \quad \text{(Machinery constraint)} \qquad (7)$$

$$a_{41}x_1 + a_{42}x_2 + a_{43}x_3 + a_{44}x_4 \leq b_4 \quad \text{(Weedicide constraint)} \qquad (8)$$

$$a_{51}x_1 + a_{52}x_2 + a_{53}x_3 + a_{54}x_4 \leq b_5 \quad \text{(Solid fertilizer constraint)} \qquad (9)$$

$$a_{61}x_1 + a_{62}x_2 + a_{63}x_3 + a_{64}x_4 \leq b_6 \quad \text{(Liquid fertilizer constraint)} \qquad (10)$$

$$a_{71}x_1 + a_{72}x_2 + a_{73}x_3 + a_{74}x_4 \leq b_7 \quad \text{(Solid insecticide constraint)} \qquad (11)$$

$$a_{81}x_1 + a_{82}x_2 + a_{83}x_3 + a_{84}x_4 \leq b_8 \quad \text{(Liquid insecticide constraint)} \qquad (12)$$

$$a_{91}x_1 + a_{92}x_2 + a_{93}x_3 + a_{94}x_4 \leq b_9 \quad \text{(Tree seedling constraint)} \qquad (13)$$

$$x_1, x_2, x_3, x_4 \geq 0 \quad \text{(Non - negativity variables)} \qquad (14)$$

The objective function is to maximize the annual revenue of crop production in the Farming Unit of UiTM Perlis where c_1, c_2, c_3 and c_4 are coefficients of the decision variable which is the annual revenue per acre for rubber, calamansi, harumanis and kelapa matag, respectively. In this study, the decision variable is land allocation area for rubber, calamansi, harumanis and kelapa matag, which is indicated by x_1, x_2, x_3 and x_4, respectively. On the other hand, b_i are the amount of resources involved in this study such as land area, labour hours, machinery, weedicides, solid fertilizers, liquid fertilizers, solid insecticides, liquid insecticides and seedlings where $i = 1, 2, 3, 4, 5, 6, 7, 8$ and 9 indicates i for land area, labour hours, machinery, weedicides, solid fertilizers, solid insecticides, liquid insecticides and seedlings, respectively. a_{ij} are the amount of resources used per acre for the crop mentioned in this study.

2.4 Model Solving

Tables 1 and 2 show the amount of resources used and the annual revenue gained per acre for rubber, calamansi, *harumanis* and *kelapa matag*, respectively, in the year 2013 at the Farming Unit of UiTM Perlis. The amount of resource allocation per acre and the annual revenue gained per acre were obtained by using the traditional method.

The amount of annual revenue gained and the amount of resources used per acre of each crop as in Tables 1 and 2 were used to formulate the linear programming model as shown below:

Table 1 Resources used per acre and total available resources

	Rubber, x_1	Calamansi, x_2	Harumanis, x_3	Kelapa Matag, x_4	Total resources available
Land area (acre)	1	1	1	1	25.8
Labour hour (h)	134.67	55.38	75.13	92.57	2392
Machinery (unit)	1	1	1	1	26
Weedicide (l)	2.67	8.31	4.43	0.86	83.7
Solid fertilizer (kg)	146.33	30.77	50.17	285.71	3495
Liquid fertilizer (l)	0	0	2.39	0	27.5
Solid insecticide (kg)	0	1.85	0.37	0	6.6
Liquid insecticide (l)	0	1.69	1.11	0	14.97
Seedling (unit)	146	38	69	155	2806

Table 2 Annual revenue per acre

Annual revenue (RM)	Rubber, x_1	Calamansi, x_2	Harumanis, x_3	Kelapa Matag, x_4
Annual revenue per acre	1637.73	382.46	6934.17	504.97

The Optimization of Crop Production: A Case Study ...

$$\text{Max } Z = 1637.73x_1 + 382.46x_2 + 6934.17x_3 + 504.97x_4 \tag{15}$$

subject to

$$x_1 + x_2 + x_3 + x_4 \leq 25.8 \quad \text{(Land constraint)} \tag{16}$$

$$134.67x_1 + 55.38x_2 + 75.13x_3 + 92.57x_4 \leq 2392 \quad \text{(Labour hr constraint)} \tag{17}$$

$$x_1 + x_2 + x_3 + x_4 \leq 26 \quad \text{(Machinery constraint)} \tag{18}$$

$$2.67x_1 + 8.31x_2 + 4.43x_3 + 0.86x_4 \leq 83.7 \quad \text{(Weedicide constraint)} \tag{19}$$

$$146.33x_1 + 30.77x_2 + 50.17x_3 + 285.71x_4 \leq 3495 \quad \text{(Solid fertilizer constraint)} \tag{20}$$

$$2.39x_3 \leq 27.5 \quad \text{(Liquid fertilizer constraint)} \tag{21}$$

$$1.85x_2 + 0.37x_3 \leq 6.6 \quad \text{(Solid insecticide constraint)} \tag{22}$$

$$1.69x_2 + 1.11x_3 \leq 14.97 \quad \text{(Liquid insecticide constraint)} \tag{23}$$

$$146x_1 + 38x_2 + 69x_3 + 155x_4 \leq 2806 \quad \text{(Tree seedling constraint)} \tag{24}$$

$$x_1, x_2, x_3, x_4 \geq 0 \quad \text{(Non-negativity variables)} \tag{25}$$

In order to solve the linear programming problem by using the simplex method, the above linear programming model is converted to a model with slack variables. The s_1, s_2, s_3, s_4, s_5, s_6, s_7, s_8 and s_9 are slack resources such as land area labour hours, machinery, weedicides, solid fertilizers, liquid fertilizers, solid insecticides, liquid insecticides, and seedlings, respectively. The model is as below:

$$\text{Max } Z = 1637.73x_1 + 382.46x_2 + 6934.17x_3 + 504.97x_4$$
$$+ 0s_1 + 0s_2 + 0s_3 + 0s_4 + 0s_5 + 0s_6 + 0s_7 + 0s_8 + 0s_9$$

subject to

$$x_1 + x_2 + x_3 + x_4 + s_1 = 25.8 \quad \text{(Land constraint)} \tag{26}$$

$$134.67x_1 + 55.38x_2 + 75.13x_3 + 92.57x_4 + s_2 = 2392 \quad \text{(Labour hour constraint)} \tag{27}$$

$$x_1 + x_2 + x_3 + x_4 + s_3 = 26 \quad \text{(Machinery constraint)} \tag{28}$$

$$2.67x_1 + 8.31x_2 + 4.43x_3 + 0.86x_4 + s_4 = 83.7 \quad \text{(Weedicide constraint)} \tag{29}$$

$$146.33x_1 + 30.77x_2 + 50.17x_3 + 285.71x_4 + s_5 = 3495 \quad \text{(Solid fertilizer constraint)} \tag{30}$$

$$2.39x_3 + s_6 = 27.5 \quad \text{(Liquid fertilizer constraint)} \tag{31}$$

$$1.85x_2 + 0.37x_3 + s_7 = 6.6 \quad \text{(Solid insecticide constraint)} \tag{32}$$

$$1.69x_2 + 1.11x_3 + s_8 = 14.97 \quad \text{(Liquid insecticide constraint)} \tag{33}$$

$$146x_1 + 38x_2 + 69x_3 + 155x_4 + s_9 = 2806 \quad \text{(Tree seedling constraint)} \tag{34}$$

$$x_1, x_2, x_3, x_4, s_1, s_2, s_3, s_4, s_5, s_6, s_7, s_8, s_9 \geq 0 \quad \text{(Non-negativity variables)} \tag{35}$$

Since this study involved more than two decision variables, it employed the simplex method to solve the problem. The application of the simplex method in linear programming was assisted by the computer software, QM for Windows.

3 Case Study Results

By using the traditional method, the Farming Unit had allocated 6 acres, 1.3 acres, 11.5 acres and 7 acres of land for rubber, calamansi, *harumanis* and *kelapa matag*, respectively. In order to maximize the revenue gained by these selective crops, linear programming suggests that the Farming Unit should allocate an optimal area of land such as 11.34 and 11.51 acres of land for rubber and *harumanis*, respectively. From the results, the land allocation area for rubber does show much difference while the land allocation area for *harumanis* has slightly increased compared to the current practice. However, by applying the optimization process of linear programming, it is suggested that calamansi and *kelapa matag* should be omitted in their system theoretically due to the reason of unprofitability. This can be revealed by the value of 0 acre in Table 3.

Hence, it can be concluded that the best combination of crops which will contribute to the maximization of annual revenue is rubber and *harumanis* while calamansi and *kelapa matag* do not contribute much revenue in selective crops production at the Farming Unit of UiTM Perlis.

Based on the results in Table 4, only labour hours and liquid fertilizers are fully utilized. However, land, machinery, weedicides, solid fertilizers, solid insecticides, liquid insecticides and seedlings are not fully utilized to achieve maximum annual revenue.

Slack resource is commonly known as unused resource. It demonstrates that 2.95 acres of land are not fully utilized out of the overall available area of land for

Table 3 Result of land allocation area

Crops	Value (acre)
Rubber, x_1	11.34
Harumanis, x_3	11.51
Calamansi, x_2	0
Kelapa Matag, x_4	0

Table 4 Slack values of resources

Resources	Slack value
Land	2.951 acres
Machinery	3 units
Weedicides	2.44 l
Solid fertilizers	1257.94 kg
Solid insecticides	2.34 kg
Liquid insecticides	2.20 l
Seedlings	356 units
Labour hour	0 h
Liquid fertilizer	0 l

selected crops. Besides, there are three units of unused machinery and 2.44 l of unused weedicide in the optimal plan. Other than that, 1257.94 kg of solid fertilizers, 2.34 kg of solid insecticides and 2.20 l of liquid insecticides are unused resources in the optimal plan of maximization of revenue for selected crops. Lastly, there are slack values of about 356 units of seedlings which should not be planted. However, it does not mean that all of the 356 units are unused resources as it is impractical to identify which trees are unproductive.

Table 5 illustrates the total amount of resources to be allocated for rubber and *harumanis* in order to achieve the maximum revenue. It shows that only 22.85 acres of land, 2391.91 h of labour, 23 units of machinery, 81.27 l of weedicides, 2237.06 kg of solid fertilizers, 27.5 l of liquid fertilizers, 4.26 kg of solid insecticides, 12.78 l of liquid insecticides and 2450 units of seedlings are allocated to these two crops. Thus, it revealed that proper allocation of resources can lead to reducing inventory costs such as the cost of buying, ordering and storing as well as avoiding wastage at the same time.

Tables 6 and 7 show the comparison of land allocation area and annual revenue is obtained by using the traditional and linear programming methods. The land

Table 5 Total amount of resources which should be allocated

Resources	Rubber, x_1	Harumanis, x_3	Total resources allocated	Total resources available
Land area (acre)	11.34	11.51	22.85	25.80
Labour hour (h)	1527.16	864.75	2391.91	2392
Machinery (unit)	11	12	23	26
Weedicide (l)	30.28	50.99	81.27	83.70
Solid fertilizer (kg)	1659.38	577.46	2237.06	3495
Liquid fertilizer (l)	0	27.50	27.50	27.50
Solid insecticide (kg)	0	4.26	4.26	6.6
Liquid insecticide (l)	0	12.78	12.78	14.97
Tree seedling (unit)	1656	794	2450	2806

Table 6 Comparison of land allocation (in acre) by using the traditional and linear programming methods

Crops	Traditional	Linear programming
Rubber	6.00	11.34
Calamansi	1.30	0
Harumanis	11.50	11.51
Kelapa matag	7.00	0

Table 7 Annual revenue obtained by using the traditional and linear programming methods

Traditional	Linear programming
RM93601.40	RM98362.90

allocation obtained by applying the linear programming method contributes a higher annual revenue compared to the allocation of land by using the traditional method. The annual revenue increased by 5% from the same piece of land if the Farming Unit applies the method of linear programming in allocating their land for each crop planted.

4 Post-optimality Analysis

In this study, post-optimality analysis is done in order to assist the Farming Unit to do experiment with the changes in value of input parameters such as the coefficient of decision variables which is revenue of each crop production per acre and total available resources.

Tables 8 and 9 show the result of post-optimality analysis generated by QM for Windows. The ranging for the value of coefficient of objective function and resources constraints allows the analysis on sensitivity of the model towards changes in value of coefficient of objective function and amount of available resources which is the right-hand side value. The coefficients of objective function are the annual revenues of selected crops production per acre. Theoretically, within the value of lower bound and upper bound, the optimal solution is still considered as optimal value in maximization annual of revenue.

Table 8 Result of post-optimality analysis

Variable	Lower bound	Original value	Upper bound
Rubber (RM)	930.04	1637.73	12429.45
Calamansi (RM)	$-\infty$	382.46	673.48
Harumanis (RM)	913.66	6934.17	∞
Kelapa Matag (RM)	$-\infty$	504.97	1125.75

Table 9 Result of post-optimality analysis

Constraint	Lower bound	Original value	Upper bound
Land area (acre)	22.85	25.80	∞
Labour hour (h)	864.47	2392	2515.17
Machinery (unit)	22.85	26	∞
Weedicide (l)	81.26	83.70	∞
Solid fertilizer (kg)	2237.06	3495	∞
Liquid fertilizer (l)	0	27.50	29.48
Solid insecticide (kg)	4.26	6.60	∞
Liquid insecticide (l)	12.77	14.97	∞
Seedling (unit)	2449.98	2806	∞

In the aspect of annual revenue of crops selected in objective function, it indicates the value of lower bound for calamansi and *kelapa matag* as negative infinity. However, the upper bound value for calamansi and *kelapa matag* is RM673.48 and RM1125.75, respectively. The negative values of the lower bound for calamansi and *kelapa matag* are logical because in optimal plan suggested by the linear programming method; both crops are not suggested for the study because the revenues are too low.

Nevertheless, upper bound value for *harumanis* is given as positive infinity. In addition, the value of positive infinity obtained is due to the area of land allocated by traditional method and linear programming method was almost the same as the difference of area was only about 0.01 acre. However, the value of upper bound given for rubber is RM12429.45. Meanwhile, the lower bound values for rubber and *harumanis* are RM930.04 and RM913.66, respectively.

The lower bound values for land area, machinery, weedicide, solid fertilizer, solid insecticide, liquid insecticide and seedling are 22.85 acres, 22.85 units which are approximate to 23 units, 81.26 l, 2237.06 kg, 4.26 kg, 12.77 l and 2449.98 units which are approximate to 2450 units, respectively. On the other hand, the upper bound values for resources mentioned previously are given as positive infinity. This positive infinity value of upper bound for the resources proved that these resources are not fully utilized in maximization of selected crops' production. Thus, any additional unit to these resources will worth nothing as it will not affect the optimal solution drastically.

However, the values of upper bound for labour hour and liquid fertilizer are 2515.17 h and 29.48 l, respectively. Moreover, the lower bound values for labour hour and liquid fertilizer are 864.47 h and 0 l, respectively. The values of upper and lower bound for labour hour and liquid fertilizer are generated since both of the resources are fully utilized in maximization of revenue in optimal plan.

In post-optimality analysis, the value of negative infinity given for lower bound values of calamansi and *kelapa matag* illustrates that the revenue of calamansi and *kelapa matag* can decrease to any amount without affecting the optimal solution extremely. Furthermore, upper bound value which is given as positive infinity for revenue of *harumanis* and amount of total available resources for land area, machinery, weedicide, solid fertilizer, solid insecticide, liquid insecticide and seedling indicates that the value can increase to any level without affecting the optimal solution. In addition, if the revenues and total available resources change within the lower and upper bound values, the optimal solution might slightly increase or decrease, but it will not affect optimal solution severely and the optimal solution will remain as optimal in maximization of annual revenue selected for crops' production in the Farming Unit.

5 Conclusions

Agriculture was once a vital aspect in human life in order to sustain life. However, due to the rapid advancement of economic growth, it has turned into one of the economic contribution factors to a country. In general, there are two vital aspects that need to be considered when dealing with agricultural activities such as the production of output and the costs incurred. Farmers have used the traditional method which depends on past experience to allocate their land area for crops planted. However, the traditional method is outdated since it may lead to wastage of resources and higher costs incurred.

This study was done at the Farming Unit of UiTM Perlis. There are three objective setups for this study, that is, to allocate areas of land for selected crops which will maximize their annual revenue, to identify a combination of crops that will maximize annual revenue and to compare the annual revenue obtained by the traditional and linear programming methods. In addition, this study will help the Farming Unit to allocate specific areas of land for the crops involved in order to improve the amount of annual revenue obtained. Apart from that, in the field of knowledge, a wider understanding in the application of the linear programming method can be learned as it is not just applicable in the industrial sector, but also in the farming sector.

In this study, four types of crops were involved, namely rubber, calamansi, *harumanis* and *kelapa matag*. They were subjected to nine resource constraints, which are land area, labour hours, machinery, weedicides, solid and liquid fertilizers, solid and liquid insecticides and seedlings. Hence, this study demonstrated that the linear programming method is a practical and efficient way to allocate limited land for selected crops planted, leading to the maximization of annual revenue obtained by the Farming Unit.

In this study, all the objective setups were successfully achieved. The first objective is to find the best allocation for area of land that will maximize the annual revenue. Based on the result obtained, the linear programming method suggested that 11.34 and 11.51 acres of land are allocated for rubber and *harumanis*, respectively, as both of the crops contribute high revenue to the Farming Unit. However, zero acre of land was allocated for calamansi and *kelapa matag*. This indicates that both crops should be omitted from the study due to their unprofitability. In aspect resources, it was illustrated that only two resources were fully used in the optimum plan, which are labour and liquid fertilizers. Meanwhile, other resources such as land area, machinery, weedicides, solid fertilizers, solid and liquid insecticides as well as seedlings were not fully utilized.

On the other hand, the second objective that is to find the optimal combination of crops was also achieved in this study. It revealed that the best combination of crops which should be planted by the Farming Unit is rubber and *harumanis*. This is because calamansi and *kelapa matag* were omitted from the farming system since calamansi and *kelapa matag* did not contribute to higher annual revenue for the Farming Unit. In the final analysis, by conducting this study, a comparison of the annual revenue obtained by using the traditional and linear programming methods was made. It proved that the annual revenue obtained by the linear programming method is superior by about 5% compared to the annual revenue obtained by the traditional method.

6 Recommendations

This study should be further improved in order to obtain better results which will be more beneficial to the Farming Unit. The first recommendation is to apply a method of goal programming to farm management. This is because goal programming is not only focused on one goal but it also focuses on maximization of output and minimization of cost of the resources constraints. It will help the Farming Unit to achieve multiple goals at one time (Sen and Nandi 2012). Other than goal programming, the method of fuzzy linear programming is also efficient to be implemented in the optimization of farm management. In fuzzy linear programming, the optimal solution is given as a trapezoidal fuzzy number and set up as a ranking fuzzy number (Masoud 2013). However, both the methods suggested for improving the research were not applied due to the limitation of the time frame in conducting the research.

Lastly, it is recommended to study the comparison between the revenue obtained by the optimization process whereby it involves long-term and short-term crops planted in the Farming Unit. Examples of long-term crops are rubber and coconut, while examples of short-term crops are papaya and vegetables. The study involving the optimization process for both types of crops is interesting since it enables the Farming Unit to determine which type of crops will contribute to higher profit and incur less cost. Apart from that, the resources required by both types of crops are slightly different due to the nature of the crops themselves.

In short, it is crystal clear that the method of linear programming is more practical as compared to the traditional method. It will help the farmers to allocate their land and other resources in a proper way which will maximize their revenue within the amount of available resources. Thus, in order to improve this current study, several recommendations are suggested such as to apply the method of goal programming, fuzzy linear programming as well as to compare the revenue obtained by short-term and long-term crops by the optimization process.

References

Abdus, S., Abedullah, M. A., & Shahzad, K. (2010). Economics of conventional and partial organic farming systems and implications for resource utilization in Punjab (Pakistan). *Pakistan Economic and Social Review, 48*(2), 245–260.

Ibrahim, H. (2007). Determining optimal maize-based enterprise in Soba local government area of Kaduna State, Nigeria. *Journal of Agriculture, Food, Environment and Extension, 6*(2), 1–5.

Issa, P., Habib, P., Majid, D., Mehdi, D., & Abolfazi, G. (2011). Performance evaluation of agricultural inputs in crops production of Zahak Payame Noor University agricultural faculty farm by linear programming. *Journal of Development and Agricultural Economics, 3*(3), 107–112.

Majeke, F., & Majeke, J. (2010). A farm resource allocation problem: Case study of small scale-commercial farmers in Zimbabwe. *Journal of Sustainable Development in Africa, 12*(2), 315–320.

Majeke, F., Majeke, J., Mufandaedza, J., & Shako, M. (2013). Modelling a small farm livelihood system using linear programming in Bindura, Zimbabwe. *Research Journal of Management Sciences, 2*(5), 20–23.

Masoud, S. (2013). The simplex method for solving fuzzy number linear programming problem with bounded variables. *Journal of Basic and Applied Scientific Research, 3*(3), 618–625.

Oni, N. O., Osuntoki, N. B., Rahaman, A., & Amao, O. D. (2013). Profit maximization among dry season vegetable farmers. *African Journal of Mathematics and Computer Science Research, 6* (4), 72–76.

Sen, N., & Nandi, M. (2012). A goal programming approach to rubber-tea intercropping management in Tripura. *Asian Journal of Management Research, 3*(1), 178–183.

Sharma, D. K., Jana, R. K., & Gaur, A. (2007). Fuzzy goal programming for agricultural land allocation problems. *Yugoslav Journal of Operations Research, 17*(1), 31–42.

Tanko, L., Baba, K. M., & Adenji, O. B. (2011). Analysis of the competitiveness of mono-crop and mixed crop enterprises in farming system of smallholder farmers in Niger State, Nigeria. *International Journal of AgriScience, 1*(6), 344–355.

Zira, B. D., & Ghide, A. A. (2013). Illustrative optimal portfolio selection for agroforestry crops: A linear programming approach. *International Journal of Management and Social Sciences Research, 2*(8), 139–141.

Preliminary Study on Food Preferences of Common Palm Civet (*Paradoxurus hermaphroditus*) in Captivity

Zakirah Zaki, Tun Firdaus Azis, Syafiqah Rahim and Sarina Mohamad

Abstract Common Palm Civet (*Paradoxurus hermaphroditus*) is one of the nocturnal mammals distributed in South and Southeast Asia which facing several threats and one of the least studied mammals in Malaysia. The objective of this experiment is to identify the food preferences of this species based on different parameter, which are color, taste, size, and odor. Two civets with different gender had been placed in two different captivities and were fed with papayas (*Carica papaya*) and pears (*Pyrus L.*). The experiment had been done for 40 days. The result showed that the male civet preferred fruits with characteristics of largest size, blue color, sweet taste, and strong odor. In contrast, female civet preferred fruits with the largest size, purple color, sweet and salty taste, and medium odor. This study will be beneficial to the farmers and authority of wildlife in implementing pest control for trapping civets, and it will be guidance for the zookeepers and the civets' owners to know the best type of foods for civets' diet.

Keywords Common palm civet · *Paradoxurus hermaphroditus* · Frugivore · Food preferences

---

Z. Zaki (✉) · T. F. Azis · S. Rahim · S. Mohamad
Universiti Teknologi Mara Cawangan Perlis, Arau, Malaysia
e-mail: zakirahzaki_95@yahoo.com

T. F. Azis
e-mail: firdausbinazis@gmail.com

S. Rahim
e-mail: nursyafiqahrahim@gmail.com

S. Mohamad
e-mail: sarin618@perlis.uitm.edu.my

© Springer Nature Singapore Pte Ltd. 2018
R. Saian and M. A. Abbas (eds.), *Proceedings of the Second International Conference on the Future of ASEAN (ICoFA) 2017 – Volume 2*,
https://doi.org/10.1007/978-981-10-8471-3_67

1 Introduction

Common Palm Civet is also known as 'Toddy Cat' or its scientific name *Paradoxurus hermaphroditus*, which is one of the least studied mammals as it is strictly nocturnal and highly secretive in nature (Krisnakumar et al. 2002). The civet is a nocturnal mammal, primarily omnivore but can be classified as highly frugivore and mainly arboreal (Choudhury 2015). This species feeds mostly on fruits matter, followed by animal matter and has the highest feeding on non-plant products during summer season (Krishna et al. 2015). It plays crucial roles in tropical forest ecosystems, as predator, prey, and seed disperser, particularly large-sized seeds (Mallick 2006). They usually hiding under the rooftops, abandoned houses and usually, they will immediately leave their places if their territories had been disturbed (Raj and Sharma 2013). The civet also had been known as 'pest' by the poultry farmers (Jothish 2011). Traits influencing the civet preferences including fruit color, size, thickness, pulp, number of seeds, and persistence of fruits on plants (Mudappa et al. 2010). Fruits feed by civets were mostly smaller than 1 cm, multi-seeded pulpy berries and drupes with moderate to higher water content, along with several large fruits more than 2 cm (Mudappa et al. 2010). By using Brown Palm Civet (*Paradoxurus jerdoni*), the dominants fruits consumed were small-seeded drupes and berries usually blue and green in color, with moderate to high water content in the pulp (Mudappa et al. 2010). The aim of this experiment is to identify the color, taste, odor, and size of fruits that were preferred by the Common Palm Civets (*P. hermaphroditus*).

2 Materials and Methods

Subjects that had been used are male Common Palm Civet and female Common Palm Civet. Lists of raw materials that had been used are dyes (red, blue, yellow, and purple), papayas (*Carica papaya*), and pears (*Pyrus L.*), salt, vinegar, sugar, water, and banana flavoring. The methods that had been used for this experiment were preparation of captivity, preparation of fruits, feeding civets with fruits in captivity, data collection, and statistical analysis.

Firstly for preparation of captivity, two captivities had been clean up without any leftover food, feces, and dirt. It is to ensure the civets to feel comfortable with the environment of the captivity. A male and female civet had been placed into two separate captivities.

Secondly, it is continued with preparation of fruits. Pears had been used for color, taste, and odor testing while papayas had been used for size testing. The pears had been cut into 2 cm^3 for 15 cubes while the papaya had been cut into 1, 3, and 4 cm^3 and for 15 cubes. For color testing, the cubes had been soaked into different dyes, while for odor testing the cubes had been soaked into banana flavoring. For taste testing, the cubes had been soaked into sugar, salt, and vinegar solution. For size testing, the cubes had been cut into different sizes, 1, 3, and 4 cm^3.

Thirdly, it is to feed civets with fruits in captivity. Each parameter had been observed for 10 days starting at 8 p.m. until 12 a.m.

Lastly, it collects data. The final weight of color, taste, and odor had been measured while for size parameter, the total weight consumed had been measured. The reading had been recorded after 12 a.m.

3 Result

The result has been observed and collected for 40 days. The total weight of fruits consumed (size) and final weight (color, taste, and odor) for both civets had been weighted and recorded. The percentage of mean for food preferences had been constructed into Figs. 1, 2, 3 and 4.

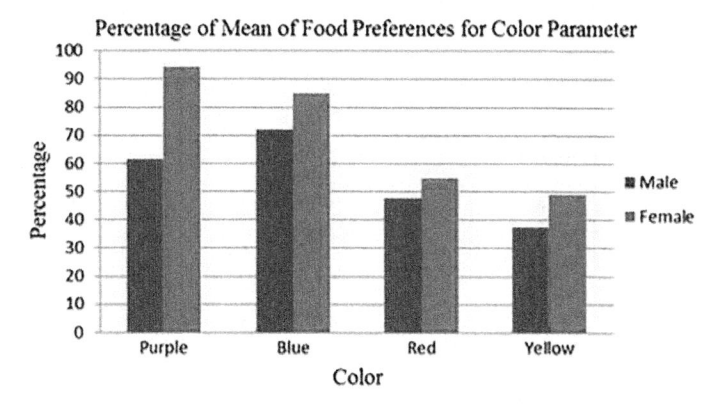

Fig. 1 Percentage mean for color

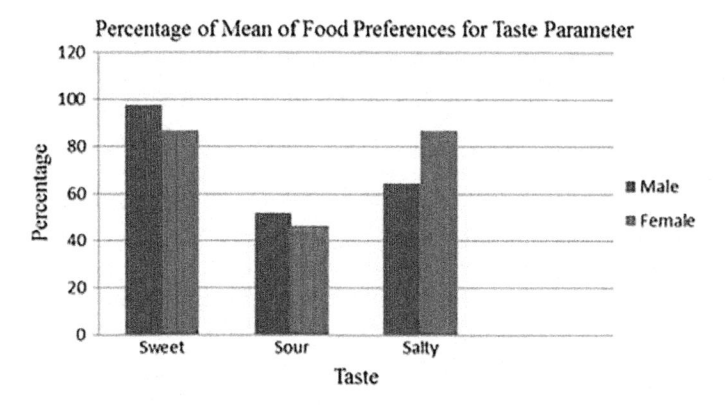

Fig. 2 Percentage mean for taste

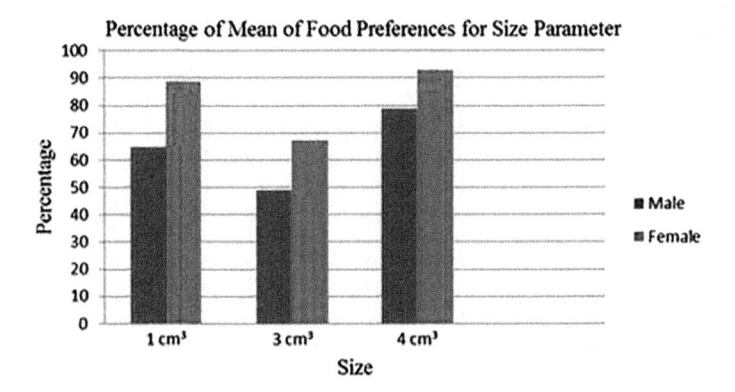

Fig. 3 Percentage mean for size

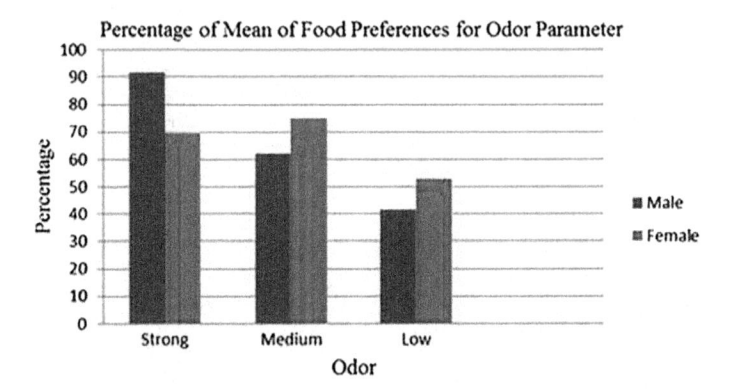

Fig. 4 Percentage mean for odor

4 Discussion

Both male and female civets chose fruits that had been cut with the largest size, 4 cm^3 with percentage for male civet is 78.73% while female civet is 92.73%. For both genders, there is a significant difference between the sizes. Usually the ability of frugivore animals to handle, swallow, and process a given fruit in an efficient way is depending to the fruit size consumed. Usually, large-seeded fruits such as family Lauraceae and Palmae will be consumed by frugivorous birds which are largely confined to large-bodied species (Wheelwright 1985). It should correspond to the frugivore's body size and crucially depends on the mouth size and the gap width (Jordano 1999). Thus, the civets preferred to eat the fruits with the largest size as their jaws are fit to grind and swallow the fruits. Based on the observation, when the civets grind the smallest fruits, 1 cm^3, the fruits can be easily slipped out through their teeth and make them lose attention to eat.

For odor, the male civet had chosen fruits with strong odor, 91.40%, while female civet had chosen medium odor, 74.80%. There is a significant difference shown by the male civet, while female civet shown no significant difference between odors. When an animal encountered an attractive odor, it will spend more time in investigating the odor compared to the less attractive odor. Thus, it shows more preference toward the medium and strongest odor compared to the low odor. Each animal exhibits different odor preference and the odor's pleasantness usually based on result of different learning process, environment culture, and life experience (Mandairon et al. 2009), and it is highly influenced by various levels of plasticity (Bensafi et al. 2007; de Araujo et al. 2005; Herz 2003). Specifically, representation of hedonic tone of smell may be modified and does not fix by learning and experience in both animals and humans (Wilson and Stevenson 2006). The banana flavor had been used in this experiment because both civets had been raised by eating banana since young. Thus, the low odor resembled as young banana, while strong odor resembled as ripe banana.

For color, the male civet preferred blue color, 72%, while the female civet preferred purple color, 94%. Both civets show significant difference between the colors. Previous study by Mudappa (2010) stated that the civets (*P. jerdoni*) choose the fruits with blue and green color (Mudappa et al. 2010). Each mammal has their own cones of color; thus, it influences their optical vision of seeing the color of objects around them. Most mammals can be classified of having a type of blindness called deuteranopia, or known as green-blind (Klein 2016). Based on the experiment, it can be concluded that the civets can see blue and purple color against the other color and tend to mix up green, brown, and red (Klein 2016). Thus, the male civet has a high tendency in choosing fruits with blue color. Furthermore, blue is a basic component of purple color, thus the optical vision of the civet reflecting the purple color as blue. It corresponds to the female civet in choosing purple fruits against other color.

For taste, the male civet preferred the sweet taste, 97.6%, while female civet preferred both sweet and salty taste, 86.8%. Both civets show significant difference between the tastes. Most animals are usually seeking and consuming foods which are similar sweet-tasting to humans (Breslin and Spector 2008). The civet has abundance of type 1 taste receptor, Tas1r2/Tas1r3. Tas1r2 gene together with Tas1r3 gene encodes for Tas1r2 and Tas1r3 receptor which enables the civet to detect the sweet taste. As the civet preferred fruits as their diet, thus it has abundance of sweet taste receptor. It is contrast to other Carnivores, such as cat, dog and lion which unable to detect the sweet taste (Jiang 2012). Next, the female civet preferred the salty taste too. Salty taste is usually related to the ripeness of fruits (Beauchamp and Jiang 2015). Some species consumed the salt, sodium, an essential mineral although there is no apparent need for them. For the plant eater and fruits eater, they need to consume more NaCl and its contrast to carnivore who does not perceive NaCl as the herbivore did since the meat-eater retained sufficient Na+ (Beauchamp and Jiang 2015). For some herbivores and omnivores, they need to find additional sodium in the environment, as they lack of ability to store of this essential electrolyte in the body (Breslin and Spector 2008). The sodium is one of the essential ions for living organism, and none of the ions can replace it as the animals usually loss the ion constantly by various secretory and excretory processes. (Breslin et al. 1993). The taste of salt in mammals can be divided into two categories: preferences of taste by the amount of salt that containing solutions and the sensitivity of salt responses to the channel blocker amiloride (Breslin et al. 1993). The least preference of taste by the male and female civet is the sour taste. The expression of PKD2L1, a type of TRP ion channel, proposed to function as a component of the acid-sensing machinery which in charge to characterize the sour-sensing taste, TRCs (LopezJimenez et al. 2006). Usually, the sour and bitter taste acts as a warning against the consumption of potentially harmful and/or poisonous chemicals (Breslin et al. 1993).

Due to the limitation of sample, this experiment is using only two civets which had been raised in different captivity since young by Miss Zakirah Zaki. Civets as we know are nocturnal mammals which actively seeking for food at night and starting to be active at dusk till dawn. Thus, there will be some problems in using civets in zoo's captivity as we need to feed and enter the zoo's captivity at night for 40 days and we need permission and zoo's guard every time when we want to enter the captivity at night. The zoo's civets will be placed together in a big captivity, and thus it will be difficult to ensure that the food is not mixed and not being eaten by different civets in the same captivity. With the obstacles arise, we had used two civets with different gender which had been raised in different captivity since young.

Both of the civets which had undergone the experiment are civets which can be classified as 'tame.' They had been raised by human, and they had a good interaction with human and their diet had been well-cared since they were little. Thus, the behavior and preferences of food might be different with the wild civets in the jungle. The animals adjusted their behavior in order to cope with the environment, which resulting phenotypic and genotypic divergence between wild and captive animals (Darwin 1868; Lickliter and Ness 1990). Based on the experiment, food preferences by civets in captivity are matching with the previous studies of food preferences by wild civets.

5 Conclusion

For the experiment that had been done for 40 days, it can be concluded that the civet chooses the fruits with the largest size and strongest smell, dark colors such as blue and purple with sweet and salty taste. The preferences of food based on the parameters, color, size, odor, and taste are similar between genders.

Acknowledgements This work is partially supported by my supervisor, Tun Mohd Firdaus, my lecturer, Nur Syafiqah Rahim and Sarina Mohamad, and I am thankful for their help and guidance that had been given to me since the day I started my project. They have given full supports and information for my writing to be accomplished. To be included, I am thankful to my parents for helping in terms of financial for my project to be done. I will be glad for receiving any of comments and suggestions from the reader in order to improve my writing in the future.

References

Beauchamp, G., & Jiang, P. (2015). Comparative biology of taste: Insights into mechanism and function. *Flavour, 4*(9), 1–3.

Bensafi, M., Rinck, F., Schaal, B., & Rouby, C. (2007). Verbal cues modulate hedonic perception of odors in 5-year-old children as well as in adults. *Chemical Senses, 32*, 855–862.

Breslin, P. A., Kaplan. J. M., Spector, A. C., Zambito, C. M., & Grill, H. J. (1993). *Lick rate analysis of sodium taste state*.

Breslin, P. A. S., & Spector, A. C. (2008). Mammalian taste receptor. *Current Biology*, (4), 148–155.

Choudhury, K. B. D., Anil, Deka, Munmun, Sarma, & Jiten, R. (2015). Phenotypic morphometric study on an adult common palm civet (*Paradoxurus hermaphrodites*). *Journal of Agriculture and Veterinary Science, 8*(4), 37–38.

Darwin, C. R. (1868). *The variation of animals and plants under domestication*. Baltimore: Johns Hopkins University Press.

de Araujo, I. E., Rolls, E. T., Velazco, M. I., Margot, C., & Cayeux, I. (2005). Cognitive modulation of olfactory processing. *Neuron, 46*, 671–679.

Herz, R. S. (2003). The effect of verbal context on olfactory perception. *Journal of Experimental Psychology: General, 132*, 595–606.

Jiang, P. (2012). Major taste loss in carnivorous mammals. *PNAS, 109*(13), 4956–4961.

Jordano, P. (1999). *Fruits and frugivory*. Estación Biológica de Doñana, C.S.I.C. Apdo. 1056, E-41080 Sevilla, Spain.

Jothish, P. S. (2011). Diet of the common palm civet *Paradoxurus hermaphroditus* in a rural habitat in Kerala, India, and its possible role in seed dispersal. *Small Carnivore Conservation, 28*, 10–11.

Klein, T. (2016). Confessions of a deuteranope. EPN, 47, 16–18. Retrived at April 11, 2016 from http://dx.doi.org/10.1051/epn/2016501.

Krishna, C. M., Kumar, A., Ray, P. C., Sarma, K, & Deka, J. (2015). *Inverstigating the foraging patterns and distribution of nocturnal frugivores with special focus on conservation threats in Namdapha National Park, Arunachal Pradesh, India*. Rufford Small Grants Program, United Kingdom.

Krishnakumar, H., Balasubramaniam, N. K., & Balakrishnan, M. (2002). Sequential pattern of behavior in the common palm civet, *Paradoxurus hermaphroditus* (Pallas). *International Journal of Comparative Physiology, 15*(14), 303–311.

Lickliter, R., & Ness, J. W. (1990). Domestication and comparative physiology: Status and strategy. *Journal of Comparative Psychology, 104*, 211–218.

LopezJimenez, N. D., Cavenagh, M. M., Sainz, E., CruzIthier, M. A., Battey, J. F., & Sullivan, S. L. (2006). Two members of the TRPP family of ion channels, Pkd1l3 and Pkd2l1, are coexpressed in a subset of taste receptor cells. *Journal of Neurochemistry, 2006*(98), 68–77.

Mallick, J. (2006). Necessity of civet conservation. Retrieved August 18, 2016 from https://www.researchgate.net/publication/200457462.

Mandairon, N., Poncelet, J., Bensafi, M., & Didier, A. (2009). Humans and mice express similar olfactory preferences. *PLoS ONE, 4*(1), 1–5.

Mudappa, D., Kumar, A., & Chellam, R. (2010). Diet and fruit choice of the brown palm civet, *Paradoxurus jerdoni*, a Viverrid Endemic to the Western Ghats rainforest, India. *Tropical Conservation Science, 3*(3), 282–300.

Raj, B., & Sharma, P. (2013). *Hand-rearing the common palm civet, Paradoxurus hermaphroditus* (pp. 1–4). Rehabber's Den, 1–4.

Wheelwright, N. T. (1985). Fruit size, gape width, and the diets of fruit-eating birds. *Ecology, 66*, 808–818.

Wilson, D., & Stevenson, R. (2006). *Learning to smell*. Baltimore: Johns Hopkins UP.

The Influence of Root Zone Temperature Manipulation on Strawberry Yields in the Tropics

Mohd Ashraf Zainol Abidin, Desa Ahmad, Ahmad Syafik Suraidi and Josephine Tening Pahang

Abstract The study was conducted to assess the quality of strawberry (*Fragaria x Ananassa*) cultivated in tropical areas. There are two treatments set up in this study. The root zone temperature (RZT) in Treatment 1 was controlled around 25 °C (±2 °C), while Treatment 2 was left to experience fluctuating RZT varying between 27 and 30 °C. Both treatments were left to experience fluctuating ambient temperature, varying from 21 to 34 °C. Samples of fruit from tropical highland areas were obtained and used as a Control Treatment in assessing the quality of strawberry produced. Though the Control Treatment obtained better results in terms of fruit size, lowering the RZT in Treatment 1 significantly increased the fruit's diameter and weight up to 8.85 and 21.60%, respectively, compared to Treatment 2. About 70.00% of the strawberries produced in Treatment 1 were of marketable size (>5 g) compared to 46.67% of the strawberries produced in Treatment 2. Treatment 1 and Treatment 2 produced strawberries that were significantly sweeter compared to those produced in Control Treatment. More than 70.00% of the strawberries produced in both treatments were above the average sweetness of the Festival variety (Brix index: >8°Brix), compared to approximately 50.00% of the strawberries produced in the Control Treatment. In conclusion, lowering the RZT was

---

M. A. Z. Abidin (✉) · J. T. Pahang
Faculty of Plantation and Agrotechnology, Universiti Teknologi MARA Perlis Branch, 02600 Arau, Perlis, Malaysia
e-mail: ashrafzainol@perlis.uitm.edu.my

J. T. Pahang
e-mail: josephine992@perlis.uitm.edu.my

D. Ahmad
Department of Biological and Agricultural Engineering, Faculty of Engineering, Universiti Putra Malaysia, 43400 UPM Serdang, Selangor, Malaysia
e-mail: desa@upm.edu.my

A. S. Suraidi
Mechanization and Automation Research Centre, MARDI Headquarters, P.O. Box 12301, 50774 Kuala Lumpur, Malaysia
e-mail: syafik@mardi.gov.my

found to produce better fruit yield for strawberry cultivation in the tropics, thus may provide an alternative solution that can comply with the growing strawberry demand.

Keywords Root zone temperature manipulation · RZT · Strawberry Tropics

1 Introduction

The term tropics refer to areas that undergo hot and wet climate continuously and are also free from frost. About 35% of the world experiences this climate including Malaysia. To be specific, Malaysia is located in the equatorial regions where air convergence is typical and rainfall is frequent; thus, constantly experiences wet conditions (Webster and Wilson 1998). The country is divided into Peninsular and East Malaysia and is mostly surrounded by the sea. This influences the climatological properties in Malaysia. The ambient temperature varies from 21 to 34 °C and has a yearly mean of around 27 °C (Razak and Roff 2011). However, agricultural highland areas such as Cameron Highlands, Kundasang and Lojing experience low ambient temperature all the year round. For example, the ambient temperature in Cameron Highlands fluctuates around 25 °C and may drop below 12 °C (Eisakhani and Malakahmad 2009).

The most popular fruit cultivated in Cameron Highlands is strawberry, also known as *Fragaria x Ananassa* sp. (Mohd Ridzuan et al. 2011) due to its pleasant taste, aroma and high nutritive values (Folta et al. 2010). Strawberries can be grown between 15 and 25 °C, at an elevation between 1000 and 1500 m above sea level (Uselis et al. 2008). The crops require day- and night-time temperatures of around 20–28 °C and 12–18 °C, respectively. This range of ambient temperature can be achieved at tropical highland areas over 1400 m above sea level in Malaysia (Mohd Ridzuan et al. 2011).

The production of this popular high-value crop (Folta et al. 2010) is expected to rise due to a significant number in demand (Mohd Ridzuan et al. 2011). However, the supply of strawberries will probably be limited due to the limited arable land in highland areas, coupled with the government's strict policy against the opening of new agricultural areas in the highlands (Barrow et al. 2009). Meanwhile, lowland areas do not provide a suitable climate for strawberry cultivation due to the tropical weather that affects the soil temperatures, extreme rainfall and the threats from insect pests (Mat Sharif 2006).

The technology that controls the environment in a greenhouse can meet the suitable crop growth requirements for temperate crops (Al-Shamiry et al. 2006) such as strawberries and tomatoes, so that they can be grown optimally in lowland areas. Tropical greenhouses experience extreme ambient temperature without the controlled system due to the high intensity of solar received, which may stunt crop growth (Chen et al. 2011). Hence, the system becomes more important (Mattas et al. 1997) in cultivating temperate crops in the tropics. However, temperate crop cultivation inside the greenhouse equipped with the environmental control system demands high cost of production, especially the initial cost, thus reducing the profit (Ahmad Syafik et al. 2010).

Manipulating the root zone temperature (RZT) to optimize crop requirements would increase crop productivity. The effect of RZT is greater on root growth, especially in the early stages of crop development. Research has shown that the crop roots are more sensitive to temperature fluctuations when compared to the crop shoots (Mohammud et al. 2012). The RZT manipulation showed significant results: increase in vegetative grow, stem diameter, internodes, root volume, nutrient absorption, fruit number, fruit qualities (Mohammud et al. 2012), leaf number, production of a greater leaf area and dry weight of cultivated crops (Dodd et al. 2000). In certain vegetative crops, the RZT manipulation caused a large increase in shoot dry weight and fruit development when light was not a limiting factor. Fortunately, light source is not a limiting factor in Malaysian climate (Mat Sharif 2006).

An alternate idea is to manipulate the RZT under a rain shelter structure in tropical lowland areas to cultivate the strawberries. This leaves the ambience to experience natural ventilation with no required energy and maintenance input (Al-Shamiry and Ahmad 2010). Generally, most of the temperate crops are less tolerant to the high ambient temperatures (Mat Sharif 2006) experienced in tropical regions. This is a huge challenge because the ambient temperatures in tropical lowland areas are normally very high compared to highland areas. With the normal temperature for growth media in lowland areas varying between 27 and 30 °C (Mohammud et al. 2011) and without exact references on growth media temperature pattern in Cameron Highlands, the RZT manipulation in providing a suitable environment to grow strawberries in lowland areas outside the control environment greenhouse is crucial. However, based on the ambient temperature in Cameron Highlands, the growth media temperature was assumed to vary around 25 °C (± 2 °C). Hence, the study was conducted to assess the quality of strawberries grown in tropical lowland areas with and without the root zone manipulation.

2 Materials and Methods

2.1 Plant Material and Growth Conditions

The study was conducted at Mechanization and Automation Research Centre, Malaysian Agricultural Research and Development Institute (MARDI) in Serdang, Selangor, Malaysia, having a 2° 59′N latitude, 101° 42′E longitude and at an altitude 37.8 m above sea level. There were two different treatments set up for this study. Both treatments were placed under open rain shelter, thus were left to experience fluctuating ambient temperature as shown in Fig. 1. The root zone

Fig. 1 Experimental site

temperature (RZT) in Treatment 1 was manipulated to vary around 25 °C (±2 °C), while Treatment 2 was left to experience fluctuating RZT. The pillow polybag was used as a growing container for crop cultivation, and the growth media used was cocopeat. The strawberries from *Festival* variety were grown in both treatments. Treatment 1 and Treatment 2 were irrigated using fertigation system at the same frequency besides undergoing similar crop management practice.

2.2 Physiological Measurements

The parameter such as fruit diameter (mm), weight (g) and Brix index (°Brix) were measured and analysed to assess the quality of strawberries grown for different temperatures. Ninety samples of fruits were randomly collected from each treatment, making the total number of samples tested up to 180 fruits. The samples of fruits from Cameron Highlands were also collected and used as a standard for comparison (Control Treatment).

The digital vernier calliper was used to measure the fruit diameter, while to determine the fruit weight, the digital weighing scales were used. The strawberries were then squeezed to obtain the juice samples. Those samples were then tested using a portable refractometer to measure their Brix index.

2.3 Statistical Analysis

An assessment process focused on the quality of the fruit. The assessment was done by comparing the data obtained in Treatment 1, Treatment 2 and Control Treatment using ANOVA test. Besides, Pearson's Correlation test was also performed to identify the relation between fruit properties with RZT. The statistical analyses were solved with the help of Statistical Package for Social Sciences (SPSS) 21.0 software.

3 Result and Discussions

3.1 Comparison of Fruits Properties

Table 1 shows the difference in fruit quality between the treatments. The two properties of fruit quality measured in this study were size and Brix index. The size properties were divided into diameter and weight. Fruit diameter recorded in Control Treatment was 25.11 mm, followed by 22.15 mm and 20.18 mm, respectively, for Treatment 1 and Treatment 2. Though Treatment 1 had a fruit diameter that was 11.83% smaller than Control Treatment, it obtained 8.85% better fruit diameter compared to Treatment 2. Besides, Treatment 1 recorded the smallest value of standard deviation. The values of fruit diameter were less spread out when compared to Treatment 2 and Control Treatment. The mean values in all treatments were significantly different from each other.

Even though Treatment 1 recorded fruit weight value that was 31.62% less than Control Treatment, the results were 21.6% better than Treatment 2. The mean values in Treatment 1, Treatment 2 and Control Treatment were 6.25 g, 4.90 g and 9.14 g, respectively. Again, Treatment 1 recorded the smallest value of standard deviation, thus recording less spread out values of fruit weight compared to other treatments.

For Brix index, Treatment 2 (9.30°Bx) recorded a higher mean value compared to Treatment 1 (9.24°Bx) and Control Treatment (8.43°Bx). Though the value was not significantly different than the value obtained in Treatment 1, it was significantly higher than that of Control Treatment. Control Treatment obtained a lower

Table 1 Fruit qualities for various treatments

Properties	Treatment	Max	Min	Mean	SD
Size (diameter), mm	1	29.50	15.90	22.14^b	2.85
	2	31.70	10.70	20.18^c	4.77
	Control	33.27	18.56	25.11^a	3.02
Size (weight), g	1	15.40	3.30	6.25^b	2.04
	2	10.30	1.00	4.90^c	2.52
	Control	180	3.40	9.14^a	3.07
Brix index, °Bx	1	13.80	5.10	9.24^a	1.81
	2	15.30	6.40	9.30^a	1.69
	Control	13.80	4.70	8.43^b	2.53

*Different letters indicate significant differences at the 0.05 level

Table 2 Marketable size and average sweetness of fruit for various treatments

Treatment	Quantity (no)		Marketable percentage (%)	Above average sweetness >8°Bx	Below average sweetness <8°Bx	Above sweetness percentage (%)
	Marketable size >5 g	Non-marketable size <5 g				
1	63	27	70.00[b]	66	24	73.33[a]
2	42	48	46.67[c]	68	22	75.55[a]
Control	85	5	94.44[a]	43	47	47.78[b]

*Different letters indicate significant differences at the 0.05 level

Brix index value compared to Treatment 1 and Treatment 2. Overall, Treatment 1 recorded 0.65% lower and 8.77% higher mean values than Treatment 2 and Control Treatment, respectively.

Based on the marketable size of strawberry (>5 g) described by Yuan et al. (2004), and average sweetness for *Festival* variety (8°Bx) described by Mohd Ridzuan et al. (2011), the ANOVA test was performed to classify the fruit yield (Table 2). Though Control Treatment produced 94.44% marketable size of strawberry, the manipulation of root zone temperature (RZT) in Treatment 1 was found to increase up to 23.33% marketable size of strawberry when compared to Treatment 2. Besides, lowland areas (major tropic areas) produced strawberry fruit that was significantly sweeter than the average sweetness for *Festival* variety as compared to the highland areas production. Though Treatment 2 produced fruit that has a higher percentage of average sweetness, only 46.67% of the strawberries produced by this treatment were of marketable size.

3.2 Pearson's Correlation Analysis

Pearson's correlation test was done to identify the correlation between fruit properties with RZT. The RZT in Control Treatment was not recorded in this study; hence, no fruit data from this treatment was involved in the correlation test. From Table 3, the strawberry's diameter and weight were found to be highly correlated with each other, with the R value equal to 0.932. Both properties were also insignificantly in correlation with the Brix index. However, correlations were

Table 3 Pearson's correlation test between fruit properties with RZT

	Diameter	Weight	Brix
Weight	0.932**	–	–
Brix	0.001	0.008	–
RZT	−0.242**	−0.284**	0.016

**Correlation is significant at the 0.01 level (2-tailed)

obtained when comparing to the RZT. For both parameters, the R values are −0.242 and −0.284, respectively. In contrast, Brix index also did not give any significant value in correlation with RZT.

3.3 Overview of Strawberry Assessment

Strawberries produced in Control Treatment were much larger and heavier compared to those produced in Treatment 1 and Treatment 2 due to the fact that the fruit's diameter and weight were greatly influenced by each other. In terms of the sweetness of the strawberries, Treatment 2 recorded the highest percentage of Brix index, followed by Treatment 1 and Control Treatment. The ambient temperature indirectly affected the fruit's diameter and weight (Palencia et al. 2013). As mentioned earlier, the ambient temperature between tropic highland and lowland areas is significant. The fruit data obtained show good argumentation with previous studies. The increase in ambient temperature also led to high respiration rate; thus, it may accelerate fruit senescence (Nunes et al. 1995). This may have contributed to the improper development of fruit. However, the result shows that the fruit growth increase with decreasing RZT. In contrast, the ambient temperature directly affected the Brix index, while the RZT had no clear effect on the Brix index cultivated in lowland areas. As an example of a RZT situation, though Treatment 2 produced higher values of Brix index compared to Treatment 1 due to the warmer RZT, the difference was insignificant. However, both treatments produced significantly higher values of Brix index than Control Treatment. Yet, the quality of the fruit was not only determined by the level of sweetness. Factors such as diameter, weight and others also play an important role in the overall quality assessment (Ruiz-Altisent et al. 2010). The sweetness of the strawberries is very important for food processing such as making jam, juice and sweet candy (Hafiz 2015). Thus, strawberry cultivation in the tropics specifically in Malaysia lowland areas could be an alternative way to meet the food processing demand.

4 Conclusion

The fruit diameter means in Treatment 1, Treatment 2 and Control Treatment were 22.14, 20.18 and 25.11 mm, respectively. The mean fruit weights for the treatments were 6.25, 4.90 and 9.14 g separately. The mean Brix index values were 9.24, 9.30 and 8.43%, respectively, for all the treatments. The root zone temperature (RZT) manipulation in Treatment 1 significantly increased the fruit diameter up to 8.85% and reduced its standard deviation value compared to Treatment 2. Treatment 1 also caused a significant increase in the fruit weight by around 21.6% compared to Treatment 2. For marketable size of strawberry (>5 g), Treatment 1, Treatment 2 and Control Treatment produced 70.00, 46.67 and 94.44%, respectively. Even though the size properties (diameter and weight) of Treatment 1 and Treatment 2 were lower than strawberries cultivated in Control Treatment, both treatments produced significantly sweeter strawberries (Brix index). In fact, Treatment 1 and Treatment 2 produced fruits that were 73.33 and 75.55% above the average sweetness of Festival variety (Brix index: >8°Brix), compared to 47.78% produced by Control Treatment. It can be seen that lowering the RZT in tropical areas increases the strawberry yield.

Acknowledgements This work is partially supported by Universiti Putra Malaysia through the Putra Grant and was conducted at the Malaysian Agriculture Research and Development Institute. All the designs and development of root zone cooling (RZC) system belong to Malaysian Agriculture Research and Development Institute. The authors are also grateful and acknowledge the helpful comments and suggestions of the reviewers, which have improved the presentation of this paper.

References

Ahmad Syafik, S. S, Mohd Faisal, H., & Mohd Hafiz, H. (2010). Effect of growth media cooling using chilled water system for tomato growth to the temperature and humidity under controlled environment structures in lowland. In *Malaysian Science and Technology Conference*, Petaling Jaya, Selangor.

Al-Shamiry, F. M. S., & Ahmad, D. (2010). Development and validation of a mathematical model for ventilation rate in crop protection structures. *Pertanika Journal of Science and Technology, 18*(December 2009), 111–120.

Al-Shamiry, F. M. S., Shariff, A. R. M., Rezuwan, K., Ahmad, D., Janius, R., & Mohamad, M. Y. (2006). Microclimate inside tunnel-roof and jack-roof tropical greenhouses structures. In *International Symposium on Greenhouse, Environmental Controls and In-house Mechanization for Crop Production in Tropics and Sub-tropics* (pp. 179–184). Cameron Highlands: Acta Horticulturae (ISHS) 710.

Barrow, C. J., Chan, N. W., & Tarmiji, M. (2009). Issues and challenges of sustainable agriculture in the Cameron Highlands. *Malaysian Journal of Environmental Management, 10*(2), 89–114.

Chen, C., Shen, T., & Weng, Y. (2011). Simple model to study the effect of temperature on the greenhouse with shading nets. *African Journal of Biotechnology, 10*(25), 5001–5014.

Dodd, I. C., He, J., Turnbull, C. G., Lee, S. K., & Critchley, C. (2000). The influence of supra-optimal root-zone temperatures on growth and stomatal conductance in *Capsicum annuum* L. *Journal of Experimental Botany, 51*(343), 239–248.

Eisakhani, M., & Malakahmad, A. (2009). Water quality assessment of Bertam river and its tributaries in Cameron highlands, Malaysia. *World Applied Sciences Journal, 7*(6), 769–776.

Folta, K. M., Clancy, M. A., Chamala, S., Brunings, A. M., Dhingra, A., Gomide, L., et al. (2010). A transcript accounting from diverse tissues of a cultivated strawberry. *The Plant Genome Journal*.

Hafiz, M. (2015). *Characteristic of fruit quality for strawberry-based food processing*. Personal Communication. Taman Agro Al-Mashoor, Cameron Highlands.

Mat Sharif, I. (2006). Design and development of fully controlled environment greenhouse for the production of selected temperate crops in lowland tropics. In *International Symposium on Greenhouse, Environmental Controls and In-house Mechanization for Crop Production in Tropics and Sub-tropics* (pp. 127–134). Cameron Highlands: Acta Horticulturae (ISHS) 710.

Mattas, K., Bentes, M., Paroussi, G., & Tzouramani, I. (1997). Assessing the economic efficiency of a soilless culture system for off-season strawberry production. *HortScience, 32*(6), 1126–1129.

Mohammud, C. H., Illias, M. K., Zaulia, O., Ahmad Syafik, S. S., & Angelina H. M. Y. (2012). BSE-501: Effect of rhizosphere cooling on tomato crop performance under controlled environment structure. In *Malaysia International Conference on Trends in Bioprocess Engineering (MICOTriBE) 2012*, pp. 1–8.

Mohammud, C. H., Rohazrin, A. R., Mohd Yusof, A., & Abd Jamil, Z. (2011). Performance of ventilation and cooling system on in-house environment in controlled environment structure. *Journal of Tropical Agriculture and Food Science, 39*(2), 267–278.

Mohd Ridzuan, M. S., Mohammad Abid, A., Saiful, Z. J., Roslina, A., Ahmad Tarmizi, S., Mahamud, S., et al. (2011). *Manual of strawberry cultivation technology using fertigation system* (1st ed.). Kuala Lumpur: Mardi Publication.

Nunes, M. C. N., Brecht, J. K., Sargent, S. A., & Morais, A. M. M. B. (1995). Effects of delays to cooling and wrapping on strawberry quality (cv. Sweet Charlie). *Food Control*.

Palencia, P., Martínez, F., Medina, J. J., & López-Medina, J. (2013). Strawberry yield efficiency and its correlation with temperature and solar radiation. *Horticultura Brasileira, 31*, 93–99.

Razak, S. A. B. D., & Roff, M. N. M. (2011). Status and potential of urban and peri-urban agriculture in Malaysia. In *Food & Fertilizer Technology Center for the Asian and Pacific Region*, Taipei, pp. 121–134.

Ruiz-Altisent, M., Ruiz-Garcia, L., Moreda, G. P., Lu, R., Hernandez-Sanchez, N., Correa, E. C., et al. (2010). Sensors for product characterization and quality of specialty crops—A review. *Computers and Electronics in Agriculture, 74*, 176–194.

Uselis, N., Lanauskas, J., Zalatorius, V., Duchovskis, P., Brazaitytė, A., & Urbonavičiūtė, A. (2008). Evaluation of the methods of soil cultivation growing dessert strawberries in beds. *Sodininkystė Ir Daržininkystė, 27*(2), 295–305.

Webster, C. C., & Wilson, P. N. (1998). *Agriculture in the tropics* (3rd ed.). London: Blackwell Science Ltd.

Yuan, B., Sun, J., & Nishiyama, S. (2004). Effect of drip irrigation on strawberry growth and yield inside a plastic greenhouse. *Biosystems Engineering, 87*(2), 237–245.

Detection of Class 1 Integron and Antibiotic Resistance Genes in *Aeromonas Hydrophila* Isolated from Freshwater Fish

Hamdan Ruhil Hayati, Mohd Daud Hassan, Ong Bee Lee, Hamid Nur Hidayahanum, A. Mohamed Nora Faten, Manaf Sharifah Raina, Tan Li Peng and Nik Mohd Fauzi Nik Nur Fazlina

Abstract The increasing incidence of antibiotic resistance strains in *Aeromonas hydrophila* is due to extensive antimicrobial usage by human and fish farms for treatment and disease prevention. Multiple antibiotic resistance phenotypes among bacteria have been shown to be attributed to integrons. Forty strains of *A. hydrophila* isolated from freshwater fish were investigated for antibiotic sensitivity and plasmid profiling. Ten antibiotics were used, namely penicillin G, cephalotin, florfenicol, streptomycin, kanamycin, erythromycin, ampicillin, gentamicin, oxytetracycline and tetracycline. Polymerase chain reaction was carried out to detect integrase genes *Int1*, *Int2* and *Int3*, gene cassette array, integron-associated aadA, sul1 and qac1 genes, streptomycin resistance genes *strA-strB*, β-lactamase resistance genes *blaTEM* and *blaSHV*, and tetracycline resistance genes *tetA-E* and *tetM*. As a result, *A. hydrophila* was sensitive to erythromycin, florfenicol, kanamycin and oxytetracycline, while the bacteria were resistant to cephalotin, gentamycin and penicillin G. In this present study, 12 out of 40 isolates contain plasmid size ranging from 6 to 23 kb. The *intI*1 gene was detected in 50% (20/40) of *A. hydrophila* strains, but no isolates contain intI2 and intI3. No gene cassette was

---

H. R. Hayati (✉) · O. B. Lee · T. L. Peng · N. M. F. N. N. Fazlina
Faculty of Veterinary Medicine, Universiti Malaysia Kelantan,
16100 Kota Bharu, Kelantan, Malaysia
e-mail: ruhilhayati1982@gmail.com

M. D. Hassan · H. N. Hidayahanum · A. M. N. Faten
Faculty of Veterinary Medicine, Universiti Putra Malaysia,
43400 Serdang, Selangor, Malaysia
e-mail: hassanmd@upm.edu.my

M. S. Raina
Faculty Plantation and Agrotechnology, Universiti Teknology MARA,
96400 Mukah, Sarawak, Malaysia
e-mail: sharifahraina@uitm.edu.my

detected from all the *A. hydrophila* isolates. Different tetracycline resistance genes (tetA and tetC) and aadA genes were detected. The detection of antibiotic resistance genes in *A. hydrophila* has public health and environmental concern.

Keywords Antibiotic resistance · Integron · *A. hydrophila*

1 Introduction

Huys et al. (2005) stated that wide use of antibiotics to treat bacterial infections and incorporation of subtherapeutic dose of antibiotics into feeds for cultured organism resulted in a global increase in antibiotic resistance among pathogenic bacteria. Rattanachaikunsopon and Phumkhachorn (2009) affirmed that accumulation of antibiotics in fish can be harmful to the environment as well as consumers. Such adverse effects of antibiotics lead to problems in aquaculture; therefore, only few a have been approved. Many countries refuse to import cultured products in which antibiotics were used. Therefore, these problems have prompted scientists to search for an alternative to replace antibiotics in controlling diseases in aquaculture.

The problem is more serious in developing countries, where antibiotics are used widely. The use of antibiotics is the most important factor in amplifying the level of resistance in a given reservoir (Vivekanandhan et al. 2002). Multiple antibiotic resistance (MAR) among *Aeromonas hydrophila* strains has been reported from many parts of the world (Vivekanandhan et al. 2002; Taylor 2003; Akinbowale et al. 2007, Kathleen et al. 2016). Under these circumstances, it will be worthwhile to find out the prevalence of antibiotic resistance of the *Aeromonas* strains that may be considered as an emerging pathogen and to identify the high-risk source.

Taylor (2003) stated that two ways by which bacterial strains develop resistance are the mutation of chromosomal genes and the transference of extrachromosomal genes through plasmids. Chromosomal mutation occurs via outside pressures of selection, and both its development and spread within a population are slow. Plasmids are extrachromosomal DNA that can be transferred directly between bacteria through conjugation. They often contain genetic material called *R* factors that encode for antimicrobial resistance. Plasmid-mediated resistance can occur rapidly within a bacterial population and can be passed between different species of bacteria. Watanabe et al. (1971) described the first *R* factor in plasmids from *Aeromonas liquifaciens* isolated from fish in Japan. Similarly, *R* factors have been described in *Aeromonas salmonicida* (Starliper and Cooper 1998), *A. hydrophila* (Shotts et al. 1976), *Edwardsiella tarda* (Aoki and Kitao 1981) and *Edwarsiella ictaluri* (Starliper et al. 1993). These bacteria can infect both food fish and ornamental fish. Therefore, plasmid-mediated resistance is a major concern in all fish production areas.

Plasmid, transposons and integrons can carry antibiotic-resistant genes and can be transferred horizontally between micro-organisms (Essen-Zandbergen et al.

2007; Sarria-Guzman et al. 2014). Integrons are unusual DNA elements which include genes encoding a site-specific DNA recombinase, a DNA integrase and an adjacent site at which a wide variety of antibiotic resistance and other genes are found as inserts (Hall and Stokes 1993). To date, five different classes of mobile integrons have been identified. The classes were distinguished by variations in the sequence of the encoded integrase (40–58% identity). Classes 1 and 2 are the most common in multiple antibiotic-resistant bacteria, and they are of concern due to their mobilization capacity which enables spread within and between species (Nguyen et al. 2014). Class 1 integrons are formed by a gene coding for an integrase (*Int*1) that corresponds to the 5′ conserved segment (5′CS); a variable region size where cassettes are located; a 3′ conserved segment (3′CS) which contains a sulphonamide resistance gene (*sul*1); a quaternary ammonium compound resistance sequence (*qac*EΔ1) (Koelman et al. 2001; Sarria-Guzaman et al. 2014). Although integrons are not mobile, they are often found associated with mobile genetic elements such as plasmids and transposons (Ndi and Barton 2011). Isolates originating from aquatic sources have also been reported to carry integrons and other genes that code for resistance determinants (Rhodes et al. 2000; Jacobs and Chenia 2007; Nawaz et al. 2010; Lukkana et al. 2012).

The objectives of this study were to investigate antibiotic sensitivity and plasmid profiling of *A. hydrophila* and then to determine the type of integrons and to identify the antibiotic resistance determinant associated with *A. hydrophila*.

2 Materials and Methods

2.1 *Bacterial Isolation and Identification*

Species of freshwater fish used in this study were *Oreochromis mossambicus* (Black tilapia), *Oreochromis* sp. (Red hybrid tilapia), *Scortum barcoo* (Jade perch), *Puntius gonionotus* (Javanese carp), *Leptobarbus hoevenii* (River carp), *Pangasius pangasius* (River catfish), *Anabas testudineus* (Climbing perch), *Clarias gariepinus* (African catfish) and *Cichlasoma* sp. (Flowerhorn). Fish weighing an average of 300–500 g and showing clinical signs of skin haemorrhages and fins rot were sampled from three fish farms in Selangor, Malaysia, during disease outbreaks. Samples of the gills, kidney, liver and spleen tissues were aseptically streaked onto TSA (LabM, UK) and incubated at 30 °C for 24 h. Representative colonies were chosen and re-streaked on fresh medium. Pure cultures were obtained in the same media. All isolates were then identified using API20E® (bioMerieux, France) with a combination of Gram staining, oxidase, catalase and O/129 sensitivity. 16S rRNA PCR assay was used to confirm the isolates of *A. hydrophila* (Chu and Lu 2005).

2.2 Antibiotic Sensitivity Tests

The Kirby–Bauer method was used to determine the antibiotic sensitivity patterns of isolates. Antibiotics tested were ampicillin (10 μg), cephalotin (10 μg), erythromycin (15 μg), florfenicol (30 μg), gentamicin (10 μg), kanamycin (30 μg), oxytetracycline (30 μg), tetracycline (30 μg), streptomycin (10 μg) and penicillin G (10 μg) (Oxoid, England) (Akinbowale et al. 2007).The results of inhibition zones were interpreted as sensitive (S), intermediary sensitive (I) and resistance (R) according to the reference to the standard provided by the National Committee for Clinical Laboratory Standards (CLSI 2016).

2.3 Multiple Antibiotic Resistance (MAR) Index

Multiple antibiotic resistance (MAR) index of the present isolates against the tested antibiotics was calculated based on the following formula: MAR index = $X/(Y \times Z)$ where X = total cases of antibiotic resistance; Y = total number of antibiotic used in the study; Z = total number of isolates. A MAR index value of equal or less than 0.2 was defined as antibiotics that were seldom or never used.

2.4 Plasmid Profiling

The plasmid profiling was done using Plasmid Extraction Kit (Yeastern Biotech, Taiwan) according to manufacturer's protocol. Plasmid was visualized following electrophoresis in 1% agarose gel, ethidium bromide staining and UV illumination.

2.5 DNA Extraction and Purification

DNeasy Blood and Tissue Kit was used to extract and purify DNA in this study according to manufacturer's protocol. To check the purity of DNA, the samples were run on 1% agarose gel, ethidium bromide staining and UV illumination.

2.6 Detection of Integrons, Integron-Associated Genes and Antibiotic Resistance Genes

Detection of integrons and the resistance genes present in them was performed using PCR amplification with the specific primers listed in Table 1. All PCR amplification were carried out in a Mastercycler (Eppendorf, Germany). Assays were carried out in 50 μL volumes containing 5 μL of 10 × PCR buffer [100 mm

Table 1 List of primers used in this study

Primer	Nucleotide sequence(5'-3')	Product size (bp)	References
Int1-F	CAGTGGACATAAGCCTGTTC	160	Koelman et al. (2001)
Int1-R	CCCGAGGCATAGACTGTA		
Int2-F	GTAGCAAACGAGTGACGAAATG	788	Mazel et al. (2000)
Int2-R	CACGGATATGCGACAAAAAGGT		
Int3-F	GCCTCCGGCAGCGACTTTCAG	979	Mazel et al. (2000)
Int3-R	ACGGATCTGCCAAACCTGACT		
hep58	TCATGGCTTGTTATGACTGT	Variable	White et al. (2000)
hep59	GTAGGGCTTATTATGCACGC		
sul1-F	CTTCGATGAGACCCGGCGGC	436	Sundstrom et al. (1988)
sul1-R	GCAAGGCGGAAACCCGCGCC		
qacED1-F	ATCGCAATAGTTGGCGAAGT	250	Stokes and Hall (1989)
qacED1-R	CAAGCTTTTGCCCATGAAGC		
aadA-F	GAGAACATAGCGTTGCCTTGGTCG	198	Sunde and Norstrom (2005)
aadA-R	GCGCGATTTTGCCGGTTA		
strA-strB-F	TTGAATCGAACTAATAT	1640	Han et al. (2004)
strA-strB-R	CTAGTATGACGTCTGTCG		
blaTEM-F	ATGAGTATTCAACATTTCCG	867	Rasheed et al. (1997)
blaTEM-R	CTGACAGTTACCAATGCTTA		
blaSHV-F	GGTTATGCGTTATATTCGCC	867	Rasheed et al. (1997)
blaSHV-R	TTAGCTTTGCCAGTGCTC		
tetA-F	GTAATTCTGAGCACTGTCGC	956	Schmidt et al. (2001a, b)
tetA-R	CTGCCTGGACAACATTGCTT		
tetB-F	CTCAGTATTCCAAGCCTTTG	535	Schmidt et al. (2001a, b)
tetB-R	CTAAGCACTTGTCTCCTGTT		
tetC-F	TCTAACAATGCGCTCATCGT	588	Schmidt et al. (2001a, b)
tetC-R	GGTTGAAGGCTCTCAAGGGC		
tetD-F	ATTACACTGCTGGACGCGAT	1070	Schmidt et al. (2001a, b)
tetD-R	CTGATCAGCAGACAGATTGC		
tetE-F	GTGATGATGGCACTGGTCAT	1198	Schmidt et al. (2001a, b)
tetE-R	CTCTGCTGTACATCGCTCTT		
tetM-F	GTTAAATAGTGTTCTTGGAG	650	Aarestrup et al. (2000)
tetM-R	CTAAGATATGGCTCTAACAA		

KCl, 20 mm MgSO$_4$.7H$_2$O, 200 mM Tris–HCl (pH 9.0), 1% Triton X-100, 100 mM (NH$_4$)$_2$SO$_4$ and 1 mg/ml BSA], 1.0 µL of 10 mM dNTPs mix, 1 µL of each primer stock solution (10 µM) and 0.5 µL of YEA Taq DNA polymerase (2.5 U/µl). Prepared DNA extract was added 3 µL to provide the DNA template. Appropriate volume of sterile dH$_2$O was added to make up the volume to 50 µL. All PCRs were subjected to amplification according to cycling parameter suggested by manufacturer's protocol (Yeastern Biotech, Taiwan): an initial denaturation step at 94 °C for 5 min; denaturation step at 94 °C for 30 s; annealing step at (Primer Tm—5 °C) for 30 s; extension step at 72 °C for 1 min and final extension at 72 °C for 7 min. PCR amplicons were analysed by electrophoresis on 1.5% agarose gel, and a 100-bp ladder was used as the molecular size marker.

3 Results

3.1 Bacterial Isolation and Identification

A total of 40 isolates of *A. hydrophila* were isolated from freshwater fishes (Table 2). All isolates were straight rod, Gram-negative and motile. The biochemical tests showed positive results for ONPG, ADH, LDC, citrate utilization, indole production, gelatinase and Voges–Proskauer. Acid was produced from mannitol. Besides, negative results were obtained from ODC, H$_2$S, urease, TDA. The isolates were further confirmed by 16S rRNA PCR (Fig. 1).

Table 2 *Aeromonas hydrophila* strains isolated from freshwater fishes

Isolates (No. and serial code)	Host species	Comments
AHBT1, AHBT2, AHBT3, AHBT4, AHBT5, AHBT6	Black tilapia (*Oreochromis mossambicus*)	Kidney and liver
AHRT7, AHRT8, AHRT9, AHRT10	Red hybrid tilapia (*Oreochromis* spp.)	Kidney and spleen
AHJP11, AHJP12, AHJP13, AHJP14, AHJP15, AHJP16, AHJP17, AHJP18, AHJP19, AHJP20	Jade perch (*Scortum barcoo*)	Skin, kidney and spleen
AHJC21, AHJC22, AHJC23, AHJC24	Javanese carp (*Puntius gonionotus*)	Kidney
AHRP25, AHRP26	River carp (*Leptobarbus hoevenii*)	Kidney
AHRC27, AHRC28, AHRC29, AHRC30	River catfish (*Pangasius pangasius*)	Kidney
AHCP31, AHCP32, AHCP33	Climbing perch (*Anabas testudineus*)	Kidney
AHAC34, AHAC35, AHAC36, AHAC37, AHAC38	African catfish (*Clarias gariepinus*)	Kidney
AHFH39, AHFH40	Flowerhorn (*Cichlasoma* sp.)	Kidney

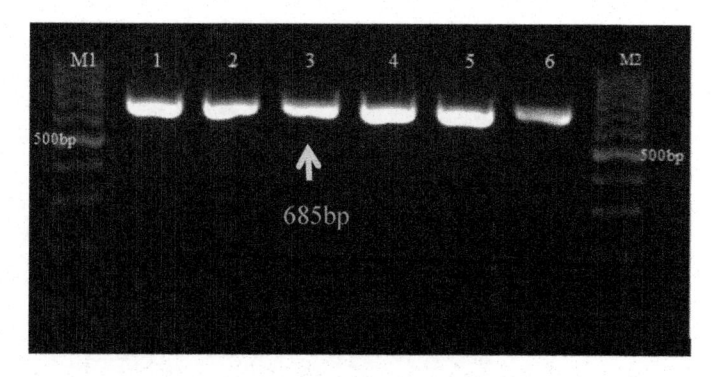

Fig. 1 16S rDNA. M1, M2: 100 bp marker; Lane 1: *A. hydrophila* isolated from Black tilapia (*O. mossambicus*, AHBT2); Lane 2: *A. hydrophila* isolated from Jade perch (*Scortumbarcoo*, AHJP12); Lane 3: *A. hydrophila* isolated from Javanese carp(*P. gonionotus*, AHJC22); Lane 4: *A. hydrophila* isolated from River carp(*L. hoevenii*, AHRP26); Lane 5: *A. hydrophila* isolated from River catfish(*P. pangasius*, AHRC28); Lane 6: *A. hydrophila* isolated from Climbing perch(*A. testudineus*, AHCP32)

3.2 Antibiotic Sensitivity and MAR Index

All isolates of *A. hydrophila* were sensitive to erythromycin, florfenicol, kanamycin and oxytetracycline. Some isolates of *A. hydrophila* also showed sensitivity towards other antibiotics such as ampicillin (10%), streptomycin (80%) and tetracycline (60%). On the other hand, the isolates were resistant to cephalotin, gentamycin and penicillin G. Furthermore, multiple antibiotic resistance (MAR) index for the bacterial isolates ranged from 0.4 to 0.5. The current results indicated that the *A. hydrophila* in these farmed fish might have been indiscriminately and continuously exposed to those antibiotics during their culturing stages (Table 3).

Table 3 MAR index and resistance profiles of *A. hydrophila* from freshwater fish

Resistance profile	Isolates with similar patterns	MAR index
CE10, GM10, PG10, AM10, TE30	AHBT1, AHBT2, AHBT3, AHBT4, AHBT5, AHBT6, AHJP11, AHJP12, AHJP13, AHJP14, AHJP15, AHJP16, AHJP17, AHJP18, AHJP19, AHJP20	0.5
CE10, GM10, PG10, S10	AHRT7, AHRT8, AHRT9, AHRT10	0.4
CE10, GM10, PG10, AM10	AHJC21, AHJC22, AHJC23, AHJC24, AHRP25, AHRP26, AHCP31, AHCP32, AHCP33, AHAC34, AHAC35, AHAC36, AHAC37, AHAC38, AHFH39, AHFH40	0.4
CE10, GM10, PG10, AM10, S10	AHRC27, AHRC28, AHRC29, AHRC30	0.5

Keys Ampicillin (10 μg); CE10: Cephalotin (10 μg); E15: Erythromycin (15 μg); FFC30: Florfenicol (30 μg); GM10: Gentamicin (10 μg); K30: Kanamycin (30 μg); OT30: Oxytetracycline (30 μg); PG10: Penicillin G (10 μg); S10:Streptomycin (10 μg); TE30: Tetracycline (30 μg)

Fig. 2 Integron detection. Lane 1: 100 bp marker; Lane 2: *A. hydrophila* isolated from Black tilapia (*O. mossambicus*, AHBT2); Lane 3: *A. hydrophila* isolated from Jade perch (*S. barcoo*, AHJP12); Lane 4: *A. hydrophila* isolated from Javanese carp (*P. gonionotus*, AHJC22); Lane 5: *A. hydrophila* isolated from Flowerhorn (*Cichlasoma* sp, AHFH40)

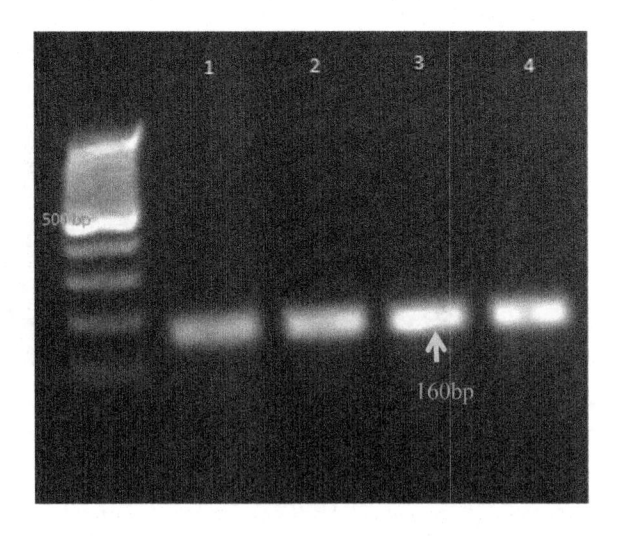

3.3 Detection of Plasmid Profiles, Integrons, Integron-Associated Genes and Antibiotic Resistance Genes

Twelve out of 40 isolates of *A. hydrophila* contained plasmid with size ranging from 6 to 23 kb. Only, 1–2 bands of plasmid were isolated. The *intI*1 gene was detected in 20/40 (50%) of *A. hydrophila* strains, but no *intI*2 and *intI*3 were detected (Fig. 2). The *aadA* was found in 5/40 (12.5%) in *A. hydrophila*, while both of tetracycline resistance genes, *tet*A and *tet*C, were found in 16/40 (40%) of *A. hydrophila* isolates. The *strA-strB*, bla$_{TEM}$ and bla$_{SHV}$ genes were not detected in any of the isolates.

4 Discussion

Antibiotics have been mixed with feed for oral administration for treatment and prevention of bacterial infections in aquaculture, and drugs in the same classes have been used for medical treatment in humans. A particular concern is that improper and overuse of antibiotics can lead to development and distribution of antimicrobial resistance among aquatic bacterial pathogens, including *A. hydrophila* that could transfer to humans (Lukkana et al. 2012).

Motile *Aeromonas* spp. was regularly found in the intestine and the gills of freshwater-cultured fish. This genus readily develops single or multiple antibiotic resistance, suggesting that this genus might be used as good indicators of antimicrobial resistance in the freshwater aquaculture environment (Nguyen et al. 2014). In this study, *A. hydrophila* isolates displayed high levels of resistance to ampicillin, penicillin G and cephalothin possibly due to the high intrinsic β-lactam resistance of this genus that are enhanced by an active efflux mechanism and/or the synergism between outer membrane impermeability or secondary resistance mechanisms such as β-lactamases or antibiotic efflux pumps (Janda and Abbott 2010; Nguyen et al. 2014).

This present study also was in agreement with Belem-Costa and Cyrino (2006), where lack of effectiveness of amoxicillin and tetracycline to *A. hydrophila* isolated from Pacu, *Piaractus mesopotamicus*. The author also stated that the isolates were susceptible to chloramphenicol, gentamicin, kanamycin and streptomycin. Son et al. (1997) also reported about *A. hydrophila* isolated from *Telapia mossambica* were resistance to tetracycline. Scoaeris et al. (2008) and Pandove et al. (2011) also reported *A. hydrophila* isolated from drinking water and surface water was resistant to ampicillin which in concordance with presents study.

In the present study, isolates of *A. hydrophila* contained plasmid size ranging from 6 to 23 kb. Radu et al. (2003) observed plasmids with sizes ranging from 2.7 to 15.7 kb in six isolates of *A. hydrophila* isolated from skin lesions of common freshwater fishes. Another study by Shome et al. (2001) stated that all nine strains of A. *hydrophila* isolated from ulcerative, septicemic and dropsy infections of freshwater fishes showed single large plasmid sized 23.13 kb. This variation in a number of plasmids reported by various workers within the *Aeromonas* genus, its different species and even in subspecies may be attributed to the source, environment or unique metabolic diversity of host bacterium harbouring the plasmid (Shome and Shome 1999; Shome et al. 2001).

The occurrence of integrons in fish-farming environments is well known. A previous study done by Ndi and Barton (2011) detected class 1 integron in 28/90 (31%) isolates of *Aeromonas* spp. from rainbow trout farms in Australia. Nawaz et al. (2010) were able to detect class 1 integron in 48% of *Aeromonas veronii* isolates from catfish. Jacobs and Chenia (2007) reported *Aeromonas spp.* isolates from tilapia, trout and koi aquaculture systems in South Africa harbour class 1 integrons with gene cassettes *ant(3″)*Ia, *aac(6′)*Ia, *dhfr1*, *oxa2a* and/or *pse1*. The occurrence of class 1 integrons with the *dhfr*, ant (3″) 1a and *catB2* genes among motile aeromonads from a fish-farming environment in Denmark (Schmidt et al. 2001a).

The study of integrons with various genes occurring in combination or as single gene cassettes in *Aeromonas* spp. from aquaculture sources from various geographical locations has been described (Labee-Lund and Sorum 2001; Henriques et al. 2006; Ndi and Barton 2011). However, this study showed that no gene cassette was detected in all bacterial isolates. This finding was in agreement with Rosser and Young (1999) and Schmidt et al. (2001a) which found integrons with no apparent gene cassette array reported from aquatic environments. So, it is not surprising that some of our isolates had no gene cassettes detected. It might be the 3′ end integron not amplified in this study. The study done by Sarria-Guzman et al. (2014) only 16 out of 20 *int*1 positive in *Aeromonas spp.* yielded the 800 bp *qacEΔ1/sul1* amplification product corresponding to the 3′ conserved region. Several attempts were done by the author to amplify the 3′ end on the remaining strains that did not show this amplicon. Another research done by Moura et al. (2012) demonstrated the presence of incomplete integrons, lacking the 3′ end, confirmed this absence by their sensitivity to sulfamethoxazole. Post et al. (2007) suggested strategies to search Tn402-like sequence, often associated with class 1 integrons.

In this study, amplicons of 150 bp were detected in *int1*. Lukkana et al. (2012) stated that amplicons of 150 bp are indicative of empty integrons. This finding also was reported previously in aquatic bacterial pathogen by Schmidt et al. Integron loses resistance cassettes in the absence of selection pressure exerted by antibiotics (Sarria-Guzman et al. 2014). The presence of empty integron indicates the possibility to capture cassettes resistant to antibiotics. Integron may play an important role in mediating gene transfer between micro-organism. This versatility has contributed to the quick adaptation of bacteria to different ecological niches (Rowe-Magnus and Mazel 2002)

In total, we found *aad*A gene in 12.5% (5/40) *A. hydrophila* which accounts for streptomycin-resistant isolates. In Australia, streptomycin was previously used in food-producing animals. The fish farms could have been contaminated by surface run-off from other agricultural activities, or streptomycin resistance could be an indication of past off-label use (Ndi and Barton 2011). Both tetracycline resistance genes, *tet*A and *tet*C were found in 16/40 (40%) of *A. hydrophila* isolates. The result was in agreement with Ndi and Barton (2011); the author detected *tet*A and *tet*C in *Aeromonas spp*. from rainbow trout farms in Australia. Noticeably, different distribution of *tet* genes found in the isolates found in the Australian marine tropical where *tet*M, *tet*E, *tet*A and *tet*D were the most common determinant (Akinbowale et al. 2007). The simultaneous occurrence of two *tet* genes was not correlated with the antibiotic sensitivity test. The *strA-strB*, bla_{TEM} and bla_{SHV} genes were not detected in any of the isolates. The β-lactam resistance phenotype observed in our strains may be due to other beta-lactamase classes not investigated in this study (Ndi and Barton 2011).

Schmidt et al. (2001a) stated that tetracycline resistance gene are frequently part of transposon, which are able to change their location within the cell, and thus achieve increased mobility, for an instance, by inserting into conjugative plasmids. Even though class 1 integrons are transposition defective, they are often plasmid borne as they are mobilizable in association with a functional transposon or by transposition proteins supplied in 'trans'. Thus, their presence might indicate whether horizontal gene transfer has occurred among the aeromonads found in and around the sampled fish farm, and associated antibiotic resistance genes could explain some of the observed resistance patterns. Sarria-Guzaman et al. (2014) stated that multidrug-resistant strains do not always have integrons. These multidrug-resistant strains without integron might contain extrachromosomal genetic plasmid or transposon that carries antibiotic resistance genes. For example, tetracycline resistance is associated generally with the presence of transposon, such as Tn1721 and Tn10 found in plasmid (Bello-Lopez et al. 2012; Ross et al. 2013).

5 Conclusion

In conclusion, constant monitoring should be conducted to compile more information on antibiotic sensitivity of *Aeromonas spp.* and other known aquatic bacteria species in order to avoid the development of antibiotic resistance superbug. The increase trend of antibiotics-resistant aeromonads will lead to treatment challenge in Aeromonas infection. Although, there has been little or no previous investigation of resistance determinants, this study recognized that high MAR index in *A. hydrophila* carrying resistance genes and integrons. The occurrence of class 1 integron, the different tetracycline resistance genes and *aad*A genes in Malaysian aquaculture is similar to what is seen in countries where antibiotics are known to be used in aquaculture. Detection of antibiotic resistance determinant in fish pathogenic bacteria has a serious impact on public health when consumer eats the fish and the farm waste drained into the river and goes into the aquatic environment.

Acknowledgements This research was supported by RUGS grant (no: 05-01-09-073RU/91736) funded by Universiti Putra Malaysia (UPM) and FRGS grant (no: R/FRGS/A06.00/00300/001/2015/000281) funded by Ministry of Higher Education.

References

Aarestrup, F. M., Agerso Gerner-Smidt, Y., Madsen, M., & Jensen, L. B. (2000). Comparison of microbial resistance phenotypes and resistance genes in *Enterococcus faecalis* and *Enterococcus faecium* from humans in the community, broilers and pigs in Denmark. *Diagnostic Microbiology and Infectious Disease, 37,* 127–137.

Akinbowale, O. L., Peng, H., & Barton, M. (2007). Diversity of tetracycline resistance genes in bacteria from aquaculture sources in Australia. *Journal of Applied Microbiology, 103,* 2016–2025.

Aoki, T., & Kitao, T. (1981). Drug resistance and transferable R plasmids in *Edwardsiella tarda* from fish culture ponds. *Fish Pathology, 15,* 277–281.

Belém-Costa, A., & Cyrino, J. E. P. (2006). Antibiotic resistence of *Aeromonas hydrophila* isolated from *Piaractus mesopotamicus* (Holmberg, 1887) and *Oreochromis niloticus* (Linnaeus, 1758). *Scientia Agricola, 63*(3), 281–284.

Bello-Lopez, J. M., Vazquez-Ocampo, N. M., Fernandez-Rendion, E., & Curiel-Quesada, E. (2012). Inability of some *Aeromonas hydrophila* strains to act as recipients of plasmid pRASa in conjugal transfer experiments. *Current Microbiology, 64,* 332–337.

Chu, W. H., & Lu, C. P. (2005). Multiplex PCR assay for the detection of pathogenic *Aeromonas hydrophila. Journal of Fish Diseases, 28,* 437–441.

CLSI. (2016). *Performance standards for antimicrobial susceptibility testing.* 9th Informational Supplement, Vol. 26, No. 1. CLSI document M100-S26. Pennsylvania, USA.

Hall, R. M., & Stokes, H. W. (1993). Integrons: Novel DNA elements which capture genes by site-specific recombination. *Genetica, 90,* 115–132.

Han, H. S., Koh, Y. J., Hur, J. S., & Jung, J. S. (2004). Occurrence of the *strA-str*B streptomycin resistance genes in *Pseudomonas* species isolated from kiwifruit plants. *The Journal of Microbiology, 42,* 365–368.

Henriques, I. S., Fonseca, F., Alves, A., Saavedra, M. J., & Correia, A. (2006). Occurrence and diversity of integrons and β-lactamase genes among ampicillin-resistant isolates from estuarine waters. *Research in Microbiology, 157,* 938–947.

Jacobs, L., & Chenia, H. Y. (2007). Characterization of integrons and tetracycline resistance determinants in *Aeromonas* spp. isolated from South African aquaculture systems. *International Journal of Food Microbiology, 114,* 295–306.

Janda, J. M., & Abbott, S. L. (2010). The genus *Aeromonas*: Taxonomy, pathogenicity, and infection. *Clinical Microbiology Reviews, 23,* 35–73.

Kathleen, M. M., Samuel, L., Felecia, C., Reagan, E. L., Kasing, A., Lesley, M., et al. (2016). Antibiotic resistance of diverse bacteria from aquaculture in Borneo. *International Journal of Microbiology, 2016,* 1–9.

Koelman, J. G. M., Stoof, J., Van Bijl Der, M. W., Vandenbroucke Grauls, C. M. J. E., & Savelkoul, P. H. M. (2001). Identification of epidemic strains of *Acinetobacter baumannii* by integrase gene PCR. *Journal of Clinical Microbiology, 39,* 8–13.

Labee-Lund, T. M., & Sorum, H. (2001). Class I integrons mediated antibiotic resistance in the fish pathogen *Aeromonas salmonicida* world-wide. *Microbial Drug Resistance, 7,* 263–272.

Lukkana, M., Wongtavatchai, J., & Chuanchuen, R. (2012). Class 1 Integrons in *Aeromonas hydrophila* Isolates from Farmed Nile Tilapia (*Oreochromisnilotica*). *Journal of Veterinary Medical Science, 74*(4), 435–440.

Mazel, D., Dychinco, B., Webb, V. A., & Davies, J. (2000). Antibiotic resistance in the ECOR collection: Integrons and identification of a novel *aad* gene. *Antimicrobial Agents and Chemother, 44,* 1568–1574.

Nawaz, M., Khan, S. A., Khan, A. A., Sung, K., Tran, Q., Kerdahi, K., et al. (2010). Detection and characterization of virulence genes and integrons in *Aeromonas veronii* isolated from catfish. *Food Microbiology, 27,* 327–331.

Ndi, O. L., & Barton, M. D. (2011). Incidence of class 1 integron and other antibiotic resistance determinants in *Aeromonas* spp. from rainbow trout farms in Australia. *Journal of Fish Diseases, 34,* 589–599.

Nguyen, H. N. K., Van, T. T. H., Nguyen, H. T., Smooker, P. M., Shimeta, J., & Coloe, P. J. (2014). Molecular characterization of antibiotic resistance in *Pseudomonas* and *Aeromonas* isolates from catfish of the Mekong Delta, Vietnam. *Veterinary Microbiology, 171,* 397–405.

Pandove, G., Sahota, P. P., Achal, V., & Vikal, Y. (2011). Detection of *Aeromonas hydrophila* in water using PCR. *American Water Works Association Journal, 103,* 59–65.

Post, V., Recchia, G. D., & Hall, R. M. (2007). Detection of gene cassette in Tn402-like class 1 integrons. *Antimicrobial Agents and Chemotherapy, 57,* 3467–3468.

Radu, S., Ahmad, N., Ling, F. H., & Reezal, A. (2003a). Prevalence and resistance to antibiotics for *Aeromonas* species from retail fish in Malaysia. *International Journal of Food Microbiology, 81,* 261–266.

Rasheed, J. K., Jay, C., Metchock, B., Berkowitz, F., Weigel, L., Crellin, J., et al. (1997). Evolution of extended-spectrum beta-lactam resistance (SHV-8) in a strain of *Escherichia coli* during multiple episodes of bacteremia. *Antimicrobial Agents and Chemother, 41,* 647–653.

Rattanachaikunsopon, P., & Phumkhachorn, P. (2009). Prophylactic effect of *Andrographis paniculata* extracts against *Streptococcus agalactiae* infection in Nile tilapia (*Oreochromis niloticus*). *Journal of Bioscience and Bioengineering, 107*(5), 579–582.

Rhodes, G., Huys, G., Swings, J., Mcgann, P., Hiney, M., Smith, P., et al. (2000). Distribution of oxytetracycline resistance plasmids between aeromonads in hospitals and aquaculture environments: implication of Tn1721 in dissemination of the tetracycline resistance determinant tet A. *Applied and Environmental Microbiology, 66,* 3883–3890.

Ross, J. A., Ellis, M. J., Hossain, S., & Haniford, D. B. (2013). Hfa restructive RNA-IN and RNA-OUT and facilitates antisense pairing in the Tn10/7810 system. *RNA, 19*(5), 670–684.

Rosser, S. J., & Young, H. K. (1999). Identification and characterization of class 1 integrons in bacteria from an aquatic environment. *Journal of Antimicrobial Chemotherapy, 44,* 11–18.

Rowe-Magnus, D. A., & Mazel, D. (2002). The role of integrons in antibiotics resistance genes capture. *International Journal of Medical Microbiology, 292,* 115–125.

Sarria-Guzmán, Y., Lopez-Ramirez, M. P., Chàvez-Romero, Y., Ruiz-Romero, E., Dendooven, L., & Bello-Lopez, J. M. (2014). Identification of Antibiotic resistance cassettes in class 1 integrons in *Aeromonas spp.* Strains isolated from fresh fish (*Cyprinuscarpio* L.). *Current Microbiology, 68,* 581–586.

Schmidt, A. S., Bruun, M. S., Dalsgaard, I., & Larsen, J. L. (2001a). Incidence, distribution and spread of tetracycline resistance determinants and integron-associated antibiotic resistance genes among motile aeromonads from fish farming environment. *Applied and Environmental Microbiology, 67,* 5675–5682.

Schmidt, A. S., Bruun, M. S., Larsen, J. L., & Dalsgaard, I. (2001b). Characterisation of class 1 integrons associated with R-plasmids in clinical *Aeromonas salmonicida* isolates from various geographical areas. *Journal of Antimicrobial Chemotherapy, 47,* 735–743.

Scoaris, D. D. O., Colacite, J., Nakamura, C. V., Ueda-Nakamura, T., de Abreu Filho, B. A., & Filho, B. P. D. (2008). Virulence and antibiotic susceptibility of *Aeromonas spp.* isolated from drinking water. *Antonie Van Leeuwenhoeck, 93*(1–2), 111–122.

Shome, R., & Shome, B. R. (1999). Study of virulence factors of *A. hydrophila* isolates causing acute abdominal dropsy and ulcerative diseases in Indian major carps. *Indian Journal of Fisheries, 46,* 133–140.

Shome, R., Shome, B. R., Gaurand, K., & Senani, S. (2001). Plasmids of *Aeromonas hydrophila* isolated from freshwater fishes. *Indian Journal of Fisheries, 48*(1), 109–110.

Shotts, E. B., Jr., Vanderwork, V. L., & Campbell, L. M. (1976). Occurrence of R factors associated with *Aeromonas hydrophila* isolates from aquarium fish and waters. *Journal of the Fisheries Board of Canada, 33,* 737–740.

Son, R., Rusul, G., Sahilah, A. M., Zainuri, A., Raha, A. R., & Salmah, I. (1997). Antibiotic resistance and plasmid profile of *Aeromonas hydrophila* isolates from cultured fish, Telapia (*Telapia mossambica*). *Letters in Applied Microbiology, 24,* 479–482.

Starliper, C. E., & Cooper, R. K. (1998). Biochemical and conjugation studies of Romet-resistant strains of *Aeromonas salmonicida* from Salmonid rearing facilities in the Eastern United States. *Journal of Aquatic Animal Health, 10,* 221–229.

Starliper, C. E., Cooper, R. K., Shotts, E. B., Jr., & Taylor, P. W. (1993). Plasmid-mediated Romet resistance of *Edwarsiella ictaluri*. *Journal of Aquatic Animal Health, 1,* 1–8.

Stokes, H. W., & Hall, R. M. (1989). A novel family of potentially mobile DNA elements encoding site specific gene integration functions: Integrons. *Molecular Microbiology, 3,* 11669–11683.

Sunde, M., & Norstrom, M. (2005). The genetic background for streptomycin resistance in *Escherichia coli* influences the distribution of MICs. *Journal of Antimicrobial Chemotherapy, 56,* 87–90.

Sundstrom, L., Radstrom, P., Swedberg, G., & Skold, O. (1988). Site-specific recombination promotes linkage between trimethoprim and sulphonamide resistance genes. Sequence characterisation of *dhfrV* and *sulI* and a recombination active locus of Tn 21. *Molecular and General Genetics, 213,* 191–201.

Taylor, P. W. (2003). Multiple antimicrobial resistance in a chronic bacterial infection of Koi carp. *North American Journal of Aquaculture, 65,* 120–125.

Van Essen-Zandbergen, A., Smith, H., Veldmanand, K., & Mevius, D. (2007). Occurrence and characteristics of class 1, 2 and 3 integrons in *Escherichia coli, Salmonella* and *Campylobacter spp.* in the Netherlands. *Journal of Antimicrobial Chemotherapy, 59,* 746–750.

Vivekanandhan, G., Savithamani, K., Hatha, A. A. M., & Lakshmanaperumalsamy, P. (2002). Antibiotic resistance of *Aeromonas hydrophila* isolated from marketed fish and prawn of south India. *International Journal of Food Microbiology, 76,* 165–168.

Watanabe, T. A., Ogata, Y., & Egusa, S. (1971). R factors related to fish culturing. *Annals of the New York Academy Science, 182,* 383–410.

White, P. A., Mciver, C. J., Deng, Y., & Rawlinson, W. D. (2000). Characterisation of two new gene cassettes, *aadA5* and *dfrA17*. *FEMS Microbiology Letters, 182,* 265–269.

Relationship of Pre-competition Anxiety and Cortisol Response in Individual and Team Sport Athletes

Jamilah Ahmad Radzi, Sarina Md Yusuf, Nurul Hidayah Amir and Siti Hannariah Mansor

Abstract This present study investigated pre-competition anxiety by using both psychological and physiological measures of anxiety among individual and team sport athletes. Malaysian males backup athletes ($N = 52$) participated in this study. Competitive State Anxiety Inventory (CSAI-2) and salivary cortisol samples were obtained from participants across three different times which at one week, one day and one hour prior to the competition. Unfortunately, the results indicated that there was no significant correlation of CSAI-2 subscale scores with cortisol response in individual sports athletes across the three different times, whereas in team sports athletes, a significant moderate negative correlation was noted in self-confidence and cortisol response at one week prior to the competition and significant strong, positive relationship between somatic anxiety and cortisol response. Future studies should use physiological measure and psychological measure at the same time prior a highly competitive situation in measuring anxiety on sport performance.

Keywords Cortisol · Individual sports athletes · Team sports athletes Anxiety

---

J. A. Radzi (✉) · S. Md Yusuf · N. H. Amir · S. H. Mansor
Faculty of Sports Science and Recreation, Universiti Teknologi MARA,
UiTM Perlis, Arau, Malaysia
e-mail: jamilaharadzi@perlis.uitm.edu.my1

S. Md Yusuf
e-mail: sarin864@salam.uitm.edu.my2

N. H. Amir
e-mail: hidayah131@perlis.uitm.edu.my

S. H. Mansor
e-mail: sitihannariah@perlis.uitm.edu.my

1 Introduction

Competition may also influence the emotion of athletes which can result in increased blood pressure and heart rate (Mitchell et al. 2005). In brief, feelings of tension, thinking of upcoming events in their mind, nervousness, worrying and involving in physiological changes such as increased in heart rate are common responses for the athletes prior to the competition (Hackfort and Spielberger 1989). All of these are common conditions which are known as anxiety (Cerin 2003; Hackfort and Spielberger 1989; Jarvis 2006; Kais and Raudsepp 2005; Martens et al. 1990; Wiggins 1998). In other words, competition can trigger emotional state of anxiety (Jarvis 2006) and this condition usually occurs in every athlete when involving with competition. Such condition is called pre-competition anxiety (Cox 2012). Anxiety would reduce cognitive ability, and impaired attention or perception of individual (Murphy 2005). In addition, the individual will also be overwhelmed with self-doubts, fear of failing and lose of self-image (Cox 2012; Jarvis 2006).

There are two types of anxiety such as state anxiety and trait anxiety. State anxiety involves feeling of apprehension, tension, fear and increase in physiological arousal. This is an immediate emotional state response to specific situation. State anxiety also consists of somatic and cognitive anxiety. Somatic anxiety is closely related to physiological aspects of anxiety which involve physical symptoms such as rapid heartbeat, shortness of breath and muscular tension (Leunes and Nation 2002; Martens et al. 1990). Another component of state anxiety is cognitive anxiety which refers to worry and emotional distress for upcoming events (Martinent et al. 2010). Trait anxiety is involved with an experience of anxiety over a long period of time towards the stressful environments (Filaire et al. 2001). It is also similar to a personality variable (Jarvis 2006; Wann 1997). Some of the Malaysian athletes are not able to compete under pressure. For instance, Chong Wei Feng, a Malaysian single badminton player admitted he was a bit nervous, his focus was disrupted and that he had failed to control the pressure in the third set which led to his loss against Rajiv Ouseph from England at BWF World Badminton Championships 2014 ("Nervousness led to loss," 2014). Therefore, competitive anxiety and the level of pre-competition anxiety are the overriding factors that can decrease and affect athletes' performance (Cox 2012; Esfahani and Soflu 2010).

Several factors have been found to influence anxiety and sport performance. Sport categories such as individual sport and team sport play an important role in anxiety (Cerin 2003; Han et al. 2006; Martens et al. 1990; Martin and Hall 1997; Zeng 2003). Previous literature reported that state anxiety were higher in individual sport compared to team sport athletes (Craft et al. 2003; Han et al. 2006; Martens et al. 1990; Martin and Hall 1997; Zeng 2003). However, self-confidence was found lower in individual than team sport athletes (Martens et al. 1990; Zeng 2003). It seems that cognitive and somatic competitive anxiety is affected by social context of sport competition that is competing as individual versus team event (Martens et al. 1990;

Martin and Hall 1997). Yet, it seems that researchers have investigated the sports without looking into the other factors of sport classification like contact and non-contact, subjective and objective sports when determining the anxiety level among individual sport athletes.

Still, most of studies often measured competitive anxiety using the self-report inventory such as Competitive State Anxiety Inventory-2 (CSAI-2). Thus, physiological changes like hormones response should be used in related above studies to have better understanding and verify the anxiety levels prior to the competition. One of the hormones that is cortisol is released from adrenal cortex which from the hypothalamic-pituitary adrenal axis (HPAA) and is released in response related to psychological state (Filaire et al. 2009; Alix-Sy et al. 2007) like anxiety (Chiodo et al. 2011). Recent studies showed cortisol response was related with pre-competition anxiety (Filaire et al. 2009; Salvador et al. 2003). According to Filaire et al. (2009) a significant positive relationship also was found between cortisol and both somatic and cognitive anxiety in competition. This shows that when psychological measure like cognitive anxiety was affected, and cortisol hormones were also affected at the same time. There are two methods can be used to measure cortisol hormones that are using blood samples or salivary samples. However, sample of saliva is the best method because it is proven to measure the cortisol response as a non-invasive method and it is easier to researcher for data collection (Kim et al. 2010; Gatti and De Polo 2010; McKay et al. 1997). Cortisol was found increases prior to competition in individual sports (Filaire et al. 2001, 2009; Kim et al. 2010; Luopos et al. 2008; Salvador et al. 2003). So, regarding a lack of studies on salivary cortisol response among Malaysian athletes, which may give a good marker in understanding of anxiety levels, and this study, therefore, will examine both psychological and physiological measures of anxiety levels as the competition approaches.

2 Methodology

2.1 Sample

Fifty-two ($N = 52$) Malaysian backup athletes voluntarily participated in this present study. The athletes were the best young athletes in Malaysia whom competed at junior national and international level competitions. The participants comprised of male athletes whose aged ranging from 16 to 23 years old. Individual sports were selected from squash, table tennis, lawn ball, bowling and track and field athletes. Team sports were selected from volleyball, cricket and petanque. Participants were drawn from non-contact, objective sport (Martens et al. 1990; Mellalieu et al. 2004) and low static demand sports only (Mitchell et al. 2005).

2.2 Instrumentation

Psychological measures: Competitive State Anxiety Inventory-2 (CSAI-2) that developed by Martens et al. (1990) was used to measure pre-competition cognitive anxiety, somatic anxiety and self-confidence of the participants. The CSAI-2 has 27 items, with nine items in each subscale and arranged in a 4-point Likert scale format. The response scale requires the participants to rate the intensity of each symptom on a scale of 1 (not at all) to 4 (very much), so resulting in scores ranging from 9 to 36 for each subscale.

Physiological measures: Salivary cortisol response was determined using the Salimetrics Salivary Cortisol Enzymes Immunoassay Kits.

2.3 Data Collection

Participants were briefly explained on the purpose of present study and all procedures involved in the study. Present study was approved by UiTM research ethical committee. Informed consent form was obtained from participants at one week prior to the competition to waive any accidental occurs during the test that does not related with the test procedures. At the same day, CSAI-2 questionnaire was later administered to the participants and was collected after participants finished answering the questionnaire. The standardized instructions were used to administer the CSAI-2 questionnaire, which were antisocial desirability instructions as suggested by Martens et al. (1990) Salivary cortisol was collected immediately after participants returned the questionnaire. CSAI-2 and saliva samples collection procedure were repeated in the same manner for one day and one hour prior to the competition. All the measurements were obtained before the beginning of any warm-up activity in order to minimize any effect of exercise or physical activity on cortisol levels (Salvador et al. 2003).

Participants were advised not to take any food or drinks for 30 min prior and no brushing of teeth two hours before collection session. Besides, participants were asked to refrain from taking alcohol the night before or the day of saliva collection. All of these guidelines (Salimetrics 2010; Gatti and De Palo 2010) were given to participants at the first meeting and were reminded to participants at pre-saliva collection day. On the day of collection, participants were asked to rinse their mouth with water before accumulate the saliva (Filaire et al. 2007) to avoid any contamination in their mouth. Participants were asked to tilt their head forward and spit their saliva directly into the plastic tubes (Salvador 2003). The samples were immediately kept in the ice and frozen at or below -20 °C within 4 h after the collection. All the samples were labelled and were sent to laboratory to analyse using Salimetrics Salivary Cortisol Enzymes Immunoassay Kit.

3 Findings and Discussion

3.1 Results

Means and standard deviation of the variables examined in the present study are presented in Table 1. The results of the data showed that the mean of cognitive anxiety and somatic anxiety scores at three different times (one week, one day and one hour prior to the competition) were higher in individual sports athletes compared to team sports athletes. But, the mean of self-confidence scores at one week and one day prior to the competition was higher in team sports compared to individual sports. On the other hand, only the mean of self-confidence scores at one hour prior to the competition was higher in individual sports compared to team sports. In terms of cortisol response, the result demonstrated that the mean of cortisol response at one week, one day and one hour prior to the competition was higher in team sports compared to individual sports.

Pearson product-moment correlation coefficient was applied to test the relationship between CSAI-2 scores and salivary cortisol response among individual and team sports athletes.

The correlations between CSAI-2 subscale scores and cortisol response in individual sports athletes are presented in Table 2. Unfortunately, the results indicated that there is no significant correlation of CSAI-2 subscale scores and cortisol response in individual sports athletes across the three different times.

Meanwhile, in team sport athletes, there is only strong, positive relationship $r = 0.710$, $n = 26$, $(p < 0.05)$ was noted between cognitive anxiety scores and somatic anxiety scores were found. The correlations between CSAI-2 subscale scores and cortisol hormone in team sports athletes are presented in Table 2. Result in the team sport athletes indicated that there is a significant relationship between somatic anxiety and cortisol response at one week prior to the competition. A strong, positive relationship, $r = 0.590$, $n = 26$, $(p < 0.05)$ was noted. In addition, the result also

Table 1 Mean and standard deviation (SD) of variables examined in individual and team sports athletes

Variables	One week before		One day before		One hour before	
	Mean	SD	Mean	SD	Mean	SD
Individual Sports						
Cognitive	20.92	6.33	21.65	5.87	21.69	6.82
Somatic	17.08	4.24	17.65	4.17	18.73	5.39
Self-confidence	26.04	4.60	25.73	4.49	25.54	4.49
Cortisol	0.132	0.066	0.101	0.072	0.130	0.089
Team Sports						
Cognitive	18.92	5.15	19.73	4.57	20.81	4.77
Somatic	15.65	4.58	17.38	4.74	17.46	5.42
Self-confidence	26.73	5.23	25.81	4.57	24.31	4.70
Cortisol	0.170	0.094	0.120	0.137	0.155	0.089

Table 2 Relationship between CSAI-2 subscale scores and cortisol response in individual and team sport athletes

	Cortisol hormone		
	Time 1	Time 2	Time 3
Individual sports			
Cognitive anxiety			
Time 1	−0.169		
Time 2		0.021	
Time 3			−0.125
Somatic anxiety			
Time 1	−0.066		
Time 2		−0.121	
Time 3			0.008
Self-confidence			
Time 1	−0.373		
Time 2		−0.027	
Time 3			−0.051
Team sports			
Cognitive anxiety			
Time 1	0.25	−0.194	
Time 2			−0.173
Time 3			
Somatic anxiety			
Time 1	0.590		
Time 2		−0.074	
Time 3			−0.133
Self-confidence			
Time 1	−0.469		
Time 2		−0.448	
Time 3			−0.15

indicated that there is a significant relationship between self-confidence and cortisol response at one week prior to the competition. A moderate negative correlation, $r = -0.469$, $n = 26$, ($p < 0.05$) was noted. There is a significant relationship between self-confidence and cortisol response at one day prior to the competition. A moderate negative correlation, $r = -0.448$, $n = 26$, ($p < 0.05$) was noted.

3.2 Discussion

No significant correlation of CSAI-2 subscale score on salivary cortisol response in individual sports was found in the present study. The finding is consistent with Chiodo et al. (2011); Luopos et al. (2008); Salvador et al. (2003) and Strahler et al.

(2010), which found no significant association for state anxiety and salivary cortisol. The similar result may be due to the selection of male athletes in the study. Cortisol response did not directly correlate to somatic anxiety but to cognitive anxiety in competition (McKay et al. 1997). This result, however, differed with Fillaire et al. (2001), who reported a significant positive relationship of salivary cortisol on both somatic anxiety and cognitive anxiety. The study utilized judo athletes, a contact sport, where the probability of physical harm from an opponent increase in contact sport may influence both psychological and physiological in a similar manner.

In team sports athletes, there was a strong, positive relationship between somatic anxiety scores with cortisol response at one week prior to the competition in team sports athletes. As suggested by Fillaire et al. (2001), psychological arousal is linked to and acts on the hypo-pituitary-adrenal axis (HPAA), which is an increase in somatic anxiety that will increase the cortisol response. Similarly, a significant positive relationship between cortisol and somatic was found prior soccer game (Alix-sy et al. 2008). The possible explanation for the similar finding is due to males' participants and the used of saliva to measure cortisol response. However, the result disagreed with findings from Alexander (2009) who have reported no correlation between cortisol response and somatic anxiety in basketball players. It should be noted that the possible reason is the use of both genders in the study, which may lead to the different findings. No relationship between cortisol and somatic anxiety was found in another study (Haneishi et al. 2007). The possible reason for the distinct findings is due to the used of females' soccer player, whereas in this present study, males athletes were employed. It is noted that participants were chewed paraffin to facilitate the salivary flow. It might influence the cortisol response as its stimulated citric acid (Kirschbaum and Hellhammer 2000).

In addition, the present study found a moderate negative correlation between self-confidence scores and cortisol response at one week and one day prior to the competition. However, the result disagreed with findings from Alexander (2009) who have reported no correlation between cortisol response and self-confidence in basketball players. It seems that cortisol response and state anxiety are interrelated in team sport only.

4 Conclusion and Future Recommendation

The result from this study may be influenced by the social context of sport competition, either participating as individual or team sports athletes. An individual athlete involved with evaluations of individual performance compared to team sports triggers a higher anxiety response prior to a competition (Han et al. 2006). It can be noted that in individual sport, the mistakes done cannot be blamed on others (Martens et al. 1990). Cognitive anxiety has been found positively correlated to somatic anxiety in both sports categories. Cognitive anxiety is also strongly associated with self-confidence in individual sports athletes. It is therefore can be concluded that cognitive anxiety is essential in both sports categories.

However, the result of cortisol response was only found to be positively correlated to somatic anxiety in team sports athletes only. Meanwhile, negative correlation was found between self-confidence and cortisol response in team sports athletes but not in individual sports athletes.

In response to the present studies, it would be advisable to measure anxiety prior or closer to the final of the competition, which might produce higher level of anxiety. Similar study at international competition would be valuable to assess the psychological sign and symptoms of Malaysian athletes in competing between local competitions compared to international competition.

References

Alix-Sy, D., Le Scanff, C., & Filaire, E. (2007). Psychophysiological responses in the pre-competition period in elite soccer players. *Journal of Sports Science and Medicine, 7*(4), 446–454.

Alix-sy, D., Le Scanff, C., & Filaire, E. (2008). Psychophysiological responses in the pre-competition period in elite soccer players. *Journal of Sports Science and Medicine, 7*, 446–454.

Alexander, D. (2009). *Psychophysiological effects of precompetition anxiety on basketball performance* (Unpublished master's thesis). California State University, Fullerton.

Cerin, E. (2003). Anxiety versus fundamental emotions as predictors of perceived functionality of pre-competitive emotional states, threat, and challenge in individual sports. *Journal of Applied Sport Psychology, 15*, 223–238.

Chiodo, S., Tessitore, A., Cortis, C., Cibelli, G., Lupo, C., Ammendolia, A., et al. (2011). Stress-related hormonal and psychological changes to official youth Taekwondo competitions. *Scandinavian Journal of Medicine and Science in Sports, 21*(1), 111–119.

Cox, R. H. (2012). *Sport psychology: Concepts and applications* (7th ed.). New York: McGraw Hill.

Craft, L. L., Magyar, T. M., Becker, B. J., &, & Feltz, D. L. (2003). The relationship between the competitive state anxiety inventory-2 and sport performance: A meta analysis. *Journal of Sport and Exercise Psychology, 25*, 44–65.

Esfahani, N., & Soflu, G. H. (2010). The comparison of pre-competition anxiety and state anger between female and male volleyball players. *World Journal of Sport Sciences, 3*(4), 237–242.

Filaire, E., Alix, D., Ferrand, C., & Verger, M. (2009). Psychophysiological stress in tennis players during the first single match of a tournament. *Psychoneuroendocrinology, 34*, 150–157.

Filaire, E., Filaire, M., & Scanff, C. L. (2007). Salivary cortisol, heart rate and blood lactate during a qualifying trial and an official race in motorcycling competition. *Journal of Sports Medicine and Physical Fitness, 47*(4), 413–417.

Filaire, E., Sagnol, M., Ferrand, C., Maso, F., & Lac, G. (2001). Psychophysiological stress in judo athletes during competitions. *The Journal of Sports Medicine and Physical Fitness, 41*(2), 263–268.

Gatti, R., & De Palo, E. F. (2010). An update: Salivary hormones and physical exercise. *Scandinavian Journal of Medicine and Science in Sports, 2*, 1–13.

Hackfort, D., & Spielberger, C. D. (1989). *Anxiety in sports: An International Perspective (No Title)*. United States of America: Hemisphere Publishing Corporation.

Han, D. H., Kim, J. H., Lee, Y. S., Bae, S. J., Bae, S. J., Kim, H. J., et al. (2006). Influence of temperament and anxiety on athletic performance. *Journal of Sports Science and Medicine, 5*, 381–389.

Haneishi, K., Fry, A. C., Moore, C. A., Schilling, B. K., Yuhua, L., & Fry, M. D. (2007). Cortisol and stress responses during a game and practice in female collegiate soccer players. *Journal of Strength and Conditioning Research, 21*(2), 583–588.

Jarvis, M. (2006). *Sport psychology: A student's handbook*. New York: Routledge.

Kais, K., & Raudsepp, L. (2005). Intensity and direction of competitive state anxiety, self confidence and athletic performance. *Journal of Kinesiology, 1,* 13–20.

Kim, K.-J., Park, S., Hoi, K. K., Tae-Won, J., Dong-Ho, P., & Kim K.-B. (2010). Salivary cortisol and immunoglobulin a responses during golf competition vs. Practice in elite male and female junior golfers. *Journal of Strength and Conditioning Research, 24*(3), 852–858.

Kirschbaum, C., & Hellhammer, D, H. (2000). Salivary cortisol. *Encyclopedia of stress, 3,* 379–383.

Leunes, A., & Nation, J. R. (2002). *Sport psychology* (3rd ed.). United States of America: Wadsworth Group.

Luopos, D., Fotini, M., Barkoukis, V., Tsorbatzoudis, H., Grouios, G., & Taitzoglou, I. (2008). Psychological and physiological changes of anxiety prior a swimming competition. *The Open Sport Medicine Journal, 2,* 41–46.

Martens, R., Vealey, R. S., & Burton, D. (1990). *Competitive anxiety in sport*. United States of America: Human Kinetics Publisher.

Martin, K. A., & Hall, C. R. (1997). Situational and intrapersonal moderator of sport competition state anxiety. *Journal of Sport Behavior, 20,* 435–446.

Martinent, G., Ferrand, C., Guillet, E., & Gautheur, S. (2010). Validation of the French version of the competitive state anxiety inventory-2 revised (CSAI-2R) including frequency and direction scales. *Psychology of Sport and Exercise, 11,* 51–57.

Mckay, J. M., Selig, S., Carlson, J., & Morris, T. (1997). Psychological stress in elite golfers during practice and competition. *The Australian Journal of Science and Medicine in Sport, 29*(2), 55–61.

Mellalieu, S. D., Hanton, S., & O'Brien, M. (2004). Intensity and direction dimensions of competitive anxiety as a function of sport type and experience. *Journal of Science & Medicine in Sport, 14,* 326–334.

Mitchell, J. H., Haskel, W., Snell, P., & Van Camp, S. P. (2005). Task force 8: Classification of sports. *Journal of American College of Cardiology, 45,* 1364–1367.

Murphy, S. (2005). *The sport psychology handbook*. United States of America: Human Kinetics.

Nervousness led to loss. (2014, August 28). Retrieved September 7, 2014, from http://www.sinarharian.com.my/sukan/gugup-punca-tewas-1.311761.

Salimetrics. (2010). Salivary cortisol enzyme immunoassay kit manual. United State of America: State College, PA.

Salvador, A., Suay, F., González-Bono, E., &, & Serrano, M. A. (2003). Anticipatory cortisol, testosterone and psychological responses to judo competition in young men. *Psychoneuroendocrinology, 28*(3), 364–375.

Strahler, K., Ehrlenspiel, F., Heene, M., & Brand, R. (2010). Competitive anxiety and cortisol awakening response in the week leading up to a competition. *Psychology of Sport and Exercise, 11*(2), 148–154.

Wann, D. L. (1997). *Sport Psychology*. United States of America: Prentice Hall.

Wiggins, M. S. (1998). Anxiety intensity and direction: Preperformance temporal patterns and expectations in athletes. *Journal of Applied Sport Psychology, 10,* 201–211.

Zeng, H. Z. (2003). The differences between anxiety and self-confidence between team and individual sports college varsity athletes. *International Sports Journal Winter, 7*(1), 28–34.

Relationship Between Muscle Architecture and Jumping Abilities Among Recreationally Active Men

Ali Md Nadzalan, Nur Ikhwan Mohamad, Jeffrey Low Fook Lee and Chamnan Chinnasee

Abstract The objective of this study is to determine the relationships between muscle architectures and jumping abilities. Thirty recreationally active men (mean age = 22.21 ± 1.59 years old) were recruited as study participants and underwent testing for muscle architecture and jumping abilities. Ultrasonography method was used to measure fascicle length, pennation angle and muscle thickness of biceps femoris (BF) and vastus lateralis (VL) for muscle architecture variables. Vertical jump and standing long jump were tested for jumping abilities. Pearson Correlation was used to determine the relationship between all muscle architectures and the jumping abilities tested. Results showed that the muscle thickness and pennation angle of BF and the muscle thickness of VL were shown to be positively correlated while the fascicle length of VL was shown to be negatively correlated with vertical jump performance. Besides that, muscle thickness of BF and the pennation angle of VL and BF were shown to be positively correlated with standing long jump performance. Results demonstrated the advantages of having thicker and more pennated muscle fascicles in enhancing jumping abilities.

Keywords Muscle · Architecture · Muscle thickness · Pennation angle Fascicle length · Jump

---

A. M. Nadzalan (✉) · N. I. Mohamad · J. L. F. Lee
Faculty of Sports Science and Coaching, Universiti Pendidikan Sultan Idris, Tanjung Malim, Perak, Malaysia
e-mail: ali.nadzalan@fsskj.upsi.edu.my

N. I. Mohamad
e-mail: nur.ikhwan@fsskj.upsi.edu.my

J. L. F. Lee
e-mail: jeffrey@fsskj.upsi.edu.my

C. Chinnasee
Faculty of Health and Sports Science, Thaksin University Papayom, Mueang Songkhla, Phatthalung, Thailand
e-mail: chamnan21@hotmail.com

1 Introduction

Muscle architecture is the physical arrangement of muscle fibers. It is important to understand the relationships of muscle architectures and movement efficiency as several previous studies have shown how different muscle architectures could be advantages to different movement or sporting performances (Abe et al. 2000; Earp et al. 2010, 2011; Kumagai et al. 2000).

Among muscle architecture parameters that could be examined includes fascicle length, pennation angle, and muscle thickness. Fascicle length is the distance of fascicle from aponeurosis to another aponeurosis (Fig. 1). It is thought that greater fascicle length represents either longer sarcomeres or more sarcomeres in line (Earp et al. 2010). Sarcomere is the contractile element in the muscles. Greater length of contractile element will enable faster contraction velocity and more force that can be applied at an increasing velocity (Sacks and Roy 1982).

Pennation angle is the direction of fascicle form the aponeurosis to another aponeurosis (Fig. 1). Greater pennation angle will cause a cross-sectional area of muscle to have more number of fibers, and therefore increase the muscle ability to produce more force (Manal et al. 2006).

Muscle thickness is measured from the muscle aponeurosis to its another aponeurosis (Fig. 1). Muscle thickness is related to muscle size, as muscle thickness increases, muscle size also increases. Increases of muscle thickness have been shown with resistance training and are highly correlated with the ability to produce greater force (Seynnes et al. 2007).

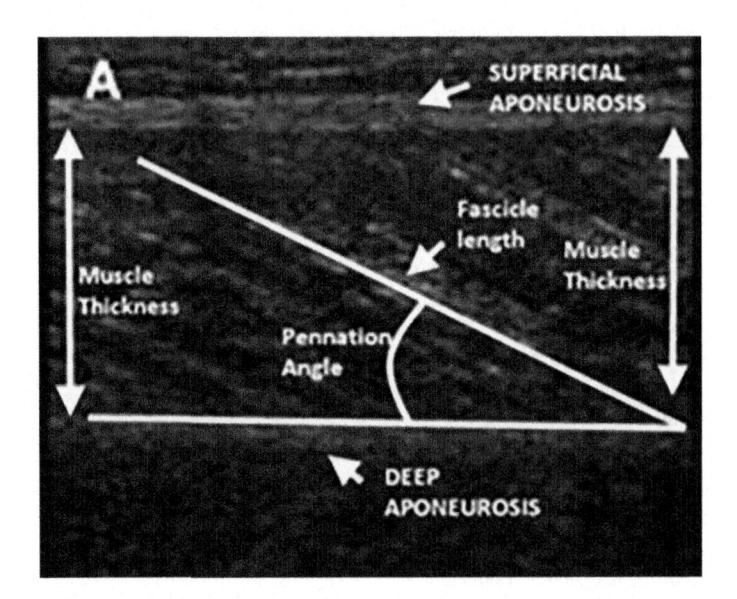

Fig. 1 Muscle architecture of vastus lateralis

Kumagai et al. (2000) in their study found faster trained sprinters were shown to have lesser pennation angles at lateral gastrocnemius (LG), medial gastrocnemius (MG), and vastus lateralis (VL) but greater muscle thickness at LG compared to slower runners. Besides that, Abe et al. (2000) in their study have found that trained sprinters were shown to have longer and thicker fascicles along with lesser pennation angles in the LG, MG, and VL muscles compared to trained distance runners. In contrast, shorter fascicles with greater muscle thickness and pennation angle were more found to be more beneficial for jumping (Earp et al. 2010). These contrast findings suggested that the muscle architectures were different based on the training adopted by the individuals.

Understanding the muscle architecture advantages to specific movement performances is important as training could be developed or planned in order to attain the specific muscle architecture adaptation. The effects of training on muscle architectures had been shown in several previous studies (Blazevich 2006; Blazevich et al. 2007; Duclay et al. 2009; Earp 2013; Earp et al. 2010, 2014). As such, resistance training has been shown to increased muscle thickness and pennation angle of muscles that been trained (Blazevich 2006; Blazevich et al. 2007; Bloomquist et al. 2013; Duclay et al. 2009; Earp 2013; Earp et al. 2014; Seynnes et al. 2007).

It is our interest to study the relationship of lower body muscle architectures and jumping performances as it can be seen that jumping is one of the most performed movements in sport. In sports, jumping is performed vertically and horizontally. Previous study has found that jumping performances are improved with greater muscle thickness, greater pennation angle, and shorter fascicles (Earp et al. 2010). However, Earp et al. (2010) only focused on the vertically directed jumping and the study was conducted on trained men. Lack of information existed on the relationship of muscle architecture, and the jumping abilities measured both vertically and horizontally. Therefore, it was the aim of this study to determine the relationship between the lower body muscle architectures and vertical jump and standing long jump performances.

2 Methodology

2.1 Participants

Thirty recreationally active, resistance untrained men (mean age = 22.21 ± 1.59 years old) were recruited as study participants. Participants had no medical problems and not consuming any performance enhancing supplementation. Participants were screened prior to testing using PAR Q. Each participant had read and signed an informed consent for testing and training approved by the Universiti Pendidikan Sultan Idris and the Thaksin University Ethics Committee (CODE E 060/2559).

Each participant self-reported that they were familiar with jumping movement but had never involved in any systematic physical training.

2.2 Vertical Jump

Vertical jump height was assessed as an indicator of vertical jump abilities using a vertical jump equipment (Vertec, USA). Participants' standing height with one arm fully extended upward was taken to set the lowest vane. Participants were required to jump up and touch the highest possible vane. Participants were allowed to swing their arms and bend their knees as to simulate the common movement performed in sports and daily life settings. The difference between standing height and jumping height was taken as the vertical jump score. Three trials were given, and the greatest height was taken as the score.

2.3 Standing Long Jump

Standing long jump distance was assessed as an indicator of horizontal jumping abilities. Participants started with standing behind a line marked on the standing long jump mat (Trident, Malaysia) with feet slightly apart. Participants were allowed to swing their arms and bend their knees to provide forward drive. Participants were asked to jump as far as possible, landing on both feet without falling backward. Three trials were given, and the greatest jump distance was taken as the standing long jump score.

2.4 Muscle Architectures Analyses

B-mode ultrasonography (F37, Aloka, Ltd, Tokyo, Japan) was used to measure muscle thickness, fascicle length, and pennation angle of VL and BF on the participant's self-reported dominant side. The ultrasound probe was positioned longitudinally and was maintained with equal contact pressure during all muscle architectures measurements (Klimstra et al. 2007). The muscle architectures of VL were done while the participant lying supine with leg straight (Pang and Ying 2006; Rutherford and Jones 1992) while the BF muscle architectures were determined while the participant lying prone with leg straight (Fredberg et al. 2008; Ward et al. 2009). All measurements were taken while the leg was in resting position. Fascicle length was calculated as the muscle thickness divided by the sine of pennation angle (Alegre et al. 2006) (Eq. 1). All ultrasound measurement was performed by the same technician. Figure 1 shows the example of image taken during ultrasonography. Three consecutive images were analyzed and averaged.

$$\text{Fascicle length} = \text{Muscle thickness}/\sin(\text{pennation angle}) \qquad (1)$$

2.5 Statistical Analyses

Descriptive statistics were used to measure the descriptive data and mean score. Pearson Correlation was used to determine the relationships between muscle architectures and the jumping abilities. Statistical significance was accepted at an α-level of $p \leq 0.05$. All statistical analyses were conducted using SPSS version 23 (IBM, New York, USA).

3 Results

Table 1 shows the physical characteristics of participants involved in this study.

Table 2 shows the muscle architecture (muscle thickness, pennation angle, and fascicle length) of the VL, VM, RF, and BF.

Table 3 shows the mean and standard deviation of vertical jump height and standing long jump distance score of the participants.

Table 4 shows the correlation analysis of muscle architectures and the vertical and standing long jump score.

Table 1 Physical characteristics of participants

Age (years)	Body mass (kg)	Height (cm)
22.21 ± 1.59	68.53 ± 3.26	170.46 ± 4.25

Table 2 Muscle architecture of participants

Muscles	Muscle thickness (cm)	Pennation angle (°)	Fascicle length (cm)
VL	2.38 ± 0.08	18.46 ± 2.40	7.60 ± 0.75
BF	2.52 ± 0.14	15.25 ± 1.16	9.61 ± 0.23

Table 3 Vertical jump and standing long jump scores

	Vertical jump (cm)	Standing long jump
Score	42.30 ± 1.56	2.43 ± 0.05

Table 4 Correlation analysis of muscle architectures vertical jump and standing long jump

Muscle	Architectures	Vertical jump	Standing long jump
		r	r
VL	MT	0.332	0.267
	PA	0.587**	0.365*
	FL	−0.560**	−0.312
BF	MT	0.661***	0.606***
	PA	0.599***	0.523**
	FL	−0.092	−0.001

*Data is significant at $p < 0.05$
**Data is significant at $p < 0.01$
***Data is significant at $p < 0.001$

4 Discussion

This study was conducted to examine the relationship between lower body muscle architectures and jumping abilities measured by vertical jump height and standing long jump distance. Results showed that the muscle thickness and pennation angle of BF and the pennation angle of VL were shown to be positively correlated while the fascicle length of VL was shown to be negatively correlated with vertical jump performance. Besides that, muscle thickness of BF and the pennation angle of VL were shown to be positively correlated with standing long jump performance.

In line with Earp et al. (2010), this study found thicker muscles were positively correlated with jumping abilities. This again demonstrated the ability of thicker muscles in producing greater force during task application (Seynnes et al. 2007).

Besides that, the pennation angle of almost all muscles investigated was shown to be positively correlated with both vertical jump and standing long jump performances. This condition could be due to the greater pennation that will allow more fibers that can be fit within a given cross-sectional area. This will then increase the physiological cross-sectional area of the muscle allowing for greater force to be developed (Blazevich et al. 2007).

Findings in the current study also showed participants with shorter fascicle lengths had more ability to performed better during jumping movement. Earp et al. (2010) suggested that this condition might be explained by the behavior of the longer fascicles that have more potential places of fascicle disruption that can contribute to higher instability of muscle fascicles.

The current findings of jumping abilities were found to be different from previous studies of sprint performances. While previous study demonstrated that faster 100-m sprinters had longer fascicle length with lesser pennation angle in the gastrocnemius (Kumagai et al. 2000), current findings showed the contrast where shorter fascicle lengths with greater pennation angle will allow for higher and further jump distance. Current findings were in line with findings of Earp et al. (2010) where greater thickness, more pennated, and shorter fascicles were correlated with jumping abilities. However, despite the differences of pennation angle and fascicle length, it should be noted that this study along with several previous studies found the advantages of having thicker muscles for running velocity and jump performances (Abe et al. 2000; Earp et al. 2010, 2011; Kumagai et al. 2000).

The knowledge on the influence of muscle architectures to specific movement is important as coaches and athletes could plan a more specific training program in order to attain specific adaptations to the muscle architectures.

5 Conclusion

Overall, the findings of this study found that relationship existed between lower body muscle architecture and jumping abilities. It is suggested for the individuals to have thicker, more pennated, and shorter fascicle of lower body muscle in enhancing jumping abilities. Training could be planned to make specific changes in muscle architecture to allow for better jumping movement abilities.

Acknowledgements The authors are gratefully acknowledged the helpful comments and suggestions of the reviewers, which have improved the presentation.

References

Abe, T., Kumagai, K., & Brechue, W. F. (2000). Fascicle length of leg muscles is greater in sprinters than distance runners. *Medicine and Science in Sports and Exercise, 32*(6), 1125–1129.

Alegre, L. M., Jiménez, F., Gonzalo-Orden, J. M., Martín-Acero, R., & Aguado, X. (2006). Effects of dynamic resistance training on fascicle length and isometric strength. *Journal of Sports Sciences, 24*(5), 501–508.

Blazevich, A. J. (2006). Effects of physical training and detraining, immobilisation, growth and aging on human fascicle geometry. *Sports Medicine, 36*(12), 1003–1017.

Blazevich, A. J., Cannavan, D., Coleman, D. R., & Horne, S. (2007). Influence of concentric and eccentric resistance training on architectural adaptation in human quadriceps muscles. *Journal of Applied Physiology, 103*(5), 1565–1575.

Bloomquist, K., Langberg, H., Karlsen, S., Madsgaard, S., Boesen, M., & Raastad, T. (2013). Effect of range of motion in heavy load squatting on muscle and tendon adaptations. *European Journal of Applied Physiology, 113*(8), 2133–2142.

Duclay, J., Martin, A., Duclay, A., Cometti, G., & Pousson, M. (2009). Behavior of fascicles and the myotendinous junction of human medial gastrocnemius following eccentric strength training. *Muscle and Nerve, 39*(6), 819–827.

Earp, J. E. (2013). The influence of external loading and speed of movement on muscle-tendon unit behaviour and its implications for training. *Dissertation*. Australia: Edith Cowan University.

Earp, J. E., Kraemer, W. J., Cormie, P., Volek, J. S., Maresh, C. M., Joseph, M., et al. (2011). Influence of muscle-tendon unit structure on rate of force development during the squat, countermovement, and drop jumps. *The Journal of Strength & Conditioning Research, 25*(2), 340–347.

Earp, J. E., Kraemer, W. J., Newton, R. U., Comstock, B. A., Fragala, M. S., Dunn-Lewis, C., et al. (2010). Lower-body muscle structure and its role in jump performance during squat, countermovement, and depth drop jumps. *The Journal of Strength & Conditioning Research, 24*(3), 722–729.

Earp, J. E., Newton, R. U., Cormie, P., & Blazevich, A. J. (2014). The influence of loading intensity on muscle–tendon unit behavior during maximal knee extensor stretch shortening cycle exercise. *European Journal of Applied Physiology, 114*(1), 59–69.

Fredberg, U., Bolvig, L., Andersen, N. T., & Stengaard-Pedersen, K. (2008). Ultrasonography in evaluation of Achilles and patella tendon thickness. *Ultraschall in der Medizin, 29*(1), 60–65.

Klimstra, M., Dowling, J., Durkin, J. L., & MacDonald, M. (2007). The effect of ultrasound probe orientation on muscle architecture measurement. *Journal of Electromyography and Kinesiology, 17*(4), 504–514.

Kumagai, K., Abe, T., Brechue, W. F., Ryushi, T., Takano, S., & Mizuno, M. (2000). Sprint performance is related to muscle fascicle length in male 100-m sprinters. *Journal of Applied Physiology, 88*(3), 811–816.

Manal, K., Roberts, D. P., & Buchanan, T. S. (2006). Optimal pennation angle of the primary ankle plantar and dorsiflexors: Variations with sex, contraction intensity, and limb. *Journal of Applied Biomechanics, 22*(4), 255.

Pang, B. S., & Ying, M. (2006). Sonographic measurement of Achilles tendons in asymptomatic subjects variation with age, body height, and dominance of ankle. *Journal of Ultrasound in Medicine, 25*(10), 1291–1296.

Rutherford, O., & Jones, D. (1992). Measurement of fibre pennation using ultrasound in the human quadriceps in vivo. *European Journal of Applied Physiology and Occupational Physiology, 65*(5), 433–437.

Sacks, R. D., & Roy, R. R. (1982). Architecture of the hind limb muscles of cats: Functional significance. *Journal of Morphology, 173*(2), 185–195.

Seynnes, O. R., de Boer, M., & Narici, M. V. (2007). Early skeletal muscle hypertrophy and architectural changes in response to high-intensity resistance training. *Journal of Applied Physiology, 102*(1), 368–373.

Ward, S. R., Eng, C. M., Smallwood, L. H., & Lieber, R. L. (2009). Are current measurements of lower extremity muscle architecture accurate? *Clinical Orthopaedics and Related Research, 467*(4), 1074–1082.

Metabolic Cost of Continuous Body Weight Circuit Training with Aerobic-Based Exercise Interval for Muscle Strength and Endurance on Young Healthy Adults

Nur Ikhwan Mohamad, Raiza Sham Hamezah and Ali Md Nadzalan

Abstract Metabolic cost and cardiorespiratory responses produced by a body weight circuit training program were assessed for the purpose of this study. Fourteen recreationally active males aged 22.57 ± 1.70 years old with body weight of 64.53 ± 8.79 kg and height of 1.68 ± 0.04 m were recruited. A body weight circuit training program was prescribed to all participants. The body weight resistance circuit was performed in a continuous mode, with a combination of aerobic-based exercise (cycling) each body weight exercises. Portable desktop metabolic analyzer with a face mask (mouthpiece) was used to assess cardiorespiratory responses during the performance. Lower post-exercise blood pressure level was recorded, with significant differences observed between the pre- and post-exercise data ($p = 0.018$). Heart rate response significantly increased after performing the exercise protocol ($p = 0.001$). Peak oxygen consumption recorded during the performance was 24.36 ± 2.53 ml/kg/min, peak amount of oxygen extracted by the lung from the air was 16.85 ± 0.45%, maximum heart rate recorded was 156.29 ± 12.65 bpm, and respiratory frequency was 34.39 ± 5.18 l/min. Metabolic cost based on energy expenditure recorded was 468.71 ± 66.84 kcal (estimated per hour), and the Metabolic Equivalent of Task (MET) recorded was 6.86 ± 0.77. Albeit lesser total duration, continuous mode used to ensure the metabolic cost recorded was in accordance with other previous findings related to circuit-based training. Body weight circuit training with active recovery in between set is an excellent way of developing and maintaining overall physical fitness level.

N. I. Mohamad (✉) · A. M. Nadzalan
Faculty of Sports Science and Coaching, Universiti Pendidikan
Sultan Idris, Tanjung Malim, Malaysia
e-mail: nur.ikhwan@fsskj.upsi.edu.my

A. M. Nadzalan
e-mail: ali.nadzalan@fsskj.upsi.edu.my

R. S. Hamezah
Everlyfit Studio, Tanjung Malim, Perak, Malaysia
e-mail: raiza.sham@yahoo.com

© Springer Nature Singapore Pte Ltd. 2018
R. Saian and M. A. Abbas (eds.), *Proceedings of the Second International Conference on the Future of ASEAN (ICoFA) 2017 – Volume 2*,
https://doi.org/10.1007/978-981-10-8471-3_72

Keywords Metabolic cost · Circuit weight training · Oxygen consumption Metabolic equivalent of task

1 Introduction

Body weight circuit training is one of the earliest modes of strength training. The main objectives of body weight circuit training include the development of muscular strength, muscular endurance, and cardiorespiratory power, apart from obviously proven body composition-related objectives (Paoli et al. 2010). As the implementation of body weight circuit training requires minimal equipment, it has become one of the most popular modes of training, especially in schools or community healthy lifestyle programs. Many commercial boot camps also utilize high-intensity body weight circuit training as their main training approach.

Strength training alone for as short as six weeks, with or without external load, has been shown to increase muscle fiber size (Friedmann-Bette et al. 2010; Ronnestad et al. 2010), strength (Roig et al. 2009, Karavirta et al. 2011), and endurance (Aagaard et al. 2011; Broekmans et al. 2011). With the use of own body weight as the resistance or loading, body weight circuit training appears to be performed in a constant load mode. However, this supposedly constant load still has certain degrees of variety during performance. The load or resistance provided by the body changes with the type and technique of exercises performed. For example, a modified push-up with bent knees will have a lower exercise resistance (much easier) compared to a normal push-up with straight legs in terms of loading.

Selection and order of exercises within any exercise programs have shown to produce different stimulus and adaptations to the body (Simao et al. 2012b). Manipulating other exercise variables such as time under tension (tempo) will further influence the physiological and mechanical effects of the exercise program (Gentil et al. 2006; Tran et al. 2006; Simao et al. 2012a). This is one of the reasons why different types of circuit training produce different metabolic cost when assessed. This study examined the metabolic cost produced by a body weight circuit training in combination with an aerobic-based exercise (cycling) in between exercise.

2 Methodology

2.1 Experimental Approach to the Problem

This acute profiling study was one of a series of studies performed in order to determine the effects of concurrent strength training on metabolic responses and adaptations. A circuit training exercise program consisted of one set of body weight squat, push-up, lunges, assisted chin-up and back-arch exercise was developed for

Table 1 Prescribed circuit exercise program

Exercise	Repetitions	Set	Duration (min)
Cycle	–	–	5
Squat	maximum	1	1
Cycle	maximum	1	2
Push-up	maximum	1	1
Cycle	maximum	1	2
Lunges	maximum	1	1
Cycle	maximum	1	2
Assisted chin-up	maximum	1	1
Cycle	maximum	1	2
Back-arch	maximum	1	1
Cycle	maximum	1	2

the purpose of this study. The exercise program prescribed to all participants is shown in Table 1. Repetition maximum means participants performed as many repetitions as possible of a single exercise with proper technique within the 60 s duration. All exercise techniques were based on National Strength & Conditioning (NSCA) recommendations (Baechle and Earle 2008).

2.2 Participants

Fourteen recreationally active males aged 22.57 ± 1.70 years of age with mean body weight of 64.53 ± 8.79 kg and height of 1.68 ± 0.04 m were recruited for this study. Participants' mean body mass index (BMI) was 22.81 ± 2.47 kg/m^2. Each participant had adequate strength training experience of more than three months at the time of data collection and actively participated in physical activities (recreational sports or exercise) 2–3 times a week. Detailed information on the research and potential risks or discomfort that may occur during the experiment was relayed to each participant prior participation. Informed consent letters from the participants were obtained before commencing the study. This study has been approved to be conducted by the UPSI Research and Management Centre (research approval code 2013-0111-106-01).

2.3 Materials

An upright spin bike (Maxx, Fitness Concept, Malaysia) was used for aerobic-based exercise. An Olympic bar (OB86, Body Solid Inc., Illinois, USA) placed on top of the safety bar in a specific position inside the power cage (GPR 378, Body Solid Inc., Illinois, USA) was used for modified chin-up exercise. All other body weight exercises were performed on an exercise mat. During exercise, a metabolic analyzer

(Fitmate pro, Cosmed, Italy) attached to a face mask was worn by the participant to measure the cardiorespiratory responses along with metabolic cost of the exercise protocol. Rate of perceived exertion (RPE) was obtained from each participant using an A4 size print of Borg Scale of RPE. Participant's body weight and height were measured using a digital body weight scale (HN-283, Omron, Kyoto, Japan) and custom-made body height scale using commercially available long fiber glass measuring tape (Aero, China). A digital wrist blood pressure monitor (HEM-6200, Omron, Kyoto, Japan) was used to assess pre- and post-exercise blood pressure level. Environmental conditions were recorded using a Wet Bulb Globe Temperature (WBGT-103 Heat Stroke Checker, Japan).

2.4 Procedures

The participants were required to attend two sessions. Each participant attended the sessions separately (one participant each time). The first session was to familiarize the participants on the study and the exercises involved. The exercise techniques were then demonstrated and the participants' exercise techniques were assessed, corrected, and standardized during the session. All familiarization and testing session were completed in a single day (morning session between 8.00 a.m. and 11.00 a.m.). All sessions were conducted in the Physical Conditioning Laboratory of the university. Room temperature recorded during the time of data collection was 22.3 oC, with ambient temperature (Ta) of 25.5 oC, relative humidity (RH) was 48.5%, and globe temperature was (Tg) was 28.4 oC.

The second session was conducted 72 h later. Although no high-intensity physical exertion was performed during the first session, the three days' rest in between sessions was provided in order to allow the participants proper rest. The participants were advised to refrain from other excessive physical activities, which might affect the outcome of the study.

The participants performed a standardized warm-up activity on a cycle ergometer for 5–7 min until the heart rate response reached the warm-up heart rate zone. A standardized dynamic stretching was then performed prior the exercise protocol.

Participants were then being assisted by the researcher to wear the face mask which is connected to the metabolic analyzer. Once the metabolic analyzer was calibrated and the breathings are being measured correctly, participants were instructed to mount on a spin bike and start cycling at a moderate cadence. At the start of the exercise, the metabolic analyzer was activated to start collecting data. All exercise repetitions were counted and the duration of each exercise was monitored using a stopwatch by a research assistant, (i.e., a qualified fitness trainer) who monitored the performance of the participants. Once started, the participants performed all of the exercises required with 2 min of moderate cadence cycling in between each exercise until the last exercise. The exercise protocol ended with a final 2 min of cycling (Fig. 1).

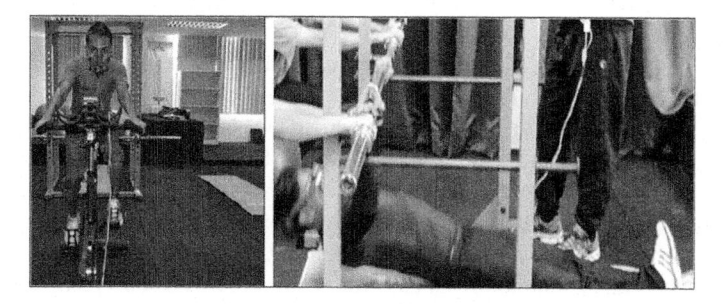

Fig. 1 Exercise area and equipment setup

2.5 Data Analysis

Raw data for all variables of interest (energy expenditure, oxygen consumption, minutes of ventilation, respiratory frequency, fraction of oxygen in expired air, exercise heart rate) during the exercise performance were collected by the metabolic analyzer, and calculation was made automatically for certain variables. The raw data were then printed out and transferred into an excel sheet for further statistical analysis. Similar procedure was done with the blood pressure raw data which were collected via a wrist digital blood pressure monitor.

2.6 Statistical Analyses

Means and standard deviations for each variable of interest were calculated and presented as the cardiorespiratory profile of the prescribed exercise program. Normality test was performed to statistically determine normal distribution of the data. Paired sample t-tests were used to compare overall cardiorespiratory data obtained with cardiorespiratory responses recorded within the anaerobic threshold time. All statistical analyses were analyzed using Statistical Package for the Social Science (SPSS) software version 17.0. Level of statistical significance was set at 0.05.

3 Results

As indicated in Table 2, lower post-exercise blood pressure level was recorded, with significant differences observed between the pre- and post-exercise data ($p = 0.018$). Heart rate response significantly increased after performing the exercise protocol ($p = 0.001$).

In terms of during the performance cardiorespiratory responses (Table 2), peak oxygen consumption recorded during the performance was 24.36 ± 2.53 ml/kg/min, peak amount of oxygen extracted by the lung from the air was 16.85 ± 0.45%, maximum heart rate recorded was 156.29 ± 12.65 bpm, and respiratory frequency was 34.39 ± 5.18 1/min. At the same time, it was found that metabolic cost based on energy expenditure recorded was 468.71 ± 66.84 kcal and the Metabolic Equivalent of Task (MET) recorded was 6.86 ± 0.77. Comparisons were also made as in Table 3 between cardiorespiratory responses recorded during the whole performance duration versus cardiorespiratory responses recorded during the measured anaerobic threshold duration. Significant differences were found in all of the variables, with higher value were observed in peak overall performance compared to peak value during the measured anaerobic threshold duration.

Table 2 Comparisons of pre- and post-exercise protocol blood pressure and heart rate responses

Variables	Pre	Post	% Difference	Significance (p-value)
Blood pressure (systolic) (mmHg)	126.14 ± 12.55	118.93 ± 12.86	5.72	0.018*
Blood pressure (diastolic) (mmHg)	81.86 ± 13.70	80.00 ± 16.59	2.27	0.709
Heart rate (bpm)	80.57 ± 13.89	117.79 ± 19.65	31.60	0.001*

Table 3 Exercise responses (outside anaerobic threshold time) and measured anaerobic threshold to the prescribed exercise protocol

Variables	Overall peak	Measured anaerobic threshold	% Difference	Significance (p-value)
Time in peak (minutes)	9.01 ± 5.35	10.95 ± 5.70	17.72	0.372
VO_2 (ml/Kg/min)	24.36 ± 2.53	19.61 ± 3.95	19.50	0.001*
FeO_2 (%)	16.85 ± 0.45	16.53 ± 0.38	1.90	0.012*
Heart rate max (bpm)	156.29 ± 12.65	143.64 ± 15.55	8.09	0.005*
Energy expenditure (kcal)	468.71 ± 66.84	381.21 ± 104.02	18.67	0.001*
Respiratory frequency (Rf) (1/min)	34.39 ± 5.18	30.40 ± 6.99	11.60	0.026*
Metabolic Equivalent of Task (MET)	6.86 ± 0.77	–	–	–
Session RPE (6–20 scale)	12.79 ± 1.63	–	–	–

4 Discussion

Post-exercise blood pressure (Table 2) was found to be lower than the readings recorded prior exercise, with the systolic level significantly low (post-exercise hypotension). In normal conditions, this was an acceptable condition as studies have shown that immediately post-exercise blood pressure level will be much lower compared with pre-exercise level (Pescatello et al. 1991; Syme et al. 2006; De Salles et al. 2010). At first the assumption was that apart from normal regulatory effect for recovery after exercise, the lower blood pressure might also be due to the effect of excess post-exercise oxygen consumption (EPOC). However, previous study has suggested that this is not the cause due to different time course of action (Williams et al. 2005). Systolic blood pressure was an indicator of the blood flow in the arteries. The harder the exercise, the more blood will be delivered in the arteries, thus higher systolic blood pressure level. The diastolic blood pressure was an indicator of pressure in the arteries, when the heart was relaxed (after each pumping out). In order to allow more oxygen delivered to the whole body, the blood vessels will be more relaxed (for wider diameter, more blood flow). This resulted in the lower systolic level. This result suggests one of the benefits of this type of exercise is lower blood pressure level, which can be used as an anti-hypertensive exercise. Similar study on circuit-based exercise also yields similar result (Paoli et al. 2013).

Energy expenditure for three circuits with 10 exercises per circuit recorded in one of the earliest studies averaged 539.7 kcal/h (about 9.0 kcal/min) (Wilmore et al. 1978). Our study recorded peak energy expenditure of 468.71 ± 66.84 kcal for ∼9 min, which means approximately 52 kcal/min. Differences in energy expenditure observed between the two studies can be attributed to the work-to-rest ratio period differences used. While our study used continuous mode (60 s work period, 0 s rest period), Wilmore et al. (1978) used interval mode for their study (30 s work: 15 s rest), allowing more rest and reducing the overall intensity (Wilmore et al. 1978).

Table 3 indicated significant different between the peak energy expenditure measured during the full period of exercise versus energy expenditure measured only during the duration of measured anaerobic threshold. Similar responses were also observed in terms of oxygen consumption (VO_2), heart rate maximum (HRM), and amount of oxygen extracted by the lung from the air (FeO_2). Simple explanation that can be given here is that the difference was due to the different energy source utilized. Cumulative effect of the whole session involved altogether the aerobic energy system, while anaerobic metabolism was the main contributor of energy during each of the body weight resistance exercise. Anaerobic energy system was the main source of energy during resistance-based exercise (Gorostiaga et al. 2010).

While the result of this study can be used to predict positive longitudinal effect from circuit training, one of the latest longitudinal studies found that the effect of circuit weight training is dependent on individuals (Kang et al. 2012). The study incorporated 10 exercises with 30 s duration for each performance. Similar to this study, Kang and colleagues incorporated one aerobic-based exercise in between of each strength exercises. Their study produced mixed findings, with some participants experiencing significant improvement in terms of body composition and other fitness benefits, others did not. If current protocol be implemented in a similar condition with their study (Kang et al. 2012) (with three sets per session, three times per week for 12 weeks), the volume of training will differ significantly, as our protocol incorporated longer exercise duration per set (60 s). What we would like to highlight here is that longer exercise duration means longer time the muscles will be under tension, which subsequently will have different metabolic, hormonal, and neural effects. Time muscle under tension has a direct consequence on metabolic cost (Toigo and Boutellier 2006), where longer submaximal lifting times had a greater energy expenditure (Scott 2012). Rates of protein synthesis had also been shown to be increased linearly with increase in time muscle under tension (Burd et al. 2012).

In terms of types of circuit training, it was shown that body weight circuit training as used in this study, while producing many positive physiological effect, it may be still not as good as the combination of cardio and external load circuit training (Monteiro et al. 2008). Previous study has shown that due to low load used, adaptation (enlargement) of muscle fibers was minimal (Harber et al. 2004).

5 Practical Applications

Body weight circuit strength training is still relevant as an excellent way of developing and maintaining overall physical fitness level. Inclusion of active recovery such as cycling in between sets is suggested in order to enhance metabolic cost of the training. Body weight circuit training as used in this study is also suggested to be used as part of a weight-loss program. For high-performance athletes, body weight circuit training is an excellent start toward a more intense strength training program with the use of external load.

Acknowledgements This study was funded by Sultan Idris Education University under the University Research Grant 2013. Authors would like to thank Syed Shahbudin Syed Omar, Andie Ujang, and Nicholas Garaman Abas for data collection assistance.

References

Aagaard, P., Andersen, J. L., Bennekou, M., Larsson, B., Olesen, J. L., Crameri, R., et al. (2011). Effects of resistance training on endurance capacity and muscle fiber composition in young top-level cyclists. *Scandinavian Journal of Medicine & Science in Sports, 21,* e298–e307.

Baechle, T. R., & Earle, R. W. (2008). *Essentials of strength training and conditioning.* New York: Human Kinetics.

Broekmans, T., Roelants, M., Feys, P., Alders, G., Gijbels, D., Hanssen, I., et al. (2011). Effects of long-term resistance training and simultaneous electro-stimulation on muscle strength and functional mobility in multiple sclerosis. *Multiple Sclerosis Journal, 17,* 468–477.

Burd, N. A., Andrews, R. J., West, D. W. D., Little, J. P, Cochran, A. J. R., Hector, A. J., et al. (2012) Muscle time under tension during resistance exercise stimulates differential muscle protein sub-fractional synthetic responses in men. *The Journal of Physiology, 590,* 351–362.

De Salles, B. F., Maior, A. S., Polito, M., Novaes, J., Alexander, J., Rhea, M., et al. (2010) Influence of rest interval lengths on hypotensive response after strength training sessions performed by older men. *The Journal of Strength & Conditioning Research, 24,* 3049–3054.

Friedmann-Bette, B., Bauer, T., Kinscherf, R., Vorwald, S., Klute, K., Bischoff, D., et al. (2010). Effects of strength training with eccentric overload on muscle adaptation in male athletes. *European Journal of Applied Physiology, 108,* 821–836.

Gentil, P., Oliveira, E., & Bottaro, M. (2006). Time under tension and blood lactate response during four different resistance training methods. *Journal of Physiological Anthropology, 25,* 339–344.

Gorostiaga, E. M., Navarro-Amezqueta, I., Cusso, R., Hellsten, Y., Calbet, J. A. L., Guerrero, M., et al. (2010). Anaerobic energy expenditure and mechanical efficiency during exhaustive leg press exercise. *PLoS ONE, 5,* e13486.

Harber, M. P., Fry, A. C., Rubin, M. R., Smith, J. C., & Weiss, L. W. (2004). Skeletal muscle and hormonal adaptations to circuit weight training in untrained men. *Scandinavian Journal of Medicine & Science in Sports, 14,* 176–185.

Kang, H. J., Lee, Y., Park, D. S., & Kang, D. H. (2012). Effects of 12-week circuit weight training and aerobic exercise on body composition, physical fitness, and pulse wave velocity in obese collegiate women. *Soft Computing, 16,* 403–410.

Karavirta, L., Häkkinen, A., Sillanpää, E., García-López, D., Kauhanen, A., Haapasaari, A., et al. (2011). Effects of combined endurance and strength training on muscle strength, power and hypertrophy in 40–67-year-old men. *Scandinavian Journal of Medicine & Science in Sports, 21,* 402–411.

Monteiro, A. G., Alveno, D. A., Prado, M., Monteiro, G. A., Ugrinowitsch, C., Aoki, M. S., et al. (2008). Acute physiological responses to different circuit training protocols. *Journal of Sports Medicine and Physical Fitness, 48,* 438–442.

Paoli, A., Pacelli, F., Bargossi, A. M., Marcolin, G., Guzzinati, S., Neri, M., et al. (2010). Effects of three distinct protocols of fitness training on body composition, strength and blood lactate. *Journal of Sports Medicine and Physical Fitness, 50,* 43–51.

Paoli, A., Pacelli, Q. F., Moro, T., Marcolin, G., Neri, M., Battaglia, G., et al. (2013). Effects of high-intensity circuit training, low-intensity circuit training and endurance training on blood pressure and lipoproteins in middle-aged overweight men. *Lipids in Health and Disease, 12,* 131.

Pescatello, L. S., Fargo, A. E., Leach, C. N., & Scherzer, H. H. (1991). Short-term effect of dynamic exercise on arterial blood pressure. *Circulation, 83,* 1557–1561.

Roig, M., Brien, K. O., Kirk, G., Murray, R., Mckinnon, P., Shadgan, B., et al. (2009). The effects of eccentric versus concentric resistance training on muscle strength and mass in healthy adults: A systematic review with meta-analysis. *British Journal of Sports Medicine, 43,* 556–568.

Ronnestad, B. R., Hansen, E. A., & Raastad, T. (2010). Effect of heavy strength training on thigh muscle cross-sectional area, performance determinants, and performance in well-trained cyclists. *European Journal of Applied Physiology, 108,* 965–975.

Scott, C. B. (2012). The effect of time-under-tension and weight lifting cadence on aerobic, anaerobic, and recovery energy expenditures: 3 submaximal sets. *Applied Physiology, Nutrition and Metabolism, 37,* 252–256.

Simao, R., Figueiredo, T., Leite, R. D., Jansen, A., & Willardson, J. M. (2012a). Influence of exercise order on repetition performance during low-intensity resistance exercise. *Research in Sports Medicine, 20,* 263–273.

Simao, R., Salles, B., Figueiredo, T., Dias, I., & Willardson, J. (2012b). Exercise order in resistance training. *Sports Medicine, 42,* 251–265.

Syme, A. N., Blanchard, B. E., Guidry, M. A., Taylor, A. W., Vanheest, J. L., Hasson, S., et al. (2006). Peak systolic blood pressure on a graded maximal exercise test and the blood pressure response to an acute bout of submaximal exercise. *American Journal of Cardiology, 98,* 938–943.

Toigo, M., & Boutellier, U. (2006). New fundamental resistance exercise determinants of molecular and cellular muscle adaptations. *European Journal of Applied Physiology, 97,* 643–663.

Tran, Q., Docherty, D., & Behm, D. (2006). The effects of varying time under tension and volume load on acute neuromuscular responses. *European Journal of Applied Physiology, 98,* 402–410.

Williams, J. T., Pricher, M. P., & Halliwill, J. R. (2005). Is postexercise hypotension related to excess postexercise oxygen consumption through changes in leg blood flow? *Journal of Applied Physiology, 98,* 1463–1468.

Wilmore, J. H., Parr, R. B., Ward, P., Vodak, P. A., Barstow, T. J., Pipes, T. V., et al. (1978). Energy cost of circuit weight training. *Medicine and Science in Sports, 10,* 75–78.

Effects of Body Weight Interval Training on Fitness Components of Primary Schoolgirls

Nurul Afiqah Bakar, Erie Zuraidee Zulkifli and Mohd Nidzam Jawis

Abstract Physical inactivity has been associated with increase of risks factor and can contribute to the development of many diseases and became a major consequence prevalence of overweight and obesity in children in which higher among girls. Hence, this study aims to investigate the effects of body weight interval training (BWIT) on fitness components of primary schoolgirls. A total of 43 schoolgirls involved in this study and divided into two groups; BWIT group followed the intervention programme during physical education classes. The control group; physical education (PE) followed the school's syllabus during physical education classes. The total duration of the intervention programme was 12 sessions, 35 min session^{-1}, with 3 times per week^{-1}, lasting for 4 weeks. Fitness components measured were anthropometrical and body composition, cardiorespiratory fitness, muscular fitness and flexibility. ANOVA with repeated measures was used to determine the differences in the parameters over time between groups. As a result, anthropometrical and body composition, cardiorespiratory fitness and muscular fitness were not significantly different between BWIT and PE groups ($p > 0.05$). However, there was significant difference for flexibility ($p = 0.011$). When compared pre- and post-intervention, BWIT showed significant improvement in cardiorespiratory fitness, muscular fitness and flexibility ($p < 0.05$) while PE group did not. The present findings were concluded that BWIT did not portray superior effects than another type of physical activity. Nevertheless, BWIT brought about positive effects to primary schoolgirls. Improvement can be made to BWIT in near future to

N. A. Bakar (✉)
Fakulti Sains Sukan Dan Rekreasi, Universiti Teknologi MARA
Cawangan Perlis, Kampus Arau, 02600 Arau, Perlis, Malaysia
e-mail: nurulafiqahbakar@gmail.com

E. Z. Zulkifli · M. N. Jawis
School of Health Sciences, Health Campus, Universiti Sains Malaysia,
16150 Kota Bharu, Kelantan, Malaysia
e-mail: erie@usm.my

M. N. Jawis
e-mail: nidzam@usm.my

be able to observe more distinctive improvement in physical fitness within groups and between groups.

Keywords Body weight interval training · Body weight exercise · Interval training Children · Physical activity

1 Introduction

Physical activity (PA) is an integral component of a healthy lifestyle (Strong et al. 2005). In recent years, the health status of young generation has become a growing topic of interest. Indeed, concern has been expressed that a large proportion of children are insufficiently active to gain health benefits (Boddy et al. 2012; Tremblay et al. 2010). Moreover, physical inactivity has been associated with an increase of risk factors, which often persist into adulthood and can contribute to the development of non-communicable diseases (Lee et al. 2012). As one of the consequences of physical inactivity, overweight and obesity remain a major health-related problem, especially in children. It has also becoming a global issue affecting various countries across population including children. Similarly, nationwide surveys in Malaysia have also revealed increasing trend in prevalence of overweight and obesity shown in the National Health and Morbidity Surveys (NHMS).

Childhood is a key stage in the development of health behaviours. Despite the well-established health benefits of physical activity, many children are not meeting the current PA recommendations (Ekelund et al. 2011), or the proportion is declining in most developing countries (Cavill et al. 2001; Tzotzas et al. 2011). Prolonged physical activities (i.e. >30 min) are contrary to a child's pattern of spontaneous exercise, which mainly comprises short intermittent efforts with high interval (Bailey et al. 1995; Van Sluijs et al. 2007).

As the nature of children to be more intermittent, interval training may be one of the best forms of exercise approach rather than the recommended continuous exercise. Interval training consists of exercise bouts intersperse by rest interval period between the sets. It also involves alternating short periods of moderate-to-intense exercise with less intense exercise in the same session. The advantage of interval training with moderate-to-high intensity apparently consumes less time than continuous prolonged endurance training, while producing comparable beneficial adaptations.

Despite the promising evidence supporting interval training in adults, there is limited research on youth, specifically in the key age group of children. Furthermore, most of the interval training in previous studies the modes of exercises had used the treadmill and/or cycle ergometer. There were limited studies that used body weight as a resistance for interval training. Due to the lack of information on body weight interval training (BWIT), the aim of this study is to provide some knowledge on body weight interval training perhaps in the approach of moderate-to-high-intensity activities, which is appropriate to gain health benefits. As for this reason, body weight interval training was used in this study as a medium of PA, which may create a pattern of regular behaviour to participate in physical activity.

2 Methods

The present study employed a pre- and post-intervention study design. Baseline data were collected prior the intervention programmes. Participants were divided into two groups. The intervention group followed the intervention programme during physical education classes, which was body weight interval training (BWIT). The control group followed the school's syllabus on physical activities during physical education classes. The total duration of the intervention programme was 12 sessions, 35 min per session, 3 times per week, lasting for 4 weeks.

The sample size was calculated by the G*Power Software (Version 3.1.9.2). Type 1α error for two-tailed test was set at 0.05 while the power of the study was set at 0.80. The effect size was set at 0.25, and the calculated sample size was 34 participants with 17 participants in each group. A 20% dropout rate was considered and thus the total of 43 participants was recruited in this study.

A total of 43 schoolgirls at year four and year five (10–11-year-olds) was involved in this study. The participants were randomly recruited from two primary schools situated in Gombak and Hulu Selangor districts of Selangor. They were divided in two groups: intervention group (body weight interval training) and control group (physical education). Throughout the study period, participants were not allowed to participate in any additional exercise programme.

The intervention programme lasted 35 min per session and consists of warm-up and cool-down and 7 activities in between (Fig. 1). Activities were done in the interval manner for 25 min with moderate-to-vigorous intensity (64–90% of maximal heart rate) determined and monitored by Polar Heart Rate monitors. There was an active rest interval of 40 s between the activities. The interval training session was made interesting by music to avoid boredom. All participants followed the exercise movements led by the instructor. Researcher made sure the participants are performing the movement correctly and adhered to the intensity and activities of the interval training.

Physical fitness tests were carried out on the participants. Participants with approved consent from parents had to fill up the participant information prior to the tests. For the fitness testing session, participants were required to wear sports attire. One day prior to each test day, the participants were advised to take some light food and drink one hour before the test and do not participate in vigorous exercise 24 h prior to test session. Warm-up exercises (10 min) were conducted prior to the test session to avoid any risk of injury during the test followed by short briefing of the test procedure which was given to the participants.

The pre-test data were obtained before the intervention programme began. Post-test were used to collect the final data from the participants after 12 sessions of

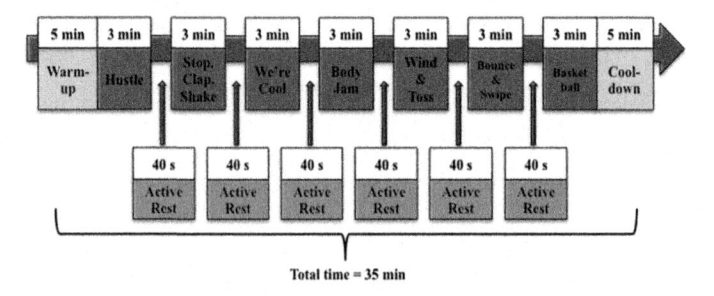

Fig. 1 Intervention programme

intervention programme. The parameters tested were (1) anthropometrical and body composition: height, weight, body mass index, waist circumference, body fat percentage, (2) cardiorespiratory fitness: one-mile run, (3) muscular fitness: standing broad jump, modified sit-up, modified push-up and (4) flexibility: sit-and-reach.

Analysis of Variance (ANOVA) with repeated measures was used to determine the differences in the anthropometrical and fitness parameters over time between groups. One-way analysis of variance was performed to determine the significance of differences between groups at pre- and post-test. The Statistical Package for Social Sciences (SPSS) version 22 was used for the statistical analysis. All values are presented as mean ± standard deviations (SDs). The accepted level of significance was set at $p < 0.05$. Results were reported as mean ± standard deviations.

3 Results

Anthropometrical and body composition results of present study are summarised in Table 1. It was found that there were significant differences on height ($p < 0.001$), weight ($p = 0.005$) and body fat percentage ($p < 0.001$), respectively, at pre-test compared to the post-test within the group. However, there were no significant differences in mean height, weight, body mass index (BMI) and body fat percentage between two groups.

ANOVA with repeated measures was used within each group indicated that there was an improvement of waist circumference (WC) at pre- and post-test (61.5 ± 11.2 versus 61.1 ± 11.1 cm), respectively. However, the result was no significant (Table 1). As for waist circumference, there was a significant difference at pre- and post-test (56.3 ± 6.3 versus 57.2 ± 6.4 cm) in control group, respectively, with $p = 0.001$.

Results indicated that there was no significant difference of body fat percentage (BF%) from 22.4 ± 7.9% (pre-test) to 22.6 ± 7.8% (post-test) with $p = 0.322$ after 12 sessions of specific training in experimental group (Table 4.1). However, there was a significant difference of body fat percentage in control group which shows an increased in measurement from pre-test to post-test (20.9 ± 4.6% versus 21.7 ± 4.9%) with $p < 0.001$, respectively.

Table 1 Anthropometrical and body composition characteristics

	Control group (PE) n = 20	Experimental group (BWIT) n = 23	P-value
Age (years)	11	11	
Height (cm)			
Pre-test	142.4 ± 5.8	139.8 ± 5.6	0.135
Post-test	143.3 ± 6.0	140.8 ± 6.0[a]	0.171
P-value	0.094	0.017	
Weight (kg)			
Pre-test	34.4 ± 7.1	35.4 ± 9.1	0.676
Post-test	35.0 ± 7.2[b]	36.3 ± 9.3[b]	0.615
P-value	0.001	0.001	
BMI (kg m^{-2})			
Pre-test	16.8 ± 2.7	18.1 ± 4.5	0.286
Post-test	17.0 ± 2.7	18.3 ± 4.5	0.264
P-value	0.566	0.378	
WC (cm)			
Pre-test	56.3 ± 6.3	61.5 ± 11.2	0.072
Post-test	57.2 ± 6.4[b]	61.1 ± 11.1	0.174
P-value	0.001	0.133	
BF%			
Pre-test	20.9 ± 4.6	22.4 ± 7.9	0.455
Post-test	21.7 ± 4.9[b]	22.6 ± 7.8	0.688
P-value	0.001	0.322	

Notes Values are means ± SD
[a]Significantly different from control group ($p < 0.05$)
[b]Significantly different from pre-test value ($p < 0.05$)

Table 2 shows parameters for cardiorespiratory component measured. From the results, it demonstrates that no significant difference was found between groups ($p = 0.394$) on one-mile run (OMR) time. However, there was a significant interaction between time and group ($p < 0.001$) on OMR time ($p < 0.001$). There was a significant improvement (14.8 ± 2.0 versus 13.2 ± 1.8 min) at pre- and post-test, respectively, ($p < 0.001$) in experimental group, whereas no significant difference of OMR time in control group at pre- and post-test (13.5 ± 1.7 versus 13.7 ± 1.4 min), respectively, ($p = 0.394$).

Predicted maximal oxygen uptake (VO$_{2max}$) results indicated no significant difference of VO$_{2max}$ ($p = 0.571$) compared to pre- and post-test, and no significance difference between groups ($p = 0.305$). In the experimental group, there was no significant difference of VO$_{2max}$ from 41.9 ± 3.8 ml kg^{-1} min^{-1} (pre-test) to 41.9 ± 4.0 ml kg^{-1} min^{-1} (post-test) with $p = 0.899$. In the control group, there was slightly impaired but no significant difference of VO$_{2max}$ at pre-test (43.0 ± 2.3 ml kg^{-1} min^{-1}) compared to post-test (42.9 ± 2.4 ml kg^{-1} min^{-1}).

Table 2 Cardiorespiratory parameters

	Control group (PE)	Experimental group (BWIT)	P-value
OMR time (min)			
Pre-test	13.5 ± 1.7	14.8 ± 2.0[a]	0.022
Post-test	13.7 ± 1.4	13.2 ± 1.8[b]	0.334
P-value	0.363	0.001	
Predicted VO$_{2max}$ (ml kg^{-1} min^{-1})			
Pre-test	43.0 ± 2.3	41.9 ± 3.8	0.275
Post-test	42.9 ± 2.4	41.9 ± 4.0	0.346
P-value	0.511	0.899	

Notes Values are means ± SD
[a]Significantly different from control group ($p < 0.05$)
[b]Significantly different from pre-test value ($p < 0.05$)

Table 3 shows the muscular fitness parameters. Results indicated a significant difference of standing broad jump (SBJ) test score between groups ($p = 0.046$) and when compared before and after intervention ($p < 0.001$). There was a significant interaction between time and group ($p < 0.001$) on standing broad jump test during the testing. There was a significant improvement in experimental group (BWIT) and control group (PE) after 12 sessions of intervention programmes. Experimental group showed an improvement from pre-test (110.3 ± 19.7 cm) to post-test (124.9 ± 18.6 cm) with $p < 0.001$. Control group showed increment from pre-test to post-test (126.9 ± 16.7 cm versus 130.7 ± 17.7 cm) with $p = 0.031$.

Another muscular parameter was sit-up test. Results indicated a significant difference of time (pre-test—post-test) with $p < 0.001$. When compared between groups, there was no significant difference ($p = 0.548$). There was a significant interaction between time and group on sit-up test during the testing ($p < 0.001$). In the experimental group (BWIT), there was a significant difference of sit-up test score in repetitions before (19.0 ± 6.6) and after intervention (25.4 ± 5.4) with $p < 0.001$. In control group (PE), although there was an improvement at pre-test (22.9 ± 3.9 reps) compared to post-test (23.2 ± 3.5 reps) but it was not statistically significant with $p = 0.750$.

Push-up test data ran with ANOVA with repeated measures indicated a significant difference between pre- and post-test ($p < 0.001$) but no significant difference when compared between groups ($p = 0.497$). There was a significant interaction between time and group on push-up test during the testing ($p < 0.001$). In the experimental group (BWIT), there was a significant difference of push-up test score in repetitions before (22.8 ± 5.9) and after intervention (29.0 ± 7.1) with $p < 0.001$. In the control group (PE), although there was an improvement from pre-test (24.1 ± 3.6 reps) to post-test (25.6 ± 3.7 reps), it was not statistically significant ($p = 0.053$).

Table 3 Muscular fitness parameter

	Control group (PE)	Experimental group (BWIT)	P-value
SBJ (cm)			
Pre-test	126.9 ± 16.7	110.3 ± 19.7[a]	0.005
Post-test	130.7 ± 17.7[b]	124.9 ± 18.6[b]	0.300
P-value	0.031	0.001	
one-min modified sit-up (reps)			
Pre-test	22.9 ± 3.9	19.0 ± 6.6[a]	0.024
Post-test	23.2 ± 3.5	25.4 ± 5.4[b]	0.119
P-value	0.750	0.001	
one-min modified push-up (reps)			
Pre-test	24.1 ± 3.6	22.8 ± 5.9	0.423
Post-test	25.6 ± 3.7	29.0 ± 7.1[b]	0.065
P-value	0.053	0.001	

Notes Values are means ± SD
[a]Significantly different from control group ($p < 0.05$)
[b]Significantly different from pre-test value ($p < 0.05$)

Table 4 Flexibility parameter

	Control group (PE)	Experimental group (BWIT)	P-value
Sit-and-reach (cm)			
Pre-test	29.5 ± 4.5	24.5 ± 4.6[a]	0.001
Post-test	29.8 ± 4.7	26.1 ± 4.5[a, b]	0.011
P-value	0.149	0.001	

Notes Values are means ± SD
[a]Significantly different from control group ($p < 0.05$)
[b]Significantly different from pre-test value ($p < 0.05$)

Table 4 shows flexibility parameter. Results indicated a significant difference between pre- and post-test ($p < 0.001$) and significant difference when compared between groups ($p = 0.004$) on the sit-and-reach test. There was a significant interaction between time and group ($p = 0.003$) during the testing. In the experimental group (BWIT), there was a significant difference of sit-and-reach test score measured in centimetres before and after 12 sessions of intervention programme (24.5 ± 4.6 versus 26.1 ± 4.5), respectively, with $p < 0.001$. In the control group (PE), there was no significant difference of sit-and-reach test with $p = 0.149$ from pre-test (29.5 ± 4.5 cm) to post-test (29.8 ± 4.7 cm).

4 Discussion

The present study found no significant improvement in cardiorespiratory fitness when compared with control for predicted VO_{2max} in 10- and 11-year-olds involved in the intervention programme. This finding was consistent with findings of the previous studies (Gilliam et al. 1980; McKeag 1991; Rowland 1985; Ignico and Mahon 1995). Physiologically, prepubescent individuals exhibit little or no significant effect of improvement in cardiorespiratory fitness. However, there is significant improvement in OMR time (Table 2). This finding was aligned with findings from previous studies. From the age of 8- to 14-year-olds, children in the period of growth and development, the maturation of children may affect the one-mile run time but the cardiorespiratory fitness does not increase to the same magnitude (Pate and Shephard 1989). This was probably due to the increased running economy, leg strength and metabolic fitness occurring with growth and maturation. For that instance, the time for one-mile run performance could improve without subsequent improvements in aerobic fitness (Bouchard et al. 1994).

In the present study, the result showed that there were no differences in standing broad jump distance, one-min modified sit-up and one-min modified push-up repetitions between body weight interval training (BWIT) group and control group after 12 sessions of intervention (Table 3). This indicated that 12 sessions of body weight interval training are not superior to other type of physical activity for muscular fitness. Based on Colchico et al. (2000), no difference in muscular parameters might be due to insufficient sessions and duration of intervention exposure towards the children. However, there was significant improvement in three muscular parameters for BWIT group when compared pre-test to post-test.

In the present study, the flexibility was significantly improved after intervention in the intervention group when compared to control. The main reason for the improvement in flexibility was that the activities designed included warm-up and cool-down procedures on each session. It was suggested that children physical fitness can be improved during the childhood years and favour training frequency of twice per week for children who participated in an introductory strength training programme. Each session lasted approximately 30–40 min, and the session ended with 5 min of stretches and cooling down activities (Faigenbaum et al. 2002). This significantly improved in sit-and-reach in the experimental group in the present study was not surprising. Girls progressively become more flexible from childhood to adolescent (Wilkinson et al. 1996).

Apparently, the reason for many parameters has no significant difference was the intensity of the BWIT. In the present study, field tests were used as the assessments tools. Not an objective and direct measurement test of the cardiorespiratory parameter only predicted maximal oxygen uptake (VO_{2max}) was obtained. Furthermore, it is speculated that the non-significant findings are due to the duration of the study that was not sufficient to induce such significant improvement of the intervention programme. It is well known that recommended exercise needs 6 weeks and above to impose beneficial effects. As for intervention period, 12 sessions, 3 times per week lasted for 4 weeks of intervention might be a little bit short to cause positive changes in the fitness components measured. Thus, future studies should consider increasing the intensity and period of intervention.

5 Conclusion

It appears that body weight interval training (BWIT), which was used in the present study, showed positive effects that may reflect through improvements in fitness components of 10- to 11-year-old schoolgirls. One-mile run time (cardiorespiratory fitness), standing broad jump, one-min modified sit-up, one-min modified push-up (muscular fitness) and sit-and-reach (flexibility) improved after intervention. However, BWIT did not portray superior effects than another type of physical activity (PA) such as in physical education (PE) classes. Body weight interval training (BWIT) was an applicable approach to promoting PA among schoolgirls. Improvement can be made to the BWIT in near future to be able to observe more distinctive improvement in physical fitness within groups and between groups.

Acknowledgements Special thanks to participants, parents, schools and Ministry of Education Malaysia, for their cooperation and commitment given throughout this study.

References

Bailey, R. C., Olson, J. O. D. I., Pepper, S. L., Porszasz, J. A. N. O. S., Barstow, T. J., & Cooper, D. M. (1995). The level and tempo of children's physical activities: an observational study. *Medicine and Science in Sports and Exercise, 27*(7), 1033–1041.

Boddy, L. M., Thomas, N. E., Fairclough, S. J., Tolfrey, K., Brophy, S., Rees, A., et al. (2012). ROC generated thresholds for field-assessed aerobic fitness related to body size and cardiometabolic risk in schoolchildren. *PLoS One, 7*(9), e45755.

Bouchard, C., Shephard, R. J., Stephens, T., & Sutton, J. R. (1994). *Exercise, fitness and health*. Champaign: Human Kinetics.

Cavill, N., Biddle, S., & Sallis, J. F. (2001). Health enhancing physical activity for young people: statement of the United Kingdom expert consensus conference. *Pediatric Exercise Science, 13*(1), 12–25.

Colchico, K., Zybert, P., & Basch, C. E. (2000). Effects of after-school physical activity on fitness, fatness, and cognitive self-perceptions: a pilot study among urban, minority adolescent girls. *American Journal of Public Health, 90*(6), 977.

Ekelund, U., Tomkinson, G., & Armstrong, N. (2011). What proportion of youth are physically active? Measurement issues, levels and recent time trends. *British Journal of Sports Medicine, 45*(11), 859–865.

Faigenbaum, A. D., Milliken, L. A., Loud, R. L., Burak, B. T., Doherty, C. L., & Westcott, W. L. (2002). Comparison of 1 and 2 days per week of strength training in children. *Research Quarterly for Exercise and Sport, 73*(4), 416–424.

Gilliam, T. B., Freedson, P. S., Geenen, D. L., & Shahraray, B. (1980). Physical activity patterns determined by heart rate monitoring in 6–7 year-old children. *Medicine and Science in Sports and Exercise, 13*(1), 65–67.

Ignico, A. A., & Mahon, A. D. (1995). The effects of a physical fitness program on low-fit children. *Research Quarterly for Exercise and Sport, 66*(1), 85–90.

Lee, I. M., Shiroma, E. J., Lobelo, F., Puska, P., Blair, S. N., Katzmarzyk, P. T., & Lancet Physical Activity Series Working Group. (2012). Effect of physical inactivity on major non-communicable diseases worldwide: An analysis of burden of disease and life expectancy. *The Lancet, 380*(9838), 219–229.

McKeag, D. B. (1991). The role of exercise in children and adolescents. *Clinics in Sports Medicine, 10*(1), 117–130.

Pate, R. R., & Shephard, R. J. (1989). Characteristics of physical fitness in youth. *Perspectives in Exercise Science and Sports Medicine, 2*, 1–46.

Rowland, T. W. (1985). Aerobic response to endurance training in prepubescent children: A critical analysis. *Medicine and Science in Sports and Exercise, 17*(5), 493–497.

Strong, W. B., Malina, R. M., Blimkie, C. J., Daniels, S. R., Dishman, R. K., Gutin, B., et al. (2005). Evidence based physical activity for school-age youth. *The Journal of Pediatrics, 146*(6), 732–737.

Tremblay, M. S., Colley, R. C., Saunders, T. J., Healy, G. N., & Owen, N. (2010). Physiological and health implications of a sedentary lifestyle. *Applied Physiology, Nutrition and Metabolism, 35*(6), 725–740.

Tzotzas, T., Evangelou, P., & Kiortsis, D. N. (2011). Obesity, weight loss and conditional cardiovascular risk factors. *Obesity Reviews, 12*(5), e282–e289.

Van Sluijs, E. M., McMinn, A. M., & Griffin, S. J. (2007). Effectiveness of interventions to promote physical activity in children and adolescents: Systematic review of controlled trials. *BMJ, 335*(7622), 703.

Wilkinson, S., Williamson, K. M., & Rozdilsky, R. (1996). Gender and fitness standards. *Women in Sport and Physical Activity Journal, 5*(1), 1–25.

Effect of Annealing Temperature on Structural Properties of $Ba_{0.9}Er_{0.1}TiO_3$ Thin Films

Zeen Vee Ooi and Ala'Eddin Ahmad Saif

Abstract In this work, $Ba_{0.9}Er_{0.1}TiO_3$ thin films have been synthesized on SiO_2/Si substrate using solgel method. The effect of annealing temperature on crystalline structure and surface morphology of the films has been studied via X-ray diffraction (XRD) and atomic force microscopy (AFM). XRD patterns reveal the crystalline structure with tetragonal phase for the films annealed at 700 °C and above. The lattice volume shows an increasing trend for the fabricated thin films with the increased annealing temperature. On the other hand, grain size obtained via AFM micrographs is increased along the increased annealing temperature which forms a larger grain during its crystallization process. The larger grain formation could be a better contribution to the alteration of other properties, but the formation of the immediate or secondary phases at 720 °C and above could also deteriorate the device performance. Among the samples, $Ba_{0.9}Er_{0.1}TiO_3$ thin film that annealed at 700 °C shows the phase-pure crystalline $BaTiO_3$ structure with no other phases, suggesting that 700 °C is the optimum annealing temperature for this type of material in our case.

Keywords Solgel · $Ba_{0.9}Er_{0.1}TiO_3$ · XRD · AFM · Annealing temperature

1 Introduction

Barium titanate, $BaTiO_3$-based lead-free ferroelectric materials have attracted numerous attentions in microelectronic and optic devices fabrication for their unique characteristics, such as high dielectric properties, thermal stability, ferroelectric properties, electro-optic coefficient, and nonlinear optical properties (Chen

---

Z. V. Ooi (✉) · A. A. Saif
School of Microelectronic Engineering, University Malaysia Perlis,
Pauh Putra Campus, 02600 Arau, Perlis, Malaysia
e-mail: zeenveeooi@gmail.com

A. A. Saif
e-mail: alaeddinsaif@gmail.com

et al. 2012a, b; Kaźmierczak-Bałata et al. 2013; Mitic et al. 2010; Woldu et al. 2015; Zhang et al. 2011). Much of the efforts were given to improve the properties of $BaTiO_3$ by introducing transition metal oxides and rare-earth oxides as a dopant into its lattice (Dang et al. 2012; Gomes et al. 2016; Qi et al. 2014). Among them, erbium (Er^{3+}), which possesses good optical properties, can be a potential candidate for optoelectronic applications (Chen et al. 2012a, b; García-Hernández et al. 2013). Viviani et al. reported the enhancement of the tetragonality of the $BaTiO_3$ with Er^{3+} dopant (Viviani et al. 2002). Besides that, Mitic et al. revealed that Er^{3+} ions can increase the dielectric constant of $BaTiO_3$ ceramic (Mitic et al. 2010). In addition, Yang et al. revealed that the good luminescence properties of Er-doped $BaTiO_3$ make it a potential candidate for optical devices (Yang et al. 2004).

In nano-material processing, the structural behavior plays an important role that affects the materials' properties. However, the structural behavior in thin film can be affected by several fabrication parameters, especially annealing temperature, which could lead to materials' property alteration (Krishnan and Munroe 2013; Rosenberger et al. 1992). Yang et al. showed that the Er-doped $BaTiO_3$ is crystallized at 650 °C. However, García-Hernández illustrated a better crystallinity for Er-doped $BaTiO_3$ film that is annealed at 700 °C (García-Hernández et al. 2013). Nonetheless, Chen et al. observed the single crystalline Er-doped $BaTiO_3$ at 750 °C (Chen et al. 2012a, b). Therefore, it is essential to study the effect of annealing temperature on the structural properties of $Ba_{0.9}Er_{0.1}TiO_3$ thin films on SiO_2/Si substrate to gain more fundamental understanding.

In this work, $Ba_{0.9}Er_{0.1}TiO_3$ solution was prepared using solgel method, spun-coated on SiO_2/Si substrate, and annealed at different temperatures. The prepared films were tested using X-ray diffraction, XRD and atomic force microscopy, AFM for the crystalline structure and surface morphology, respectively. The lattice parameters, roughness parameters, and grain size were extracted from the analyzed results.

2 Experimental Detail

2.1 Solution Preparation

$Ba_{0.9}Er_{0.1}TiO_3$ solution has been prepared from the initial materials, i.e., barium acetate, erbium acetate, and titanium (IV) isopropoxide, using solgel technique. In order to dissolve these materials, acetic acids and 2-methoxyethanol have been employed for acetates and titanium (IV) isopropoxide, respectively. A specific amount of acetates have been mixed with the preheated acetic acid to form Ba–Er solution. Then, this solution has been refluxed at 120 °C for 2 h. At the same time, stoichiometric amount of titanium (IV) isopropoxide has been added and stirred with 2-methoxyethanol at room temperature to form Ti solution. Lastly, both solutions were mixed and stirred at room temperature, followed by reflux process at 120 °C for 1 h.

2.2 Sample Preparation

Three $Ba_{0.9}Er_{0.1}TiO_3$ thin films have been fabricated on SiO_2/Si substrate via spin-coating method. The substrates have been cleaned with acetone, isopropyl alcohol, and rinsed with deionized water, followed by bake-dry process to eliminate the moisture surface. The solution then has been deposited on the entire surface of the substrate with the spin rate at 5000 rpm for 20 s, followed by baking process at 200 °C for all deposited samples to vaporize solvent and improve adhesion. The deposition and baking steps have been repeated until desired thickness has been obtained. Finally, all samples have been annealed separately at 700, 720, and 740 °C for 1 h to form crystalline structure.

2.3 Characterization

The crystalline structure of the fabricated $Ba_{0.9}Er_{0.1}TiO_3$ thin films has been studied via XRD (D2 Phaser, Bruker) that employed CuKα radiation, which operated at 30 kV and 10 mA. The patterns of the thin films have been collected at Bragg angle, 2θ from 20° to 60° with the scan step of 0.02° and the scan step time of 96 s. The surface characteristics of the films have been intensively observed using AFM (SPA400, SII Nanotechnology, Inc.) which operated in tapping mode. Two evaluation areas, i.e., 1×1 μm² and 10×10 μm², have been chosen for grain and roughness investigation, respectively. Gwyddion software has been employed to measure grain size and analyst roughness parameters.

3 Results and Discussions

3.1 XRD Analysis

Figure 1 shows the XRD results for $Ba_{0.9}Er_{0.1}TiO_3$ thin films that annealed at 700, 720, and 740 °C. The results show that the films are well-crystallized with perovskite structure after the annealing process, where the perovskite peaks are matched with the JCPDS 05-0626. However, it is found that the (001) or (100) peak, which should be appeared at the Bragg angle, 2θ around 22.2° for $Ba_{0.9}Er_{0.1}TiO_3$ thin film, is associated with the peak contributed by the sample holder. Besides, one can observe that the existence of the secondary phases due to higher annealing

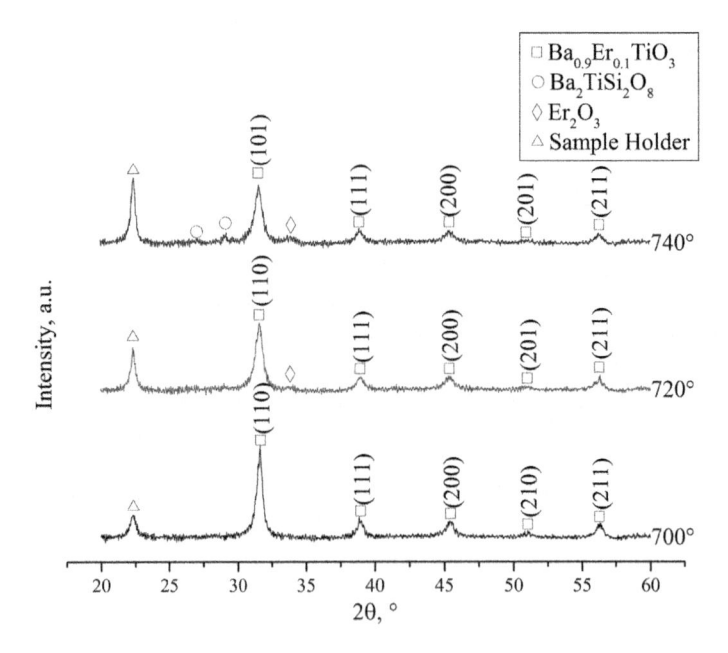

Fig. 1 XRD patterns of $Ba_{0.9}Er_{0.1}TiO_3$ thin films on SiO_2/Si substrate that annealed at various temperatures

temperature. For the films that annealed at 720 °C, Er_2O_3 has been detected. Meanwhile, with further increment in annealing temperature, Er_2O_3 with an addition phase, $Ba_2TiSi_2O_8$, also known as fresnoite, have been observed in the pattern.

The fresnoite is attributed to the interdiffusion among the $Ba_{0.9}Er_{0.1}TiO_3$ layer and SiO_2 layer at high-temperature heating process (Stawski et al. 2012). In the work from Stawski et al., they observed the similar phase above 750 °C in solgel $BaTiO_3$ thin film on Pt/Ti/Si substrate. They found that at high temperature, the Si will start to diffuse into $BaTiO_3$ precursor, which leads to the formation of the secondary phases. On the other hand, a small relatively low-intensity count of Er_2O_3 phase is detected (marked with diamonds), which indicates that the Er^{3+} dopant is not completely diffused into the $BaTiO_3$ lattice at 720 °C. This could be attributed to the distortion of the $BaTiO_3$ lattice structure by the Si, which the Si starts to diffuse into $BaTiO_3$ for the formation of fresnoite. From the result, it shows that film that annealed with 700 °C exhibits single phase with perovskite structure and no intermediate or secondary phases, which indicates that 700 °C could be the most suitable annealing temperature for $Ba_{0.9}Er_{0.1}TiO_3$ thin film in our work.

From Fig. 1, it can be noticed that as the annealing temperature increased, the perovskite peaks are mostly shifted toward lower angle. This shifting could be attributed to the formation of the byproduct, Er_2O_3, which indicating the less Er^{3+} ions substituted at $BaTiO_3$ host lattice at higher annealing temperature. On the other hand, such shifting also leads to the changes of the lattice orientation for all annealed films, which can be observed particularly at the $2\theta \approx 31.60°$ and $51.05°$.

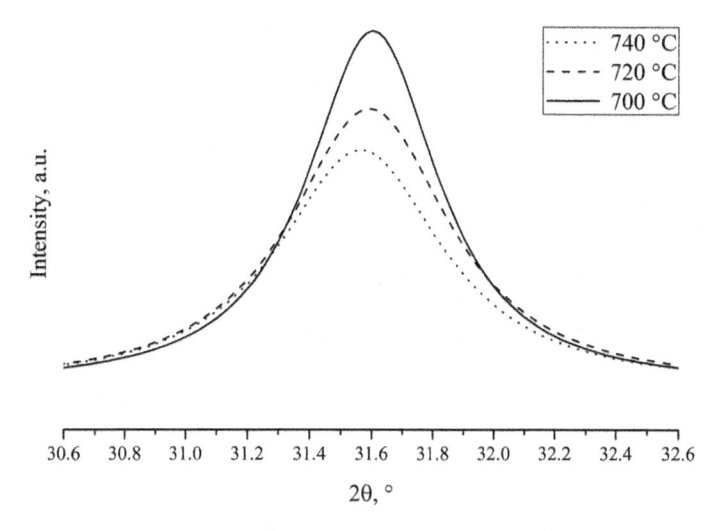

Fig. 2 Enlarged (101) (110) peaks for $Ba_{0.9}Er_{0.1}TiO_3$ thin films

Figure 2 shows the enlarged (101) (110) peak for $Ba_{0.9}Er_{0.1}TiO_3$ thin films which annealed at different temperatures. From the figure, one can notice the peak is shifted toward lower angle as a result of increased annealing temperature, which could indicate that the formation of a larger spacing between miller planes, which in turn can form larger lattice size. Besides, it is found that the intensity of the peak is reduced for higher annealing temperature. This shows that the crystallinity of the peak is reduced at higher temperature for this type of material.

In order to gain more knowledge on this behavior, lattice parameters, such as lattice cell lengths, a, c, and unit cell volume, V of the films have been evaluated from the XRD patterns using equations as followed (Cullity and Stock 2001).

$$\frac{1}{d^2} = \frac{h^2 + k^2}{a^2} + \frac{l^2}{c^2} \quad (1)$$

$$V = a^2 c \quad (2)$$

where h, k, and l are the Miller indices and d is the interplanar spacing. The spacing can be determined using Bragg equation given by

$$\lambda = 2d \sin \theta \quad (3)$$

The lattice parameters of the $Ba_{0.9}Er_{0.1}TiO_3$ thin films that annealed at various temperatures are listed in Table 1. The obtained lattice values show that the prepared films are crystallized with tetragonal structure at room temperature. In addition, it is found that $Ba_{0.9}Er_{0.1}TiO_3$ lattices are expanding at higher

Table 1 Lattice parameters of $Ba_{0.9}Er_{0.1}TiO_3$ films annealed at various temperatures

T_{anneal} [°C]	a [Å]	c [Å]	V [Å³]	a/c	Structure
700	4.0005	4.0066	64.1219	1.0015	Tetragonal
720	4.0019	4.0039	64.1225	1.0005	Tetragonal
740	4.0005	4.0101	64.1798	1.0024	Tetragonal

temperature, which could be attributed to the formation of Er_2O_3 within the films, with less Er^{3+} contents distributed in the $BaTiO_3$ host lattice, where Ba^{2+} ions (1.35 Å) own larger ionic radius than Er^{3+} ions (0.96 Å).

3.2 AFM Analysis

Figure 3 shows the three-dimensional AFM micrographs for $Ba_{0.9}Er_{0.1}TiO_3$ thin films that are captured with 10×10 μm² scanned area. From the images, the samples are free from crack and pinhole, which gives a good quality surface.

Figure 4 presented the two-dimensional AFM microimages that are scanned at 1×1 μm² evaluation area for $Ba_{0.9}Er_{0.1}TiO_3$ thin films. It can be seen that the grains are distributed in uniform, fine, and dense manner.

In order to gain more information for the structural effect, the roughness parameters have been extracted statically from recorded AFM micrographs. The average roughness, R_a, and root mean square roughness, R_q, are used to describe the average and the square root distribution of the surface height, respectively. Generally, the value of R_q is always greater than R_a (Gadelmawla et al. 2002). On the other hand, maximum peak and valley, R_t, refers to the overall roughness of the evaluated film surface. Skewness roughness, R_{sk}, is used to explain the symmetrical behavior of the profile about the mean line. Negative R_{sk} value demonstrates the valleys, scratches, or pits are dominant over the scanned area, while positive R_{sk} value shows that the peaks are dominant over the observed area (Tudose et al. 2007). Roughness kurtosis, R_{ku}, is used to evaluate the sharpness of the probability

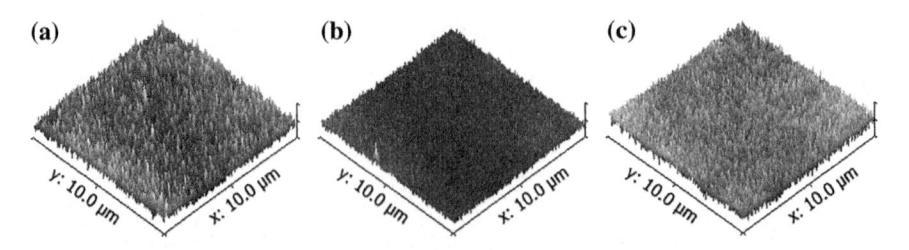

Fig. 3 Three-dimensional AFM surface micrographs for $Ba_{0.9}Er_{0.1}TiO_3$ thin films annealed at **a** 700 °C, **b** 720 °C, and **c** 740 °C

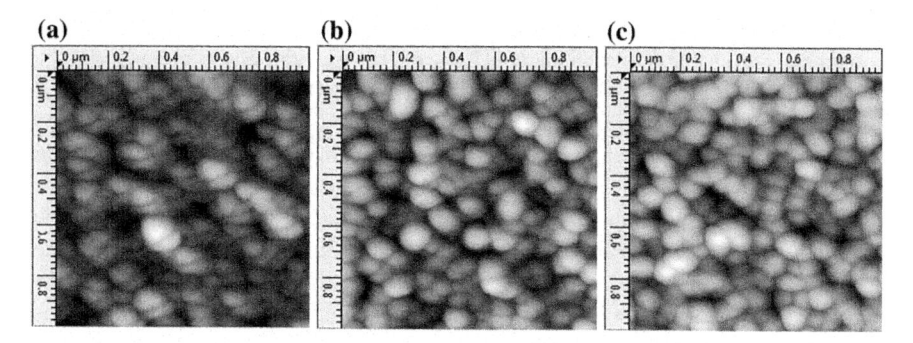

Fig. 4 Two-dimensional AFM grain distribution micrographs for $Ba_{0.9}Er_{0.1}TiO_3$ thin films annealed at **a** 700 °C, **b** 720 °C, and **c** 740 °C

Table 2 Roughness parameters and grain size of $Ba_{0.9}Er_{0.1}TiO_3$ films annealed at various temperatures

T_{anneal} [°C]	R_a [nm]	R_q [nm]	R_t [nm]	R_{sk} [nm]	R_{ku} [nm]	Grain size [nm]
700	2.86	3.62	33.07	0.438	0.441	77.57
720	3.38	4.28	65.43	0.167	0.821	83.44
740	3.66	4.61	39.68	0.0255	0.103	84.89

density of the profile. If the R_{ku} value is lower than three, the surface of film is said to be relatively flat, else the surface owns spiky behavior.

Table 2 summarized the roughness parameters and grain size of $Ba_{0.9}Er_{0.1}TiO_3$ thin films which annealed at various temperatures. The result shows that the surface roughness and grain size values of films are slightly greater at higher temperature. For the R_{sk}, it shows that the peaks are dominant at all the annealed films. On the other hand, the R_{ku} values below three indicate that all films have a relatively flat surface, which shows that the solution has been well-spread during the spin-coating process.

From the results, it shows that the nuclei began to grow and form into crystalline grains at 700 °C. At higher temperature, the larger grains are formed over the entire films. This could be attributed to the higher atom mobility with increasing temperature that can cause a more effective recrystallization of the material, which in turn leads to the formation of larger grains (Supasai et al. 2010).

4 Conclusion

In this work, solgel perovskite $Ba_{0.9}Er_{0.1}TiO_3$ thin films have been successfully fabricated on SiO_2/Si substrate and annealed at different temperatures. The diffraction patterns recorded using XRD illustrates that $Ba_{0.9}Er_{0.1}TiO_3$ thin films exhibit perovskite structure at room temperature. The calculated lattice volume showed an increasing trend from 6.412 to 6.418 nm with the increment in the annealing temperature from 700 to 740 °C. The micrographs that obtained via AFM show dense, smooth, uniformly distributed grains over the surface for all annealed samples. The measure grain size demonstrated a growing trend from 77.57 to 84.89 nm as the annealing temperature increased from 700 to 740 °C. The surface roughness is increased with an increase of annealing temperature, which could be attributed to the grain growth effect during the heat treatment. Furthermore, positive skewness was obtained for all samples, specifying that the peaks are dominant over the surface. Meanwhile, the roughness kurtosis with value less than three was achieved for all samples, showing that the surface is relatively flat, where the solution is well-spread over the entire surface. Among the samples, it can be seen that the optimum annealing temperature of $Ba_{0.9}Er_{0.1}TiO_3$ thin films is at 700 °C, where a pure perovskite structure with no intermediate or secondary phases is obtained. This suggests that 700 °C is the most suitable temperature in our case for $Ba_{0.9}Er_{0.1}TiO_3$ thin films.

Acknowledgements This work is partially supported by Fundamental Research Grant Scheme (FRGS) with the Grant number 9003-00479, funded by the Ministry of Higher Education (KPT). The authors also gratefully acknowledge the helpful comments and suggestions of the reviewers, which have improved the presentation.

References

Chen, L., Liang, X., Long, Z., & Wei, X. (2012a). Upconversion photoluminescence properties of Er^{3+}-doped $Ba_xSr_{1-x}TiO_3$ powders with different phase structure. *Journal of Alloys and Compounds, 516*, 49–52. https://doi.org/10.1016/j.jallcom.2011.11.121.

Chen, L., Wei, X. H., & Fu, X. (2012b). Effect of Er substituting sites on upconversion luminescence of Er^{3+}-doped $BaTiO_3$ films. *Transactions of the Nonferrous Metals Society of China, 22*, 1156–1160. https://doi.org/10.1016/S1003-6326(11)61299-5.

Cullity, B. D., & Stock, S. R. (2001). *Elements of X-ray diffraction*. Reading, Massachusetts Book Chapter.

Dang, N. V., Nguyen, H. M., Chuang, P. Y., Thanh, T. D., Lam, V. D., Lee, C. H. et al. (2012). Structure of $BaTi_{1-x}Fe_xO_{3-del}$ multiferroics using X-ray analysis. *Chinese Journal of Physics, 50*, 262–270. http://psroc.org/cjp/download.php?type=paper&vol=50&num=2&page=262. Accessed May 31, 2017.

Gadelmawla, E. S., Koura, M. M., Maksoud, T. M. A., Elewa, I. M., & Soliman, H. H. (2002). Roughness parameters. *Journal of Materials Processing Technology, 123*, 133–145. https://doi.org/10.1016/S0924-0136(02)00060-2.

García-Hernández, M., García-Murillo, A., de Carrillo-Romo, F. J., de Morales-Ramírez, Á. J., Meneses-Nava, M.A., Gonzalez-Penguelly, B. et al. (2013). Effect of starting materials on the morphological and optical properties of Er doped $BaTiO_3$ nanocrystalline films. *Material Transaction, 54*, 806–810. https://doi.org/10.2320/matertrans.m2012313.

Gomes, M. A., Lima, Á. S., Eguiluz, K. I. B., & Salazar-Banda, G. R. (2016). Wet chemical synthesis of rare earth-doped barium titanate nanoparticles. *Journal Materials Science, 51*, 4709–4727. https://doi.org/10.1007/s10853-016-9789-7.

Kaźmierczak-Bałata, A., Bodzenta, J., Krzywiecki, M., Juszczyk, J., Szmidt, J., & Firek, P. (2013). Application of scanning microscopy to study correlation between thermal properties and morphology of BaTiO$_3$ thin films. *Thin Solid Films, 545*, 217–221. https://doi.org/10.1016/j.tsf.2013.08.007.

Krishnan, P. S. S. R., & Munroe, P. R. (2013). Influence of fabrication conditions on the ferroelectric polarization of barium titanate thin films. *Journal of Asian Ceramic Societies, 1*, 149–154. https://doi.org/10.1016/j.jascer.2013.04.001.

Mitic, V. V., Nikolic, Z. S., Pavlovic, V. B., Paunovic, V., Miljkovic, M., Jordovic, B., et al. (2010). Influence of rare-earth dopants on barium titanate ceramics microstructure and corresponding electrical properties. *Journal of the American Ceramic Society, 93*, 132–137. https://doi.org/10.1111/j.1551-2916.2009.03309.x.

Qi, Y., Zhang, L., Jin, G., Wan, Y., Tang, Y., Xu, D., et al. (2014). UV-visible spectra and conductive property of Mn-doped BaTiO$_3$ and Ba$_{0.93}$Sr$_{0.07}$TiO$_3$ ceramics. *Ferroelectrics, 458*, 64–69. https://doi.org/10.1080/00150193.2013.849995.

Rosenberger, J., Nass, R., & Schmidt, H. (1992). Crystallization behaviour of barium titanate thin films. In *European materials research society monographs* (pp. 343–349). Elsevier, Oxford. https://doi.org/10.1016/b978-0-444-89344-4.50041-0.

Stawski, T. M., Vijselaar, W. J. C., Göbel, O. F., Veldhuis, S. A., Smith, B. F., Blank, D. H. A., et al. (2012). Influence of high temperature processing of sol–gel derived barium titanate thin films deposited on platinum and strontium ruthenate coated silicon wafers. *Thin Solid Films, 520*, 4394–4401. https://doi.org/10.1016/j.tsf.2012.02.029.

Supasai, T., Dangtip, S., Learngarunsri, P., Boonyopakorn, N., Wisitsoraat, A., & Hodak, S. K. (2010). Influence of temperature annealing on optical properties of SrTiO$_3$/BaTiO$_3$ multilayered films on indium tin oxide. *Applied Surface Science, 256*, 4462–4467. https://doi.org/10.1016/j.apsusc.2010.01.072.

Tudose, I. V., Horváth, P., Suchea, M., Christoulakis, S., Kitsopoulos, T., & Kiriakidis, G. (2007). Correlation of ZnO thin film surface properties with conductivity. *Applied Physics A, 89*, 57–61. https://doi.org/10.1007/s00339-007-4036-3.

Viviani, M., Buscaglia, M. T., Buscaglia, V., Mitoseriu, L., Testino, A., & Nanni, P. (2002). Electrical properties of Er-doped BaTiO$_3$ ceramics for PTCR applications. In *Proceedings of the 13th IEEE International Symposium on Applications of Ferroelectrics, 2002. ISAF 2002* (pp. 103–106). https://doi.org/10.1109/isaf.2002.1195881.

Woldu, T., Raneesh, B., Sreekanth, P., Ramana Reddy, M. V., Philip, R., & Kalarikkal, N. (2015). Size dependent nonlinear optical absorption in BaTiO$_3$ nanoparticles. *Chemical Physics Letters, 625*, 58–63. https://doi.org/10.1016/j.cplett.2015.02.020.

Yang, X., Guo, H., Chen, K., Zhang, W., Lou, L., Yin, M. (2004). Er^{3+} doped BaTiO$_3$ optical-waveguide thin films elaborated by sol-gel method. *Journal of Rare Earths, 22*, 36–39. http://cj.re-journal.com/en/guokanshow.asp?id=4218. Accessed May 31, 2017.

Zhang, Y., Hao, J., Mak, C. L., & Wei, X. (2011). Effects of site substitutions and concentration on upconversion luminescence of Er^{3+}-doped perovskite titanate. *Optics Express, 19*, 1824–1829. https://doi.org/10.1364/OE.19.001824.

The Effect of Zinc Addition on the Characteristics of Sn–2.0Ag–0.7Cu Lead-Free Solders

Ramani Mayappan and Amirah Salleh

Abstract Sn–Pb solder was a popular solder in electronics industry. Due to the negative effects from the usage of that solder, researchers start to find a replacement for the Sn–Pb solder. The new solder should maintain the same or better properties than the Sn–Pb solder. In this study, a newly developed solder which is Sn–2.0Ag–0.7Cu with the addition of 0.5, 1.0, 1.5, and 3.0 wt% of Zn was studied. These solders were prepared via powder methodology method and several characterizations were done on them. The 1.0 wt% Zn solder shows the lowest melting point of 222.30 °C and the 0 wt% Zn shows the highest value of 225.65 °C. The 1.0 wt% Zn solder shows the lowest Cu_6Sn_5 intermetallic thickness value of 1.58 µm and reasonable joint strength. The presence of 1.0 wt% Zn in the Sn–2.0Ag–0.7Cu solder improves the solder properties.

Keywords Intermetallic · Solder · Sn–Ag–Cu · Cu_6Sn_5

1 Introduction

Solders are widely used to link chips to their packaging substrates in flip chip technology as well as in surface mount technology (Zeng and Tu 2002). Since the electronic components are getting more complicated, solder becomes the most important mechanism in electronic devices because it provides thermal, electrical, and mechanical continuity. Soldering is a metallurgical method that uses a filler metal, the solder, with a melting point below 425 °C (Manko 1979). Solders are important in the

---

R. Mayappan (✉)
Faculty of Applied Science, Universiti Teknologi MARA, Perlis Campus, 02600 Arau, Perlis, Malaysia
e-mail: ramani@perlis.uitm.edu.my

A. Salleh
Faculty of Applied Science, Universiti Teknologi MARA, 40450 Shah Alam, Selangor, Malaysia
e-mail: amirahsalleh90@gmail.com

assembly and interconnection of the silicon die or chip. The characteristics and the properties of solders are crucial to the integrity of a solder joint (Abtew and Selvaduray 2000). The electrical and mechanical properties of the solder joints have become an important issue in electronics industry (Amin et al. 2014). Sn–Pb solders for metal interconnections have a long history and continue to provide many benefits, such as ease of handling, low melting temperatures, good workability, ductility, and excellent wetting on Cu and its alloys (Suganuma 2001). However, Sn–Pb solder has some drawbacks in terms of toxicity. The usage of Pb in solder is hazardous toward human health and environment. In electronic industry, the Pb from the disposal of electronic products is harmful. Preventing Pb from being sent to landfills and other waste disposal sites is already implemented in Japan country. The substitutes of this Sn–Pb solder were studied by many researchers (Shalaby 2015; Sobhy et al. 2016; Geipel et al. 2017). Ternary and higher sequences solders are most likely based on the binary eutectic Sn–Ag, Sn–Cu, Sn–Zn, or Sn–Bi alloys. The most encouraging and recommended substitution of Sn–Pb solder is Sn–Ag–Cu (SAC) solder and to be more specific is the Sn–2.0Ag–0.7Cu. SAC has good mechanical properties and soldering ability. The downside of this SAC solder is its high melting temperature and the thicker intermetallic layer. To improve the performance of the SAC solder, a fourth element, Zn, was added into the Sn–2.0Ag–0.7Cu to improve the characteristics of the solder.

Reaction of solder and substrate will create an intermetallic compounds (IMC) layer at the interface of the solder and substrate. The IMC layer shows the metallurgic bonding between the solder and the substrate interface. The formation of IMC layer means that the solder has a good wetting characteristic. But the excessive thickness of IMC layer is harmful to the reliability of the solder joint since IMC is brittle in nature. The IMC layer thickness will increase due to current heating during the usage of the components. Hence, the intermetallic layer growth will affect the joint strength of the solder joint. To retard the growth of intermetallic compound layer, the fourth element was added. Zn has attracted the attention of many researchers. The minor addition of Zn was believed to suppress and apparently slow down the growth of intermetallics (Song et al. 2010; Kotadia et al. 2010; Jee et al. 2007). Furthermore, it was reported that the addition of 1.0 wt% of Zn in the Sn–3.0Ag–0.5Cu solder reduces the melting temperature, enhances the mechanical strength, and refines the solder microstructure (Wang et al. 2007). Study from Xu et al. found that the addition of Zn in Sn–3.5Ag eutectic solder can prompt the formation of Cu_5Zn_8 intermetallic and restrain the formation of the Cu_6Sn_5 intermetallic. Therefore, in this study, several wt% of Zn were added into the Sn–2.0Ag–0.7Cu. The influence of adding Zn into Sn–2.0Ag–0.7Cu was studied, and the formation on intermetallic thickness, the joint strength, and the melting temperature was reported and discussed.

2 Methodology

2.1 Solder Preparation

In this study, the Sn–2.0Ag–0.7Cu–xZn (x = 0, 0.5, 1.0, 1.5, 3.0) was used as solder. The solder was prepared using the powder metallurgical method. Each element with the desired amount of solder powder (solder range: 5–15 µm) was pre-weighted. The mixture then was mixed and blended by using roll mill machine for about 2 h with the speed of 200 rpm. After the process of mixing and blending, the samples were weighed for 1 g then compacted by using hydraulic press at 14 MPa of pressure in order to form a pellet. The pellets were sintered for 2 h at the temperature of 150 °C in an argon atmosphere. A similar mixing method was used in our previous study (Mayappan et al. 2014).

Melting points of the Sn–2.0Ag–0.7Cu–xZn solders were investigated using differential scanning calorimeter (Jade DSC Perkin-Elmer). The solder samples were heated to 250 °C at heating rate of 10 °C/min. The entire scanning was carried out under inert nitrogen atmosphere.

For the study of interfacial reactions, the solder was placed on the Cu substrate and melted to 250 °C for one minute. Furthermore, the samples were isothermally soldered for 15 min (for x = 0.5Zn), 30 min (for x = 1.0Zn), 45 min (for x = 1.5Zn), and 60 min (for x = 3.0Zn). After soldering, the sample was cross-sectioned and mounted using epoxy resin. Polishing was done with silicon carbide (SiC) papers until 2000 grit and final polishing was done with 1.0 and 0.3 µm silica powders, and etched for 1 s to reveal the intermetallic microstructure. The Cu/solder interface was then analyzed under Scanning Electron Microscope (SEM), and the average thickness of the intermetallic was calculated by using ImageJ software.

2.2 Joint Strength Analysis

Two copper strips were soldered together at 250 °C for different time depending to their Zn content as mentioned above. The two ends of the strips were pulled using tensile machine. The values for the maximum stress were recorded.

3 Results and Discussions

3.1 Melting Temperature Results

Figure 1 shows the melting temperature results on the Sn–2.0Ag–0.7Cu solder with the addition of different percentage of Zn. Low melting point is one of the requirements for the newly developed solder to substitute the Sn–Pb solder. The

Fig. 1 Melting temperature variation with the addition of Zn

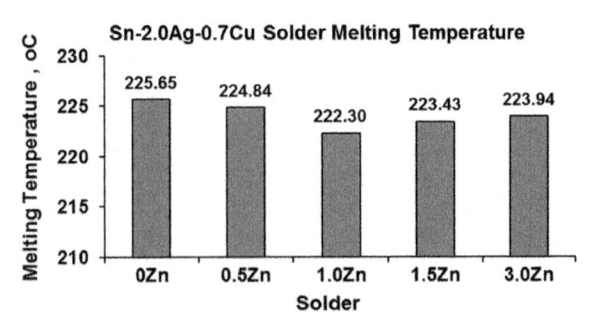

results show that the addition of small amount of Zn lowers the melting temperature with 1.0 wt% gave the lowest value.

3.2 Intermetallic Morphology Results

During soldering, the interfacial interactions between a molten solder (Sn–2.0Ag–0.7Cu–xZn) and the Cu substrate result in the formation and growth of intermetallic (IMC) phases at the substrate/solder interface. When the solder reached its melting point, it will start to melt and react with the Cu substrate. Initially, the Cu substrate will partially dissolve into the molten solder alloy upon contact. As this dissolution progresses, the liquid solder becomes supersaturated (close to the interface) with the dissolved metal substrate, leading to the subsequent formation and growth of intermetallic compounds. The intermetallic compound layer(s) will continuously evolve during the lifetime of the solder joint due to current heating.

Figure 2 shows the intermetallic formation for the Sn–2.0Ag–0.7Cu–1.0Zn solder after soldering for 1 and 30 min. As reported in our previous report (Mayappan et al. 2014), only a single Cu_6Sn_5 intermetallic was formed after

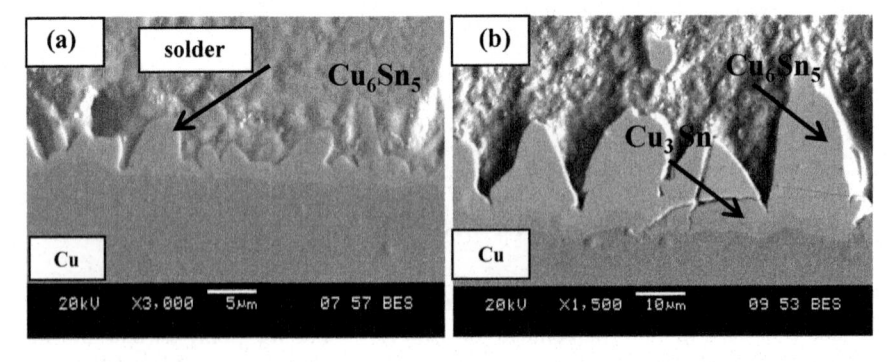

Fig. 2 Intermetallic formation **a** after 1 min soldering **b** after 30 min soldering for Sn–2.0Ag–0.7Cu–1.0Zn solder

Fig. 3 X-ray diffraction pattern for the solder interface

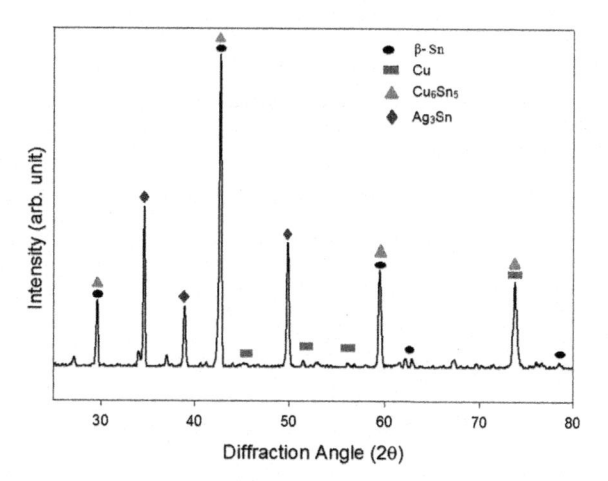

soldering for 1 min. After prolong soldering time, a second layer near the Cu substrate was observed. This layer was identified as Cu_3Sn. The Cu_3Sn phase has a darker contrast due to its lower atomic number (backscattered). The formation of the Cu_3Sn layer may be attributed decreasing amount of Cu atoms in the Cu substrate. This leads to the formation of Cu_3Sn phase which requires lesser Cu.

The XRD analysis was carried out on the solder joint. The solder joint was etched to remove Sn on the solder system. Figure 3 shows the XRD analysis result. The formation of Cu_6Sn_5 intermetallic can be confirmed from this result. The Cu_3Sn was not detected in the XRD result since Cu_3Sn intermetallic was located below the Cu_6Sn_5 layer. So, only Cu_6Sn_5 intermetallic can be detected.

3.3 Intermetallic Thickness Results

Figure 4 shows the Cu_6Sn_5 intermetallic thickness variations with Zn addition. The 1.0 wt% Zn is optimum in retarding the intermetallic growth since it gives the lowest thickness after soldering for one minute.

Fig. 4 Intermetallic thickness after soldering for 1 min

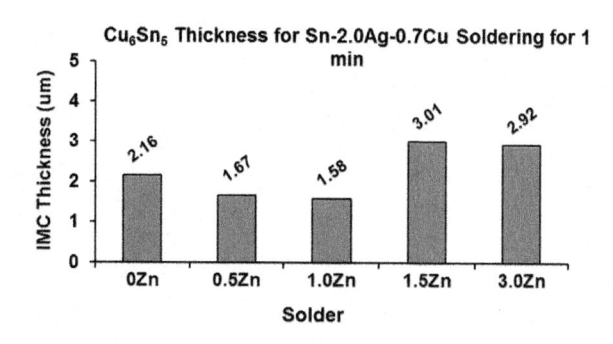

Figure 5 shows the Cu_6Sn_5 intermetallic thickness with different amount of Zn after soldering at different time. Clearly, the presence of Zn in the Sn–2.0Ag–0.7Cu solder retarded the Cu_6Sn_5 intermetallic.

Figure 6 shows the Cu_3Sn intermetallic thickness after soldering for different time with different Zn wt%. Generally, the addition of Zn has retarded the growth of Cu_3Sn intermetallic. According to Yu and Duh, the growth of Cu_3Sn was attributed to the rapid diffusion of Cu atoms from substrate into the Cu_6Sn_5. Furthermore, they emphasized that this diffusion resulted in the formation of Kirkendall voids near the Cu_3Sn/Cu interface. This explains how Zn reduces the intermetallic thickness of solder alloys (Park and Arróyave 2011).

After the Cu atoms arrive at the interface of Cu_3Sn/Cu_6Sn_5 by diffusion through the grain boundaries of the Cu_3Sn layer, the following interfacial reaction happens:

$$Cu_6Sn_5 + 9Cu \rightarrow 5Cu_3Sn \qquad (1)$$

By this reaction, Cu_6Sn_5 is converted to Cu_3Sn at the interface. Because of this reaction, the amount of Cu atoms that can diffuse to the interface of Cu_6Sn_5/solder is greatly reduced. As the result, Cu_3Sn grows rapidly with temperature and time by consuming Cu_6Sn_5 at the interface of Cu_3Sn/Cu_6Sn_5. The growth of Cu_6Sn_5 on the solder side mainly depends on the availability of Cu atoms in the solder. Since most of the Cu atoms in the bulk solder have been taken to form Cu_6Sn_5 particles in the eutectic structure, the amount of free Cu atoms that can diffuse to the solder/Cu_6Sn_5 interface is very small, greatly limiting the growth of Cu_6Sn_5 on the solder side.

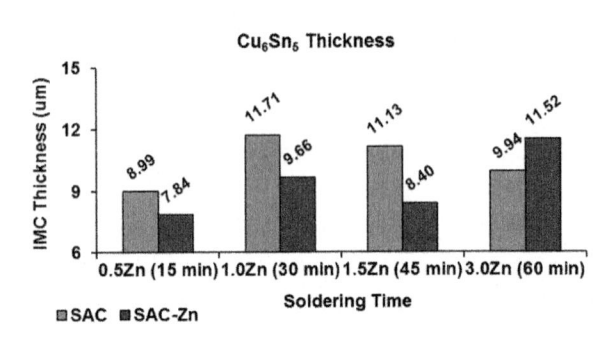

Fig. 5 Cu_6Sn_5 Intermetallic thickness after soldering for different time

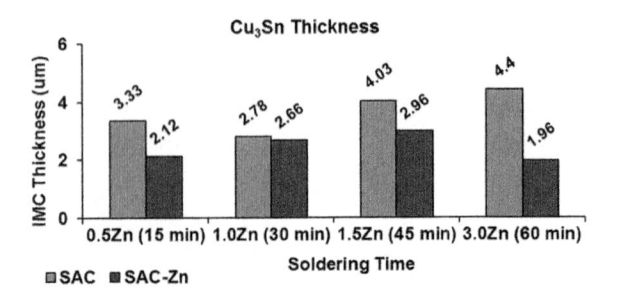

Fig. 6 Cu_3Sn Intermetallic thickness after soldering for different time

Fig. 7 Maximum stress after soldering for 1 min

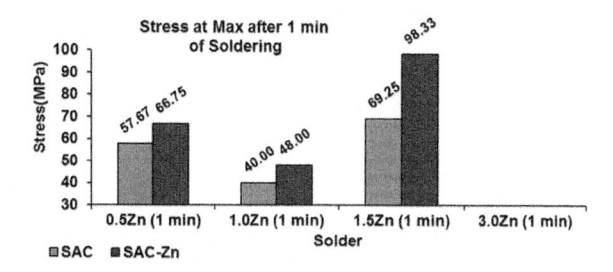

Therefore, during thermal aging, the Cu$_3$Sn layer expended on both sides, resulting in the shifting of the Cu/Cu$_3$Sn interface toward the Cu$_6$Sn$_5$. Peng et al. (2007) reported that because of this the Cu$_6$Sn$_5$ intermetallic grew in much slower rate compared to Cu$_3$Sn intermetallic.

The significantly depressed growth of Cu$_6$Sn$_5$ and Cu$_3$Sn phases for the SAC–Zn/Cu reaction couple can be attributed to the effect of Zn addition in SAC solder matrix on the migration of Sn atoms toward the solder/Cu$_6$Sn$_5$ interface (Wang et al. 2006).

3.4 Tensile Strength Results

Figure 7 shows the stress at maximum for the stress–strain curve when the sample was tested under Instron. After soldering for 1 min, the solder joints with Zn addition show higher stress value. This may due the refine morphology of the scallop Cu$_6$Sn$_5$ intermetallic.

4 Conclusion

In this study, the Sn–2.0Ag–0.7Cu–*x*Zn (*x* = 0, 0.5, 1.0, 1.5, 3.0) was successfully prepared using powder metallurgy method. The thickness of intermetallic, joint strength, melting temperature of the solders were investigated. The formation of Cu$_6$Sn$_5$ intermetallic was observed at 1 min aging for Sn–2.0Ag–0.7Cu and Sn–2.0Ag–0.7Cu–*x*Zn. Meanwhile, there were two layers of intermetallic formed when aging at prolong time which were Cu$_6$Sn$_5$ and Cu$_3$Sn layers. Based on the results, the intermetallic thickness for the Sn–2.0Ag–0.7Cu–*x*Zn was smaller than Sn–2.0Ag–0.7Cu. The retardation of the intermetallic layer indicates that Zn is an excellent retarding agent. Next, the stress strength of the solders decreased with the increased of aging time and the melting temperature of the solders were lower with the addition of zinc. Based on the results, it can be concluded that the 1.0 wt% Zn is the best percentage for the Sn–2.0Ag–0.7Cu solder (Fig. 8).

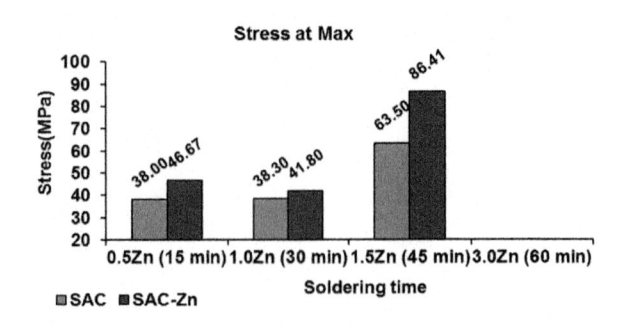

Fig. 8 Maximum stress after soldering for different time

Acknowledgements This study is financially supported by The Minister of Higher Education (MOHE) for the Grant 600-RMS/FRGS 5/3 (104/2015). Authors also thank Universiti Teknologi MARA and UniMAP for facilities. Authors also thank Nurul Alia Bt Azami, Nur Ilina Bt Zikri, Farah Nasuha Bt Aziz, and Fasihah Bt Pondi for their work reported in this paper.

References

Abtew, M., & Selvaduray, G. (2000). Lead-free solders in microelectronics. *Journal of Materials and Science Engineering R, 27,* 94–141.

Amin, N. A. A. M., Shnawah, D. A., Said, S. M., Sabri, M. F. M., & Arof, H. (2014). Effect of Ag content and the minor alloying element Fe on the electrical resistivity of Sn–Ag–Cu solder Alloy. *Journal of Alloys and Compounds, 599,* 114–120.

Geipel, T., Moeller, M., Walter, J., Kraft, A., & Eitner, U. (2017). Intermetallic compounds in solar cell interconnections: Microstructure and growth kinetics. *Solar Energy Materials and Solar Cells, 159,* 370–388.

Jee, Y. K., et al. (2007). Effect of Zn on the intermetallic formation and reliability of Sn–3.5Ag solder on a Cu pad. *Journal of Materials Research, 22,* 1879–1887.

Kotadia, H., et al. (2010). Reactions of Sn–3.5Ag-based solders containing Zn and Al additions on Cu and Ni(P) substrates. *Journal of Electronic Materials, 39,* 2720–2731.

Manko, H. H. (1979). *Solder and soldering* (2nd ed.). New York: McGraw-Hill.

Mayappan, R., Yahya, I., Ghani, N. A. A., & Hamid, H. A. (2014). The effect of adding Zn into the Sn–Ag–Cu solder on the intermetallic growth rate. *Journal of Materials Science: Materials in Electronics, 25,* 2913–2922.

Park, M. S., & Arróyave, R. (2011). Computational investigation of intermetallic compounds (Cu_6Sn_5 and Cu_3Sn) growth during solid-state aging process. *Computational Materials Science, 50,* 1692–1700.

Peng, W., et al. (2007). Effect of thermal aging on the interfacial structure of SnAgCu solder joints on Cu. *Microelectronics Reliability, 47,* 2161–2168.

Shalaby, R. M. (2015). Indium, choromium and nickel-modified eutectic Sn–0.7 wt% Cu lead-free solder rapidly solidified from molten state. *Journal of Materials Science: Materials in Electronics, 26,* 6625–6632.

Sobhy, M., El-Refai, A. M., & Fawzy, A. (2016). Effect of graphene oxide nano-sheets (GONSs) on thermal, microstructure and stress-strain characteristics of Sn-5 wt% Sb-1 wt% Ag solder alloy. *Journal of Materials Science: Materials in Electronics, 27,* 2349–2359.

Song, H. Y., et al. (2010). Effects of Zn addition on microstructure and tensile properties of Sn–1Ag–0.5Cu alloy. *Materials Science and Engineering A, 527,* 1343–1350.

Suganuma, K. (2001). Advances in lead-free electronics soldering. *Journal of Current Opinion in Solid State And Materials Science, 5,* 55–64.

Wang, F. J., et al. (2006). Depressing effect of 0.2 wt% Zn addition into Sn–3.0Ag–0.5Cu solder alloy on the intermetallic growth with Cu substrate during isothermal aging. *Journal of Electronic Materials, 35,* 1818–1824.

Wang, F. J., et al. (2007). Intermetallic compound formation at Sn–3.0Ag–0.5Cu–1.0Zn lead-free solder alloy/Cu interface during as-soldered and as-aged conditions. *Journal of Alloys and Compounds, 438,* 110–115.

Zeng, K., & Tu, K. N. (2002). Six cases of reliability study of Pb-free solder joints in electronic packaging technology. *Journal of Materials Science and Engineering R, 38,* 55–105.

Influence of Grain Size on the Isothermal Oxidation of Fe–40Ni–24Cr Alloy

Noraziana Parimin, Esah Hamzah and Astuty Amrin

Abstract The influence of grain size on the oxidation of Fe–40Ni–24Cr alloy was studied in this work. The present paper focuses on the isothermal oxidation behaviour at 700 °C. Solution treatment at three different temperatures, namely 950, 1050 and 1150 °C, was applied to Fe–40Ni–24Cr alloy to alter the average grain size of the samples. The results showed that the average grain size increased with increase in solution treatment temperature. Optical microscopy, scanning electron microscopy (SEM) and energy dispersive X-ray spectroscopy (EDS) were employed in this study to analyse the oxidation behaviour of solution-treated samples. The solution-treated samples were subjected to oxidation experiment under isothermal conditions for 500 h. The oxide scales formed during oxidation were generally complex, and their morphologies and structure were influenced by the alloy structure and expose conditions and environment. The kinetics of oxidation followed the parabolic law which represents diffusion-controlled oxide growth rate. After 500-h exposure, all alloy displayed protective oxidation with the fine grain samples exhibiting the lowest weight gain, hence the superior oxidation resistance. Smaller grain size improves the protective oxidation behaviour by enhancing exfoliation resistance and reducing oxidation rate.

Keywords Fe–40Ni–24Cr alloy · Isothermal oxidation · Grain size

---

N. Parimin (✉)
School of Material Engineering, Universiti Malaysia Perlis, 02600 Arau, Perlis, Malaysia
e-mail: noraziana@unimap.edu.my

E. Hamzah
Faculty of Mechanical Engineering, Universiti Teknologi Malaysia, 81310 Skudai, Johor, Malaysia
e-mail: esah@fkm.utm.my

A. Amrin
UTM Razak School of Engineering & Advanced Technology, UTM Kuala Lumpur, 54100 Kuala Lumpur, Malaysia
e-mail: astuty@ic.utm.my

1 Introduction

Nickel (Ni)-based superalloys are used in nuclear reactors, electrical resistance heaters, gas turbines, petrochemical, aerospace and heat-treating industries, due to their favourable strength and excellent resistance to oxidation (and many other aggressive environments) at elevated temperatures (Barnard et al. 2010). Ni-based superalloys have the ability to form protective surface oxide scales at high temperatures that provide them with resistance to further high-temperature corrosion (Barnard et al. 2010; Fulger et al. 2009). Based on its advantages, these alloys are widely used in many applications as high-temperature structural materials. HR-120 (Fe–40Ni–24Cr) alloy is a general-purpose engineering material for applications that require resistance to heat and corrosion. The alloy is a solid-solution-strengthened heat-resistant alloy that provides excellent strength at elevated temperature combined with very good resistance to carburizing and sulfidizing environments. Its oxidation resistance is comparable to other widely used Fe–Ni–Cr materials, such as alloys 330 and 800H (HAYNES International Ins). The composition of HR-120 alloy is a face-centred cubic solid solution with a high degree of metallurgical stability. HR-120 alloy is a standard material of construction for various types of thermal processing equipment such as industrial heating applications, industrial furnace and heat-treating operations. In industrial heating applications, the alloy is used for wire mesh furnace belts and basket liners. In industrial furnaces, the alloy is used for radiant tubes, muffles, retorts and cast link belt pins. Heat treating operations include bar frame heat treating baskets and heat treating fixtures.

HR-120 alloy was used in the petrochemical industries that typically contain a small amount of Nb and Ti as an alloying element (Tari et al. 2009; Tan et al. 2008a; Ray et al. 2003; Kaya et al. 2002). This alloy is strengthened by carbides precipitation such as MC carbides composed of NbC, TiC and (Nb, Ti)C (Sustaita-Torres et al. 2012; Borjali et al. 2012; Piekarski 2010; Dehmolaei et al. 2008; Ribeiro et al. 2003; de Almeida et al. 2002; Piekarski 2001). Additionally, addition of Al and Si may form an intermetallic phase subjected to ageing treatment, in order to induce precipitation hardening effect, such as $Ni_3(Al, Ti)$, $Ni_3(Nb, Ti, Al)$, $Ni_{16}Nb_6Si_7$ (Tan et al. 2008a; Kaya et al. 2002; Sustaita-Torres et al. 2012; Piekarski 2010; Dutta 2009). These materials experience tremendously harsh conditions at various temperatures and environment upon service. Under these conditions, the materials undergo compositional change and develop an oxide layer due to oxidation at high temperature.

HR-120 alloy exhibits good resistance to oxidizing environments and can be used at temperatures up to 1205 °C. It is the ability of alloys to form protective surface oxide scales at high temperatures that endow them with resistance to further high-temperature corrosion (Barnard et al. 2010). Under the aforementioned applications, HR-120 alloys encounter elevated temperatures, thermal cycling and/or long service life. In order to survive with severe environment applications, these alloys must develop a thermodynamically stable protective oxide layer with excellent corrosion resistance and also good resistance to oxide exfoliation. One way to mitigate oxide exfoliation is by refining the grain size of the alloy structure by means of heat treatment processes.

Recent researcher has studied the technique to mitigate oxide exfoliation by grain alteration such as shot-peening (Tan et al. 2008b), grain boundary engineering (Tan et al. 2006, 2008a, 2011, 2013a, b; Xu et al. 2012; Nie et al. 2010), laser surface treatment (Voisey et al. 2006; Zhua et al. 1995; De et al. 1994; Stokes et al. 1989; Stott et al. 1987) and surface treatment (Hänsel et al. 2003; Gheno et al. 2012). While grain refinement methods have never been covered for HR-120 alloy, this method was successful for other materials such as intermetallic alloy (Zheng et al. 2009; Yang et al. 2001; Niu et al. 2003; Perez 2002; Perez and Adeva 1998), alloy 617 (Jo et al. 2010), stainless steel (Peng et al. 2005) and superalloys (Klein et al. 2014; Geng et al. 2012).

Studies on isothermal and/or cyclic high-temperature oxidation behaviour by heat treatment processes of HR-120 (Fe–40Ni–24Cr) alloy are limited in the literature and, thus, are the subject of this research. The appropriate scenario(s) of the materials in this study under subsequently detailed oxidizing conditions will be evaluated by means of weight change due to isothermal oxidation, as well as oxide morphology.

2 Methods

The materials used in this study were commercial HAYNES HR-120 alloy, supplied by Haynes International, Inc., with the measured chemical compositions (in wt%): 40.45 Ni, 24.11 Cr, 0.05 C, 0.08 Al, 0.03 Ti, 0.44 Si, 0.7 Mn, 0.01 P, 0.11 Cu, 0.25 Mo, 0.001 B, 0.17 Co, 0.44 Nb, 0.05 W, and balance Fe. A test sample was cubical coupons with nominal dimensions of $10 \times 10 \times 3$ mm. The as-received alloy was subjected to heat treatment by means of different solution treatment temperature, namely 950, 1050 and 1150 °C for 3 h followed by water quench. The different solution treatment temperatures were chosen based on the findings by other researchers (Cai et al. 2003; Zhang et al. 2008; Noraziana 2015), recorded that solution treatment from 950 to 1200 °C was provide beneficial grain size alteration for Fe–Ni–Cr alloy. These samples are denoted as solution-treated 950 °C (ST950), solution-treated 1050 °C (ST1050) and solution-treated 1150 °C (ST1150). The microstructure of solution-treated samples was characterized using optical microscopy. The grain size of solution-treated sample was measured using linear intercept methods. The solution-treated samples were ground to P600 grit surface finish and clean with acetone, and the dimension of the sample was measured. All samples were weight before and after the oxidation experiment to measure the weight change using Metler AT400 analytical balance with sensitivity of 0.1 mg. The isothermal oxidation tests were investigated by means of discontinuous testing at 700 °C up to 500 h. The surface morphology of oxidized samples was characterized by scanning electron microscopy (SEM) and energy dispersive X-ray spectroscopy (EDX).

3 Results and Discussion

3.1 Microstructure of Solution-Treated Samples

Figure 1 shows the optical micrographs of three solution-treated samples. The microstructure of all samples shows a matrix austenite phase. Results indicate that increase in solution treatment temperature increases the grain size of the samples. ST950 samples show the finer grain size that is 27.27 μm followed by ST1050 samples with average grain size of 32.85 μm. ST1150 samples indicate the coarser grain size that is 37.13 μm.

3.2 Kinetic of Oxidation

Figure 2 illustrates the weight changes of the three solution-treated samples which were oxidized at 700 °C for 500 h. The curves indicate the trend of the weight gain as a function of exposure time for ST950, ST1050 and ST1150 samples, respectively. ST950 exhibited the lowest weight gain out of the three samples, hence the superior oxidation resistance, while ST1150 samples exhibited the highest weight gain. The oxidation kinetics for all samples follows a parabolic rate law which represents diffusion-controlled oxide growth rate. The weight gain data of the ST950 sample was oxidized very rapidly at the beginning, but after a few hours of exposure, the rate decreased to a very low value recorded a small amount of increment in weight gain. The fine grain samples indicate a lower oxidation rate that shows the fine grains promotes the higher grain boundary area that was act as a diffusion path of ion movement across the metal-gas surface. The higher ion diffusion provides rapid oxide layers growths that protect the alloys surface. Similar trends occurred for ST1050 samples with rapid oxidation at the beginning.

Fig. 1 Optical micrograph of three solution-treated samples: **a** ST950, **b** ST1050 and **c** ST1150

Fig. 2 Weight gain data for three solution-treated samples oxidized at 700 °C

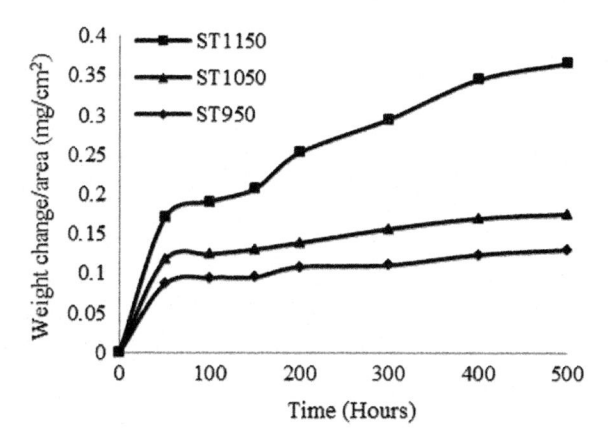

However, after exposing to a longer time, the behaviour of the ST1050 samples is significantly different from the ST950 sample, and it gains significantly more weight due to the oxidation.

In contrast, the weight gain of coarse grain size gradually increased due to oxide growth during exposure to working temperature. The ST1150 sample recorded a high weight gain at the beginning, followed by a steady weight gain until 150 h. The weight gain gradually increased until 200 h of time exposed at working temperature, which then sharply increased until 400 h of exposure. After 400 h of exposure, the oxidation rate mildly decreased due to the possible occurrence of minor oxide spallation or oxide cracking which will be discussed later. This higher weight gain is due to the possible continuous oxide growth to develop a protective layer at the alloy surface.

3.3 Surface Morphology of Oxidized Samples

Oxide morphology of ST950 sample oxidized at 500 h and analysed with SEM corresponding SEI is shown in Fig. 3. The low magnification SEI of Fig. 3a shows a continuous layer of oxide with distinct facets and discrete oxide particle. Overgrow of discrete oxide particle was enriched in the element of Nb and O which is possibly composed of Nb-rich oxide particle. In addition, the surface morphology displays two different regions of oxides, which are rough region (highlighted black in Fig. 3b) where oxide particle randomly distributed and flat region (highlighted brown in Fig. 3c). Further EDX analysis of the oxide particle at A indicated the enrichment of element O and Cr, with a minor presence of element Mn, Fe and Ni. This observation suggested that at the rough region was mainly composed of Cr rich oxide with slightly disperse spinel phase. Furthermore, EDX analysis on area B indicates that the presence of element O, Cr, Fe, Ni with a smaller percentage of Mn was detected. This observation indicates the development of mixed oxide composed of spinel, haematite and Cr-rich oxides phase as identified in the phase analysis.

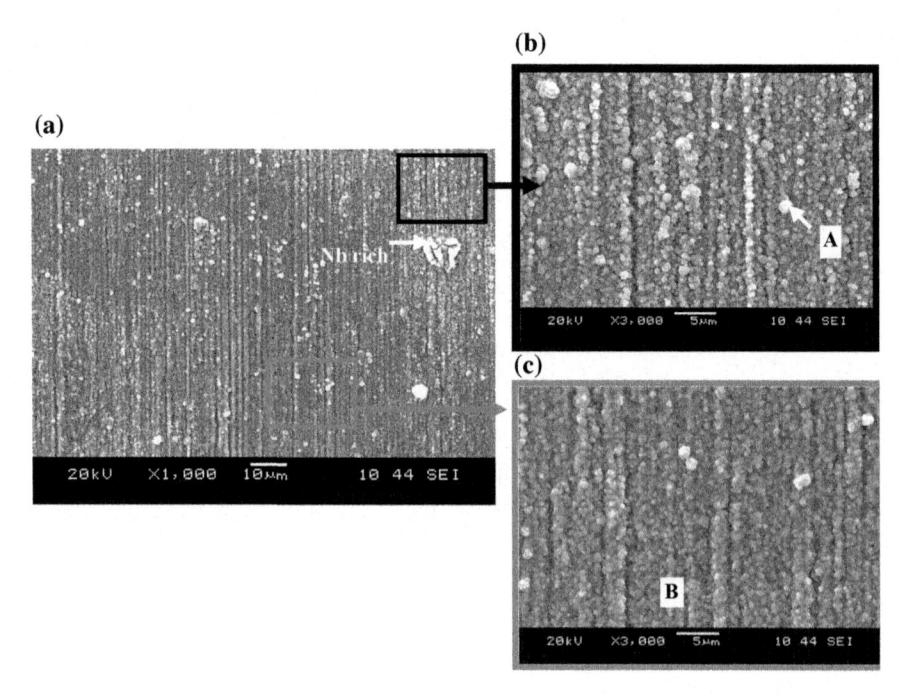

Fig. 3 **a** Low and **b**, **c** high magnification SEM images of the surface morphology of ST950 samples for 500 h (Color figure online)

The formation of these oxide layers was suggested that it may be controlled by the refinement of the grain structure that provides higher grain boundary area which increased the diffusion path for ion diffusion across the oxide scale to develop a rapid protective oxide layer.

The surface morphologies of ST1050 sample oxidized at 500 h are shown in Fig. 4. The low magnification SEI of Fig. 4a shows a serration and discrete oxide islands at alloy matrix. The discrete oxide island indicates a big area of oxide growth at isolated area (highlighted red in Fig. 4c) analyse with EDX at C indicated the enrichment of element Cr and O, suggested composes of Cr oxides. Observation at the matrix area indicates the presence of elements O, Cr, Fe, Ni and Mn as analyse with EDX at area D. The observation identifies the formation of mixed oxide phases at that area. In addition, the Nb-rich oxides also observed on the alloy surface and on overgrow Cr-rich oxides. Nb-rich oxide also identified distributed along the surface, which grows as isolated oxide particle as shown in Fig. 4b.

The surface morphologies of ST1150-oxidized samples at 500 h are shown in Fig. 5a. Serration and undulation surfaces were observed on low magnification SEI as shown in Fig. 5a. EDX analysis at the area *E* indicates the enrichment of Cr and O, which suggested the formation of Cr-rich oxides at the region, while EDX analysis

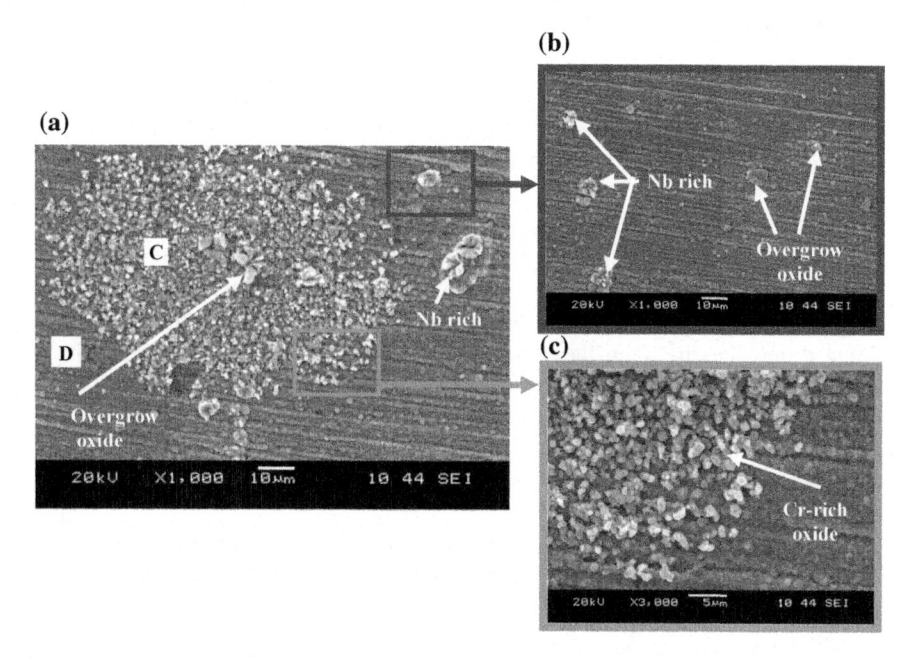

Influence of Grain Size on the Isothermal Oxidation... 783

Fig. 4 **a**, **b** Low and **c** high magnification SEM images of the surface morphology of ST1050 samples for 500 h (Color figure online)

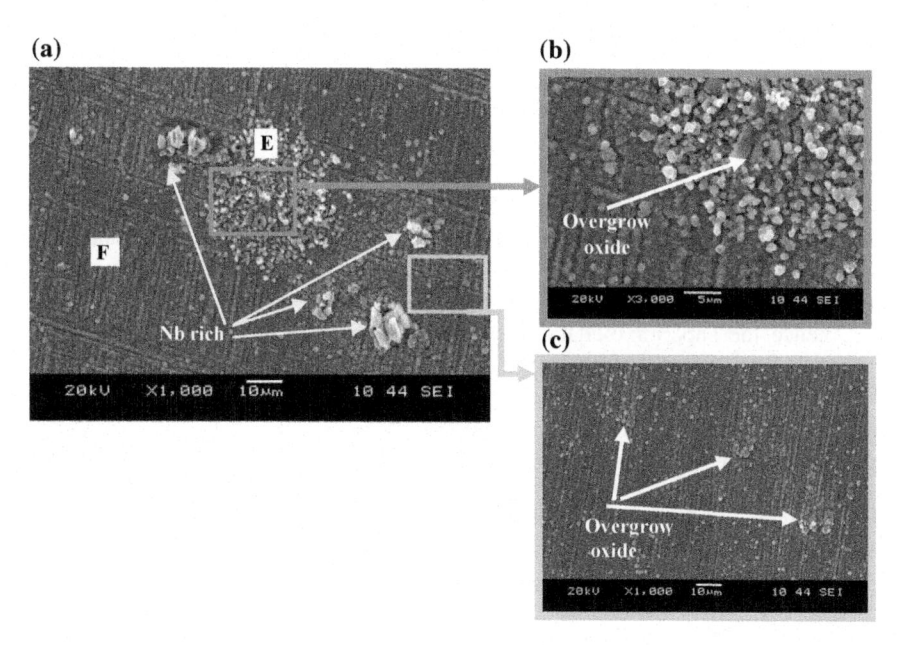

Fig. 5 **a**, **c** Low and **b** high magnification SEM images of the surface morphology of ST1150 samples for 500 h (Color figure online)

at the area *F* indicates the presence of elements O, Cr, Fe, Ni and Mn, which suggested the formation of mixed oxide composed of spinel, haematite and Cr-rich oxide phase. The isolated Nb-rich oxide particles also noticeable overgrow at the alloy surface. Another observed phenomenon indicates the separate oxide islands with evidence of crack induced by the overgrown oxide particle as highlighted green in Fig. 5b.

Nb was added to the alloy system to improve the mechanical properties of the alloy. However, the Nb addition resulted in a large fraction of Nb-rich precipitates which encouraged the occurrences of pitting (Tan et al. 2008a). Therefore, the pitting may be associated with the Nb-rich precipitate due to the massive compositional differences between the matrix and precipitate. The extensive growth of oxide scale tends to create the small difference in volume thermal expansion coefficients between the oxides and metals (Tan et al. 2008b). Cracks induce the oxide to grow faster as it develops the unprotective surface. This phenomenon will continuously happen until the stable protective oxide layer develops at the alloy surface.

This feature supports the oxidation kinetics trend for ST1150 sample as discussed earlier that shows the slight decrease in the weight change after 400-h exposure due to the minor oxide spallation induced by crack of overgrown oxides. In addition, Fig. 5c displays the overgrown oxide island formed on the sample surface as highlighted in yellow.

4 Conclusion

Solution treatment was applied to Fe–40Ni–24Cr alloy to alter the grain size of the alloy to improve protective oxidation behaviour. Solution-treated samples were oxidized at 700 °C for 500 h. It was found that the fine and coarse grain size was produced by the solution treatment on Fe–40Ni–24Cr alloys. The oxidation of all samples tends to follow the parabolic rate law which represents diffusion-controlled oxide growth rate. The oxidation rate of the fine grain (ST950) was slower compared to the coarse grain (ST1050 and ST1150). ST950 exhibited the lowest weight gain, hence the superior oxidation resistance. The oxide morphology formed on oxidized ST1050 and ST1150 samples is mostly uniform oxide scale with overgrow of discrete oxide particle composed of Nb-rich oxide. The morphological surfaces of the ST1150 samples indicate an evidence of crack induced by the overgrown oxide particle, resulting in the high oxidation rate.

Acknowledgements The authors would like to thank the Ministry of Higher Education Malaysia for the research fund under the Fundamental Research Grant Scheme (FRGS) (Project No. FRGS/1/2016/TK05/UNIMAP/02/4).

References

Barnard, B. R., Liaw, P. K., Buchanan, R. A., & Klarstrom, D. L. (2010). Affects of applied stresses on the isothermal and cyclic high-temperature oxidation behavior of superalloys. *Materials Science and Engineering A, 527*(16–17), 3813–3821.

Borjali, S., Allahkaram, S. R., & Khosravi, H. (2012). Effects of working temperature and carbon diffusion on the microstructure of high pressure heat-resistant stainless steel tubes used in pyrolysis furnaces during service condition. *Materials and Design, 34,* 65–73.

Cai, D., Mei, Y., Pulin, N., & Wenchang, L. (2003). Influence of solution treatment temperature on mechanical properties of a Fe–Ni–Cr alloy. *Materials Letters, 57*(24–25), 3805–3809.

de Almeida, L. H., Ribeiro, A. F., & Le May, I. (2002). Microstructural characterization of modified 25Cr–35Ni centrifugally cast steel furnace tubes. *Materials Characterization, 49*(3), 219–229.

De Damborenea, J., Lopez, V., & Vázquez, A. J. (1994). Improving high-temperature oxidation of Incoloy 800H by laser cladding. *Surface and Coatings Technology, 70,* 107–113.

Dehmolaei, R., Shamanian, M., & Kermanpur, A. (2008). Microstructural characterization of dissimilar welds between alloy 800 and HP heat-resistant steel. *Materials Characterization, 59*(10), 1447–1454.

Dutta, R. S. (2009). Corrosion aspects of Ni–Cr–Fe based and Ni–Cu based steam generator tube materials. *Journal of Nuclear Materials, 393*(2), 343–349.

Fulger, M., Ohai, D., Mihalache, M., Pantiru, M., & Malinovschi, V. (2009). Oxidation behavior of Incoloy 800 under simulated supercritical water conditions. *Journal of Nuclear Materials, 385*(2), 288–293.

Geng, S., Qi, S., Zhao, Q., Ma, Z., Zhu, S., & Wang, F. (2012). Effect of columnar nano-grain structure on the oxidation behavior of low-Cr Fe–Co–Ni base alloy in air at 800 °C. *Materials Letters, 80,* 33–36.

Gheno, T., Monceau, D., & Young, D. J. (2012). Mechanism of breakaway oxidation of Fe–Cr and Fe–Cr–Ni alloys in dry and wet carbon dioxide. *Corrosion Science, 64,* 222–233.

Hänsel, M., Boddington, C. A., & Young, D. J. (2003). Internal oxidation and carburisation of heat-resistant alloys. *Corrosion Science, 45*(5), 967–981.

Jo, T. S., Kim, S. H., Kim, D.-G., Park, J. Y., & Do, Kim Y. (2010). Effects of grain refinement on internal oxidation of Alloy 617. *Journal of Nuclear Materials, 402*(2–3), 162–166.

Kaya, A. A., Krauklis, P., & Young, D. J. (2002). Microstructure of HK40 alloy after high temperature service in oxidizing/carburizing environment: I. oxidation phenomena and propagation of a crack. *Material Characterization, 49*(1), 11–21.

Klein, L., von Bartenwerffer, B., Killian, M. S., Schmuki, P., & Virtanen, S. (2014). The effect of grain boundaries on high temperature oxidation of new γ'-strengthened Co–Al–W–B superalloys. *Corrosion Science, 79,* 29–33.

Nie, S. H., Chen, Y., Ren, X., Sridharan, K., & Allen, T. R. (2010). Corrosion of alumina-forming austenitic steel Fe–20Ni–14Cr–3Al–0.6Nb–0.1Ti in supercritical water. *Journal of Nuclear Materials, 399*(2–3), 231–235.

Niu, Y., Cao, Z. Q., Gesmundo, F., Farnè, G., Randi, G., & Wang, C. L. (2003). Grain size effects on the oxidation of two ternary Cu–Ni–20 wt% Cr alloys at 700–800 °C in 1 atm O_2. *Corrosion Science, 45*(6), 1125–1142.

Noraziana, P. (2015). *Effect of solution treatment on high temperature oxidation of Fe–33Ni–19Cr and Fe–40Ni–24Cr alloys* (Ph.D. Thesis, Universiti Teknologi Malaysia).

Peng, X., Yan, J., Zhou, Y., & Wang, F. (2005). Effect of grain refinement on the resistance of 304 stainless steel to breakaway oxidation in wet air. *Acta Materialia, 53*(19), 5079–5088.

Perez, P. (2002). Influence of the alloy grain size on the oxidation behaviour of PM2000 alloy. *Corrosion Science, 44,* 1793–1808.

Perez, P., & Adeva, P. (1998). Influence of exposure time and grain size on the oxidation behaviour of a PM Ni_3Al alloy at 635 °C. *Corrosion Science, 40*(4), 631–644.

Piekarski, B. (2001). Effect of Nb and Ti additions on microstructure, and identification of precipitates in stabilized Ni–Cr cast austenitic steels. *Materials Characterization, 47*(3–4), 181–186.

Piekarski, B. (2010). The influence of Nb, Ti, and Si additions on the liquidus and solidus temperatures and primary microstructure refinement in 0.3C–30Ni–18Cr cast steel. *Materials Characterization, 61*(9), 899–906.

Ray, A. K., Sinha, S. K., Tiwari, Y. N., Swaminathan, J., Das, G., Chaudhuri, S., et al. (2003). Analysis of failed reformer tubes. *Engineering Failure Analysis, 10*(3), 351–362.

Ribeiro, A. F., De Almeida, L. H., Fruchart, D., & Bobrovnitchii, G. S. (2003). Microstructural modifications induced by hydrogen in a heat resistant steel type HP-45 with Nb and Ti additions. *Journal of Alloys and Compounds, 357*, 693–696.

Stokes, P. S. N., Stott, F. H., & Wood, G. C. (1989). The influence of laser surface treatment on the high-temperature oxidation of Cr_2O_3-forming alloys. *Materials Science and Engineering A, 121*, 549–554.

Stott, G. C., Bartlett, F. H., & Wood, P. K. N. (1987). The influence of laser surface treatment on the high temperature oxidation of Cr_2O_3—forming Alloys. *Materials Science and Engineering, 88*, 163–169.

Sustaita-Torres, I. A., Haro-Rodríguez, S., Guerrero-Mata, M. P., de la Garza, M., Valdés, E., Deschaux-Beaume, F., et al. (2012). Aging of a cast 35Cr–45Ni heat resistant alloy. *Materials Chemistry and Physics, 133*(2–3), 1018–1023.

Tan, L., Sridharan, K., & Allen, T. R. (2006). The effect of grain boundary engineering on the oxidation behavior of INCOLOY alloy 800H in supercritical water. *Journal of Nuclear Materials, 348*(3), 263–271.

Tan, L., Ren, X., Sridharan, K., & Allen, T. R. (2008a). Corrosion behavior of Ni-base alloys for advanced high temperature water-cooled nuclear plants. *Corrosion Science, 50*(11), 3056–3062.

Tan, L., Ren, X., Sridharan, K., & Allen, T. R. (2008b). Effect of shot-peening on the oxidation of alloy 800H exposed to supercritical water and cyclic oxidation. *Corrosion Science, 50*(7), 2040–2046.

Tan, L., Allen, T. R., & Yang, Y. (2011). Corrosion behavior of alloy 800H (Fe–21Cr–32Ni) in supercritical water. *Corrosion Science, 53*(2), 703–711.

Tan, L., Allen, T. R., & Busby, J. T. (2013a). Grain boundary engineering for structure materials of nuclear reactors. *Journal of Nuclear Materials, 441*(1–3), 661–666.

Tan, L., Busby, J. T., Chichester, H. J. M., Sridharan, K., & Allen, T. R. (2013b). Thermomechanical treatment for improved neutron irradiation resistance of austenitic alloy (Fe–21Cr–32Ni). *Journal of Nuclear Materials, 437*(1–3), 70–74.

Tari, V., Najafizadeh, A., Aghaei, M. H., & Mazloumi, M. A. (2009). Failure analysis of ethylene cracking tube. *Journal of Failure Analysis and Prevention, 9*(4), 316–322.

Voisey, K. T., Liu, Z., & Stott, F. H. (2006). Inhibition of metal dusting of alloy 800H by laser surface melting. *Applied Surface Science, 252*(10), 3658–3666.

Xu, P., Zhao, L. Y., Sridharan, K., & Allen, T. R. (2012). Oxidation behavior of grain boundary engineered alloy 690 in supercritical water environment. *Journal of Nuclear Materials, 422*(1–3), 143–151.

Yang, S., Wang, F., & Wu, W. (2001). Effect of microcrystallization on the cyclic oxidation behavior of β-NiAl intermetallics at 1000 °C in air. *Intermetallics, 9*, 741–744.

Zhang, J. F., Tu, Y. F., Xu, J., Zhang, J. S., & Zhang, J. L. (2008). Effect of solid solution treatment in microstructure of Fe–Ni based high strength low thermal expansion alloy. *Journal of Iron and Steel Research, 15*(1), 75–78.

Zheng, H. Z., Lu, S. Q., & Huang, Y. (2009). Influence of grain size on the oxidation behavior of $NbCr_2$ alloys at 950–1200 °C. *Corrosion Science, 51*(2), 434–438.

Zhua, S. M., Wang, L., Li, G. B., & Tjongb, S. C. (1995). Laser surface alloying of Incoloy 800H with silicon carbide: microstructural aspects. *Materials Science and Engineering A, 201*, 5–7.

Performance Measurement of Small and Medium Enterprises (SMEs) in Malaysia: Implications for Enhanced Competitive Advantage Toward the Implementation of TQM

Wan Ahmad Yusmawiza Wan Yusoff, Naim Ben Ali and M. Boujelbene

Abstract This paper discusses the performance measurement (PM) of small to medium enterprises (SMEs) and focuses on manufacturing industry in Malaysia. This paper shows the current performance situation and provides guidance for companies', especially manufacturing sector in Malaysia by considering the implementation of Total Quality Management (TQM). The method used in this paper is by doing the survey and analysis of the data collection as well as to implement the objective of this project. A total of 80 survey SMEs manufacturing industries in Malaysia participated in this survey. The companies are measured based on the number of year established and the number of employee with each TQM aspect. The result concludes that the companies in SME manufacturing sector still need to improve in order to implement the government target of SME improvement.

Keywords Small and medium enterprises · Performance measurement
Total quality management · Management system · Productivity

W. A. Y. W. Yusoff (✉)
Manufacturing and Materials Department International Islamic University,
Kulliyyah of Engineering, Selangor, Malaysia
e-mail: yusmawiza@iium.edu.my

N. B. Ali
Photovoltaic and Semiconductor Materials Laboratory, El-Manar University-ENIT,
PO. Box 37 Le Belvedere, 1002 Tunis, Tunisia

M. Boujelbene
College of Engineering, Industrial Engineering Department, University of Hail,
Ha'il, Kingdom of Saudi Arabia

© Springer Nature Singapore Pte Ltd. 2018
R. Saian and M. A. Abbas (eds.), *Proceedings of the Second International Conference on the Future of ASEAN (ICoFA) 2017 – Volume 2*,
https://doi.org/10.1007/978-981-10-8471-3_77

1 Introduction

Malaysian SMEs can be grouped into three categories: micro, small, or medium. These groupings are decided based on either the numbers of people a business employs or on the total sales or revenue generated by a business in a year. Malaysian SMEs are a vital component of the country's economic development. A small and medium enterprise (SME) in manufacturing and manufacturing-related services (MRS) is an enterprise with full-time employees not exceeding 150 or with annual sales turnover not exceeding RM25 million. For more detail, SME can be classified as in Table 1.

1.1 Problem Statement

Currently, the SMEs in Malaysia are trying to improve their companies as to implement the aims of Malaysia's Government. The Government aims to improve the productivity in SME sector in Malaysia competitively with other countries over the world. This study aims to contribute to the Total Quality Management (TQM) literature by assessing on the TQM practices and organizational performances of SMEs with and without ISO 9000 certification in the developing economy of Malaysia. The focus is to examine whether research models developed for industrialized countries are applicable in the Malaysian region. In addition, this study also develops a framework of quality measurements based on prescriptive, conceptual, practitioner, and empirical literatures to generically benchmark the quality measurements of TQM for the use of SME manufacturing in Malaysia. The analysis of current performance measurement in SME development processes has been described briefly in Table 1. It shows the important sample that covered all the dimension of performance, and few exhibited properties that also mapped to the characteristics of performance measures and to the requirements of an effective development process.

The uses of characteristics are to evaluate existing performance measurement system, enable strategic objective identification, and enable performance measure development, which have been mentioned by Radam et al. (2008).

Table 1 Definition of SMEs in Malaysia

Size	Manufacturing and MRS	
Micro	Less than 5 employees	Less than RM250,000
Small	Between 5 and 50 employees	Between RM250,000 and less than RM10 million
Medium	Between 51 and 150 employees	Between RM10 million and RM25 million

Hashim (2000)

The other theoretical models that are focused in a strategic performance measurement system are quality, flexibility, time, and finance. These terms have been successfully covered by Ballantyne and Brignall (1994); Hashim (2000). The performance measure in SMEs manufacturing industry also can be related with using of Total Quality Management (TQM). Therefore, to illustrate the importance of an effective development process for introducing new systems into SMEs, a case study on the development of TQM in SMEs was studied by Eddy (1998); Ong et al. (2010). While there are commitments to TQM practices, opinion has been divided on its implementation with some researchers supporting the idea of combination with ISO 9000 certification as a first step toward TQM by Kumar et al. (2008). They stressed about identification of critical characteristics of performance measures, use a survey to establish whether SMEs measure performance strategically, and use a case study to investigate whether the process identified is appropriate within a SME context which has been recognized by Hudson et al. (2001). According to Lee-Mortimer (2006); Hudson et al. (2001), performance measurement also needs to define multiple data collection methods that minimize the threats to validity and reliability of information and define process that underlines the fact that SMEs place equal attentions on both the financial and non-financial measures.

1.2 Total Quality Management (TQM)

Performance measurement can be achieved when there are efforts contributed toward it. However, the plan and implementation should be done in a systematic manner to ensure its efficiency. Total Quality Management is one of the systems that should be incorporated to achieve the objective of performance measurement.

TQM also can be defined as the part of the overall management system that includes the aspects such as personal and organizational structure, planning activities, practices, procedure, processes, and resources for developing, implementing, achieving, reviewing, and maintaining the manufacturing performance. In this case, the research is done on performance measurement of SME industry that specializes in manufacturing industry. TQM is identified to use as a tool to do survey on manufacturing performance within SME industries in Malaysia. TQM approach to long-term success views continuous improvement in all aspects of an organization as a journey and not as a short-term destination. It aims to radically transform the organization through progressive changes in the attitudes, practices, structures, and systems. TQM transcends the "product quality" approach, involves everyone in the organization, and encompasses its every function administration, communications, distribution, manufacturing, marketing, planning, training. To implement successfully TQM, most organizations must take a major paradigm shift. In many situations, the conventional "solving a crisis is success". However, the TQM wisdom is, "not having a crisis is success." The latter is much harder to measure successful organizations especially in SMEs manufacturing industry, and prevention is the established culture and philosophy.

2 Methodology

2.1 Design Questionnaire

The questionnaire study is one of the ways to get the information needed. It surveys the experiences of the manufacturers in managing their company toward achieving better performance in manufacturing sector. From the survey, the understanding or prediction of human behavior or conditions related to this issue can be made. This type of survey consists of several stages to be implemented before getting the yield.

2.2 Data Processing

Data processing is the most critical part in survey research. All the data received from the respondents should be kept properly, and the date of received must be recorded. For this project, the sample data that collected will be analyzed by using SPSS (Statistical Package for Social Science) after the distribution phase is completed. The data obtained from the questionnaire will be set into SPSS and Microsoft Excel 2007. The advantages using SPSS are as follows: It will enable the user to score and analyze the questionnaire data immediately and in various ways while working on scores and carrying out calculations.

2.3 Analyzing Data Using SPSS and Microsoft Office Excel 2007

The data obtained from the questionnaire will be set into SPSS and Microsoft Office Excel 2007. This SPSS features will be used to analyze data to perform several tests. The test is comparison test. These tests will distinguish whether the questions are reliable and acceptable or not. From the test performed, the data will be analyzed or interpreted and some discussion and recommendation will be given according to the results.

3 Results

There are some objectives to measure the data collected and simplify in the simple word:

1. To analyze the relationship by the number of employees and year of established between each of key performance indicators.
2. To clarify the percentage of performances from each of key performance indicator.

A total of 80 surveys components were sent out to SME manufacturing industry which is producing various type of material of product from July 2010 until October 2010. Three methods were used to approach respondent. Forty questionnaires were sending via e-mail, 20 through post, and another 20 by door-to-door. There are 27 replies that were received. All the replies were maintained in hard copy. The total response rate was 33.75%. From this rate of response percent, the respondent consists of some of the following industries: electrical and electronics industry of 25.9%; 14.8% of paper, printing, and publishing industry; 11.1% of metal product industry; 7.4% of machinery and engineering industry; 7.4% of non-metallic material industry; 3.7% of plastic industry; 3.7% of textile industry; and others/services industry of 26%. From the survey, it is found that the percentage of micro-industry is 40.7%, small industry is 11.1%, and 48.2% of respondent can be classified as medium industries.

3.1 The Relationship Between Manufacturing Indicator and No. of Employee

Table 2 and Fig. 1 show that the lowest mean is scrap or rework that scored only from 1 to 1.67 or can be considered as poor than other criterion, while the highest are on-time delivery rate and human resource practice which are contributed from mean of 3.64 to 5.0.

Table 2 Manufacturing indicator versus number of employees

No. of employees		Reduction in manufacturing cycle time	On-time delivery rate	Scrap/rework as a percentage of sales	Performance (3-year growth): customer base	Human resource practices
5–19	Mean	2	3.64	1.64	3.27	3.64
	N	11	11	11	11	11
20–49	Mean	2.33	3.67	1.67	3	3.67
	N	3	3	3	3	3
50–99	Mean	1.6	4.3	1.4	4	4.3
	N	10	10	10	10	10
100–150	Mean	1	5	1	5	5
	N	3	3	3	3	3
Total	Mean	1.78	4.04	1.48	3.7	4.04
	N	27	27	27	27	27

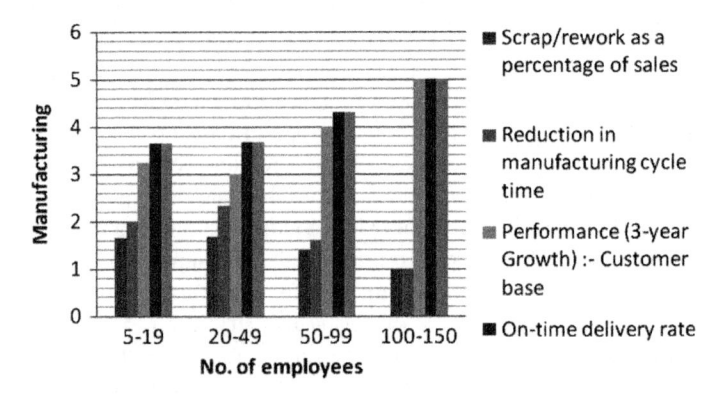

Fig. 1 Manufacturing indicator versus number of employee

3.2 The Relationship Between Manufacturing Indicator and Year of Established

Table 3 and Fig. 2 depict the lowest mean scored by scrap or rework that is from mean of 1 to 1.8, while the highest mean or very good level is contributed by human resource practices that scored from 3.58 to 5.0. Thus, the manufacturing indicator aspect still should be more empowered especially for scrap and rework criterion.

Table 3 Manufacturing indicator of year of established

Year of established		Reduction in manufacturing cycle time	On-time delivery rate	Scrap/ rework as a percentage of sales	Performance (3-year growth): customer base	Human resource practices
<1995	Mean	1.00	5.00	1.00	5.00	5.00
	N	3	3	3	3	3
1996–2000	Mean	1.14	4.71	1.00	4.43	4.71
	N	7	7	7	7	7
2001–2005	Mean	2.40	3.60	1.80	3.00	3.60
	N	5	5	5	5	5
>2006	Mean	2.08	3.58	1.75	3.25	3.58
	N	12	12	12	12	12
Total	Mean	1.78	4.04	1.48	3.70	4.04
	N	27	27	27	27	27

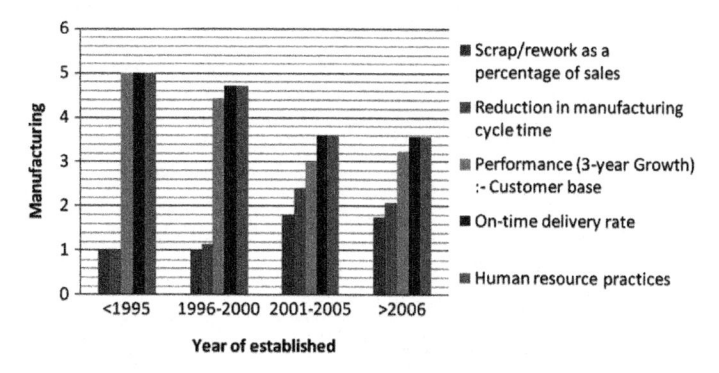

Performance Measurement of Small and Medium Enterprises ... 793

Fig. 2 Manufacturing indicator versus year of established

3.3 Relationship Between Manufacturing Management and Number of Employee

Table 4 shows the level of manufacturing management criterion between numbers of employee that is fairly in good level which contributes from total value of mean up to 3.82. As shown in Fig. 3, the manufacturing management criterion is effectively done for the companies that have number of employee from 100 to 150 compared to companies that only have number of employee from 50 to 99 and 20 to 49. As shown in Table 5, the implementation of manufacturing management criterion between the numbers of year established is fairly in good level that contributes the total mean value up to 3.82 but zero defects, cost reduction, and saving scored the average level mean value of 3 and 3.24. Figure 4 shows the manufacturing management criterion is effectively done for the companies that have long time established which is from year 1995 and below compared to the company established from 2006 and above.

Table 4 Manufacturing management versus number of employee

Number of employees		QC problem solving technique	5s awareness	Zero defect	Standardization (S.O.P/S.O.C)	Cost reduction and saving
20–49	Mean	3.75	3.75	3.00	3.25	3.25
	N	4	4	4	4	4
50–99	Mean	3.70	3.30	2.90	3.50	3.00
	N	10	10	10	10	10
100–150	Mean	4.33	4.67	3.33	4.33	4.00
	N	3	3	3	3	3
Total	Mean	3.82	3.65	3.00	3.59	3.24
	N	17	17	17	17	17

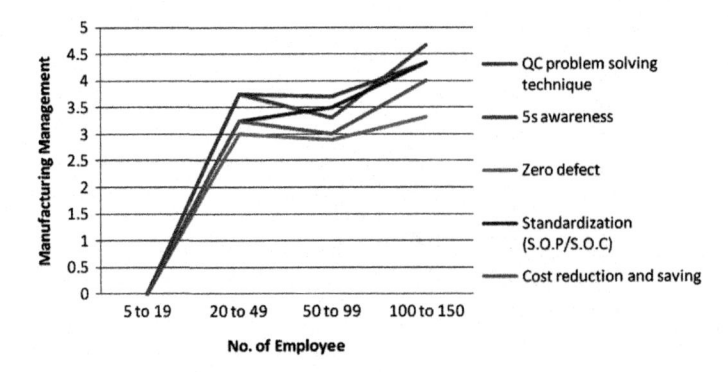

Fig. 3 Graph of manufacturing management versus number of employee

Table 5 Manufacturing management versus year of established

Year of company being established		QC problem-solving technique	5s awareness	Zero defect	Standardization (S.O.P/S.O.C)	Cost reduction and saving
1995 or earlier	Mean	4.33	4.67	3.33	4.33	4.00
	N	3	3	3	3	3
1996–2000	Mean	3.88	3.50	3.00	3.63	3.13
	N	8	8	8	8	8
2001–2005	Mean	3.60	3.40	3.00	3.20	3.00
	N	5	5	5	5	5
2006 or above	Mean	3.00	3.00	2.00	3.00	3.00
	N	1	1	1	1	1
Total	Mean	3.82	3.65	3.00	3.59	3.24
	N	17	17	17	17	17

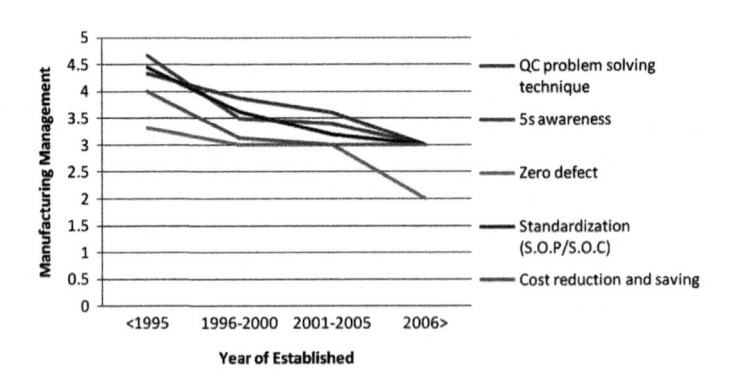

Fig. 4 Graph of manufacturing management versus year of established

3.4 Relationship Between Manufacturing Management and Year of Established

According to Table 5, the implementation of manufacturing management criterion between the numbers of year established is fairly in good level that contributes the total mean value up to 3.82 but zero defects, cost reduction, and saving scored the average level mean value of 3 and 3.24. As shown in Fig. 4, the manufacturing management criterion is effectively done for the companies that have long time established which is from year 1995 and below compared to the company established from 2006 and above. Table 6 indicate the highest percent of good and very good of 74.1% is performance (3-year Growth): Customer base while reduction in manufacturing cycle time and scrap/rework as a percentage of sales give the lowest percentage of 11.1%. Table 7 shows the percentage of good level on manufacturing management criteria. QC problem solving technique gives the highest score which is 70.6% and the lowest is Zero defect aspect which is scored 0.0%.

4 Discussions

The concept of Total Quality Management has been used as a guideline to measure the performance level of this industry. The paradigm of TQM impressed about the elements such as developing a corporate quality policy, designing and implementing a quality system, and also developing measures for capturing quality costs

Table 6 Manufacturing indicator

KPI for manufacturing	Percentage (%)
Reduction in manufacturing cycle time	11.1
On-time delivery rate	70.3
Scrap/rework as a percentage of sales	11.1
Performance (3-year growth): customer base	74.1
Human resource practices	70.3

Table 7 Manufacturing management

Manufacturing management	Percentage (%)
QC problem-solving technique	70.60
5s awareness	41.20
Zero defect	0.00
Standardization (S.O.P/S.O.C)	47.10
Cost reduction and saving	23.50

and benefits from improved quality. In addition, the TQM concept also touched about benchmarking of business and improvement of managerial and technical processes. Besides that, the selected respondents have been made. It is important in order to evaluate the level of performance of SME manufacturing industries, especially in TQM aspects. Thus, based on TQM approaches, on the questionnaire, the respondents need to give rating value from 1 to 5 that represent whether it is very poor, average, or excellent according to the level of implementation of their company. Based on the study analysis, the measurement of performance level of respondents is focusing on two factors: The first factor is how the number of year established by a company can be related to those TQM aspects, and the second factor is how the number of employee by a company can be illustrated in terms of those TQM parameters. As shown in Table 4 and Fig. 3, the overall manufacturing management criterion is fairly in good level. However, the zero defect achievement as shown in Table 5 does not contribute any percentage of good level and still considered in average level. It seriously needs to improve by reducing the mistakes during production in order to achieve the standard level of production quality. However, the relationship between number of year established and manufacturing management criterion is inversely proportional. It is because, as the number of year operating by the companies increases the manufacturing management criteria decreases. By the way, the relationship by number of employee and manufacturing management criterion as shown in Table 5 and Fig. 4, the relationship is directly proportional as the number of employees increases, the manufacturing management criterion also increases. The biggest problem encountered by manufacturers in achieving better performance level is lack of commitment from management side. Besides that, there are still other problems encountered such an unstable companies' management system, financial problems, and problem related to employees. The strength of company also effected by how long has been the company established. For the new manufacturer company, they usually do not have enough experience to counter the weaknesses. The employees also consist of various backgrounds. Thus, the difference level of education, knowledge, and skills also contributes factors that affected performance level. Furthermore, the experts also will demand for higher salary, which is hardly afforded by the non-well-established company. However, most of the respondents believe that all problems can be overcome effectively if there are full commitment and integration between employers and employees within a company. Thus, as recommendation, the employer should encourage the employee and staff about the all management skill awareness as discussed before by providing training course to each employee. The training course can be conducted by program such as campaign, talk, and seminar. The higher level of the company should be given enough information and knowledge about Total Quality Management (TQM) aspects. The staff should know how to implement it in order to gain the better performance level toward company.

5 Conclusion

The objectives of the project are considered achieved since the results achieved. From the result obtained, it can be concluded that the performance level in quality management aspect among manufacturer of SME industry in Malaysia is in fair condition. There are still a lot of spaces need to be improved to help them being more aware on the importance of manufacturing performance. It can also be concluded that the level of performance increases as the number of year of companies established and number of employee increased. The establishment of TQM within company enables the integration of difference aspects of needed skills and also led to waste minimization of a company. It also enables the manufacturer to satisfy the level of standard of quality system or ISO 9001 that practically contributes toward high-quality performance.

Acknowledgements This research is partially funded by the Research Endowment Fund, IIUM. We would like to give special thanks to Engr. Muhammad Fauzan Noraini and Engr. Mohd Norazrul Ismail, SME Corp. Malaysia, El-Manar University Le Belvedere 1002-Tunis, Tunisia, for their endless help in the design, development, and implementation of this program.

References

Ballantyne, J., & Brignall, S. (1994). *A taxonomy of performance measurement frameworks* (Research Paper No. 135). Warwick: Warwick Business School.

Eddy, D. M. (1998). Performance measurement: Problems and solutions. *Journal of Health Affairs, 17*(4), 7–25.

Hashim, M. K. (2000). Business strategy and performance in Malaysian SMEs: A recent survey. *Malaysian Management Review, 35*(2), 1–10.

Hudson, M., Smart, A., & Bourne, M. (2001). Theory and practice in SME performance measurement systems. *Journal of IJOPM, 21*(8), 1096.

Kumar, V., DeGrosbois, D., Choisne, F., & Kumar, U. (2008). Performance measurement by TQM adopters. *The TQM Journal, 20*(3), 209–222.

Lee-Mortimer, A. (2006). A lean route to manufacturing survival. *Assembly Automation, 26*(4), 104.

Ong, J. W., Ismail, H., & Yeap, P. F. (2010). Malaysian small and medium enterprises: The fundamental problems and recommendations for improvement. *Journal of Asia Entrepreneurship and Sustainability, 6*(1), 39.

Radam, A., Abu, M. L., & Abdullah, A. M. (2008). Technical efficiency of small and medium enterprise in Malaysia: a stochastic frontier production model. *International Journal of Economics and Management, 2*(2), 395–408.

The Effect of Ethanol on Drying Defects of Oil Palm Trunk (OPT)

Suhaida Azura, Ahmad Fauzi Awang,
Shaikh Abdul Karim Yamani Zakaria, Junaiza Ahmad Zaki
and Izzah Azimah Noh

Abstract Wood-based industries in Malaysia are dependent on resources from natural forests and forest plantations. The oil palm trunk has created the massive biomass waste. The high consumption of these resources increases the costs of product development and expansion. One of the alternative resources to overcome the high dependency on local timber is the biomass waste from the oil palm industry which can be turned into value-added products, but oil palm trunk (OPT) is reported to be difficult to dry due to its extremely high green moisture content. Thus, this study aimed to determine the susceptibility of oil palm lumber (OPL) to drying defect and to compare the drying properties of OPL that is soaked with ethanol to OPL without ethanol. The analysis was performed on 40 samples obtained from the bottom part of the OPT. The samples were subjected to treatment with the following concentrations of ethanol liquid—65, 75, and 85%. The results showed that drying using ethanol produced less defect compared to drying without ethanol. The different percentages of ethanol concentration depicted that 85% ethanol concentration is better than 65 and 75% because the defects showed significant improvement at the higher level.

Keywords Oil palm trunk · Oil palm lumber · Defect · Quick drying test Ethanol concentration

---

S. Azura (✉) · I. A. Noh
Faculty of Applied Science, Universiti Teknologi Mara, Shah Alam,
Selangor, Malaysia
e-mail: suhaidaazura90@yahoo.com

I. A. Noh
e-mail: izzahnoah@gmail.com

A. F. Awang · S. A. K. Y. Zakaria · J. A. Zaki
Faculty of Applied Science, Universiti Teknologi Mara Pahang,
Bandar Tun Abdul Razak Jengka, Pahang, Malaysia
e-mail: ahmad_fauzi@pahang.uitm.edu.my

S. A. K. Y. Zakaria
e-mail: syamani@yahoo.com

J. A. Zaki
e-mail: jun_dis@yahoo.com

1 Introduction

Oil palm (*Elaeis guineensis Jacq.*) is an agricultural plant with non-wood fibres lignocellulosic material originated from West Africa and is widely cultivated in Malaysia for its fruits that produced oil (Ahmad Fauzi et al. 2013). The Malaysian Timber Industry Board (MTIB) reported that in 1990 the oil palm industry in Malaysia experienced rapid growth where it covered 1.7 million hectares and this figure increased to 5.6 million hectares in 2015 (Aminah 2017).

Oil palm is a monocotyledon species with great variation in density values at different parts of its stem. Hence, the utilisation of oil palm lumber (OPL) should emphasise on its various properties that are related to low density (Srivaro et al. 2014). In most countries, it is prohibited to burn the trunks after felling, even though the practice was common until recently. The advantages of the oil palm lumber (OPL) are that it is cheap, light, and available in large amounts (Razak et al. 2008).

Oil palm trunk (OPT) possesses the largest biomass with varied moisture content and density. However, to date there are very few studies being done on its application in the wood industry.

Nevertheless, the commercialisation of OPT as a biomass resource presents several challenges (Anis et al. 2007; Edi Suhaimi et al. 2008). One in particular is the economic life cycle of an oil palm is only between 25 and 30 years, and after this period, OPT is no longer considered to be valuable commercially (Othman et al. 2009). When it exceeded the aforementioned duration, OPT is converted into OPL as an alternative to sawn timber. Therefore, this study aimed to determine the susceptibility of OPL to drying defect and to compare between the drying properties of OPL that is soaked with ethanol to OPL without ethanol treatment.

2 Materials and Methods

The bottom portion of OPT aged 30 years old was collected from an oil palm plantation at Felda Jengka 19, in Pahang. The diameter of the trunk was approximately 40–50 cm. The OPT was sawn to the specific lumber size using the band saw. Subsequently, a straight-line ripsaw was used to rip it into the final size of the intended samples which was 25 mm thick × 100 mm wide × 200 mm long.

The Terazawa's quick drying test (QDT) method was adopted to dry the samples in this study (Terazawa and Tsutsumoto 1976; Terazawa 1965; Yusoff and Othman 2016). The samples were soaked in ethanol liquid with different concentrations that were 65, 75, and 85%, while the control sample was not applied with ethanol. The whole drying process took between 5 and 7 days to achieve 12% moisture content.

After the drying process, the defects were classified into three groups, end check, honeycomb and deformation, which followed the Terazawa's quick drying defect. The 5-point Likert scale was implemented to rate the defect level where the scale of 1 indicated a good sample that was free from defects to the rating of 5 that indicated that the sample had severe defect. Each parameter required 10 samples, and thus, a total of 40 samples were gathered for the analysis. Samples were labelled as D1 for end check, D2 for honeycomb and D3 for deformation. Figure 1a shows the ranting

(a)

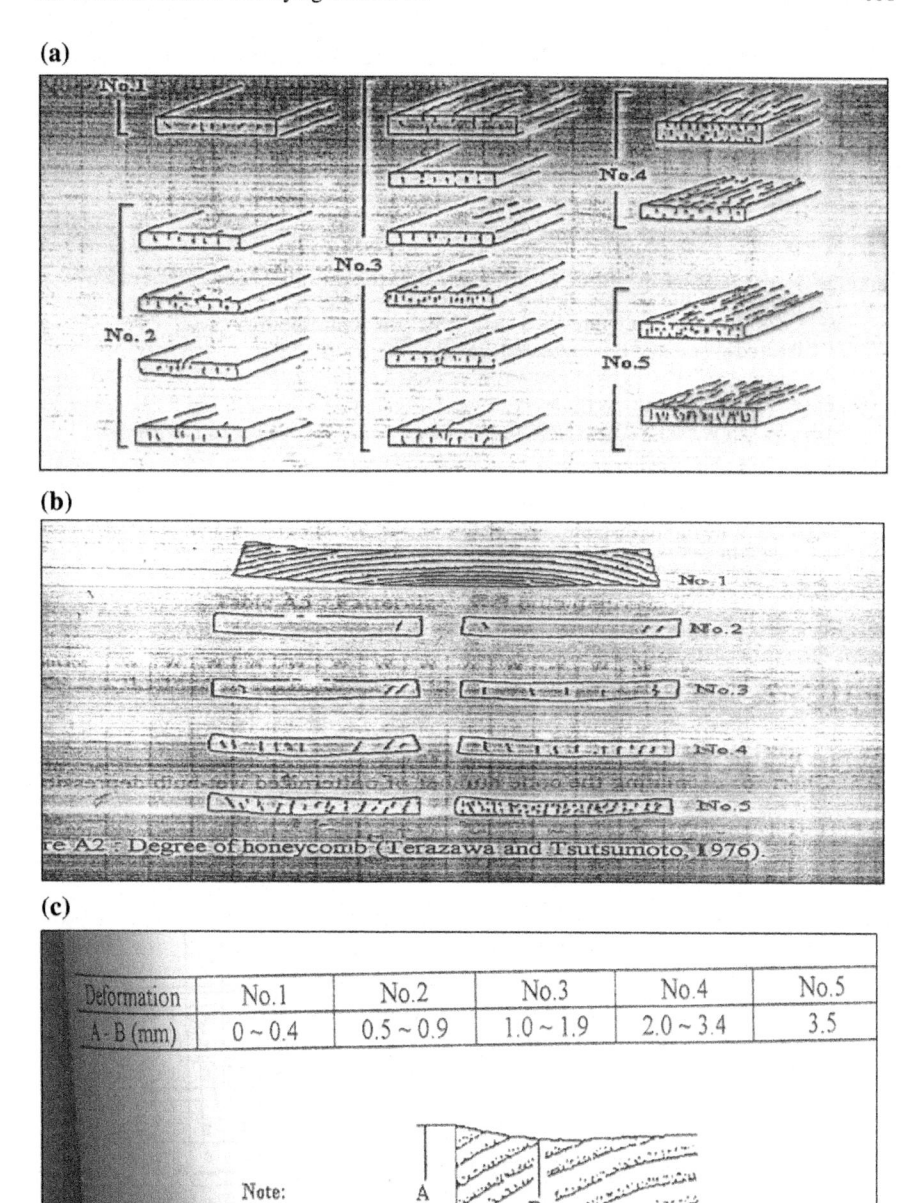

(b)

(c)

Deformation	No.1	No.2	No.3	No.4	No.5
A - B (mm)	0 ~ 0.4	0.5 ~ 0.9	1.0 ~ 1.9	2.0 ~ 3.4	3.5

Note: A - B (mm)

Fig. 1 **a** Ranting defects for end check, **b** ranting defects for honeycomb, **c** ranting defects for deformation

defects for end check, Fig. 1b shows the ranting defects for honeycomb, and Fig. 1c shows the ranting defect for deformation.

3 Results and Discussion

The data were analysed using SPSS to determine the significant levels between the untreated and treated methods. Table 1 shows the result for the treated OPL at different drying defects on end check, honeycomb and deformation. Notably, the mean rating for the end check defects between treated OPL and untreated OPL was significant at 6.40. Moreover, the honeycomb defects had a mean rating of 5.00, whereas the deformation defects were highly significant with its mean rating of 27.74.

Table 2 shows the F-value at different ethanol concentrations. The mean value for end check defects was not significant at 2.25 followed by the honeycomb defects which were significant with a mean rating of 3.86. In addition, the deformation defects are not significant with a mean rating of 2.10.

With the data in Tables 1 and 2, the total mean rating of defects was indicated to be lower for the treated OPL as compared to the untreated OPL (refer to Fig. 1).

Figure 2 illustrates that the untreated OPL had higher drying defects as compared to the treated OPL. The highest mean rating for the untreated OPL was 3.10, while both the mean rating for end check and honeycomb defects were 1.50. On the contrary, the OPL at end check defect recorded the lowest drying defects with a mean rating of 1.07. Moreover, the mean rating for honeycomb defect was 1.10 and the deformation defect was 1.77. This shows that the defect on end check, honeycomb and deformation had significant difference between treated and untreated of OPL.

Comstock and Cote Jr. (1968) and Ahmad Fauzi et al. (2012) observed that the application of ethanol as a solvent in the drying process using oven dry method can minimise the time required for drying as well as the defects on wood. Essentially, ethanol can stabilise and lower the surface tension, avoiding the structure from collapsing. Nonetheless, deformation can still occur after the soaking process but at a different rate compared to the conventional way of soaking without ethanol.

Table 1 F-value for treatment (treated and untreated) of OPL

Source	Df ($n - 1$)	Check	Honeycomb	Deformation
Treatment	1	6.40	5.00	27.74

Table 2 F-value of treated OPL at different ethanol concentrations

Source	Df ($n - 1$)	Check	Honeycomb	Deformation
Concentration		2.25	3.86	2.10

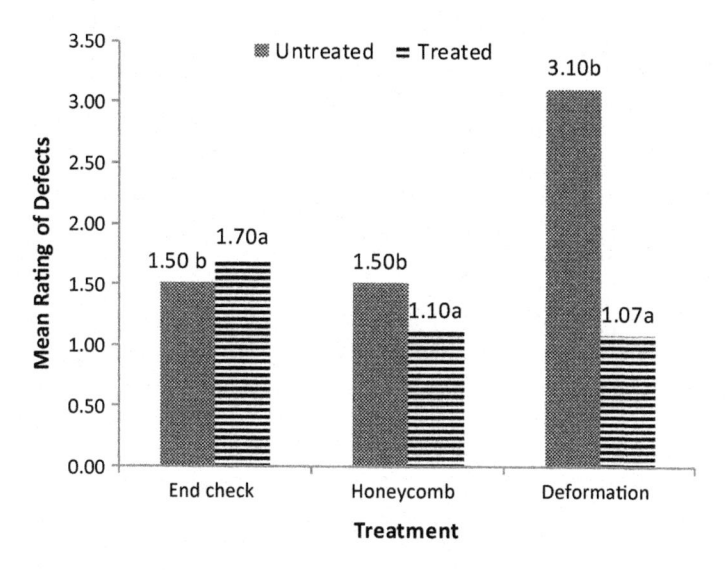

Fig. 2 Mean values of defects for untreated OPL and treated OPL

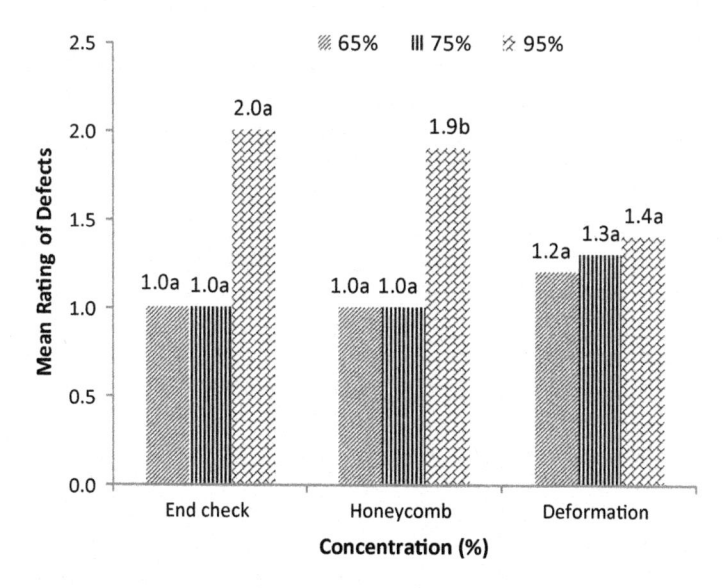

Fig. 3 Mean rating of defects for different ethanol concentrations

Figure 3 depicts no significant difference between the end check and honeycomb defects in ethanol concentrations of 65 and 75% because they attained the same mean value which was 1.00. Additionally, end check and honeycomb defects at 85% ethanol concentration had mean values of 1.20 and 1.30, respectively. Apart from that, the deformation defects demonstrated that the drying of OPL in 65%

ethanol concentration had a mean value of 2.00. On the contrary, the mean value for deformation defects at 75% ethanol concentration was 1.90, whereas 85% ethanol concentration recorded a mean value of 1.40.

These results showed that 85% ethanol concentration was better than 65 and 75% ethanol concentrations because the mean values of three different drying defects did not vary significantly. Accordingly, Comstock and Cote Jr. (1968) found that drying in 85% ethanol concentration at high temperature did not cause the pit border to be aspirated, thus resulting in faster drying and lower drying defects.

Apart from that, Petty and Puritch (1970) stated that 5 days are required for drying when 85% ethanol concentration was used. High concentration of ethanol increased the number of pit membrane pores per conducting and hence decreased the drying defect. Notably, an average of 27,000 pit membrane pores per conducting tracheid were presented on solvent-dried wood as opposed to only 600 on air-dried wood. Pits aspiration in softwood can be reduced or prevented by means of certain organic liquids. It is probable that the reduced pit aspiration caused by drying from organic liquids is due to the lower surface tension of the liquids compared to water. The relatively rapid drying rates obtained in solvent seasoning are no doubt related to this (Ellwood and Ecklund 1961). According to Fauzi et al., the ethanol can help to drying method for the material is shorter in drying time with minimal defects. Ethanol liquid can stabilize the vessel and fibre structure, and this might hinder the collapse and deformation occurs after soaking process, compared to normal oven dry without ethanol (Yusoff and Othman 2016). With particular interests to industry is that by using the ethanol vapours, high-temperature drying can be used, thus the wood could be dried at a fast-drying rate (Pang 2006).

Figure 4 shows the drying defects of end check, honeycomb and deformation that occur on OPL.

From the *scanning electron microscopy* (SEM) image analysis in Fig. 5b, ethanol liquid at 85% concentration can stabilize the vessel and fibre structure and this might hinder the collapse, and deformation occurs after soaking process, compared to normal oven dry without ethanol (Fig. 5a). From the observation, most of the penetration occurred through the parenchyma cells. According to Tomimura (1992), parenchyma in OPT is soft and less strength compared to vascular bundles. Besides, parenchyma is attached with the centre core which easily allows the penetration of the chemical. Conversely, the structure vascular bundles of OPT are

Fig. 4 Drying defect of end check, honeycomb and deformation

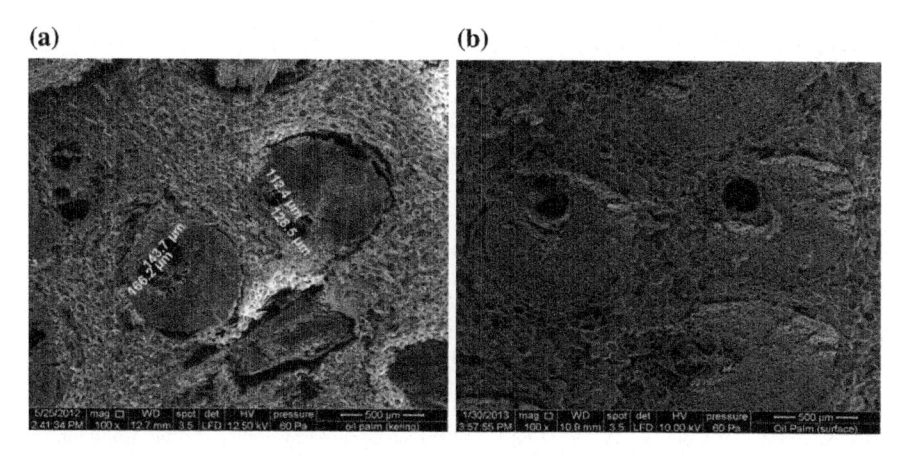

Fig. 5 *Scanning electron microscopy* (SEM) analysis of OPL. **a** Untreated OPL, **b** treated OPL

hard and the density area is rich in vascular bundles, like the central region which is slightly rich in parenchyma and very soft. This suggests that the structure can easily collapse and deform to abnormal shape. From the microscopic inspection (Fig. 5b), it is evident that the ethanol spread well over the surfaces and cell wall. It is possible that ethanol links react with hydroxyl groups in the OPT to make the structure more harden and would reduce the lignin content and crystalline of cellulose and increase surface area (Alam et al. 2005).

4 Conclusion

This study observed the OPL treated with ethanol and untreated OPL, and three types of defects were detected which were end check, honeycomb and deformation. Consequently, this study discovered that OPT treated with ethanol had lower defects and minimised drying time. At high ethanol concentration, the Margo strands were prevented from collapsing, and thus, the pit was not aspirated. This contributed to a faster and more efficient method of drying OPL. Furthermore, the drying defects were significantly reduced at high ethanol concentration of 85%.

Further studies to investigate the drying properties of OPT using ethanol solvent are recommended. These studies may explore the optimum ethanol concentration and use different OPT portions such as the top, middle and bottom stems. Apart from that, various parameters may also be applied.

Acknowledgements Praise to the Almighty Allah the most gracious and the most merciful, for His blessing that gives us health either physical or mental to finish this research. Our appreciation is extended to the financial support of Dana Kecemerlangan (011000120020) UiTM Pahang. We are blessed with having very dear friends who stood by us throughout our study and research writing.

References

Ahmad Fauzi, O., Edi Suhaimi, B., Zaidon, A., Shaikh Abdul Karim, Y., Saadiah, S., & Shafie, A. (2012). Analysis drying defect of oil palm trunk. In *Proceedings of 4th International Symposium of Indonesian Wood Research Society*, Makassar (pp. 44–49).

Ahmad Fauzi, O., Shaikh Abdul Karim, Y., Saadiah, S., Shafie, A., Junaiza, A. Z., Nur Hannani, A. L., et al. (2013). Study on drying oil palm trunk with ethanol. Prosiding Konferensi Akademik Universiti Teknologi MARA. Bukit Gambang Resort City, Pahang, (pp. 127–133).

Alam, M. Z., Muhammad, N., & Mahmat, M. E. (2005). Production of cellulase from oil palm biomass assubstrate by solid state bioconversion. *American Journal of Applied Sciences, 2*.

Aminah, S. (2017). "Kayu sawit tahan lasak," Kosmo, January 3, 2017.

Anis, M., Kamarudin, H. & Soon, L. W. (2007). Challenges in drying of oil palm wood. In *PIPOC2007 Proceedings of the International Palm Oil Congress: Empowering Change*. Kuala Lumpur.

Comstock, G. L., & Cote, W. A., Jr. (1968). Factors affecting permeability and pit aspiration in coniferous sapwood. *Wood Science and Technology, 2*, 279–291.

Edi Suhaimi, B., Mohd Hamami, S., & H'ng, P. S. (2008). Anatomical characteristics and utilization of oil palm wood. In T. Nobuchi & S. Mohd Hamami (Eds.), *The formation of wood in tropical forest trees—A challenge from the perspective of functional wood anatomy* (pp. 161–178). Serdang: Universiti Putra Malaysia.

Ellwood, E. L., & Ecklund, B. A. (1961). Treatment to improve wood permeability as an approach to the drying problem. Wood technologiest, forest products laboratory, Richmond, California.

Othman, S., Nurjannah, S., Rokiah, H., Lili Hanum, M. Y., Razak, W., Nor Yuziah, M. Y., et al. (2009). Evaluation on the suitability of some adhesives for laminated veneer lumber from oil palm trunks. *Journal of Material and Design, 30*(9), 3572–3580.

Pang, S. (2006). Using methanol and ethanol vapours as drying media for producing bright colour wood in drying of radiate pine. In *15th International Drying Symposium (IDS 2006)*. Budapest, Hungary, August 20–23, 2006.

Petty, J. A., & Puritch, G. S. (1970). The effects of drying on the structure and permeability of the wood of *Abies grandis*. *Wood Science and Technology, 4*, 140–154.

Razak, W., Sulaiman, O., Nurjannah, S., Rokiah, H., Lili Hahum, M. Y., Nor Yuziah, M. Y., et al. (2008). Evaluation on the suitability of some adhesives for laminated veneer lumber from oil palm trunks. *Journal of Materials and Design, 30*(9), 3572–3580.

Srivaro, S., Chaowana, P., Matan, N., & Kyokong, B. (2014). Lightweight sandwich panel from oil palm core and rubberwood veneer face. *Journal of Tropical Forest Science, 26*, 50–57.

Terazawa, S. (1965). *An easy method for the determination of wood drying schedule*. Tokyo: Wood Technological Association of Japan.

Terazawa, S., & Tsutsumoto, T. (1976). *Wood drying*. Tokyo: Wood Technological Association of Japan.

Tomimura, Y. (1992). Chemical characteristics and utilization of oil palm trunk. *JARQ, 25*, 283–288.

Yusoff, N. M., & Othman, A. F. (2016). A method for drying monocotyledon lumber. I.P.C.O Malaysia. Malaysia, Abdul Rahman R.

The Effect of Fly Ash and Bottom Ash Pile in Problematic Soil Due to Liquefaction

Mohd Ikmal Fazlan Rozli, Juhaizad Ahmad, Mohd Asha'Ari Masrom, Syahrul Fithri Senin and Abdul Samad Abdul Rahman

Abstract Generally, a construction structure is located on a by soil preferably a good and sound soil properties. However due to limited location of a good and sound soil properties, engineers were challenge by constructing a structure on a problematic soil condition. Mostly, the engineer will be facing building a structure on a loose sandy soil and worst it has a higher chance of liquefaction. This sandy soil can cause the problem to the structure of the building especially if the bearing capacity of the soil is not adequate. For example, the settlement, cracking will happen due to the earthquakes phenomena and worst-case scenario the building will have to undergo a liquefaction attack aftershock. Thus, to overcome this serious problem, the soil properties must be improved by a suitable method such as soil stabilization. The objective of this research is to assess the effectiveness of using fly ash and bottom ash on soil stabilization in liquefaction condition. In this research, piling was used to increase the density of the soil. This will help to reduce the number of settlement during liquefaction process. The amount of cement (5% of soil weight) and fly ash with bottom ash is used with different percentages (10, 20, 30, 40%). The sand will be mixed with the bottom ash and fly ash in different percentages. After conducting a series of tests, it is concluded that 30% of bottom and fly ash give a better result in reducing the settlement value than other mixture.

Keywords Bottom ash · Fly ash · Soil liquefaction · Pile

M. I. F. Rozli (✉) · J. Ahmad · M. A. Masrom · S. F. Senin
Faculty of Civil Engineering, Universiti Teknologi MARA,
13500 Permatang Pauh, Penang, Malaysia
e-mail: ikmal601@ppinang.uitm.edu.my

A. S. A. Rahman
Faculty of Civil Engineering, Universiti Teknologi MARA,
40450 Shah Alam, Selangor, Malaysia
e-mail: kempass@hotmail.com

© Springer Nature Singapore Pte Ltd. 2018
R. Saian and M. A. Abbas (eds.), *Proceedings of the Second International Conference on the Future of ASEAN (ICoFA) 2017 – Volume 2*,
https://doi.org/10.1007/978-981-10-8471-3_79

1 Introduction

Nowadays, there is a lot of high-rise building construction being built in all over the world, for example PETRONAS Twin Tower in Malaysia and Burj Khalifa in Dubai. Normally in this big and megastructure project, the engineers will be facing different types of soil condition in different places but in the same site construction. That is not all, due to the unique plateau of our earth, the non-seismic zone country now is experiencing almost regular seismic activity. This brutal disaster has destroyed so many buildings and claimed many of human lives. The condition of aftershock can be as brutal as during shocking activity. What happens usually aftershock is that the water contained under the loose soil will rise up to the surface, thus creating a mud flood or known as soil liquefaction. Soil liquefaction can be described as natural phenomenon when saturated or partially saturated soil suddenly loses its strength and also its stiffness. This phenomenon occurs because of the applied stress. Regularly, earthquakes shaking or other sudden change in anxiety condition will cause the soil to a fluid structure condition. Soil liquefaction also can be considered as one of the major causes of instability to the buildings after construction and also structures during earthquakes (Huang and Yu 2013). To counter that several methods like ground improvement and modification technique can be used for soft soil rehabilitation (Sa'adon 2009). Recently, stabilizing methods using fly ash and bottom ash are gaining acceptances in geotechnical field. This is because fly ash and bottom ash is more cheap and environmental-friendly.

2 Liquefaction of Soil

Normally, earthquake will cause soil liquefaction to happen as shown in Fig. 1. Soil liquefaction is defined as a phenomenon wherein a mass of soil loses its large percentage of shear resistance due to subjected loading in either monotonic or cyclic forms (Terzaghi and Peck 1967). This phenomenon occurs in because of the applied stress. Normally, earthquake shaking or other sudden change in stress condition will causes the soil to become like a liquid form. Soil liquefaction also can be considered as one of the major causes of instability to the buildings after construction and also structures during earthquakes (Huang and Yu 2013). When liquefaction occurs, the shear strength of the soil will decrease to nearly zero. This will cause the soil to lose its strength and unable to support the foundation and give a negative impact to the whole structure such as buildings. The building could sink into the ground. Although earthquake shaking gives more damage to the structure, liquefaction also will give damage to underground pipeline, harbour facilities and roads or highway surface (Perkins 2001). Normally, the earthquakes phenomena will cause the soil liquefaction to occur. This liquefaction occurs is coming from the pressure of the earthquake disaster. The liquefaction normally happens on saturated soil. This soil is completely filled with a lot of water. This earthquake disaster that causes a rapid big scale of shaking makes the sand grains compress the spaces that

Soil liquefaction

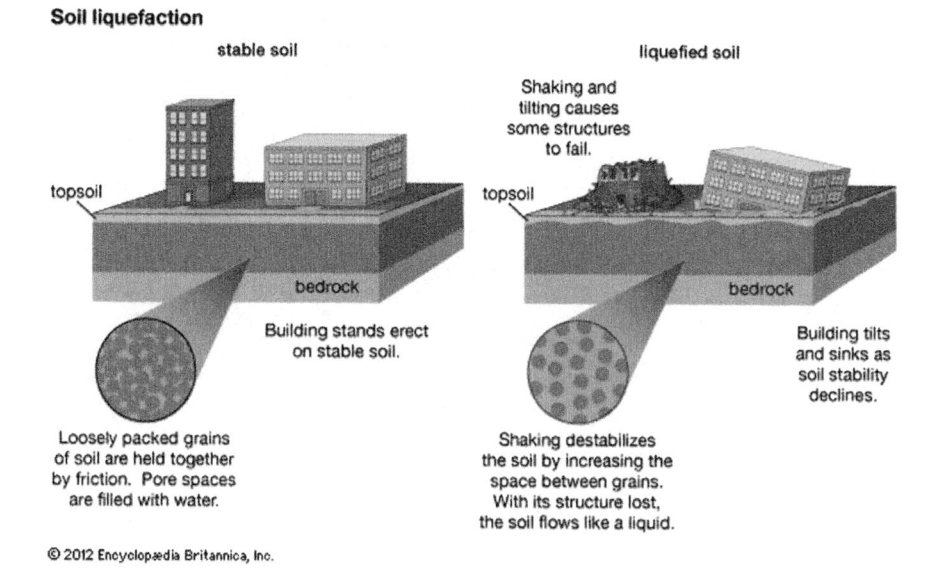

Fig. 1 Behaviour of soil during loading and unloading of specimen (Britannica Encyclopedia 2012)

were filled with water, but suddenly the water will pushing it back and the sand particles will float in the water (Marto 2016).

3 Fly Ash

Fly ash is one of the natural products from the coal combustion process. The fly ash is a material that is nearly same as volcanic ash. These elements mainly consist of iron, silica and Alumina (Upadhyay 2007). The volume of fly ash would increase as the demand for power increases. This substance will make the concrete stronger and durable than normal concrete made with Portland cement. Fly ash particles are almost totally spherical in shape. This will allow them to flow freely in any mixtures. That capability is one of the properties making fly ash a desirable admixture for concrete. Nowadays, fly ash has been used in many infrastructure projects, such as roads, highways, building and bridges. This can help to reduce the cost. Fly ash can be used as an alternative to conventional materials in the construction of geotechnical and geo-environmental infrastructure (Makusa 2012). Figure 2 shows a sample of fly ash.

Fig. 2 Fly ash

4 Bottom Ash

Bottom ash is the coarser component of coal ash, comprising about 10% of the waste. Rather than floating into the exhaust stacks, it settles to the bottom of the power plant's boiler. Bottom ash also makes a useful construction material. The European Coal Combustion Products Association estimates the use of bottom ash in the construction industry at 46% and the use of fly ash at 43%. Bottom ash applications include filler material for structural applications and embankments, aggregate in road bases, sub-bases, pavement, and lightweight concrete products, as feedstock in the production of cement. These particles are quite porous and look like volcanic lava. If the bottom ash is competent, it can be used as lightweight aggregate which can be mixed in concrete to form concrete blocks. The concrete blocks will be much lighter and just as strong. The chemical makeup of fly and bottom ash varies significantly and is dependent on the source and composition of the coal being burned. Figure 3 shows a sample of bottom ash.

Fig. 3 Bottom ash

5 Methodology

5.1 Soil Classification

The sandy soil was used as the main sample in this study. For this soil classification, two (2) tests were conducted, namely sieve analysis test and particle density test. The sandy soil was collected at a location in Butterworth. The sieve analysis test will be conducted on this sandy soil to determine its properties. The sieve analysis was conducted by sieving the soil sample by using sieve pan. The particle density test is used to determine the mass of the soil sample. Figure 4b and 4b depicts the apparatus for sieve and penetration test, respectively.

5.2 Shaking Table Test

The mechanical testing will be conducted in this research to determine the settlement of the soil sample and the condition of the weight applied. The shaking table will simulate the actual earthquake wave. The reaction of soil sample can be determined by setting the parameters into three conditions. The condition is only sand with loading applied, sand with formation of groundwater table and with loading applied. The third conditions are with the present of the piling full with

Fig. 4 a Sieve apparatus, **b** penetration test apparatus

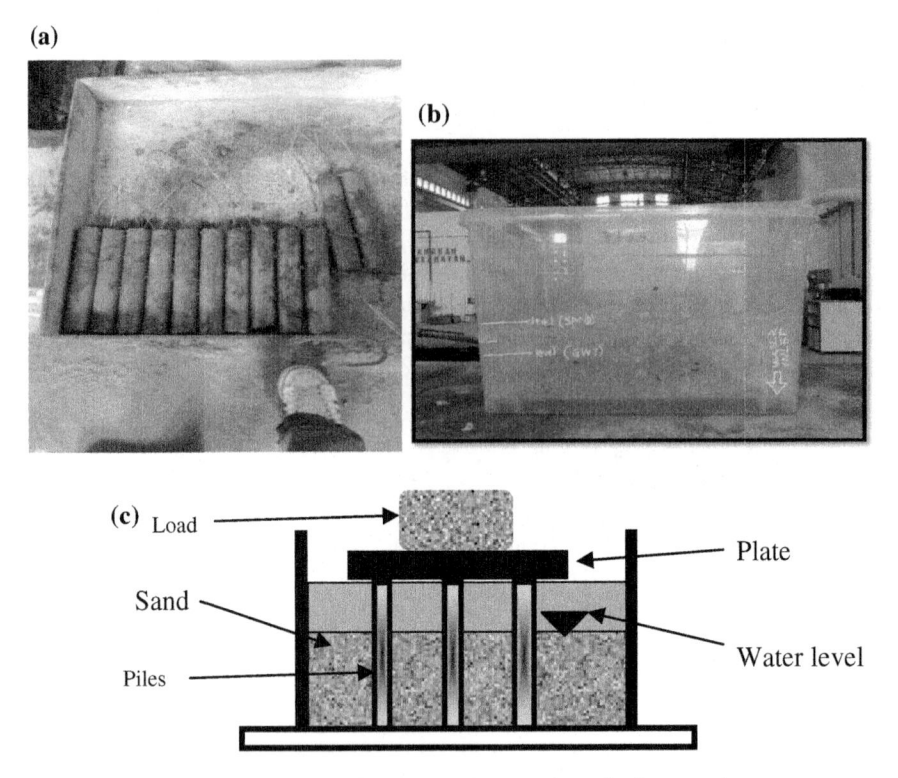

Fig. 5 a Sample of pile, b soil liquefaction container, c schematic diagram of test

concrete mixes with bottom ash and fly ash into the sand with load is applied. Figure 5a shows the sample of pile used in the test. Figure 5b shows the containment used to replicate the condition of soil liquefaction. Figure 5c shows the schematic diagram of the soil liquefaction set-up.

6 Results

6.1 Soil Classification

After conducting a number of tests, it is concluded that the condition of soil is as per researcher expected. It is not a well-graded sand; thus, the sample is applicable to this research. The results were as shown in Fig. 6. By using some theoretical calculation, it is further found out that the coefficient of gradient (Cg) of the soil sample is 3.6, thus categorized it in not well-graded sand.

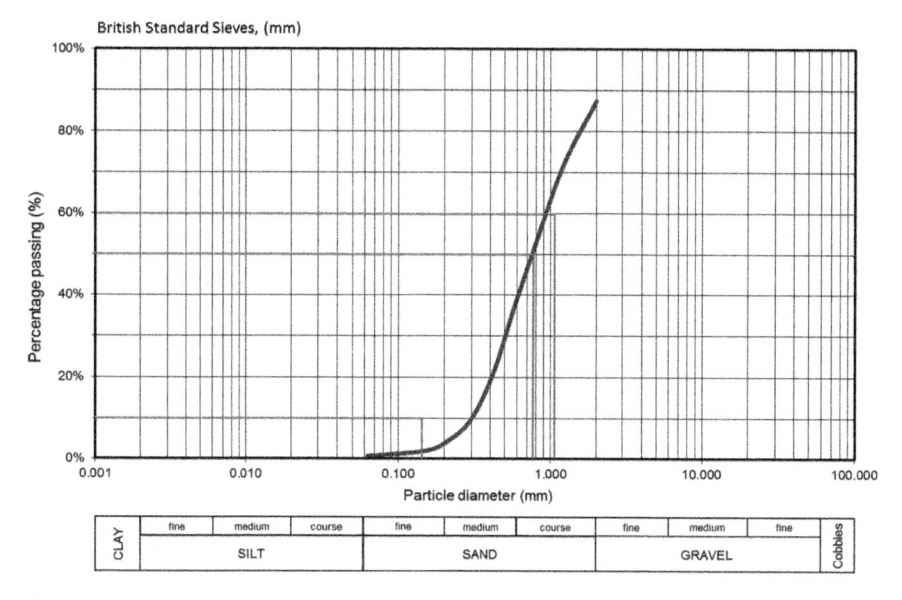

Fig. 6 Sieve graph analysis

6.2 Shaking Table Test

From this test, six points were considered as point that will be measured the settlements of the soil during liquefaction happen. There are third conditions that are considered when the result was taken. Piling A consists of 10% fly ash and bottom ash with cement. Piling B consists of 20% fly ash and bottom ash with cement. Piling C consists of 30% fly ash and bottom ash with cement. Piling D consists of 40% fly ash and bottom ash with cement. The result for settlement is shown in the table.

The first condition is with presence of loading but without the formation of piling and water table called as control variable. The second situation is with the presence of piling but without presence of groundwater table. The third situation is loading with the presence of piling and groundwater table. For the first conditions, the settlements that happen were around 19 mm.

For the second conditions, the settlement value when piling A is applied during testing is about 16 mm. The second value when piling B is tested is 12 mm. The settlement value when piling C is applied is 10 cm. For piling D, the value of settlement is 8.5 mm.

The third conditions, when piling A is applied is 16 mm. The settlement's value when piling B is applied is 13.4 mm. The settlements for piling C tested are 10 mm. The value of settlement is decreasing due to the increasing strength of the piling. The value of piling D is 13 mm. All the results are best viewed in Fig. 7.

	A (10%FA+BA)	B (20%FA+BA)	C (30%FA+BA)	D (40%FA+BA)
Control Condition (without piling and water)	19			
Second Condition (Piling and without water)	16	12	10	8.5
Third Condition (Piling and water)	16	13.4	10	13

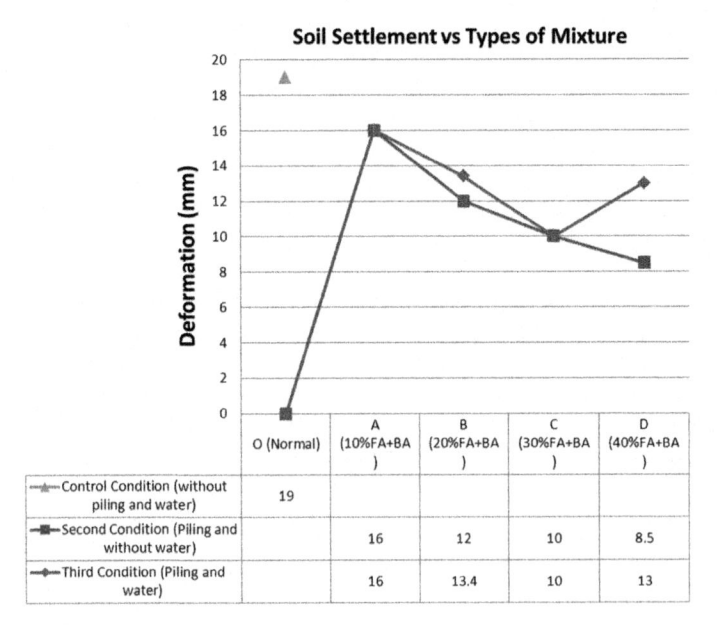

Fig. 7 Graph of soil settlement

7 Discussion and Conclusion

From the data collected in shaking table test, it showed that the settlement values obtained for 30% of fly ash and bottom ash are the lowest than other percentage. From this statement, it can be concluded that the percentage of ash plays an important role in the piling admixture for resisting settlement due to liquefaction. The maximum bearing capacity of soil could be achieved when the correct value of ash is added into mixture. Besides that, at 30% of ash used in mixture, in condition where without water table, the settlement value is 10 mm. This showed the lowest value of settlement from all other percentages of admixtures. Based on the result obtained, it can concluded from two (2) experimental works for sieve analysis test and particle density test has proved that this soil is a poorly graded SAND.

Acknowledgements The authors would like to thank the Research Management Institute, University Teknologi MARA, Malaysia, and the Ministry of Higher Education, Malaysia, for the funding [Ref. No 600-RMI/RAGS 5/3 (179/2014)] this research work under Research Acculturation Grant Scheme (RAGS). The authors also gratefully acknowledge the helpful comments and suggestions of the reviewers to improve the content of this paper. Nevertheless, the authors want to express their gratitude to the technicians of Heavy Structures Laboratory, Faculty of Civil Engineering, UiTM for conducting this research work successfully.

References

Britannica Encyclopedia. (2012). Retrieved December 11, 2006 from https://global.britannica.com/science/soil-liquefaction.

Huang, Y., & Yu, M. (2013). Review of soil liquefaction characteristics during major earthquakes of the twenty-first century. *Natural Hazards, 65*(3), 2375–2384.

Makusa, G. P. (2012). *Soil stabilization methods and material in engineering practice* (E. a. N. r. e. Department of Civil, Division of Mining and Geotechnical Engineering, Trans.). Sweden: Luleå University of Technology, Luleå.

Marto, A. (2016). Liquefaction potential of Nusajaya City. Faculty of Civil Engineering, Universiti Teknologi Malaysia.

Perkins, J. B. (2001). The real dirt on liquefaction—A guide to the liquefaction hazard in future earthquakes affecting the San Francisco bay area, Association of Bay Area Governments.

Sa'adon, M. F. B. (2009). *Determination of basic soil properties and shear strength of Pekan soft clay* (F. o. C. E. Resources, Trans.). Universiti Malaysia Pahang.

Terzaghi, K., & Peck, R. B. (1967). *Soil mechanics in engineering practice*. New York: Wiley.

Upadhyay, A., (2007). Characterization and utilization of fly ash. (Bachelor of Technology), National Institute of Technology Rourkela, Orissa-769008.

A Method for the Full Automation of Euler Deconvolution for the Interpretation of Magnetic Data

Nuraddeen Usman, Khiruddin Abdullah and Mohd Nawawi

Abstract The conventional Euler deconvolution methodology requires the user to choose appropriate structural index (SI) in order to compute the source parameters, and this makes the operation tedious and time consuming. To solve the mentioned problem, a method based on Euler's homogeneity relation for full automation of magnetic data interpretation is presented in this paper. The technique uses multiple linear regression (MLR) methodology to estimate background, horizontal coordinate (x_0 and y_0), depth and structural index (SI) simultaneously is prepared and used for this task. The technique involves the use of first-order derivatives, independent of analytic signal (AS) and the derivatives are computed directly from the total field grid. It is fast means of magnetic data interpretation and easy to implement.

Keywords Euler deconvolution · Multiple linear regression · Structural index Automatic interpretation

1 Introduction

Geophysical techniques have been applied to investigate the subsurface of the earth in order to explore geological structures of economic interest (in most cases) in areas of hydrology, solid minerals, hydrocarbons, engineering, geothermal studies, geo-hazard assessment, geochemical and environmental studies (Loke et al. 2013;

N. Usman (✉) · K. Abdullah · M. Nawawi
School of Physics, Universiti Sains Malaysia, 11800 Gelugor, Penang, Malaysia
e-mail: nu14_phy055@usm.my

K. Abdullah
e-mail: khirudd@usm.my

M. Nawawi
e-mail: mnawawi@usm.my

Yang et al. 2015). With the aid of the invasive geophysical techniques (Gerovska and Araúzo-Bravo 2003; Gerovska et al. 2010; Cooper 2014; Cooper and Whitehead 2016; Salem et al. 2007), it is possible to determine the horizontal and vertical positions of concealed metallic objects in the near vicinity of the earth's surface in addition to the delineation of deep-seated structures. The conventional Euler deconvolution technique (Thompson 1982; Reid et al. 1990; Mushayandebvu et al. 2004; Ugalde and Morris 2010; Barbosa and Silva 2011; Oruç and Selim 2011; Chen et al. 2014) requires the input of appropriate SI. However, the concept of SI applies to some simple shapes only and the geology of the earth's subsurface is very complex. Attempts were also made to automate the Euler technique (Keating and Pilkington 2004; Salem et al. 2007; Stavrev 1997; Gerovska and Arouzo-Bravo 2003; Gerovska et al. 2010) but each method suffers some drawback.

To solve the mentioned problem, a technique (based on Euler deconvolution relation) that does not require the input of SI is presented in this paper. Euler deconvolution technique has been applied extensively in delineating geologic boundaries (Hsu et al. 1996; Ugalde and Morris 2010; Barbosa and Silva 2011), and locating geothermal sources or hot springs (Nouraliee et al. 2015), and they are usually combined with other geophysical methods to ensure enhanced interpretation of the geology of subsurface.

2 Methodology

Euler deconvolution method is based on Euler's homogeneity relationship, it is first initiated to solve 2D magnetic field by Thompson (1982). Euler deconvolution equation is normally used in order to find the source location (x_0, y_0, z_0),

$$(x - x_0)\frac{\partial F}{\partial x} + (y - y_0)\frac{\partial F}{\partial y} + (z - z_0)\frac{\partial F}{\partial z} = N(B - F) \quad (1)$$

where x, y, z are the observation point coordinates; x_0, y_0 and z_0 are the source locations; $\frac{\partial F}{\partial x}, \frac{\partial F}{\partial y}, \frac{\partial F}{\partial z}$ are the potential derivatives; N is the structural index; B is the background of field F (Thompson 1982). Equation 1 forms the basis for the methodology used in this research.

The multiple linear regression (MLR) method has been extensively applied in statistics to establish relationships among multiple variables (more than two) by fitting a straight line to the observed data. MLR model can be expressed as (Levine et al. 2001):

$$Y_i = \beta_0 + \beta_1 X_{1i} + \beta_2 X_{2i} + \cdots + \beta_k X_{ki} + \varepsilon_i \quad (2)$$

where β_0 is the intercept of Y_i; β_1, β_2, β_k denote the slope of regression lines for each variable, respectively; ε_i is the error term and Y_i is the dependent variable. To apply the MLR on 3D Euler deconvolution technique, Eq. 1 can be written in this form (Eq. 3):

$$x\frac{\partial F}{\partial x} + y\frac{\partial F}{\partial y} + z\frac{\partial F}{\partial z} = NB + x_0\frac{\partial F}{\partial x} + y_0\frac{\partial F}{\partial y} + z_0\frac{\partial F}{\partial z} - NF \quad (3)$$

Equation 3 can be solved using MLR method where $Y_i = x\frac{\partial F}{\partial x} + y\frac{\partial F}{\partial y} + z\frac{\partial F}{\partial z}$, $\beta_0 = NB, \beta_1 = x_0, \beta_2 = y_0, \beta_3 = z_0, \beta_4 = N, X_1 = \frac{\partial F}{\partial x}, X_2 = \frac{\partial F}{\partial y}, X_3 = \frac{\partial F}{\partial z}$ and $X_4 = -F$.

All β's are the coefficients that need to be solved and all X_i values are the independent variables, in which the values are known from the data. Assuming that Eq. 3 is linear, the equation could solve 5 unknowns using MLR: N, B, x_0, y_0 and z_0 (position of source). The equation can be written in the matrix form in order to be solved in linear system.

$$\begin{bmatrix} \frac{\partial F}{\partial x_1} & \frac{\partial F}{\partial y_1} & \frac{\partial F}{\partial z_1} & N \\ .. & .. & .. & .. \\ .. & .. & .. & .. \\ \frac{\partial F}{\partial x_n} & \frac{\partial F}{\partial y_n} & \frac{\partial F}{\partial z_n} & N \end{bmatrix} \begin{bmatrix} x_0 \\ y_0 \\ z_0 \\ B \end{bmatrix} = \begin{bmatrix} x_1\partial F/\partial x_1 & y_1\partial F/\partial x_1 & z_1\partial F/\partial x_1 & NF_1 \\ .. & .. & .. & .. \\ x_n\partial F/\partial x_n & x_n\partial F/\partial y_n & x_1\partial F/\partial z_n & NF_n \end{bmatrix} \quad (4)$$

2.1 Buried Concrete Wall Model

This model (Fig. 1) is simulated to estimate the magnetic response of the field model used in this study, as such the modelling parameters of the former are the same as that of the latter. The data of this model is synthesized using the equation given by Bhattacharyya (1964). The inclination and declination of the model were 67.3° and −1°, respectively; the magnetization intensity was 0.1632 A/m. The depth to top of the targets/objects was 0.6 m and oriented horizontally, the width and height of the objects were 0.45 and 0.4 m, respectively, the objects were 4-m long. The inversion (Eq. 3) was carried out using the window size 5. The filtering parameters were 5 and 0.4 for the convolution window and deviation of structural index, respectively. The acceptable regression error was 50% and threshold of analytic signal was 14.

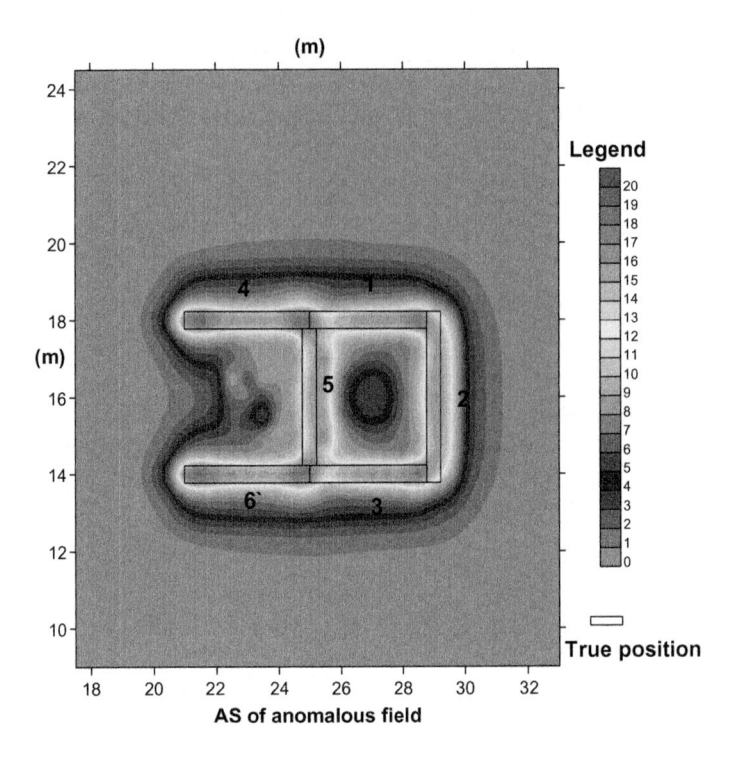

Fig. 1 Analytic signal of magnetic field (buried concrete wall model)

2.2 Field Model

The total magnetic field intensity is obtained from one of the two test sites of Near Surface Geophysics Group of the Geological Society of London located at Leicester University. The data and other relevant information can be retrieved at http://www.nsgg.org.uk/test-sites/. It is located at latitude 52° 36′ 27″ N and longitude 1° 5′ 12″ W Southmeads road, Oadby, Leicester. The site can be characterized as gentle hill-top with top layer of about 0.3-m thick. The boulder clay material is distributed throughout the site; it is 16–18 m thick underlined by liassic clays and limestone. One of the areas selected from the site (Area 4), consisting of six concrete walls with different composition is used for this study. The targets are 4-m long (along box 2) and 8-m wide (along boxes 3 and 6) in the north and east direction, respectively; it is designed to simulate magnetic response of buried concrete wall (Fig. 2). The depth to bottom of the targets/objects was 1 m and oriented horizontally, the width and height of the objects were 0.45 and 0.4 m, respectively (Chambers et al. 2002).

A trench was built and fills with some building materials such as modern concrete, corbelled brick, aggregate, stone, peat and sand. The buried objects (Fig. 2) are non-reinforced casted concrete (1), engineering bricks (2), stone blocks (3), aggregate of concrete and engineering block (4), sand used for building purpose (5), moulding material containing peat (decay vegetable) and leap (6).

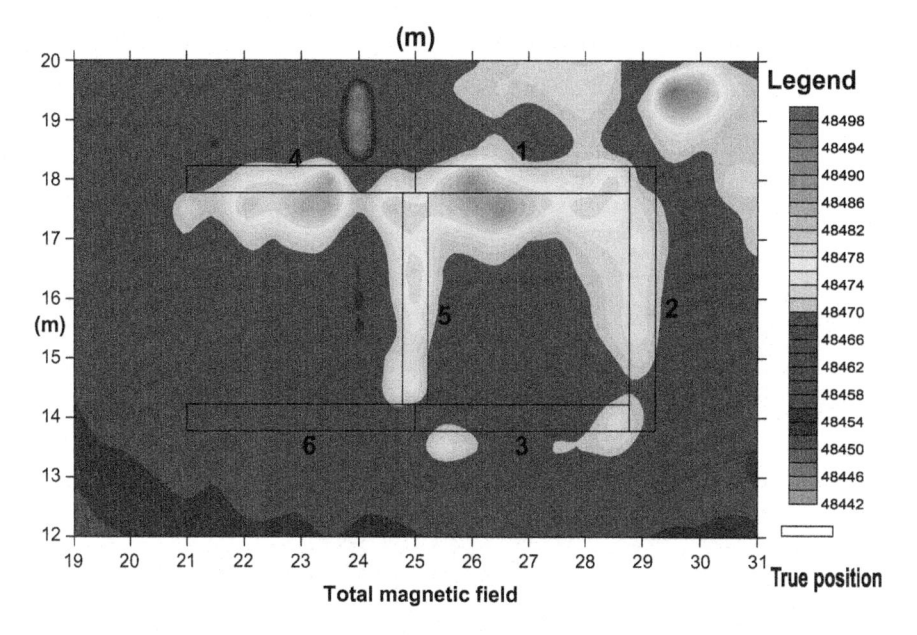

Fig. 2 Total magnetic field intensity of the field model

The data were obtained using two sets of sensor that recorded both the total field of lower and upper console, and the gradiometer reading simultaneously. The distance between the lower console and the ground surface was 30 cm above the ground; and the station spacing was 0.5 m. The total magnetic field data is presented as contour map (Fig. 2). The vertical and horizontal derivatives were computed from the total magnetic field intensity data, analytic signal is also computed. The data is inversed using Eq. (3) using the window size of 5. The output of the inversion was filtered using the convolution window of 5 and deviation of structural index was 0.4. The error involved in the estimate of unknown parameters for both real and synthetic data can be obtained using Eq. (5), where ρ_i and $\hat{\rho}_i$ are the true and mean estimated parameters.

$$\text{Error} = \left|\frac{\rho_i - \hat{\rho}_i}{\rho_i}\right| \times 100 \qquad (5)$$

3 Result and Discussion

3.1 Buried Concrete Wall Model

The inversion result of synthetic model parameters (Table 1) has low error when compared with the true parameters especially on position estimates. The average value of the depth solutions obtained coincided with depth to the top of the target

Table 1 Inversion result of synthetic model after filtering (number of solutions in the parenthesis)

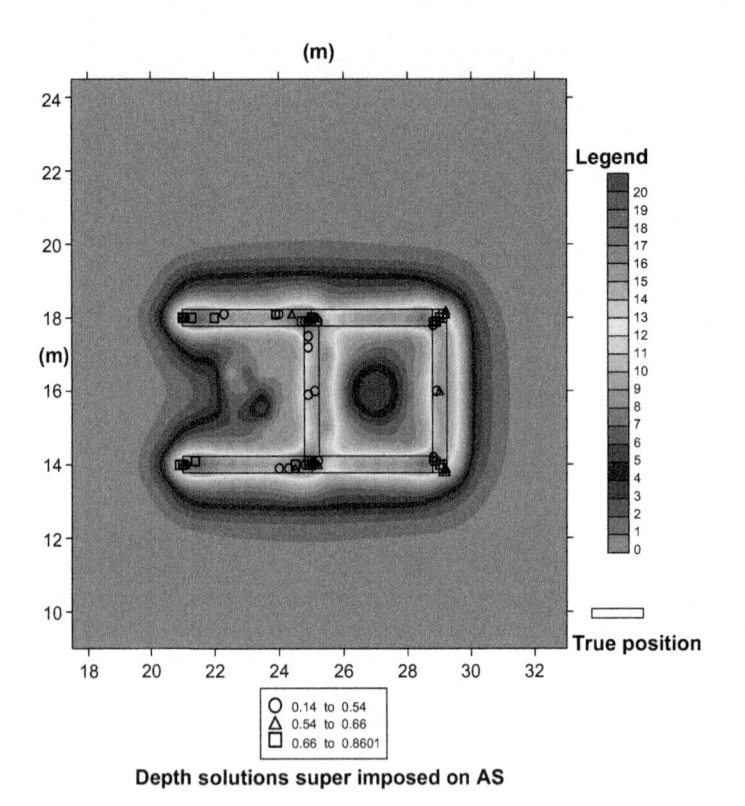

Structure	x (m)	x_0 (m)	y (m)	y_0 (m)	z (m)	z_0 (m)	N	N_0
Concrete wall (97)	21–28.775	20.9–29.2	13.775–18.225	13.8–18.2	0.6	0.14–0.86	–	0.1–2.27

Fig. 3 Depth solution of concrete wall model superimposed on analytic signal

with error of about 2.4%. For the structural index, the average value indicated that the target is more of dike structure. This result has demonstrated that the present technique can be used to delineate structures in geotechnical investigation with reliable position and depth estimates. The estimated depth values are superimposed on AS and presented in Fig. 3. In normal practice, reduction to the pole (RTP) is applied to the magnetic data (Thompson 1982) prior to the use of Euler deconvolution. However, in this research, the new technique estimated the unknown parameters with good accuracy without applying reduction to the pole

3.2 Field Model

The result obtained after the filtering indicated that the horizontal position in x-direction varies from 20.8 to 29.2 m while the estimate along y-direction is from 13.3 to 18.8 m (Fig. 4). The true horizontal extent of the model along x- and y-directions was 8.225 and 4.45 m, respectively. The error recorded for the estimation of horizontal extent (x_0 and y_0) of the target along x is smaller (2.1%) than along y (23%) (Table 2). According to Breiner (1973), the maximum acceptable error for the estimate of locations is 30%. The estimated depth obtained after the filtering varies from 0.31 to 2.18 m (Table 2 and Fig. 4), and the average value is 0.87 m. The average depth solutions (0.87 m) coincided with depth to the centre of the buried targets which is 0.8 m (depth to the bottom and depth to the top of the target were 1 and 0.6 m, respectively) with the error of about 9%. The shallow depth solutions are due to filling material used to cover the target while the deeper sources were partly due to concrete used to build the trench. The majority of the solutions obtained are due to the buried targets, that is why the average depth value is closer to the actual depth to the centre of the target. Because of the unknown SI value of the target (concrete wall) in the geophysical literature and presence of many sources involved in this model, the interpretation of estimated structural index tends to be ambiguous. The average SI of the target obtained is about 0.4, and this value can be attributed to contact between the ends of targets considering the theoretical value of contact is zero.

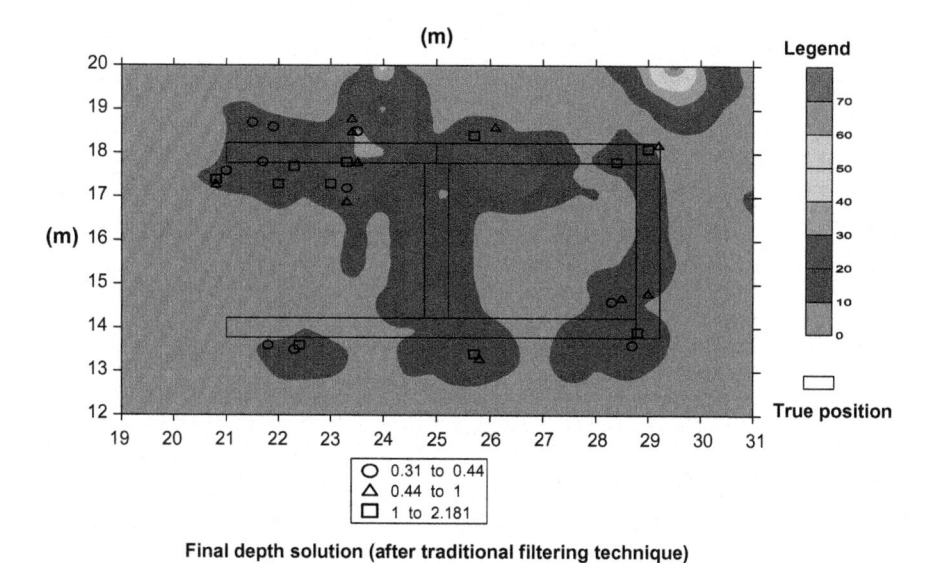

Fig. 4 Depth solutions of buried walls model superimposed on AS

Table 2 Inversion result of field model after filtering (number of solutions in the parenthesis)

Structure	x (m)	x_0 (m)	y (m)	y_0 (m)	z (m)	z_0 (m)	N	N_0
Field model (31)	21–28.775	20.8–29.2	13.775–18.225	20.8–29.2	0.6–1 (centre = 0.8)	0.31–2.18 (mean = 0.87)	–	(mean = 0.4)

4 Conclusion

This research demonstrated that the new Euler deconvolution technique can be applied to engineering/geotechnical investigation and obtain a reliable result. The estimation of position coordinates and SI simultaneously assisted in the interpretation of different magnetic sources in the construction site. The theoretical modelling validated the effectiveness of this technique to be used in any location without reduction to the pole/equator or pseudo-gravity. In addition to the application of Euler deconvolution to the large data set, it can also be used to analyse few data points obtained from small area. It is fast means of magnetic data interpretation and easy to implement.

References

Barbosa, V. C., & Silva, J. B. (2011). Reconstruction of geologic bodies in depth associated with a sedimentary basin using gravity and magnetic data. *Geophysical Prospecting, 59*(6), 1021–1034.

Bhattacharyya, B. K. (1964). Magnetic anomalies due to prism-shaped bodies with arbitrary polarization. *Geophysics, 29*(4), 517–531.

Breiner, S., (1973). Applications manual for portable magnetometers.

Chambers, J., Ogilvy, R., Kuras, O., Cripps, J., & Meldrum, P. (2002). 3D electrical imaging of known targets at a controlled environmental test site. *Environmental Geology, 41*(6), 690–704.

Chen, Q., Dong, Y., Cheng, G. S., Han, L., Xu, H. H., & Chen, H. (2014). Interpretation of fault system in the Tana Sag, Kenya, using edge recognition techniques and Euler deconvolution. *Journal of Applied Geophysics, 109*, 150–161.

Cooper, G. R. (2014). Using the analytic signal amplitude to determine the location and depth of thin dikes from magnetic data. *Geophysics, 80*(1), J1–J6.

Cooper, G. R., & Whitehead, R. C. (2016). Determining the distance to magnetic sources. *Geophysics, 81*(2), J25–J34.

Gerovska, D., & Araúzo-Bravo, M. J. (2003). Automatic interpretation of magnetic data based on Euler deconvolution with un-prescribed structural index. *Computers & Geosciences, 29*(8), 949–960.

Gerovska, D., Araúzo-Bravo, M. J., Stavrev, P., & Whaler, K. (2010). MaGSoundDST—3D automatic inversion of magnetic and gravity data based on the differential similarity transform. *Geophysics, 75*(1), L25–L38.

Hsu, S. K., Sibuet, J. C., & Shyu, C. T. (1996). High-resolution detection of geologic boundaries from potential-field anomalies: An enhanced analytic signal technique. *Geophysics, 61*(2), 373–386.

Keating, P., & Pilkington, M. (2004). Euler deconvolution of the analytic signal and its application to magnetic interpretation. *Geophysical Prospecting, 52*(3), 165–182.

Levine, D. M., Ramsey, P. P., & Smidt, R. K. (2001). *Applied statistics for engineers and scientists: Using Microsoft Excel and Minitab*. London: Pearson.

Loke, M. H., Chambers, J. E., Rucker, D. F., Kuras, O., & Wilkinson, P. B. (2013). New developments in the direct-current geoelectrical imaging method. *Journal of Applied Geophysics, 95*, 135–156.

Mushayandebvu, M. F., Lesur, V., Reid, A. B., & Fairhead, J. D. (2004). Grid Euler deconvolution with constraints for 2D structures. *Geophysics, 69*(2), 489–496.

Nouraliee, J., Porkhial, S., Mohammadzadeh-Moghaddam, M., Mirzaei, S., Ebrahimi, D., & Rahmani, M. R. (2015). Investigation of density contrasts and geologic structures of hot springs in the Markazi Province of Iran using the gravity method. *Russian Geology and Geophysics, 56*(12), 1791–1800.

Oruc, B., & Selim, H. (2011). Interpretation of magnetic data in the Sinop area of Mid Black Sea, Turkey, using tilt derivative, Euler deconvolution, and discrete wavelet transform. *Journal of Applied Geophysics, 74*, 194–204.

Reid, A. B., Allsop, J. M., Granser, H., Millett, A. T., & Somerton, I. W. (1990). Magnetic interpretation in three dimensions using Euler deconvolution. *Geophysics, 55*(1), 80–91.

Salem, A., Williams, S., Fairhead, D., Smith, R., & Ravat, D. (2007). Interpretation of magnetic data using tilt-angle derivatives. *Geophysics, 73*(1), L1–L10.

Stavrev, P. Y. (1997). Euler deconvolution using differential similarity transformations of gravity or magnetic anomalies. *Geophysical Prospecting, 45*(2), 207–246.

Thompson, D. T. (1982). EULDPH: A new technique for making computer-assisted depth estimates from magnetic data. *Geophysics, 47*(1), 31–37.

Ugalde, H., & Morris, W. A. (2010). Cluster analysis of Euler deconvolution solutions: New filtering techniques and geologic strike determination. *Geophysics, 75*, L61–L70.

Yang, J., Agterberg, F. P., & Cheng, Q. (2015). A novel filtering technique for enhancing mineralization associated geochemical and geophysical anomalies. *Computers & Geosciences, 79*, 94–104.

Enhancing Ride Comfort of Quarter Car Semi-active Suspension System Through State-Feedback Controller

Muhamad Amin Zul Ifkar Mohd Fauzi, Fitri Yakub,
Sheikh Ahmad Zaki Shaikh Salim, Hafizal Yahaya,
Pauziah Muhamad, Zainudin A. Rasid, Hoong Thiam Toh
and Mohamad Sofian Abu Talip

Abstract The objective of this study is to simulate the road disturbance toward suspension in quarter car system. Suspension consists of the system of springs, shock absorbers, and linkages that connects a vehicle to its wheel and allows relative motion between the car body and the wheel. This paper shows the mathematical modeling in order to design the quarter car suspension system using Simulink and MATLAB software. The work shows the effect of suspension travel in quarter car system toward road profile by using state-feedback controller. The state-feedback controller's purpose is to decrease the continuous damping in suspension system. The inconsistency condition of the road is the main element that affects the ride comfort which is in this paper represented by different heights of road profile. In suspension principles, the road wheels and vehicle body produce vertical forces which are rotational motions. Therefore, state-feedback controller

---

M. A. Z. I. M. Fauzi · F. Yakub (✉) · S. A. Z. S. Salim · H. Yahaya
P. Muhamad · Z. A. Rasid · H. T. Toh
Malaysia-Japan International Institute of Technology, Universiti Teknologi
Malaysia, Jalan Sultan Yahya Petra, 54100 Kuala Lumpur, Malaysia
e-mail: mfitri.kl@utm.kl.my

S. A. Z. S. Salim
e-mail: sheikh.kl@utm.kl.my

H. Yahaya
e-mail: hafizal.kl@utm.kl.my

P. Muhamad
e-mail: pauziah.kl@utm.kl.my

Z. A. Rasid
e-mail: arzainudin.kl@utm.kl.my

H. T. Toh
e-mail: ththiam.kl@utm.kl.my

M. S. A. Talip
Faculty of Engineering, University of Malaya, 50603 Kuala Lumpur, Malaysia
e-mail: sofian_abutalip@um.edu.my

© Springer Nature Singapore Pte Ltd. 2018
R. Saian and M. A. Abbas (eds.), *Proceedings of the Second International Conference on the Future of ASEAN (ICoFA) 2017 – Volume 2*,
https://doi.org/10.1007/978-981-10-8471-3_81

must be able to reduce body deflection caused by road disturbance to achieve the ride comfort of driver and passengers. The results show that the proposed controller is capable of reducing the vibration of suspension after experiencing the bumps with different heights.

Keywords Ride comfort · Quarter car model · Semi-active suspension Suspension travel · State-feedback controller · PID controller

1 Introduction

For the past centuries, people used wooden wheel as their support to the body of vehicle with less comfort to the driver and passengers. Nowadays, technologies have increased rapidly, and a lot of researches have been conducted on the car suspension system that enhances the level of satisfaction to the driver and passengers. The structure of a car suspension system separates the car body and the wheel. The system is connected with springs, shock absorbers, and linkages that connects the car body to the wheel which allows relative motion between the two parts of the suspension system (Zhang et al. 2012) and (Sathishkumar et al. 2014).

The rotational motions are needed in every car suspension system. The roles of each motion are important to achieve the ride comfort to driver and passengers. In this paper, the quarter car suspension system understudied has only one relative motion which is the vertical motion that also affects the ride comfort of the driver and passengers. Car suspension system can be categorized into three types of suspension which are the passive, semi-active, and active suspension system (Pekgokgoz et al. 2010). A lot of researches demonstrated that suspension system provides considerable measurements which are contributing to the performance of car's braking and handling (Alvarez-Sanchez 2013).

Most of conventional suspension systems use the passive suspension in commercial vehicles that make the ride comfort to be affected by the road condition (Mou et al. 2015). The semi-active suspension system absorbs the conventional spring element of passive suspension but uses a controllable damper to reduce the vibration of the suspension after the presence of road disturbances. The active suspension system is able to store, dissipate, and apply energy to the suspension system (Ghazaly et al. 2015). Unfortunately, it is difficult for the passive suspension system to decrease the vibration, where a soft spring will allow for excessive oscillation, while a hard spring causes passenger discomfort due to road inconsistencies (Yakub et al. 2016a, b).

This research uses PID controller and state-feedback controller in the control analysis of a quarter car semi-active suspension system. The main objective of controller's design is to reduce the vibration and damping in semi-active suspension system model that are due to road disturbance. The controller must be able to get the best transient response in the quarter car semi-active suspension system. Therefore, the ride comfort can be achieved through the ability of controller.

This paper will be structured as follows. Section 2 describes the mathematical model of quarter car semi-active suspension system. The simulation will be done in MATLAB and Simulink software, and the study of response of the system with the controllers will be conducted. Next in Sect. 3, suitable controller is designed which is state feedback in order to reduce the vibration that occurs in the passive suspension system based on the previous simulations that have been conducted. The results are discussed in Sect. 4. Last but not least, Sect. 5 describes the conclusion and the possible future works.

2 Model of Quarter Car

The model of quarter car in Fig. 1 includes the linear motion of the sprung mass, M_2, which represents the car body with passengers, and the unsprung mass M_1 which corresponds to the mass of the wheel and suspension. The disturbance input w is the road profile, and the x_2 represents the positions of the sprung of the mass, whereas x_1 is the position of the unsprung mass. The dampers are b_2 and b_1, whereas the springs are k_2 and k_1. To design a controller, every subsystem is involved in ensuring the ride comfort of passengers and driver. Therefore, the

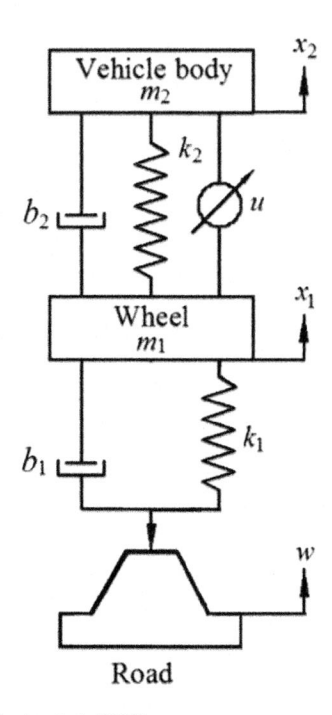

Fig. 1 Quarter car model (Florin et al. 2013)

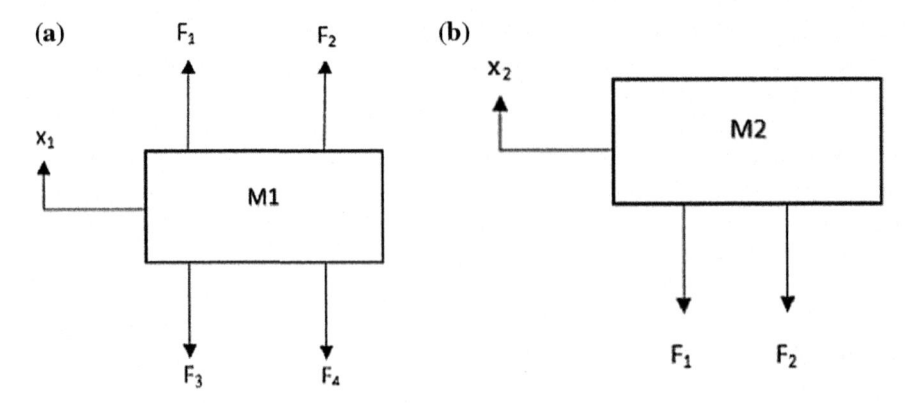

Fig. 2 Free body diagram **a** sprung mass and **b** unsprung mass

controller will be designed to meet this criteria need. The aim of the mathematical modeling is to get a state-space representation of the quarter car model.

The free body diagram of the quarter car model can describe all the contact forces acting on the body of the vehicle and the wheels where the mathematical equations of motion can be obtained. Therefore, the equations will be transferred to the MATLAB and Simulink software in order to be simulated. To derive the mathematical model above, free body diagram is needed such as shown in Fig. 2.

From the free body diagrams, the relationship between the force and displacement is

$$F_1 = b_2(\dot{X}_2 - \dot{X}_1) \tag{1}$$

$$F_2 = k_2(x_2 - x_1) \tag{2}$$

From the body diagram shown in Fig. 2a, the equation can be extracted and rewritten as in Eqs. 1 and 2 by substituting the equation obtained from Eqs. 3 and 4

$$M_i \ddot{X}_2 = -F_1 - F_2 \tag{3}$$

$$M_2 \ddot{X}_2 = -b_2(\dot{X}_2 - \dot{X}_1) - k_2(x_2 - x_1) \tag{4}$$

For unsprung wheel mass, M_1 is generated the body force x_1 and related suspension springs, damper and controller related forces F_1, F_2, F_3, and F_4. The equations formed (Eqs. 5–8) are extracted from the body diagram shown in Fig. 2b.

$$F_3 = b_1(\dot{X}_1 - w) \tag{5}$$

$$F_4 = k_1(x_1 - w) \tag{6}$$

From the free body diagram shown in Fig. 2b, the equation can be extracted and rewritten as in Eqs. 1, 2, 5, and 6 by substituting the equation obtained from Eqs. 7 and 8.

$$M_1\ddot{X}_1 = F_1 + F_2 - F_3 - F_4 \tag{7}$$

$$M_1\ddot{X}_1 = b_2(\dot{X}_2 - \dot{X}_1) + k_2(x_2 - x_1) - b_1(\dot{X}_1 - w) - k_1(x_1 - w) - u \tag{8}$$

The state-space representation of the controlled system of quarter car model in Fig. 1 can be formalized as the following:

$$\ddot{X} = Ax + Bu \tag{9}$$

$$y = Cx + Du \tag{10}$$

where the state vector x is defined as in the following:

$$x = \begin{bmatrix} x_1\dot{X}_1 & x_2 - x_1 & \dot{X}_2 - \dot{X}_1 \end{bmatrix}^T \tag{11}$$

where x_1 is the car body displacement, $\dot{X}_1$ is the car body velocity, $x_2 - x_1$ is the suspension deflection, and $\dot{X}_2 - \dot{X}_1$ is the suspension velocity. The state-space matrices are defined as in the following (Popovic et al. 2011) (Table 1, Fig. 3):

$$A = \begin{bmatrix} 0 & 1 & 0 & 0 \\ \frac{-b_2 b_1}{M_2 M_1} & 0 & \left[\frac{b_2}{M_2}\left(\frac{b_2}{M_2} + \frac{b_2}{M_1} + \frac{b_1}{M_1}\right)\right] & -\frac{b_2}{M_2} \\ \frac{b_1}{M_1} & 0 & -\left[\frac{b_2}{M_2} + \frac{b_2}{M_1} + \frac{-k_1}{M_1}\right] & 1 \\ \frac{k_1}{M_1} & 0 & -\left[\frac{k_2}{M_2} + \frac{k_2}{M_1} + \frac{k_1}{M_1}\right] & 0 \end{bmatrix} \tag{12}$$

$$B = \begin{bmatrix} 0 & 0 \\ \frac{1}{M_1} & \frac{b_2 b_1}{M_2 M_1} \\ 0 & -\frac{b_1}{M_1} \\ \left[\frac{1}{M_2} + \frac{1}{M_1}\right] & -\frac{k_1}{M_1} \end{bmatrix} \tag{13}$$

$$C = \begin{bmatrix} 0 & 0 & 1 & 0 \end{bmatrix} \tag{14}$$

$$D = \begin{bmatrix} 0 & 0 \end{bmatrix} \tag{15}$$

Table 1 Passenger vehicle parameter (Pekgokgoz et al. 2010)

Description	Units	Values
Body (sprung) mass	M_2 (kg)	2500
Axle (unsprung) mass	M_1 (kg)	320
Suspension stiffness	k_2 (N/m)	80,000
Suspension stiffness	k_1 (N/m)	500,000
Tire damping	b_2 (Ns/m)	350
Tire damping	b_1 (Ns/m)	15,020

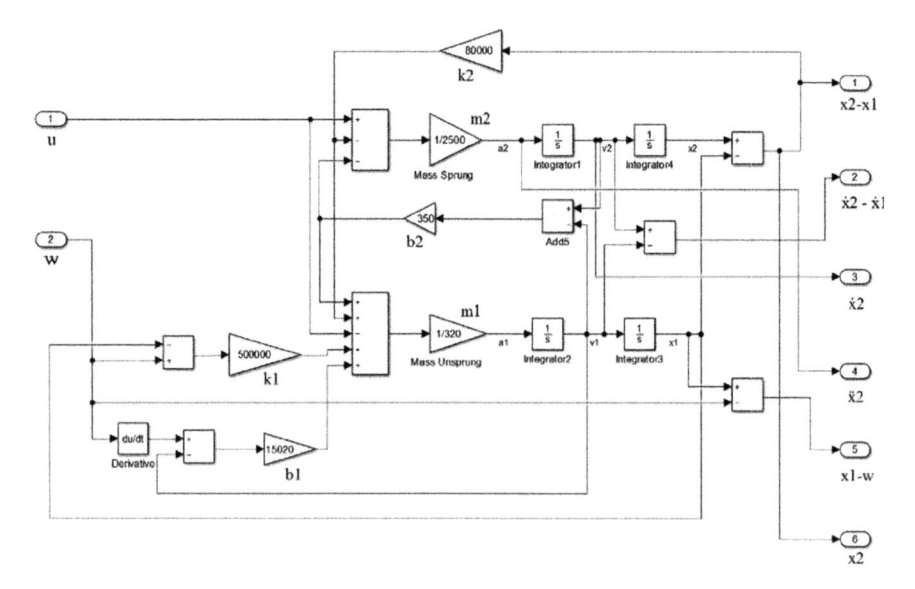

Fig. 3 Simulink block diagram of the quarter car system

3 Controller Design

In this section, the controller is divided into two parts which are the PID and the state-feedback controllers.

3.1 PID Controller

A lot of researches had been conducted using PID controller in quarter car model semi-active suspension system. The PID controller is the most-used feedback control design. PID is a short form for proportional–integral–derivative, i.e., the three terms operating on the error signal to produce a control signal. The PID controllers are the most well-established class of control systems. However, they cannot be used in several more complicated cases such as in the multiple input and multiple output systems. In order to minimize the vibration in a suspension system, the PID equation is used as stated below (Du et al. 2012):

$$u(t) = K_p e(t) + K_i \int_0^t e(\tau) d\tau + K_d \, de(t)/dt \qquad (16)$$

From the mathematical equation above, K_p is denoted as the proportional gain, K_i is the integral coefficients, K_d is the derivative coefficients, and the error signal is

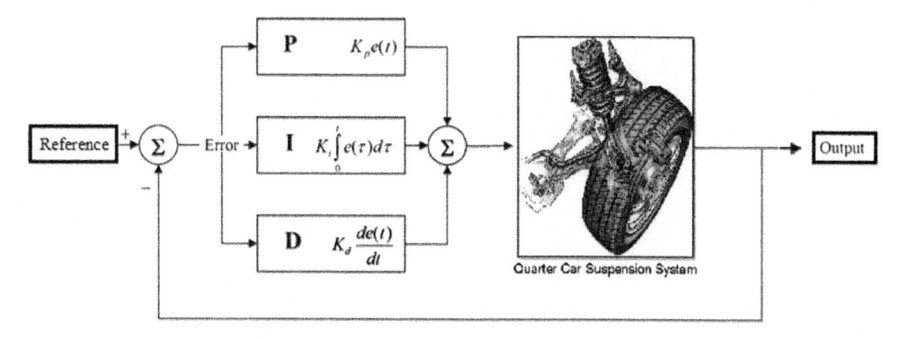

Fig. 4 PID controller system

represented by *e*. For this study, the trial-and-error method is chosen to see the characteristic and function of P (proportional), I (integral), and D (derivative). Next, several tunings and the simulations have been done in MATLAB and Simulink software until the reasonable response is obtained (Fig. 4).

3.2 State-Feedback Controller

State-feedback controller is designed to get the desirable closed-loop performance in terms of both transient and steady-state response characteristics. The arbitrary closed-loop eigenvalue placement via state-space feedback can be achieved if the open-loop state equation is controllable (Yakub et al. 2016a, b). The open-loop system understudied which is the plant of the system is presented by the linear time-invariant state equation stated as below (Du et al. 2012):

$$\dot{X}(t) = Ax(t) + Bu(t) \tag{17}$$

$$y(t) = Cx(t) \tag{18}$$

The result of state-feedback control law is stated as below:

$$u(t) = -Kx(t) + r(t) \tag{19}$$

where K is the constant state-feedback gain matrix ($m \times n$) that gives the closed-loop state equation with the desired performance characteristics.

The value of K as the state-feedback gain will be used in the quarter car semi-active suspension in state-feedback controller as shown as below (Fig. 5).

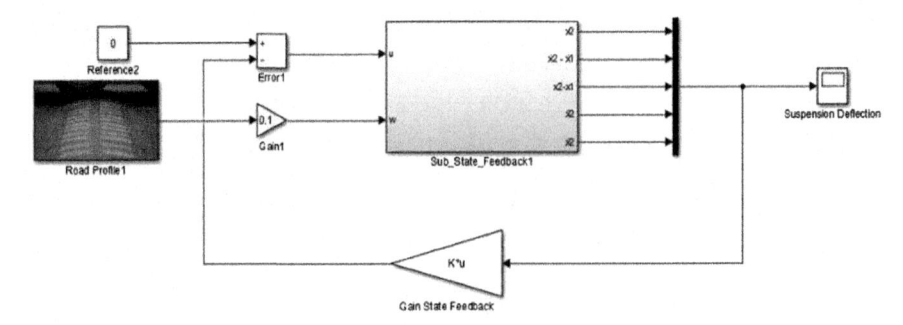

Fig. 5 Simulink block diagram of state-feedback controller design

4 Result and Discussion

The simulation has been run using MATLAB and Simulink software. The parameters that will be observed are focused on the suspension deflection. The goal of the results is to achieve the shortest time for the vibration to settle down and minimize the vibration in these three parameters. The PID controller simulation is from the MATLAB and Simulink software, and the values of gain K_P, K_i, and K_D can be tuned by PID tuner. The value of K_P is 1664245.42, K_i is 1248152.87, and K_D is 416051.36. For the state-feedback controller design and the state-feedback gain vector, K can be obtained by using Eqs. (17) to (19) as shown below:

$$K = [2.5e2 \quad 4.3e6 \quad 6.4e9 \quad 3.5e3 \quad 9.5e6]^T \tag{20}$$

The road profile uses different heights of bump as shown in Fig. 6.

The road profile consists of 0.05 m bump, 0.03 m bump, and 0.04 m bump. All bumps have the same width according to the time. The simulation is done by comparing three controllers which are the passive controller corresponding to the uncontrolled system, the PID controller, and the state-feedback controller of the

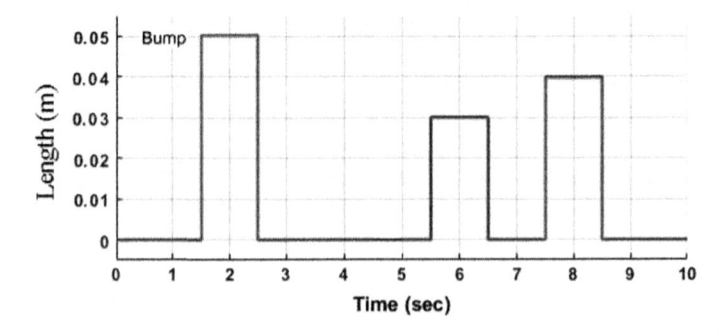

Fig. 6 Road profile

quarter car suspension system. Figure 7 shows the passive controller result, whereas Fig. 8 shows the comparison between the effects of the PID controller and state-feedback controller of suspension travel in the quarter car semi-active suspension. In Fig. 7, the uncontrolled system which uses the passive controller takes approximately 45 s times to reach zero stability.

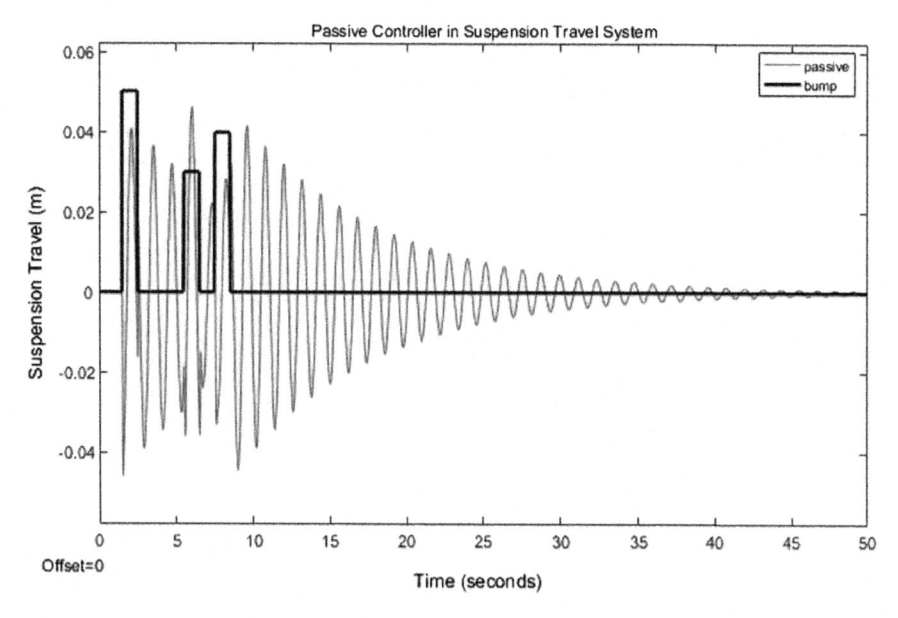

Fig. 7 Passive controller in suspension travel

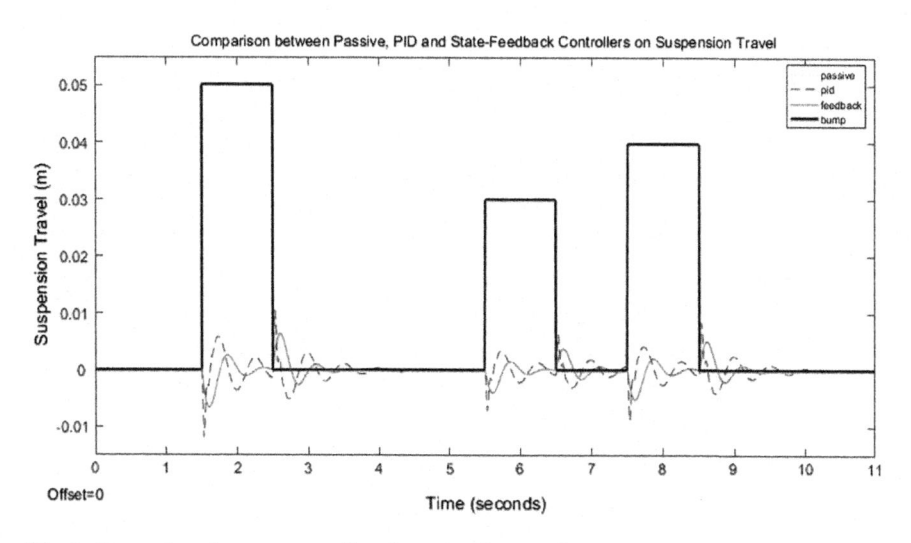

Fig. 8 Comparison between controllers in suspension travel

Figure 8 shows that the comparison in suspension travel corresponds to PID controller and state-feedback controller. Both of controllers have good settling times where the PID controller took less than three second, whereas state-feedback controller took less than two second to reach zero stability. The vibration of suspension travel after the second bump cannot achieve zero stability completely due to the distance between second bump and third bump which is close to each other. However, the vibration is able to be reduced after the first and third bumps with acceptable settling time of around 0.8 s.

By comparing the three controllers, the PID controller and state-feedback controller provide very much better response in the quarter car semi-active suspension system. It also clearly shows that the state feedback can give faster response time in settling time compared to the PID controller. Simulation result demonstrates the proposed state-feedback controller is capable of reducing the vibration of suspension after the car hits bumps with different heights of bump. Therefore, the potential to improve ride comfort of passengers and driver was developed using the state-feedback controller, and its performance was examined. The MATLAB and Simulink software is shown to have capability in handling the control design and simulation for suspension travel using different types of controller.

5 Conclusion

A lot of researches have been conducted to increase the ride comfort of passengers and driver of car due to conditions on the road. There are a lot of controllers available to solve this problem. In this study, it helps to simulate the ability of controllers in semi-active suspension system toward road disturbances such as bump in reducing the vibration of suspension travel in suspension system. The state-feedback controller shows the fastest settling time compared to the passive controller and the PID controller which lead to increase in the ride comfort of passengers and driver. Suspension travel and suspension vibration effects have been significantly improved by using the state-feedback controller as compared to the passive controller and the PID controller. In the future, the potential to increase the stability and the ride comfort of passengers and driver using the intelligent-based method of controller will be studied and implemented in this research.

Acknowledgements The authors would like to thank Dr. Hatta Ariff and Dr. Shamsul Sarip for their comments and advice. This work was financially supported through research grant of Universiti Teknologi Malaysia under Vot number 11H67. The authors also gratefully acknowledge the helpful comments and suggestions of the reviewers, which have improved the presentation.

References

Alvarez-Sánchez, E. (2013). A quarter-car suspension system: Car body mass estimator and sliding mode control. *Procedia Technology, 7,* 208–214.

Du, H., Li, W., & Zhang, N. (2012). Integrated seat and suspension control for a quarter-car with driver model. *IEEE Transactions on Vehicular Technology, 61*(9), 3893–3908.

Florin, A., Ioan-Cozmin, M. R., & Iliana, P. (2013). Pasive suspension modeling using matlab, quarter car model, input signal step type. *TECNOMUS, 3,* 258–263.

Ghazaly, N. M., Ahmed, A. E. N. S., Ali, A. S., & El-Jaber, G. T. A. (2015). PID controller of active suspension system for a quarter car model. *International Journal of Advances in Engineering & Technology, 8*(6), 899–909.

Mou, R., Hou, L., Jiang, Y., Zhao, Y., & Wei, Y. (2015). Study of automobile suspension system vibration characteristics based on the adaptive control method. *International Journal of Acoustics & Vibration, 20*(2), 101–106.

Pekgokgoz, R. K., Gurel, M. A., & Bilgehan, M. (2010). Active suspension of cars using fuzzy logic controller optimized by genetic algorithm. *International Journal of Engineering and Applied Science, 2*(4), 38–54.

Popovic, V., Vasic, B., Petrovic, M., & Mitic, S. (2011). System approach to vehicle suspension system control in CAE environment. *Strojniški vestnik-Journal of Mechanical Engineering, 57* (2), 100–109.

Sathishkumar, P., Jancirani, J., John, D., & Manikandan, S. (2014). Mathematical modelling and simulation quarter car vehicle suspension. *International Journal of Innovative Research in Science, Engineering and Technology, 3*(1), 1280–1283.

Yakub, F., Muhamad, P., Thiam, T. H., Fawazi, N., Sarip, S., Ali, M. S. M, Zaki, S. A. (2016a) Enhancing vehicle ride comfort through intelligent based control. In *2016 IEEE International Conference on Automatic Control & Intelligent Systems,* (pp. 66–71), October 22, Shah Alam, Malaysia.

Yakub, F., Lee, S., & Mori, Y. (2016b). Comparative study of MPC and LQC with disturbance rejection control for heavy vehicle rollover prevention in an inclement environment. *Journal of Mechanical Science and Technology, 30*(8), 3835–3845.

Zhang, Z., Cheung, N. C., Cheng, K. W. E., Xue, X., & Lin, J. (2012). Direct instantaneous force control with improved efficiency for four-quadrant operation of linear switched reluctance actuator in active suspension system. *IEEE Transactions on Vehicular Technology, 61*(4), 1567–1576.

Heavy Metals Concentration in Water Convolvulus (*Ipomoea aquatica* and *Ipomoea reptans*) and Potential Health Risk

Siti Nuur Ruuhana Saidin, Farah Ayuni Shafie,
Siti Rohana Mohd Yatim and Rodziah Ismail

Abstract Water convolvulus (*Ipomoea aquatica* and *Ipomoea reptans*) or their Malay name "*kangkung*" is tropical, semiaquatic, and fast-growing vegetables belonging to the morning glory family. They are among the green leafy vegetables widely consumed by Malaysian population. However, excessive use of fertilizer and pesticides in "*kangkung*" cultivation and contaminated source of irrigation may cause harm to the human health. Samples of *Ipomea* were collected from local wet market and supermarket, and digested and analyzed by Atomic Absorption Spectrophotometer (AAS). There were significant differences of heavy metal concentration for zinc, copper, manganese, and nickel in both species. The average concentration of Fe, Mn, Zn, Cu, and Ni in *I. aquatica* was found to be 270.20, 91.14, 32.95, 3.99, and 1.79 mg/kg, respectively, while Fe, Mn, Zn, Cu, and Ni in *I. reptans* was 275.97, 212.63, 87.33, 18.56, and 3.14 mg/kg, respectively. The daily intake of aforementioned heavy metals was within the safe limit values under Malaysia Food Regulations 1985 and World Health Organization's provisional tolerable daily intake. Subsequently, the Health Risk Index (HRI) for daily intake of Fe, Zn, Cu, and Ni was less than 1 except for Mn (HRI > 1) indicating possible health effect upon consumption. The use of fertilizers and pesticides should be monitored, and regular sampling of the vegetables is recommended to minimize the heavy metal uptake by water convolvulus, thus reducing human exposure.

1 Introduction

Vegetables contain both essential (copper, zinc, manganese, cobalt) and nonessential or toxic metals (cadmium, arsenic, chromium, mercury, lead). Heavy metals are non-biodegradable, persistent environmental contaminants and easily accumulate in the edible parts of leafy vegetables (Mapanda et al. 2005). Prolonged

---

S. N. R. Saidin · F. A. Shafie · S. R. M. Yatim · R. Ismail (✉)
Department of Environmental Health and Safety, Faculty of Health Sciences,
Universiti Teknologi MARA Selangor, Puncak Alam Campus,
42300 Selangor, Malaysia
e-mail: rodziah_fsk@salam.uitm.edu.my

consumption of heavy metals-contaminated vegetables may lead to the chronic accumulation in kidney and liver and subsequently cause disturbance to numerous human biochemical processes which lead to cardiovascular, nervous, kidney, and bone diseases (Suruchi and Pankaj 2011).

Several studies have shown heavy metal contamination in water convolvulus (Khairiah et al. 2014; Zarcinas et al. 2004; Rai and Sinha 2001; Marcussen et al. 2008; Li et al. 2014; Maimon et al. 2009; Kananke et al. 2014), but assessment on potential health risk from consumption is overlooked. Therefore, this study aims to evaluate the potential of heavy metal (Fe, Mn, Zn, Cu, Ni) accumulation in water convolvulus as well as to determine the human health risk associated with it.

2 Methodology

2.1 Sampling

This cross-sectional study was conducted in Changlun, Kedah. Changlun is a main border town in Kedah, Malaysia, in the district of Kubang Pasu, 42 km north from Alor Setar. A total of 32 samples of water convolvulus (*Ipomoea aquatica* and *Ipomoea reptans*) samples were randomly selected from 12 wet markets and supermarkets. Freshly harvested water convolvulus (500 g) was taken and placed in polyethylene bags. The geographical location of Changlun local wet market and supermarket is 6.4308°N, 100.4297°E and 6.4353°N, 100.4303°E, respectively.

2.2 Sample Preparation and Digestion

Water convolvulus samples were washed with distilled water to remove soil and dust particles. The edible portion of sample was weighed and chopped into small pieces. The samples were then oven-dried for 24 h at 80 °C. Then, the dried samples were powdered and stored in polyethylene container. After that, the powdered samples were weighted using digital weighing scale and digested using hotplate acid to destruct organic matters in samples by using wet digestion method (US EPA Method 3050A).

2.3 Heavy Metal Analysis

Heavy metal concentration was analyzed by using AA Analyst 400 Perkin Elmer Flame Atomic Absorption Spectrophotometer (FAAS). The measurement was made using a hollow cathode lamp of Fe, Mn, Zn, Cu, and Ni at respective wavelengths. The data of heavy metal analysis from FAAS was converted from mg/L or ppm to mg/kg to determine the exact concentration present in the samples. Concentration of metals in all samples was calculated (Likuku and Obuseng 2015) as shown in Eq. 1.

$$\text{Concentration (mg/kg)} = \frac{\text{Concentration (mg/l)} \times \text{Volume, V}}{\text{Sample mass, M (kg)}} \quad (1)$$

where concentration (mg/l) is the concentration of metal in digested solution from FAAS; volume, V (l), is the final volume after sample digestion (100 ml or 0.1 l); and sample mass, M (kg), is the mass of the sample used in acid digestion (2 g or 2×10^{-3} kg).

2.4 Daily Intake of Metals

The Daily Intake of Metals (DIM) was calculated in accordance with Likuku and Obuseng (2015) as shown in Eq. 2.

$$\text{DIM} = \frac{\text{Concentration (mg/kg)} \times \text{DIR (kg/day)}}{\text{BW (kg)}} \quad (2)$$

where concentration (mg/kg) is the metal concentration, DIR (kg/day) is the daily intake rate, and BW(kg) is the average body weight.

The daily intake rate of water convolvulus was calculated by converting the food frequency to amount of food intake by using formula adapted from Wessex Institute of Public Health. The amount of food intake was calculated as shown in Eq. 3.

$$\begin{aligned}\text{Amount of food (g/day)} = \ &\text{Frequency of intake} \times \text{Serving size}\\ &\times \text{ total numbers of servings}\\ &\times \text{ weight of food in one serving}\end{aligned} \quad (3)$$

2.5 Health Risk Index (HRI)

The Food Frequency Questionnaire used in the survey participated by 282 respondents was adapted from the US National Cancer Institute (2008). The survey consisted of three parts: demographic profile of respondents, water convolvulus consumption information, and public awareness on heavy metal contamination in water convolvulus. The Health Risk Index for exposure to selected heavy metals through ingestion of water convolvulus was estimated from ratio of Daily Intake of Metals (DIM) (mg/kg/day) to the oral reference dose (RfD) (Likuku and Obuseng 2015). The RfD values for Fe, Mn, Zn, Cu, and Ni were, respectively, 0.7, 0.014, 0.3, 0.04, and 0.02 in mg/kg/day (Likuku and Obuseng 2015; Chauhan and Chauhan 2014).

3 Analysis and Findings

3.1 Level of Heavy Metal Concentration in Water Convolvulus

The lead and cadmium concentration were not detected in *I. aquatica* and *I. reptans*. Zinc, iron, copper, manganese, and nickel were found in both *I. aquatica* and *I. reptans* (Fig. 1). These metals were also present in relatively high amount in the vegetables taken from small vegetable farms in Terengganu and Kelantan, Malaysia (Khairiah et al. 2014). Mean concentration of metal found in water convolvulus around Peninsular Malaysia was 0.8, 0.18, and 3.9 for copper, nickel, and zinc in mg/kg, respectively (Zarcinas et al 2004). High amounts of manganese, zinc, copper, and iron are due to the necessity of these essential elements in plant metabolic activity such as photosynthesis and their significance in the early stages of seedling growth (Khairiah et al. 2014; Ismail et al. 2005).

3.2 Comparison of Heavy Metals Concentration Between Ipomoea aquatica *and* Ipomoea reptans

Independent *t-test* was conducted to compare the mean concentration of zinc for the two types of water convolvulus. There was a significant difference in zinc concentration in *I. aquatica* and *I. reptans* ($p = 0.000$). Mann–Whitney U test was conducted for iron, copper, manganese, and nickel concentration in *I. aquatica* and *I. reptans*. There was no significant difference in iron concentration between *I. aquatica* and *I. reptans* ($p = 0.473$). For concentration of copper, manganese, and

Fig. 1 Heavy metals concentration difference in *Ipomoea aquatica* and *Ipomoea reptans*

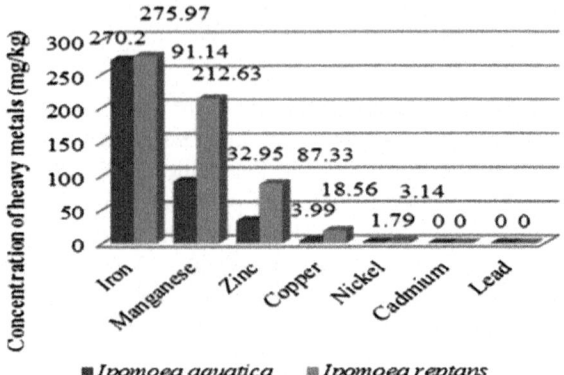

nickel, there was a significant difference between *I. aquatica* and *I. reptans* since the *p*-value is < 0.001.

The differences of heavy metal concentration in the two types of water convolvulus might be because of the differences of heavy metal concentration found in soil, irrigation water as well as air in the vicinity of the cultivation area. The use of irrigation water that may have been contaminated by industrial or municipal effluents, sewage sludge, and chemicals might also influence the level of contamination (Singh and Kumar 2006). Heavy metal uptake, accumulation, exclusion, deposition on foliage, and retention efficiency are among the mechanisms enabling the presence of heavy metals on the plant. As plants are the base in the food chain, the concern is the biomagnification of heavy metals in higher order of organism such as animals and human beings.

3.3 Compliance with Legal Requirements

The Daily Intake of Metals (DIM) for zinc, iron, copper, manganese, and nickel in both types of *Ipomoea* does not exceed the maximum permitted proportion by Malaysia Food Regulations 1985 and recommended safe limit of Food and Agriculture Organization (FAO) and World Health Organization (WHO) based on average adult body weight of 60 kg (Table 1). Noteworthy, national and international limit for the selected heavy metals is also within the permissible level, and thus it does not call for concern for human exposure upon consumption.

3.4 Health Risk Assessment

The Health Risk Index (HRI) for daily intake of zinc, iron, copper, and nickel for male and female consumption was less than 1. HRI ratio of less than 1 shows no obvious risk from the contaminants over a lifetime of exposure (Gupta et al. 2013). However, the HRI for manganese exposure for both types of water convolvulus was

Table 1 Daily intake of metals (DIM) for *Ipomoea aquatica* and *Ipomoea reptans*

Heavy metal	Male		Female		Malaysia Food Regulations 1985 (mg/kg)	FAO/WHO recommended safe limit (mg/day)
	DIM *I. aquatica* (mg/day)	DIM *I. reptans* (mg/day)	DIM I. aquatica (mg/day)	DIM I. reptans (mg/day)		
Zn	0.04	0.1	0.04	0.11	–	60
Fe	0.33	0.28	0.38	0.32	–	45
Cu	0	0.02	0	0.03	–	3
Mn	0.09	0.22	0.1	0.25	–	11
Ni	0	0	0	0	–	1.4

more than 1. The HRI ratio for *I. aquatica* is 6.43 for male and 7.14 for female, while the HRI for *I. reptans* is 15.71 for male and 17.86 for female. When HRI ratio is more than 1, the contaminants may produce an adverse effect and immediate intervention programs should be introduced and implemented (Gupta et al. 2013). Nervous system disturbance, behavior change, and fertility problems were among the adverse effects observed in animals nourished with abnormal amounts of manganese.

4 Conclusion

Heavy metal concentration of zinc, iron, copper, manganese, and nickels in *I. reptans* was significantly higher than *I. aquatica*. The Daily Intake of Metals (DIM) for zinc, iron, copper, manganese, and nickel concentration was lower than the recommended limits specified in Malaysia Food Regulations 1985 and FAO/WHO. The main concern is the possible health risk associated with the exposure to high level of manganese. Regular monitoring of heavy metal contamination in vegetables by the Ministry of Health and Department of Agriculture should be conducted to ensure the level of heavy metals in water convolvulus does not exceed the maximum permitted proportion to prevent excessive accumulation to the widely consumed green leafy vegetables in Malaysia.

References

Chauhan, G., & Chauhan, P. U. K. (2014). Human health risk assessment of heavy metals via dietary intake of vegetables grown in wastewater irrigated area of Rewa, India. *International Journal of Scientific and Research Publications, 4*(9), 1–9.

Gupta, S., Jena, V., Jena, S., Davi, N., Matic, N., Radojevi, D., et al. (2013). Assessment of heavy metal contents of green leafy vegetables. *Croatian Journal of Food Science and Technology, 5*(2), 53–60.

Ismail, B. S., Farihah, K., & Khairiah, J. (2005). Bioaccumulation of heavy metals in vegetables from selected agricultural areas. *Bulletin of Environmental Contamination and Toxicology, 74*(2), 320–327.

Kananke, T., Wansapala, J., & Gunaratne, A. (2014). Heavy metal contamination in green leafy vegetables collected from selected market sites of Piliyandala area, Colombo District, Sri Lanka. *American Journal of Food Science and Technology, 2*(5), 139–144.

Khairiah, J., Saad, B. S., Habibah, J., Salem, N., Semail, A., & Ismail, B. S. (2014). Heavy metal content in soils and vegetables grown in an Inland Valley of Terengganu and a River Delta of Kelantan, Malaysia. *Research Journal of Environmental and Earth Sciences, 6*(6), 307–312.

Li, Y., Wang, H., Wang, H., Yin, F., Yang, X., & Hu, Y. (2014). Heavy metal pollution in vegetables grown in the vicinity of a multi-metal mining area in Gejiu, China: total concentrations, speciation analysis, and health risk. *Environmental Science and Pollution Research, 21*, 12569–12582.

Likuku, A. S., & Obuseng, G. (2015). Health risk assessment of heavy metals via dietary intake of vegetables irrigated with treated wastewater around Gaborone, Botswana. In *International*

Conference on Plant, Marine and Environmental Sciences (PMES-2015). Kuala Lumpur, Malaysia.

Maimon, A., Khairiah, J., Ahmad Mahir, R., Aminah, A., & Ismail, B. S. (2009). Comparative accumulation of heavy metals in selected vegetables, their availability and correlation in lithogenic and nonlithogenic fractions of soils from some agricultural areas in Malaysia. *Advances in Environmental Biology, 3*(3), 314–321.

Mapanda, F., Mangwayana, E. N., Nyamangara, J., & Giller, K. E. (2005). The effects of long-term irrigation using water on heavy metal contents of soils under vegetables. *Agriculture, Ecosystem and Environment, 107,* 151–156.

Marcussen, H., Joergensen, K., Hoilm, P. E., Brocc, D., Simmons, R. W., & Dalsgaard, A. (2008). Element contents and food safety of water spinach (*Ipomoea aquatica* Forssk.) cultivated with wastewater in Hanoi, Vietnam. *Environmental Monitoring and Assessment, 139,* 77–91.

Rai, U. N., & Sinha, S. (2001). Distribution of metals in aquatic edible plants: Trapa Natans (Roxb.) Makino and *Ipomoea aquatica* Forsk. *Environmental Monitoring and Assessment, 70,* 241–252.

Singh, S., & Kumar, M. (2006). Heavy metal load of soil, water and vegetables. *Environmental Monitoring and Assessment, 120,* 79–91.

Suruchi, & Pankaj, K. (2011). Assessment of heavy metal contamination in different vegetables grown in and around urban areas. *Research Journal of Environmental Toxicology, 5*(3), 162–179.

US Cancer Institute. (2008). Usual dietary intakes: NHANES food frequency questionnaire. Retrieved from https://epi.grants.cancer.gov/diet/usualintakes/ffq.html.

Zarcinas, B. A., Fauziah, I., Mclaughlin, M. J., & Cozens, G. (2004). Heavy metals in soils and crops in Southeast Asia. 1. Peninsular Malaysia. *Environmental Geochemistry and Health, 26,* 343–357.

Analyzing Throughput in a Smartphone-Based Grid Computing

Alif Faisal Ibrahim, Muhammad Amir Alias and Syafnidar Abdul Halim

Abstract Grid computing is defined as professionally controlled, efficient, powerful, and large-scale distributed system for tasks sharing over several clustered computers in various platforms, different network bandwidths, operating systems, and computing abilities to create a virtual supercomputer for high-throughput computing. A variation of the grid computing is the smartphone-based grid system. Smartphones can be linked together to form a grid and be utilized to contribute in research and development that requires vast computational power. It is estimated that by 2020, there will be 6.1 billion users of smartphones. In our research, we developed a smartphone-based grid computing and then analyzed the throughput as the number of clients connecting to it increases. The throughput was measured using the software Iperf for Android. The result from the experiments shows that as the number of clients in the grid increases, the throughput of the grid decreases slightly. There were many factors that contribute to the slight decrement of the throughput such as the processing power of the server and the number of clients that is not significant enough to stimulate a change.

Keywords Grid computing · Supercomputing · Mobile grid · Smartphone-based grid

A. F. Ibrahim (✉) · M. A. Alias · S. A. Halim
Faculty of Computer and Mathematical Sciences,
Universiti Teknologi MARA, Shah Alam, Malaysia
e-mail: aliffaisal@perlis.uitm.edu.my

M. A. Alias
e-mail: amiralassad@gmail.com

S. A. Halim
e-mail: syafnidar@perlis.uitm.edu.my

© Springer Nature Singapore Pte Ltd. 2018
R. Saian and M. A. Abbas (eds.), *Proceedings of the Second International Conference on the Future of ASEAN (ICoFA) 2017 – Volume 2*,
https://doi.org/10.1007/978-981-10-8471-3_83

1 Introduction

Grid computing is defined as professionally controlled, efficient, powerful, and large-scale distributed system for tasks sharing over several clustered computers in various platforms, different network bandwidths, operating systems, and computing abilities to create a virtual supercomputer for high-throughput computing (Fedak 2010; Foster et al. 2001; Shi et al. 2011; Zhu et al. 2006). A grid system functions as coordinators of resources due to its decentralized control environment; it conveys nontrivial quality of service (QoS) and uses open, general-purpose standards, protocols, and interfaces (Foster et al. 2001; Sadashiv and Kumar 2011). Grid computing can maximize computing and data resources, pool them for large computing workloads, share them across networks, and allow cooperation among different devices.

A variation of the grid computing is the smartphone-based grid system. Smartphones can be linked together to form a grid and be utilized to contribute in research and development that requires vast computational power. It is estimated that by 2020, there will be 6.1 billion users of smartphones (Lunden 2015). However, these smartphones are usually idle and not turned off even when in sleep mode. Hence, it creates a waste in resources when the smartphones could be optimally utilized by creating a grid computing.

According to Behera and Tripathy (2014), as the computer grid increases in size and gets more complex, the tendency for the system to fail also increases. Therefore, the performance of the grid computing is vital to ensure jobs are completed within the required time. The performance of the grid can be measured in terms of bandwidth, latency, delay, or throughput. However, in our research, we focused on the throughput. Throughput is the number of message or work sent in a period of time. In our research, we developed a smartphone-based grid computing and then analyzed the throughput of the grid as the number of clients connecting to it increases.

2 Grid Computing

Grid computing is the act of sharing tasks over multiple computers which collaborates to create a virtual supercomputer for high-throughput computing (Fedak 2010). Resources such CPU cycles, data storage, applications, and network bandwidth are utilized, coordinated, and shared through grid computing without the requirement to buy new hardware (Bertis et al. 2009; Opitz et al. 2008). Grid computing is also defined as an aggregation of variably located resources into an expanded distributed system (Katsaros and Polyzos 2007). A grid is a system that consists of several connected computers or any devices even though these resources are located at different locations. Gengan et al. (2014) explained that a grid is formed by connecting heterogeneous devices and hardware to do very specific jobs. If these specific jobs require high processing capabilities and storage, it can be achieved in a grid environment (Sathyan and Rijas 2009).

The variant of grid computing includes computational grid, desktop grid, wireless grid, and smartphone-based grid. Computational grid is a combination of hardware and software that will provide a complex and huge computational power that is dependable, pervasive, and inexpensive (Foster et al. 2001). The desktop grid is an aggregation of desktop computers into a grid. The aim of the desktop grid architecture is to utilize the idle processing power of the desktop. A wireless grid is the combination of heterogeneous wireless devices into a grid with the goal of providing computational resources to application or to do complex jobs. A smartphone-based grid computing is where multiple smartphones are connected together in a network that will provide solution to big and complicated problem (Phan et al. 2002).

2.1 Grid Implementation

Grid is created through clustering processes and complex middleware such as Globus Toolkit (Marosi et al. 2010) and Berkeley Open Infrastructure for Network Computing (BOINC) (Black and Bard 2011; BOINC n.d.) that require tremendous effort to set up and need to be maintained professionally by the institution or grid owner in order to utilize large-scale computing and storage infrastructure (Kalochristianakis et al. 2012).

2.2 Importance of Network Throughput to Grid Computing

Reliable information on network throughput could enhance the total processing time for job execution in the grid since assigning large number of jobs to sites with sufficient computational power but insufficient network throughput could cause the job to be delayed (Lee et al. 2011). Data throughput in smartphone-based grid application would be a major bottleneck issue, and task processing would be affected by the issue. Therefore, real-time execution of job is important to avoid delays which would affect the performance and integrity of the grid.

3 Experimentation Setup

Figure 1 shows the smartphone-based grid topology that was implemented. Four smartphones and a server connected wirelessly to the grid network using mobile hot spot. BOINC middleware was installed in the server to monitor the clients in the grid and also for jobs implementations. NativeBOINC is an Android-based middleware installed in smartphones which allow it to integrate with the grid. Iperf was used in the experiment to measure the throughput in the smartphone-based grid.

Four experiments were conducted to analyze the throughput of the smartphone-based grid as the number of clients connecting to it increases. Initially, the first experiment analyzed the throughput when only a client was connected to the grid. Then, Iperf command was executed from the client and the throughput was recorded. The throughput collected from the first experiment was used as the control throughput. The experiment was repeated with different number of clients which are 2, 3, and 4 clients, respectively. Table 1 summarizes the method for the experiment.

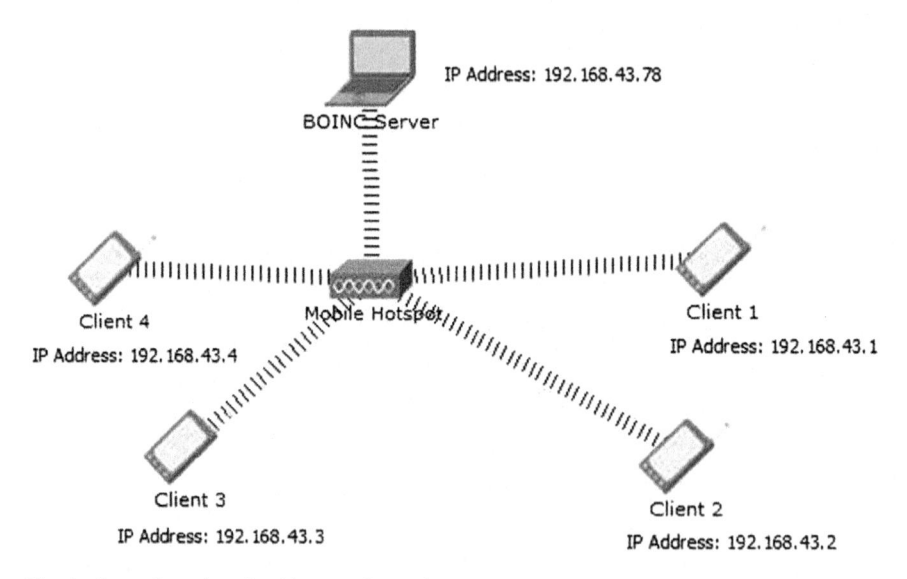

Fig. 1 Smartphone-based grid network topology

Table 1 Experimentation method

Experiment	Number of clients	Steps of experimentation
1(control)	1	Connect the client(s) to the smartphone-based grid
2	2	Run the Iperf software on the client and the server
3	3	Record the throughput
4	4	Repeat each experiment for three (3) times and calculate the average throughput

4 Results and Discussion

Table 2 shows the average throughput for the experiment. The average throughput for experiment one was 45.3 Mb/s. For experiment two, the average throughput was 44.3 Mb/s. While for experiment three, the average throughput was 44.2 Mb/s. Finally for experiment four, the average throughput was 43.5 Mb/s. Based on the control data from experiment one which is 45.3 Mb/s, we can identify that the throughput decreases as the number of clients connected to the smartphone-based grid increases. This correlates with the theory that throughput is influenced by the number of traffic in the network. In this case with the increasing number of clients, the traffic load also increases. Therefore, the throughput decreases.

Figure 2 shows the illustration of the throughput decrement. In this graph, we can see that the trend was clear. As the number of clients increases, the throughput decreases. The graph's scale was small; therefore, we may see a sharp decline in the throughput when the number of clients increases. The difference between the average throughputs of experiment one and experiment two was 1.0 Mb/s. This difference is small and not as significant to the performance of the grid. Furthermore, we can also see a much smaller margin of difference between experiments two and three which was 00.1 Mb/s. This was due to the idea that as the grid had already managed two devices, from connecting two clients in the previous experiment, the grid server was able to manage one more additional device without needing more resources. Then, another slight difference was shown from the graph as the throughput falls another 1.3 Mb/s from connecting three clients to

Table 2 Average throughput of smartphone-based grid

Experiment	Number of clients	Average throughput (Mb/s)
1(control)	1	45.3
2	2	44.3
3	3	44.2
4	4	43.5

Fig. 2 Average throughput in smartphone-based grid

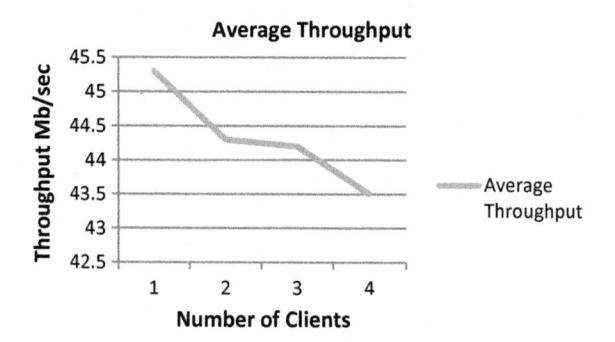

connecting four clients. This quite large difference compared to the previous experiments shows that as we add more devices to the grid, resources in the grid were shared and that this decreases the effective data rate of the grid as more traffic was generated. Although from the graph the difference between the average throughputs of each experiment was shown clearly. But, the difference was insignificant enough to actually trouble the network performance of the grid.

5 Conclusion

Grid computing is the idea of aggregating heterogeneous computing resources to execute complex jobs. These complex jobs are processing-intensive jobs that require high processing capabilities. Network performance is important to a grid as it dictates the efficiency of grid. The higher the network performance, the more efficient the grid is. The experiment that was conducted to evaluate the network performance of the smartphone-based grid shows that as the number of clients in the grid increases, the throughput of the grid decreases slightly. There were many factors that contribute to the slight decrement of the throughput such as the processing power of the server and the number of clients that is not significant enough to stimulate a change. A suggestion for improvement, future researcher should increase the number of clients in smartphone-based grid in order to significantly affect the performance of the server itself. This is important so that we can identify the maximum resources that can be aggregated at one time. By understanding this threshold, we can also provide solutions to such aggregating servers so that their processing power can be combined and used to create even bigger grid and connect more variety of devices in the grid.

Acknowledgements The authors gratefully acknowledge the helpful comments and suggestions of the reviewers, which have improved the presentation.

References

Behera, I., & Tripathy, C. R. (2014). Performance modelling and analysis of mobile grid computing systems. *International Journal of Grid and Utility Computing, 5*(1), 11–20. https://doi.org/10.1504/ijguc.2014.058244.

Bertis, V., Bolze, R., Desprez, F., & Reed, K. (2009). From dedicated grid to volunteer grid: Large scale execution of a bioinformatics application. *Journal of Grid Computing, 7*(4), 463–478.

Black, M., & Bard, G. (2011, 21–23 September). *SAT over BOINC: An application-independent volunteer grid project.* Paper presented at the Grid Computing (GRID), 2011 12th IEEE/ACM International Conference on BOINC, About BOINC. Retrieved October 25, 2015 from http://boinc.berkeley.edu/.

Fedak, G. (2010). Recent advances and research challenges in desktop grid and volunteer computing. In F. Desprez, V. Getov, T. Priol, & R. Yahyapour (Eds.), *Grids, P2P and Services*

Computing (pp. 171–185). Boston, MA: Springer. Retrieved from http://link.springer.com/chapter/10.1007/978-1-4419-6794-7_14.

Foster, I., Kesselman, C., & Tuecke, S. (2001). The anatomy of the grid: Enabling scalable virtual organizations. *International Journal of High Performance Computing Applications, 15*(3), 200–222.

Gengan, D., Schoeman, M. A., & Poll, J. A. V. D. (2014). *An ant-based mobile agent approach to resource discovery in grid computing.* Paper presented at the Proceedings of the Southern African Institute for Computer Scientist and Information Technologists Annual Conference 2014 on SAICSIT 2014 Empowered by Technology, Centurion, South Africa.

Kalochristianakis, M. N., Georgatos, F., Gkamas, V., Kouretis, G., & Varvarigos, E. (2012). Deploying LiveWN grids in the greek school network. *Journal of Grid Computing, 10*(2), 237–248.

Katsaros, K., & Polyzos, G. C. (2007). *Towards the realization of a mobile grid.* Paper presented at the Proceedings of the 2007 ACM CoNEXT conference, New York.

Lee, C., Abe, H., Hirotsu, T., & Umemura, K. (2011). *Predicting network throughput for grid applications on network virtualization areas.* Paper presented at the Proceedings of the first international workshop on Network-aware data management, Seattle, Washington, USA.

Lunden, I. (2015). 6.1B smartphone users globally by 2020, overtaking basic fixed phone subscriptions. Retrieved October 16, 2015 from http://techcrunch.com/2015/06/02/6-1b-smartphone-users-globally-by-2020-overtaking-basic-fixed-phone-subscriptions/#.lajzkhf:RPIH.

Marosi, A. C., Balaton, Z., Kacsuk, P., & Drótos, D. (2010). SZTAKI desktop grid: Adapting clusters for desktop grids. In F. Davoli, N. Meyer, R. Pugliese, & S. Zappatore (Eds.), *Remote instrumentation and virtual laboratories* (pp. 133–144).

Opitz, A., König, H., & Szamlewska, S. (2008). What does grid computing cost? *Journal of Grid Computing, 6*(4), 385–397.

Phan, T., Huang, L., & Dulan, C. (2002). *Challenge: Integrating mobile wireless devices into the computational grid.* Paper presented at the Proceedings of the 8th Annual International Conference on Mobile Computing and Networking, Atlanta, Georgia, USA.

Sadashiv, N., & Kumar, S. M. D. (2011). *Cluster, grid and cloud computing: A detailed comparison. 2011 6th international conference on computer science education (ICCSE)* (pp. 477–482). Presented at the 2011 6th International Conference on Computer Science Education (ICCSE).

Sathyan, J., & Rijas, M. (2009). *Job management in mobile grid computing.* Paper presented at the Proceedings of the 7th International Conference on Advances in Mobile Computing and Multimedia, Kuala Lumpur, Malaysia.

Shi, X., Jin, H., Wu, S., Zhu, W., & Qi, L. (2011). Adapting grid computing environments dependable with virtual machines: Design, implementation, and evaluations. *The Journal of Supercomputing, 66*(3), 1152–1166.

Zhu, T., Wu, Y., & Yang, G. (2006). Scheduling divisible loads in the dynamic heterogeneous grid environment. In *Proceedings of the 1st International Conference on Scalable Information Systems, InfoScale '06.* New York, NY, USA: ACM.

UAV/Drone Zoning in Urban Planning: Review on Legals and Privacy

Norzailawati Mohd Noor, Intan Zulaikha Mastor and Alias Abdullah

Abstract The use of drones or unmanned aerial vehicles (UAVs) in commercial applications has the potential to dramatically alter several industries, and, in the process, change our attitudes and behaviors regarding their impact on our daily life. This paper attempts to review a legal and privacy in urban planning context regards to the use of unmanned aerial vehicle (UAV)/drones provided by international and local approached. It is including studied an impacts of these existing policies to the usage of drones in the urban planning context. The review consists of comparing an existing law, prohibition, restriction, and guideline in drone operation to preserve safety and security of the people, property, and environment. Few items need to be measure in designing the zones such as safety, security, and privacy based on the technical and airspace aspects. Thus, by evaluating existing laws and regulations practiced in countries around the world, it will assist in designing drone zoning especially in Malaysia which able to manage urban planning practices and to regulate a general guidelines of zoning drone based on banned/prohibited, restricted, and allowed/controlled zones within the urban context and ensuring the livability and resiliency of cities.

Keywords Drones/UAV · Zoning · Privacy · Law and guidelines
Urban planning

1 Introduction

The advent of new and emerging technologies has broad economic, social, and personal impacts. Drones or robotic planes which are also defined as unmanned aerial vehicles (UAVs), unmanned aircraft systems (UAS), and remotely piloted aircraft (RPA) are rapidly developed over past decades vastly for military and

N. M. Noor (✉) · I. Z. Mastor · A. Abdullah
Department of Urban and Regional Planning, Kulliyah Architecture and Environmental Design, International Islamic University Malaysia, Selangor, Malaysia
e-mail: norzailawati@iium.edu.my

© Springer Nature Singapore Pte Ltd. 2018
R. Saian and M. A. Abbas (eds.), *Proceedings of the Second International Conference on the Future of ASEAN (ICoFA) 2017 – Volume 2*,
https://doi.org/10.1007/978-981-10-8471-3_84

civilian purposes because of its low-cost monitoring solutions (Ivosevic et al. 2015). Drones are thus becoming increasingly important in the fields of science, technology and society. Traditionally, When it first emerge, drones are use for the military purpose in World War I and yet over time, especially these few decades drone usage was introduced in the non-military sectors such as atmospheric research, earth and weather observations, and also policing remote sensing (Kennington and Berger 2014). The vast participation in the application of drones in various aspects, there are in need of laws and regulations for the drone users to comply to ensure the public safety especially in urban area. Urban area is the center of activities, and public is free to have their own activities' including flying a drone for their own purpose with not aware that it can be disrupt and trespassing a virtual boundary between urban community. However, lack of clear regulation to follow, drones has been freely hovering in urban spaces and public areas causing discomfort and controversial reactions by the oversight agencies involved (Rao et al. 2016; Schlag 2012).

The UAV/drone in Malaysia is expending in terms of its application not only for government agenda but also spreads into civilian fields such as plantation, natural resources management, safety and security, and also for leisure purpose. The rules and regulation designed for drone/UAVs rely on few aspects that control the usage such as, the surrounding condition, type and size of aircraft, categories of users and safety and security, especially unclear border zone for drones to fly in urban areas. Thus, bring this paper focus on a related but slightly different phenomenon: the emergence of commercial drones in urban area that requires zones and specific laws to be designed in protecting human privacy and safety.

2 Methodology

Understanding a subject as extensive as relationship between technology and society requires a suitably broad approach; thus, we performed a discourse analysis of various documents in order to investigate how various stakeholders perceived legal on drones. Discourse is comprehensive concept that includes any practice by which individuals imbue reality with meaning (Gee 2014). In the field of commercial drones, we considered various stakeholders such as governmental regulatory organizations, judicial bodies, research institutes, public policy organizations, drone manufacturers, technology developers, service providers, news organizations, insurance companies, non-profits acting in public interest, activist for privacy, activist for and against drones, public and private establishment drones users, and individual users. We used an aggregate corpus of seventy articles published 2000–2016, accessed through academic and non-academic databases and search engines and considered the text produced by these stakeholders as the baseline, and use content analysis to make inference. The text was then divided and coded based on origin, purpose, and content. Finally, the inferences were classified into schemes that represented the facets of specific stakeholder and one that represented the society response.

2.1 Legal Aspects of Drone Application: A Theoretical

Regulations for military and civilian drone applications differ considerably. Military drones, which have existed for a while, have regulations developed over time to cover only a limited set of activities' in specific and controlled air space. The broad applications of civilian drones, and their relatively small impact if compromised, have appreciable clogged and therefore delayed regulations specific urban space and private environment. Department of Civil Aviation (DCA) Malaysia stated that UAVs operating in Malaysia must meet or exceed the safety and operational standards as the manned aircraft. The civil drones, which not exceed 20 kg, are prohibited in controlled airspace or within aerodrome traffic zone without permission of air traffic controller. Civil drones can only be flown within the operator's line of sight, and they may not use the onboard cameras as their primary medium of sight. They are restricted to airspeed of 100 mph, can be flown at daytime, and when there is minimum of three miles of weather visibility. Most importantly, they must yield the right of way to any other aircrafts irrespective of whether it is manned or unmanned (Rao et al. 2016).

Some specific regulation in drones; such as United States (US) impose a Federal drone law stated that there are no enforceable regulations that apply to the general public that prohibit or restrict all aircraft from flying within certain parameters. In Section 336 Special Rule for Model Aircraft (2012), defined model aircraft as unmanned aircraft that is capable of sustained flight in the atmosphere, flown within visual line of sight of the person operating the aircraft and flown for hobby or recreational purposes (Sachs 2015). Federal rules prohibit any aircraft from operating in the flight restricted zone around our nation's capital without specific approval, which includes all unmanned aircraft. The airspace around Washington, D.C., is more restricted than in any other part of the country (Sudekum 2014).

2.2 Overview on Drone/UAV Laws

Currently, France holds the largest number of drone operators in Europe with over 1600 companies and has one of the most advanced laws regulating the use of civilian drones. They imposed in capital city of Paris to be classified as a strict no-fly zone. Small civilian drones are banned from areas such as nuclear facilities, which are protected by a no-fly zone that covers a 2.5 km radius and a height of 1,000 m buffer parameters (MLV Drone 2015).

Transport Canada regulated a UAV/drone regulation with sets a clear line between "unmanned aerial vehicles" which is for commercial use and "model aircraft" which is another category for recreational use. The definition of a model aircraft is it must have less than 35 kg; it is individually owned not under any company for commercial use; and are not use for seeking profits. The aircrafts must meet these conditions in order to be considered as recreational vehicle, making it subject to lower inspection by the authority. Aircraft, which do not meet these criteria, is considered under the category of "unmanned aerial vehicles," and it is required for Special Flight Operations certificates for its operations (Transport Canada 2015).

UAV/drone laws in UK are similar to the current policies of the USA, which it is more as guidelines rather than a comprehensive set of regulations. Unmanned Aircraft System Operations in UK Airspace Guidance (CAP 722) claim jurisdiction over UAV use in the UK. Under this legislation, UAVs are divided based on their use that both require permits. The weight limit for UAVs under category of "small unmanned aircraft" in UK is 20 kg (De Castella 2014). Thus, the aircraft more likely only requires a minor "Permit to Fly" classification, which is relatively easy to acquire, but does limit in terms of airspace to fly and altitude limit. In this case, rural area is more likely suitable for its flight operation.

Meanwhile, Singapore imposed a UAV/drone regulation through Civil Aviation Authority of Singapore (CAAS) with launching online portal for drone operators permit application and activity permit is required for flying drones that weigh more than 7 kg for any purpose, business, or recreation (Kok 2015). Those who fly drones for business purposes are required to apply for both permits regardless of the weight of the aircraft. Recreation or research drones do not require a permit if the weight of the aircraft is less than 7 kg (Civil Aviation Authority of Singapore 2015). However, an activity permit is required if the unmanned aircraft is using a restricted or dangerous airspace or area within 5 km of a military base regardless of operating height. If drones are flown indoors at a private residence or indoor area and the flying does not affect the general public at all then there is no permits are required or its operation (Civil Aviation Authority of Singapore 2015).

Brazil imposed at national level using UAVs to patrol its borders (Agencia Nacional de Aviacao Civil 2015). In 2013, a company, XMobots, has been permitted to fly the first civil drone for purpose of monitoring Jirau dam (Stochero 2013). Thailand regulates this application into two categories: sports and research purposes, and personal use (Barrow 2015), which users are needed to secure prior permissions and submit a flight plan. An exception is made for drones used by the film industry, which is considered alongside the latter category rather than under the former category.

Meanwhile, Cambodia put a serious legal action, after a series of incidents involving hobby and commercial drones, Phnom Penh's City Hall put an official ban on using drones in Phnom Penh's airspace in April 2015 (Parameswaran 2015). Drones are only allowed to fly only with permission of City Hall on a pre-arranged flight path. Tourists are still allowed to use UAVs outside of the city, but are advised to exercise caution when flying drones in heavily populated tourist areas such as Angkor Wat. There are no known commercial drone ventures in Cambodia at this time (Sovuthy and Ho 2015). Vietnam's Law on Civil Aviation, Article 81 Section 2b, states: "The Ministry of National Defense shall grant flight permission to Vietnamese and foreign military aircrafts operating civil flights in Viet Nam and to unmanned aircrafts." (Tuoi Tre News 2015). Recently, the ministry of defense wants all drone-flyers, personal or commercial use, to have license in order to take flight (Defense World Bureau 2015). The legislation, however, is unclear as to how it regulates smaller civilian unmanned aircraft used in the country.

Indonesia introduced regulations that will limit the areas in which civilians may use unmanned aircrafts. Licenses are needed to be acquired and new regulations lays out restrictions on where drones can fly, and also requires drone users to register their flight plans with the local civil navigation authority. The Philippines' drone regulations were implemented in June 2014 with imposing a licenses for the operators attached with evidence on experience and passed a training courses and passenger being imposed with clearance and pay a luxury tax of 100,000 Philippines Pesos or approximately $2200 (Calleja 2014; Andrade 2015).

Based on the overview, we can conclude most of the country are aware on the development of UAV/drone usage and future risk to the human, so that many efforts have been done although it yet to success for overall; however, the specific regulations for the specific areas and purposes should be develop especially in urban area and effect to human activities'.

3 Analysis and Findings

3.1 Societal Impact of UAV/Drones

The common societal impact can be focus on the safety and privacy of urban resident. According to Rao et al. (2016), his analysis found that there are three broad classes of issues in drones that need a further attention. There are a safety and security, conception of civilian airspace and privacy and ownership. Safety, the freedom from harm is basic human right that are guaranteed and protected by constitution of most nations. Currently, the use of drones in civilian airspace has triggered concerns are directed toward both the technology and the user. Concerns regarding the technology center around the battery life lift capacity, airworthiness, and reliability of the drones. The primary critics with the flying of commercial drones over public space are that small mistake could result in crashes that threaten the health, well-being, and property of public.

3.1.1 Privacy and Ownership

The issue rises on the airspace over private property and standards and expectation for its protection. In a public space such as a park or on a street, the reasonable expectation of privacy does not apply. Therefore, since a person is present in a public place, there is also not legal basis to make a claim of a breach of their privacy. The research about people's privacy concerns is diverse and contradictory in terms of theory (for instance, identifies 15 different theories of privacy online contexts), methods (Van Zoonen 2014 discuss the usage of experiments, survey, qualitative interviews and document analysis) and outcomes in particular with respect to the (lack of) influence of age gender and other socio-demographic features on privacy concerns. The same argument also extends up to an extent, to private property that is visible from public spaces.

However, these laws assume that sight is confined to the eye level. Drones disrupt the expectations of reasonable privacy since they are operated in a public place, yet can capture images and sound from that are not traditionally available in public. This gap in the law allows for the possibility of unwarranted surveillance without fear of repercussion. Current privacy laws state that it is illegal to record the interior of a home or owned building, even if the camera is placed outside. This creates uncertainty since even if the drone is being flown within eyesight and over the private property of the operator; there is the possibility of being in violation of privacy laws since it provides a monitoring capability that is not yet legitimated by the law.

4 Conclusion and Future Outlooks

The rapid evolution of drones for civilian applications has created several challenges; regulatory, safety, privacy, security, and the uncertain landscape for new business models. This paper shows, in constructing guidelines for the application of UAV/drones in Malaysia, that there is in need of clear definition and category of users, types and air space the drone flown, in designing an area to be banned/prohibited, restricted, or allowed/free. Majority of the guidelines and regulations applied in many countries are implemented by taking into account all three items as mentioned above; the user, the aircraft, and the air space. The general aspects such as type of users able to define on what type of drones they are using. Thus, the detail specifications of the drones are able to portray the ability and compatibility of the drone, where the legislators able to find out whether the drone is safe to be flown in different type of air space. In terms of regulation concerning the air space, some country has very specific area where the drone is prohibited, restricted, and allowed. In terms of zoning the drone area, it is crucial to identify the type of land use and its activity. Some zones are fixed, such as buildings and gathering square. But there is country that very detail in defining the air space. The temporary prohibited area, for example, are not fixed, it is based on events and occasions that held in that particular area in some certain timeframe. Above all, those law and legislation concerning the applications of UAVs/drones are applied based on the safety and security factors. There are pros and cons come along with UAV technology, but with a best designed guidelines and zoning for its applications will give more positive impacts and open to new findings in urban planning in particular, and reduce the negative impacts and gray area in legislation statutes. As the population of civilian drones and their users expands globally, the risk of accidents both digital and physical is destined to multiply. The future success of civilian drones depends on the ability of varied stakeholders to reconsider how this emerging technology platform can be best harnessed to serve the broad interest of society.

Acknowledgements The authors greatly acknowledge the Ministry of Higher Education Malaysia for a research grant on Fundamental Research Grant Scheme (FRGS 15-1850426), International Islamic University of Malaysia for providing facilities and support in this study. Authors sincerely thank all referees for their suggestions to improve the manuscript.

References

Agencia Nacional de Aviacao Civil. (2015). *PSURs—Aircraft Systems Remotely piloted*. Retrieved August 25, 2015 from Agencia Nacional de Aviacao Civil: http://www2.anac.gov.br/rpas/.

Andrade, J. I. (2015) *Gov't set to clip wings of drones—Philippine Daily Inquirer—Philippine UAV News*. Retrieved from Philippines UAV Review: http://philippinedrones.blogspot.my/2015/04/govt-set-to-clip

Barrow, R. (2015). Restriction flying drone in Thailand. Richard Barrow writing about Thailand in social media. http://www.richardbarrow.com/2015/01/restrictions-on-flying-drones-in-thailand/ retrieved on 12/7/2017.

Calleja, N. P. (2014). *Drones must be registered, their 'pilots' licensed*. Retrieved from Phillipine Daily Inquirer: http://newsinfo.inquirer.net/623268/drones-must-be-registered-their-pilots-licensed.

Civil Aviation Authority of Singapore. (2015). *Flying of Unmanned Aircraft*. Retrieved June 5, 2015 from CAAS: http://www.caas.gov.sg/caas/en/ANS/unmanned-aircraft.html.

De Castella, T. (2014, December 9). *Where you can and can't fly a drone*. Retrieved April 2015from BBC News: http://www.bbc.com/news/magazine-30387107.

Defence World Bureau. (2015). *Vietnam MoD wants licence for Drone-Flyers*. Retrieved August, 2015 from Defence World.Net: http://www.defenseworld.net/news/13740/Vietnam_MoD_Wants_Licence_For_Drone_Flyers.

Gee, J. P. (2014). *An introduction to discourses analysis: Theory and method*. London: Routledge.

Ivosevic, B., Han, Y.-G., Cho, Y., & Kwon, O. (2015). The use of conservation drones in ecology and wildlife research. *Journal of Ecology and Environment, 38*(1), 113–118.

Kennington, W. L., & Berger, M. (2014). *Why land use lawyers care about the law of unmanned system*. Minnesota: Thomson Reuters.

Kok, L. M. (2015). *CAAS launches online portal for drone permit applications in Singapore*. Retrieved from The Straits Times Singapore: http://www.straitstimes.com/singapore/caas-launches-online-portal-for-drone-permit-applications-in-singapore.

MLV Drone (2015). *France—Rules for flying recreational drones*. Retrieved from MLVDRONE: http://www.mlvdrone.fr/rules-for-flying-recreational-drones/.

Parameswaran, P. (2015). *Cambodia Bans Drones*. Retrieved from The Diplomat: http://thediplomat.com/2015/02/cambodia-bans-drones/.

Rao, B., Gopi, A. G., & Maione, R. (2016). The societal impact of commercial drones. *Technology in Society, 45*, 83–90.

Sachs, P. (2015). *Current US Drone Law*. Retrieved from Drone Law Journal: http://dronelawjournal.com/.

Schlag, S. (2012) New privacy battle: How the expanding use of drones continues to erode our concept of privacy and privacy rights. *Pitt. J. Tech. L. Pol'y.*

Sovuthy, K., & Ho, T.-K. (2015). *Chinese Tourists Arrested for Flying Drone Near Royal Palace.* Retrieved from The Cambodian Daily: https://www.cambodiadaily.com/archives/chinese-tourists-arrested-for-flying-drone-near-royal-palace-87357/.

Stochero, T. (2013). *ANAC grants 1st authorization for private and civil drone fly in Brazil.* Retrieved from Paraiba Valley and Region: http://g1.globo.com/sp/vale-do-paraiba-regiao/noticia/2013/05/anac-concede-1-autorizacao-para-drone-particular-e-civil-voar-no-brasil.html.

Sudekum, B. (2014). *Don't Fly Drones Here.* Retrieved from Mapbox: https://www.mapbox.com/blog/dont-fly-here/.

Transport Canada. (2015). *Flying a drone or an unmanned air vehicle (UAV) for work or research.* Retrieved from Transport Canada: http://www.tc.gc.ca/eng/civilaviation/standards/general-recavi-uav2265.htm?WT.mc_id=21zwi.

Tuoi Tre News. (2015). *Users need permission to fly camera drones: Vietnam defense ministry.* Retrieved from Tuoi Tre News: http://tuoitrenews.vn/society/29767/vietnam-users-need-permission-to-fly-camera-drones-defense-ministry.

Van Zoonen, L. (2014). *What do users want from their future means of identity management?* (Final Report.) http://imprintsfutures.org/assets/images/pdfs/End%20report%20IMPRINTS.pdf.

Indigenous and Produce Vegetable Consumption in Selangor, Malaysia

Nur Filzah Aliah and Emmy Hainida Khairul Ikram

Abstract The discovery of increased health-protecting properties of nutrient bioactive compounds in vegetables found in several studies has been strongly connected to the recommendation of consuming higher intake of vegetables as one of the vital dietary components. Malaysian indigenous and produce vegetables are among of the plants that possess variety of the health benefit compounds which will alleviate micronutrient-related deficiencies. The main purpose of this study is to compare the consumption pattern of indigenous and produce vegetables in rural and urban setting. A cross-sectional survey was conducted in both study area, Shah Alam and Tanjung Karang. A stratified random sampling was employed. About 213 households were selected, 91 from the rural area in Tanjung Karang, Kuala Selangor, and 122 from the urban area in Shah Alam. Quantitative data collected include the household demographic profile and socio-economic characteristics, availability, accessibility, diversity and consumption of Malaysian indigenous and preferences of vegetables. Moreover, data are collected on knowledge on health-protecting benefits related to test the relation with consumption of indigenous and commercial vegetables in both settings. There is no significant difference in factors that influence the consumption pattern of indigenous and produce vegetables except for availability of vegetables sold in both study areas. Unavailability of indigenous vegetables sold contributes to low intake frequency. Small production and less market demand due to unpopularity of indigenous vegetables are the reasons for the lack of availability in both study areas.

Keywords Consumption · Indigenous · Vegetables

N. F. Aliah (✉) · E. H. K. Ikram
Faculty of Health Sciences, Universiti Teknologi MARA, Puncak Alam Campus, 43200 Bandar Puncak Alam, Selangor, Malaysia
e-mail: filzahaliah@gmail.com

E. H. K. Ikram
e-mail: emmy4546@puncaksalam.uitm.edu.my

1 Introduction

Being a country with multi-ethnic and multicultural populations, the diversity of Malaysian foods has become one of the main attraction and symbol of life in Malaysia. Eating trends have evolved according to the passage of time, and recently, people are adapting to current eating trends to cope with rapid and busy living. Furthermore, the rapid growth of Malaysia food industry and 24-h food vendors contributes unhealthy eating trends and lifestyle. In the recent National Health and Morbidity Survey (Ministry of Health Malaysia 2015), the main highlights were given to the prevalence of fruit and vegetables consumption among Malaysian. The survey found that only 6.0% of Malaysian adults consumed 5 serving or more fruits or vegetables per day which is very low. From NHMS survey, it was found that almost half of Malaysian adults 18 years and above have hypercholesterolaemia (47.7%), 17.5% having diabetes and 30.3% having hypertension. Low consumption of vegetables, poor nutritional and diet choices, lack of knowledge and difficulties of obtaining healthy foods are several other factors that contribute to unhealthy lifestyle and eventually lead to the development of chronic diseases such as cardiovascular disease, cancer, diabetes and others (Langat 2014). Socio-economic and psychological characteristics influence the individuals' preferences on type and amount of vegetable consumption (Yeh et al. 2016).

In the current eating trends, it can be said that vegetables and fruits are the least food group eaten by people (Mad Nasir et al. 2010). The consumption pattern of vegetables is influenced by (1) availability (Yeh et al. 2016), (2) awareness level (Kendzierski et al. 2015), (3) socio-economic background (Patrick et al. 2005; Zimmer and Prachuabmoh 2012) and (4) preferences (Farragher et al. 2016). These studies suggested that low consumption of vegetables, poor nutritional and diet choices, lack of knowledge and difficulties of obtaining healthy foods are several other factors that contribute to unhealthy lifestyle and eventually lead to the development of chronic diseases such as cardiovascular disease, cancer, diabetes and others (Langat 2014). A cross-sectional study was conducted by comparing the consumption of indigenous vegetables and produce vegetables in urban and rural areas. Indigenous vegetables are defined as crops from which the tender leaves, stems and petioles are harvested and used in the preparation of vegetables. These crops are subdivided into roots or tubers and leafy (Department of Agriculture Forestry and Fisheries 2009). Produce vegetables are defined as crops that commonly sold by supermarkets, farmer's market and greengrocers (Morgan 1991). The conceptual framework involved independent and dependent variables which are types of vegetables and consumption pattern of indigenous vegetables and produced vegetables. In this study, affecting the relationship between independent and dependent variables was the confounding variables which are the availability and accessibility, preferences, socio-economic status, level of awareness and food behaviour pattern.

2 Methods

2.1 Study Design

We chose to conduct a cross-sectional study in urban and rural areas. The areas chosen for this study are categorized into rural and urban area. Rural is defined as any area with population less than 10,000 people having agriculture and natural resources in which its population either clustered, linear or scattered. Meanwhile, urban is defined as gazette areas with their adjoining built-up areas, which had a combined population of 10,000 or more at the time of the Census 2010 or the special development area that can be identified, which at least had a population of 10,000 with at least 60% of population (aged 15 years and above) were involved in non-agricultural activities. Tanjung Karang in Kuala Selangor area was chosen as the rural setting and Shah Alam for urban setting.

2.2 Sampling Method

Stratified random sampling was employed as sampling method for this study as two different settings are used. Stratified random sampling is a method of sampling that involves division of population into different subgroups. The sample is then selected from each subgroup by selecting an equal number of elements from each subgroup. The sample size was determined by using Eq. 1:

$$ss^0 = \frac{Z^{02^0} * (p) * (1-p)}{c^{02}} \quad (1)$$

where $Z = Z$ value (e.g. 1.96 for 95% confidence level); p = percentage picking a choice, expressed as decimal (0.5 used for sample size needed); c = confidence interval, expressed as decimal (e.g. 0.1 = ± 1).

Total 192 households were chosen from both settings, and samples chosen were approached physically with sets of questionnaire and information sheet explaining the purpose of this study at the first page. Besides that, consent form was also attached. Participation is fully voluntary and no money reimbursement was made. Inclusion criteria for samples in this study are permanent residents who are currently living in Kuala Selangor and Shah Alam area and are able to understand the questionnaire provided. The exclusion criteria for samples in this study are non-resident of the study area and samples who have allergy towards any kind of vegetables.

2.3 Study Procedure

This study was done through observation and questionnaire. Observation was made on availability of both types of vegetables at supermarkets, market, minimarts in the study areas. A total of six supermarkets and three markets were observed in Shah Alam area. Meanwhile in Tanjung Karang, one supermarket, two markets and one minimarket were observed. Different modes are chosen for questionnaire administration: (1) online survey, and (2) interviews, depending on participant's convenience. This study used self-administered online survey form and interview administration to distribute questionnaire. Interview-administered approach was used to distribute questionnaire only when email approach is not applicable in certain situation.

2.4 Research Tools

1. *Questionnaire on Vegetables Consumption Pattern*
   Questionnaire used for this study is self-developed and consists of five sections: section one contains the socio-demographic data of samples, section two is about sample's common knowledge regarding benefits of vegetables consumption, section three is about the availability of vegetables in sample's residence area, section four is about the sample's preferences and frequency of vegetables consumption, and section five is about the behaviour and readiness of samples towards vegetables consumption. Different sections of the questionnaire are obtained from various established food and behaviour questionnaires.
2. *Observation on Availability of Vegetables Sold*
   We went to places where residents of study areas usually go to purchase vegetables. The availability of the vegetables sold to the resident was observed. The farmers or greengrocers were also interviewed on what kind of vegetables consumers buy, the highest purchasing type of vegetables and factors influenced consumers purchasing level.
   Content Validity
   In developing survey form used in this study, the validation procedures of instrument tested were only at the first phase which includes only facet and content validity. The survey form has gone through expert's judgment (lecturers) and gained several critics in terms of appearance, relevance and representatives of its elements.
   Data Analysis
   Data collected were analysed by using Statistical Package of Social Science (SPSS) version 21.0. Types of statistical analysis used are chi-square and descriptive analysis. Data collection procedures are summarized in Fig. 1.

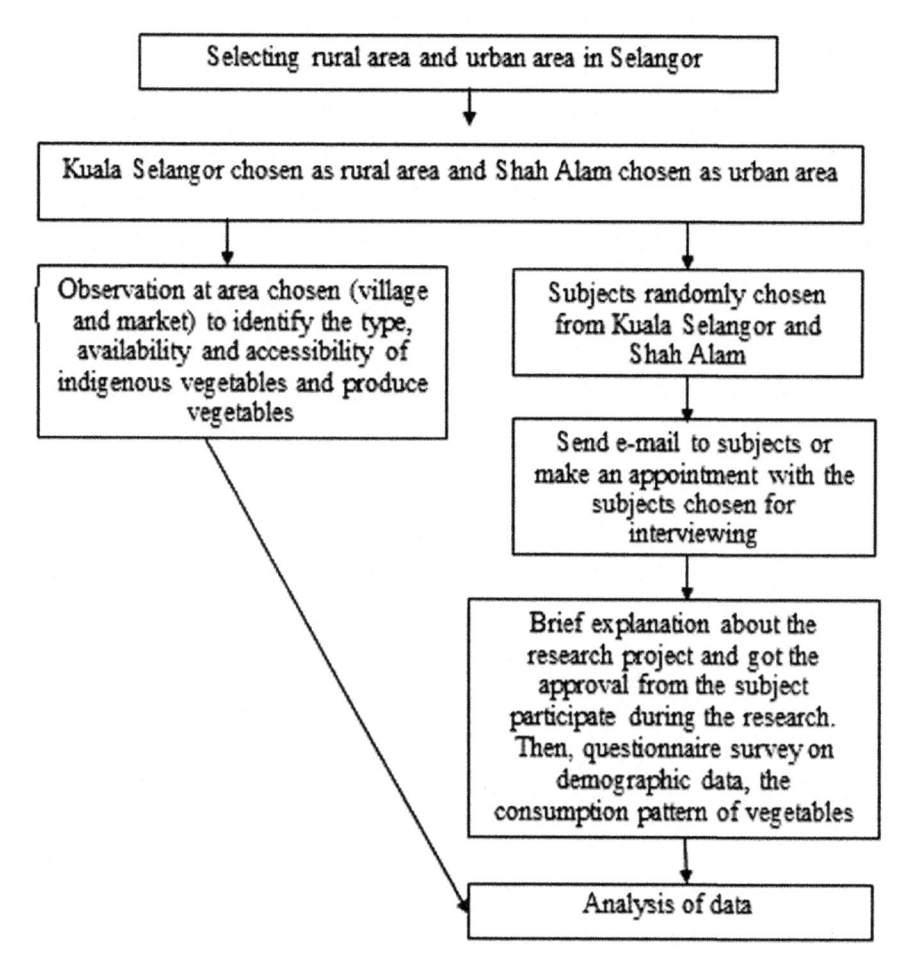

```
┌─────────────────────────────────────────────────┐
│ Selecting rural area and urban area in Selangor │
└─────────────────────────────────────────────────┘
                         ↓
┌───────────────────────────────────────────────────────────────┐
│ Kuala Selangor chosen as rural area and Shah Alam chosen as urban area │
└───────────────────────────────────────────────────────────────┘
           ↓                                    ↓
┌─────────────────────────────┐    ┌─────────────────────────────┐
│ Observation at area chosen  │    │ Subjects randomly chosen    │
│ (village and market) to     │    │ from Kuala Selangor and     │
│ identify the type,          │    │ Shah Alam                   │
│ availability and            │    └─────────────────────────────┘
│ accessibility of indigenous │                 ↓
│ vegetables and produce      │    ┌─────────────────────────────┐
│ vegetables                  │    │ Send e-mail to subjects or  │
└─────────────────────────────┘    │ make an appointment with the│
                                   │ subjects chosen for         │
                                   │ interviewing                │
                                   └─────────────────────────────┘
                                                ↓
                                   ┌─────────────────────────────┐
                                   │ Brief explanation about the │
                                   │ research project and got the│
                                   │ approval from the subject   │
                                   │ participate during the      │
                                   │ research. Then, questionnaire│
                                   │ survey on demographic data, │
                                   │ the consumption pattern of  │
                                   │ vegetables                  │
                                   └─────────────────────────────┘
                                                ↓
                                   ┌─────────────────────────────┐
                                   │      Analysis of data       │
                                   └─────────────────────────────┘
```

Fig. 1 Data collection procedure

3 Results

3.1 Subject Characteristics

There was a good response rate (127%, $n = 122$ from 96 households) of urban survey population and (94.7%, $n = 91$ from 96 households) from rural survey population with total 213 respondents completing the survey form as compared to total 192 households chosen for the study. 57.3% of total respondent came from urban area and 42.7% came from rural area. 34.7% from respondents secured more than RM3000 for household income per month, followed by 31% of them securing less than RM1000 per month. 90.1% of them received tertiary level of education followed by 8.9% of them receiving secondary level and 0.5% at primary level of education. The list of data can be referred in Table 1.

Table 1 Subject characteristics

Characteristics	Total (N = 213)		Men (n = 46)		Women (n = 167)	
	n	(%)	n	(%)	n	(%)
Residence area						
Urban	122	57.3	29	23.8	93	76.2
Rural	91	42.7	17	18.7	74	81.3
Term of resident						
<1 year	13	6.1	1	7.7	12	92.3
1–3 years	36	16.9	10	27.8	26	72.2
3–5 years	27	12.7	11	40.7	16	59.3
>5 years	137	64.3	24	17.5	113	82.5
Age						
13–19 years	8	3.8	0	0	8	100
20–39 years	191	89.7	43	22.5	148	77.5
40–64 years	14	6.6	3	21.4	11	78.6
>65 years	0	0	0	21.6	0	78.4
Race						
Malay	211	99.1	46	21.6	165	78.2
Chinese	1	0.5	0	0	1	100
Indian	0	0	0	0	0	0
Others	1	0.5	0	0	1	100
Religion						
Islam	211	99.1	46	21.8	165	78.2
Buddha	1	0.5	0	0	1	100
Hindu	0	0	0	0	0	0
Christian	1	0.5	0	0	1	100
Others	0	0	0	0	0	0
Marital status						
Single	186	87.3	37	19.9	149	80.1
Married	27	12.7	9	33.3	18	66.7
Divorced	0	0	0	0	0	0
Number of person per household						
1–2 person	16	7.5	5	31.3	11	68.8
3–5 person	73	34.3	16	21.9	57	78.1
6–7 person	70	32.9	11	15.7	59	84.3
>7 person	54	25.4	14	25.9	40	74.1
Type of diets						
Normal	212	99.5	46	21.7	166	78.3
Vegan	1	0.5	0	0	1	100
Vegetarian	0	0	0	0	0	0
Lacto-ovo	0	0	0	0	0	0
Lacto	0	0	0	0	0	0

(continued)

Table 1 (continued)

Characteristics	Total (N = 213)		Men (n = 46)		Women (n = 167)	
	n	(%)	n	(%)	n	(%)
Education level						
None	1	0.5	0	0	1	100
Primary education	1	0.5	0	0	1	100
Secondary education	19	8.9	7	36.8	12	63.2
Tertiary education	192	90.1	39	20.3	153	79.7
Household income						
<RM 1000	66	31	17	25.8	49	74.2
RM 1000–RM 2000	38	17.8	11	28.9	27	71.1
RM 2000–RM 3000	35	16.4	6	17.1	29	82.9
>RM 3000	74	34.7	12	16.2	62	83.8
Groceries expenses per month						
<RM 100	19	8.9	3	15.8	16	84.2
RM 100–RM 300	93	43.7	21	22.6	72	77.4
RM 300–RM 500	55	25.8	7	12.7	48	87.3
>RM 500	46	21.6	15	32.6	31	67.4

3.2 Preferences of Vegetables Intake

A total of 26 indigenous vegetables and 27 produce vegetables were evaluated for sample's preferences. Indigenous vegetables and produce vegetables are categorized according to their shapes and texture. To date, there is no report published on consumption pattern of indigenous vegetables in Malaysia. Therefore, the results from this study for these vegetables would provide new information about the consumption pattern of these vegetables (Fig. 2).

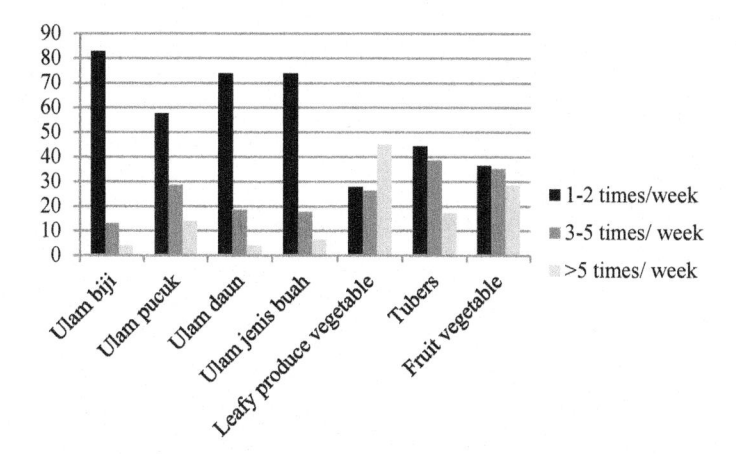

Fig. 2 Intake frequency for different categories of vegetables

Table 2 Preferences of vegetables intake

Vegetables category		Preferences			
		Highest (%)		Lowest (%)	
Indigenous vegetables	Indigenous pea, ulam biji	Bitter bean	52.10	Dogfruit	46
	Indigenous shoot, ulam pucuk	Fern shoot	78.90	Papaya shoot	61
	Indigenous leaves, ulam daun	Ulam raja	57.80	Papaya Flower	51.60
	Indigenous fruit vegetables, ulam jenis buah	Young papaya	64.80	Bitter Chinese melon	61
Produce vegetables	Leafy produce vegetables	Salad	91.10	Celery	23
	Tubers	Potato	96.70	Lotus root and Radish	52.60
	Fruit vegetables	Cucumber	78.90	Bitter melon	62

Among these vegetables, the highest intake frequency of indigenous vegetables is once a week (Table 2). Contrary to produce vegetables, the distribution of intake frequency is well distributed throughout the week. For leafy produce vegetables, the highest intake frequency is twice intake per week, followed by five intake and everyday intake per week. For tubers, the highest intake frequency is once a week, followed by thrice and twice a week. Moreover, for fruit vegetable, the highest intake frequency is twice and thrice a week, followed by once a week and four intakes per week.

3.3 Knowledge on Vegetables in Both Residency Areas

All items in this section ask about general knowledge regarding benefit of vegetables to health. Most of the respondent with mean of 97.9% ($n = 209$) answered the questionnaire correctly. Three other questions in the section are about general knowledge regarding vegetables and with mean 80.9% ($n = 172$) of respondent answer correctly. Chi-square analysis indicated that there are no significant relationships between residency areas with level of knowledge regarding benefits of vegetables to health ($p < 0.05$).

3.4 Preferences and Availability of Vegetables

Chi-square analysis indicated that there is no significant value to prove the strong correlation between these two variables except for two items questioning the sources of vegetables the survey population consumed with score ($p = 0.029$) and the reason behind unavailability of vegetables with score ($p = 0.006$).

Based on observation done in both study areas, there are a lot of access and availability of produce vegetables such as in supermarkets, daily wet market and minimarts on daily basis. Different for indigenous vegetables, most of it rarely sold in supermarkets and daily wet market in urban areas but it can be found sold only on farmers' market day or known to the local as Pasar Tani. Meanwhile for rural areas, half of the indigenous vegetables listed in this study were available in wet market or weekend market nearby the study area.

3.5 Behaviour and Readiness Towards Vegetable Consumption

This section asks on the tendency to consume vegetables under certain circumstances such as different environment, taste and perception. Furthermore, preferences of how the vegetables are served for consumption were also asked. The analysis indicated that there is no statistical significance between food behaviour and availability of vegetables in studied areas ($p = 0.008$).

4 Discussion

The consumption of some underutilized indigenous vegetables has been restricted by lack of knowledge regarding health-protecting benefits of the plant. Level of education is no longer a barrier to vegetable consumption pattern among Selangor rural and urban area. This is comparable to several studies, for instance, a study done on Portuguese adults which found that educational attainment was more frequently associated with food intake compared to income. In several studies, it shows that low income is associated with a poorer dietary quality. Household income did not significantly influence the monthly grocery expenses as the expenses depend on the amount of person in a household. Many studies that tried to find out correlations between nutrition knowledge and dietary intake failed to prove the statistical significance resulting in more curiosity to find out the relevancy of correlation between nutrition knowledge and dietary intake. The result in this study shows that there is no significant relationship between the level of knowledge regarding benefits of vegetables and intake frequency among population in both areas in Selangor. This may due to majority of folks in both areas received tertiary level of education. This is contrary to a study on relationship between nutrition knowledge and food intake that support the relationship between nutrition knowledge in promoting healthier diet intake. Availability of vegetables in particular area is crucial to the influence of consumption pattern because most of the population purchased vegetables in nearby supermarket or grocery store. However, the availability of indigenous and produce vegetables sold in supermarkets or grocery store is different. Through observation, most indigenous vegetables rarely sold commercially in supermarket or daily wet market and only can be found on farmer's market day. This may due to small production of these crops and also less market demand. Less market demand may be driven by unpopularity of indigenous vegetables among Malaysian citizens.

5 Conclusion

It can be concluded that intake frequency of both types of vegetables was not influenced by the difference in residency areas. Other factors tested in this study such as household income, level of education, nutrition knowledge, preferences and availability and food behaviour show no statistical significance between rural and urban setting. It is suggested that the development of the survey form used in this study is continued until the last phase in future research to produce more valid and reliable data that contribute to a bigger scope of statistical analysis. It is also recommended that the subject of either Malaysian indigenous vegetables or any indigenous vegetables existed worldwide is further explored in future research in terms of consumption pattern, nutrient contents and their benefit in combating diseases.

Acknowledgements This work is supported by Universiti Teknologi MARA (600-IRMI/DANA5/3/LESTARI 0091/2016). The authors also gratefully acknowledge Norazmir Md Nor and Khairil Anuar Isa for expert guidance on the questionnaire and SPSS.

References

Department of Agriculture Forestry and Fisheries, R. of S. A. (2009). Indigenous food crops. *Agriculture, forestry and fisheries*.

Farragher, T., Wang, W. C., & Worsley, A. (2016). The associations of vegetable consumption with food mavenism, personal values, food knowledge and demographic factors. *Appetite, 97*, 29–36. https://doi.org/10.1016/j.appet.2015.11.005.

Kendzierski, D., Ritter, R. L., Stump, T. K., & Anglin, C. L. (2015). The effectiveness of an implementation intentions intervention for fruit and vegetable consumption as moderated by self-schema status. *Appetite, 95*, 228–238. https://doi.org/10.1016/j.appet.2015.07.007.

Langat, P. (2014). Vegetables : A comparative study of agricultural and pastoral.

Mad Nasir, S., Jinap, S., Alias, R., Abdul Ghariff, R., Sheng, T. Y., & Ahmad Hanis Izani, A. H. (2010). Food consumption trend; Transforming issues into. *Journal of Agribusiness Marketing*.

Ministry of Health Malaysia. (2015). National Health Morbidity Survey (NHMS).

Morgan, R. (1991). Vegetable storage in root cellars and basements in Alaska. *University of Alaska, Cooperative Extension Service (USA)*.

Patrick, H., Nicklas, T. A., Hughes, S. O., & Morales, M. (2005). The benefits of authoritative feeding style: caregiver feeding styles and children's food consumption patterns. *Appetite, 4*(2), 243–249.

Yeh, M.-C., Glick-Bauer, M., & Wechsler, S. (2016). Fruit and vegetable consumption in the United States: Patterns, barriers and federal nutrition assistance programs. *Fruits, Vegetables, and Herbs, Bioactive*, pp 411–422.

Zimmer, Z., & Prachuabmoh, V. (2012). Comparing the socioeconomic status—Health gradient among adults 50 and older across rural and urban areas of Thailand in 1994 and 2007. *Social Science and Medicine, 74*(12), 1921–1928.

3D-MSWT: An Alternative Tool in Developing Students' Understanding in Landform Interpretation Course

Ruwaidah Borhan, Azran Mansor and Nur Hanim Ilias

Abstract Student's engagement and understanding is becoming vital important for development of high-quality teaching and learning experience. Generally accepted that this become a challenge in the outcome-based learning environment. Traditionally, learning the landform and topography in Landscape Architecture subject needed the student to be able to understand and visualize the types of different terrain in the 2D aspect. This paper revealed the treatment when learning in a different approach by using 3D-modelling sculpture in water tank method (3D-MSWT). An experimental research has been carried out by using both quantitative and qualitative approaches. A simple random sampling technique is used to the Landscape Architecture Diploma students in second semester at UiTM Cawangan Perak ($n = 35$) to clue the findings. This method has shown a significant increase in the students' engagement and understanding of the topic compared to the conventional method.

Keywords Landform · Topography · 3D-modelling sculpture · Water tank

1 Introduction

The traditional teaching method in delivering landform and earthworks subject in Landscape Architecture course is arguable whether it is still relevant to optimize the students' understanding. This study is aligned with government agenda to promote more innovation practice at aim to optimize student's learning potential in classroom. The alternative method in learning the landform interpretation on 2D and 3D

R. Borhan (✉) · A. Mansor · N. H. Ilias
Universiti Teknologi MARA, Shah Alam, Malaysia
e-mail: aidah866@perak.uitm.edu.my

A. Mansor
e-mail: azran973@perak.uitm.edu.my

N. H. Ilias
e-mail: nurha048@perak.uitm.edu.my

© Springer Nature Singapore Pte Ltd. 2018
R. Saian and M. A. Abbas (eds.), *Proceedings of the Second International Conference on the Future of ASEAN (ICoFA) 2017 – Volume 2*,
https://doi.org/10.1007/978-981-10-8471-3_86

aspect is tested. The objective of this research is to investigate the relationship between the alternative learning methods towards students' understanding on the landform interpretation, and secondly to evaluate the student's perception on the alternative learning method. The methodology of this research includes the development of objectives, data collection and sampling, establishing model and simulation, quantitative and qualitative data analysis, and finally conclusion and recommendation. This research uses two methods, namely the quantitative (through questionnaire before and after the method is adopt) and qualitative (data obtained from observation and interviews of respondents) study. Data analysis is crucial to enable researchers understand more on student's feedback and result towards the alternative learning method. At the end of this research, the finding is able to reveal the differences between using a 3D-modelling sculpture in water tank method (3D-MSWT) towards students' understanding in learning landform interpretation and how this alternative method has brought different towards students' understanding of the subject. This research is important as it may be able to give impact to students' understanding level towards subject matter as well as a part of the innovation to teaching and learning experience by the students.

2 Literature Review

An era of learning patterns of face-to-face method has evolved in parallel with the transforming of the new technology that's acquired the learners been more creative and critical thinker. The learning scope covers a very broad domain, namely cognitive (thinking and mind), affective (feeling and emotion) and psychomotor (physical member body). According to the study by Barnet and Ceci, the core of educational policy is transferring of learning across content from 2D to 3D which are the context be able in development of abstract thinking and in particular development of a flexible representation. Moreover, seeing the sculpture or target material in depth in a moving realistic format can make more fun and excitement towards the learners therefore it gives more effective and legible explanation towards the subject (Dalgarno et al. 2010). Through the study by Hill, learning occurs when experience leads to a relatively permanent change in knowledge or behaviour. According to the study by Alias and Zainuddin (2005), their finding shows that dealing with the new method or technology can open mindedness among the students which leads to the positive attitudes in learning environment. In additional, Slavin suggests that learning as a change in an individual is a result of experience. People began to learn from birth, and all learning is closely linked to the experience.

The 3D model sculpture is one of the alternative ways and practice in delivering the content and context to the learners by representing the relationship of spatial and visual through experience. It is been agreed by Frisby (1980), who stated that 3D perspective is an alternative method of representing spatial relationship which requires more inferences to derive depth information about visual scene than stereoscopic presentation and is generally agreed to be less effective in conveying spatial relationship and depth information. Thru 3D images, students have better understanding of the relationship study, and 3D images also help in increasing student's achievement were reported study by Ribaupiere and Wilson (2012). Moreover, thru the research by Umar et al. (2011), 3D learning tools are very effective in accessing the students to have a better understanding in the learning process. Meanwhile, according to Barnett and Ceci's framework studies, transferring of learning from 2D sources allows for specific manipulation of modality (from 2D to 3D and vice versa), physical context (different experimenter, different room), temporal context (immediate v. delay), task difficulty (simple v. complex tasks), memory demands (practice, repetition, delay) and perceptual and linguistic cues.

However, based on Dalgarno et al. (2010), heightened interest in 3D learning tools has not been accompanied at the same level by empirical research designed to test their impact on teaching and learning. Thus, there is a persistent and perceived need expressed in the educational literature for more such research to illuminate the effects of 3D environments on learning outcomes. O'Byrne et al. found that the use of realistic interactive anatomy modelling is of special value for kinaesthetic learners and is supportive of self-learning, a process which requires active and deliberate cognitive engagement on the learners' part. This is an opportunity to be investigated on landform interpretation in landform and earthworks subject.

3 Methodology

This experimental research is carried out by using both quantitative and qualitative approaches. The quantitative part is explained and analysed with statistical procedures. The data is obtained from the survey questionnaire's feedbacks from Landscape Architecture Diploma students in second semester at UiTM Cawangan Perak ($n = 35$) through simple random sampling technique. On the other hand, the qualitative approach is focused on explanatory meaning and description towards the data that are collected. Data is obtained from the literature review, interview and observation to evaluate the opinion, view, perception of the students based on their experience of implementing the 3D-modelling sculpture in water tank method (3D-MSWT) in learning the subject.

First stage (Literature Review): The statement of issues and problems is developed. Second stage (Data Collection and Sampling): Structured questionnaire is administered in the context of face-to-face and formal interviews. They are carried out after using the 3D-MSWT to the same group of respondent. This incorporated with a series of observation towards respondent. Third stage (3D-MSWT's method): The prototype of 3D-MSWT is developed based on the previous background study and the literature review at aim to improve students' understanding by encourage respondent kinaesthetic involvement and engagement during learning the landform interpretation subject. Fourth stage (Modelling and Simulation): Respondents undergone a simulation experience of the 3D-MSWT's method in learning the landform interpretation subject. This is ended up with a set of survey questionnaire to be answered to measure the level of their understanding after using this method. Fifth stage (Analysing of Quantitative and Qualitative Data): The quantitative data is analysed by using SPSS computer software application (descriptive statistics and correlation analysis) while the qualitative data by using the SWOT technique to understand in depth the scenario and help the researchers to elaborate and support the statistical result. Sixth stage (Conclusion): The conclusion and recommendation out of the result is made.

4 Result and Discussion

The questionnaire contains three main components, namely (a) background of respondent, (b) perception towards 3D-MSWT method (IEN) and (c) importance (IMP). The two main constructs were the main variables to clue the student motivation and understanding in the landform and earthwork subject. Every item in the questionnaire is followed by five choices of answers using the Likert Scale. Choices of response range from (1) Highly Disagree to (5) Highly Agree for the IEN and IMP dimensions. The validation and confirmation of all constructs were done using Exploratory Factor Analysis (EFA). EFA is used to gather information about the interrelationship among a set of variables. The result for the level of reliability was found by calculating the Cronbach's Alpha. The dimensions of the construct have a good reliability value as the Cronbach's Alpha value exceeds 0.60. The results indicated that the Alpha values for perception of Interesting/Enjoyable (IEN) = 0.94 and Perception of Importance (IMP) = 0.91. These results of Alpha value for all construct and dimensions achieved good Alpha reliability levels as shown in Table 1.

Table 2 shows the percent agree for each construct. Data represent that students are motivated in using the 3D-MSWT method in landform and earthwork class where the per cent agree for each construct is more that 50%. They agreed that the session using 3D-MSWT was interesting and enjoyable (73.8%) and importance (65.8%).

The t-test analysis conducted (Table 3) shows that there are no statistically different between male and female students' motivation during teaching and learning session using 3D-MSWT method. Both are equally motivated in the same way.

Table 1 Cronbach's Alpha value for all constructs

Constructs	Item	Description of items	Corrected item-total correlation	Reliability (Cronbach's Alpha)
Interesting/ enjoyable (IEN)	1	I can easily understand the topic with this teaching aid	0.86	0.94
	2	I can easily relate the 3D aspect of landform and earthwork to their contour line	0.81	
	3	I can easily explain the landform physical characteristic with this teaching aid	0.75	
	4	I have increased my confidence level in landform and modifying contour line related subject	0.79	
	5	I enjoyed learning the subject with this teaching aid	0.76	
	6	The 3D-MSWT method has positively affected my motivation to learn this subject	0.76	
Importance (IMP)	7	I feel this teaching aid is very important to have in landform and earthwork class	0.76	0.91
	8	I feel this teaching aid is a must have in landform and earthwork class	0.75	
	9	I feel this teaching aid is very important to ensure student understanding of the landform characteristics well	0.72	
	10	I would highly recommend to use 3D-MSWT as teaching aid in understanding landform and contour interpretation class	0.79	

Table 2 Inner motivation of students for each construct

Construct	Items	% Agree
Interest/enjoyment	1, 2, 3, 4, 5, 6	73.8
Importance	7, 8, 9, 10	65.8

$N = 35$

Table 3 Difference in motivation for male and female students

Pair	N	Mean	Std. dev	T	df	Sig. 2 tailed
Male	17	4.90	1.07	0.022	33	0.982
Female	18	4.90	0.85			

$p > 0.05$

The feedbacks from students in the experimental group regarding the using of 3D-MSWT method in landform and earthwork class are as follows:

1. These activities enable them to engage with interactive learning in class.
2. 3D-MSWT method is straightforward and easy to use.
3. It is effective in helping them to understand the landform interpretation concept faster.

Few excerpts of positive comments by the experimental student are: 'It is so easy to visualize on how landform is associated with the earth contour line', 'It is easy for me to understand the concept of landform modification' and 'It is so easy to read the conversion between 2D drawings to 3D landform build and vice versa. These comments clearly support that 3D-MSWT method can certainly increased student motivation and student understanding on landform interpretation and earthwork concept in class. In other words, the treatment conducted to the subjects of the experimental group was successful to reinforce their understanding landform interpretation and earthwork subject.

5 Conclusion

It can be concluded from the research findings that the study group showed a significant improvement on confidence and understanding of the subject matter after treatment is conducted. It shows that the 3D-MSWT method was successfully reinforced understanding of second semester of Landscape Architecture students in UiTM Cawangan Perak. This is because the 3D-MSWT method was used as teaching aid to promote the interactive collaboration learning managed to enhance the understanding on landform interpretation. Therefore, the teaching method by using 3D-MSWT method as a teaching aid has a high positive impact on the Landscape Architecture students about landform interpretation and earthworks subject.

Acknowledgements The research reported in this article was supported by grant from the Universiti Teknologi MARA (Grant No: 600-IRMI/DANA 5/3//ARAS (0087/2016)). The authors also gratefully acknowledge the helpful comments and suggestions of the reviewers, which have improved the presentation.

References

Alias, N. A., & Zainuddin, A. M. (2005). Innovation for better teaching and learning: Adopting the learning management system. *Malaysian Online Journal of Instructional Technology, 2*(2), 27–40.

Dalgarno, B., et al. (2010). What are the learning affordances of 3-D virtual environments? *British Journal of Educational Technology, 41*(1), 10–32.

Frisby, J. (1980). *Seeing: Illusion, brain, and mind*. Oxford: Oxford United Press.
Ribaupierre, S., & Wilson, T. D. (2012). Construction of a 3-D anatomical model for teaching temporal lobectomy. *Computers in Biology and Medicine, 42,* 692–696. https://doi.org/10.1016/j.compbiomed.2012.03.005.
Umar, R. S., et al. (2011). Menggunakan Animasi di dalam Instruksi Khas untuk Kanak-Kanak Disleksia. Jurnal Teknologi Pendidikan Malaysia, Jilid 1, Nombor 2.

Analysis of Weld Line Movement for Hot Runner System

Saiful Bahri Mohd Yasin, Zuliahani Ahmad, Sharifah Nafisah Syed Ismail, Noor Faezah Mohd Sani and Zafryll Amir Zulkifly

Abstract In the plastic industry, high-performance structural and fully furnished characteristics of products are important scales that need to be considered by reducing and avoiding the defects occurred. One of the common defects occurred in the molding process is the present of weld line. Hence, the purpose of this study was to investigate the formation of weld lines and their movement regarding the temperature variations in hot runner system by using computer-aided engineering (CAE) software. The hot runner system was designed to have two injection locations in which the temperature at one of the injection location was fixed at 270 °C and the other was set in the range of 270–190 °C with reduction interval of 10 °C. The result shows that the weld line movements relocated far from the center point with increasing in averagely 1.2-mm-interval for the decrement of temperature and the final weld line distance changed to 10.22 mm at 80 °C (higher temperature reduction). While, the weld line appearance remains to have low visibility due to the melt temperature difference was found to range ±20 °C.

Keywords Melt filling · Hot runner system · Mold temperature and weld lines and using computer-aided engineering (CAE)

S. B. M. Yasin (✉) · Z. Ahmad · S. N. S. Ismail · Z. A. Zulkifly
Polymer Technology Department, Faculty of Applied Sciences, Universiti Teknologi Mara Cawangan Perlis Kampus Arau, Arau, Perlis, Malaysia
e-mail: sbmyasin@gmail.com

Z. Ahmad
e-mail: zuliahani@perlis.uitm.edu.my

S. N. S. Ismail
e-mail: sh.nafisah86@yahoo.com

Z. A. Zulkifly
e-mail: zafryllamir95@yahoo.com

N. F. M. Sani
Faculty of Applied Sciences, Universiti Teknologi MARA,
Perak Branch, Tapah Campus, 35400 Tapah Road, Perak, Malaysia
e-mail: faezah125@uitm.edu.my

1 Introduction

Nowadays, the need for the polymer-based products is quite increasing in which the applications cover almost the field areas. The polymer-based products are well-known for their outstanding performances especially in delivering the high strength properties and flexibility (Uygunoglua et al. 2015). These advantages give the priority for the polymer product to become a leader as material selection, thus replacing other materials such as metal, glass, ceramic, and wood. However, various factors can affect the polymer properties, especially regarding its strength. The main factor is the problem during manufacturing process, in which the production of defect on the product surface can affect both mechanical and appearance properties of the product (Zhi-yun et al. 2009; Brahimi et al. 1994).

One of the common defects occurred in the molding process is the present of weld line on the product surface. Weld line can be divided into two types which are cold weld line or also known as knit line and hot weld line or also known as meld line (Dzulkipli and Azuddin 2017). It is actually a thin hair-like crack that is produced when two separate melt fronts or streams met each other and recombined during the materials filled into the mold cavity (Onken and Hopmann 2016). As the consequence, the unsuitable emerging position of this weld line on the product surface can affect the aesthetic value of the product as it is one of the visible marks on the product surface and the most destructive that it can reduce the material strength at this position.

The weld line can cause the vicinity area to become weak regarding certain reasons such as lowering the molecular entanglement and disturb the molecular orientation (Morelli et al. 2017). For these reasons, it is crucial to know how the weld line is formed and the way to eliminate or relocate it. The weld line is produced due to certain factors such as multiple gating being used, presence of flow obstacle in the mold, different part thickness and jetting (Selden 1997; Wu and Liang 2005). The present of weld line becomes more critical issue if it is situated at the weak area or area that will be subjected to force and at the surface appearance of the product. The action to remove the weld line is quite impossible, therefore the alternative way can be taken by relocating or moving away the present of weld line from emerging at the product that promote high stress during application (failure area) or at the area that need to appear smooth (Kagitci and Tarakcio 2016).

Based on the previous study, in order to relocate the position of weld line, previous remedies such as changing the gate location, maintaining the nominal wall thickness, and providing venting system near the weld line's area had been done. However, in the application of hot runner system, an alternative to move the location of weld line to the less sensitive area had become a new introduction as there is limited research about this finding.

In this experiment, the present of weld line is able to be detected by using CAE simulation. In this simulation, the idea of using hot runner system is proposed in order to study the relationship between the setup temperature differences and the movement of the weld line. As the consequence, the weld line is expected to be positioned in the non-failure area.

2 Methodology

2.1 Material Description

High-density polyethylene (HDPE) plastic was selected to be used in this simulation which has 120 mm (length) × 60 mm (width) × 3 mm (thickness). Based on the design, this plastic product has hollow section in the middle with fillet radius at edge of the product. HDPE has the processing parameter of melt temperature 234.5 °C with range of minimum 193 °C and maximum 274 °C. The mold temperature for HDPE is found to be 39 °C with range of minimum 4 and 49 °C. The mechanical properties of the material for both elastic modulus first and second principles are 1240 MPa. The poisons ratio is found to be 0.426. The shear modulus of HDPE material is 434 MPa. The rheological properties in terms of melt-mass flow rate (MFR) are 20 g/10 min at temperature 190 °C with load 2.16 kg. The HDPE melt density is 0.73043 g/cm^3, while the solid density is 0.95163 g/cm^3.

2.2 Process Analysis

The analysis was done by observing the weld line formation when the temperature of both runners was set at the same values. The result shows the weld lines formed at the center of the product, which is indicated as at an origin point, O. This point is considered as a weak section due to thin part dimension and almost high exposure to the applied force. The origin point has also become one of the areas that has significant appearance on the product surface. Therefore, the analysis was carried out to prove the difference in hot runner temperatures that lead to induce the movement of the weld line in order to give a better placement for the weld lines formation, far from the point O, or closer to product end to avoid the fracture toward the product when the force is being applied and to maintain high strength of the product. The movements of weld line to the invisible area will improve product strength and furnished characteristics of product. This project theory can be visualized in Fig. 1. The direction of weld lines is expected to move toward the lower temperature runner (LR), while at the right region (RR), a fixed temperature was set at the control condition (270 °C). The analysis had been done by using Autodesk MoldFlow Adviser software.

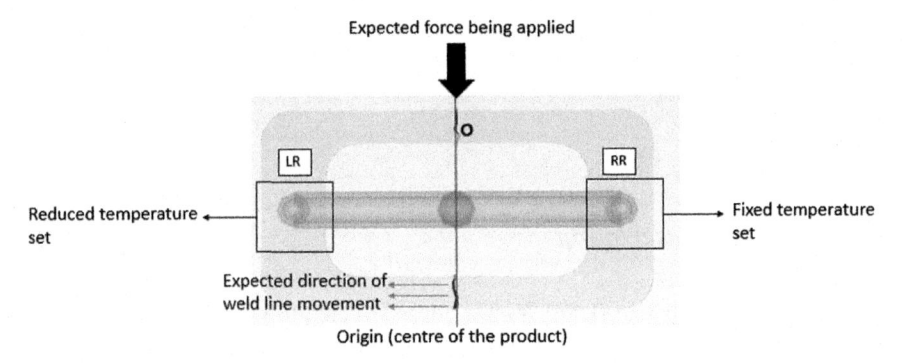

Fig. 1 Weld line formation at the center of the product

3 Result and Discussion

The weld line movement from eight different analyses and one control condition (both runners with same temperature) for the hot runner system was observed and recorded as illustrated in Table 1.

The following figures show the temperature setting in both LR and RR and the resulting distance of weld line formation from point **O**.

Figure 2 shows the first modeling simulated from the CAE software for the control condition. The temperatures for each runner were set at 270 °C, since there are no temperatures different, thus the material started to melt and flow at the same rates. As the consequence, the flow fronts were found to meet each other at the origin point (0 mm from the origin, 60 mm from the product end). Next, the temperature of LR is reduced in the interval of 10 °C, in which temperature is reduced in range from 260 to 190 °C as shown in Figs. 3, 4, 5, 6, 7, 8, 9, and 10, while the temperature of RR is maintained at 270 °C.

Table 1 Weld line distance from origin against hot runner temperature

Hot runner temperature setting (°C)	Temperature difference (°C)	Distance of weld line from origin (mm)
LR (270); RR (270)	0	0
LR (260); RR (270)	10	2.71
LR (250); RR (270)	20	3.39
LR (240); RR (270)	30	4.04
LR (230); RR (270)	40	5.89
LR (220); RR (270)	50	6.17
LR (210); RR (270)	60	8.27
LR (200); RR (270)	70	9.18
LR (190); RR (270)	80	10.22

Analysis of Weld Line Movement for Hot Runner System 887

Fig. 2 LR (270); RR (270)

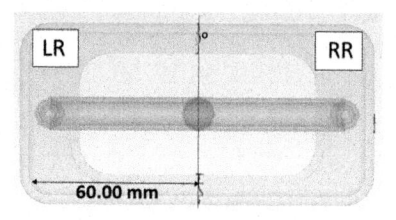

Fig. 3 LR (260); RR (270)

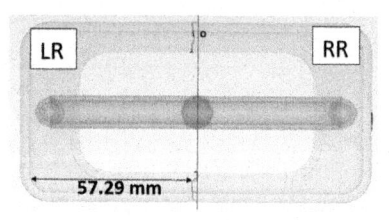

Fig. 4 LR (250); RR (270)

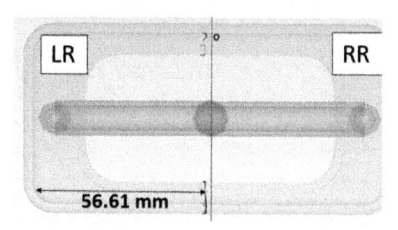

Fig. 5 LR (240); RR (270)

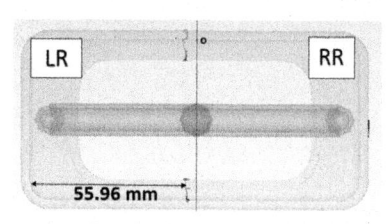

Fig. 6 LR (230); RR (270)

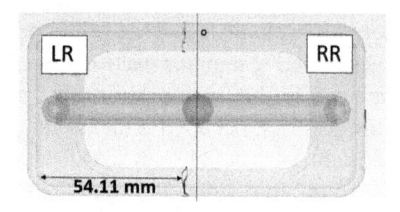

Fig. 7 LR (220); RR (270)

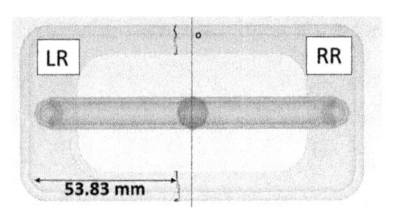

53.83 mm

Fig. 8 LR (210); RR (270)

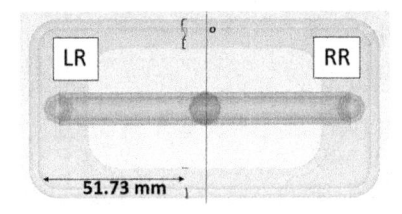

51.73 mm

Fig. 9 LR (200); RR (270)

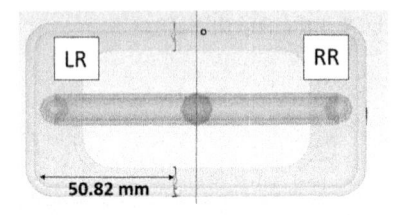

50.82 mm

Fig. 10 LR (190); RR (270)

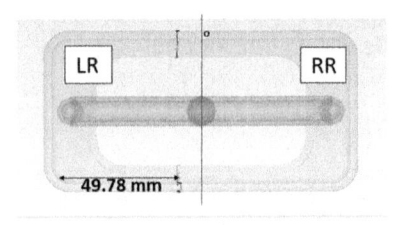

49.78 mm

Based on these results, it shows that there is a significant change in the distance between weld lines and the origin point, where the distance of weld line changed to averagely 1.2 mm for each of the temperature difference. This is due to the hot runner with high temperature that will easily melt the materials, and the material will flow first occupying the product, while the low-temperature runner provides a resistant in material melt and flow into the cavity. Thus, it demonstrates that the both flow fronts met at the low-temperature runner system and produced the weld line at this area.

Based on the final result, the distance of weld line is increased to 10.22 mm at 80 °C (higher temperature reduction) as illustrated in Fig. 10. The distance of weld lines was increased to 10.22 mm and able to move away from the force area or failure area. In addition, the weld line will be expected to have moderately low visibility on the product surface due to the melt temperature for the LR (at 190 °C) and RR (at 270 °C) which were recorded at 234.6 and 238.6 °C, respectively, as shown in Fig. 11. Thus, an invisible weld line that occurred due to the melt temperatures was found to have the tolerance to range ±20 °C, which means around to 234.5 °C (the processing parameter of melt temperature).

Figure 12 shows a plotted graph to present the relationship between the decrement of the temperature difference at the left hot runner (LR) and distance of weld line change from the origin point, **O**. The trend shows that the relationship is directly proportional in which the higher the temperature difference, the longer distance of weld line from point **O**. From this analysis, it shows that the application of hot runner system is able to move the weld line position to non-failure area. This situation is significant to overcome low structural strength at the center part.

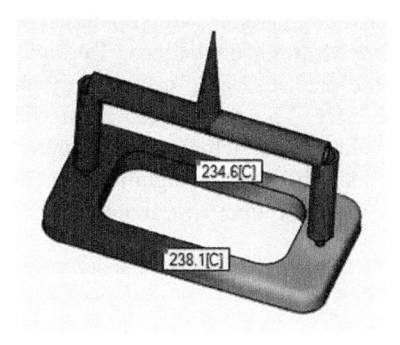

Fig. 11 Temperature difference

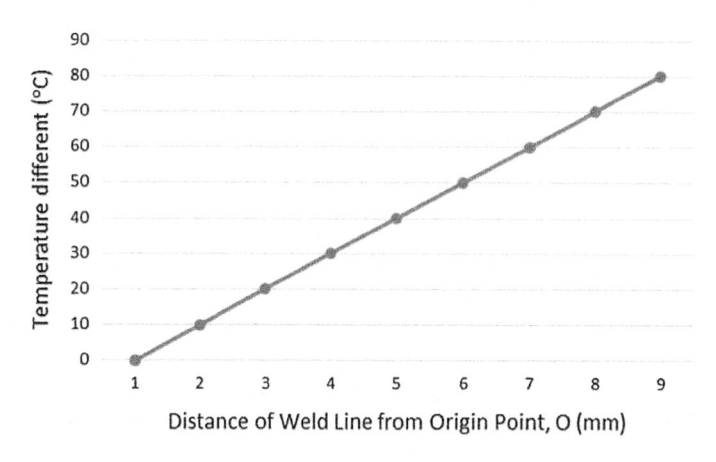

Fig. 12 Temperature difference against distance of weld line from origin point, O

4 Conclusion

As conclusion, the weld line analyses were successfully done by using CAE software. It shows that the application of hot runner system was able to be analyzed in order to study the movement of weld lines before starting the manufacturing process. Based on the result, the weld line formation can be adjusted by varying, in terms of reducing the temperature of LR (ranging from 260 to 190 °C) by interval of 10 °C while maintaining the temperature of RR at 270 °C. The weld line movements show a trend of moving far from point O with increasing in averagely 1.2-mm-interval for the decrement of temperature and the final weld line distance changed to 10.22 mm at 80 °C (higher temperature reduction). While, the weld line appearance remain to have low visibility due to the melt temperature difference was found to range ±20 °C.

The recommendations can be taken into account by proposing the real experimental test in order to make comparison with simulation result, where CAE simulation is just only capable of doing predictions. Through real testing, the result produced will become more precise. Next, further study on the visibility of weld line produced also can be done. The process can be preceded through simulation and real test in which the value of melt temperature difference above 20 °C can lead to the divisibility of weld line. This study is significant to check the visibility of the weld line that also affects in the product appearance.

References

Brahimi, B., Ait-kadi, A., & Ajji, A. (1994). Weld lines and mechanical properties of injection molded polyethylene/polystyrene/copolymer blends. *Polymer Engineering and Science, 34*(15).

Dzulkipli, A. A., & Azuddin, M. (2017). Study of the effects of injection molding parameter on weld line formation. *Advances in Material & Processing Technologies Conference,* 663–672.

Kagitci, C. Y., & Tarakcio, N. (2016). The effect of weld line on tensile strength in a polymer composite part. *The International Journal of Advanced Manufacturing Technology,* 1125–1135.

Morelli, C. L., de Sousa, J. A., & Pouzada, A. S. (2017). Assessment of weld line performance of PP/talc moldings produced in hot runner injection molds. *Polymer Engineering and Sciences*.

Onken, J., & Hopmann, C. (2016). Prediction of weld line strength in injection-moulded parts made of unreinforced amorphous thermoplastics. *International Polymer Science and Technology,* 574–580.

Selden, R. (1997). Effect of processing on weld line strength in five thermoplastics. *Polymer Engineering and Science, 37*(1).

Uygunoglua, T., Gunes, I., & Brostow, W. (2015). Physical and mechanical properties of polymer composites with high content of wastes including boron. *Materials Research*.

Wu, C. -H., & Liang, W. -J. (2005). Effects of geometry and injection-molding parameters. *Polymer Engineering and Science*.

Zhi-yun, Y., Yu-mei, D., Peng-cheng, X., & Wei-min, Y. (2009–04). Weld line defect in injection molding. *Development of Visual Device for Injection Molding Process*.

Ethnobotanical Study on Plant Materials Used in Malay Traditional Post-partum Bath (*Mandi Serom*) Among Malay Midwives in Kedah

Nur Illani Abdul Razak, Rashidi Othman and Josephine Tening Pahang

Abstract A study was carried out on the traditional knowledge of plants used in *mandi serom*, a traditional post-partum bath among the Malays in the state of Kedah, Malaysia. In Malay culture, *mandi serom* is an essential treatment after childbirth. Information was obtained from 15 Malay midwives (*bidan kampung*) through interviews, and also by observing and participating in their activities during each visit. A total of 40 species of plants were collected during the botanical surveys. The species are dominated by trees, followed by zingibers, herbaceous, shrubs, and climbers. Plant parts most commonly used are leaves, whole plant, bark, seeds, bulb, flowers, roots, and gall. Only three species are similarly used by all the midwives which are *Cymbopogon nardus* (*serai wangi*), *Lawsonia inermis* (*inai*), and *Pandanus amaryllifolius* (*pandan*). The majority of Malay women in Kedah put their trust in midwives to conduct *mandi serom* during the post-partum confinement period. The plants are used to rid the body of odour, for spiritual cleansing, for hygienic purposes, and to ward off mystical forces known as *makhluk halus* in Malay culture. They believe that without proper formulation from the midwife, *mandi serom* may not be healing and effectual. This is the first systematic study of Malay post-partum bath (*mandi serom*) in northern Malaysia. This study helps to preserve traditional knowledge of Malay midwifery practice and protect Malay natural heritage, and at the same time, these new ethnobotanical records can be subjected to clinical studies and serve as a guideline for women health care in Malaysia.

N. I. A. Razak · J. T. Pahang
Faculty of Plantation and Agrotechnology, Universiti Teknologi Mara,
Perlis Branch, Arau Campus, 02600 Arau, Perlis, Malaysia
e-mail: illanirazak@perlis.uitm.edu.my

R. Othman (✉)
Herbarium Unit, Department of Landscape Architecture Kulliyyah of Architecture and Environment Design, International Institute for Halal Research and Training (INHART), International Islamic University Malaysia, 53100 Kuala Lumpur, Malaysia
e-mail: rashidi@iium.edu.my

Keywords Malay midwifery · Traditional bath · Ethnobotany Traditional knowledge · Medicinal plants · Malay culture

1 Introduction

Historically, midwives have always been around to help women give birth using natural procedures by involving medicinal plants. Such traditional procedures were the primary form of treatment among traditional peoples with limited access to biomedicine. Midwives from various ethnic groups all over the country use diverse groups of plants in their practice of midwifery to treat many health issues pertaining to fertility problems, birth control, pregnancy, parturition, post-partum care, neonatal care, and primary health care of women, infants, and children (Nigenda et al. 2006; Coe 2008). Their knowledge of medicinal plants has played a major role in pre- and post-natal care in many rural and urban areas (Whitaker 2003; Gollin 2004; Hamilton 2004; McNeely 2005; Green et al. 2006).

Medicinal plants have a significant role during pregnancy, birth, and post-partum care in many rural areas of the world. Plants used in women's health-related conditions, such as female fertility, menopause, abortion, menorrhoea, birth control, pregnancy, birth (parturition), post-partum (puerperium), lactation, infant and children care, have been documented for various ethnic groups (Singh et al. 1984; Bourdy and Walter 1992; Jain et al. 2004; Ticktin and Dalle 2005; Lamxay et al. 2011; Kim and Lean 2013; Torri 2013; Sein 2013). However, as stated by Pfeiffer and Butz many researches on these plants often focus on the knowledge of male traditional healers, while missing the wealth of knowledge held by women.

In the Malay community, midwives are commonly known as *bidan, bidan kampung,* or *mak bidan.* Traditional midwifery is an exclusive female occupation compared to many other traditional healers in Malay folk medicine such as clairvoyant *(nujum),* shaman *(bomoh),* and medicine man *(dukun)* which tend to be dominated by men. Her concern is not merely with the actual birth itself but also with the care of the mother from pregnancy through to the forty-fourth post-partum day. Not only trusted in childbirth, her duty also concerns weddings, pregnancies, pre- and post-natal care, nursing and weaning, infant and childcare, family planning, menstrual cycle conditions, abortion, treatment muscular aches, and pain and bone setting (Chen 1981).

Baths are used extensively in traditional healing. *Mandi serom* is a term commonly used by Malay locals referring to traditional post-partum bath during the 44 days of confinement period. This traditional post-partum bathing serves to freshen the body with the use of aromatic leaves in bath water and at the same time helps to heal the woman internally. During *mandi serom,* the woman must sit on a stool while bathing. Standing up and squatting is strictly prohibited. The use of aromatic herbs in *mandi serom* is believed to remove foul body odours from lochia, freshen the body and mood, make the body warm, and expel wind from the body (Barakbah 2007; Hassan 2007; Jamal et al. 2011).

The bathing practice involves a wide array of plants normally prepared by the *bidan kampung* (midwife). The midwife will gather the plants for the preparation of mandi serom, or sometimes family members of the new mother will take the responsibility to collect plants as instructed by the midwife. One of the interesting parts in Malay traditional post-partum care is that practices lightly vary from one location to another. The herbal formulations used in each particular location are usually based on local plants found in that area (Hasan 2007; Ishak 2012).

The knowledge of plant utilisation in traditional midwifery has long been upheld. Midwives have emerged as one of the most important persons in village life and the most commonly available herbalist. However, today their status is not as popular as back then due to reliance on modern medicine. There are lots more to discover about Malay traditional medicine, and the effort should be pursued.

2 Material and Method

Fifteen Malay traditional villages in three districts of Kedah which are Kubang Pasu, Kota Setar, and Pokok Sena were chosen as study areas for folk ethnobotanical survey of medicinal plants. The selected villages are traditional Malay villages with forests and paddy fields nearby. The herbal medicine practised by the villagers is predominantly plant based that grows naturally in the surrounding habitats and cultivated plants. All verbal information was obtained through face-to-face interviews with 15 Malay midwives (*bidan kampung*) guided by a predetermined set of questions during each visit (Martin 1995). Plant samples and plant parts collected during the fieldwork were preserved as voucher specimens. Each plant voucher specimen was collected according to standard practice (Martin 1995; Alexiades and Sheldon 1996; Alexiades 1996). They were identified with the help of published authentic literature such as Werner (2002), Samy, Sugumaran and Lee (2005), Min et al. (2006), and Zakaria and Mohd (2010). To facilitate cross-checking of plant species, the specimens were identified through various floristic records or secondary data such as books, Internet, and Forest Research Institute of Malaysia (FRIM) herbarium, Kepong, in addition to previous research studies and journals to ascertain the nomenclature as further detailed by Bandaranayake (1998). Unidentifiable voucher specimens were numbered and brought to FRIM herbarium for further examination.

3 Results and Discussion

A total of 40 plant species were documented during the surveys by the Malay traditional midwives in *mandi serom* as shown in Table 1. The 40 plant species of medicinal plants correspond to 26 families. The plant habitats are dominated by trees, followed by zingibers, herbaceous, shrubs, and climbers. Among the different

Table 1 List of plant species used during *mandi serom* by 15 traditional Malay midwives in Kedah

Scientific name	Local name	Family	Plant habitat	Parts used
Allium sativum	Bawang putih	Alliaceae	Herbaceous	Bulb
Alpinia galanga	Lengkuas	Zingiberaceae	Zingiber	Leaves
Averrhoa bilimbi	Belimbing buluh	Oxalidaceae	Tree	Leaves
Barringtonia racemosa	Putat	Lecythidaceae	Tree	Leaves
Blumea balsamifera	Capa	Asteraceae	Herbaceous	Leaves
Centella asiatica	Pegaga	Apiaceae	Herbaceous	Leaves
Cinnamomum iners	Daun teja	Lauraceae	Tree	Leaves
Cinnamomum zeylanicum	Kulit kayu manis	Lauraceae	Tree	Bark
Citrus hystrix	Limau purut	Rutaceae	Shrub	Leaves
Curcuma longa	Kunyit	Zingiberaceae	Zingiber	Leaves
Cymbopogon nardus	Serai wangi	Gramineae	Herbaceous	Leaves
Elephantopus scaber	Tutup bumi	Asteraceae	Herbaceous	Whole plant
Entada spiralis	Sintok	Leguminosae	Climber	Root
Etlingera elatior	Kantan	Zingiberaceae	Zingiber	Leaves
Euphorbia tirucalli	Tulang-tulang	Euphorbiaceae	Shrub	Leaves
Flemingia strobilifera	Meringan	Leguminosae	Shrub	Leaves
Hibiscus rosa-sinensis	Bunga raya putih	Malvaceae	Shrub	Leaves
Kaempferia galanga	Cekur	Zingiberaceae	Zingiber	Whole plant
Lawsonia inermis	Daun inai	Lythraceae	Tree	Leaves
Quercus infectoria	Manjakani	Fagaceae	Tree	Gall
Mangifera spp	Mangga telur	Anacardiaceae	Tree	Leaves
Michelia × alba	Bunga chempa putih	Magnoliaceae	Tree	Leaves
Micromelum pubescens	Cemumar	Rutaceae	Tree	Leaves
Mimosa pudica	Semalu	Fabaceae	Shrub	Whole plant
Mimusops elengi	Bunga tanjung	Sapotaceae	Tree	Leaves, flowers
Morinda citrifolia	Mengkudu kecil	Rubiaceae	Tree	Leaves
Musa paradisiaca	Pisang kelat	Musaceae	Zingiber	Bark
Pandanus amaryllifolius	Pandan	Pandanaceae	Herbaceous	Leaves
Peronema canescens	Sungkai	Verbenaceae	Tree	Leaves
Pimpinella anisum	Jintan manis	Apiaceae	Herbaceous	Seed
Piper betle	Sireh	Piperaceae	Climber	Leaves
Pluchea indica	Beluntas	Asteraceae	Herbaceous	Leaves
Premna serratifolia	Bebuas	Verbenaceae	Herbaceous	Leaves
Punica granatum	Delima	Lythraceae	Shrub	Leaves
Syzygium cumini	Jambu keling	Myrtaceae	Tree	Leaves
Trigonella foenum-graecum	Halba	Leguminosae	Herbaceous	Seed

(continued)

Table 1 (continued)

Scientific name	Local name	Family	Plant habitat	Parts used
Vitex pubescens	Halban	Verbenaceae	Tree	Leaves
Vitex trifolia	Lemuni	Lamiaceae	Tree	Leaves
Zingiber cassumunar	Bonglai	Zingiberaceae	Zingiber	Whole plant
Zingiber officinale var rubrum	Halia bara	Zingiberaceae	Zingiber	Leaves

plant parts, leaves (71%) are the most frequently used in *mandi serom* preparation and administered as external medicine, followed by whole plant (10%), bark (5%), seeds (5%), bulb (3%), flowers (2%), roots (2%), and gall (2%). Only three species are commonly used by all the interviewed midwives which are *Cymbopogon nardus (serai wangi), Lawsonia inermis (inai)*, and *Pandanus amaryllifolius (pandan)*. The selections of plant species also vary between each midwife. Different midwives used different combinations of plants during the *mandi serom* rituals. Surprisingly, not one of them used the exact same combination and amount of plants for *mandi serom* even though they are living in the same state. Their knowledge is somehow different from one another in terms of plant usage selections and applications; however, the significance of *mandi serom* practice is the same.

The results suggest that the selection of plant composition is mediated by the availability of plant species in Alor Setar, as well the ancestry knowledge of herbal plants among the Malay traditional midwives. This study also indicates large variations among midwives in terms of the total number of species used, plant application, restriction during the bath, and selection group of plants in *mandi serom*. There is a need for more studies on similar practices with different localities to obtain more accurate scenarios in the variation of medicinal plant usage by Malay traditional midwives.

4 Conclusion

The documentation of plant selections in *mandi serom* shows the thriving of traditional knowledge among Malay traditional midwives *(bidan kampung)* in Alor Setar, Kedah, who depend mainly on native plant species. The study also shows the selection of plants relies on the availability of plants existed in that particular area which presents its own identity based on cultural value and environmental conditions.

This current ethnobotanical field survey carried out among Malay traditional midwives living around Alor Setar, Kedah, reveals that many medicinal plants are still used by the Malay population in the sampled areas. Since many plant species are suggested as potential resources in Malay midwifery practice, this should encourage further research in ethnomedicine. The current data will expand the genetic resources obtainable in the area of research and signify a potential source of natural products for woman health care. The preservation of these plant species is the gateway towards developing efficacious remedies for women during post-partum period. There is greater ability to discover and extend the knowledge into pharmacological studies while conserving our landscape to ensure the legacy of Malay culture is well preserved. It is important not only to record such ethnomedical knowledge and conduct further studies but also to take steps to conserve these medicinal plants before they are lost to humankind forever.

Acknowledgements The authors would like to thank the Malay midwives who were willing to contribute to the study towards the development of knowledge on medicinal plants and preservation of Malay traditional knowledge.

References

Alexiades, M. N. (1996). Collecting ethnobotanical data: An introduction to basic concepts and techniques. *Advances in Economic Botany, 10,* 53–96.
Alexiades, M. N., & Sheldon, J. W. (Eds.). (1996). *Selected guidelines for ethnobotanical research: A field manual.* New York: The New York Botanical Garden Press.
Bandaranayake, W. M. (1998). Traditional and medicinal uses of mangroves. *Mangroves and Salt Marshes, 2*(1), 137.
Barakbah, A. (2007). Ensiklopedia Perbidanan Melayu. *Utusan Publication and Distributors Sdn. Bhd.*
Bourdy, G., & Walter, A. (1992). Maternity and medicinal plants in Vanuatu I. The cycle of reproduction. *Journal of Ethnopharmacology, 37*(3), 179–196.
Chen, P. C. (1981). Traditional and modern medicine in Malaysia. *Social Science & Medicine. Part A: Medical Psychology & Medical Sociology, 15*(2), 127–136.
Coe, F. G. (2008). Rama midwifery in Eastern Nicaragua. *Journal of Ethnopharmacology, 117*(1), 136–157.
Gollin, L. (2004). Subtle and profound sensory attributes of medicinal plants among the Kenyah leppo' ke of east Kalimantan, Borneo. *Journal of Ethnobiology, 2*(4), 173–201.
Green, G., Bradby, H., Chan, A., & Lee, M. (2006). "We are not completely Westernised": Dualmedical systems and pathways to health care among Chinese migrant women in England. *Social Science and Medicine, 6*(2), 1498–1507.
Hamilton, A. C. (2004). Medicinal plants, conservation and livelihoods. *Biodiversity and Conservation, 13*(5), 1477–1517.
Hasan, Z. (2007). *Beauty is beyond skin deep: Traditional treatments for women.* Serdang, Kuala Lumpur, Malaysia: Malaysian Agricultural Research and Development Institute, MARDI.
Ishak, N. (2012). *A study on plant materials selection in traditional Malay midwifery practices as potential softscape elements in Malay garden.* Kuala Lumpur: International Islamic University Malaysia.
Jamal, J. A., Ghafar, Z. A., & Husain, K. (2011). Medicinal plants used for postnatal care in Malay traditional medicine in the Peninsular Malaysia. *Pharmacognosy Journal, 3*(24), 15–24.
Jain, A., Katewa, S. S., Chaudhary, B. L., & Galav, P. (2004). Folk herbal medicines used in birth control and sexual diseases by tribals of Southern Rajasthan, India. *Journal of Ethnopharmacology, 90*(1), 171–177.
Kim, S. L., & Lean, K. S. (2013). Herbal medicines: Malaysian women's knowledge and practice. In *Evidence-based complementary and alternative medicine.*
Lamxay, V., de Boer, H. J., & Björk, L. (2011). Traditions and plant use during pregnancy, childbirth and postpartum recovery by the Kry ethnic group in Lao PDR. *Journal of Ethnobiology and Ethnomedicine, 7*(14), 1–15.

Martin, G. J. (1995). *Ethnobotany: A methods manual. People and plants conservation manual*. London: Chapman and Hall.
McNeely, J. A. (2005). Biological and cultural diversity: The double helix of sustainable development. In J. T. Arnason, P. M. Catling, E. Small, P. T. Dang, & J. D. H. Lambert (Eds.), *Biodiversity and health; focusing research to policy* (pp. 3–9). Ottawa: NRC Press.
Min, B. C., Omar-Hor, K., & Chow Lin, O. Y. (2006). *1001 garden plants in Singapore*. Singapore: National Parks Board.
Nigenda, G., Ruiz, J. A., & Bejarano, R. (2006). University-trained nurses in Mexico: An assessment of educational atrition and labor wastage. *Salud Pública de México, 48*(1), 22–29.
Samy, J., Sugumaran, M., & Lee, K. L. (2005). In K. M. Wong (Ed.), *Herbs of Malaysia: An introduction to the medicinal, culinary, aromatic and cosmetic use of herbs*. Times Editions.
Sein, K. K. (2013). Beliefs and practices surrounding postpartum period among Myanmar women. *Midwifery, 29*(11), 1257–1263.
Singh, Y. N., Ikahihifo, T., Panuve, M., & Slatter, C. (1984). Folk medicine in Tonga. A study on the use of herbal medicines for obstetric and gynaecological conditions and disorders. *Journal of Ethnopharmacology, 12*(3), 305–329.
Ticktin, T., & Dalle, S. P. (2005). Medicinal plant use in the practice of midwifery in rural Honduras. *Journal of Ethnopharmacology, 96*(1), 233–248.
Torri, M. C. (2013). Perceptions and uses of plants for reproductive health among traditional midwives in Ecuador: Moving towards intercultural pharmacological practices. *Midwifery, 29*(7), 809–817.
Werner, R. (2002). *Medicines in malay villages*, Vol. 2. University of Malaya Press.
Whitaker, E. D. (2003). The idea of health: History, medical pluralism, and the management of the body in Emilia-Romagna, Italy. *Medical Anthropology Quarterly, 17*(3), 348–375.
Zakaria, M., & Mohd, M. A. (2010). *Traditional Malay medicinal plants*. Kuala Lumpur: Institut Terjemahan Negara Malaysia.

Meiofaunal Responses to Azoic Sediment in a Sandbar-Regulated Estuary in the East Coast of Peninsular Malaysia

Rohayu Ramli, Zaleha Kassim and Muhammad Akmal Roslani

Abstract Series of field experiments were conducted in a remote estuarine creek of the Mengabang Telipot River (05°24.860″N, 103°5.266″E) on April 2012 and November 2012 to investigate the colonisation of the meiofauna on the azoic sediment. A total of 25 bottles (modified holes on the cap) filled with azoic sediment were deposited during low tide. Five replicates were retrieved on day 1, 4, 7, 10 and 13 days post-placement. Five replicates of sediment control were also taken to provide baseline information on the resident meiofaunal community. On each sampling occasion, the physico-chemical parameters such as temperature, salinity, dissolved oxygen and pH value were measured in situ. Both experiments in April and November 2012 recorded three major meiobenthic taxa, namely Nematoda, Gnathostomulida and Copepoda. The one-way ANOVA result had proved that the density of the meiofauna found in April 2012 had significant different with the azoic sediment at most stage of the experiment ($p < 0.05$). Meanwhile, in November 2012, only the densities of copepods and nematodes were found statistically significant different during the experiment period (One-way ANOVA, $p < 0.05$). Environmental factors, substrate characteristics, the availability of the food and behaviour of individual taxa are very crucial in determining the consistency of the new recruitment to survive, recolonise and therefore reproduce in the new substrate.

Keywords Colonisation · Azoic sediment · Meiofauna

R. Ramli (✉) · M. A. Roslani
Faculty of Applied Science, Universiti Teknologi MARA Cawangan Perlis, Arau, Perlis, Malaysia
e-mail: rohayuramli.85@gmail.com

M. A. Roslani
e-mail: areqmal125@gmail.com

Z. Kassim
Kulliyyah of Science, International Islamic University Malaysia, Kuantan, Pahang, Malaysia
e-mail: drzack@iium.edu.my

1 Introduction

Colonisation of sediments by meiofauna is generally a rapid procedure. Dispersion, transport and settling are the processes which lead to the colonisation which obvious in marine environments (Alve 1999). Recruitment by water column processes is the common medium for the meiofauna to disperse and colonise new area (Giere 1993). According to Thistle et al. (1995), most of the meiofaunal taxa lack of pelagic larvae and they are known to enter the water column either passively through suspended or actively through swimming. Drift, suspension and emergence are the main transport components for the colonisation process for most taxa (Giere 1993). Vertically, upward movement within the substrate also was identified as another important mechanism in colonisation especially for interstitial species which do not have swimming abilities such as nematodes, foraminifera and oligochaetes (Bo et al. 2006). These mechanisms including larval colonisation are important in controlling the meiofauna settlement process in the new area (Köhler et al. 2008).

Many colonisation studies have concluded that meiofauna are able to rapidly colonise new habitat (natural or artificial substrates) or disturbed areas (De Troch et al. 2005; Atilla and Fleeger 2000) especially by the opportunistic colonisers. Successful colonisation was derived from the factors of habitat complexity, food availability and refuge from the predators (Atilla and Fleeger 2000). Relatively in Malaysia, little attention has been paid to the influence of colonisation on the population dynamics of meiofauna in any mesocosm or field experiment. The colonisation experiment was designed to provide information on rate of meiofaunal colonisation and mode of meiofaunal to move into the new habitat. More specifically, this study was aimed in investigating the colonisation rate of the meiofauna on the azoic sediment and mode of colonisation in short-term period of time, as well as the relationships of possible factors in structuring community in the colonisation process.

2 Methods

Series of field experiments were conducted in a remote estuarine creek of the Mengabang Telipot River (05°24.860″N, 103°5.266″E) on April 2012 and November 2012 (Fig. 1). Mengabang Telipot is a river with a sandbar-regulated estuary in the east coast of peninsular Malaysia (Mei et al. 2012). This chosen area is an estuary ecosystem which consists of *Rhizophora apiculata*, *Nypa fruticans* and *Lumnitzera racemosa* (black mangrove) trees. The estuary receives inflows from the artificial stream and experiencing significant fluctuations which largely depends on the sandbar-opened and sandbar-closed events with regard on the tidal fluctuations and rainfall.

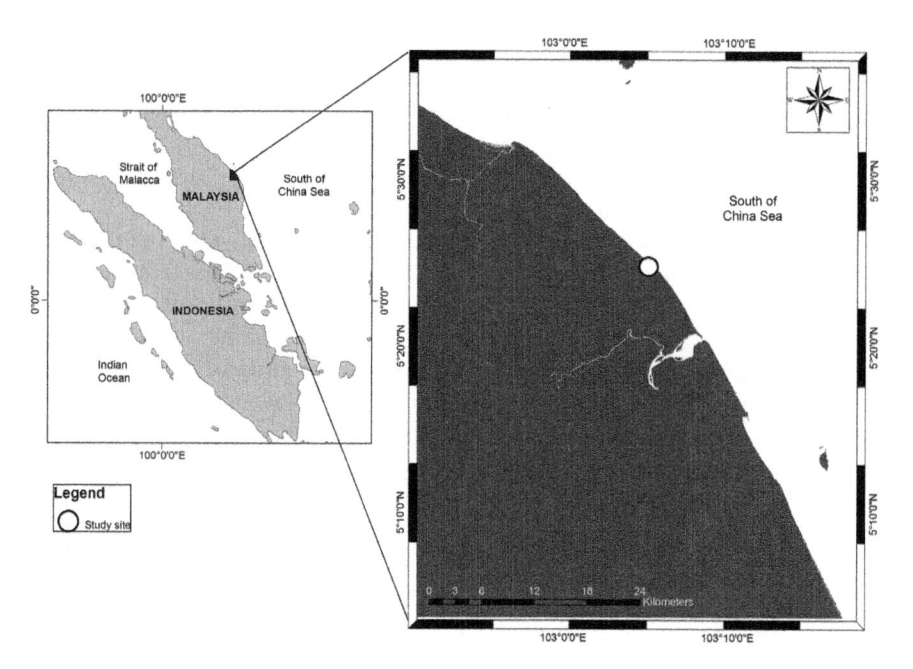

Fig. 1 Map showing the study site

About 2 kg of surface sediments were collected in the study area. After collection, they were combusted in the laboratory muffle furnace at 500 °C for 6 h in order to obtain azoic sediment. The specimen bottles (4 cm diameter) with modified holes on the cap were used as experimental tubes. The position of the location was taken by using the portable GPS instrument. A total of 25 bottles filled with azoic sediment were deposited during low tide. The bottles were embedded into the sediment with the distance 1 cm from each other and 2 cm from the sediment surface to allow the immigration either by crawling or burrowing infaunal organism.

Five replicates were retrieved on day 1, 4, 7, 10 and 13 days post-placement. Five replicates of sediment control were also taken within experimental area to provide baseline information on the resident meiofaunal community. On each sampling occasion, the physico-chemical parameters such as temperature, salinity, dissolved oxygen and pH value were measured in situ. Three replicates of sediment samples for particle (grain) size analysis were collected once during the experimental period. Same procedures were applied for both experiments in April and November.

Upon arrival in the laboratory, faunal extraction were done by decanting the substrates through 500 and 63 µm sieves with tap water for 5–6 times. The materials that retained on 63 µm sieve were transferred into 5% buffered formalin with the addition of the Rose bengal stain (McIntyre and Warwick 1984). Major meiofaunal taxa were identified and enumerated under a dissecting microscope. The meiobenthos was identified into taxa group by referring to Higgin and Thiels (1988). The mean values and the standard deviation also calculated from the samples analysis. Determination of the particle grain size analysis was done by using the laser-diffraction analyser, Malven Master Sizer 2000. The Udden/Wentworth scale (Wentworth 1922) was adopted to classify the characteristics of the particle size (Eleftheriou and McIntyre 2005).

The density of meiobenthos was expressed in number of individuals (N) per unit area (10 cm^2) based on the mean of five replicates. The rates of the colonisation also were determined by the abundance of meiofauna on the azoic sediment compared to the field samples during the experiment. The mode of colonisation was determined by the general morphologies as well as behavioural characteristics of the meiofauna. Experimental effects on the densities of major meiofaunal taxa were determined by using univariate analysis based on a one-way ANOVA. Independent Samples t-test was used to compare the mean of physico-chemical parameters between the two months.

3 Results

3.1 Environmental Variables

April 2012 is in the inter-monsoon season while November 2012 is in the north-east monsoon season. The physico-chemical parameter readings recorded in the study area during all the experiment periods were showed fluctuation pattern especially in salinity. There are significance differences between the physico-chemical parameter readings between experiment in April 2012 and November 2012 (Independent Samples t-test, $p < 0.05$).

In April 2012, the mean temperature was lowest on day seven (28.3 °C) while highest temperature was recorded on day 13 (34.5 °C). The highest mean of salinity was on day four (30.2 ppt) and the lowest mean recorded for salinity was 14.0 ppt on day ten. The highest mean of pH was 8.2 on the 13th day while the lowest mean recorded was 7.7 on the first day. The highest mean for DO (mg/L) was recorded also on day 13 (4.0 mg/L) while the lowest mean was 2.7 mg/L which was on the first day. The bottom was composed of 80% of sand.

In November 2012, the highest mean temperature was on day four (30.8 °C) while the lowest was recorded on day ten (28.1 °C). The highest mean of salinity was on day seven (11.4 ppt) and the lowest mean recorded for salinity is 1.8 ppt on day ten. The salinity values in this experiment period is continue to decrease as the rainfall in heavy and frequent. The highest mean of pH is 7.4 on day four while the lowest mean recorded is 7.1 on the first day. Similar trend also observed for the mean of DO (mg/L) which highest on the day four (4.2 mg/L) while the lowest mean is 3.3 mg/L which on the first day. Sedimentary analysis indicates that the bottom is composed of 72% of sand.

3.2 Meiofauna Responses Towards the Azoic Sediment

Experiment in April 2012 recorded three major meiobenthic taxa, namely Nematoda, Gnathostomulida and Copepoda. Other groups were found in low numbers was classified as others. Only two major meiofauna groups were recorded in field samples which apparently show the highest density of nematodes accounting for 95% of the total meiofauna present in the study area. The remaining proportion was occupied by small numbers of copepods and other meiofauna groups. Colonisation of the azoic sediment also dominated by the nematodes (49%), copepods and copepods nauplii made up of 34 and 16%, respectively, followed by 1% of the Gnathostomulida group. Figure 2 depicts the recolonisation trends of the meiofauna in the azoic sediment treatment compared to field samples.

Morphologically, nematodes were absent of any swimming legs and defined as slower coloniser compared than copepods. However, in the present study, nematodes were abundant in the azoic sediment (Fig. 2a). The one-way ANOVA result also proved that the density of the nematode has significant different with the azoic sediment at most of the stage of the experiment ($p < 0.05$). Although the colonisation showed a decreasing trend after the fourth day, surprisingly their abundance

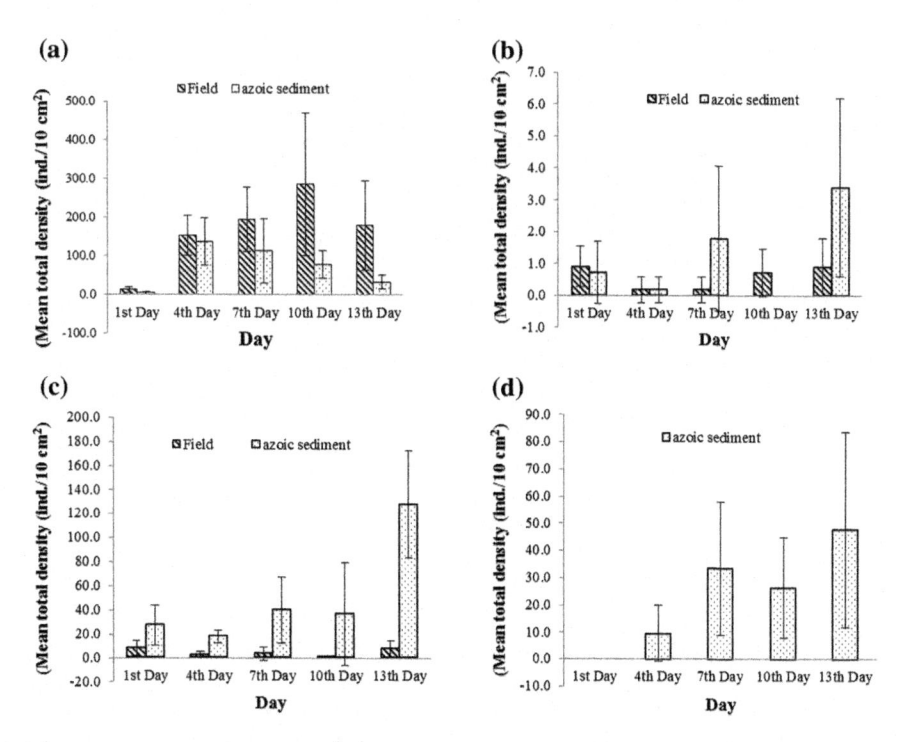

Fig. 2 Mean density (ind./10 cm^2) of meiofauna that was found in the field samples and colonising the azoic sediment in April 2012 **a** Nematodes, **b** Gnathostomulids, **c** Copepods and **d** Copepod nauplii

remained higher than copepods. The nematodes abundant are high at the initial of the experiment and gradually decreased over time probably due to the high emigration and little or no probability of recolonisation.

Azoic sediment was successfully colonised by the copepods even though there are poor copepods distribution in the field samples throughout the experiment period (Fig. 2c). Faster colonisation rate by copepods was recorded and the number is increasing at the end of the experiment followed by the present of copepods nauplii (Fig. 2d). The colonisation rate of copepods nauplii clearly showed the same pattern as the colonisation rate of the adults. The one-way ANOVA result also showed that azoic sediment had effected on the density of the copepods and their nauplii at most of the stage of the experiment ($p < 0.05$). Only small numbers of Gnathostomulids were recorded colonising the azoic sediment (Fig. 2b). Nevertheless, the one-way ANOVA result also revealed that this substrate affected the density of the gnathostomulids at all stage of the experiment ($p < 0.05$).

Contrary to the previous experiment results, the colonisers from the three major taxa Nematoda, Gnathostomulida and Copepoda were found colonising the azoic sediment with very low density. In November 2012, field samples still dominating by the nematodes accounting for 72% of the total meiofauna followed by Copepoda group (18%). Gnathostomulids group was found to contribute for 5% of the total meiofauna while only 1% of copepods nauplii were recorded compared to the previous experiment which found absent in the study area. The remaining proportion scored by the other meiofauna groups. Figure 3 shows the recolonisation trends of the meiofauna in the azoic sediment treatment compared to field samples.

The azoic sediment still dominated by the colonisation of nematodes (Fig. 3a). There were high proportion of nematodes recorded (73%) compared to the 49% which was recorded in the previous experiment. However, the colonisation rate of nematodes was kept decreasing after the first day until the end of the experiment. The copepods number was clearly dropped in this month (Fig. 3c). However, the present of copepod nauplii in the azoic sediment is proportional to the density of the copepods in both series of the experiments (Fig. 3d). Gnathostomulids were found colonise the azoic sediment with the same pattern during the previous experiment (Fig. 3b). The result from one-way ANOVA test also showed that only the densities of copepods and nematodes were found statistically significant different during the experiment period.

4 Discussion

Mengabang Telipot estuary has been challenged by the changes of seasons which largely influenced the key components in structuring the fauna community. The estuarine ecosystem was integrated with the ebb-flood system creating highly variable on physical and chemical conditions. The sandbar-opened event facilitates the inflow and outflow of the water through the estuary while sandbar-closed event helps to retain the water flow hence increase the water level. The tidal currents and

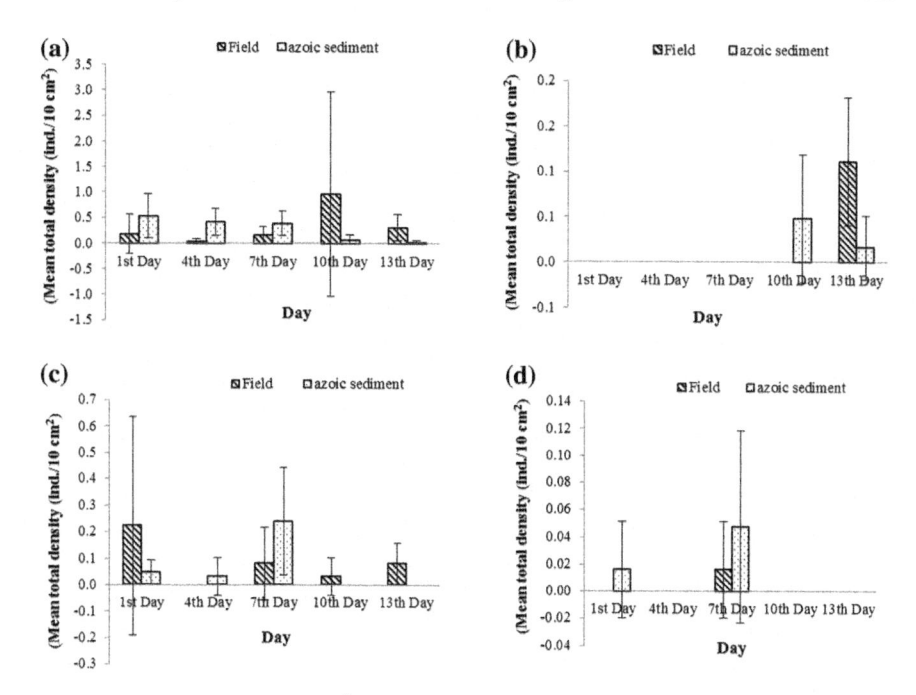

Fig. 3 Mean density (ind./10 cm^2) of meiofauna that was found in the field samples and colonising the azoic sediment in November 2012 **a** Nematodes, **b** Gnathostomulids, **c** Copepods and **d** Copepod nauplii

the riverine flows drive the complex currents which are important in the ecological processes in the estuary (Molles 2013). These currents responsible in the transportation of the organism, the circulation to renew the nutrients and oxygen and also facilitate in removing the waste (Friedhelm and Sabine 2006).

High copepods density was found colonised the azoic sediment. This condition suggesting that copepods, being good swimmers, could freely immigrate and emigrate into the new substrate. This is also proved in the previous studies that copepods are excellent early colonisers and preferred to the azoic sediment (Danovaro and Mirto 2004; Atilla and Fleeger 2000). Moreover, the present of the copepod nauplii proving that the copepods had utilised the azoic sediment as breeding ground. According to Giere (1993), most of the copepods and nauplii were restricted to the oxidised, uppermost centimetre of sediment. This surficial distribution also enhancing the probability of passive suspension from the current action and thus encourage the recolonisation process. The copepods might preferentially colonise the new habitat because of the more available space which has not impacted by the silt (Freese and Arthur 2006). Moreover, the azoic sediment which used in this experiment consists of more than 70% sand which is favoured by the copepod especially harpacticoid copepod. This result indicates that sediment grain size is also crucial for grazers. Favourable sedimentary condition increases the ability of harpacticoid to migrate and foraging the food.

According to Danovaro and Mirto (2004) and Sun and Fleeger (1994) in contrast to copepods, nematodes are generally poor swimmers and slower colonisers of azoic sediments. Although the colonisation rates showed the decreasing trend, colonisation of the nematodes was relatively higher than the copepods. Nematodes generally prefer high organic content sediment and rich in food supply (Atilla and Fleeger 2000; Gee and Somerfield 1997). However, the suitability of organic content and food also is important in controlling the distribution of nematodes. In the present study, the factors of the size of azoic sediment (coarse sand) and the absent of any organic matter (food source) for the nematodes consumption might lead the nematodes to become constantly leaving the azoic sediment.

Nematodes also most rapidly colonised the new habitat via suspended transport (Rohayu and Zaleha 2016). Nematodes species that live at the sediment surface are faster to colonise compared to nematodes species which live deeper in the sediment (Ullberg and Olafsson 2003). In contrast, copepods are highly mobile which can colonise new substrate rapidly (Atilla et al. 2003). The ability to rapidly colonise free space seems to be species-specific, with the sequence of colonisation succession related to the lifestyle and motility of the different harpacticoid species (Callens et al. 2011). Moreover, several behavioural and morphological aspects are required to colonise new habitat. The body shape also always related to an interstitial and burrowing lifestyle. These lifestyles and morphological characteristics could be the primary determinant to these harpacticoid copepods species in the dispersal either by infaunal (by burrowing) or active transport (by swimming) to colonise the new substrates (Rohayu and Zaleha 2016).

5 Conclusion

Meiofauna of different taxa/groups were capable to colonise the new substrates over short period of time. The colonisation pattern is relying on the abilities of meiofauna to disperse either by infaunal burrowing or by active and passive transport. Higher abundance of meiofauna in the study area may also increase the potential colonisers, hence encourage the colonisation of the meiofaunal on the new substrates. Nevertheless, the colonisation process also restricted by a number of factors. Unfavourable environmental condition such as low salinity and hydrologic event such as high water flow may influence the potential colonist to colonise the new substrate. The substrate characteristics, the availability of the food in the new substrate and behaviour of an individual taxa are very crucial in determining the establishment of the new recruitment to survive, recolonise and therefore continue the generation by reproduce in the new substrate. Lack of practical information about some of those factors could limit the persistence of the findings and thus, more studies are needed to rectify this shortcoming. Successful colonisation on the new substrates revealed that the substrates can be tools to identify the survival rate between taxa of the meiofauna.

References

Alve, E. (1999). Colonization of new habitats by benthic foraminifera: A review. *Earth Science Reviews, 46*, 167–185.

Atilla, N., & Fleeger, J. (2000). Meiofaunal colonization of artificial substrates in an estuarine embayment. *PSZN: Marine Ecology, 21*, 69–83.

Atilla, N., Wetzel, M. A., & Fleeger, J. W. (2003). Abundance and colonization potential of artificial hard substrate-associated meiofauna. *Journal of Experimental Marine Biology and Ecology, 287*, 273–287.

Bo, T., Cucco, M., Fenoglio, S., & Malacarne, G. (2006). Colonisation patterns and vertical movements of stream invertebrates in the interstitial zone: A case study in the Apennines, NW Italy. *Hydrobiologia, 568*, 67–78.

Callens M., Gheerardyn, H., Ndaro, S. G. M., De Troch, M., & Vanreusel, A. (2011). Harpacticoid copepod colonization of coral fragments in a tropical reef lagoon (Zanzibar, Tanzania). *Journal of the Marine Biological Association of the United Kingdom*, 1–11. Marine Biological Association of the United Kingdom.

Danovaro, R., & Mirto, S. (2004). Meiofaunal colonisation on artificial substrates: A tool for biomonitoring the environmental quality on coastal marine ecosystem. *Marine Pollution Bulletin, 48*, 919–926.

De Troch, M., Vandepitte, L, Raes, E. S. M., & Vincx, M. (2005). A field colonization experiment with meiofauna and seagrass mimics effect of time, distance and leaf surface area. *Marine Biology*, 1–14.

Eleftheriou, A., & McIntyre, A. (2005). *Methods for the study of marine benthos*. 3rd Edn. Blackwell Science Ltd., 416 p.

Freese, S., & Arthur, V. B. (2006). Effects of flow on meiofauna colonization in artificial streams and reference sites within the Illinois River, Arkansas. *Hydrobiologia, 571*, 169–180.

Friedhelm, G., & Sabine S. (2006). Estuaries and soft bottom shores. In G. Friedhelm, H. T. Kris, P. M. Paciencia, & M. Josef (Eds.), *Ecology of insular Southeast Asia. The Indonesian Archipelago* (pp. 215–228). First edition. The Netherlands: Elsevier, Amsterdam.

Gee, J. M., & Somerfield, P. J. (1997). Do mangrove diversity and leaf litter decay promote meiofaunal diversity? *Journal of Experimental Marine Biology and Ecology, 218*, 13–33.

Giere, O. (1993). *Meiobenthology* (p. 328). Berlin Heidelberg: Springer-Verlag.

Higgins, R. P., & Thiel, H. (1988). *Introduction to the study of meiofauna*. Washington, DC: Smithsonian Institution Press.

Köhler, G. V., Laudien, J., Knott, J., Velez, J., & Sahade, R. (2008). Meiobenthic colonisation of soft sediments in arctic glacial Kongsfjorden (Svalbard). *Journal of Experimental Marine Biology and Ecology, 363*, 58–65.

McIntyre, A. D., & Warwick, R. M. (1984). Meiofauna techniques. In A. D McInTyre, & N. A. Holme (Eds.), *Methods for the study marine benthos* (pp. 217–244). IBP Handbook 16 Oxford: Blackwell Scientific Publications.

Mei, K. K., Edlic, S., Suhaimi, S., & Norhayati, M. T. (2012). Sandbar-regulated hydrodynamic influences on river hydrochemistry at Mengabang Telipot River, Peninsular Malaysia. *Environmental Monitoring Assessment, 184*, 7653–7664.

Molles, J. M. C. (2013). *Ecology, concepts and applications* (567 p). 6th edition. Published by Mc Graw-Hill Companies Inc.

Rohayu, R., & Zaleha, K. (2016). Colonisation of meiofauna: Selectivity of substrates. *Journal of Applied Environmental and Biological Sciences, 6*(12), 133–137.

Sun, B., & Fleeger, J. W. (1994). Field experiments on the colonization of meiofauna into sediment depressions. *Marine Ecology Progress Series, 110*, 167–175.

Thistle, D., Weatherly, G. L., & Ertman, S. C. (1995). Shelf harpacticoid copepods do not escape into the seabed during winter storms. *Journal of Marine Research, 53*, 847–863.
Ullberg, J., & Olafsson, E. (2003). Effects of biological disturbance by Monoporeia affinis (Amphipoda) on small-scale migration of marine nematodes in low-energy soft sediments. *Marine Biology 143*: 867–874.
Wentworth, C. K. (1922). A scale of grade and class terms for clastic sediments. *Journal of Geology, 30*, 377–392.

Coastal Deposit Characteristic Influenced by Terrestrial Organic Matter and Its Sedimentary Structure at Jangkang Beach, Bengkalis District, Riau Province—Indonesia

Yuniarti Yuskar, Tiggi Choanji, Dewandra Bagus EP, Adi Suryadi and Rani A. Ramsof

Abstract Jangkang Beach is located in the Deluk village, Bengkalis District, Riau Province, Indonesia. This beach is overgrown by mangrove plants and peat. Its presence resulting varying thickness amount of terrestrial organic material deposited to the shore. Purpose of this research is to determine the characteristics of coastal sediments that influenced by the distribution of terrestrial organic materials and its sedimentary structures. The method is used for field survey that consists of drilling core at five locations, lateral measured section, and granulometric analysis at laboratory. This research resulted in that coastal sediments consist of medium to fine sand, silt, clay, and organic material. The terrestrial organic material (TOM) deposited at Jangkang Beach has size ±35 cm which is wooden to finely sized (less than 2 mm) with thickness 5–15 cm. Distribution of this material located on littoral zone (intertidal) and number of TOM decreased seaward with increasingly more fine wood. This TOM was deposited above ripple mark sedimentary structure. Ripple mark has slope 100–480 with symmetrical shape. Other sedimentary structures found are bioturbation structure (burrow, track, trail), lenses of TOM, cross lamination, and parallel lamination.

Keywords Terrestrial organic matter (TOM) · Ripple mark · Coastal deposit Bengkalis Island

Y. Yuskar (✉) · T. Choanji · D. B. EP · A. Suryadi · R. A. Ramsof
Department of Geological Engineering, Faculty of Engineering, Universitas Islam Riau, Jl. Kaharuddin Nasution No 113, Marpoyan, Riau 28284, Indonesia
e-mail: yuniarti_yuskar@eng.uir.ac.id

1 Introduction

Large amounts of terrestrial organic carbon are annually transported from the continents to the oceans mainly by fluvial transport or, in lower amounts, by aeolian dust (Hopmans et al. 2004). Estuaries and shelf seas cover only 10% of the total seafloor, but they bury more than 90% of marine sedimentary organic matter owing to high terrestrial inputs and marine primary productivity (Xing et al. 2011).

The organic matter preserved in marine sediments contains varying contributions of terrestrial and marine source inputs, an issue relevant along coastal margins where most carbon burial occurs (Hedges et al. 1997). Terrestrial organic matter deposited in marine sediments might then be expected to undergo less efficient re-mineralization and therefore be preferentially buried (Burdige 2005). Understanding the fate of terrestrial organic matter (TOM) in marine sediments is of importance for a number of reasons. In part, this interest stems from the observation that marine organic matter is broadly considered to be more reactive than terrestrial organic matter (Burdige 2005).

Sedimentary structures which are common on the coast are ripple marks structure. One of the most common modified types of ripple includes those in which superimposition of a secondary set on an earlier set takes place. In many instances, subdued secondary crests (symmetrical, angular) are found to be oriented at right angles to the primary lunate ripple set (Sarkar 1981).

Jangkang Beach located in the northern island of Bengkalis, Riau Province, and dealing with the Straits (Fig. 1). The Strait of Malacca is located between the east coast of Sumatra Island in Indonesia and the west coast of Peninsular Malaysia and is linked with the Strait of Singapore at its southeast end. Three smaller straits are found within this waterway: the Bengkalis Strait, located between the islands of Bengkalis and Sumatra; the Rupat Strait, between Rupat and Sumatra; and the Johore Strait, situated between the southern tip of Peninsular Malaysia and the north coast of Singapore (Thia-eng et al. 2000). It causes the formation of currents and

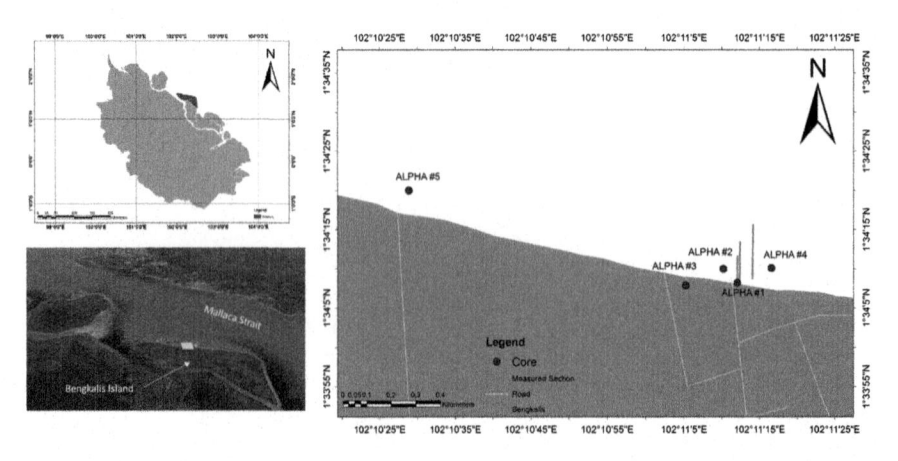

Fig. 1 Location map of Jangkang Beach, Bengkalis District, Riau, Indonesia

strong waves. As a result, terrestrial organic matter derived from unstable materials such as thick peat along the coast continues to be eroded, transported, and deposited into the sea. This study focuses on the deployment of terrestrial organic matter deposited and its sedimentary structures along the coast of Jangkang (Fig. 1).

Stratigraphy of the study area is composed of rocks that include surface deposits which are Young Superficial Deposit (Qh) and Older Superficial Deposit (Qp). Young Superficial Deposit consists of clays, silts, clean gravel, vegetation rafts, peat swamps and Older Superficial Deposits consist of clays, silts, clayey gravels, vegetation rafts.

2 Methodology

The methodology used here consists of several steps and measurement, starting from surveying the location and marking the drilling position (Yuskar and Choanji 2017). Drilling was conducted using hand auger on 2-m depth at five locations. Four locations of drilling located at Jangkang Beach, while one location at Deluk Beach, continued with making lateral cross section/lateral measured section of zone affected by tides. The lateral cross section is conduct from the sea tide limit up to lowest position of the seawater. The samples of data were taken from every 10 m. The location of the drilling location and lateral measured section can be seen in Fig. 1.

Granulometric analysis at laboratory conducts using mesh sieve with the size in 2.38, 1.19, 0.6, 0.297, 0.149, 0.074, 0.04, and 0.02 cm for five drilling core and lateral measured section samples. This analysis will measure dominant grain size, sorting, skewness, and kurtosis (Table 1).

Table 1 Availability of research data

Data	Name of data	Location	Coordinate Long	Lat	Depth	Length
Core	Alpha #1	Jangkang Beach	01°34′08.21″	102°11′12.00″	200 cm	–
	Alpha #2	Jangkang Beach	01°34′10.03″	102°11′10.15″	200 cm	–
	Alpha #3	Jangkang Beach	01°34′07.88″	102°11′05.10″	200 cm	–
	Alpha #4	Jangkang Beach	01°34′10.13″	102°11′16.57″	200 cm	–
	Alpha #5	Jangkang Beach	01°34′20.00″	102°10′28.86″	200 cm	
Measure section	Line 1	Jangkang Beach	01°34′08.6″	102°11′14″	–	90 m
	Line 2	Jangkang Beach	01°34′08.6″	102°11′12″		180 m
	Line 3	Jangkang Beach	01°34′07.45″	102°11′12.34″		210 m

3 Result

3.1 Core Analysis

In cores of sediment from five locations, three types of sediment found are clay, silt, fine sand. All of these sediments contain terrestrial organic matter (TOM) with varying percentages. The sediment structure found is parallel lamination, cross-lamination, lenses of TOM, bioturbation (burrow, track, trail). Based on analysis and physical description of 200-m sediment cores, characteristics of sediment obtained as follows:

1. Laminated silt with fine TOM
2. Mud with fine TOM
3. Fine sand with fine TOM
4. Laminated mud with lenses of TOM
5. Cross-laminated silt with TOM (Fig. 2).

3.2 Lateral Measured Section Analysis

The lateral measure section analysis was conducted at three sections laterally from the tidal boundary to the low tide with length 90, 180, and 210 m. Data resulted are sediments on the surface and trenching at some point with depth 30–50 cm. From analysis data, it is found that the change of sediments characteristic is influenced by the amount of TOM deposited, current, wave, activity of the organism, and the position of the beach against the river or the trench carries TOM to the shore.

Based on lateral measured section, sediment characteristics are as follows:

1. Mud with large—coarse wood (>10 cm)
2. Mud with coarse—medium wood (10–2 cm)
3. Mud with medium—fine wood (2–0.02 cm)
4. Woody material (TOM supported—fine material) with bioturbation structure

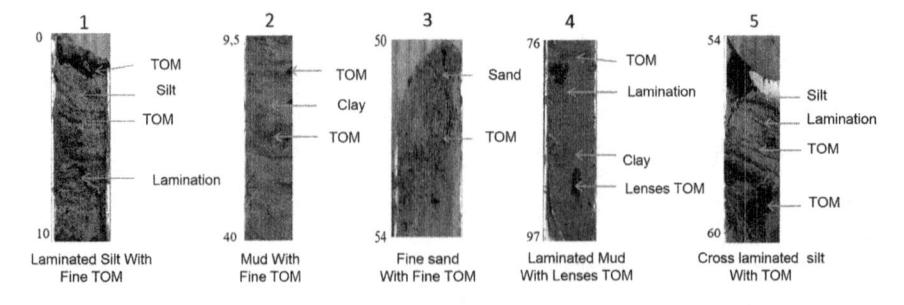

Fig. 2 Coastal Sediment Facies based on core data at research area

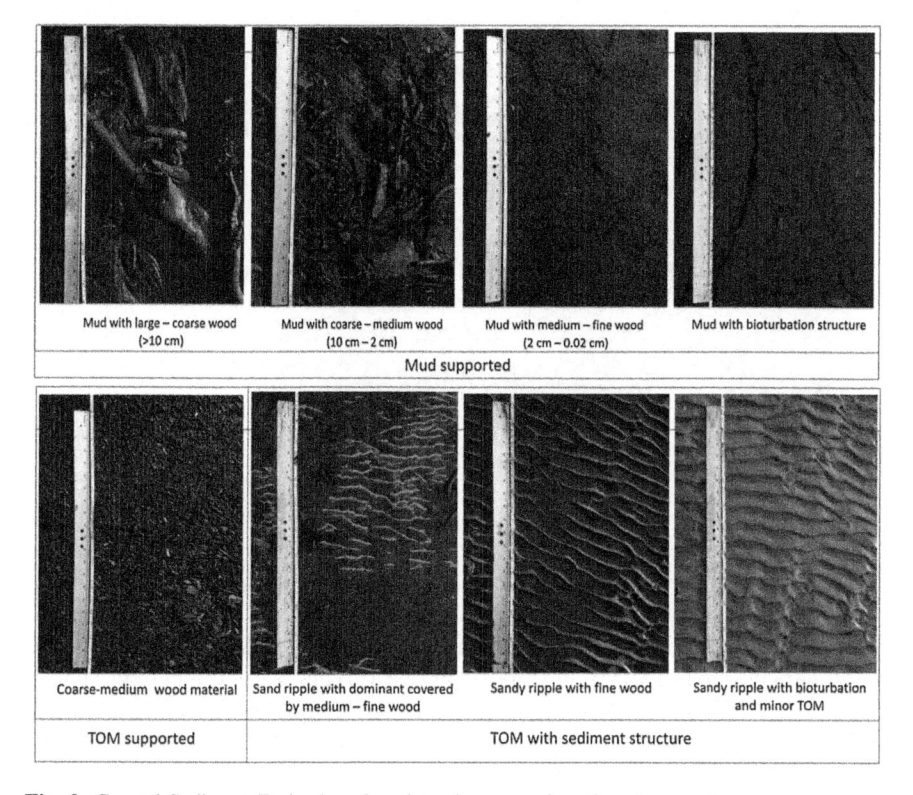

Fig. 3 Coastal Sediment Facies based on lateral measured section at research area

5. Woody material (TOM supported—coarse material)
6. Sand ripple with dominant covered by medium—fine wood
7. Sand ripple with fine wood
8. Sandy ripple with bioturbation and minor TOM (Fig. 3).

Large to coarse particle size of TOM commonly found at the end of river or trench that facing beach current and gradually become fining toward the sea. Meanwhile, from the trenching mud or clay only as lenses at sand layer. It is shown that clay deposited when there stagnant water at coastal during sea water resending and then it will covered by sand material during tide up process. Based on its geometry characteristics, ripple mark structure is symmetrical in fine sand to very fine sand with TOM content and the angle of the ripple varies from 10° to 48° (Fig. 4).

Ripples can form in sand with grain diameters between 0.06 and 0.7 mm (but see later for a revised limit). They can be generated by waves, currents or both together, very small waves and weak currents are unable to move sand, so cannot create or modify ripples, very large waves or strong currents wash out ripples, when the strengths of the waves and currents diminish, the ripples can be slow to respond,

Fig. 4 Ripple mark at line #3

the orientation of the ripples is aligned with the forcing waves or current, and changes (with a time-lag) as the forcing direction changes, ripples can be flattened by biological action in a few hours (Soulsby et al. 2012). Since ripples on the seabed are constantly varying in response to the varying wave and current conditions, the prediction of the rate of response of the ripples is just as important as prediction of the equilibrium geometry. Effectively, the ripples are constantly trying to catch up with the driving conditions (Soulsby et al. 2012). Internally, the ripples, irrespective of their symmetry, are often characterized by unidirectional bundles of foresets consisting of rhythmically alternating sand and mud laminae. The sets of cross-laminae may be complexly organized with planar or curved erosional boundaries separating them. In many instances, internal structures typical of wave ripples are also noted (Sarkar 1981).

3.3 Granulometric Analysis

Analysis on four surface sample points obtained mean value 0.286, standard deviation 0.534 with good grain sorting, skewness value 0.534 (very fine skewness) and very platykurtic. The calculation results that the current work on grain size from 0.01 to 99.1 is suspension currents that carry a very small sediment material, and the grain size from 99.1 to 100 resulting in the change of current from the suspension current into traction current which is a very strong (Fig. 5).

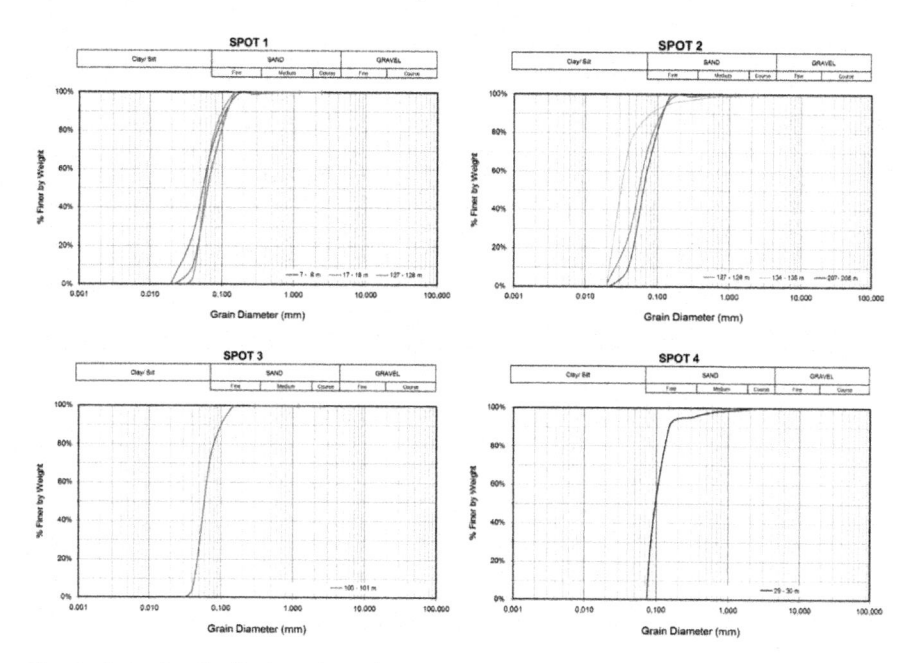

Fig. 5 Grain size distribution of sample data at research area

4 Conclusion

The coastal sediments have varying TOM content influenced by the coastal position of rivers or trench that carries TOM in coast, currents and waves that cause coastal abrasion, and organism activity. Sediments deposited on the beach are mud, silt, and fine sand with the number of TOM decreasing toward the sea. TOM has a size of large (>10 cm) to fine (<2 mm). The dominant sedimentary structures are ripple mark, cross-lamination, parallel lamination, lenses of TOM, and bioturbation (burrow, track, and trail).

Acknowledgements This work is supported by Department of Geological Engineering, Universitas Islam Riau. Authors also gratefully acknowledge the helpful comments and suggestions of the reviewers, which have improved the paper and presentation.

References

Burdige, D. J. (2005). *Burial of terrestrial organic matter in marine sediments: A re-assessment, 19,* 1–7. https://doi.org/10.1029/2004GB002368.
Hedges, J. I., Keil, R. G., & Benner, R. (1997). *What happens to terrestrial organic matter in the ocean?, 27*(5).

Hopmans, E. C., Weijers, J. W. H., Schefuß, E., Herfort, L., Schouten, S., & Damste, J. S. S. (2004). *A novel proxy for terrestrial organic matter in sediments based on branched and isoprenoid tetraether lipids, 224,* 107–116. https://doi.org/10.1016/j.epsl.2004.05.012.

Sarkar, S. (1981). Ripple marks in intertidal lower bhander sandstone (Late Proterozoic), Central India: A morphological analysis. *Sedimentary Geology, 29,* 241–282.

Soulsby, R. L., Whitehouse, R. J. S., & Marten, K. V. (2012). Prediction of time-evolving sand ripples in shelf seas. *Continental Shelf Research, 38,* 47–62. https://doi.org/10.1016/j.csr.2012.02.016.

Thia-eng, C., Gorre, I. R. L., Ross, S. A., & Regina, S. (2000). *The Malacca Straits, 41*(0).

Xing, L., Zhang, H., Yuan, Z., Sun, Y., & Zhao, M. (2011). Terrestrial and marine biomarker estimates of organic matter sources and distributions in surface sediments from the East China Sea shelf. *Continental Shelf Research, 31*(10), 1106–1115. https://doi.org/10.1016/j.csr.2011.04.003.

Yuskar, Y., & Choanji, T. (2017). Uniqueness deposit of sediment on floodplain resulting from lateral accretion on tropical area: Study case at Kampar River. *Indonesia, 2*(1), 14–19.

Influence of Grain Size on Isothermal Oxidation of Fe–33Ni–19Cr Alloy

Zahraa Zulnuraini, Noraziana Parimin and Izzat Mohd Noor

Abstract This research study describes the influence of grain size on isothermal oxidation of Fe–33Ni–19Cr alloy. The Fe–33Ni–19Cr alloy was undergone heat treatment of three different temperatures, namely 1000, 1100, and 1200 °C for 2 h of soaking time followed by water quench to vary the grain size of alloy. This alloy was ground by using several grit of sandpaper as well as weighed by using analytical balance and measured by using vernier calipers before oxidation test. The heat-treated Fe–33Ni–19Cr alloy was isothermally oxidized at 800 °C for 150 h. The characterization of oxidized samples was carried out using optical microscope, scanning electron microscope equipped with energy dispersive X-ray and X-ray diffraction. The result shows that increasing the heat treatment temperature will increase the average grain size. Fine grain size promotes the higher boundary area which acts as an ion diffusion path across the metal–gas interface. Heat-treated sample at 1000 °C with fine grain structure shows minimum weight gain and lower oxidation rate compared to alloys heat treated at 1100 and 1200 °C. Alloy heat treated at 1000 °C shows continuous oxide layer formed on the surface, while alloy heat treated at 1200 °C indicates oxide spallation. Besides, phase analysis shows that the oxidized sample formed several oxide phases.

Keywords Fe–33Ni–19Cr · Isothermal oxidation · Grain size · Heat treatment

---

Z. Zulnuraini · N. Parimin (✉) · I. M. Noor
School of Materials Engineering, Universiti Malaysia Perlis,
02600 Kangar, Perlis, Malaysia
e-mail: noraziana@unimap.edu.my

Z. Zulnuraini
e-mail: zaraieza12@gmail.com

I. M. Noor
e-mail: izzatpp@gmail.com

1 Introduction

Superalloy is the development of material with great mechanical property, elevated temperature oxidation, and corrosion resistant needed for elevated temperature service. Superalloys have an austenitic face-centered cubic (FCC) crystallization matrix and are able to deform at elevated temperature. Superalloy is generally classified within three dominant classes which are nickel based, iron based, and cobalt based. The development of superalloys is mainly enhanced by the industrial service where their high performance, great formability, and weldability are needed generally by aviation, aerospace, and nuclear industries (Khanna 2002).

Ni-based superalloys are characterized by their excellent high-temperature mechanical like high-temperature strength, creep resistance and fatigue life, and corrosion resistance. These alloys are largely used for high-temperature application, such as gas turbines in aerospace and power-generating industry. Ni-base superalloys derive their strength primarily from precipitation of γ' phase within grains and additionally by carbide precipitation at grain boundaries. The size, distribution, morphology, and composition of microstructural features like γ' and carbides are essential to obtain the desired properties (Joseph et al. 2017).

Cr_2O_3 has crack-free scale properties that bear the cost of their corrosion resistance to elevated temperature alloy. This oxide, however, becomes vulnerable when it came to cracking, especially during high-temperature cycling. Oxidation formations are susceptible to surface treatment and microstructure. Protective layer is formed through oxide-forming elements, chromium, silicon, and aluminum (Cao et al. 2016).

This project focused on the influence of grain size on isothermal oxidation of Fe–33Ni–19Cr alloy. The analysis is based on the heat treatment effect on grain size and oxidation test of the sample in high-temperature surrounding.

2 Experimental

Material used in this research is Fe–33Ni–19Cr alloy. It is a nickel-based alloy to be oxidized in this research as it possesses great oxidation and corrosion resistance in high-temperature oxidation. There are various characteristic of this Fe–33Ni–19Cr alloy which is good mechanical strength, high-temperature resistance, high creep, and rupture properties. The full compositions of Fe–33Ni–19Cr alloy (in wt%): 32.51 Ni, 18.90 Cr, 0.078 C, 0.534 Al, 0.315 Si, 0.556 Mn and balance Fe.

Plate shape material is cut into dimension of approximately 10 mm × 10 mm × 3 mm. The samples were grind by using 600grit SiC paper. All samples are weighed before and after oxidation test.

Heat treatment was done on three different temperatures, namely 1000, 1100, and 1200 °C for 2 h followed by water quench. These heat-treated samples denoted as HT1000, HT1100, and HT1200 for samples heat treated at 1000, 1100, and 1200 °C, respectively. Discontinuous oxidation test was done at 800 °C for 150 h.

The surface morphology of oxidized sample was characterized by using scanning electron microscope (SEM) equipped with energy dispersive X-ray (EDX). The phase analysis was identified by using X-ray diffraction (XRD) technique.

3 Results and Discussion

3.1 Microstructure of Heat-Treated Fe–33Ni–19Cr Alloy

The metallographic of etched non-oxidized Fe–33Ni–19Cr alloy was analyzed under optical microscope to determine the microstructure. Figure 1 indicated the growth of grain size in every different heat treating temperatures.

The micrograph from Fig. 1 was used to determine the average grain size number. The average grain size of alloy was increases as increasing in heat treatment temperature. HT1000 developed the fine grain structure which is 71.8 µm, whereas HT1200 develops the coarse grain structure which is 124.4 µm.

3.2 Oxidation Kinetic

For the oxidation kinetics, the curve had been calculated for the heat-treated samples of 1000, 1100, and 1200 °C that undergo oxidation at 800 °C. By each and every samples of the heat-treated temperature, the kinetic changes were calculated by dividing the weight changes of the sample with surface area. Figure 2 shows the graph of weight changes/surface area as a function of time. From the graph plotted, the trend of the weight changes value was increasing along with increasing oxidation time. HT1200 sample recorded a high weight change value compared to HT1100 and HT1000. This sample indicated a high weight gain pattern showing a high oxidation rate, whereas the fine grain sample of HT1000 shows the lowest weight changes denoted a low oxidation rate.

Figure 3 was plotted to identify the parabolic rate constant for all three heat-treated samples oxidized at 800 °C. Theoretically, oxidation rate is determined by the value of K_p where higher value of K_p leads to higher oxidation rate and vice

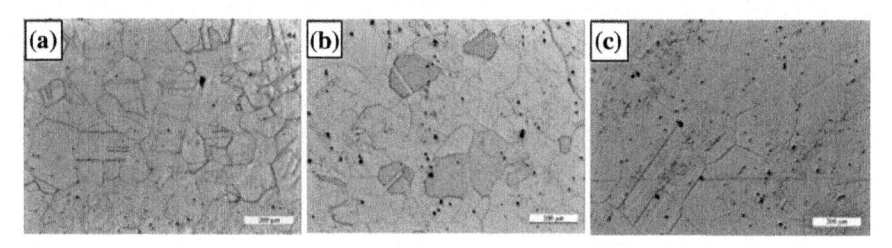

Fig. 1 Microstructure of heat-treated Fe–33Ni–19Cr alloy with magnification 50×: **a** HT1000, **b** HT1100, **c** HT1200

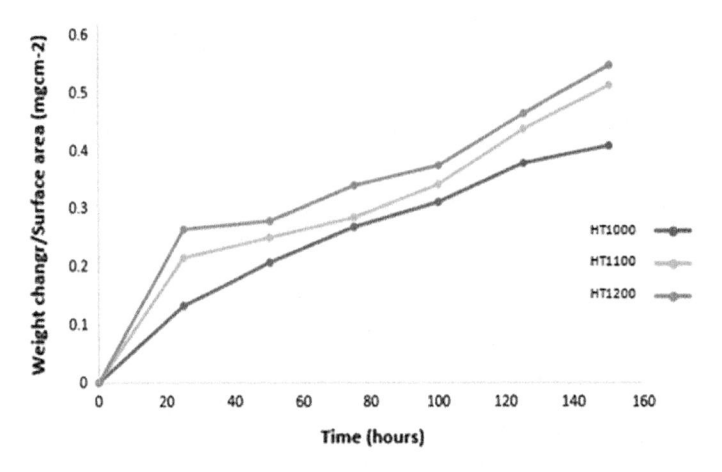

Fig. 2 Isothermal oxidation kinetic of Fe–33Ni–19Cr alloy at 800 °C

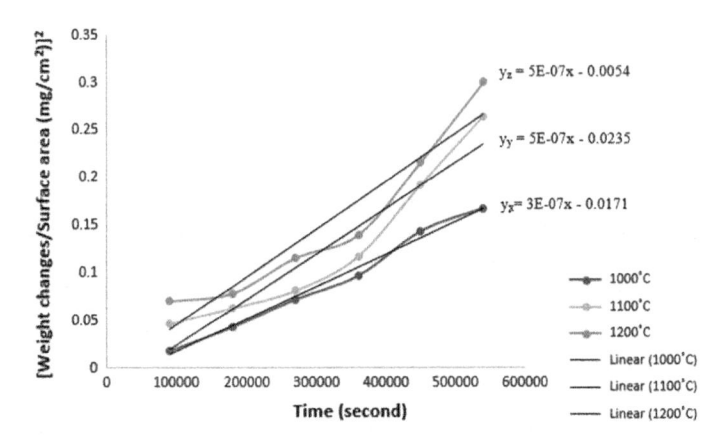

Fig. 3 Graph of square of weight changes indicating parabolic rate constant

versa. This value is determined by comparing the equation $x^2 = K_p t + c$ which is obtained from the square of weight changes/surface area as a function of time. Base on the plotted graph, the K_p value for HT1000, HT1100, and HT12000 was 3×10^{-7}, 5×10^{-7}, and 5×10^{-7} mg^{2}cm^{-4}s^{-1}, respectively. This result indicated that the higher the heat-treated temperature of the sample with coarse grain structure, the higher the oxidation rate. This result was also supported by Tan et al. in his study where he stated that the Fe–33Ni–19Cr alloy sample showed weight gains, ascending with increasing temperature (Tan et al. 2008).

3.3 Phase Analysis

From Fig. 4, the oxide thickness can be regulated based on the peak of austenite and oxides. Based on the result obtained, all three samples after oxidation contain austenite and other oxide phase. The existence of austenite features the base metal of Fe–33Ni–19Cr alloy. The oxide phase detected in all denoted samples was Cr_2O_3, TiO_2, $(Ti_{0.97}Cr_{0.03})O_2$, Fe_3O_4, and spinel ($MnCr_2O_4$, $MnFe_2O_4$, $FeCr_2O_4$, $NiCr_2O_4$, and $NiFe_2O_4$) (Niewolak et al. 2016).

3.4 Surface Morphology Analysis

SEM was used in order to determine the surface morphology of the oxidized samples. Fig. 5 shows SEM image of oxidized sample at 50 h. Figure 5a shows a low magnification of SEM HT1000 which was 1000× indicated an uneven oxide layer forms in the matrix. In Fig. 5b, high magnification of 3000× HT1000 produces more

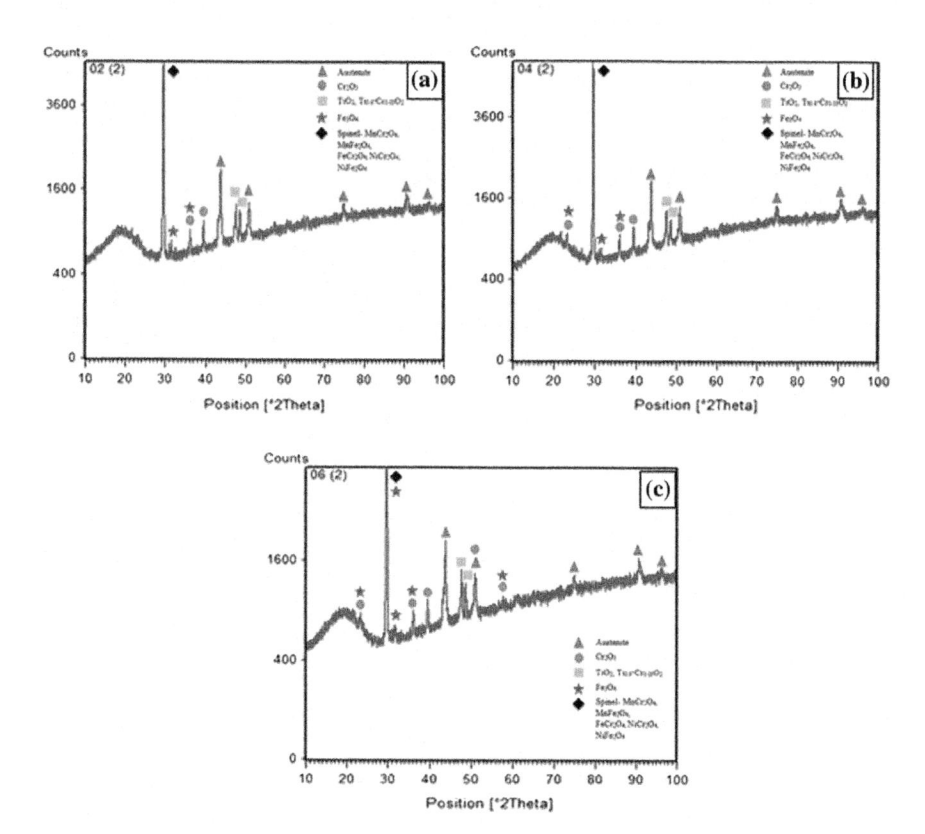

Fig. 4 XRD result for oxidized samples heat treated at three different temperatures **a** HT1000, **b** HT1100, **c** HT1200

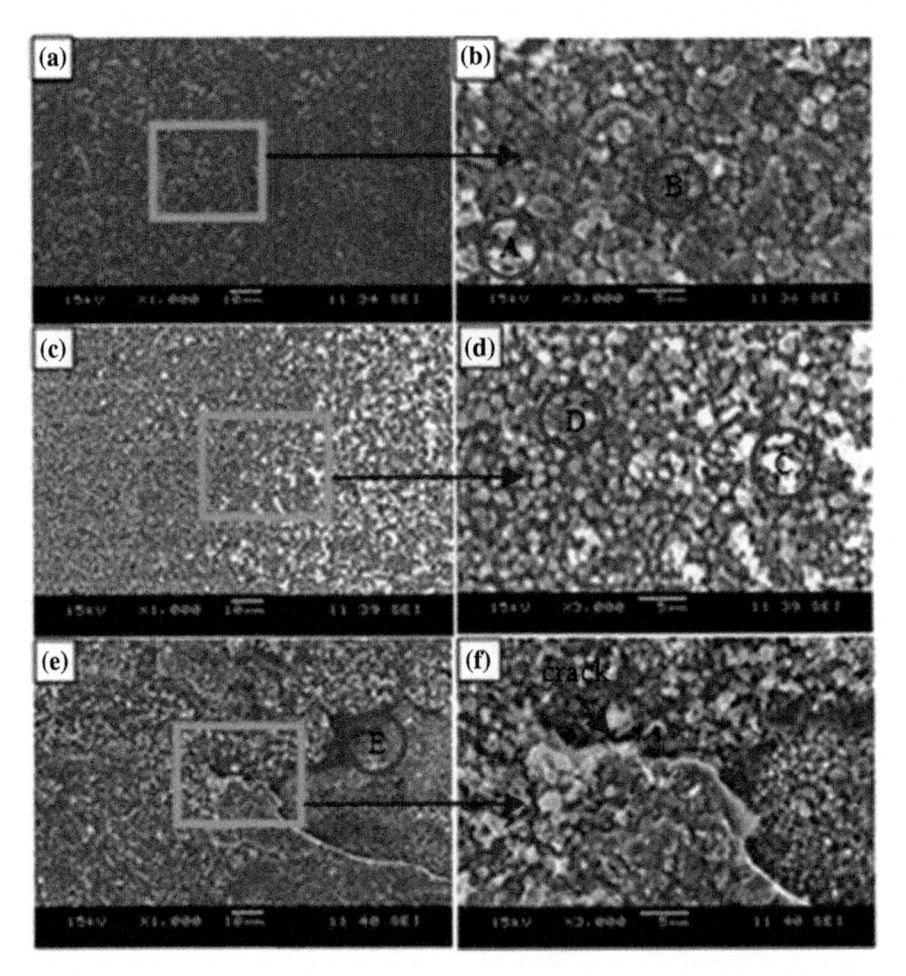

Fig. 5 SEM image of Fe–33Ni–19Cr alloy oxidized at 800 °C for 50 h: **a** HT1000 at 1000×, **b** HT1000 at 3000×, **c** HT1100 at 1000×, **d** HT1100 at 3000×, **e** HT1200 at 1000×, **f** HT1200 at 3000×

close-up image of uneven oxide particle where light region (A) and dark region (B) can be seen clearly. Both of these regions refer to the oxide scales that grew up gradually as the temperature increased, where light region was much denser layer, while dark region was less dense layer form on the surface of the sample. Figure 5c of HT1100 sample shows a low magnification of SEM which was 1000× indicated the formation of compact oxide layer in the matrix throughout the 50 h exposure of time. As the exposure time increase, uniform oxide scale was formed at the surface of the samples. The SEM image on Fig. 5d (HT1100) with a high magnification of 3000× shows compact oxide particle formed on the surface. The different of the oxide layer can be seen where (C) were much denser layer and (D) were less dense layer. Figure 5e shows a low magnification of SEM of HT1200 sample which was

1000× indicated porous oxide structure with evidence of crack. This oxide scale has great tendency to spall. Figure 5f of HT1200 shows a high magnification of SEM which was 3000× with a more close-up spall oxide structure (E) and crack that promoted higher oxidation rate due expanded area. This observation indicated that a fresh oxide layer will continuously form on the alloy surface to seal the unprotected area and crack. This mechanism will cause increase in weight change due to the formation of new oxide scale. This observation was related with the oxidation kinetic of this samples that showed higher oxidation rate and weight changes as shown in Fig. 2.

Figure 6 shows SEM image of oxidized samples at 100 h. Figure 6a shows a low magnification of SEM of HT1000 sample which was 1000× indicated the

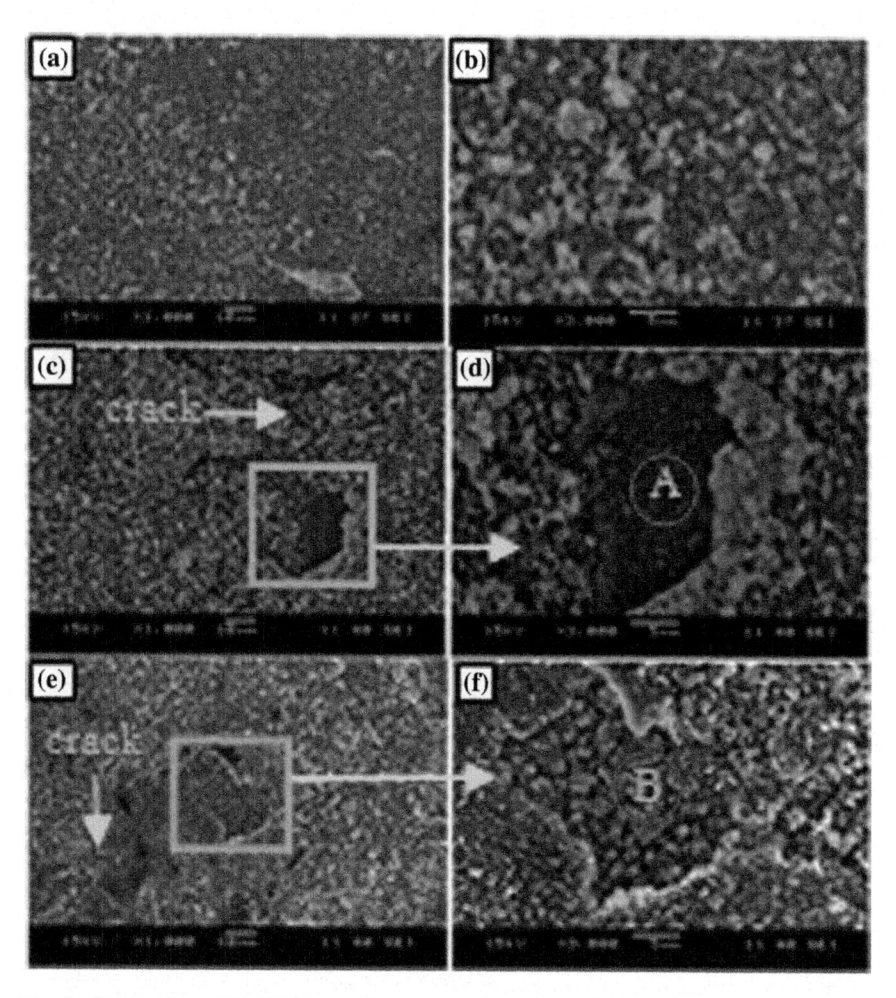

Fig. 6 SEM image of Fe–33Ni–19Cr alloy oxidized at 800 °C for 100 h: **a** HT1000 at 1000×, **b** HT1000 at 3000×, **c** HT1100 at 1000×, **d** HT1100 at 3000×, **e** HT1200 at 1000×, **f** HT1200 at 3000×

continuous growth of oxide layer in the matrix. The oxide particles are located mainly at metal grain boundaries (Niewolak et al. 2016). Figure 6b (HT1000) shows a high magnification of SEM which was 3000× where much denser layer and less dense layer of oxide scale can be seen clearly.

Figure 6c shows a low magnification of SEM of HT1100 sample which was 1000× indicated the continuous growth of compact oxide layer in the matrix. Spallation oxide layer was spotted at the bottom right of the image. A crack can also be seen on top of the spall area. This crack was probably because of the broken oxide layer and additional oxide particles formed between the crack surface. Figure 6d (HT1100) shows a high magnification of SEM which was 3000× producing a close-up image of the spallation oxide layer (A). The oxide spallation will reveal an exposed area on the surface that promotes the formation of new oxide scale. This new oxide scale was contributed to the increasing weight gain recorded on the oxidation kinetics (Fig. 2) for this sample.

Figure 6e shows a low magnification of SEM of HT1200 sample which was 1000× indicated a more uniform oxide layer on the sample surface. Convoluted scales with lots of cracks appeared and scale spallation occurred. This finding was reported by other researcher (Patriarca et al. n.d.; Wang and Lou 1990). Figure 6f (HT1200) shows a high magnification of SEM which was 3000× showed the formation of porous oxide structure started to decrease due to the formation of fresh oxide scale on the pores and cracked area where development of compact oxide structure occur.

Spallation of the oxide layer (B) can also be seen. This oxide spallation reveals an exposed area on the surface that promotes the formation of new oxide scale. This new oxide scale was contributed to the increasing weight gain recorded on the oxidation kinetics (Fig. 2) for this sample.

4 Conclusion

This project was focusing on the influence of grain size on isothermal oxidation of Fe–33Ni–19Cr alloy. Following the results obtained, the main conclusion of this research can be listed as follows:

1. The heat treatment process of Fe–33Ni–19Cr alloy has varied the grain size of Fe–33Ni–19Cr alloy. Increasing the heat treatment temperature will increase the grain size.
2. The kinetic of oxidation of heat-treated Fe–33Ni–19Cr alloy obeyed parabolic rate law. Fine grain sample, which is alloy heat treated at 1000 °C shows lower oxidation rate, whereas alloy heat treated at 1200 °C shows higher oxidation rate.
3. Phase analysis of oxidized heat-treated Fe–33Ni–19Cr alloy was identified using XRD methods. All samples indicated austenitic phase, and several oxide phases

were formed consisting of Cr_2O_3, TiO_2, $(Ti_{0.97}Cr_{0.03})O_2$, Fe_3O_4, and spinel $(MnCr_2O_4, MnFe_2O_4, FeCr_2O_4, NiCr_2O_4,$ and $NiFe_2O_4)$.
4. The oxide morphology formed on oxidized sample is mostly uniform oxide scale. Oxidized sample of heat treatment at 1000 °C shows continuous oxide scale, whereas oxidized sample of heat treatment at 1100 and 1200 °C indicated a formation of crack and evidence of oxide spallation area.

Acknowledgements The authors would like to thank the Ministry of Higher Education, Malaysia, for the research fund under the Fundamental Research Grant Scheme (FRGS) (Project No. FRGS/1/2016/TK05/UNIMAP/02/4).

References

Khanna, A. S. (2002). *Introduction to high temperature oxidation and corrosion.* ASM international.

Joseph, C., Persson, C., & Hörnqvist, M. (2017). Crossmark. *Materials Science & Engineering A, 679*(October 2016), 520–530. http://doi.org/10.1016/j.msea.2016.10.048.

Cao, J., Zhang, J., Chen, R., Ye, Y., & Hua, Y. (2016). Materials characterization high temperature oxidation behavior of Ni-based superalloy GH202. *Materials Characterization, 118,* 122–128. https://doi.org/10.1016/j.matchar.2016.05.013.

Tan, L., Ren, X., Sridharan, K., Allen, T. R. (2008). Effect of shot-peening on the oxidation of alloy 800H exposed to supercritical water and cyclic oxidation. *Corrosion Science 50,* 2040–2046.

Niewolak, L., Zurek, J., Menzler, N. H., Grüner, D., Quadakkers, W. J., Zurek, J., … Quadakkers, W. J. (2016). *Materials at High Temperatures Oxidation and reduction kinetics of iron and iron based alloys used as storage materials in high temperature battery,* 3409 (April).

Patriarca, P., Slaughter, G. M., & Maxwell, W. A. (n.d.). *Corrosion of incoloy 800 and nickel base alloy weldments in steam.*

Wang, F., & Lou, H. (1990). Oxidation behaviour and scale morphology of normal—grained CoCrAl. *Materials Science and Engineering, 129,* 279–285.

Differentiation of Displacement Factor for Stiff and Soft Clay in Additional Modulus of Subgrade Reaction of Nailed-Slab Pavement System

Anas Puri

Abstract Displacement factor was proposed to use in calculation of additional modulus of subgrade reaction. The moduli are used in designing the Nailed-slab Pavement System. A curve of displacement factor was proposed by Puri (2017) for soft clay, and the inverse one was recommended by Hardiyatmo (2011b) for stiff clay. This paper is aimed to develop displacement factor for stiff clay based on Hardiyatmo data and compared to Puri's curve. Results show that there is no significant differentiation of displacement factor between both clay consistencies. It can be concluded that the curve of displacement factor for soft clay from Puri's curve can be used for stiff clay.

Keywords Rigid pavement · Nailed-slab system · Subgrade · Displacement factor · Clay

1 Introduction

Hardiyatmo (2008) introduced a new proposed method that is called Nailed-slab System. It was developed from the pavement of the *Sistem Cakar Ayam Modifikasi* (CAM) by changing the cylindrical foundation with short micropiles. Hardiyatmo (2008) conducted several studies on a nailed-slab under dynamic loads, and studies on vertical loadings were done by Hardiyatmo 2008, 2011a), Nasibu (2009), Dewi (2009), Taa (2010), Somantri (2013), and Puri et al. (2011a, b, 2012a, b, 2013a, b, c, d). Nailed-slab System due to tension loading was studied by Puri et al. (2015), and Puri (2016). Hardiyatmo (2011a) proposed an analysis method for determining the additional modulus of subgrade reaction. The additional modulus of subgrade reaction is the additional modulus developed by a pile. Meanwhile, the modulus of subgrade reaction is the modulus considered from a slab. Puri et al. (2012a)

A. Puri (✉)
Department of Civil Engineering, Universitas Islam Riau, Pekanbaru, Indonesia
e-mail: anaspuri@eng.uir.ac.id

© Springer Nature Singapore Pte Ltd. 2018
R. Saian and M. A. Abbas (eds.), *Proceedings of the Second International Conference on the Future of ASEAN (ICoFA) 2017 – Volume 2*,
https://doi.org/10.1007/978-981-10-8471-3_92

modified the Hardiyatmo method by considering the tolerable deflection or allowable deflection of a pavement slab (δ_a) as an approach to safety construction.

Hardiyatmo (2011b) used the inverse of the displacement factor which developed from full-scale test in stiff clay. The displacement factor is the ratio between the relative displacement between the pile and soil (δ_0) and the displacement of the pile head (δ_p). Since the δ_p is assumed to be similar with slab deflection (δ_s), then this factor is written as

$$\alpha = \frac{\delta_0}{\delta_s} \qquad (1)$$

Puri (2017) proposed a curve of displacement factor (δ_0/δ_s) for soft clay. In this paper, the curve of the δ_0/δ_s ratio for stiff clay based on Hardiyatmo data (2011b) will be developed and will be compared to Puri's curve (2017).

Hardiyatmo (2011a) proposed Eq. (2) in determining the additional modulus of subgrade reaction (Δk). The relative displacement between the pile and soil is considered.

$$\Delta k = \frac{\delta_0 A_s}{\delta_s^2 s^2} (a_d c_u + p_0 K_d \tan \varphi_d) \qquad (2)$$

where δ_0: relative displacement between pile and soil (m), δ_s: deflection of surface of slab (m), A_s: surface area of pile shaft (m^2), s: pile spacing (m), a_d: adhesion factor (non-dimensional), c_u: undrained cohesion (kN/m^2), p_o: average effective overburden pressure along pile (kN/m^2), K_d: coefficient of lateral earth pressure in pile surroundings (non-dimensional), and ϕ_d: soil internal friction angle (°).

Figure 1 shows the curve of the inverse of the displacement factor (δ_s/δ_0) based on the full-scale test of a single pile in stiff clay. The pile and slab were connected by bolts. The pile diameter was 20 cm, and the length of the pile varied between 1.0 and 2.0 m.

Figure 2 shows the curve of displacement factor (δ_0/δ_s) for soft clay according to Puri (2017). This curve was based on full-scale test on single pile Nailed-slab System in soft clay. The slab and pile were connected monolithically.

Fig. 1 Relationships of δ_s/δ_0 ratio versus slab deflection (Hardiyatmo, 2011b)

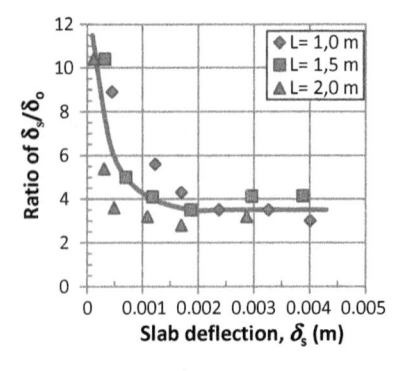

Fig. 2 Curve of displacement factor (α) for soft clay (Puri 2017)

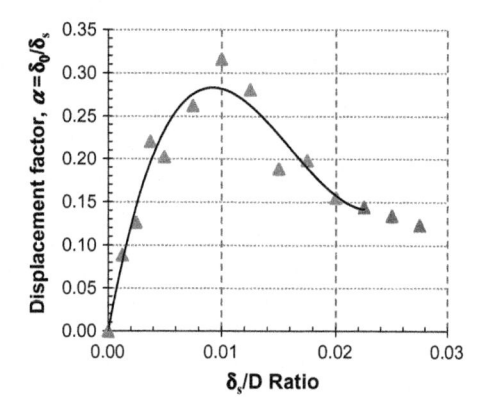

The additional modulus of subgrade reaction (Δk) is used in determining the equivalent modulus of subgrade reaction (k′) as given as Eq. (3) (Hardiyatmo 2011a; Dewi 2009; Puri et al. 2012a):

$$k' = k + \Delta k \qquad (3)$$

where k: modulus of subgrade reaction from plate load test (kN/m^3) and Δk: additional modulus of subgrade reaction due to pile installation under slab (kN/m^3), the modulus of subgrade reaction from a plate load test (k) is usually taken by using a circular plate, and it should be corrected to the slab shape of the nailed slab. The secant modulus is recommended.

2 Methodology

The object of this research is the inverse of displacement factor from Hardiyatmo (2011b) as shown in Fig. 1 and the curve of displacement factor from Puri (2017) as in Fig. 2. Hardiyatmo (2011b) used data from Dewi (2009) to develop Fig. 1. Dewi (2009) conducted Nailed-slab System with single pile by using reinforced concrete. The used slab was circular slab with 1.00 m in diameter and 0.15 m in thickness. There was no lean concrete under the slab. The used pile was 0.20 m in diameter and 1.00, 1.50, and 2.00 m in length variation, respectively. Slabs and piles were connected by using some bolts. The Nailed-slab was constructed in stiff clay. The unconfined compression strength q_u of clay was 120.28 kN/m^2. The characteristic strength of concrete was 21.97 MPa.

Puri (2017) also conducted Nailed-slab System with single pile by using reinforced concrete. The used slab was rectangular slab with 1.20 m width and 0.15 m in thickness. There was 0.05 m lean concrete under the slab. The used pile was 0.20 m

Table 1 Clay properties

No.	Parameter	Unit	Stiff clay Dewi (2009)	Soft clay Pur et al. (2013c)
1	Specific gravity, G_s	–	2.45	2.55
2	Consistency limits:			
	- Liquid limit, LL	%	88.46	80.50
	- Plastic limit, PL	%	24.43	28.48
	- Shrinkage limit, SL	%	n/a	9.34
	- Plasticity index, PI	%	56.05	59.98
3	Water content, w	%	24.60–36.10	54.87
4	Fines content	%	92.93	92.29
5	Sand content	%	7.30	6.89
6	Gravel content	%	0.41	0
7	Bulk density, γ	kN/m^3	17.00	16.32
8	Dry density, γ_d	kN/m^3	12.52	10.90
9	Undrained shear strength, s_u	kN/m^2	60.14	20.14
10	CBR	%	n/a	0.83
11	Soil classification:			
	- AASHTO	–	A-7-6	A-7-6
	- USCS	–	CH	CH

in diameter and 1.50 m in length. Slabs and piles were connected monolithically. The Nailed-slab was constructed in soft clay. The unconfined compression strength q_u of clay was 40.28 kN/m^2. The characteristic strength of concretes was 29.21, 17.4, and 14.5 MPa for slab, pile, and lean concrete, respectively. Table 1 shows clay properties from both researchers.

The displacement factor for stiff clay will be defined by using Fig. 1. The value of δ_s/δ_0 will be inversed become displacement factor ($\alpha = \delta_0/\delta_s$) and compared to Puri (2017).

3 Results and Discussion

Displacement factor for stiff clay based on Hardiyatmo (2011b)—only for 1.50 m pile length—is shown in Fig. 3 and combined to Puri (2017). It is seen that there is no differentiation between both soil consistency up to 0.01 in δ_s/D ratio. Significant differentiation came up after 0.01 in δ_s/D ratio. Figure 4 shows the curve of displacement factor for all data and is compared to Puri (2017). It is can be concluded for both cases that there is no effect of soil consistency. Although there are some differents between both cases, Puri (2017) used lean concrete, larger, and

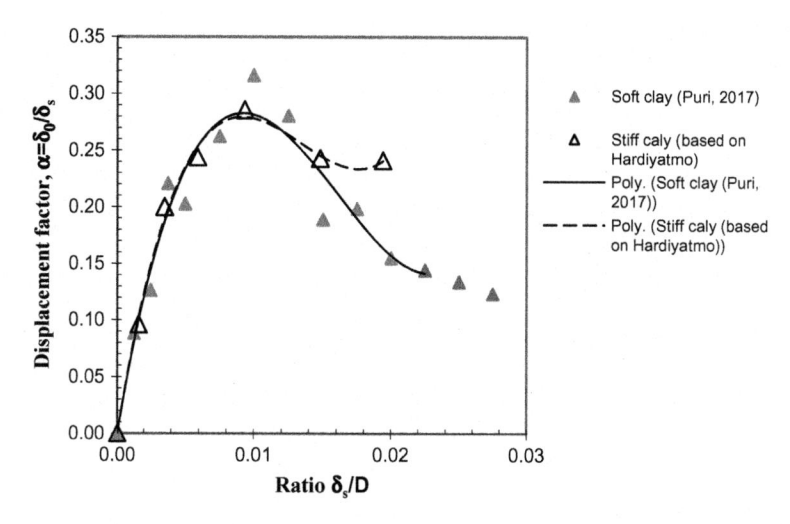

Fig. 3 Displacement factor for soft and stiff clay

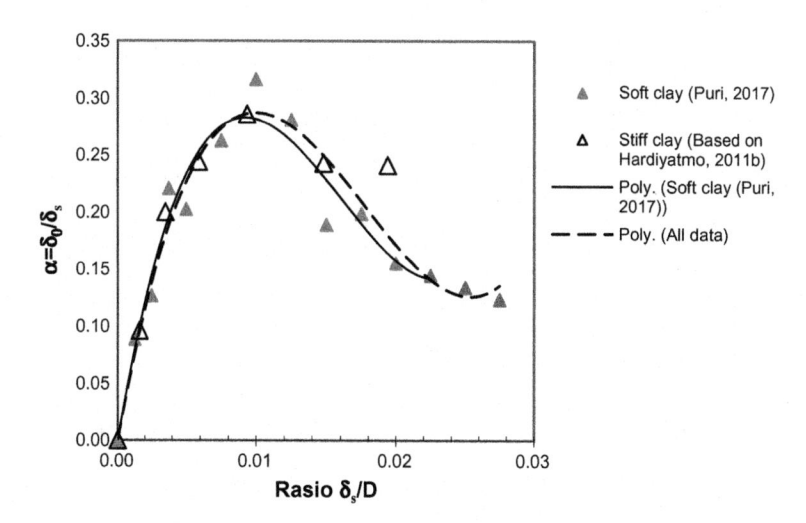

Fig. 4 Displacement factor for all data

rectangular slab while Hardiyatmo (2011b) used circular and smaller slab without lean concrete. And there was also differed on slab-pile connection type. These factors can be neglected, because of the relative displacement between pile and soil was response of these conditions.

According to Fig. 4, it can be concluded that the curve of displacement factor for soft clay from Puri (2017) can be used for stiff clay.

4 Conclusion

Curve of displacement factor for stiff clay was developed and compared to displacement factor for soft clay. It seems that there is no significant differentiation between both clay consistencies. It can be concluded that the curve of displacement factor for soft clay from Puri (2017) can be used for stiff clay. In this case, effect of lean concrete, slab shape and dimension, and slab-pile connection type were neglected, because of the relative displacement between pile and soil was response of these conditions.

Acknowledgements This work is partially supported by Universitas Islam Riau. The author also gratefully acknowledges the helpful comments and suggestions of the reviewers, which have improved the presentation.

References

Dewi, D. A. (2009). *Study on effect of single pile due to the value of equivalent modulus of subgrade reaction from full-scale loading tests (Master's Thesis)*. Graduate Program Gadjah Mada University, Yogyakarta, Indonesia.

Hardiyatmo, H. C. (2008). Nailed-slab system for reinforced concrete slab on rigid pavement. In *Proceedings of the National Seminar on Appropriate Technology for Handling Infrastructures* (pp. M1–M7), 12 April, MPSP JTSL FT UGM, Yogyakarta, Indonesia.

Hardiyatmo, H. C. (2011a). Method to analyze the deflection of the nailed slab system. *IJCEE-IJENS, 11*(4), 22–28.

Hardiyatmo, C. H. (2011b). *Designing of pavement roads and soil investigation: Flexible pavement, rigid pavement, modified chicken foot foundations, nailed-slab system*. Yogyakarta, Indonesia: Gadjah Mada University Press.

Nasibu, R. (2009). *Study on modulus of subgrade reaction due to effect of pile attached under plate (loading test on full scale) (Master's Thesis)*. Yogyakarta, Indonesia: Graduate Program Gadjah Mada University.

Puri, A. (2016). Behavior of uplift resistance of single pile row nailed-slab pavement system on soft clay sub grade. In *Proceeding of the 3rd Asia Future Conference (AFC)*, Kitakyushu, Japan, 29 Sept–3 Oct 2016.

Puri, A. (2017). Developing the curve of displacement factor for determination the additional modulus of sub grade reaction on nailed-slab pavement system. *International Journal of Technology, 1*, 1117–1126. ISSN 2086-9614.

Puri, A., Hardiyatmo, C. H., Suhendro, B., & dan Rifa'i, A. (2011a). Experimental study on deflection of slab which reinforced by short friction piles in soft clay. In *Proceedings of the 14th Annual Scientific Meeting (PIT) HATTI* (pp. 317–321), Yogyakarta, 10–11 Feb, Indonesia.

Puri, A., Hardiyatmo, H. C., Suhendro, B., & dan Rifa'i, A. (2011b). Contribution of wall barrier to reduce the deflection of nailed-slab system in soft clay. In *Proceedings of the 9th Indonesian Geotech. Conference and 15th Annual Scientific Meeting (KOGEI IX & PIT XV) HATTI* (pp. 299–306), Jakarta, 7–8 Dec, Indonesia.

Puri, A., Hardiyatmo, H. C., Suhendro, B., & dan Rifa'i, A. (2012a). Determining additional modulus of subgrade reaction based on tolerable settlement for the nailed-slab system resting on soft clay. *IJCEE-IJENS, 12*(3), 32–40.

Puri, A., Hardiyatmo, H. C., Suhendro, B., & dan Rifa'i, A. (2012b). Application of the additional modulus of subgrade reaction to predict the deflection of nailed-slab system resting on soft clay due to repetitive loadings. In *Proceedings of Pertemuan Ilmiah Tahunan ke-16 (PIT) HATTI* (pp. 217–222), Jakarta, 4 Dec, Indonesia.

Puri, A., Hardiyatmo, H. C., Suhendro, B., & dan Rifa'i, A. (2013a). Pile spacing and length effects due to the additional modulus of subgrade reaction of the nailed-slab system on the soft clay. In *Proceedings of the 13th International Symposium on Quality in Research (QiR)* (pp. 1032–1310), Yogyakarta, 25–28 June, Indonesia.

Puri, A., Hardiyatmo, H. C., Suhendro, B., & dan Rifa'i, A. (2013b). Deflection analysis of nailed-slab system which reinforced by vertical wall barrier under repetitive loadings. In *Proceedings of the 6th Civil Engineering Conference in Asian Region (CECAR6)* (pp. TS6-10–TS6-11), Jakarta, 20–22 Aug, Indonesia.

Puri, A., Hardiyatmo, H. C., Suhendro, B., & dan Rifa'i, A. (2013c). Application of the method of nailed-slab deflection analysis on full scale model and comparison to loading tests. In *Proceedings of Konferensi Nasional Teknik Sipil ke-7 (KoNTekS7)* (pp. G201–G211), Universitas Negeri Sebelas Maret, Surakarta, 24–26 October, Indonesia.

Puri, A., Hardiyatmo, H. C., Suhendro, B., & dan Rifa'i, A. (2013d). Behavior of fullscale nailed-slab system with variation on load positions. In *Proceedings of the 1st International Conference on Development Infrastructure (ICID)* (pp. 26–36), UMS, Solo, 1–3 Nov, Indonesia.

Puri, A., Hardiyatmo, H. C., Suhendro, B., & dan Rifa'i, A. (2015). Pull out test of single pile row nailed-slab system on soft clay. In *Proceeding of the 14th International Conference on Quality in Research (QiR)* (pp. 63–68), Universitas Indonesia, Lombok, 10–13 Aug.

Somantri, A. K. (2013). *Kajian Lendutan Pelat Terpaku pada Tanah Pasir dengan Menggunakan Metode Beam on Elastic Foundation (BoEF) dan Metode Elemen Hingga (Master's Thesis)*. Graduate Program Gadjah Mada University.

Taa, P. D. S. (2010). *Effects of installation of group pile due to slab uplift of nailed-slab resting on expansive subgrade (Master's Thesis)*. Yogyakarta, Indonesia: Graduate Program Gadjah Mada University.

Decision Support System of Vegetable Crop Stipulation in Lowland Plains

Ause Labellapansa, Ana Yulianti and Arif Zalbiahdi

Abstract Agriculture is a technology pattern that requires energy to flow, process, change, and produces energy. Rengat is an agrarian sub-district located in Riau Province, Indonesia, with the main livelihood of the population comes from agriculture industry. For onfarm activities, many cultivated commodities are vegetables and fruits. This leads to the fact that stipulation of suitable vegetable crops in lowland areas is crucial. To provide solutions, a decision support system is needed to determine suitable vegetable crops in lowland areas. The decision support system process is done by using Naive Bayesian Classification method. Methodology used in this research requires interviews and data collecting directly from the Agriculture Food Crops and Horticulture Agent, Indragiri Hulu Regency. If farmers want to access decision support systems to determine the suitable vegetable crops in lowland areas, farmers are required to register to the Agent. This decision support system can facilitate farmers in determining the suitable vegetable crops in lowland plains.

Keywords Vegetable crops · Naive Bayesian Classification · Agro-industry

1 Preliminary

Vegetable plants are source for vitamins and minerals or common term for food of plant origin that usually contains high water content and consumed in a fresh state or after being processed minimally. Vegetables are needed for some health benefits. The contents of various vitamins, carbohydrates, and minerals in vegetables cannot be substituted with staple foods (Kanisius 1992).

A good temperature range for lowland vegetable growth is greater than in highland vegetables. Actually, the plants will grow better at average monthly

---

A. Labellapansa (✉) · A. Yulianti · A. Zalbiahdi
Program Studi Teknik Informatika, Fakultas Teknik, Universitas Islam Riau,
Jl. Kaharudin Nasution no. 113, Pekanbaru, Indonesia
e-mail: ause.labella@eng.uir.ac.id

temperature of 21 °C and above. The average temperature for optimum growth is at 26–28.5 °C. If the minimum average temperature is lower than 10 °C, plant growth will be disrupted. The vegetables well grown in the lowlands are chili, tomato, long bean, eggplant, caisim, cucumber, kale, garlic, onion, spinach, lettuce, leek, basil, winged, pumpkin, and bitter gourd.

There is a science field of artificial intelligence that is able to assist farmers in determining the appropriate vegetable crops in the lowlands. This decision support system is a systematic approach to the essence of a problem, the gathering of facts, the mature determination of the alternatives encountered, and the taking of actions which, according to the calculations, are the most appropriate action (Suryadi et al. 2000). The decision support system developed in this research uses Naive Bayesian Classification method.

2 Literature Review

Several previous studies have been conducted both to determine suitable vegetable crops in lowland plains and to research of the similarity on methods used. In a study conducted by Habibullah et al. (2014), decision support system was developed to determine the suitable agricultural land types for the cultivation of fruit crops using the similarity method. The steps in his research include data collection method of data which obtained from the literature studies, and then knowledge representation will be done on data that have been obtained to be used as a knowledge base on decision support systems. Similarity method is used to find solutions based on the conclusion of the similarity on existing objects, and then the greatest similarity value will be selected as a solution. From the results of system testing, it is said that the use of similarity method on the system has provided good results on appropriate fruit based on land criteria.

Research conducted by Utami Ferry Hari (2015), a system was developed for Agent of Counseling for Agriculture, Farming, and Fisheries to determine soil fertility level using Naive Bayes algorithm in data mining. The steps in his research include data collection methods from the literature studies, then knowledge representation is done for the data already obtained, and then finds the probability value of each criterion. Naive Bayes method is used to find the solution for the conclusion based on the largest probability value. From the results of system testing, it is said that the utilization of Naive Bayes method on the system has given good results to determine fertile or infertile produce on ground based on the criteria entered.

Lubis research (Lubis 2013) has developed a decision support system for determining the feasibility of agricultural areas using SAW method. The steps in his research such as data collection methods from literature studies, then knowledge representation will be done on the data obtained and to find the weight value for each attribute. SAW method is used to find the solution of conclusion based on the value of weight value for each attribute. Based on the results of system testing, it is said that the use of SAW method on the system has given good results to produce a decent area for agriculture.

3 Research Methodology

This research is divided into several stages that are used to develop decision support system, starting from problem identification or analysis, knowledge acquisition, knowledge representation, development and testing.

3.1 Knowledge Acquisition

Knowledge acquisition is a process to collect data knowledge of a problem. This study uses two types of data obtained through different ways. The data used is primary data which taken directly through the source which in this case is Agriculture Food Crops and Horticulture Agent Indragiri Hulu, Riau, Indonesia. The primary data collected in this research is done through interviews. There are ten kinds of criteria obtained from the acquisition, namely soil type, soil texture, soil pH, temperature, rainfall, soil fertility, soil moisture, altitude, wind speed, and sunlight. As for vegetable crops as many as 393 data of vegetable crops were available.

3.2 Knowledge Representation

After the data collection process is done, next is data representation into the knowledge base to describe in forms of design so becomes systematic. The representation of knowledge made into the decision support system determines the appropriate vegetable crops in this lowland plains uses a table consisting of 12 tables.

3.3 Implementation in Naive Bayesian

Naive Bayesian Classification is a statistical classification that can be used to predict the probability of membership of a class. Probability is a value used to measure the degree of occurrence of a random event. Probability is generally a probability that will happen (Kusrini et al. 2009). It is important to know that Naive Bayesian Classification method requires a number of clues to determine what class is suitable for the analyzed sample. Therefore, the method of Naive Bayesian Classification is adjusted as follows:

$$P(C|F1...Fn) = \frac{P(C)P(F1...Fn|C)}{P(F1...Fn)} \quad (1)$$

where C variable represents the class, while the $F1$ variable... Fn represents the characteristics of the clues needed to perform the classification. Then the formula explains that probability of entering a sample of certain characteristics in class C (Posterior) is the probability of the emergence of class C (prior to the entry of the sample, often called the prior), multiplied by probability of characteristics of the sample in class C (also called likelihood) divided by the probability of appearance of characteristics of samples globally (also called evidence). Therefore, the above formula can also be written simply as follows:

$$\text{Posterior} = \frac{\text{prior} \times \text{likelihood}}{\text{evidence}} \quad (2)$$

Evidence values are always fixed for each class in a single sample. For classification with continuous data, the Gauss density formula is used:

$$P(X_i = \chi_i | Y = y_j) = \frac{1}{\sqrt{2\pi}\sigma_{ij}} e^{-\frac{(\chi_i - \mu_{ij})^2}{2\sigma_{ij}^2}} \quad (3)$$

where P: opportunities, e: 2,718282, π: 3,14, X_i: attribute to I, x_i: the attribute value to I, Y: classes are in search, y_i: subclass Y is in search, μ: (mean), states the mean of all attributes, σ: (standard deviation), declares a variant of all attributes.

The flow of the Naive Bayesian Classification method can be seen in Fig. 1: (Saleh Alfa 2015):

1. Read training data
2. Calculate amount and probability, but if numerical data then:

   a. Find the mean and standard deviation values of each parameter which is numerical data. The equation used to calculate the average value of the count (mean) can be seen as follows:

   $$\mu = \frac{\sum_{i=1}^{n} \chi_i}{n} \quad (4)$$

   or

   $$\mu = \frac{\chi_1 + \chi_2 + \chi_3 + \cdots \chi_n}{n} \quad (5)$$

Fig. 1 Method of Naive Bayesian Classification

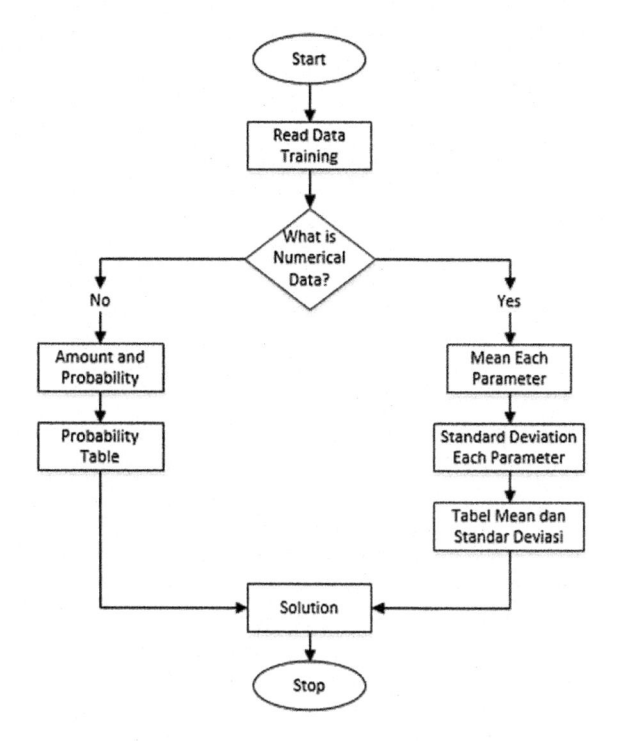

where

- μ average count (mean)
- χ_i sample value to $-i$
- n the number of samples and equations to calculate the standard deviation value (standard deviation) can be seen as follows:

$$\sigma = \sqrt{\frac{\sum_{i=1}^{n}(\chi_i - \mu)^2}{n-1}} \qquad (6)$$

where

- σ standard deviation
- χ_i value x to $-i$
- μ average count
- n   number of samples

However, if the *i*th attribute is continuous, then $P(\chi_i|C)$ is estimated with the Gauss density function.

$$f(x) = \frac{1}{\sqrt{\frac{2\pi}{\sigma}}} e^{\frac{-(x-\mu)^2}{2\sigma^2}} \quad (7)$$

with μ = mean, and σ = standard deviation.

$$f(x) = \frac{1}{\sqrt{2\pi\sigma}} e^{\frac{-(x-\mu)^2}{2\sigma^2}} \quad (8)$$

b. Find probabilistic values by calculating the appropriate amount of data from the same category divided by the amount of data in that category.

3. Get values in mean tables, standard deviations, and probabilities.
4. The solution is then generated.

Table 1 presents six training data from two types of plants, namely Capri and Cabbage with different values of type of soil, soil texture, soil pH, temperature, rainfall, soil fertility, soil moisture, elevation place, wind velocity, and sunlight. There are 393 of training data.

4 Results and Discussion

Based on the following data in Table 1, a farmer wants to plant crops in a field with criteria: Soil Type: **Alluvial**, Soil Texture: **Friable**, Soil pH: **Medium**, Temperature: **Low**, Rainfall: **Medium**, Soil Fertility: **Height**, Soil Humidity: **Low**, Elevation Place: **Medium**, Wind Speed: **Low**, Sunshine: **Medium**. Which vegetable crops fit the criteria.

The probability value of each criterion is obtained from the training data in Table 1. The probability value of each criterion is as follows:

The Probability of Soil Type Criteria is shown in Table 2. Soil type for Capri consists of organosol = 0, humus = 1, alluvial = 2 while the soil type for Cabbage consists of organosol = 2, humus = 0, alluvial = 1. The Probability of Soil Texture Criteria is shown in Table 3.

The Probability of the Land pH Criteria is shown in Table 4. The Temperature Criteria Probability is shown in Table 5. The Probability of Rainfall Criteria is shown in Table 6. The Probability of Soil Fertility Criteria is shown in Table 7. The Probability of Soil Humidity Criteria is shown in Table 8. The Probability of Place

Table 1 Data training table

Type of soil	Soil texture	Soil pH	Temperature	Rainfall	Soil fertility	Soil moisture	Elevation place	Wind velocity	Sunlight
Vegetable crops: Capri									
Alluvial	Loose	Height	Medium	Medium	Height	Low	Medium	Low	Low
Alluvial	Loose	Medium	Low	Low	Height	Medium	Medium	Medium	Medium
Humus	Soft	Medium	Low	Medium	Medium	Low	Low	Low	Medium
Vegetable crops: Cabbage									
Alluvial	Look	Medium	Low	Height	Low	Height	Low	Height	Height
Organasol	Loose	Medium	Low	Medium	Height	Low	Low	Low	Height
Organasol	Look	Medium	Low	Height	Low	Height	Medium	Height	Medium

Elevation Criteria is shown in Table 9. The Probability of Wind Speed Criteria is shown in Table 10. The Probability of Sunlight Criteria is shown in Table 11. The Probability of Vegetable Criteria is shown in Table 12.

Temperature: **Low**, Rainfall: **Medium**, Soil Fertility: **Height**, Soil Humidity: **Low**, Altitude Place: **Medium**, Wind Speed: **Low**, Sunshine: **Medium**. Then it can be calculated:

Table 2 Probability of soil type criteria

Type of soil	Number of events		Probability	
	Capri	Cabbage	Capri	Cabbage
Peat	0	2	0/3	2/3
Humus soil	1	0	1/3	0/3
Friable soil	2	1	2/3	1/3
Amount	3	3	1	1

Table 3 Probability of soil texture criteria

Soil texture	Number of events		Probability	
	Capri	Cabbage	Capri	Cabbage
Soft	1	0	1/3	0/3
Look	0	2	0/3	2/3
Loose	2	1	2/3	1/3
Amount	3	3	1	1

Table 4 Probability of soil pH criteria

Soil pH	Number of events		Probability	
	Capri	Cabbage	Capri	Cabbage
Low	0	0	0/3	0/3
Medium	2	3	2/3	3/3
Height	1	0	1/	1/3
Amount	3	3	1	1

Table 5 Probability of temperature criteria

Temperature	Number of events		Probability	
	Capri	Cabbage	Capri	Cabbage
Low	2	3	2/3	3/3
Medium	1	0	1/3	0/3
Height	0	0	0/3	0/3
Amount	3	3	1	1

Table 6 Probability of rainfall criteria

Rainfall	Number of events		Probability	
	Capri	Cabbage	Capri	Cabbage
Low	1	0	1/3	0/3
Medium	2	1	2/3	1/3
Height	0	2	0/3	2/3
Amount	3	3	1	1

Table 7 Probability of soil fertility criteria

Soil fertility	Number of events		Probability	
	Capri	Cabbage	Capri	Cabbage
Low	0	2	0/3	2/3
Medium	1	0	1/3	0/3
Height	2	1	2/3	1/3
Amount	3	3	1	1

Table 8 Probability of soil humidity criteria

Soil moisture	Number of events		Probability	
	Capri	Cabbage	Capri	Cabbage
Low	2	1	2/3	1/3
Medium	1	0	1/3	0/3
Height	0	2	0/3	2/3
Amount	3	3	1	1

Table 9 Probability of site elevation criteria

Elevation place	Number of events		Probability	
	Capri	Cabbage	Capri	Cabbage
Low	1	2	1/3	2/3
Medium	2	1	2/3	1/3
Height	0	0	0/3	0/3
Amount	3	3	1	1

Table 10 Wind speed criteria probability

Wind velocity	Number of events		Probability	
	Capri	Cabbage	Capri	Cabbage
Low	2	1	2/3	1/3
Medium	1	0	1/3	0/3
Height	0	2	0/3	2/3
Amount	3	3	1	1

Table 11 Probability of sunlight criteria

Sunlight	Number of events		Probability	
	Capri	Cabbage	Capri	Cabbage
Low	1	0	1/3	0/3
Medium	2	1	2/3	1/3
Height	0	2	0/3	2/3
Amount	3	3	1	1

Table 12 Probability of vegetable criteria

Vegetable crops	Number of events		Probability	
	Capri	Cabbage	Capri	Cabbage
Amount	3	3	1	1

$$\text{Capri Likelihood} = \frac{2}{3} \times \frac{2}{3} \times \frac{2}{3} \times \frac{2}{3} \times \frac{2}{3} \times \frac{2}{3} \times \frac{2}{3} \times \frac{2}{3} \times \frac{2}{3} \times \frac{2}{3} \times \frac{3}{6}$$
$$= 0.666 \times 0.666 \times 0.666 \times 0.666 \times 0.666$$
$$\times 0.666 \times 0.666 \times 0.666 \times 0.666 \times 0.666 \times 0.5$$
$$= 0.008671$$

$$\text{Cabbage Likelihood} = \frac{1}{3} \times \frac{1}{3} \times \frac{3}{3} \times \frac{3}{3} \times \frac{1}{3} \times \frac{1}{3} \times \frac{1}{3} \times \frac{1}{3} \times \frac{1}{3} \times \frac{1}{3} \times \frac{3}{6}$$
$$= 0.333 \times 0.333 \times 1 \times 1 \times 0.333 \times 0.333$$
$$\times 0.333 \times 0.333 \times 0.333 \times 0.333 \times 0.333 \times 0.5$$
$$= 7.62079E - 05$$

The probability value can be calculated by normalizing the likelihood so that the number of values obtained = 1.

$$\text{Capri Probability} = \frac{0.008671}{0.008671 + 7.62079E - 05} = 0.991287746$$

$$\text{Cabbage Probability} = \frac{7.62079E - 05}{7.62079E - 05 + 0.008671} = 0.008712254$$

So it can be concluded that plants suitable for agriculture with criteria: Soil Type: Alluvial, Soil Texture: Gembur, pH Soil: Medium, Temperature: Low, Rainfall: Medium, Soil Fertility: Height, Soil Humidity: Low, Elevation Place: Wind Speed: Low, Sunlight: Medium, i.e., Crop Capri with probability value 0.991287746.

Testing of decision support system is done by using *black box* and by testing by Head of Agriculture Agent. Based on the results of testing by the Head of Agent, the results of manual calculations were obtained and systems provide the same conclusion.

5 Conclusion

The conclusions in this study are:

1. In this system there are previous data from vegetable crops based on the criteria, so that the output will be obtained in the form of values close to 1 or = 1.
2. This system can facilitate the Agency of Agriculture and farmers in determining the appropriate vegetable crops in the lowlands by using more detailed data.

References

Habibullah et al. (2014). Sistem pendukung keputusan penentuan kesesuain jenis lahan pertanian untuk budidaya tanaman buah-buahan menggunakan metode similarity berbasis web. *Jurnal Sarjana Teknik Informatika, 2*(2).

Kanisius. (1992). *Petunjuk praktis bertanam sayuran.* Yogyakarta: Kanisius.

Kusrini, et al. (2009). *Algoritma data mining.* Yogyakarta: Andi.

Lubis, E. (2013). Sistem pendukung keputusan penentuan kelayakan daerah pertanian menggunakan metode saw. *Pelita Informatika Budi Darma,* V(3).

Saleh, A. (2015). Implementasi metode klasifikasi naive bayes dalam memprediksi besarnya penggunaan listrik rumah tangga. *Citec Journal, 2*(3).

Suryadi, et al. (2000). *Sistem pendukung keputusan suatu wacana structural edealisasi dan implementasi konsep pengambilan keputusan.* Bandung: PT Remaja Rosdakarya.

Utami Feri Hari. (2015). Penentuan tingkat kesuburan tanah di balai penyuluhan pertanian perikanan dan kehutanan dengan menggunakan algoritma naive bayes dalam data mining. *Riau Journal of Computer Science, 1*(1), 27–38.

Geology and Geochemistry Analysis for Ki Index Calculation of Dompak Island Granite Bauxites to Determine the Economical Mineral

Catur Cahyaningsih, Arrachim Maulana Putera, Gayuh Pramukti and Mohammad Murtaza Sherzoy

Abstract The research area is geographically located at coordinates 00° 52′ 00″N and 104° 33′ 00″E, administratively belongs to the region Dompak Island, District of Bukit Bestari, Bintan Regency. The method used is field orientation, geochemistry using X-ray fluorescence (XRF) and work studio. Based on XRF, data can calculate of the index Ki index. The results of the analysis concluded that the dominant presence of the compounds Al_2O_3 with the amount of 39.11%, Fe_2O_3 with amount 32.88%, SiO_2 with amount 5.63% and TiO_2 with amount 2894%. Ki index bauxite in research area classified into high quality to low, economical enough to do the exploration.

Keywords Bauxite · Ki index · XRF · Economical · Mineral

1 Introduction

Bauxite resources are economic concentrations of aluminum, formed from alteration and chemical weathering of alumosilicate-rich parent rocks (Ahmadnejad et al. 2017; Alex and Kumar 2017; Argyraki et al. 2017; Buccione et al. 2016; Guevara et al. 2017; Han et al. 2017; Zhu et al. 2016; Wang et al. 2015, 2016). Bauxite ($Al_2O_3 \cdot 2H_2O$) has octahedral crystal system, consisting of 35–65% Al_2O_3, 2–10% SiO_2, 2–20% Fe_2O_3, 1–3% TiO_2, and 10–30% H_2O. As the alumina ore, bauxite contains at least 35% Al_2O_3, SiO_2 5%, 6% Fe_2O_3, and 3% TiO_2. Bauxite is formed from rocks that have high levels of aluminum, iron levels and slightly lower

---

C. Cahyaningsih (✉) · A. M. Putera · G. Pramukti
Geological Engineering Department, Islamic University of Riau, Jl.
KH Nasution no. 113 Perhentian Marpoyan, Pekanbaru, Riau 28284, Indonesia
e-mail: caturcahyaningsih@eng.uir.ac.id

M. M. Sherzoy
Academy of Sciences of Afghanistan, Sher Ali Khan Watt, Shari-e-naw,
894, Kabul, Afghanistan
e-mail: murtaza_sherzoy2000@yahoo.com

© Springer Nature Singapore Pte Ltd. 2018
R. Saian and M. A. Abbas (eds.), *Proceedings of the Second International Conference on the Future of ASEAN (ICoFA) 2017 – Volume 2*,
https://doi.org/10.1007/978-981-10-8471-3_94

Table 1 Distribution grade bauxite reserves

Kelas cadangan	Al₂O₃	SiO₂
A	>50%	0%
B	48–50%	0–13%
C	<48%	>13%

levels of free quartz. At the time of chemical weathering of rocks, it undergoes chemical elements silica (Si) dissolved and loose from the crystal as well as some elements of Iron. Alumina, titanium, and mineral oxidation are concentrated as sludge residue. Rocks that can meet the requirements are, among others, nephelin, syenite, claystone/shalestone. It will undergo a process rock lateritization (temperature exchange process continuously so that the rocks undergo weathering) (Deng et al. 2017; Valeton 1972; Cahyaningsih 2016; Chengrong et al. 2016; Croymans et al. 2017; Zamanian et al. 2015; Wang et al. 2015; Shi et al. 2016; Rao et al. 2016).

Deposition of bauxite in the area of Bintan discovered in 1924 and the first advantage is a Dutch company, *NVN Nederlansch Indies Bauxiet Exploitatie Maatschapij* (NVNIBEM). The distribution of minerals of bauxite (alumina clay) spreads widely in Bintan Island and the surrounding area. Bauxite in the area of research is the result of weathering of granite rocks which are the bedrock of Bintan Island, is generally spread on the morphology of the plains up with ramps that allow intensive weathering process can take place. Table 1 shows the class division of bauxite reserves.

Potential distribution of bauxite is quite likely in the subdistrict of East Bintan, on the mainland and islands—islands in the vicinity, is a mining area and most of the former bauxite mine (Retty Dwi Kisnawati 2016; Fatriadi et al. 2017).

2 Regional Geology

Regional stratigraphy of the study area is an old granite Trias. Reddish-greenish-gray granite, coarse grained, consist of feldspar, quartz, hornblende and biotite, generally mineral textured primer and form a broad pluton exposed batholiths in Batam Island and Bintan Island, the result of weathering and peneplain process generates economic minerals such as bauxite deposits, based on the location and mineralogical composition, granite inni grouped into several plutons Granite Pluton Kawal in Bintan and Granite Pluton Nongsa in Batam, shown in Fig. 1.

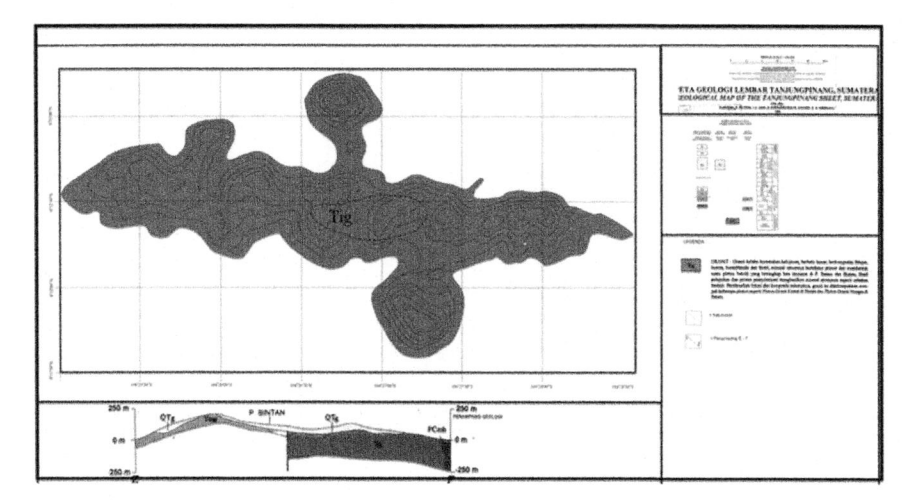

Fig. 1 Map showing the regional geology Dompak Island, Bintan Regency, Riau Islands

3 Methods

The method used in this study consisted of three methods: field orientation, geochemical, studio. Field orientation method is to record against the outcrop and collect bauxite sample. The method used is geochemical analysis using X-ray fluorescence (XRF) machine. Studio method is an activity creating a geological map and index Ki calculation. A total of 27 samples of bauxite in the area of research in analysis using XRF to find the percentage of content of Aluminum, Iron, Silica, and Titanium in bauxite samples were collected by using the grid method.

4 Results

Results of the study area consist of geology, stratigraphy, geochemistry, and Ki index calculation.

4.1 Geology and Stratigraphy of Study Areas

Based on the results of research area will not find granite outcrops of Triassic identify. Granite above precipitated unconformity Triassic Formation Goungon deposited in Pliocene–Pleistocene period, shown in Fig. 2.

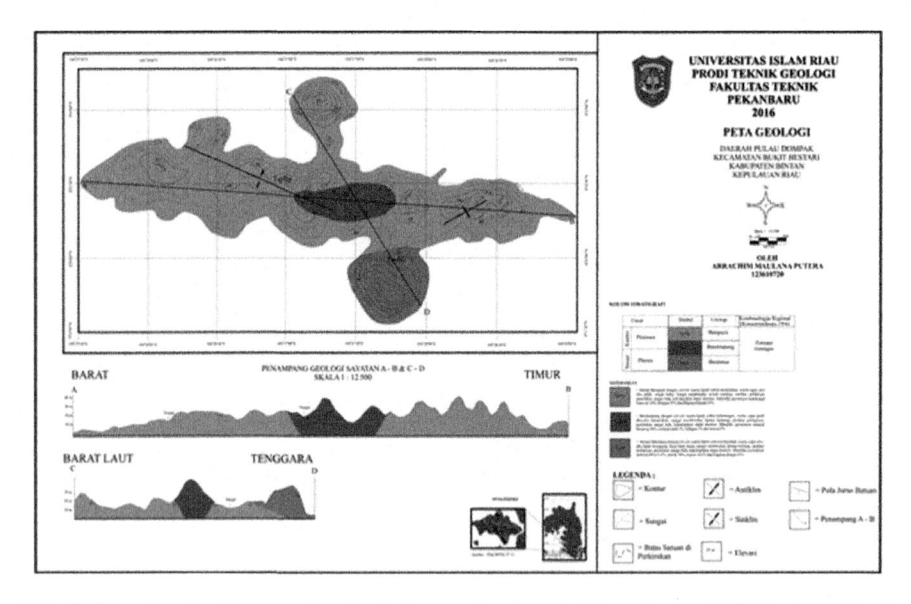

Fig. 2 Map showing the area of geological research area consisting of several lithologies units of the rock research area is divided into three, namely: Quartz Wacke Sandstone unit, Siltstone unit, and Mudstone unit

1. Quartz Wacke Sandstone unit

Quartz Wacke Sandstone unit shows characteristics of colors weathered reddish brown and fresh color gray-white, large grain very subtle, form of grain rounded, pack closed, sorting good, permeability moderate, compactness is quite soft and non-carbonated. Have a percentage minerals of 15% quartz, feldspar 35% and 35% rock fragments classified as Quartz Wacke.

2. Unit siltstone

Siltstone unit shows the characteristics of weathered brown color and fresh color gray, silt grain size, grain shape rounded, closed containers, sorting good, moderate permeability, compactness can be kneaded and non-carbonated. Percentage of minerals feldspar 0.4%, matrix 50%, 14.6% quartz, opaque minerals 10%, and 35% rock fragments that can be classified as the siltstone in the south with a percentage of 20% in the area of research.

3. Unit mudstone

Mudstone unit showing the characteristics of a brownish-yellow color and a fresh white color pinkish gray, large grain clay, grain shape is very rounded, enclosed containers, sorting is very good, poor permeability, compactness can be kneaded and non-carbonated. Has a 96% percentage of clay minerals, mineral opaque 1%, 1% and quartz feldspar 2% can be classified as mudstone. Distribution in the middle with a percentage of 20% in the area of research.

4.2 Geochemistry and Ki Index Calculation

Results of the XRF analyze which shows the composition of the chemical elements in the bauxite areas of research showing the chemical content by percentage as follows: Al_2O_3 with the amount of 39.11%, Fe_2O_3 with the amount of 32.88%, SiO_2 with amount of 5.63% and TiO_2 in the amount of 2.894%. Based on the XRF analyze, there are several differences in the chemical composition of the different elements in each lithology in the area of research.

Elements of Al_2O_3 to dominate in the siltstone that is by the percentage amount of 58.66%, element Fe_2O_3 to dominate in sandstone that is by the percentage amount of 68.68%, an element of SiO_2 dominate the claystone is by the percentage amount of 8.65%, an element of TiO_2 dominate the siltstone that is with the percentage amount of 1.8% shown in Fig. 3.

Minerals bauxite research area has an average content of aluminum, iron, silica, and titanium that have economic potential. The bauxite mineral economics can be calculated using an index Ki. The results of the analysis of Ki index calculation (Fig. 4) found that sample 2 has a low-level types bauxite despite having a value of Al 15.39%. Sample 4 has the type of bauxite high level with a value of 58.66% Al. Sample 11 had a kind of a high level of bauxite with a value of 38.05% Al. Sample 16 has a low-level types of bauxite with a value of 20.63% Al. Sample 21 has a low-level types of bauxite with a value of 44.97% Al. Of the percentages by classification included into the reserve grade type B is quite economical to do exploration and exploitation activities in the research area.

Fig. 3 Differences in chemical content of bauxite: Al2O3, Fe₂O₃, SiO₂, and TiO₂ in each rock unit of research area

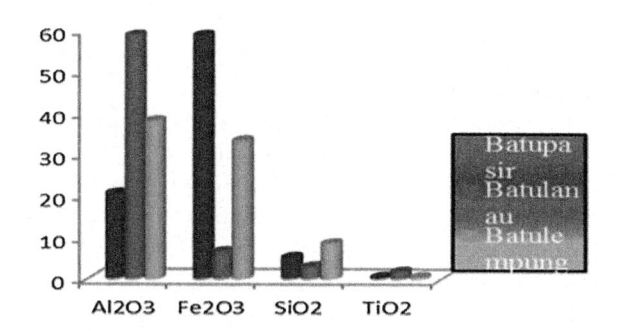

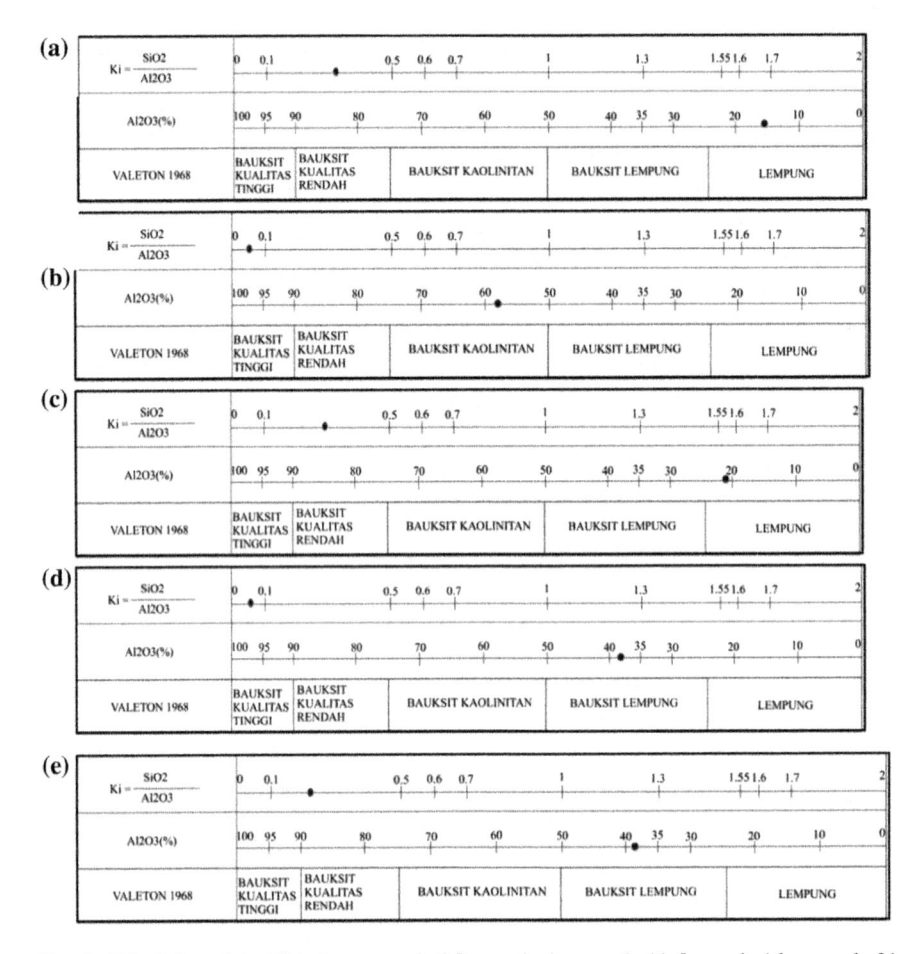

Fig. 4 Calculation of the Ki index **a** sample 2 **b** sample 4 **c** sample 11 **d** sample 16 **e** sample 21

5 Conclusion

Geological regional research area can be three units of rocks that Quartz Wacke Sandstone unit, Siltstone unit and Mudstone unit, Conformable stratigraphic conditions at the time of Pliosesen–Plistosen comparable with Gaungon Formation. Geochemical studies of bauxite region contain Al_2O_3 with the amount of 39.11%, Fe_2O_3 with the amount of 32.88%, SiO_2 with the amount of 5.63%, and TiO_2 with the amount of 2.894%. Ki index from 0 to 0.5 with the classification of high quality to low quality is bauxite. Type deposits in the study area included in the reserve grade type B are quite economical for exploration.

Acknowledgements Thanks are given to Department of Energy and Mineral Resources Riau Islands Province which has provided support and Universitas Islam Riau.

References

Ahmadnejad, F., Zamanian, H., Taghipour, B., & Zarasvandi, A. (2017). Mineralogical and geochemical evolution of the Bidgol bauxite deposit, the Zagros Mountain Belt, Iran: Implications for ore genesis, rare earth elements fractionation and parental affinity. *Ore Geology Reviews*. Elsevier BV. https://doi.org/10.1016/j.oregeorev.2017.04.006.

Alex, T. C., Kumar, R. (2017). International journal of mineral processing surface and bulk activation of a siliceous bauxite during attrition milling. *International Journal of Mineral Processing, 160*, 32–38. Elsevier BV. https://doi.org/10.1016/j.minpro.2017.01.006.

Argyraki, A., Boutsi, Z., Zotiadis, V. (2017). Towards sustainable remediation of contaminated soil by using diasporic bauxite: Laboratory experiments on soil from the sulfide mining village of Stratoni, Greece. *Journal of geochemical Exploration*. Elsevier BV. https://doi.org/10.1016/j.gexplo.2017.03.007.

Buccione, R., Mongelli, G., Sinisi, R., & Boni, M. (2016). Relationship between geometric parameters and compositional data: A new approach to karst bauxites exploration. *Journal of Geochemical Exploration*. Elsevier BV. https://doi.org/10.1016/j.gexplo.2016.08.002.

Cahyaningsih, C. (2016). Hydrology analysis and rainwater harvesting effectiveness as an alternative to face water crisis in the old village Bantan-Riau Bengkalis District. *Journal of Dynamics, 1*(1), 27–30. https://doi.org/10.21063/JoD.2016.V1.1.27-30.

Chengrong, H., Chuan, C., Xiaofei, W., & Yiwei, L. (2016). Acid transformation of bauxite residue: Conversion of its alkaline characteristics. *Journal of Hazardous Materials*. Elsevier BV. https://doi.org/10.1016/j.jhazmat.2016.10.073.

Croymans, T., Schroeyers, W., Krivenko, P., Kovalchuk, O., Pasko, A., Hult, M., Marissens, G., Lutter, G., & Schreurs, S. (2017). Radiological characterization and evaluation of the high volume of bauxite residue of alkali activated concretes. *Journal of Environmental Radioactivity, 168*, 21–29. Elsevier Ltd. https://doi.org/10.1016/j.jenvrad.2016.08.013.

Deng, B., Li, G., Luo, J., Ye, Q., Liu, M., Peng, Z., & Jiang, T. (2017). Enrichment of Sc_2O_3 and TiO_2 from bauxite ore residues. *Journal of Hazardous Materials*. Elsevier BV. https://doi.org/10.1016/j.jhazmat.2017.02.022.

Fatriadi, R., Asteriani, F., & Cahyaningsih, C. (2017). Effectiveness of the national program for community empowerment (PNPM) for infrastructure development accelerated and geoplanology in District of Marpoyan Peace, Pekanbaru. *Journal of Geoscience, Engineering, environment, and Technology, 2*(1), 53–63.

Guevara, H. P. R., Ballesteros, F. C., Vilando, A. C., Daniel, M., de Luna, G., & Lu, M. (2017). Recovery of oxalate from bauxite wastewater using homogeneous fluidized-bed granulation process. *Journal of Cleaner Production*. Elsevier BV. https://doi.org/10.1016/j.jclepro.2017.03.172.

Han, Y., Ji, S., Lee, P., & Oh, C. (2017). Bauxite residue simultaneous neutralization with mineral carbonation using atmospheric CO_2. *Journal of Hazardous Materials, 326*, 87–93. Elsevier BV. https://doi.org/10.1016/j.jhazmat.2016.12.020.

Rao, C., Mermans, J., Blanpain, B., Pontikes, Y., Binnemans, K., & Van Gerven, T. (2016). Selective recovery of rare earths from bauxite residue by combination of sulfation, roasting and leaching. *Minerals Engineering, 92*, 151–159. Elsevier Ltd. https://doi.org/10.1016/j.mineng.2016.03.002.

Retty Dwi Kisnawati, S. (2016). Separation of bauxite alumina on residue (red mud) originating from Riau with sodalime sintering method, *5*(2), 160–163.

Shi, B., Qu, Y., & Li, H. (2016). The caused by alkaline bauxite residue in leaves of Atriplex canescens. *Ecological Engineering*, 6–11. Elsevier BV. https://doi.org/10.1016/j.ecoleng.2016.10.008.

Valeton. (1972). Bauxites.

Wang, Q., Deng, J., Liu, X., Zhao, R., & Cai, S. (2015). Provenance of late carboniferous bauxite deposits in the North China Craton: New constraints on marginal arc construction and accretion processes. *Gondwana Research*. International Association for Gondwana Research. https://doi.org/10.1016/j.gr.2015.10.015.

Wang, S., Li, X., Wang, S., Li, Q., Chen, C., Feng, F., & Chen, Y. (2016). Three-dimensional orebody modeling and intellectualized longwall mining for bauxite stratiform deposits. *Transactions of Nonferrous Metals Society of China. The Nonferrous Metals Society of China*, 26(10), 2724–2730. https://doi.org/10.1016/S1003-6326(16)64367-4.

Zamanian, H., Ahmadnejad, F., & Zarasvandi, A. (2015). *Mineralogical and geochemical investigations of the Mombi bauxite deposit, the Zagros Mountains of Iran*. Chemie der Erde - Geochemistry: Elsevier GmbH. https://doi.org/10.1016/j.chemer.2015.10.001.

Zhu, F., Liao, J., Xue, S., Hartley, W. Zou, Q., & Wu, H. (2016). Science of the total environment evaluation of aggregate following microstructures in natural regeneration characterized as bauxite residue by synchrotron-based X-ray micro-computed tomography. *Science of the Total Environment, 573*, 155–163. Elsevier BV. https://doi.org/10.1016/j.scitotenv.2016.08.108.

Accuracy of Algorithm C4.5 to Study Data Mining Against Selection of Contraception

Des Suryani, Ause Labellapansa and Eka Marsela

Abstract Family planning is an attempt to infuse or plan the number and spacing of pregnancy using contraception. Contraception is a method or tool used to prevent pregnancy. Arrangements can be made with the use of contraceptives such as pills, spirals, implants. Patients may consult with the midwife or physician in choosing a contraceptive that is suitable or convenient to use. But in reality there are still many patients, especially patients at Clinic Pratama Hasanah Pekanbaru, who still hesitate in choosing contraceptives in accordance with the patient's body condition. For that, writer was interested to do study data of contraception usage to patient at Clinic Pratama Hasanah Pekanbaru to get pattern of decision tree and its level of accuracy. Based on this, it is necessary to evaluate the data collection of contraceptive usage to determine the pattern of contraceptive selection. The data mining process uses the classification method with the Decision Tree C4.5 algorithm. The attributes used in determining the pattern of selection of contraceptives consist of nine regular attributes, i.e., age, term of usage, menstrual cycle, just married, just give birth, breastfeeding, already having offspring, health problems, and have more than four children, whereas the attribute label/class that is contraceptive used. This study resulted in recognition of contraceptive pattern selection with accuracy of 93.15% (excellent classification). With this level of accuracy, it will help the midwife to direct the patient in the selection of contraceptives based on the resulting pattern.

Keywords Data mining · Decision Tree C4.5 · Classification · Contraception

---

D. Suryani (✉) · A. Labellapansa · E. Marsela
Department of Information Technology, Universitas Islam Riau,
Pekanbaru, Indonesia
e-mail: des.suryani@eng.uir.ac.id

1 Introduction

Information technology is growing so rapidly and touching on all sides of life. In order for human activities more easily and efficiently, many researchers do collaboration between computer science disciplines and other disciplines. One of the collaborations took place between computer science and medical science to choose a contraceptive tool for family planning (FP) acceptors.

Contraception is a method or tool used to prevent pregnancy. In other words, KB is planning the number of families. Arrangements can be made with the use of contraceptives such as pills, syringes, spirals. Patients can consult a physician and then choose a contraceptive device that is suitable or convenient to use.

However, despite consultation with a doctor, the choice of methods or contraceptives is not an easy thing. This is because effects that affect the body will not be known before the contraceptive is used. Each method has advantages and disadvantages. Despite the advantages and disadvantages, all contraceptives are available; women still find it difficult to control fertility safely, effectively, with acceptable methods. It is not surprising that women feel that the use of contraception is sometimes problematic and forced to choose contraceptives that do not match the adverse consequences of or not using the FP method at all.

In Clinic Pratama Hasanah Pekanbaru, the patient in choosing contraceptives especially with consultation with midwife or nurse who will identify by giving question to patient.

Based on the above background, the authors are interested to analyze the data of contraceptive device selection to Clinic Pratama Hasanah Pekanbaru patients to get the pattern of decision tree and the level of accuracy. The method used is to apply the method of classification with algorithm C4.5. With a better level of accuracy, it will help midwives and the public to know the appropriate contraceptive and appropriate body condition.

2 The C4.5 Algorithm for Building Decision Trees

Planning in having children is one of the important things in a family. To anticipate the occurrence of pregnancy, couples usually do family planning (FP) program. Some contraceptives used for the family planning program include pills, syringes, spirals. But determining contraceptives is not an easy task.

Usually, couples should consult a doctor- or midwife-related contraceptive to be used. However, this does not mean this way without risk, because doctors do the diagnosis based on the knowledge they have. In one condition, a midwife or doctor may be mistaken when suggesting an appropriate contraceptive for a woman. As a result, women feel that the use of contraception is sometimes problematic because they have to choose a method that does not match.

The selection of contraceptives in Clinic Pratama Hasanah Pekanbaru is done by way of the society must fill the registration form, then nurse will identify by giving question to patient. From the results of this identification, the doctor will advise contraceptives that are considered appropriate to the condition of the patient. Although the midwife is an expert in this field, in reality they also have limited memory and stamina work. This could have made a mistake in identifying the appropriate contraceptive device. But until now, there are still many people who do not fit with the method or contraceptives because the situation and condition of the body of each individual will continue to change.

Nugroho (2014) have used the k-nearest neighbor algorithm in the classification method. This research successfully applied k-nearest neighbor algorithm to calculate and give result of a recommendation class to choose contraception method.

Trisnawarman and Erlysa (2007) designed a decision support system that can assist in determining appropriate methods or devices of contraception and in accordance with the situation and condition of the body of each user.

Kamaludin (2012) has created a decision support system for choosing contraceptives using the simple additive weighting method where Kamaludin uses age, cost, side effect, duration, tool efficacy, menstrual history, potential complications, health benefits, and health status.

Differences with the research made are to dig the data in a training data in the form of contraceptive device selection data with age attributes, duration of use, menstrual period, new marriage, new childbirth, breastfeeding, having offspring, health problems, children more than four, and label attributes of contraceptive devices while the results of analysis in the form of contraceptive pattern selection.

In general, the C4.5 algorithm for building decision trees is as follows:

1. Select an attribute as root
2. Create a branch for each value
3. For the case in the branch
4. Repeat the process for each branch until all the cases on the branch have the same class.

An attribute is selected as a root based on the highest gain value of the attributes. To calculate the gain used, the formula is shown in Eq. (1):

$$\text{Gain}(S, A) = \text{Entropy}(S) - \sum_{i=1}^{n} \frac{|Si|}{|S|} * Entropy(Si) \quad (1)$$

While the calculation of entropy value can be seen in the following Eq. (2):

$$\text{Entropy}(S) = \sum_{1}^{n} -pi * \log_2 pi \quad (2)$$

where

S   The Set of Cases
A   Features
N   Number of partitions S
P   The proportion of Si to S

3   Process Modeling and Knowledge Outcomes

Data of clinical contraceptive use Pratama Hasanah Pekanbaru in 2015 and 2016. Data performed data cleaning. Then the data is done to determine the selection of data to be analyzed. Selection result data is data that has been validated. In this case amounted to 146 records such as Table 1. The data is continued to the process of data transformation; that is, the data is converted into the appropriate format for processing in data mining. The result of data transformation process (dataset) is tested using C4.5 algorithm through RapidMiner software. The results of this test obtained the decision tree as in Fig. 2.

After experimenting with data processing using Decision Tree C4.5 algorithm to data of contraceptive used tool counted 146 record with nine regular attributes, i.e., age, term of usage, menstrual cycle, just married, just give birth, breastfeeding, already having offspring, health problems, and having more than four children, whereas the attribute label/class that is contraceptive used.

3.1   Modeling Process

Data mining process to data set is processed by using algorithm C4.5 through application of RapidMiner Studio version 7.3 as in Fig. 1.

3.2   Knowledge Outcomes

Based on the modeling process is generated knowledge in the form of a pattern of decision tree which can be seen in Fig. 2.

Table 1   Number of patients in selection of contraceptives

No.	Contraceptive used	Number of patients
1	Spiral	47
2	Pill	59
3	Implant	26
4	Sterile	14

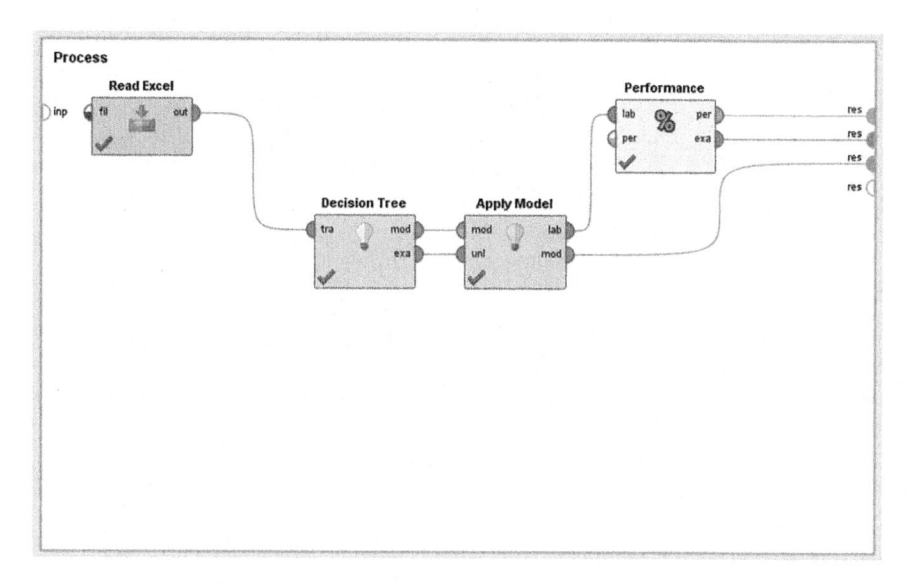

Accuracy of Algorithm C4.5 to Study Data Mining Against ... 959

Fig. 1 Applying the model

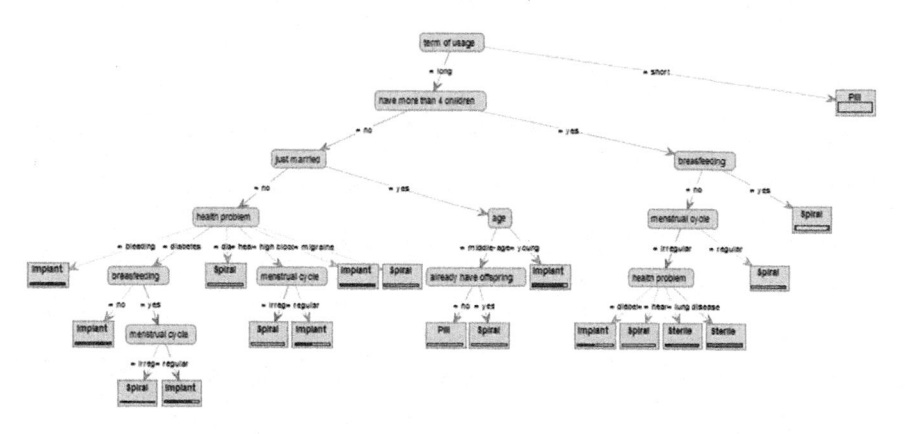

Fig. 2 Tree resulting from C4.5 algorithm

The decision tree in Fig. 2 can be spelled out in the form of rule as follows:

term of usage = long
| have more than 4 children = no
| | just married = no
| | | health problem = bleeding: Implant {Implant = 4, Pill = 0, Spiral = 0, Sterile = 0}
| | | health problem = diabetes
| | | | breastfeeding = no: Implant {Implant = 2, Pill = 0, Spiral = 0, Sterile = 0}

| | | | breasfeeding = yes
| | | | | menstrual cycle = irregular: Spiral {Implant = 1, Pill = 0, Spiral = 2, Sterile = 0}
| | | | | menstrual cycle = regular: Implant {Implant = 4, Pill = 0, Spiral = 1, Sterile = 0}
| | | health problem = dizzy: Spiral {Implant = 0, Pill = 0, Spiral = 5, Sterile = 0}
| | | health problem = heart disease
| | | | menstrual cycle = irregular: Spiral {Implant = 0, Pill = 0, Spiral = 5, Sterile = 0}
| | | | menstrual cycle = regular: Implant {Implant = 1, Pill = 0, Spiral = 1, Sterile = 0}
| | | health problem = high blood pressure: Implant {Implant = 3, Pill = 0, Spiral = 0, Sterile = 0}
| | | health problem = migraine: Spiral {Implant = 0, Pill = 0, Spiral = 2, Sterile = 0}
| | just married = yes
| | | age = middle-aged
| | | | already have offspring = no: Pill {Implant = 0, Pill = 6, Spiral = 0, Sterile = 0}
| | | | already have offspring = yes: Spiral {Implant = 0, Pill = 0, Spiral = 6, Sterile = 0}
| | | age = young: Implant {Implant = 8, Pill = 0, Spiral = 1, Sterile = 0}
| have more than 4 children = yes
| | breasfeeding = no
| | | menstrual cycle = irregular
| | | | health problem = diabetes: Implant {Implant = 1, Pill = 0, Spiral = 1, Sterile = 0}
| | | | health problem = dizzy: Spiral {Implant = 0, Pill = 0, Spiral = 2, Sterile = 0}
| | | | health problem = heart disease: Sterile {Implant = 1, Pill = 0, Spiral = 0, Sterile = 4}
| | | | health problem = lung disease: Sterile {Implant = 0, Pill = 0, Spiral = 0, Sterile = 8}
| | | menstrual cycle = regular: Spiral {Implant = 0, Pill = 0, Spiral = 4, Sterile = 1}
| | breasfeeding = yes: Spiral {Implant = 1, Pill = 0, Spiral = 16, Sterile = 1}
term of usage = short: Pill {Implant = 0, Pill = 53, Spiral = 1, Sterile = 0}

3.3 The Accuracy Level

The accuracy level obtained from the 146 train data records used in the Decision Tree C4.5 modeling can be seen in Fig. 3.

accuracy: 93.15%

	true Implant	true Pill	true Spiral	true Sterile	class precision
pred. Implant	23	0	4	0	85.19%
pred. Pill	0	59	1	0	98.33%
pred. Spiral	2	0	42	2	91.30%
pred. Sterile	1	0	0	12	92.31%
class recall	88.46%	100.00%	89.36%	85.71%	

Fig. 3 Performance vector

Based on Fig. 3, out of 146 record numbers there are 23 correctly predicted records of choosing Implant, 59 records are predicted to correctly select Pill, 42 records are predicted to correctly choose Spiral, and 12 records are correctly predicted to choose Sterile with accuracy = (23 + 59 + 42 +12)/146 = 93.15%. So that, the result gives accuracy rate obtained from data processing, i.e., 93.15% which can be classified excellent.

4 Conclusion

Based on the results of the study and modeling process using Decision Tree C4.5 algorithm conducted on the data of contraceptive use in Clinic Pratama Hasanah Pekanbaru can be drawn conclusion as follows:

1. Data selection of contraceptives can be used as training data to generate knowledge by using the method of classification.
2. The results of training data study of contraceptive usage tools can form knowledge in the form of decision tree models that transform data into decision trees that represent the rules.
3. Accuracy rate obtained from data processing reaches 93.15% which can be classified excellent so that this result can be a reference for midwife in advised patient in choosing contraception.

Acknowledgements This research was partially supported by Universitas Islam Riau.

References

Kamaluddin (2012). Sistem Pendukung Keputusan Dalam Pemilihan Alternatif Alat Kontrasepsi Menggunakan Simple Additive Weighting. *Jurnal Jurusan Teknik Informatika UIN SGD Bandung*.

Nugroho, C. G., Nugroho, D., & Fitriasih, S. H. (2014). Sistem Pendukung Keputusan Untuk Pemilihan Metode Kontrasepsi Pada Pasangan Usia Subur Dengan Algoritma K-Nearest Neighbour (KNN). *Jurnal Ilmiah SINUS*. ISSN:1693–1173.

Trisnawarman D., Erlysa W. (2007). Sistem Penunjang Keputusan Pemilihan Metode/Alat Kontrasepsi. Gematika. *Jurnal Manajemen Informatika*, *9*(1).

Home Monitoring System Based on Cloud Computing Technology and Object Sensor

Evizal Abdul Kadir, Apri Siswanto and Ari Yulian

Abstract High mobility of Indonesian residence is increasing by the time, especially for the people who live in urban areas. Most of housing and room are leave without any guidance and very risk to the theft and violation. In this research propose, a home monitoring system based on object sensor and cloud computing to keep the information before forward to owner. House monitoring system used cloud computing technology using Webcam motion detection used for the detection system and Raspberry Pi 3 as processor to store data as buffer before send to cloud, as well as webcam and motion movement. Every section for image storage media when image captured then store in google drive that can be call any time and also cloud computing and notification that is using email or push messaging. Results show that moving object can be detect while object under position up to 5 m, an alert send to owner for notification that something happen in the house. With this system will help residence more safe and reduce risk to leave house.

Keywords Cloud computing · Monitoring · Object sensor · Raspberry pi 3

1 Introduction

In today's modern era, the level of mobility of Indonesian society is increasing, especially for people who live in urban areas. The busy work of the day causes the interaction at home to be reduced. In addition to the many job matters, one of the cultures that are already inherent in Indonesian society is to return to their home-

---

E. A. Kadir (✉) · A. Siswanto · A. Yulian
Department of Information Technology, Faculty of Engineering,
Universitas Islam Riau, Jl. Kaharuddin Nasution No. 113, Pekanbaru
Riau 28284, Indonesia
e-mail: evizal@eng.uir.ac.id

town or commonly called "Mudik" during public holidays or during school holidays (Rido and Faldana 2014).

High population density in urban areas and job vacancies are getting less while the need for food and accommodation should remain fulfilled for the sake of survival. To meet the needs, sometimes humans take action outside the norms of humanity, committing criminal acts such as cases of theft that many occur lately. In addition to the factors of fulfillment of needs, other factors that make a person commit a crime that is because of the opportunity.

At the time of leaving the house in a state of empty and with the many cases of theft that occurred, would cause a fear for the owner of the house. The most common precautions used are with a security guard or a Closed Circuit Television (CCTV) camera installed in every corner of the room. Security officers who are expected to be the number one solution to home security or implementation cannot be said to be effective as most security guys sit in their posts alone, while the use of CCTV devices that can record and display real-time room conditions that are considered more effective cost but not least in the cost of installation and maintenance (Rido and Faldana 2014).

The other side, technological progress is increasing rapidly, so almost every individual already have gadgets, whether it be a smartphone or tablet. Internet technology is no less rapid, using the Internet Service Provider (ISP) and also provides a variety of package options at an affordable price and supported by Third Generation (3G) or even Fourth Generation (4G) networking technology, making internet access is getting faster in accessing streaming video content.

One of the Internet technologies that became the center of attention in recent years is cloud computing technology. Cloud computing is a combination of the use of computer technology (computing) in a network with Internet-based development (cloud) that has the function to run programs or applications through computers connected at the same time. Cloud computing technology makes the Internet a central server for managing data and user applications. Cloud computing also provides data storage capacity large enough, so users can store a lot of data on a large scale and can be reopened wherever and whenever.

From some backgrounds above, it is necessary to build a system capable of monitoring a room that is easy to access on all gadget devices, cost-effective, and practical. This system uses a wireless device and Webcam as a medium of shooting or object as well as cloud computing as an online image storage media. It is hoped that with the technology of merging the two technologies, this system will become more practical and easy to access anywhere.

2 Literature Review

Research conducted by (Rido and Faldana 2014), entitled "Monitoring System Home Based on Cloud Computing Technology". The research discusses the making of home monitoring system using Webcam by integrating with cloud computing technology as storage media. Differences with this research that will be located on the motion detection using Webcam, raspberry Pi and motioneye with notification to Email.

Another study is a room monitoring system, conducted by Setiawan (2013) entitled "The Design of a Room Monitoring System Using OpenWrt-based Webcams." The study discusses the creation of a room security system using the TP-LINK 3420 router with OpenWrt firmware. Features include Webcam to record the state of the room, soundcard as an alarm output and notification based on security level, i.e., security level as notification in the form of alarm, security level two notification in the form of alert and SMS, and security level three notification in the form of SMS and email to user. Differences with research that will be the author of the adoption of storage with cloud technology also use motion detection and notification.

Another study was a Web-based home observer, conducted by (Zul Ihsan and Widyawan 2013) entitled "Architecture of Web-Based Monitoring System Using IP Camera." The study discussed about making home monitoring system by using Internet network and also using IP Camera. This monitoring system requires an architectural design in accordance with the conditions of the home network in general. Also use four monitoring architecture conditions that can be applied in home internet user network.

3 Methodology

Cloud computing is a client-server model, where resources such as servers, storage, networks, and software can be used as a service that users can remotely access at any time. Users can enjoy the various services provided by the cloud computing provider, without requiring too much help from technicians or support from the provider. Figure 1 shows the illustration of the cloud computing diagram to be used.

Cloud computing is a combination of the use of computer technology (computing) in a network with Internet-based development (cloud) that has the function to run programs or applications through computers connected at the same time, but not all connected via the Internet using cloud computing. Cloud-based computer system technology is a technology that makes the Internet as a central server to manage data and user applications. This technology allows users to run programs without installation and allows users to access their personal data through computers with Internet access.

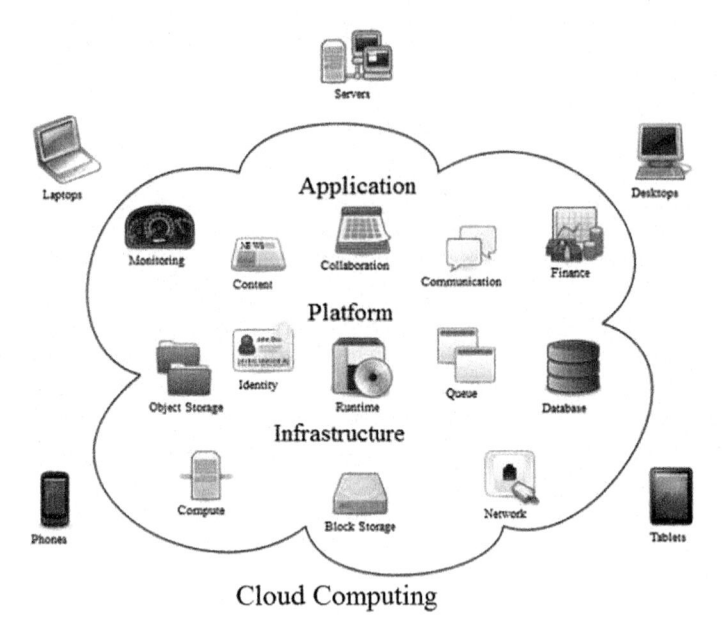

Fig. 1 Cloud computing illustration diagram

Raspberry Pi 3

Raspberry Pi 3 is a minicomputer board; the size of a credit card can be called single-board computer (SBC), and raspberry Pi 3 is already equipped with Wi-fi and Bluetooth. The specifications are as follows: Broadcom BCM2837 processor 64-bit quad-core 1.2 GHz, 1 GB RAM, MCM43438 Wi-Fi and low-energy Bluetooth, general purpose input/output (GPIO) 40 pins, 4 USB 2.0 ports, ethernet 10/100 Mbps, 4-pole stereo and video output ports, HDMI port, CSI camera port to connect to RPi camera, DSI display port to connect to RPi touch screen, microSD slot. For a minicomputer with limited hardware specifications, then for the operating system is also lightweight type. Figure 2 shows the Raspberry Pi3 board that will be used to process a moving object image.

Raspberry Pi 3 is a very flexible platform; there are many things that can be done with Raspberry Pi 3, i.e.:

1. Media Learning Programming
   Raspberry Pi already has interpreters and compilers from various programming languages such as Python, Java, and C++.
2. General Purpose Computing
   Raspberry Pi can be used as a computer as it connects to a monitor and adjusts its graphical display through a Web browser.
3. Media Center

Fig. 2 Propose Raspberry Pi 3 Model B

Raspberry has an HDMI port and audio/video. Raspberry Pi can be easily connected to the monitor. This advantage is supported by the power of Raspberry Pi processor is enough to play full screen high definition video and also in Raspberry Pi itself already has XBMC (media player) that supports as a kind of media file format.

Motion Object Detector

Motioneye is a Web interface for the use of security cameras called motion for Raspberry Pi. Motioneye is a development of previous versions of motionpie to be compatible with several different versions of Raspberry Pi. Motioneye is created by Callin Crisan to develop the use of motion created by Kanneth Larvsen. Motioneye has Web interface with responsive design, user and security password, streaming MJPG, motion detection with JPEG and AVI file output, browsing and downloading media files via Web, and camera settings (Crisan 2014).

PYTHON Programming

Python is a dynamic object-oriented programming language; it can be used for a variety of software development. Python provides strong support for integration with other programming languages and other tools. And also, python can run on many platforms or operating systems like Windows, Linux or platform, Mac OS X, OS/2, Amiga, and python have also been ported into Java virtual machine and .NET.

Python is distributed under Open Source Initiative (OSI) open source licenses, so python is free to use and also free to use, and there are some python features:

1. Python is powerful and fast
   Python users often use batteries included phrases to describe standard libraries. The library covers everything from non-syncing processing to compressed files. Python itself is a collection of very good modules and can handle practically every problem domain.
2. Python plays well with others
   Python can integrate with component object model (COM), .NET, and common object request broker architecture (CORBA) objects. If we use Java libraries, Jython is a Python implementation for Java Virtual Machine.
3. Python runs everywhere
   Python is available for many widely used operating systems, such as Windows, Unix or Linux, OS/2, Mac, Amiga, and many other operating systems. There is also a python version running on .NET, Java virtual machine.

4 Result and Discussion

In home monitoring systems based on cloud computing technology using a webcam with motion detection that after implementation, will be tested against the system. System testing is a process to verify that all elements of the system have been integrated and functioning correctly. Home-based monitoring system cloud computing technology uses a Webcam with motion detection is built and designed based on user roles, which will use this system can to perform monitoring or monitoring from a distance by looking at the capture results of cloud computing. Home monitoring system can be accessed by using Web browser (Fig. 3).

Main view form of the display is generated from the camera that will be displayed on the Web. Figure 4 shows the implementation of the main display form process.

```
sudo wget goo.gl/hRdgZP
```

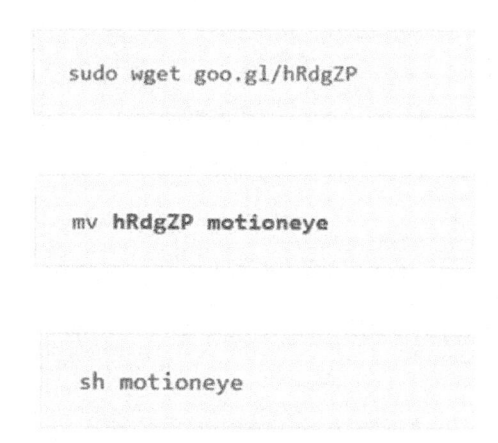

```
mv hRdgZP motioneye
```

```
sh motioneye
```

Fig. 3 Script instruction of motion program

Fig. 4 Form tampilan utama

Capture report form in motion is used to do temporary storage in the report in motion, and there is a feature to download images and can be delete images. Figure 5 shows the implementation of the capture report form process in motion.

The capture report form on google drive is used to capture storage of movements and store the capture in google drive by online. Figure 6 shows the implementation of the capture report form process in google drive.

Once the system has been successfully established, the system will be tested. Testing is required to ensure the system is running as expected. One way of testing is by using black box testing. This test focuses on the functions that exist in the

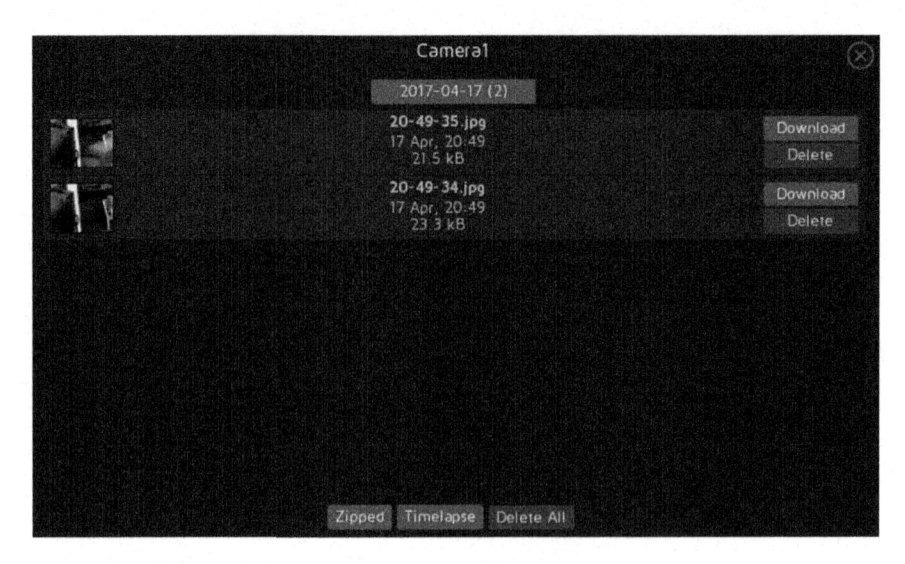

Fig. 5 Report form based on motion sensor captured

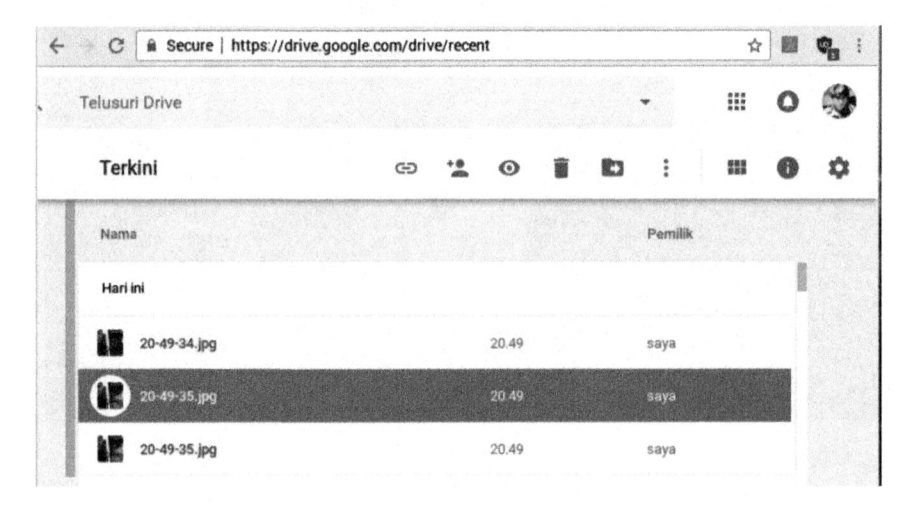

Fig. 6 Report form based on captured in google drive

program without having to know how the function is made whether in accordance with the expected results. If the expected results match the test results, this means the system corresponds to a predetermined goal. If not in accordance with the expected, then the system will be reviewed and performed repairs in accordance with existing errors. Here is an analysis of the results of black box testing. Table 1 shows that the results of capture distance movement and affects the success of the process of capture processing.

Table 1 Motion object based on capture in camera

No	Distance (m)	Captured moving object	Results
1	0.6		Terdeteksi Pergerakan
2	1		Terdeteksi Pergerakan
3	2		Terdeteksi Pergerakan
4	3		Terdeteksi Pergerakan
5	4		Terdeteksi Pergerakan
6	5		Terdeteksi Pergerakan

5 Conclusion

Based on the analysis and discussion of home-based monitoring system of cloud computing technology using Webcam with motion detection, then with this home monitoring system can help users or house owners to know the circumstances in the room by looking at capture from google drive. With this system can help provide solutions in home security and homeowners feel comfortable when leaving home because there is already a webcam that has been installed in the room if there is automatic movement of capture and will be uploaded to google drive.

Acknowledgements This work is partially supported by Universitas Islam Riau; the authors also gratefully acknowledge the helpful comments and suggestions of the reviewers, which have improved the presentation.

References

Crisan, C. (2014). http://www.howtoembed.com/projects/raspberry-pi/95-motioneye-with-raspberry- pi. diakses 14 Januari 2017.

Rido, A., Faldana, R. (2014). Sistem Monitoring Rumah Berbasis Teknologi Cloud Computing. *SESINDO 2014*.

Setiawan, A. (2013). Rancang Bangun Sistem Monitoring Ruangan Menggunakan Webcam Berbasis Openwrt (Doctoral dissertation, UIN SUNAN KALIJAGA).

Sofana, I. (2012). *Cloud Computing Teori dan Praktik (Open Nebula, VMware, dan Amazon AWS)*. Bandung: Informatika.

Zul Ihsan, M., & Widyawan, W. (2013). Arsitektur Sistem Pemantau Rumah Bebasis Web dengan Mengunakan IP Camera, Reserchgate.

Geological Mapping of Silica Sand Distribution on the Muda Island and Ketam Island, Estuary of Kampar River, Indonesia

Husnul Kausarian, Tiggi Choanji, Detri Karya, Evizal Abdul Kadir and Adi Suryadi

Abstract Silica sand is produced by weathered igneous rock where transported and deposited in the area such as banks of a river, lake, or beach. The study area of the silica sand distribution is on Muda Island and Ketam Island, Pelalawan Regency, Riau Province, Indonesia. Silica sand source deposits in this area are influenced by the northern part of the estuary of Kampar River which is dealing with the straits of Melaka that rich with sand sediments reserve. Melaka Strait in the northwest part of this estuary transports the sand materials from the Indian Ocean, while in the northeast part the sand sediment materials were obtained from the South China Sea and Riau Islands. Silica sand deposition process occurs due to the estuary system which is a bore tidal system. Landsat 8 that has been taken from June 2016 shows the significance of the sandbars development that almost thrives on the whole estuary of the Kampar River and the spread almost evenly from Muda Island and Ketam Island. From the result of laboratory testing using X-ray fluorescence (X-RF) for the sample of Muda Island contains the silica compound with the percentage is 92% and for the sample from Ketam Island, the percentage compound of silica sand in the sand

---

H. Kausarian (✉) · T. Choanji · A. Suryadi
Department of Geological Engineering, Universitas Islam Riau,
Jalan Kaharuddin Nasution No. 113, Pekanbaru, Riau 28284, Indonesia
e-mail: husnulkausarian@eng.uir.ac.id

T. Choanji
e-mail: tiggich@eng.uir.ac.id

A. Suryadi
e-mail: adisuryadi@eng.uir.ac.id

D. Karya
Faculty of Economy, Universitas Islam Riau, Jalan Kaharuddin Nasution No. 113, Pekanbaru, Riau 28284, Indonesia
e-mail: de.ka87@yahoo.co.id

E. A. Kadir
Department of Information Engineering, Universitas Islam Riau, Jalan Kaharuddin Nasution No. 113, Pekanbaru, Riau 28284, Indonesia
e-mail: evizal@eng.uir.ac.id

sample is 90.5%. The result from X-ray diffraction (X-RD) analysis in the sample of Muda Island shows the high peak is quartz and miciocline compounds with the peak count of 3000 cps and for the silica sand sample from Ketam Island, the high peak are muscovite and quartz which has more than 32,000 cps.

Keywords Silica sand · Estuary · Kampar River · X-ray fluorescence X-ray diffraction

1 Introduction

Silica sand is one of the minerals are relatively abundant in Indonesia. This is possible due to the condition of Indonesia which almost half in the form of acid igneous rocks as a source of the material forming the silica sand (Wicaksono 2012). Silica sand found in many coastal areas of rivers, lakes, beaches, and most of the shallow sea. Silica sand (quartz sands) formed by weathering of acid igneous rocks (Brown 2000; Kausarian et al. 2013) such as granite, gneiss, or other igneous rock containing the main mineral quartz (Rikke et al. 2010).

The result of the rock weathering is a process of sedimentation which will be transported and carried by water or wind as the sedimentation agents and then deposited on the bank of a river, lake or beach as the deposit/sedimentation environment (Holland and Elmore 2008; Carter 2013). Because the deposit numbers are quite large and look white along the banks of a river, lake, or beach, then in Indonesia, the famous name of this sand is white sand. This study aims to conduct an inventory and determine the potential (characterization and utilization) of silica sand resources in the area of Muda Island and Ketam Island, Pelalawan, Riau Province, Indonesia, as the basic information of the potential that can be used in the presence of silica sand in this area (Kausarian et al. 2016, 2017; Konstantinos 2014; Nugrahanti et al. 2014).

The location of Muda Island and Ketam Island (Fig. 1) is located in the estuary of Kampar River, Pelalawan, Riau Province, Indonesia as a part of Central Sumatra Basin (Heidrick and Aulia 1993; Dawson et al. 1997; Pubellier and Morley 2014; Rafat and Navneet 2011) which are located at 0° 17′ 5.82″N and 102° 52′ 28.46″E for Muda Island and 0° 20′ 5.73″N and 102° 57′ 35.51″E for Ketam Island.

2 Details of Experiments

Geological Mapping and Sample Collecting

The geological mapping has been carried out thoroughly to determine the distribution of silica sand contained in Muda Island and Ketam Island like observation point plotting, observation outcrops of silica sand, and sand sampling. Based on the field

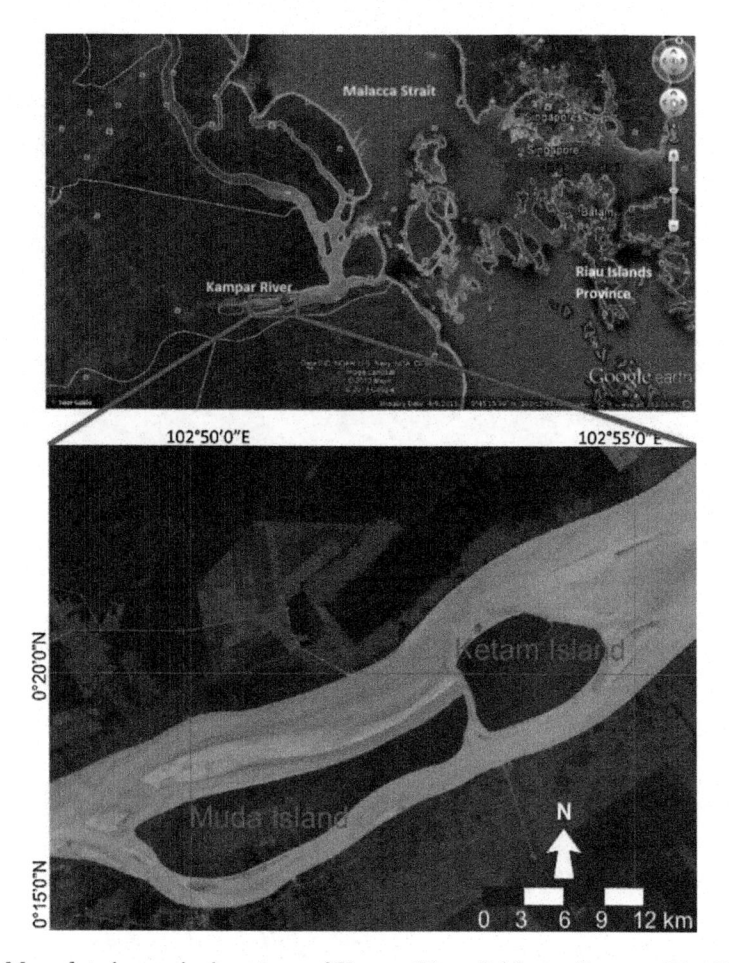

Fig. 1 Map of study area in the estuary of Kampar River, Pelalawan Regency, Riau Province, Indonesia

observation (Lubis et al. 2017), determined observation points are ten locations, where seven observation points located in the area of Muda Island and four points in the Ketam Island (Fig. 2). The sampling of silica sand has been carried out at each point of observations (Fig. 3).

Laboratory Experimental

After sampling point at the represent locations, these samples were taken to the laboratory for further testing to determine the content and character of silica sand (Kausarian et al. 2016). The results of the laboratory test are needed to determine

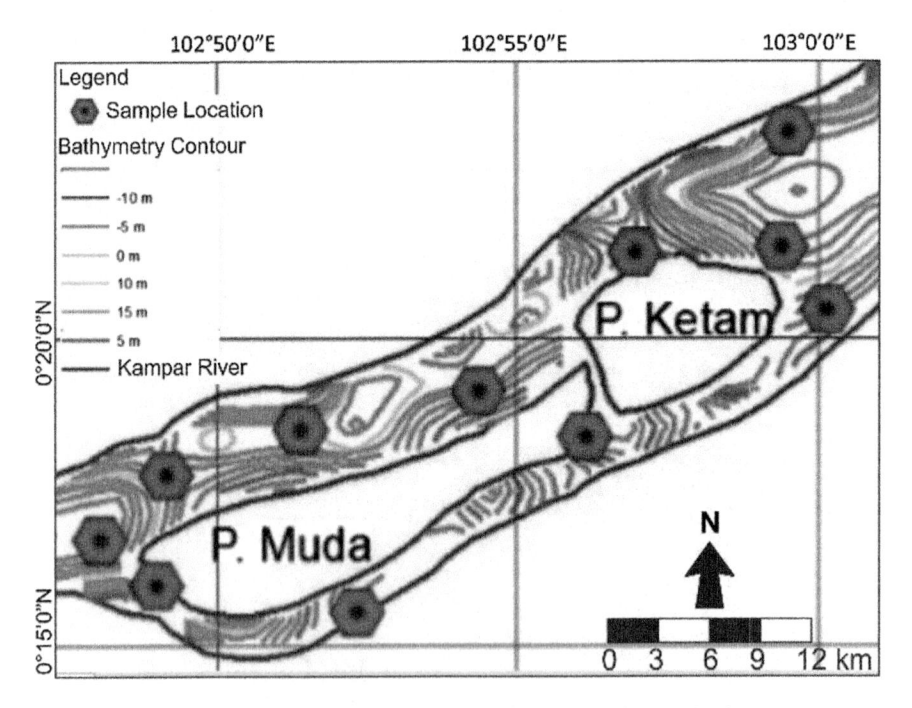

Fig. 2 Sample locations were collected on the Muda Island and Ketam Island

Fig. 3 Samples of silica sand from Muda Island (left) and Ketam Island (right)

the chemical analysis of silica sand samples to determine the types of elements, physical properties, and the percentage of the corresponding elements as well as the usefulness of the silica sand. The laboratory testing that was used is X-ray fluorescence (X-RF) (Murthy and Rao 2016) and X-ray diffraction (X-RD) (Ghalya et al. 1994), this testing is focused on determining the chemical composition, mineral deposits, the percentage composition, and others of compounds/minerals inside of silica sand. Another laboratory test is grain size for the particle of silica sand. This analysis will help to find the character of the silica sand. The size will give the explanation of the origin and the disturbing process that happens to the silica sand, for the explanation of the uniformity of the grain size will give the interpretation for distance of transportation process from the source/origin to the sedimentation area in this current time.

3 Result and Discussion

Silica Sand Sedimentation Process

Silica sand is a sediment deposit comes from rocks that contain silicon dioxide (quartz SiO_2) such as granite, rhyolite, and granodiorite. Silica sand deposit occurs after the process of transportation, sorting, and sedimentation. Therefore, deposit of silica sand in nature is never found in a pure state. Natural quartz grains, in general, are mixed with clay, feldspar, magnetite, ilmenite, limonite, pyrite, mica (biotite), zircon and hornblende and organic material from plants, and so on. In this area, water controls the process of transportation which caused the sandstone becomes increasingly subtle and relatively purer. The pollutant material generally gives color to the silica sand, so that the color produced can be shown the degree of purity. Sources of silica sand contained in these areas come not from in situ rock or soil from the Muda Island and Ketam Island because these areas are consisting of peat soil which has high carbon content, making it impossible to produce the deposit of silica sand. Silica sand source in this area comes from the Riau Islands Province which is located in front of the estuary of Kampar River. Riau Islands Province consists of igneous rock that potentially was weathered by the main agent sedimentology which is seawater and produces silica sand.

The main formations that made the estuary of Kampar River are young surface sediment formation (Qh) and old surface sediment formation (QP). The northern part of the estuary of Kampar River is dealing with the straits of Melaka which rich with sand sediments reserve, while the southern part is the land area. Melaka strait obtains sand sediment supply by its current. Melaka Strait in the northwest part of this estuary transports the sand materials from the Indian Ocean, while in the northeast part the sand sediment materials were obtained from the South China Sea and Riau Islands. Estuary of Kampar River which is located directly in front of this

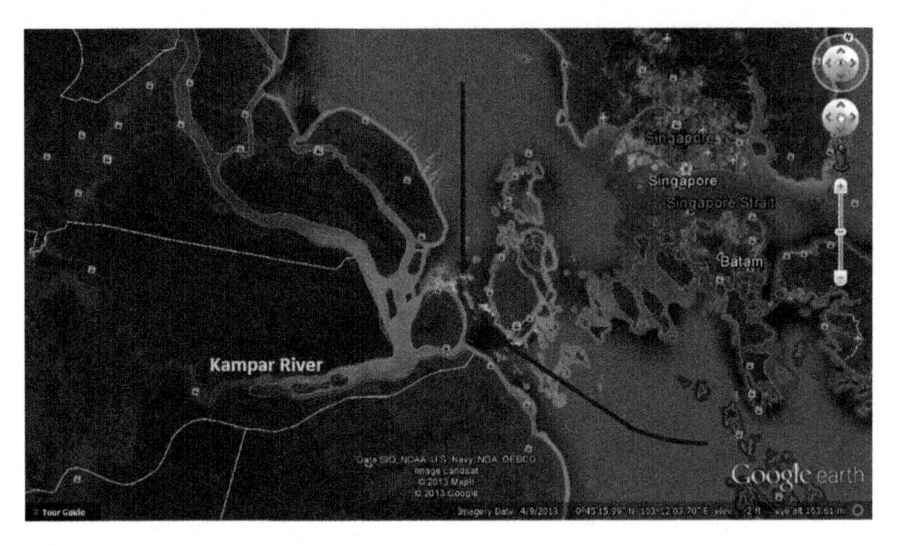

Fig. 4 Current of Melaka Strait comes to the open wide of the estuary of Kampar River (shown by red arrows)

strait gets the sand sources in large numbers. It is caused by the movement of the currents that carries the source of these sediments into the estuary of Kampar River, where the morphology of this estuary shaped as a very open and wide and allows the deposition of the large sand sources in this area (Fig. 4).

Silica sand deposition process occurs due to the estuary system which is a bore tidal system. The significant influence from the tidal wave has occurred throughout the Kampar River, and the waters in the downstream area bring the large sand source and subsequent sedimentation occurs around the Muda and Ketam Island. The sand materials brought by the tidal wave and deposited when the current speed conditions weakened due to the convergence of tidal current from the river wave.

Satellite Data Interpretation

The results of satellite image processing from Landsat 8 in 2016 (Fig. 5) also support the information of silica sand distribution in this region which shows the significance of the sandbars development that almost thrives on the whole estuary of the Kampar River and spread almost evenly from Muda Island and Ketam Island. This is evident from the emergence of sandbars recorded by satellite imagery, which shows the appearance of the sand appears up to the surface and not submerged by the water.

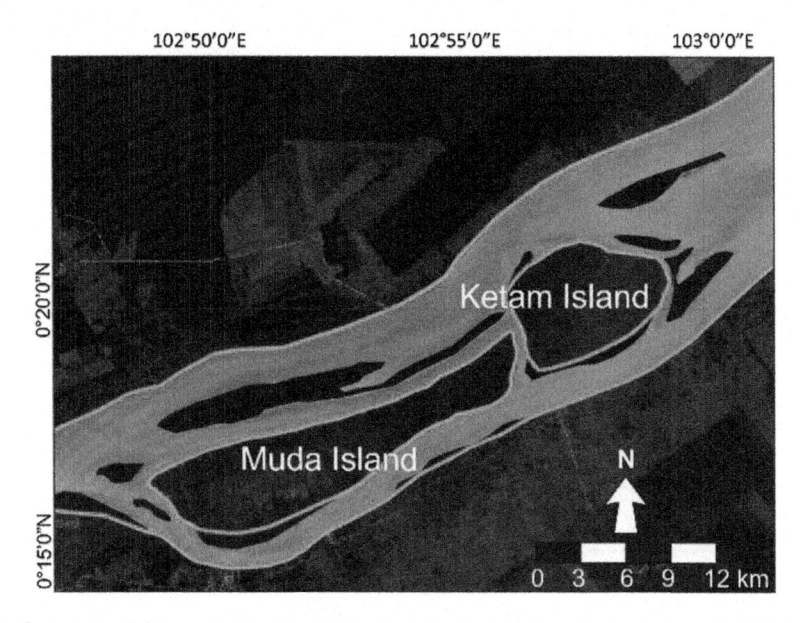

Fig. 5 Landsat 8 image shows the distribution of silica sandbar (red color) on the Muda and Ketam Islands on June 2016

Table 1 Result of silica sand compound percentage from the sample of Muda Island and Ketam Island

Compound	Percentage (%)	
	Muda Island	Ketam Island
MgO	0.5	0.54
Al_2O_3	5.491	5.571
SiO_2	91.847	90.47
P_2O_5	0.686	1.288
Cl	0.002	0.021
K_2O	0.874	1.016
CaO	0.144	0.266
Ti	0.109	0.169
V	0	0
Cr	0.001	0.003
Mn	0.001	0.003
Fe_2O_3	0.269	0.48
Zn	0.001	0.001
As	0	0
Rb	0.004	0.007
Sr	0.002	0.003
Y	0.001	0.003
Zr	0.01	0.057
Ag	0.057	0.101
Pb	0.001	0.001
Eu	0.001	0.004
Re	0	0

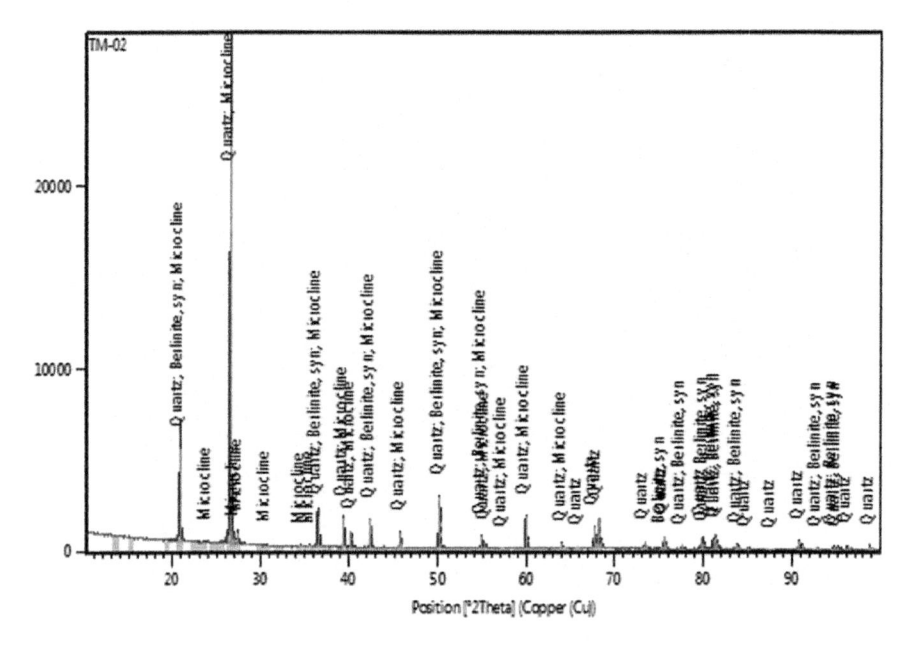

Fig. 6 X-ray diffraction (X-RD) result shows the abundance of quartz and miciocline compounds as the high peak in the sample of Muda Island

Laboratory Test Result

From the result of laboratory testing using X-ray fluorescence (X-RF) for the sample of Muda Island (Table 1) has an abundance of compounds such as SiO_2, TiO_2, Al_2O_3, Fe_2O_3, MnO, MgO, CaO, Na_2O, K_2O, P_2O_5. The result shows the compound of silica (SiO_2) is the highest percentage compared to other compounds. From X-RF result, the silica content in this location showed 92% and followed by mineral/aluminum compound is 5.5% and potassium compound is 0.7% as the major minerals. The results of these percentages prove large silica content. The result from X-ray diffraction (X-RD) analysis in the sample of Muda Island (Fig. 6) shows the high peak is quartz and miciocline compounds. The compound miciocline and quartz has a peak count of 3000 cps.

The result of laboratory X-RF (see Table 1) for the Ketam Island sample shows silica (SiO_2) is the highest percentage 90.5%, followed by the aluminum compound is 5.5% and potassium compound is 1.3% as major minerals. The result of X-RD analysis (Fig. 6) in the Ketam Island sample shows muscovite and quartz compounds are the high peak in the calculation of the points using X-RD which is more than 32,000 cps.

4 Conclusion

The study of silica sand distribution in the area of Muda Island and Ketam Island in the estuary of Kampar River shows around these islands completely surrounded by silica sand deposit. Distribution is seen very clearly based on the result of geological mapping that has been done and supported by the interpretation result of Landsat 8 satellite image that was taken in June 2016. The sand deposits are proven as silica sand based on the result of laboratory tests of X-RF and X-RD which shows the content of silica is very high in this sand. The results expected from this research of silica sand distribution is being a source reference data for the government in the effort to develop the potential areas that could be optimized for the processing of silica sand as the mining industry and others.

Acknowledgements This work is partially supported by Badan Penelitian dan Pengembangan Provinsi Riau. The authors also gratefully acknowledge the helpful comments and suggestions of the reviewers, which have improved the presentation.

References

Brown, J. R. (2000). Chapter 12 Sands and Green Sand. *Foseco Ferrous Foundryman's Handbook* (Eleventh Edition), 146–166.

Carter, R. W. G. (2013). *Coastal environments: An introduction to the physical, ecological, and cultural systems of coastlines*. Academic Press.

Dawson, W. C., Yarmanto, Sukanta, U., Kadar, D., & Sangree, S. B. (1997). *Regional sequence stratigraphic correlation Central Sumatra*. Rumbai: PT Caltex Pacific Indonesia.

Ghalya, A. E., Ergüdenlera, A., & Lauferb, E. (1994). Study of agglomeration characteristics of silica mineral-straw ash mixtures using scanning electronic microscopy and energy dispersion X-ray techniques. *Bioresource Technology, 48*(2), 127–134.

Heidrick, T. L., Aulia, K. (1993). A structural and tectonic model of the coastal plains block, Central Sumatra Basin, Indonesia. *The Pertamina Chevron and Texaco Proceeding*, 285–317.

Holland, K. T., & Elmore, P. A. (2008). A review of heterogeneous sediments in coastal environments. *Earth-Science Reviews, 89*(3), 116–134.

Kausarian, H., Mursyidah, & Sugeng, W. (2013). Silica mineral potency of Bukit Pelintung as base material of solar cell. *Journal of Ocean, Mechanical and Aerospace—Science and Engineering, 2*, 20–24.

Kausarian, H., Sumantyo, J. T. S., Kuze, H., Karya, D., & Panggabean, G. F. (2016). Silica Sand Identification using ALOS PALSAR Full Polarimetry on The Northern Coastline of Rupat Island, Indonesia. *International Journal on Advanced Science, Engineering and Information Technology, 6*(5), 568–573.

Kausarian, H., Sri Sumantyo, J. T., Kuze, H., Aminuddin, J., & Waqar, M. M. (2017). Analysis of polarimetric decomposition, backscattering coefficient, and sample properties for identification and layer thickness estimation of silica sand distribution using L-band synthetic aperture radar. *Canadian Journal of Remote Sensing, 43*, 95–108.

Kausarian, H., Sumantyo, J. T. S., Kuze, H., Karya, D., & Wiyono, S. (2006). The origin and distribution of silica mineral on the recent surface sediment area, Northern Coastline of Rupat Island, Indonesia. *ARPN Journal of Engineering and Applied Sciences, 12*(4), 980–989.

Konstantinos, I. V., Georgios, C., Spiridon, P., & Nikolas, P. B. (2014). Market developments and industrial innovative applications of high purity quartz refines. *Procedia Economics and Finance, International Conference on Applied Economics, ICOAE, 2014,* 624–633.

Lubis, M. Z., Taki, H. M., Anurogo, W., Pamungkas, D. S., Wicaksono, P., Aprilliyanti, T. (2017). Mapping the distribution of potential land drought in Batam Island using the integration of remote sensing and geographic information systems (GIS). *IOP Conference Series: Earth and Environmental Science, 98*(1), 012012.

Murthy, I. N., & Rao, J. B. (2016). Investigations on physical and chemical properties of high silica mineral, Fe-Cr slag and blast furnace slag for foundry applications. *Procedia Environmental Sciences, 35,* 583–596.

Nugrahanti, A., Guntoro, A., Fathaddin, M. T., & Djohor, D. S. (2014). The impact of the production of neighbour wells on well productivity in a shale gas reservoir. *IIUM Engineering Journal, 15*(1), 41–53.

Pubellier, M., & Morley, C. K. (2014). The basins of Sundaland (SE Asia): Evolution and boundary conditions. *Marine and Petroleum Geology, 58,* 555–578.

Rafat, S., & Navneet, C. (2011). Use of silicon and ferrosilicon industry by-products (silica fume) in cement paste and mortar. *Resources, Conservation and Recycling, 55*(8), 739–744.

Rikke, W., Henrik, F., Afsoon, M. K., Johan, B. S., Jesper, S., & Mette, L. K. P. (2010). Development of early diagenetic silica and quartz morphologies—Examples from the Siri Canyon, Danish North Sea. *Sedimentary Geology, 228*(34), 151–170.

Wicaksono, N. (2012). Survei Potensi Pasir Kuarsa di Daerah Ketapang Propinsi Kalimantan Barat. *Jurnal Sains dan Teknologi Indonesia, 11*(2), 126–132.

Expert System Diagnoses on Degenerative Diseases Using Bayes Theorem Method

Ause Labellapansa, Ana Yulianti and Islahudin

Abstract Degenerative disease is the process of decreasing functions of cells or organs of body that generally occurs in old age. Nowadays, with a high level of stress, lifestyle and poor diet makes degenerative diseases suffered by many young people at productive age and will certainly increase the high degenerative disease sufferers. Prevention of degenerative diseases is necessary to do. One of possible ways is to use artificial intelligence science with help of computers. Expert systems that are part of artificial intelligence are developed to diagnose degenerative diseases. This expert system is designed using forward chaining and Bayes theorem method to deal with uncertainty issues that arise. The steps in this research are done by conducting interviews with specialists in internal diseases (degenerative diseases) and by observing hospital to collect medical record data of patient's diagnosis. There are as many as 6 diseases diagnosed with 42 symptoms. After patient's diagnoses entered into the application, probability value of a disease and each probability of symptoms comparison will be calculated. After testing is done, this expert system has a success rate of 90%.

Keywords Forward chaining · Degenerative disease · Expert system
Bayes theorem

1 Preliminary

Degenerative disease is a medical term used to describe a disease that arises due to the process of decreasing functions of cells in body, i.e., from the previous normal state to a worse condition. Degenerative disease occurs due to the aging process, but

---

A. Labellapansa (✉) · A. Yulianti · Islahudin
Fakultas Teknik, Universitas Islam Riau,
Jl. Kaharudin Nasution no. 113, Pekanbaru, Indonesia
e-mail: ause.labella@eng.uir.ac.id

© Springer Nature Singapore Pte Ltd. 2018
R. Saian and M. A. Abbas (eds.), *Proceedings of the Second International Conference on the Future of ASEAN (ICoFA) 2017 – Volume 2*,
https://doi.org/10.1007/978-981-10-8471-3_98

now degenerative disease started to affect younger people, especially those living in urban areas.

Some of the main causes are the wrong diet and lifestyle changes due to urbanization and modernization, such as alcohol consumption, cigarettes smoking, stress, not enough sports activities, often sleep late at night, and so forth. The effects of degenerative diseases will usually show indications of weakening of immune systems and possibly result in death. Diseases that are categorized as degenerative diseases are: diabetes mellitus, hypertension, coronary heart, stroke, cancer, osteoporosis, uric acid, rheumatoid arthritis, and atherosclerosis (Ananta 2009).

Based on the above explanation, early prevention of degenerative diseases should be done. This research uses the help of computers by utilizing the branch of artificial intelligence expert system to perform early diagnosis of degenerative diseases. The expert system is a computer program that can adopt experts' knowledge into a computer by combining knowledge and an inference system to match the ability to solve a problem like an expert (Ignizio 1991).

Expert system which is developed uses forward chaining and Bayes theorem method to deal with uncertainties that arise. Based on knowledge acquisition from specialist doctors, medical record data and literature books, knowledge of six types of degenerative disease, i.e., hypertension, diabetes mellitus, coronary heart, stroke, dyslipidemia, and osteoarthritis with 42 kinds of symptoms, are obtained.

2 Literature Review

Several previous studies have been conducted both in conducting degenerative diseases diagnoses as well as research in terms of similarity of methods used. Dhani and Yamasari (2014) designed and built expert system to diagnose degenerative diseases. The research resulted in a rule-based expert system design that is stored into a database. To deal with the uncertainty, this expert system uses the Dempster–Shafer method. From the results of system testing, it can be concluded that the Dempster–Shafer method can be used to diagnose degenerative diseases well based on the symptoms felt by patient.

Research conducted by Anggraini et al. (2014) builds a diagnostic system for children's ear and nose and throat (ENT) diseases on Android-based platform. The system is built using forward chaining data analysis method. As for the uncertainty, it uses Bayes theorem method. This study produces an expert system that can diagnose seven types of ENT diseases. In this diagnosis process, symptoms perceived are entered to the system and probability value of a disease and each symptoms probability comparison will be calculated.

Johan and Fadlil (2013) developed an expert system to identify giant prawns diseases by using Bayes theorem method. In his research, it is said that the application of Bayes theorem method can perform the process of diagnosis of prawn shrimp disease on a desktop-based application. On this Bayes theorem method, the symptoms that inputted to the system will calculate level of probabilities against disease and sought the highest level of probability of disease as a result of diagnosis. Results are issued by the system of the level of certainty of diagnosis, definitions, as well as suggested solutions.

3 Research Methodology

This research is divided into several stages starting with identifying problems, knowledge acquisition, knowledge representation to create rules, application development, and application testing. Out of the knowledge acquisition, there are 6 kinds of degenerative diseases with 42 kinds of symptoms which were obtained. The acquisition of medical record data is obtained from 50 data used to find the probability value. Knowledge is represented in the form of decision tables.

3.1 Knowledge Acquisition

Knowledge acquisition is a process to collect data knowledge of a problem. This study uses two types of data which were obtained through several different ways. The primary data used is taken directly through interviews with one of the doctors in one of Pekanbaru hospitals along with medical record data. This study also obtained information from medical books in the library, research journals, and other literature that support this research.

3.2 Knowledge Representation

After data collection process has been completed, representation of the knowledge base and rule base is next, then encoded, organized, and described in another systematic design. The representation of knowledge for this expert system uses decision tables, and then, a rule is formed.

4 Results and Discussion

Table 1 is a decision table of degenerative diseases that contains relations between symptoms and diseases. From this decision, table rules are created for this expert system. The number of illnesses consists of 6 diseases and as many as 42 symptoms (Tables 2 and 3).

Table 1 Degenerative disease decision table

Disease code	Symptom code										
	G1	G2	G3	G4	G5	G6	G7	G8	G9	G10	G11
P1	X	X	X	X	X	X					
P2	X		X	X	X	X	X	X	X	X	X
P3				X	X	X					
P4	X		X		X	X				X	
P5	X										
P6										X	

Disease code	Symptom code									
	G12	G13	G14	G15	G16	G17	G18	G19	G20	G21
P1		X					X	X		
P2	X							X		
P3		X					X	X		X
P4		X	X	X	X	X	X			
P5		X					X		X	X
P6									X	

Disease code	Symptom code									
	G22	G23	G24	G25	G26	G27	G28	G29	G30	G31
P1	X	X	X	X	X	X				
P2	X		X	X	X	X	X	X	X	X
P3				X	X	X				
P4	X		X		X	X				X
P5	X									
P6										X

Disease code	Symptom code										
	G32	G33	G34	G35	G36	G37	G38	G39	G40	G41	G42
P1			X					X	X		
P2	X	X						X			
P3			X	X	X	X	X	X	X		
P4			X					X			
P5			X					X		X	
P6										X	X

Table 2 List of symptoms names on degenerative disease

No	Symptom code	Symptom name	No	Symptom code	Symptom name
1	G1	Dizzy	22	G22	Out of breath
2	G2	Buzzing ears	23	G23	Fast heartbeat
3	G3	The eye view is blurred	24	G24	Body sweats a lot
4	G4	Easily tired when doing various activities	25	G25	Chest discomfort when exercising or moving
5	G5	Nausea even vomiting	26	G26	Joint pain
6	G6	Head aches	27	G27	Stiffness in joints
7	G7	Frequent urination especially at night	28	G28	Inflammation of the joints
8	G8	Often feel thirsty	29	G29	Weak muscles
9	G9	Excessive hunger or eating a lot	30	G30	Cracking sounds of each joint
10	G10	Tingling	31	G31	Lisp
11	G11	Scratched or cut that takes long time to heal	32	G32	Eyes are hard to open
12	G12	Easy to drowsy	33	G33	Missing balance
13	G13	Weight loss	34	G34	Difficult to swallow
14	G14	Bloated	35	G35	Seizures
15	G15	Bowel obstruction	36	G36	Often fainted
16	G16	Diarrhea	37	G37	Face flushed
17	G17	Blood derived from the anus or rectum	38	G38	Swelling of joints
18	G18	Loss of appetite	39	G39	Abdominal pain
19	G19	Fever	40	G40	Neck pain
20	G20	Calf pain when walking	41	G41	Arm paralysis
21	G21	Chest pain	42	G42	Sudden confusion

Table 3 List of degenerative disease names

No	Disease code	Name of disease	No	Disease code	Name of disease
1	P1	Hypertension	4	P4	Stroke
2	P2	Diabetes mellitus	5	P5	Dyslipidemia
3	P3	Coronary heart	6	P6	Osteoarthritis

4.1 Handling Uncertainty

In this research, the Bayes theorem method is used to handle the uncertainty of the expert system. The concept of the Bayes theorem is to look for the highest probability of a disease which can be expressed in equation (Ananta 2009).

$$P(H|E) = \frac{P(E|H) \cdot P(H)}{P(E)} \quad (1)$$

With:

P(H|E) : Probability hypothesis H if given evidence E
P(E|H) : The probability of the appearance of the evidence E, if the hypothesis H is true
P(H) : The probability of hypothesis H regardless of any evidence
P(E) : Probability evidence E

To obtain such probability, the probability value of the disease regardless of any symptom can be calculated by the process of Eq. 2.

$$P(\text{Disease } X) = \frac{\text{Number of Cases of Disease Patients } X}{\text{Number of Cases}} \quad (2)$$

Furthermore, to calculate the probability of symptoms of the disease can be done in the process of Eq. 3.

$$P(\text{Symptoms } Y | \text{Disease } X) = \frac{\text{Number of Cases of Symptoms } Y \text{ on Illness } X}{\text{Number of Cases of Disease Patients } X} \quad (3)$$

In this expert system, the probability value is taken from the existing case data. Given an example of calculation using Bayes theorem which answer is obtained from choosing symptoms that appear on the system to the user, the user selects symptoms by choosing radio button on **G7** symptoms (frequent urination especially at night), **G8** symptoms (often feeling thirsty or thirsty), symptoms **G9** (excessive hunger or eating a lot), and symptoms of **G14** (flatulence).

Based on the user's answer and Table 1, the rules are met which are: The four symptoms are found only in P2 (symptoms G7, G8, and G9) and P3 (symptoms G14). So the probability search for Bayes theorem is only in P2 and P3 disease.

Diabetes Mellitus Disease (P2):

Symptoms G7: Frequent urination especially at night

$$P(P2|G7) = \frac{P(G7|P2) * P(P2)}{\begin{array}{c}P(G7|P1) * P(P1) + P(G7|P2) * P(P2) + \\ P(G7|P3) * P(P3) + P(G7|P4) * P(P4) + \\ P(G7|P5) * P(P5) + P(G7|P6) * P(P6)\end{array}}$$

$$P(P2|G7) = \frac{0.9 * 0.2}{0 * 0.25 + 0.9 * 0.2 + 0 * 0.17 + 0.46 * 0.13 + 0 * 0.15 + 0.3 * 0.1}$$
$$= \frac{0.18}{0.18} = 1$$

Symptoms G8: Often feel thirsty

$$P(P2|G8) = \frac{P(G8|P2) * P(P2)}{\begin{array}{c}P(G8|P1) * P(P1) + P(G8|P2) * P(P2) + \\ P(G8|P3) * P(P3) + P(G8|P4) * P(P4) + \\ P(G8|P5) * P(P5) + P(G8|P6) * P(P6)\end{array}}$$

$$P(P2|G8) = \frac{0.85 * 0.2}{0 * 0.25 + 0.85 * 0.2 + 0 * 0.17 + 0.46 * 0.13 + 0 * 0.15 + 0.3 * 0.1}$$
$$= \frac{0.17}{0.17} = 1$$

Symptoms G9: Excessive hunger or eating a lot

$$P(P2|G9) = \frac{P(G9|P2) * P(P2)}{\begin{array}{c}P(G9|P1) * P(P1) + P(G9|P2) * P(P2) + \\ P(G9|P3) * P(P3) + P(G9|P4) * P(P4) + \\ P(G9|P5) * P(P5) + P(G9|P6) * P(P6)\end{array}}$$

$$P(P2|G9) = \frac{0.85 * 0.2}{0 * 0.25 + 0.85 * 0.2 + 0 * 0.17 + 0 * 0.13 + 1 * 0.15 + 0.4 * 0.1}$$
$$= \frac{0.17}{0.17} = 1$$

Symptoms G014: Flatulence

$$P(P2|G14) = 0$$

Then, based on Tables 1 and 4, probability of P2 disease which is identified by choosing all the symptoms that belong to the rule of disease P2 is 7.4. Then,

Table 4 Cases data of degenerative disease

Case code	Symptoms	Results diagnose disease	Case code	Symptoms	Results diagnose disease
K1	G13, G4, G1, G6, G2	Hypertension	K51	G22, G18, G16, G13, G23, G21, G17, G15	Coronary heart
K2	G39, G18, G4, G36, G13, G2	Hypertension	K52	G14, G25, G21, G15, G35, G23, G16	Coronary heart
K3	G33, G19, G13, G2, G32, G18, G3	Hypertension	K53	G25, G22, G19, G17, G13, G24, G21, G18, G14	Coronary heart
K4	G39, G32, G19, G4, G33, G23, G5, G2	Hypertension	K54	G22, G19, G16, G14, G24, G21, G18, G15, G6	Coronary heart
K5	G39, G33, G3, G36, G4, G2	Hypertension	K55	G23, G21, G14, G5, G35, G24, G22, G15, G6, G25	Coronary heart
K6	G39, G19, G4, G23, G6, G1	Hypertension	K56	G23, G21, G18, G16, G35, G22, G19, G17, G14	Coronary heart
K7	G36, G13, G5, G19, G6, G3	Hypertension	K57	G25, G23, G21, G14, G4, G24, G22, G18, G6	Coronary heart
K8	G36, G18, G1, G39, G23, G4	Hypertension	K58	G23, G21, G15, G6, G24, G22, G16, G14, G5	Coronary heart
K9	G39, G23, G5, G2, G32, G6, G4	Hypertension	K59	G14, G6, G25, G22, G19, G17, G13, G35, G24, G21, G18	Coronary heart
K10	G39, G32, G5, G3, G33, G18, G4, G1	Hypertension	K60	G25, G23, G21, G6, G35, G24, G22, G14	Coronary heart
K11	G36, G23, G18, G4, G33, G19, G5, G1	Hypertension	K61	G35, G24, G22, G17, G14, G25, G23, G21, G15	Coronary heart
K12	G5, G39, G18, G6, G4, G36, G13	Hypertension	K62	G23, G21, G14, G24, G22, G16, G4, G25	Coronary heart
K13	G23, G4, G39, G6, G3	Hypertension	K63	G38, G35, G33, G29, G10, G5, G37, G34, G31, G27, G6	Stroke

(continued)

Table 4 (continued)

Case code	Symptoms	Results diagnose disease	Case code	Symptoms	Results diagnose disease
K14	G36, G3, G1, G23, G2	Hypertension	K64	G38, G34, G31, G18, G3, G37, G33, G29, G13, G1	Stroke
K15	G32, G19, G39, G23, G3	Hypertension	K65	G38, G35, G31, G6, G37, G34, G29, G5	Stroke
K16	G39, G23, G13, G5, G1, G33, G18, G6, G2	Hypertension	K66	G6, G37, G34, G31, G10, G3, G38, G35, G33, G29	Stroke
K17	G19, G2, G39, G23, G6, G1, G32	Hypertension	K67	G38, G35, G33, G29, G1, G37, G34, G31, G5	Stroke
K18	G39, G18, G5, G33, G6, G4	Hypertension	K68	G38, G35, G33, G29, G10, G37, G34, G31, G27, G6	Stroke
K19	G36, G6, G2, G39, G23, G4, G1	Hypertension	K69	G29, G18, G37, G34, G31, G27, G5, G38, G35, G33	Stroke
K20	G33, G23, G4, G1, G32, G6, G2	Hypertension	K70	G37, G34, G31, G18, G10, G38, G35, G33, G29, G13	Stroke
K21	G36, G4, G2, G39, G6, G3, G1	Hypertension	K71	G37, G34, G31, G18, G3, G38, G35, G33, G29, G13	Stroke
K22	G2, G32, G19, G13, G3, G36, G23, G18, G5	Hypertension	K72	G34, G31, G27, G6, G38, G35, G33, G29, G10, G37	Stroke
K23	G39, G23, G3, G1, G33, G18, G2	Hypertension	K73	G37, G34, G31, G13, G38, G35, G33, G29, G3	Stroke
K24	G39, G23, G6, G4, G33, G13, G5, G2	Hypertension	K74	G37, G34, G31, G6, G38, G35, G33, G29	Stroke
K25	G39, G23, G6, G3, G36, G19, G5, G1	Hypertension	K75	G38, G35, G33, G29, G18, G10, G37, G34, G31, G27, G13	Stroke
K26	G3, G19, G9, G7, G1, G10, G8	Diabetes mellitus	K76	G41, G21, G18, G42, G22, G20	Dyslipidemia

(continued)

Table 4 (continued)

Case code	Symptoms	Results diagnose disease	Case code	Symptoms	Results diagnose disease
K27	G12, G10, G7, G5, G11, G9, G6, G4	Diabetes mellitus	K77	G41, G20, G1, G42, G21, G13	Dyslipidemia
K28	G11, G8, G5, G9, G7, G1	Diabetes mellitus	K78	G41, G21, G1, G42, G22, G20	Dyslipidemia
K29	G11, G9, G5, G12, G10, G8, G4	Diabetes mellitus	K79	G42, G22, G18, G1, G41, G20, G13	Dyslipidemia
K30	G10, G7, G3, G12, G8, G4	Diabetes mellitus	K80	G42, G22, G20, G41, G21, G18	Dyslipidemia
K31	G19, G8, G6, G9, G7, G1	Diabetes mellitus	K81	G20, G42, G21, G18, G41	Dyslipidemia
K32	G12, G9, G7, G5, G11, G8, G6, G3	Diabetes mellitus	K82	G42, G22, G20, G13, G41, G21, G18, G1	Dyslipidemia
K33	G10, G7, G12, G8, G3	Diabetes mellitus	K83	G41, G20, G42, G22, G18	Dyslipidemia
K34	G12, G10, G8, G1, G19, G11, G9, G7	Diabetes mellitus	K84	G41, G18, G42, G20	Dyslipidemia
K35	G11, G7, G4, G12, G9, G6, G19	Diabetes mellitus	K85	G42, G22, G1, G41, G20	Dyslipidemia
K36	G11, G9, G7, G1, G10, G8, G6	Diabetes mellitus	K86	G42, G21, G18, G41, G20, G13	Dyslipidemia
K37	G12, G8, G6, G19, G9, G7, G5	Diabetes mellitus	K87	G42, G22, G20, G1, G41, G21, G18	Dyslipidemia
K38	G12, G9, G7, G5, G3, G11, G8, G6, G4	Diabetes mellitus	K88	G42, G22, G20, G41, G21, G13	Dyslipidemia
K39	G8, G6, G4, G9, G7, G5, G1	Diabetes mellitus	K89	G41, G21, G18, G42, G22, G20, G13	Dyslipidemia
K40	G19, G11, G9, G7, G12, G10, G8, G4	Diabetes mellitus	K90	G42, G22, G20, G41, G21	Dyslipidemia
K41	G11, G7, G5, G12, G8, G6, G1	Diabetes mellitus	K91	G40, G28, G26, G30, G27, G20	Osteoarthritis
K42	G11, G8, G6, G12, G9, G7	Diabetes mellitus	K92	G26, G30, G27, G10, G40, G28	Osteoarthritis

(continued)

Table 4 (continued)

Case code	Symptoms	Results diagnose disease	Case code	Symptoms	Results diagnose disease
K43	G12, G9, G5, G1, G19, G11, G6, G3	Diabetes mellitus	K93	G30, G27, G10, G40, G28, G26	Osteoarthritis
K44	G4, G1, G11, G8, G6, G3, G9, G7	Diabetes mellitus	K94	G40, G28, G20, G30, G26	Osteoarthritis
K45	G12, G9, G7, G5, G11, G8, G6, G3	Diabetes mellitus	K95	G30, G27, G40, G28, G26	Osteoarthritis
K46	G25, G23, G21, G15, G5, G24, G22, G16, G14, G4	Coronary heart	K96	G30, G27, G20, G40, G28, G26	Osteoarthritis
K47	G35, G23, G21, G14, G24, G22, G15, G4	Coronary heart	K97	G40, G28, G26, G30, G27	Osteoarthritis
K48	G24, G22, G16, G14, G25, G23, G21, G15	Coronary heart	K98	G28, G26, G40, G27	Osteoarthritis
K49	G24, G22, G13, G5, G23, G21, G6	Coronary heart	K99	G40, G28, G10, G30, G26	Osteoarthritis
K50	G24, G22, G14, G6, G25, G23, G21, G13, G5	Coronary heart	K100	G40, G28, G26, G30, G27, G20	Osteoarthritis

$$\frac{1+1+1+0}{7.4} = 0.40 * 100\% = 40\%$$

Coronary Heart Disease (P3):
Symptoms G7: Frequent urination especially at night

$$P(\text{P3}|\text{G7}) = 0$$

Symptoms G8: Often feel thirsty

$$P(\text{P3}|\text{G8}) = 0$$

Symptoms G9: Excessive hunger or eating a lot

$$P(P3|G9) = 0$$

Symptoms G14: Flatulence

$$P(P1|G7) = \frac{P(G14|P3) * P(P3)}{\begin{array}{c} P(G14|P1) * P(P1) + P(G14|P2) * P(P2) + \\ P(G14|P3) * P(P3) + P(G14|P4) * P(P4) + \\ P(G14|P5) * P(P5) + P(G14|P6) * P(P6) \end{array}}$$

$$P(P3|G14) = \frac{0.88 * 0.17}{0 * 0.25 + 0 * 0.2 + 0.88 * 0.17 + 0.46 * 0.13 + 0 * 0.15 + 0.3 * 0.1}$$
$$= \frac{0.15}{0.15} = 1$$

Then, based on Tables 1 and 4, known probability of P3 disease which is identified by choosing all the symptoms that include the rule of disease P3 is 9.1. Then,

$$\frac{0+0+0+1}{9.1} = 0.109 * 100\% = 10.9\%$$

The results of Bayes theorem's calculations concluded that patient was diagnosed with diabetes mellitus disease with a confidence level of 0.40 or 40%. The system will then provide an explanation ranging from natural symptoms, diagnosis results, confidence levels, and solutions in accordance with the disease they suffer.

4.2 System Testing Based on Rule/Diagnose Rules with Experts

Expert system testing is done by using black box in addition to standard gold testing by experts directly. Based on the test results from experts, the conclusion of both experts diagnosis and system diagnosis provides the same conclusion.

5 Conclusion

The conclusions in this study are as follow:

1. Expert system with Bayes theorem uncertainty method can be applied to diagnose degenerative diseases in patients based on symptoms felt by the patient.

2. From the results of system testing, expert system of degenerative disease diagnoses with Bayes theorem method has a success rate of 90% and the percentage of the usage of this system is at 88.85% which interpreted very good to be implemented.

References

Ananta. (2009). *Waspada gejala penyakit mematikan jantung koroner dengan tiga penyakit yang berkaitan hipertensi, diabetes mellitus, dan stroke*. Jakarta: Tugu Publisher.
Anggraini, D. et al. (2014). Diagnosa penyakit telinga hidung dan tenggorokan (THT) pada anak dengan menggunakan sistem pakar berbasis mobile android. *Jurnal Coding Sistem Komputer Universitas Tanjungpura, 2*(2), 8–14. ISSN: 2338-493x.
Dhani, S. R., & Yamasari, Y. (2014). Rancang bangun sistem pakar untuk mendiagnosa penyakit degeneratif. *Jurnal Manajemen Informatika, 3*(2), 17–25.
Ignizion, P. J. (1991). *Introduction to expert systems: the development and implementation of rule-based expert systems*. United States of America: McGraw-Hill Inc.
Johan, W. M., & Fadlil, A. (2013). Sistem pakar untuk mengidentifikasi penyakit udang galah dengan metode theorem bayes. *Jurnal Sarjana Teknik Informatika, 1*(1), 8–14. e-ISSN: 2338-5197.

The Analysis Factor to Determine Modern Store Location in Pekanbaru City

Puji Astuti, Febby Asteriani, Yoghi Kurniawan and Idham Nugraha

Abstract The development of modern stores in Indonesia influenced the development of modern stores in many cities. Pekanbaru becomes an attractive destination for retail businesses. It had the potential to grow in terms of trade. This indication is shown by the proliferation of modern shops in several locations in Pekanbaru. There are many factors that affect the modern shops owner to determining location of the modern shops. These factors related to accessibility, population, land prices, building rental rates, competition, physical condition of the land, infrastructure, and the provision of space. This study used a deductive approach with quantitative analysis techniques. At first, the theory of the modern store locations is collected and carried stabilization theory, by distributing questionnaires to the respondents. Then, the results of questionnaires are processed by factor analysis using the program Predictive Analytics Software (PASW) Statistics 18. The results of data processing by factor analysis showed that the initial factor of 8 or 20 observed variables reduces to 4 new factors with 18 variables that factor into the spread. The new factor composed of factor 1 (number and density of the population), factor 2 (availability of pedestrian path/pedestrian), factor 3 (soil type), and factor 4 (communication network availability).

Keywords Modern stores · Analysis factors · Minimarket · Supermarket

P. Astuti (✉) · F. Asteriani · Y. Kurniawan · I. Nugraha
Department of Urban Planing, Faculty of Engineering, Universitas Islam Riau,
Jl. Kaharuddin Nasution, Pekanbaru, Riau 28284, Indonesia
e-mail: pujiastutiafrinal@eng.uir.ac.id

F. Asteriani
e-mail: febbyastriani@eng.uir.ac.id

© Springer Nature Singapore Pte Ltd. 2018
R. Saian and M. A. Abbas (eds.), *Proceedings of the Second International Conference on the Future of ASEAN (ICoFA) 2017 – Volume 2*,
https://doi.org/10.1007/978-981-10-8471-3_99

1 Introduction

Modern stores located in Pekanbaru are started up since 1993. The service system imposed in minimarkets and supermarkets is self-service, where visitors are given the freedom to select the desired item and then bring it to the cashier to pay. The type of goods sold at minimarkets and supermarkets is the everyday needs of society. Secondary and tertiary sectors contribute more dominant than the primary sector, and the economic structure of the Pekanbaru City supported by trade, hotels, and restaurants amounted to 3,180,369 (billion rupiah) with the distribution percentage of 32.23%. Based on data from the Integrated Service Agency and Department of Industry and Trade of the Pekanbaru City, modern stores scattered throughout the Pekanbaru City are officially registered in the year 2012–2013 amounted to 60 modern stores. Distribution of modern stores in the Pekanbaru City is dominated by minimarket 49 stores and supermarket 11 stores.

Modern stores have spread in various regions. The entry of the giant retailer adds to a long list of modern shops there. Increasing this facility does provide a variety of options to the community to meet their needs. However, existence modern stores in many locations can lead to various problems, such as competition, accessibility, and inefficiency of land. The growth of modern stores in Pekanbaru feared could damage competition from a variety of existing trading facilities, as well as raising fears of monopoly of land in the surrounding area. The business locations cause problems such as congestion due to the accessibility of such business activities as well as the lack of adequate vehicle parking. The most important thing in setting up the trading business is being able to determine the location factor in the progress and success of a business and can provide benefits for businesses. In addition, the business location selection decision or modern retail trade shops will be focused on maximizing profits. In general, the ideal location for activities/trading business is the location with the potential for the commercial area, easily accessible, the potential for growth, interception of competitors, and the location of economic (Sopiah 2008). Some of the problems arising from modern stores in the city of Pekanbaru are:

1. Competition: Modern shop in an area can take place between the members of modern stores or between modern shop with other trading facilities such as traditional markets and shopping centers. This competition can give a positive or negative impact. One of the positive impacts is the growing economy of a region, while the negative impact to be considered is the competition with the traditional markets which are a group of small merchants, where it can be deadly small traders.
2. Accessibility: The modern stores around the site will affect the accessibility of the modern shop. That cause arises problem of traffic jams as a result of the activities of a modern store. And the parking space available is inadequate for consumer vehicles, as well as the loading and unloading of goods.
3. Inefficiency of land: These modern stores that are in a particular region can monopolize the land in the surrounding area.

Indonesian Presidential Regulation No. 112 of 2007 explained the characteristics of modern stores as follows:

1. The floor area of the modern store sales can be explained as follows:

    a. Minimarket, less than 400 m^2.
    b. Supermarket, 400–5000 m^2.
    c. Hypermarket, over 5000 m^2.
    d. Department Store, over 400 m^2.
    e. Grocery, over 5000 m^2.

2. The type of goods sold at modern stores can be explained as follows:

    a. Minimarket, supermarkets, and hypermarkets sell retail consumer goods, especially food products and other household products;
    b. Department store sells consumer goods in retail clothing and accessories products primarily to the arrangement of items based on gender and/or age level of the consumer; and
    c. Grocery wholesale selling consumer goods.

According to Ma'ruf (2006), there are several factors in considering the selection of the location or point of outlets: foot traffic, vehicle traffic, the parking facilities, public transport, the composition of the store, the location of the establishment of outlets, and terms and conditions of use of the room.

2 Research Methodology

This study uses a deductive approach with quantitative analysis techniques. Its aims to identify the factors that determine or be considered the owner/manager in the selection of the modern store locations in the city of Pekanbaru. The scope of areas in this study is in the administrative area of Pekanbaru, which consists of 11 districts: District of Pekanbaru Kota, Limapuluh District, Senapelan District, Sukajadi District, Rumbai District, Rumbai Pesisir District, Tenayan Raya District, Bukit Raya District, Marpoyan Damai District, Tampa District, and Payung Sekaki District.

Modern stores in the scope of this study, a minimarket and a supermarket, are registered on the relevant agencies in the city of Pekanbaru. The scope of the discussion in this study is related to the variables to be considered in choosing the location of modern stores such as minimarkets and supermarkets. Variables that will be used in the analysis to determine the factors in the site selection of modern shops in Pekanbaru are shown in Table 1.

Data obtained from field observations and the results of questionnaires to the owner/manager of modern stores such as minimarkets and supermarkets regarding their votes in an election modern store locations. Questionnaire distributed questionnaire types of options. This type of measurement used is itemized rating scale. This type of measurement provides a selection in according to this study used four types of options (Asteriani 2005). Respondents were asked to give an assessment.

Table 1 Selected variables

No	Variables		Symbol
1	Accessibility	Distance	
		Proximity to the residential area	X1
		Proximity to the business center	X2
		Proximity to the main roads	X3
		Transportation	
		Crossed by public transport	X4
		Availability vehicle parking	X5
		Availability of pedestrian	X6
		The physical condition of the road	
		Road surface quality	X7
		Width of the road	X8
2	Population	Population density	X9
		Population income level	X10
3	Price of the land	Price of the land	X11
4	Building rent	Building rent	X12
5	Completion	Size of competitor	X13
		Proximity to the competitor	X14
6	Land condition	Topography	X15
		Type of soil	X16
7	Infrastructure	Availability of the electricity network	X17
		Availability of the communication network	X18
8	Condition of the space	General rules of space utilization	X19
		Intensity of the space conditions	X20

Source Result of analysis, 2013

Assessment is provided comprising: 1 = Strongly Specifies (SS); 2 = Not Determine (ND); 3 = Specifies (S); 4 = Very Specifies (VS).

Sampling technique used in this research is purposive sampling. Based on the analysis used in this study, there is no minimum sample size acceptable in the factor analysis, factor analysis as a general rule, the number of respondents is at least three times the number of variables (Simamora 2005).

Due to the variables in this study amounted to 20 variables and using a ratio of 1: 3, the number of respondents is 60 people. From the secondary data have identified as many as 60 modern stores in the city of Pekanbaru and has been determined to be the respondent. Of the total respondents, 60% or 36 respondents are owners and 40% or 24 respondents were managers of minimarkets or supermarkets that are on the assessment of the questionnaires distributed.

The approach in this study uses quantitative analysis techniques. The main characteristics important factor of the analysis is the ability to reduce the existing data, so that will facilitate analysis. By providing a series of correlation coefficients for a class of variables, factor analysis technique allowed us to see whether some of the major patterns of relationship that allows data can be compiled or reduced to a class of factors smaller.

This factor analysis using the program Predictive Analytics Software (PASW) Statistics 18. Stages of the factor analysis are as follows:

1. Formulate the problem: Determine the factors which are to be taken into consideration by the owners/managers in selecting the location of modern stores in Pekanbaru.
2. Establish a correlation matrix: that factor analysis can be done, should be correlated variables analyzed. If the correlation coefficient is too small, then the factor analysis cannot be done, or by removing the variable with the coefficient is small. Ideally, the original variables are correlated with each other and can be correlated with a factor, as a new variable, derived from the original variables. PASW program provides a test indicator to the value of the correlation coefficient between the members of variables and factors, through the Bartlett's test of sphericity and the value of Kaiser–Meyer–Olkin (KMO). The Bartlett's test of sphericity used to test that the variables are correlated in the population. While the index KMO compares the magnitude of the correlation coefficient was observed with the magnitude of the correlation coefficient partial. KMO value which is small shows that the correlation between pairs of variables cannot be explained by other variables and factor analysis cannot be done. Ideally, the factor analysis KMO index must be greater than 0.5.
3. Extraction of factors; at this stage of the process will be the core of the analysis of factors, namely the extraction of the set of variables that exist (KMO > 0.5) thus forming one or more factors. There are two methods in this process, namely principal components analysis (PCA) and common factor analysis (CFA). Both of these methods can be selected, in particular for calculating scales/coefficient score factors. In the principal components analysis (PCA) are used with the justification that the principal is the one things that many factors must specify a minimum, taking into account the maximum variance in the data used in subsequent multivariate analysis. Instead common factor analysis (CFA) is used with justification that the main purpose identifies the underlying dimensions and variance common interest. In this study, PCA will be used.
4. Rotation factor; In the rotation factor, factor matrix is transformed into simple matrix making it easier to interpret. There are two methods in the rotation factor, which is orthogonal and oblique rotation. Orthogonal where the axis of rotation is the rotation of a straight to maintained upright (Angle 90°). Many used methods are varimax rotation procedure. This procedure is orthogonal methods that seek to minimize the number of high loading variables on one factor, thus facilitate the creation of factors interpretation. Produce orthogonal factor rotation that is not correlated with each other. While oblique rotation in which the rotation axis is not maintained to be perpendicular to each other (angle 90°) and factors are not correlated. This method can be used, if the factors in the population strongly correlate. In this study, will be using the varimax rotation.

5. Interpretation of factors: Interpretation can be made easier with the surrogate variables, where variables have a high loading factors on each others, then the variable representing each factors.

3 Result and Discussion

Analysis factors is basically used to reduce the data, which is a process to summarize a number of variables to be fewer and named it as a factor. This study uses 20 variables that is a consideration in the selection of the modern store locations in Pekanbaru.

1. Defining issues

The initial step in the factor analysis is to formulate the problem. This step is done by determining the variables as many as 20 variables (Table 1).

2. Creating correlation matrix

Based on the results of data processing through the program PASW Statistic18 in Kaiser-Meyer-Olkin (KMO) table and Barlett Test of sphericity. It is known that the KMO value of $0.744 > 0.5$ and statistical values. Bartlett test of sphericity amounted to 823.457 with 0.000. Significance values are less than 0.05. This means that the analysis can be continued.

Then further analyzed by looking at the value Measure of Sampling Adequacy (MSA). If each variable has a value MSA > 0.5, then that variable is valid. If there is a variable with the MSA of less than 0.5, then the variable is not valid, so removed and re-calculation of the variable. MSA value from the table obtained one variable that MSA value is less than 0.5, namely: general provisions variable space utilization ($\times 19$) with a value of 0.500. Then, these variables are issued and performed recalculation (Table 2).

Then, proceed again with the MSA and see the value of each variable. It is known that variable has a value of 19 MSA > 0.5 make that variable is valid. For the MSA for each variable can be seen in Table 3.

Table 2 Kaiser–Meyer–Olkin (KMO) and Bartlett test of sphericity

Kaiser–Meyer–Olkin Measure of Sampling Adequacy	0.767		
Bartlett's test of sphericity	Approx. chi-square	773.558
	Df	171
	Sig.	0.000

Source Result of analysis, 2013

Table 3 Values for each MSA variable

No	Variables	MSA	Description
1	Proximity to the residential area (X1)	0.808	Valid
2	Proximity to the business center (X2)	0.760	Valid
3	Proximity to the main roads (X3)	0.768	Valid
4	Crossed by public transportation (X4)	0.796	Valid
5	Availability of vehicle parking (X5)	0.810	Valid
6	Availability of pedestrian (X6)	0.779	Valid
7	Road surface quality (X7)	0.757	Valid
8	Width of the road (X8)	0.879	Valid
9	Population density (X9)	0.787	Valid
10	Population income level (X10)	0.767	Valid
11	Price of the land (X11)	0.773	Valid
12	Building rent (X12)	0.815	Valid
13	Size of competitor (X13)	0.787	Valid
14	Proximity to the competitor (X14)	0.723	Valid
15	Topography (X15)	0.803	Valid
16	Type of soil (X16)	0.574	Valid
17	Availability of the electricity network (X17)	0.658	Valid
18	Availability of the communication network (X18)	0.631	Valid
19	Intensity of the space conditions (X20)	0.659	Valid

Source Result of analysis, 2013

1. Determining of total factor/factor extraction

Total factor is determined from eigenvalue. The bigger the coefficient eigenvalue a factor means the representative in representing a number of variables. Factors known representative is the factor with eigenvalue greater than or equal to 1. By using the method of PCA four (4) factors are obtained with eigenvalue $\geq$ 1, factor of 1, which is a major factor that has eigenvalues of 6.902, a factor of 2 has eigenvalues amounted to 2.802, a factor of 3 has eigenvalues of 2.218, and a factor of 4 has eigenvalues of 1.302 (Table 4).

Table 4 Factors with eigenvalue $\geq$ 1

Factor	Eigenvalues	Percent of variance	Cumulative percent
1	6902	36,326	36,326
2	2802	14,749	51,076
3	2218	11,674	62,750
4	1302	6852	69,603

Source Result of analysis, 2013

2. Factor rotation

Factor rotation is the simplification of the matrix factor, which has a structure that is quite difficult to interpret. To facilitate interpretation of factors, matrix factor is transformed into simpler matrices with the rotation of factors.

Variables shown in Table 5 are as follows:

- X1 (proximity to the residential area) has a loading factor (correlation) of 0.711, the factor of 1. Because the strongest correlation is at a factor of 1, then the variable X1 (proximity to the residential area) became a member of one factor.
- X2 (proximity to the business center) has a loading factor (correlation) of 0.556, on factor of 2. Because the strongest correlation is at a factor of 2, then the variable X2 (proximity to the business center) become a member of a factor of 2.
- X3 (proximity to the main roads) has a loading factor (correlation) of 0.778, the factor of 1. Because the strongest correlation is at a factor of 1, then X3 (proximity to the main roads) became a member of factor 1.
- X4 (crossed by public transportation) has a loading factor (correlation) of 0.594, the factor of 4. Because the strongest correlation is at a factor of 4, then the variable X4 (crossed by public transportation) became a member of factor 4.

Table 5 Rotated component matrix

Variables	Component			
	1	2	3	4
Proximity to the settlement area (X1)	0.711	0.141	0.190	0.263
Proximity to the business center (X2)	0.275	0.556	0.169	0.232
Proximity to the main roads (X3)	0.778	0.007	0.116	0.198
Crossed by public transportation (X4)	0.436	0.270	0.107	0.594
Availability of vehicle parking (X5)	0.689	0.251	0.007	0.223
Availability of pedestrian (X6)	0.065	0.845	0.105	0.269
Road surface quality (X7)	0.325	0.200	0.400	0.725
Width of the road (X8)	0.418	0.204	0.410	0.442
Population density (X9)	0.866	0.014	0.113	0.247
Population income level (X10)	0.610	0.529	−0.389	−0.024
Price of the land (X11)	0.303	0.250	0.804	0.036
Building rent (X12)	0.320	0.819	−0.164	−0.134
Size of competitor (X13)	0.738	0.308	0.294	0.084
Proximity to the competitor (X14)	0.439	0.658	−0.012	−0.363
Topography (X15)	0.048	0.581	0.466	0.154
Type of soil (X16)	0.012	−0.046	0.850	−0.001
Availability of the electricity network (X17)	0.575	−0.229	−0.309	0.408
Availability of the communication network (X18)	0.242	−0.030	−0.148	0.811
Intensity of the space conditions (X20)				

Source Result of analysis, 2013

- X5 (availability of vehicle parking) has a loading factor (correlation) of 0.689, the factor of 1. Because the strongest correlation is at a factor of 1, then the variable X5 (availability of vehicle parking) be a member of a factor 1.
- X6 (availability of pedestrian) has a loading factor (correlation) of 0.845, on factor of 2. Because the strongest correlation is at a factor of 2, then the variable X6 (availability of pedestrian) be a member of a factor of 2.
- X7 (road surface quality) has a loading factor (correlation) of 0.725, the factor of 4. Because the strongest correlation is at a factor of 4, then the variable X7 (road surface quality) be a member of a factor of 4.
- X8 (width of the road) has a loading factor (correlation) of 0.442, the factor of 4. Because the strongest correlation is at a factor of 4, then the variable X8 (width of the road) be a member of a factor of 4.
- X9 (population density) has a loading factor (correlation) of 0.866, the factor of 1. Because the strongest correlation is at a factor of 1, then the variable X9 (population density) be a member of a factor 1.
- X10 (population income level) has a loading factor (correlation) of 0.610, the factor of 1. Because the strongest correlation is at a factor of 1, then the variable X10 (population income level) be a member of one factor 1.
- X11 (price of the land) has a loading factor (correlation) of 0.804, the factor of 3. Since the strongest correlation is at a factor of 3, then the variable X11 (price of the land) be a member of a factor of 3.
- X12 (building rent) has a loading factor (correlation) of 0.819, on factor of 2. Because the strongest correlation is at a factor of 2, then the variable X12 (building rent) be a member of a factor of 2.
- X13 (size of competitor) has a loading factor (correlation) of 0.738, the factor of 1. Because the strongest correlation is at a factor of 1, then the variable X13 (size of competitor) be a member of a factor 1.
- X14 (proximity to the competitor) has a loading factor (correlation) of 0.658, on factor of 2. Because the strongest correlation is at a factor of 2, then the variable X14 (proximity to the competitor) be a member of a factor of 2.
- X15 (topography) has a loading factor (correlation) of 0.581, on factor of 2. Because the strongest correlation is at a factor of 2, then the variable X15 (topography) be a member of a factor of 2.
- X16 (type of soil) has a loading factor (correlation) of 0.850, the factor of 3. Since the strongest correlation is at a factor of 3, then the variable X16 (type of soil) be a member of a factor of 3.
- X17 (availability of the electricity network) has a loading factor (correlation) of 0.575, the factor of 1. Because the strongest correlation is at a factor of 1, then the variable X17 (availability of the electricity network) be a member of a factor 1.
- X18 (availability of the communication network) has a loading factor (correlation) of 0.811, the factor of 4. Because the strongest correlation is at a factor of 4, then the variable X18 (availability of the communication network) be a member of a factor of 4.

- X20 (intensity of the space conditions) has a loading factor (correlation) of 0.680, on factor of 2. Because the strongest correlation is at a factor of 2, then the variable X20 (intensity of the space conditions) be a member of a factor of 2.

3. Interpretation factor

Interpretation is done by grouping variable factors that have a loading factor of at least 0.50. Variables with the loading factor under 0.50 were excluded from the model. In Table 6, the road width variables incorporated in the first factor are excluded, because of the loading factor under 0.50.

Furthermore, 18 of these variables are spread into four factors: factor I, factor II, factor III, and factor IV with 69.603% of the total variance. It showed that it was able to explain the factors considered to determine or be a consideration in picking the modern store locations in Pekanbaru. The four factors are named according to the variables grouped scattered upon these factors. Factor analysis did not specify the name of each of the factors and concepts for factors produced. The name and

Table 6 Interpretation factor

No	Variables	Factor	Eigenvalue	Loading factor	Percent of variance
1	Population density (X9)	FACTOR I	6902	0.866	36,326
2	Proximity to the main roads (X3)			0.778	
3	Size of competitor (X13)			0.738	
4	Proximity to the residential area (X1)			0.711	
5	Availability of vehicle parking (X5)			0.689	
6	Population density (X9)			0.610	
7	Availability of the electricity network (X17)			0.575	
8	Availability of pedestrian (X6)	FACTOR II	2802	0.845	14,749
9	Building rent (X12)			0.819	
10	Intensity of the space conditions (X20)			0.680	
11	Proximity to the competitor (X14)			0.658	
12	Topography (X15)			0.581	
13	Proximity to the business center (X2)			0.556	
14	Type of soil (X16)	FACTOR III	2218	0.850	11,674
15	Price of land (X11)			0.804	
16	Availability of the communication network (X18)	FACTOR IV	1302	0.811	6852
17	Road surface quality (X7)			0.725	
18	Crossed by public transportation (X4)			0.594	
19	Width of the roads (X8)			0.442	

Source Result of analysis, 2013

concept or meaning of each factor is determined based on the theory of a surrogate, whereas variable which has high loading factors on each of the factors, then variable represents the name of each factors.

From Table 6, it has identified factors in the selection of the modern store locations in Pekanbaru. The utilization condition of space is removed from the model; it means that it is not considered by the owners/managers in selecting a business location minimarket and supermarket. Due to the lack of clarity regarding the dissemination or use directives is to the community, especially businesses minimarket and a supermarket in Pekanbaru. Meanwhile, the variable width of the road is also not considered, because its conditions on the location of modern store business observed are sufficient for the survival of businesses run by entrepreneurs of minimarkets and supermarkets. The results of data processing by analysis factor also identified a group of variables that are primary or dominant factor in the site selection minimarket and a supermarket in Pekanbaru. The main factor is a factor I, wherein the variables belonging to the first factor are things more decisive or more to be considered in choosing a store location modern than other variables belonging to the factor II, factor III, and factor IV.

4 Conclusion

The factors that determine the conduct of the election modern store locations in Pekanbaru are:

1. Factor I consists of seven variables: population density, proximity to the main roads, size of competitors, proximity to the residential areas, availability of vehicle parking lot, income level, and availability of the electricity networks.
2. Factor II consists of six variables: availability of pedestrian, building rent, provision of space intensity, proximity to competitors, topography, and proximity to the business center.
3. Factor III consists of two variables: type of soil and land prices.
4. Factor IV consists of three variables: availability of the communication network, quality of the road surface, and crossed by public transport.

Acknowledgements This work is partially supported by Urban and Regional Planning Department, Faculty of Engineering, Islamic University of Riau.

References

Asteriani, F. (2005). *Analisis Peringkat Faktor-Faktor Pemilihan Lokasi Ruko dari Sudut Pandang Pengguna dan Pengembang Ruko di Kota Pekanbaru.* Yogyakarta: Universitas Gajah Mada.
Badan Pusat Statistik Kota Pekanbaru. Pekanbaru Dalam Angka 2013.

Badan Pusat Statistik Kota Pekanbaru. Pekanbaru Dalam Angka 2014.
Dinas Pasar Kota Pekanbaru. (2013). *Ekspose Profil Dinas Pasar*.
Ma'ruf, H. (2006). *Pemasaran Ritel*. Jakarta: Gramedia.
Peraturan Mentri Nomor: 53/M-Dag/Per/12/2008. Pedoman Penataan dan Pembinaan Pasar Tradisional. Dinas Perindustrian dan Perdagangan Republik Indonesia, Jakarta.
Peraturan Presiden Republik Indonesia Nomor 112 tahun 2007 tentang Penataan dan Pembinaan Pasar Tradisional Pusat Perbelanjaan, dan Toko Modern. Jakarta.
Simamora, B. (2005). *Analisis Multivariat Pemasaran*. Jakarta: Gramedia.
Sopiah, S. (2008). *Manajemen Bisnis Ritail*. Andi, Yogyakarta.

Landslide Hazard Map Using Aster GDEM 30m and GIS Intersect Method in Tanjung Alai, XIII Koto Kampar Sub-District, Riau, Indonesia

Tiggi Choanji, Idham Nugraha, Muhammad Sofwan and Yuniarti Yuskar

Abstract In this paper, we are trying to produce landslide hazard map by using field data and geospatial technology using GIS at Tanjung Alai, XIII Koto Kampar Sub-District, Kampar, Riau Province, Indonesia, which has several spots of landslide location. Susceptibility analysis has involved the location of the potential slope failures and hazard zonation map near the road. We derived geology, slope, aspect, soil type, structural lineament, and rainfall accumulation parameters per year from Aster GDEM 30m. Analysis from six parameters shows that topography ranges from 45 to 506 m. Geology consists of four formations exposed in the research area. Slopes are ranging from 15% to more than 50%. For aspect, there are nine classified aspects/directions of slope in this research area. Soil type is divided into clay and sand. Structural lineament trending northwest–southeast rainfall shows average 567–808 mm. From all combined factor maps, the result showing landslide hazard index ranged from very low to very high. So this derived map hopefully will give suggestion for the best plan for mitigation and disaster management plan in the area.

Keywords Landslide · Aster GDEM 30m · Geology · Hazard · Mitigation

T. Choanji (✉) · Y. Yuskar
Department of Geological Engineering, Universitas Islam Riau,
Riau, Indonesia
e-mail: tiggich@eng.uir.ac.id

I. Nugraha · M. Sofwan
Department of Urban and Regional Planning, Universitas Islam Riau,
Riau, Indonesia
e-mail: idham.nugraha@eng.uir.ac.id

1 Introduction

Landslide is defined as the movement of a mass of rock debris or earth down the slope (Cruden 1991). The term 'Landslide' encompasses events such as ground movement, rock falls, and failures of slopes, topples, slides, spreads, and flows such as debris flows, mudflows, or mudslides (Varnes 1994). Gravity acting on a steep slope is the primary reason for the landslides. Tanjung Alai area has most viewable landscape with hilly and lake dam surrounding the area. But also, it holds potential disaster due to the combination of factors such as heavy rainfall, steep slopes, lack of vegetation or deforestation, rugged topography, incompetent geological formation, and structurally fragmented rock. The correlation between associated factors that cause landslide and landslide areas can be allocated from the connection between areas without past landslide and landslide-related parameters (Dahal and Dahal 2017). In order to provide the landslide hazard map, this method was implemented using GIS techniques (Regmi et al. 2010).

2 Condition of the Area

Tanjung Alai area is located at XIII Koto Kampar Sub-District, Kampar District, Riau, Indonesia (Fig. 1). The research area is located near Barisan Mountain and belongs to Granite Formation, Bohorok Formation, Sihapas Formation, and Tanjung Pauh Member. Granite Formation on the area consists of foliated, partly gneisses granite. Bohorok Formation consists of wackes, conglomeratic wackes,

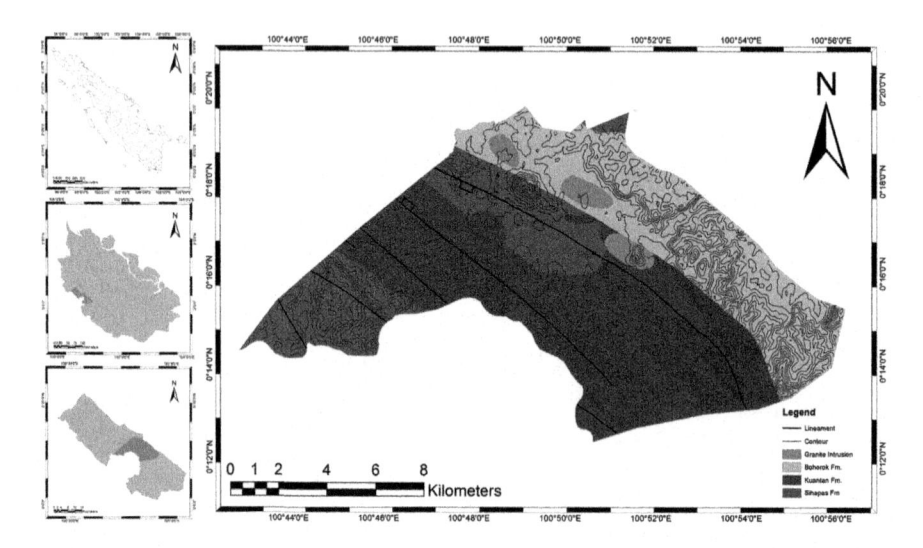

Fig. 1 Geological map at Tanjung Alai, Kampar, Riau

and turbidite deposits. Sihapas Formation consists of conglomeratic sandstones and siltstones (Clarke et al. 1982). A Tanjung Pauh Member consists of chlorite carbonate schist that is strongly lineated. The research area ranges from 45 to 506 m from mean sea level.

Structural geology development in the area was trending northwest–southeast (Putra and Choanji 2016; Choanji 2016). It consists of normal fault and anticline which have same direction with regional tectonic develop in the research area.

The large total of the research area is 163.84 km^2. The mean annual precipitation ranges from 567 to 808 mm. Most of the slopes face east and west, and its location is near the road. The slope gradient generally increases with increase in elevation (Jayanthi et al. 2016). Because of the road location that accessing two province where the food supply is needed for each other and the cut slope considered as heavily joint slope due to fault system on that area (Choanji and Indrajati 2016; Putra and Choanji 2016; Choanji 2017; Choanji and Indrajati 2016), hazard analysis is necessary to do for the sustainability development between these two provinces.

3 Methodology

The main step for landslide hazard mapping is data collection and preparation of a spatial database from which relevant factors can be extracted (Ciampalini et al. 2016). The main feature of this method is comparing the possibility of landslide occurrence with observed landslides.

Based on field survey, various causative factors were identified, including geology, structural lineament, slope, slope aspect, mean annual rainfall, soil type. All of these factors can be zoned by using Geographic Information Systems (Wu and Ke 2016). These thematic maps were prepared by using Aster GDEM 30m, and also data were collected at 2014 from BMKG (Badan Meteorologi, Klimatologi, dan Geofisika). All of thematic map were prepared using GIS software ArcGIS 10.1.

In this research, six parameters were used for hazard analysis. By using intersect analysis, we combined all of parameter data into a single thematic map by the method which uses Eq. 1. This method is an associative operation where all the data combine in one union.

$$A \cap B = \{x : x \in A \wedge x \in B\} \quad (1)$$

Then, all parameters convert into raster and result in landslide hazard index (LHI) (Dahal and Dahal 2017; Guzzetti et al. 2012). For susceptibility score, we use Indonesia National Standard for Preparation and Determination of Vulnerability Zones of Land Movement from PVMBG, 2015 (Amri et al. 2016) (Table 1).

Table 1 Parameter for landslide hazard index (LHI)

No.	Data	Parameter	Class	Class value	Score	weight score
1	Aster GDEM 30m	Slope	15–30%	1	0.25	0.3
			30–50%	2	0.5	
			50–70%	3	0.75	
			>70%	4	1	
2		Direction of slope (Aspect)	Flat	0	0	0.05
			North–Northwest	1	0.125	
			West	2	0.25	
			West–Northeast	3	0.375	
			Northeast	4	0.5	
			Southwest	5	0.625	
			East	6	0.75	
			Southeast	7	0.875	
			South	8	1	
3	Geology	Type of rock	Alluvial	1	0.33	0.2
			Sediment	2	0.667	
			Volcanic	3	1	
4		Distance from fault	>400 m	1	0.2	0.05
			300–400 m	2	0.4	
			200–300 m	3	0.6	
			100–200 m	4	0.8	
			0–100 m	5	1	
5	Soil type	Soil type	Sand	1	0.3	0.1
			Sandy clay	2	0.666	
			Clay	3	1	
6	Hydrology	Annual cumulative rainfall	<2000 mm	1	0.33	0.2
			2000–3000 mm	2	0.667	
			>3000 mm	3	1	

4 Result and Discussion

4.1 Slope

From the result of the thematic map of slope (Fig. 2a) in this research area ranged from 15% up to more than 70%, with categories class large from 1.33 km^2 at slope more than 70%, 4.225 km^2 at slope 50–70%, 37.666 km^2 at slope 30–50%,

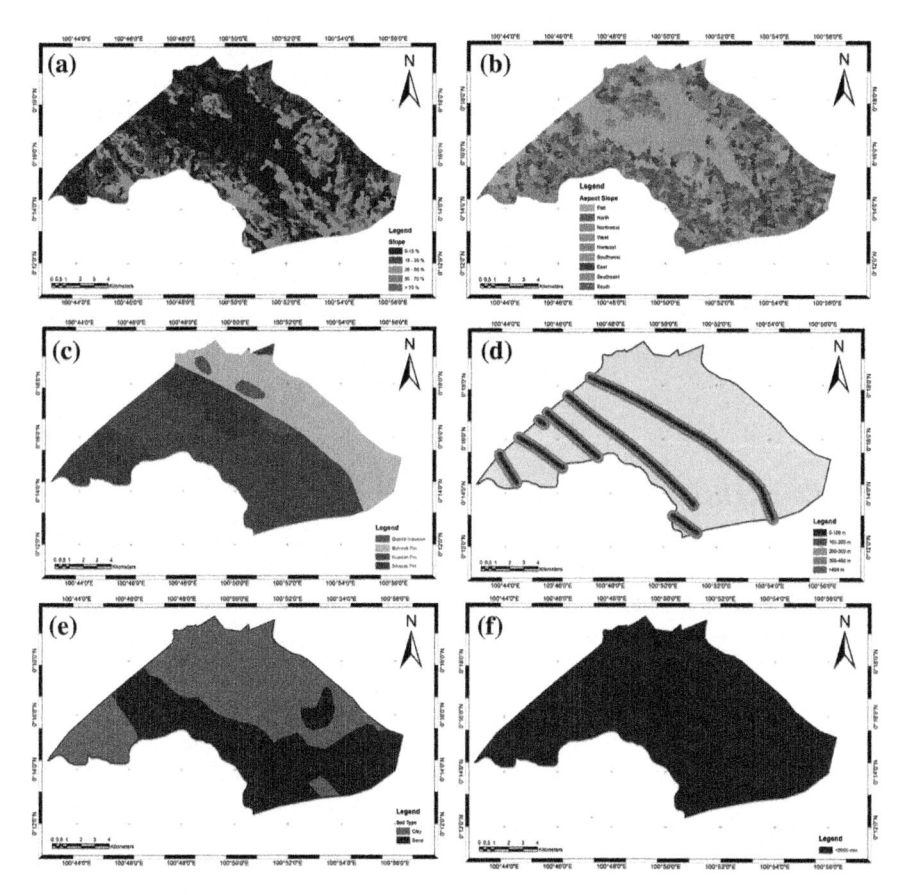

Fig. 2 Thematic map **a** slope, **b** aspect of slope, **c** geology map, **d** distance to fault, **e** soil type, **f** annual cumulative rainfall

36.68 km² at slope 15–30%, and 82.68 km². Each category will be classified into four classes and multiplied by score weight. And the result is visualized on Fig. 2a.

4.2 Aspect

The aspect of thematic map (Fig. 2b) is categorized into nine aspects/directions of slope collected from all directions. The results of this aspect are that 545 areas are flat, 762 areas are facing north direction, 518 areas are facing northwest direction,

389 areas are facing west direction, 324 areas are facing northeast direction, 634 areas are facing southwest direction, 437 areas are facing east direction, 304 areas are facing southeast direction, and 288 areas are facing south direction. The result of this analysis is visualized on Fig. 2b.

4.3 Geology

The geological map (Fig. 2c) in this research area consists of four formations, which are Granite Intrusion, Bohorok Formation, Kuantan Formation, and Sihapas Formation.

4.4 Fault Distance

The fault map distance (Fig. 2d) in this research area consists of several lineaments of normal fault that buffered into four class of categories, which are 0–100 m, 100–200 m, 200–300 m, 300–400 m, and more than 400 m.

4.5 Soil Type

Soil type map (Fig. 2e) in this research area consists of two types of soil which are clay and sand, so only two classes are developed in this research area.

4.6 Rainfall Annual Cumulative

The rainfall annual cumulative map (Fig. 2f) in research area was referred from 2014 BMKG data of annual rainfall, based on four locations, and interpolated it with inverse distance weighted method, so the result in this area is categorized into single class <2000 mm of annual rainfall cumulative.

After all of six maps resulted, we conduct the intersection method to classify the LHI values resulted from all thematic maps. As a result, from six factor maps, slope has the highest contribution to the landslide hazard index and then geology, aspect, soil type, fault distance, and rainfall annual cumulative.

So based on LHI map, there are 2.89% on very high category, 14.4% on high category, 31.38% on medium category, 34.36% on low category, and 16.52% on very low category (Fig. 3).

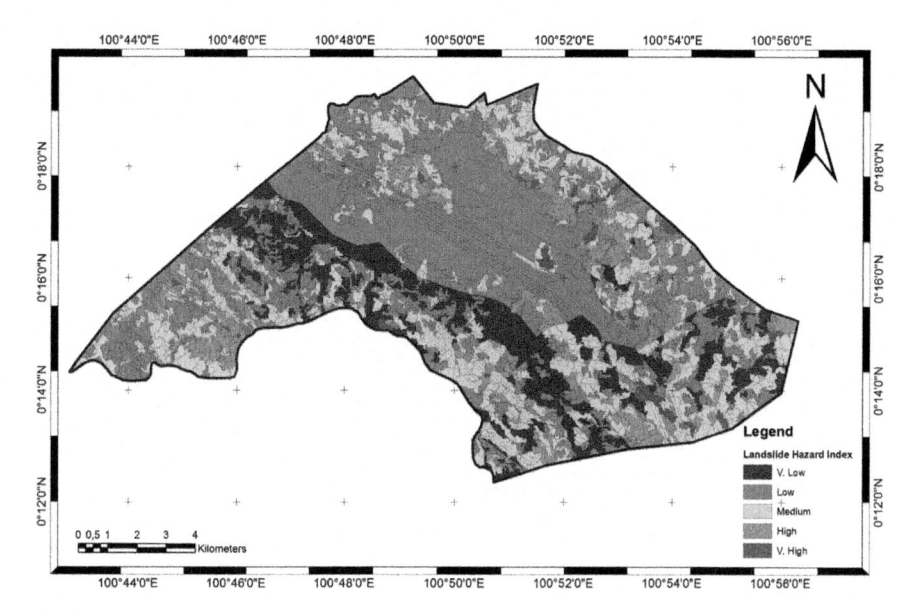

Fig. 3 Landslide hazard index map

5 Conclusion

So the conclusion of this paper is drawn as follows:

- Landslide hazard mapping is essential in delineating landslide-prone areas by collecting all the factor maps that are categorized into several classes of hazard.
- By using landslide hazard index, it will make easier to identify location with high dangerous potential hazard.
- Based on LHI map, the research area at Tanjung Alai is categorized dominantly from low to medium. So, by overlaying it with land use and roadmap, it will give a best contribution for delineating hazard position.

Acknowledgements This work is supported by Department of Geological Engineering and Department of Urban and Regional Planning of Universitas Islam Riau. The authors also gratefully acknowledge the helpful comments and suggestions of the reviewers, which have improved the article and the presentation.

References

Amri, M. R., Yulianti, G., Yunus, R., & Wiguna, S. (2016). *Risiko Bencana Indonesia*, (p. 140).
Choanji, T. (2016). Indikasi Struktur Patahan Berdasarkan Data Citra Satelit dan Digital Elevation Model (DEM) di Sungai Siak, Daerah Tualang dan Sekitarnya Sebagai Pertimbangan Pengembangan Pembangunan Wilayah. *Jurnal Saintis, 16*(2), 22–31.

Choanji, T. (2017). Slope analysis based On SRTM digital elevation model data: study case on Rokan IV Koto area and surrounding. *Journal of Dynamics, 1*(2).

Choanji, T., & Indrajati, R. (2016). Analysis of structural geology based on satellite image and geological mapping on Binuang Area, Tapin Region, South Kalimantan. In *Geosea XIV and 45th IAGI Annual Convention 2016 (GIC 2016)* (Vol. 45).

Ciampalini, A., Raspini, F., Bianchini, S., Lagomarsino, D., & Moretti, S. (2016). *A landslide susceptibility map of the Messina Province (Sicily, Italy)* (pp. 657–661).

Clarke, M. C. G., Kartawa, W., Djunuddin, A., Suganda, E., & Bagdja, M. (1982). *Peta Geologi Lembar Pekanbaru*. Bandung.

Cruden, D. M. (1991). A simple definition of a landslide. *Bulletin of the International Association of Engineering Geology, 43*(1), 27–29. https://doi.org/10.1007/BF02590167.

Dahal, B. K., & Dahal, R. K. (2017). Landslide hazard map: Tool for optimization of low-cost mitigation. *Geoenvironmental Disasters, 4*(1), 8. https://doi.org/10.1186/s40677-017-0071-3.

Guzzetti, F., Mondini, A. C., Cardinali, M., Fiorucci, F., Santangelo, M., & Chang, K. T. (2012). Landslide inventory maps: New tools for an old problem. *Earth-Science Reviews, 112*(1–2), 42–66. https://doi.org/10.1016/j.earscirev.2012.02.001.

Jayanthi, J., Naveen Raj, T., & Suresh Gandhi, M. (2016). *Identification of landslide-prone areas using remote sensing techniques in Sillahallawatershed, Nilgiris District, Tamilnadu, India* (pp. 1947–1952).

Putra, D. B. E., & Choanji, T. (2016). Preliminary analysis of slope stability in Kuok and surrounding areas. *Journal of Geoscience, Engineering, Environment, and Technology, 1*(1), 41–44. https://doi.org/10.24273/jgeet.2016.11.5.

Regmi, N. R., Giardino, J. R., Vitek, J. D., & Dangol, V. (2010). Mapping landslide hazards in western Nepal: Comparing qualitative and quantitative approaches. *Environmental and Engineering Geoscience, 16*(2), 127–142. https://doi.org/10.2113/gseegeosci.16.2.127.

Varnes, D. J. (1994). Landslide hazard zonation: A review of principle and practice. *Bulletin of the International Association of Engineering Geology*. UNESCO.

Wu, Y., & Ke, Y. (2016). Landslide susceptibility zonation using GIS and evidential belief function model. *Arabian Journal of Geosciences, 9*(17). https://doi.org/10.1007/s12517-016-2722-1.